The SAGE Handbook of

Multilevel Modeling

The SAGE Handbook of

Multilevel Modeling

Edited by
Marc A. Scott,
Jeffrey S. Simonoff and
Brian D. Marx

SAGE

Los Angeles | London | New Delhi
Singapore | Washington DC

SAGE

Los Angeles | London | New Delhi
Singapore | Washington DC

SAGE Publications Ltd
1 Oliver's Yard
55 City Road
London EC1Y 1SP

SAGE Publications Inc.
2455 Teller Road
Thousand Oaks, California 91320

SAGE Publications India Pvt Ltd
B 1/I 1 Mohan Cooperative Industrial Area
Mathura Road
New Delhi 110 044

SAGE Publications Asia-Pacific Pte Ltd
3 Church Street
#10-04 Samsung Hub
Singapore 049483

Editor: Jai Seaman
Assistant editor: Anna Horvai
Production editor: Ian Antcliff
Copyeditor: Richard Leigh
Proofreader: Derek Markham
Indexer: David Rudeforth
Marketing manager: Ben Griffin-Sherwood
Cover design: Wendy Scott
Typeset by: C&M Digitals (P) Ltd, Chennai, India
Printed in Great Britain by: Henry Ling Limited, at the Dorset Press, Dorchester, DT1 1HP

First published 2013

Library of Congress Control Number is Available

British Library Cataloguing in Publication data

A catalogue record for this book is available from the British Library

ISBN 978-0-85702-564-7

Contents

Notes on Contributors

Sudipto Banerjee received an MS and PhD in Statistics from the University of Connecticut. Prior to this he received a BS (Honors) from Presidency College and an MStat from the Indian Statistical Institute, both in Calcutta (now called Kolkata), India. He is currently a tenured Professor of Biostatistics in the School of Public Health, University of Minnesota Twin Cities. His research focuses upon statistical modeling and analysis of geographically referenced datasets, Bayesian statistics (theory, methods, and applications), statistical computing/software, and the melding of numerical/physical models with field data from industrial hygiene. He has published over 60 peer-reviewed journal articles, several book chapters, and has co-authored a book titled *Hierarchical Modeling and Analysis for Spatial Data* (Chapman & Hall/CRC, 2009). He has overseen the development of several spatial software packages within the R statistical framework. In 2009 he was honored with the Abdel El Shaarawi Award from The International Environmetrics Society – accorded to a young investigator (below the age of 40) every year who has made outstanding contributions to the field of environmetrics. Sudipto is also an elected member of the International Statistical Institute and is an inductee of the Pi chapter of the Delta Omega National Honor Society.

Gerard van Breukelen is Associate Professor at the Department of Methodology and Statistics, Maastricht University, the Netherlands. He obtained his master's degree (*cum laude*) and PhD in mathematical psychology and psychometrics. He now specializes in the design and analysis of intervention studies in psychology and health sciences, in particular sample sizes for multilevel experiments and confounding in nonrandomized studies. He has been advisor to several PhD projects in optimal design, including that of his present co-author, and he is author and co-author to numerous publications in statistics, psychology, health sciences, and medicine. His teaching ranges from logistic regression and factor analysis to multilevel analysis and structural equation modeling.

Ryan P. Browne is Assistant Professor at the Department of Mathematics and Statistics at the University of Guelph. Ryan holds degrees in statistics from the University of Waterloo: BMath 2004, MMath 2006, and PhD 2009. Based on his PhD work, Ryan and his advisors were awarded the 2011 W.J. Youden Award in Interlaboratory Testing by the American Statistical Association. Ryan's current research focus is on model-based clustering and classification. In addition, he is interested in measurement models, specifically in assessing the quality of a measurement system, which was the focus of his PhD thesis.

Sean P. "Jack" Buckley is Commissioner of the National Center for Education Statistics (NCES). He is currently on leave from New York University, where he is Associate Professor of Applied Statistics. He also served previously as Deputy Commissioner of NCES from 2006 to 2008. Buckley is known for his research on school choice, particularly charter schools, and on statistical methods for public policy. He has also taught statistics and education policy at Georgetown University, Boston College, and the State University of New York at Stony Brook. Buckley also spent five years in the U.S. Navy as a surface warfare officer and nuclear reactor engineer, and he worked in the intelligence community as an analytic methodologist. He holds an AB in Government from Harvard and an MA and PhD in political science from SUNY Stony Brook.

Brian S. Caffo received his PhD from the University of Florida's Department of Statistics in 2001. He is Professor in the Department of Biostatistics at Johns Hopkins University. He has worked in the areas of statistical computing, categorical data analysis, and imaging data. He has a particular focus in the analysis of biological heterogeneity in brain imaging data for basic science and clinical research. He has developed statistical analysis methods for the analysis of MRI, functional MRI, Diffusion Tensor Imaging (DTI), EEG, electrocorticography (ECoG) and nuclear medicine imaging data. With Dr. Crainiceanu, he leads the Statistical Methods and Applications for Research in Technology (SMART, www.smart-stats.org) research group at Johns Hopkins. The group, boasting a diverse collection of faculty, postdoctoral fellows, and students, is leading efforts in the analysis of high-dimensional longitudinal and multilevel functional data, especially in the field of medical imaging.

Gerda Claeskens is Professor of Statistics affiliated with the research center OR and Business Statistics and with the Leuven Statistics Research Center of the KU Leuven, Belgium. Her research topics include lack-of-fit and goodness-of-fit testing, model selection, model averaging, and semi- and nonparametric estimation. She received her PhD degree from Limburgs Universitair Centrum in 1999, and previously held positions at TU Eindhoven, the Australian National University, and Texas A&M University. She co-authored the 2008 book *Model Selection and Model Averaging* (Claeskens and Hjort, Cambridge University Press, 2008), and is the author of numerous journal publications on the abovementioned research topics.

Ciprian M. Crainiceanu received his PhD in Statistics from Cornell University in 2003 and is an Associate Professor of Biostatistics at Johns Hopkins University. He is a specialist in functional, longitudinal, and nonparametric methods. His main expertise is in the area of massive and complex observational studies obtained from the integration of new measurement technologies and classical observational studies. His scientific interests are centered on biosignals such as electroencephalograms (EEG), multimodality Magnetic Resonance Imaging (MRI), and human activity measurements obtained from wearable devices. Together with Dr Caffo, Dr Crainiceanu leads the Statistical Methods for Analysis of Research Technologies (SMART, www.smart-stats.org) research group. Dr Crainiceanu is also a specialist in measurement error modeling; he co-authored the second edition of the monograph *Measurement Error in Nonlinear Models: A Modern Perspective.*

Robert Crouchley obtained a PhD in Mathematics (Statistics) from Imperial College, London, in 1991. His supervisor was Professor Sir David Cox. Rob is Professor of Applied Statistics in the Department of Economics, Lancaster University, and was (2002–2012) Director of the Lancaster University Management School (LUMS) Centre for e-Science; he has directed research projects on the development and deployment of grid-enabled statistical computing software. Rob also lectures on econometrics. He has published many applied research papers as well as writing and editing several books on statistical modeling. He is co-author of the book *Multivariate generalized Linear Mixed Models Using R* (Chapman & Hall/CRC, 2011) with Daman Berridge. He currently represents the UK's Royal Statistical Society on the Government Statistical Service Methodology Advisory Committee. His research interests include statistical modeling, He development of database management, and grid computing systems.

Marijtje A.J. van Duijn is Associate Professor of Statistics in the Department of Sociology, University of Groningen. Her research interests are in the development and application of random effects (multilevel) models for discrete or complex data, such as longitudinal, grouped, or social network data, and often inspired by collaboration with social scientists. Together with Mark Huisman, she contributed two chapters to the *Sage Handbook on Social Network Analysis* (SAGE, 2011), one on statistical models, the other on software for social network analysis.

Craig Enders, PhD, is an Associate Professor in the Quantitative Psychology concentration in the Department of Psychology at Arizona State University, where he teaches graduate-level courses in missing data analyses, multilevel modeling, and longitudinal modeling. The majority of his research focuses on analytic issues related to missing data analyses and multilevel modeling. His book, *Applied Missing Data Analysis*, was published by Guilford Press in 2010.

Christel Faes is Professor of Statistics in the Interuniversity Institute for Biostatistics and Statistical Bioinformatics at the Hasselt University in Belgium. She received an MS degree in Mathematics (2000) from the University of Hasselt and a PhD degree in Biostatistics (2004) from the Hasselt University. Her research interests include statistical methods for correlated, multivariate, clustered, and spatial data, with applications in infectious diseases, veterinary epidemiology, and non-clinical studies. She serves as associate editor for *Biometrics*.

Ludwig Fahrmeir is Professor Emeritus for Statistics at the Department of Statistics, Ludwig-Maximilians-University of Munich. He obtained both his PhD and his Habilitation degree at the Technical University of Munich. After a short period as a visiting professor at the University of Dortmund, he was appointed full professor at the University of Regensburg in 1978. He returned to his hometown, Munich, in 1991 and became Professor at the Department of Statistics of the LMU. From 1995 to 2006 he served as the speaker of the collaborative research center "Statistical Analysis of Discrete Structures, with applications in Econometrics and Biometrics," funded by the German National Science Foundation. His research interests cover asymptotic theory, state space models and semiparametric regression, longitudinal and spatial data analysis, Bayesian statistics, and a broad range of applications such as credit risk, development economics, and neuroscience. Becoming Professor Emeritus in 2010, he has increased free time to enjoy family life, skiing, traveling and guitar playing.

Andrew O. Finley received a PhD in Natural Resources Science and Management and an MS in Statistics from the University of Minnesota. He is currently an Assistant Professor in the Departments of Forestry and Geography at the Michigan State University. His research interests lie in developing methodologies for monitoring and modeling environmental processes, Bayesian statistics, spatial statistics, and statistical computing. Andrew received the 2009 American Statistical Association Section on Statistics and the Environment's Young Investigator Award.

Garrett Fitzmaurice is Professor of Psychiatry (Biostatistics) at the Harvard Medical School and Professor in the Department of Biostatistics at the Harvard School of Public Health. He is a Fellow of the American Statistical Association and a member of the International Statistical Institute. He has served as Associate Editor for *Biometrics*, the *Journal of the Royal Statistical Society, Series B*, and *Biostatistics*; currently, he is Statistics Editor for the journal *Nutrition*. His research and teaching interests are in methods for analyzing longitudinal and repeated measures data. A major focus of his methodological research has been on the development of statistical methods for analyzing repeated binary data and for handling the problem of attrition in longitudinal studies. Much of his collaborative research has concentrated on applications to mental health research, broadly defined. He co-authored the textbook *Applied Longitudinal Analysis*, 2nd edition (Wiley, 2011) and co-edited the book *Longitudinal Data Analysis* (Chapman & Hall/CRC Press, 2008).

Andrzej Gałecki is a Research Professor in the Division of Geriatric Medicine, Department of Internal Medicine, and Institute of Gerontology at the University of Michigan Medical School, and is Research Scientist in the Department of Biostatistics at the University of Michigan School of Public Health. He earned his MSc in Applied Mathematics (1977) from the Technical University of Warsaw, Poland, and an MD (1981) from the Medical University of Warsaw. In 1985 he earned a PhD in Epidemiology from the Institute of Mother and Child Care in Warsaw (Poland). He is a member of the Editorial Board of the *Open Journal of Applied Sciences*. Since 1990, Dr Galecki has collaborated with researchers in gerontology and geriatrics. His research interests lie in the development and application of statistical methods for analyzing correlated and over-dispersed data. He developed the SAS macro NLMEM for nonlinear mixed effects models, specified as a solution to ordinary differential equations. He also proposed a general class of variance–covariance structures for the analysis of multiple continuous dependent variables measured over time. This methodology is considered to be one of first approaches to joint models for longitudinal data.

Dani Gamerman earned a degree in Mechanical Engineering from the Military Engineering Institute, Rio de Janeiro, in 1980, an MSc in Statistics from the Instituto Nacional de Matemática Pura e Aplicada (IMPA) in 1983, and a PhD in Statistics from the University of Warwick in 1987. He has been Professor of Statistics at UFRJ since 1996, where he supervises MSc and PhD students. He has been an invited lecturer at various scientific meetings in Brazil and abroad. He has been a visiting lecturer of a few Universities in Brazil and abroad, and *colaborador honorífico* of Universidade Rey Juan Carlos in Madrid. He is author of the books *Monte Carlo Markov Chain: Stochastic Simulation for Bayesian Inference*, published by Chapman and Hall in 1997 (1st edition) and 2006 (2nd edition, with Hedibert F. Lopes) and *Statistical Inference: An Integrated Approach* (with Helio S. Migon), published by Arnold in 1999. He has published papers in many statistical journals, including *Journal of the Royal Statistical Society, Series B* and *Biometrika*. His current research interests include dynamic models, item response theory, spatial statistics, survival analysis, stochastic simulation, econometrics, and Bayesian inference.

Helena Geys is Professor of Statistics in the Interuniversity Institute for Biostatistics and Statistical Bioinformatics at the University of Hasselt in Belgium, and Associate Director of Nonclinical Statistics at Janssen Pharmaceutical Companies of Johnson & Johnson. She holds a master's degree in Mathematics from the University of Antwerp and a PhD degree in biostatistics from the University of Hasselt. She published methodological work on clustered non-normal data with applications in developmental toxicity, risk assessment, spatial epidemiology, surrogate marker validation, and pseudo-likelihood inference. She served on the editorial panel of the Archives of Public Health, Belgium, was associate editor for the *Journal of Agricultural, Biological and Environmental Statistics* (*JABES*), and was a member of the *JABES* management committee. She has authored or co-authored more than 50 publications in major statistical and epidemiological journals and is one of the editors of a book on topics in modeling clustered data.

Jeff Gill is Professor of Political Science, Professor of Biostatistics, and Professor of Surgery (Public Health Sciences) at Washington University in St. Louis. He does extensive work in the development of Bayesian hierarchical models, nonparametric Bayesian models, elicited prior development from expert interviews, as well as in fundamental issues in statistical inference. He has extensive expertise in statistical computing, and Markov chain Monte Carlo (MCMC) tools in particular. His current theoretical work centers on new hybrid algorithms for product partition clustering, Bayesian meta-modeling, and Polya-tree mixture models. Current applied work includes energetics and cancer, long-term mental health outcomes from children's exposure to war, pediatric head trauma, and assessment of transplant centers.

Harvey Goldstein was formerly Professor of Statistical Methods at the Institute of Education from 1997–2005. He is currently Professor of Social Statistics at the University of Bristol. He has been a member of the Council of the Royal Statistical Society, and chair of its Educational Strategy Group. He is currently chair of the technical advisory committee for the RSS Centre for Statistical Education. He was awarded the RSS Guy medal in Silver in 1998 and was elected a fellow of the British Academy in 1997. He has been the principal applicant on several major ESRC-funded research projects since 1981. His current major research interest is in the methodology of multilevel modeling. His recent book, *Multilevel Statistical Models* (Wiley, 2011, 4th edition) is the standard reference text in this important area of statistical data analysis. Most recently he has helped to develop efficient methods for handling missing data in multilevel models and procedures for unbiased and efficient record linkage of large datasets.

Masataka Harada is a postdoctoral research scientist at New York University's Center for the Promotion of Research Involving Innovative Statistical Methodology (PRIISM), where he is developing methods of sensitivity analysis, some of which (Imbens' Sensitivity Analysis, -isa-, and Generalized Sensitivity Analysis, -gsa-) are available as Stata ado-files on the Statistical Software Components (SSC) website. He specializes in econometrics and political science. His main interest is causal inference of political theories using quasi-experimental methods. His recent work includes the quasi-experimental investigation of strategic use of debts in Southern States in the U.S. when the Voting Rights Act was enacted in 1965. He received an MA in Social Science and PhD in Public Policy from the University of Chicago.

Larry V. Hedges is the Board of Trustees Professor of Statistics and Social Policy and a faculty fellow of the Institute for Policy Research at Northwestern University. Dr Hedges is a national leader in the fields of educational statistics and evaluation. His research focuses broadly on the development and application of statistical methods for the social, medical, and biological sciences. He is best known for his work on developing statistical methods for meta-analysis. Widely published, Hedges has authored or co-authored numerous journal articles and books, including *Statistical Methods for Meta-Analysis: A Practical Guide to Modern Methods of Meta-Analysis* (with I. Olkin, Academic Press, 1985) and *The Handbook of Research Synthesis* (with H. Cooper, Russell Sage Foundation, 1993). Dr Hedges is a member of the National Academy of Education and the Society of Multivariate Experimental Psychology and a fellow of the American Statistical Association and the American Psychological Association.

Jennifer Hill is an Associate Professor of Applied Statistics and Co-director of the Center for the Promotion of Research Involving Innovative Statistical Methodology (PRIISM) at the Steinhardt School for Culture, Education, and Human Development at New York University. Hill earned her PhD in Statistics from Harvard University, followed by a postdoctoral fellowship in Child and Family Policy at Columbia University's School of Social Work. Hill's research for many years has negotiated the intersection between statistical methodology and research in the social and behavioral sciences, as well as educational policy. She is interested in methods and study designs that allow researchers to go beyond making purely associational observations to having the capacity to potentially answer causal questions. In particular, she focuses on situations in which it is difficult or impossible to perform traditional randomized experiments, or when even seemingly pristine study designs are complicated by missing data or hierarchically structured data. Hill has published in a variety of leading journals including *Journal of the American Statistical Association*, *American Political Science Review*, *American Journal of Public Health*, and *Developmental Psychology*.

Ardo van den Hout is a lecturer at the Department of Statistical Science at University College, London, UK. He has previously worked in the Medical Research Council Biostatistics Unit, Cambridge, UK. His primary research areas are multi-state survival models and longitudinal data analysis, with applications in epidemiology and biostatistics.

Joop Hox is full professor of Social Science Methodology at the department of Methodology and Statistics of the Faculty of Social Sciences at Utrecht University. As Methodology chair, he is responsible for the research, development, and teaching carried out at the faculty in the field of social science methods and techniques. His research interests are multilevel modeling and data quality in surveys. He has been invited speaker at several conferences and is chair of the Amsterdam Multilevel conferences, held every two years. He is editor of the journal *Methodology*, a founding member of the European Association of Methodology (EAM), and past editor of the EAM book series.

Göran Kauermann is Professor of Statistics at the Ludwig-Maximilian-University (LMU) Munich, Germany. After receiving his PhD in statistics at the Technical University Berlin, Germany, in 1994 he spent a year as visiting scholar at the University of Chicago, USA. He became Assistant Professor at the LMU in Munich, Germany, in 1998, and worked at the University of Glasgow, Scotland, as Senior Lecturer from 2000 to 2003. In 2003 he was appointed as full Professor of Statistics at Bielefeld University, Germany. Professor Kauermann is currently Editor of *AStA – Advances in Statistical Analysis* and from 2005 to 2013 he was Chair of the Deutsche Arbeitsgemeinschaft Statistik, a consortium of all German statistical societies. His research focuses on nonparametric models and penalized regression in generalized linear and generalized mixed models.

Thomas Kneib is Professor of Statistics at the Department of Economics, University of Göttingen, and speaker of the interdisciplinary Centre for Statistics at the University of Göttingen. He received his PhD in statistics in 2006 from the Department of Statistics, University of Munich. During his time as postdoctoral researcher, from 2006 until 2009, he was visiting Professor for Applied Statistics at the Faculty of Mathematics and Economics, University of Ulm, and substitute Professor for Statistics at the Department of Economics, University of Göttingen. In 2009 he became Professor for Applied Statistics at the department of mathematics, University of Oldenburg, before he moved to Göttingen in 2011. His research interests are semiparametric regression, spatial statistics, quantile and expectile regression, Bayesian statistics, and statistical learning techniques.

Torben Kuhlenkasper is Junior Professor of Applied Econometrics at Goethe University Frankfurt, Germany, and a Research Fellow of the Hamburg Institute of International Economics (HWWI). He graduated in Business Administration and Economics at Bielefeld University, Germany, in 2006 and 2007, where he also received his PhD in 2011. Before being appointed as a Junior Professor in 2012 he worked as a Senior Economist at HWWI in the field of empirical economic research. His research focuses on the application of non- and semiparametric regression models in econometrics, especially in labor economics and its related fields.

Nan Laird is the Harvey V. Fineberg Professor of Biostatistics at the Harvard School of Public Health. Dr. Laird has contributed to methodology in many different fields, including meta-analysis, statistical genetics, and longitudinal data. She is a co-author of the book *Applied Longitudinal Analysis* with Garrett Fitzmaurice and James Ware, and is a co-author of the book *Fundamentals of Modern Statistical Genetics* with Christoph Lange. She is the recipient of many awards and prizes, including Fellow of the American Statistical Association, and the Samuel Wilks Award in 2011.

Stefan Lang is Professor for Applied Statistics at the Department of Statistics, University of Innsbruck. He is editor of *Advances in Statistical Analysis* (ASTA) and associate editor of *Statistical Modeling*. He received his PhD in statistics in 2001 from the Department of Statistics, University of Munich. He became Professor of Statistics at the University of Leipzig in 2005. He finally moved to the University of Innsbruck in October 2006. His research interests cover Bayesian statistics, semiparametric regression, spatial statistics, multilevel models, and applications in ecology, marketing science, development economics, etc.

Wei Liu received a BSc in Mathematics and an MSc in Statistics from Northeast Normal University in China. Then she got another MSc and a PhD in Statistics from Memorial University of Newfoundland and the University of British Columbia in Canada, respectively. Since 2007, she has been an Assistant Professor in the Department of Mathematics and Statistics at York University in Toronto, Canada. Her current research interests include longitudinal data, missing data, measurement errors, mixed effects models, and order-restricted statistical inference.

Kimberly S. Maier is an Assistant Professor of Measurement and Quantitative Methods and an affiliate of the Educational Policy program in the College of Education at Michigan State University. Dr Maier's research focuses on the development of statistical methods for complex data structures, with an emphasis on the development and application of extensions to multilevel models for policy research. Currently, Maier is studying the application of multilevel item response theory to education achievement measures and attitudinal surveys using Bayesian techniques. Maier's methodological work as author or co-author appears in journals such as *Journal of Educational and Behavioral Statistics* and *Structural Equation Modeling: A Multidisciplinary Journal*.

Brian D. Marx is a full professor in the Department of Experimental Statistics at Louisiana State University. His main research interests include P-spline smoothing, ill-conditioned regression problems, and high-dimensional chemometric applications. He is currently serving as coordinating editor for the journal *Statistical Modelling* and is past chair of the Statistical Modelling Society. He has also taught as a visiting professor at Stanford University, University of Munich, and Utrecht University. He is co-author of the forthcoming (2013) book *Regression: Models, Methods, and Applications*, with Ludwig Fahrmeir, Thomas Kneib, and Stefan Lang.

Charles E. McCulloch is Professor of Biostatistics in the Division of Biostatistics in the Department of Epidemiology and Biostatistics at the University of California, San Francisco. He is a Fellow of the American Statistical Association, and author (with John M. Neuhaus and Shayle R. Searle) of *Generalized, Linear, and Mixed Models*, now in its second edition.

Paul D. McNicholas is a Professor and University Research Chair in Computational Statistics at the Department of Mathematics and Statistics at the University of Guelph. Prior to taking a position at Guelph, he completed a PhD in Statistics, an MSc in High Performance Computing, and an MA in Mathematics at Trinity College Dublin, Ireland. His research interests are in computational statistics; with a focus in classification and clustering using mixture models.

Mirjam Moerbeek is Associate Professor, Department of Methods and Statistics, Utrecht University, the Netherlands. She graduated with a master's degree in biometrics (*cum laude*) and wrote a PhD thesis on the design and analysis of experiments with multilevel data. She specializes in statistical power analysis and the optimal design of experiments, in particular for multilevel and longitudinal data. She is joint organizer of courses, conferences, and symposia on optimal design, multilevel analysis, and longitudinal data analysis. She has received prestigious research grants from the Netherlands Organization for Scientific Research to build her own research group and to supervise PhD students.

Geert Molenberghs is Professor of Biostatistics at Universiteit Hasselt and Katholieke Universiteit Leuven in Belgium. He received a BS degree in Mathematics (1988) and a PhD in biostatistics (1993) from Universiteit Antwerpen. He has published on surrogate markers in clinical trials, and on categorical, longitudinal, and incomplete data. He was Joint Editor of *Applied Statistics* (2001–2004), Co-editor of *Biometrics* (2007–2009), and Co-editor of *Biostatistics* (2010–2012). He was President of the International Biometric Society (2004–2005), received the Guy Medal in Bronze from the Royal Statistical Society and the Myrto Lefkopoulou award from the Harvard School of Public Health. Geert Molenberghs is founding director of the Center for Statistics. He is also the director of the Interuniversity Institute for Biostatistics and Statistical Bioinformatics (I-BioStat). Jointly with Geert Verbeke, Mike Kenward, Tomasz Burzykowski, Marc Buyse, and Marc Aerts, he has authored books on longitudinal and incomplete data, and on surrogate marker evaluation. Geert Molenberghs received several Excellence in Continuing Education Awards from the American Statistical Association for courses at Joint Statistical Meetings.

James E. Monogan III is an Assistant Professor in the Department of Political Science at the University of Georgia. He received his PhD from the University of North Carolina at Chapel Hill in 2010. His primary research interests are in political methodology and American politics. Within political methodology, he focuses on the analysis of geographically referenced and time-dependent data. In American politics, he studies the quality of representation in the United States, public opinion, elections, and state politics and policy. His research has been published in the *Journal of Politics*, *State Politics & Policy Quarterly*, *Publius: The Journal of Federalism*, and the *Journal of Theoretical Politics*.

Jeffrey S. Morris is a Professor and Deputy Chair in the Department of Biostatistics at the University of Texas M.D. Anderson Cancer Center. He received his PhD in Statistics from Texas A&M University in 2000. His main research contributions include developing new methods for complex, high-dimensional object data such as functional and quantitative image data, bioinformatics methods for high-throughput genomic and proteomic technologies, and collaborative medical and biological research in cancer. His methods have been applied to a variety of data, including carcinogenesis biomarker, actigraphy, microarray, copy number, fMRI, proteomics, and EEG data. His work has been recognized with a number of prestigious awards including the Harvard University Myrto Lefkopoulou Invited Lecture and the Noether Young Investigator Award.

John M. Neuhaus is Professor of Biostatistics in the Division of Biostatistics in the Department of Epidemiology and Biostatistics at the University of California, San Francisco. He is a Fellow of the American Statistical Association, and author (with Charles E. McCulloch and Shayle R. Searle) of *Generalized, Linear, and Mixed Models*, now in its second edition.

Vicente Núñez-Antón is Professor of Statistics, Department of Econometrics and Statistics, University of the Basque Country UPV/EHU, Bilbao, Spain. He received his undergraduate degree in Electronic Engineering (1987) from ESPOL (Ecuador), and a Master in Theoretical Statistics (1989) and a PhD in Statistics (1993) from The University of Iowa (USA). His research interests include longitudinal data analysis, survival analysis, nonparametric estimation methods, Bayesian methods, goodness-of-fit testing and health-related quality of life (HRQoL) methods. He is one of the founders and one of the main researchers of the BIOSTATNET Network in Biostatistics. He has authored or co-authored four books and more than 50 scholarly articles in peer-reviewed journals. He is an elected member of the International Statistical Institute. He was associate editor for *Applied Statistics*, and is currently associate editor for *Statistical Modelling* and *METHODOLOGY – European Journal of Research Methods for the Behavioral and Social Sciences*.

David Rindskopf is Distinguished Professor of Educational Psychology and Psychology at the City University of New York Graduate School. He is a Fellow of the American Statistical Association and the American Educational Research Association, Past President of the Society for Multivariate Experimental Psychology, and Editor of the *Journal of Educational and Behavioral Statistics*. His research interests are categorical data, latent variable models, and multilevel models. Current projects include: (i) showing how people subconsciously use complex statistical methods to make decisions in everyday life, (ii) introducing floor and ceiling effects into logistic regression to model response probabilities constrained to a limited range, (iii) using multilevel models to analyze data from single case designs.

Adam Sagan, PhD is Associate Professor of Marketing Research at Cracow University of Economics, Cracow, Poland. His main research interests lie in two fields of marketing research. The first is the analysis of consumer behavior and market segmentation based on means–end chains theory and laddering interviews. The second field of interest is the application of latent variable and path models in the analysis of buyer–seller interactions and measurement of the dimensions of value for the customer in relationship marketing. He is author of the handbook of marketing research (in Polish) and has published papers in *Advances in Data Analysis* and *Journal of Targeting, Measurement and Analysis for Marketing*.

Rens van de Schoot graduated *cum laude* for the research master's degree in Development and Socialization of Children and Adolescents at Utrecht University in The Netherlands. He obtained his PhD, also *cum laude*, on informative hypotheses and Bayesian statistics at the department of Methods and Statistics. Currently, he is Assistant Professor at Utrecht University and Extraordinary Professor at the North-West University in South Africa. Besides his research on how to directly evaluate expectations and the labor market position of PhD candidates, he collaborates with many developmental researchers. He is co-editor of the *European Journal of Developmental Psychology*, president of the young researchers union of the European Association of Developmental Psychology, and he is vice-chair for the scientific committee of the Dutch Institute for Psychologists (NIP).

Marc A. Scott is an Associate Professor of Applied Statistics at New York University's Steinhardt School of Culture, Education, and Human Development, and co-directs the Center for the Promotion of Research Involving Innovative Statistical Methodology (PRIISM). He received his PhD in Statistics from New York University and spent the next two years as a Senior Research Associate at Columbia University's Institute on Education and the Economy. Scott's research primarily involves the development of statistical models for longitudinal data, including models for sequence analysis and event histories. This methodological work has been motivated through collaborations with policy researchers on topics such as wage inequality and immobility, educational attainment, child development and achievement, disease progression in respiratory illness and reliability of sleep studies. Recent methodological papers have appeared in *Statistical Science*, *Sociological Methodology*, *Journal of Educational and Behavioral Statistics*, *Journal of the Royal Statistical Society Series C*, and *Statistics in Medicine*.

Patrick E. Shrout (PhD, 1976, University of Chicago) is Professor of Psychology at NYU, where he teaches statistics and measurement courses to graduate and undergraduate students. Prior to moving to NYU, he was on the biostatistics faculty at Columbia University. His methodological interests are on inferences that can be made from non-experimental data. His substantive interests are stress, social support and coping in intimate relationships, and in cross-cultural studies in psychiatric epidemiology. He is Past-President of both the American Psychopathological Association and the Society of Multivariate Behavioral Research. Shrout has been elected Fellow of the American Association for the Advancement of Science, American Statistical Association, the American Psychological Association (Divisions 5 and 8), the Association for Psychological Science, and the Society of Experimental Social Psychology. He is currently an Associate Editor of *Psychological Methods*.

Marina Silva Paez earned a degree in Statistics from Universidade Federal do Rio de Janeiro (UFRJ) in 1998, MSc in Statistics from UFRJ in 2000, and PhD in Statistics from UFRJ in 2004. She was a research student at Lancaster University from September 2003 to August 2004, and has been Associate Professor of Statistics at UFRJ since 2005. She supervises MSc and PhD students, and has published papers in statistical journals, including *Environmetrics* and *Environmental and Ecological Statistics*, mainly on the subject of space-time Bayesian models. She has been an invited speaker at many scientific meetings in Brazil. Her current research interests include space-time models, spatial statistics, dynamic models, item response theory, and Bayesian inference.

Jeffrey S. Simonoff is Professor of Statistics in the Department of Information, Operations and Management Sciences of the Leonard N. Stern School of Business at New York University. He is a Fellow of the American Statistical Association, a Fellow of the Institute of Mathematical Statistics, and an Elected Member of the International Statistical Institute. He has been at NYU Stern since receiving his PhD in Statistics from Yale University in 1980. He has authored or co-authored more than 90 articles and five books on the theory and applications of statistics, including *A Casebook for a First Course in Statistics and Data Analysis* with S. Chatterjee and M.S. Handcock (1995), *Smoothing Methods in Statistics* (1996), *Analyzing Categorical Data* (2003), *Nonprofit Trusteeship In Different Contexts* with R. Abzug (2004), and *Handbook of Regression Analysis* with S. Chatterjee (2013). His research interests include computer-intensive statistical methodology, modern regression methods, categorical data, smoothing methods, and applications of statistics to business, industrial, and scientific problems.

Russell Steele is an Associate Professor in the Department of Mathematics and Statistics at McGill University. He also holds a position as an investigator at the Jewish General Hospital in the Division of Clinical Epidemiology. His primary statistical methodological interests lie in the areas of methods for analyzing data with missing values and Bayesian model selection, although he is more broadly interested in statistical applications. He has a broad range of substantive interests in medicine, publishing work in rheumatology, sports medicine, and design and interpretation of meta-analyses. In particular, he has published several papers in high-impact rheumatology journals with other members of the Canadian Scleroderma Research Group over the last seven years. Russell has also collaborated with scientists in geography and the biological sciences on problems in multilevel modeling.

Brian D.M. Tom is an Investigator Scientist at the UK Medical Research Council Biostatistics Unit and an Affiliated Lecturer at the Department of Pure Mathematics and Mathematical Statistics of the University of Cambridge. He has previously worked as a Consulting Statistician and a Project Statistician within the University of Cambridge's Department of Public Health and Primary Care. His research areas are quite varied, ranging from epidemiology to biostatistics and bioinformatics. On the methodological side, he has particular interests in bias and efficiency, causal and dynamic modeling, event history and survival analysis, integrative genomics, and longitudinal data analysis. On the applied side, he has worked and is working in various chronic and acute disease areas, such as Rheumatology, Hepatitis C and Pandemic Influenza.

Zachary Townsend is a research affiliate at New York University's Center for the Promotion of Research Involving Innovative Statistical Methodology (PRIISM). Zac's research interests include causal inference in policy analysis, and the use of multilevel modeling in education and criminal justice contexts. As a research affiliate, he works closely with the PRIISM faculty on research projects. Zac received a MPA from New York University's Wagner School of Public Service, where he focused on public finance and statistical methods, and an AB in Applied Mathematics–Economics and Public Policy from Brown University.

Geert Verbeke has published extensively on various aspects of mixed models for longitudinal data analyses, about which he co-authored and co-edited several text books (*Springer Lecture Notes* 1997; *Springer Series in Statistics* 2000 and 2005; Chapman & Hall/CRC 2009). He held visiting positions at the Gerontology Research Center and the Johns Hopkins University (Baltimore, MD), was International Program Chair for the International Biometric Conference in Montreal (2006), Joint Editor of the *Journal of the Royal Statistical Society, Series A* (2005–2008), and Co-editor of *Biometrics* (2010–2012). He served and serves on a variety of committees of the International Biometric Society, is an elected Fellow of the American Statistical Association and Elected Member of the International Statistical Institute. He was elected international representative in the Board of Directors of the American Statistical Association (2008–2010). Geert Verbeke has earned Excellence in Continuing Education Awards in 2002, 2004, 2008, and 2011 for short courses taught at the Joint Statistical Meetings of the American Statistical Association. He received the International Biometric Society Award for the best *Biometrics* paper in 2006, and received accreditation as a professional statistician by the American Statistical Association (2010–2016).

Jeroen K. Vermunt is a full professor in the Department of Methodology and Statistics at Tilburg University, the Netherlands. His research is on methodologies of social, behavioral, and biomedical research, with a special focus on latent variable models and techniques for the analysis of categorical, multilevel, and longitudinal data sets. He has widely published on these topics in statistical and methodological journals and has also co-authored many articles in applied journals in which these methods are used to solve practical research problems. He is the co-developer (with Jay Magidson) of the Latent GOLD software package. In 2005, Vermunt was awarded the Leo Goodman award by the Methodology Section of the American Sociological Association.

Sharon L. Weinberg is Professor of Applied Statistics and Psychology and former Vice Provost for Faculty Affairs at New York University. She received an AB degree in mathematics and a PhD degree in psychometrics and research design methodology from Cornell University. Dr Weinberg is the author of over sixty articles, books, and reports on statistical methodology, statistical education, evaluation, and on such applied areas as clinical and school psychology, special education, and higher education. She is the recipient of several major grants from Federal agencies, including the National Science Foundation, the National Institute of Drug Abuse, and the Office of Educational Research and Improvement. Her current textbook, *Statistics Using SPSS: An Integrative Approach*, co-authored with former graduate student Sarah Knapp Abramowitz, and published by Cambridge University Press, is in its second edition. She is co-editor of *Diversity in American Higher Education: Toward a More Comprehensive Approach*, published in 2011 by Routledge Press. She is currently on the Editorial Board of *Educational Researcher.*

Brady T. West is a Research Assistant Professor in the Survey Methodology Program, located within the Survey Research Center at the Institute for Social Research on the University of Michigan–Ann Arbor (U-M) campus. He also serves as a Statistical Consultant at the U-M Center for Statistical Consultation and Research (CSCAR). He earned his PhD from the U-M Program in Survey Methodology in 2011. Before that, he received an MA in Applied Statistics from the U-M Statistics Department in 2002, being recognized as an Outstanding First-year Applied Masters student. His current research interests include the implications of measurement error in auxiliary variables and survey paradata for survey estimation, survey nonresponse, interviewer variance, and multilevel regression models for clustered and longitudinal data. He is the lead author of a book comparing different statistical software packages in terms of their mixed effects modeling procedures (*Linear Mixed Models: A Practical Guide using Statistical Software*, Chapman Hall/CRC Press, 2006), with a second edition currently being written, and he is a co-author of a second book entitled *Applied Survey Data Analysis* (with Steven Heeringa and Pat Berglund), which was published by Chapman and Hall in April 2010.

Andrew J. Womack is a Postdoctoral Research Associate in the Department of Statistics at the University of Florida. His research focuses on objective Bayesian model selection, Bayesian hierarchical modeling, and cluster analysis. He is also interested in statistical applications related to gene expression data, signal processing, machine learning, and social science research. Andrew earned a BS from the University of Kansas and a PhD in Mathematics from Washington University in St. Louis.

Lang Wu received a BSc from East China Normal University, an MSc from Tulane University, and a PhD in Statistics from the University of Washington at Seattle. He then joined Harvard School of Public Health as a postdoctoral fellow. Since 2000, he has been an Assistant Professor, Associate Professor, and full Professor in the Department of Statistics at the University of British Columbia in Vancouver, Canada. His current research interests include longitudinal data analysis, missing data and measurement error problems, multivariate one-sided hypothesis testing, and join models. His book *Mixed Effects Models for Complex Data* (Chapman & Hall/CRC, 2009) provides an overview of mixed effects models, including nonlinear mixed effects models and generalized linear mixed models for longitudinal data and multilevel data.

Jin-Ting Zhang is currently an Associate Professor, Department of Statistics and Applied Probability, National University of Singapore, and a visiting fellow in the Department of Operation Research and Financial Engineering, Princeton University, USA. He earned his PhD in 1999 at the Department of Statistics, University of North Carolina at Chapel Hill, USA under the supervision of Professors Jianqing Fan and James Steve Marron. He has published two books and a number of research papers in top journals. His research interests include longitudinal data analysis, functional data analysis, high-dimensional data analysis, nonparametric smoothing, and heteroscedastic ANOVA and MANOVA, among others.

Dale L. Zimmerman is Robert V. Hogg Professor, Department of Statistics and Actuarial Science, University of Iowa. He received his PhD in Statistics from Iowa State University in 1986. His research interests include spatial statistics, longitudinal data analysis, multivariate analysis, mixed linear models, and environmental statistics. He has authored or co-authored two books and more than 70 scholarly articles in peer-reviewed journals. He was named a Fellow of the American Statistical Association in 2001, and he received the Distinguished Achievement Award from the Section on Statistics and Environment of the American Statistical Association in 2007. He is currently an associate editor for *Journal of the Royal Statistical Society, Series B* and for *Annals of Applied Statistics*.

Preface

Multilevel models (MLMs) are statistical models that, broadly speaking, are characterized by complex patterns of variability, usually focusing on nested structures of, for example, students in schools, animals in litters, longitudinal measurements of individuals or countries, and so on. Such models have a long history, being associated with early work on design and analysis of experiments, repeated measures, and Bayesian models. However, until recently, researchers could not ask or answer many important questions: often the data were too expensive to collect and analysis was hampered by limited computational resources. As late as the mid-1990s, there were only a few software packages for fitting multilevel models. Now, an applied researcher has many software choices, but given that each package was developed by researchers who posed their questions in slightly different frameworks and disciplinary contexts, there are major differences in notation, data structures, and so forth, associated with these packages.

Different approaches have led to similar but distinct strands in the literature to the point at which we have many different names for different forms or formulations of MLMs, such as random effects models, random coefficient models, and hierarchical linear models. The list is much larger if we examine subfields such as panel data models, which are often referred to as longitudinal models, pooled cross-sectional and time-series models, or growth curve models. Mixed effect, random effect, and fixed effect models all have a specific meaning in certain disciplines, but the distinction tends to confuse rather than clarify their purpose. Again, these distinctions often reflect the historical development of the framework, be it biology, psychology, sociology, policy science, market or econometric research, and so forth.

This volume is premised on the notion that researchers across and within a variety of disciplines have much to learn from one another, particularly in the area of multilevel modeling, where many related ideas converge. Moreover, the ways in which these ideas differ from each other are equally illuminating. By examining the common threads and subtle differences between methods that we commonly lump together under the rubric "multilevel modeling," we hope that the applied researcher will recognize where her problem fits in this arena, and perhaps even discover that the answer to a slightly reformulated question is more informative for her research. We expect that sharing information across disciplinary boundaries will lead to significant gains in productivity as well as open up new areas of inquiry. We note, however, that this is too often hampered by jargon, notation or more simply from a lack of communication and awareness of what has already been done in other fields. This volume seeks to break down disciplinary boundaries by organizing and framing the types of questions addressed by different constituencies so that they form a coherent body of knowledge, while still maintaining the unique contributions of each. In essence, our goal is to deconstruct the vast and disconnected literature with the goal of reconstructing it in this thorough and unified handbook.

Even with a substantial literature in this area, the answers to methodological questions that are fairly well-established for methods such as regression are much less definitely addressed for MLMs. For example, in regression, we have a fairly clear set of approaches to missing data, graphics and diagnostic methods, and matters of robustness and model selection. For MLMs, each of these poses a fairly unique challenge. For example, there is a substantial literature on pattern missingness that is crucial to the analysis of MLMs, while multiple imputation techniques for MLMs are less well developed and neither approach has fully penetrated mainstream software packages. This volume will gather the most current innovations in such areas and establish through both theory and practical example how to approach the variety of technical concerns that arise.

This volume is aimed at the Masters/Ph.D. level researcher, whether a statistician or a researcher in an applied field. We hope that it appeals to researchers who might have some exposure to the basics of multilevel modeling (as can be found in the many excellent introductory texts about MLMs that are available), but who are facing new research problems or who want to branch out to new fields, and can benefit from the volume's broad coverage, discussion of philosophical issues and controversies in the field, and detailed attention to applications in different fields. We also hope that it can serve researchers who wish to understand where their discipline-specific models fit in the broader universe of methods and models. We reiterate that each discipline and approach provides unique insights and perspectives.

We are pleased to have a truly impressive set of authors of chapters. They include many of the leading lights in the field, and many are fundamental contributors to the topics about which they have written. They bring a broad perspective to the material, representing eleven countries and four continents, coming from academia, research institutions, industry, and government, with backgrounds and experience in fields including biostatistics, economics, education, marketing, political science, psychology, and statistics.

Material related to the book can be found at its associated website, which provides access to the data sets used in the book, examples of computer code for performing analyses in the book, and an errata list. The reader is encouraged to visit http://people.stern.nyu.edu/jsimonof/MultilevelHandbook/.

We would like to thank Patrick Brindle for bringing this project to us, and thank Anna Horvai and him for their help in bringing it to fruition. We would also like to thank the authors of the chapters for sharing their time and expertise cheerfully and willingly. We would like to thank Jeff Gill, Mark Handcock, Jennifer Hill, Thomas Kneib, Charles McCulloch, Geert Molenberghs, John Neuhaus, Ardo van den Hout, and Geert Verbeke in particular for their editorial assistance in the preparation of this volume. We are hopeful that readers will benefit from reading this book, and are especially confident regarding that, given how much we have learned just from editing it. Finally, we would like to thank our families for their love and support.

Marc A. Scott
New York, New York

Jeffrey S. Simonoff
New York, New York

Brian D. Marx
Baton Rouge, Louisiana

October, 2012

Multilevel Modeling

Jeffrey S. Simonoff
New York University, USA

Marc A. Scott
New York University, USA

Brian D. Marx
Louisiana State University, USA

1 INTRODUCTION

In this chapter we provide some introductory background for the chapters to come. We start with discussion of how multilevel modeling can be viewed as a natural generalization of three different types of models: regression, analysis of variance, and time series models. We then note how algorithmic advances and the introduction of statistical software has had a particularly strong effect on the development of multilevel modeling. This is followed by a description of the design and goals of the book, and we close with some discussion of what is not covered here, including future possibilities, applications, and opportunities.

2 REGRESSION, ANOVA, AND TIME SERIES MODELS

Regression analysis is undoubtedly the most widely used statistical method. The standard regression model

$$y_i = \beta_0 + \beta_1 x_{1i} + \cdots + \beta_p x_{pi} + \varepsilon_i \quad (1)$$

with $\boldsymbol{\varepsilon}$ forming a random sample from a Gaussian population with mean 0 and variance σ^2, despite its simplicity, is a remarkably powerful and flexible tool for data summary and prediction. Least-squares-based methods are well understood, and possess well-known optimality properties. Of course, when the assumptions underlying (1) do not hold, these favorable characteristics do not necessarily hold, and the model and estimation scheme need to be changed. One such situation is when observations arise in a *nested* or *clustered* fashion.

Such data occur often in practice. For example, in an educational setting, information and responses might be available at the level of each student, but an experimental condition might be administered at the classroom level. Since all of the students within a class are exposed to the same classroom conditions (that is, students are nested within classrooms), it would be expected that the errors ε_i for students from the same classroom would be correlated with each other (reflecting unmodeled differences between classrooms, such as the effect of the teacher), a violation of the standard regression model assumptions.

Such nesting need not only occur just at one level, as students can be nested within classrooms, which are nested within schools, which are nested within school districts, and so on. Further, in a situation like this the importance of each level of nesting could be different for different users. Effects at the level of an individual student are likely to be of greater interest to students and teachers, but from a policy point of view relationships at the school or district level might be a more important focus. What is needed is a way to generalize (1) to allow for this nested structure.

Consider now a simple experimental setting, where the response of interest is a numerical measure of patient health such as blood pressure level. The purpose of the experiment is to explore the relationship between blood pressure and several types of physical and mental exercise, such as yoga, meditation, aerobic exercise, and so on. In this situation a natural representation of the relationships in the data is the one-way analysis of variance (ANOVA) model

$$y_{ij} = \mu + \alpha_j + \varepsilon_{ij}, \tag{2}$$

where y_{ij} is the response for the ith observation in the jth exercise group (for example, yoga) and α_j is the *fixed effect* of being in the jth group. This is, in fact, a special case of the regression model (1), with the predictors being effect codings that code the groups (Chatterjee and Simonoff, 2013, Section 6.3.1). It is well known that for this model the best estimated expected response for any member of the jth group is $\overline{y}_j$, the mean response for observations from that group.

Imagine instead a situation where a specific treatment regimen is applied to patients in different treatment facilities, with each treatment facility defining a group. In this situation it is reasonable to view the set of facilities as constituting a random sample from the population of treatment facilities. A model that is consistent with this is

$$y_{ij} = \mu + u_j + \varepsilon_{ij}, \tag{3}$$

where u_j is the *random effect* of being in the jth group (treatment facility), and is a Gaussian random variable with mean 0 and variance σ_u^2 that is independent of $\boldsymbol{\varepsilon}$.

The key distinction between models (2) and (3) is that the fixed effects model (2) has only one source of variability (ε_{ij}, with variance σ^2), while the random effects model (3) has two *components of variance* (ε_{ij} and u_j, with variances σ^2 and σ_u^2, respectively). These two sources of variability reflect the variability between observations that are in the same group, measured by the within-group variance σ^2, and the variability between observations that are in different groups, measured by the between-group variance σ_u^2. A single number that summarizes the relative importance of the two variance components is the *intraclass correlation* (ICC)

$$\rho = \frac{\sigma_u^2}{\sigma_u^2 + \sigma^2},$$

which is the proportion of variability accounted for by the variability between groups. It is also equal to the correlation between two observations within the same group, reinforcing the correspondence of variability between groups and correlation within groups.

It can be shown that in this situation unbiased estimates of $\mathbf{u}$ satisfy

$$\hat{u}_j \approx \frac{\left(\frac{n_j}{\sigma_u^2+\sigma^2}\right)\overline{y}_j + \left(\frac{1}{\sigma_u^2}\right)\overline{y}}{\frac{n_j}{\sigma_u^2+\sigma^2} + \frac{1}{\sigma_u^2}} = \frac{n_j\rho\overline{y}_j + \overline{y}}{n_j\rho + 1} \tag{4}$$

(Gelman and Hill, 2007, p. 253). Note that, in practice, estimates of the variance components are necessary to construct these estimates. Equation (4) shows that the random effects model provides a sensible compromise between the fixed effects strategy of treating each group separately (yielding $\overline{y}_j$ as the predicted y for all members of the jth group) and the null strategy of assuming that there are no group effects at all (yielding the overall mean $\overline{y}$ as the predicted y for all observations). Gelman and Hill (2007, Section 12.2) refer to this "borrowing strength" from other groups as partial pooling, reflecting a compromise between no pooling and complete pooling. The compromise is "sensible" because of the form of the weights in (4). In particular, groups with larger sample sizes are given more weight, and the smaller ρ is (and hence the smaller the between-group variability is relative to the total variability), the more relative weight is given to complete pooling. As Robinson (1991) notes, these notions are related to the concept of regression to the mean, as they imply that the best estimate of a characteristic of an offspring is based on parental averages that are then regressed (shrunk) towards the population mean.

The weakness of model (3) is that it does not allow for the possibility of more complex fixed-effects structure or more complex random-effects structure. What is needed is a way to generalize (3) to allow for these more complex possibilities.

Another way the assumptions of the standard regression model (1) can be violated is if there is a time ordering to the observations, resulting in autocorrelation of the errors. That would, of course, require the application of time series methodology, of which there is a vast literature.

A more challenging situation is when data have both a cross-sectional and time series structure; that is, when individual units are repeatedly sampled at different points in time. Such data are called *longitudinal data*. They are a special case of *repeated measures data*, in which the repeated measurements for an individual unit are of the response at different time points (rather than, for example, under different experimental conditions). In economic data the term *panel data* is typically used for this type of situation.

Since the measurements at different time points are taken on the same individual, the issues of an induced correlation of observations within each individual noted earlier (measured by the ICC ρ) still apply. In addition to this, however, the possibility of autocorrelation of errors within an individual must also be considered. This suggests the use of multiple time series models, such as are described in Lütkepohl (2005), including vector autoregressive and moving average processes, cointegrated processes, multivariate ARCH and GARCH processes, among others.

The weakness of these models is that they are designed for the situation where there are a relatively small number of series, measured at a relatively large number of time points. Longitudinal data are typically the opposite situation—a relatively large number of individuals measured at a relatively small number of time points. What is needed are models that are designed for the latter situation. The models need to be flexible enough to handle when subject-level inference is of less interest and time effects are less important (as is often true in longitudinal data), when subjects might constitute the entire population (implying the use of fixed rather than random effects), and complex time structures reflecting, for example, the occurrence of seasonal and shock effects (as is often true for economic panel data).

All three of these scenarios (generalizing regression, generalizing random effects ANOVA, and generalizing time series models) lead directly to the use of *multilevel models*. The natural generalization of models (1), (2), and (3) is

$$y_{ij} = \beta_0 + \beta_1 x_{1ij} + \cdots + \beta_p x_{pij} + u_{0j} + \varepsilon_{ij}, \tag{5}$$

the *random intercepts model* (since $\beta_{0j} \equiv \beta_0 + u_{0j}$ defines different random intercepts for each group). This can be generalized further by allowing for different types of time-related structure in the joint distributions of $\mathbf{u}$ or $\boldsymbol{\varepsilon}$ (or both). The random intercepts model (5) can also be generalized to allow for more complex random effects, such as random slopes and multiple levels of nesting.

Since multilevel models generalize regression, ANOVA, and time series models, lessons from each of those classes of models still apply. In particular, preliminary graphical examination of data, diagnostic measures and tests of assumptions, transformation of variables, and so on, all have their role in multilevel data analysis. Each of these issues will be discussed in future chapters.

3 THE DEVELOPMENT OF MULTILEVEL MODELING

Many of the succeeding chapters discuss the historical development of different types of multilevel model, so we will only briefly provide some context here. More detailed discussion of this can also be found in de Leeuw and Kreft (1986), Engel (1990), and Robinson (1991). As the discussion in the previous section would suggest, the different ways that multilevel models generalize more basic models corresponds to their adoption at different times in different subject areas.

The earliest examples come from the use of random effects in agricultural experiments. The original fundamental work is due to R.A. Fisher, described in Chapter 7 of the 1925 edition of his *Statistical Methods for Research Workers* (Fisher, 1925). Eisenhart (1947) built on this work by explicitly considering the distinction between fixed and random effects. The best linear unbiased predictor (shrinkage) technique was introduced by Henderson (1950) in the context of animal breeding. Lindley and Smith (1972) demonstrated the close connection of this technique to Bayes and Empirical Bayes methods, and Efron and Morris (1975) discussed the connection to James–Stein estimation.

Models with random parameters have long been a part of the econometric treatment of panel data. Early examples include Hildreth and Houck (1968), Swamy (1970, 1971), Hsiao (1974, 1975), Cooley and Prescott (1976), and Amemiya (1978). MaCurdy (1981) explicitly discussed the connection between multiple time series models and the analysis of panel data.

The recognition that the use of regression models that ignore hierarchical structures can lead to incorrect inferences in Aitkin et al. (1981) had a profound effect on educational research in the 1980s. The naturally hierarchical structure of educational data noted in the previous section resulted in a good deal of research on the application of the *hierarchical linear model* to educational data, including Burstein (1980), Goldstein (1985, 1987), Aitkin and Longford (1986), de Leeuw and Kreft (1986), Raudenbush and Bryk (1986), Raudenbush (1988), and Bryk and Raudenbush (1992).

Laird and Ware (1982) extended earlier work on growth curves (Rao, 1965; Fearn, 1975) and repeated measures (Harville, 1977) to linear random effects models for longitudinal data. This work has provided the basis for many extensions, including to binary data (Stiratelli et al., 1984), nonlinear models (Lindstrom and Bates, 1990), missing and censored data (Brown, 1990), nonparametric regression (Rice and Wu, 2001), and regression trees (Hajjem et al., 2011; Sela and Simonoff, 2012).

Advances in computing in general, and computer software packages in particular, have had a very strong effect on the development of multilevel modeling. The early ANOVA-based algorithms required balanced designs; adaptations to unbalanced data lose the benefit of exact distributional results and can lead to negative variance estimates (Searle, 1971, 1987). Harville (1977) discussed maximum likelihood and restricted maximum likelihood estimation. Laird and Ware (1982) described the use of the EM (expectation–maximization) algorithm of Dempster et al. (1977) to calculate these estimates, and the HLM software package introduced in 1988 (Bryk et al., 1988) was based on this approach. The uses of iterative generalized least squares (Goldstein, 1986) and Newton–Raphson and Fisher scoring algorithms (Jennrich and Schluchter, 1986) to conduct likelihood-based inference also date from this time, with the former method being the basis of ML2 (Rasbash et al., 1989) and its successors such as MLwiN (Rasbash et al., 2004), the latter method being the basis of BMDP 5V (Schluchter, 1988). The availability of these packages greatly expanded the applicability of multilevel models in the late 1980s.

In the 1990s and 2000s these packages (which were specifically designed for multilevel models, other than BMDP 5V) were joined by general-purpose packages with modules or functions designed for multilevel modeling, including SAS, R, SPSS, and Stata. These functions were eventually expanded to include coverage of nonlinear mixed models and generalized linear mixed models for non-Gaussian data. Chapter 26 discusses these and other packages in more detail.

The computational revolution in Bayesian analysis brought about by Markov Chain Monte Carlo methods has also had a strong effect on multilevel modeling, which is unsurprising given the natural Bayesian interpretation of random effect distributions as prior distributions. Several packages include these capabilities, including MLwiN and R.

4 THE DESIGN AND GOALS OF THIS BOOK

The chapters are divided into four parts, the first three of which involve aspects of model development and the last of which emphasizes applications. The subject of model development is viewed in a very general way, including, in addition to methodology, aspects of the history of multilevel modeling, design and data collection, and software. Some chapters highlight innovative cutting-edge methodology while also acknowledging and addressing current controversies in the field. To the extent possible, the chapters utilize a common set of notation, and a chapter is dedicated to establishing the connections and equivalence of the most common notations used in practice.[1] The applications chapters highlight both generality and generalizability, thereby encouraging different communities of researchers to learn from the research programs of others; that is, the centerpiece of each of the chapters is the underlying methodology.

Part I of the Handbook, *Multilevel Model Specification and Inference*, is a self-contained overview of multilevel modeling, which includes topics of key interest to practitioners, such as design considerations and causal inference, that are sometimes available only through more specialized outlets. The overall framework for multilevel modeling is provided in Chapter 1, including a concise introduction to different types of multilevel model and their associated terminology. This is followed by a chapter on notation, which establishes the rules and rationale guiding notation one finds in practice, and discusses different contexts that would recommend one notational form over another. Nearly all chapters utilize a form discussed in this chapter, and when they deviate, specific rationale is provided.[2] Frequentist and Bayesian approaches to estimation and inference are developed next in Chapters 3 and 4, respectively.

Chapters 5 and 6 are two perspectives on what at first appear to be unrelated issues: placing distributional assumptions on group effects and centering predictors. The former emphasizes the modeling assumptions and their implications for interpretation, wherein it is revealed that between-group relationships and their within-group analog are implicitly tied to these assumptions through estimation choices. The chapter ends with a discussion of centering that potentially unifies what has often been viewed as a contentious topic. This chapter is followed by a full chapter on centering, which exposes the underlying complexity of the choice with an emphasis on the implicit research questions that result. The overarching theme is whether effects can be characterized as between- or within-subjects, tying the discussion to the issues presented in the prior chapter. Chapter 7 discusses how the standard approaches to model selection and measures of fit require significant modification in the realm of multilevel modeling, particularly in the area of accounting for degrees of freedom. Examples provide further insight as to what may be appropriate in different multilevel contexts (e.g., nested versus panel data).

At this point in the presentation, the reader has most of the tools needed to model Gaussian outcomes, so Chapter 8 provides an overview of the extension of generalized linear models to include multilevel structure. Following this is a chapter on longitudinal data models. It provides a thorough introduction to these models, including the somewhat unique considerations of time-series modeling and loss to follow-up. It also includes a discussion of generalized linear mixed models and estimation using generalized estimating equations.

Returning to the theme of re-examining standard approaches and assumptions, Chapter 10 discusses a variety of error structures that fill the spectrum between simpler compound symmetry (or random intercept) models and fully unstructured covariance. The first section ends with a chapter on design considerations and one on causal inference. Optimal design is especially important in grouped or multilevel data, where the group sizes may be somewhat small, and this chapter covers important designs such as the cluster randomized trial. As many research questions involve binary outcomes, this situation is included in the discussion. Chapter 12 tackles the challenge of causal inference and is a thorough overview of the additional considerations one must confront with observational as well as randomized studies in the multilevel setting. This chapter builds on material from Chapter 5.

The next part of the Handbook develops methodology for handling variations and extensions of multilevel models. While some of the chapters may appear to be quite specialized, many practitioners are trying to draw inference from increasingly complex data structures and the relationships they represent. In other words, practitioners increasingly find themselves facing new types of multilevel data and their research questions require new methodology. Chapter 13 on multilevel functional data analysis (MFDA) is an excellent example of the "outcome" variable being more complex than a continuous scalar measurement. In MFDA the outcome is a function, perhaps indexed by time or space. In a similar vein, Chapter 21 extends multilevel models to accommodate multiple measures on the same subject when they are nested, by group or time. The chapter notably does not limit the outcomes to be Gaussian. Just as outcomes can be more complex, the models themselves may be; the natural extension of linear models to nonlinear models can also incorporate random effects and thus multilevel structure, and this is the topic of Chapter 14. Perhaps the more commonly known nonlinear multilevel models, generalized linear mixed models (GLMMs), are discussed next; these were introduced in Part I of the Handbook, but now estimation methods are given in greater detail. In most discussions of GLMMs, binary and count outcomes are developed quite thoroughly, while categorical outcomes with more than two levels are not. Yet multilevel categorical data are quite common among practitioners, so we dedicate the subsequent chapter to this topic exclusively.

Chapters 17–20 extend features of multilevel models that some practitioners take for granted. For example, while linear models can capture a variety of mean structures, nonparametric approaches allow one to more closely capture an underlying relationship. Semiparametric models for the mean may be more realistic and may improve the modeling of covariance as well. Chapter 17 develops a framework for nonparametric extensions of mean and covariance structures. Chapter 18 develops related semiparametric models for the mean using a roughness penalty approach that is a multilevel analog of generalized additive models (Hastie and Tibshirani, 1990). Chapter 19 returns to the linear model, but now allows the predictor effects to vary with time, all within the multilevel framework. Chapter 20 extends the class of random effects distributions available to multilevel modelers using a latent class approach. The simplest way to characterize this extension is, "what if you wished to draw your random effects from a mixture of Gaussian densities?" This chapter merges ideas from latent growth curve modeling (Jones and Nagin, 2007) and model-based clustering (Banfield and Raftery, 1993) to form a coherent framework for more realistic longitudinal data modeling.

Having established both the general framework and extensions of multilevel models, Part III of the Handbook covers model fitting and specification issues. For example, Chapter 22 begins with a general discussion of robustness and then covers essential topics such as robust standard error estimates, bootstrapping, and using Bayesian approaches (e.g., replacing normal distributions with t- or Cauchy distributions). All of these ideas are developed with the particular features of multilevel modeling in mind. Missing data are quite common in applied statistical analyses, and Chapter 23 provides a comprehensive review of approaches to dealing with missingness. After defining the types of missingness that may occur, the authors discuss historically common approaches such as last observation carried forward to multiple imputation and then pattern mixture models. Next, Chapter 24 offers a detailed discussion of the assumptions in multilevel models and how they can be assessed. The chapter emphasizes important graphical techniques and clearly describes the limitations inherent in some approaches. Chapter 25 offers a unique perspective on a substantive concern—endogeneity—and an estimation technique known as Generalized Estimating Equations (GEE). The perspective offered, that GEE is potentially robust with respect to this issue, is examined closely via a simulation study. The chapter has ties to many chapters in the book, including those emphasizing generalized linear models, the choice between fixed and random effects, and causality (Chapters 15, 5, and 12, respectively).

We close this part of the Handbook with a practical discussion of software packages in Chapter 26. In particular, the authors emphasize the connection between estimation methods and software to implement them, as well as specialty software for more recently introduced multilevel models. For example, there is discussion of generalized additive mixed models, models for dyads or social networks, and implementation of complex survey weights, to name a few.

Part IV of the Handbook contains selected applications that illustrate domains of inquiry making novel use of multilevel models. For example, Chapter 27 describes how meta-analysis clearly benefits from multilevel modeling approaches; what is less apparent is that the "level one units" consist of study findings and the within-study uncertainty, but not the within-study subject responses. In Chapter 28, the author presents an overview of the ways in which policy analysts utilize multilevel models, and points out several considerations that are both common in that field and commonly overlooked. Examples include micronumerosity and spatial correlation. Coverage of these topics is broad enough that this chapter serves as a useful springboard for delving more deeply in chapters covering fixed versus random effects, causality, and spatial modeling (Chapters 5, 12 and 31, respectively). Chapter 29 offers a broader overview than its predecessor, providing insight into the opportunities and pitfalls that accompany multilevel modeling in the social and behavioral sciences. This chapter might be read immediately after the first few chapters in the Handbook.

The last four chapters are similar in spirit to Part II of the Handbook in that they involve extending an established methodology to accommodate multilevel data and structures. Chapter 30 discusses the frailty model in survival analysis—perhaps one of the earliest extensions of an existing class of models to handle a form of nesting; data of this nature are quite common in the medical sciences, but the method has broad applicability. In some problems, the unit of analysis is a location in space (e.g., weather). When multiple measures are collected at specific points, we have grouped data, but observations across groups are correlated due to spatial proximity. Multilevel models for such data must thus include models for several forms of covariance. Chapter 31 discusses these, complementing the general treatment of multivariate outcomes given in Chapter 21. An independent area of research of its own, structural equation models are particularly useful for capturing relationships between latent variables, making them popular in the behavioral sciences. In Chapter 32, our market research application, the author develops the multilevel structural equation model and applies it to preference data. Similar to spatial data models, network and relational data models represent complex correlation structure, but they must represent relationships on graphs, which is distinct from nearly every other topic discussed in this book (one way to see this is to consider network data to be a realization of an adjacency matrix of relations). Chapter 33 applies multilevel models for friendship networks of students nested in classrooms. These last four chapters are excellent examples of how the addition of multilevel structures improves our understanding of complex human behaviors.

5 WHAT ISN'T IN THIS BOOK

Although we have made a genuine effort to leave "no stone unturned," there remain some multilevel modeling topics that are either not fully covered or completely absent from this handbook, and this section surveys some of these missing topics. Broadly speaking, these coverage gaps can be delineated into two camps: ones due to space limitations and areas of future research.

We start with "big data," which has become a gigantic issue in nearly all areas of statistics, and multilevel modeling is no exception. With such data, computational issues clearly become increasingly important, especially in combination with complex models. As a case in point, Strandén and Lidauer (1999) have addressed real time and applied challenges in solving large mixed models on continuous evaluation of dairy cattle with random regression, which requires both a fast algorithm and solution method. These authors cleverly used conjugate gradient iterative methods, efficiently calculating the multiplication of a vector with a matrix in (three) reordered steps. Generalized linear mixed models (GLMM) require additional attention to computation speed. Ver Hoef and London (2010) considered pseudo-GLMMs, with many binary repeated measures on each subject, that have computational time that increases linearly with the number of observations. These authors impressively tackled computational issues for such binary time series, containing over 100 fixed effects, 50 random effects parameters, and 1.5×10^5 observations. In other related work, Alam (2010) developed a two-step pseudo-likelihood estimation technique for the GLMM that is more computationally efficient than conventional algorithms. This author specifically addressed models having random effects that are (weakly) correlated between groups, while presenting a binary response application to Swedish credit default data. More recently, Schelldorfer and Bühlmann (2011) developed a lasso-type approach useful for variable screening in high-dimensional GLMMs. The approach is based on an ℓ_1-penalized algorithm, and is implemented in the R package `glmmlasso`. Also related to rich data are problems that contain intensive longitudinal data. A portion of Chapter 13 is dedicated to such intensive data, where in some cases outcomes take the form of a curve for each subject. We find commonality in all of the above research efforts: efficient computation in rich multilevel modeling settings.

Another approach useful for handling "big data," especially when (generalized) linear mixed models struggle to track the data structure, is the use of regression trees. Tree-based models are based on a recursive partitioning of the data and have strengths and popularity for several reasons: they are more adept at (automatically) capturing nonadditive or interactive structure, can also easily handle a mix of numeric covariates and factors, while being invariant to (monotone) re-expressions of the response. Extensions have been put forth into hierarchical settings, using the EM algorithm. Hajjem et al. (2011) developed a mixed effects regression tree for nested data within clusters. Sela and Simonoff (2012) developed methodology that combines the structure of mixed effects models for both longitudinal and clustered data, with the flexibility of tree-based estimation methods. The strength of using a tree-based approach becomes apparent in this setting as this mixed effect approach, for example, can handle unbalanced clusters, allows clusters to be split, and can incorporate random effects and also observation-level covariates.

More in the spirit of "enormous data," Chapter 33 of this handbook develops multilevel models for social networks. It is natural to wonder how such statistical models can be used and extended with massive network data, such as that from social media platforms like Facebook and Twitter. Although the models described in Chapter 33 have demonstrated effectiveness in describing and predicting smaller networks, their complexity will grow exponentially for networks with millions of individuals and potentially hundreds of millions of ties between those individuals.

Another area that is not extensively covered here is *marginal structural models*, which are an alternative to structural nested models, and thus relevant to our handbook. Using observational data, Robins et al. (2000) developed a new class of causal models for the (consistent) estimation of the causal effect of, e.g., a time-dependent exposure in the presence of time-dependent covariates, which may additionally be simultaneously confounders and intermediate variables. Chapters 5 and 12 of this handbook do contain some discussion of marginal structural models, but not in any great detail.

We do not cover problems related to *small area estimation* (SAE) in multilevel models in this volume. Essentially, the SAE problem is to estimate unknown parameters (e.g., the mean or quantiles) for areas in which only small (or no) samples are available. Precision of estimation is also of interest in this context. Pfeffermann (2012) has compiled a contemporary review containing important developments in SAE during the last decade, and he considers both design-based and model-dependent approaches (the latter with both frequentist and Bayesian methods). A selection of papers that are specifically related to small area estimation in the multilevel models setting include Pfeffermann et al. (1998), Feder et al. (2000), and Pfeffermann et al. (2008). An issue related to SAE is that of weighting for complex survey design. Rabe-Hesketh and Skrondal (2006) nicely presented the topic of multilevel modeling of complex survey data, and the references therein contain, more generally, a broad coverage of the role of sampling weights.

Lastly, coverage of measurement error (quantitative variables) and classification error (categorical variables) is absent from this handbook. Fox and Glas (2002) devoted an entire chapter on the modeling of measurement error in structural multilevel models, which is a nice overview and also provides many references. Somewhat more recently, Goldstein et al. (2008) developed models that adjusted for measurement errors in normally distributed predictors and response variables and categorical predictors with misclassification errors. Using Markov Chain Monte Carlo, the models allowed for hierarchical data structure and for correlations among the errors and misclassifications. See also Chapter 14 of Goldstein (2011).

NOTES

1 Some chapters are more technically sophisticated than others, sometimes relying heavily on matrix notation. The reader should consult Chapter 2 which, in addition to introducing that notation, provides some references for deeper understanding and further examples.

2 A good example of such deviation is Chapter 12 on causal inference, which utilizes the potential outcomes notation discussed in Rubin (1978), modified to incorporate a multilevel model framework.

REFERENCES

Aitkin, M., Anderson, D., and Hinde, J. (1981) 'Statistical Modeling of Data on Teaching Styles', *Journal of the Royal Statistical Society, Series A*, 144: 419–61.

Aitkin, M. and Longford, N. (1986) 'Statistical Modeling Issues in School Effectiveness Studies', *Journal of the Royal Statistical Society, Series A*, 149: 1–43.

Alam, M.M. (2010) 'Feasible Estimation of Generalized Linear Mixed Models (GLMM) with Weak Dependency Between Groups', `http://oru.diva-portal.org/smash/get/diva2:389313/FULLTEXT01`, accessed September 14, 2012).

Amemiya, T. (1978) 'A Note on a Random Coefficients Model', *International Economic Review*, 19: 793–6.

Banfield, J.D. and Raftery, A.E. (1993) 'Model-based Gaussian and Non-Gaussian Clustering', *Biometrics*, 49: 803–21.

Brown, C.H. (1990) 'Protecting Against Nonrandomly Missing Data in Longitudinal Studies', *Biometrics*, 46: 143–55.

Bryk, A.S. and Raudenbush, S.W. (1992) *Hierarchical Linear Models*. Newbury Park, CA: Sage.

Bryk, A., Raudenbush, S., Seltzer, M., and Congdon, R. (1988) *An Introduction to HLM: Computer Program and Users Manual*. Chicago: University of Chicago Dept. of Education.

Burstein, L. (1980) 'The Analysis of Multi-Level Data in Educational Research and Evaluation', *Review of Research in Education*, 8: 158–233.

Chatterjee, S. and Simonoff, J.S. (2013) *Handbook of Regression Analysis*. Hoboken, NJ: John Wiley and Sons.

Cooley, T.F. and Prescott, E.C. (1976) 'Estimation in the Presence of Stochastic Parameter Variation', *Econometrica*, 44: 167–84.

de Leeuw, J. and Kreft, I.G.G. (1986) 'Random Coefficient Models for Multilevel Analysis', *Journal of Educational Statistics*, 11: 57–85.

Dempster, A.P., Laird, N.M., and Rubin, D.B. (1977) 'Maximum Likelihood with Incomplete Data via the E-M Algorithm', *Journal of the Royal Statistical Society, Series B*, 39: 1–38.

Efron, B. and Morris, C. (1975) 'Data Analysis Using Stein's Estimator and Its Generalizations', *Journal of the American Statistical Association*, 70: 311–9.

Eisenhart, C. (1947) 'The Assumptions Underlying the Analysis of Variance', *Biometrics*, 3: 1–21.

Engel, B. (1990) 'The Analysis of Unbalanced Linear Models With Variance Components', *Statistica Neerlandica*, 44: 195–219.

Fearn, T. (1975) 'A Bayesian Approach to Growth Curves', *Biometrika*, 62: 89–100.

Feder, M., Nathan, G., and Pfeffermann, D. (2000) 'Multilevel Modeling of Complex Survey Longitudinal Data with Time Varying Random Effects', *Survey Methodology*, 26: 53–65.

Fisher, R.A. (1925) *Statistical Methods for Research Workers*. Edinburgh: Oliver and Boyd.

Fox, J.-P. and Glas, C.A.W. (2002) 'Modeling Measurement Error in a Structural Multilevel Model', in Marcoulides, G.A. and Moustaki, I. (eds), *Latent Variable and Latent Structure Models*, London: Lawrence Erlbaum Associates, 245–69.

Gelman, A. and Hill, J. (2007) *Data Analysis Using Regression and Multilevel/Hierarchical Models*. Cambridge: Cambridge University Press.

Goldstein, H. (1985) *Multilevel Statistical Models*. New York: Halstead Press.

Goldstein, H. (1986) 'Multilevel Mixed Model Analysis Using Iterative Generalized Least Squares', *Biometrika*, 73: 43–56.

Goldstein, H. (1987) *Multilevel Models in Education and Social Research*. Oxford: Oxford University Press.

Goldstein, H. (2011) *Multilevel Statistical Models*, 4th ed. Chichester: John Wiley and Sons.

Goldstein, H., Kounali, D., and Robinson, A. (2008) 'Modeling Measurement Errors and Category Misclassifications in Multilevel Models', *Statistical Modeling*, 8: 243–61.

Hajjem, A., Bellavance, F., and Larocque, D. (2011) 'Mixed Effects Regression Trees for Clustered Data', *Statistics and Probability Letters*, 81: 451–9.

Harville, D.A. (1977) 'Maximum Likelihood Approaches to Variance Component Estimation and to Related Problems', *Journal of the American Statistical Association*, 72: 320–40.

Hastie, T.J. and Tibshirani, R.J. (1990) *Generalized Additive Models*. Boca Raton, FL: Chapman & Hall/CRC.

Henderson, C.R. (1950) 'Estimation of Genetic Parameters (abstract)', *Annals of Mathematical Statistics*, 21: 309–10.

Hildreth, C. and Houck, J.P. (1968) 'Some Estimators for a Linear Model with Random Coefficients', *Journal of the American Statistical Association*, 63: 584–95.

Hsiao, C. (1974) 'Statistical Inference for a Model With Both Random Cross-Sectional and Time Effects', *International Economic Review*, 15: 12–30.

Hsiao, C. (1975) 'Some Estimation Methods for a Random Coefficients Model', *Econometrica*, 43: 305–25.

Jennrich, R.I. and Schluchter, M.D. (1986) 'Unbalanced Repeated-Measures Models With Structured Covariance Matrices', *Biometrics*, 42: 805–20.

Jones, B.L. and Nagin, D.S. (2007) 'Advances in Group-Based Trajectory Modeling and an SAS Procedure for Estimating Them', *Sociological Methods Research*, 35: 542–71.

Laird, N.M. and Ware, J.H. (1982) 'Random Effects Models for Longitudinal Data', *Biometrics*, 38: 963–74.

Lindley, D.V. and Smith, A.F.M. (1972) 'Bayes Estimates for the Linear Model', *Journal of the Royal Statistical Society, Series B*, 34: 1–41.

Lindstrom, M.J. and Bates, D.M. (1990) 'Nonlinear Mixed Effects Models for Repeated Measures Data', *Biometrics* 46: 673–87.

Lütkepohl, H. (2005) *New Introduction to Multiple Time Series Analysis*. Berlin: Springer.

MaCurdy, T.E. (1981) 'Multiple Time-Series Models Applied to Panel Data', *NBER Working Paper No. 646* (http://www.nber.org/papers/w0646, accessed August 27, 2012).

Pfeffermann, D., Skinner, C., Goldstein, H., Holmes, D., and Rasbash, J. (1998) 'Weighting for Unequal Selection Probabilities in Multilevel Models (with discussion)', *Journal of the Royal Statistical Society, Series B*, 60: 23–56.

Pfeffermann, D., Terryn, B., and Moura, F. (2008) 'Small Area Estimation Under a Two-Part Random Effects Model with Application to Estimation of Literacy in Developing Countries', *Survey Methodology*, 34: 233–47.

Pfeffermann, D. (2012) 'New Important Developments in Small Area Estimation', *Statistical Science*, 27: to appear.

Rabe-Hesketh, S. and Skrondal, A. (2006) 'Multilevel Modeling of Complex Survey Data', *Journal of the Royal Statistical Society, Series A*, 169: 805–27.

Rao, C.R. (1965) 'The Theory of Least Squares When the Parameters Are Stochastic and its Application to the Analysis of Growth Curves', *Biometrika*, 52: 447–58.

Rasbash, J., Prosser, R., and Goldstein, H. (1989) *ML2: Software for Two-Level Analysis*. London: Institute of Education, University of London.

Rasbash, J., Steele, F., Browne, W.J., and Prosser, B. (2004) *A User's Guide to MLwiN, Version 2.0*. Bristol, UK: Centre for Multilevel Modeling.

Raudenbush, S.W. (1988) 'Education Applications of Hierarchical Linear Models: A Review', *Journal of Educational Statistics*, 12: 85–116.

Raudenbush, S.W. and Bryk, A.S. (1986) 'A Hierarchical Model for Studying School Effects', *Sociology of Education*, 59: 1–17.

Rice, J.A. and Wu, C.O. (2001) 'Nonparametric Mixed Effects Models for Unequally Sampled Noisy Curves', *Biometrics*, 57: 253–9.

Robins, J.M., Hernán, M., and Brumback, B. (2000) 'Marginal Structural Models and Causal Inference in Epidemiology', *Epidemiology*, 11: 550–60.

Robinson, G.K. (1991) 'That BLUP Is a Good Thing: Estimation of Random Effects', *Statistical Science*, 6: 15–51.

Rubin, D.B. (1978) 'Bayesian Inference for Causal Effects: The Role of Randomization', *The Annals of Statistics* 6: 34–58.

Schelldorder, J. and Bühlmann, P. (2011) 'GLMM-Lasso: An Algorithm for High-Dimensional Generalized Linear Mixed Models Using ℓ_1-Penalization', (http://arxiv.org/abs/1109.4003, accessed September 6, 2012).

Schluchter, M.D. (1988) *BMDP 5V: Unbalanced Repeated Measures: Models with Structured Covariance Matrices*. BMDP Statistical Software.

Searle, S.R. (1971) *Linear Models*. New York: John Wiley and Sons.

Searle, S.R. (1987) *Linear Models for Unbalanced Data*. New York: John Wiley and Sons.

Sela, R.J. and Simonoff, J.S. (2012) 'RE-EM Trees: A Data Mining Approach for Longitudinal and Clustered Data', *Machine Learning*, 86: 169–207.

Stiratelli, R., Laird, N., and Ware, J.H. (1984) 'Random-Effects Models for Serial Observations With Binary Response', *Biometrics*, 40: 961–71.

Strandén, I. and Lidauer, M. (1999) 'Solving Large Mixed Linear Models Using Preconditional Conjugate Gradient Iteration', *Journal of Dairy Science*, 82: 2779–87.

Swamy, P.A.V.B. (1970) 'Efficient Inference in a Random Coefficient Regression Model', *Econometrica*, 38: 311–23.

Swamy, P.A.V.B. (1971) *Statistical Inference in Random Coefficients Regression Models*. New York: Springer.

Ver Hoef, J.M. and London, J.M. (2010) 'Fast Computing of Some Generalized Linear Mixed Pseudo-Models with Temporal Autocorrelation', *Computational Statistics*, 25: 39–55.

PART I

Multilevel Model Specification and Inference

1

The Multilevel Model Framework

Jeff Gill
Washington University, USA

Andrew J. Womack
University of Florida, USA

1.1 OVERVIEW

Multilevel models account for different levels of aggregation that may be present in data. Sometimes researchers are confronted with data that are collected at different levels such that attributes about individual cases are provided as well as the attributes of groupings of these individual cases. In addition, these groupings can also have higher groupings with associated data characteristics. This *hierarchical structure* is common in data across the sciences, ranging from the social, behavioral, health, and economic sciences to the biological, engineering, and physical sciences, yet is commonly ignored by researchers performing statistical analyses. Unfortunately, neglecting hierarchies in data can have damaging consequences to subsequent statistical inferences.

The frequency of nested data structures in the data-analytic sciences is startling. In the United States and elsewhere, individual voters are nested in precincts which are, in turn, nested in districts, which are nested in states, which are nested in the nation. In healthcare, patients are nested in wards, which are then nested in clinics or hospitals, which are then nested in healthcare management systems, which are nested in states, and so on. In the classic example, students are nested in classrooms, which are nested in schools, which are nested in districts, which are then nested in states, which again are nested in the nation. In another familiar context, it is often the case that survey respondents are nested in areas such as rural versus urban, then these areas are nested by nation, and the nations in regions. Famous studies such as the American National Election Studies, Latinobarometer, Eurobarameter, and Afrobarometer are obvious cases. Often in population biology a hierarchy is built using ancestral information, and phenotypic variation is used to estimate the heritability of certain traits, in what is commonly referred to as the "animal model." In image processing, spatial relationships emerge between the intensity and hue of pixels. There are many hierarchies that emerge in language processing, such as topic of discussion, document type, region of origin, or intended audience. In longitudinal studies, more complex hierarchies emerge. Units or groups of units are repeatedly observed over a period of time. In addition to group hierarchies, observations are also grouped by the unit being measured. These models are extensively used in the medical/health sciences to model the effect of a stimulus or treatment regime conditional on measures of interest, such as socioeconomic status, disease prevalence in the environment, drug use, or other demographic information. Furthermore, the frequency of data at different levels of aggregation is increasing as more data are generated from geocoding, biometric monitoring, Internet traffic, social networks, an amplification of government and corporate reporting, and high-resolution imaging.

Multilevel models are a powerful and flexible extension to conventional regression frameworks. They extend the linear model and the generalized linear model by incorporating levels directly into the model statement, thus accounting for aggregation present in the data. As a result, all of the familiar model forms for linear, dichotomous, count, restricted range, ordered categorical, and unordered categorical outcomes are supplemented by adding a structural component. This structure classifies cases into known groups, which may have their own set of explanatory variables at the group level. So a hierarchy is established such that some explanatory variables are assigned to explain differences at the individual level and some explanatory variables are assigned to explain differences at the group level. This is powerful because it takes into account correlations between subjects within the same group as distinct from correlations between groups. Thus, with nested data structures the multilevel approach immediately provides a set of critical advantages over conventional, flat modeling where these structures emerge as unaccounted-for heterogeneity and correlation.

What does a multilevel model look like? At the core, there is a regression equation that relates an outcome variable on the left-hand side to a set of explanatory variables on the right-hand side. This is the basic individual-level specification, and looks immediately like a linear model or generalized linear model. The departure comes from the treatment of some of the coefficients assigned to the explanatory variables. What can be done to modify a model when a point estimate is inadequate to describe the variation due to a measured variable? An obvious modification is to treat this coefficient as having a distribution as opposed to being a fixed point. A regression equation can be introduced to model the coefficient itself, using information at the group level to describe the heterogeneity in the coefficient. This is the heart of the multilevel model. Any right-hand side effect can get its own regression expression with its own assumptions about functional form, linearity, independence, variance, distribution of errors, and so on. Such models are often referred to as "mixed," meaning some of the coefficients are *modeled* while others are *unmodeled*.

What this strategy produces is a method of accounting for structured data through utilizing regression equations at different hierarchical levels in the data. The key linkage is that these higher-level models are describing *distributions* at the level just beneath them for the coefficient that they model as if it were itself an outcome variable. This means that multilevel models are highly symbiotic with Bayesian specifications because the focus in both cases is on making supportable distributional assumptions.

Allowing multiple levels in the same model actually provides an immense amount of flexibility. First, the researcher is not restricted to a particular number of levels. The coefficients at the second grouping level can also be assigned a regression equation, thus adding another level to the hierarchy, although it has been shown that there is diminishing return as the number of levels goes up, and it is rarely efficient to go past three levels from the individual level (Goel and DeGroot 1981, Goel 1983). This is because the effects of the parameterizations at these super-high levels gets washed out as it comes down the hierarchy. Second, as stated, any coefficient at these levels can be chosen to be modeled or unmodeled and in this way the mixture of these decisions at any level gives a combinatorially large set of choices. Third, the form of the link function can differ for any level of the model. In this way the researcher may mix linear, logit/probit, count, constrained, and other forms throughout the total specification.

1.2 BACKGROUND

It is often the case that fundamental ideas in statistics hide for a while in some applied area before scholars realize that these are generalizable and broadly applicable principles. For instance, the well-known EM algorithm of Dempster, Laird, and Rubin (1977) was pre-dated in less fully articulated forms by Newcomb (1886), McKendrick (1926), Healy and Westmacott (1956), Hartley (1958), Baum and Petrie (1966), Baum and Eagon (1967), and Zangwill (1969), who gives the critical conditions for monotonic convergence. In another famous example, the core Markov chain Monte Carlo (MCMC) algorithm (Metropolis et al. 1953) slept quietly in the *Journal of Chemical Physics* before emerging in the 1990s to revolutionize the entire discipline of statistics. It turns out that hierarchical modeling follows this same storyline, roughly originating with the statistical analysis of agricultural data around the 1950s (Eisenhart 1947, Henderson 1950, Scheffé 1956, Henderson et al. 1959). A big step forward came in the 1980s when education researchers realized that their data fit this structure perfectly (students nested in classes, classes nested in schools, schools nested in districts, districts nested in states), and that important explanatory variables could be found at all of these levels. This flurry of work focused on the *hierarchical linear model* (HLM) and was developed in detail in works such as Burstein (1980), Mason et al. (1983), Aitkin and Longford (1986), Bryk and Raudenbush (1987), Bryk et al. (1988), De Leeuw and Kreft (1986), Raudenbush and Bryk (1986), Goldstein (1987), Longford (1987), Raudenbush (1988), and Lee and Bryk (1989). These applications continue today as education policy remains an important empirical challenge. Work in this literature was accelerated by the development of the standalone software packages `HLM`, `ML2`, `VARCL`, as well as incorporation into the `SAS` procedure `MIXED`, and others. Additional work by Goldstein (notably 1985) took the two-level model and extended it to situations with further nested groupings, non-nested groupings, time series cross-sectional data, and more. At roughly the same time, a series of influential papers and applications grew out of Laird and Ware (1982), where a random effects model for Gaussian longitudinal data is established. This Laird–Ware model was extended to binary outcomes by Stiratelli, Laird, and Ware (1984) and GEE estimation was established by Zeger and Liang (1986). An important extension to non-linear mixed effects models is presented in Lindstrom and Bates (1988). In addition, Breslow and Clayton (1993) developed quasi-likelihood methods to analyze generalized linear mixed models (GLMMs).

Beginning around the 1990s, hierarchical modeling took on a much more Bayesian complexion now that stochastic simulation tools (e.g. MCMC) had arrived to solve the resulting estimation challenges. Since the Bayesian paradigm and the hierarchical reliance on distributional relationships between levels have a natural affinity, many papers were produced and continue to be produced in the intersection of the two. Computational advances during this period centered around customizing MCMC solutions for particular problems (Carlin et al. 1992, Albert and Chib 1993, Liu 1994, Hobert and Casella 1996, Jones and Hobert 2001, Cowles 2002). Other works focused on solving specific applied problems with Bayesian models: Steffey (1992) incorporates expert information into the model, Stangl (1995) develops prediction and decision rules, Cohen et al. (1998) model arrest rates, Zeger and Karim (1991) use GLMMs to study infectious disease, Christiansen and Morris (1997) build on count models hierarchically, Hodges and Sargent (2001) refine inference procedures, Pauler et al. (2001) model cancer risk, and Pettitt et al. (2006) model survey data from immigrants. Recently, Bayesian prior specifications in hierarchical models have received attention (Hadjicostas and Berry 1999, Daniels and Gatsonis 1999, Gelman 2006, Booth et al. 2008). Finally, the text by Gelman and Hill (2007) has been enormously influential.

A primary reason for the large increase in interest in the use of multilevel models in recent years is due to the ready availability of sophisticated general software solutions for estimating more complex specifications. A review of software available for fitting these models is presented in Chapter 26 of this volume. For basic models, the `lme4` package in `R` works quite well and preserves `R`'s intuitive model language. Also, `Stata` provides some support through the `XTMIXED` routine. However, researchers now routinely specify generalized linear multilevel models with categorical, count, or truncated outcome variables. It is also now common to see non-nested hierarchies expressing cross-classification, mixtures of nonlinear relationships within hierarchical groupings, and longitudinal considerations such as panel specifications and standard time-series relations. All of this provides a rich, but sometimes complicated, set of variable relationships. Since most applied users are unwilling to spend the time to derive their own likelihood functions or posterior distributions and maximize or explore these forms, software like `WinBUGS` and its cousin `JAGS` are popular (Bayesian) solutions (`Mplus` is also a helpful choice).

1.3 FOUNDATIONAL MODELS

The development of multilevel models starts with the simple linear model specification for individual i that relates the outcome variable, y_i, to the systematic component, $x_i\beta_1$, with unexplained variance falling to the error term, ε_i, giving:

$$y_i = \beta_0 + x_i\beta_1 + \varepsilon_i, \qquad (1.1)$$

which is assumed to meet the standard Gauss–Markov assumptions (linear functional form, independent errors with mean zero and constant variance, no relationship between x_i and errors). The normality of the errors is not a necessary assumption for making inferences since standard least squares procedures produce an estimate of the standard error (Amemiya 1985, Ravishanker and Dey 2002), but with reasonable sample size and finite variance the central limit theorem applies. For maximum likelihood results, which produce the same estimates as does least squares, the derivation of the estimator begins with the assumption of normality of population residuals. See the discussion on pages 583–6 of Casella and Berger (2001) for a detailed derivation.

1.3.1 Basic Linear Forms, Individual-Level Explanatory Variables

How does one model the heterogeneity that arises because each i case belongs to one of $j = 1, \ldots, J$ groups where $J < n$? Even if there does not exist explanatory variable information about these J assignments, model fit may be improved by binning each i case into its respective group. This can be done by loosening the definition of the single intercept, β_0, in (1.1) to J distinct intercepts, β_{0j}, which then groups the n cases, giving them a common intercept with other cases if they land in the same group. Formally, for $i = 1, \ldots, n_j$ (where n_j is the size of the jth group):

$$y_{ij} = \beta_{0j} + x_{ij}\beta_1 + \varepsilon_{ij},$$

where the added j subscript indicates that this case belongs to the jth group and gets intercept β_{0j}. The β_{0j} are group-specific intercepts and are usually given a common normal distribution with mean β_0 and standard deviation σ_{u_0}. The overall intercept β_0 is referred to as a *fixed effect* and the difference $u_{0j} = \beta_{0j} - \beta_0$ is a *random effect*. Subscripting the u with 0 denotes that it relates to the intercept term and distinguishes it from other varying quantities to be discussed shortly. Since the β_1

Figure 1.1 Varying Intercepts

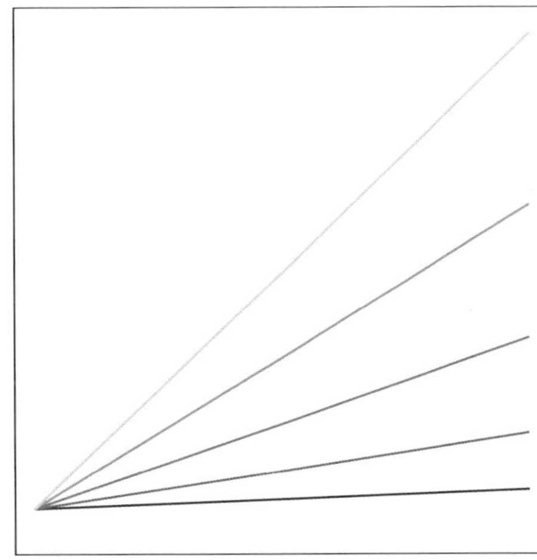

Figure 1.2 Varying Slopes

coefficient is not indexed by the grouping term j, this is still constant across all of the $n = n_1 + \cdots + n_J$ cases and evaluated with a standard point estimate. This model is illustrated in Figure 1.1, which shows that while different groups start at different intercepts, they progress at the same rate (slope). This model is sufficiently fundamental that it has its own name, the *varying-intercept* or *random-intercept* model.

In a different context, one may want to account for the groupings in the data, but the researcher may feel that the effect is not through the intercept, where the groups start at a zero level of the explanatory variable x, having reason to believe that the grouping affects the slopes instead: as x increases, group membership dictates a different change in y. So now loosen the definition of the single slope, β_1, in (1.1) to account for the groupings according to:

$$y_{ij} = \beta_0 + x_{ij}\beta_{1j} + \varepsilon_{ij},$$

where the added j subscript indicates that the n_j cases in the jth group get slope β_{1j}. The intercept now remains fixed across the cases in the data and the slopes are given a common normal distribution with mean β_1 and standard deviation σ_{u_1}. In this situation, β_1 is a *fixed effect* and the difference $u_{1j} = \beta_{1j} - \beta_1$ is a *random effect*, and subscripting the u with 1 denotes that these random effects relate to the slope as opposed to the intercepts. This is illustrated in Figure 1.2 showing divergence from the same starting point for the groups as x increases. This model is also fundamental enough that it gets its own name, the *varying-slope* or *random-slope* model.

Suppose the researcher suspects that the heterogeneity in the sample is sufficiently complex that it needs to be modeled with both a varying-intercept and a varying-slope. This is a simple combination of the previous two models and takes the form:

$$y_{ij} = \beta_{0j} + x_{ij}\beta_{1j} + \varepsilon_{ij},$$

where membership in group j for case ij has two effects, one that is constant and one that differs from others with increasing x. The vectors (β_{0j}, β_{1j}) are given a common multivariate normal distribution with mean vector (β_0, β_1) and covariance matrix $\mathbf{\Omega}_u$. The vector of means (β_0, β_1) is the *fixed effect* and the vectors of differences $(u_{0j}, u_{1j}) = (\beta_{0j} - \beta_0, \beta_{1j} - \beta_1)$ are the *random effects*. A synthetic, possibly exaggerated, model result is given in Figure 1.3. Not surprisingly, this is called the *varying-intercept, varying-slope* or *random-intercept, random-slope* model. Notice from the simple artificial example in

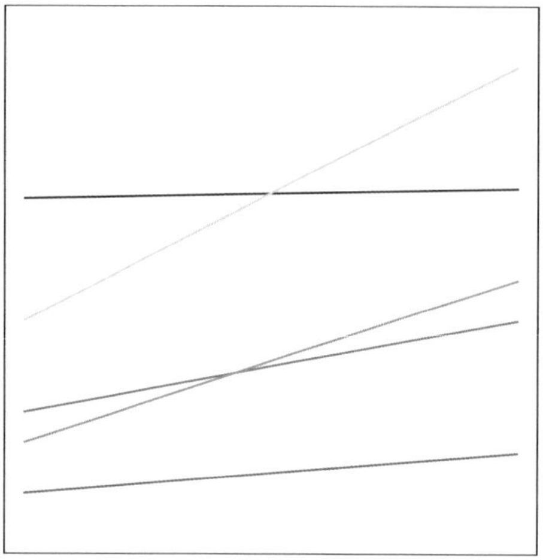

Figure 1.3 Varying Intercepts and Slopes

the figure that it is already possible to model quite intricate differences in the groups for this basic linear form.

Before moving on to more complicated models, a clear explanation of what is meant by *fixed effects* and *random effects* is necessary. *Fixed effects* are coefficients in the mean structure that are assumed to have point values, whereas *random effects* are coefficients that are assumed to have distributions. For example, in the models thus far considered the *random effects* have been assumed to have a common Gaussian distribution. A distinction must be made, then, for when group-level effects are assumed to be *fixed effects* or *random effects*. In the random-intercepts model, for example, the β_{0j} can be assumed to be point values instead of being modeled as coming from a distribution, as *fixed effects* as opposed to *random effects*. This distinction is quite important and modifies both the assumptions of the models as well as the estimation strategy employed in analyzing them.

1.3.2 Basic Linear Forms, Adding Group-Level Explanatory Variables

The bivariate linear form can be additively extended on the right-hand side to include more covariates, which may or may not receive the grouping treatment. A canonical *mixed* form is one where the intercept and the first q explanatory variables have coefficients that vary by the $j = 1, \ldots, J$ groupings (for a total of $q + 1$ modeled coefficients), but the next p coefficients, $q + 1, q + 2, \ldots, q + p$, are fixed at the individual level. This is given by the specification:

$$\begin{aligned} y_{ij} = \; & \beta_{0j} + x_{1i}\beta_{1j} + \cdots + x_{qi}\beta_{qj} \\ & + x_{(q+1)i}\beta_{q+1} \\ & + \cdots + x_{(q+p)i}\beta_{q+p} + \varepsilon_{ij}, \end{aligned}$$

where membership in group j for case ij has $q+1$ effects. The vectors of group-level coefficients $(\beta_{0j}, \ldots, \beta_{qj})$ are given a common distribution, which for now will be assumed to be Gaussian with mean vector $(\beta_0, \ldots, \beta_q)$ (the *fixed effects*) and covariance matrix $\mathbf{\Omega}_u$. As before, the vectors of differences $\mathbf{u}_j = (\beta_{0j} - \beta_0, \ldots, \beta_{qj} - \beta_q)$ are the *random effects* for this model.

An important aspect of these models, which greatly facilitates their use in generalized linear models and nonlinear models, is writing them in a hierarchical fashion. The model is written as a specification of a regression equation at each level of the hierarchy. For example, consider the model where

$$y_{ij} = \beta_{0j} + x_{1ij}\beta_{1j} + x_{2ij}\beta_2 + \varepsilon_{ij}, \quad (1.2)$$

where there are two random coefficients and one fixed coefficient. The group-level coefficients can be written as

$$\beta_{0j} = \beta_0 + u_{0j} \qquad \beta_{1j} = \beta_1 + u_{1j} \quad (1.3)$$

and the u_{0j}, u_{1j} appear as errors at the second level of the hierarchy. Assumptions are then made about the distributions of the individual-level errors (ε_{ij}) and group-level errors (u_{0j}, u_{1j}). For example, a common set of assumptions in linear regression is ε_{ij} are independent and identically distributed (iid) with common variance, and (u_{0j}, u_{1j}) is bivariate normal with mean 0 and covariance matrix $\mathbf{\Omega}$.

The model given in (1.2), (1.3) does not impose explanatory variables at the second level, given by the J groupings, since $\mathbf{u}_j \sim N_q(\mathbf{0}, \mathbf{\Omega})$ by assumption. Until explanatory variables at this second level are added, there is not a modeled reason for the differences in the group-level coefficients. The fact that one can model the variation at the group level is a key feature of treating them as *random effects* as opposed to *fixed effects*. Returning to the linear model with two group-level coefficients and one individual-level coefficient in (1.2), $y_{ij} = \beta_{0j} + x_{1ij}\beta_{1j} + x_{2ij}\beta_2 + \varepsilon_{ij}$, model each of the two group-level coefficients with their own regression equation and index these by $j = 1$ to J:

$$\begin{aligned} \beta_{0j} &= \beta_0 + \beta_3 x_{3j} + u_{0j} \\ \beta_{1j} &= \beta_1 + \beta_4 x_{4j} + u_{1j}, \end{aligned} \tag{1.4}$$

and the group-level variation is modeled as depending on covariates. The explanatory variables at the second level are called *context-level* variables, and the idea of *contextual specificity* is that of the existence of legitimately comparable groups. These context-level variables are constant in each group and are subscripted by j instead of ij to identify that they only depend on group identification.

Substituting the two definitions in (1.4) into the individual-level model in (1.2) and rearranging produces:

$$\begin{aligned} y_{ij} &= (\beta_0 + \beta_3 x_{3j} + u_{0j}) \\ &\quad + (\beta_1 + \beta_4 x_{4j} + u_{1j})x_{1ij} \\ &\quad + \beta_2 x_{2ij} + \varepsilon_{ij} \\ &= (\beta_0 + \beta_1 x_{1ij} + \beta_2 x_{2ij} + \beta_3 x_{3j} \\ &\quad + \beta_4 x_{4j} x_{1ij}) \\ &\quad + (u_{0j} + u_{1j} x_{1ij} + \varepsilon_{ij}) \end{aligned} \tag{1.5}$$

for the ijth case. The composite fixed effects now have a richer structure, accounting for variation at both the individual and group levels. In addition, the error structure is now modeled as being due to specific group-level variables. In this example, the composite error structure, $(u_{1j}x_{1ij} + u_{0j} + \varepsilon_{ij})$, is heteroscedastic since it is conditioned on levels of the explanatory variable x_{1ij}. This composite error shows that uncertainty is modified in a standard linear model (ε_{ij}) by introducing terms that are correlated for observations in the same group. Though it seems that this has increased uncertainty in the data, it is just modeling the data in a fuller fashion. This richer model accounts for the hierarchical structure in the data and can provide a significantly better fit to observed data than standard linear regression.

It is important to understand the exact role of the new coefficients. First, β_0 is a universally assigned intercept that all i cases share. Second, β_1 gives another shared term that is the slope coefficient corresponding to the effect of changes in x_{1ij}, as does β_2 for x_{2ij}. These three terms have no effect from the multilevel composition of the model. Third, β_3 gives the slope coefficient for the effect of changes in the variable x_{3j} for group j, and are applied to all individual cases assigned to this group. It therefore varies by group and not individual. Fourth, and surprisingly, β_4 is the coefficient on the interaction term between x_{1ij} and x_{4j}. But though no interaction term was specified in the hierarchical form of the model this illustrates an important point. Any hierarchy that models a slope on the right-hand side imposes an interaction term if this hierarchy contains group-level covariates. While it is easy to see the multiplicative implications from (1.5), it is surprising to some that this is an automatic consequence.

In the Laird–Ware form of the model, the fixed effects are separated from the random effects as in (1.5). In this formulation, the covariates on the random effects are represented by zs rather than xs in order to distinguish the two. This structure can be written (in matrix form) for group j as $\mathbf{y}_j = \boldsymbol{X}_j\beta + \boldsymbol{Z}_j\mathbf{u}_j + \varepsilon_j$. Here, $\mathbf{y}_j$ is the vector of observations, $\boldsymbol{X}_j$ is the fixed effects design matrix, and $\boldsymbol{Z}_j$ is a the random effects design matrix for group j. The vector $\boldsymbol{\beta}$ is the vector of fixed effects, which are assumed to have point values, whereas $\mathbf{u}_j$ is the vector of random effects for group j which are modeled by a distributional assumption. Finally, ε_j is a vector of individual-level error terms for group j. From this formulation, it is apparent that multilevel models can be expressed in a single-level expression, although this does not always lead to a more intuitive expression.

1.3.3 The Model Spectrum

In language that Gelman and Hill (2007) emphasize, multilevel models can be thought of as sitting between two extremes that are available to the researcher when groupings are known: *fully pooled* and *fully unpooled*. The fully pooled model treats the group-level variables as individual variables, meaning that group-level distinctions are ignored and these effects are treated as if they are case-specific. For a model with one explanatory variable measured at the individual level (x_1) and one measured at the group level (x_2), this specification is:

$$y_i = \beta_0 + x_{1i}\beta_1 + x_{2i}\beta_2 + \varepsilon_i.$$

In contrast to (1.5), there is no u_{0j} modeled here. This is an assertion that the group distinctions do not matter and the cases should all be treated homogeneously, ignoring the (possibly important) variation between categories. At the other end of the spectrum is a set of models in which each group can be treated as a separate dataset and modeled completely separately:

$$y_{ij} = \beta_{0j} + x_{ij}\beta_{1j} + \varepsilon_{ij},$$

for $j = 1, \ldots, J$. Note that the group-level predictor x_2 does not enter into this equation because $x_{2i}\beta_2$ is constant within a group and therefore subsumed into the intercept term. Here there is no second level to the hierarchy and the βs are assumed to be fixed parameters, in contrast to the distributional assumptions made in the mixed model. The fully unpooled approach is the opposite distinction from the fully pooled approach and asserts that the groups are so completely different that it does not make sense to associate them in the same model. In particular, the values of slopes and intercept from one group have no relationship to those in other groups. Such separate regression models clearly overstate variation between groups, making them look more different than they really should be.

Between these two polar group distinctions lies the multilevel model. The word "between" here means that groups are recognized as different, but because there is a single model in which they are associated by common individual-level fixed effects as well as distributional assumptions on the random effects, the resulting model therefore compromises between full distinction of groups and the full ignoring of groups. This can be thought of as partial-pooling or semi-pooling in the sense that the groups are collected together in a single model, but their distinctness is preserved.

To illustrate this "betweeness", consider a simple varying-intercepts model with no explanatory variables:

$$y_{ij} = \beta_{0j} + \varepsilon_{ij}, \tag{1.6}$$

which is also called a *mean model* since β_{0j} represents the mean of the jth group. If there is an assumption that $\beta_{0j} = \beta_0$ is constant across all cases, then this becomes the fully pooled model. Conversely, if there are J separate models each with their own β_{0j} which do not derive from a common distribution, then it is the fully unpooled approach. Estimating (1.6) as a partial pooling model (with Gaussian distributional assumptions) gives group means that are a weighted average of the n_j cases in group j and the overall mean from all cases. Define first:

$\bar{y}_j$	fully unpooled mean for group j
$\bar{y}$	fully pooled mean
σ_0^2	within-group variance (variance of the ε_{ij})
σ_1^2	between-group variance (variance of the $\bar{y}_j$)
n_j	size of the jth group.

Then an approximation of the multilevel model *estimate* for the group mean is given by:

$$\hat{\beta}_{0j} = \frac{\frac{n_j}{\sigma_0^2}\bar{y}_j + \frac{1}{\sigma_1^2}\bar{y}}{\frac{n_j}{\sigma_0^2} + \frac{1}{\sigma_1^2}}. \tag{1.7}$$

This is a very revealing expression. The estimate of the mean for a group is a weighted average of the contribution from the full sample and the contribution from that group, where the weighting depends on relative variances and the size of the group. As the size of arbitrary group j gets small, $\bar{y}_j$ becomes less important and the group estimate borrows more strength from the full sample. A zero size group, perhaps a hypothesized case, relies completely on the full sample size, since (1.7) reduces to $\hat{\beta}_{0j} = \bar{y}$. On the other hand, as group j gets large, its estimate dominates the contribution from the fully pooled mean, and is also a big influence on this fully pooled mean. This is called the *shrinkage* of the mean effects towards the common mean. In addition, as $\sigma_1^2 \to 0$, then $\hat{\beta}_{0j} \to \bar{\mathbf{y}}$, and as $\sigma_1^2 \to \infty$, then $\hat{\beta}_{0j} \to \bar{\mathbf{y}}_j$. Thus, the group effect which is at the heart of a multilevel model, is a balance between the size of the group and the standard deviations at the individual and group levels.

1.4 EXTENSIONS BEYOND THE TWO-LEVEL MODEL

Multilevel models are not restricted to linear forms with interval-measured outcomes over the entire real line, nor are they restricted to hierarchies which contain only one level of grouping or nested levels of grouping. The stochastic assumptions at each level of the hierarchy can be made in any appropriate fashion for the problem being modeled. This added flexibility of the MLM provides a much richer class of models and captures many of the models used in modern scientific research.

1.4.1 *Nested Groupings*

The generalization of the mixed effects model to nested groupings is straightforward and is most easily understood in the hierarchical framework of (1.2), (1.3) as opposed to the single equation of (1.5).

Consider the common case of survey respondents nested in regions, which are then nested in states, and so on. The individual level comprises the first hierarchy of the model and captures the variation in the data that can be explained by individual-level covariates. In this example, the outcome of interest is measured support for a political candidate or party, with covariates that are individualized such as race, gender, income, age, and attentiveness to public affairs. The second level of the model in this example is immediate region of residence, and this comes with its own set of covariates including rural/urban measurement, crime levels, dominant industry, coastal access, and so on. The third level is state, the fourth level is national region, and so on. Each level of the model comes with a regression equation where the variation in intercepts or slopes that are assumed to vary do so with the possible inclusion of group-level covariates.

Consider a three-level model with individual-level covariate x_1, level-two group covariate x_2, and level-three covariate x_3. The data come as y_{ijk}, indicating the ith individual in the jth level two group which is contained in the kth level three group. In the previous example, i represents survey respondents, j represents immediate region, and k represents state. Allowing both varying-intercepts and varying-slopes in the regression equation at the individual level, gives:

$$y_{ijk} = \beta_{0jk} + \beta_{1jk}x_{1ijk} + \varepsilon_{ijk},$$

where the ε_{ijk} are assumed to be independently and normally distributed. At the second level of the model, there are separate regression equations for the intercepts and slopes:

$$\begin{aligned}\beta_{0jk} &= \beta_{0k} + \beta_{2k}x_{2jk} + u_{0jk}\\ \beta_{1jk} &= \beta_{1k} + \beta_{3k}x_{2jk} + u_{1jk},\end{aligned}$$

where the vectors of (u_{0jk}, u_{1jk}) are assumed to have a common multivariate normal distribution. At the third level of the model, there are separate regression equations for the intercepts and slopes:

$$\begin{aligned}\beta_{0k} &= \beta_0 + \beta_4 x_{3k} + u_{3k}\\ \beta_{2k} &= \beta_2 + \beta_5 x_{3k} + u_{4k}\end{aligned}$$

$$\beta_{1k} = \beta_1 + \beta_6 x_{3k} + u_{5k}$$
$$\beta_{3k} = \beta_3 + \beta_7 x_{3k} + u_{6k},$$

where the vectors of level three residuals $(u_{3k}, u_{4k}, u_{5k}, u_{6k})$ are assumed to have a common multivariate normal distribution. In analogy to (1.5), this model includes eight fixed effects parameters capturing the intercept (β_0), the three main effects ($\beta_1, \beta_2, \beta_4$), the three two-way interactions ($\beta_3, \beta_5, \beta_6$), and the three-way interaction of the covariates (β_7) as well as a rich error structure capturing the nested groupings within the data. Just from this simple framework, extensions abound. For example, since the level two residuals are indexed by both j and k, a natural relaxation of the model is to let the distribution of $\mathbf{u}_{jk}$ depend on k and then bring these distributions together at the third level. Alternatively, one can specify the model such that the level three covariate only affects intercepts and not slopes, or that slopes and intercepts vary at level two but only intercepts vary at level three, both of which are easy modifications in this hierarchical specification.

1.4.2 Non-Nested Groupings

In order to generalize to the case of non-nested groupings, consider data with two different groupings at the second level. In an economics example, imagine modeling the income of individuals in a state who have both an immediate region of residence and an occupation. These workers are then naturally grouped by the multiple regions and the jobs, where these groups obviously are not required to have the same number of individuals: there are more residents in a large urban region than a nearby rural county, and one would expect more clerical office workers than clergymen, for example. This is non-nested in the sense that there are multiple people in the same region with the same *and* different jobs. Represent region of residence with the index r and occupation with the index o, letting iro refer to the ith individual who is in both the rth region class and the oth occupation class. Of course, such an individual does not necessarily exist—there may be no ranchers in New York City. A regression equation with individual-level covariate x and intercepts which vary with both groupings is given by:

$$y_{iro} = \beta_0 + x_{iro}\beta_1 + u_r + u_o + \varepsilon_{iro}, \quad (1.8)$$

where the random effects u_r have one common normal distribution and the random effects u_o have a different common normal distribution. To add varying slopes to (1.8), simply modify the equation to be

$$y_{iro} = \beta_0 + x_{iro}\beta_1 + u_{0r} + u_{0o} + u_{1r}x_{iro} + u_{1o}x_{iro} + \varepsilon_{iro},$$

and make appropriate distributional assumptions about the random effects.

The addition of a random effect u_{ro} to (1.8) that depends on both region and occupation would give this model three levels: the individual level, the level of intersections of region and occupation, and the level of region or occupation. The second level of the hierarchy would naturally nest in both of the level three groupings. There are many ways to extend the MLM with crossed groupings to take into account complicated structures that could generate observed data. The key to effectively using these models in practice is to consider the possible ways in which different groupings can affect the outcome variable and then include these in appropriately defined regression equations.

1.4.3 Generalized Linear Forms

The extension to generalized linear models with non-Gaussian distributional assumptions at the individual level is also straightforward. This involves inserting a link function between the outcome variable and the additive-linear form based on the right-hand side with the explanatory variables. For a two-level model, this means modeling the linked variable as:

$$\eta_{ij} = [\beta_{0j} + x_{1ij}\beta_{1j}] + \cdots + x_{qij}\beta_{qj} + x_{(q+1)ij}\beta_{q+1} + \cdots + x_{(q+p)ij}\beta_{q+p},$$

where this is then related to the conditional mean of the outcome, $\eta_{ij} = g(E[y_{ij}|\boldsymbol{\beta}_j])$, where there is conditioning on the vector of group-level coefficients $\boldsymbol{\beta}_j$. This two-level model is completed by making an assumption about the distribution of $\boldsymbol{\beta}_j$. In contrast to (1.5), the stochastic components of the model at the individual and group levels cannot simply be added together because the link function g is only linear in the case of normally distributed data. Thus, the random effects are difficult or impossible to "integrate out" of the likelihood, and the marginal likelihood of the data cannot be obtained in closed form.

To illustrate this model, consider a simplified logistic case. Suppose there are outcomes from the same binary choice at the individual level, a single individual-level covariate (x_1), and a single group-level covariate (x_2). Then the regression equation for varying-intercepts and varying-slopes is given by:

$$\begin{aligned} p(y_{ij} = 1|\beta_{0j}, \beta_{1j}) &= \text{logit}^{-1}(\beta_{0j} + x_{1ij}\beta_{1j}) \\ &= 1 - \left(1 + e^{\beta_{0j} + x_{1ij}\beta_{1j}}\right)^{-1} \\ \beta_{0j} &= \beta_0 + \beta_2 x_{2j} + u_{0j} \\ \beta_{1j} &= \beta_1 + \beta_3 x_{2j} + u_{1j}. \end{aligned}$$

Assume that the random effects $\mathbf{u}_j$ are multivariate normal with mean $\mathbf{0}$ and covariance matrix $\boldsymbol{\Omega}$. The parameters to be estimated are the coefficients in the mean structure (the fixed effects), and the elements of the covariance matrix $\boldsymbol{\Omega}$. Notice that the distribution at the second level is given explicitly as a normal distribution whereas the distribution at the first level is implied by the assumption of having Bernoulli trials. It is common to stipulate normal forms at higher levels in the model since they provide an easy way to consider the possible correlation of the random effects. However, it is important to understand that the assumption of normality at this level is exactly that, *an assumption*, and thus must be investigated. If evidence is found to suggest that the random effects are not normally distributed, this assumption must be relaxed or fixed effects should be used.

Standard forms for the link function in $g(E[y_{ij}|\boldsymbol{\beta}_j])$ include probit, Poisson (log-linear), gamma, multinomial, ordered categorical forms, and more. The theory and estimation of GLMMs is discussed in Chapter 15, and the specific case of qualitative outcomes is discussed in Chapter 16. Many statistical packages (see Chapter 26) are available for fitting a variety of these models, but as assumptions are relaxed about distributional forms or more exotic generalized linear models are used, one must resort to more flexible estimation strategies.

1.5 VOCABULARY

An unfortunate consequence of the development of multilevel models in disparate literatures is that the vocabulary describing these models differs, even for the exact same specification. The primary confusion is between the synonymous terms of *multilevel model* or *hierarchical model* and the descriptions that use *effects*. Varying-coefficients models, either intercepts or slopes, are often called *random effects models* since they are associated with distributional statements like $\beta_{0j} \sim N(\beta_0 + \beta_1 x_{0j}, \sigma_u^2)$. A related term is *mixed models*, meaning that the specification has both modeled and unmodeled coefficients. The term *fixed effects* is unfortunately not used as cleanly as implied above, with different meanings in different particular settings. Sometimes this is applied to unmodeled coefficients that are constant across individuals, or "nuisance" coefficients that are uninteresting but included by necessity in the form of control variables, or even in the case where the data represent a population instead of a sample.

The meanings of "fixed" and "random" can also differ in definition by literature (Kreft and De Leeuw 1988; Section 1.3.3, Gelman 2005). The obvious solution to this confusion is to not worry about labels but to pay attention to the implications of *subscripting* in the described model. These specifications can be conceptualized as members of a larger multilevel family where indices are purposely *turned on* to create a level, or *turned off* to create a point estimate.

1.6 CASE STUDY: PARTY IDENTIFICATION IN WESTERN EUROPE

As an illustration, consider 23,355 citizens' feeling of alignment with a political party in ten Western European countries[1] taken from the Comparative Study of Electoral Systems (CSES) for 16 elections from 2001 to 2007. The natural hierarchy for these eligible voters is: district, election, and country (some countries held more than one parliamentary election during this time period). The percentage of those surveyed who felt close to one party varies from 0.29 (Ireland 2002) to 0.65 (Spain 2004).

Running a logistic regression model on these data using individual-, district-, and country-level covariates as though they are individual-specific (fully pooled) requires dramatically different ranges for the explanatory variables to produce reliable coefficients. Since Western European countries do not show such differences in individual-level covariates and the country-level covariates do not vary strongly with the outcome variable (correlation of -0.25), the model needs to take into account higher-level variations. This is done by specifying a hierarchical model to take into account the natural groupings in the data.

The CSES dataset provides multiple ways to consider hierarchies through region and time. Respondents can be nested in voting districts, elections, and countries. Additionally, one could add a time dynamic taking into account that elections within a single country have a temporal ordering. Twelve of

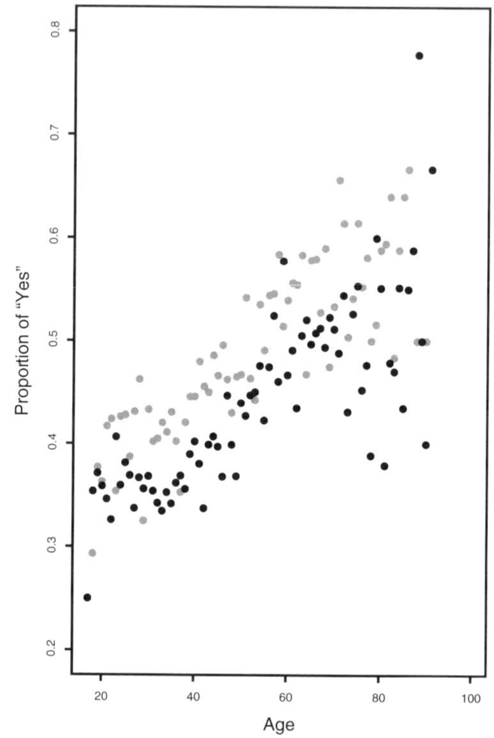

Figure 1.4 Empirical Proportion of "Yes" Versus Age, by Gender

the elections considered belong to groupings of size two (grouped by country), and in four countries there was a single election. If a researcher expects heterogeneity to be explained by dynamics within and between particular elections, the developed model will be hierarchical with two levels based on districts and elections. In Figure 1.4 this is shown by plotting the observed fraction of "Yes" answers for each age, separated by gender (gray dots for men, black dots for women), for these elections. Notice that in the aggregated data women are generally less likely to identify with a party for any given age, even though identification for both men and women increases with age.

The outcome variable for our model is a dichotomous measure from the question "Do you usually think of yourself as close to any particular political party?" coded zero for "No" and one for "Yes" (numbers

B3028, C3020_1). Here attention is focused on only a modest subset of the total number of possible explanatory variables. The individual-level demographic variables are: `Age` of respondent in years (number 2001); `Female`, with men equal to zero and women equal to one (number 2002); the respondent's income quintile labeled `Income`; and the respondent's municipality coded `Rural/Village`, `Small/Middle Town`, `City Suburb`, `City Metropolitan`, with the first category used as the reference category in the model specification (numbers B2027, C2027). Subjects are nested within electoral districts with a district-level variable describing the number of seats elected by proportional representation in the given district. Additionally, these districts are nested within elections with an election-level variable describing the effective number of parties in the election. The variable `Parties` (number 5094) gives the effective number of political parties in each country, and `Seats` (number 4001) indicates the number of seats contested in each district of the first segment of the legislature's lower house where the respondent resides. Further details can be found at `http://www.cses.org/varlist/varlist_full.htm`.

For an initial analysis, ignore the nested structure of the data and simply analyze the dataset using a logistic generalized linear model. The outcome variable is modeled as

$$y_i|p_i \sim \text{Bern}(p_i) \qquad \log\left(\frac{p_i}{1-p_i}\right) = \mathbf{x}_i\boldsymbol{\beta},$$

where $\mathbf{x}_i$ is the vector of covariates for the ith respondent and $\boldsymbol{\beta}$ is a vector of coefficients to be estimated. The base categories for this model are men for `Female`, the first income quantile, and `Rural/Village` for region. Table 1.1 provides this standard logistic model in the first block of results.

For the second model, analyze a two-level hierarchy: one at district level and one at the election level, represented by random intercept contributions to the individual level. The outcome variable is now modeled as:

$$\begin{aligned}
y_{ijk}|p_{ijk} &\sim \text{Bern}(p_{ijk}) \\
\log\left(\frac{p_{ijk}}{1-p_{ijk}}\right) &= \beta_{0jk} + \mathbf{x}_{ijk}\beta \\
\beta_{0jk} &= \beta_{0k} + \beta_{\texttt{Seats}} \\
&\quad \times \texttt{Seats}_{jk} + u_{0jk} \\
\beta_{0k} &= \beta_0 + \beta_{\texttt{Parties}} \\
&\quad \times \texttt{Parties}_k + u_{0k} \\
u_{0jk} &\sim \mathcal{N}(0, \sigma_d^2) \\
u_{0k} &\sim \mathcal{N}(0, \sigma_e^2).
\end{aligned}$$

Therefore, `Seats` and `Parties` are predictors at different levels of the model. Since `Parties` is constant in an election, it predicts the change in intercept at the election level of the hierarchy. `Seats` is constant in districts but varies within an election, so it predicts the change in intercept at the district level. In addition, all of the district-level random effects have a common normal distribution which does not change depending on election, and the election-level random effects have a different common normal distribution. The three levels of the hierarchy are evident and stochastic processes are introduced at each level: Bernoulli distributions at the data level and normal distributions at the district and election level.

The multilevel model fits better by several standard measures, with a difference in AIC (Akaike information criterion) of 525 and a difference in BIC (Bayesian information criterion) of 509 in favor of the multilevel model. This shows that the multilevel model fits the data dramatically better than the GLM. As a description of model fit, also consider the percent correctly predicted with the naïve criterion (splitting predictions at the arbitrary 0.5 threshold). The standard GLM gives 57.367% estimated correctly, whereas the multilevel model gives 60.522%.

In Table 1.1 there are essentially no differences in the coefficient estimates between the two models for `Age`, `Female`, `Income Level 4`, `Income Level 5`, and

Table 1.1 Contrasting Specifications, Voting Satisfaction

	Standard Logit GLM			Random Intercepts Version		
	Estimate	Std Error	*z*-score	Estimate	Std Error	*z*-score
Intercept	−0.3946	0.0417	−9.46	−0.1751	0.0944	−1.855
`Age`	0.0156	0.0008	19.25	0.0168	0.0008	20.207
`Female`	−0.2076	0.0267	−7.76	−0.2098	0.0272	−7.709
`Income Level 2`	0.1178	0.0425	2.77	0.0350	0.0436	0.801
`Income Level 3`	0.2282	0.0427	5.35	0.1474	0.0439	3.357
`Income Level 4`	0.2677	0.0443	6.04	0.2468	0.0453	5.454
`Income Level 5`	0.2388	0.0451	5.30	0.2179	0.0466	4.673
`Small/Middle Town`	0.0665	0.0392	1.70	−0.0822	0.0429	−1.916
`City Suburb`	0.1746	0.0431	4.05	−0.0507	0.0500	−1.014
`City Metropolitan`	0.1212	0.0359	3.38	0.0636	0.0417	1.525
`Parties`	−0.0408	0.0142	−2.87	−0.1033	0.0846	−1.222
`Seats`	0.0047	0.0010	4.82	0.0027	0.0019	1.403
Residual Deviance	31608 on 23343 df			31079 on 23341 df		
Null Deviance	32134 on 23354 df			31590 on 23352 df		
				$\sigma_d = 0.2402$, $\sigma_e = 0.32692$		

`Small/Middle Town`. However, notice the differences between the two models for the coefficient estimates of `Income Level 2`, `Income Level 3`, `City Suburb`, `City Metropolitan`, `Parties`, and `Seats`. In all cases where observed differences exist, the standard generalized linear model gives more reliable coefficient estimates (smaller standard errors), but this is misleading. Once the correlation structure of the data is taken into account, there is more uncertainty in the estimated values of these parameters.

To further evaluate the fit of the regular GLM, consider a plot of the ranked fitted probabilities from the model against binned estimates of the observed proportion of success in the reordered observed data. Figure 1.5 indicates that although the model does describe the general trend of the data, it misses some features since the point-cloud does not adhere closely to the fitted line underlying the assumption of the model. The curve of fitted probabilities only describes 47% of the variance in these empirical proportions of success. This suggests that there are additional features of the data, and it is possible to capture these with the multilevel specification. The fitted probabilities of the multilevel model describe 71% of the variation in the binned estimates of the observed proportion of success.

After running the initial GLM, one could have improperly concluded that the number of seats in a district, the effective number of political parties, and city type significantly influenced the response. However, these variables in this specification are mimicking the correlation structure of the data, which was more effectively taken into account through the multilevel model framework, as evidenced by the large gains in predictive accuracy. This is also apparent taking the binned empirical values and breaking down their variance in various ways. To see that the random effects have a meaningful impact on data fit, compare how well the fixed effects and the full model predict the binned values. Normally, a researcher would be happy if the group-level standard deviations were of similar size to the residual standard deviation, as small group effects relative to the residual effect indicate that the grouping is not effective in the model specification. Use of the binned empirical values mimics this kind of analysis for a generalized linear mixed model. The variance of the binned values minus the fixed effects is 0.0111

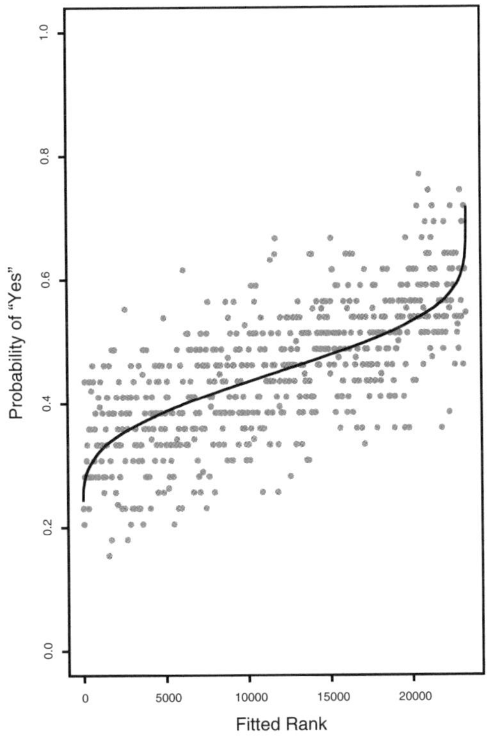

Figure 1.5 Probability of "Yes" Versus Fitted Rank

and the variance of the binned values minus the fitted values from the full model is 0.0060, indicating that a significant amount of variation in the data that is indeed captured by the random effects.

1.6.1 *Computational Considerations*

Multilevel models are more demanding of statistical software than more standard regressions. In the case of the Gaussian–Gaussian multilevel model such as the one-way random effects model, it is possible to integrate out all levels of the hierarchy to be left with a likelihood for $\mathbf{y}$ which only depends on the fixed effects and the induced covariance structure. Maximum likelihood (or restricted maximum likelihood) estimates for the coefficients on the fixed effects as well as the parameters of the variance components can be computed.

Moving beyond the simple Gaussian–Gaussian model greatly complicates estimation considerations. This includes both relaxations of the assumptions of the linear mixed effects models and complications arising from nonlinear link functions in generalized linear mixed models. A sophisticated set of estimation strategies using approximations to the likelihood have been developed to overcome these difficulties and are discussed in Chapter 3. Alternatively, the Bayesian paradigm offers a suite of MCMC methods for producing samples from the posterior distribution of the model parameters. These alternatives are discussed in Chapter 4.

1.7 SUMMARY

This introduction to multilevel models provides an overview of a class of regression models that account for hierarchical structure in data. Such data occur when there are natural levels of aggregation whereby individual cases are nested within groups, and those groups may also be nested in higher-level groups. It provides a general description of the model features that enable multilevel models to account for such structure, demonstrates that ignoring hierarchies produces incorrect inferential statements in model summaries, and has illustrated that point with a simple example using a real dataset.

Aitkin and his co-authors (especially, 1981, 1986) introduced the linear multilevel model in the 1980s, concentrating on applications in education research since the hierarchy in that setting is obvious: students in classrooms, classrooms in schools, schools in districts, and districts in states. These applications were all just linear models and yet they substantially improved fit to the data in educational settings. Since this era more elaborate specifications have been developed for nonlinear outcomes, non-nested hierarchies, correlations between hierarchies, and more. This has been a very active area of research both theoretically and in applied settings. These developments are described in detail in subsequent chapters.

Multilevel models are flexible tools because they exist in the spectrum between fully pooled models, where groupings are ignored, and fully unpooled models, where each group gets its own regression statement. This means that multilevel models recognize both commonalities within the cases and differences between group effects. The gained efficiency is both notational and substantive. The notational efficiency occurs because there are direct means of expressing hierarchies with subscripts, nested subscripts, and sets of subscripts. This contrasts with messy "dummy" coding of group definitions with large numbers of categories. Multilevel models account for individual- versus group-level variation because these two sources of variability are both explicitly taken into account. Since all non-modeled variation falls to the residuals, multilevel models are guaranteed to capture between-group variability when it exists. These forms are also a convenient way of estimating separately, but concurrently, regression coefficients for groups. The alternative is to construct separate models whereby between-group variability is completely lost. In addition, multilevel models provide more flexibility for expressing scientific theories, which routinely consider settings where individual cases are contained in larger groups, which themselves are contained in even larger groups, and so on. Moreover, there are real problems with ignoring hierarchies in data. The resulting models will have the wrong standard errors on group-affected coefficients since fully pooled results assume that the apparent commonalities are results of individual effects. This problem also spills over into covariances between coefficient estimates. Thus, the multilevel modeling framework not only respects the data, but provides better statistical inference which can be used to describe phenomena and inform decisions.

NOTE

1 These are: Switzerland, Germany, Spain, Finland, Ireland, Iceland, Italy, Norway, Portugal, and Sweden.

REFERENCES

Aitkin, M., Anderson, D., and Hinde, J. (1981) 'Statistical Modeling of Data on Teaching Styles', *Journal of the Royal Statistical Society, Series A*, 144: 419–61.

Aitkin, M. and Longford, N. (1986) 'Statistical Modeling Issues in School Effectiveness Studies', *Journal of the Royal Statistical Society, Series A*, 149: 1–43.

Albert, J.H. and Chib, S. (1993) 'Bayesian Analysis of Binary and Polychotomous Response Data', *Journal of the American Statistical Association*, 88: 669–79.

Amemiya, T. (1985) *Advanced Econometrics*. Cambridge, MA: Harvard University Press.

Baum, L.E. and Eagon, J.A. (1967) 'An Inequality with Applications to Statistical Estimation for Probabilistic Functions of Markov Processes and to a Model for Ecology', *Bulletin of the American Mathematical Society*, 73: 360–3.

Baum, L.E. and Petrie, T. (1966) 'Statistical Inference for Probabilistic Functions of Finite Markov Chains', *Annals of Mathematical Statistics*, 37: 1554–63.

Booth, J.G., Casella, G., and Hobert, J.P. (2008) 'Clustering Using Objective Functions and Stochastic Search', *Journal of the Royal Statistical Society, Series B*, 70: 119–39.

Breslow, N.E. and Clayton, D.G. (1993) 'Approximate Inference In Generalized Linear Mixed Models', *Journal of the American Statistical Association*, 88: 9–25.

Bryk, A.S. and Raudenbush, S.W. (1987) 'Applications of Hierarchical Linear Models to Assessing Change', *Psychological Bulletin*, 101: 147–58.

Bryk, A.S., Raudenbush, S.W., Seltzer, M., and Congdon, R. (1988) *An Introduction to HLM: Computer Program and User's Guide*. 2nd edn. Chicago: University of Chicago Department of Education.

Burstein, L. (1980) 'The Analysis of Multi-Level Data in Educational Research and Evaluation', *Review of Research in Education*, 8: 158–233.

Carlin, B.P., Gelfand, A.E., and Smith, A.F.M. (1992) 'Hierarchical Bayesian Analysis of Changepoint Problems', *Applied Statistics*, 41: 389–405.

Casella, G. and Berger, R.L. (2001) *Statistical Inference*. 2nd edn. Belmont, CA: Duxbury Advanced Series.

Christiansen, C.L. and Morris, C.N. (1997) 'Hierarchical Poisson Regression Modeling', *Journal of the American Statistical Association*, 92: 618–32.

Cohen, J., Nagin, D., Wallstrom, G., and Wasserman, L. (1998) 'Hierarchical Bayesian Analysis of Arrest Rates', *Journal of the American Statistical Association*, 93: 1260–70.

Cowles, M.K. (2002) 'MCMC Sampler Convergence Rates for Hierarchical Normal Linear Models: A

Simulation Approach', *Statistics and Computing*, 12: 377–89.

Daniels, M.J. and Gatsonis, C. (1999) 'Hierarchical Generalized Linear Models in the Analysis of Variations in Health Care Utilization', *Journal of the American Statistical Association*, 94: 29–42.

De Leeuw, J. and Kreft, I. (1986) 'Random Coefficient Models for Multilevel Analysis', *Journal of Educational Statistics*, 11: 57–85.

Dempster, A.P., Laird, N.M., and Rubin, D.B. (1977) 'Maximum Likelihood from Incomplete Data via the EM Algorithm', *Journal of the Royal Statistical Society, Series B*, 39: 1–38.

Eisenhart, C. (1947) 'The Assumptions Underlying the Analysis of Variance', *Biometrics*, 3: 1–21.

Gelman, A. (2005) 'Analysis of Variance: Why It Is More Important Than Ever', *Annals of Statistics*, 33: 1–53.

Gelman, A. (2006) 'Prior Distributions for Variance Parameters in Hierarchical Models', *Bayesian Analysis*, 1: 515–33.

Gelman, A. and Hill, J. (2007) *Data Analysis Using Regression and Multilevel/Hierarchical Models*. Cambridge: Cambridge University Press.

Goel, P.K. (1983) 'Information Measures and Bayesian Hierarchical Models', *Journal of the American Statistical Association*, 78: 408–10.

Goel, P.K. and DeGroot, M.H. (1981) 'Information About Hyperparameters in Hierarchical Models', *Journal of the American Statistical Association*, 76: 140–7.

Goldstein, H. (1985) *Multilevel Statistical Models*. New York: Halstead Press.

Goldstein, H. (1987) *Multilevel Models in Education and Social Research*. Oxford: Oxford University Press.

Hadjicostas, P. and Berry, S.M. (1999) 'Improper and Proper Posteriors with Improper Priors in a Poisson-Gamma Hierarchical Model', *Test*, 8: 147–66.

Hartley, H.O. (1958) 'Maximum Likelihood Estimation From Incomplete Data', *Biometrics*, 14: 174–94.

Healy, M. and Westmacott, M. (1956) 'Missing Values in Experiments Analysed on Automatic Computers', *Journal of the Royal Statistical Society, Series C*, 5: 203–6.

Henderson, C.R. (1950) 'Estimation of Genetic Parameters', *Biometrics*, 6: 186–7.

Henderson, C.R., Kempthorne, O., Searle, S.R., and von Krosigk, C.M. (1959) 'The Estimation of Environmental and Genetic Trends From Records Subject To Culling', *Biometrics*, 15: 192.

Hobert, J.P. and Casella, G. (1996) 'The Effect of Improper Priors on Gibbs Sampling in Hierarchical Linear Mixed Models', *Journal of the American Statistical Association*, 91: 1461–73.

Hodges, J.S. and Sargent, D.J. (2001) 'Counting Degrees of Freedom in Hierarchical and Other Richly Parameterized Models', *Biometrika*, 88: 367–79.

Jones, G.L. and Hobert, J.P. (2001) 'Honest Exploration of Intractable Probability Distributions via Markov Chain Monte Carlo', *Statistical Science*, 16: 312–34.

Kreft, I.G.G. and De Leeuw, J. (1988) 'The Seesaw Effect: A Multilevel Problem?', *Quality and Quantity*, 22: 127–37.

Laird, N.M. and Ware, J.H. (1982) 'Random-effects Models for Longitudinal Data', *Biometrics*, 38: 963–74.

Lee, V.E. and Bryk, A.S. (1989) 'Multilevel Model of the Social Distribution of High School Achievement', *Sociology of Education*, 62: 172–92.

Lindstrom, M.J. and Bates, D.M. (1988) 'Newton-Raphson and EM Algorithms for Linear Mixed-Effects Models for Repeated-Measures Data', *Journal of the American Statistical Association*, 83: 1014–22.

Liu, J.S. (1994) 'The Collapsed Gibbs Sampler in Bayesian Computations with Applications to a Gene Regulation Problem', *Journal of the American Statistical Association*, 89: 958–66.

Longford, N.T. (1987) 'A Fast Scoring Algorithm for Maximum Likelihood Estimation in Unbalanced Mixed Models With Nested Random Effects', *Biometrika*, 74: 817–27.

Mason, W.M., Wong, G.Y., and Entwistle, B. (1983) 'Contextual Analysis Through the Multilevel Linear Model', in S. Leinhardt (ed.), *Sociological Methodology 1983–1984*. Oxford: Blackwell, 72–103.

McKendrick, A.G. (1926) 'Applications of Mathematics to Medical Problems', *Proceedings of the Edinburgh Mathematical Society*, 44: 98–130.

Metropolis, N., Rosenbluth, A.W., Rosenbluth, M.N., Teller, A.H., and Teller E. (1953) 'Equation of State Calculations by Fast Computing Machines', *Journal of Chemical Physics*, 21: 1087–91.

Newcomb, S. (1886) 'A Generalized Theory of the Combination of Observations So As to Obtain the Best Results', *American Journal of Mathematics*, 8: 343–66.

Pauler, D.K., Menon, U., McIntosh, M., Symecko, H.L., Skates, S.J., and Jacobs, I.J. (2001) 'Factors Influencing Serum CA125II Levels in Healthy Postmenopausal Women', *Cancer Epidemiology, Biomarkers & Prevention*, 10: 489–93.

Pettitt, A.N., Tran, T.T., Haynes, M.A., and Hay, J.L. (2006) 'A Bayesian Hierarchical Model for Categorical Longitudinal Data From a Social Survey of Immigrants', *Journal of the Royal Statistical Society, Series A*, 169: 97–114.

Raudenbush, S.W. (1988) 'Education Applications of Hierarchical Linear Models: A Review', *Journal of Educational Statistics*, 12: 85–116.

Raudenbush, S.W. and Bryk, A.S. (1986) 'A Hierarchical Model for Studying School Effects', *Sociology of Education*, 59: 1–17.

Ravishanker, N. and Dey, D.K. (2002) *A First Course In Linear Model Theory*. New York: Chapman & Hall/CRC.

Scheffé, H. (1956) 'Alternative Models for the Analysis of Variance', *Annals of Mathematical Statistics*, 27: 251–71.

Stangl, D.K. (1995) 'Prediction and Decision Making Using Bayesian Hierarchical Models', *Statistics in Medicine*, 14: 2173–90.

Steffey, D. (1992) 'Hierarchical Bayesian Modeling With Elicited Prior Information', *Communications in Statistics–Theory and Methods*, 21: 799–821.

Stiratelli, R., Laird, N., and Ware, J.H. (1984) 'Random-Effects Models for Serial Observations with Binary Response', *Biometrics*, 40: 961–71.

Zangwill, W.I. (1969) *Nonlinear Programming: A Unified Approach*. Englewood Cliffs, NJ: Prentice-Hall.

Zeger, S.L. and Karim, M.R. (1991) 'Generalized Linear Models With Random Effects: A Gibbs Sampling Approach', *Journal of the American Statistical Association*, 86: 79–86.

Zeger, S.L. and Liang, K.-Y. (1986) 'Longitudinal Data Analysis for Discrete and Continuous Outcomes', *Biometrics*, 42: 121–30.

2

Multilevel Model Notation—Establishing the Commonalities

Marc A. Scott
New York University, USA

Patrick E. Shrout
New York University, USA

Sharon L. Weinberg
New York University, USA

2.1 INTRODUCTION

Multilevel models (MLMs) have broad applicability, having been usefully applied to problems in the biological sciences, social policy, and health sciences, among others. In most of the disciplines there are textbooks that begin with the common language of that discipline and then introduce the multilevel framework. Many of these texts have been particularly influential, and the number of new texts continues to grow, with a rough grouping into those that emphasize analysis of longitudinal data and those that emphasize nested social groups. Some representative texts are those by Bryk and Raudenbush (1992); Diggle, Heagerty, Liang, and Zeger (2002); Fitzmaurice, Ware, and Laird (2004); Gelman and Hill (2007); Goldstein (1995); Hox (2002); Longford (1993); Singer and Willett (2003); and Snijders and Bosker (1999).

Although the various texts cover many overlapping concepts and techniques, they often describe these methods using very different notation. The goals of this chapter are to review core concepts in multilevel models, to illustrate by example the ways in which different families of notation represent these core concepts, and to establish commonalities among the different families of notation so as to deepen our understanding of them. The different sets of notation we address are by no means exhaustive, but they do capture the bulk of what is used in practice. Obviously we cannot cover all variations, and so we focus on those that are influential in major software implementations of multilevel models. These include HLM and MLwiN, the MIXED procedures in SAS and SPSS, xtmixed in STATA and lme and lmer in the R environment. See Chapter 26 for discussion of these and related software packages.

2.2 AN EMPIRICAL EXAMPLE

Before introducing and reviewing the different approaches to notation, we begin with a simple multilevel example to illustrate the techniques. Suppose that educational researchers were interested in exploring the role of gender and economic disadvantage among students in the early stages of high school in the U.S. A gap between the performance of boys and girls on math and English has been examined quite closely in the literature (Hyde and Mertz, 2009; Rampey, Dion, and Donahue, 2009). Boys tend to score higher in math, and whereas girls tend to score higher in examinations of English, the latter gap diminishes somewhat by high school. The researchers might wonder if the tendency for males to outperform females in math also extends to social studies, and whether that tendency is larger in schools whose students have fewer economic resources. Social studies achievement gaps have been examined much less frequently in the U.S. education literature.

We explore these questions using data on a standardized test of History, Citizenship, and Geography that was administered to a national sample of over 12,000 10th grade students in the National Educational Longitudinal Study (NELS) during 1990.[1] These students were sampled from 1225 schools, and in our example we focus on only two levels of nesting: students within school and schools within the sample. The exam score has a reasonably symmetric distribution with a mean of approximately 50 and a standard deviation of approximately 10. For simplicity of presentation, we ignore a number of other features of the NELS sampling design.[2] We also ignore the matter of missing data; the exam was not administered or the information was unavailable for about 2000 students in the sample. In our analysis we use information about the gender (SEX, coded +.5 for female, −.5 for male) of each student as well as whether each school had more than 20% of their population receiving free school lunch (FSL, coded +.5 if so, −.5 if not). We use this second variable as a proxy for whether the school was serving an economically disadvantaged student population, and under this definition about 40% of the schools were doing so.[3] These data were extracted from the public use files and, as such, may be freely obtained from the Handbook website.

In addition to testing for differences on the dependent variable between the (fixed) levels of SEX and FSL, respectively, a fundamental goal of multilevel analysis is to estimate the variability in test results that can be attributed to school effects rather than just to individual student effects, and how these components of variation change with the inclusion of predictors. To assess the variability due to school effects, a so-called unconditional means model can be fit to the data. This model posits that the score of a randomly selected student (the ith student in a randomly selected jth school) can be represented as the sum of the population mean (β_0), the effect of being in school j (u_{0j}) and the effect of being a particular student i within school j (ε_{ij}). We write the unconditional means model as $y_{ij} = \beta_0 + u_{0j} + \varepsilon_{ij}$. By convention, we assume $u_{0j} \sim N(0, \sigma^2_{u0})$ and $\varepsilon_{ij} \sim N(0, \sigma^2_{\varepsilon})$, independent of one another. In this "top-down" approach, one seeks to explain sources of variation (initially documented via an unconditional means model) by successive inclusion of predictors. As an alternative, at the other extreme, one may include all predictors in a model first and then use various model selection techniques (see Chapter 7) to determine the proper random effects to include. A middle ground, which we do not illustrate here, includes all predictors that vary at the individual level as the baseline model, including group-level predictors only in the second phase of the analysis.[4]

The unconditional means model for these NELS data, fit using restricted maximum likelihood estimation (REML; see Pinhiero and Bates, 2000, and Chapter 3 in this volume) in STATA (version 11.2), yields an estimated population test score mean of $\beta_0 = 51.07$ (se=.18), an estimated variability in test scores between schools of $\sigma^2_{u0} = 19.70$ and within schools (between students) of $\sigma_\varepsilon^2 = 77.87$. These results are given in Table 2.1, row 1, labeled Unconditional Means. We may calculate the proportion of overall variation that is due to school effects as an intra-class correlation (ICC) of 19.70/(19.70+77.87), or approximately 20% for these data. To determine whether differences on the dependent variable can be related to a known school-level characteristic and account for some of the random variation between schools, FSL is added as a predictor to the unconditional means model equation (see Table 2.1, row 2). When we entered this variable into the equation predicting test score we found that it predicted an effect of −4.36 (se=.33) for the gap between schools serving economically disadvantaged vs advantaged students. Although this difference was significant at a .05 level, it explained a somewhat small portion of the variation between schools. That estimated variability dropped to 15.29 from 19.70, and the within-school variation in scores was essentially unchanged.

Multilevel models contain two forms of stochastic variation, usually called variance components, captured in random effects (u_{0j}) and error (ε_{ij}) terms, the former representing unexplained systematic differences between groups (schools) and the latter unexplained, subject-level differences within groups (schools). Some software packages use the term random effects to refer to either of these two terms, perhaps due to their common stochastic nature. In the third model fit (Table 2.1, row 3), we included a subject-specific variable, SEX, the student's gender. The effect for gender (female − male), was −1.65 (se=.19), so females were associated with worse outcomes according to this model. This effect was significant, but its inclusion had very little impact on either of the two variance components. Table 2.1, row 4 shows the results of a slightly more complicated model that adjusts for gender, school-level economic advantage, and a cross-level interaction term between gender and economic advantage. All parameter estimates are significant; the positive effect associated with the interaction term for a female in a school with economically disadvantaged students (1.20) provides a combined effect for such females of −0.38, which is much closer to zero and surely not significant, suggesting that boys and girls in these schools fare about the same, under this model. Again, we found no noticeable change in the variance components.

At this point, our interaction model can be written,

$$\begin{aligned} y_{ij} &= \beta_0 + \beta_1 \mathrm{FSL}_j + \beta_2 \mathrm{SEX}_{ij} \\ &\quad + \beta_3 \mathrm{FSL}_j \mathrm{SEX}_{ij} + u_{0j} + \varepsilon_{ij}, \end{aligned}$$

with $u_{0j} \sim N(0, \sigma_{u0}^2)$ and $\varepsilon_{ij} \sim N(0, \sigma_\varepsilon^2)$, independent of one another. Note that FSL is indexed by j, as it varies at the school level, while SEX is indexed by ij, as it references students within schools. An important research question that is readily addressed with MLMs is whether the effect associated with females, β_2, is homogeneous across schools. In Model 4, we determined that the sex effect seemed to vary with the economic advantage of the schools, but there might be other ways in which the schools vary besides FSL. We, therefore, include an additional term in our model $u_{1j}\mathrm{SEX}_{ij}$ to allow the gender effect to vary by other unobserved school attributes different from FSL, given that FSL has already been controlled for in the model. We assume $u_{1j} \sim N(0, \sigma_{u1}^2)$ independent of the other random components of the model. The results of this model fit are given in row 5 of Table 2.1. The variance on the gender effect is substantively meaningful at 8.06, and the error variance has dropped a few units from 77.10 to 75.09. A likelihood ratio test comparing models with and without the varying gender effect suggests including the additional variance component ($p < .0001$). The interaction effect, β_3, remains significant.

One way to interpret the new variance component is that (controlling for FSL and school differences in level) the gap between females and males varies from school to school in terms of other unnamed school-specific effects with a standard deviation of

Table 2.1 Parameter Estimates for NELS School Example

	Parameter Estimates						
	Fixed Effects				Variance Components		
Model	Const	FSL	SEX	Inter	σ^2_{u0}	σ^2_{u1}	σ^2_{ε}
1: Unconditional Means	51.07 (0.18)				19.70		77.87
2: + School-Level (FSL)	50.81 (0.17)	−4.36 (0.33)			15.29		77.79
3: + Student-Level (SEX)	50.87 (0.17)	−4.32 (0.33)	−1.65 (0.19)		15.24		77.17
4: + Interaction (FSL × SEX)	50.86 (0.17)	−4.36 (0.33)	−1.58 (0.19)	1.20 (0.38)	15.21		77.10
5: + Random Slopes (SEX)	50.85 (0.17)	−4.37 (0.33)	−1.63 (0.22)	1.32 (0.44)	15.14	8.06	75.09

$\sqrt{8.06} = 2.84$. Based on Model 5, setting FSL to 0 so that we are examining the population average school, we see that a school-specific effect of $u_{1j} = -1$ would yield a gender gap of −4.47, rather than the population average for that profile, which is −1.63. Given the variance component $\sigma_{u1} = 2.84$, we would also expect the gender gap to reverse signs in some schools, for larger values of u_{1j}. The decrease in error variance is offset by a reallocation to systematic gender differences by school. Given the effect coding, the variance associated with the different gender effects by school including not only FSL, but also, as yet unnamed, other school-level attributes, is $\text{Var}(u_{0j} + u_{1j}\text{SEX}_{ij}) = \sigma^2_{u0} + 0.25\sigma^2_{u1}$, which is estimated to be 17.16, about a two unit increase over that of the prior model.

These examples include many of the typical terms used in MLMs: school-level variables (FSL), student-level variables (SEX), their interaction, and school-level random effects u_{0j}, u_{1j}, along with variance components, σ^2_{u0}, σ^2_{u1}, and σ^2_{ε}. Yet there are many different ways to express the models just presented. Table 2.2 describes these equivalent models using three distinct forms of notation. In the first column is the form we have been using, often called *composite notation*, in which the model is expressed in a form similar to regression with a complex error structure. The last column is a matrix version of this. The middle column is sometimes referred to as *level* or *hierarchical notation*, as the model is constructed in layers. Some disciplinary traditions might label the middle column the *random coefficient model* (e.g., Longford, 1993). These three approaches to model formulation employ certain naming conventions, which we represent in the table as different names for variance components or for random effects themselves. In the next sections we describe the notational differences in detail.

2.3 COMPOSITE AND HIERARCHICAL FORMS

We presented our NELS example in what has become known as composite form (see Singer, 1998), which is essentially the presentation used for mixed effects models. This can be understood as an extension of linear regression, allowing consideration of main effects, interactions, and terms representing stochastic variation. The flexibility of this notation is apparent from the example, which builds from the intercept model (Model 1) to a model with multiple explanatory variables and stochastic terms (Model 5). In these models the β coefficients represent the average effects of FSL, gender and/or their interaction, and these are typically called *fixed effects*. The u effects represent heterogeneity between the groups (school differences in our example), and these are typically called *random effects*. The

Table 2.2 Three Different Notations for the Models Discussed in the Example

	Notation type		
	Composite	Level	Matrix
1	$y_{ij} = \beta_0 + u_{0j} + \varepsilon_{ij}$ $u_{0j} \sim N(0, \sigma^2_{u0})$ $\varepsilon_{ij} \sim N(0, \sigma^2_\varepsilon)$	$y_{ij} = \beta_{0j} + \varepsilon_{ij}$ $\beta_{0j} = \gamma_{00} + u_{0j}$ $u_{0j} \sim N(0, \sigma^2_{u0})$ $\varepsilon_{ij} \sim N(0, \sigma^2_\varepsilon)$	$\boldsymbol{y} = X\boldsymbol{\beta} + Z\boldsymbol{\gamma} + \boldsymbol{\varepsilon}$ $X = \mathbf{1}_N$ $Z = \bigoplus_j \mathbf{1}_{n_j}$ $\boldsymbol{\gamma} \sim N(\mathbf{0}, G)$ $\boldsymbol{\varepsilon} \sim N(\mathbf{0}, R)$ $G = \tau_{00} I_J;\ R = \sigma^2_\varepsilon I_N$
2	$y_{ij} = \beta_0 + \beta_1 FSL_j$ $+ u_{0j} + \varepsilon_{ij}$ $u_{0j} \sim N(0, \sigma^2_{u0})$ $\varepsilon_{ij} \sim N(0, \sigma^2_\varepsilon)$	$y_{ij} = \beta_{0j} + \varepsilon_{ij}$ $\beta_{0j} = \gamma_{00} + \gamma_{01} FSL_j + u_{0j}$ $u_{0j} \sim N(0, \sigma^2_{u0})$ $\varepsilon_{ij} \sim N(0, \sigma^2_\varepsilon)$	$\boldsymbol{y} = X\boldsymbol{\beta} + Z\boldsymbol{\gamma} + \boldsymbol{\varepsilon}$ $X = (\mathbf{1}, \boldsymbol{FSL})$ $Z = \bigoplus_j \mathbf{1}_{n_j}$ $\boldsymbol{\gamma} \sim N(\mathbf{0}, G)$ $\boldsymbol{\varepsilon} \sim N(\mathbf{0}, R)$ $G = \tau_{00} \boldsymbol{I}_J;\ R = \sigma^2_\varepsilon \boldsymbol{I}_N$
3	$y_{ij} = \beta_0 + \beta_1 FSL_j + \beta_2 SEX_{ij}$ $+ u_{0j} + \varepsilon_{ij}$ $u_{0j} \sim N(0, \sigma^2_{u0})$ $\varepsilon_{ij} \sim N(0, \sigma^2_\varepsilon)$	$y_{ij} = \beta_{0j} + \beta_{1j} SEX_{ij} + \varepsilon_{ij}$ $\beta_{0j} = \gamma_{00} + \gamma_{01} FSL_j + u_{0j}$ $\beta_{1j} = \gamma_{10}$ $u_{0j} \sim N(0, \sigma^2_{u0})$ $\varepsilon_{ij} \sim N(0, \sigma^2_\varepsilon)$	$\boldsymbol{y} = X\boldsymbol{\beta} + Z\boldsymbol{\gamma} + \boldsymbol{\varepsilon}$ $X = (\mathbf{1}, \boldsymbol{FSL}, \boldsymbol{SEX})$ $Z = \bigoplus_j \mathbf{1}_{n_j}$ $\boldsymbol{\gamma} \sim N(\mathbf{0}, G)$ $\boldsymbol{\varepsilon} \sim N(\mathbf{0}, R)$ $G = \tau_{00} \boldsymbol{I}_J;\ R = \sigma^2_\varepsilon \boldsymbol{I}_N$
4	$y_{ij} = \beta_0 + \beta_1 FSL_j + \beta_2 SEX_{ij}$ $+ \beta_3 FSL_j SEX_{ij}$ $+ u_{0j} + \varepsilon_{ij}$ $u_{0j} \sim N(0, \sigma^2_{u0})$ $\varepsilon_{ij} \sim N(0, \sigma^2_\varepsilon)$	$y_{ij} = \beta_{0j} + \beta_{1j} SEX_{ij} + \varepsilon_{ij}$ $\beta_{0j} = \gamma_{00} + \gamma_{01} FSL_j + u_{0j}$ $\beta_{1j} = \gamma_{10} + \gamma_{11} FSL_j$ $u_{0j} \sim N(0, \sigma^2_{u0})$ $\varepsilon_{ij} \sim N(0, \sigma^2_\varepsilon)$	$\boldsymbol{y} = X\boldsymbol{\beta} + Z\boldsymbol{\gamma} + \boldsymbol{\varepsilon}$ $X = (\mathbf{1}, \boldsymbol{FSL}, \boldsymbol{SEX},$ $\boldsymbol{FSL} \cdot \boldsymbol{SEX})$ $Z = \bigoplus_j \mathbf{1}_{n_j}$ $\boldsymbol{\gamma} \sim N(\mathbf{0}, G)$ $\boldsymbol{\varepsilon} \sim N(\mathbf{0}, R)$ $G = \tau_{00} \boldsymbol{I}_J;\ R = \sigma^2_\varepsilon \boldsymbol{I}_N$
5	$y_{ij} = \beta_0 + \beta_1 FSL_j + \beta_2 SEX_{ij}$ $+ \beta_3 FSL_j SEX_{ij}$ $+ u_{0j} + u_{1j} SEX_{ij} + \varepsilon_{ij}$ $u_{0j} \sim N(0, \sigma^2_{u0})$ $u_{1j} \sim N(0, \sigma^2_{u1})$ $\varepsilon_{ij} \sim N(0, \sigma^2_\varepsilon)$	$y_{ij} = \beta_{0j} + \beta_{1j} SEX_{ij} + \varepsilon_{ij}$ $\beta_{0j} = \gamma_{00} + \gamma_{01} FSL_j + u_{0j}$ $\beta_{1j} = \gamma_{10} + \gamma_{11} FSL_j + u_{1j}$ $u_{0j} \sim N(0, \sigma^2_{u0})$ $u_{1j} \sim N(0, \sigma^2_{u1})$ $\varepsilon_{ij} \sim N(0, \sigma^2_\varepsilon)$	$\boldsymbol{y} = X\boldsymbol{\beta} + Z\boldsymbol{\gamma} + \boldsymbol{\varepsilon}$ $X = (\mathbf{1}, \boldsymbol{FSL}, \boldsymbol{SEX},$ $\boldsymbol{FSL} \cdot \boldsymbol{SEX})$ $Z = \bigoplus_j (\mathbf{1}, \boldsymbol{SEX}_j)$ $\boldsymbol{\gamma} \sim N(\mathbf{0}, G)$ $\boldsymbol{\varepsilon} \sim N(\mathbf{0}, R)$ $G = \bigoplus_j \begin{pmatrix} \tau_{00} & 0 \\ 0 & \tau_{11} \end{pmatrix}$ $R = \sigma^2_\varepsilon \boldsymbol{I}_N$

pattern of subscripts indicate which level of variation is being considered. When j is noted the variation is at the school level, whereas when ij is noted the variation is at the child (within-school) level.

For example, suppose we consider a variation of the models shown in Table 2.2:

$$y_{ij} = \beta_0 + \beta_1 \text{FSL}_j + \beta_2 \text{SEX}_{ij} + \beta_3 \text{FSL}_j \text{SEX}_{ij} + \varepsilon_{ij}. \quad (2.1)$$

This is like Models 4 and 5, except that it does not include random effects for the intercept or the sex effect. The variation of children within school is said to be explained by a simple function of the average effects of FSL, SEX, and their interaction. In a typical regression model we would assume that the error terms, ε_{ij}, are independent and homoscedastic, but in this context that assumption would not be credible.

In some fields it is conventional to refine (2.1) by explicitly considering between-school variation of β_0 and β_2 rather than by adding u_j terms to the model. This variation can be represented by adding subscripts to β_0 and β_2:

$$\begin{aligned} y_{ij} &= \beta_{0j} + \beta_1 \text{FSL}_j + \beta_{2j} \text{SEX}_{ij} \\ &\quad + \beta_3 \text{FSL}_j \text{SEX}_{ij} + \varepsilon_{ij}. \end{aligned} \tag{2.2}$$

We can then define:

$$\begin{aligned} \beta_{0j} &= \beta_0 + u_{0j} \\ \beta_{2j} &= \beta_2 + u_{1j} \end{aligned} \tag{2.3}$$

and, as above, assume normality and independence of the random terms u_{0j}, u_{1j}, and ε_{ij}.

Depending on your perspective, this is either a minor change or one with much deeper implications. The hierarchy embedded in the above notation may be made even more explicit by labeling the top equation (2.2) level one and labeling the second set of equations (2.3) level two. The level one equation describes a typical regression model with a single error term, but the regression parameters β_{0j} and β_{2j} in this top level suggest, by their subscript j, potential school-to-school variability in social studies test scores and in gender gap effects, respectively. To make even more explicit the multilevel nature of this type of design, one may choose to include in the level one equation only those predictors that vary at the subject level, reserving those predictors that vary at the school level to be included in the set of level two equations. The advantage of this notation convention is that it can help to enforce distinctions among levels. Using this approach, outcomes, predictors, effects, and error terms all contain subscripts that reflect the level at which they vary. Because SEX_{ij} varies at the subject level, it can be understood to belong on level one. Likewise, because FSL_j does not vary at the subject level, but at the school level, it can be included in a level two equation to explain variation at level two. The interaction between SEX_{ij} and FSL_j "crosses" levels and, as such, does not, strictly speaking, belong on level one. In the context of this interaction, FSL_j can be understood as a predictor of the parameter β_{2j} that represents the relationship between SEX_{ij} and y_{ij}. Using a variant of the notation introduced by Bryk and Raudenbush (1992) and Goldstein (1995), we reformulate (2.2) and (2.3) as follows:

Level one:

$$y_{ij} = \beta_{0j} + \beta_{1j} \text{SEX}_{ij} + \varepsilon_{ij} \tag{2.4}$$

Level two:

$$\begin{aligned} \beta_{0j} &= \gamma_{00} + \gamma_{01} \text{FSL}_j + u_{0j} \\ \beta_{1j} &= \gamma_{10} + \gamma_{11} \text{FSL}_j + u_{1j}, \end{aligned} \tag{2.5}$$

with u_{0j}, u_{1j} as before. Note that r_{ij} is sometimes used to represent the error terms in multilevel models, but we believe that this has the potential to associate error, a stochastic model component, with residual, its corresponding estimate, so we prefer and adopt the use of ε_{ij} to avoid this confusion and be consistent with usage elsewhere in the volume. Notice that the level one expression (2.4) has fewer terms than (2.2), and the coefficients in (2.5) at level two conform to a different naming convention. This change of notation completes the transformation of Model 5 from the form given in Table 2.2, column 1, to the corresponding row of column 2. We find no definitive name for this form of the model, although it is sometimes described as *level notation* for obvious reasons. We will do the same, sometimes referring to it as *hierarchical, level notation* as well. Note that the level two "outcome" β_{0j} in the column two representation of Model 5 is a random intercept varying at the school level, is "explained" in a regression model with predictor FSL_j, and has error term u_{0j}. The level two error, u_{0j}, can be understood as local to that regression sub-model, or as a random effect in the corresponding composite model that represents Model 5 in column 1 of Table 2.2. Critically, and as we illustrate by example in later sections of this chapter, how one characterizes the random effects—as controls in a regression model, as part of the dependent error structure, or as error terms in a set of sub-model regressions—is tied to how one formulates the MLM notationally and to how one interprets results in practice.

Another important difference between the hierarchical and composite notation systems is how level two effects between group variables such as FSL are interpreted. In the first equation of (2.5), the level two outcome we are trying to explain represents an intercept shift, and so one could interpret γ_{01} as the expected change in group *mean* for a unit change in group predictor FSL_j, net of all else in the model. One of the striking differences that the notation evokes is that cross-level interactions become part of sub-models for varying coefficients. Thus, γ_{11} reflects the effect of FSL_j on β_{1j}, but the latter modifies the effect of SEX_{ij} on the outcome, so this is an interaction effect in the composite notation.

The use of γ for the level two coefficients follows Bryk and Raudenbush (1992), who employ several strategies for managing coefficient names across levels. First, they use a common Greek letter for coefficients within each level and a different letter between levels. In our two-level model, level one employs βs, while level two employs γs. Further, the fixed subscripts at each level (our numbers $0, 1, 2, \ldots$) are allocated in a systematic manner, reflecting the level and equation within that level. At level one, all subscripts are allocated sequentially, as we have done above, so β_{0j} precedes β_{1j} and so forth. At level two, model coefficients are given a subscript reflecting the outcome they explain (the left-hand side of the equation), followed by a subscript allocated sequentially within that equation. Thus, γ_{00} is the constant in a level two model that explains β_{0j}, while γ_{01} would be the coefficient of the first predictor in that same model.

This level notation approach to model specification is associated with several software packages, notably HLM and MLwiN, while the composite form is closer in structure to the syntax used in SAS (PROC MIXED), STATA (xtmixed), and R (lme, lmer). Bayesian modeling packages such as WinBUGS naturally relate terms through hierarchical structures, more closely resembling the level notation. In-depth discussion of issues related to software implementation is given in Chapter 26. We note that an alternative notation adopted by Gelman and Hill (2007), with its origins in the syntax associated with the BUGS family of packages and Browne, Goldstein, and Rabash (2001), incorporates the group identification using index look-up functions, wherein $j[i]$ (read: j of i) identifies the group j to which unique index i belongs.

2.4 MATRIX NOTATION

To those familiar with matrix algebra, the matrix notation in column three of Table 2.2 provides a convenient mechanism for organizing a variety of models in a common form. For those less familiar with matrices, it may be profitable to learn about the basics so that it is possible to consult sources, such as software help files, that use this notation. One reason mathematicians often prefer this formulation is that all five models in Table 2.2 are represented by exactly the same matrix equation,

$$\boldsymbol{y} = X\boldsymbol{\beta} + Z\boldsymbol{u} + \boldsymbol{\varepsilon}. \tag{2.6}$$

To show where equation (2.6) comes from and what it means, we first consider a simpler version. Suppose the subjects of group (school) j are put in a list $\boldsymbol{y}_j = \{y_{1j}, y_{2j}, y_{3j}, \ldots, y_{n_j j}\}$, which we call data vector $\boldsymbol{y}_j$ (the convention is to view this list as a column of numbers or symbols, as in a spreadsheet). The data from all the groups can be combined in one sample vector, $\boldsymbol{y} = \{\boldsymbol{y}_1, \boldsymbol{y}_2, \boldsymbol{y}_3, \ldots, \boldsymbol{y}_J\}$, where J is the number of groups, and the vector has $N = \sum_j n_j$ elements, where n_j is the number of subjects in group j. To represent the model for group j we write $\boldsymbol{y}_j = X_j\boldsymbol{\beta} + Z_j\boldsymbol{u}_j + \boldsymbol{\varepsilon}_j$, where X_j is the design matrix for the fixed effects, and Z_j is the design matrix for the random effects. As described in the matrix Appendix to Fitzmaurice et al. (2004), $X_j\boldsymbol{\beta}$ and $Z_j\boldsymbol{u}_j$ represent matrix multiplication, which are sums of products of design values and parameter weights. For example, if we were representing Model 5 in Table 2.2, $\boldsymbol{y}_j$ would have n_j

elements, X_j would have n_j rows and four columns, and Z_j would have n_j rows and two columns. Suppose group j was economically disadvantaged (FSL = .5) and $n_j = 3$. Suppose further that the second person in the group was female (SEX = .5) and the first and third were males (SEX = −.5). The matrix equation for this group for Model 5 would be:

$$\begin{pmatrix} y_{1j} \\ y_{2j} \\ y_{3j} \end{pmatrix} = \begin{pmatrix} 1 & .5 & -.5 & -.25 \\ 1 & .5 & +.5 & +.25 \\ 1 & .5 & -.5 & -.25 \end{pmatrix} \begin{pmatrix} \beta_0 \\ \beta_1 \\ \beta_2 \\ \beta_3 \end{pmatrix} + \begin{pmatrix} 1 & -.5 \\ 1 & +.5 \\ 1 & -.5 \end{pmatrix} \begin{pmatrix} u_{0j} \\ u_{1j} \end{pmatrix} + \begin{pmatrix} \varepsilon_{1j} \\ \varepsilon_{2j} \\ \varepsilon_{3j} \end{pmatrix}.$$

This single equation represents three separate linear combinations of the parameters and design indicators plus specific error terms. When the $\boldsymbol{y}_j$ vectors are combined into the sample vector with $N = \sum_j n_j$ elements, it is necessary to construct equivalent structures for X, Z, and $\boldsymbol{u}$. For Model 5 in Table 2.2, X becomes an $N \times 4$ matrix with each of the group-specific X_j matrices being adjoined row-wise. The matrix Z in equation (2.6) becomes more complicated, as each group has its own random effect vector $\boldsymbol{u}_j$. The Z matrix has qJ columns, where q is the number of random effects in the model. We can use the direct sum operation from matrix algebra to represent the matrix as $Z = \bigoplus_j Z_j$. The $\bigoplus$ operation creates a block-wise diagonal matrix with the group-level matrices Z_j on the block. For example,

$$(Z_1 \oplus Z_2) = \begin{pmatrix} Z_1 & 0 \\ 0 & Z_2 \end{pmatrix},$$

where 0 is a matrix of zeros whose dimension depends on the dimensions of Z_1 and Z_2. Returning to equation (2.6), $\boldsymbol{y}$ is an $N \times 1$ vector of observations, X is an $N \times p$ design matrix for fixed effects, $\boldsymbol{\beta}$ is a $p \times 1$ vector of fixed effects coefficients, Z an $N \times qJ$ design matrix for modeling random effects, $\boldsymbol{u}$ is the $qJ \times 1$ vector of random effects, and $\boldsymbol{\varepsilon}$ an $N \times 1$ vector of error terms. One often assumes $\boldsymbol{u} \sim N_{qJ}(0, G)$ and $\boldsymbol{\varepsilon} \sim N_N(0, R)$, independent of one another, and from this one can see that the components of $\boldsymbol{u}$ are stacked sequentially, one group's effects after the other.

Examples of how the matrix terms would be defined for our five representative models are given in Table 2.2, column 3. When all vectors and matrices are built up from their group-level components, we have p =1, 2, 3, 4, and 4, respectively, for models 1 through 5; q is 1 in models 1 through 4, and 2 in model 5. Also note that in the general notation, I_N is $N \times N$, I_{qJ} is $qJ \times qJ$, $\mathbf{1}_N$ is the $N \times 1$ column vector of ones and $\mathbf{1}_{n_j}$ is the $n_j \times 1$ column vector of ones corresponding to group j. All models in the table assume the independence of all random effects, but this can be relaxed. The use of τ_{00} and τ_{11} for variance components in matrix notation leads to the natural generalization for dependent random effects, in which τ_{01} would represent a covariance term, as in

$$G = \bigoplus_j \begin{pmatrix} \tau_{00} & \tau_{01} \\ \tau_{01} & \tau_{11} \end{pmatrix}.$$

Predictors **FSL** and **SEX** are $N \times 1$ column vectors. $\mathbf{SEX}_j$ represents the vector of values for that predictor in school j, for all students in the school, as in our example above with three students in the school. This general matrix formulation is quite common, although sometimes the specifics vary (a different symbol for $\boldsymbol{u}$ or $\boldsymbol{\varepsilon}$, for example).

In this notation, the level of the predictor—whether it varies by group or not—is implicit, revealed only in how one might use subscripts to reference any single element of the outcome vector or design matrix. From the point of view of model specification and estimation, whether a predictor is at the subject or group level is not essential, as everything is captured in the design matrix X. Group membership is identified in columns of the design matrices (if they are part of the model) through the use of indicator variables, as discussed above.

The value in representing models through matrix notation cannot be overstated. This simpler form makes it possible to focus on other aspects of the model that we for the most part ignore in this chapter, such as estimation, inference, complex covariance structure, model predictions, diagnostics, and missing data. However, limiting one's exposure to models represented in matrix notation has the potential to obscure the very reason one develops MLMs in practice: to examine relationships between predictors and outcomes in the presence of complex sampling designs and their resultant data structures. In many of the chapters that follow, matrix notation will be the primary means by which models are developed. When appropriate, the relationship between the different forms of notation discussed here can be used to make concrete some of the more abstract presentations.

Matrix notation is typically organized following composite notation structure, and, as mentioned, subscripts differentiating the manner in which predictors may vary is difficult to embed in this notation. In what follows, we primarily contrast composite and level notation, as these have direct impact on model specification and interpretation in most software packages.

2.5 EXAMPLE: TWO TYPES OF NOTATION IN PRACTICE

To illustrate the manner in which composite and level notational forms link to statistical software packages and the implications of that correspondence, we return to our example evaluating U.S. tenth grade social studies test scores. We begin by evaluating a simpler, main effects model, which is row 3 in Table 2.2, and reproduce both the composite and level forms of notation here for completeness.

$$y_{ij} = \beta_0 + \beta_1 \text{FSL}_j + \beta_2 \text{SEX}_{ij} + u_{0j} + \varepsilon_{ij}, \tag{2.7}$$

or

Level one:

$$y_{ij} = \beta_{0j} + \beta_{1j}\text{SEX}_{ij} + \varepsilon_{ij} \tag{2.8}$$

Level two:

$$\beta_{0j} = \gamma_{00} + \gamma_{01}\text{FSL}_j + u_{0j} \tag{2.9}$$
$$\beta_{1j} = \gamma_{10},$$

with $u_{0j} \sim N(0, \sigma^2_{u_0})$ and $\varepsilon_{ij} \sim N(0, \sigma^2_\varepsilon)$, independent of one another. The two forms of notation, (2.7) and (2.8)–(2.9), use different terms, β or γ, for the same effect, which is a by-product of the way in which each set of models was built up, sequentially. This also leads to a different ordering of the terms in each type of model, but we maintain this difference to be consistent with the prior discussion and the presentation in Tables 2.1 and 2.2. We estimate the model parameters using REML with STATA and HLM software.

In STATA, the model is specified in two parts, fixed effects followed by random effects, and the grouping factor (school) provides the multilevel structure. In this example, the syntax is `xtmixed Y FSL SEX || SCH_ID:`, where `SCH_ID` is the unique identifier for the school. The fixed effects include both of our predictors, specified as one might in a regression model, with no indication of the level at which they vary. Accordingly, within a data file, the value of a school-level predictor is the same for all students who are in the same school. The absence of any predictors following the double bar and group identifier '`|| SCH_ID:`' indicates (by default) that only the intercept (denoted by STATA as _cons) potentially varies by school. In Table 2.3, the STATA output from fitting this model is organized into fixed effect estimates followed by variance components.

We note that both the sex effect and indicator for free school lunch are negative and significant. The estimates for $\hat{\sigma}_{u_0}$ (sd(_cons)) and $\hat{\sigma}_\varepsilon$ (sd(Residual)) are 3.9 and 8.8, respectively. For the social studies test, after adjusting for these two predictors, the proportion of variation that is between schools is somewhat small compared to that within schools.

Table 2.3 STATA Output for Main Effects, Random Intercept Model on Tenth Grade Social Studies Test Scores

Y	Coef.	Std. Err.	z	P>|z|	[95% Conf.	Interval]
FSL	-4.322866	.3331719	-12.97	0.000	-4.975871	-3.669862
SEX	-1.653334	.1870637	-8.84	0.000	-2.019972	-1.286696
_cons	50.86574	.1662935	305.88	0.000	50.53981	51.19167

Random-effects Parameters		Estimate	Std. Err.	[95% Conf.	Interval]
SCH_ID: Identity					
	sd(_cons)	3.903847	.1524490	3.616201	4.214373
	sd(Residual)	8.784368	.0660321	8.655897	8.914747

To specify this same model using HLM, we utilize the hierarchical, level form of the notation. HLM software has a graphical user interface (GUI) that facilitates model specification at each level. It also has the option to display the model in composite form. What also distinguishes HLM software is that predictors may be managed separately, depending on their level. So a level one predictor may be in a student-level file, in our example, while a level two predictor may be in a school-level file.[5] These files are linked through the school identifier, since for each student we know which school he or she attends. We do not reproduce the level notation generated by the HLM software. It is only slightly different from our equation (2.7), with minor substitution of symbols, to be discussed. Table 2.4 is slightly edited output from the estimation.

The STATA and HLM estimates are extremely similar, but note that the latter's presentation directly reflects the notation. For fixed effects, the sub-model for β_{0j} is given first, followed by the sub-model for β_{1j} (β is represented by `B` in the ASCII output, and γ is represented by `G`). The ordering reflects the perspective that we are estimating sub-models for latent intercepts and slopes, as they are explicitly represented through notation. The terms in each sub-model are listed *within* each latent effect, so that the FSL effect, `G01`, for `B0` is listed as the predictor following the intercept for that sub-model, `G00`. The SEX effect `G10` is the intercept in the sub-model for `B1`.

2.6 TARGETING LATENT COMPONENTS

As alluded to previously, different notations did not, for the most part, develop in a vacuum. Many variations in notation were introduced in the context of specific sets of research questions within disciplines. A good example of this is the interest, or lack thereof, in variance components, which are the parameters governing the stochastic terms in a model. While it may not be immediately apparent, notation can influence one's perspective on these components (and vice versa). In the composite form of the MLM, all of the stochastic components can be grouped together and simply viewed as complex covariance structure that corrects for the correlation inherent in the sampling design and potential heteroscedasticity. For example, the random intercept model is equivalent to the error structure known as *compound symmetry* or *exchangeable* correlation. Models with compound symmetry or other familiar forms of heterogeneous error may be fit using generalized least squares. More complex error structures are described in Chapter 10. Once a plausible form of covariance structure is included in the model,

Table 2.4 HLM Output for Equivalent Main Effects, Random Intercept Model on Tenth Grade Social Studies Test Scores

Fixed Effect	Coefficient	Standard Error	T-ratio	Approx. d.f.	P-value
For INTRCPT1, B0					
INTRCPT2, G00	50.865741	0.164830	308.596	1082	0.000
FSL, G01	-4.322833	0.328996	-13.139	1082	0.000
For SEX slope, B1					
INTRCPT2, G10	-1.653335	0.243976	-6.777	8701	0.000

Final estimation of variance components:

Random Effect		Standard Deviation	Variance Component	df	Chi-square	P-value
SCH_ID: Identity						
INTRCPT1,	u0	3.90347	15.23709	1082	3028.22466	0.000
level-1,	r	8.78440	77.16567			

standard errors for parameter estimates are properly adjusted. Some researchers do not care to explore the parameters governing this correction, as the covariance is simply a nuisance to them. See Chapter 25 for a discussion of generalized estimating equations, which for the most part takes this perspective.

In this section, we focus on variance components derived from the multilevel nature of the model. This perspective is exemplified by the common use of a baseline model that quantifies the amount of between- versus within-group variation in the outcome of interest *before* including predictors in the model. Singer and Willett (2003) suggest initially fitting an *unconditional means* model, which we presented as Model 1 in our tables. That model had a single latent variable represented by β_{0j} in level notation, with variance components, σ^2_{u0} and σ^2_{ε}, corresponding to the variances of the level two error term associated with β_{0j} and level one residual variance. Researchers often hypothesize that by adding a predictor at a particular level to a model, one will observe a reduction in the variance of the random effects associated with that level. Adding a (level two) predictor indexed by j to Model 1 can be understood as an attempt to explain group-level differences as represented by β_{0j}, and thus one would expect that σ^2_{u0} would be reduced in the new model as we are targeting that portion of the variation. Contrast this with introducing a level one predictor indexed by subscripts ij in the level one equation. Here, we would expect that an effective predictor would result in a reduction in σ^2_{ε}. These expectations as to which component of the variance is "targeted" by the inclusion of predictors at different levels follows naturally from the hierarchical, level notation. See Raudenbush (2009) for additional discussion of targeting components of variation.

The above discussion of targeting variance components is idealized. In practice, one should be prepared for more complex and nuanced scenarios, as all of the model parameters are estimated simultaneously, and thus conditioning on additional predictors changes the interpretation and potentially the impact of the original and new parameters. Nevertheless, ideally one may organize one's research questions to explain variance components with the inclusion of predictors. In Section 2.2, we provided the results for an increasingly complex set of models, but we were not explicit about how each model establishes or attempts to explain components of variation. We do this now, referring to Tables 2.1 and 2.2. Model 1 is the unconditional means model, establishing estimates $\hat{\sigma}^2_{u_0} = 19.70$ and $\hat{\sigma}^2_{\varepsilon} = 77.87$. As a reminder, this type of model contains no predictors. Model 2 introduces the school-level predictor FSL_j, which can be seen in Table 2.2, column 2, in the level two formula $\beta_{0j} = \gamma_{00} + \gamma_{01}\mathrm{FSL}_j + u_{0j}$, to explain between school differences captured in level two error term u_{0j}. The estimate of the variance of that component drops to 15.29 with FSL_j in the model, while the residual variance estimate is effectively unchanged. Model 3 introduces a student-level (or level one) predictor, SEX_{ij}, which, given its placement in the model, should explain the residual variation. The estimate of $\hat{\sigma}^2_{\varepsilon}$ drops from 77.79 to 77.17, which is a very minor change.

Model 4 allows the effect of SEX_{ij} to vary across schools, addressing the question of whether the gender gap varies as a function of a school's FSL level. Given its representation in level notation, this addition would explain the between-school variation in β_{1j}. To assess the extent to which the addition of FSL_j accounts for the variation in β_{1j}, one would first need to introduce a random effect in the level two model that would estimate that variation.[6] We forego this analysis because the addition of FSL in Model 4 did not alter the variance components from what they were in Model 3. Model 5 contains every predictor and their interaction, as well as a random effect for every level one predictor. The reduction in residual variance from 77.10 to 75.09 suggests that some of the residuals from Model 4 had additional structure within school. In some schools, the gap between males and females was larger, while in others it was smaller. This variation is captured in the $\sigma^2_{u_1}$ term, which is estimated to be 8.06.

One of the distinctions between composite and level notation is how one characterizes interaction terms. In the composite form of the model, one usually ascribes a regression interpretation: the effect of the interaction term is given, net of all fixed and random effects. If one uses level notation and refers to the sub-model equation in which the interaction effect resides, then one is likely to describe the effect in terms of the slope on the level two predictor. In our Model 5 in Table 2.2, $\beta_{1j} = \gamma_{10} + \gamma_{11}\mathrm{FSL}_j + u_{1j}$, so γ_{11} is the slope on FSL_j in a model predicting the latent variable β_{1j}. As we shall now illustrate, first with STATA and next with HLM, this descriptive difference translates to a difference in the way in which STATA and HLM present their output.

The estimates associated with Model 5 may be obtained using the following STATA command, where the variable SEX_FSL represents the product between SEX and FSL computed in advance by the researcher:

```
xtmixed Y FSL SEX SEX_FSL ||
SCH_ID: SEX
```

As a reminder, in this syntax the fixed effects are given first, followed by the `||` and then the random effects. The STATA output associated with this command is given in Table 2.5. With respect to the two random effects, as noted earlier in this chapter, sd(_cons) refers to the random effect intercept variance and sd(SEX) refers to the random effect slope variance for the SEX predictor.

In contrast to the STATA output, the output for Model 5 based on HLM, given in Table 2.6, mirrors the level notation. In particular, the interaction term is reported in this output as simply the effect of the predictor FSL_j on the latent variable β_{1j}. Accordingly, the effect `G11` is referenced as a fixed effect for the SEX slope, `B1` in the HLM output. Notably, the variance component σ_{u_1} is listed by referring to `SEX slope, u1`.

2.7 THREE-LEVEL NESTING

There are often even more complicated dependencies in the data than two-level nesting. For example, we might have children nested within classrooms and classrooms nested within schools. Composite notation can accommodate three levels of nesting

Table 2.5 STATA Output for Interaction, Random Intercept and Slopes Model on Tenth Grade Social Studies Test Scores

Y	Coef.	Std. Err.	z	P>|z|	[95% Conf.	Interval]
FSL	−4.367319	.3331512	−13.11	0.000	−5.020283	−3.714354
SEX	−1.630995	.2217208	−7.36	0.000	−2.06556	−1.196431
SEX_FSL	1.32297	.4439517	2.98 0.003	.4528407	2.193099	
_cons	50.84589	.1661933	305.94	0.000	50.52016	51.17163

Random-effects Parameters	Estimate	Std. Err.	[95% Conf.	Interval]
SCH_ID: Independent				
sd(SEX)	2.839104	.2812424	2.338089	3.447478
sd(_cons)	3.89056	.1526332	3.602616	4.201517
sd(Residual)	8.665238	.0668994	8.535105	8.797356

Table 2.6 HLM Output for Equivalent Interaction, Random Intercept and Slopes Model on Tenth Grade Social Studies Test Scores

Fixed Effect	Coefficient	Standard Error	T-ratio	Approx. d.f.	P-value
For INTRCPT1, B0					
INTRCPT2, G00	50.845968	0.164809	308.514	1082	0.000
FSL, G01	−4.367333	0.330727	−13.205	1082	0.000
For SEX slope, B1					
INTRCPT2, G10	−1.630818	0.220597	−7.393	1082	0.000
FSL, G11	1.322565	0.441995	2.992	1082	0.003

Final estimation of variance components:

Random Effect		Standard Deviation	Variance Component	df	Chi-square	P-value
SCH_ID: Identity						
INTRCPT1,	u0	3.89067	15.13729	723	2344.81255	0.000
SEX slope,	u1	2.83042	8.01128	723	894.23124	0.000
level-1,	r	8.66558	75.09223			

simply by adding a further subscript to any terms used. Level notation adds a subscript as well, but it also introduces a new level, and as such one must track which sub-model includes which predictors a bit more carefully. Random effects specified on one level are usually taken to be independent of those on other levels, but even the more basic models have somewhat complicated interpretations. For example, assume student i is in classroom j in school k, and we have a model with a single predictor SEX_{ijk} (we have essentially "inserted" a level between schools and students in our previous models). A variant of a three-level composite model specification with random intercepts and slopes could be:

$$y_{ijk} = (u_{0jk} + u_{00k} + \delta_{000}) + (u_{1jk} + u_{10k} + \delta_{100})\text{SEX}_{ijk} + \varepsilon_{ijk},$$

with the usual normal, independence assumptions on the stochastic terms. Note that we have used parameter names that differ from the usual βs so that the correspondence with hierarchical, level notation will be clearer. We have also grouped the terms according to their role modifying the intercept or slope. Thus the random intercept is $\beta_{0jk} = u_{0jk} + u_{00k} + \delta_{000}$ and the random slope $\beta_{1jk} = u_{1jk} + u_{10k} + \delta_{100}$. Hierarchical notation would break these components down in stages, with $\beta_{0jk} = \gamma_{00k} + u_{0jk}$ and then $\gamma_{00k} = \delta_{000} + u_{00k}$. An analogous set of terms exists for the random slope term.

The hierarchical model specification would, of course, categorize these stages as levels:

Level one:

$$y_{ijk} = \beta_{0jk} + \beta_{1jk}\text{SEX}_{ijk} + \varepsilon_{ijk}$$

Level two:

$$\begin{aligned}\beta_{0jk} &= \gamma_{00k} + u_{0jk}\\ \beta_{1jk} &= \gamma_{10k} + u_{1jk}\end{aligned}$$

Level three:

$$\begin{aligned}\gamma_{00k} &= \delta_{000} + u_{00k}\\ \gamma_{10k} &= \delta_{100} + u_{10k},\end{aligned}$$

with the usual distributional assumptions. First, note that we have loosely followed the approaches of Bryk and Raudenbush (1992) and Goldstein (1995), utilizing a fairly complex subscripting scheme to identify the parameters and their free indices. Level-one parameters contain one fixed subscript, as in $\beta_{0..}$ or $\beta_{1...}$ The parameters become outcomes at the next level. The error term ε_{ijk} parallels the outcome, y_{ijk}, in that only free indices are included. This structure is maintained at level two, in which the outcome β_{0jk} has a single fixed subscript 0, and the error term u_{0jk} uses parallel form. The level two coefficient γ_{00k} begins with a double subscript 00, reflecting the additional level of nesting (as compared to terms in level one). On level three, explanatory terms have triple fixed subscripts, such as 000, and error terms have only two fixed subscripts, as they must correspond to that level's outcome, which also has two fixed subscripts, suggesting further that variables at level three only display variability at level three. While the plethora of subscripts may seem awkward or cumbersome, they are useful for identifying the covariances of individual terms. At level two, the two variances may be written $\sigma^2_{u_0}$ and $\sigma^2_{u_1}$, with covariance σ_{u_0,u_1}. At level three, the corresponding terms are $\sigma^2_{u_{00}}$ and $\sigma^2_{u_{10}}$, with covariance $\sigma_{u_{00},u_{10}}$. The extra subscripts make it possible to unambiguously reference all of these terms, and we can identify the level of each component by the number of subscripts used.

Notice that, regardless of notation, there are now two sources of variation other than the error: classroom level, and school level. Additional predictors can be used to target these new variance components as well.

In STATA, with the composite notation, the additional level translates to an additional nested grouping structure, so the model that we have described, with random intercepts and slopes for the single predictor SEX_{ijk}, can be specified as:

```
xtmixed Y SEX || SCH_ID: SEX
|| CLASS_ID: SEX
```

where `CLASS_ID` is a fictitious classroom identifier. This model specification yields five variance components, two for each level of nesting (one intercept and slope each for school and classroom) and one for the residual variance. In HLM, the model is specified through the GUI, and it would correspond quite closely to the level notation used above.

2.8 LONGITUDINAL DATA

In the longitudinal setting, a single observation for a person must be related through notation to a particular time period. Because there are many persons and many periods (each called a *person-period*) any reference to a single observation must indicate which person and which period. This, too, can become more complicated when subjects are nested, as described above, and longitudinal. It has become fairly standard in longitudinal data modeling to write the outcome for a specific person-period as y_{it}, denoting subject i's tth period. If we view outcomes as a sample from a stochastic process, $Y(t)$, then this notation identifies the subject and time of observation. It is important to note that t may represent the actual time of an event or simply the occasion number. In the latter case, one would write a random intercept, random linear growth model as:

$$y_{it}=\beta_0 + \beta_1 TIME_{it} + u_{0i} + u_{1i} TIME_{it} + \varepsilon_{it},$$

and t would take on integer values as it represents occasion for subject i, and $TIME_{it}$ would be free to vary, identifying the actual time of the observation on the tth occasion. When using this notation, it is common to substitute the subscript j for t, as time is simply another index, so y_{ij} is subject i's outcome on the jth occasion. With the explicit link to time removed in the y_{ij} notation, some researchers prefer to view time as nested within subject (a repeated measures ANOVA perspective), and thus y_{ji} is used to represent the jth occasion for the ith subject. The advantage to the y_{it} notation is that it makes explicit reference to time in the outcome and so other model components indexed by t are immediately identifiable as time effects. When there are no additional levels of nesting in the model, any of these approaches will work, though we recommend the use of t in the notation.

We now employ the y_{it} notation for subject i's outcome at occasion t and rewrite the above random intercepts and slopes model, described as the *unconditional growth* model by Singer and Willett (2003). If we proceed this way, the outermost subscript, t, rather than the innermost subscript (which is standard), changes on level one. Here is the model:

Level one:

$$y_{it} = \beta_{0i} + \beta_{1i} TIME_{it} + \varepsilon_{it}$$

Level two:

$$\begin{aligned} \beta_{0i} &= \gamma_{00} + u_{0i} \\ \beta_{1i} &= \gamma_{10} + u_{1i}, \end{aligned}$$

with the usual distributional assumptions. The indices used above have an unambiguous interpretation. As long as the modeling remains strictly longitudinal, with two levels of nesting, and the index t, which clearly represents time, is used, the ordering of the subscripts is a non-issue.

When the data are nested and longitudinal, it becomes clear that there are advantages to using subscript t, as one is charged with representing both nested information and the longitudinality of the data. Say we are interested in observing student i in classroom j four times during a single school year, in which case she does not switch classrooms during the observation period. Writing the outcome as y_{tij} or y_{ijt} seems reasonable (note that the former approach is simply a more nested analog of y_{ti} notation). Either is preferable to something like $y_{ij,t}$, which does separate the two key aspects of the outcome, but could be viewed as cumbersome. Once again, the choice of subscript notation is tied to whether one tends to use the composite or hierarchical, level notation for the rest of the model specification. In composite notation, there is no compelling reason to "nest" time within subjects, so it seems either subscript notation suffices.

A nested longitudinal model with random intercepts and slopes at every level of grouping could be written in composite notation as:

$$\begin{aligned} y_{ijt}=&(u_{0ij} + u_{00j} + \delta_{000}) \\ &+ (u_{1ij} + u_{10j} + \delta_{100}) TIME_{ijt} + \varepsilon_{ijt}, \end{aligned}$$

with the usual distributional assumptions. Here, we have deliberately chosen Greek letters that would correspond closely to the hierarchical model we are about to present. Note how the subscripts modifying the intercept and slope are decreasing in the number of (literal) indices as we move from left to right.

In a hierarchical, level specification, each level represents the point at which some components vary, with level one representing the innermost layer of nesting, traditionally. In the longitudinal setting, time or occasion must vary at level one, and thus the optimal notation to reflect this places the time index first. A three-level model for nested longitudinal data with a single predictor x_{tij} other than time could thus be written:

Level one:

$$y_{tij} = \beta_{0ij} + \beta_{1ij} TIME_{tij} + \beta_{2ij} x_{tij} + \varepsilon_{tij}$$

Level two:

$$\begin{aligned} \beta_{0ij} &= \gamma_{00j} + u_{0ij} \\ \beta_{1ij} &= \gamma_{10j} + u_{1ij} \\ \beta_{2ij} &= \gamma_{20j} + u_{2ij}. \end{aligned}$$

Level three:

$$\begin{aligned}\gamma_{00j} &= \delta_{000} + u_{00j}\\ \gamma_{10j} &= \delta_{100} + u_{10j}\\ \gamma_{20j} &= \delta_{200} + u_{20j},\end{aligned}$$

with the usual distributional assumptions.

In this very general model, there are random effects for time, at the individual and classroom level, indicated by β_{1ij} and its stochastic components u_{1ij} (individual) and u_{10j} (classroom). One can also see the impetus in level notation for using a different symbol for the coefficients at each level. Some of these notational conventions translate readily into composite notation, as one can see from the correspondence above. One advantage to the composite approach is that fewer intermediary terms (in the above case, the βs and γs) are needed and there are no constraints on the placement of random effects, so their inclusion or exclusion takes place in one equation. Using either notation, it becomes clear that there may be individual effects (level or slope) as well as effects due to classrooms. In the hierarchical notation, any time-varying predictor, such as x_{tij}, or the interaction of two time-varying predictors, must be included on level one, whereas there is no such restriction for the composite form (in part, because there is only one level in the latter case).

We have observed that in longitudinal data modeling, time often has an elevated status. The above discussion underscores the importance of this perspective. However, when time is simply an index for occasion, and those occasions are referenced as $TIME_{ti}$ (the ordering *is* relevant here), one could say that t is just another index, like i or j, and that clearly occasion t is *nested* within subject i. Further, one might argue along those same lines that $TIME_{ti}$ is just a subject-varying predictor, like any other x_{ti}. Because, however, nearly every longitudinal data model has some time-trend component which may potentially differ between subjects, time, in fact, does have an elevated status in longitudinal data modeling. Moreover, given a longitudinal setting, one must pay particular attention to the issue of homogeneity across time for both predictors and error terms. Although effective model builders would make a similar set of statements about any modeling problem, for those working with longitudinal data, it is clear that this consideration has some added importance.

Matrix notation, most closely resembling the composite form, is only slightly altered to represent longitudinal data. Laird and Ware (1982) introduced notation that is fairly standard at this point, in which the grouping factor is indexed by i rather than j, but otherwise is quite similar to the structure developed in Section 2.4. A single individual's vector of observations, $\boldsymbol{y}_i$, is modeled in a manner analogous to (2.6) above:

$$\boldsymbol{y}_i = X_i\boldsymbol{\beta} + Z_i\boldsymbol{u}_i + \boldsymbol{\varepsilon}_i,$$

where X_i and Z_i are $n_i \times p$ and $n_i \times q$ design matrices consisting of predictors or a column of ones. $\boldsymbol{\beta}$ $(p \times 1)$ is taken as fixed, which in this context implies non-varying by individual, $\boldsymbol{u}_i \sim N_q(0, G_0)$, and $\boldsymbol{\varepsilon}_i \sim N_{n_i}(0, R_i)$, independent of one another. Let us say that m is the number of individuals and that $N = \sum_{i=1}^{m} n_i$ above is the total number of observations. Clearly, this model can be made equivalent to (2.6) by setting $J = m$, $\boldsymbol{y} = (\boldsymbol{y}_1', \ldots, \boldsymbol{y}_m')'$, $X = (X_1'|\cdots|X_m')'$, $Z = \bigoplus_{i=1}^{m} Z_i$, $\boldsymbol{u} = (\boldsymbol{u}_1', \ldots, \boldsymbol{u}_m')'$, $\boldsymbol{\varepsilon} = (\boldsymbol{\varepsilon}_1', \ldots, \boldsymbol{\varepsilon}_m')'$, $G = \bigoplus_{i=1}^{m} G_0$, and $R = \bigoplus_{i=1}^{m} R_i$, where $\bigoplus$ is the direct sum (block diagonal) operator.

Further details related to longitudinal data modeling are given in Chapter 9.

2.9 CROSSED EFFECTS

Crossed effects are usually expressed in composite form, and different subscripts are used for different group memberships. So, for example, a student's third grade classroom may be indexed by j and her fourth grade classroom by k. There is no reason to assume that every student in classroom j ends up in classroom k a year later, so these effects are non-nested. A simple model for this situation might be written

$$y_{ijk} = \beta_0 + \beta_1 x_{1ij} + \beta_2 x_{2ik} + u_{1j} + u_{2k} + \varepsilon_{ijk},$$

with the usual distributional assumptions, and in which the single observation y_{ijk} represents an outcome for student i who was in classrooms j and k in third and fourth grades, respectively; x_{1ij} a predictor associated with student i when she was in classroom j; x_{2ik} a predictor associated with student i when she was in classroom k; and u_{1j} and u_{2k} are effects for the classrooms in the two grades. An interaction term, u_{12jk}, can be introduced, further exploiting the crossed structure. This would be the additional effect associated with being in both classrooms j and k. Of course, the full distribution of $(u_{1j}, u_{2k}, u_{12jk})'$ remains to be specified.

The composite form of notation seems more natural in the context of crossed effects, highlighting the need for the researcher to be familiar with several forms of notation.

2.10 CONCLUSION

This chapter began by describing models at the level of individual observations and demonstrated how subscript notation could be used to organize the model components. We showed that the way in which we organize these components often reflects the research questions we wish to answer. Of course, any model formulation may be translated into one of the other forms through algebraic manipulation. Sometimes this is necessary for practical reasons, as the syntax used by different software partially reflects the discipline in which it was developed. Sometimes, a notation has the stronger effect of providing a framework for model selection and interpretation. Whatever the purpose, this chapter has provided a roadmap for understanding the differences between some common notational approaches used in MLM and the ramifications of those differences for the consideration and interpretation of effects. We hope this roadmap will be helpful to researchers in utilizing each approach to its fullest.

NOTES

1 Although NELS contains longitudinal data on individual subjects, we focus on a single time point.

2 Students are nested within classrooms, which are nested within schools, which are nested within districts and states. Some districts were sampled with higher probability, and the NELS public data set includes cluster and weighting information to allow analysts to estimate the effects in the target population.

3 The FSL measure was based on the school attended in grade 8. By grade 10, about 1% of students have moved, so we make an adjustment to this measure as discussed in a footnote in Section 2.5.

4 This last approach is common to fields that prefer what we call "level notation," to be described shortly. We thank a reviewer for making this distinction.

5 Recent versions of HLM software will construct a school-level file from the student-level version, by assuming that school-level predictors are unique within the school-level file (a single value of the predictor is chosen by the software); no such distinction is made or enforced by software that is solely based on a composite form, such as STATA. In the NELS data, FSL is assessed for each student in grade 8, so it is possible for that predictor to vary within the school attended in tenth grade; this occurred in fewer than 10% of schools, so we simply took the most frequent FSL status and assigned it to the whole school.

6 Rather than the equation $\beta_{1j} = \gamma_{10} + \gamma_{11}\mathrm{FSL}_j$ as one of the level two sub-models, we might first include the simpler sub-model equation $\beta_{1j} = \gamma_{10} + u_{1j}$, which would provide an estimate of the variance of β_{1j} (via u_{1j}) in the output.

REFERENCES

Browne, W.J., Goldstein, H., and Rabash, J. (2001) 'Multiple Membership Multiple Classification (MMMC) Models', *Statistical Modeling*, 1: 103–24.

Bryk, A.S. and Raudenbush, S.W. (1992) *Hierarchical Linear Models: Applications and Data Analysis Methods*. Thousand Oaks, CA: Sage.

Diggle, P.J., Heagerty P.J., Liang K.-Y., and Zeger S.L. (2002) *Analysis of Longitudinal Data*, 2nd edition. Oxford: Clarendon Press.

Fitzmaurice, G.M., Ware, J.H., and Laird, N.M. (2004) *Applied Longitudinal Analysis*. Hoboken, NJ: Wiley-Interscience.

Gelman, A. and Hill, J. (2007) *Data Analysis Using Regression and Multilevel/Hierarchical Models*. New York: Cambridge University Press.

Goldstein, H. (1995) *Multilevel Statistical Models*. London: Edward Arnold.

Hox, J. (2002) *Multilevel Analysis: Techniques and Applications* Mahwah, NJ: Erlbaum.

Laird, N.M. and Ware, J.H. (1982) 'Random-effects Models for Longitudinal Data' *Biometrics*, 38: 963–74.

Longford, N.T. (1993) *Random Coefficient Models*. Oxford: Oxford University Press.

Hyde, J.S. and Mertz, J.E. (2009) 'Gender, Culture, and Mathematics Performance', *Proc. Natl. Acad. Sci.*, 106: 8801–7.

Pinheiro, J.C. and Bates, D.M. (2000) *Mixed-effects Models in S and SPLUS*. New York: Springer.

Rampey, B.D., Dion, G.S., and Donahue, P.L. (2009) *NAEP 2008 Trends in Academic Progress (NCES 2009-479)*. Washington, DC: National Center for Education Statistics, Institute of Education Sciences, U.S. Department of Education.

Raudenbush, S.W. (2009) 'Targets of Influence in Hierarchical Models for Longitudinal Data', In *Longitudinal Data Analysis*, Garrett Fitzmaurice, Marie Davidian, Geert Molenberghs, and Geert Verbeke (eds). New York: Chapman and Hall, 167–90.

Singer, J.D. (1998) 'Using SAS PROC MIXED to Fit Multilevel Models, Hierarchical Models, and Individual Growth Models', *Journal of Educational and Behavioral Statistics*, 25: 323–55.

Singer, J.D. and Willett, J.B. (2003) *Applied Longitudinal Data Analysis: Modeling Change and Event Occurrence*. New York: Oxford University Press.

Snijders, T.A.B. and Bosker, R. (1999) *Multilevel Analysis: An Introduction to Basic and Advanced Multilevel Modeling*. London: Sage.

3

Likelihood Estimation in Multilevel Models

Harvey Goldstein
Centre for Multilevel Modeling,
University of Bristol, United Kingdom

3.1 A BRIEF HISTORY

Fisher (1925) introduced the notion of the *intraclass coefficient*. His basic example was a set of measurements on a sample of brother pairs. In terms of modern notation he defined the quantities:

$$\hat{\beta}_0 = \frac{1}{N}\sum_{ij} y_{ij}$$

$$\hat{\sigma}_T^2 = \frac{1}{N}\sum_{ij}\left(y_{ij} - \hat{\beta}_0\right)^2$$

$$\hat{\sigma}_u^2 = \frac{1}{m}\sum_{i}\left(y_{i1} - \hat{\beta}_0\right)\left(y_{i2} - \hat{\beta}_0\right),$$

where i, j index level 1 units (brothers) and level 2 units (pairs), and there are $m = N/2$ pairs. The implicit 2-level composite model can be written as

$$y_{ij} = \beta_0 + u_j + \varepsilon_{ij} \qquad (3.1)$$
$$u_j \sim N(0, \sigma_u^2) \quad \varepsilon_{ij} \sim N(0, \sigma_\varepsilon^2),$$

where the u_j, ε_{ij} are assumed to have independent (normal) distributions. Fisher defined the intraclass correlation as $\rho = \sigma_u^2/\sigma_T^2$, where $\sigma_T^2 = \sigma_u^2 + \sigma_\varepsilon^2$.

Fisher extended this to the case where there were greater than two level 1 units per level 2 unit, but only considered the balanced case. His basic concern was to estimate the correlation between pairs of level 1 units, rather than to find an expression for the percentage of variation at level 2, although in this simple 'variance component' model, as he pointed out, they have the same value.

Fisher's ideas were developed largely in the areas of agriculture and genetics, and a summary of what came to be known as random effects analysis of variance can be found in Scheffé (1959). These models were developed for purely hierarchical as well as cross-classified variance component data structures, including covariates, using moment estimators based on equating expected mean squares. While the computations were relatively straightforward for balanced designs, unbalanced designs led to complications with various approximations being developed.

The next major development, and arguably the start of the 'modern' era of multilevel modeling, is the paper by Lindley and Smith (1972). While ostensibly concerned with a Bayesian formulation for multiple regression, they introduced, via a prior distribution for the level 2 effects, the now familiar expressions for the fixed and random parameters and shrunken residuals. Their algorithm anticipates the IGLS algorithm described below.

Harville (1977) sets out a full maximum likelihood framework for a multilevel model, including maximum and restricted maximum likelihood estimation.

Aitkin et al. (1981) describe a random coefficient model and obtain maximum likelihood estimates using an EM algorithm. That paper was important in that it showed how a multilevel analysis could change the inferences obtained from a single-level regression model. Aitkin and Longford (1986) describe a Fisher scoring algorithm for maximum likelihood estimation (MLE) in n-level models, and Laird and Ware (1982) describe a general random coefficient model for analysis of repeated measures data. Goldstein (1986) describes the iterative generalized least squares (IGLS) algorithm with extensions to multivariate multilevel models and models with measurement errors.

Subsequent work essentially elaborated these models in terms of likelihood-based estimation, for example the use of quasi-likelihood methods for discrete response data, the further use of the EM algorithm (Bryk and Raudenbush, 1992) and extensions to structural equation modeling (McDonald and Goldstein, 1989; Skrondal and Rabe-Hesketh, 2004). By the early 1990s fairly general-purpose software was available, but only in a few specialist packages (VARCL, HLM, ML3).

In the 1990s, Bayesian methods came into their own with the application of MCMC algorithms – see Chapter 4.

3.2 THE N-LEVEL MODEL WITH RANDOM COEFFICIENTS

3.2.1 Notation

Consider the simple 2-level model (3.1). As the number of levels increases the subscripting will extend so that, for example, a 5-level model will require subscripts i,j,k,l,m. Furthermore, as we shall see later with cross-classified and especially multiple membership models, we can readily have many ways of classifying units that will lead to rather cumbersome, and for multiple membership models non-standard, subscripting. When we come to discuss these models we will outline an alternative, more general, notation.

For now, to describe the estimation algorithms we shall use a simple 2-level formulation that captures the essential steps and has straightforward generalizations to n-level models. We can write the general 2-level model with random coefficients as

$$Y = X\beta + E,$$

where $Y = \{y_{ij}\}$, $X = \{X_{ij}\}$,

$$X_{ij} = \{x_{0ij}, x_{1ij}, \ldots, x_{pij}\},$$

and

$$E = E_1 + E_2 = Z_{\varepsilon ij}\varepsilon_{ij} + Z_{uij}u_j,$$

or simply

$$Y = X\beta + Z_\varepsilon \varepsilon + Z_u u.$$

The residual matrices E_1, E_2 have expectation zero with

$$E(E_1 E_1^T) = V_{(2)1},\ E(E_2 E_2^T) = V_{(2)2}$$

$$E(E_1 E_2^T) = 0,\ V_2 = V_{2(1)} + V_{2(2)}. \quad (3.2)$$

Note that we have allowed more than one simple random (intercept) effect at level 1 to allow for the case where we have complex level 1 variance that depends on explanatory variables. In the standard model the level 1 residuals are assumed independent across level 1 units, so that $V_{2(1)}$ is diagonal with ijth element

$$\text{var}(\varepsilon_{ij}) = \sigma^2_{\varepsilon ij} = z^T_{\varepsilon ij}\Omega_\varepsilon z_{\varepsilon ij}$$

and Ω_e is the $(q_1 \times q_1)$ covariance matrix of the level 1 residuals.

The level 2 residuals are assumed independent across level 2 units, $V_{2(2)}$ is block-diagonal with j-th block

$$V_{2(2)j} = z_{uj}^T \Omega_u z_{uj},$$

and Ω_u is the $q_2 \times q_2$ covariance matrix of the level 2 residuals. The jth block of V_2 is therefore given by

$$V_{2j} = \oplus_j \sigma_{eij}^2 + V_{2(2)j}, \qquad (3.3)$$

where $\oplus$ is the direct sum operator.

For some models such as time series models, the requirement of independence among the residuals for the level 1 units is relaxed. In this case the first term on the right-hand side of (3.3) is replaced by the particular structure for $V_{2(1)}$.

3.2.2 Estimation Using Iterative Generalized Least Squares (IGLS)

For known V_2 and omitting the subscript for convenience, the generalized least squares estimate of the fixed coefficients is

$$\hat{\beta} = \left(X^T V^{-1} X\right)^{-1} X^T V^{-1} Y, \qquad (3.4)$$

with covariance matrix $(XVX)^{-1}$.

For known β, we form

$$Y^* = \tilde{Y}\tilde{Y}^T \quad \tilde{Y} = Y - X\beta,$$

and we have $E(Y^*) = V$. We now write $Y^{**} = \text{vec}(Y^*)$, where vec is the vector operator stacking the columns of Y^* underneath each other. We can now write a linear model involving the random parameters θ, that is the elements of Ω_u, Ω_e, as

$$E\left(Y^{**}\right) = Z^*\theta,$$

where Z^* is the design matrix for the random parameters.

For example, for the model given by (3.1) for a level 2 unit with 2 level 1 units, we have

$$E\begin{pmatrix} \tilde{y}_1^2 \\ \tilde{y}_1 \ \tilde{y}_2 \\ \tilde{y}_2^2 \end{pmatrix} = \sigma_u^2 \begin{pmatrix} 1 \\ 1 \\ 1 \end{pmatrix} + \sigma_e^2 \begin{pmatrix} 1 \\ 0 \\ 1 \end{pmatrix}.$$

We now carry out a generalized least squares analysis to estimate θ, namely

$$\begin{aligned} \hat{\theta} &= (Z^{*T} V^{*-1} Z^*)^{-1} Z^{*T} V^{*^{-1}} Y^{**}, \\ & V^* = V \otimes V, \qquad (3.5) \end{aligned}$$

where $\otimes$ is the Kronecker product. The covariance matrix of $\hat{\theta}$ is given by

$$(Z^{*T} V^{*-1} Z^*)^{-1} Z^{*T} V^{*-1} \text{cov}(Y^{**}) V^{*-1} Z^* \times (Z^{*T} V^{*-1} Z^*)^{-1}.$$

Now we have

$$Y^{**} = \text{vec}(\tilde{Y}\tilde{Y}^T) = \tilde{Y} \otimes \tilde{Y}.$$

Using a standard result (see, for example, Searle et al., 1992, Section 12.3), we have

$$cov(\tilde{Y} \otimes \tilde{Y}) = (V \otimes V)(I + S_N),$$

where $V \otimes V = V^*$ and S_N is the vec permutation matrix.

As Goldstein and Rasbash (1992) note, the matrix A where $Z^* = \text{vec}(A)$ is symmetric, and hence

$$\begin{aligned} V^{*-1} Z^* &= V^{-1} \otimes V^{-1} \text{vec}(A) \\ &= \text{vec}(V^{-1} A V^{-1}) \end{aligned}$$

and $V^{-1}AV^{-1}$ is symmetric so that, using a standard result, we have

$$S_N V^{*-1} Z^* = V^{*-1} Z^*.$$

After substituting in the above expression for $\text{cov}(\hat{\theta})$ we obtain

$$\text{cov}(\hat{\theta}) = 2(Z^{*T} V^{*-1} Z^*)^{-1}.$$

The iterative generalized least squares (IGLS) procedure (Goldstein, 1986) iterates between (3.4) and (3.5) using the current estimates of the fixed and random parameters. Typical starting values for the fixed parameters are those from an ordinary least squares analysis. At convergence, assuming multivariate normality, the estimates are maximum likelihood.

The IGLS procedure produces biased estimates in general for the random parameters and this can be important in small samples.

Goldstein (1989) shows how a simple modification leads to restricted iterative generalized least squares (RIGLS) or restricted maximum likelihood (REML) estimates which are unbiased. If we rewrite (3.3) using the *estimates* of the fixed parameters $\hat{\beta}$ we obtain

$$\begin{aligned} E(Y^*) &= V_2 - X\text{cov}(\hat{\beta})X^T \\ &= V_2 - X(X^T V_2^{-1} X)^{-1} X^T. \quad (3.6) \end{aligned}$$

By taking account of the sampling variation of the $\hat{\beta}$ we can obtain an unbiased estimate of V_2 by adding the second term in (3.6), the *hat matrix*, to Y^* at each iteration until convergence. In the case where we are estimating a variance from a simple random sample this becomes the standard procedure for using the divisor $n-1$ rather than n to produce an unbiased estimate.

Full details of efficient computational procedures for carrying out all these calculations are given by Goldstein and Rasbash (1992).

3.2.3 Estimation of Random Effects (Residuals)

Conditional on the parameter estimates, we can obtain estimates for the random effects, or residuals, at any level. We give details for a 2-level model and there are corresponding expressions for models with further levels.

Denote the set of m distinct residuals at level 2 by

$$u = \{u_1, \ldots, u_m\}, u_j^T = \{u_{j1}, \ldots, u_{jn_j}\},$$

where n_j is the number of level 1 units in the jth level 2 unit. Since the residuals at any level are independent of those at any other level, we have the posterior or predicted residual estimates given by

$$\hat{u}_j = E(u_j | \tilde{Y}_j V_j),$$

where $\tilde{Y} = Y - X\beta$. We consider the regression of the set of all residuals u on $\tilde{Y}$ which gives the general estimator for each level 2 unit in the form

$$\hat{u}_j = Z_j \Omega_u V_j^{-1} \tilde{Y}_j, \quad (3.7)$$

where Z_j is the matrix of explanatory variables for the random coefficients or effects for the jth level 2 unit. We obtain consistent estimators by substituting sample estimates of the parameters in (3.7). These estimates are linear functions of the responses and their unconditional covariance matrix is given by

$$\begin{aligned} &\Omega_u Z_j^T V^{-1}(V - X(X^T V^{-1} X)X^T) \\ &V^{-1} Z_j \Omega_u. \quad (3.8) \end{aligned}$$

The second term (in 3.8) derives from considering the sampling variation of the estimates of the fixed coefficients and can be ignored in large samples, and we obtain a consistent estimator by substituting parameter estimates in

$$\Omega_u Z^T V^{-1} Z \Omega_u,$$

where we drop the j suffix for convenience. If we wish to study the distributional properties of standardized residuals for diagnostic purposes then the unconditional covariance matrix (3.8) should be used to standardize the estimated residuals. If, however, we wish to make inferences about the true u, for example to construct confidence intervals or test differences, then we require the 'comparative' covariance matrix of $\hat{u} = E[(\hat{u} - u)(\hat{u} - u)^T]$. The covariance matrix is given by substituting parameter estimates in

$$\begin{aligned} &\Omega_u - \Omega_u Z^T V^{-1}(V - X(X^T V^{-1} X)X^T) \\ &V^{-1} Z \Omega_u. \quad (3.9) \end{aligned}$$

We note that no account is taken of the sampling variability associated with the estimates of the random parameters in (3.8) or (3.9). Thus, with small numbers of units, a procedure such as bootstrapping should be used to estimate these covariance matrices or a delta method approximation can be used as shown below. Skrondal and Rabe-Hesketh (2004) provide a detailed discussion.

We can derive the following form for the composite residual estimator that is computationally convenient,

$$\hat{u} = (\Omega_u^{-1} + Z^T V_1^{-1} Z)^{-1} Z^T V_1^{-1} \tilde{Y},$$

where $Z = \{Z_1, \ldots .Z_m\}$, and with the comparative covariance matrix as

$$(\Omega_u^{-1} + Z^T V_1^{-1} Z)^{-1},$$

where $V_1 = \sigma_\varepsilon^2 I_n$ and n is the total number of level 1 units. Corresponding to (3.9) we have the following additional correction factor, to be subtracted from this covariance matrix, that takes into account the sampling variation of the fixed coefficients:

$$(\Omega_u^{-1} + Z^T V_1^{-1} Z)^{-1} Z^T V_1^{-1} X (X^T V^{-1} X)^{-1} \times X^T V_1^{-1} Z (\Omega_u^{-1} + Z^T V_1^{-1} Z)^{-1}.$$

3.3 DELTA METHOD ESTIMATORS FOR THE COVARIANCE MATRIX OF RESIDUALS

Consider the case of a 2-level normal model

$$Y = X\beta + Z_\varepsilon \varepsilon + Z_u u.$$

Using a well-known equality, we have

$$\text{var}(\hat{u}|y, \beta, \theta) = E[\text{var}(\hat{u}|y, \beta, \theta)] + \text{var}[E(\hat{u}|y, \beta, \theta)], \qquad (3.10)$$

where the terms on the right-hand side of (3.10) are regarded as functions of the model parameters and evaluated at the sample estimates. The comparative covariance matrix is the first term in (3.10).

For the second term we shall use the first-order approximation derived from the Taylor expansion about $E(\hat{\theta}) = \theta$ for the covariance matrix of a function, namely

$$\text{cov}[g(\hat{\theta})] \approx \left(\frac{\partial g}{\partial \theta}\right)^T \text{cov}(\hat{\theta}) \left(\frac{\partial g}{\partial \theta}\right)\bigg|_{\hat{\theta}}.$$

In some circumstances we may wish to have a better approximation, in which case, assuming multivariate normality, we obtain the additional contribution, evaluated at the sample estimates,

$$\frac{1}{4}\left(\frac{\partial^2 g}{\partial \theta^2}\right)[2A_1 + A_2]\left(\frac{\partial^2 g}{\partial \theta^2}\right)\bigg|_{\hat{\theta}},$$

where $A_1 = \{a_{ij}^2\}$, $\text{cov}(\hat{\theta}) = \{a_{ij}\}$, $A_2 = aa^T$ with $a = \{a_{ii}\}$.

For $\hat{u}$ as a function of the random parameters θ, we have

$$d_k^T = \frac{\partial g}{\partial \theta_k} = [-\Omega_u Z^T V^{-1} \frac{\partial V}{\partial \theta_k} V^{-1} + \frac{\partial \Omega_u}{\partial \theta_k} Z^T V^{-1}]\tilde{y}$$
$$\frac{\partial \Omega_u}{\partial \theta_k} = 0 \text{ if } \theta_k \text{ not at level 2.}$$

Note that the elements of $\frac{\partial V}{\partial \theta_k}$ are just the elements of the design vector for the parameter θ_k, and that

$$\frac{\partial V^{-1}}{\partial \theta_k} = -V^{-1}\frac{\partial V}{\partial \theta_k}V^{-1}.$$

The row vector d_k has q elements, one for each residual at level 2, with $d = \{d_k\}$ a t x r_u matrix where t is the total number of random parameters. The second adjustment term in (3.10) is therefore

$$d^T \text{cov}(\hat{\theta}) d.$$

These adjustments are implemented in MLwiN (Rasbash et al., 2009).

3.3.1 An Example Using Educational Achievement Data

The dataset used to illustrate estimation for the standard normal multilevel model consists of examination and test scores for 4059 students in 65 schools in Inner London. A full account of the data is given by Goldstein et al. (1993). In our analyses, the response is an examination score taken in the final year of compulsory schooling (year 11) and the explanatory variable is a reading test score taken in year 6 just before entry to secondary (high) school. The examination and test scores are separately transformed to normality using normal equivalent deviates across the whole sample. The analyses are carried out using MLwiN.

Figure 3.1 is a scatter plot of the examination score against the reading score, and Figure 3.2 is the set of 65 separate simple regression lines, constrained to be parallel, one for each school, fitted to the data.

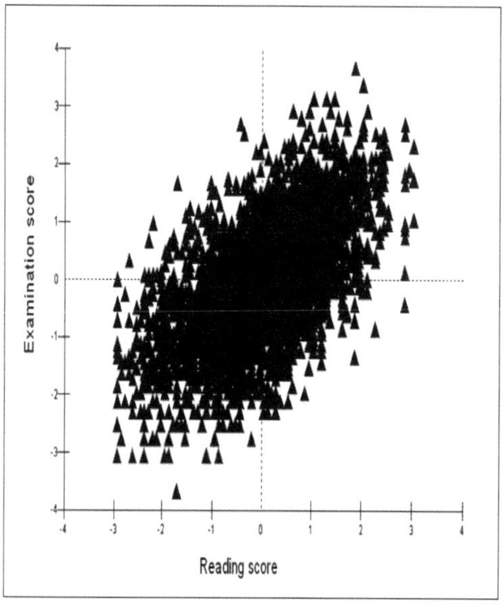

Figure 3.1 Examination Score Versus Reading Score for Educational Achievement Data

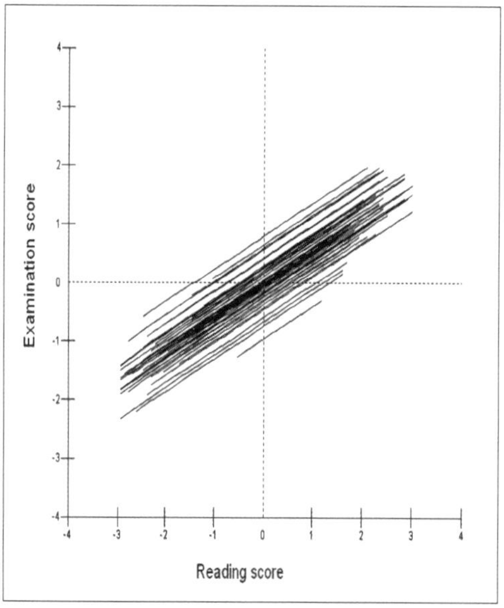

Figure 3.2 Separate Parallel Regression Lines Fitted to Examination Score Versus Reading Score Data

There is a clear relationship between the scores with a sample correlation of 0.59 and variation between schools: the null hypothesis of no difference in school intercepts yields a highly significant chi squared of 620.7 with 64 degrees of freedom.

Table 3.1 Examination Scores Related to Reading Test Scores (standard errors)

Explanatory variable	Fixed effects model	2-level model
Intercept	–	0.002 (0.040)
Reading score	0.559 (0.012)	0.563 (0.012)
Level 2 variance	–	0.092 (0.018)
Level 1 variance	0.556 (0.012)	0.566 (0.013)
−2 log likelihood	9139.8	9357.2

The model assumed in Figure 3.2 is the single-level fixed effect model where a separate intercept is fitted for each school, namely

$$y_i = \beta_0 + \beta_1 x_{i1} + \cdots + \beta_{64} x_{i64} + \varepsilon_i, \quad (3.11)$$

where x_{ij} is a dummy variable for the jth school ($j = 1 \ldots 64$) and the term for school 65 is incorporated into the intercept.

We now formulate the 2-level version of (3.11), namely

$$y_{ij} = \beta_0 + \beta_1 x_{ij} + u_j + \varepsilon_{ij}. \quad (3.12)$$

Note that, for simplicity, we omit the distributional statements when these are associated with standard normal models.

The IGLS estimates from fitting (3.12) are given in Table 3.1. We see the coefficient for the reading score is similar for the fixed effect and 2-level random effect model. From a formal point of view, the fixed effect model is a better fit to the data as judged by the AIC criterion, namely $(9139.8 + 2 * 67) - (9357.2 + 2 * 4) = -91.4$. As our models become more complex, fixed effects models will tend to provide less good fits. More importantly, they rapidly become unmanageable with increasing complexity, for example where several coefficients vary across different higher-level units. Fixed effects models also, by definition, do not allow us to explore factors that might explain the variation across higher-level units. Finally, for many scientific purposes, it is natural to view the higher-level units, as well as the level 1 units, as sampled from a (super)population of units, such as schools,

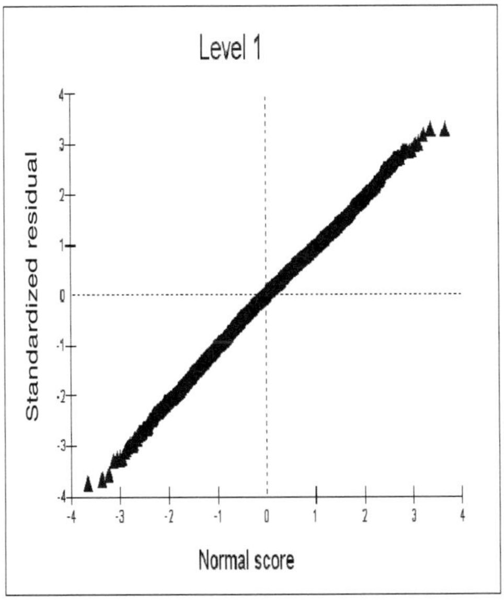

Figure 3.3 Level 1 Standardized Residual Plot

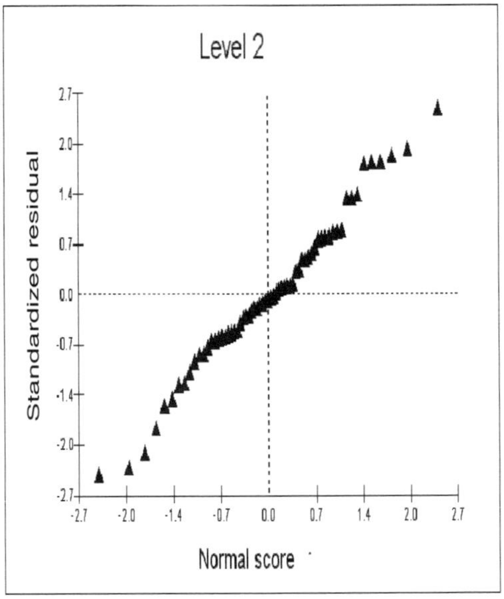

Figure 3.4 Level 2 Standardized Residual Plot

in which case a random effects formulation is implied.

Figures 3.3 and 3.4 show normal residual plots where the residuals are standardized using the diagnostic standard errors and, as expected, these are reasonably close to straight lines, indicating that the distributional assumption of normality is satisfied.

Table 3.2 shows results for models where the reading score coefficient is allowed to vary across schools and additionally where the level 1 variance becomes a function of the reading score.

As pointed out earlier with respect to (3.2), we can fit complex level 1 variance functions and our full 2-level model can be written as:

$$y_{ij}=\beta_0+\beta_1 x_{1ij}+u_{0j}+u_{1j}x_{1ij}+\varepsilon_{0ij}+\varepsilon_{1ij}x_{1ij}$$

$$\begin{pmatrix} u_{0j} \\ u_{1j} \end{pmatrix} \sim N\begin{pmatrix} \sigma^2_{u0} & \\ \sigma_{u01} & \sigma^2_{u1} \end{pmatrix}$$

$$\begin{pmatrix} \varepsilon_{0ij} \\ \varepsilon_{1ij} \end{pmatrix} \sim N\begin{pmatrix} \sigma^2_{\varepsilon 0} & \\ \sigma_{\varepsilon 01} & \sigma^2_{\varepsilon 1} \end{pmatrix},$$

so that the level 1 variance is given by $\sigma^2_{e0} + 2\sigma_{\varepsilon 01}x_{1ij} + \sigma^2_{\varepsilon 1}x_{1ij}$. In Table 3.2 we have set the term $\sigma^2_{\varepsilon 1}$ to zero since it is estimated as very small.

If we wish to compare the two models in Table 3.2 we have a conventional chi squared test statistic of 5.3 on 1 degree of freedom which has a p-value of 0.02, so that we would judge that the level 1 variance is a decreasing linear function of the reading score. If we wish to compare the model with just a single (homoscedastic) variance at level 1 with the 2-level model in Table 3.1 then we are effectively testing $\sigma^2_{u0} = \sigma_{u01} = 0$. Since, however, one of these parameters is a variance and cannot be negative, we need to adjust the usual tail area for the chi squared statistic. With 2 degrees of freedom the 5% level becomes 5.0 rather than the conventional 6.0. This is known as the Chi-bar statistic and further details are given by Goldstein (2011, Section 2.8). We see that the deviance for this test is 40.3, which is highly significant.

Figures 3.5 and 3.6 are plots of the variance functions at each level. We note that the level 2 between-school variance is a quadratic function of the reading test score.

While this modeling of variance is straightforward and can be extended to several explanatory variables, it does not constrain

Table 3.2 Examination Scores Related to Reading Test Scores with Random Coefficients. Standard errors in brackets

Explanatory variable	Random coefficient level 2	Random level 2 and complex level 1 variance
Intercept	−0.012 (0.04)	−0.012 (0.04)
Reading score	0.557 (0.02)	0.558 (0.02)
Level 2		
Intercept variance	0.090 (0.018)	0.091 (0.018)
Reading coefficient variance	0.015 (0.004)	0.014 (0.004)
Covariance	0.018 (0.007)	0.019 (0.007)
Level 1		
σ^2_{e0}	0.554 (0.012)	0.553 (0.015)
σ^2_{e1}	–	−0.015 (0.006)
σ_{e01}	–	–
−2 log likelihood	9316.9	9311.6

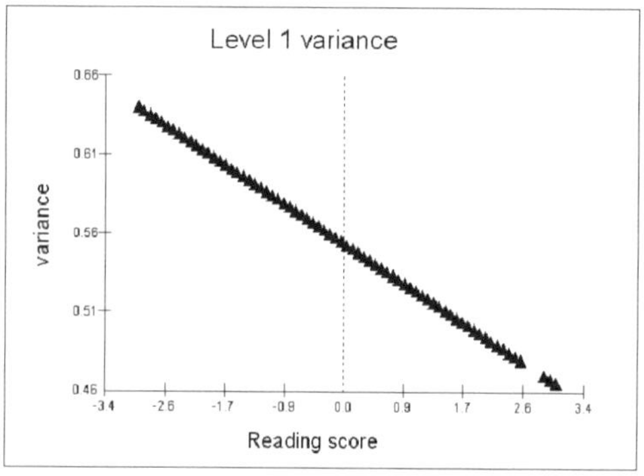

Figure 3.5 Variance Function at Level 1

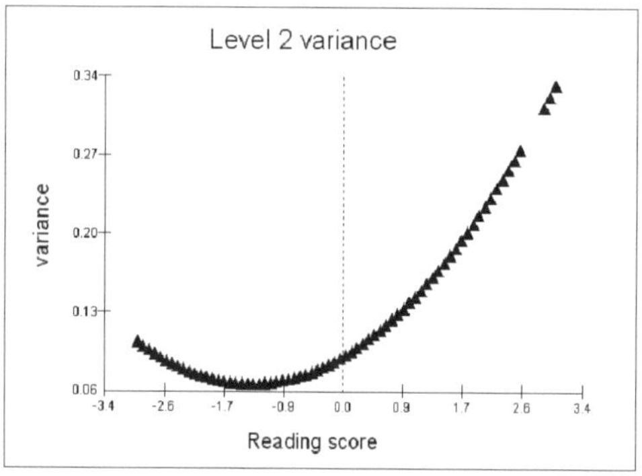

Figure 3.6 Variance Function at Level 2

the variance to be positive. A more satisfactory general formulation is to write, say the level 1 variance, as an exponential function so that in our model this would become $\exp(\sigma^2_{\varepsilon 0} + 2\sigma_{\varepsilon 01}x_{1ij} + \sigma^2_{\varepsilon 1}x_{1ij})$. An example is given in Goldstein (2010, Section 9.5).

3.3.2 Estimation Using Profile and Extended Likelihood

Consider the following 2-level random effect model with a single level 1 random intercept:

$$y = X\beta + Z_u u + \varepsilon$$
$$u \sim N(0, \Omega_u),\ \varepsilon \sim N(0, \sigma^2_\varepsilon).$$

We write twice the log likelihood, ignoring a function of the number of observations, as

$$2L(\theta, \beta) = -\log|V| - \mathrm{tr}(V^{-1}S)$$
$$= -\log|V| - (Y - X\beta)^T V^{-1}(Y - X\beta),$$

with $S = (Y - X\beta)^T(y - X\beta)$, $V = E(S)$, where θ is the vector of random parameters (variances and covariances) and $V = V(\theta) = E(\varepsilon\varepsilon^T)$. If we have a maximum likelihood (ML) estimate of β at a given value of θ then

$$2L(\theta, \hat{\beta}) = -\log|V| - \mathrm{tr}(V^{-1}S)$$
$$= -\log|V| - (Y - X\hat{\beta})^T V^{-1}(Y - X\hat{\beta}) \qquad (3.13)$$

is the profile likelihood for the random parameters θ. The IGLS algorithm alternates between maximizing (3.13) for θ and then obtaining the conditional maximum likelihood (generalized least squares) estimate of β. For given values of β the profile likelihood can be plotted for values of θ and is useful for obtaining confidence regions.

We can write the *extended likelihood*, referred to in different contexts as a penalized likelihood or an h-likelihood (Lee, Nelder, and Pawitan, 2006), that includes the actual random effects (residuals) u treated as parameters:

$$\begin{aligned} 2L(\theta,\beta,u) = &-\log|V_1| \\ &-(Y-X\beta-Zu)^T V_1^{-1}(Y-X\beta-Zu) \\ &-\log|\Omega_u| - u^T\Omega_u^{-1}u, \end{aligned} \tag{3.14}$$

where $V_1=\sigma_\varepsilon^2 I$.

If we maximize (3.14) for the random effects, given (β,θ), we obtain the usual estimator which can be written conveniently as

$$\hat{u} = (Z^T\hat{V}_1^{-1}Z+\Omega_u^{-1})^{-1}Z^T\hat{V}_1^{-1}(Y-X\hat{\beta}). \tag{3.15}$$

Now, given estimates for (θ,u), the profile likelihood for the fixed effects becomes

$$\begin{aligned} 2L(\hat{\theta},\beta,\hat{u}) = &-\log|\hat{V}_1| \\ &-(Y-X\beta-Z\hat{u})\hat{V}_1^{-1}(Y-X\beta-Z\hat{u}) \\ &-\log|\hat{\Omega}_u| - \hat{u}^T\hat{\Omega}_u^{-1}\hat{u}. \end{aligned} \tag{3.16}$$

Thus, a modification of the IGLS procedure is to iterate between calculating the fixed effects using (3.16), namely

$$\hat{\beta} = (X^T\hat{V}_1^{-1}X)^{-1}X^T\hat{V}_1^{-1}(Y-Z\hat{u}),$$

which when V_1 is diagonal is just ordinary least squares (OLS), and then calculating the random parameters from (3.13) as in the standard IGLS algorithm, and then the random effects from (3.15). In some circumstances this may be computationally convenient. Goldstein (2010, Chapter 15) gives an example in the case of a general multilevel smoothing model. Note that the OLS standard errors for the fixed coefficients based simply on (3.16) are incorrect. The standard errors should be computed using generalized least squares as described above.

3.3.3 The EM Algorithm

To illustrate the EM procedure, consider the 2-level variance components model

$$Y = X\beta + u + \varepsilon.$$

The vector of level 2 residuals is treated as missing data and the 'complete' data consists of the observed vector Y and the u_j treated as responses. The joint distribution of these, assuming Normality, is

$$\begin{pmatrix} Y \\ u \end{pmatrix} = N\left\{\begin{pmatrix} X\beta \\ 0 \end{pmatrix}, \begin{pmatrix} V & J^T\sigma_u^2 \\ \sigma_u^2 J & \sigma_u^2 I \end{pmatrix}\right\}, \tag{3.17}$$

where J is a matrix of ones. This generalizes readily to the case where there are several random coefficients. We can now derive the distribution of $u_j|Y$ as before, and we can also write down the normal log likelihood function for (3.17) with a general set of random coefficients, namely

$$\begin{aligned} \log(L) \propto &-N\log\left(\sigma_\varepsilon^2\right) - m\log|\Omega_u| \\ &-\sigma_\varepsilon^{-2}\sum_{ij}\varepsilon_{ij}^2 - \sum_j u_j^T\Omega_u^{-1}u_j. \end{aligned}$$

Maximizing this for the random parameters, we obtain

$$\begin{aligned} \hat{\sigma}_\varepsilon^2 &= N^{-1}\sum_{ij}\varepsilon_{ij}^2 \\ \hat{\Omega}_u^2 &= m^{-1}\sum_j u_j u_j^T. \end{aligned} \tag{3.18}$$

We require the expected values, conditional on the Y and the current parameters, of the terms under the summation signs in (3.18), these being the sufficient statistics. We then substitute these expected values in (3.18) for the updated random parameters. These expected values are simply the 'shrunken' residuals. With these updated values of the random parameters we can form V and hence obtain the updated estimates for the fixed parameters using generalized least squares. We note that the expected values of the sufficient statistics can be obtained using the general result for a random parameter vector θ:

$$E\left(\theta\theta^T\right) = \text{cov}(\theta) + [E(\theta)][E(\theta)]^T .$$

This is known as the E (expectation) step of the algorithm, and the computations in (3.18) the M (maximization) step. Given starting values, based upon OLS, these computations are iterated until convergence is obtained. Convenient computational formulae for computing these quantities at each iteration can be found in Bryk and Raudenbush (1992).

3.4 GENERALIZED LINEAR MODELS

So far, we have considered maximum likelihood estimation when the response is normal with an identity link function. In other cases, and in particular for generalized linear models, the (marginal) likelihood needs to be evaluated numerically, for example via quadrature or simulation (Ng et al., 2006), and a search procedure used to find the maximum. For complex models this can be time consuming and beyond a certain point becomes infeasible. In such cases approximate methods are available based, for example, upon quasi-likelihood methods, but these approximations are known to underestimate parameter values, especially where there are few level 1 units per level 2 unit. In such cases MCMC typically becomes the method of choice. This topic is taken up elsewhere in this volume and we shall not discuss it further.

3.5 MULTIVARIATE MODELS

Consider Table 3.3 where for each individual we have two responses, say height and weight. Using 2-level model notation, subscript i indexes the responses ($i = 1, 2$) and j the individual. Dummy variables Z_1, Z_2 indicate the response and X is a covariate.

Table 3.3 Bivariate Response Data Structure for Three Individuals

Individual	Response	Z_1	Z_2	X
1	y_{11} (height)	1	0	x_1
1	y_{21} (weight)	0	1	x_1
2	y_{12} (height)	1	0	x_2
2	y_{22} (weight)	0	1	x_2
3	y_{13} (height)	1	0	x_3

A bivariate normal model can be written as:

$$y_{ij} = \beta_{01} z_{1j} + \beta_{02} z_{2j} + \beta_{11} z_{1j} x_j + \beta_{12} z_{2j} x_j + u_{1j} z_{1j} + u_{2j} z_{2j}$$

$$\begin{pmatrix} u_{1j} \\ u_{2j} \end{pmatrix} \sim N\begin{pmatrix} \sigma_{u1}^2 & \\ \sigma_{u12} & \sigma_{u2}^2 \end{pmatrix}.$$

Each response then gives:

$$y_{1j} = \beta_{01} + \beta_{11} x_j + u_{1j}$$

$$y_{2j} = \beta_{02} + \beta_{12} x_j + u_{2j}.$$

Thus, by adding an extra lowest level which serves to define the multivariate structure but with no associated random effects, we can define a multivariate model to which higher levels and classifications can be added. Existing software that allows no variation to be specified at level 1 can be used and the IGLS algorithm is readily adapted. Data do not need to be balanced so that not all individuals need to be measured on all responses.

3.6 MORE COMPLEX MODELS: CROSS CLASSIFICATIONS

So far, we have considered a single higher-level random classification in our model. In many cases, however, we may wish to include several such classifications. For example, in our educational example we might suppose that, in addition to the effect of secondary school attended, the primary (elementary) school attended may affect the exam score. Each student can be classified by a particular combination of primary and secondary schools, and a simple variance components model for such a data structure can be written as

$$y_{ij} = \beta_0 + \beta_1 x_{ij} + u_{1j} + u_{2j} + \varepsilon_{ij},$$

where u_{1j}, u_{2j} refer to the secondary and primary school residuals respectively.

As mentioned earlier, it is useful to introduce a more general notation that is also more informative and flexible. An alternative indexing procedure was first introduced by Browne et al. (2001) that is quite general and flexible, and a similar notation is given by Gelman and Hill (2007). Instead of the above we write the composite model as follows. The superscript identifies the 'classification' or set of units at any given level. For convenience we adopt the convention that the level 1 superscript is omitted. The full model can then be written as:

$$y_i = \beta_0 + \beta_1 x_{ij} + u^{(2)}_{secondary(i)} + u^{(3)}_{primary(i)} + \varepsilon_i \quad (3.19)$$

$$\varepsilon_i \sim N(0, \sigma^2_\varepsilon),$$
$$u^{(2)}_{secondary(i)} \sim N(0, \sigma^2_{u(2)}),$$
$$u^{(3)}_{primary(i)} \sim N(0, \sigma^2_{u(3)}).$$

Details of an efficient algorithm to obtain maximum likelihood estimates using IGLS are given by Goldstein (2010, Appendix 12.1). In essence, if we have a cross classification such as in (3.19), one of the classifications, say primary school, is treated as a standard hierarchical classification and for the other (secondary school) we fit a set of dummy variables, one for each secondary school, with coefficients that are random at level 2, mutually uncorrelated and independent of the first classification and with variances constrained to be equal. In addition, these coefficients are constrained to have means of zero – that is, they do not enter the fixed part of the model. Such a model, and more complex ones with corresponding constraints, can be fitted readily in MLwiN. One of the drawbacks is that if the non-hierarchical classification has a large number of units then this requires large amounts of memory and can be slow. MCMC methods are much more efficient in such cases

Table 3.4 Analysis of Examination Scores by Secondary and Primary School Attended. The subscript 3 refers to primary and 2 to secondary school. Standard errors in brackets

Parameter	Estimate	Estimate
Fixed		
Intercept	5.98	5.99
Verbal Reasoning	0.16 (0.003)	0.16 (0.003)
Random		
$\sigma^2_{u(3)}$	0.27 (0.06)	–
$\sigma^2_{u(2)}$	0.011 (0.021)	0.28 (0.06)
σ^2_e	4.25 (0.10)	4.26 (0.10)

and are recommended for general use (Goldstein, 2010, Chapter 12).

The following example illustrates how the introduction of a second classification can modify inferences.

The data consist of scores on a school leaving examination obtained by 3435 students who attended 19 secondary schools cross classified by 148 primary schools in Fife, Scotland (Paterson, 1991). Before their transfer to secondary school at the age of 12, each student obtained a score on a verbal reasoning test, measured about the population mean of 100 and with a population standard deviation of 15. Table 3.4 shows the results of fitting the model given by (3.19) with and without the inclusion of the primary school classification. The response is examination score and the explanatory variable is verbal reasoning test score.

We see that most of the level 2 variation is due to the primary school attended. This may be due to the fact that primary schools are smaller and centered in small local neighbourhoods, whereas secondary schools draw students from several primary schools so that social and other differences among local neighbourhoods are averaged, with less resulting variation in exam scores.

3.7 MORE COMPLEX MODELS: MULTIPLE MEMBERSHIP

In some circumstances units can be members of more than one higher-level unit. An example is friendship patterns where at any time individuals can be members of more than one friendship group. In an educational system students may attend more than one secondary school. In all such cases we shall assume that for each higher-level unit to which a lower-level unit belongs there is a known weight (summing to 1.0 for each lower-level unit) which represents, for example, the amount of time spent in that unit. The choice of weights may be important and we might carry out a sensitivity analysis to determine how alternative choices of weights affect inferences.

A simple variance components model with multiple membership of secondary schools can be written as

$$
\begin{aligned}
&y_i = (X\beta)_i + \sum_{j \in school(i)} w_{i,j}^{(2)} u_j^{(2)} + \varepsilon_i \\
&u_j^{(2)} \sim N(0, \sigma_{u(2)}^2),\ \varepsilon_i \sim N(0, \sigma_\varepsilon^2) \\
&school(i) \in \{1, \ldots, J\},\ i = 1, \ldots, N \\
&\sum_j w_{i,j}^{(2)} = 1.
\end{aligned}
$$

To obtain maximum likelihood estimates we adopt a similar procedure to that for cross classified structures. Now we just have a single classification with a dummy variable for each level 2 unit that takes the value $w_{i,j}^{(2)}$ for units j and zero otherwise, with random coefficients constrained to have equal variances and zero fixed part contributions. As for cross classifications, MCMC methods are preferred for large data sets. The importance of taking account of multiple membership structures is that a failure to do so can lead to underestimation of higher-level variances.

Multiple membership models are also useful for spatial modeling. In this case the level 2 units are typically geographical areas and the known weights represent the relative contributions of these to the response. Often, if a level 1 unit is located in area j then this will be given a larger weight, with the surrounding areas sharing the remaining weights. A discussion is given by Goldstein (2010, Chapter 13).

3.8 DISCUSSION

This chapter has set out the basic procedures for maximum likelihood estimation in the case of linear multilevel models, including multivariate, cross classified, and multiple membership models.

For many standard models, MLE is fast and is implemented in most general-purpose as well as specialized packages. As has been indicated, however, it becomes cumbersome and even infeasible for models of even moderate complexity, especially for generalized linear multilevel models and latent variable models. In some cases, as implemented in MLwiN, approximate likelihood based methods can serve to provide useful starting values for more flexible and efficient MCMC algorithms. Likelihood or approximate likelihood based methods may also be useful in the data exploration stage where a large number of different models or model classes are being tried.

REFERENCES

Aitkin, M., Anderson, D., and Hinde, J. (1981) 'Statistical Modeling of Data on Teaching Styles', *Journal of the Royal Statistical Society, Series A*, 144: 419–61.

Aitkin, M. and Longford, N. (1986) 'Statistical Modeling Issues in School Effectiveness Studies', *Journal of the Royal Statistical Society, Series A*, 149: 1–43.

Browne, W., Goldstein, H., and Rasbash, J. (2001) 'Multiple Membership Multiple Classification (MMMC) Models', *Statistical Modeling*, 1: 103–24.

Bryk, A.S. and Raudenbush, S.W. (1992) *Hierarchical Linear Models*. Newbury Park, CA: Sage.

Fisher, R.A. (1925) *Statistical Methods for Research Workers*. London: Oliver and Boyd.

Gelman, A. and Hill, J. (2007) *Data Analysis Using Regression and Multilevel/Hierarchical Models*. New York: Cambridge University Press.

Goldstein, H. (1986) 'Multilevel Mixed Model Analysis Using Iterative Generalized Least Squares', *Biometrika*, 73: 43–56.

Goldstein, H. (1989) 'Restricted (Unbiased) Iterative Generalised Least Squares Estimation', *Biometrika*, 76: 622–3.

Goldstein, H. (2011) *Multilevel Statistical Models*. 4th edn. Chichester: Wiley.

Goldstein, H. and Rasbash, J. (1992) 'Efficient Computational Procedures for the Estimation of Parameters in Multilevel Models Based on Iterative Generalised Least Squares', *Computational Statistics and Data Analysis*, 13: 63–71.

Goldstein, H., Rasbash, J., Yang, M., Woodhouse, G., Pan, H., Nuttall, D., and Thomas, S. (1993) 'A Multilevel Analysis of School Examination Results', *Oxford Review of Education*, 19: 425–33.

Harville, D.A. (1977) 'Maximum Likelihood Approaches to Variance Component Estimation and to Related Problems', *Journal of the American Statistical Association*, 72: 320–40.

Laird, N.M. and Ware, J.H. (1982) 'Random Effects Models for Longitudinal Data', *Biometrics*, 38: 963–74.

Lee, Y., Nelder, J.A., and Pawitan, Y. (2006) *Generalized Linear Models with Random Effects: Unified Analysis via H-Likelihood*. London: Chapman and Hall.

Lindley, D.V. and Smith, A.F.M. (1972) 'Bayes Estimates for the Linear Model', *Journal of the Royal Statistical Society, Series B*, 34: 1–41.

McDonald, R.P. and Goldstein, H. (1989) 'Balanced Versus Unbalanced Designs for Linear Structural Relations in Two-Level Data', *British Journal of Mathematical and Statistical Psychology*, 42: 215–32.

Ng, E.S.W., Carpenter, J.R., Goldstein, H., and Rasbash, J. (2006) 'Estimation in Generalised Linear Mixed Models with Binary Outcomes by Simulated Maximum Likelihood', *Statistical Modeling*, 6: 23–42.

Paterson, L. (1991) *Trends in Attainment in Scottish Secondary Schools*. New York: Academic Press.

Rasbash, J., Steele, F., Browne, W., and Goldstein, H. (2009). *A User's Guide to MLwiN Version 2.10*. Bristol, Centre for Multilevel Modeling, University of Bristol.

Searle, S.R., Casella, G., and McCulloch, C.E. (1992) *Variance Components*. New York, Wiley.

Scheffé, H. (1959) *The Analysis of Variance*. New York: Wiley.

Skrondal, A. and Rabe-Hesketh, S. (2004) *Generalized Latent Variable Modeling: Multilevel, Longitudinal, and Structural Equation Models*. Boca Raton, FL: Chapman & Hall/CRC.

4

Bayesian Multilevel Models

Ludwig Fahrmeir
University of Munich (Emeritus), Germany

Thomas Kneib
University of Göttingen, Germany

Stefan Lang
University of Innsbruck, Austria

4.1 BAYESIAN INFERENCE IN A NUTSHELL

This section is a strongly condensed version of an introduction to Bayesian inference given in Fahrmeir et al. (2013) Appendix B. Other, more elaborate, introductions to Bayesian inference are presented, e.g. in O'Hagan (1994), Gelman et al. (2003), and Held and Sabanes-Bove (2012).

The fundamental difference to likelihood-based inference is that the unknown parameters in a statistical model, $\boldsymbol{\theta} = (\theta_1, \ldots, \theta_p)'$, say, are not considered as fixed, deterministic quantities but as variables with a *prior distribution*. A Bayesian model therefore consists of two parts:

- *Prior distribution:* Any (possibly subjective) information about the unknown parameter $\boldsymbol{\theta}$ is expressed by specifying a probability distribution for $\boldsymbol{\theta}$. This distribution is called a prior (distribution) of $\boldsymbol{\theta}$. The specification of a prior for $\boldsymbol{\theta}$ does not necessarily mean that the unknown parameters are actually stochastic quantities. Rather, the prior describes the degree of uncertainty about the unknown parameters prior to the statistical analysis.
- *Observation model:* The observation model specifies the conditional distribution of observable quantities, that is the random sample variables $\boldsymbol{y} = (y_1, \ldots, y_n)'$, given the parameters. The observation model corresponds to the likelihood in a frequentist perspective.

Based on the prior, denoted by $p(\boldsymbol{\theta})$, and the observation model, denoted by $p(\boldsymbol{y} \,|\, \boldsymbol{\theta})$, Bayes' theorem determines the distribution of $\boldsymbol{\theta}$ after the data are known through the statistical experiment, that is the conditional distribution of $\boldsymbol{\theta}$ given the observations $\boldsymbol{y} = (y_1, \ldots, y_n)'$. We obtain

$$\begin{aligned} p(\boldsymbol{\theta} \,|\, \boldsymbol{y}) &= \frac{p(\boldsymbol{y} \,|\, \boldsymbol{\theta})\, p(\boldsymbol{\theta})}{\int p(\boldsymbol{y} \,|\, \boldsymbol{\theta})\, p(\boldsymbol{\theta})\, d\boldsymbol{\theta}} \\ &= c \cdot p(\boldsymbol{y} \,|\, \boldsymbol{\theta})\, p(\boldsymbol{\theta}), \end{aligned}$$

with the normalizing constant $c = [\int p(\boldsymbol{y} \,|\, \boldsymbol{\theta})\, p(\boldsymbol{\theta})\, d\boldsymbol{\theta}]^{-1}$. This conditional distribution is called the *posterior (distribution)*.

Bayesian inference completely relies on the characteristics of the posterior. The usual point estimators in Bayesian inference are the posterior mean, the posterior median, and the posterior mode. The posterior mean is given by

$$\begin{aligned}\hat{\boldsymbol{\theta}} = \mathrm{E}(\boldsymbol{\theta} \mid \boldsymbol{y}) &= \int \boldsymbol{\theta}\, p(\boldsymbol{\theta} \mid \boldsymbol{y})\, d\boldsymbol{\theta} \\ &= c \cdot \int \boldsymbol{\theta}\, p(\boldsymbol{y} \mid \boldsymbol{\theta})\, p(\boldsymbol{\theta})\, d\boldsymbol{\theta}.\end{aligned}$$

Thus, the analytical or numerical determination of the posterior requires the computation of (possibly high-dimensional) integrals.

The posterior mode is the value $\hat{\boldsymbol{\theta}}$ that (globally) maximizes the posterior density, i.e.

$$\hat{\boldsymbol{\theta}} = \operatorname{argmax}_{\boldsymbol{\theta}}\, p(\boldsymbol{\theta} \mid \boldsymbol{y}) = \operatorname{argmax}_{\boldsymbol{\theta}}\, p(\boldsymbol{y} \mid \boldsymbol{\theta})\, p(\boldsymbol{\theta}).$$

The second expression shows that no integration is necessary to compute the posterior mode, because the normalizing constant is not needed.

The posterior median, that is the median of the posterior distribution, is sometimes preferred to the posterior mean because it is more robust against outliers.

Point estimators reduce the information of the posterior to a single number. In particular, it is not possible to see how precise these estimates are. For the posterior mean, a natural measure of estimator variability is the posterior variance; for the posterior median, the interquartile distance seems to be appropriate to measure its variability. In the case of the posterior mode, the curvature of the posterior at the mode, i.e. the observed Fisher information, is a natural choice. Another popular way of assessing uncertainty are Bayesian confidence intervals, so-called credible intervals or, more generally, credible regions. A region C of the parameter space is said to be a $(1-\alpha)$ credible region for $\boldsymbol{\theta}$ if

$$P(\boldsymbol{\theta} \in C \mid \boldsymbol{y}) \geq 1 - \alpha.$$

If C is an interval it is called a credible interval. In other words, a credible region contains (at least) a probability mass $1 - \alpha$ of the posterior. A credible region is a highest posterior density (HPD) region if the posterior inside the region is always greater than outside.

For a long time, the main difficulty in applying Bayesian methods has been that in many situations the posterior is not available analytically or even numerically. MCMC simulation methods, which appeared in the statistics community from the 1990s on, revolutionized Bayesian inference as they allow the drawing of random samples from even high-dimensional posteriors. Most characteristics of the posterior can then be estimated (with arbitrary precision) on the basis of the samples. For instance, the posterior mean can be estimated by the sample arithmetic mean, a 95% credible interval by the 2.5% and 97.5% empirical quantiles, etc.

The conceptually most simple MCMC algorithm is the Gibbs sampler (Gelfand and Smith, 1990). In many cases the full posterior $p(\boldsymbol{\theta} \mid \boldsymbol{y})$ of the entire parameter vector is analytically intractable, while the conditional densities of parameter blocks $\boldsymbol{\theta}_1, \ldots, \boldsymbol{\theta}_S$ of $\boldsymbol{\theta}$ are tractable analytically. We denote by $p(\boldsymbol{\theta}_s \mid \cdot)$ the conditional density of $\boldsymbol{\theta}_s$ given all other blocks $\boldsymbol{\theta}_1, \ldots, \boldsymbol{\theta}_{s-1}, \boldsymbol{\theta}_{s+1}, \ldots, \boldsymbol{\theta}_S$ and the data $\boldsymbol{y}$, and call it full conditional. Gibbs sampling consists of successively drawing random numbers from the full conditionals in each iteration of the sampler. After a burn-in phase the sampled random numbers can be considered as realizations from the marginal posteriors $p(\boldsymbol{\theta}_1 \mid \boldsymbol{y}), p(\boldsymbol{\theta}_2 \mid \boldsymbol{y}), \ldots, p(\boldsymbol{\theta}_S \mid \boldsymbol{y})$. More specifically, the Gibbs sampler generates random numbers as follows:

1. Specify initial values $\boldsymbol{\theta}_1^{(0)}, \ldots, \boldsymbol{\theta}_S^{(0)}$ and the number T of iterations. Set $t = 1$.
2. For $s = 1, \ldots, S$: Draw random numbers from the full conditionals

$$\begin{aligned}&p(\boldsymbol{\theta}_S \mid \cdot) \\ &\quad = p(\boldsymbol{\theta}_s \mid \boldsymbol{\theta}_1^{(t)}, \ldots, \boldsymbol{\theta}_{s-1}^{(t)}, \boldsymbol{\theta}_{s+1}^{(t-1)} \\ &\quad\quad , \ldots, \boldsymbol{\theta}_S^{(t-1)}, \boldsymbol{y}).\end{aligned}$$

Note that the most recent states are used in the conditioning set of parameter blocks.

3. Stop if $t = T$, otherwise set $t = t + 1$ and go to 2.

If a Gibbs sampler is not applicable because the full conditionals are not known distributions, so-called Metropolis–Hastings algorithms or generalizations must be applied. Details can be found, e.g. in Robert and Casella (2004), Gelman et al. (2003), and Gamerman and Lopes (2006).

4.1.1 Example: Bayesian Ridge Regression

We illustrate the Bayesian inference machinery with a Bayesian version of ridge regression. The modeling framework is the classical linear model $\boldsymbol{y} = \boldsymbol{X}\boldsymbol{\beta} + \boldsymbol{\varepsilon}$ with p-dimensional parameter vector $\boldsymbol{\beta} = (\beta_1, \ldots, \beta_p)'$ and Gaussian errors $\boldsymbol{\varepsilon} \sim N(\boldsymbol{0}, \sigma^2\boldsymbol{I})$. Ridge regression minimizes the penalized least squares criterion

$$\text{PLS}(\boldsymbol{\beta}) = (\boldsymbol{y}-\boldsymbol{X}\boldsymbol{\beta})'(\boldsymbol{y}-\boldsymbol{X}\boldsymbol{\beta})+\lambda\boldsymbol{\beta}'\boldsymbol{\beta}. \quad (4.1)$$

The penalty shrinks the parameters towards zero in order to reduce the variance of the least squares estimator at the cost of a (typically small) bias. The amount of penalization is governed by the parameter λ. Using a particular prior for the regression coefficients, we obtain a Bayesian version of ridge regression. More specifically, we assume a priori independent regression coefficients β_j, $j = 1, \ldots, p$, and set $\boldsymbol{\beta} \mid \tau^2 \sim N(\boldsymbol{0}, \tau^2\boldsymbol{I})$. The terminology 'Bayesian ridge regression' is justified as it can be shown that maximizing the corresponding posterior with respect to $\boldsymbol{\beta}$ is equivalent to minimizing the penalized least squares criterion (4.1) for fixed $\lambda = \sigma^2/\tau^2$.

While in the frequentist approach to ridge regression the penalty parameter λ is estimated outside of the optimization criterion, e.g. via cross validation, the Bayesian approach allows for simultaneous inference for the regression parameters and the amount of penalization measured through τ^2. This is obtained by defining an additional prior for τ^2. A convenient and flexible choice is an inverse gamma distribution $\tau^2 \sim \text{IG}(a_{\tau^2}, b_{\tau^2})$. The advantage of this specification is that the full conditional for τ^2 is again an inverse gamma distribution, allowing for straightforward MCMC inference. However, other choices for τ^2 are possible and often useful; see the discussion on the choice of τ^2 in Section 4.2.2. The Bayesian model is completed by assuming another $\text{IG}(a, b)$ prior for the overall variance parameter σ^2.

It is now straightforward to develop a Gibbs sampler that consecutively samples from the full conditionals of $\boldsymbol{\beta}$, σ^2, and τ^2. The full conditionals are given by

$$\begin{aligned}
\boldsymbol{\beta} \mid \cdot &\sim \text{N}\left(\left(\boldsymbol{X}'\boldsymbol{X} + \frac{\sigma^2}{\tau^2}\boldsymbol{I}\right)^{-1}\boldsymbol{X}'\boldsymbol{y},\right. \\
&\qquad\left.\left(\frac{1}{\sigma^2}\boldsymbol{X}'\boldsymbol{X} + \frac{1}{\tau^2}\boldsymbol{I}\right)^{-1}\right) \\
\sigma^2 \mid \cdot &\sim \text{IG}(a + \frac{n}{2}, \\
&\qquad b + \frac{1}{2}(\boldsymbol{y} - \boldsymbol{X}\boldsymbol{\beta})'(\boldsymbol{y} - \boldsymbol{X}\boldsymbol{\beta})) \\
\tau^2 \mid \cdot &\sim \text{IG}(a_{\tau^2} + \frac{p}{2}, b_{\tau^2} + \frac{1}{2}\boldsymbol{\beta}'\boldsymbol{\beta}).
\end{aligned}$$

Drawing random numbers from this well-known distribution is standard. For non-Gaussian responses, a direct Gibbs sampler is not applicable. For many distributions, e.g. logit or probit models and Poisson regression, data augmentation schemes are available that allow the use of a slightly modified Gibbs sampler: see Albert and Chib (1993) for probit models, Holmes and Held (2006) and Frühwirth-Schnatter and Frühwirth (2010) for logit models, Frühwirth-Schnatter and Wagner (2006) and Frühwirth-Schnatter et al. (2009) for Poisson and gamma regression. In situations where the data augmentation trick does not work, Metropolis–Hastings algorithms with iteratively weighted least squares proposals can be used, see Gamerman (1997) for details.

4.2 MULTILEVEL MODELS BASED ON GAUSSIAN PRIORS

4.2.1 The General Concept

Gill and Womack (Chapter 1) showed that all linear mixed or multilevel models (LMMs), including models with varying (random) intercepts and slopes as well as levels with nested and non-nested (crossed) groupings, can be cast into the unifying observation model

$$y = X\beta + \sum_{l=1}^{L} Z_l u_l + \varepsilon = \eta + \varepsilon, \quad \varepsilon \sim N(\mathbf{0}, \sigma^2 I), \tag{4.2}$$

(see Chapter 1, Section 1.4). In this general form, y is a $n \times 1$ vector of observations, β is a $p \times 1$ vector of non-varying ('fixed') unknown coefficients, and Z_l and u_l are appropriately constructed design matrices and Gaussian random effect vectors at the lth group level. Extensions to generalized linear multilevel models (GLMMs) are obtained for observations y following exponential family distributions and linking (conditional) expectations to the linear multilevel predictor in (4.2) through a link function, i.e.

$$\mathbb{E}(y) = g(\eta) \tag{4.3}$$

(see Chapter 13, Section 1.4.3). From a Bayesian perspective, (4.2) or (4.3) are observation or measurement models, conditional on unknown parameters, and have to be supplemented by specifying prior distributions for the parameters.

For the population parameters β, a proper Gaussian prior

$$\mathrm{p}(\beta) \sim N(c, C) \tag{4.4}$$

is a natural choice if corresponding (subjective) prior information is available or, as in ridge regression, if estimates should be shrunk towards c, in particular $c = \mathbf{0}$. In the latter case, $C = \sigma^2 I$ is an obvious choice. For Gaussian multilevel models with error variance σ^2, the conjugate prior $\beta \sim N(c, \sigma^2 C)$ is an alternative to (4.4).

Without prior information or aiming at shrinkage estimates, a noninformative, flat, but improper, prior

$$\mathrm{p}(\beta) \propto \text{const}, \tag{4.5}$$

is our preferred default choice for inference in a multilevel model. Theoretical results, e.g. in Sun et al. (2001) for linear mixed models, or Fahrmeir and Kneib (2009) for the class of structured additive regression models described in Section 4.2.2, and references given therein, provide sufficient conditions for propriety of posteriors, implying that MCMC simulation will work correctly. It seems tempting to approximate the noninformative prior (4.5) through a weakly informative but still proper prior $\mathrm{p}(\beta) \sim N(c, C)$, C large, to guarantee posterior propriety. As pointed out in Berger and Pericchi (2001), the use of such vague priors can be dangerous, in particular in connection with model selection based on Bayes factors; see the remark in Section 4.5. In the case of problems with the noninformative prior (4.5), objective priors may be constructed by using a small part of the data as a training set, as in the intrinsic Bayes factor approach (Berger and Pericchi, 1996), or a fraction of the data likelihood, as in the fractional Bayes factor approach (O'Hagan, 1995). Another possibility is to construct proper priors from historical data, as in Chen et al. (2003).

The Gaussian random effects distributions

$$u_l \sim N(\mathbf{0}, \Omega_l) \tag{4.6}$$

are interpreted as Gaussian priors for the parameter vectors u_l. More exactly, these priors are conditionally Gaussian, because Ω_l will also contain unknown parameters such as variances or correlations. Denoting such unknown parameters through θ_l, (4.6) is then replaced through

$$u_l \mid \theta_l \sim N(\mathbf{0}, \Omega_l(\theta_l)), \quad \theta_l \sim \mathrm{p}(\theta_l),$$

where $\mathrm{p}(\theta_l)$ denotes a (hyper-)prior. Its specific form depends on the structure of $\Omega_l(\theta_l)$; see the next section for examples. For Gaussian models with error variance σ^2, an

IG(a, b) prior (with a and b small) as in linear models is a default choice.

In an empirical Bayes approach, these 'hyperparameters' as well as the error variance σ^2 are considered as 'fixed' but unknown and are determined through ML or REML estimation. It turns out that empirical Bayes inference and likelihood-based inference in LMMs are more or less identical. Here we focus on full Bayesian inference, where additional hyperpriors are specified for hyperparameters and estimation of all parameters relies on MCMC techniques.

The generic forms (4.2) and (4.3) of multilevel models allow the specification of various forms of nested and non-nested grouping effects in (G)LMMs by appropriate definition of design matrices and assumptions on $\boldsymbol{\Omega}_l = \text{Cov}(\boldsymbol{u}_l)$. An important feature of the additive structure of the predictor is that it also allows the incorporation of random effects vectors with correlated elements instead of the conventional independent and identically distributed (iid) random effects. Thus, it is possible to extend (G)LMMs to (generalized) additive multilevel models ((G)AMMs) by incorporating Bayesian smoothing or P-splines, or to include spatial effects in the form of Gaussian (Markov) random field priors for geoadditive regression models. In the following, we consider such semiparametric multilevel models (SPMMs) and the corresponding specification of design matrices $\boldsymbol{Z}_l$ and priors for $\boldsymbol{u}_l$ in more detail.

4.2.2 *Linear and Semiparametric Bayesian Multilevel Models*

To simplify presentation and notation, we focus on specific models rather than describing them all in full generality.

(G)LMMs for longitudinal data

Let $(y_{ij}, \boldsymbol{x}_{ij}),\ j = 1, \ldots, n_i$, denote the values of a response variable y and a covariate vector $\boldsymbol{x}$ observed at times $t_{i1} < \ldots < t_{ij} < \ldots < t_{in_i}$ for individuals $i = 1, \ldots, n$ in a longitudinal study. An LMM with population effects $\boldsymbol{\beta}$ and individual-specific random effects $\boldsymbol{u}_i,\ i = 1, \ldots, n$, is defined hierarchically through the measurement model

$$y_{ij} = \boldsymbol{x}_{ij}'\boldsymbol{\beta} + \boldsymbol{z}_{ij}'\boldsymbol{u}_i + \varepsilon_{ij}, \quad \varepsilon_{ij} \sim N(0, \sigma^2). \tag{4.7}$$

The covariate vector $\boldsymbol{x}_{ij}$ includes a constant 1 for the intercept term and z_{ij} is a subvector of $\boldsymbol{x}_{ij}$, leading to random intercept models for $z_{ij} \equiv 1$ and to random slope models if z_{ij} contains further covariates. The errors are assumed to be iid, but extensions to autocorrelated errors are possible. The random effects prior is defined through

$$\boldsymbol{u}_i = (u_{i0}, \ldots, u_{im})' \overset{\text{iid}}{\sim} N(\boldsymbol{0}, \boldsymbol{Q}),$$

with unknown (elements in) $\boldsymbol{Q}$. Then the prior for the vector $\boldsymbol{u} = (\boldsymbol{u}_1', \ldots, \boldsymbol{u}_m')'$ of all random effects is

$$\boldsymbol{u} \sim N(\boldsymbol{0}, \boldsymbol{\Omega}), \quad \boldsymbol{\Omega} = \text{blockdiag}(\boldsymbol{Q}, \ldots, \boldsymbol{Q}).$$

Collecting all responses and errors in $\boldsymbol{y}$ and $\boldsymbol{\varepsilon}$ and defining $\boldsymbol{X}$ and the design matrix $\boldsymbol{Z}$ in the usual way, the LMM (4.7) can be written in the generic form (4.2):

$$\boldsymbol{y} = \boldsymbol{X\beta} + \boldsymbol{Zu} + \boldsymbol{\varepsilon}, \quad \boldsymbol{\varepsilon} \sim N(\boldsymbol{0}, \sigma^2\boldsymbol{I}).$$

We have deliberately assumed that the hierarchy contains only one ($L = 1$) further level after the data level. Extensions to models with more levels ($L > 1$) would be necessary if, for example, individuals are assigned to groups. GLMMs for non-Gaussian responses are defined in the usual way linking the linear predictor $\boldsymbol{\eta} = \boldsymbol{X\beta} + \boldsymbol{Zu}$ to $\mathbb{E}(\boldsymbol{y})$ through $g(\boldsymbol{\eta})$.

For full Bayesian inference, hyperpriors have to be assigned to the hyperparameters $\sigma^2 = \text{Var}(\varepsilon_{ij})$ and $\boldsymbol{Q} = \text{Cov}(\boldsymbol{u}_i)$. The standard choice for σ^2 is an inverse gamma prior IG(a, b),

$$\text{p}(\sigma^2) \propto \frac{1}{(\sigma^2)^{a+1}} \exp\left(-\frac{b}{\sigma^2}\right),$$

which is proper for $a > 0, b > 0$. The limiting case $a \to 0, b \to 0$ is Jeffreys' noninformative but improper prior $\text{p}(\sigma^2) \propto \sigma^{-2}$; see

Fahrmeir et al. (2012) Appendix B and particularly O'Hagan (1994) or Held and Sabanes-Bove (2012) for an introduction to Jeffreys' prior.

For an unspecified variance-covariance matrix $\boldsymbol{Q}$, the conjugate prior is an inverse Wishart distribution $\text{IW}(a, \boldsymbol{B})$, with $a \to 0$, $\boldsymbol{B} \to \boldsymbol{0}$ leading to Jeffreys' noninformative but improper prior $\text{p}(\boldsymbol{Q}) \propto |\boldsymbol{Q}|^{-\frac{m+2}{2}}$. Assuming that components $u_{i0}, \ldots, u_{im}$ of $\boldsymbol{u}_i$ are a priori independent, $\boldsymbol{Q}$ reduces to a diagonal matrix and the $\text{IW}(a, b)$ prior to the product of $\text{IG}(a_k, b_k)$ priors for $\delta_k^2 = \text{Var}(u_{ik})$, $k = 0, \ldots, m$. It seems that Jeffreys' noninformative priors are a natural choice in the absence of information about σ^2 and $\boldsymbol{Q}$. However, its use can lead to improper posteriors, with MCMC samplers unable to detect this. Therefore, 'diffuse' IG priors with small $a = b = \kappa = 0.1, 0.01$, or 0.001 have been suggested. While this choice works well for σ^2, $\text{IG}(\kappa, \kappa)$ priors can actually be quite informative in models with a small number of units or groups; see, e.g., Gelman (2006), who also suggested alternative priors such as a uniform prior on standard deviations δ_k or a folded-Cauchy density.

Because IG and IW priors are (conditionally) conjugate priors, they have the computational advantage that Gibbs sampling is possible for posteriors. A good strategy is to start with these priors, to check sensitivity with respect to hyperparameters (at least in sparse data situations), and possibly to proceed with alternative priors; see Section 4.4.

GLMMs for grouped data

In their case study on party identification in Western Europe, Gill and Womack (Chapter 1) apply a multilevel logit model to binary responses y_{ijk}, where i indexes individuals and j and k two nested groups. The hierarchical form is

$$
\begin{aligned}
y_{ijk}|\pi_{ijk} &\sim \text{Bern}(\pi_{ijk}), \\
\log\left(\frac{\pi_{ijk}}{1-\pi_{ijk}}\right) &= \beta_{0jk} + \boldsymbol{x}'_{ijk}\boldsymbol{\beta} = \eta_{ijk} \\
\beta_{0jk} &= \beta_{0k} + \boldsymbol{x}'_{jk}\boldsymbol{\alpha} + u_{jk} \\
\beta_{0k} &= \beta_0 + \boldsymbol{x}'_k\boldsymbol{\gamma} + v_k \\
u_{jk} &\overset{\text{iid}}{\sim} N(0, \sigma_u^2) \qquad (4.8) \\
v_k &\overset{\text{iid}}{\sim} N(0, \sigma_v^2).
\end{aligned}
$$

Inserting the models for β_{0jk} and β_{0k} into the measurement model yields the predictor

$$
\eta_{ijk} = \beta_0 + \boldsymbol{x}'_k\boldsymbol{\gamma} + \boldsymbol{x}'_{jk}\boldsymbol{\alpha} + \boldsymbol{x}'_{ijk}\boldsymbol{\beta} + u_{jk} + v_k. \qquad (4.9)
$$

Redefining $\boldsymbol{\beta}$ as $\boldsymbol{\beta} = (\boldsymbol{\beta}_0, \boldsymbol{\alpha}', \boldsymbol{\beta}', \boldsymbol{\gamma}')'$, setting $\boldsymbol{u}_1 = (v_1, \ldots, v_k, \ldots)'$ and $\boldsymbol{u}_2 = (u_{11}, \ldots, u_{jk}, \ldots)'$, and defining the design matrices $\boldsymbol{X}$, $\boldsymbol{Z}_1$, and $\boldsymbol{Z}_2$ appropriately, the vector $\boldsymbol{\eta}$ of all predictors in (4.9) becomes

$$
\boldsymbol{\eta} = \boldsymbol{X}\boldsymbol{\beta} + \boldsymbol{Z}_1\boldsymbol{u}_1 + \boldsymbol{Z}_2\boldsymbol{u}_2,
$$

with $\boldsymbol{u}_1 \sim N(\boldsymbol{0}, \boldsymbol{\Omega}_1 = \sigma_v^2\boldsymbol{I})$, $\boldsymbol{u}_2 \sim N(\boldsymbol{0}, \boldsymbol{\Omega}_2 = \sigma_u^2\boldsymbol{I})$.

As for LMM, $\text{IG}(\kappa, \kappa)$ priors are the most convenient hyperpriors for σ_v^2 and σ_u^2, but others may be considered as well.

If additional random slopes are introduced at the data level and the (jk) level, then the construction of $\boldsymbol{Z}_1$ and $\boldsymbol{Z}_2$ becomes more complicated and $\boldsymbol{\Omega}_1$, $\boldsymbol{\Omega}_2$ may be block diagonal instead of diagonal, but the general form of the linear predictor is retained.

GAMMs

It is often questionable whether the effect of a continuous covariate is linear or nonlinear. Generalized additive multilevel models (GAMMs) extend the linear predictor of GLMMs by adding non- or semiparametric functional effects of continuous covariates. For example, the longitudinal LMM (4.7) may be extended to

$$
\begin{aligned}
y_{ij} &= \boldsymbol{x}'_{ij}\boldsymbol{\beta} + \boldsymbol{z}'_{ij}\boldsymbol{u}_i + f(t_{ij}) + \varepsilon_{ij}, \\
&\varepsilon_{ij} \sim N(0, \sigma^2),
\end{aligned}
$$

where $f(t)$ is a nonlinear time trend, or age may be modeled through $f(\text{age})$ instead of the linear effect $\beta \cdot \text{age}$ in the multilevel model (4.9) of Gill and Womack (Chapter 1, Section 1.6).

A typical function $f_l(x_l)$ of a continuous covariate x_l can be modeled through Bayesian P-splines or smoothing splines by assuming a (truncated) basis function approximation

$$f_l(x_{il}) = \sum_{k=1}^{m} u_{lk} B_k(x_{il}),$$

where $\boldsymbol{u}_l = (u_{l1}, \ldots, u_{lm})'$ are (random) basis coefficients and $B_k(x)$, $k = 1, \ldots, m$, are B-spline basis functions. Bayesian P-splines (Lang and Brezger, 2004; Brezger and Lang, 2006) are the Bayesian version of the Eilers and Marx (1996) P-splines (see also Chapter 19 for more details on the frequentist interpretation of P-splines and their connection to mixed models), where the quadratic difference penalty is replaced through the Gaussian smoothness prior

$$\mathrm{p}(\boldsymbol{u}_l|\tau_l^2) \propto \exp\left(-\frac{1}{2\tau_l^2}\boldsymbol{u}_l'\boldsymbol{K}_l\boldsymbol{u}_l\right), \quad (4.10)$$

and $\boldsymbol{K}_l = \boldsymbol{D}'\boldsymbol{D}$, with $\boldsymbol{D}$ as the matrix of first or second differences. Then the vector $\boldsymbol{f}_l = (f_l(x_{1l}), \ldots, f_l(x_{nl}))'$ of function values is given by

$$\boldsymbol{f}_l = \boldsymbol{Z}_l\boldsymbol{u}_l,$$

where the design matrix $\boldsymbol{Z}_l$ is the matrix of B-spline basis function values. Thus, GAMM predictors are still in the general form (4.2). However, for P-splines the Gaussian prior (4.10) for $\boldsymbol{u}_l$ is (partially) improper, because the penalty matrix $\boldsymbol{K}_l$ has rank $(m-d)$, where $d = 1$ or $d = 2$ for first- or second-order difference penalties, respectively. Thus

$$\boldsymbol{u}_l \sim N(\boldsymbol{0}, \boldsymbol{\Omega}_l), \quad \boldsymbol{\Omega}_l = \tau_l^2\boldsymbol{K}_l^-,$$

where $\boldsymbol{K}_l^-$ is a generalized inverse. The variance parameter τ_l^2 acts as a (inverse) smoothing parameter, with hyperprior $\tau_l^2 \sim \mathrm{IG}(\kappa, \kappa)$. Bayesian smoothing splines or other basis function expansions can be handled in the same way by defining appropriate design and penalty matrices.

Geoadditive regression models

In diverse scientific and public data sources, spatial information about individual observations is available in addition to the covariate or group information considered so far. Let $s \in D \subseteq \mathbb{R}^2$ denote the spatial location for a typical observation y. A basic differentiation is between continuous and discrete spatial information. In the continuous case, s is represented through spatial coordinates and may vary continuously in D. In the discrete case, also referred to as area data, $s \in \{1, \ldots, S\}$ is a spatial index denoting, for example, region or area s within S regions. Let $f_{\mathrm{geo}}(s)$ denote the effect of a typical observation with location at s. Geoadditive regression models extend linear or additive predictors by including $f_{\mathrm{geo}}(s)$ in additive form. For example, in a longitudinal study with additional spatial information the predictor is now

$$y_{ij} = \boldsymbol{x}_{ij}'\boldsymbol{\beta} + \boldsymbol{z}_{ij}'\boldsymbol{u}_i + f_{\mathrm{geo}}(s_i) + \varepsilon_{ij},$$

where $s_1, \ldots, s_n$ are the spatial locations of individuals $i = 1, \ldots, n$ in regions $s = 1, \ldots, S$. Let $\boldsymbol{u}_{\mathrm{geo}} = (u_{\mathrm{geo}}(1), \ldots, u_{\mathrm{geo}}(S))$ denote the vector of spatial effects of these regions. To account for spatial correlation, a Gaussian Markov random field (GMRF)

$$\mathrm{p}(\boldsymbol{u}_{\mathrm{geo}} \mid \tau_{geo}^2) \propto \exp\left(-\frac{1}{2\tau_{\mathrm{geo}}^2}\boldsymbol{u}_{\mathrm{geo}}'\boldsymbol{K}_{\mathrm{geo}}\boldsymbol{u}_{\mathrm{geo}}\right)$$

is assumed as a prior, where $\boldsymbol{K}_{\mathrm{geo}}$ is a penalty matrix enforcing smoothness of neighboring spatial effects. The smoothing parameter τ_{geo}^2 again has an $\mathrm{IG}(\kappa, \kappa)$ prior. The vector $\boldsymbol{f}_{\mathrm{geo}} = (f_{\mathrm{geo}}(s_1), \ldots, f_{\mathrm{geo}}(s_n))$ can be expressed as

$$\boldsymbol{f}_{\mathrm{geo}} = \boldsymbol{Z}_{\mathrm{geo}}\boldsymbol{u}_{\mathrm{geo}},$$

with an incidence matrix $\boldsymbol{Z}_{\mathrm{geo}}$, and is incorporated into the geoadditive predictor

$$\eta = \boldsymbol{X\beta} + \boldsymbol{Zu} + \boldsymbol{Z}_{\mathrm{geo}}\boldsymbol{u}_{\mathrm{geo}},$$

which is again in the general form (4.2).

More details on geoadditive models, including continuous spatial information, are described in Fahrmeir and Kneib (2011, Chapter 5); compare also Chapter 31. A very good reference on spatial models and data analysis is Banerjee et al. (2003).

STAR models

Obviously, nonlinear effects $f_l(x_l)$ and spatial effects $f_{\text{geo}}(s)$ can be introduced into any multilevel predictor for nested or non-nested grouped data. In addition, varying coefficient terms $f(x_1) \cdot x_2$ or interaction terms $f(x_1, x_2)$ can be handled in an analogous manner. We call models with such different types of function appearing in additive form in the predictor *structured additive regression* (STAR) models; see Fahrmeir, Kneib, and Lang (2004), Brezger and Lang (2006), Fahrmeir and Kneib (2011), and Fahrmeir, Kneib, Lang, and Marx (2012). All STAR models have predictors of the generic form (4.2) or (4.3), with Gaussian priors $\boldsymbol{u}_l \sim N(\boldsymbol{0}, \boldsymbol{\Omega}_l)$ for random effects, but with possibly rank-deficient covariance matrices $\boldsymbol{\Omega}_l$. This allows the development of Bayesian inference within a rather unified framework; see the following section. To summarize in general form and to distinguish notationally between iid random effects for groups and correlated random effects corresponding to functions such as $f(x)$ or $f_{\text{geo}}(s)$, we extend the general linear multilevel predictor in (4.2) to the multilevel STAR predictor

$$\boldsymbol{\eta} = \boldsymbol{X}\boldsymbol{\beta} + \sum_{l=1}^{L} \boldsymbol{Z}_l \boldsymbol{u}_l + \sum_{r=1}^{R} \boldsymbol{Z}_r^f \boldsymbol{u}_r^f. \quad (4.11)$$

In (4.11), $\boldsymbol{u}_l \sim N(\boldsymbol{0}, \boldsymbol{\Omega}_l)$ are random effect vectors at the lth level for groups or clusters of size G_l, $l = 1, \ldots, L$, and $\boldsymbol{Z}_l$ are the lth-level design matrices. Correspondingly, $\boldsymbol{Z}_r^f$ and $\boldsymbol{u}_r^f$ are design matrices and coefficient vectors representing different functions f_r, such as main and interaction effects of continuous covariates, varying coefficient terms, and spatial effects. The priors for coefficients are

$$\begin{aligned} &\mathrm{p}(\boldsymbol{\beta}) \propto \text{const} \quad \text{or} \quad \boldsymbol{\beta} \sim N(\boldsymbol{c}, \boldsymbol{C}) \\ &\boldsymbol{u}_l \sim N(\boldsymbol{0}, \boldsymbol{\Omega}_l) \;\; l = 1, \ldots, L \\ &\boldsymbol{u}_r^f \sim N(\boldsymbol{0}, \tau_r^2 \boldsymbol{K}_r^-), \;\; r = 1, \ldots, R. \end{aligned}$$

Hyperpriors are inverse Wishart priors $\mathrm{IW}(a_l, \boldsymbol{B}_l)$ for $\boldsymbol{Q}_l$ or inverse gamma priors $\mathrm{IG}(a_{lk}, b_{lk})$, $k = 0, \ldots, m_l$ if $\boldsymbol{Q}_l = \operatorname{diag}(\delta_{l0}, \ldots, \delta_{lm_l})$, and inverse $\mathrm{IG}(a_r, b_r)$ priors for τ_r^2, $r = 1, \ldots, R$. For Gaussian models we assume $\sigma^2 \sim \mathrm{IG}(a, b)$ as a standard prior for the error variance.

4.2.3 Bayesian Inference

In principle, an empirical Bayes approach based on posterior mode estimation of regression coefficients $\boldsymbol{\beta}$, $\boldsymbol{u}$, and $\boldsymbol{u}^f$ is possible. Suppressing dependence on unknown hyperparameters $\boldsymbol{Q}_l$ and τ_r^2, this is equivalent to maximizing the numerator

$$\mathrm{p}(\boldsymbol{y}|\boldsymbol{\eta})\, \mathrm{p}(\boldsymbol{\beta})\, \mathrm{p}(\boldsymbol{u})\, \mathrm{p}(\boldsymbol{u}^f)$$

in Bayes' theorem for the posterior $\mathrm{p}(\boldsymbol{\beta}, \boldsymbol{u}, \boldsymbol{u}^f | \boldsymbol{y})$. Taking logarithms leads to maximizing the penalized likelihood

$$\begin{aligned} \mathrm{l}(\boldsymbol{y}|\boldsymbol{\eta}) \quad &= \quad \frac{1}{2} \sum_{l=1}^{L} \boldsymbol{u}_l' \boldsymbol{\Omega}_l^{-1} \boldsymbol{u}_l \\ &\quad - \frac{1}{2} \sum_{r=1}^{R} \frac{1}{\tau_r^2} (\boldsymbol{u}_r^f)' \boldsymbol{K}_r \boldsymbol{u}_r^f, \end{aligned}$$

where $\mathrm{l}(\boldsymbol{y}|\boldsymbol{\eta}) = \log \mathrm{p}(\boldsymbol{y}|\boldsymbol{\eta})$ is the data likelihood and the other terms are quadratic penalties. This shows the ridge-type shrinkage effect of the Gaussian priors. Hyperparameters in $\boldsymbol{\Omega}_l = \operatorname{blockdiag}(\boldsymbol{Q}_l, \ldots, \boldsymbol{Q}_l)$ and the smoothing parameters τ_r^2 are determined through (restricted) ML estimation.

For full Bayesian inference, MCMC techniques are commonly applied to simultaneously estimate regression coefficients and hyperparameters. For Gaussian multilevel STAR models $\boldsymbol{y} = \boldsymbol{\eta} + \boldsymbol{\varepsilon}$, $\boldsymbol{\varepsilon} \sim N(\boldsymbol{0}, \sigma^2 \boldsymbol{I})$, the full conditionals for regression coefficients are Gaussian. Explicit expressions for the means and covariance matrices of the full conditionals

$$\begin{aligned} &\boldsymbol{\beta}|\cdot \sim N(\boldsymbol{\mu}_\beta, \boldsymbol{\Sigma}_\beta), \\ &\boldsymbol{u}_l|\cdot \sim N(\boldsymbol{\mu}_l, \boldsymbol{\Sigma}_l), \quad \boldsymbol{u}_r^f|\cdot \sim N(\boldsymbol{\mu}_r^f, \boldsymbol{\Sigma}_r^f) \end{aligned}$$

can be found in Fahrmeir and Kneib (2011, Section 4.2.2) for the special case $L = 1$, with straightforward extensions to more levels. Full conditionals for σ^2, $\boldsymbol{Q}_l$, and τ_r^2

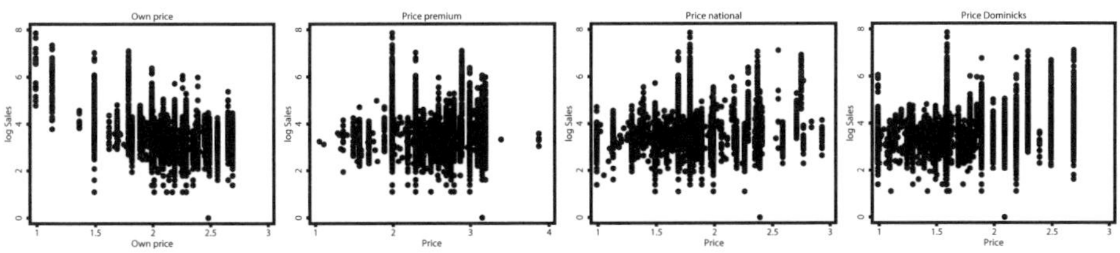

Figure 4.1 Scatter Plots Between the Log Number of Sold Packages and the Price of Tree Fresh (upper left panel) as well as Prices of Competitive Brands

are again inverse gamma or Wishart with updated parameters. The resulting Gibbs sampler for drawing from these full conditionals is an extension of Algorithm 4.2, p. 215, in Fahrmeir and Kneib (2011). The structure of the full conditional is close to that of the Gibbs sampler for Bayesian ridge regression in Section 4.1. For non-Gaussian exponential family responses, MH algorithms are needed to draw from full conditionals of regression coefficients; see Brezger and Lang (2006) and Fahrmeir and Kneib (2011, pp. 219) for the case $L = 1$. An approximate alternative that allows one to avoid simulation-based inference are integrated nested Laplace approximations as recently proposed by Rue et al. (2009).

It turns out that a hierarchical formulation of Bayesian GLMMs, GAMMs or geoadditive regression models as in (4.8) can be quite favorable for MCMC-based inference. Such hierarchical formulations and its merits for MCMC simulations are discussed in Lang et al. (2011b).

4.3 CASE STUDY: SALES OF ORANGE JUICE

We illustrate Bayesian multilevel models with supermarket scanner data from 'Dominicks Finer Foods', a major supermarket chain operating in the Chicago metropolitan area. The data were provided by the James M. Kilts Center, GSB, University of Chicago. The data include weekly unit sales and respective retail prices for different brands of orange juice (premium, national, and store brands) in 81 stores of the chain over a time span of 89 weeks. We restrict ourselves to one of the national brands, the brand 'Tree Fresh'. In the following we model the log-sales y_{ij} in store i, $i = 1, \ldots, 81$, at time t_{ij} in dependence of its own item-price p_{ij} and competitive prices $pc1_{ij}$, $pc2_{ij}$, and $pc3_{ij}$ in three quality tiers (premium brand, national brand, and Dominicks private store brand). To get an intuition about the data, Figure 4.1 shows scatter plots of log-sales against the covariates. We clearly observe an inverse relationship between unit sales and own-item price. The relationship seems to be nonlinear as the slope decreases for prices higher than about 2. The situation is less clear-cut for the impact of competitive prices on the sales of Tree Fresh: it is harder to discern any relationships in the scatter plots. Of course, economic theory suggests a positive correlation between sales and prices of competitors, which can be observed at least for the price of Dominicks. In order to investigate possible store-specific heterogeneity it is instructive to plot the log-sales price relationship for different stores. The left panel of Figure 4.2 displays log-sales against own item-price for four different stores, exhibiting a clear tendency for outlet-specific heterogeneity. Obviously, the sales overall level and also the slope of the relationship is different for the four stores. The scatter plots suggest transforming prices to log-prices in order to capture the nonlinearity in the relationships satisfactorily. We therefore estimate the two-level random effects model

$$\begin{aligned} y_{ij} &= \beta_0 + \beta_1 logp_{ij} + \beta_2 \cdot logpc1_{ij} \\ &\quad + \beta_3 logpc2_{ij} + \beta_4 logpc3_{ij} \\ &\quad + u_{i0} + u_{i1} logp_{ij} + u_{i2} logpc1_{ij} \\ &\quad + u_{i3} logpc2_{ij} + u_{i4} logpc3_{ij} + \varepsilon_{ij}, \end{aligned} \tag{4.12}$$

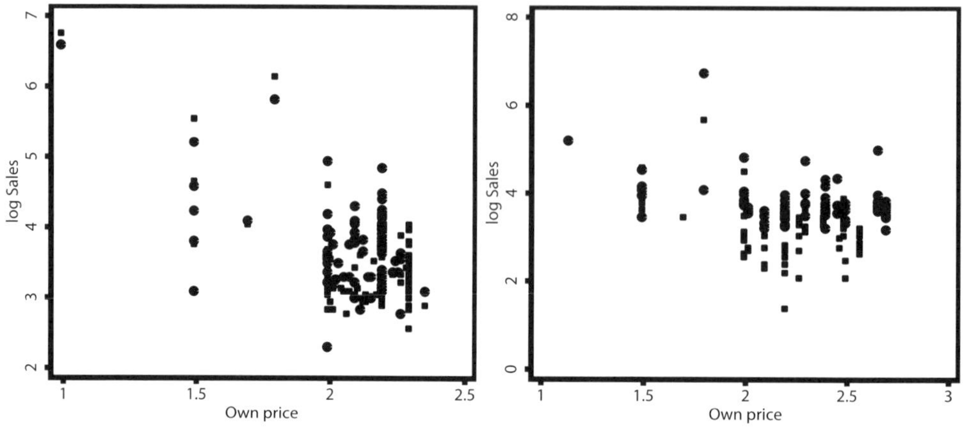

Figure 4.2 Scatter Plots Between the Log Number of Sold Packages and the Own Price in Four Different Stores. Each plot displays the log-sales versus price in two outlets using different symbols (plus signs and black dots)

Table 4.1 Posterior Means, Standard Deviations and 95% Credible Intervals for Fixed Effects and Random Effects Variances

Variable	post.mean	stdev	95% CI	re var.
intercept	4.00	0.069	[3.86,4.13]	0.254
logp	−2.54	0.062	[−2.66,−2.41]	0.177
logpc1	0.45	0.033	[0.39,0.52]	0.005
logpc2	1.11	0.033	[1.04,1.18]	0.013
logpc3	0.54	0.027	[0.49,0.59]	0.015

with a priori independent random effects given by $u_{ik} \sim N(0, \delta_k^2)$, $k = 0, \ldots, 4$. The model is estimated with the program BayesX (Brezger et al., 2005, 2007) using MCMC simulations based on 100,000 iterations after a burn-in period of 2000 iterations. Of the 100,000 samples, every 100th sample has been stored and used for posterior inference.

Table 4.1 shows the estimated posterior mean, standard deviation, and 95% credible intervals for the fixed effects. The last column of Table 4.1 shows the estimated variances δ_k^2, $k = 0, \ldots, 4$ of the corresponding outlet-specific random effects.

The estimated log-linear price effects are in agreement with economic theory. Increasing own-prices reduce sales while increasing prices of competing brands lead to increased sales for Tree Fresh. The strength of the outlet-specific random variation across stores measured through the random effects variances in the last column of Table 4.1 is quite different. While the random intercept (measuring the outlet-specific overall sales level) and the own-price random effects are comparably strong, with variances of $\delta_0^2 = 0.254$ and $\delta_1^2 = 0.177$ (compared to the posterior mean of the overall variance of $\sigma^2 = 0.19$), the random effects of cross-price effects are comparably small. A way to get an intuition about the store-specific heterogeneity is to plot the store-specific price curves. This is done for four outlets in Figure 4.3, showing considerable own-price heterogeneity.

Model comparison in hierarchical Bayesian models estimated with MCMC simulations can be done via the deviance information criterion (DIC), as proposed by Spiegelhalter et al. (2002); see also Section 4.5 for a critical discussion of the DIC in multilevel models. Comparing the full random effects model (4.12) with a model without random effects, we obtain a DIC reduction of more than 140 points (8278 for the parametric model versus 8136 for the random effects model) which is another clear indicator for the existence of store-specific heterogeneity. However, it is

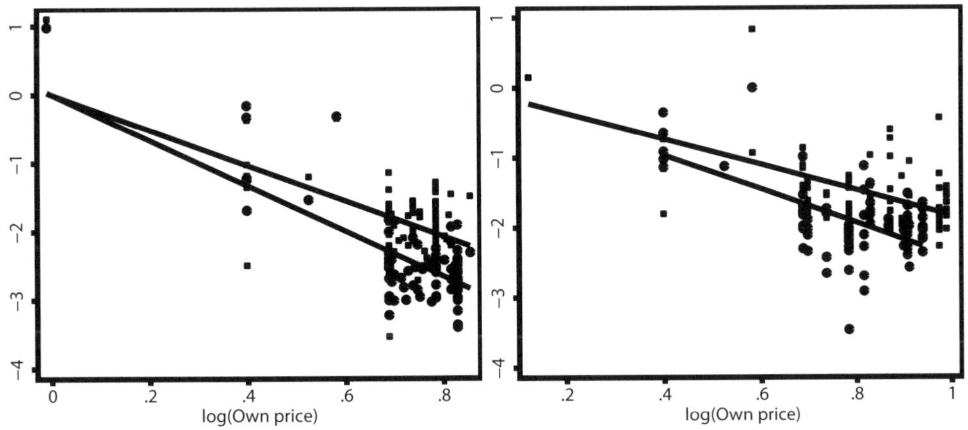

Figure 4.3 Log-Sales Versus Log-Price Curves for Four Different Outlets together with Corresponding Partial Residuals

less clear that the cross-price random effects $u_{i2}, \ldots, u_{i4}$ are really needed. Estimation without these random effects reveals a DIC of 8148, so that the full random effects model is still slightly better (in terms of the DIC) than the reduced random effects model.

A somewhat unsatisfactory aspect of our models so far is that the type of nonlinearity of the price effects is determined in advance using a log-transformation of prices. To investigate whether the log-transformation is really appropriate we can replace the linear fixed effects in (4.12) by Bayesian P-splines (compare Section 4.2.2) leading to the model

$$\begin{aligned} y_{ij} = \beta_0 &+ f_1(logp_{ij}) + f_2(logpc1_{ij}) \\ &+ f_3(logpc2_{ij}) + f_4(logpc3_{ij}) + u_{i0} \\ &+ u_{i1} logp_{ij} + u_{i2} logpc1_{ij} \\ &+ u_{i3} logpc2_{ij} \\ &+ u_{i4} logpc3_{ij} + \varepsilon_{ij}. \end{aligned} \tag{4.13}$$

We assumed a priori a monotonically decreasing effect of the own price and monotonically increasing effects of cross prices. Imposing monotonicity constraints is comparably straightforward in a Bayesian approach; see Brezger and Steiner (2008) for details. If the log-transformation is sufficient to capture the nonlinearity, the estimated curves $\hat{f}_1, \ldots, \hat{f}_4$ should be approximately linear. Figure 4.4 shows the estimated price effects. The plots reveal that the log-transformation is a reasonable approximation to the nonlinear effects, although considerable additional nonlinearity remains. The DIC of this model is 6606, which is approximately 1500 units below the strictly parametric models. This is a huge improvement in the goodness of fit and confirms our claim that there is considerable additional nonlinearity beyond the log-transformation. The size of the random effects measured through the random effects variances is in the same range as in the parametric model (4.13). To demonstrate the outlet-specific heterogeneity, Figure 4.5 shows some outlet-specific log-sales own-price curves which are now additively composed of the nonlinear log-price effect f_1 and the linear log-price random effect $u_{i1} logp_{ij}$.

A different approach to deal with nonlinearity of price effects and outlet-specific heterogeneity simultaneously has been proposed in Lang et al. (2011a). The paper assumes the model

$$\begin{aligned} y_{ij} &= \beta_0 + u_{i0} + f_1(p_{ij})(u_{i1} + 1) \\ &\quad + f_2(pc1_{ij})(u_{i2} + 1) \\ &\quad + f_3(pc2_{ij})(u_{i3} + 1) \\ &\quad + f_4(logpc4_{ij})(u_{i4} + 1) + \varepsilon_{ij}, \end{aligned}$$

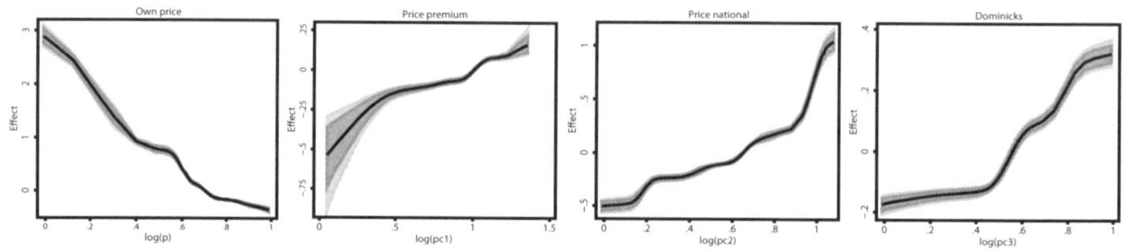

Figure 4.4 Estimated Price Effects in the Model (4.13). Shown are the posterior mean together with 80% and 95% pointwise credible intervals

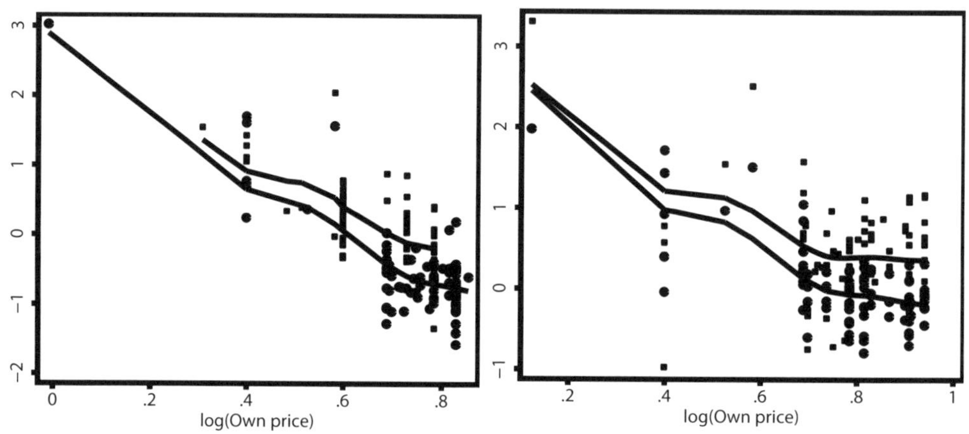

Figure 4.5 Log-Sales Versus Log-Price Curves in Model (4.13) for Four Different Outlets together with Corresponding Partial Residuals

where

- f_1 is an unknown smooth nonlinear decreasing function of the brand's own price, and $f_2, \ldots, f_4$ are unknown smooth nonlinear increasing functions of prices of competing brands;
- u_{i0} is a random store effect accounting for heterogeneity in baseline sales across stores; and
- $u_{i1} + 1, \ldots, u_{i4} + 1$ are store-specific scaling factors accommodating heterogeneity of the nonparametric own- and cross-item price effects across stores (while the overall functional forms remain constant from store to store).

More details on the approach and results are given in Lang et al. (2011a); see also Lang et al. (2011b).

4.4 EXTENSIONS TO MODELS WITH NON-GAUSSIAN PRIORS

One of the main advantages of taking a Bayesian approach to multilevel modeling is the flexibility in defining extended model types and classes by adding additional layers to the hierarchical prior structure. In the following, we will briefly discuss some of these extended models with non-Gaussian priors arising either by adding suitable hyperpriors to the variance parameters, leading to scale mixture of normal priors, or by replacing the usual, parametric iid prior with mixture distributions of different type. When using suitable prior choices involving conjugate priors, it is often possible to retain Gibbs sampling steps at several places in the MCMC algorithms. In addition, due to the modularity of MCMC algorithms, flexible approaches can easily be incorporated into the model structures discussed in Section 4.2, such as additive or geoadditive mixed models.

4.4.1 Conditionally Gaussian Priors

For simplicity, we will restrict most of the presentation in this paragraph to individual-specific random intercepts u_{i0}, but similar extensions can easily be derived for vectors of random effects. The standard assumption in Bayesian (G)LMMs is a Gaussian random effects prior u_{i0} iid $N(0, \delta_0^2)$. Although, from a Bayesian perspective, it may often be safe to assume Gaussian priors if the data are informative enough to outweigh the prior information, more flexible and/or robust alternative prior specifications are often desirable. Such priors can easily be determined by supplementing the unknown variance parameter δ_0^2 with hyperpriors $p(\delta_0^2)$. Then the prior for u_{i0} is no longer Gaussian but instead given by

$$p(u_{i0}) = \int p(u_{i0}|\delta_0^2)p(\delta_0^2)d\delta_0^2$$

when marginalizing with respect to the hyperprior. For example, the simple, hierarchical random effects prior

$$u_{0i} \mid \delta_0^2 \overset{iid}{\sim} N(0, \delta_0^2), \quad \delta_0^2 \sim IG(a, b)$$

considered in Section 4.2 in fact already induces a marginal t-distribution for the random effects. This partly explains the greater robustness of Bayesian random effects models with respect to deviations from the normal assumption for random effects, since the marginal t-prior has heavier tails than the normal distribution. A more detailed discussion of robust approaches for multilevel modeling is given in Chapter 22.

Specifying other types of hyperpriors $p(\delta_0^2)$ allows one to determine prior structures with different properties. These may, for example, aim at further improving the robustness in terms of prior distributions with heavier tails or at improving regularization properties based on priors that achieve stronger shrinkage towards zero for small random effects. For example, a marginal Laplace prior for random effects u_{i0} is obtained by the following prior hierarchy:

$$u_{i0}|\delta_0^2 \sim N(0, \delta_0^2), \quad \delta_0^2 \sim \text{Exp}(0.5\lambda^2).$$

Then the marginal distribution can be shown to be given by

$$\begin{aligned} p(u_{i0}) &= \int_0^\infty p(u_{i0}|\delta_0^2)p(\delta_0^2)d\delta_0^2 \\ &\propto \exp(-\lambda|u_{i0}|). \end{aligned}$$

Hence, adding an exponential prior to the variance allows the replacement of the t-distribution arising from the standard random effects distribution with a Laplace distribution corresponding to an absolute value penalty. The latter has the advantage that small random effects are shrunk towards zero more heavily than with a Gaussian prior, while larger random effects are assigned a smaller penalty. This behavior is, in fact, analogous to replacing ridge regularization with LASSO regularization in a frequentist regularization approach, and has mostly been considered for regularizing the 'fixed effects' part of the predictor instead of the random effects (see Park and Casella, 2008, for details), in particular with high-dimensional vectors of regression coefficients. Several extensions of the basic Bayesian LASSO have been proposed following Park and Casella (2008); see, for example, Hans (2009); Griffin and Brown (2010); Carvalho et al. (2010).

The main advantage of scale mixtures of Gaussians as random effects distributions is that the full conditional of the random intercepts u_{i0} remains Gaussian. Therefore, when computing the full conditionals for the random effects, exactly the same calculations arise as in the original Gaussian random effects model and the full conditionals obtained there remain valid. The only additional task that has to be performed is to derive the full conditionals for the hyperparameters in the scale mixture representation.

Various other types of conditionally Gaussian priors can be achieved by varying the choices on the hyperpriors in the scale mixture. For example, a prior distribution corresponding to the frequentist elastic net (Zou and Hastie, 2005) is obtained as a mixture of a Gaussian and a Laplace prior and can therefore again be cast into the conditionally Gaussian prior framework; see Fahrmeir et al. (2010) for more examples and a more detailed discussion.

4.4.2 Mixture Priors for Random Effects

The approaches discussed in the previous paragraph replace the Gaussian random effects prior with more flexible alternatives but still assume one common, parametric prior for the random effects where the specific type of distribution has to be chosen before the analysis. In this subsection, we will instead focus on prior hierarchies that enable the flexible estimation of the random effects distribution simultaneously with estimation of the random effects themselves. These approaches will be based on different types of (Gaussian) mixture distributions (see Chapter 20 for a detailed, frequentist introduction to mixture models in multilevel regression with inference based on the expectation maximization algorithm).

The most immediate choice for a mixture prior are finite mixture models

$$p(\boldsymbol{u}_i|\boldsymbol{\pi}, \boldsymbol{\phi}) = \sum_{k=1}^{K} \pi_k p(\boldsymbol{u}_i|\boldsymbol{\phi}_k),$$

where the random effects are assumed to arise from one of K mixture components $p(\boldsymbol{u}_i|\boldsymbol{\phi}_k)$ characterized by the mixture parameters summarized in $\boldsymbol{\phi} = (\boldsymbol{\phi}_1, \dots, \boldsymbol{\phi}_K)$ (possibly supplemented with additional hyperpriors), and $\boldsymbol{\pi} = (\pi_1, \dots, \pi_K)'$ is a vector of mixture weights such that $\pi_1 + \cdots + \pi_K = 1$. When assuming Gaussian mixture components, i.e. $p(\boldsymbol{u}_i|\boldsymbol{\phi}_k)$ are Gaussian densities $N(\boldsymbol{\mu}_k, \boldsymbol{\Sigma}_k)$ and $\boldsymbol{\phi}_k$ collects unknown parameters in the mean $\boldsymbol{\mu}_k$ and the covariance matrix $\boldsymbol{\Sigma}_k$, we obtain the so-called heterogeneity model (Frühwirth-Schnatter, 2006). It has the particular advantage that (similar to the previous subsection) it can be reformulated in a conditionally Gaussian fashion. Therefore, we introduce latent class indices $c_i \in \{1, \dots, K\}$, $i = 1, \dots, n$, that indicate the mixture component to which a specific observation is assigned. Then the conditional distribution of $\boldsymbol{u}_i$ given the class index c_i is Gaussian, i.e.

$$\boldsymbol{u}_i|c_i = k \sim N(\boldsymbol{\mu}_k, \boldsymbol{\Sigma}_k).$$

Including the class indices as additional unknowns in the MCMC algorithm allows one to obtain standard full conditionals as with Gaussian random effects but with separate updates for each of the mixture components (see, for example, Frühwirth-Schnatter, 2006, for such an MCMC algorithm).

Inferences in a mixture model may become unstable when considering a large number of latent classes K, which, however, may be required to obtain enough flexibility when estimating random effects densities. As a remedy, Komárek and Lesaffre (2008) propose penalized mixtures of Gaussians where a large number of mixture components is used in combination with a penalty on the mixture weights. Since a large number of components is employed, the expectations and covariance matrices can safely be assumed to be fixed if they are defined on a dense grid covering the support of the (standardized) random effects density.

Then the only remaining unknowns are the mixture weights which are assigned a random walk prior (on the logit scale) to achieve smooth random effects densities.

A completely nonparametric approach to the joint estimation of the random effects density and the random effects is obtained by specifying a hyperprior on the space of probability distributions instead of assuming the class of Gaussian mixtures a priori. The most common choice to achieve such a nonparametric random effects model are Dirichlet processes, i.e.

$$\boldsymbol{u}_i \sim G, \quad G \sim DP(\nu_0, G_0),$$

where $DP(\nu_0, G_0)$ denotes a Dirichlet process with concentration parameter $\nu_0 > 0$ and base distribution G_0. See Ferguson (1973), Blackwell and MacQueen (1973), and Antoniak (1974) for early introductions to DP modeling; Neal (2000) for details on corresponding sampling schemes; Kleinman and Ibrahim (1998) and Escobar and West (1998) for DP applications in the context of random effects modeling; and Ghosh and Ramamoorthi (2010) for a recent review. Informally, the parameters of the Dirichlet process can be explained as follows: The base distribution G_0 is a distribution that corresponds to a parametric standard assumption for the prior that shall be replaced with a nonparametric alternative. In the context of random effects models, this will typically be the Gaussian random effects prior. The concentration parameter ν_0 specifies our prior confidence in this base distribution. If ν_0 is large, the resulting Dirichlet process will concentrate most of the probability mass close to the base distribution, and we basically retain the standard model arising with G_0 as random effects prior. If, on the other hand, ν_0 is small, ample deviations from the base distribution may be encountered. In practice, a hyperprior is typically assigned to the concentration parameter to avoid too subjective choices.

A more detailed understanding of Dirichlet processes can be obtained from the constructive, stick-breaking representation introduced by Sethuraman (1994): Any $G \sim DP(\nu_0, G_0)$ can be represented as

$$G(\cdot) = \sum_{k=1}^{\infty} \pi_k \delta_{\boldsymbol{\phi}_k}(\cdot),$$

where $\delta_{\boldsymbol{\phi}_k}(\cdot)$ are Dirac measures (i.e. point masses) located at cluster-specific parameter vectors $\boldsymbol{\phi}_k$ drawn from the base distribution G_0, i.e.

$$\boldsymbol{\phi}_k \overset{\text{iid}}{\sim} G_0,$$

independently from the (random) weights π_k. The weights are generated through a *stick-breaking process*:

$$\begin{aligned}\pi_1 = v_1, \pi_k &= v_k \left(1 - \sum_{j=1}^{k-1}(1-\pi_j)\right) \\ &= v_k \prod_{j=1}^{k-1}(1-v_j),\end{aligned}$$

for $k = 2, 3, \ldots$, with

$$v_k \overset{\text{iid}}{\sim} \text{Be}(1, \nu_0).$$

Thus, realizations G of the Dirichlet process can be obtained as infinite mixtures of point masses. The locations $\boldsymbol{\phi}_k$ are an iid sample from G_0, and the weights π_k are generated through a 'stick-breaking procedure', first breaking the part $\pi_1 = v_1$ from a stick of length $1 = \pi_1 + \pi_2 + \cdots$, then breaking off the part $\pi_2 = v_2(1 - \pi_1)$ from the remaining stick of length $1 - \pi_1$, and so on.

The stick-breaking representation of Dirichlet processes has two immediate consequences. First of all, Dirichlet processes can be considered infinite mixtures and therefore provide a natural extension to the finite mixture models considered at the beginning of this section. On the other hand, all random effects measures arising as realizations of a Dirichlet process are almost surely discrete, with all probability mass concentrated on the locations $\boldsymbol{\phi}_k$, $k = 1, 2, \ldots$.

The latter property has the disadvantage that assigning a Dirichlet process prior directly to the random effects distribution induces clusters among the individuals sharing exactly the same random effects, which is often unrealistic in practice. As a consequence, the assumption of a Dirichlet process is often moved to a higher level in the prior hierarchy, yielding Dirichlet process mixture priors. For example, we may assume that groups of subjects share the same normal distribution for the random effects, but that there are differences in the groups with respect to either the mean, the covariance matrix, or both. More generally, a Dirichlet process mixture is obtained by the following prior hierarchy:

$$\begin{aligned} \boldsymbol{u}_i &\overset{\text{ind}}{\sim} p(\boldsymbol{u}_i|\boldsymbol{\phi}_i), \\ \boldsymbol{\phi}_i &\overset{\text{iid}}{\sim} G, \\ G &\sim DP(\nu_0, G_0). \end{aligned}$$

Here, the random effects are assumed to be realized independently from distributions $p(\boldsymbol{u}_i|\boldsymbol{\phi}_i)$ with individual-specific parameters $\boldsymbol{\phi}_i$. These are generated according to a probability measure that is realized from a Dirichlet process with concentration parameter ν_0 and base distribution G_0. Since the realization of the Dirichlet process is almost surely discrete, there will be ties among the individual-specific parameters $\boldsymbol{\phi}_i$ such that groups of individuals sharing the same random effects distribution arise.

More specifically, in the case of random effects, we may assume the following prior hierarchy:

$$\begin{aligned} \boldsymbol{u}_i &\overset{\text{ind}}{\sim} N(\boldsymbol{\mu}_i, \boldsymbol{\Omega}), \\ \boldsymbol{\mu}_i &\overset{\text{iid}}{\sim} G, \\ G &\sim DP(\nu_0, G_0), \end{aligned}$$

i.e. the random effects are independent Gaussian distributed with a common covariance matrix Ω but differing means $\boldsymbol{\mu}_i$ that are assigned a Dirichlet process prior. Conjugate prior choices completing this prior hierarchy are a Gaussian base distribution G_0 (possibly with additional hyperpriors on the corresponding parameters) and an inverse Wishart prior for the covariance matrix $\boldsymbol{\Omega}$; see Fahrmeir and Kneib (2011) for more details and references.

The modularity of MCMC approaches also allows for an easy combination of Dirichlet process mixtures with more general modeling components; see Heinzl et al. (2011) for an example where additive mixed models are supplemented with Dirichlet process mixture priors for random effects. Note that the stick-breaking procedure discussed above may not be the best way to facilitate Bayesian inference, and that several alternative suggestions exist (see, for example, Roy and Hobert, 2007, for more details).

4.5 REMARKS ON MODEL ASSESSMENT

Advances in Bayesian computation make it possible to fit increasingly flexible hierarchical models to multilevel data. Model assessment is desirable to choose between a number of candidate models and to check model adequacy. This has led to immense methodological research, often with competing suggestions and strategies. The classical formal approach to Bayesian model choice is based on marginal likelihoods $p(y)$, after integrating out parameters, on Bayes factors $p(y|M_1)/p(y|M_2)$ as the ratio of the marginal likelihoods of competing models M_1, M_2 out of a finite set of candidate models, including the correct one, and on posterior model probabilities $p(M_1|y)$, $p(M_2|y)$, etc. Computation of marginal likelihoods and Bayes factors is quite demanding, however, for complex hierarchical models. Frühwirth-Schnatter (2006) contains a good survey of simulation-based approximations of marginal likelihoods. Frühwirth-Schnatter and Wagner (2006) propose an auxiliary variable approach for non-Gaussian models; see also further references given therein.

A serious problem with the Bayes factor for some of the models discussed in Section 4.2 is that a part of the prior specification is improper, e.g. flat priors for fixed effects or smoothing priors for P-splines. General approaches for constructing objective proper priors have been developed for so-called intrinsic Bayes factors (Berger and Pericchi, 1996) and factorial Bayes factors (O'Hagan, 1995); see also Berger and Pericchi (2001) for a comparison.

Model selection criteria can avoid problems with improper priors. The deviance information criterion (DIC) of Spiegelhalter et al. (2002), which we used in the case study of

Section 4.3, has the big advantage that it is easily computed from standard MCMC output. As with the AIC, however, it may underpenalize model complexity. Another issue, already pointed out in Spiegelhalter et al. (2002) and the discussion accompanying this chapter, is the fact that the effective model dimension or number of parameters, included in the definition of the DIC, depends on the parameterization of the model. This can be a problem in more complex hierarchical models. It is recommended 'to specify the level of the hierarchy the model focuses on'. Obviously, problems arise with models involving latent variables and missing data. Celeux et al. (2006) discuss this with a view towards mixture models and models with random effects. For longitudinal or clustered data models with only one additional level, we consider all 'random effects', including basis function coefficients for nonlinear effects in GAMMs or STAR models, to be a focus of the model. This corresponds to the conditional DIC in Celeux et al. (2006), but the DIC should be used with caution in models with more levels because sufficient empirical experience is lacking.

Posterior predictive model checking (Gelman et al., 1996, 2003) can be implemented as a by-product of MCMC simulations and is useful for assessing model adequacy through various diagnostic tools, ranging from graphical goodness-of-fit checks to omnibus measures and custom summary statistics to assess specific aspects of a working model. Because posterior predictive model checking is based on generating replicated data via samples from the posterior, given all data, it induces conservatism and is useful for goodness-of-fit assessment, but less for truly predictive assessment. The latter could be based, in principle, on cross-validatory predictive distributions, see Gelfand (1996) for early suggestions, but computation efforts become extreme for exact leaving-one-out methods. Marshall and Spiegelhalter (2003, 2007) propose an approximate cross-validatory sampling scheme, which seems to be quite useful. More experience with complex hierarchical models is needed, however, to recommend it generally. Another possibility is minimizing predictive loss (Gelfand and Ghosh, 1998).

As for any regression model, predictor selection is an important issue. This task comprises selection of variables with fixed effects, specification of possibly nonlinear functional effects, selection of random components, and, moreover, selection of individual random effects. We refer the reader to Section 4.5.3 of Fahrmeir and Kneib (2011) for a survey; to Chen and Dunson (2003), Cai and Dunson (2006), and Frühwirth-Schnatter and Tüchler (2008) for selection of random components through choice of random effects covariance matrices; and to Frühwirth-Schnatter and Wagner (2010) for selection of individual random effects.

Finally, Gaussian mixture priors and $DP(M)$ priors as sketched in Section 4.4.2 are rather useful for investigating whether Gaussian random effects priors are adequate or too simplistic.

REFERENCES

Albert, J. and Chib, S. (1993) 'Bayesian Analysis of Binary and Polychotomous Response Data', *Journal of the American Statistical Association*, 88: 669–79.

Antoniak, C. (1974) 'Mixtures of Dirichlet Processes with Applications to Bayesian Nonparametric Problems', *Annals of Statistics*, 2: 1152–74.

Banerjee, S., Carlin, B.P., and Gelfand, A.E. (2003) *Hierarchical Modeling and Analysis for Spatial Data*. New York: Chapman & Hall / CRC.

Berger, J.O. and Pericchi, L.R. (1996) 'The Intrinsic Bayes Factor for Model Selection and Prediction', *Journal of the American Statistical Association*, 91: 109–22.

Berger, J.O. and Pericchi, L.R. (2001) 'Objective Bayesian Methods for Model Selection: Introduction and Comparison (with Discussion)', in P. Lahiri (ed.), *Model Selection*, 38: 135–207.

Blackwell, D. and MacQueen, J. (1973) 'Ferguson Distributions via Pólya Urn Schemes,' *Annals of Statistics*, 1: 353–5.

Brezger, A. and Lang, S. (2006) 'Generalized Additive Regression Based on Bayesian P-splines', *Computational Statistics & Data Analysis*, 50: 967–91.

Brezger, A. and Steiner, W. (2008) 'Monotonic Regression Based on Bayesian P-splines: An Application to Estimating Price Response Functions from

Store-level Scanner Data', *Journal of Business and Economic Statistics*, 26: 90–104.

Brezger, A., Kneib, T., and Lang, S. (2005) 'Bayesx: Analyzing Bayesian Structured Additive Regression Models', *Journal of Statistical Software*, 14: 1–22.

Brezger, A., Kneib, T., and Lang, S. (2012) *Bayesx Manuals*. Technical report, Department of Statistics, University of Munich. www.bayesx.org

Cai, B. and Dunson, D. (2006) 'Bayesian Covariance Selection in Generalized Linear Mixed Models', *Biometrics*, 62: 446–57.

Carvalho, C.M., Polson, N.G., and Scott, J.G. (2010) 'The Horseshoe Estimator for Sparse Signals', *Biometrika*, 97: 465–80.

Celeux, G., Forbes, F., Robert, C.P., and Titterington, D.M. (2006) 'Deviance Information Criteria for Missing Data Models', *Bayesian Analysis*, 1: 651–74.

Chen, M.H., Ibrahim, J., Shao, Q.M., and Weiss, R. (2003) 'Prior Elicitation for Model Selection and Estimation in Generalized Linear Mixed Models', *Journal of Statistical Planning and Inference*, 111: 57–76.

Chen, Z. and Dunson, D. (2003) 'Random Effects Selection in Linear Mixed Models', *Biometrics*, 59: 762–9.

Eilers, P.H.C. and Marx, B.D. (1996) 'Flexible Smoothing using B-splines and Penalties', *Statistical Science*, 11: 89–121.

Escobar, M.D. and West, M. (1998) 'Computing Bayesian Nonparametric Hierarchical Models', in D. Dey, P. Müller, and D. Sinha (eds), *Practical Nonparametric and Semiparametric Bayesian Statistics*. New York: Springer, 1: 22.

Fahrmeir, L. and Kneib, T. (2009) 'Propriety of Posteriors in Structured Additive Regression Models: Theory and Empirical Evidence', *Journal of Statistical Planning and Inference*, 39: 843–59.

Fahrmeir, L. and Kneib, T. (2011) *Bayesian Smoothing and Regression for Longitudinal, Spatial and Event History Data*. Oxford: Oxford University Press.

Fahrmeir, L., Kneib, T., and Lang, S. (2004) 'Penalized Structured Additive Regression for Space-time Data: A Bayesian Perspective', *Statistica Sinica*, 14: 731–61.

Fahrmeir, L., Kneib, T., and Konrath, S. (2010) 'Bayesian Regularisation in Structured Additive Regression: A Unifying Perspective on Shrinkage, Smoothing and Predictor Selection', *Statistics and Computing*, 20: 203–19.

Fahrmeir, L., Kneib, T., Lang, S., and Marx, B. (2013) *Regression: Models, Methods and Applications*. Berlin: Springer.

Ferguson, T.S. (1973) 'A Bayesian Analysis of Some Nonparametric Problems', *Annals of Statistics*, 1: 209–30.

Frühwirth-Schnatter, S. (2006) *Finite Mixture and Markov Switching Models*. Berlin: Springer.

Frühwirth-Schnatter, S. and Frühwirth, R. (2010) 'Data Augmentation and MCMC for Binary and Multinomial Logit Models', in T. Kneib and G. Tutz (eds), *Statistical Modeling and Regression Structures: Festschrift in Honour of Ludwig Fahrmeir*. Berlin: Springer Verlag.

Frühwirth-Schnatter, S. and Tüchler, R. (2008) 'Bayesian Parsimonious Covariance Estimation for Hierarchical Linear Mixed Models', *Statistics and Computing*, 18: 1–13.

Frühwirth-Schnatter, S. and Wagner, H. (2006) 'Auxiliary Mixture Sampling for Parameter-driven Models of Time Series of Small Counts with Applications to State Space Modeling', *Biometrika*, 93: 827–41.

Frühwirth-Schnatter, S. and Wagner, H. (2010) 'Bayesian Variable Selection for Random Intercept Modeling of Gaussian and Non-Gaussian Data', in Bernardo et al. (eds), *Bayesian Statistics 9*. Oxford: Oxford University Press.

Frühwirth-Schnatter, S., Frühwirth, R., Held, L., and Rue, H. (2009) 'Improved Auxiliary Mixture Sampling for Hierarchical Models of Non-Gaussian Data', *Statistics and Computing*, 19: 479–92.

Gamerman, D. (1997) 'Efficient Sampling from the Posterior Distribution in Generalized Linear Mixed Models', *Statistics and Computing*, 7: 57–68.

Gamerman, D. and Lopes, H.F. (2006) *Markov Chain Monte Carlo: Stochastic Simulation for Bayesian Inference*. London: Chapman and Hall.

Gelfand, A. (1996) 'Model Determination using Sampling Based Methods', in W.R. Gilks, S. Richardson, and D.J. Spiegelhalter (eds), *Markov Chain Monte Carlo in Practice*. Boca, Raton: Chapman & Hall/CRC, 145–61.

Gelfand, A.E. and Ghosh, S.K. (1998) 'Model Choice: A Minimum Posterior Predictive Loss Approach', *Biometrika*, 85: 1–11.

Gelfand, A.E. and Smith, A.F.M. (1990) 'Sampling Based Approaches to Calculating Marginal Densities', *Journal of the American Statistical Association*, 85: 398–409.

Gelman, A.E. (2006) 'Prior Distributions for Variance Parameters in Hierarchical Models', *Bayesian Analysis*, 1: 515–34.

Gelman, A.E., Meng, X.-L., and Stern, H.S. (1996) 'Posterior Predictive Assessment of Model Fitness via Realized Discrepancies', *Statistica Sinica*, 6: 733–60.

Gelman, A.E., Carlin, J.B., Stern, H.S., and Rubin, D.B. (2003) *Bayesian Data Analysis*. New York: Chapman and Hall.

Ghosh, J.K. and Ramamoorthi, R.V. (2010) *Bayesian Nonparametrics*. New York: Springer.

Griffin, J.E. and Brown, P.J. (2010) Inference with Normal-Gamma Prior Distributions in Regression Problems. *Bayesian Analysis*, 5: 171–88.

Hans, C. (2009) 'Bayesian Lasso Regression', *Biometrika*, 96: 835–45.

Heinzl, F., Fahrmeir, L., and Kneib, T. (2012) 'Additive Mixed Models with Dirichlet Process Mixture and P-spline Priors', *Advances in Statistical Analysis*, 1: 47–68.

Held, L. and Sabanes-Bove, D. (2012) *Applied Statistical Inference*. Berlin: Springer (forthcoming).

Holmes, C.C. and Held, L. (2006) 'Bayesian Auxiliary Variable Models for Binary and Multinomial Regression', *Bayesian Analysis*, 1: 145–68.

Kleinman, K. and Ibrahim, J. (1998) 'A Semiparametric Bayesian Approach to the Random Effects Model', *Biometrics*, 54: 921–38.

Komárek, A. and Lesaffre, E. (2008) 'Generalized Linear Mixed Model with a Penalized Gaussian Mixture as a Random Effects Distribution', *Computational Statistics & Data Analysis*, 52: 3441–58.

Lang, S. and Brezger, A. (2004) 'Bayesian P-splines', *Journal of Computational and Graphical Statistics*, 13: 183–212.

Lang, S., Steiner, W., and Wechselberger, P. (2011) *Simultaneously Accommodating Heterogeneity and Functional Flexibility in Store Sales Models* Technical report, Institute of Statistics, University of Innsbruck.

Lang, S., Umlauf, N., Wechselberger, P., Harttgen, K., and Kneib, T. (2013) *Multilevel Structured Additive Regression*. *Statistics and Computing*, to appear.

Marshall, E.C. and Spiegelhalter, D.J. (2003) 'Approximate Cross-validatory Predictive Checks in Disease Mapping Models', *Statistics in Medicine*, 22: 1649–60.

Marshall, E.C. and Spiegelhalter, D.J. (2007) 'Identifying Outliers in Bayesian Hierarchical Models: A Simulation-based Approach', *Bayesian Analysis*, 2: 409–44.

Neal, R.M. (2000) 'Markov Chain Sampling Methods for Dirichlet Process Mixture Models', *Journal of Computational and Graphical Statistics*, 9: 249–65.

O'Hagan, A. (1994) *Kendall's Advanced Theory of Statistics Vol. 2B: Bayesian Inference*. London: Arnold.

O'Hagan, A. (1995) 'Fractional Bayes Factors for Model Comparison (with Discussion)', *Journal of the Royal Statistical Society, Series B*, 57: 99–138.

Park, T. and Casella, G. (2008) 'The Bayesian Lasso', *Journal of the American Statistical Association*, 103: 681–6.

Robert, C.P. and Casella, G. (2004) *Monte Carlo Statistical Methods*. New York: Springer.

Roy, V. and Hobert, J.P. (2007) 'Convergence Rates and Asymptotic Standard Errors for Markov Chain Monte Carlo Algorithms for Bayesian Probit Regression', *Journal of the Royal Statistical Society, Series B*, 69: 607–23.

Rue, H., Martino, S., and Chopin, N. (2009) 'Approximate Bayesian Inference for Latent Gaussian Models by using Integrated Nested Laplace Approximations', *Journal of the Royal Statistical Society, Series B*, 71: 319–92.

Sethuraman, J. (1994) 'A Constructive Definition of Dirichlet Priors', *Statistica Sinica*, 2: 639–50.

Spiegelhalter, D.J., Best, N.G., Carlin, B.P., and van der Linde, A. (2002) 'Bayesian Measures of Model Complexity and Fit', *Journal of the Royal Statistical Society B*, 65: 583–639.

Sun, D., Tsutakawa, R., and He, Z. (2001) 'Propriety of Posteriors with Improper Priors in Hierarchical Linear Mixed Models', *Statistica Sinica*, 11: 77–95.

Zou, H. and Hastie, T. (2005) 'Regularization and Variable Selection via the Elastic Net', *Journal of the Royal Statistical Society Series B*, 67: 301–20.

5

The Choice Between Fixed and Random Effects

Zac Townsend
New York University, USA

Jack Buckley
National Center for Education Statistics and New York University, USA

Masataka Harada
New York University, USA

Marc A. Scott
New York University, USA

5.1 INTRODUCTION

In prior chapters, group-specific effects were assumed to be drawn from a distribution, typically Gaussian. In applied research in the social and behavioral sciences, economics, public health, public policy, and many other fields, alternatives to this choice are often made, with the most common being the "fixed effects" approach. In its most basic formulation, group-specific intercepts are modeled using indicator variables, effectively making them open parameters in the model and not assigning them a distributional form. The choice between random and fixed effects has implications for the interpretation of parameter estimates and for estimation efficiency.[1] Specifically,

- the effect estimates for each group will be different depending on the modeling choice;
- the β estimates for the predictors in the model may be different depending on the modeling choice;
- the reliability with which one can make predictions for new groups differs; and
- there is a tradeoff between efficiency and bias, with random effects being more efficient than fixed effects, but potentially biased (we defer formalizing our definition of bias to a later section).

The remainder of this chapter will discuss these options as well as a hybrid version of the two. It provides guidance as to when an applied researcher should use each model, and describes situations when neither is appropriate and other models should be considered.

We proceed by describing the two models in Section 5.2 before discussing the different assumptions, describing estimation, and giving advice on when to use each in Section 5.3. Section 5.4 outlines an example of using fixed effects and random effects with data from the National Assessment of Educational Progress, a series of large-scale assessments of schoolchildren in the United States.

5.2 THE FIXED EFFECTS AND RANDOM EFFECTS MODELS

In this chapter, we outline both random and fixed effects models. We will refer to models as fixed if they model unit-specific components in longitudinal data or group-specific components in clustered data as separate parameters, and random effects if they are drawn from a (often Gaussian) probability distribution.

We will delve into model specifics shortly, but here we mention that with grouped data there are two types of relationship that are in play: between-group and within-group. Failure to understand this interplay can result in an ecological fallacy, in which one makes inferences about the nature of units based on information about the groups containing them. In the context of multilevel models, we will show that variation naturally divides into within- and between-group components, so that potential for misinterpretation exists whenever the two types of relationship differ for a predictor—for example, when neighborhood income matters for some outcome, but relative income within that group does not. Diez Roux (2004) and Wakefield (2003) give extensive overviews (see also Gelman, 2006). We note that the standard ecological fallacy typically occurs when one does not have both unit- and group-level data available to disentangle the two types of effects.

5.2.1 *The Random Effects Model*

Consider the multilevel model

$$y_{ij} = \beta_0 + \beta_1 x_{1ij} + u_{0j} + \varepsilon_{ij}, \qquad (5.1)$$

where y_{ij} is the outcome for the ith subject in the jth group. It is also represented more explicitly as a multilevel model as

$$y_{ij} = \beta_{0j} + \beta_1 x_{1ij} + \varepsilon_{ij}, \qquad (5.2)$$

where

$$\beta_{0j} = \beta_0 + u_{0j}. \qquad (5.3)$$

Equation (5.2) is often referred to as the level one equation and (5.3) is the level two equation. In both formulations, it is usually assumed that $u_{0j} \sim N(0, \sigma^2_{u0})$ is independent of $\varepsilon_{ij} \sim N(0, \sigma^2_{\varepsilon})$. This model proposes that the values of the outcome, y_{ij}, are based on the population-level constant, β_0, the single explanatory variable, x_{1ij} and the associated parameter β_1, and a group-specific component, u_{0j}. Although the random effects model can include more complex random components such as two-way (e.g., time- and unit-specific) intercepts or multiple group-specific intercepts that have a nested structure (e.g., classrooms and schools), in this chapter we consider a single group-specific intercept to focus on the choice between random and fixed effects.

Researchers using random effects models often focus on the variance components, such as σ^2_{u0}. It is instructive, however, to examine the group-specific effects, as these serve as controls for all other effects, and thus have the potential to alter the other effect estimates. In the random effects model, when $u_{0j} \sim N(0, \sigma^2_{u0})$, given constant β_0, we have $\beta_{0j} \sim N(\beta_0, \sigma^2_{u0})$, which are group-specific intercepts. Gelman and Hill (2007, p. 257) call this a "soft constraint" that is applied to the intercepts, which shrinks the estimates of β_{0j} toward the common mean β_0, a phenomenon they call a partial pooling (see Lindley and Smith, 1972; Smith, 1973; Efron and Morris, 1975; Morris, 1983; and Kreft and de Leeuw, 1998, for some background). This term derives from the "pooled" and "unpooled" modeling approaches described in Gelman and Hill (2007) and introduced in Chapter 1. A pooled model simply applies OLS or similar estimation techniques to grouped data, *ignoring the grouping*. An unpooled model may be defined in at least two ways. First, it could be understood as separate regressions, one for each group. This would entail unique regression parameters for every predictor as well as the intercept, and small sample sizes may yield highly imprecise estimates. A more common approach is to estimate group-specific intercepts, effectively using one parameter per group, and to pool the estimates of all other predictor effects; this definition is suggested by Gelman and Hill (2007). This is equivalent to estimating the so-called fixed effects model and will be what we adopt in our subsequent discussion.

The intercepts in the random effects model are a weighted average of estimates from the two types of pooling. In our simple model (5.1), assuming the data are balanced, the random intercepts and slope coefficients are

$$\hat{\beta}_{0j} \approx \frac{n_j/\sigma_\varepsilon^2}{n_j/\sigma_\varepsilon^2 + 1/\sigma_{u_0}^2}(\bar{y}_j - \hat{\beta}_1\bar{x}_{1j}) + \frac{1/\sigma_{u_0}^2}{n_j/\sigma_\varepsilon^2 + 1/\sigma_{u_0}^2}\beta_0 \quad (5.4)$$

$$\hat{\beta}_1 = \left(\sum_i\sum_j (x_{1ij} - \lambda\bar{x}_{1j})^2\right)^{-1} \quad (5.5)$$
$$\cdot \left(\sum_i\sum_j (x_{1ij} - \lambda\bar{x}_{1j})(y_{ij} - \lambda\bar{y}_j)\right),$$

where

$$\lambda = 1 - \sqrt{\frac{\sigma_\varepsilon^2}{\sigma_\varepsilon^2 + n_j\sigma_{u_0}^2}},$$

n_j is the number of observations within a group (group size), $\bar{y}_j$ and $\bar{x}_{1j}$ are within-group means of the outcome variable and the explanatory variable, and $\sigma_{u_0}^2$ and σ_ε^2 are variances of y, u_0, and ε, respectively (Gelman and Hill, 2007, p. 258; Afshartous and de Leeuw, 2005, pp. 112–13; Wooldridge, 2010, p. 287).[2] In order to utilize these formulas to get estimates of the coefficients, one must replace the variances and the intercept with their estimates.

In Equation (5.4), the group-specific residuals $(\bar{y}_j - \beta_1\bar{x}_{1j})$ employ the β_1 from the no-pooling model (or fixed effects model) and the overall intercept (β_0) is obtained from the complete pooling model (or OLS). Likewise, the slope estimate in Equation (5.5), $\hat{\beta}_1$, reduces to that of the fixed effects model when $\lambda = 1$ and that of OLS when $\lambda = 0$. That is, two weight parameters, $(n_j/\sigma_\varepsilon^2)/(n_j/\sigma_\varepsilon^2 + 1/\sigma_{u_0}^2)$ and λ, serve as a measure of the shrinkage toward the unpooled estimators. When group-level effects are estimable with great precision, $\sigma_{u_0}^2$ becomes large relative to σ_ε^2, and these weight parameters approach one, very little shrinkage occurs, and the random and fixed effects estimates are nearly identical. Conversely, when $\sigma_{u_0}^2$ becomes relatively small, these two weight parameters approach zero, and shrinkage to the population-level effect (ignoring group) becomes nearly complete.

When the assumptions in the random effects model hold, these above estimates are more efficient than their fixed effects counterparts, and this is one of the key reasons for choosing such models. The assumptions, however, may be more or less plausible in different applications, such as randomized experiments versus observational studies. See Chapter 12 for further discussion.

5.2.2 Fixed Effects Model

If one wants to estimate group-specific intercepts without imposing a distribution or utilizing information from the other groups, one may choose to estimate a fixed effects model. The model can be represented in a similar way to equation (5.1), but with a modification to the group effects:

$$y_{ij} = \beta_0 + \beta_1 x_{1ij} + \sum_{j'=2}^{J} u_{0j'} I\left\{j = j'\right\} + \varepsilon_{ij}, \quad (5.6)$$

with the indicator function $I\left\{j = j'\right\}$ tracking the $J - 1$ group-level effects (group 1 taken as the reference category). The implication of this reformulation is that the group effects u_{0j} may be correlated with the other predictor (or predictors, more generally). In situations in which the orthogonality of the random effects and other predictors is implausible, fixed effects models are one alternative.

Another way to characterize this model in the context of multilevel models is to describe the random effects distribution as having infinite variance, $\sigma_{u0}^2 = \infty$, so the model does not borrow any information across groups.[3] Although, in this sense, the fixed effects model is a special case of random effects, our simplest fixed effects model is not really a multilevel model in the traditional use of the term, as the group level has been reduced to a set of indicators and their corresponding effects.

A tradeoff that may not be immediately apparent is that one cannot include any group-level explanatory variables, such as each group's average income, due to collinearity with the group fixed effect. If one is interested in variance components and how they change with the addition of predictors at group and subject level, then one is severely hampered by the fixed effects approach. However, one may not have this type of interest.

Fixed effects models allow group-specific components u_{0j} to be correlated with other covariates x_{1ij}, which is certainly a less restrictive approach. This has led some researchers to interpret these components as controls that proxy for any omitted variable that is "fixed" or constant for the entire group. However, the form of "omitted variable" that may be captured this way is quite restrictive (see Baltagi, 2005; Wooldridge, 2010, for some discussion of omitted variable bias and how it relates to this model; see also Chapter 12). At a minimum, it is clear that the omitted confounder must be time- (or group-) constant.

Of course, one may estimate (5.6) using OLS, but in practice, a de-meaning approach, to be described in the next section, is usually employed. Given the estimators of random effects model, the derivation of fixed effects estimators is straightforward. By setting $\sigma_{u0}^2 = \infty$ in (5.4) and (5.5),

$$\hat{\beta}_{0j} = \bar{y}_j - \hat{\beta}_1 \bar{x}_{1j}, \text{ where}$$

$$\hat{\beta}_1 = \left(\sum_i \sum_j (x_{1ij} - \bar{x}_{1j})^2 \right)^{-1} \cdot \left(\sum_i \sum_j (x_{1ij} - \bar{x}_{1j})(y_{ij} - \bar{y}_j) \right).$$

These fixed effects estimators are often called the within estimators because they do not attempt to explain between-group differences, and are solely based on *deviations* of x_{1ij} and y_{ij} from the respective estimated group means $\bar{x}_{1j}$ and $\bar{y}_j$. Fixed effect methods control for all stable characteristics of the groups in the study. In a longitudinal study, this would mean that fixed effects control for time-invariant characteristics of each individual, such as personal history before the observation period.

While we have briefly noted some of the differences between the two estimators, our next section will provide more specific details of those estimation procedures and then discuss the relative merits or conditions favoring each of the techniques.

5.3 A COMPARISON OF THE TWO ESTIMATORS

This section describes the assumptions behind the fixed and random effects models, outlines the estimation techniques of both models and provides general advice regarding when to use and not use each technique. We then describe some hybrid models developed that satisfy the "random effects assumption" while still offering some of the benefits of fixed effects models.

5.3.1 The Assumptions of Each Model

Wooldridge (2010) states that the fixed effects model must satisfy the strict exogeneity assumption, $E(\varepsilon_{ij}|x_{1i'j}, u_{0j}) = 0$ for all values of i' given any i, to obtain an unbiased fixed effects estimator in the presence of any unobserved group-specific confounders. That is, ε_{ij} is not only independent of its contemporaneous covariates but also should be independent of the covariates that belong to the same group ($x_{1i'j}$ such that $i' \neq i$). If the condition on ε_{ij} is not met in the fixed effects model, then potential unobserved omitted variables could bias the estimator for x_{1ij}'s effect.

In addition to this strong exogeneity assumption, the random effects model assumes conditional mean independence of group-specific effects u_{0j} given all covariates x_{1ij} (Wooldridge, 2010). Formally,

$$\mathrm{E}(u_{0j}|x_{1ij}) = \mathrm{E}(u_{0j}) = 0.$$

This assumption is also referred to as orthogonality or, somewhat less formally, no correlation between u_{0j} and x_{1ij} ($\mathrm{Cov}(u_{0j}, x_{1ij}) = 0$). Usually it is further assumed that $u_{0j} \sim N(0, \sigma^2_{u0})$. This is perhaps the fundamental difference between the fixed and random effects approaches: the latter assumes that u_{0j} are orthogonal to ε_{ij} and to any predictors x_{1ij} in the model, while the former implicitly allows correlation with predictors.

This same concern of bias under violation of assumptions exists in the random effects model, but the additional assumption that the random effects are orthogonal to explanatory variables may be harder to justify, particularly where causal inference is the goal. However, in practice the researcher may wish to trade a small amount of bias for efficiency, and we reiterate that the hybrid models we present in Section 5.3.5 provide a useful compromise approach. Causal modeling for multilevel models requires a much deeper framework than we present in this chapter—see Chapter 12 of this volume for further discussion of assumptions in multilevel models for causal inference.

The differences in the above assumptions—and their implications—has led some researchers to frame the choice between random and fixed effects models as a bias/variance trade-off. Random effects models may be biased (for the within-group effect) when the model assumptions fail to hold, but they are more efficient than fixed effects models when they do. The bias is often framed as a form of omitted variable bias (see Bafumi and Gelman, 2006). Specifically, if there is a group-constant confounder c omitted from the model, and it is correlated with another predictor, such as x_{1ij}, (and the outcome), then the corresponding parameter β_1 will be biased when c is omitted. One of motivations for using grouped data is that we can recover the (within-group) β_1 that would have been estimated had we included the confounder c. With respect to recovering this parameter, fixed effects models are unbiased and random effects models are efficient (this is the setup for the Hausman Test). However, this statement applies only to the basic forms of these models; in Section 5.3.5 we show that centering can be used to achieve the same goal for random effects models.

5.3.2 Estimating Each Model

Fixed effects estimation can be conducted either by group de-meaning both the dependent and explanatory variables using the *within* transformation (sometimes called the fixed effects transformation), or by adding a set of dummy variables for each group j and using OLS, as previously described. To perform the *within* transformation, we can average equation (5.6) for each group, and then subtract the new group means from the original, obtaining:

$$\begin{aligned} y_{ij} - \bar{y}_j &= \left(\beta_0 + \beta_1 x_{1ij} + u_{0j} + \varepsilon_{ij}\right) \\ &\quad - \left(\beta_0 + \beta_1 \bar{x}_{1j} + u_{0j} + \bar{\varepsilon}_j\right) \\ &= \beta_1(x_{1ij} - \bar{x}_{1j}) + (\varepsilon_{ij} - \bar{\varepsilon}_j). \end{aligned} \tag{5.7}$$

Thus, the de-meaning of the original equation ensures that de-meaned covariates are orthogonal to any group effects, u_{0j}. It also implicitly controls for group effects, or partials them out of the model. The *within* nature of the effects is evident from this formulation, for now a one-unit change in the de-meaned covariate reflects change relative to one's group. Chapter 6 discusses this concept in some detail. The estimation of equation (5.7) with OLS yields fixed effects estimators.[4]

Alternatively, fixed effects estimators can be obtained by adding a set of group indicators as explanatory variables and applying OLS to equation (5.6). These two methods are equivalent; the interested reader can consult Davidson and MacKinnon (1993) and Wooldridge (2010).

Random effects models may be estimated in a variety of ways, including feasible generalized least squares (FGLS), maximum-likelihood estimation (MLE), or Bayesian approaches with MCMC. See Chapters 3 and 4 of this book, Hsiao (2003), and Gelman and Hill (2007) for an overview of potential approaches to estimation. Here, we introduce a step-by-step approach to calculate random effects estimators given by Johnston and DiNardo (1997).[5] Each step illuminates important properties of random effects models, but the main idea is to find the weight parameter, λ. Random effects estimators are partial-pooling estimators situated between no pooling (or fixed effects) and complete pooling (or OLS) estimators, and λ determines the proportion of these two quantities.

The first step is to find no-pooling and complete-pooling estimators. This can be easily done by estimating (5.6) or (5.7) and by estimating (5.1) without a group-specific component (u_{0j}) using OLS. Second, using the residuals obtained in the first step, we estimate σ_ε^2 and $\sigma_{u_0}^2$ using the following formulae:

$$\hat{\sigma}_\varepsilon^2 = \frac{\hat{\varepsilon}'_{FE}\hat{\varepsilon}_{FE}}{n(n_j - k - 1)}$$
$$\hat{\sigma}_{u_0}^2 = \frac{\hat{\varepsilon}'_{OLS}\hat{\varepsilon}_{OLS}}{n - k} - \frac{\hat{\sigma}_\varepsilon^2}{n_j},$$

where $\hat{\varepsilon}_{FE}$ is a residual vector from the no-pooling model, $\hat{\varepsilon}_{OLS}$ is that from the complete-pooling model, and k is the number of explanatory variables. Now, we have all the information required to calculate λ. Remember that the fixed effects model is defined to be the same as the unpooled model mentioned above, while the final step for this approach to estimating a random effects model is simply to regress $y_{ij} - \hat{\lambda}\bar{y}_j$ on $x_{1ij} - \hat{\lambda}\bar{x}_{1j}$. To repeat, this regression is identical to that of the fixed effects model when $\lambda = 1$, and that of OLS when $\lambda = 0$. Because $0 \leq \lambda \leq 1$, one way to conceptualize the random effects estimator is as the regression of *partially* de-meaned variables.

Another way of conceptualizing and then estimating random effects models is through a matrix representation of predictors and error structure and vector notation for outcomes. This requires specification of a complex covariance structure, in which some between-element correlation must be modeled. Typically, observations are assumed independent across groups, but not within groups, resulting in a block-diagonal structure. An example of a classic random intercepts model is given in Chapter 2, row 4 of Table 2.2; in matrix notation, using the terms X, Z, β, G, and R to capture the fixed and random effects designs, the fixed effects parameters, the covariance of the random effects, and the covariance of the errors, respectively, the outcome vector $\mathbf{y} \sim N(X\beta, ZGZ'+R)$. Estimation is usually made using the method of maximum likelihood, but Bayesian approaches are also common. As a by-product of this notation, we can express the best linear unbiased predictor (BLUP; see Robinson, 1991) of the random effect vector for group j as:

$$\begin{aligned}\hat{\mathbf{u}}_j &= E(\mathbf{u}_j|Y_{\cdot j}, X_{\cdot j}, Z_{\cdot j}, \hat{\beta}) \qquad (5.8)\\ &= GZ_{\cdot j}'(Z_{\cdot j}'GZ_{\cdot j} + R)^{-1}(Y_{\cdot j} - X_{\cdot j}\hat{\beta}),\end{aligned}$$

where "$\cdot j$" refers to that portion of the vector or matrix associated with group j (over all associated i). In the case of a random intercept model, Z, G, and R have simpler structure, so this becomes:

$$\begin{aligned}\hat{u}_{0j} &= E(u_{0j}|Y_{\cdot j}, X_{\cdot j}, Z_{\cdot j}, \hat{\beta}, \hat{\sigma}_u^2, \hat{\sigma}_\varepsilon^2)\\ &= \hat{\sigma}_u^2(\hat{\sigma}_u^2 Z_{\cdot j}'Z_{\cdot j} + \hat{\sigma}_\varepsilon^2 I)^{-1}(Y_{\cdot j} - X_{\cdot j}\hat{\beta}).\end{aligned}$$

While most of our discussion assumes some interest in estimating these effects, we reiterate that random effects models estimate the parameters *governing* the effects, while fixed effects models incorporate them through group-specific parameters. The BLUPs are post hoc "best guesses" of the group effects, which, under likelihood theory, have asymptotic distributions, and thus known precision. A new group's effect along with its precision could thus be estimated using (5.8). In the case of fixed effects, it is possible to define an analogous prediction for a group effect from residualized outcomes, but the lack of model assumptions offers no shrinkage of that prediction, nor an estimate of its precision that utilizes information from all groups.

5.3.3 The Debate Between Fixed and Random Effects

Whenever a textbook introduces the random and fixed effects models, it discusses which method is to be used when. However, readers are sometimes perplexed by the wide range of prescriptions proposed by different textbooks. The goal of this subsection is to provide some guidance on the opinions and prescriptions encountered throughout the literature. Here we provide three criteria, namely satisfaction of the assumptions, compatibility with the quantities of interest, and trade-off between bias reduction and efficiency, on which researchers can base their judgment in selecting a model.

The first criterion is whether the selected method satisfies the required assumptions. We have pointed out that the random effects assumption, or conditional independence between the covariates and group-specific components, may be unrealistic. An advantage to the fixed effects model is that it allows for group-specific components to be correlated with the covariates. However, when the random effects assumption is satisfied, random effects models are more efficient. As we see in the next section, the Hausman test takes advantage of this property to adjudicate between the choice of fixed or random effects.

Although the fixed effects model requires fewer assumptions than the random effects model, this does not necessarily mean the fixed effects model is always more robust against violation of its assumptions. In particular, the fixed effects model is known to be more susceptible to several violations of the regression assumptions and other problems typical of regression modeling. For example, the impact of measurement errors in covariates is amplified in fixed effects models (Johnston and DiNardo, 1997). A drawback of the fixed effects model is its susceptibility to the violation of strict exogeneity. As discussed in Section 5.3.1, under the assumption of strict exogeneity, the error terms must be orthogonal to any values of the covariates that belong to the same group. Typically, this assumption is violated when there is an endogenous covariate in panel data or when a subject's outcome may affect another subject's outcome *within the same group*. To be fair, the random effects model must also satisfy the strict exogeneity assumption. However, because random effects estimators are derived from both within-group and between-group variations, the latter of which is unaffected by the strict exogeneity assumption, the violation has the potential to bias fixed effects estimators more (Johnston and DiNardo, 1997).

The second criterion is whether the quantities of interest are estimable with the fixed effects model. Specifically, the quantities of interest obtained from a fixed effects model are limited in two ways. First, fixed effects models cannot estimate the effect of a variable that has no within-group variation because fixed effects subsume all observed and unobserved group-specific variation. For example, one cannot estimate the effect of race or gender with the panel data of repeated individual observations because race and gender do not change over time. Also, with grouped data that contain students' academic performance, one cannot use classroom fixed effects if the students receive treatment on a classroom basis. Further, even if explanatory variables have some within-group variance, they are often imprecisely estimated in fixed effects models due to the additional degrees of freedom used (Ashenfelter, et al., 2003).

As alluded to previously, fixed effects models have a limited capability to make predictions about the distribution of an outcome variable at different levels. For example, with the random effects model, a researcher whose interest is the effect of extended class hours on students' test scores can predict the distribution of the test scores at classroom, school, or population level. With the random effects model, a researcher can incorporate the information from random intercepts at different levels into the predictive distribution of the outcome variable and this extends to new observations and groups. Although it is possible to estimate fixed effects for new groups in an ad hoc manner in fixed effects models, the precision of those estimates is not readily available, nor does any parameter directly reflect the overall between-group (and, in more nested models, between-level) variation. This observation has led some researchers to frame the tradeoff between fixed and random effects as: random effects are samples from a population while fixed effects represent the entire population of interest or conditional inference on the available groups in the sample (see, for example, Mundlak, 1978). The basis for this characterization is classical presentations of ANOVA; we have shown that the implications of the choice on parameter estimates requires the practitioner to consider additional implications of the choice.

The final criterion is the tradeoff between how much bias can be reduced and how much efficiency is lost with the fixed effects model. Fixed effects models add group-specific indicators as explanatory variables, controlling for time-constant, homogeneous group-specific confounding, including that due to unobservables, provided the underlying assumptions hold. This means that, to return to the example of panel data on individuals, confounding variables such as personal or family background before the observation period and any personal attributes such as race, gender, or genetics are controlled for, up to the validity of the model assumptions, primary among which is the assumption that group effects have no heterogeneity within that group (e.g., no change over time in the panel data setting). In grouped data, e.g., observations on students in classrooms, a fixed effects model would control for quality of teachers, family background, and any experiences that classmates might share. It is the promise of controlling for such confounders, and the expectation that they are always lurking, that has led some researchers to endorse primarily fixed effects modeling with multilevel data (see, for example, Allison, 2005).

However, the contribution to bias reduction with the fixed effects model must also be evaluated in terms of its loss of efficiency. Fixed effects models lose their efficiency, particularly in the following two instances. First, for a large number of groups and a small number of within-group observations, the number of degrees of freedom consumed by the fixed effects will be large. For example, if a researcher has panel data that consist of 100 group observations ($J = 100$), the introduction of a fixed effects model reduces the degree of freedom by 99. Or, in the panel case, if the number of time periods is two, the fixed effects model halves the degrees of freedom. The tails of the t-distribution become thicker as the degrees of freedom decreases, which implies that the coefficients are estimated with greater uncertainty. Goldstein (2003) advises using the fixed effects model when there are a few groups and moderately large numbers of units in each. Snijders and Bosker (1999) also advise using a random effects model if the number of groups is more than ten. Second, when explanatory variables have little within-group variation, they are highly correlated with the fixed effects, and fixed effects estimators will be inefficient (Bartels, 2008). In a simple example, suppose that a researcher is interested in estimating a policy effect using the panel data of 50 states in the U.S. If only 10 states changed the policy during the observed period, the fixed effects estimators are calculated based on these 10 states, ignoring 40 states in the data because fixed effects subsume all time-invariant covariates.

To make the choice between fixed and random effects, a researcher needs to consider all of what we have discussed, such as the plausibility of the random effects assumption and strict exogeneity, the need to include group-level predictors, efficiency, degrees of freedom, the presence and degree of measurement error in the data, and the amount of within-group variance.

5.3.4 *Comparing Fixed Effects and Random Effects: The Hausman Test*

Although most of the criteria for the choice between fixed and random effects models require subjective judgments, test statistics have been developed for some of them. The Hausman test (Hausman, 1978) is most frequently used to check the validity of the random effects assumption, namely the conditional independence between group-specific intercepts (u_{0j}) and covariates ($\mathbf{x}_{ij}$). Briefly speaking, the Hausman test is a comparison between the parameter estimates of the fixed effects and random effects models, and supports the random effects assumption if the difference between two parameters is sufficiently small.

To see the logic, let us show the formal definition of the test statistic, H, below:

$$H = (\hat{\beta}_{RE} - \hat{\beta}_{FE})'(\hat{V}_{FE} - \hat{V}_{RE})^{-1} (\hat{\beta}_{RE} - \hat{\beta}_{FE}),$$

where $H \sim \chi^2$ with k degrees of freedom (the number of time-varying coefficients in both models), $\hat{\beta}$ denotes a vector of estimated coefficients, $\hat{V}$ denotes a variance-covariance matrix of estimators, and the subscripts RE and FE denote the random or fixed effect estimator as the source of the estimates. We test the null hypothesis that the results of the two estimators are not different. When the random effects assumption holds, both the fixed and random effects estimators are consistent. However, when the assumption does not hold, only the fixed effects estimator is consistent, and the random effects estimator is biased. Moreover, when the assumption holds, the fixed effects estimator is inefficient compared to the random effects model. As the violation of the assumption becomes less serious, the difference in point estimates between two parameters decreases in the numerator, while the difference in variances increases in the denominator. Thus, the size of H is interpreted as the extent of the deviation from the assumption.

It must be noted, however, that the Hausman test does not always provide a definitive answer. One problem is that although statistical significance of the test statistic, H, is fairly reliable evidence for the bias in the random effects estimates, statistical insignificance is not necessarily evidence for unbiasedness. This is because the Hausman test relies on the consistency of the two estimators. As (5.4) and (5.5) indicate, fixed and random effects estimators converge only if the group size (n_j) and/or the within-group variance (σ^2_{u0}) is large. Thus, in a small finite sample and/or with covariates that have little within-group variance, these two estimators may not significantly differ from each other even when the random effects assumption is violated, and the Hausman test cannot distinguish between the satisfaction of the random effects assumption and the lack of convergence.[6] A number of studies have been conducted to improve the performance of the Hausman test. Ahn and Low (1996) refined the Hausman test to better detect the violation when the composite error term ($u_{0j} + \varepsilon_{ij}$) contains little within-group variance (σ^2_{u0}). Bole and Rebec (2004) show that a bootstrap version of the Hausman statistics marginally improved its performance when the violation is minimal. To date, none has been established as a definitive alternative to the Hausman test.

5.3.5 *The Benefits of Both Models: Hybrid Models*

The classic back and forth of fixed versus random effects is that a researcher should have less concern about bias when using fixed effects, while random effects are more efficient. However, several interesting compromises have been proposed. The two most prominent approaches are correlated random effects (for example, Mundlak, 1978 and Chamberlain, 1982) and adaptive centering (Raudenbush, 2009). Several other recent papers (Gelman, 2006; Bartels, 2008; Leyland, 2010) also propose various solutions to the correlation between predictors and random effects that allow researchers to more comfortably estimate varying parameters.

The literature has accurately redefined the random effects assumption outlined above as one of cluster confounding. When only a single, unadjusted version of a predictor is included in a model, standard random effects approaches implicitly assume that the within- and between-group effects are equal for changes in any component of $\mathbf{x}_{ij}$ (Zorn, 2001; Skrondal and Rabe-Hesketh, 2004; Rabe-Hesketh and Everitt, 2006; Bartels, 2008). We can decompose any particular x_{1ij} into different between- and within-group effects, which we can call x_{1j}^B and x_{1ij}^W, and the result is that β_1 is a weighted average of the two processes. Bafumi and Gelman (2006) label this a type of omitted variable bias, and go on to propose group mean centering as a workaround solution. This is a form of the hybrid model described in Allison (2005).

In the case of one predictor, a hybrid model would be

$$\begin{aligned} y_{ij} &= \beta_0 + \beta_{W1}(x_{1ij} - \bar{x}_{1j}) \\ &\quad + \beta_{G1}(\bar{x}_{1j}) + u_{0j} + \varepsilon_{ij}, \end{aligned}$$

where β_{W1} captures the within-group effect of the predictor x_1, β_{G1} captures the between-group effect, and u_{0j} is a random effect. All other terms, such as the group means $\bar{x}_{1j}$, are as they have previously been defined, as are the distributional assumptions on the stochastic terms. The within-group effect should be the same as that obtained via the fixed effects estimator.[7] Two things are unique about this approach: first, we can include group-constant predictors, such as $\bar{x}_{1j}$, in the model. In fact, any group-constant predictors may be included without affecting β_{W1}, as $(x_{1ij}-\bar{x}_{1j})$ is orthogonal to any group-constant predictor. Second, the random intercept variance $\sigma^2_{u_0}$ measures the remaining between-group variation in level, and inference for this parameter is obtainable. In the fixed-effects setting, say, using group-specific indicators, one cannot include additional group-specific predictors, and the inferential framework for assessing between-group variance is somewhat ad hoc (we describe this in the example). In this sense, the hybrid method has all of the advantages of both random- and fixed-effects approaches. Allison (2005), drawing in particular on Neuhaus and Kalbfleisch (1998), describes how this model may be implemented in the SAS Statistical Programming language, and notes that a Wald test of the hypothesis $\beta_{W1} - \beta_{G1} = 0$ provides a Hausman-like test of whether random effects models provide the same coefficients of interest as the fixed effects models.

The above approach is also quite similar to Mundlak's (1978) formulation, and simply adds the mean for each time-varying covariate. This identifies the endogeneity concern as a result of attempting to model two processes in one term (see also Snijders and Bosker, 2012, p. 56; Berlin, Kimmel, Ten Have, and Sammel, 1999). Others similarly argue for unit characteristics that are correlated with the mean (Clark, Crawford, Steele, and Vignoles, 2010). Bartels (2008) and Leyland (2010) are more interpretable reformulations of Mundlak (1978). Proponents of fixed effects either call this method a compromise approach or a form of the Hausman test to choose between fixed and random effects (Hsiao, 2003, pp. 44–50; Allison, 2009, pp. 33–5; Wooldridge, 2010; Greene, 2011; and others). In truth, they are essentially the same model as those argued for by Bafumi and Gelman (2006), Gelman and Hill (2007), and throughout the multilevel models literature: random effects models with additional time-invariant predictors, which address the exogeneity assumption.

We also note that in fields such as psychology and education, group mean centering has a long history (e.g., Bryk and Raudenbush,

1992; Raudenbush and Bryk, 2002), but the connections to the econometric fixed-effects approach have only been made more recently (Allison, 2009). Lastly, the practical researcher should always check the plausibility of the assumptions behind the modeling, no matter the name, and realize that not all random effects models place the same demands on the random effects assumption.

5.3.6 Other Extensions

If a researcher is interested in the causal effects of (potentially time-dependent) treatments on time-varying regressors, neither fixed nor random effects may be appropriate (see Sobel, 2012). The methodological challenge arises because the effects of later treatments may themselves be outcomes of earlier instances of the treatment. Structural nested models, proposed in Robins (1993, 1994, 1997), were designed to estimate the effect of time-varying treatments, while marginal structural models (Robins, 1998; Robins, Hernán, and Brumback, 2000; Robins and Hernán, 2009) are a more recently developed alternative approach. This issue is discussed in Chapter 12 as well.

5.4 AN EXAMPLE: PROFICIENCY STANDARDS AND ACADEMIC ACHIEVEMENT

To illustrate the fixed and random effects approaches, we consider an open question in U.S. education policy: are more rigorous or stringent performance standards on state assessments associated with improved academic outcomes? Although states in the U.S. develop their own assessments and set their own proficiency standards, the Federal government assesses student performance for state-representative samples using a series of common tests known as the National Assessment of Educational Progress or NAEP.[8]

The National Center for Education Statistics publishes a series of studies that use the NAEP to place the states' performance or proficiency standards on a common metric—the NAEP scale for a given subject and grade level (NCES, 2011). This enables the creation of a panel dataset at the state level that includes states' performance on NAEP, states' proficiency levels or the "cutscores" that they set on their own state assessments mapped onto the NAEP scale, and an additional set of time-varying covariates measuring various student demographics and other factors shown to be related to academic outcomes. These data can be used to estimate the relationship between the stringency of states' performance standards and the average academic achievement of their students.

Specifically, the models estimated in this section condition the state average score on NAEP grade 4 mathematics (the outcome measure) on the following covariates:

- the state proficiency or performance standard for its state grade 4 mathematics test mapped to the NAEP scale (the policy variable of interest);
- the percentage of students in the state who are eligible for free or reduced price school lunch as a measure of poverty;
- the percentage of disabled students;
- the percentage of English language learners;
- the percentage of Black students;
- the percentage of Hispanic students;
- the percentage of Asian students; and
- the percentage of Native American or Alaska Native students in each state.

NAEP is given at the state level every other year, but state test mapping data—the policy variable of interest—are currently only available for academic years 2004–5, 2006–7, and 2008–9. Thus, for these examples, $T = 3$. In the following models, dichotomous indicators or fixed effects for time periods are also included to control for any common trends among the states during this time period. Also, not all data are available for all states in each time period, but these data are assumed to be missing completely at random (Little and Rubin, 2002; see Chapter 23 of this volume).

Table 5.1 Estimation Results: Random Effects Model

Variable	Coefficient	(Std Err.)
lagged standard	0.058	(0.024)
% free and reduced lunch	−0.086	(0.035)
% disabled	0.000	(0.001)
% ELL	0.001	(0.001)
% Black	−0.212	(0.041)
% Hispanic	−0.115	(0.039)
% Asian	−0.119	(0.051)
% Am.Ind./Al.Nat.	−0.300	(0.108)
2008	0.061	(0.363)
2010	1.278	(0.451)
Intercept	236.495	(5.577)
N	132	
σ_{u0}	2.743	
$\chi^2_{(10)}$	85.127	

Finally, because it is theoretically unlikely that a change in state standards would immediately translate into a change in academic achievement, the mapped state proficiency standard is included in each specification below as lagged one time period (two years).

5.4.1 *The Random Effects Model*

If we are willing to assume that the unobservables are uncorrelated with the other covariates, we can estimate the model:

$$\text{NAEP}_{it} = \mathbf{x}_{it}\beta + \delta\text{Standard}_{i,t-1} + u_{0i} + \varepsilon_{it},$$

with the random effects estimator, assuming that $u_{0i} \sim \text{Normal}(0, \sigma^2_{u0})$. The results are presented in Table 5.1. The key coefficient of policy interest, δ, is positive and statistically significant at conventional levels, implying a positive relationship between stringency of academic performance standards and student achievement. The between-states variance component is estimated to be $\hat{\sigma}_{u0} = 2.743$.

5.4.2 *The Fixed Effects Approach*

An alternative means of estimating the linear unobserved effects model is by means of the fixed effects transformation (time de-meaning) to eliminate the time-invariant unobservables:

$$\begin{aligned}\text{NAEP}_{it} - \overline{\text{NAEP}_i} = \\ &(\mathbf{x}_{it} - \bar{\mathbf{x}}_i)\beta \\ &+ \delta(\text{Standard}_{i,t-1} - \overline{\text{Standard}_i}) \\ &+ \varepsilon_{it} - \bar{\varepsilon}_i.\end{aligned}$$

Here, we see a substantively different result (Table 5.2). The estimated coefficient for the policy variable of interest, lagged state proficiency standard, is much smaller in magnitude and no longer statistically significant at conventional levels ($p = .175$, two-tailed). We note that an estimate of the between-states variation captured in this fixed effects specification is much larger, at $\hat{\sigma}_{u0} = 9.936$. This standard deviation is based directly on the 50 state fixed effects implicitly estimated in this model (values not shown). Inference for this variance component is made indirectly through an F-test on the underlying fixed effects. It is highly significant in this case, but implies something about the states in the sample—at least one has non-zero deviation from the mean outcome level, net of all else—rather than a property of all states in the population (at other historical periods, for example). This value is substantially larger than that estimated in the random effects model, suggesting either that there is additional state-level variation that should be modeled, or that fixed effects models are overestimating the differences we would observe in this population under some sampling mechanism.

5.4.3 *The Hausman Test*

So which is the correct specification? Although it does not provide a definitive answer, the Hausman test may be useful to the applied researcher to guide the choice between the FE and RE models, assuming unobserved covariates. For the example data,

$$\begin{aligned}H &= (\hat{\beta}_{RE} - \hat{\beta}_{FE})'(\hat{V}_{FE} - \hat{V}_{RE})^{-1} \\ &\quad(\hat{\beta}_{RE} - \hat{\beta}_{FE}) \\ &= 49.78.\end{aligned}$$

Table 5.2 Estimation Results: Fixed Effects Model

Variable	Coeff.	(Std Err.)
lagged standard	0.032	(0.023)
% free and reduced lunch	0.059	(0.037)
% disabled	0.0004	(0.001)
% ELL	0.001	(0.001)
% Black	−0.257	(0.305)
% Hispanic	0.037	(0.234)
% Asian	−0.896	(0.387)
% Am.Ind./Al.Nat.	0.087	(0.385)
2008	−0.117	(0.349)
2010	0.135	(0.556)
Intercept	237.953	(7.537)
N	132	
σ_{u0}	9.936	
$F_{(59,72)}$	2.676	

Since $H \sim \chi^2$ with $k = 10$ degrees of freedom (recall, the number of time-varying coefficients in both models), the null hypothesis that the difference in estimated coefficients between the two models is not systematic is rejected at $p < .0005$. Thus, the results of the Hausman test suggest that the fixed effects model may be accounting for important unobserved heterogeneity that may otherwise be biasing the observed relationships.

5.4.4 *The Hybrid Model*

Alternatively, as noted in Section 5.3.5 above, one may instead fit a "hybrid" model of the fixed and random effects approaches:

$$\begin{aligned}\text{NAEP}_{it} = (\mathbf{x}_{it} - \bar{\mathbf{x}}_i)\beta_w \\ + \bar{\mathbf{x}}_i\beta_g + \delta_w(\text{Standard}_{i,t-1} \\ - \overline{\text{Standard}_i}) \\ + \delta_g\overline{\text{Standard}_i} + u_{0i} + \varepsilon_{it}.\end{aligned}$$

The results of this model, presented in Table 5.3, are informative. The hybrid model yields the same coefficients for within-state effects as the fixed effects model, and very similar precision. The variance component σ_{u0} has been reduced to 2.712, a small drop from the random effects model's findings, reflecting the additional explanatory

Table 5.3 Estimation Results: Hybrid Model

Variable	Coeff.	(Std Err.)
centered lagged standard	0.032	(0.023)
centered % free/reduced lunch	0.058	(0.037)
centered % disabled	0.000	(0.001)
centered % ELL	0.001	(0.001)
centered % Black	−0.254	(0.305)
centered % Hispanic	0.045	(0.234)
centered % Asian	−0.902	(0.386)
centered % Am.Ind./Al.Nat.	0.091	(0.384)
centered 2008	−0.120	(0.348)
centered 2010	0.135	(0.555)
mean lagged standard	0.164	(0.041)
mean % free/reduced lunch	−0.412	(0.050)
mean % disabled	−0.009	(0.004)
mean % ELL	0.002	(0.003)
mean % Black	0.017	(0.046)
mean % Hispanic	−0.021	(0.035)
mean % Asian	−0.109	(0.042)
mean % Am.Ind./Al.Nat.	−0.172	(0.094)
mean 2008	2.869	(5.474)
mean 2010	−2.135	(4.139)
Intercept	222.121	(9.988)
N	132	
σ_{u0}	2.712	
$\chi^2_{(20)}$	216.599	

power of group-specific predictors. Group-level effects can be contrasted with within-group effects through the construction of a contextual effect, whose substantive interpretation is described in Raudenbush and Bryk (2002). A Wald test for the difference between any two such between/within effects is analogous to the Hausman test, and in our example, for lagged standard, the difference between 0.032 and 0.164 has an associated $\chi_1^2 = 7.80$, with $p = 0.005$, which results in the same inferential conclusion as the prior Hausman test at conventional significance levels. Note that the estimates from the hybrid model are sensitive to the choice of predictors, just as in any regression, be it using fixed or random effects for groups.

5.5 CONCLUSION

The literature is full of advice, often conflicting, about when to use fixed or random effects models. Adding to this confusion, random effects models have several extensions which relax the random effects assumption, including the long history of hybrid models outlined above. Standard random effects models can obtain estimators that are potentially biased if the random effects assumption is violated, but fixed effect models come with an inability to account for any time-invariant variables, difficulty in making out-of-sample predictions, and statistical inefficiency. While the Hausman test is often noted as a tool for evaluating bias in the random effects model, Clark and Linzer (2012) have shown through simulations that it may not be reliable. Ultimately, evaluating the balance of bias and variance in the two techniques is not a simple task, but applied researchers should not immediately shy away from using a random effects model because of the bias concerns, as the associated variance reduction may be worthwhile. Our example above serves a useful guide for other potential analyses in which one fits fixed effects, random effects, and hybrid models. By exploring the relationship between all three estimates, researchers can more completely understand the associations and potential effects of some forms of confounding.

NOTES

1 Editors' note: In this chapter, the underlying model is effectively the same throughout, but by changing assumptions about model parameters and, more importantly, by changing the estimator, different goals are achieved. These competing goals are the subject of much debate in the applied statistics literature; this chapter describes the assumptions underpinning various estimators and reframes some of the debate in terms of between- and within-group predictor effects.

2 $\hat{\beta}_{0j}$ is only approximately equal to the right-hand side of equation (5.4) because the actual fitting techniques are slightly more complex. Statistical programs fit the models through either Bayesian inference or an augmented least squares technique (see Gelman and Hill, 2007, sections 18.3–18.4 for detailed descriptions on both techniques and extensions for more complex multilevel models). In the first technique, one uses an iterative algorithm that alternatively estimates the particular intercepts, and the means and variances, called the hyperparameters, which in our case above includes σ^2_{u0}, σ^2_{ε}, and β_0 (Lindley and Smith, 1972; Efron and Morris, 1975). In the second technique, one develops a weighted least squares regression (Afshartous and de Leeuw, 2005, pp. 112–13).

3 This is why fixed effects are sometimes called "unmodeled effects" and random effects are called "modeled effects" (Bafumi and Gelman, 2006). Additionally, see Gelman (2005) or Gelman and Hill (2007, p. 245) for a discussion of all the conflicting different meanings of the term "fixed effects". Here, we mean the fifth definition in that list.

4 The degrees of freedom associated with the demeaning of the outcomes require an adjustment to the standard error of the estimated effects. The reader may also be concerned that from a multilevel perspective, these models ignore the within-group correlation between observations. However, the demeaning process accounts for a group-constant effect, which captures a specific form of within-group correlation, analogous to first differencing with pre/post designs.

5 The following account applies to balanced data where the group size is the same for all groups. Unbalanced data require some adjustments.

6 Specifically, if the within variation is small, the fixed effects estimates may not be asymptotically normal and this may invalidate the Hausman test (Hahn, Ham, and Moon, 2011).

7 Little has been written about the relative efficiency of this compared to the fixed effects estimator, but our own simulations suggest that the precision is comparable in singly-nested models.

8 For more information on these assessments, including subjects and grade levels tested, see `http://nces.ed.gov/nationsreportcard/`.

REFERENCES

Afshartous, D. and de Leeuw, J. (2005) 'Prediction in Multilevel Models', *Journal of Educational and Behavorial Statistics*, 30: 109–39.

Ahn, S.C. and Low, S. (1996) 'A Reformulation of the Hausman Test for Regression Models with Pooled Cross-Section-Time-Series Data', *Journal of Econometrics*, 71: 309–19.

Allison, P.D. (2005) *Fixed Effects Regression Methods for Longitudinal Data Using SAS*. Cary, NC: SAS Publishing.

Allison, P.D. (2009) *Fixed Effects Regression Models*. London: Sage.

Ashenfelter, O., Levine, P.B., and Zimmerman, D.J. (2003) *Statistics and Econometrics: Methods and Applications*. New York: J. Wiley.

Bafumi, J. and Gelman, A. (2006) 'Fitting Multilevel Models When Predictors and Group Effects Correlate', Paper presented at the annual meeting of the Midwest Political Science Association, Chicago, IL.

Baltagi, B. (2005) *Econometric Analysis of Panel Data*, 3rd edition. New York: Wiley.

Bartels, B.L. (2008) 'Beyond "Fixed Versus Random Effects": A Framework for Improving Substantive and Statistical Analysis of Panel, Time-Series Cross-Sectional, and Multilevel data', Political Methodology Conference. Ann Arbor, MI. Available at `http://home.gwu.edu/~bartels/cluster.pdf` [accessed 26 August 2012].

Berlin, J.A., Kimmel, S.E., Ten Have, T.R., and Sammel, M.D. (1999) 'An Empirical Comparison of Several Clustered Data Approaches Under Confounding Due to Cluster Effects in the Analysis of Complications of Coronary Angioplasty', *Biometrics*, 55: 470–6.

Bole, V.A. and Rebec, P. (2004) 'Bootstrapping the Hausman Test in Panel Data Models'. Available at SSRN: `http://ssrn.com/abstract=628321` [accessed 26 August 2012].

Bryk, A.S. and Raudenbush, S.W. (1992) *Hierarchical Linear Models in Social and Behavioral Research: Applications and Data Analysis Methods*, 1st edition. Newbury Park, CA: Sage Publications.

Chamberlain, G. (1982) 'Multivariate Regression Models for Panel Data,' *Journal of Econometrics*, 1: 5–46.

Clark, T.S. and Linzer, D.A. (2012) 'Should I Use Fixed Or Random Effects?', Unpublished working paper. Available at `http://userwww.service.emory.edu/~tclark7/randomeffects.pdf` [accessed 26 August 2012].

Clarke, P., Crawford, C., Steele F., and Vignoles, A. (2010) 'The Choice Between Fixed and Random Effects Models: Some Considerations for Educational Research', CMPO working paper: University of Bristol. Available at `http://www.bristol.ac.uk/cmpo/publications/papers/2010/wp240.pdf` [accessed 26 August 2012].

Davidson, R. and MacKinnon, J.G. (1993) *Estimation and Inference in Econometrics*. New York: Oxford University Press.

Diez Roux, A.V. (2004) 'The Study of Group-Level Factors in Epidemiology: Rethinking Variables, Study Designs, and Analytical Approaches', *Epidemiologic Reviews*, 26: 104–11.

Efron, B. and Morris, C. (1975) 'Data Analysis Using Stein's Estimator and its Generalizations', *Journal of the American Statistical Association*, 70: 311–19.

Gelman, A. (2005) 'Analysis of Variance – Why it is More Important than Ever', *Annals of Statistics*, 33: 1–53.

Gelman, A. (2006) 'Multilevel Modeling: What It Can and Can't Do', *Technometrics*, 48: 241–51.

Gelman, A. and Hill, J. (2007) *Data Analysis Using Regression and Multilevel/Hierarchical Models*. New York: Cambridge.

Goldstein, H. (2003) *Multilevel Statistical Models*, 3rd edition. New York: Wiley.

Greene, W. (2011) 'Fixed Effects Vector Decomposition: A Magical Solution to the Problem of Time-invariant Variables in Fixed Effects Models?', *Political Analysis* 19: 135–46.

Hahn, J., Ham, J.C., and Moon, H.R. (2011) 'The Hausman Test and Weak instruments', *Journal of Econometrics*, 160: 289–99.

Hausman, J.A. (1978) 'Specification Tests in Econometrics', *Econometrica*, 46: 1251–71.

Hsiao, C. (2003) *Analysis of Panel Data*, 2nd edition. New York: Cambridge University Press.

Johnston, J. and DiNardo, J. (1997) *Econometric Methods*. New York: McGraw Hill Co.

Kreft, I.G. and de Leeuw, J. (1998) *Introducing Multilevel Modeling*. London: Sage Publications.

Leyland, A.H. (2010) 'No Quick Fix: Understanding the Difference Between Fixed and Random Effect Models', *Journal of Epidemiology and Community Health*, 64: 1027–8.

Lindley, D.V. and Smith, A.F.M. (1972) 'Bayes Estimates for the Linear Model', *Journal of the Royal Statistical Society, Series B (Methodological)*, 34: 1–41.

Little, R.J.A. and Rubin, D.B. (2002) *Statistical Analysis With Missing Data*, 2nd edition. New York: Wiley.

Morris, C.N. (1983) 'Parametric Empirical Bayes Confidence intervals', in *Scientific Inference, Data Analysis and Robustness*. Box, G.E.P., Leonard, T., and Wu, C.F. (eds). New York: Academic Press, 25–50.

Mundlak, Y. (1978) 'Pooling of Time-Series and Cross-Section Data', *Econometrica*, 46: 69–85.

National Center for Education Statistics (2011) *Mapping State Proficiency Standards On to the NAEP Scales: Variation and Change in State Standards for Reading and Mathematics, 2005–2009, (NCES 2011–458)*. Institute of Education Sciences, U.S. Department of Education, Washington, DC.

Neuhaus, J.M. and Kalbfleisch, J.D. (1998) 'Between and Within Cluster Covariate Effects in the Analysis of Clustered Data', *Biometrics*, 54: 726–31.

Rabe-Hesketh, S. and Everitt, B.S. (2006) *A Handbook of Statistical Analyses Using STATA*, 4th edition. Boca Raton, FL: Chapman and Hall/CRC.

Raudenbush, S.W. (2009) 'Adaptive Centering With Random Effects: An Alternative to the Fixed Effects Model for Studying Time-Varying Treatments in School Settings', *Education Finance and Policy*, 4: 468–91.

Raudenbush, S.W. and Bryk, A.S. (2002) *Hierarchical Linear Models*, 2nd edition. Thousand Oaks: Sage Publications.

Robins, J.M. (1993) 'Analytic Methods for HIV Treatment and Cofactor Effects', in *AIDS Epidemiology – Methodological Issues* (pp. 213–29). Ostrow, D.G. and Kessler, R. (eds). New York: Plenum Publishing.

Robins, J.M. (1994) 'Correcting for Non-Compliance in Randomized Trials Using Structural Nested

Mean Models', *Communications in Statistics*, 23: 2379–412.

Robins, J.M. (1997) 'Causal inference from Complex Longitudinal Data', in *Latent Variable Modeling and Applications to Causality. Lecture Notes in Statistics (120)* (pp. 69–111). Berkane, M. (ed.). New York: Springer Verlag.

Robins, J.M. (1998) 'Marginal Structural Models', *1997 Proceedings of the American Statistical Association, Section on Bayesian Statistical Science*, 1–10.

Robins, J.M. and Hernán, M. (2009) 'Estimation of the Causal Effects of Time-Varying Exposures', in *Longitudinal Data Analysis*, Fitzmaurice, G., Davidian, M., Verbeke, G., and Molenberghs G. (eds). Boca Raton, FL: Chapman and Hall, 553–97.

Robins, J.M., Hernán, M.A., and Brumback, B. (2000) 'Marginal Structural Models and Causal inference in Epidemiology', *Epidemiology*, 11: 550–60.

Robinson, G.K. (1991) 'That BLUP Is A Good Thing: The Estimation of Random Effects', *Statistical Science*, 6: 15–32.

Skrondal, A. and Rabe-Hesketh, S. (2004) *Generalized Latent Variable Modeling: Multilevel, Longitudinal and Structural Equation Models*. Boca Raton: Chapman and Hall.

Smith, A.F.M. (1973) 'A General Bayesian Linear Model', *Journal of the Royal Statistical Society, Series B (Methodological)*, 35: 67–75.

Snijders, T.A.B. and Bosker, R. (1999) *Multilevel Analysis: An Introduction to Basic and Advanced Multilevel Modeling*. London: Sage.

Snijders, T.A.B. and Bosker, R. (2012) *Multilevel Analysis: An Introduction to Basic and Advanced Multilevel Modeling*, 2nd edition. London: Sage.

Sobel, M.E. (2012) 'Does Marriage Boost Men's Wages?: Identification of Treatment Effects in Fixed Effects Regression Models for Panel Data', *Journal of the American Statistical Association*, 107: 521–9.

Wakefield, J.C. (2003) 'Sensitivity Analyses for Ecological Regression', *Biometrics*, 59: 9–17.

Wooldridge, J. (2010) *Economic Analysis of Cross Section and Panel Data*. Boston: MIT Press.

Zorn, C. (2001) 'Estimating Between- and Within-Cluster Covariate Effect, With an Application to Models of International Disputes', *International Interactions*, 27: 433–45.

6

Centering Predictors and Contextual Effects

Craig K. Enders

Arizona State University, USA

6.1 INTRODUCTION

Multilevel regression models enjoy widespread use in many disciplines. These models are attractive because they provide researchers with a mechanism for analyzing data sets where individuals are nested within higher-order organizational units (i.e., clusters). Examples of multilevel data structures are abundant in the literature and include situations where students are nested within classrooms or schools, clients are nested within therapists, employees are nested within workgroups or companies, and individuals are nested within family units, to name a few. Unlike single-level regression models, multilevel models allow researchers to examine explanatory variables at all levels of the data hierarchy.

Specifying a multilevel model involves a number of nuances that are not germane to single-level regression analyses. For example, this chapter is devoted to the topic of centering predictor variables. In single-level models, centering often has no material impact on parameter estimates. It is only in models with higher-order (e.g., interaction, quadratic) effects that centering matters, and even then the impact of centering is relegated to the lower-order terms (Aiken and West, 1991). In contrast, centering usually has an important impact on multilevel model parameter estimates and the substantive interpretation of those estimates. To further complicate matters, researchers must choose between two centering approaches, grand mean centering and group mean centering. With few exceptions, these methods are not equivalent and can produce different parameter estimates and model fit. The purpose of this chapter is to describe the differences between grand and group mean centering and provide researchers with guidelines for implementing these procedures.

The organization of this chapter is as follows. I begin by describing grand and group mean centering in the context of a cross-sectional model with two levels. Next, I show that the approaches differentially partition variability and I explain how the presence or absence of between-cluster variation in a level-1 predictor impacts the interpretation of the parameter estimates. Having established the differences between the centering approaches, I discuss a so-called contextual model where grand and group mean centering produce equivalent estimates. A central premise of this chapter is that theoretical or substantive considerations should guide centering decisions. To this end, I discuss centering choices for several prototypical research goals. Finally, I conclude with a brief discussion of centering in three-level models.

Throughout the chapter, I use an artificial data set that is loosely based on a math problem-solving intervention described by Montague, Enders, and Dietz (2011). The data set mimics a scenario where researchers randomly assign 40 schools, each with 25 students, to an intervention and a control condition (i.e., a cluster-randomized design). In addition to a binary level-2 intervention code, the data set contains a number of student-level covariates, including a lunch assistance indicator, math self-concept scores, standardized math scores, and pre-test problem-solving scores. The outcome variable is end-of-year math problem-solving test scores. I use variables from this data set to illustrate concepts throughout the chapter. For readers that are interested in duplicating the illustrative analyses, the data set and the computer programs are available on the Handbook's companion website.

6.2 OVERVIEW OF CGM AND CWC

Unlike single-level regression models, researchers employing multilevel models can choose between two centering options, grand mean centering and group mean centering (also referred to as centering within context). To maintain consistency with previous research (e.g., Enders and Tofighi, 2007; Kreft, de Leeuw, and Aiken, 1995), I refer to these approaches as CGM and CWC, respectively. Although CGM and CWC are the principal centering approaches for multilevel models, deviating predictors around a meaningful constant (e.g., the mean of a normative sample, a clinically meaningful cutoff) is also an option, as is using the raw scores. I do not consider these alternatives because they share the same properties as CGM and produce identical estimates, with the exception of the intercept coefficient.

CGM expresses scores as deviations from the grand mean. To illustrate, consider a random intercept model with a single individual-level (i.e., level-1) predictor variable. The CGM regression equation is

$$y_{ij} = \beta_0 + \beta_1 \left(x_{ij} - \bar{x}\right) + u_{0j} + \varepsilon_{ij}, \quad (6.1)$$

where y_{ij} and x_{ij} are the outcome and the predictor variable scores for case i in cluster j, respectively, β_0 and β_1 are the intercept and slope coefficients, respectively, $\bar{x}$ is the grand mean, u_{0j} is a level-2 error that allows the intercepts to vary across clusters, and ε_{ij} is the within-cluster error for case i. In contrast, CWC expresses the level-1 predictor as deviations from the cluster means. The corresponding CWC model is

$$y_{ij} = \beta_0 + \beta_1 \left(x_{ij} - \bar{x}_j\right) + u_{0j} + \varepsilon_{ij}, \quad (6.2)$$

where $\bar{x}_j$ is the average x score in cluster j, and all other terms have the same meaning as (6.1).

The previous equations show that CGM deviates scores around a constant value, whereas CWC deviates scores around cluster-specific values. To illustrate the impact of these two approaches, reconsider the artificial data example. Suppose that the researcher uses standardized achievement test scores to predict end-of-year math problem solving. Figure 6.1 shows a scatterplot of the raw data from three different schools. The symbols with solid fill denote the cluster means. Notice that both variables have a non-zero intraclass correlation because the means vary across clusters. In particular, the horizontal separation of the solid symbols represents between-cluster variation on x.

Figure 6.2 shows the scatterplot after applying CGM to the predictor variable. The plot illustrates three important points. First, expressing each score as a deviation from the grand mean relabels the x values but has no effect on the data structure. The fact that the

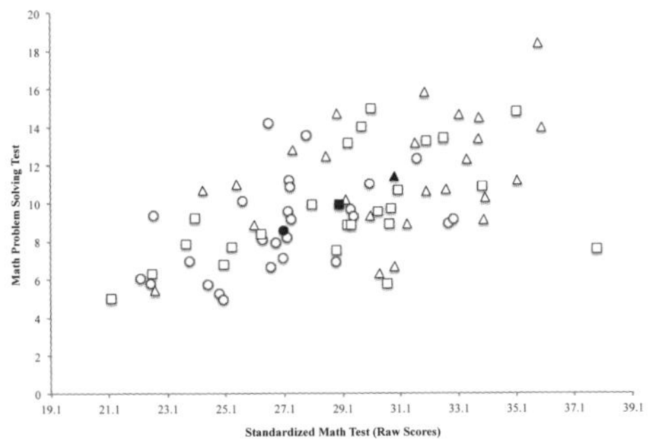

Figure 6.1 Raw Score Scatterplot of Standardized Achievement and Problem-Solving Scores from Three Different Schools. The symbols with solid fill denote the cluster means. The horizontal separation of the solid symbols represents between-cluster variation on x

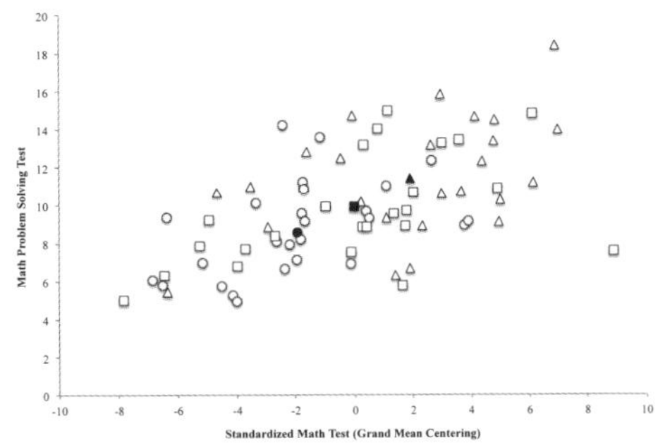

Figure 6.2 CGM Scatterplot of Standardized Achievement and Problem-Solving Scores from Three Different Schools. The symbols with solid fill denote the cluster means. The plot is identical to that in Figure 6.1 except for the score values on the horizontal axis. In particular, the horizontal separation of the solid symbols (i.e., the between-cluster mean differences) is identical to that in Figure 6.1

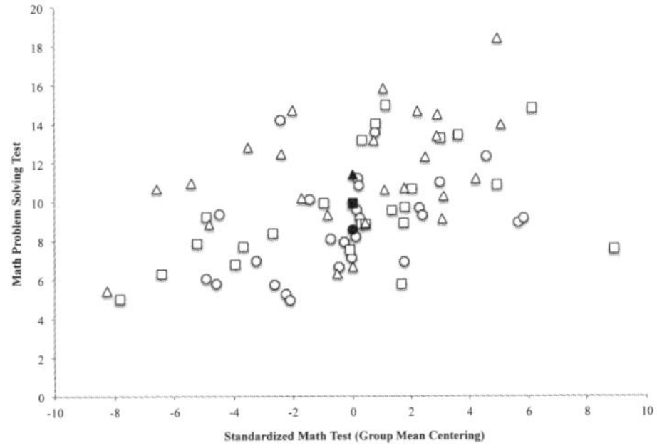

Figure 6.3 CWC Scatterplot of Standardized Achievement and Problem-Solving Scores from Three Different Schools. The symbols with solid fill denote the cluster means. The averages fall on the same vertical plane because all x **means equal zero under CWC**

CGM and raw score scatterplots are identical except for the score values on the horizontal axis illustrates this point. Second, CGM has no effect on the interpretation of the score values. Both the raw score and the deviation score metrics quantify a student's overall standing on the predictor, such that a student with a high raw score necessarily has a high deviation score, and a student with a low raw score has a low deviation score. Third, CGM retains between-cluster mean differences on the predictor variable and thus has no impact on the intraclass correlation. To verify this point, notice that the horizontal separation of the cluster means (i.e., the solid symbols) is identical in the CGM and raw score scatterplots.

Figure 6.3 shows the scatterplot after applying CWC to the predictor. Again, the scatterplot illustrates three important points. First, notice that the CWC and raw score scatterplots are markedly different. Expressing each score as a deviation from its cluster mean alters the rank ordering of the x values and fundamentally changes the data structure. Second, CWC changes the interpretation of the score values. Whereas the raw score metric quantifies a student's overall standing on the predictor, the deviation score metric reflects a student's standing relative to other members of the same cluster. Consequently, a student with a relatively high achievement score would have a positive deviation score if she attended a school with a low achievement average, but she could have negative deviation score if she attended a school with a high achievement average. As I discuss later in the chapter, the cluster-dependent nature of CWC deviation scores is important to consider when implementing this approach. Finally, notice that CWC eliminates between-cluster mean differences on the predictor variable, such that the intraclass correlation equals zero. As seen in the figure, the solid symbols fall on the same vertical plane, owing to the fact that CWC subtracts the cluster means from the data.

The previous figures demonstrate that CGM and CWC rescale a predictor variable in fundamentally different ways. Perhaps not surprisingly, the two centering approaches can produce different parameter estimates. To illustrate this outcome, I fit the random intercept models from (6.1) and (6.2) to the artificial data set. Table 6.1 gives the parameter estimates from the two analyses. Notice that CGM and CWC produced identical estimates (within rounding error) of the intercept coefficient and the level-1 error variance (i.e., $\hat{\beta}_0$ and $\hat{\sigma}^2_\varepsilon$, respectively), but different estimates of the regression slope and the intercept variance (i.e., $\hat{\beta}_1$ and $\hat{\sigma}^2_{u_0}$, respectively); the slope estimates differed by approximately nine-tenths of a standard error unit, and the intercept variance estimates differed by a much larger amount. Furthermore, the models were not equivalent and produced different fit; the deviance values for CGM and CWC were 4455.803 and 4483.201, respectively. I use the estimates in Table 6.1 throughout the subsequent sections to highlight important differences between the two approaches. As an aside, I used the full information maximum likelihood estimator in the SPSS mixed modeling procedure for all data analysis examples. The fractional degrees of freedom values in the tables are a product of the Satterthwaite approximation in SPSS.

6.3 WHY DO CGM AND CWC DIFFER?

The discrepancies observed in Table 6.1 stem from the fact that CGM and CWC differentially partition variability in the predictor. In a multilevel data set, a level-1 predictor variable potentially contains two sources of variability. Specifically, the total deviation between a score and the grand mean is comprised of a deviation between the score and the cluster mean and a deviation between the cluster mean and the grand mean, as follows:

$$\begin{aligned} x_{ij} - \bar{x} &= \left(x_{ij} - \bar{x}_j\right) + \left[\bar{x}_j - \bar{x}\right] \\ &= \mathrm{W}_x + \mathrm{B}_x. \end{aligned} \tag{6.3}$$

Equation (6.3) is identical to ANOVA in the sense that the total deviation reflects both within-group and between-group variation; the equation denotes these components as W_x and B_x, respectively. The same decomposition applies to the outcome variable,

$$y_{ij} - \bar{y} = \left(y_{ij} - \bar{y}_j\right) + \left[\bar{y}_j - \bar{y}\right] = \mathrm{W}_y + \mathrm{B}_y.$$

In the multilevel framework, the cluster means (i.e., $\bar{x}_j$) are level-2 variables that can correlate with the level-2 component of the outcome variable (i.e., $\bar{y}_j$). Consequently, multilevel data structures not only contain two sources of variation, but they also contain two sources of association: W_x can correlate with W_y, and B_x can correlate with B_y. This concept is fundamental to understanding the differences between CGM and CWC. Equation (6.3) shows that CGM yields a composite predictor variable that contains both within- and between-cluster variation. By extension, the total association between x and y can reflect a within-cluster correlation (i.e., W_x with W_y), a between-cluster correlation (i.e., B_x with B_y), or both. Returning to Figure 6.2, the CGM scatterplot clearly shows a positive association within each cluster, such that higher standardized math scores predict higher problem-solving scores. In addition, the solid symbols in the figure show a positive correlation between the cluster means, $\bar{x}_j$ and $\bar{y}_j$. In this particular example, the level-1 and level-2 associations have the same sign (they need not be the same), but the strength of the relationship is different at the two levels.

Unlike CGM, CWC removes between-group variability from the predictor variables, leaving only within-cluster (i.e., level-1) variation. Rearranging the terms from (6.3) and

Table 6.1 CWC and CGM Random Intercept Model Parameter Estimates

Parameter	Method	Est.	*SE*	*t*	*df*
Intercept (β_0)	CWC	10.730	0.356	30.159	40
	CGM	10.730	0.250	42.882	38.531
Level-1 Slope (β_1)	CWC	0.328	0.020	16.646	960
	CGM	0.345	0.019	17.804	998.572
Intercept Variance ($\sigma^2_{u_0}$)	CWC	4.882	1.132	4.312	N/A
	CGM	2.323	0.571	4.070	N/A
Error Variance (σ^2_{ε})	CWC	4.537	0.207	21.909	N/A
	CGM	4.540	0.207	21.892	N/A

Note. Variance components use a *z* rather than a *t* reference distribution.

isolating the within-cluster deviation illustrates this point:

$$\left(x_{ij} - \bar{x}_j\right) = x_{ij} - \bar{x} - \left[\bar{x}_j - \bar{x}\right] = \mathrm{W}_x. \tag{6.4}$$

This equation shows that CWC yields a predictor variable that contains only within-cluster variation. By extension, the total association between x and y can reflect only within-cluster correlation (i.e., W_x with W_y) because the within-cluster portion of x is orthogonal to the between-cluster portion of y, by definition. Returning to Figure 6.3, the CWC scatterplot shows a positive association within each cluster, such that higher standardized math scores predict higher problem-solving scores. However, because the cluster means are constant (i.e., all $\bar{x}_j = 0$), the correlation between $\bar{x}_j$ and $\bar{y}_j$ is necessarily zero.

The previous discussion demonstrates that CGM produces a composite predictor variable that contains both within-cluster and between-cluster variation, whereas CWC produces a variable that contains only within-cluster variation. The differential treatment of the between-cluster variation leads to two important conclusions: CGM yields a level-1 predictor that can correlate with both level-1 and level-2 variables, whereas CWC yields a level-1 predictor that is orthogonal to all level-2 variables. These conclusions have important implications for model specification and they provide the basis for the centering recommendations that I outline later in the chapter. The differential treatment of between-cluster variation also has an important impact on the numeric estimates and substantive interpretation of multilevel model parameters.

6.4 THE IMPACT OF CENTERING ON THE INTERPRETATION OF PARAMETER ESTIMATES

Table 6.1 demonstrates that CGM and CWC can produce different parameter estimates and standard errors. Although not immediately obvious from the table, the two sets of estimates differ in their substantive interpretation. In this section, I show that the treatment of the between-cluster (i.e., level-2) variation is responsible for these differences.

6.4.1 Interpretation of CGM Parameters

To understand the CGM parameter estimates, consider the level-1 (i.e., within-cluster) regression model with a single predictor variable:

$$y_{ij} = \beta_{0j} + \beta_{1j}\left(x_{ij} - \bar{x}\right) + \varepsilon_{ij}. \tag{6.5}$$

Taking the expectation of (6.5) within each cluster yields

$$E\left(y_j\right) = \beta_{0j} + \beta_{1j}\left(\bar{x}_j - \bar{x}\right),$$

and rearranging terms to isolate the intercept gives the following expression:

$$\beta_{0j} = E\left(y_j\right) - \beta_{1j}\left(\bar{x}_j - \bar{x}\right). \tag{6.6}$$

In words, (6.6) says that the CGM intercept equals the expected value (i.e., the outcome mean) for cluster j minus an adjustment term that depends on the slope coefficient and the difference between $\bar{x}_j$ and the grand mean. Readers may recognize that the right side of (6.6) is identical to the expression for an adjusted mean in ANCOVA. Consequently, β_{0j} represents the adjusted outcome mean for cluster j (i.e., the mean that would result after equating $\bar{x}_j$ to the grand mean).

Defining β_{0j} as an adjusted mean has implications for interpreting the intercept variance. To illustrate, consider the composite regression model that follows from (6.5):

$$\begin{aligned} y_{ij} &= \beta_0 + \beta_1(x_{ij} - \bar{x}) \\ &\quad + u_{0j} + u_{1j}(x_{ij} - \bar{x}) + \varepsilon_{ij}. \end{aligned} \tag{6.7}$$

The β_0 coefficient is the average β_{0j} value (i.e., the average adjusted mean), and u_{0j} is an error term that captures cluster-level heterogeneity in the intercepts; conceptually, u_{0j} is a mean-centered version of β_{0j}. Researchers routinely assume that the intercept error terms are normally distributed with a mean of zero and a variance equal to σ^2_{u0}. Given the interpretation of β_{0j}, it follows that the CGM estimate of σ^2_{u0} quantifies the variability in the adjusted means (i.e., the remaining variability in the outcome means after removing the influence of $\bar{x}_j$).

Turning to the regression slope, recall from the previous section that CGM yields a predictor variable that contains both within- and between-cluster variation. The fact that x is a composite of two sources of variability implies that the total association between x and y can reflect a within-cluster correlation (i.e., W_x with W_y), a between-cluster correlation (i.e., B_x with B_y), or both. When the intraclass correlation of x equals zero, the β_1 coefficient in (6.7) is a pure estimate of the average within-cluster regression of y on x. However, when the intraclass correlation is non-zero, β_1 is a weighted average of the within- and between-cluster slopes, where the magnitude of the intraclass correlation determines the weighting (Raudenbush and Bryk,

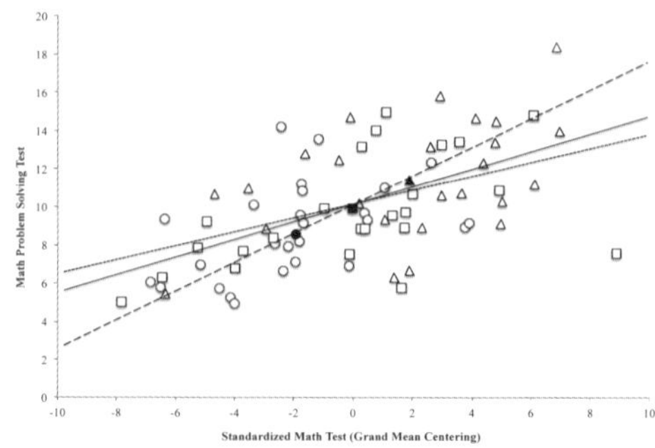

Figure 6.4 CGM Scatterplot of Standardized Achievement and Problem-Solving Scores from Three Different Schools. The dotted line represents the average within-cluster (i.e., level-1) regression, and the dashed line denotes the between-cluster (i.e., level-2) regression. The solid line corresponds to the CGM regression coefficient, which estimates the total association between *x* and *y*. Because CGM blends two associations into a single estimate, its regression slope is stronger than the level-1 regression but weaker than the level-2 regression

2002, pp. 135–9). To illustrate this point, Figure 6.4 shows the CGM scatterplot for the three clusters in Figure 6.2. The dotted line represents the average within-cluster regression, and the dashed line denotes the between-cluster regression of $\bar{y}_j$ on $\bar{x}_j$. The solid line corresponds to the total association between x and y, which is the CGM estimate of β_1 from (6.7). Notice that the CGM regression slope is stronger than the within-cluster regression but weaker than the between-cluster regression. This follows from the fact that β_1 is a weighted average of the two slopes.

Returning to (6.7), u_{1j} is an error term that captures between-cluster differences in the regression slopes. Consistent with the intercept errors, researchers routinely assume that the slope error terms are normally distributed with a mean of zero and a variance equal to σ^2_{u1}. The CGM estimate of σ^2_{u1} quantifies the between-cluster variance of the regression slopes. As an aside, Raudenbush and

Bryk (2002, pp. 145–9) noted that the CGM estimate of $\sigma^2_{u_1}$ tends to be smaller than that of CWC when x has substantial between-cluster variation. The reason for the numeric difference is somewhat technical and stems from the fact that CGM defines the level-1 intercept as an adjusted mean. When the range of x values for a particular cluster does not include the grand mean, the adjustment term in (6.6) extrapolates beyond the data and yields an unreliable estimate of that cluster's β_{0j} value. Maximum likelihood estimation (more accurately, the EM algorithm) compensates for this noise by augmenting β_{0j} with a constant term that depends on the average intercept (i.e., an empirical Bayes estimate). Infusing the intercepts with a constant term restricts their variability, resulting in a phenomenon that the multilevel literature refers to as shrinkage. The covariance between the cluster-specific intercepts and slopes (a parameter that follows from assuming normally distributed errors) allows this range restriction to propagate to the slope variance, thereby reducing the magnitude of the slope heterogeneity relative to CWC.

6.4.2 Interpretation of CWC Parameters

To understand the CWC parameter estimates, consider the level-1 (i.e., within-cluster) regression model with a single predictor variable,

$$y_{ij} = \beta_{0j} + \beta_{1j}\left(x_{ij} - \bar{x}_j\right) + \varepsilon_{ij}.$$

Taking the expectation of this equation within each cluster gives

$$E\left(y_j\right) = \beta_{0j} + \beta_{1j}\left(\bar{x}_j - \bar{x}_j\right) = \beta_{0j}. \quad (6.8)$$

Equation (6.8) shows that the CWC intercept is simply the expected value (i.e., the outcome mean) for cluster j. Unlike CGM, the CWC estimate does not adjust for and does not depend on between-cluster differences on x.

Perhaps not surprisingly, defining β_{0j} as a cluster mean leads to a different interpretation of the intercept variance. The composite regression model that follows from (6.8) is

$$\begin{aligned} y_{ij} &= \beta_0 + \beta_1\left(x_{ij} - \bar{x}_j\right) + u_{0j} \\ &\quad + u_{1j}\left(x_{ij} - \bar{x}_j\right) + \varepsilon_{ij}. \end{aligned} \quad (6.9)$$

As before, the β_0 coefficient is the average β_{0j} value (i.e., the grand mean), and u_{0j} is an error that captures cluster-level heterogeneity in the intercepts. Given the previous definition of β_{0j}, it follows that the CWC estimate of $\sigma^2_{u_0}$ quantifies the variability in the outcome means. Because CWC does not remove the effect of $\bar{x}_j$ from the outcome variable means, the CWC estimate of the intercept variance is necessarily larger than that of CGM (to the extent that $\bar{x}_j$ predicts $\bar{y}_j$). Further, the CWC estimate of $\sigma^2_{u_0}$ is usually identical (or quite similar) to the corresponding estimate from an unconditional model.

Turning to the regression slope, recall from the previous discussion that CWC yields a predictor variable that contains only within-cluster variation. Because the intraclass correlation of x equals zero after applying CWC, the β_1 coefficient in (6.9) is the pooled within-cluster regression of y on x. It is important to reiterate that CWC requires a relative score interpretation, such that the value of x depends only on other members of the cluster. Consequently, the CWC slope coefficient reflects the expected change in y for a one-unit *within-cluster* increase on x. Finally, the CWC estimate of $\sigma^2_{u_1}$ quantifies the between-cluster variance in the regression slopes.

6.5 DATA ANALYSIS EXAMPLE 1

To illustrate the differences between CGM and CWC parameter estimates, I used the math problem-solving data to estimate three random intercept models. Model 1 was an unconditional model with no predictor,

$$y_{ij} = \beta_0 + u_{0j} + \varepsilon_{ij}, \quad (6.10)$$

where y_{ij} is the math problem-solving score for student i in school j. Model 2 added level-1 standardized test scores (i.e., x) as a predictor of math problem solving. The CWC model was

$$y_{ij} = \beta_0 + \beta_1 \left(x_{ij} - \bar{x}_j\right) + u_{0j} + \varepsilon_{ij}, \quad (6.11)$$

and the corresponding CGM model was

$$y_{ij} = \beta_0 + \beta_1 \left(x_{ij} - \bar{x}\right) + u_{0j} + \varepsilon_{ij}. \quad (6.12)$$

Model 3 included standardized test scores as a level-1 predictor and the school-level test score means as a level-2 predictor. The CWC model was

$$\begin{aligned} y_{ij} &= \beta_0 + \beta_1(x_{ij} - \bar{x}_j) \\ &\quad + \beta_2(\bar{x}_j - \bar{x}) + u_{0j} + \varepsilon_{ij}, \quad (6.13) \end{aligned}$$

and the corresponding CGM model was

$$y_{ij} = \beta_0 + \beta_1(x_{ij} - \bar{x}) + \beta_2(\bar{x}_j - \bar{x}) + u_{0j} + \varepsilon_{ij}. \quad (6.14)$$

Note that I centered the cluster means at the grand mean in (6.13) and (6.14) to facilitate the interpretation of the intercept.

Table 6.2 gives the parameter estimates from the five models. The unconditional model in (6.10) produced just three parameter estimates: the grand mean, $\hat{\beta}_0 = 10.730$, the intercept variance, $\hat{\sigma}^2_{u_0} = 4.829$, and the level-1 error variance, $\hat{\sigma}^2_{\varepsilon} = 5.846$. These values serve as useful benchmarks for understanding the remaining models.

Next, consider the models with a single level-1 predictor, (6.11) and (6.12). Both models produced an identical estimate of the grand mean, $\hat{\beta}_0 = 10.730$. In contrast, the slope coefficients differed by approximately nine-tenths of a standard error unit, $\hat{\beta}_{1_{\text{CWC}}} = 0.328$ and $\hat{\beta}_{1_{\text{CGM}}} = 0.345$. Because CWC eliminated the between-cluster variability from the predictor variable, its slope estimates the average within-cluster regression of math problem solving on standardized test scores. From a substantive perspective this means that, within a typical school, a one-point increase in standardized test scores is associated with a 0.328 increase in problem-solving scores, on average. On the other hand, CGM retained the level-2 variation in the predictor, so its slope quantifies the overall regression of math problem solving on standardized test scores. That is, across all schools, a one-point increase in test scores is associated with a 0.345 increase in problem-solving scores, on average. The CGM estimate is somewhat ambiguous because it combines the within- and between-school regression coefficients. In this particular example, the strong between-cluster (i.e., level-2) association effectively pulled the CGM regression line away from the within-cluster slope (see Figure 6.4).

Considering the variance components, the CWC intercept variance was considerably larger than that of CGM: $\hat{\sigma}^2_{u_0} = 4.882$ versus $\hat{\sigma}^2_{u_0} = 2.323$, respectively. Again, the differences stem from the composition of the predictor variable. The CWC model could not explain between-cluster variation in the outcome because it included a pure within-cluster predictor. Consequently, the CWC intercept variance estimate had the same interpretation and the same numeric value (within estimation error) as that of the unconditional model. In contrast, the CGM model included a composite predictor that explained a substantial portion of the between-cluster variation in the problem-solving means. Consequently, the CGM intercept variance was substantially smaller than that of CWC because it quantifies the variation that remains after accounting for standardized test score means (i.e., the variability in the adjusted models). Finally, both models produced an identical estimate of the level-1 error variance because this parameter is unaffected by centering decisions.

Finally, consider the models that added the standardized math means as a level-2 predictor, (6.13) and (6.14). The intercept coefficients were unaffected by the additional predictor because I centered the cluster means. To understand the regression slopes, recall that CWC yields a level-1 predictor that is orthogonal to level-2 predictors, such that the within-cluster deviations in (6.13) were uncorrelated with the cluster means. Consequently, $\hat{\beta}_1$ estimates the pooled within-cluster slope, and $\hat{\beta}_2$ estimates the

Table 6.2 CWC and CGM Parameter Estimates from Data Analysis 1

Parameter	Method	Est.	*SE*	*t*	*df*
	Model 1: Unconditional Model (Equation (6.10))				
Intercept (β_0)	N/A	10.730	0.356	30.159	40
Intercept Variance ($\sigma^2_{u_0}$)	N/A	4.829	1.132	4.265	N/A
Error Variance (σ^2_ε)	N/A	5.846	0.267	21.909	N/A
	Model with a Level-1 Predictor (Equations (6.11) and (6.12))				
Intercept (β_0)	CWC	10.730	0.356	30.159	40
	CGM	10.730	0.250	42.882	38.531
Level-1 Slope (β_1)	CWC	0.328	0.020	16.646	960
	CGM	0.345	0.019	17.804	998.572
Intercept Variance ($\sigma^2_{u_0}$)	CWC	4.882	1.132	4.312	N/A
	CGM	2.323	0.571	4.070	N/A
Error Variance (σ^2_ε)	CWC	4.537	0.207	21.909	N/A
	CGM	4.540	0.207	21.892	N/A
	Contextual Effect Model (Equations (6.13) and (6.14))				
Intercept (β_0)	CWC	10.730	0.166	64.645	40
	CGM	10.730	0.166	64.645	40
Level-1 Slope (β_1)	CWC	0.328	0.020	16.646	960
	CGM	0.328	0.020	16.646	960
Level-2 Slope (β_2)	CWC	0.853	0.071	11.991	40
	CGM	0.525	0.074	7.108	46.373
Intercept Variance ($\sigma^2_{u_0}$)	CWC	0.921	0.247	3.734	N/A
	CGM	0.921	0.247	3.734	N/A
Error Variance (σ^2_ε)	CWC	4.537	0.207	21.909	N/A
	CGM	4.537	0.207	21.909	N/A

between-cluster regression of the math problem-solving means on the standardized test score means. In contrast, the CGM model was comprised of correlated predictors because the deviation scores in (6.14) contained the cluster means. As a result, the CGM estimate of $\hat{\beta}_1$ is a partial regression coefficient that reflects the unique contribution of standardized math scores, controlling for the cluster means. In a similar vein, $\hat{\beta}_2$ quantifies the incremental contribution of the standardized test score means above and beyond the level-1 scores.

Turning to the variance components, CWC and CGM produced identical estimates of $\hat{\sigma}^2_{u_0}$ and $\hat{\sigma}^2_\varepsilon$. Although the models differentially partitioned the variability in standardized achievement scores, the right side of (6.13) and (6.14) fundamentally include the same effects. Consequently, the intercept variance estimates quantify the residual variation in the problem-solving means, and the level-1 error variance estimates quantify the residual variation in the problem-solving scores.

6.6 CONTEXTUAL MODELS

The literature often refers to (6.13) and (6.14) as contextual or compositional models (Firebaugh, 1978). These models have a long history in the literature and are useful for assessing the relative magnitude of a predictor variable at level-1 and level-2. Willms (1986) described a classic example of a contextual effect that occurs when the mean socioeconomic status of a school (i.e., $\bar{x}_j$) influences student achievement above and beyond student-level socioeconomic status (i.e., x_{ij}). Contextual models are unique because CWC and CGM produce equivalent solutions, such that the coefficients from one model are algebraic transformations of the other. In the seminal work on this topic, Kreft et al. (1995) derived the following linkages between CGM and CWC regression coefficients:

$$\begin{aligned} \beta_{0_{\mathrm{CWC}}} &= \beta_{0_{\mathrm{CGM}}} - \beta_{1_{\mathrm{CGM}}}\bar{x} \\ \beta_{1_{\mathrm{CWC}}} &= \beta_{1_{\mathrm{CGM}}} \\ \beta_{2_{\mathrm{CWC}}} &- \beta_{1_{\mathrm{CWC}}} = \beta_{2_{\mathrm{CGM}}}. \end{aligned} \tag{6.15}$$

Note that Kreft et al. (1995) considered the situation where the level-2 means are in raw score form. When the cluster means are centered at the grand means as in (6.13) and (6.14), the intercept coefficients require no transformation and $\beta_{0_{\mathrm{CWC}}} = \beta_{0_{\mathrm{CGM}}}$. The relationships in (6.15) hold in models with random intercepts and slopes, although the variance components are equivalent only in random intercept models.

To illustrate the equivalence of CWC and CGM, reconsider the estimates in the bottom panel of Table 6.2. Notice that the equalities outlined by Kreft et al. (1995) hold exactly:

$$\begin{aligned} \beta_{0_{\mathrm{CWC}}} &= \beta_{0_{\mathrm{CGM}}} = 10.730 \\ \beta_{1_{\mathrm{CWC}}} &= \beta_{1_{\mathrm{CGM}}} = 0.328 \\ \beta_{2_{\mathrm{CWC}}} - \beta_{1_{\mathrm{CWC}}} &= 0.853 - 0.328 \\ &= \beta_{2_{\mathrm{CGM}}} = 0.525. \end{aligned}$$

Because the analysis models included only random intercepts, the variance components were also identical. The equivalence of CWC and CGM in this situation implies that researchers can employ either centering approach to address contextual model research questions. However, it is important to point out that the models produce different substantive interpretations. In particular, the CWC estimate of β_2 quantifies the level-2 regression of the outcome means on the predictor means, whereas the corresponding CGM coefficient estimates the *difference* between the level-1 and level-2 regressions.

The contextual model warrants an important caveat before proceeding. Researchers sometimes believe that introducing the cluster means as a level-2 predictor eliminates the differences between CWC and CGM. This conclusion is generally limited to the very simple models in (6.13) and (6.14) and does not extend to the vast majority of analyses found in published research articles. As an example, a model with an interaction effect involving a level-1 variable (e.g., a cross-level interaction) can produce complex combinations of within- and between-cluster variation, and including the variable means does not fully disentangle these effects. As a general rule, introducing the cluster means as a level-2 predictor does not nullify the need for centering.

6.7 CENTERING RECOMMENDATIONS

The previous analyses illustrated that the differences between CGM and CWC are not simply cosmetic. Rather, centering affects both the numeric values and the substantive interpretation of the parameter estimates. A central premise of this chapter is that theoretical or substantive considerations should guide centering decisions. In this section, I discuss centering choices for five common research goals: (a) evaluating contextual effects; (b) assessing the influence of level-2 variables, including between-cluster interactions; (c) evaluating the influence of level-1 variables; (d) assessing interaction effects involving level-1 predictors, including within- and cross-level interactions; and (e) estimating the influence of level-2 variables while controlling for level-1 covariates. Although these scenarios are far from exhaustive, they provide a general framework for thinking about centering decisions that readily generalizes to more complex analyses.

As an aside, I limit the subsequent discussion to the use of multilevel models for explanation rather than prediction. In the latter situation, using the raw score predictors is probably the most convenient option because it preserves between-cluster mean differences and because it facilitates interpretation of the predicted values (e.g., describing a predicted score as a function of a set of raw scores is arguably more meaningful than doing so with a set of deviation scores). Interested readers can consult Afshartous and de Leeuw (2005) for an overview of prediction in multilevel models.

6.7.1 Contextual Models

Recall from the previous discussion that the goal of a contextual analysis is to compare the within- and between-cluster regression coefficients. As seen in (6.13) and (6.14), the simplest form of a contextual model includes x and $\bar{x}_j$ as level-1 and level-2 predictors, respectively. Kreft et al. (1995) thoroughly studied this model and showed that CWC and CGM produce equivalent solutions, such that the coefficients from one model are algebraic transformations of the other (see (6.15)). The variance estimates are also equivalent in random intercept models but not in random slope models. Given that CWC and CGM are equivalent, personal preference is the only reason to choose one method over the other. The CGM model is slightly more convenient because it yields a direct test of the contextual effect, whereas the CWC model requires a contrast between two regression coefficients. However, this is not a limiting factor because most multilevel software packages can test the difference between two coefficients.

6.7.2 Level-2 Predictors

Centering decisions for level-2 variables are straightforward because CWC is not applicable; by definition, the values of level-2 variable are constant within a given cluster. Consequently, centering at a constant (e.g., CGM) or leaving the variables on their raw metrics is the only option. When a model includes only main effects, this decision is largely one of personal preference because centering affects only the intercept coefficient. When a model includes higher-order effects such as a between-cluster interaction between two level-2 variables, well-established recommendations from the multiple regression literature can guide centering decisions. For example, Aiken and West (1991) recommended centering the explanatory variables at their grand means and forming a product term from the centered predictors. This procedure is ideally suited for multilevel analyses with interactions between level-2 variables.

6.7.3 Level-1 Predictors

Researchers routinely use multilevel models to describe relationships among level-1 variables. This is arguably the most difficult scenario that I address in this section because centering decisions generally require a theoretical or a substantive rationale. In particular, the scaling of a predictor variable should align with a researcher's conceptualization of the corresponding construct. Klein, Dansereau, and Hall's (1994) work from the organizational research literature provides a useful framework for thinking about this issue. According to these authors, level-1 constructs reflect individual differences *independent* of cluster membership or individual differences *within* clusters. When a construct is independent of cluster membership, the interpretation of score values does not depend on other members of the same group (i.e., an individual's absolute standing on the construct is what matters). CGM is appropriate for this situation because it maintains absolute score differences among all level-1 units. In contrast, a construct that reflects within-cluster differences requires a relative score interpretation (i.e., an individual's standing relative to others in the group is what matters). This type of construct is often associated with a so-called frog pond effect (Davis, 1966). Frog pond effects arise from social comparisons whereby individuals judge themselves relative to others in their peer group. For example, Marsh and Hau (2003) describe a scenario where a capable student who attends a school with high average achievement could have poor academic self-concept because his achievements are below average relative to those of his classmates (i.e., the student is a small frog in a big pond). CWC is ideal for frog pond effects because it yields a variable that reflects relative score differences within each cluster.

To illustrate the application of CGM and CWC, consider an educational study that examines the association between standardized test scores and special education placement in a sample of students nested within schools. If the influence of achievement on special education placement were independent of school membership, a low-scoring student would have the same chance of being placed in special education, regardless of the school that she attended. This situation could occur, for example, if teachers used an objective cutoff score as the basis for all special education referrals. CGM is appropriate for this situation because a student's overall standing on the standardized test influences special education placement and the performance of other students in the school does not play a role. In contrast, if the influence of achievement on special education placement were context-dependent, a student's placement in special education would depend on the school he attended. Such would be the case if teachers judged a student's performance relative to his classmates or to other students in the school (i.e., a frog pond effect). Notice that absolute achievement level is not the determining factor here because a low-scoring student would be more likely to be placed in special education if he attended a school with high average test scores (Hibel, Farkas, and Morgan, 2010). In this situation, CWC is appropriate because a student's relative standing to others in the group influences special education placement.

As a second example, consider an occupational stress study that examines the association between employee workload and psychological well-being in a sample of employees nested within workgroups. If the influence of workload on stress were independent of cluster membership, two employees that work the same number of hours would have the same well-being scores, on average, regardless of the workgroup to which they belonged. This situation could arise if heavy workload negatively impacts psychological well-being because an employee has less time to spend with family, has fewer opportunities to pursue leisure time activities, etc. Again, CGM is appropriate because an employee's judgment about his absolute workload influences well-being. In contrast, if the impact of workload on well-being were dependent on cluster membership, two employees that work the same number of hours but belong to different workgroups would have dissimilar well-being scores. This situation could occur if an employee that works more hours than others in her group has lower psychological well-being because she perceives the situation as unfair. Notice that the absolute number of work hours per week is not a determining factor in this example because perceptions of unfairness could arise in groups with a low average workload. CWC is appropriate because it expresses workload relative to other members of the same workgroup.

The previous examples underscore the point that level-1 centering decisions generally require a theoretical framework or a substantive rationale. In the absence of such information, the data structure may provide some insight into the appropriate centering approach. Klein et al. (1994) argued that constructs that are independent of cluster membership (i.e., situations where CGM is appropriate) should not produce a contextual effect. Although a number of study-specific characteristics can influence the magnitude of the between-cluster regression (e.g., measurement properties, data collection design, statistical power), estimating contextual models for the level-1 predictors may be a useful exploratory step. Constructs that require a relative score interpretation (i.e., situations where CWC is appropriate) necessarily require between-cluster mean differences (Bliese and Jex, 2002; Klein et al., 1994). Consequently, estimating intraclass correlations for the level-1 predictors may provide insight into the utility of a frog pond model, although the presence of between-group variability does not preclude CGM. As an aside, CGM and CWC will provide nearly identical estimates when there are no contextual effects or when the intraclass correlations for the predictor variables equal zero.

Finally, although it may seem unnatural to center a set of dummy or effect codes, the decision to use CGM or CWC also extends to nominal predictors (Enders and Tofighi, 2007). To illustrate, consider an educational study that examines the association between free and reduced lunch assistance (0 = no free or reduced lunch, 1 = received lunch assistance) and standardized test scores in a sample of students nested within schools. Like continuous predictors, a binary variable can contain within- and between-cluster variability. In this example, the cluster mean quantifies the proportion of students that receive lunch assistance in a particular school, and the level-2 variation reflects between-school differences in this proportion. When choosing between CGM and CWC, a researcher must again decide whether the absolute or relative score value is important. For CGM (or raw score dummy coding) to make sense, a researcher must believe that the impact of economic disadvantage is independent of school influence (e.g., a student from a low-income family would attain the same level of achievement, regardless of the school that he attended). In contrast, for CWC to make sense, the researcher would necessarily expect that economic disadvantage is context-dependent construct (e.g., a student from a low-income family would be at a greater disadvantage if he attended a school with a high average income).

6.7.4 *Interaction Effects Involving Level-1 Predictors*

Researchers routinely use multilevel models to examine moderation (i.e., interactive) effects where the magnitude of the relationship between two level-1 variables depends on the value of a third variable (e.g., the regression of y on x depends on a third variable z). In ordinary least squares regression, grand mean centering the predictor variables and computing a product term from the centered scores is standard practice (Aiken and West, 1991). Multilevel models are more complex because they allow for within-level interactive effects (i.e., the moderator is another level-1 variable) and cross-level interactive effects (i.e., the moderator is a level-2 variable). In some respects, centering decisions for interactive models are no different from those for models with only level-1 main effects. As before, CGM is appropriate for constructs that call for an absolute score interpretation, and CWC is preferred for constructs that require a relative score interpretation.

Although substantive concerns should ideally guide centering decisions, applying CGM to interactive models requires caution because it potentially confounds two or more interaction effects (Enders and Tofighi, 2007; Hofmann and Gavin, 1998; Raudenbush, 1989). To illustrate, consider a cross-level interaction model where the level-1 regression of y on x depends on the values of a level-2 variable w. The random intercept model is

$$\begin{aligned} y_{ij} &= \beta_0 + \beta_1 \left(x_{ij} - \bar{x}\right) \\ &\quad + \beta_2 w_j + \beta_3 \left(x_{ij} - \bar{x}\right) w_j + u_{0j} + \varepsilon_{ij}, \end{aligned} \tag{6.16}$$

where the β_3 coefficient quantifies the cross-level interaction effect. To understand the potential for confounding, consider the composition of the product term in (6.16),

$$\begin{aligned} \left(x_{ij} - \bar{x}\right) w_j &= \left[\left(x_{ij} - \bar{x}_j\right) + \left(\bar{x}_j - \bar{x}\right)\right] w_j \\ &= \left(x_{ij} - \bar{x}_j\right) w_j + \bar{x}_j w_j. \end{aligned}$$

Notice that the CGM product term is a composite variable that contains a cross-level product and a between-cluster product involving w_j and $\bar{x}_j$. The composite nature of the CGM product term raises the possibility that the β_3 coefficient in (6.16) may reflect an ambiguous combination of interaction effects. Hofmann and Gavin (1998) demonstrated this point, showing that a non-zero between-group interaction can produce a significant cross-level interaction, even when no such effect exists in the population. Note that CWC does not suffer from this problem because its product term reflects only a cross-level interaction.

The potential confounding becomes more complex when the CGM model includes a within-level interaction effect between a pair of level-1 variables. To illustrate, consider a random intercept model where the level-1 regression of y on x depends on the values of another level-1 variable z:

$$\begin{aligned} y_{ij} &= \beta_0 + \beta_1 \left(x_{ij} - \bar{x}\right) + \beta_2 \left(z_{ij} - \bar{z}\right) \\ &\quad + \beta_3 \left(x_{ij} - \bar{x}\right) \left(z_{ij} - \bar{z}\right) + u_{0j} + \varepsilon_{ij}. \end{aligned} \tag{6.17}$$

Again, the composition of the product term demonstrates the problem:

$$\begin{aligned} &\left(x_{ij} - \bar{x}\right) \left(z_{ij} - \bar{z}\right) \\ &\quad = [(x_{ij} - \bar{x}_j) + (\bar{x}_j - \bar{x})][(z_{ij} - \bar{z}_j) \\ &\quad\quad + (\bar{z}_j - \bar{z})] \\ &\quad = (x_{ij} - \bar{x}_j)(z_{ij} - \bar{z}_j) \\ &\quad\quad + (\bar{x}_j - \bar{x})(z_{ij} - \bar{z}_j) \\ &\quad\quad + (x_{ij} - \bar{x}_j)(\bar{z}_j - \bar{z}) \\ &\quad\quad + (\bar{x}_j - \bar{x})(\bar{z}_j - \bar{z}). \end{aligned} \tag{6.18}$$

This expression shows that the CGM product term is potentially a composite of four interaction effects: (a) the moderating effect of z on the association between x and y (i.e., the within-level interaction); (b) the moderating influence of $\bar{x}_j$ on the association between z and y (i.e., a cross-level interaction); (c) the moderating effect of $\bar{z}_j$ on the association between x and y (i.e., a cross-level interaction); and (d) the moderating influence of $\bar{z}_j$ on the association between $\bar{x}_j$ and $\bar{y}_j$ (i.e., the between-cluster interaction).

The composite nature of the CGM product term threatens the validity of the β_3 estimate in (6.16) and (6.17) and opens the door for erroneous substantive conclusions. Estimating exploratory models that include additional terms may provide some protection against this problem. For example, expanding (6.16) gives the following model:

$$\begin{aligned} y_{ij} = &\beta_0 + \beta_1 \left(x_{ij} - \bar{x}\right) + \beta_2 w_j \\ &+ \beta_3 \left(x_{ij} - \bar{x}\right) w_j + \beta_4 \bar{x}_j \\ &+ \beta_5 \bar{x}_j w_j + u_{0j} + \varepsilon_{ij}. \end{aligned} \tag{6.19}$$

Because the level-1 and level-2 predictors are potentially correlated, β_4 quantifies the net contribution of $\bar{x}_j$ (i.e., the difference between the within- and between-cluster regressions), and β_5 quantifies the net contribution of the between-cluster interaction above and beyond that of the cross-level interaction (i.e., the difference between the cross-level and the between-level interaction effects). Similarly, expanding (6.17) gives the model

$$\begin{aligned} y_{ij} = &\beta_0 + \beta_1 \left(x_{ij} - \bar{x}\right) + \beta_2 \left(z_{ij} - \bar{z}\right) \\ &+ \beta_3 \left(x_{ij} - \bar{x}\right) \left(z_{ij} - \bar{z}\right) + \beta_4 \bar{x}_j \\ &+ \beta_5 \bar{z}_j + \beta_6 \bar{x}_j \left(z_{ij} - \bar{z}\right) \\ &+ \beta_7 \bar{z}_j \left(x_{ij} - \bar{x}\right) + \beta_8 \bar{x}_j \bar{z}_j + u_{0j} + \varepsilon_{ij}. \end{aligned} \tag{6.20}$$

Consistent with (6.19), the additional coefficients in (6.20) (i.e., β_4 through β_8) quantify the net influence of the between-cluster effects above and beyond the within-cluster predictors. Performing a joint significance test of these additional parameters may identify unexpected contextual effects or confounded interaction effects.

6.7.5 *Level-1 Covariates*

In the previous examples, the level-1 predictors were central to the substantive research goals. Level-1 centering decisions also arise when researchers want to assess the influence of level-2 variables while controlling for level-1 covariates. This situation is quite different because centering choices do not require a substantive rationale. Rather, CGM is virtually always preferable. To illustrate, consider a random intercept model with a single level-2 predictor and a single level-1 covariate,

$$y_{ij} = \beta_0 + \beta_1 x_{ij} + \beta_2 w_j + u_{0j} + \varepsilon_{ij}. \tag{6.21}$$

The model in (6.21) routinely arises in cluster-randomized designs where the researcher assigns entire groups (e.g., classrooms or schools) to experimental conditions. For example, in the context of the math problem-solving data, w_j could represent a binary level-2 intervention code (e.g., 0 = comparison school, 1 = intervention school), and x_{ij} could represent any number of individual-level covariates (e.g., pre-test scores, standardized achievement test scores, the lunch assistance indicator).

To begin, consider the behavior of CWC. Recall from the previous discussion that CWC yields level-1 scores that are orthogonal to level-2 variables. This implies that w_j and x_{ij} are uncorrelated and the level-2 coefficient is not adjusted for the covariate. In fact, applying CWC to (6.16) would produce a β_2 coefficient that is identical to that from a model that omits x entirely. In contrast, CGM yields level-1 scores that are correlated with level-2 variables (e.g., w_j is correlated with the level-2 component of the covariate, $\bar{x}_j$). Because the predictors are correlated, the CGM estimate of the β_1 coefficient reflects the net influence of the level-2 predictor after partialling out the level-1 covariate.

Recasting the composite regression equation as separate level-1 and level-2 models further illustrates the difference between CWC and CGM. The level-1 regression equation is

$$y_{ij} = \beta_{0j} + \beta_{1j}x_{ij} + \varepsilon_{ij},$$

and the level-2 equations are

$$\begin{aligned} \beta_{0j} &= \beta_0 + \beta_2 w_j + u_{0j} \\ \beta_{1j} &= \beta_1. \end{aligned} \qquad (6.22)$$

Recall from (6.8) that CWC yields a β_{0j} value that corresponds to the mean of cluster j. With this interpretation in mind, the CWC estimate of the β_2 coefficient in (6.22) quantifies the influence of w_j on the outcome means. Notice that this interpretation makes no reference to x. In contrast, (6.6) shows that the β_{0j} term from a CGM model corresponds to the adjusted mean for cluster j. By extension, the CGM estimate of β_2 reflects the influence of w_j on means that have been adjusted for between-cluster differences on x. In sum, CGM controls for level-1 variables, and CWC does not. For this reason, CGM is the appropriate choice when the research goal is to assess the influence of level-2 variables while controlling for level-1 covariates.

6.7.6 Summary

Table 6.3 summarizes the centering recommendations from this section. Three of the five scenarios are relatively straightforward because CGM is the only appropriate method (centering level-2 predictors, controlling for level-1 covariates) or because either CGM or CWC are appropriate (contextual models). Situations where level-1 predictors are the analytic focus—either as main effects or as part of an interaction—are the most challenging because centering decisions generally require a substantive rationale. Choosing between CGM and CWC largely depends on whether a researcher conceptualizes the construct as independent of or dependent on cluster membership. In the absence of a strong conceptual model, estimating contextual models for the level-1 predictors may be a useful exploratory step because constructs that require an absolute score interpretation should not produce a contextual effect (Klein et al., 1994).

6.8 DATA ANALYSIS EXAMPLE 2

To illustrate the previous centering recommendations, I fit a random intercept model to the problem-solving data set. My primary analytic goal was to determine whether academic self-concept moderated the influence of the level-2 intervention variable (0 = comparison school, 1 = intervention school). The analysis also included three student-level covariates: free or reduced lunch status, standardized math achievement test scores, and

Table 6.3 Summary of Centering Recommendations

Research Goal	CGM	CWC
Evaluate contextual effects	Always applicable. The level-2 coefficient quantifies the contextual effect. Equivalent to CWC in random intercept models.	Always applicable. The difference between the level-1 and level-2 coefficients quantifies the contextual effect. Equivalent to CGM in random intercept models.
Assess level-2 predictors	Always applicable. Use guidelines from the single-level regression literature to guide level-2 centering decisions.	Not applicable because level-2 scores are constant within each cluster.
Evaluate level-1 predictors	The construct reflects individual differences independent of cluster membership. An individual's absolute standing on the construct is what matters.	The construct reflects individual differences within clusters. An individual's standing relative to other members of the same cluster is what matters.
Assess interaction effects involving level-1 predictors	The construct reflects individual differences independent of cluster membership. An individual's absolute standing on the construct is what matters. Potential for confounded interaction effects.	The construct reflects individual differences within clusters. An individual's standing relative to other members of the same cluster is what matters.
Control for level-1 covariates	Always applicable. Level-2 coefficients are adjusted for cluster differences on the covariate, as in ANCOVA.	Not appropriate because level-2 coefficients are not adjusted for cluster differences on the covariate.

pre-test problem-solving scores. The random intercept model is as follows:

$$\begin{aligned} probsolving_{ij} = {} & (\beta_0 + \beta_1 frlunch_{ij}) \\ & + (\beta_2 stanmath_{ij}) \\ & + (\beta_3 pretest_{ij}) \\ & + (\beta_4 selfconcept_{ij}) \\ & + (\beta_5 txgrp_j) \\ & + (\beta_6 selfconcept_{ij})(txgrp_j) \\ & + (u_{0j} + \varepsilon_{ij}), \end{aligned} \tag{6.23}$$

where β_4 and β_5 are the main effects of self-concept and intervention condition, respectively, and β_6 is the cross-level interaction term.

In line with my previous recommendations, I applied CGM to the covariates, including the binary lunch assistance indicator; this decision did not require a theoretical rationale because my only goal was to control for these variables. In contrast, academic self-concept required careful consideration because it was integral to the substantive question. Marsh and colleagues (e.g., see Marsh and Hau, 2003) argued that academic self-concept is a relativistic construct that forms when students compare their academic achievements to those of their peer group (i.e., a frog pond effect). For example, a capable student who attends a school with high average achievement could have a poor self-concept because he perceives himself as below average relative to his peers, whereas the same student could have a positive self-concept if he attends a school where his peers have lower achievement, on average. This theoretical framework suggests that CWC is appropriate because the resulting self-concept scores depend only on other students from the same school. Finally, although I could have centered the level-2 predictor at its grand mean, I used the raw dummy code; CGM would have converted the dummy variable to an effect code with values of $-.5$ and $.5$. This decision defined the intercept as the adjusted mean for the comparison schools.

Table 6.4 gives the parameter estimates from the analysis. After controlling for the level-1 covariates, neither self-concept nor its interaction with the intervention variable were

Table 6.4 Parameter Estimates from Data Analysis 2

Parameter	Est.	*SE*	*t*	*df*	*p*
Intercept (β_0)	9.814	0.249	39.439	35.918	<.001
Lunch Assistance Slope (β_1)	−0.120	0.159	−0.752	962.897	0.452
Achievement Slope (β_2))	0.238	0.023	10.531	995.856	<.001
Pre-Test Slope (β_3)	0.339	0.037	9.255	991.127	<.001
Self-Concept Slope (β_4)	0.004	0.030	0.144	961.119	0.886
Intervention Slope (β_5)	1.833	0.352	5.207	35.953	<.001
Interaction Slope (β_6)	0.016	0.042	0.372	955.545	0.710
Intercept Variance ($\sigma^2_{u_0}$)	1.069	0.292	3.655	N/A	<.001
Error Variance (σ^2_ε)	4.215	0.193	21.857	N/A	<.001

Note. Variance components use a *z* rather than a *t* reference distribution.

significant predictors of post-test problem-solving scores, $\hat{\beta}_4 = .004$ and $\hat{\beta}_6 = .016$, respectively. However, the intervention effect was significant, such that intervention schools scored approximately 1.8 points higher, on average, $\hat{\beta}_5 = 1.832$. Although self-concept and its interaction were not significant, it is important to reiterate that CWC requires a relative score interpretation. For example, the $\hat{\beta}_4$ coefficient reflects the expected change in problem-solving scores for a one-unit within-school increase in self-concept.

6.9 EXTENSIONS: THREE-LEVEL MODELS

Thus far, I have focused on the role of centering in two-level models. To date, no methodological studies have investigated the role of centering in three-level models; this is an important area of future research. Although the scope of this chapter is necessarily limited to two-level models, a brief discussion of three-level models is warranted. To provide a context for the subsequent discussion, consider a research scenario where students are nested within schools at level-2, and schools are nested within districts at level-3. That is, unit *i* is nested within cluster *j*, and cluster *j* is nested within cluster *k*.

Level-1 predictors are the most complex variables in a three-level model because they potentially contain within-group variability and two sources of between-group variability. The following equation partitions the total deviation between a score and the grand mean into the three components

$$x_{ijk} - \bar{x} = \left(x_{ijk} - \bar{x}_{jk}\right) + \left[\bar{x}_{jk} - \bar{x}_k\right] + \{\bar{x}_k - \bar{x}\} = \mathrm{W}_1 + \mathrm{B}_2 + \mathrm{B}_3, \qquad (6.24)$$

where W_1 represents within-group variability among level-1 units, B_2 denotes between-group variability among level-2 units, and B_3 denotes between-group variability among level-3 units. Returning to the math problem-solving example, consider standardized achievement test scores. The variation in this variable reflects within-school differences among students (i.e., W_1), between-group differences in the school means (i.e., B_2), and between-group differences in the district means (i.e., B_3). Because the left side of (6.24) is consistent with grand mean centering, it follows that CGM produces a composite predictor variable that contains all three sources of variation. Consistent with the conclusions from a two-level model, this implies that the predictor can correlate with level-1, level-2, and level-3 variables. Retaining all three sources of variability (and thus three sources of association) may be desirable in some situations, whereas isolating lower-level components may be necessary in other cases.

Applying CWC to level-1 variables becomes somewhat more complex in three-level models because you can deviate scores around the level-2 (e.g., school) means or around the level-3 (e.g., district) means.

Consistent with (6.4), deviating scores around the level-2 means removes all between-group variability from the predictor variable, leaving only level-1 variation. Consequently, this approach produces a predictor variable that can correlate only with other level-1 predictors. A second option is to deviate the level-1 predictor around the level-3 (e.g., district) means. This approach mimics the behavior of CGM in a two-level model because it yields a composite predictor variable that contains two sources of variation, as shown below,

$$\begin{aligned} x_{ijk} - \bar{x}_k &= (x_{ijk} - \bar{x}_{jk}) + [\bar{x}_{jk} - \bar{x}_k] \\ &= \mathrm{W}_1 + \mathrm{B}_2. \end{aligned}$$

As you might expect, centering at the level-3 means yields scores that can correlate with level-1 and level-2 variables.

Next, consider level-2 predictor variables (e.g., school size). In a two-level model, centering level-2 variables was not an issue because CGM was the only option. However, CGM and CWC are now both applicable because a level-2 predictor potentially contains two sources of variability. The following equation partitions the total deviation between a level-2 score w_{jk} and the grand mean into the two components

$$w_{jk} - \bar{w} = [w_{jk} - \bar{w}_k] + \{\bar{w}_k - \bar{w}\} = \mathrm{W}_2 + \mathrm{B}_3, \tag{6.25}$$

where W_2 denotes within-group variability among level-2 units, and B_3 denotes between-group variability among level-3 units. Returning to the math problem-solving example, consider a level-2 predictor such as school size. The variation in this variable reflects within-district differences among schools (i.e., W_2) and between-district differences in the school size means (i.e., B_3). Because the left side of (6.25) is consistent with grand mean centering, it follows that CGM produces a composite predictor variable that retains both sources of variation. Consistent with the conclusions from a two-level model, this implies that the level-2 predictor can correlate with level-2 and level-3 variables. In contrast, centering the level-2 predictor at the level-3 means eliminates the between-group variation, leaving only within-cluster variability. Doing so creates a level-2 variable that correlates only with other level-2 variables.

The previous discussion illustrates the complexities that arise with three-level models. Adding a third level to the analysis creates myriad analytic possibilities, and thoroughly addressing centering decisions could easily fill an entire chapter. Nevertheless, my earlier centering recommendations do imply a few general principles. For example, if a lower-level variable (e.g., a level-1 or a level-2 predictor) is the analytic focus, choosing between CGM and CWC largely depends on whether a researcher conceptualizes the construct as independent of or dependent on cluster membership. This principle becomes even more important when the lower-level variable contributes to an interaction effect because the presence of higher-level variability results in a complicated composite interaction term that may not have a clear interpretation. In contrast, when a variable serves as a covariate, retaining the higher-level variation is usually desirable because doing so partials out the effect of that variable. In every situation, considering the variables with which a lower-level predictor can correlate may be a useful exercise when considering different centering options. Again, these are just a few preliminary recommendations. Additional research is needed to fully clarify the role of centering in three-level models.

6.10 CONCLUSIONS

This chapter examined the important role of centering predictor variables in multilevel models. The overarching goal of the chapter was to delineate the differences between grand and group mean centering and to provide researchers with guidelines for implementing these procedures. With few exceptions, CGM and CWC are not equivalent and can produce different numeric estimates as well as a different substantive interpretation. The fact that the two approaches partition the variability in different ways explains these discrepancies. Specifically, CGM produces a composite predictor variable that contains both within-cluster and between-cluster variation, whereas CWC produces a variable that contains only within-cluster variation. This variance partitioning (or lack thereof in the case of CGM) affects the associations among the analysis variables, such that CGM yields a level-1 predictor that can correlate with both level-1 and level-2 variables, whereas CWC produces a level-1 predictor that is orthogonal to all level-2 variables. The associations among the analysis variables have important implications for model specification and provided the basis for the centering recommendations that I outlined in the chapter. Although my recommendations are far from exhaustive, they provide a general framework for thinking about centering decisions that readily generalizes to more complex analyses, including three-level models.

REFERENCES

Afshartous, D. and de Leeuw, J. (2005) 'Prediction in Multilevel Models', *Journal of Educational and Behavioral Statistics*, 30: 109–39.

Aiken, L.S. and West, S.G. (1991) *Multiple Regression: Testing and Interpreting Interactions*. Newbury Park, CA: Sage.

Bleise, P.D. and Jex, S.J. (2002) 'Incorporating a Multilevel Perspective into Occupational Stress Research: Theoretical, Methodological, and Practical Implications', *Journal of Occupational Health Psychology*, 7: 265–76.

Davis, J.A. (1966) 'The Campus as a Frog Pond: An Application of Theory of Relative Deprivation to Career Decisions for College Men', *American Journal of Sociology* 72: 17–31.

Enders, C.K. and Tofighi, D. (2007) 'Centering Predictor Variables in Cross-sectional Multilevel Models: A New Look at an Old Issue', *Psychological Methods*, 12: 121–38.

Firebaugh, G. (1978) 'A Rule for Inferring Individual-level Relationships from Aggregate Data', *American Sociological Review*, 43: 557–72.

Hibel, J., Farkas, G., and Morgan, P.L. (2010) 'Who is Placed in Special Education?' *Sociology of Education*, 83: 312–32.

Hofmann, D.A. and Gavin, M.B. (1998) 'Centering Decisions in Hierarchical Linear Models: Implications for Research in Organizations', *Journal of Management*, 24: 623–41.

Klein, K.J., Dansereau, F., and Hall, R.J. (1994) 'Levels Issues in Theory Development, Data Collection, and Analysis', *The Academy of Management Review*, 19: 195–229.

Kreft, I.G.G., de Leeuw, J., and Aiken, L.S. (1995) 'The Effect of Different Forms of Centering in Hierarchical Linear Models', *Multivariate Behavioral Research*, 30: 1–21.

Marsh, H.W. and Hau, K-T. (2003) 'Big-fish-little-pond Effect on Academic Self-concept: A Cross-cultural (26-country) Test of the Negative Effects of Academically Selective Schools', *The American Psychologist*, 58: 364–76.

Montague, M., Enders, C., and Dietz, S. (2011) 'The Effects of Cognitive Strategy Instruction on Middle School Students' Math Problem Solving', *Learning Disability Quarterly*, 34: 262–72.

Raudenbush, S.W. (1989) 'A response to Longford and Plewis', *Multilevel Modeling Newsletter*, 1: 8–11.

Raudenbush, S.W. and Bryk, A.S. (2002) *Hierarchical Linear Models: Applications and Data Analysis Methods* (2nd edition). Thousand Oaks, CA: Sage.

Willms, J.D. (1986) 'Social Class Segregation and its Relationship to Pupils' Examination Results in Scotland', *American Sociological Review*, 51: 224–41.

7

Model Selection for Multilevel Models

Russell Steele

McGill University, Canada

7.1 INTRODUCTION

One of the more challenging aspects of modern applied problems is specifying the statistical model in a principled, rigorous, and ultimately defensible manner. For standard statistical analyses, such as multiple linear or logistic regression, one can rely on an established philosophy and practices to select models. Although there are various approaches to basic model specification and selection, the principles and differences amongst methods are well understood and can be easily explained, even to non-statisticians. However, only some of the fundamental principles from simpler models can be readily transported to the context of multilevel modeling, and the usual criteria and strategies for model selection must be adapted.

The purpose of this chapter is to provide a broad picture of the issues surrounding model selection and specification for multilevel models, focusing on the most commonly used approaches for model selection. A comprehensive treatment of the subject would require significantly more space than is possible in this chapter, so instead the focus will be on developing a sturdy framework for readers that will allow them to interpret other existing and future advances in model selection. We choose here to focus on percentage variance explained (R^2) as a measure of model fit and the use of common information criteria (AIC and BIC), as these are the most popular approaches to model selection. It is important to understand that no one criterion can be, in some sense, considered "best", as all criteria have relative strengths and weaknesses and illustrating these differences is an underlying motive for this chapter.

The chapter begins with two motivating examples from rheumatic disease epidemiology that are typical of a multilevel setting. A brief review of model selection for multiple linear regression models then follows. The next section discusses the critical tensions that arise when trying to apply those simple ideas to the multilevel model framework. The remainder of the section focuses on particular model selection criteria and the mechanics of multilevel model selection. Section 7.5 then uses both data examples to illustrate the similarities and differences amongst the various criteria. The chapter concludes with a discussion of other approaches and current directions for research.

7.2 MOTIVATING EXAMPLE: LONGITUDINAL LUNG FUNCTION IN SYSTEM SCLEROSIS PATIENTS

Systemic (system) sclerosis (or scleroderma) is a rare but debilitating auto-immune disease that results in fibrotic organ tissue. System sclerosis (SSc) affects primarily women, and the median age of onset is approximately 55. Progression of the disease usually begins in the extremities, with hardening of skin tissue in the fingers and hands, although more severe forms of the disease progress into the trunk and fibrosis of major organ systems.

In this chapter, we consider modeling the physical function of scleroderma patients, as measured by the Health Assessment Questionnaire Disability Index (HAQ) (Bruce and Fries, 2003). The HAQ asks patients about eight categories of function: dressing, arising, eating, hygiene, walking, reach, grip, and outside activity, and yields an overall score between 0 and 3.

We will examine the data in two ways. First, we analyze only the baseline visit data, i.e. the data for each subject at study entry for 535 subjects from 13 of the centers. The question of interest is the relationship of the Scleroderma Disease Activity Index (SDAI) (Valentini et al., 2001; Della Rossa et al., 2001) to patient function, controlling for possible confounding. We hypothesize that high levels of disease activity should lead to lower function, which may depend on the type of disease (limited or diffuse) and could be confounded by age, disease duration, and/or gender.

The multilevel problem is that patient baseline visits are nested within recruiting centers, a situation where we have a small number of clusters (centers) relative to the total number of patients. The observed association between activity and function could be confounded if there is a heterogeneity in the level of disease activity between centers, particularly due to the highly variable number of subjects per center. Figure 7.1 shows plots of the activity score vs the HAQ function score, separated by center, with limited disease patients as the grey circles and diffuse patients as the black circles.

We also present longitudinal data for the HAQ. In this case, we have clustering of measurements within subjects (in addition to the clustering by center), so that the number of visits per subject is small compared to the number of patients. Note that we will not address the issue of patient drop-out in this chapter, although we have addressed it elsewhere (Schnitzer et al., 2011). Similarly, for purposes of brevity and simplicity, in this example we will ignore the clustering of patients within centers. Figure 7.2 shows the longitudinal disability for 421 SSc patients from the CSRG registry who have been observed for at least two annual visits and have complete baseline data and observed HAQ scores at all visits (with a total of 1530 follow-up visits across all patients).

The goal of the second analysis is to characterize the longitudinal pattern of disability for SSc patients while adjusting for the baseline visit patient function. Figure 7.2 presents additional challenges in choosing statistical models. The trajectories for the patients appear to generally be increasing (i.e. increasing disability or decreasing function over time), although there is a large amount of heterogeneity in the observed slopes.

All of the above considerations are complicated by a need for a systematic way to decide amongst the various models that has valid and, just as importantly, interpretable statistical properties. We will explore these issues, as well as others, in the remainder of the chapter.

7.3 OVERVIEW OF MODEL SELECTION IN THE CONTEXT OF MULTIPLE LINEAR REGRESSION MODELS

In this section, we review the statistical properties of standard model selection criteria and strategies for multiple linear regression models. We use the multiple linear regression model to develop intuition about what

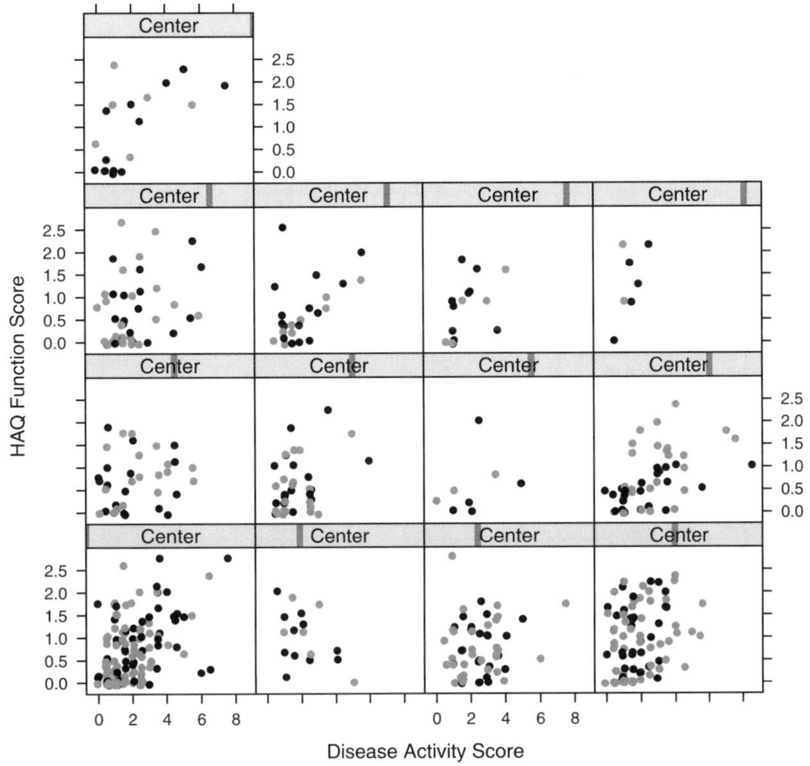

Figure 7.1 Plot of HAQ vs SDAI by Center for 549 Subjects with Complete Baseline Data from the Canadian Scleroderma Research Group Patient Cohort. Grey circles indicate limited disease patients, dark circles indicate diffuse disease patients

might constitute a reasonable model selection approach for multilevel models. For more detailed information about variable selection procedures for the standard multiple linear regression model, we refer readers to modern linear regression texts, e.g. Montgomery et al. (2001) or Weisberg (2005).

In the case of linear regression, the set of possible models generally corresponds to a set of linear models with different design matrices. Usually, it is assumed that the outcome variable, $\mathbf{y}$, can be modeled as:

$$\mathbf{y} = X\boldsymbol{\beta} + \boldsymbol{\varepsilon},$$

where $\mathbf{y}$ is an $(n \times 1)$ vector, X is an $n \times p$ design matrix, $\boldsymbol{\beta}$ is a $(p \times 1)$ vector of coefficients and $\boldsymbol{\varepsilon}$ is an $(n \times 1)$ vector of error terms. We additionally assume here that $\boldsymbol{\varepsilon} \sim N_n(0, \sigma^2 I_{n\times n})$, where $I_{n\times n}$ is the $(n\times n)$ identity matrix. We emphasize that for a different design matrix, X, the coefficients $(\boldsymbol{\beta})$ will be of varying dimension and meaning and the error variance (σ^2) will be different in meaning and value.

Squared error fit criteria are often used to evaluate models, most notably the coefficient of determination (R^2) and adjusted R^2. The R^2 criterion estimates the proportion of the overall variance of the outcome variable, $\mathbf{y}$, that is explained by the regressor variables, and can be written as:

$$R^2 = \frac{(\mathbf{y} - \hat{\mathbf{y}})'(\mathbf{y} - \hat{\mathbf{y}})}{(\mathbf{y} - \bar{\mathbf{y}})'(\mathbf{y} - \bar{\mathbf{y}})}.$$

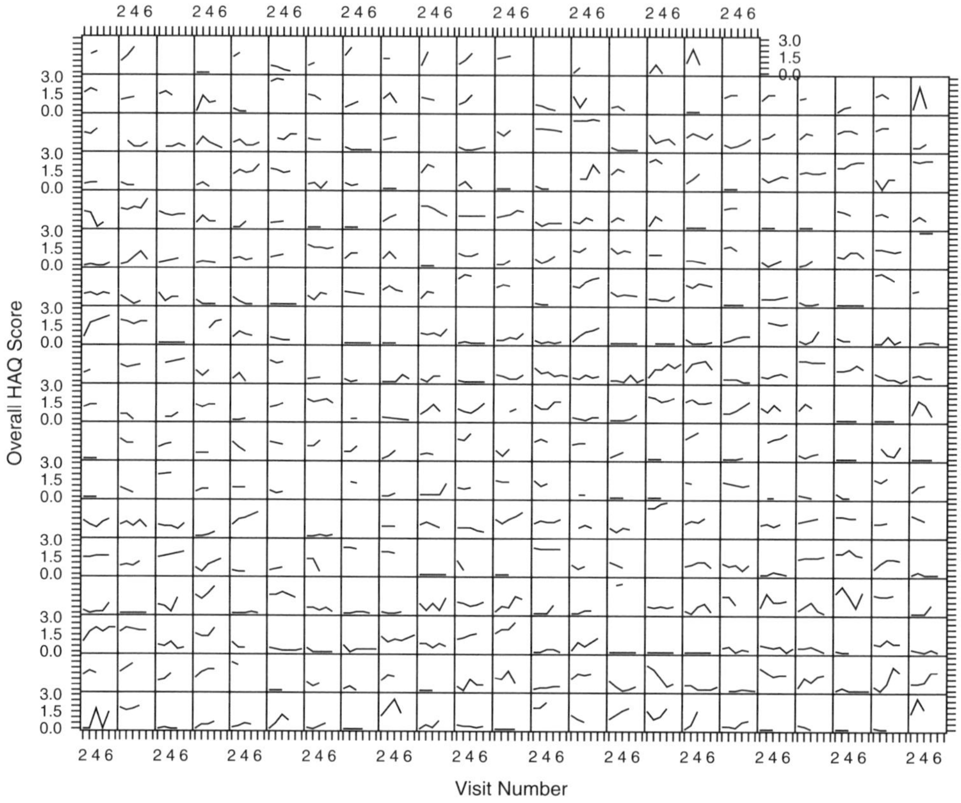

Figure 7.2 Longitudinal HAQ Scores from Canadian Scleroderma Research Group Registry

It is well known that R^2 cannot be decreased in linear models by adding columns to the design matrix, and is likely to increase. R^2_{adj} is frequently used as an alternative measure and can be written as

$$R^2_{\text{adj}} = 1 - (1 - R^2) \times \frac{n-1}{n-p}.$$

Therefore, for small- to moderate-sized data sets, R^2_{adj} will penalize models with more covariates through the increase in p in the denominator, so that it is possible for R^2_{adj} to decrease with an increase in the number of columns of the design matrix. In practice, it is usually recommended to choose the most parsimonious model that maintains a sufficiently large value of either R^2 or R^2_{adj}.

One can also utilize the likelihood as a measure of fit. In multiple linear regression models, the deviance is intimately related to the residual squared error of the model. Under an assumption of normal errors, the maximum likelihood estimators for $\boldsymbol{\beta}$ and σ^2 minimize the deviance

$$\begin{aligned} D(\hat{\boldsymbol{\beta}}, \hat{\sigma}^2) &= n \log 2\pi \\ &+ n \log \frac{(\mathbf{y} - \hat{\mathbf{y}})'(\mathbf{y} - \hat{\mathbf{y}})}{n} + n. \end{aligned}$$

When comparing two nested models, the more complex model will always have a deviance at least as small as the deviance of the less complex model. One can use asymptotic results for standard likelihood ratio tests or the distribution of the residual sum of squares to test for statistically significant decreases in deviance. However, two other criteria are far more commonly used when utilizing the deviance as a measure of the fit of the model:

the Akaike Information Criterion (AIC) and the Bayesian or Schwarz information criterion (BIC or, sometimes, SIC). These two criteria are penalized deviance criteria and penalize in different ways for the complexity of the model, where $AIC = D(\boldsymbol{\beta}, \sigma^2) + 2k$ and $BIC = D(\boldsymbol{\beta}, \sigma^2) + k \log n$, and k is the number of free parameters in the model.

The justification for the AIC comes from the original paper by Akaike (1973), where he showed that for regular statistical models (which includes the class of linear regression models), when the true model is not among the candidate models, the model in a class of models that obtains the minimum AIC, asymptotically, will be the model that has minimum Kullback–Leibler distance to the true model. Hence, the AIC tends to perform very well with respect to choosing models for prediction, because it tends to favor models that fit the observed data well, even if the models under consideration may not contain all of the true covariates that generated the data.

There are several justifications for using the BIC, although the most commonly stated justification comes from Bayesian inference. The BIC can be justified as being approximately equal to the negative log-integrated likelihood of a statistical model for large sample sizes or under a particular choice of prior distribution for the parameters of the model (Schwarz, 1978). The BIC yields very precise estimates of the log-integrated likelihood under a particular class of priors (Kass and Wasserman, 1995). The integrated likelihood is important in Bayesian inference as the posterior probability of a model given the observed data is proportional to the product of $p(y)$ and the prior probability assigned to the model. Choosing models according to minimum BIC should, therefore, choose models that have the largest posterior model probability in the class of models considered under reasonable prior distributions (Kass and Raftery, 1995).

Burnham and Anderson present a far more nuanced and detailed comparison of the two criteria in sections 6.3 and 6.4 of their text (Burnham and Anderson, 2002) than what appears above. In particular, they make clear that small values of the AIC will yield models that have fitted values that are close to the fitted values of the data-generating model as the sample size increases, regardless of whether the true model is within the class of models being considered or not. For very large n, choosing models with small values of the BIC, on the other hand, will select the model with the right dimension (e.g. in most regression examples this would be the right number of predictors) if the true model is contained in the set of all possible models, or the model with the smallest dimension that provides fitted values that are close to the fitted values of the data generating model. Their most salient point is that when the true model is not amongst those models under consideration, the "best" target model for AIC can vary with the sample size, but the "best" model for the BIC does not depend on the sample size (although the BIC's ability to stably detect this "best" model may require large samples).

7.4 EXTENDING STANDARD MODEL SELECTION CRITERIA TO MULTILEVEL MODELS

We now consider the potential pitfalls in trying to naïvely extend the criteria discussed previously to the context of multilevel models. Assume we are choosing amongst a class of multilevel models, where each can be written as:

$$\mathbf{y} = X\boldsymbol{\beta} + Z\mathbf{u} + \boldsymbol{\varepsilon}, \qquad (7.1)$$

where Z is now the $(n \times q)$ design matrix for modeling covariance amongst elements of $\mathbf{y}$s and $\mathbf{u}$ is the $(q \times 1)$ vector of random effect parameters, where the columns of X and Z vary between models. This general formulation of the problem allows for different models to differ in their fixed effect and/or random effect design matrices and allows for nested and non-nested sets of models.

Challenges arise when attempting to apply intuition and standard criteria from standard linear models to multilevel models. The primary issues that we will address are:

- choosing amongst the marginalized, REML, or conditional versions of the likelihood
- choosing a reference model in order to interpret goodness-of-fit
- determining the correct sample size (or error degrees of freedom)
- counting the number of model parameters (or model degrees of freedom), and
- accounting for violations of basic assumptions required by the criteria.

7.4.1 R^2 and R^2_{adj}

Extending even a notion as simple as R^2 to the multilevel setting proves challenging. Snijders and Bosker (1994) give simple examples of where the idea of R^2 as a measure of proportional reduction in the error variance can fail, particularly in the case of unbalanced designs. Two general issues arise when trying to formulate a measure of percentage variance explained for linear mixed models: a) how to account for the contribution of the random effect design matrix, and b) the choice of a null reference model to represent the total amount of variability.

Snijders and Bosker (1994) redefine the R^2 measure for mixed models in terms of the reduction in *prediction error*, rather than error for the observations. In a standard multiple linear regression model, these two ideas are equivalent: the true error variance is the mean-squared prediction error for the linear regression model. However, these things are not equivalent in a multilevel model. Instead, one must consider different prediction errors at different levels, where the different prediction errors depend critically on the level of the prediction. For example, if focused on prediction within-individual (level-1 prediction), then one should consider prediction at the level of the individual observations of $\mathbf{y}$. In this case, the prediction error at the first level, conditional on $\mathbf{z_i}$, can be written as $Var(y_i - \mathbf{x_i}\boldsymbol{\beta}|\mathbf{z_i}) = \mathbf{z_i}Var(\mathbf{u})\mathbf{z_i}^t + \sigma^2$, where y_i is the ith element of $\mathbf{y}$, and $\mathbf{x_i}$ and $\mathbf{z_i}$ are the ith rows of X and Z, respectively. However, this calculation is conditional on a particular row of the random effects design matrix, so Snijders and Bosker (1994) recommend using

$$\begin{aligned}Var(y_i - \mathbf{x_i}\boldsymbol{\beta}) = \text{tr}(&Var(\mathbf{u})Var(\mathbf{z_i}))\\ &+ E(\mathbf{z_i})Var(\mathbf{u})E(\mathbf{z_i})^t\\ &+ \sigma^2,\end{aligned}$$

where $Var(\mathbf{z_i})$ is the variance-covariance matrix for the vectors $\mathbf{z_i}$ sampled from the population and $E(\mathbf{z_i})$ is the mean for those vectors. Note here that the interpretation becomes difficult if the design matrix is from a designed experiment, rather than an observational study, because $\mathbf{z_i}$ would be fixed rather than random. Now a reference variance must be chosen, and the authors suggest that, for level-1 testing, the correct variance is the marginal variance of $\mathbf{y}$, which can be estimated with either the sample variance or the variance estimated from a null model with only intercept terms in X and Z. Therefore, their R^2 value at the first level, by plugging estimates for all expectations, variance-covariance matrices, and σ^2, is:

$$R^2_{\text{SB}-1} = 1 - \frac{tr(Var(\mathbf{u})Var(\mathbf{z_i})) + E(\mathbf{z_i})Var(\mathbf{u})E(\mathbf{z_i})^t + \sigma^2}{Var(\mathbf{y})}.$$

Estimates for the terms in the numerator should come from fitting the linear mixed model of interest. They suggest estimating the denominator from a model using only a constant intercept term and a random intercept term (obviously only relevant to two-level multilevel models with only one clustering variable), rather than using the observed sample variance of $\bar{y}$, as $\hat{\sigma_0}^2 + \hat{\tau}_0^2$ where $\hat{\sigma}_0^2$ is the estimated error variance and $\hat{\tau}_0^2$ is the estimated variance of the random intercept terms from the null model.

Snijders and Bosker assert that the primary level of interest for fixed effects inference is at the second level, i.e. the prediction of the *mean* at the second level, not the individual observation itself. To simplify ideas, the authors assume that there is the same number (v) of level-1 units within each unit of level-2. Then the best linear predictor for the mean of the response variable at the second level for a randomly selected unit j, say $\bar{y}_j$, is equal to $\bar{\mathbf{x}}_{\mathrm{j}}\boldsymbol{\beta}$, where the $\bar{\mathbf{x}}_{\mathrm{j}}$ is the average of the X_i vectors for unit j. Again conditioning on Z, one obtains that the mean squared prediction error of $\bar{\mathbf{x}}_{\mathrm{j}}\boldsymbol{\beta}$ is $Var(\bar{y}_j - \bar{\mathbf{x}}_{\mathrm{j}}\boldsymbol{\beta}|Z_{\cdot j}) = \bar{\mathbf{z}}_{\cdot \mathrm{j}} Var(\mathbf{u})\bar{\mathbf{z}}^t_{\cdot j} + \frac{\sigma^2}{v}$ (where '$\cdot j$' represents the rows in group j) because of the averaging over the level-1 units of level-2 unit j. Again, taking the expectation with respect to Z, one obtains

$$Var(\bar{y}_j - \bar{\mathbf{x}}_{\mathrm{j}}\boldsymbol{\beta}|Z_{\cdot j}) = \mathrm{tr}(Var(\mathbf{u})Var(\bar{\mathbf{z}}_{\cdot \mathrm{j}})) + E(\bar{\mathbf{z}}_{\cdot j})Var(\mathbf{u})E(\bar{\mathbf{z}}_{\cdot j})^t + \frac{\sigma^2}{v}.$$

Note that $Var(\bar{\mathbf{z}})$, the variance-covariance matrix for the column means of Z, will also depend on v here. When the data are unbalanced, i.e. there is an unequal number of level-1 units per level-2 unit, the authors suggest using either a "representative" value of v or a data-based estimate of v, such as the harmonic mean. Here, the null model is considered with respect to the level-2 predicted mean, $\bar{y}_j$, estimated under a model with only intercept terms in X and Z. The analogous null variance would then be $\hat{\tau}_0^2 + \hat{\sigma}_0^2/v$, where v is chosen as mentioned previously. This leads to their criterion for variance explained at the second level:

$$R^2_{SB-2} = (\hat{\tau}_0^2 + \hat{\sigma}_0^2/v)^{-1}[\mathrm{tr}(Var(\mathbf{u})Var(\bar{\mathbf{z}}_{\cdot j})) + E(\bar{\mathbf{z}}_{\cdot j})Var(\mathbf{u})E(\bar{\mathbf{z}}_{\cdot j})^t + \sigma^2/v)]$$

Edwards et al. (2008) and Orelien and Edwards (2008) discuss several limitations of the approaches of Snijders and Bosker (1994), including the restriction to homogeneous variance between level-2 units and an assumption that the within-person covariance matrix is $\sigma^2 I_{n_i \times n_i}$. Edwards et al. (2008) propose, for choosing amongst models with different fixed effects only, an R^2 statistic based on the relationship between R^2 and the overall model F-statistic in the multiple linear regression model. The authors note that for the multiple linear regression model, one can write the F-statistic for comparing a model with $p-1$ predictors to the null model with only a fixed effect intercept as:

$$F = \frac{R^2}{p-1} \times \frac{r}{1-R^2},$$

where R^2 is the R^2 value for the model with $p-1$ predictors and $r = n - \mathrm{rank}(X)$. Solving for R^2 gives

$$R^2 = \frac{(p-1)F/r}{1+(p-1)F/r}.$$

Edwards et al. (2008) extend this idea to the linear mixed model, using the F-statistic from a Wald test of the appropriate set of model coefficients in the above expression, yielding a multilevel R^2 statistic, referred to in the remainder as R^2_{E}. Users must choose the denominator degrees of freedom and the choice of the null model for the Wald comparison. Edwards et al. (2008) discuss the several choices available for denominator degrees of freedom for Wald tests of this form, suggesting that the Kenward–Roger degrees of freedom should be preferred. The choice of the null model is similar to the idea suggested by Snijders and Bosker (1994) in that the null model is allowed to vary depending on the analysis. However, the R^2_{E} statistic assumes a null reference model for the F-test has only an intercept column in the fixed effects design matrix X but contains the same random effects design matrix Z and assumed correlation structure for $\boldsymbol{\varepsilon}$ as the non-null model. Edwards et al. (2008) argue that, for testing for fixed effects, methods that utilize a null model without the same random effects and correlation structure as the target model cannot separate an increase in explanatory power that is due solely to the fixed effects from an increase due to changing the covariance structure of $\mathbf{y}$.

7.4.2 Deviance-Based Measures of Fit

As in the previous subsection, certain challenges arise when attempting to modify the usual multiple linear regression deviance based criteria for use in linear mixed models. The first challenge that arises is the choice of likelihood. The most commonly used choice is the marginal likelihood of the data, with the random effects parameters integrated out. For the linear mixed model (7.1), the marginal likelihood can be written as

$$LL_{\mathrm{MARG}}(\boldsymbol{\beta}, \Sigma) = -\frac{1}{2}(\mathbf{y} - X\boldsymbol{\beta})^t (\Sigma)^{-1}(\mathbf{y} - X\boldsymbol{\beta}) - \frac{n}{2}\log(2\pi) - \frac{1}{2}\log|\Sigma|, \quad (7.2)$$

where $\Sigma = Var(\varepsilon) + ZVar(\mathbf{u})Z^t$.

Harville (1974) instead suggested the use of the restricted (or error) likelihood to correct for the underestimation of parameters in Σ, as discussed in Chapter 3. The SAS statistical package's PROC MIXED (SAS Institute Inc., 2008) uses a different REML likelihood from Harville (1974):

$$LL_{\mathrm{REML}}(\Sigma) = -\frac{1}{2}(\mathbf{y} - X\boldsymbol{\beta}(\Sigma))^t(\Sigma)^{-1}(\mathbf{y} - X\boldsymbol{\beta}(\Sigma)) - \frac{n-p}{2}\log(2\pi) - \frac{1}{2}\log|\Sigma| - \frac{1}{2}\log|X^{'}\Sigma X|, \quad (7.3)$$

where now the original fixed effects coefficients, $\boldsymbol{\beta}$, have been eliminated from the formulation, instead written as functions of the parameters of Σ, $\boldsymbol{\beta}(\Sigma) = [X'(\Sigma)^{-1}X]^{-1}X'(\Sigma)^{-1}\mathbf{y}$.

Although it performs better for estimation of the underlying parameters of Σ, using the REML likelihood to compare models with different fixed effects structures is awkward, because different fixed effect design matrices correspond to different constraints on the ys. It's also important to note that for the REML likelihood we only have $n - p$ linearly independent observations and the number of free parameters is smaller because we have eliminated the free parameters for the p fixed effect coefficients. Therefore, generally the recommendation is that one should not compare REML likelihoods with different fixed effects (see, for example, Wolfinger, 1993). However, Gurka (2006) does compare models with different REML likelihoods using a Bayesian justification. The argument uses a result from Harville (Harville, 1974) that expresses the REML likelihood as a Bayesian marginal likelihood where the fixed effects parameters have been integrated out with respect to a particular, data-dependent prior distribution. Therefore, the reader should understand two things. First, comparing REML likelihoods with different choices of fixed effects cannot be well justified from a non-Bayesian perspective and is highly non-standard. Second, even if one does so, it is only under a particular choice of a data-dependent prior for the fixed effects parameters, which may or may not be reasonable from the perspective of the researcher. Finally, one can never compare ML values from one model to REML values of another.

Vaida and Blanchard (2005) posited that marginalizing the likelihood (either Eq. (7.2) or Eq. (7.3)) was not appropriate for inference about cluster-specific parameters (such as the random effect parameters). Instead, they advocated the use of the conditional likelihood:

$$LL_{\mathrm{COND}}(\boldsymbol{\beta}, u, \sigma^2) = -\frac{n}{2}\log(2\pi) - \frac{1}{2}\sum_{i=1}^{I}\log|V| - \frac{1}{2}(\mathbf{y} - [X\boldsymbol{\beta} + Z\mathbf{u}])^t (V)^{-1} \times (\mathbf{y} - [X\boldsymbol{\beta} + Z\mathbf{u}]),$$

where V is the assumed covariance matrix for $\boldsymbol{\varepsilon}$. Their motivation derives from similar ideas in Snijders and Bosker (1994) in that they differentiate between different levels of prediction. Vaida and Blanchard (2005) suggest that the choice is between population prediction (in which the marginal likelihood is of interest) or cluster-level prediction (in which the conditional likelihood is of interest).

The marginal likelihood version of Akaike's original version of the AIC for a particular model m is

$$AIC_{\text{MARG}} = -2LL_{\text{MARG}}(\hat{\boldsymbol{\beta}}, \hat{\Sigma}) + 2(p+k),$$

and the REML version is

$$AIC_{\text{REML}} = -2LL_{\text{REML}}(\hat{\Sigma}) + 2k,$$

where k is the number of parameters to be estimated in Σ. For example, in a model with only random intercept terms (i.e. Z is a column of 1s) and a diagonal covariance matrix for $\boldsymbol{\varepsilon}$, k is equal to 2. Because most major software packages use REML as the default likelihood approach, it is likely that most researchers are using the REML form of the AIC. However, the issue is far from settled (Gurka, 2006).

Vaida and Blanchard (2005) instead proposed an alternative version of the AIC based on the conditional likelihood and called it the Conditional AIC, referred to here as the CAIC. Note that the abbreviation CAIC is also used for the consistent AIC (Bozdogan, 1987), a different criterion that has appeared in the literature. Counting the number of free parameters in the conditional likelihood becomes subtly difficult. Clearly, the parameters $\mathbf{u}$ are not similar in kind to either the p fixed effect parameters or the k parameters that are used in the construction of Σ, as the random effect parameters are not completely "free". With an assumption of a normal distribution on the $\mathbf{u}$, these parameters are restricted in some sense and this restriction makes typical asymptotic arguments difficult (for a more detailed explanation, see Pauler, 1998). A related theoretical issue is that these are not population-level parameters, so the usual asymptotics will not apply.

Vaida and Blanchard (2005) use an estimate of the number of effective degrees of freedom of the mean model from Hodges and Sargent (2001), $\rho = \text{tr}(H_1)$, where H_1 is the matrix that projects the observed data into the space of fitted values from the linear mixed model. The approximate form of the conditional AIC criterion of Vaida and Blanchard (2005), then, is:

$$CAIC = -2LL_{\text{COND}}(\hat{\boldsymbol{\beta}}, \hat{\mathbf{u}}, \hat{\sigma}^2) + 2(\rho + 1),$$

where ρ is an estimate of the number of parameters for the mean structure of the data and 1 is added for the error variance parameter. Note that Vaida and Blanchard (2005) derived their criterion under the assumption that $Var(\mathbf{u})$ is known (which is why parameters in $Var(\mathbf{u})$ do not appear in the penalty). Unfortunately, in most cases this would need to be estimated. The authors state that deriving the correct value for ρ for estimated $Var(\mathbf{u})$ would be extremely complicated and give arguments for why plugging an estimate into the calculation of H_1 would yield reasonable estimates. Greven and Kneib (2010) show that a correction to the conditional AIC proposed by Liang et al. (2008) fixes the problem, but point out that the correction is computationally expensive. Greven and Kneib (2010) propose a different correction that is less computationally expensive and performs well in simulated data examples, and we use their correction in our examples.

The Bayesian Information Criterion (BIC) provides an obvious alternative to the AIC for mixed models. Note that the BIC and AIC both use the log-likelihood of the model, merely with different penalties ($2p$ for the AIC versus $p \log n$ for the BIC). So, one must also make a decision about which of the different versions of the likelihood (marginal, REML, or conditional) to use with the BIC. The critical difference between the AIC and BIC with regards to the likelihood is one of interpretation. Recall that model selection via the BIC can be justified as an approximation to model selection using posterior probabilities of the model. Versions of the BIC using both the marginal and REML likelihoods can be justified with respect to Laplace approximations, with similar accounting of parameters, for choosing amongst fixed effect models with a common random effects structure, assuming that the maximum likelihood estimates for all parameters are not along the boundary (for example, random effects variance components suitably far from zero). Issues with regards to counting parameters for the BIC are similar to those when using the AIC.

The key additional difficulty in formulating the BIC for mixed models is choosing the correct sample size for the $\log n$. Wolfinger (1993) uses the number of unconstrained observed values, n or $n - p$, in his description of using the BIC for both marginal and REML log-likelihood model selection. However, Pauler (1998) argues that the penalty should correspond to the order of the determinant of the Fisher information for the parameter vector of interest in an exponential family model, in accordance with the original work by Schwarz (1978). However, the order of the determinant of the Fisher information will not necessarily be $p \log n$ for linear mixed models, so the approximation may not be accurate. Kass and Raftery (1995) also note that the choice of sample size for general uses of the Schwarz approximation to the integrated likelihood can be difficult.

Pauler (1998) instead developed a version of the BIC for selecting amongst mixed models in the presence of random effects in the common situation where $\mathbf{y}$ can be decomposed into J $(n_j \times 1)$ sub-vectors, $\mathbf{y_j}$, which are mutually independent (for example, n_j repeated measurements on the jth subject for J independently sampled subjects or clusters). Her adjustment to the BIC was justified via a Laplace approximation and a particular form of prior distribution (a modification of the unit-information prior of Kass and Wasserman (1995)). The basic premise was that the fixed effect parameters could be divided into two groups: those for whom the respective columns of X were proportional to some column of Z, and those where this was not the case, i.e. fixed effects with associated random effects and fixed effects with no associated random effects. Pauler proves that, under certain regularity assumptions, the BIC penalty should be $\sum_{j=1}^{p+1} \log(N_j)$, where $N_j = J$ if a fixed effect has an associated random effect and $N_k = n$ if it does not. In the case of multiple linear regression with no random effects, the penalty term then reduces to the usual $p \log n$. Pauler et al. (1999) proposed a more complex version of the BIC to specifically address the problem of selecting amongst models with differing choices for Z, but we do not have space to discuss it here.

7.5 ILLUSTRATION IN TWO APPLICATIONS

We now examine several of the model selection approaches discussed previously in the context of data from the Canadian Scleroderma Research Group cohort. Two potential uses of the registry data were discussed in the introduction. The first was a cross-sectional analysis that potentially needed to adjust for between-subject correlation within different recruiting centers. The second described a problem in longitudinal data analysis that would require the adjustment of within-subject correlation in addition to the subjects being nested within centers. We will look at each problem in detail in the next two subsections.

7.5.1 Cross-Sectional Analysis of the Association Between Disease Activity and Physical Function

In this cross-sectional analysis, we examine the relationship of patient disease activity with patient physical function, adjusting for potential confounding and for possible clustering of the association by recruiting center. The primary outcome measure for this analysis is the baseline measurement of the Health Assessment Questionnaire Disability Index (HAQ), which ranges from 0 to 3. Although the scores themselves are restricted in range, we will use linear and linear mixed models with normal errors for illustration.

The primary covariate of interest is the Valentini Scleroderma Disease Activity Index (SDAI). We also adjust for potential confounding variables, i.e. variables that

Table 7.1 Parameter Estimates for the 18 Models for the Cross-Sectional Data. Standard errors are in ()s

Model		Fixed Effect Coefficients				$\sqrt{\text{Var. Comp.}}$		
		Intercept	SDAI	Duration	Duration: SDAI	Error	Intercept	SDAI
A1	ML	0.34 (0.14)	NA	−0.002 (0.003)	NA	0.66	NA	NA
	REML	0.34 (0.14)	NA	−0.002 (0.003)	NA	0.66	NA	NA
A2	ML	0.34 (0.15)	NA	−0.003 (0.003)	NA	0.66	0.08	NA
	REML	0.34 (0.15)	NA	−0.003 (0.003)	NA	0.66	0.09	NA
A3	ML	0.40 (0.14)	NA	−0.0008 (0.003)	NA	0.62	0.37	0.14
	REML	0.40 (0.14)	NA	−0.0009 (0.003)	NA	0.62	0.38	0.14
B1	ML	0.03 (0.14)	0.14 (0.02)	−0.0002 (0.003)	NA	0.63	NA	NA
	REML	0.03 (0.14)	0.14 (0.02)	−0.0002 (0.003)	NA	0.63	NA	NA
B2	ML	0.03 (0.15)	0.14 (0.02)	−0.0006 (0.003)	NA	0.62	0.10	NA
	REML	0.04 (0.15)	0.14 (0.02)	−0.0005 (0.003)	NA	0.63	0.11	NA
B3	ML	0.04 (0.15)	0.14 (0.02)	−0.0004 (0.003)	NA	0.62	0.12	0.03
	REML	0.05 (0.15)	0.14 (0.02)	−0.0003 (0.003)	NA	0.62	0.14	0.03
C1	ML	−0.05 (0.15)	0.18 (0.03)	0.007 (0.006)	−0.004 (0.002)	0.63	NA	NA
	REML	−0.05 (0.15)	0.18 (0.03)	0.007 (0.006)	−0.004 (0.002)	0.63	NA	NA
C2	ML	−0.05 (0.15)	0.18 (0.03)	0.008 (0.005)	−0.004 (0.002)	0.62	0.10	0.62
	REML	−0.04 (0.15)	0.18 (0.03)	0.007 (0.005)	−0.004 (0.002)	0.62	0.11	NA
C3	ML	−0.04 (0.15)	0.18 (0.03)	0.007 (0.005)	−0.004 (0.002)	0.62	0.11	0.01
	REML	−0.03 (0.16)	0.18 (0.03)	0.007 (0.005)	−0.004 (0.002)	0.62	0.13	0.02

could be considered causes of both the level of disease activity and patient physical function. We only consider subject-level confounding variables in this analysis and the covariates we adjust for include gender, age, and the scleroderma classification of the patient (limited vs diffuse disease) and disease duration.

With respect to the clustering of the data, patient data is collected from 13 recruiting centers. It is possible that there is heterogeneity in not only overall levels of function between centers, but also differences in the association between disease activity and function amongst centers.

All of these substantive considerations lead to a variety in the models that could be considered. We will consider the following fixed effects structures for the data:

(A) Null model: Adjusting only for potential confounders (gender, age, disease duration, disease type);
(B) All confounders from the null model including the SDAI as a covariate; and
(C) All confounders from the null model and including an interaction between SDAI and disease duration (along with the main effects).

We also consider the following random effects design matrices:

1 No random effects terms, i.e. fitted multiple linear regression models ignoring clustering by center;
2 Random intercept, clustering by center;
3 Random intercept and slope for disease activity again with clustering by center, assuming an unstructured covariance matrix for the random effect pairs (i.e. giving three covariance parameters).

We will refer to models by the letter that indicates their fixed effect structure (A, B, or C) and a number for their assumed random effects structure (1, 2, or 3). For example, we will refer to the model with potential confounders, SDAI with no interaction term, and only random intercepts by center as model B1. Summaries of the parameter estimates for all models using both marginal likelihood and REML likelihood are contained in Table 7.1.

Tables 7.2 and 7.3 contain the values of the criteria for both versions of REML and the marginal likelihood for all models and criteria. First we assess the model fit for all of the

Table 7.2 Estimates of Percentage Variance Explained According to Various Approaches for the Cross-Sectional Data Example. Maximum values for R^2 per row are in bold

	A1	A2	A3	B1	B2	B3	C1	C2	C3
$R^2_{\text{SB}-1}$	5.4	5.8	4.7	13.7	13.6	13.6	14.2	14.1	**14.2**
R^2_{E}	5.4	21.5	17.9	13.7	32.3	**52.6**	14.2	27.7	51.1
$R^2_{\text{SB}-2}$	NA	**−2.7**	−182.7	NA	−10.3	−14.9	NA	−9.9	−11.1

Table 7.3 Deviance-Based Criteria Results for the Cross-Sectional Data Example. Optimal values per row are in bold. Note that 1000 has been subtracted from each of the criterion values

		A1	A2	A3	B1	B2	B3	C1	C2	C3
AIC	Marginal LL	86.39	86.28	52.49	39.62	36.82	40.59	38.26	**35.50**	39.44
	REML	108.3	107.54	73.95	66.05	**62.34**	65.86	73.00	69.33	73.15
AIC_{VB}	Cond. LL	86.39	84.23	37.81	39.62	32.82	37.74	38.26	**31.49**	41.72
BIC	Marginal LL	112.08	116.26	91.03	**69.60**	71.08	83.41	72.52	74.04	86.54
	REML-2	112.57	116.08	91.04	**70.31**	70.88	82.94	77.27	77.86	90.23
BIC_{Paul}	Marginal LL	112.08	112.54	78.66	69.60	**67.36**	71.05	72.52	70.32	74.18
	REML-1	75.62	75.42	42.96	26.39	23.25	27.90	22.16	**19.04**	24.00
	REML-2	112.57	112.38	79.92	70.31	**67.17**	71.83	77.27	74.16	79.11

models using R^2. The R^2 criteria act quite differently than what is normally observed for R^2 criteria in the usual linear models setting. Note that R^2 can decrease as fixed and/or random effects are added to models. The $R^2_{\text{SB}-1}$ and R^2_{E} criteria all ostensibly measure the extent to which the fixed effects are explaining the variability of the outcome measure, although the interpretations of the criteria vary. Both R^2 measures indicate better fits for the models with SDAI or the SDAI × Duration interaction than for the models without SDAI at all. The R^2_{E} seems to yield the most interpretable values of the R^2 criterion in this case. Recall that the Edwards criterion measures the extent to which the fixed effects explain variance conditional on a particular random effects structure. As noted previously, R^2_{E} indicates a better fit for Model B2 over A2 and C2, but this general pattern holds for the random slope model as well (Model B3 is preferred to Models C3 and A3). The $R^2_{\text{SB}-2}$ criterion is negative for all models, which is likely due to the small amount of between-center variability in intercepts and slopes.

We will now step readers through potential strategies that could be used to choose models via the two deviance-based criteria. Diggle (1988) proposed a three-step procedure that relied on the likelihood ratio test for selecting models in a repeated measures setting. The first step requires a preliminary selection of the fixed effects design matrix using standard multiple linear regression. The recommendation was to choose a model that was deliberately less parsimonious in order to avoid any problems with potential confounding. Then, using the residuals from the ordinary least squares analysis, one chooses a form for the random effect design matrices. Finally, formal testing proceeded using generalized likelihood-ratio tests for the fixed effect design matrix and maximum likelihood estimation for all model parameters. Wolfinger (1993) extended this strategy to the general mixed model using model selection criteria.

First, we use a standard multiple linear regression model (ignoring the correlation within centers) to choose a suitably complex provisional fixed effects mean structure. Given our particular hypotheses of interest, we will simply use fixed effects structure C. Next, a REML model selection approach is used to choose amongst potential covariance structures for the data, conditional on the provisional mean structure, i.e. we compare models C1, C2, and C3. It's important to remember that the usual likelihood calculation for model C1 done by standard statistical packages would not be applicable here if we are comparing likelihood-based criteria for model C1 to the REML criterion of C2 and C3. We must, instead, calculate the REML likelihood for model C1, which contains additional terms not included in the standard linear model likelihood. If we look at the REML criteria for these three models in Table 7.3, we see that the random intercept model (C2) is preferred if we use the AIC based on REML, but the fixed effect only model (C1) is preferred if we use the usual REML BIC, due to the small differences in likelihood amongst all three models and the severity of the penalty for the extra parameters of the random effects models. It is perhaps a bit surprising that the REML likelihoods between models C2 and C3 are not dramatically different, given the non-negligible size of the random intercept standard deviation (0.11) relative to the overall variability in the data (for example, the standard deviation over all outcomes is 0.67). In fact, most researchers would use the random intercept model because it corresponds to their notions of heterogeneity amongst different recruiting centers, even if there is only modest evidence that the center variable is important to the model. We see here the challenge in choosing random effects structures according to a mixture of objective procedures and subjective external knowledge about the data.

Returning to the Diggle–Wolfinger (DW) approach, assume for now that we choose the random intercept model because it has the best value of the AIC amongst C1, C2, and C3 and it corresponds to our prior intuition about the clustering. Now, we select an optimal model amongst the various fixed effects models with a random intercept structure (A2, B2, and C2). The DW prescription at this point is to use marginal maximum likelihood criteria (rather than REML). The standard AIC values for this problem (based now on the marginal likelihood model) are 1086.4, 1036.8, and 1035.5 respectively, which suggests that B2 and C2 are significantly better models than A2, with C2 as either slightly preferred to B2 or not conclusively better (depending on one's interpretation of the differences in AIC). The AIC values based on the REML likelihood (using the number of parameters equal to the number of covariance parameters) are 1107.5, 1062.3, and 1069.3, indicating clear evidence in favor of model B2 over the other two models. We see in this example, as in Gurka (2006), that the REML and marginal likelihood analyses can lead to different orderings of models.

The ordering of the conditional AIC values (Greven and Kneib, 2010) give similar inference to that of the marginal likelihood AICs, with values of 1084.06, 1032.71, and 1031.38. The estimated number of parameters used as the penalty in the conditional AIC calculation with the maximum likelihood estimates were 9.7, 12.1, and 13.2. We see that these are slightly higher than what would be expected from naive counting of fixed effect and covariance parameters (7, 8, and 10 respectively).

The naive calculation BIC using the marginal likelihood for B2 (1071.1) is the best of the three models, although the value for C2 is close (1074.0), compared to a much worse value for A2 (1116.3). The Pauler (1998) versions of the BIC within this comparison do not change the model preference, because the number of random effect parameters is the same across models A2, B2, and C2.

To summarize the deviance criteria results in the DW framework, we note that our AIC and BIC results track the ambiguous conclusion of the Wald test between models B2 and C2 for the significance of the interaction term in C2 (which, using a hypothesis test, yields a p-value of 0.068). All versions of the AIC and BIC clearly differentiate A2 as a suboptimal model, but disagree about preferring either B2 or C2 to the other.

Note that the DW approach avoids some of the more delicate model selection issues that were discussed earlier. For example, the Pauler (1998) version of the BIC (with different penalties for fixed effect terms with and without associated random effects) yields the same inference as the standard version because the random effects terms are fixed in the model. In fact, most of the issues with the marginal criteria discussed so far are rendered moot by conditioning on the structure of the random effects. However, particularly with respect to the deviance-based criteria, there is no need to follow the DW prescription and one can instead optimize over all nine models simultaneously.

We see some heterogeneity in model selection optimizing deviance-based criteria across all models, depending on which versions are used, and even within version, depending on which likelihood is used. The standard AIC measures using the marginal likelihood favors Model C2 (including SDAI fixed effect and random intercepts and slopes), whereas the AIC utilizing the REML likelihood favors Model B2 (including SDAI and only random intercepts). The Conditional AIC of Greven and Kneib chooses the Model C2 with random intercepts and slopes and the interaction between the SDAI and disease duration.

The BIC criteria give quite different pictures of what the preferred model should be. The standard version of the BIC with the marginal likelihood or REML actually prefers Model B1 (SDAI slope, no random effects), gravely penalizing the additional parameters needed to model the random effects. The Pauler (1998) version of the BIC, however, penalizes the random effects terms significantly less (only log(13) instead of log(535) in the marginal likelihood case), so that it chooses Model B2 for marginal and REML likelihoods.

Taking the example as a whole and summarizing over all choices of criterion and likelihood, it appears that the preferred model in this case would indeed be Model B2, with modest evidence in favor of Model C2.

7.5.2 Longitudinal Analysis of Physical Function

We now present model selection results for the longitudinal physical function data presented in Section 7.2. Recall that the goal of the longitudinal analysis is to characterize the progression of disability in scleroderma patients over time using the annual visit data. We again consider nine models, with different fixed effects structures denoted by A, B, and C:

(A) Adjusting only for baseline HAQ score;
(B) Allowing for a trend in time (in number of visits) and adjusting for baseline HAQ DI score;
(C) Allowing for an interaction between time (number of visits) and Baseline HAQ DI score;

and random effects design matrices:

1 No random effects terms, i.e. fitted multiple linear regression models ignoring clustering by patient;
2 Random intercept, clustering by patient;
3 Random intercept and slope of time with clustering by patient, assuming an unstructured covariance matrix for the random effect pairs (i.e. giving three covariance parameters).

It's important to emphasize that the longitudinal analysis is of a very different nature than the cross-sectional analysis of the previous subsection. In particular, the cross-sectional data contained a small number of clusters (13 centers) with a moderate to large number of observations per cluster. In contrast, the longitudinal data contains a large number of clusters (409 patients), but only a small number of observations per subject (an average of approximately three annual follow-up visits per patient, with a maximum of six). Table 7.4 summarizes the parameter estimates across the nine models of interest, both for the ML and REML parameterizations.

Table 7.5 summarizes the results of the goodness of fit criteria for the nine longitudinal models of interest. In general, we see that much higher percentages of the variability in longitudinal HAQ scores between patients are explained by the models compared to the cross-sectional example. With respect to the choice of fixed effects, the $R^2_{\mathrm{SB}-1}$ criterion only eliminates Model A1 (no temporal adjustment and no random effects terms) from consideration. Otherwise, it indicates that the other eight models explain extremely similar amounts of variability in the response.

Table 7.4 Parameter Estimates for the 18 Models for the Longitudinal Data. Standard errors are in ()s

		Fixed Effect Coefficients				$\sqrt{\text{Var. Comp.}}$		
Model		Intercept	Base HAQ	Time	Time: Base HAQ	Error	Intercept	Time
A1	ML	0.20 (0.02)	0.81 (0.02)	NA	NA	0.46	NA	NA
	REML	0.20 (0.02)	0.81 (0.02)	NA	NA	0.46	NA	NA
A2	ML	0.20 (0.03)	0.82 (0.03)	NA	NA	0.29	0.35	NA
	REML	0.20 (0.03)	0.82 (0.03)	NA	NA	0.29	0.35	NA
A3	ML	0.16 (0.03)	0.83 (0.03)	NA	NA	0.26	0.34	0.09
	REML	0.16 (0.03)	0.83 (0.03)	NA	NA	0.26	0.34	0.09
B1	ML	0.13 (0.04)	0.81 (0.02)	0.020 (0.009)	NA	0.46	NA	NA
	REML	0.13 (0.04)	0.81 (0.02)	0.020 (0.009)	NA	0.46	NA	NA
B2	ML	0.09 (0.04)	0.82 (0.03)	0.032 (0.006)	NA	0.29	0.36	NA
	REML	0.09 (0.04)	0.82 (0.03)	0.032 (0.006)	NA	0.29	0.36	NA
B3	ML	0.06 (0.03)	0.83 (0.03)	0.039 (0.007)	NA	0.26	0.32	0.09
	REML	0.06 (0.03)	0.83 (0.03)	0.039 (0.007)	NA	0.26	0.32	0.09
C1	ML	0.07 (0.05)	0.89 (0.05)	0.037 (0.013)	−0.022 (0.013)	0.46	NA	NA
	REML	0.07 (0.05)	0.89 (0.05)	0.037 (0.013)	−0.022 (0.013)	0.46	NA	NA
C2	ML	0.05 (0.04)	0.87 (0.04)	0.043 (0.009)	−0.016 (0.009)	0.29	0.35	NA
	REML	0.05 (0.04)	0.87 (0.04)	0.043 (0.009)	−0.016 (0.009)	0.29	0.36	NA
C3	ML	0.04 (0.04)	0.85 (0.04)	0.045 (0.011)	−0.009 (0.011)	0.26	0.32	0.09
	REML	0.04 (0.04)	0.85 (0.04)	0.045 (0.011)	−0.009 (0.011)	0.26	0.32	0.09

The R^2_{E} criterion, at first glance, performs somewhat erratically, with a significant decrease in explanatory value of models B2 and C2 relative to the other models. In particular, models B2 and C2 suffer in comparison with model A2, which does not allow for a change in disability over time at either the fixed or random effect level. A closer examination showed that this decrease is completely due to the use of the recommended Kenward–Roger degrees of freedom for calculating R^2. The estimated denominator degrees of freedom for models A2 and A3 are 520.1 and 509.0, respectively. However, for models B2 and C2, the estimated denominator degrees of freedom are 1019.7 and 1630.5. The larger estimated degrees of freedom for these two models drives down the value of the Edwards version of the R^2. The estimated denominator degrees of freedom for random slope models B3 and C3 are 559.8 and 557.6, respectively, which is why the Edwards R^2 values for those two models are on a par with the value for model A2.

The $R^2_{\text{SB}-2}$ criterion indicates that a great deal of level-2 variability is being explained by the models with random effects, even though the reference model in this case is the same model used for the Xu criterion. However, similar to the $R^2_{\text{SB}-1}$ criterion, no model is clearly favored amongst the six random effects models, although the models with random intercepts only seem to perform slightly better than the models with random slopes and intercepts.

The results for the various deviance-based criteria are given in Table 7.6. Unlike the cross-sectional data example, the versions of the AIC and BIC agree fairly convincingly on the ordering of the models. All versions of the AIC and BIC, regardless of the choice of likelihood, substantially prefer models with random slopes and intercepts (i.e. models A3, B3, and C3) compared to models either without random effects or with only a random intercept. The criteria based on the marginal likelihood and the REML-2 likelihood lead to a choice of model B3, which has fixed effects for the baseline HAQ DI and Time, but no interaction. For the AIC criteria, the choice of B3 over C3 is less conclusive, due to less of a penalty for the extra fixed effect parameter

Table 7.5 Estimates of Percentage Variance Explained According to Various Approaches for the Longitudinal Data Example. Maximum values for R^2 per row are in bold

	A1	A2	A3	B1	B2	B3	C1	C2	C3
$R^2_{\text{SB}-1}$	45.8	59.4	58.2	58.2	59.5	58.6	58.3	**59.6**	58.6
R^2_{E}	45.8	62.1	**65.5**	58.2	46.2	64.1	58.3	35.0	64.5
$R^2_{\text{SB}-2}$	NA	**66.8**	62.7	NA	66.6	63.1	NA	66.6	63.2

Table 7.6 Deviance-Based Criteria Results for the Longitudinal Data Example. Optimal values per row are in bold. Note that 1150 has been subtracted from each of the criterion values

		A1	A2	A3	B1	B2	B3	C1	C2	C3
AIC	Marginal LL	806.1	170.5	63.29	802.9	144.7	**38.78**	802.0	143.5	40.14
	REML	815.4	177.9	70.91	817.8	158.5	**52.38**	821.7	162.9	58.88
AIC_{VB}	Cond. LL	806.1	–247.7	–480.1	802.9	–285.5	–502.0	802.0	–287.0	– **500.2**
BIC	Marginal LL	822.1	191.8	95.28	824.2	171.3	**76.11**	828.8	175.5	82.81
	REML	820.8	188.6	92.2	823.1	169.1	**73.7**	827.1	173.6	80.2
BIC_{Paul}	Marginal LL	822.1	190.6	91.41	824.2	170.0	**72.24**	828.8	174.2	78.94
	REML	820.8	187.3	88.37	823.1	167.8	69.84	827.1	172.3	76.34

compared to the BIC criteria. The conditional AIC criterion very strongly favors Model C3 to all others.

The lack of substantial differences amongst versions of the criteria for this example is most likely driven by two factors. First, there is a large number of independent clusters in this example (409 patients compared to only 13 centers in the cross-sectional data). The larger sample size likely leads to more stable performance from the AIC and BIC criteria, due to the fact that both have their basis in asymptotic approximations that are more likely to hold with a large number of independent clusters. The other reason the picture is quite different in this example is the extent of between-subject heterogeneity in both mean level of disability and the slope. In the cross-sectional example, there was very little estimated heterogeneity between clusters, which meant that random effects variance and covariance parameters were often very near zero, which again calls into question the validity of the basic asymptotic approximations. The longitudinal data showed much clearer evidence of heterogeneity, and so the parameter estimates were further from the boundary and the asymptotic basis for the usual criteria are more likely to hold. Note that in this problem the DW approach would have led to the same conclusions, as the random intercept and random slope models were so clearly favored by all of the deviance-based criteria for the most complex fixed effect models (C1, C2, and C3).

7.6 DISCUSSION

Model selection for multilevel modeling is materially more challenging than in the usual linear model context. The most important challenges result from ambiguity in choosing amongst marginal or conditional likelihoods, potential null reference models, and seemingly reasonable choices for both the number of model parameters and the sample size (or error degrees of freedom). The different versions of standard criteria presented in this chapter address one or more of these concerns. The two examples represented two typical extremes of the usual multilevel model analyses. The cross-sectional example contained data on a small number of independent clusters containing a moderate to large number of repeated observations. The longitudinal example, however, provided a context with a large number of independent clusters with a small number of repeated observations per cluster. Considerably greater heterogeneity of model selection across criteria was observed in the cross-sectional example, due to both the small number of independent clusters and the smaller extent of the heterogeneity between clusters.

Vaida and Blanchard (2005) discuss the relationship between the effective number of degrees of freedom proposed and the estimate used in the DIC criterion of Spiegelhalter et al. (2002), pointing out that there is a close relationship between the two. In fact, worked examples for the linear mixed model in Spiegelhalter et al. (2002) and Greven and Kneib (2010) indicate that these effective numbers of degrees would be exactly the same in the case of a known variance-covariance matrix for the model residuals, known random effects variance parameters, and uniform priors on the fixed effects coefficients. The estimates of the effective dimensionality would possibly differ under informative priors for the fixed effects coefficients and very likely would differ in the case of unknown model variance and covariance parameters. In the case of the DIC, the estimates of effective dimensionality of the model will depend on the choice of prior.

As noted by Gurka (2006), one of the likely reasons that there do not exist well-established linear mixed model selection approaches is the inconsistency with which familiar criteria are calculated by the most-used statistical packages. His comments related particularly to deviance-based criteria within SAS, although the same issues arise with Stata (StataCorp, 2009), SPSS (SPSS Inc., 2010), and the R statistical package (R Development Core Team, 2011). No package calculates all of the criteria discussed in this paper, and even the criteria that are able to be obtained are often not calculated in the same way. For example, the lme4 library in the R statistical package includes the fixed effect parameters in the AIC and BIC penalties for the REML likelihood, whereas SAS PROC MIXED does not. At the time of writing there does not currently exist a validated R package to calculate Kenward–Roger degrees of freedom. However, progress is being made in this direction. Note that software for mixed models is also addressed in Chapter 26 of this text.

We also note that in the last 15 years, penalized regression approaches have become very popular for performing stepwise model selection in linear mixed models.

Rather than favor one particular approach to variable selection in the multilevel model context, practitioners should instead use different approaches and compare the sets of best models according to a variety of criteria. When presenting results for multilevel analyses, it is likely to be much more informative to examine how relevant substantive conclusions change depending on the model selected rather than to select a single model which is considered optimal according to a single version of a single criterion.

REFERENCES

Akaike, H. (1973) 'Information Theory and an Extension of the Maximum Likelihood Principle' in *Second International Symposium on Information Theory*, Volume 1. New York: Springer Verlag, 267–81.

Bozdogan, H. (1987) 'Model Selection and Akaike's Criterion (AIC): The General Theory and Analytical Extensions', *Psychometrika*, 52: 345–70.

Bruce, B. and Fries, J.F. (2003) 'The Stanford Health Assessment Questionnaire: A Review of Its History, Issues, Progress, and Documentation', *The Journal of Rheumatology*, 30: 167–78.

Burnham, K.P. and Anderson, D.R. (2002) *Model Selection and Multimodel Inference: A Practical Information-theoretic Approach*. New York: Springer.

Della Rossa, A., Valentini, G., Bombardieri, S., Bencivelli, W., Silman, A.J., D'Angelo, S., Cerinic, M.M., Belch, J.F., Black, C.M., Becvar, R., Bruhlman, P., Cozzi, F., Czirják, L., Drosos, A.A., Dziankowska, B., Ferri, C., Gabrielli, A., Giacomelli, R., Hayem, G., Inanc, M., McHugh, N.J., Nielsen, H., Scorza, R., Tirri, E., van den Hoogen, F.H., and Vlachoyiannopoulos, P.J. (2001) 'European Multicentre Study to Define Disease Activity Criteria for Systemic Sclerosis. II. Identification of Disease Activity Variables and Development of Preliminary Activity Indexes', *Annals of the Rheumatic Diseases*, 60: 592–8.

Diggle, P.J. (1988) 'An Approach to The Analysis of Repeated Measurements', *Biometrics*, 44: 959–71.

Edwards, L.J., Muller, K.E., Wolfinger, R.D., Qaqish, B.F., and Schabenberger, O. (2008) 'An R^2 Statistic for Fixed Effects in The Linear Mixed Model', *Statistics in Medicine*, 27: 6137–57.

Greven, S. and Kneib, T. (2010) 'On the Behaviour of Marginal and Conditional AIC in Linear Mixed Models', *Biometrika*, 97: 773–89.

Gurka, M.J. (2006) 'Selecting the Best Linear Mixed Model Under REML', *The American Statistician*, 60: 19–26.

Harville, D. (1974) 'Bayesian Inference for Variance Components Using Only Error Contrasts', *Biometrika*, 61: 383–5.

Hodges, J.S. and Sargent, D.J. (2001) 'Counting Degrees of Freedom in Hierarchical And Other Richly-parameterised Models', *Biometrika*, 88: 367–79.

Kass, R. and Raftery, A. (1995) 'Bayes Factors', *Journal of the American Statistical Association*, 90: 773–95.

Kass, R. and Wasserman, L. (1995) 'A Reference Bayesian Test for Nested Hypotheses and Its Relationship to the Schwarz Criterion', *Journal of the American Statistical Association*, 90: 928–34.

Liang, H., Wu, H., and Zou, G. (2008) 'A Note on Conditional AIC for Linear Mixed-Effects Models', *Biometrika*, 95: 773–8.

Montgomery, D., Peck, E., Vining, G., and Vining, J. (2001) *Introduction to Linear Regression Analysis*. New York: Wiley.

Orelien, J. and Edwards, L. (2008) 'Fixed-effect Variable Selection in Linear Mixed Models Using R2R2 Statistics', *Computational Statistics and Data Analysis*, 52: 1896–907.

Pauler, D. (1998) 'The Schwarz Criterion and Related Methods for Normal Linear Models', *Biometrika*, 85: 13–27.

Pauler, D., Wakefield, J., and Kass, R. (1999) 'Bayes Factors and Approximations for Variance Component Models', *Journal of the American Statistical Association*, 94: 1242–53.

R Development Core Team (2011) *R: A Language And Environment For Statistical Computing.*

SAS Institute Inc. (2008) *SAS/STAT 9.2 User's Guide*. Cady, NC: SAS Institute Inc.

Schnitzer, M., Hudson, M., Baron, M., and Steele, R. (2011) 'Disability in Systemic Sclerosis – A Longitudinal Observational Study', *The Journal of Rheumatology*, 38: 685–92.

Schwarz, G. (1978) 'Estimating the Dimension of a Model', *The Annals of Statistics*, 6: 461–4.

Snijders, T.A.B. and Bosker, R.J. (1994) 'Modeled Variance in Two-Level Models', *Sociological Methods and Research*, 22: 342–63.

Spiegelhalter, D.J., Best, N.G., Carlin, B.P., and van der Linde, A. (2002) 'Bayesian Measures Of Model Complexity And Fit', *Journal of the Royal Statistical Society: Series B*, 64: 583–639.

SPSS Inc. (2010) *PASW Statistics 19*.

StataCorp (2009) *Stata Statistical Software: Release 11*.

Vaida, F. and Blanchard, S. (2005) 'Conditional Akaike Information For Mixed-effects Models', *Biometrika*, 92: 351–70.

Valentini, G., Della Rossa, A., Bombardieri, S., Bencivelli, W., Silman, A.J., D'Angelo, S., Cerinic, M.M., Belch, J.F., Black, C.M., Bruhlmann, P., Czirják, L., De Luca, A., Drosos, A.A., Ferri, C., Gabrielli, A., Giacomelli, R., Hayem, G., Inanc, M., McHugh, N.J., Nielsen, H., Rosada, M., Scorza, R., Stork, J., Sysa, A., van den Hoogen, F.H., and Vlachoyiannopoulos, P.J. (2001) 'European Multicentre Study To Define Disease Activity Criteria For Systemic Sclerosis. I. Clinical And Epidemiological Features Of 290 Patients From 19 Centres', *Annals of the Rheumatic Diseases*, 60: 585–91.

Weisberg, S. (2005) *Applied Linear Regression*. New York: Wiley.

Wolfinger, R. (1993) 'Covariance Structure Selection In General Mixed Models', *Communications in Statistics – Simulation and Computation*, 22: 1079–106.

8

Generalized Linear Mixed Models–Overview

Geert Verbeke and Geert Molenberghs
Interuniversity Institute for Biostatistics and Statistical Bioinformatics
Katholieke Universiteit Leuven and Universiteit Hasselt, Belgium

8.1 INTRODUCTION

In applied sciences, one is often confronted with the collection of *correlated data* or otherwise hierarchical data. This generic term embraces a multitude of data structures, such as multivariate observations, clustered data, repeated measurements, longitudinal data, and spatially correlated data. In particular, studies are often designed to investigate changes in a specific outcome which is measured repeatedly over time in the participating persons. This is in contrast to cross-sectional studies where the response of interest is measured only once for each individual. Longitudinal studies are conceived for the investigation of such changes, together with the evolution of relevant covariates.

A very important characteristic of data to be analysed is the type of outcome. Methods for Gaussian longitudinal data form no doubt the best-developed and most advanced body of research (Verbeke and Molenberghs, 1997; 2000); the same is true for software implementation. This is natural, since the special status and the elegant properties of the normal distribution simplify model building and ease software development. A number of software tools, such as the SAS procedure MIXED, the S-PLUS function `lme`, and MLwiN have been developed in this area. However, categorical (nominal, ordinal, and binary) and discrete outcomes are also very prominent in statistical practice. For example, in many surveys regarding educational testing or quality of life, as well as in behavioural observations, responses are often scored on binary or ordinal scales. Generalized linear models, traditionally used to analyse cross-sectional discrete outcomes, can be extended in many ways to accommodate longitudinal data. In Diggle et al. (2002), there is a lot of emphasis on generalized estimating equations (Liang and Zeger, 1986), which is discussed in Chapters 15 and 25. As an alternative, generalized linear mixed models have been proposed and/or studied by, for example, Stiratelli, Laird, and Ware (1984), Wolfinger and O'Connell (1993), and Breslow and Clayton (1993). Fahrmeir and Tutz (2001) devote an entire book to generalized linear models for multivariate settings. In this chapter we will provide a general introduction to generalized linear mixed models, and illustrate them in the context of discrete data.

In correlated settings, each unit (respondent, cluster, person, patient, ...) typically has a *vector* $\boldsymbol{y}$ of responses (here, we will use the terms 'person' or 'individual' to indicate study units). This leads to several, generally nonequivalent, extensions of univariate models. In a *marginal model*, marginal distributions for each component of $\boldsymbol{y}$ are modeled based on a set X of predictor variables. The correlation among the components of $\boldsymbol{y}$ can then be captured either by adopting a fully parametric approach or by means of working assumptions, such as in the semiparametric approach of Liang and Zeger (1986). Alternatively, in a *mixed-effects model*, the predictor variables X are supplemented with a vector $\boldsymbol{u}$ of subject-specific (or random) effects, conditional upon which of the components of $\boldsymbol{y}$ are usually assumed to be independent. This does not preclude that more elaborate models are possible if residual dependence is detected (Longford, 1993). Finally, a *conditional model* describes the distribution of the components of $\boldsymbol{y}$, conditional on X but also conditional on (a subset of) the other components of $\boldsymbol{y}$. A discussion of conditional models can be found in Molenberghs and Verbeke (2005, Chapter 11).

This chapter, which will focus on mixed-effects models, is organized as follows. In Section 8.2, our notation and the generalized linear mixed model (GLMM) will be introduced, and some special cases discussed. In Section 8.3 we will briefly discuss estimation and inference within the GLMM framework. In Section 8.4 we will explain why care is needed for the correct interpretation of the parameters in generalized linear mixed models. We will continue with two examples, one with a binary outcome and one with a count outcome, presented in Section 8.5. Finally, some concluding remarks will be presented in Section 8.6.

8.2 NOTATION AND MODEL FORMULATION

Let y_{ij} denote the jth measurement available for the ith cluster, $i = 1, \ldots, m$, $j = 1, \ldots, n_i$, where clusters can represent families in surveys, study participants in longitudinal studies, classrooms in educational testing experiments, litters in toxicology studies, etc. Observe that this notation is different from notation used elsewhere in this book, where y_{ij} represents the measurement for observation i in cluster j. Our notation is more conventional in the literature on longitudinal data and clustered data, while the other notation stems from the multilevel context. Also, the notation used here is consistent with the notation of Chapter 15 about estimation and inference in GLMMs. Further, let $\boldsymbol{y}_i$ denote the corresponding vector of all measurements for cluster i. Our general model assumes that $\boldsymbol{y}_i$ (possibly appropriately transformed) satisfies

$$\boldsymbol{y}_i|\boldsymbol{u}_i \sim F_i(\boldsymbol{\zeta}, \boldsymbol{u}_i), \qquad (8.1)$$

i.e., conditional on $\boldsymbol{u}_i$, $\boldsymbol{y}_i$ follows a prespecified distribution F_i, possibly depending on covariates, and parameterized through a vector $\boldsymbol{\zeta}$ of unknown parameters, common to all clusters. Further, $\boldsymbol{u}_i$ is a q-dimensional vector of cluster-specific parameters, called random effects, assumed to follow a so-called mixing distribution G which may depend on a vector $\boldsymbol{\psi}$ of unknown parameters, i.e., $\boldsymbol{u}_i \sim G(\boldsymbol{\psi})$. The $\boldsymbol{u}_i$ reflect the between-cluster heterogeneity in the population with respect to the distribution of $\boldsymbol{y}_i$.

Different specifications of F_i and G lead to specific models. One approach is to leave G completely unspecified and to use non-parametric maximum likelihood (NPML, Böhning, 1999) estimation, which maximizes the likelihood over all possible distributions G. The resulting estimate $\widehat{G}$ is then always discrete with finite support. Depending on the context, this is not always believed to be a realistic reflection of the true heterogeneity between units. One therefore often assumes G to be of a specific parametric form. Most models assume the random-effects distribution G to be normal, leading to the classical mixed model, but other choices are possible as well. In Section 8.2.1 we will first focus on the special case of the linear mixed model for Gaussian outcomes. Afterwards, in Section 8.2.2, the generalized linear mixed model will be introduced. The linear model is very illustrative as it shows very explicitly how the various parameters in the model can be interpreted, but also how the cluster-specific parameters $\boldsymbol{u}_i$ induce an association structure between the measurements within clusters.

8.2.1 *Linear Mixed Models*

When Gaussian hierarchical data are considered, a general, and very flexible, class of parametric models is obtained from introducing random effects $\tilde{\boldsymbol{u}}_i$ in the multivariate linear regression model. Suppose that a specific outcome y is observed repeatedly over time for a set of persons, and suppose that the individual trajectories are of the type shown in Figure 8.1. Obviously, a linear regression model with intercept and linear time effect seems plausible to describe the data of each person separately. However, different persons tend to have different intercepts and different slopes. One can therefore assume that the outcome y_{ij}, measured at time t_{ij}, satisfies $y_{ij} = \tilde{u}_{i0} + \tilde{u}_{i1}t_{ij} + \varepsilon_{ij}$. Assuming the vector $\tilde{\boldsymbol{u}}_i = (\tilde{u}_{i0}, \tilde{u}_{i1})'$ of person-specific parameters to be bivariate normal with mean $(\beta_0, \beta_1)'$ and 2×2 covariance matrix D, and assuming the error terms ε_{ij} to be normal as well, this leads to a so-called linear mixed model. In practice, one will often formulate the model as

$$y_{ij} = (\beta_0 + u_{i0}) + (\beta_1 + u_{i1})t_{ij} + \varepsilon_{ij},$$

with $\tilde{u}_{i0} = \beta_0 + u_{i0}$ and $\tilde{u}_{i1} = \beta_1 + u_{i1}$, and the new random effects $\boldsymbol{u}_i = (u_{i0}, u_{i1})'$ assumed to have mean zero.

The above model can be viewed as a special case of the general linear mixed model which assumes that the outcome vector $\boldsymbol{y}_i$ follows a multivariate normal distribution, with mean vector $X_i\boldsymbol{\beta} + Z_i\boldsymbol{u}_i$ and some covariance matrix Σ, and which further assumes that the random effects $\boldsymbol{u}_i$ follow a (multivariate) normal distribution, i.e., it is assumed that the n_i-dimensional vector $\boldsymbol{y}_i$ satisfies

$$\begin{aligned} \boldsymbol{y}_i|\boldsymbol{u}_i &\sim N(X_i\boldsymbol{\beta} + Z_i\boldsymbol{u}_i, \Sigma_i), && (8.2)\\ \boldsymbol{u}_i &\sim N(\mathbf{0}, D), && (8.3) \end{aligned}$$

where X_i and Z_i are $(n_i \times p)$ and $(n_i \times q)$ matrices of known covariates, $\boldsymbol{\beta}$ is a p-dimensional vector of regression parameters, called the fixed effects, D is a general $(q \times q)$ covariance matrix, and Σ_i is a $(n_i \times n_i)$ covariance matrix depending on i only through its dimension. Note that this model satisfies our general definition of a mixed model, with F_i in (8.1) defined in (8.2) and with G as in (8.3). Often, it is assumed that $\Sigma_i = \sigma^2 I_{n_i}$, with I_{n_i} being the identity matrix of dimension n_i. One then assumes that, conditionally on the cluster-specific parameters $\boldsymbol{u}_i$, the outcomes y_{ij} are independent. This assumption is therefore referred to as the *conditional independence assumption*. More details about model formulation in this context can be found in Chapter 2.

8.2.2 *The Generalized Linear Mixed Model*

The generalized linear mixed model is the most frequently used mixed-effects model for discrete outcomes (see also Chapter 16). As an example, consider a study in which a specific test is taken repeatedly over time by a set of children. The outcome y_{ij} taken at time (age) t_{ij} is of binary type, i.e., pass/fail. Continuing the use of subject-specific regression models, a logistic model could be proposed in this case, i.e., one could assume y_{ij} to be Bernoulli distributed with success probability π_{ij} satisfying

$$\begin{aligned} \text{logit}(\pi_{ij}) &= \ln\left[\frac{\pi_{ij}}{1-\pi_{ij}}\right]\\ &= (\beta_1 + u_i) + \beta_2 t_{ij}. && (8.4) \end{aligned}$$

Hence, we have a logistic regression model for each person separately. Through the person-specific intercepts u_i, the model allows all

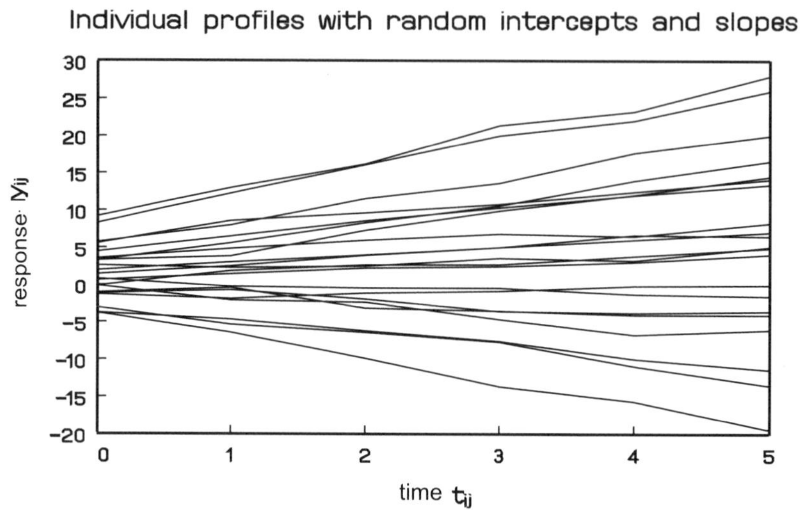

Figure 8.1 Hypothetical Example of Continuous Longitudinal Data which can be Well Described by a Linear Mixed Model with Random Intercepts and Random Slopes. The thin lines represent the observed subject-specific evolutions. The bold line represents the population-averaged evolution

persons to be different with respect to their ability to pass the tests. As before, the u_i will be called random effects, and will be assumed to be normally distributed with mean zero. A graphical representation of this model is given in Figure 8.2. As for the linear model, person-specific slopes could have been included as well.

A general formulation is as follows. Conditionally on random effects $\boldsymbol{u}_i$, it is assumed that the elements y_{ij} of $\boldsymbol{y}_i$ are independent, with density function of an exponential family form

$$f_i(y_{ij}|\boldsymbol{u}_i) = \exp[(y_{ij}\eta_{ij} - a(\eta_{ij}))/\phi + c(y_{ij}, \phi)], \qquad (8.5)$$

with a and c functions and ϕ an overdispersion parameter, and further with mean $\mathrm{E}(y_{ij}|\boldsymbol{u}_i) = a'(\eta_{ij}) = \mu_{ij}(\boldsymbol{u}_i)$ and variance $\mathrm{Var}(y_{ij}|\boldsymbol{u}_i) = \phi a''(\eta_{ij})$, and where, apart from a link function h, a linear regression model with parameters $\boldsymbol{\beta}$ and $\boldsymbol{u}_i$ is used for the mean, i.e.,

$$h(\boldsymbol{\mu}_i(\boldsymbol{u}_i)) = X_i\boldsymbol{\beta} + Z_i\boldsymbol{u}_i. \qquad (8.6)$$

Note that the linear mixed model is a special case, with identity link function. As in the linear model, the random effects $\boldsymbol{u}_i$ are assumed to follow a multivariate normal distribution with mean $\mathbf{0}$ and covariance D. Usually, the canonical link function is used, i.e., $h = a'^{-1}$, such that $\boldsymbol{\eta}_i = X_i\boldsymbol{\beta} + Z_i\boldsymbol{u}_i$. For example, for binomial data, the logit link is canonical. For Poisson data, the log link is canonical. The advantage of this choice is a simplified form for the score equations for which, in many cases, fast and stable fitting algorithms can be constructed. Note also that when a link function other than the identity link is used, the fixed effects need to be interpreted in a way which is consistent with the link. For example, if a logit link is used, as in (8.4), the slope β_2 is the change in the log odds of success associated with a one-unit change in the predictor, holding all other effects in the model fixed; if a log link is used, $\exp(\beta)$ is the multiplicative change in the mean of the outcome associated with a one-unit change in the predictor, holding all other effects in the model fixed.

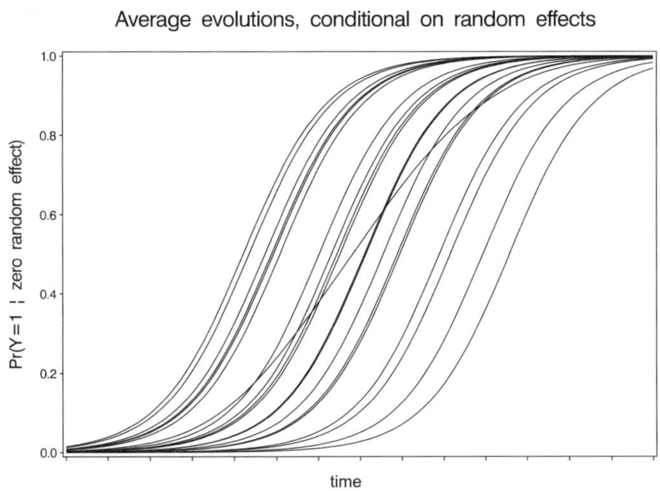

Figure 8.2 Graphical Representation of a Random-Intercepts Logistic Model. The thin lines represent the subject-specific logistic regression models. The bold line represents the population-averaged evolution

8.3 ESTIMATION AND INFERENCE

In general, unless a fully Bayesian approach is followed (see, e.g., Gelman et al., 1995), inference is based on the marginalized model for $\boldsymbol{y}_i$ which is obtained from integrating out the random effects over their distribution $G(\boldsymbol{\psi})$. Let $f_i(\boldsymbol{y}_i|\boldsymbol{u}_i)$ and $g(\boldsymbol{u}_i)$ denote the density functions corresponding to the distributions F_i and G, respectively. As indicated before, $g(\boldsymbol{u}_i)$ will often be the density of the $N(\boldsymbol{0}, D)$ distribution. The marginal density function of $\boldsymbol{y}_i$ is

$$f_i(\boldsymbol{y}_i) = \int f_i(\boldsymbol{y}_i|\boldsymbol{u}_i)g(\boldsymbol{u}_i)d\boldsymbol{u}_i, \qquad (8.7)$$

which depends on the unknown parameters $\boldsymbol{\zeta}$ (in F_i) and $\boldsymbol{\psi}$ (in G). Assuming independence of the units, estimates $\widehat{\boldsymbol{\zeta}}$ and $\widehat{\boldsymbol{\psi}}$ can be obtained from maximizing the likelihood function built from (8.7).

Depending on F_i and G, the integration in (8.7) may or may not be possible analytically. For example, it can be easily shown that the marginal distribution for $\boldsymbol{y}_i$ under the linear mixed model defined by (8.2) and (8.3) equals

$$\boldsymbol{y}_i \sim N(X_i\boldsymbol{\beta}, V_i = Z_i D Z_i' + \Sigma_i).$$

The mean vector equals $X_i\boldsymbol{\beta}$, and is parameterized as in classical (multivariate) regression models. The covariance matrix V_i has a very specific parameterization, and illustrates that the variability observed in the outcomes $\boldsymbol{y}_i$ can be partially ascribed to variability between subjects (the random-effects covariance D) and variability within subjects (the error covariance Σ_i).

In most other mixed models, the marginal distribution (8.7) can no longer be derived analytically, and approximations are required. Proposed solutions are based on Taylor series expansions of $f_i(\boldsymbol{y}_i|\boldsymbol{u}_i)$, on approximations of the data, on numerical approximations of the integrals, or on applications of the EM algorithm (Dempster, Laird, and Rubin, 1977). Some of these techniques were or will be discussed in Chapters 3, 4, and 15. Other references include Lavergne and Trottier (2000), Pinheiro and Bates (1995), Verbeke and Molenberghs (2000), and Molenberghs and Verbeke (2005).

While interest is often primarily in estimation and inference for the parameters in the marginal distribution (8.7), one is sometimes also interested in obtaining predictions for the random effects $\boldsymbol{u}_i$ as well. They reflect between-person variability, which makes them helpful for detecting special profiles (i.e., outlying individuals) or groups of individuals showing extraordinary behavior. Such predictions are also needed whenever there is interest in predicting person-specific profiles. Inference for the random effects is often based on their so-called posterior distribution $f_i(\boldsymbol{u}_i|\boldsymbol{y}_i)$, given by

$$f_i(\boldsymbol{u}_i|\boldsymbol{y}_i) = \frac{f_i(\boldsymbol{y}_i|\boldsymbol{u}_i)\, g(\boldsymbol{u}_i)}{\int f_i(\boldsymbol{y}_i|\boldsymbol{u}_i)\, g(\boldsymbol{u}_i)\, d\boldsymbol{u}_i}, \quad (8.8)$$

in which the unknown parameters $\boldsymbol{\zeta}$ and $\boldsymbol{\psi}$ are replaced by their estimates obtained earlier from maximizing the marginal likelihood. The mean or mode corresponding to (8.8) can be used as point estimates for $\boldsymbol{u}_i$, yielding the so-called empirical Bayes (EB) estimate.

Since fitting of generalized linear mixed models is based on maximum likelihood principles, inferences for the parameters are readily obtained from classical maximum likelihood theory. Indeed, assuming the fitted model is appropriate, the obtained estimators are asymptotically normally distributed with the correct values as means, and with the inverse Fisher information matrix as covariance matrix. Hence, asymptotic Wald-type tests, comparing standardized estimates to the standard normal distribution, can be easily performed. Alternatively, asymptotic likelihood ratio tests and score tests can be used as well. One complication often ignored in practice is that estimation of parameters in covariance structures such as the random-effects covariance D occurs under the restriction of non-negative definiteness for the resulting covariance structure. This implies that null hypotheses involving such parameters are often on the boundary of the parameter space. As a result, classical maximum likelihood theory no longer applies, and none of the classical testing procedures (Wald, likelihood ratio, score) remain valid. A typical example is the test for the need of a specific random effect. Consider a model as in Section 8.2.1 with random intercepts and slopes, and consider testing whether random slopes are needed in the model. Under the alternative hypothesis, D is a 2×2 non-negative definite matrix, which reduces to a non-negative scalar under the null hypothesis. Clearly, the null hypothesis is on the boundary of the alternative parameter space as it requires the random-slopes variance to be zero. Stram and Lee (1994, 1995) have shown that, in this case, the asymptotic null distribution for the likelihood ratio test statistic is a mixture of a χ^2_1 and a χ^2_2, with equal probability $1/2$, rather than the standard χ^2_2 one would expect under classical likelihood theory. In general, the asymptotic null distribution for the likelihood ratio test statistic for testing a null hypothesis which allows for q correlated random effects versus an alternative of $q+1$ correlated random effects is a mixture of a χ^2_q and a χ^2_{q+1}, with equal probability $1/2$. For more general settings, e.g., comparing models with q and $q+q'$ ($q' > 1$) random effects, the null distribution is a mixture of χ^2 random variables (Shapiro, 1988; Raubertas, Lee, and Nordheim, 1986), the weights of which can only be calculated analytically in a number of special cases. Similar results can be derived for the score test. Building upon Silvapulle and Silvapulle (1995) and Shapiro (1988), Verbeke and Molenberghs (2003) have shown that the score test is asymptotically equivalent to the likelihood ratio test, and that the same mixtures of chi-squared distributions appear as asymptotic null distributions.

8.4 INTERPRETATION OF REGRESSION PARAMETERS IN MIXED MODELS

The common aspect in generalized linear mixed models is the fact that all of these models can be interpreted as person-specific statistical models, i.e., models which contain cluster-specific parameters, and therefore require an interpretation conditionally on the cluster. More specifically, the vector $\boldsymbol{\beta}$ of regression parameters needs to be interpreted conditionally on the random effect $\boldsymbol{u}_i$. This can be indicated explicitly using the notation $\boldsymbol{\beta}^{\mathrm{RE}}$. In many practical situations, however, one is interested in studying 'average outcomes', such as average success probabilities, and how these averages depend on known covariates. In general, such population-averaged profiles require the fitting of marginal models which, in principle, can contain the same covariates X_i. The corresponding vector $\boldsymbol{\beta}^{\mathrm{M}}$ of regression parameters then has a marginal interpretation. It should be strongly emphasized that, except in very specific cases, $\boldsymbol{\beta}^{\mathrm{RE}}$ and $\boldsymbol{\beta}^{\mathrm{M}}$ are not identical, and have a different interpretation. This will be further explained in the following sections, first for linear mixed models, then for generalized linear mixed models.

8.4.1 Linear Mixed Models

As discussed before, the linear mixed model can be interpreted as a linear regression model for the vector $\boldsymbol{y}_i$ of repeated measurements for each cluster separately, where some of the regression parameters are specific (random effects, $\boldsymbol{u}_i$), while others are not (fixed effects, $\boldsymbol{\beta}$). Note that the assumption $\mathrm{E}(\boldsymbol{u}_i) = \boldsymbol{0}$ implies that the mean of $\boldsymbol{y}_i$ still equals $X_i\boldsymbol{\beta}$, such that the fixed effects in the mixed-effects model (8.2) can also be interpreted marginally. Hence, in this case, $\boldsymbol{\beta}^{\mathrm{RE}} = \boldsymbol{\beta}^{\mathrm{M}}$. This is a very important and useful property in the linear mixed model setting, that does not carry over to more general settings. Thus, in this context, not only do the fixed effects reflect the effect of changing covariates within specific clusters, they also measure the marginal effect in the population of changing the same covariates.

8.4.2 Generalized Linear Mixed Models

Since the generalized linear mixed model (8.5) and (8.6) specifies the conditional distribution of y_{ij}, conditional on the random effects $\boldsymbol{u}_i$, it only directly yields the conditional expectation $\mathrm{E}(y_{ij}|\boldsymbol{u}_i)$ and additional calculations are needed to obtain the population average response $\mathrm{E}(y_{ij})$. More specifically, integration is needed over the distribution G of random effects:

$$\mathrm{E}(\boldsymbol{y}_i) = \int \mathrm{E}(y_{ij}|\boldsymbol{u}_i) g(\boldsymbol{u}_i) d\boldsymbol{u}_i. \tag{8.9}$$

The only reason why in linear mixed models the regression parameters also have a population-average interpretation is that the link function h is the identity link, implying that the conditional mean $\mathrm{E}(y_{ij}|\boldsymbol{u}_i)$ is a linear function of the random effects $\boldsymbol{u}_i$. This no longer holds for other mixed models.

As an illustration, consider the logistic mixed model for binary outcomes. Using the logit link, the average conditional on the random effects, equals

$$\begin{aligned}\mathrm{E}(y_{ij}|\boldsymbol{u}_i) &= P(y_{ij} = 1|X_i, Z_i, \boldsymbol{\beta}, \boldsymbol{u}_{ii}) \\ &= \frac{\exp(X_i\boldsymbol{\beta} + Z_i\boldsymbol{u}_i)}{1 + \exp(X_i\boldsymbol{\beta} + Z_i\boldsymbol{u}_i)}.\end{aligned}$$

The population averaged mean, implied by the above generalized linear mixed model, is obtained from integrating over the random effects, i.e.:

$$\begin{aligned}\mathrm{E}(y_{ij}) &= \mathrm{E}[P(y_{ij} = 1|X_i, Z_i, \boldsymbol{\beta}, \boldsymbol{u}_i)] \\ &= \mathrm{E}\left[\frac{\exp(X_i\boldsymbol{\beta} + Z_i\boldsymbol{u}_i)}{1 + \exp(X_i\boldsymbol{\beta} + Z_i\boldsymbol{u}_i)}\right] \\ &\neq \frac{\exp(X_i\boldsymbol{\beta})}{1 + \exp(X_i\boldsymbol{\beta})}.\end{aligned}$$

Note that the implied population average is *not* of a logistic form any more. In general, $h^{-1}(X_i\boldsymbol{\beta}) \neq \mathrm{E}\left[h^{-1}(X_i\boldsymbol{\beta} + Z_i\boldsymbol{u}_i)\right]$, for an arbitrary link function $h(\cdot)$, unless in specific cases such as the identity link used in linear mixed models. As discussed by Neuhaus, Kalbfleisch, and Hauck (1991), the parameter estimates obtained from fitting the logistic mixed model are typically larger in absolute value than their marginal counterparts that would be obtained from directly fitting a marginal model with mean model

$$\mathrm{E}(y_{ij}) = \frac{\exp(X_i\boldsymbol{\beta})}{1 + \exp(X_i\boldsymbol{\beta})}.$$

However, one should not refer to this phenomenon as *bias* since the two sets of parameters target different scientific questions.

Nevertheless, in some cases an approximate relationship can be derived between the marginal and mixed-effects model parameters. Consider the specific case of the binomial model for binary data, with the logit canonical link function, and where the only random effects are intercepts u_i. As before, let $\boldsymbol{\beta}^{\mathrm{RE}}$ represent the vector of regression parameters in the mixed model. It can then be shown that the *implied* marginal mean $\boldsymbol{\mu}_i = \mathrm{E}(y_{ij})$ satisfies $h(\boldsymbol{\mu}_i) \approx X_i \boldsymbol{\beta}^{\mathrm{M}}$, with

$$\boldsymbol{\beta}^{\mathrm{M}} = [c^2 \mathrm{Var}(u_i) + 1]^{-1/2} \boldsymbol{\beta}^{\mathrm{RE}},$$

in which c equals $16\sqrt{3}/15\pi$. Hence, although the parameters $\boldsymbol{\beta}$ in the generalized linear mixed model have no marginal interpretation, they do show a strong relation to their marginal counterparts. Note that we do, indeed, obtain marginal parameters which are smaller in absolute value than the mixed model counterparts, with larger differences for larger random-intercepts variances. In Figure 8.2, this is reflected in the less steep increase in the population-averaged evolution when compared to all subject-specific evolutions.

Note that, because the random-effects approach results in a marginal likelihood, thereby completely specifying the distribution of $\boldsymbol{y}_i$, it is possible to derive the marginal average trends in the data. As is indicated in (8.9), this requires averaging the conditional means over the random effects $\boldsymbol{u}_i$. In most cases, this is not possible analytically. Numerical integration methods can be used, but it is often much easier to use numerical averaging by sampling a large number M of random-effects vectors $\boldsymbol{u}_i$ from their fitted distribution $N(\boldsymbol{0}, \widehat{D})$, and to estimate $\mathrm{E}(\boldsymbol{y}_i)$ by

$$\begin{aligned}\widehat{\mathrm{E}}(\boldsymbol{y}_i) &= \frac{1}{M}\sum_i^M \widehat{\mathrm{E}}(\boldsymbol{y}_i|\boldsymbol{u}_i) \\ &= \frac{1}{M}\sum_i^M h^{-1}(X_i\widehat{\boldsymbol{\beta}}^{\mathrm{RE}} + Z_i\boldsymbol{u}_i).\end{aligned}$$

This can be calculated for a fine grid of covariate values X_i, such that a graphical representation for average trends can be obtained. As stated before, the marginal average $\mathrm{E}(y_{ij})$ is in general not of the same parametric form as the conditional mean $\mathrm{E}(y_{ij}|\boldsymbol{u}_i)$. Hence, the averaging over the random effects will not yield formal estimates for the elements in $\boldsymbol{\beta}^{\mathrm{M}}$ that would be estimated when fitting marginal models. It can only provide estimates of $\mathrm{E}(y_{ij})$ for specific values for the covariates, or a plot of population-averaged trends. This will be further illustrated in Section 8.5.

8.5 CASE STUDIES

8.5.1 A Binary Outcome: The Toenail Data

As an example of a logistic mixed model, we analyse data from a randomized, double-blind, parallel group, multicentre study, for the comparison of two oral treatments for toenail dermatophyte onychomycosis. Patients with a clinical diagnosis of toe onychomycosis confirmed by a positive direct microscopy and a positive culture for dermatophytes at a central laboratory were randomly assigned to treatment A or treatment B. After a 12-week treatment period, there was a follow-up period of 36 weeks. Patients returned to the hospital at months 0 (baseline), 1, 2, 3, 6, 9, and 12. More details can be found in De Backer et al. (1996, 1998). The outcome of interest here is the severity of the infection, coded as 0 (not severe) or 1 (severe). The question of interest is whether the probability of having a severe infection decreases over time, and whether that evolution is different for the two treatment groups. A summary of the number of patients in the study at each time-point and the number of patients with severe infections is given in Table 8.1.

This data set has been analysed extensively in Molenberghs and Verbeke (2005), who proposed a logistic random-intercepts model with normally distributed random effects. More specifically, it is assumed that the probability π_{ij} for subject i to have a severe infection at time t_{ij} satisfies model (8.4) with intercept and slope $\beta_{\mathrm{A}0}$ and $\beta_{\mathrm{A}1}$ if the patient belongs

Table 8.1 Toenail Data: Number and Percentage of Patients with Severe Toenail Infection, for each Treatment Arm Separately

	Group A			Group B		
	# severe	# patients	percentage	# severe	# patients	percentage
Baseline	54	146	37.0%	55	148	37.2%
1 month	49	141	34.7%	48	147	32.6%
2 months	44	138	31.9%	40	145	27.6%
3 months	29	132	22.0%	29	140	20.7%
6 months	14	130	10.8%	8	133	6.0%
9 months	10	117	8.5%	8	127	6.3%
12 months	14	133	10.5%	6	131	4.6%

Table 8.2 Toenail Data: Parameter Estimates (Standard Errors) Obtained from Fitting a Logistic Random-Intercepts Model and from a GEE Analysis Assuming Unstructured Working Correlation Matrix

	GLMM	GEE	
Parameter	Estimate $\widehat{\beta}^{\mathrm{RE}}$ (s.e.)	Estimate $\widehat{\beta}^{\mathrm{M}}$(s.e.)	Ratio $\widehat{\beta}^{\mathrm{RE}}/\widehat{\beta}^{\mathrm{M}}$
Intercept group A (β_{A0})	−1.63 (0.44)	−0.72 (0.17)	2.27
Intercept group B (β_{B0})	−1.75 (0.45)	−0.65 (0.17)	2.69
Slope group A (β_{A1})	−0.40 (0.05)	−0.14 (0.03)	2.87
Slope group B (β_{B1})	−0.57 (0.06)	−0.25 (0.04)	2.22
Random intercepts s.d. (τ)	4.02 (0.38)		

to treatment group A and with intercept and slope β_{B0} and β_{B1} otherwise. The random effects u_i are assumed normal with mean 0 and standard deviation τ. The results reported in Table 8.2 have been obtained by using adaptive Gaussian quadrature with 50 quadrature points for the numerical integration to approximate the likelihood contributions (8.7).

There seems to be evidence for much between-subject heterogeneity in the population. Indeed, the random-intercepts variance is estimated to be $\widehat{\tau}^2 = 16.16$. This implies that the approximate relation between the subject-specific regression parameters $\boldsymbol{\beta}^{\mathrm{RE}}$ and the population-average regression parameters $\boldsymbol{\beta}^{\mathrm{M}}$ is given by yielding $[c^2\widehat{\tau}^2 + 1]^{1/2} = 2.56$, implying a huge difference between subject-specific effects and average trends. This is also graphically illustrated in Figure 8.3. Panel (a) shows the predicted evolution for an 'average' subject in treatment group A and treatment group B, respectively, where 'average' subjects are subjects with random effect equal to zero, i.e., $u_i = 0$. These evolutions are given by:

$$P(y_{ij}=1|u_i=0)=\begin{cases}\dfrac{\exp(-1.63-0.40t_{ij})}{1+\exp(-1.63-0.40t_{ij})}\\[2ex]\dfrac{\exp(-1.75-0.57t_{ij})}{1+\exp(-1.75-0.57t_{ij})},\end{cases}$$

for Treatments A and B, respectively. Panel (b) in Figure 8.3 shows the population-average trends for both treatment groups, obtained from

$$P(Y_{ij}=1)=\begin{cases}\mathrm{E}\left[\dfrac{\exp(-1.63+u_i-0.40t_{ij})}{1+\exp(-1.63+u_i-0.40t_{ij})}\right]\\[2ex]\mathrm{E}\left[\dfrac{\exp(-1.75+u_i-0.57t_{ij})}{1+\exp(-1.75+u_i-0.57t_{ij})}\right],\end{cases}$$

for Treatments A and B, respectively, in which the expectation was approximated using the sampling-based procedure described in Section 8.4.2, with $M = 1000$. Obviously, the evolutions in both panels are very different. On a logit scale, the slopes of the evolutions in panel (a) are approximately 2.56 times larger than the slopes of the evolutions in panel

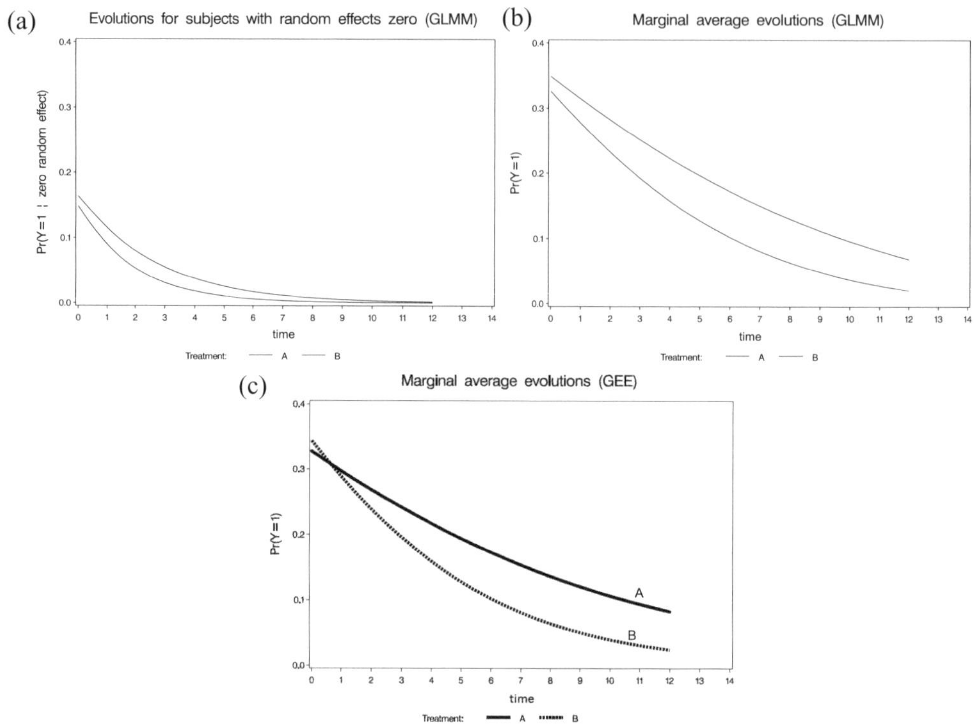

Figure 8.3 Toenail Data: Treatment-Specific Evolutions. (a) Evolutions for an 'average' person, i.e., with $u_i = 0$. (b) Average evolution as obtained from marginalizing the GLMM. (c) Average evolution estimated directly from fitting a marginal model using GEE with unstructured working correlation

(b). Note that the fitted population-average probabilities are close to the observed proportions of severe infections summarized in Table 8.1. Note also that, in linear mixed models, the evolutions of 'average' subjects would be equal to the population-average evolutions, implying identical panels (a) and (b) in the corresponding graphs.

Note that, although the sampling-based approximation to the population-average trends yields a graphical representation of the evolution over time of the average probability for a severe infection, estimates of the parameters in the marginal logistic counterpart of the model can only be obtained from directly fitting a marginal model, for example, using generalized estimating equations (GEE, Liang and Zeger, 1986). As an illustration, we performed a GEE analysis with unstructured working correlation assumption, and with the same logistic model as before but, of course, with the random effects omitted. The parameter estimates and associated standard errors (corrected for possible misspecification of the association structure) are also included in Table 8.2. The ratio between each parameter estimate and its logistic mixed-model counterpart is also presented in the last column of the table. All ratios are between 2.22 and 2.87, somewhat in the neighborhood of the predicted approximate ratio 2.56. The fitted marginal model equals

$$P(y_{ij} = 1) = \begin{cases} \frac{\exp(-0.72-0.14t_{ij})}{1+\exp(-0.72-0.14t_{ij})} \\ \frac{\exp(-0.65-0.25t_{ij})}{1+\exp(-0.65-0.25t_{ij})}, \end{cases}$$

for Treatments A and B, respectively. A graphical representation of these average evolutions is presented in panel (c) of Figure 8.3, which shows evolutions very similar to those obtained from marginalizing the fitted logistic mixed model.

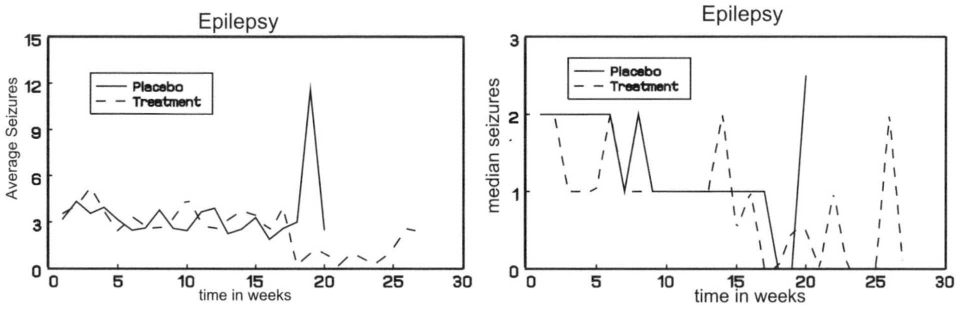

Figure 8.4 Epilepsy Data: Average and Median Evolutions Over Time

8.5.2 The Epilepsy Data

As a second illustration of the use of generalized linear mixed models, we analyse data from a randomized, double-blind, parallel group multicenter study for the comparison of placebo with a new anti-epileptic drug (AED), in combination with one or two other AEDs. The study is described in full detail in Faught et al. (1996). The randomization of epilepsy patients took place after a 12-week baseline period that served as a stabilization period for the use of AEDs, and during which the number of seizures were counted. After that period, 45 patients were assigned to the placebo group, 44 to the active (new) treatment group. Patients were then measured weekly. Patients were followed (double-blind) during 16 weeks, after which they were entered into a long-term open-extension study. Some patients were followed for up to 27 weeks. The outcome of interest is the number of epileptic seizures experienced during the last week, i.e., since the last time the outcome was measured. The key research question is whether or not the additional new treatment reduces the number of epileptic seizures. As a summary of the data, Figure 8.4 shows average and median evolutions. The unstable behaviour can be explained by the presence of extreme values, but is also the result of the fact that very few observations are available at some of the time-points, especially past week 20. This is also reflected in Table 8.3, which shows the number of measurements at a selection of time-points. Note the serious drop in number of measurements past the end of the actual double-blind period, i.e., past week 16.

Table 8.3 Epilepsy Data: Number of Measurements Available at a Selection of Time-Points, for Both Treatment Groups Separately

	# Observations		
Week	Placebo	Treatment	Total
1	45	44	89
5	42	42	84
10	41	40	81
15	40	38	78
16	40	37	77
17	18	17	35
20	2	8	10
27	0	3	3

Let y_{ij} denote the number of epileptic seizures patient i experiences during week j of the follow-up period. Furthermore, as before, t_{ij} is the time-point at which y_{ij} has been measured, $t_{ij} = 1, 2, \ldots,$ until at most 27. Following Molenberghs and Verbeke (2005), a Poisson mixed model will be used, with the canonical logarithmic link function and with subject-specific intercepts as well as slopes. More specifically, we assume that

$$Y_{ij}|\boldsymbol{u_i} \sim \text{Poisson}(\lambda_{ij}),$$

$$\log(\lambda_{ij}) = \begin{cases} (\beta_0 + u_{i0}) + (\beta_1 + u_{i1})t_{ij} \\ (\beta_0 + u_{i0}) + (\beta_2 + u_{i1})t_{ij}, \end{cases} \quad (8.10)$$

Table 8.4 Epilepsy Study: Parameter Estimates (Standard Errors) for a Poisson Mixed Model with Random Intercepts and Slopes

Effect	Parameter	Estimate (s.e.)
Common intercept	β_0	0.7739 (0.1293)
Slope placebo	β_1	−0.0245 (0.0096)
Slope treatment	β_2	−0.0193 (0.0097)
Variance of intercepts	d_{11}	1.2859 (0.2231)
Variance of slopes	d_{22}	0.0024 (0.0006)
Correlation rand. eff.	ρ	−0.3349 (0.1318)

for placebo and treated groups, respectively, where the random effects $\boldsymbol{u}_i = (u_{i0}, u_{i1})'$ are assumed to be normally distributed with mean vector $\mathbf{0}$ and 2×2 covariance matrix D, parameterized with the random-intercepts variance d_{11}, the random-slopes variance d_{22}, and the correlation ρ between intercepts and slopes. We assume the same fixed intercept for the two groups. This reflects our prior belief that, due to the randomization, the initial values are equally distributed in both treatment groups.

The results reported in Table 8.4 have been obtained by using adaptive Gaussian quadrature with 10 quadrature points for the numerical integration to approximate the likelihood contributions (8.7). A comparison of the slopes β_1 and β_2 yields a non-significant result ($p > 0.6$). Note that these fixed effects have a subject-specific interpretation again. Hence, a comparison of β_1 and β_2 aims at testing whether a specific subject receiving placebo would experience a different evolution under the active treatment.

Panel (a) in Figure 8.5 shows the predicted evolution for an 'average' subject in both treatment groups, respectively. These evolutions are now given by:

$$\lambda_{ij}|\boldsymbol{u}_i = 0 = \begin{cases} \exp[0.7739 - 0.0242t_{ij}] \\ \quad \text{placebo} \\ \\ \exp[0.7739 - 0.0191t_{ij}] \\ \quad \text{treated.} \end{cases}$$

Panel (b) in Figure 8.5 shows the population-average trends for both treatment groups, obtained from

$$\lambda_{ij} = \begin{cases} \mathrm{E}\{\exp[(0.7739 + u_{i0}) \\ \quad +(-0.0242 + u_{i1})t_{ij}]\} \\ \mathrm{E}\{\exp[(0.7739 + u_{i0}) \\ \quad +(-0.0191 + u_{i1})t_{ij}]\} \end{cases}$$

for placebo and treated groups, respectively, in which the expectation was again approximated using the sampling-based procedure described in Section 8.4.2, with $M = 1000$. This time, sets (u_{i0}, u_{i1}) need to be sampled from a bivariate normal distribution with mean $\mathbf{0}$ and covariance matrix equal to the fitted random-effects covariance matrix (see Table 8.4)

$$D = \begin{pmatrix} 1.2859 & -0.0185 \\ -0.0185 & 0.0024 \end{pmatrix}.$$

As before, we observe that an average subject shows an evolution which is completely different from the average evolution, due to the non-linear link function used to model the conditional expectation of the outcome values as a function of time. Note also that the fitted population-average curves show some curvature, even though the linear predictor in (8.10) is linear in time.

8.6 CONCLUSIONS

One of the most flexible classes of model for the analysis of multilevel data is the generalized linear mixed model (GLMM). Not only does the GLMM account for the correlation in a natural way by explicitly separating out variability between clusters and variability within clusters, it also allows modeling flexible data structures such as highly

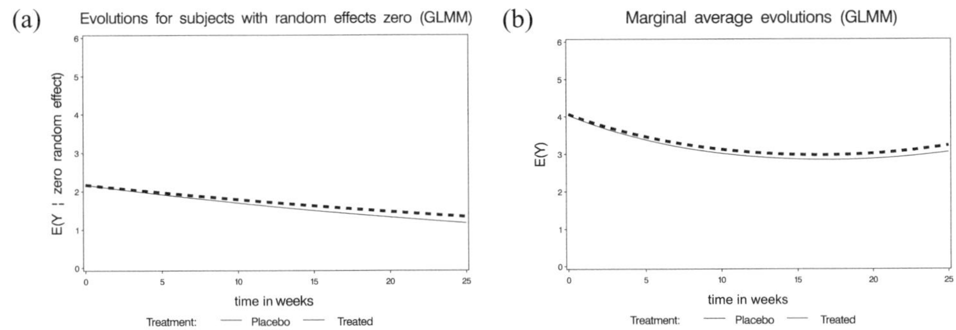

Figure 8.5 Epilepsy Study: Treatment-Arm Specific Evolutions. (a) Evolutions for an 'average' person, i.e., with $u_i = 0$. (b) Average evolution as obtained from marginalizing the GLMM

unbalanced longitudinal data in which outcomes are measured a different number of times for the various study participants and/or in which measurements are taken at subject-specific time-points. We have provided a general overview on GLMMs, including model formulation, parameter interpretation, and some general issues about model fitting. Starting from the special case of the linear mixed model (LMM) for Gaussian outcomes, we have illustrated how latent random effects imply particular association structures in the data. Afterwards, the GLMM was introduced in full generality, with a brief discussion of issues about parameter estimation and inference. More details can be found in Chapters 3 and 15. Since the GLMMs are formulated conditionally on subject-specific parameters, parameters have, in general, a subject-specific interpretation rather than a population average interpretation. This has been discussed in full detail in the context of the logistic random-intercepts model. Finally, two case studies have been presented, in which a logistic and a Poisson mixed model were used for the analysis of binary data and counts, respectively. All statistical analyses performed in this chapter were performed using the SAS software, which includes three procedures (MIXED, GLIMMIX, and NLMIXED) for the fitting of mixed models. The GEE results presented in this chapter (the toenail example in Section 8.5) were performed using the SAS procedure GENMOD. Programs and outputs can be found in Verbeke and Molenberghs (2000) and Molenberghs and Verbeke (2005), while datasets are available through the authors' web pages: `www.ibiostat.be/software`.

REFERENCES

Böhning, D. (1999) *Computer-Assisted Analysis of Mixtures and Applications: Meta-analysis, Disease Mapping and Others*, Number 81 in Monographs on Statistics and Applied Probability. London: Chapman & Hall/CRC.

Breslow, N.E. and Clayton, D.G. (1993) 'Approximate Inference in Generalized Linear Mixed Models', *Journal of the American Statistical Association*, 88: 9–25.

De Backer, M., De Keyser, P., De Vroey, C., and Lesaffre, E. (1996) 'A 12-week Treatment for Dermatophyte Toe Onychomycosis: Terbinafine 250 mg/day vs. Itraconazole 200 mg/day – A Double-blind Comparative Trial', *British Journal of Dermatology*, 134: 16–17.

De Backer, M., De Vroey, C., Lesaffre, E., Scheys, I., and De Keyser, P. (1998) 'Twelve Weeks of Continuous Oral Therapy for Toenail Onychomychosis Caused by Dermatophytes: A Double-blind Comparative Trial of Terbinafine 250 mg/day versus Itraconazole 200 mg/day', *Journal of the American Academy of Dermatology*, 38: S57–S63.

Dempster, A.P., Laird, N.M., and Rubin, D.B. (1977) 'Maximum Likelihood from Incomplete Data via the EM Algorithm', *Journal of the Royal Statistical Society, Series B*, 39: 1–38.

Diggle, P.J., Heagerty, P., Liang, K.-Y., and Zeger, S.L. (2002) *Analysis of Longitudinal Data*. Oxford: Clarendon Press.

Fahrmeir, L. and Tutz, G. (2001) *Multivariate Statistical Modeling based on Generalized Linear Models*, 2nd Edition. New York: Springer-Verlag.

Faught, E., Wilder, B.J., Ramsay, R.E., Reife, R.A., Kramer, L.D., Pledger, G.W., and Karim, R.M. (1996) 'Topiramate Placebo-controlled Dose-ranging Trial in Refractory Partial Epilepsy using 200-, 400-, and 600-mg Daily Dosages', *Neurology*, 46: 1684–90.

Gelman, A., Carlin, J.B., Stern, H.S., and Rubin, D.B. (1995) *Bayesian Data Analysis*. New York: Chapman & Hall.

Lavergne, C. and Trottier, C. (2000) 'Sur l'estimation dans les Modèles Linéaires Généralisés à Effets Aléatoires', *Revue de Statistique Appliquée*, 48: 49–67.

Liang, K.-Y. and Zeger, S.L. (1986) 'Longitudinal Data Analysis using Generalized Linear Models', *Biometrika*, 73: 13–22.

Longford, N.T. (1993) *Random Coefficient Models*. Oxford: Clarendon Press.

Molenberghs, G. and Verbeke, G. (2005) *Models for Discrete Longitudinal Data*. New York: Springer.

Neuhaus, J.M., Kalbfleisch, J.D., and Hauck, W.W. (1991) 'A Comparison of Cluster-specific and Population-averaged Approaches for Analyzing Correlated Binary Data', *International Statistical Review*, 59: 25–30.

Pinheiro, J.C. and Bates, D.M. (1995) 'Approximations to the Log-likelihood Function in the Nonlinear Mixed-effects Model', *Journal of Computational and Graphical Statistics*, 4: 12–35.

Raubertas, R.E., Lee, C.I.C., and Nordheim, E.V. (1986) 'Hypothesis Tests for Normal Means Constrained by Linear Inequalities', *Communications in Statistics – Theory and Methods*, 15: 2809–33.

Shapiro, A. (1988) 'Towards a Unified Theory of Inequality Constrained Testing in Multivariate Analysis', *International Statistical Review*, 56: 49–62.

Silvapulle, M.J. and Silvapulle, P. (1995) 'A Score Test against One-sided Alternatives', *Journal of the American Statistical Association*, 90, 342–9.

Stiratelli, R., Laird, N.M., and Ware, J.H. (1984) 'Random-effects Models for Serial Observations with Binary Responses', *Biometrics*, 40, 961–71.

Stram, D.A. and Lee, J.W. (1994) 'Variance Components Testing in the Longitudinal Mixed Effects Model', *Biometrics*, 50: 1171–7.

Stram, D.A. and Lee, J.W. (1995) 'Correction to: Variance Components Testing in the Longitudinal Mixed Effects Model', *Biometrics*, 51: 1196.

Verbeke, G. and Molenberghs, G. (1997) *Linear Mixed Models in Practice: A SAS-oriented Approach*, Number 126 in Lecture Notes in Statistics. New York: Springer.

Verbeke, G. and Molenberghs, G. (2000) *Linear Mixed Models for Longitudinal Data*. New York: Springer.

Verbeke, G. and Molenberghs, G. (2003) 'The Use of Score Tests for Inference on Variance Components', *Biometrics*, 59, 254–62.

Wolfinger, R.D. and O'Connell, M. (1993) 'Generalized Linear Mixed Models: A Pseudo-likelihood Approach', *Journal of Statistical Computation and Simulation*, 48: 233–43.

9

Longitudinal Data Modeling

Nan M. Laird
Harvard School of Public Health

Garrett M. Fitzmaurice
Harvard School of Public Health & McLean Hospital

Data collected in a longitudinal study consist of repeated measures made on individuals over time. In this chapter we show how longitudinal data can be put into the multilevel framework, and we also highlight important distinctions between longitudinal and general multilevel data. Longitudinal data can be considered a special case of 2-level data where measurement occasions (the level 1 units) are clustered within individuals (the level 2 units). Unlike many multilevel designs, there is a natural ordering of the measurement occasions that has important implications for longitudinal analysis. In some instances longitudinal data have more than two levels, e.g., observations from a multicenter longitudinal clinical trial can be regarded as multilevel data having 3 levels, with repeated measurement occasions (level 1 units) clustered within individuals (level 2 units) clustered within clinics (level 3 units). This chapter explores the connections and distinctions between models for longitudinal data analysis and multilevel models. Special features of longitudinal data analysis, such as missing data, dropout or lost to follow-up, and modeling within-subject covariances are discussed, as a lead-in to other chapters where these topics are considered in detail. Additionally, we discuss extensions of longitudinal models to discrete outcomes. Three examples involving very different types of longitudinal studies, clinical trials, epidemiological cohort studies, and studies of dynamic economic behaviors are discussed.

9.1 INTRODUCTION

The use of multilevel models extends quite naturally to encompass models for data collected in longitudinal studies. Here we refer to a longitudinal study as one where measurements of the same subjects are taken repeatedly through time. Whereas it is often reasonable to assume subjects are independent, the repeated measures on the same subject are emphatically not assumed to be independent observations. The primary objective of a longitudinal study is measuring changes in the response over time. Often we wish to study change as a function of an intervention (e.g., drug treatment), an event (e.g., onset of menarche), or individual characteristics (e.g., gender, exposures). The structure of longitudinal data can be considered a special case of multilevel data, where subjects are the level 2 units, and the repeated measures are the level 1 units. Longitudinal data share the common feature of all two-level data structures that measurements on lower-level units (within a higher-level unit) are more similar or positively correlated. An important difference is that, with longitudinal data, the measures at the lower level are ordered by time. This has some important implications in terms of modeling the covariance structure of the level 1 units. A related feature of longitudinal studies is the occurrence of missingness and dropout; with repeated measures, missing data points and dropout can seriously complicate the analysis.

Designs for longitudinal studies can differ substantially in terms of number and spacing of repeated measures and their balance over individuals in the study. With longitudinal data, one can take a multilevel modeling approach via the introduction of random effects at the higher level, or one can choose to bypass the 2-level structure of the data and use a so-called *marginal modeling* approach. This is useful when the goal is to study overall effects, and not within-level effects. In some cases, the methods and inferences are similar, but in other settings they are not.

In general, longitudinal studies include two main types of predictor: predictors whose values do not change throughout the duration of the study, and predictors whose values change over time. The former are referred to as time-invariant predictors and common examples include gender and fixed experimental treatments or interventions. The latter are referred to as time-varying predictors and examples include time since baseline and environmental exposures.

In this chapter we will first consider three examples of longitudinal studies which illustrate the potential complexities. We then discuss the broad class of linear mixed effects models which is frequently used to analyze longitudinal data and show how it relates to multilevel models. We show how marginal models are formulated for longitudinal data and point to other chapters considering related

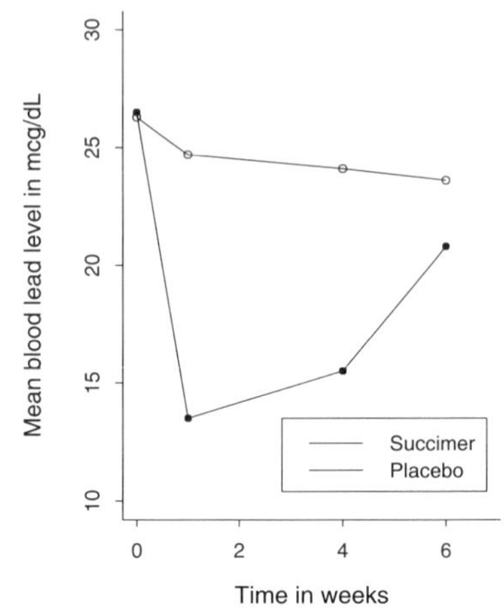

Figure 9.1 Mean Blood Lead Levels at Baseline (Week 0), Week 1, Week 4, and Week 6 in the Succimer and Placebo Groups

topics. Because the longitudinal data models we discuss can be viewed as special cases of the multilevel model, estimation and inference for these models are discussed in other chapters (see Chapter 3). We also consider models for longitudinal data when the outcomes are discrete (e.g., binary responses or counts). Finally, we conclude the chapter with a brief discussion of issues raised by missing data in longitudinal studies.

9.1.1 Case Studies

The first example we consider is the Treatment of Lead-Exposed Children (TLC) Trial (Fitzmaurice et al., 2011). This study was a randomized controlled trial, typical of many clinical trials designed to study whether therapeutic agents can effect a change in level of some health outcome. In the TLC trial, an oral chelating treatment called Succimer is compared to placebo in children with high blood lead levels. A common protocol for such trials is to take measurements on all subjects at baseline, before randomization, then at specified intervals following randomization until the end of specified follow-up. Although the TLC trial lasted for three years, and involved repeated treatment interventions, we will discuss only the first six weeks of follow-up here. Change in blood lead level at week 6 following the initiation of treatment (with an oral chelating agent or a placebo) is our primary endpoint. In addition to baseline and week 6, the design called for all subjects to be measured at weeks 1 and 4 following randomization. When the primary endpoint does not utilize the intermediate repeated measures, in our case measures taken at weeks 1 and 4, these measures might be ignored in the analysis. However, such measurements are useful in the event that subjects are lost to follow-up. The mean response profiles for the two groups randomized to Succimer and placebo are presented in Figure 9.1. This plot is produced by simply calculating the mean response at each occasion, within each treatment group, and joining adjacent means with a series of line segments.

Here, the randomized subjects can be thought of as determining the level 2 units; the four repeated measures are the level 1 units. The active treatment group (treatment with the oral chelating agent, Succimer) or the control (treatment with a placebo) is a covariate which is time invariant, i.e., an individual's treatment status does not change over time. In the TLC trial, the treatment was administered only once during the first six weeks, at randomization. In other cases, the treatment might be administered at multiple times during the study. In some cases, these more complex designs can be analyzed using the same general approach we discuss here.

If there were no subjects who had missing data or were lost to follow-up in the first six weeks, the TLC trial can be considered a balanced and complete design. It is balanced in the sense that all subjects are measured at the same set of occasions, and the data are complete. While most clinical trials are designed to be balanced, incompleteness, due to either intermediate missingness, dropout, or loss to follow-up is often inevitable. In the case where the design calls for the same measurement protocol for each subject, we would consider the study balanced, but incomplete if some observations are missing. This is an important consideration for both the strategy for modeling and analyzing the data. In some instances, as in the next example, the subjects can have the correct number of observations, but times of measurement may be subject-specific; alternatively, measurements may be mistimed. This raises a different type of complication for the analysis.

The TLC used a design common in clinical trials, and there are many variations on this theme. For example, we may have dichotomous outcomes, or counts, measured repeatedly on individuals; methods for longitudinal analyses of discrete outcomes are considered in Section 9.3. We can have a cross-over design, where the treatment changes within subject; in a cross-over study treatment is a time-varying predictor whose values at any occasion are determined a priori by study design. Often in epidemiological studies there will be planned missingness, e.g., in rotating panel designs, only selected subjects will be measured on different occasions. The next example illustrates a very different type of study which is inherently unbalanced.

The MIT Growth and Development Study (Fitzmaurice et al., 2011) was an observational study designed to characterize how measures of body fat change in girls before and after menarche. While it is accepted that body fat increases about the time of menarche, of interest was whether the increase persisted up to four years after. Pre-menarche, non-obese girls were selected into the study, and measured annually until four years after menarche, or the end of follow-up. The time of menarche was recorded for each participant. While the design is almost balanced with respect to calendar time (with the exception of the final measurement, the design is balanced), what is of interest is time relative to menarche. That is, time is rescaled so that a value of zero corresponds to menarche, a positive value indicates time after menarche, and a negative value denotes time before menarche. Since menarche is a random event, differing for each girl, the resulting sample design

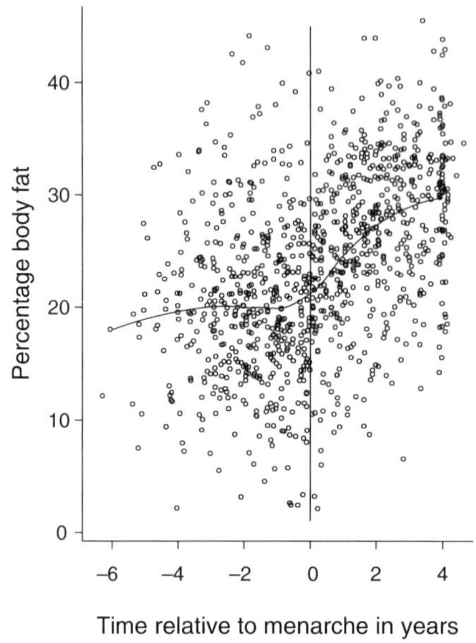

Figure 9.2 Time Plot of Percent Body Fat Against Time, Relative to Age of Menarche (in Years), with *Lowess* Smoothed Curve

is quite unbalanced. Although most girls had four measurements after menarche, the number of measures taken before was not fixed by design. Figure 9.2 illustrates all of the data points plotted against time relative to menarche. Note the cluster of measures at four years post menarche is by design, but the other measures are scattered randomly.

The third example is from the National Longitudinal Survey of Youth (NLSY). These longitudinal surveys, conducted by the U.S. Department of Labor's Bureau of Labor Statistics, are designed to collect data over time on the labor market experiences of various subgroups of the U.S. population. These surveys are particularly important for economists interested in the study of employment behavior of individuals in labor markets. Our example focuses on a sample of 545 full-time working males who completed their schooling by 1980 and who were followed annually from 1980 to 1987 (Vella and Verbeek, 1998). The question of interest is the effect of trade union membership on wage rate determination for young men. Trade union membership was determined by whether or not an individual had his annual wage set in a collective bargaining agreement. This predictor varied over time indicating movement in and out of union membership (approximately 50% of the sample of men changed union membership status at least once during follow-up). Additional variables collected in this study included education (years of schooling), labor market experience, marital status, annual hours worked, location, and type of occupation. In assessing the union effect on wages, it is important to control or adjust for these latter characteristics. Also, note that in this example trade union membership is both time-varying and *stochastic*, in the sense that its values at any measurement occasion cannot be accurately predicted because they are governed by a random mechanism. As a result, we must carefully consider the relationship between wages at any year and subsequent union membership status in the following year. For example, workers who receive lower wages (after adjusting for other characteristics) may be more likely to seek union employment in subsequent years; in that case, union membership status is said to be *endogenous* to the determination of wages. In general, when the predictor of interest is time-varying and stochastic, far greater care is needed in modeling and interpreting its relationship to the response variable.

Finally, we remark that the models that we consider in the following sections can also be used for cluster correlated data, where different measures on the same subject are of interest. For example, in the Connecticut Child Survey (Fitzmaurice et al., 2011) of behavioral outcomes in school children, a random sample of school children were drawn for behavioral assessment as well as use of school and community health services. A measure of behavioral assessment was obtained from the parent and from the teacher. These two measures can be considered correlated level 1 responses, with the children being the level 2 subjects. Here the level 1 responses are not ordered by time, but they are distinctly different in interpretation.

9.2 A COMMON STRUCTURE FOR LONGITUDINAL STUDIES

We first introduce some notation that we will use in the rest of the chapter. Laird and Ware (1982) introduced a notation that is very commonly used in longitudinal studies, where y_{ij} denotes the outcome measurement taken on the ith subject at the jth occasion. However, we note that this notation reverses the subscripting order and conflicts with the usual notation for multilevel models, whereby i indexes the level 1 units, and j indexes the level 2 units. Therefore, to avoid confusion, and to distinguish the notation from that used in other chapters in the book, we use y_{it} to denote the outcome measure, where i indexes the subject (level 2 unit), and t indexes the measurement number (level 1 unit). This is the notation that is most widely used in the statistical literature on longitudinal and time series analysis (where t is used to denote time). Notice that, in balanced designs, we could choose to let t index time itself (e.g., $t = 1$ means time 1, $t = 2$ means time 2, etc.), but this is not workable in unbalanced designs where different people are measured at different times. Hence, we let t index measurement number rather than time or occasion; we use the terms time or occasion somewhat interchangeably to denote the specification of measurements in the design. Thus t varies from 1 to n_i, where n_i is the total number of observations made on the ith subject. In the TLC trial a subject measured at all four occasions has $t = 1, 2, 3, 4$ (corresponding to measurements obtained at weeks 0, 1, 4, and 6), whereas a subject missing baseline has $t = 1, 2, 3$ (corresponding to measurements obtained at weeks 1, 4, and 6). Adopting this notation means that it will be necessary to explicitly introduce time in the model for the mean response and the covariance matrix. However, this provides us with a flexible notation for handling missing data and unbalanced designs.

A general linear mixed model for y_{it} can be written as

$$y_{it} = \boldsymbol{x}'_{it}\boldsymbol{\beta} + z'_{it}\boldsymbol{u}_i + \varepsilon_{it}, \qquad (9.1)$$

where $\boldsymbol{\beta}$ is the $p \times 1$ vector of fixed regression coefficients, $\boldsymbol{x}_{it}$ is the $p \times 1$ vector of predictors for the ith subject at the tth occasion, $\boldsymbol{u}_i$ is the $q \times 1$ vector of random effects for the ith subject at level 2, z_{it} is the corresponding design vector for the random effects, and ε_{it} is a random error term. Several important assumptions are commonly made for this mixed model. First, we assume that $\boldsymbol{u}_i$ and ε_{it} are independent and that both have mean zero. The assumption of zero means for the random effects implies that an ordinary linear regression model for the mean of y_{it} holds:

$$E(y_{it}|\boldsymbol{x}_{it}) = \mu_{it} = \boldsymbol{x}'_{it}\boldsymbol{\beta}, \qquad (9.2)$$

when averaged over the distribution of the random effects. That is, the general linear mixed model for y_{it} in equation (9.1) also implies a linear regression model for the marginal mean (averaging over the random effects) of y_{it}. Here, $\boldsymbol{\beta}$ models the fixed effects, i.e., the time effects, as well as the effects of interventions or covariates, and their interactions. In this chapter we will focus on the specification of the random error terms, since modeling the mean response is not fundamentally different from choosing an ordinary regression model for the mean response over time. Also, note that the model for the mean of a subject ($\boldsymbol{x}'_{it}\boldsymbol{\beta}$), can be selected independently from the model for the variance-covariance structure. We focus instead on models for the random effects and their variance components. A consequence of the zero mean assumption is that ordinarily the elements of z_{it} must be a subset of the elements of $\boldsymbol{x}_{it}$. The reason for this is not immediately obvious, but stems from the fact that $z'_{it}\boldsymbol{u}_i$ can then be usefully interpreted as an error defined with respect to a mean curve. So, for example, if u_i is scalar and $z_{it} = 1$ for all t, then the model for y_{it} must include an intercept term. It would not be realistic to assume the overall intercept is zero if individuals deviate randomly from this.

The $z_{it}'u_i$ term can be interpreted in several ways, e.g., it measures the deviation of the ith subject from the overall mean, or it provides a device for introducing correlation among the repeated measures on a single subject. This latter remark follows from the fact that any two observations on the ith subject, say t and t', will be correlated because they share the same random effects. The definition of z_{it} will be important in interpreting these random effects, as we discuss below.

The second assumption made is that u_i and ε_{it} are independent of each other for every i and every t, and the $u_i, u_j, \varepsilon_{it}, \varepsilon_{jt}$ are also independent for all $i \neq j$. This is an important assumption of the linear mixed model, namely that the random effects stemming from the different levels are independent. An alternative class of models takes the u_is to be fixed effects in the model (see, for example, Chapter 10 of Wooldridge, 2010).

Incorporating all of the observations on the ith subject into a vector of length n_i, denoted by y_i, and all the corresponding rows x_{it} into an $n_i \times p$ matrix X_i, and likewise for the $n_i \times q$ matrix Z_i, we can write

$$y_i = X_i\beta + Z_i u_i + \varepsilon_i,$$

where ε_i is an $n_i \times 1$ column vector with elements $\varepsilon_{i1}, \varepsilon_{i2}$, etc. Although we can continue to concatenate vectors in order to have a single large model for all the observations, there is not a great advantage in doing this for longitudinal data with only two levels, independent subjects, and repeated measures on subjects.

The variance-covariance structure is modeled by making assumptions on the ε_{it} or on the u_i, or both. In particular, the general form,

$$\text{Var}(y_i) = Z_i \Omega Z_i' + \Sigma_i, \qquad (9.3)$$

follows from assuming u_i and ε_i are independent of each other, with $\text{Var}(u_i) = \Omega$ and $\text{Var}(\varepsilon_i) = \Sigma_i$. Notice that Ω is not subscripted by i; rather, we assume the variance of the random subject effects is the same for all subjects. If desired, we can allow Ω to depend on level 2 covariates of the subjects, but we do not pursue this additional flexibility here. In contrast, Σ_i will depend upon i via the index n_i unless all subjects are measured at the same set of occasions. This variance-covariance structure given by (9.3) is very general, and in fact too general to be very useful, since without some restrictions on Ω and/or Σ_i, we can easily overparameterize the variance-covariance matrix, resulting in non-identifiability of the model parameters. We now consider possible structures of Ω and Σ_i that are commonly used in longitudinal designs. In considering these structures, it is useful to keep in mind that what we choose to use will depend in large measure on the design: the balance, timing and number of repeated occasions, etc. Of course, our goal is to have a model which can accommodate many different data sets found in practice.

9.2.1 Unrestricted Variance Models for Longitudinal Data

The least restrictive model for $\text{Var}(y_i)$ in longitudinal studies can be used in the setting where the design is balanced so that subjects are all measured on the same set of times (usually a relatively small number of times). By least restrictive, we mean that there are no assumptions about how the variances and covariances change over time. There are actually two ways to achieve such a model. Consider a subject measured on all occasions so that $n_i = T$, where T denotes the maximum number of measurement times. One approach is to set $\text{Var}(u_i) = 0$ so that

$$\text{Var}(y_i) = \Sigma, \qquad (9.4)$$

where Σ is now an arbitrary $T \times T$ variance-covariance matrix. This variance structure allows the variance to change over time as well as arbitrary correlations for all possible pairs

Table 9.1 Estimated Covariance Matrix (with Correlations Reported Below the Diagonal) for the Blood Lead Levels at Baseline, Week 1, Week 4, and Week 6 For the Children from the TLC Trial

Covariance Matrix			
25.2	19.1	19.7	22.2
0.57	44.3	35.5	29.7
0.57	0.78	47.4	30.6
0.58	0.58	0.58	58.7

of times. This model can easily accommodate missing observations by setting

$$\text{Var}(\boldsymbol{y}_i) = \Sigma_i,$$

where Σ_i is an $n_i \times n_i$ matrix constructed by removing the rows and columns of Σ that correspond to the missing observation. We can also characterize Σ_i as

$$\Sigma_i = \text{I}_i \Sigma \text{I}_i',$$

where I_i is an $n_i \times T$ matrix constructed by starting with a $T \times T$ identity matrix and removing any row corresponding to a missing observation. For example, in the TLC study, a subject with a missing baseline measurement would have

$$\text{I}_i = \begin{pmatrix} 0 & 1 & 0 & 0 \\ 0 & 0 & 1 & 0 \\ 0 & 0 & 0 & 1 \end{pmatrix}.$$

This variance-covariance model is quite convenient for the TLC study, with only four occasions of measurement, common to all subjects.

Alternatively, this same variance structure is attained when we set $\Sigma_i = 0$ so that

$$\text{Var}(\boldsymbol{y}_i) = Z_i \Omega Z_i'.$$

Again assuming a balanced design we let Ω be an arbitrary $T \times T$ variance-covariance matrix, and Z_i be the $T \times T$ identity matrix if no observations are missing. Thus $\text{Var}(\boldsymbol{y}_i) = \Omega$. If $n_i < T$, we take $Z_i = \text{I}_i$. We can think of the vector of random subject effects, $\boldsymbol{u}_i$, as giving the deviation of each subject from the overall mean at each time point. This illustrates the trade-off in structuring Ω versus Σ_i.

This unrestricted covariance structure was used in analysis of the blood lead level data from the TLC trial (see Table 9.1). Notice that the estimated variance of the blood lead levels doubles over the course of the trial. Some degree of heterogeneity of the variability over time is not uncommon in longitudinal studies. In Table 9.1, five of the six pairwise correlations among the repeated measures are quite similar in magnitude; however, this pattern cannot be expected in general. Ordinarily, the pairwise correlations among repeated measures tend to decrease with increasing time separation.

9.2.2 *Variance Component Models for Longitudinal Data*

When T is large, and/or the measurements are unstructured, as is often the case in multilevel settings, the unrestricted approach described in Section 9.2.1 is unsatisfactory. The simplest approach, and one of the most popular, is to let Ω and Σ_i each be scalars so that each has only one variance component (denoted here by σ_u^2 and σ_ε^2, respectively). When i indexes level 2 subjects in the longitudinal study, we take $q = 1$ and z_i to be an $n_i \times 1$ vector of ones. We can think of u_i, the ith subject's single random effect, as representing an underlying (or latent) level of response that persists over time and influences all repeated measurements on that subject. Here, σ_u^2 is a scalar that captures subject-to-subject variability in the u_is. To model the level 1 variance, a simple structure is often assumed:

$$\Sigma_i = \sigma_\varepsilon^2 \text{I}_{n_i},$$

where I_{n_i} is an $(n_i \times n_i)$ identity matrix. This implies that $\text{Var}(\varepsilon_{it})$ is constant over time, and $\text{Cov}(\varepsilon_{it}, \varepsilon_{it'}) = 0$ for $t \neq t'$. Thus we can write

$$y_{it} = \boldsymbol{x}'_{it}\boldsymbol{\beta} + z'_{it}u_i + \varepsilon_{it},$$

so that the assumptions imply that

$$\text{Var}(y_{it}) = \sigma_u^2 + \sigma_\varepsilon^2$$

and

$$\text{Cov}(y_{it}, y_{it'}) = \sigma_u^2, \text{ for all } i \text{ and } t \neq t'.$$

One justification for this correlation structure is it assumes that, conditional on a specific subject, repeated observations taken on that subject are independent. While this may be reasonable in many situations, for example, when the level 2 variable is a school or a classroom and level 1 refers to teachers or students, it is rarely useful in longitudinal studies because the assumption that variances and covariances are constant throughout the period of the study is generally unrealistic. The model remains popular, however, because in balanced designs, estimation and testing can be greatly simplified.

More useful structures for $\text{Var}(\boldsymbol{y}_i)$ are obtained by enriching either Ω or Σ_i. One popular approach is to let Z_i denote the 'design on time'; for example, if we assume a linear trend to model change over time, we might assume each level 2 unit has their own subject-specific intercept and slope. Here, the $(n_i \times 2)$ design matrix Z_i would be comprised of the vector of ones (for the subject-specific intercepts) and the vector of times (for the subject-specific slopes) corresponding to the n_i observations on subject i. For example, a child measured at weeks 0, 1, and 6 in the TLC trial would have

$$Z_i = \begin{pmatrix} 1 & 0 \\ 1 & 1 \\ 1 & 6 \end{pmatrix},$$

and $\boldsymbol{u}_i$ would be the intercept and slope used to determine the ith subject's deviation from the overall linear trend. Here, Ω would be a 2×2 matrix giving the variance of the random subject coefficients, the random intercepts and slopes, as well as their covariance. The overall time trends (when averaged over subjects) would be modeled in $X_i\boldsymbol{\beta}$ and would depend upon treatment group, and possibly other covariates. We can then assume that, conditional on an individual's own trajectory, $X_i\boldsymbol{\beta} + Z_i\boldsymbol{u}_i$, the repeated deviations are independent, i.e.,

$$\Sigma_i = \sigma_\varepsilon^2 I_i,$$

as previously defined. With this model, both $\text{Var}(y_{it})$ and $\text{Cov}(y_{it}, y_{it'})$ will depend upon the variance components and on the elements of z_{it} and $z_{it'}$. This model can be made more complex if desired by adding additional columns to Z_i which include additional level 1 covariates. For example, Donnelly et al. (1995) discuss a model for longitudinal growth in lung function where the subject-specific trajectories depend not only on age of measurement, but also height at the tth measurement. However, parameters in Ω may be poorly estimated if the model is too large.

A different approach to enriching Ω and Σ_i is to leave Z_i as a vector of ones, and Ω as a scalar, but to add correlation structure to Σ_i. This approach has been taken by Diggle (1988); also, see Diggle et al. (2002). Some care is required when attempting to enrich both Ω and Σ_i. As was mentioned earlier, this can lead to identifiability problems due to an overparameterized variance-covariance matrix. For example, in many longitudinal studies there may be insufficient information in the data at hand to support separate estimation of randomly varying intercepts and slopes in addition to a complex serial correlation structure (e.g., autoregressive) for the within-subject errors. A more detailed discussion of these issues, and alternative models for the covariance, can be found in Chapter 10.

Finally, to highlight some of the main ideas discussed so far, we consider a longitudinal analysis of the data from the MIT Growth and Development study. A linear mixed model approach was taken to fit a linear spline, with

Table 9.2 Estimated Regression Coefficients (Fixed Effects) and Standard Errors for the Percent Body Fat Data

Variable	Estimate	SE	Z
Intercept	21.361	0.565	37.84
Time	0.417	0.157	2.65
$(\text{Time})_+$	2.047	0.228	8.98

a knot at time zero (here denoting menarche) to the measurements of percent body fat over time. Recall that the main question of interest was whether the slope on time changes after menarche, hence using a linear spline makes it easy to address this question. Specifically, we assumed that each girl has a piecewise linear spline growth curve with a knot at the time of menarche. That is, each girl's growth curve can be described with an intercept and two slopes, one slope for the changes in response before menarche, another slope for the changes in response after menarche. We fit the following linear mixed effects model:

$$\begin{aligned} y_{it} &= \beta_0 + \beta_1 t_{it} + \beta_2 (t_{it})_+ + u_{0i} \\ &\quad + u_{1i} t_{it} + u_{2i} (t_{it})_+ + \varepsilon_{it}, \end{aligned}$$

where t_{it} denotes the time, relative to menarche (i.e., before or after menarche), of the tth measurement on the ith subject (with $t_{it} = 0$ at menarche), and $(t_{it})_+ = t_{it}$ if $t_{it} > 0$ and $(t_{it})_+ = 0$ if $t_{it} \leq 0$. Here, $\boldsymbol{x}_{it} = \boldsymbol{z}_{it} = [1, t_{it}, (t_{it})_+]'$, $\boldsymbol{u}_i \sim N_3(0, \Omega)$, and $\boldsymbol{\varepsilon}_i \sim N_{n_i}(0, \Sigma_i)$, where $\Sigma_i = \sigma_\varepsilon^2 \mathrm{I}_{n_i}$.

Table 9.2 gives the fixed effects estimate of the overall mean curve, the intercept and two slopes, before (β_1) and after menarche ($\beta_1 + \beta_2$). Notice that this approach allows us to estimate mean percent body fat at menarche, even though few if any girls were actually measured then. Based on the magnitude of the estimate of β_2, relative to its standard error, there is a significant difference between the slopes before and after menarche. The estimate of the population mean pre-menarcheal slope is 0.42 and indicates that the annual rate of body fat growth is less that 0.5%. In contrast, the estimate of the population mean post-menarcheal slope is 2.46 (0.417 + 2.047) and indicates that the annual rate of body fat growth is approximately 2.5%, almost six times higher than the corresponding rate in the pre-menarcheal period.

The matrix Ω has six parameters, three parameters for the variances of the random intercepts and slopes, and three parameters for their covariances. Estimates of the variance and covariance parameters are given in Table 9.3. Much of the subject-to-subject variability is contained in the intercept. The covariance between a subject's intercept and post-menarche slope is negative, suggesting that girls with higher body fat at menarche tend to gain less after menarche. The model allows us to calculate values of the variances and covariances for measures of percent body fat at arbitrary times; this is illustrated in Table 9.4. Here, we see a slight decline in variance over time; in addition, the strength of the correlation declines with increasing time separation. The latter feature of the correlation is not uncommon in longitudinal studies.

9.3 MODELS FOR DISCRETE LONGITUDINAL DATA

In the previous section we considered models for longitudinal data that are appropriate when the response variable, y_{it}, is numeric and has an approximate normal distribution. However, in many applications the longitudinal response is not continuous. For example, the response recorded on individuals may be binary, ordinal, or a count. When the longitudinal response is discrete rather than

Table 9.3 Estimated Covariance of the Random Effects and Standard Errors for the Percent Body Fat Data

Parameter	Estimate	SE
$\mathrm{Var}(u_{0i})$	45.941	5.739
$\mathrm{Var}(u_{1i})$	1.631	0.433
$\mathrm{Var}(u_{2i})$	2.750	0.964
$\mathrm{Cov}(u_{0i},u_{1i})$	2.526	1.219
$\mathrm{Cov}(u_{0i},u_{2i})$	−6.110	1.873
$\mathrm{Cov}(u_{1i},u_{2i})$	−1.751	0.598
$\mathrm{Var}(\varepsilon_{it}) = \sigma^2$	9.473	0.544

continuous, the linear models for changes in the mean response discussed in Section 9.2 are not appealing. Instead, generalized linear models (Nelder and Wedderburn, 1972) can be extended for analyzing discrete longitudinal data, where a suitable non-linear transformation of the mean response is assumed to be linear in the covariates, X_i.

In the univariate setting, generalized linear models (see, for example, McCullagh and Nelder, 1989) provide a unified class of models for regression analysis of observations on a discrete or continuous response. Broadly speaking, generalized linear models have been extended in two distinct ways to handle the correlation among longitudinal responses. The first extension leads to a class of regression models called 'marginal models'. The term 'marginal' is used in this context to emphasize that the model for the mean response depends only on X_i and not on any random effects. By analogy with the linear models of the previous section, marginal models set $\mathrm{Var}(\boldsymbol{u}_i) = 0$; they allow for dependence among the repeated measures via a so-called 'working' covariance matrix. Marginal models provide a straightforward extension of generalized linear models to longitudinal data and they have frequently been applied in the biomedical and health sciences. The second extension of generalized linear models to longitudinal data accounts for the correlation among the repeated measures via the introduction of a vector of random effects, $\boldsymbol{u}_i$. In these models, the mean response depends on both X_i and $\boldsymbol{u}_i$, via a known link function. These models are known as *generalized linear mixed effects models* (GLMMs) and they extend in a natural way the conceptual approach represented by the linear mixed effects models discussed in Section 9.2. Before discussing these two broad classes of models, we must emphasize that they differ not only in approach for accounting for the correlation among longitudinal data, but also in terms of regression parameters with distinct interpretations. As a result, somewhat greater care is required in the choice of model for discrete longitudinal data.

9.3.1 Marginal Models

In our discussion of marginal models we use notation similar to that introduced in Section 9.2. Specifically, we let y_{it} denote the response variable for the ith subject on the tth measurement occasion. The response variable can be continuous, binary, ordinal, or a count; the notation does not distinguish between the different types of response.

A key feature of marginal models for longitudinal data is that they separately model the mean response and the covariance among the repeated responses. In a marginal model, the goal is to make inferences about the former, whereas the latter is sometimes regarded as a nuisance characteristic of the data that must be accounted for to make correct inferences. Specifically, a marginal model for

Table 9.4 Estimated Marginal Correlations (on the Off-Diagonals) Among Repeated Measures of Percent Body Fat Between Four Years Pre- and Post-Menarche, with Estimated Variances along the Main Diagonal

Time in years (relative to menarche)								
−4	−3	−2	−1	0	1	2	3	4
61.3	0.82	0.78	0.71	0.61	0.60	0.57	0.52	0.47
0.82	54.9	0.81	0.76	0.70	0.68	0.64	0.60	0.54
0.78	0.81	51.8	0.80	0.76	0.74	0.71	0.66	0.60
0.71	0.76	0.80	52.0	0.81	0.79	0.76	0.71	0.64
0.61	0.70	0.76	0.81	55.4	0.81	0.78	0.73	0.66
0.60	0.68	0.74	0.79	0.81	49.1	0.79	0.76	0.70
0.57	0.64	0.71	0.76	0.78	0.79	44.6	0.77	0.74
0.52	0.60	0.66	0.71	0.73	0.76	0.77	41.8	0.76
0.47	0.54	0.60	0.64	0.66	0.70	0.74	0.76	40.8

longitudinal data has the following three-part specification:

1 The conditional mean of each response, $E\left(y_{it}|\boldsymbol{x}_{it}\right) = \mu_{it}$, is assumed to depend on the covariates through a known link function

$$g\left(\mu_{it}\right) = \boldsymbol{x}'_{it}\boldsymbol{\beta}.$$

2 The conditional variance of each y_{it} is assumed to depend on the mean according to

$$\text{Var}\left(y_{it}|\boldsymbol{x}_{it}\right) = \phi\, v\left(\mu_{it}\right),$$

where $v\left(\mu_{it}\right)$ is a known 'variance function' (i.e., a known function of the mean, μ_{it}) and ϕ is a scale parameter that is either fixed and known or may need to be estimated.

3 The conditional within-subject dependence among the repeated responses is assumed to be a function of an additional set of association parameters, $\boldsymbol{\alpha}$ (and may also depend upon the means, μ_{it}); e.g., elements of $\boldsymbol{\alpha}$ might represent the pairwise correlations or log odds ratios among the repeated responses, depending on the type of response.

This three-part specification of a marginal model makes the extension of generalized linear models to longitudinal data apparent. The first two parts of the marginal model correspond to the standard generalized linear model (although with no explicit distributional assumptions about the responses). The third component, incorporating the time dependence among the repeated responses from the same individual, represents the main extension to longitudinal data.

To fix ideas, we consider two examples of marginal models using this three-part specification. For the first example suppose that y_{it} is a binary response, taking values of 0 (denoting 'failure') or 1 (denoting 'success'). For example, in the Muscatine Coronary Risk Factor (MCRF) study, a longitudinal survey of school-age children in Muscatine, Iowa (Woolson and Clarke, 1984; Lauer et al., 1997), a measure of obesity was obtained biennially from 1977 to 1981. The longitudinal outcome variable in this study is the binary indicator of whether a child is obese at each occasion. The goal of the study was to examine the development and persistence of obesity, a known risk factor for coronary disease, in children (see Fitzmaurice et al., 2011, for a detailed presentation of the results of a longitudinal analysis of these data).

Suppose it is of interest to relate changes in $\mu_{it} = E(y_{it}|\boldsymbol{x}_{it}) = \Pr(y_{it} = 1|\boldsymbol{x}_{it})$ over time to the covariates. An example of a marginal model for y_{it} is given by the following three-part specification:

1 The mean of y_{it} is related to the covariates by a logit link function,

$$\log\left(\frac{\mu_{it}}{1-\mu_{it}}\right) = \boldsymbol{x}'_{it}\boldsymbol{\beta}.$$

2 The variance of each y_{it} depends on the mean response,

$$\text{Var}\,(y_{it}|\boldsymbol{x}_{it}) = \mu_{it}(1-\mu_{it}),$$

with fixed $\phi = 1$.

3 The within-subject dependence among the repeated binary responses is assumed to have an unstructured pairwise log odds ratio pattern,

$$\log \text{OR}\,(y_{is}, y_{it}|\boldsymbol{x}_{is}, \boldsymbol{x}_{it}) = \alpha_{st}, \quad \text{for } s \neq t,$$

where

$$\text{OR}\,(y_s, y_t) = \frac{\Pr(y_s = 1, y_t = 1)\Pr(y_s = 0, y_t = 0)}{\Pr(y_s = 1, y_t = 0)\Pr(y_s = 0, y_t = 1)}.$$

The marginal model specified above is a logistic regression model where the within-subject association is specified in terms of pairwise log odds ratios rather than pairwise correlations. For binary data, the odds ratio is a more natural measure of dependence and, unlike the correlation, is far less constrained by the marginal probabilities, μ_{it}.

For the second example, suppose that y_{it} is a count and it is of interest to relate changes in the expected count (or expected rate) over time to the covariates. For example, in a placebo-controlled clinical trial of patients with leprosy at the Eversley Childs Sanitarium in the Philippines (Snedecor and Cochran, 1967), patients were randomized to either of two antibiotics or to a placebo. Baseline and follow-up measurements of the number of leprosy bacilli at six sites of the body where the bacilli tend to congregate were recorded for each subject. The outcome variable in this study is the total count of the number of leprosy bacilli at the six sites at each occasion (see Fitzmaurice et al., 2011, for a detailed analysis of these data).

An example of a marginal model for y_{it} is given by the following three-part specification:

1 The mean of y_{it} is related to the covariates through a log link function,

$$\log\,(\mu_{it}) = \boldsymbol{x}'_{it}\boldsymbol{\beta}.$$

2 The variance of each y_{it}, given the effects of the covariates, depends on the mean response,

$$\text{Var}\,(y_{it}|\boldsymbol{x}_{it}) = \phi\,\mu_{it},$$

where ϕ is a scale parameter to be estimated.

3 The within-subject dependence among the repeated responses is assumed to have an unstructured pairwise correlation pattern,

$$\text{Corr}\,(y_{is}, y_{it}|\boldsymbol{x}_{is}, \boldsymbol{x}_{it}) = \alpha_{st}, \quad \text{for } s \neq t.$$

The marginal model specified above is a log-linear regression model, with an extra-Poisson variance assumption to allow for potential *overdispersion*. In many applications, count data have variability that far exceeds that predicted by the Poisson distribution; this phenomenon is often referred to as overdispersion or extra-Poisson variability. A scale factor ϕ (with $\phi \neq 1$) is included in the specification of the variance to allow for overdispersion. Finally, the within-subject association is specified in terms of an unstructured pairwise correlation pattern. We note that an unstructured correlation matrix is more appealing in studies with a balanced design and a relatively small number of repeated measures.

These two examples of marginal models demonstrate how the specification of the three components of a marginal model might differ according to the type of response variable. In principle, any suitable link function can be chosen, and alternative assumptions about the variances and within-subject associations can be made. In general, the choices for the three components of a marginal model will reflect statistical and subject-matter considerations (see Chapter 13 of Fitzmaurice et al., 2011, for numerous examples of the application of marginal models to longitudinal studies). Note that for the case of a continuous response, a marginal model can be specified in terms of equations (9.2) and (9.4) in Section 9.2.

Next, we consider estimation of the marginal model parameters. In general, the three-part marginal model specification does not determine the joint distribution of $\boldsymbol{y}_i$; an exception is the linear models considered in Section 9.2 where the multivariate normal distribution for $\boldsymbol{y}_i$ is completely specified by the mean vector and the variance-covariance matrix. Consequently, maximum likelihood (ML) estimation of the parameters in the marginal model requires additional assumptions about the joint distribution of $\boldsymbol{y}_i$. This presents two alternative ways to proceed. In principle, the three-part marginal model specification can be extended by making full distributional assumptions about the vector of responses, $\boldsymbol{y}_i$. However, there are a number of difficulties with the application of likelihood-based marginal models for the analysis of discrete longitudinal data. The first is the lack of a convenient joint distribution for discrete multivariate responses, in particular, one having similar properties to the multivariate normal distribution. Paradoxically, the most general joint distribution for discrete longitudinal data requires specification of many higher-order moments despite the fact that there is, in some sense, substantially less information in the discrete data case. Second, for many joint distributions for discrete longitudinal data there is no closed form expression for the joint probabilities in terms of the marginal model parameters; this complicates maximization of the likelihood. Third, due to the curse of dimensionality, likelihood-based approaches face difficulties with sparseness of data as the number of repeated measures increases. Finally, many of the aforementioned difficulties are exacerbated by the fact that ML estimates of $\boldsymbol{\beta}$ are sensitive to misspecification of the higher-order moments. That is, for many proposed likelihood-based methods, if the marginal model for the mean response has been correctly specified but the model for the higher-order moments has not, then the ML estimator of $\boldsymbol{\beta}$ is badly biased.

These difficulties surrounding ML estimation have led to the recognition that it is possible to estimate the regression parameters, $\boldsymbol{\beta}$, without making full distributional assumptions about $\boldsymbol{y}_i$. The avoidance of distributional assumptions leads to a method of estimation known as *generalized estimating equations* (GEE). The GEE approach, advocated in the biostatistical literature by Liang and Zeger (1986), is a natural extension of the quasi-likelihood approach (Wedderburn, 1974) for generalized linear models to the multivariate response setting. The essential idea behind the GEE approach is to extend quasi-likelihood methods by incorporating additional nuisance parameters for the covariance matrix of the vector of responses. For example, given a model for the pairwise correlations among the responses, the corresponding covariance matrix can be constructed as the product of the standard deviations and correlations,

$$V_i = A_i^{\frac{1}{2}} \mathrm{Corr}(\boldsymbol{y}_i) A_i^{\frac{1}{2}},$$

where A_i is a diagonal matrix with $\mathrm{Var}(y_{it}) = \phi\, v\,(\mu_{it})$ along the diagonal, and $\mathrm{Corr}(\boldsymbol{y}_i)$ is a correlation matrix (here assumed to be a function of $\boldsymbol{\alpha}$). Note that because the variances are a function of the means (and perhaps a scale parameter, ϕ), the GEE approach incorporates additional parameters $\boldsymbol{\alpha}$ for $\mathrm{Corr}(\boldsymbol{y}_i)$ rather than for $\mathrm{Cov}(\boldsymbol{y}_i)$. In the GEE approach, V_i is referred to as a 'working' covariance matrix to distinguish it from the true underlying covariance matrix of $\boldsymbol{y}_i$. The term 'working' in this context acknowledges uncertainty about the assumed model for the variances and within-subject associations. Because the GEE depend on both $\boldsymbol{\beta}$ and $\boldsymbol{\alpha}$, an iterative two-stage estimation procedure is required; this has been implemented in many widely available software packages.

The GEE approach has a number of appealing properties for estimation of the regression parameters in marginal models. First, in many longitudinal designs the GEE estimator of $\boldsymbol{\beta}$ is almost efficient when compared to the maximum likelihood estimator. For example, it can be shown that the GEE has a similar expression to the generalized least squares (GLS) estimator of $\boldsymbol{\beta}$ in linear models (identity link function) for continuous responses that are assumed to have a multivariate normal distribution (see Chapter 3 for details on estimation for linear mixed models). Specifically, with balanced data the GEE estimator of $\boldsymbol{\beta}$ in linear models is identical to the GLS estimator. Second, the GEE estimator has a very appealing robustness property, yielding a consistent estimator of $\boldsymbol{\beta}$ even if the within-subject associations among the repeated measures have been misspecified. It only requires that the model for the mean response be correct.

Although the GEE approach yields a consistent estimator of $\boldsymbol{\beta}$ under misspecification of the within-subject associations, the usual model-based standard errors obtained under the misspecified model for the within-subject association are not valid. However, valid standard errors for the resulting estimator $\widehat{\boldsymbol{\beta}}$ can be obtained using the empirical or so-called 'sandwich' estimator of $\text{Cov}(\widehat{\boldsymbol{\beta}})$ (Cox, 1961; Huber, 1967; White, 1982; Royall, 1986). The 'sandwich' estimator is also robust in the sense that, with sufficiently large samples, it provides valid standard errors when the assumed model for the covariances among the repeated measures is not correct. We note that the 'sandwich' estimator can also be used to obtain variance estimates in the linear model setting to guard against possible misspecification of the variance structure.

In summary, marginal models are a very natural way to extend generalized linear models to longitudinal responses. Marginal models specify a generalized linear model for the longitudinal responses at each occasion but also include a model for the within-subject association among the responses. A crucial aspect of marginal models is that the mean response and within-subject association are modeled separately. For other than linear models, this separation of the modeling of the mean response and the dependence among responses has important implications for interpretation of the regression parameters. In particular, the regression parameters, $\boldsymbol{\beta}$, in the marginal model have so-called *population-averaged* interpretations. That is, they describe how the mean response in the population is related to the covariates. For example, regression parameters in a marginal model might have interpretation in terms of contrasts of the changes in the transformed mean responses in sub-populations (e.g., different treatment or exposure groups). For the last 20 years, the GEE approach has been the dominant method for estimation of marginal models for longitudinal data. The GEE method is semi-parametric, in that the estimating equations are derived without fully specifying the joint distribution of $\boldsymbol{y}_i$. The GEE method provides consistent estimators of the marginal regression parameters, $\boldsymbol{\beta}$, under the weak condition that only the marginal model for the mean response is correctly specified.

9.3.2 Generalized Linear Mixed Models

In the previous section, we discussed how marginal models can be considered an extension of generalized linear models that *directly* incorporate the within-subject association among the repeated measurements. In a certain sense, marginal models account for the *consequences* of the correlation among the repeated measures but do not provide any explanation for its potential source. An alternative approach for accounting for the within-subject association, and one that provides a source for the within-subject association, is via the introduction of random effects in the model for the mean response. Following the same basic ideas as in linear mixed effects models, generalized linear models can be extended to longitudinal data by allowing a subset of the regression coefficients to vary randomly from one individual to another. These models are known as *generalized linear mixed (effects) models* (GLMMs), and they extend in a natural way the conceptual approach represented by the linear mixed effects models discussed in Section 9.2.

In GLMMs the model for the mean response is conditional upon both measured covariates, X_i, and unobserved random effects, $\boldsymbol{u}_i$; it is the inclusion of the latter that induces correlation among the repeated responses marginally, when averaged over the distribution of the random effects. However, as we discuss later, with non-linear link functions, the introduction of random effects has implications for the interpretation of the 'fixed effects' regression parameters.

The generalized linear mixed model can be formulated using the following two-part specification:

1 Given a $q \times 1$ vector of random effects $\boldsymbol{u}_i$, the y_{it} are assumed to be conditionally independent and to have exponential family distributions with conditional mean depending on both fixed and random effects,

$$g\{E(y_{it}|\boldsymbol{u}_i)\} = \boldsymbol{x}'_{it}\boldsymbol{\beta} + z'_{it}\boldsymbol{u}_i,$$

for some known link function, $g(\cdot)$. The conditional variance is assumed to depend on the conditional mean according to $\text{Var}(y_{it}|\boldsymbol{u}_i) = \phi\, v\{E(y_{it}|\boldsymbol{x}_{it}, \boldsymbol{u}_i)\}$, where $v(\cdot)$ is a known variance function and ϕ is a scale parameter that may be known or may need to be estimated.

2 The random effects, $\boldsymbol{u}_i$, are assumed to be independent of the covariates $\boldsymbol{x}_{it}$, and to have a multivariate normal distribution, with zero mean and $q \times q$ covariance matrix Ω.

These two components completely specify a broad class of generalized linear mixed models. In principle, the conditional independence assumption in the first component is not necessary, but is commonly made. Similarly, any multivariate distribution can be assumed for the $\boldsymbol{u}_i$; in practice, however, it is common to assume that the $\boldsymbol{u}_i$ have a multivariate normal distribution.

Next, consider an example of a GLMM using this two-part specification. For this example suppose that y_{it} is a binary response, taking values of 0 (denoting 'failure') or 1 (denoting 'success'). An example of a GLMM for y_{it} is given by

$$\text{logit}\{E(y_{it}|\boldsymbol{u}_i)\} = \boldsymbol{x}'_{it}\boldsymbol{\beta} + u_i,$$

where, given u_i, the y_{it} are assumed to be conditionally independent Bernoulli random variables, and u_i has a normal distribution with zero mean and constant variance. This model is often referred to as the simple 'logit-normal model' and is very similar in spirit to the random intercept model for continuous outcomes discussed in Section 9.2.

Although this model is simple, it is not so straightforward to fit because maximum likelihood estimation requires maximization of the marginal likelihood. The marginal likelihood requires integration over the distribution of the random effect and cannot be evaluated in a closed form. The normal distribution assumption for the random effects in generalized linear mixed models leads, in general, to intractable likelihood functions; an exception is the linear mixed model for continuous data.

Because no simple analytic solutions are available, Stiratelli, Laird, and Ware (1984) proposed an approximate method of estimation for the logit-normal model based on empirical Bayes ideas. This led to the development of a general approach for fitting GLMMs known as penalized quasi-likelihood (PQL); see, for example, Schall (1991), Breslow and Clayton (1993), and Wolfinger (1993). Despite its generality, PQL can often yield quite biased estimators of the variance components, which in turn leads to biased estimators of $\boldsymbol{\beta}$, especially for longitudinal binary data. This has motivated research on bias corrections (e.g., Breslow and Lin, 1995) and on more accurate approximations (e.g., Raudenbush, Yang, and Yosef, 2000). An appealing alternative to these approximate methods is the use of numerical integration. In particular, adaptive Gaussian quadrature with the numerical integration centered around the empirical Bayes estimates of the random effects, permits maximization of the marginal likelihood with any desired degree of accuracy (e.g., Anderson and Aitkin, 1985; Hedeker and Gibbons, 1994, 1996). Adaptive Gaussian quadrature is especially appealing for longitudinal data where the dimension of the random effects is often relatively low. See Chapters 8 and 15 for a more detailed discussion of estimation in GLMMs. See Chapter 14 of Fitzmaurice et al. (2011) for illustrative examples of the application of GLMMs to data from longitudinal studies.

Finally, we consider the interpretation of the fixed effects $\boldsymbol{\beta}$ in GLMMs. These regression parameters are said to have 'subject-specific' interpretations in terms of changes in the transformed mean response for an individual. That is, to interpret a component of $\boldsymbol{\beta}$ we must consider a unit change in the corresponding covariate while holding fixed $\boldsymbol{u}_i$ (and the remaining covariates). The most natural way to hold $\boldsymbol{u}_i$ fixed is to focus on the conditional mean response of a given individual. In the longitudinal setting, this interpretation of the components of $\boldsymbol{\beta}$ is most natural when the covariate is time-varying. (Alternatively, $\boldsymbol{\beta}$ can be interpreted in terms of the comparison of individuals with different values of the covariate but with the same values for $\boldsymbol{u}_i$.) Except in certain special cases, the fixed effects, $\boldsymbol{\beta}$, are not directly comparable to the regression parameters in a marginal model. The implied model for the marginal means can only be obtained by integrating over the distribution of the random effects. This involves taking an expectation of a non-linear function of $\boldsymbol{u}_i$:

$$\begin{aligned}\mu_i &= E(\boldsymbol{y}_i|X_i)\\ &= E\{E(\boldsymbol{y}_i|X_i,\boldsymbol{u}_i)\}\\ &= E\{g^{-1}(X_i\boldsymbol{\beta}+Z_i\boldsymbol{u}_i)\}\\ &= \int_{-\infty}^{\infty} g^{-1}(X_i\boldsymbol{\beta}+Z_i\boldsymbol{u}_i)f(\boldsymbol{u}_i)d\boldsymbol{u}_i,\end{aligned} \tag{9.5}$$

where $f(\boldsymbol{u}_i)$ is the joint probability density function for $\boldsymbol{u}_i$. Even though $E(\boldsymbol{u}_i)=0$, the marginal mean, $E(\boldsymbol{y}_i|X_i)$, given by equation (9.5) does not, in general, have a closed-form expression and, moreover,

$$E(\boldsymbol{y}_i|X_i) \neq g^{-1}(X_i\boldsymbol{\beta}),$$

for any $\boldsymbol{\beta}$. That is, marginalization of a GLMM does not yield a generalized linear model for the marginal means.

9.3.3 Some Remarks on Marginal versus Mixed Effects Models

In the longitudinal setting, marginal models and generalized linear mixed models are widely used for the analysis of discrete data. Although these two classes of models are conceptually similar to the corresponding linear models discussed in Section 9.2, there is an important distinction. With discrete data, due to the non-linear link functions that are commonly adopted, the fixed effects $\boldsymbol{\beta}$ in generalized linear mixed models are not comparable to the regression parameters $\boldsymbol{\beta}$ in marginal models. The lack of comparability reflects the distinct targets of inference associated with these two classes of model. The fixed effects in GLMMs describe the effects of covariates on changes in an individual's response over time while the regression parameters in marginal models describe the effects of covariates on changes in the population mean response over time. The linear models discussed in Section 9.2, with covariance expressed either in terms of equation (9.3) or (9.5), are an exception. With an identity link function, the regression parameters have the same interpretation in both classes of model. That is, marginalization of the linear mixed effects model yields:

$$\begin{aligned}E(\boldsymbol{y}_i|X_i) &= E\{E(\boldsymbol{y}_i|X_i,\boldsymbol{u}_i)\}\\ &= E(X_i\boldsymbol{\beta}+Z_i\boldsymbol{u}_i)=X_i\boldsymbol{\beta}.\end{aligned}$$

Simply stated, in the linear mixed effects model $\boldsymbol{\beta}$ has a marginal interpretation because the average of the linear rates of change over time for individuals is the same as the linear rate of change over time in the population mean response. In contrast, for any *non-linear* function of $\boldsymbol{u}_i$, say $g^{-1}(X_i\boldsymbol{\beta}+Z_i\boldsymbol{u}_i)$,

$$E\{g^{-1}(X_i\boldsymbol{\beta}+Z_i\boldsymbol{u}_i)\} \neq g^{-1}(X_i\boldsymbol{\beta}).$$

Recognizing that 'subject-specific' and 'population-averaged' effects are different, and address distinct scientific questions, the choice between marginal and generalized linear mixed models for longitudinal data must be made on subject-matter grounds.

Interestingly, much of the multilevel literature on the analysis of discrete data is dominated by the use of generalized linear mixed effects models. Conceptually, it may seem natural to account for the intra-cluster correlations at different levels via the introduction of random effects at different levels. However, for discrete data, this does raise subtle issues concerning the interpretation of the fixed effect regression parameters and questions about what is the relevant target of inference; see Hubbard et al. (2010) for a review of the assumptions behind, and the type of inference provided by, marginal and mixed effects models in the multilevel data setting.

9.4 MISSING DATA

An important issue for longitudinal data analysis that has received somewhat more attention than in the analysis of multilevel data is the consideration of missing data. Missing data are a common and challenging problem in longitudinal studies. Indeed, missing data are almost the rule, not the exception, in longitudinal studies. For example, study participants do not always appear for a scheduled observation, or they may simply leave the study before its completion. When some observations are missing, the data are necessarily unbalanced over time, since not all individuals have the same number of repeated measurements obtained at a common set of occasions. However, to distinguish missing data in a longitudinal study from other kinds of unbalanced data, such data sets are often referred to as being 'incomplete'. This distinction is important and emphasizes the fact that an intended measurement on an individual could not be obtained.

When longitudinal data are incomplete, there are implications for their analysis that go beyond whether a particular statistical method can handle unbalanced longitudinal data. First, with missing data there must necessarily be some loss of information. However, besides causing inefficiency, a greater concern in many settings is that missing data can introduce bias in the estimates of change over time. As a result, when longitudinal data are incomplete, the reasons for any missingness must be carefully considered. Some types of missing data are relatively benign and do not complicate the analysis; others are not and can potentially introduce bias in the estimates of the regression parameters.

All methods for longitudinal data analysis implicitly make assumptions about the reasons for any missingness, often referred to as the *missing data mechanism*. The missing data mechanism can be thought of as a model that describes the probability that a response is observed or missing at any occasion. There is an important distinction between mechanisms that are referred to as *missing completely at random* (MCAR) and *missing at random* (MAR). In particular, the distinction between them determines the appropriateness of likelihood-based inference under assumptions about the joint distribution for the responses and GLS/GEE methods that do not require assumptions about the joint distribution.

Longitudinal data are said to be MCAR when the probability that responses are missing is unrelated to either the specific values that, in principle, should have been obtained (the *missing* responses) or the set of observed responses. The key feature of MCAR is that the observed data can be considered a random sample of the complete data; consequently, any method of analysis that produces valid inferences in the absence of missing data will also yield valid inferences when missing data are MCAR. In contrast to MCAR, data are said to be MAR when missingness depends on the set of observed responses, but is unrelated to the missing responses. Put another way, if subjects are stratified on the basis of similar values for the observed responses, missingness is simply the result of a chance mechanism. However, when data are MAR an analysis restricted to the 'completers' (i.e., those with no missing data) is not valid because the 'completers' are a biased sample. While ML estimation of $\boldsymbol{\beta}$ is valid when data are MAR it requires that the entire joint distribution has been correctly specified. For example, with the linear models discussed in Section 9.2, this requires correct specification of not only the regression model for the mean response but also the model for the covariance among the responses. In a sense, ML estimation allows the missing values to be validly 'predicted' or 'imputed' using the observed data and a correct model for the joint distribution of the responses.

There is a third type of missing data mechanism that is referred to as *not missing at random* (NMAR) or 'non-ignorable' missingness. This arises when the probability that responses are missing is related to the specific values that should have been obtained. When longitudinal data are NMAR, almost all standard methods of longitudinal analysis are not valid and yield biased estimates of mean response trends. When missingness is thought to be NMAR, it is important to carefully assess the sensitivity of inferences to a variety of plausible assumptions concerning the missingness process. A more detailed discussion concerning the assumptions about the missing data mechanism, and their implications for analysis, can be found in Chapter 23.

Although methods for handling missing data have been a dominant theme in longitudinal data analysis, the problem has not received the same degree of attention in the multilevel data setting. Incomplete data commonly arise in multilevel data settings, e.g., missing data on children within classrooms or missing patient data within clinics; however, because the cluster sizes (e.g., classroom or clinic sizes) are random the extent of missingness in the data at hand is somewhat less transparent. Nonetheless, missing data in the multilevel setting has the potential to introduce bias whenever units (at any level) with missing data differ in important ways from those with complete data.

9.5 SUMMARY

This chapter has provided an introduction to the analysis of longitudinal data, using both classical approaches for balanced designs and multilevel models for unbalanced designs. We show how both approaches can be extended to binary or count data. Although it is clear that repeated measures on independent subjects can form the level 1 (repeated measures) and level 2 (subjects) units in a multilevel design, there are several features which set longitudinal data apart from general multilevel data. First, the time-ordering of longitudinal data will be an important feature in most longitudinal studies. For example, in simple longitudinal clinical trial designs such as that used for the TLC trial, the treatment effects on change from baseline are of primary interest. These can be modeled using standard regression models, but the variance-covariance structure will often not be stable over time. In cross-over designs, subjects are given different treatments in different time periods and the primary interest is the treatment comparison. Here, the treatment comparison is a between-time-point contrast, but the contrast used depends upon the order that treatment is given. In the cross-over study setting, there are several approaches to modeling the mean response depending on whether we choose to adjust for so-called carry-over effects, but generally we assume variances and covariances are stable over time, and a multilevel variance component model is often used. In general, the regression models used for the mean responses will have similar structure for longitudinal and multilevel data, but the covariance structures often differ.

A second important feature of longitudinal data is that missing responses or dropout may occur. While the multilevel framework allows us to incorporate level 2 units with missing observations quite naturally, both from a modeling and a computational standpoint, the thornier question of inference arises since missingness and dropout can result in badly biased estimates of $\boldsymbol{\beta}$. There is substantial statistical literature on missing data and dropout. While it is clear that missingness causes loss of information, the effect on bias in parameter estimates or tests depends upon the mechanism causing the missingness as well as the method of handling the missing data in the analysis. This topic will be taken up in greater detail in Chapter 23.

ACKNOWLEDGMENTS

Section 9.3 of this chapter is based in part on Chapter 1 of *Longitudinal Data Analysis* by Fitzmaurice, Davidian, Verbeke, and Molenberghs (2009). The writing of this chapter was supported by grant MH 054693 from the National Institutes of Health (NIH).

REFERENCES

Anderson, D.A. and Aitkin, M. (1985) 'Variance Components Models with Binary Response: Interviewer Variability', *Journal of the Royal Statistical Society, Series B*, 47: 203–10.

Breslow, N.E. and Clayton, D.G. (1993) 'Approximate Inference in Generalized Linear Mixed Models', *Journal of the American Statistical Association*, 88: 9–25.

Breslow, N.E. and Lin, X. (1995) 'Bias Correction in Generalized Linear Models with a Single Component of Dispersion', *Biometrika* 82: 81–91.

Cox, D.R. (1961) 'Tests of Separate Families of Hypotheses', in *Proceedings of the Fourth Berkeley Symposium on Mathematics, Probability and Statistics: Vol. 1*. Berkeley: University of California Press, 105–23.

Diggle, P.J. (1988) 'An Approach to the Analysis of Repeated Measurements', *Biometrics*, 44: 959–71.

Diggle, P.J., Heagerty, P.J., Liang, K.-Y., and Zeger, S.L. (2002) *Analysis of Longitudinal Data*, 2nd edition. Oxford: Clarendon Press.

Donnelly, C.A., Laird, N.M., and Ware, J.H. (1995) 'Prediction and Creation of Smooth Curves for Temporally Correlated Longitudinal Data', *Journal of the American Statistical Association*, 90: 984–9.

Fitzmaurice, G., Davidian, M., Verbeke, G., and Molenberghs, G. (2009) *Longitudinal Data Analysis*. Boca Raton, FL: Chapman and Hall/CRC Press.

Fitzmaurice, G.M., Laird, N.M., and Ware, J.H. (2011) *Applied Longitudinal Analysis*, 2nd edition. New Jersey: Wiley.

Hedeker, D. and Gibbons, R.D. (1994) 'A Random-Effects Ordinal Regression Model for Multilevel Analysis', *Biometrics*, 50: 933–44.

Hedeker, D. and Gibbons, R.D. (1996) 'MIXOR: A Computer Program for Mixed-Effects Ordinal Regression Analysis', *Computer Methods and Programs in Biomedicine*, 49: 157–76.

Hubbard, A.E., Ahern, J., Fleischer, N.L., Van der Laan, M., Lippman, S.A., Jewell, N., Bruckner, T., and Satariano, W.A. (2010) 'To GEE or Not to GEE: Comparing Population Average and Mixed Models for Estimating the Associations Between Neighborhood Risk Factors and Health', *Epidemiology*, 21: 467–74.

Huber, P.J. (1967) 'The Behavior of Maximum Likelihood Estimates Under Nonstandard Conditions', in *Proceedings of the Fifth Berkeley Symposium on Mathematical Statistics and Probability: Vol 1*. Berkeley: University of California Press, 221–33.

Laird, N.M. and Ware, J.H. (1982) 'Random Effects Models for Longitudinal Data', *Biometrics*, 38: 963–74.

Lauer, R.M., Clarke, W.R., and Burns, T.L. (1997) 'Obesity in Childhood: The Muscatine Study', *Acta Pædiatrica Scandinavica*, 38: 432–7.

Liang, K.-Y. and Zeger, S.L. (1986) 'Longitudinal Data Analysis Using Generalized Linear Models', *Biometrika*, 73: 13–22.

McCullagh, P. and Nelder, J. (1989) *Generalized Linear Models*, 2nd edition. London: Chapman and Hall/CRC Press.

Nelder, J.A. and Wedderburn, R.W.M. (1972) 'Generalized Linear Models', *Journal of the Royal Statistical Society, Series A*, 135: 370–84.

Raudenbush, S.W., Yang, H.-L., and Yosef, M. (2000) 'Maximum Likelihood for Generalized Linear Models with Nested Random Effects Via High-Order, Multivariate Laplace Approximation', *Journal of Computational and Graphical Statistics*, 9: 141–57.

Royall, R.M. (1986) 'Model Robust Confidence Intervals Using Maximum Likelihood Estimators', *International Statistical Review*, 54: 221–6.

Schall, R. (1991) 'Estimation in Generalized Linear Models with Random Effects', *Biometrika*, 78: 719–27.

Snedecor, G.W. and Cochran, W.G. (1967) *Statistical Methods*, 6th ed. Ames, IA: Iowa State Press.

Stiratelli, R., Laird, N.M., and Ware, J.H. (1984) 'Random Effects Models for Serial Observations with Binary Response', *Biometrics*, 40: 961–71.

Vella, F. and Verbeek, M. (1998) 'Whose Wages Do Unions Raise? A Dynamic Model of Unionism and Wage Rate Determination for Young Men', *Journal of Applied Econometrics*, 13: 163–83.

Wedderburn, R.W.M. (1974) 'Quasi-likelihood Functions, Generalized Linear Models, and the Gauss-Newton Method', *Biometrika*, 61: 439–47.

White, H. (1982) 'Maximum Likelihood Estimation Under Mis-Specified Models', *Econometrica*, 50: 1–26.

Wolfinger, R. (1993) 'Laplace's Approximation for Nonlinear Mixed Models', *Biometrika*, 80: 791–5.

Wooldridge, J. (2010) *Econometric Analysis of Cross Section and Panel Data*, 2nd edition. Cambridge, MA: MIT Press.

Woolson, R.F. and Clarke, W.R. (1984) 'Analysis of Categorical Incomplete Longitudinal Data', *Journal of the Royal Statistical Society, Series A*, 147: 87–99.

10

Complexities in Error Structures Within Individuals

Vicente Núñez-Antón
The University of the Basque Country, UPV/EHU, Spain

Dale L. Zimmerman
The University of Iowa, USA

10.1 INTRODUCTION

Many types of data have a hierarchical or clustered structure. For example, trees growing in the same area or descended from the same genetic line tend to be more alike than trees chosen at random from the population at large. Individuals or subjects in a study may be further nested within geographical areas. Multilevel data structures also arise in longitudinal studies where the same individual's responses over time are correlated with each other (see also Chapter 9). This is the more natural setting where within-individual correlation or covariance structures that can model specific data behaviors arise. Some of the characteristics that need to be appropriately modeled are stationary or nonstationary variances and correlations, and serial correlation. Additional issues that need to be appropriately addressed are the complexity of the structures and the possible use of specific random effects models, as is more common in some multilevel modeling approaches.

Although the majority of covariance complexities in multilevel models appear in the longitudinal context, there are other modeling possibilities that may need to be considered. Very simple within-individual error structures, such as compound symmetry, may be used outside the longitudinal situation, such as for clustered data, and these may be extended to allow for heteroscedasticity. However, most research on within-individual error structures has been carried out in the longitudinal data setting and, therefore, it makes sense to study its complexities in this context.

The choice of a model for the within-individual error structure should account for model flexibility and model parsimony. In this chapter we concentrate on these issues and describe and motivate the different modeling possibilities for these error structures. We do not deal here with model selection or model fitting issues, mainly because these are issues to be described in more detail in the specific context of multilevel modeling in Chapters 7, 9 and 15. Among the possible models used for longitudinal-type error covariance structures, we try to obtain the right balance between flexibility and parsimony, where the unstructured covariance model is the most flexible and least parsimonious, and models such as the autoregressive, structured antedependence, Wiener, random effects, and compound symmetric models are the least flexible and most parsimonious. Intermediate models, such as the Toeplitz, unstructured antedependence, and Huynh–Feldt models will also be described. Finally, the appropriate use of the aforementioned models in a longitudinal data setting will be illustrated with simulated and real data applications.

10.2 GENERAL MODEL: HYPOTHESIS AND NOTATION

We suppose that the observations follow the standard linear regression model given by:

$$\boldsymbol{y} = X\boldsymbol{\beta} + Z\boldsymbol{u} + \boldsymbol{\varepsilon}, \qquad (10.1)$$

where $\boldsymbol{y}$ is an $n \times 1$ vector of observations, X is an $n \times p$ design matrix, $\boldsymbol{\beta}$ is a $p \times 1$ vector of fixed coefficients, Z an $n \times q$ design matrix for modeling covariance, $\boldsymbol{u}$ a $q \times 1$ vector of random effects, and $\boldsymbol{\varepsilon}$ a vector of error terms. We assume that $\boldsymbol{u} \sim N_q(0, \Omega)$ and $\boldsymbol{\varepsilon} \sim N_n(0, \Sigma)$ are independent of one another. The focus of this chapter will be on the modeling of the covariance matrix Σ. For longitudinal data (see also Chapter 9), Laird and Ware (1982) have proposed a variant of model (10.1) where a single individual's vector of observations, $\boldsymbol{y}_i$, is modeled as:

$$\boldsymbol{y}_i = X_i\boldsymbol{\beta} + Z_i\boldsymbol{u}_i + \boldsymbol{\varepsilon}_i, \qquad (10.2)$$

where X_i and Z_i are $n_i \times p$ and $n_i \times q_i$ design matrices consisting of predictors or a column of ones, $\boldsymbol{\beta}$ is taken as fixed (i.e., constant for all individuals), $\boldsymbol{u}_i \sim N_{q_i}(0, \Omega_i)$, $\boldsymbol{\varepsilon}_i \sim N_{n_i}(0, \Sigma_i)$, and the $\boldsymbol{u}_i$s and $\boldsymbol{\varepsilon}_i$s are mutually independent. If we have m individuals and write $n = \sum_{i=1}^{m} n_i$ for the total number of observations, this model is equivalent to model (10.1) if we set $\boldsymbol{y} = (\boldsymbol{y}_1', \ldots, \boldsymbol{y}_m')'$, $X = (X_1', \ldots, X_m')'$, $Z = \oplus Z_i$, $\boldsymbol{u} = (\boldsymbol{u}_1', \ldots, \boldsymbol{u}_m')'$, $q = \sum_{i=1}^{m} q_i$, $\boldsymbol{\varepsilon} = (\boldsymbol{\varepsilon}_1', \ldots, \boldsymbol{\varepsilon}_m')'$, $\Omega = \oplus\Omega_i$, and $\Sigma = \oplus\Sigma_i$, where $\oplus$ is the direct sum (block diagonal) operator. It is also assumed that the covariance models that generate the Ω_is and Σ_is are the same across subjects; thus, if all subjects were observed at exactly the same times, then all the Ω_is would be equal and so would the Σ_is.

As mentioned above, we use model (10.1) to illustrate the issues related to the choice of a model for the within-subject error structure, and to motivate the reasons why they should account for model flexibility and model parsimony. For simplicity and because it does not add anything new to the focus of this chapter, we have decided to use the notation in model (10.1) and not that in model (10.2), described in Laird and Ware (1982), and specifically adapted for longitudinal data.

Goldstein (2003, Chapter 5) has indicated that there is a need to consider the modeling of the commonly existing within-subject autocorrelation, proposing several models, such as autoregressive, polynomials or more data-specific correlation models. Jones (1990) has brought out the possibility of partially confounding the within-subject serial correlation (i.e., correlation that is a monotonic decreasing function of time-lag), common in this type of data, with the random between-subject effects usually modeled in the second term in equation (10.1). Moreover, he mentioned that, in real data situations, it is often not clear if there is some type of serial correlation, random subject effects, or both. Thus, models with random coefficients, serial correlation, or both should be fitted and the best-fitting model can be selected with the use of some model selection criteria, such as Likelihood Ratio Tests (LRT) or Akaike's (1974) Information Criterion (AIC). For interested readers, Chapter 7 provides additional details on the issue of model interpretation, specification, and selection in the specific context of multilevel models, and Chapter 26 deals with the issue of software for fitting mixed-effects and multilevel models. Wolfinger (1996) presented a unified discussion of a useful collection of heterogeneous covariance matrix structures for the within-subject covariance matrix. In addition, he discussed issues pertaining to estimation, testing, and model selection, as well as the fitting possibilities for most of these models in commercial software. Zimmerman and Núñez-Antón (2001 and 2010) proposed the parametric multivariate regression modeling approach to jointly modeling the mean and within-subject covariance structure, analyzing the advantages it has over the classical analysis of variance and general multivariate approaches. They proposed to postulate, fit, evaluate, and compare parametric models for the data's mean structure and covariance structure with the use of graphical and descriptive tools as well as some model selection fitting criteria, such as LRT, AIC, and Schwarz's (1978) Bayesian Information Criterion (BIC). In the following sections we will introduce some general ideas that are important when modeling both the mean and within-subject covariance matrix.

10.3 MEAN MODELING

We start with some general ideas and recommendations about the mean structure, where our focus will be on the modeling of the first term (i.e., the fixed effects) on the right-hand side of equation (10.1). The choice for the mean structure depends, just as in regression analysis, on model parsimony (i.e., a simple or complex mean model), ease of interpretation, specific characteristics of the data being considered, and the flexibility and objectives of the inferences to be carried out by the researcher. One should also take into account both the scientific objectives and the design of the study. Very commonly, the use of an additive mean model such as the one in model (10.1), which could depend on the time at which measurements were taken and on additional covariates, is reasonable. One initial approach is to compute and examine the sample means at each measurement time for the different treatments being compared, and similarly to compute sample correlations between the response and each covariate at each measurement time. In addition, and after fitting the selected initial mean model, a plot of the fitted residuals versus time and/or other covariates would confirm, or indicate the required modifications to, the initial mean model. A well-known and helpful graphical diagnostic related to these summary statistics is the profile plot, which helps identify how the mean changes over time for the different treatments or groups in the data. Profile plots can be easily obtained using most statistical software packages; in the widely used statistical package *R*, for example, the relevant function is *matplot*. The profile plot is just a graph of the response values for each of the subjects in the study along time, or for the different treatments or groups under study. The mean profile for the whole group of subjects under that specific treatment or for the complete data set may be included in the plot for comparison with the profiles of individual subjects.

Researchers working in the area have motivated and illustrated the use of the aforementioned tools, including the profile plot, as a very helpful graphical tool for mean modeling purposes (see, e.g., Pourahmadi, 2001; Diggle et al., 2002; Weiss, 2005; or Zimmerman and Núñez-Antón, 2010). To illustrate this, we consider the data from a study of cattle growth reported by Kenward (1987), where the data are weights (in kg) for 30 cattle receiving one of two possible treatments (labeled henceforth as treatments A and B) for internal parasites, recorded on 11 occasions. Figure 10.1 shows the profile plots for treatments A and B in this data set, and Table 10.1 lists the corresponding sample means. It is clear that the mean weights for the two treatments tend to increase with time, and that this increase does not seem to be linear. Table 10.1 also indicates that mean cattle weights are different for the two treatments. All of the aforementioned information suggests the use of an unstructured mean model (i.e., one parameter for each time point) that includes a treatment effect or a polynomial model in time that also includes a treatment effect. Models including time-by-treatment interactions or nonlinear mean models could also be considered (see, e.g., Cepeda-Cuervo and Núñez-Antón, 2009).

Table 10.1 Sample Means (in kg) and Sample Variances for Cattle Weights under Treatments A and B in the Cattle Data

Time	Treatment A		Treatment B	
(in days)	Mean	Variance	Mean	Variance
0	226.2	105.6	224.6	105.3
14	230.3	155.1	227.9	108.4
28	246.9	165.2	243.5	147.1
42	265.6	184.9	262.5	198.5
56	281.2	243.0	276.4	217.7
70	294.9	283.8	290.1	250.4
84	304.7	306.6	299.2	248.2
98	312.9	340.7	317.7	234.1
112	315.1	389.2	319.7	287.0
126	324.1	470.1	326.9	404.7
133	325.5	444.6	320.5	598.6

10.4 COVARIANCE MODELING

As in Section 10.3, we will start with some general ideas that may help the practitioner decide on suitable error structures within individuals that could be considered for the specific data set being analyzed. We should also mention that we will focus on the modeling of the within-individual covariance structure Σ for the errors in equation (10.1). Some of the main issues that should be considered when modeling this covariance structure are the existence or not of stationarity in variance and/or correlation and, in addition, if there is serial correlation or not in the data. The answer to each of these questions influences the modeling possibilities for the data set under study. Therefore, and before describing and motivating some of the most commonly used within-subject covariance model structures, the initial question is how this can be assessed or appropriately evaluated.

As in the previous section, the profile plot, as well as the sample variances and correlations can also be used to evaluate both the stationarity in variance and the existence of serial correlation. The sample variances reported in Table 10.1 and the profile plots in Figure 10.1 provide clear evidence that variances tend to increase with time, indicating that stationarity in variance may not be a reasonable assumption for this data set. In addition, we see only very modest crossing of individual cow's profiles over time: the heaviest and lightest cattle at the beginning of the study tend to remain among the heaviest and the lightest for both treatment groups. This consistent behavior within subjects can usually be considered as a clear sign of a large positive correlation for observations taken on the same subject. Table 10.2 includes the within-subject sample correlations for cattle in treatments A and B. These reported correlation values confirm the aforementioned large positive correlations, as well as the fact that correlations decrease as the elapsed time between measurements increases (serial correlation). This latter fact cannot be easily ascertained from the profile plot, but can be clearly observed in the sample correlation matrices reported in Table 10.2 as a mostly monotonic decrease of correlations within columns as one moves away from the main matrix diagonal.

Two well-known and very helpful graphical diagnostics related to these summary statistics are the ordinary scatterplot matrix (OSM) and the Partial Regression-on-Intervenors Scatterplot Matrix (PRISM) (Zimmerman, 2000). The OSM is a two-dimensional array of pairwise scatterplots of standardized responses (or residuals), and it represents the graphical

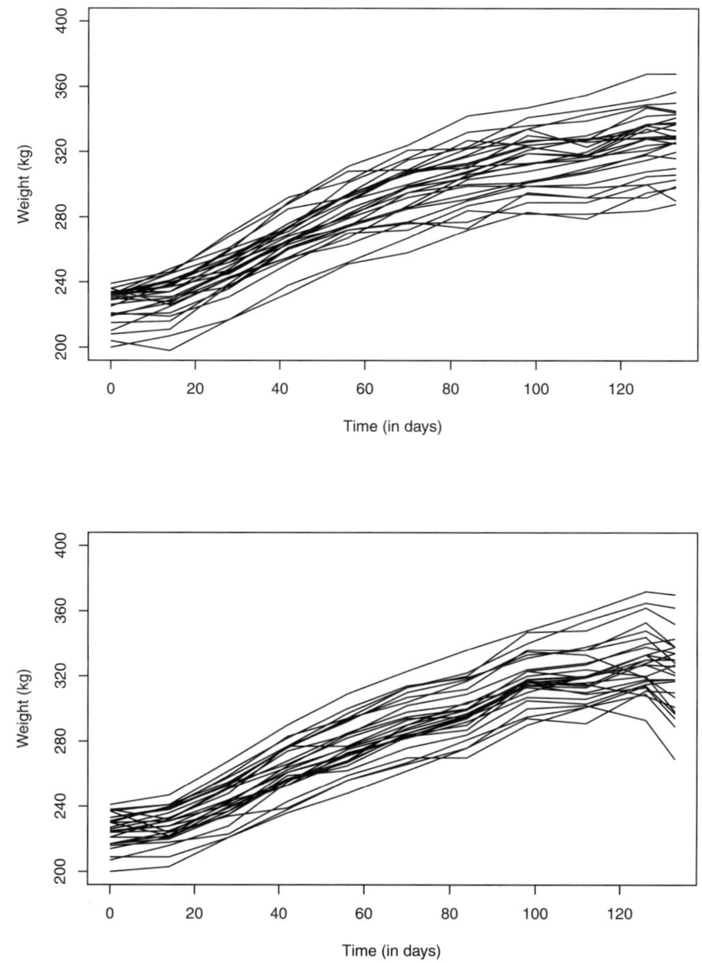

Figure 10.1 Profile Plots for the Cattle Data Treatments Groups A (top panel) and B (lower panel). The thicker line indicates the overall mean profile for this group. Reprinted with permission from Zimmerman, D.L. and Núñez-Antón, V. (2010) ***Antedependence Models for Longitudinal Data*****. Boca Raton, FL: Chapman & Hall/CRC**

equivalent of the sample correlation matrix. The PRISM is a two-dimensional array of certain partial regression plots. More specifically, it is the graphical equivalent of a matrix of certain partial correlations, where the plot in row j and column k ($j \geq k$) of the array is the partial regression plot of standardized response (or residual) variables at times k and $(j + 1)$, adjusted for all standardized responses (or residuals) at intervening times $k + 1, \dots, j$. Random scatter in the (j, k)th plot indicates that the responses (or residuals) at times k and $(j + 1)$ are conditionally independent, given the intervening responses (or residuals), whereas departures from random scatter indicate conditional dependence. A layout of the PRISM is shown in Figure 10.2. As will be illustrated in later sections (see, e.g., Section

Table 10.2 Sample Correlations for Cattle Weights Under Treatments A (upper triangle) and B (lower triangle) in the Cattle Data

1.0	.82	.76	.66	.64	.59	.52	.53	.52	.48	.48
.86	1.0	.91	.84	.80	.74	.63	.67	.60	.58	.55
.83	.94	1.0	.93	.88	.85	.75	.77	.71	.70	.68
.68	.89	.93	1.0	.94	.91	.83	.84	.77	.73	.71
.67	.84	.88	.95	1.0	.94	.87	.89	.84	.80	.77
.66	.84	.87	.95	.98	1.0	.93	.94	.90	.87	.83
.61	.78	.82	.91	.94	.97	1.0	.93	.93	.88	.86
.63	.81	.84	.92	.92	.95	.95	1.0	.97	.94	.92
.63	.80	.79	.90	.93	.95	.93	.96	1.0	.96	.96
.48	.65	.67	.78	.78	.82	.76	.78	.83	1.0	.98
.44	.57	.62	.73	.68	.74	.71	.71	.75	.92	1.0

10.4.5), the PRISM can also be used to detect the possible existence of serial and/or stationary correlations, a fact that may help decide if a first-order autoregressive process (see Section 10.4.3) or an antedependence model (see Section 10.4.5) should be considered.

The OSM can be easily obtained in most statistical software packages such as, for example, with the function *splom* in the *lattice* package of *R*. The PRISM can also be readily obtained with simple programming in *R*. Interested readers can obtain such programs from the second author's webpage: `homepage.stat.uiowa.edu/~dzimmer/R-functions-for-AD/`. Additional graphical diagnostics that could also be used for informal model identification are described in Jones and Rice (1992), Segal (1994), and Pourahmadi (2002). An alternative graphical tool that can only be used and related to the autocorrelation function under stationarity conditions is the variogram (see, e.g., Diggle et al., 2002, as well as Chapter 31). Given that the OSM and PRISM can be used for very general conditions, here we only concentrate and illustrate the usefulness of these more general and commonly used graphical diagnostic tools for covariance model selection.

To illustrate the usefulness of the OSM and PRISM graphical diagnostic plots, we show these plots for the treatment A group of the cattle data set (see Figure 10.3). The OSM indicates that correlations between responses tend to decrease as a function of elapsed time but remain non-negligibly positive even for the longest elapsed time. This suggests that models with constant correlation or banded covariance structures may not be appropriate for this treatment group. The OSM also indicates that same-lag correlations tend to increase as the experiment progresses, calling into question the use of stationary correlation models, such as the autoregressive models or Toeplitz covariance models to be introduced in Sections 10.4.3 and 10.4.4, respectively (this can also be seen from the information provided in Table 10.2), but at the same time supporting the use of nonstationary models such as antedependence models, to be introduced in Section 10.4.5 (Zimmerman and Núñez-Antón, 2010). In the PRISM, all scatterplots show random scatter except those on the first diagonal and one on the second diagonal (i.e., the one in row 7, column 6). These results suggest that models having some type of conditional independence for observations should be considered. In this specific case, antedependence models of order at most two (because the PRISM shows random scatter plots for the third and higher diagonals) could be a good model option to try to fit for this data set.

We now describe some of the most commonly used covariance models that can be considered for modeling the error structure within individuals and also discuss issues related to model flexibility and model

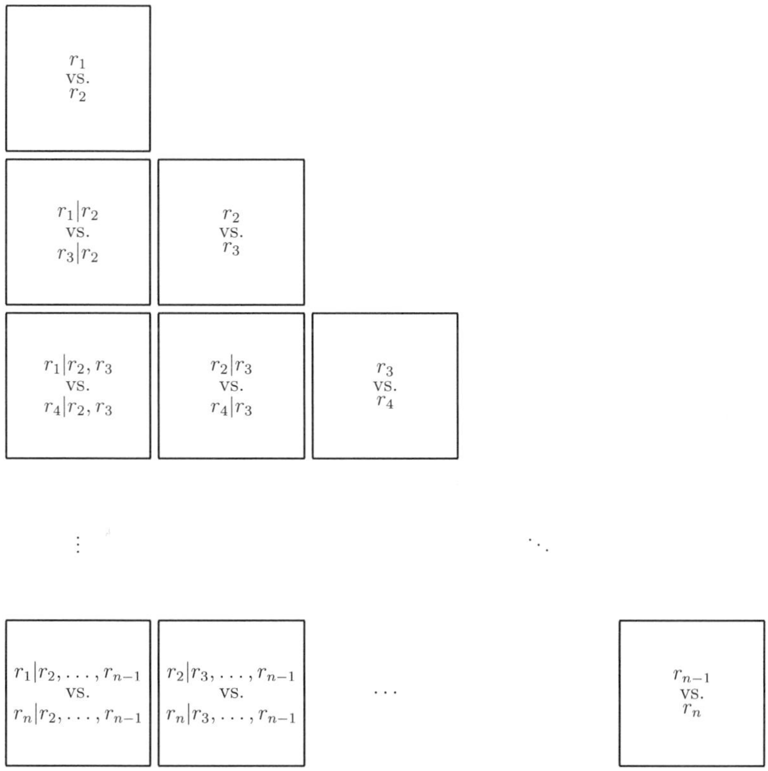

Figure 10.2 Partial Regression-on-Intervenors Scatterplot Matrix (PRISM) Layout for Standardized Responses (or Residuals), denoted here by *r*

parsimony and how a covariance model can strike the right balance between them. Readers interested in alternative covariance models that may be more commonly used in other areas can additionally consult the research reported in Hand and Crowder (1996), Verbeke and Molenberghs (2001), Davis (2002), Diggle et al. (2002), Fitzmaurice, Laird, and Ware (2004), or Weiss (2005).

10.4.1 Random Coefficients Models

Random coefficients (RC) models were originally introduced by Rao (1959) and have become very popular, mainly because of their intuitive interpretability and their close relation to Bayesian techniques (see, e.g., Laird and Ware, 1982; Reinsel, 1982; or Rutter and Elashoff, 1994). The general random coefficients model can be written as in equation (10.1), where it is assumed that $\Omega = (\omega_{ij})$ is a positive definite unstructured matrix, and that $\Sigma = \sigma^2 I$, $\sigma^2 > 0$. Under the assumptions for model (10.1), we have that the covariance matrix of $\boldsymbol{y}$ is given by $V = Z\Omega Z' + \sigma^2 I$. Special cases include the linear random coefficients (LRC) and quadratic random coefficients (QRC) models. In the linear case, $Z = [\mathbf{1}, \boldsymbol{t}]$, with $\mathbf{1}$ being an $n \times 1$ vector of ones and $\boldsymbol{t} = (t_1, \ldots, t_n)'$ an $n \times 1$ vector containing the times at which measurements were taken. This model is also equivalent to the standard hierarchical linear model used in educational statistics (Goldstein, 1987; Bryk and Raudenbush, 1992). In the quadratic case, $Z = [\mathbf{1}, \boldsymbol{t}, (t_1^2, t_2^2, \ldots, t_n^2)']$.

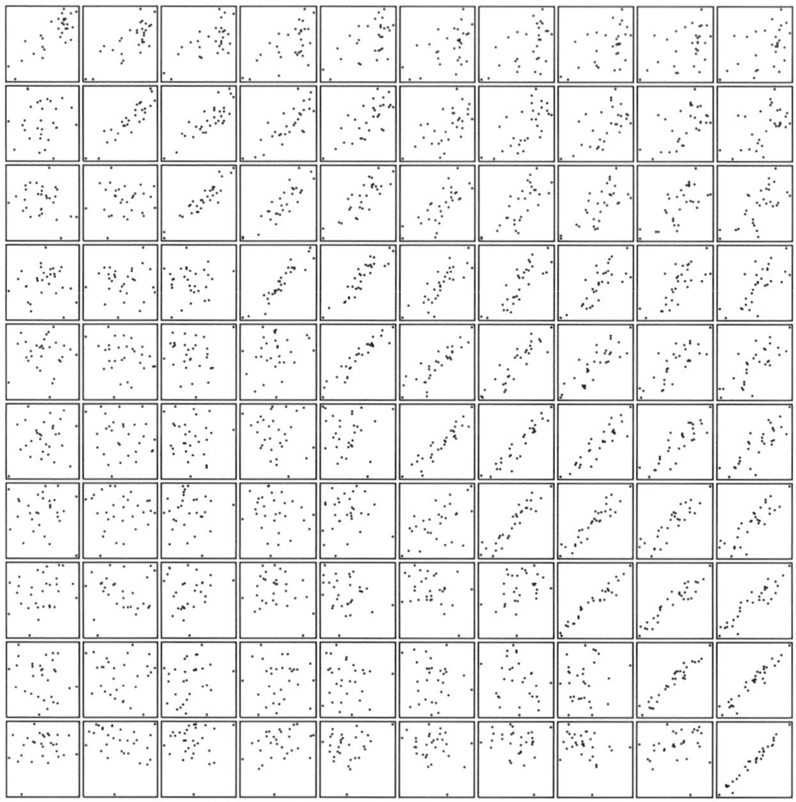

Figure 10.3 Partial Regression-on-Intervenors Scatterplot Matrix (PRISM) Plot (lower triangle, including main diagonal) and Ordinary Scatterplot Matrix (OSM) Plot (upper triangle, including main diagonal) for Cattle Data Treatment Group A

As will be seen in Section 10.4.3, the compound symmetry structure is equivalent to the case $Z = \mathbf{1}$. Another property of the whole class of random coefficients models is that they tend to produce conservative estimates of the standard errors of mean-model parameters, when compared with more complicated covariance structures (Lange and Laird, 1989). This is especially true when the only nonrandom coefficient in the model is the overall intercept.

Three characteristics of random coefficients models are their lack of flexibility, their applicability to situations in which the measurement times are unequally spaced, and their parsimony (i.e., the number of parameters is unrelated to the number of measurement times), so that it has $O(1)$ parameters. Moreover, the lack of flexibility shown by random coefficients models specified in their current form can be alleviated by considering, for example, fractional polynomials or spline-based structures in the random effects (see, e.g., Verbyla et al., 1999; or Ruppert, Wand, and Carroll, 2003, and references therein). In addition, if the random effects structure is supplemented with a sufficiently versatile error covariance matrix, potentially rich structures result. However, there are situations where this will not suffice and a richer covariance structure modeling should be incorporated in the marginal covariance matrix (see, e.g., Zimmerman and Núñez-Antón, 2001).

Another very interesting feature is that random coefficients models have not usually been considered as a way of parametric covariance modeling, probably because of their main motivation as varying regressions across subjects rather than being considered as a way of modeling within-subject similarity. Nevertheless, they may be regarded as parametric covariance structures that allow for nonconstant variances and nonstationary correlations of particular forms, a fact that does not appear to be widely known by practitioners or researchers using these models. For example, the linear random coefficients model structure for a subject with observations taken at equally-spaced time points $t_i = i$, $i = 1, \ldots, n$, is given by $V = [v_{ij}]_{i=1,\ldots,n}^{j=1,\ldots,n}$, so that

$$v_{jj} = \sigma^2 + \omega_{00} + 2\omega_{01}j + \omega_{11}j^2,$$

$$v_{jk} = \frac{\omega_{00} + \omega_{01}(j+k) + \omega_{11}jk}{d_j d_k},$$

where $d_j = \sqrt{\sigma^2 + \omega_{00} + 2\omega_{01}j + \omega_{11}j^2}$ and $d_k = \sqrt{\sigma^2 + \omega_{00} + 2\omega_{01}k + \omega_{11}k^2}$. This structure is very flexible, allowing, for example, for increasing or decreasing variances; and correlations, some of which are negative while others are positive. However, it precludes the variance from being a concave-down function of time, and it precludes the variances from being constant if the same-lag correlations are not. These characteristics suggest that these models may not be appropriate for the cattle data.

10.4.2 Independent Increments

In analyzing a longitudinal data set it may be reasonable to assume that errors follow a simple random walk covariance model. The independent increments (I-I) structure (see, e.g., Louis, 1988) models this type of process. In this case, if we assume that $n = 4$, we have:

$$\Sigma = \begin{pmatrix} \sigma_{11} & \sigma_{11} & \sigma_{11} & \sigma_{11} \\ \sigma_{11} & \sigma_{11}+\sigma_{22} & \sigma_{11}+\sigma_{22} & \sigma_{11}+\sigma_{22} \\ \sigma_{11} & \sigma_{11}+\sigma_{22} & \sigma_{11}+\sigma_{22}+\sigma_{33} & \sigma_{11}+\sigma_{22}+\sigma_{33} \\ \sigma_{11} & \sigma_{11}+\sigma_{22} & \sigma_{11}+\sigma_{22}+\sigma_{33} & \sigma_{11}+\sigma_{22}+\sigma_{33}+\sigma_{44} \end{pmatrix}.$$

A similar but more parsimonious model is the Wiener (WI) covariance structure, which may be viewed as a continuous-time analogue of the random walk model. For this structure, we have that $\sigma_{jk} = \sigma^2 \min(t_j, t_k), \sigma^2 > 0$, so that:

$$\Sigma = \sigma^2 \begin{pmatrix} t_1 & t_1 & t_1 & t_1 \\ t_1 & t_2 & t_2 & t_2 \\ t_1 & t_2 & t_3 & t_3 \\ t_1 & t_2 & t_3 & t_4 \end{pmatrix}.$$

As can be seen, this model applies only when the variances are increasing over time. Directly fitting the I-I structure is an alternative to differencing the data and assuming a homogeneous covariance structure. In addition, this model allows for increasing variances (linearly or nonlinearly) and also for nonstationary correlations between equidistant measurements that, in general, can be modeled so that they increase (nonlinearly) over time. These characteristics suggest that the independent increments model may be appropriate for the cattle data.

10.4.3 Compound Symmetry and First-Order Autoregressive Models

Compound symmetry (CS) and first-order autoregressive (AR(1)) models are homogeneous covariance structures; that is, the variance along the main diagonal is constant. However, they differ in the way the covariance (or correlation) is modeled. For the CS model, the covariance (or correlation) remains constant, whereas for the AR(1) model, it decreases exponentially. If we assume that $n = 4$, we have that the corresponding covariance structures for the CS and AR(1) models are, respectively:

$$\sigma^2 \begin{pmatrix} 1 & \rho & \rho & \rho \\ & 1 & \rho & \rho \\ & & 1 & \rho \\ & & & 1 \end{pmatrix}$$

and

$$\sigma^2 \begin{pmatrix} 1 & \rho & \rho^2 & \rho^3 \\ & 1 & \rho & \rho^2 \\ & & 1 & \rho \\ & & & 1 \end{pmatrix}.$$

The CS model assumes stationary variances and correlations and, in addition, it also assumes that the correlation is the same for all pairs of observations, which sometimes is a very strong assumption in the context of longitudinal or multilevel modeling. A real example for the use of the compound symmetry model corresponds to data on a cohort of adult men in Boston reported by D'Agostino et al. (1995). The study consisted of 2280 men aged 21–80 years at the time of entry, and it was a longitudinal study of aging initiated in 1963 and conducted at the Veterans' Administration Outpatient Clinic in Boston, Massachusetts. Study participants returned for repeat physical examination every five years if their age was under 52 and every three years if it was over 52 years. A detailed smoking history was obtained from the subjects at the time of each examination. Subjects were then classified at each visit as current, former, or never smokers. Measurements on the tracings of forced vital capacity maneuvers (i.e., spirometric data) were obtained for each subject. In this specific data set, for never smokers, the compound symmetry model provided a significantly better fit than the first-order autoregressive model, with an estimated correlation of about $\hat{\rho} = 0.66$.

The AR(1) model also assumes stationary variances and correlations. However, in this case, correlations show a banded structure, where correlations between equidistant observations remain constant and correlations decrease as the elapsed time between measurements increases. A real example illustrating the adequacy of the first-order autoregressive model was provided by Goldstein (2003), who reported on a sample of 26 boys each having their height measured on nine occasions between the ages of 11 and 14 years (Harrison and Brush, 1990). The measurements were taken approximately three months apart. The estimated autocorrelation coefficient was $\hat{\rho} = 0.19$, and the AR(1) model fitted significantly better than the alternative covariance models considered therein, such as the compound symmetry or random effects models.

In addition, some of the characteristics described in Section 10.4 when constructing the OSM and PRISM plots for the cattle data, as well as the sample correlation values reported in Table 10.2, indicate that the CS model may not be appropriate for this data set, but that the AR(1) model may. Moreover, the sample variance values reported in Table 10.1, where the variance is clearly nonstationary, indicate that neither model is appropriate. This is why, nonstationary-variance (heterogeneous) versions of these models (i.e., the CSH and ARH(1) models) have also been proposed in the literature (see, e.g., Wolfinger, 1996). As can be seen, the CS and AR(1) models have only two parameters and, thus, they are very parsimonious models, having only $O(1)$ parameters. Their heterogeneous counterparts, the CSH and ARH(1), models have $O(n)$ parameters. With regard to model flexibility, it is clear that the heterogeneous versions are more flexible than the homogeneous ones because they allow for nonstationary variances. However, they retain the parsimony of the constant or banded correlation structures, which are sufficiently flexible to fit some real data sets quite well.

Another example that may be useful to illustrate the applicability of some of the aforementioned covariance structures is the speech recognition data reported by Tyler et al. (1988). The data consist of scores (percentage of correct responses) on a sentence test administered under audition-only conditions to groups of individuals wearing one of two types of cochlear implants, labeled here as A and B. Implants were surgically implanted about five to six weeks prior to being electrically connected to an external speech processor. Subjects were bilaterally deaf, thus baseline values were all zero. Twenty subjects received implant A and 21 received implant B. Measurements were originally scheduled at 1, 9, 18, and 30 months after connection.

Table 10.3 Sample Means and Variances for Percentage Test Scores for Cochlear Implants A and B in the Speech Recognition Data

Time	Implant A		Implant B		Pooled
(in months)	Mean	Variance	Mean	Variance	Variance
1	28.52	419.3	19.40	371.5	395.4
9	49.14	652.1	38.24	547.5	599.8
18	53.16	657.3	45.00	496.8	577.1
30	58.99	658.2	51.10	553.8	606.0

Some subjects did not show up for one or more of their scheduled follow-ups, so some data are missing. Figure 10.4 shows the profile plots for the cochlear implant groups A and B in this data set, and Table 10.3 gives the corresponding sample means. Both mean profiles increase over time, with the largest increase occurring from the first to the second measurement after implantation. In addition, mean profiles suggest that audition tends to improve as measurement times increase and that it also tends to remain constant at the end of the study, or even decrease for some individuals. That is, the rate of improvement in hearing does not seem to be linear in time. All of the above suggests that a quadratic or cubic mean model in time, or even an unstructured mean model for each implant group, may be appropriate, and that individuals with implant A hear better, on average, than those with implant B.

The variances do not appear to increase substantially over time. However, the sample variances reported in Table 10.3 provide some evidence that the variance of the first-time measurements is slightly smaller than those of measurements at the remaining times and, thus, cast doubt on a strict assumption of stationarity in variances. Note that the pooled variance is also included here because all of the authors that have analyzed this data set concluded that the same covariance structure can be assumed for the two types of implant. In addition, and as with the cattle data, we see little crossing of individuals' hearing measurements over time, which is a sign of the existence of large positive correlations among observations taken on the same subject. Table 10.4 includes the pooled within-subject sample correlations for this data. These confirm the aforementioned large positive correlations, as well as the fact that correlations decrease as the elapsed time between measurements increases; that is, there is clear evidence of the existence of serial correlation. Figure 10.5 features the OSM supporting this claim. In addition, correlations in the same diagonal tend to increase slowly as the study progresses (i.e., audition behavior tends to become more consistent over time). All of the plots in the PRISM (see Figure 10.5) except those on the main diagonal are random, indicating conditional independence. Therefore, autoregressive or antedependence models of order one (see Section 10.4.5) may be appropriate for these data.

Table 10.4 Sample Correlations for Percentage Test Scores for the Speech Recognition Data

t=1	1			
t=9	0.85	1		
t=18	0.70	0.90	1	
t=30	0.64	0.87	0.95	1

10.4.4 Toeplitz Models

The homogeneous (TOEP) and heterogeneous (TOEPH) Toeplitz models generalize the AR(1) and ARH(1) models, respectively. In the Toeplitz models, correlations

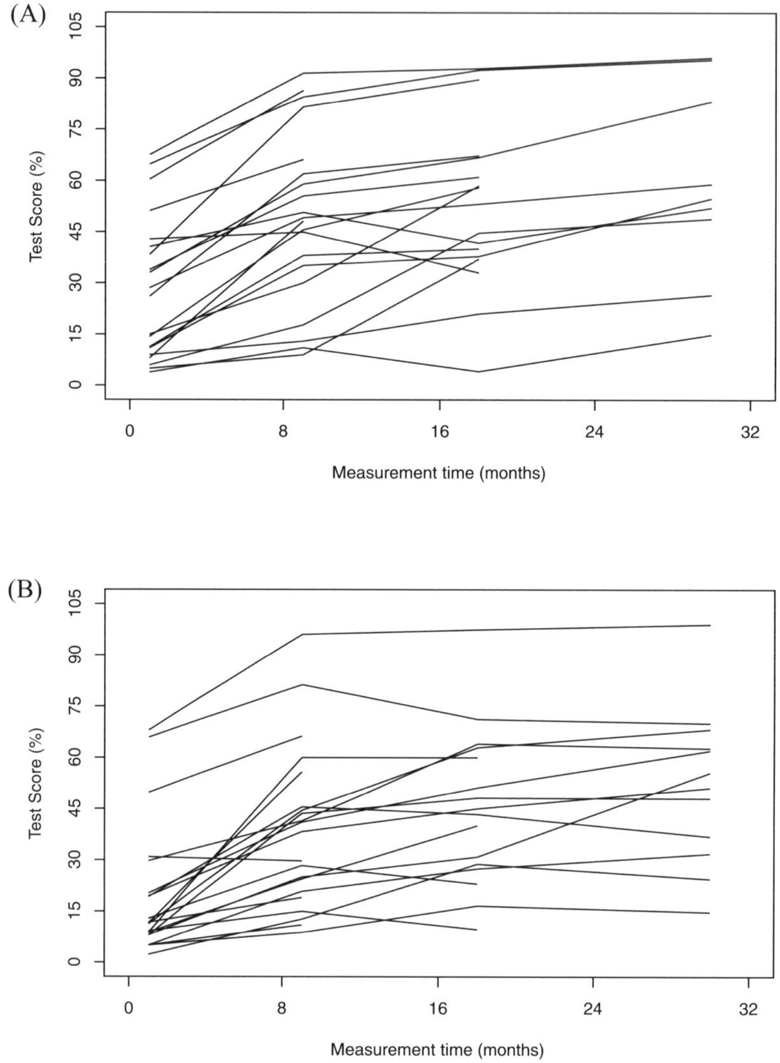

Figure 10.4 Profile Plot for the Speech Recognition Data Cochlear Implants Groups A (top panel) and B (lower panel). The thicker line indicates the overall mean profile for this group. Reprinted with permission from Zimmerman, D.L. and Núñez-Antón, V. (2010) ***Antedependence Models for Longitudinal Data. Boca Raton, FL: Chapman & Hall/CRC***

are not assumed to decrease in an exponential way but, instead, they are assumed to vary in an unstructured manner (i.e., no specific model or parameterization is imposed on them). However, these models maintain the somewhat restrictive stationarity assumption for correlations. To illustrate, if $n = 4$, the covariance structures for the TOEP and TOEPH models are, respectively:

$$\sigma^2 \begin{pmatrix} 1 & \rho_1 & \rho_2 & \rho_3 \\ & 1 & \rho_1 & \rho_2 \\ & & 1 & \rho_1 \\ & & & 1 \end{pmatrix}$$

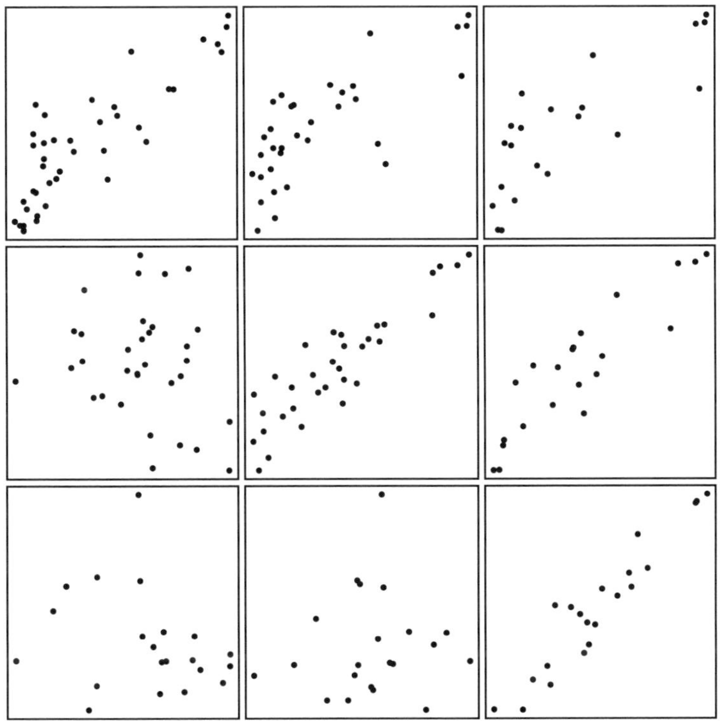

Figure 10.5 Partial Regression-on-Intervenors Scatterplot Matrix (PRISM) Plot (lower triangle, including main diagonal) and Ordinary Scatterplot Matrix (OSM) Plot (upper triangle, including main diagonal) for the Speech Recognition Data

and

$$\begin{pmatrix} \sigma_{11} & \sqrt{\sigma_{11}\sigma_{22}}\rho_1 & \sqrt{\sigma_{11}\sigma_{33}}\rho_2 & \sqrt{\sigma_{11}\sigma_{44}}\rho_3 \\ & \sigma_{22} & \sqrt{\sigma_{22}\sigma_{33}}\rho_1 & \sqrt{\sigma_{22}\sigma_{44}}\rho_2 \\ & & \sigma_{33} & \sqrt{\sigma_{33}\sigma_{44}}\rho_1 \\ & & & \sigma_{44} \end{pmatrix}.$$

As can be seen, the TOEP and TOEPH models have n and $2n-1$ parameters; that is, $O(n)$ parameters, and, thus, they are not very parsimonious models. With regard to model flexibility, it is clear that the heterogeneous version is more flexible than the homogeneous one because it allows for nonstationary variances. However, both versions retain a banded correlation structure.

As we have seen in Section 10.4, the OSM and PRISM for the cattle treatment A data (see Figure 10.3), as well as the sample variance and correlation values reported in Tables 10.1 and 10.2, indicate that the TOEPH model may be appropriate for this data set. The clear nonstationarity in variance in this data set argues against the use of a TOEP model. With regard to the speech recognition data, the OSM and PRISM (see Figure 10.5), as well as the sample variance and correlation values reported in Tables 10.3 and 10.4, indicate that the heterogeneous Toeplitz model, TOEPH, may be a model worth considering but, again, the clear nonstationarity in variance advises against the use of a TOEP model. However, both data sets also exhibit nonstationary correlations, which calls into question the usefulness of the

TOEPH model. Moreover, as will be seen in Section 10.4.5, antedependence models may also be a good choice to consider for the speech recognition data.

It is worth mentioning that a simpler version of the Toeplitz models can also be considered for situations in which subdiagonals in the covariance matrix for values larger than k are negligible. For example, for the previous case, the TOEPH(2) model can be written as:

$$\begin{pmatrix} \sigma_{11} & \sqrt{\sigma_{11}\sigma_{22}}\rho_1 & \sqrt{\sigma_{11}\sigma_{33}}\rho_2 & 0 \\ & \sigma_{22} & \sqrt{\sigma_{22}\sigma_{33}}\rho_1 & \sqrt{\sigma_{22}\sigma_{44}}\rho_2 \\ & & \sigma_{33} & \sqrt{\sigma_{33}\sigma_{44}}\rho_1 \\ & & & \sigma_{44} \end{pmatrix}.$$

However, these models may not be appropriate for the cattle or the speech recognition data sets because of their persistent nonzero correlations. Zimmerman and Núñez-Antón (2010) reported the analysis of a fruit fly mortality data set in which a heterogeneous Toeplitz model of moderate order was one of the best-fitting models for the within-individual covariance structure.

10.4.5 Unstructured and Structured Antedependence Models

Multivariate normal errors $\varepsilon_1, \ldots, \varepsilon_n$ are antedependent of order s (AD(s)) if ε_j and ε_{j+k+1}, conditioned on the intervening errors $\varepsilon_{j+1}, \ldots, \varepsilon_{j+k}$, are independent, for all $j = 1, \ldots, n-k-1$ and all $k \geq s$. The original definition of AD models is due to Gabriel (1962), but a formal parametric model was not proposed therein. An equivalent definition for the AD(s) model, but parametrically specified, can be written as:

$$\varepsilon_j = \sum_{k=1}^{s^*} \phi_{jk}\varepsilon_{j-k} + \nu_j (j = 2, \ldots, n), \tag{10.3}$$

where $s^* = \min(s, j-1)$ and the ν_js are independent normal random variables with mean zero and variances $\sigma_j^2 > 0$, which could be time dependent. From the parametrically specified AD(s) model in (10.3), we can clearly see that AD models generalize stationary AR models. That is, AD models allow for the existence of serial correlation but, in addition, they can have nonstationary variances and correlations. Particular cases of interest are the AD($n-1$) model, which is equivalent to the unstructured model (see Section 10.4.7) and the AD(0) model, which is equivalent to the heterogeneous independent model (i.e., $\Sigma = \text{diag}(\sigma_{11}, \ldots, \sigma_{nn})$).

We refer to model (10.3) as an unstructured AD model of order s, where by unstructured we mean that the ϕ_{jk} and σ_j^2 parameters are not written or further parameterized as functions of a smaller number of parameters. An AD(s) model has $(s+1)(2n-s)/2$ parameters. In addition, Gabriel (1962) proposed a likelihood ratio test that allows us to determine the order of antedependence for a given data set. Equation (10.3) is an autoregressive specification of an AD(s) model, that is, a parametric specification as a function of the autoregressive coefficients $\{\phi_{jk}: j = 2\ldots, n; k = 1, \ldots, s^*\}$ and innovation variances $\{\sigma_j^2: j = 1, \ldots, n\}$ (the ν_js being the innovations). The elements of Σ may be determined from the autoregressive parameters and innovation variances by recursive use of the Yule–Walker equations. For example, for an AD(2) model, we have that:

$$\begin{aligned} \sigma_{11} &= \sigma_1^2 \\ \sigma_{21} &= \phi_{21}\sigma_1^2 \\ \sigma_{22} &= \phi_{21}^2\sigma_1^2 + \sigma_2^2 \\ \sigma_{j,j-k} &= \phi_{j1}\sigma_{j-1,j-k} + \phi_{j2}\sigma_{j-2,j-k} \\ &\quad + \sigma_j^2 I_{\{k=0\}} (k = 0, 1, \ldots, \\ &\quad j-1\ ;\ j = 3, \ldots, n). \end{aligned}$$

A second, equivalent, way of parameterizing the AD(s) model is the covariance specification, for which we parametrically specify the variances and covariances (or correlations) in the model. It can be easily shown that variances and covariances in the first s subdiagonals (and superdiagonals) of the covariance matrix of an AD(s) model remain unstructured, whereas the remaining covariances are completely determined by these values. As an example, we can write the covariance matrix for the AD(1) model for $n = 4$, which has seven parameters, as:

$$\begin{pmatrix} \sigma_{11} & \sqrt{\sigma_{11}\sigma_{22}}\rho_1 & \sqrt{\sigma_{11}\sigma_{33}}\rho_1\rho_2 & \sqrt{\sigma_{11}\sigma_{44}}\rho_1\rho_2\rho_3 \\ & \sigma_{22} & \sqrt{\sigma_{22}\sigma_{33}}\rho_2 & \sqrt{\sigma_{22}\sigma_{44}}\rho_2\rho_3 \\ & & \sigma_{33} & \sqrt{\sigma_{33}\sigma_{44}}\rho_3 \\ & & & \sigma_{44} \end{pmatrix}. \tag{10.4}$$

In this case, elements of the main and first subdiagonal (and superdiagonal) are determined by $\sigma_{11}, \sigma_{22}, \sigma_{33}$, and σ_{44}, and ρ_1, ρ_2, and ρ_3, the seven unstructured parameters in the model. Here, $\rho_j \equiv \rho_{j,j+1}$. In general, for $s > 1$, correlations in the $s+1, \ldots, n-1$ subdiagonals of Σ will depend on the correlations of the first s subdiagonals. However, this dependence is much more complicated than the multiplicative structure observed in (10.4). In addition, when $s > 1$ the parameter space generally cannot be stated in terms of simple constraints on individual correlations.

A third way of specifying an AD(s) model is in terms of the corresponding precision matrix, Σ^{-1}. For an AD(s) model, elements on the main diagonal and the first s subdiagonals (and superdiagonals) of Σ^{-1} are unstructured, but the remaining elements are zero (Gabriel, 1962; Zimmerman et al., 1998; or Zimmerman and Núñez-Antón, 2010). Interested readers can find additional details and results on these specifications for AD models in Zimmerman and Núñez-Antón (2010).

From the above, it is clear that unstructured AD models have more parameters (i.e., are more flexible) than particular cases of it, such as, for example, stationary AR models. For example, when $n = 10$, the AD(1) and AD(2) models have 19 and 27 parameters, respectively, whereas the most commonly used stationary models have considerably fewer parameters. This raises the question of whether there are alternative or simpler AD models that are flexible enough to adequately model nonstationarity of correlations or handle unequal spacing of measurement times and, at the same time, are sufficiently parsimonious that the number of parameters does not increase with the number of measurement times. This question led Núñez-Antón and Woodworth (1994) and Zimmerman and Núñez-Antón (1997) to introduce classes of structured AD (SAD) models. The basic idea consists of parameterizing the autoregressive coefficients, correlations, or partial correlations (depending on whether an autoregressive, covariance, or precision matrix specification is used) of the AD(s) model, so that they follow a Box–Cox power function of time, and parameterizing the innovation variances, variances, or partial variances (again depending on the specification) as polynomial or step functions of time. The Box–Cox transformation function is given by

$$f(t;\lambda) = \begin{cases} (t^\lambda - 1)/\lambda & \text{if } \lambda \neq 0 \\ \log t & \text{if } \lambda = 0. \end{cases} \tag{10.5}$$

The classes of autoregressive and covariance-specified SAD models of order s are as follows:

Autoregressive specification (SAD-AE):

$$\begin{aligned} \phi_{jk} &= \phi_k^{f(t_j;\lambda_k)-f(t_{j-k};\lambda_k)} \\ &\qquad (j = s+1, \ldots, n;\ k = 1, \ldots, s), \\ \sigma_j^2 &= \sigma^2 g(t_j;\boldsymbol{\psi}) \ \ (j = s+1, \ldots, n). \end{aligned} \tag{10.6}$$

Here, g is a parametrically specified function having few parameters (e.g., a low-order polynomial function).

Covariance specification (SAD-CE):

$$\begin{aligned} \rho_{j,j-k} &= \rho_k^{f(t_j;\lambda_k)-f(t_{j-k};\lambda_k)} \\ &\qquad (j = k+1, \ldots, n; k = 1, \ldots, s), \\ \sigma_{jj} &= \sigma^2 g(t_j;\boldsymbol{\psi})(j = 1, \ldots, n). \end{aligned} \tag{10.7}$$

The constraints associated with (10.7) are $\rho_k > 0$, $\sigma^2 > 0$, and $\{\boldsymbol{\psi}: g(t_j;\boldsymbol{\psi}) > 0\}$, and those associated with (10.6) are $\phi_k > 0$, $\sigma^2 > 0$, and $\{\boldsymbol{\psi}: g(t_j;\boldsymbol{\psi}) > 0\}$. Exponentiation of the ϕ_ks and the ρ_ks in (10.6) and (10.7) by the Box–Cox form of f given by (10.5) prescribes that the autoregressive coefficients of order k (i.e., the ϕ_ks in the SAD-AE specification above) and the correlations on the kth subdiagonal of the correlation matrix (i.e., the ρ_ks in the SAD-CE specification above) are monotonic increasing if $\lambda_k < 1$, monotonic decreasing if $\lambda_k > 1$, or constant if $\lambda_k = 1$ $(k = 1, \ldots, s)$. That is, f effects a nonlinear deformation upon the time axis such that correlations between measurements equidistant in the deformed scale are constant and, thus, stationarity in correlation in the transformed scale is obtained.

An SAD model has fewer parameters than an AD model of the same order. For example, if g is a quadratic function of time, the SAD-AE(1) and SAD-AE(2) would have 6 and 10 parameters, respectively, and the corresponding SAD-CE(1) and SAD-CE(2) would have 5 and 7 parameters, respectively. There is also another parsimonious class of SAD models proposed by Pourahmadi (1999), which takes same-diagonal autoregressive coefficients and logarithms of innovation variances to be low-order polynomial functions of time.

As we have seen in Section 10.4, the OSM and PRISM for the treatment A cattle data (see Figure 10.3), as well as the sample variance and correlation values reported in Tables 10.1 and 10.2, indicate that AD models may be appropriate for this data set. That is, models having some type of conditional independence for observations should be considered. In this specific case, antedependence models of orders at most two (because the PRISM shows random scatter plots only for the third and higher diagonals) could be a good model option to try to fit for this data set. This is supported by observing that nonstationarity in both variances and correlations are exhibited by this data set, a characteristic easily modeled by some type of antedependence model. Readers interested in the final selected covariance models for the treatment A cattle data may wish to consult the complete results reported in Chapter 9 of Zimmerman and Núñez-Antón (2010), where the best-fitting covariance models, based on OSM, PRISM, AIC, BIC, and LRT criteria, were a covariance or marginally specified structured AD model of order two that assumed a cubic in time variance function and a Box–Cox power law model for the correlations. Competing models were all structured or unstructured versions of AD models, as well as their special cases, such as, for example, homogeneous or heterogeneous autoregressive models.

With regard to the speech recognition data, the OSM and PRISM (see Figure 10.5), as well as the sample variance and correlation values reported in Tables 10.3 and 10.4, indicate that AD models of order one or SAD models of the same order should be considered. In this case, it seems that there is one variance for the first time point and a different variance for the remaining ones. Thus, considering a step variance function may be appropriate in this case. Correlations seem to increase along diagonals, indicating nonstationarity in correlations and motivating the use of some type of antedependence model for this data set. Readers interested in the final selected covariance models for these data may verify that the best-fitting covariance model, based on OSM, PRISM, AIC, BIC, and LRT criteria, was a covariance or marginally specified structured AD model of order one, with a step variance function and a Box–Cox power law model for the correlations (Zimmerman and Núñez-Antón, 2010, Chapter 9). Competing models were all structured or unstructured versions of AD models, as well as their special cases, such as, for example, homogeneous or heterogeneous autoregressive models. As expected, in these data sets, random coefficients models did not provide a good fit. For more details on additional model selection criteria, readers can consult Chapter 7.

In order to further illustrate the usefulness of the OSM and PRISM to discriminate and suggest possible models to fit to data, Figures 10.6 and 10.7 include the OSM and PRISM plots for 100 realizations simulated from AR(1) and AD(1) processes, respectively, over six time points. Figure 10.6 suggests the existence of serial correlation and also of stationary correlations for equally spaced observations. Plots out of the main diagonal in the PRISM exhibit random scatter, suggesting that only models including a first order dependence should be considered. In the same way, Figure 10.7 also suggests the existence of serial correlation but, contrary to what we

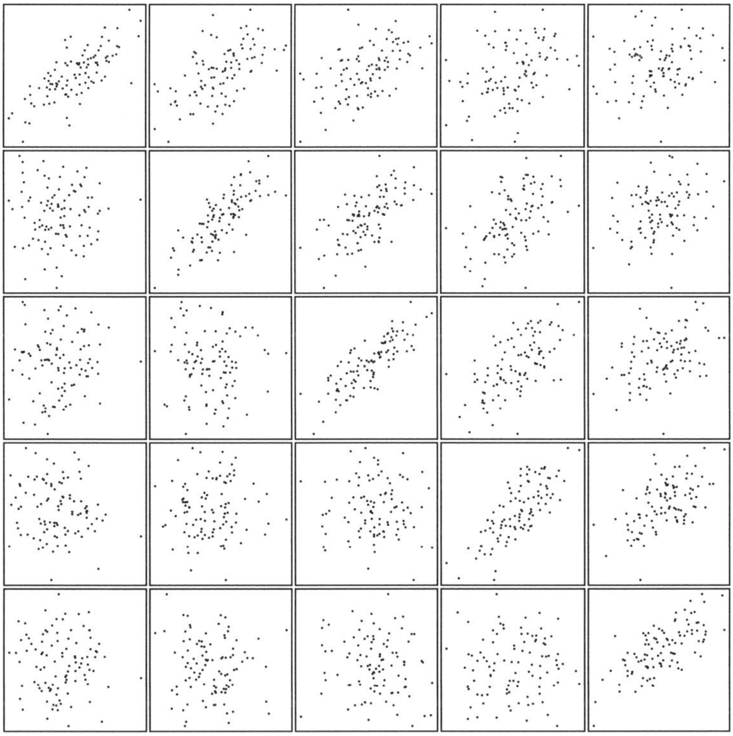

Figure 10.6 Partial Regression-on-Intervenors Scatterplot Matrix (PRISM) Plot (lower triangle, including main diagonal) and Ordinary Scatterplot Matrix (OSM) Plot (upper triangle, including main diagonal) for 100 Realizations of a First Order Autoregressive Process with $\rho = 0.80$ Over Six Time Points

have just observed in Figure 10.6 for the AR(1) process, this correlation is not stationary and it actually increases along the diagonals. This is clear evidence for the existence of nonstationarity in correlation that could very well be modeled by AD or SAD models but, at the same time, advises against the use of stationary autoregressive models. Off-diagonal plots in the PRISM exhibit random scatter, suggesting that only models including a first order dependence should be considered.

As we have already mentioned above, AD models have $(s+1)(2n-s)/2$ parameters; that is, $O(n)$ parameters, whereas the number of parameters in the SAD models do not depend on the number of observations; that is, $O(1)$ parameters. Therefore, SAD models are more parsimonious than AD models. With regard to model flexibility, it is clear that AD and SAD models are capable of modeling general behaviors exhibiting nonstationarity in variances and correlations in a real data situation. Thus, they can be very useful models when these characteristics are observed in the data set.

10.4.6 Factor Analytic Models

Factor analytic (FA) models arise from factor analysis, a very important statistical technique in fields such as educational statistics and psychometrics (McDonald, 1985; or Mulaik, 2009). The general factor analytic form (see, e.g., Everitt, 2009; Mulaik, 2009; or Afifi et al., 2011) is:

$$\Sigma = \Lambda\Lambda' + D,$$

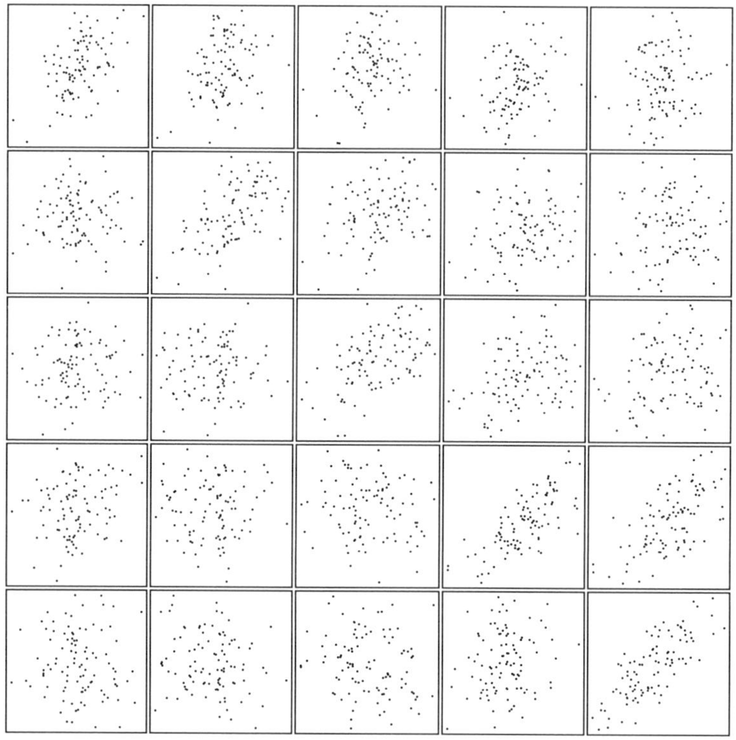

Figure 10.7 Partial Regression-on-Intervenors Scatterplot Matrix (PRISM) Plot (lower triangle, including main diagonal) and Ordinary Scatterplot Matrix (OSM) Plot (upper triangle, including main diagonal) for 100 Realizations of an Antedependence Model of Order $s = 1$, with $\phi_{i,i-1} = 0.3 + 0.1i$, Over Six Time Points

where Λ is an $n \times k$ matrix and D is an $n \times n$ diagonal matrix. The k columns or elements of Λ are called factor loadings and the non-null elements of D are called specific variances. This model can be motivated by indicating that the portion of the total variance modeled in Λ represents variability attributable to some underlying common factors, whereas the remaining portions are unique (for more details, see Mardia et al., 1979, Chapter 9; Anderson, 1984, Chapter 14; or Seber, 1984, Chapter 5). In the context of longitudinal or multilevel data analysis, factor analytic models provide another useful and distinct way of modeling heterogeneity. We only consider here the first-order factor analytic model, where all columns of Λ except the first are equal to zero. This model is known as the first-order FA (i.e., FA(1)) model because it only models one common factor. Moreover, a special case of the first-order factor analytic model is the random coefficients models in Section 10.4.1. The FA(1) model also assumes that the specific variances in D are all equal, so that for $n = 4$, this model can be written as:

$$\Sigma = \begin{pmatrix} \lambda_1 \\ \lambda_2 \\ \lambda_3 \\ \lambda_4 \end{pmatrix} \begin{pmatrix} \lambda_1 \\ \lambda_2 \\ \lambda_3 \\ \lambda_4 \end{pmatrix}' + \begin{pmatrix} \sigma^2 & & & \\ & \sigma^2 & & \\ & & \sigma^2 & \\ & & & \sigma^2 \end{pmatrix}.$$

As was the case for some of the models mentioned in previous sections, there is a heterogeneous version of the FA(1) model, the FAH(1) model, for which the specific variances are not assumed to be equal. Obviously, the FA(1) model is more parsimonious than the FAH(1) model. Higher-order FA structures are also possible, although in this case constraints must be placed on a few of the parameters in order to fix the rotation of the factors (for more details on this, see, Jennrich and Schluchter, 1986). For $n = 4$, the FAH(1) model can be written as:

$$\Sigma = \begin{pmatrix} \lambda_1 \\ \lambda_2 \\ \lambda_3 \\ \lambda_4 \end{pmatrix} \begin{pmatrix} \lambda_1 \\ \lambda_2 \\ \lambda_3 \\ \lambda_4 \end{pmatrix}' + \begin{pmatrix} \sigma_1^2 & & & \\ & \sigma_2^2 & & \\ & & \sigma_3^2 & \\ & & & \sigma_4^2 \end{pmatrix}.$$

As can be seen, the FA(1) and FAH(1) models have $n + 1$ and $2n$ parameters; that is, $O(n)$ parameters, and, thus, they are not very parsimonious models. With regard to model flexibility, it is clear that these models can provide another useful and distinct way of modeling heterogeneity, both in variance and correlation. Thus, they are, in general, flexible covariance structures allowing one to model nonstationary variances. In addition, both versions can generate covariances that allow for non-banded correlation structures and, thus, are very flexible covariance models to be used in real data set problems. As an example, let us consider the FA(1) model in which $n = 4$, $\Lambda = (8, 6, 4, 2)'$, and $\sigma^2 = 4$, so that we have:

$$\Sigma = 4 \cdot \begin{pmatrix} 1 & 0.92 & 0.87 & 0.69 \\ & 1 & 0.85 & 0.67 \\ & & 1 & 0.63 \\ & & & 1 \end{pmatrix}.$$

This very simple example, as well as the general properties these models have, indicate that they may be appropriate for the cattle and the speech recognition data sets.

10.4.7 Unstructured Covariance Model

The unstructured covariance model is the extreme case of parametrically modeling the error's covariance structure. Its use is commonly limited to the cases where measurements for different subjects are taken at a common set of time points (as in the cattle data), but, in general, it is very difficult, if not impossible, to consider this structure for cases where observation times are random, or otherwise varying from subject to subject (as in the speech recognition data). In the latter case, and because this structure does not depend on the actual observation times, if it does provide a good fit, a random coefficients model may make sense. In the unstructured covariance model, no structure is imposed on any of the elements of the covariance matrix, so that this model has $n(n+1)/2$ variance and covariance parameters or, equivalently, $n(n + 1)/2$ variance and correlation parameters to be estimated. Given that it represents a completely general model for the error covariance structure within individuals, it does not require that observations be equally spaced. In this model, the parameter space for Σ to be positive definite cannot be expressed as linear inequality restrictions on individual parameters. The unstructured model for $n = 4$ would have 10 parameters and can be written as:

$$\Sigma = \begin{pmatrix} \sigma_{11} & \sigma_{12} & \sigma_{13} & \sigma_{14} \\ & \sigma_{22} & \sigma_{23} & \sigma_{24} \\ & & \sigma_{33} & \sigma_{34} \\ & & & \sigma_{44} \end{pmatrix}.$$

The main advantage of the unstructured covariance model structure is its complete flexibility; that is, it represents the most flexible covariance model. However, this model presents several disadvantages, among which it has a large, i.e., $O(n^2)$, number of parameters, so that it is the least parsimonious covariance model structure. Furthermore, there are computational difficulties of estimation when subjects are not observed at the same times.

10.5 CONCLUSIONS AND FINAL RECOMMENDATIONS

We have introduced the parametric multivariate regression modeling approach for jointly modeling the mean and covariance structures, where these two components are to be appropriately selected. This chapter has mainly centered on issues related to modeling of the within-individual covariance structure for the errors. Additional details on model interpretation, specification, and selection in the context of longitudinal data can be found in Chapters 7, 8, and 9. We have recommended and illustrated the use of graphical diagnostics, such as the profile plot, OSM, and PRISM, as well as sample mean, variance, and correlation values as an effective way of selecting reasonable and plausible candidates for the mean and covariance structure models. In addition, we have introduced and motivated the appropriateness of the most commonly used covariance structure models in practice, including their use in real data situations, as well as in simulations.

We have also noted that some specific characteristics that often need to be appropriately modeled are the presence of stationary or nonstationary variances and correlations, as well as the existence of serial correlation. Moreover, an additional issue that needs to be appropriately addressed is the possibility of considering simple or more complex structures, or the possible use of specific random effects models, as is more common in some multilevel modeling approaches. We have concentrated on these issues, as well as the issue of trying to obtain the right balance between the model's flexibility and parsimony. Along these lines, it is clear that an extremely flexible model will lead to inefficient estimation, and a very parsimonious model will lead to a poor fit.

In the past, many authors did not pay enough attention to the model's error covariance structure within individuals and, thus, assumed that existing models, such as the compound symmetry, random coefficients, or unstructured models were the ones to be used. As we have seen in previous sections, the compound symmetry and unstructured models represent two extremes, with the parsimonious (with $O(1)$ parameters) and inflexible compound symmetry model often not fitting the data very well and the unparsimonious (with $O(n^2)$ parameters) and very flexible unstructured model often being inefficient. In the mean time, intermediate models have been proposed, so the existing gap between the two extremes has been filled with new covariance modeling proposals. Some of these models have been introduced here. Their main purpose, besides modeling the characteristics present in the data set under study, is to strike the right balance between model flexibility and parsimony. Those that have $O(n)$ parameters, such as the heterogeneous first-order autoregressive, Toeplitz, factor analytic, and, especially, the unstructured antedependence models, seem to be the most successful at achieving this balance. In the examples presented in this chapter, structured antedependence models were those providing the best fit based on goodness-of-fit criteria, such as AIC or BIC, specific tests between nested models, such as the LRT tests, or on graphical diagnostic tools, such as the OSM and PRISM plots.

ACKNOWLEDGMENTS

Research supported by Ministerio Español de Ciencia e Innovación, FEDER, and Departamento de Educación del Gobierno Vasco (UPV/EHU Econometrics Research Group) under research grants MTM2007-60112, MTM2010-14913, and IT-334-07, US12/09, UFI11/03, and IT-642-13.

REFERENCES

Afifi, A., May, S., and Clark, V.A. (2011) *Practical Multivariate Analysis*, 5th Edition. Boca Raton, FL: Chapman & Hall/CRC.

Akaike, H. (1974) 'A New Look at the Statistical Model Identification', *IEEE Transactions on Automatic Control*, 19: 716–23.

Anderson, T.W. (1984) *An Introduction to Multivariate Statistical Analysis*, 2nd Edition. New York: Wiley.

Bryk, A.S. and Raudenbush, S.W. (1992) *Hierarchical Linear Models*. Newbury Park, CA: Sage Publications, Inc.

Cepeda-Cuervo, E. and Núñez-Antón, V. (2009) 'Bayesian Modeling of the Mean and Covariance Matrix in Normal Nonlinear Models', *Journal of Statistical Computation and Simulation*, 79: 837–53.

D'Agostino, R.B., Sparrow, D., Weiss, S., and Rosner, B. (1995) 'Longitudinal Models for Analysis of Respiratory Function', *Statistics in Medicine*, 14: 2205–16.

Davis, C.S. (2002) *Statistical Methods for the Analysis of Repeated Measurements*. New York: Springer.

Diggle, P.J., Heagerty, P.J., Liang, K.-Y., and Zeger, S.L. (2002) *Analysis of Longitudinal Data*, 2nd Edition. New York: Oxford University Press.

Everitt, B.S. (2009) *Multivariable Modeling and Multivariate Analysis for the Behavioral Sciences*. Boca Raton, FL: Chapman & Hall/CRC.

Fitzmaurice, G.M., Laird, N.M., and Ware, J.H. (2004) *Applied Longitudinal Analysis*. New York: Wiley.

Gabriel, K.R. (1962) 'Ante-dependence Analysis of an Ordered Set of Variables', *Annals of Mathematical Statistics*, 33: 201–12.

Goldstein, H. (1987) *Multilevel Models in Educational and Social Research*. New York: Oxford University Press.

Goldstein, H. (2003) *Multilevel Statistical Model*, 3rd Edition. London: Arnold.

Hand, D. and Crowder, M. (1996) *Practical Longitudinal Data Analysis*. London: Chapman & Hall.

Harrison, G.A. and Brush, G. (1990) 'On Correlations between Adjacents Velocities and Accelerations in Longitudinal Growth Data', *Annals of Human Biology*, 17: 55–7.

Jennrich, R. and Schluchter, M.D. (1986) 'Unbalanced Repeated Measures Models with Structured Covariance Matrices', *Biometrics*, 42: 805–20.

Jones, R.H. (1990) 'Serial Correlation or Random subject Effects?', *Communications in Statistics – Simulation and Computation*, 19: 1105–23.

Jones, M.C. and Rice, J.A. (1992) 'Displaying the Important Features of Large Collections of Similar Curves', *The American Statistician*, 46: 140–5.

Kenward, M.G. (1987) 'A Method for Comparing Profiles of Repeated Measurements', *Applied Statistics*, 36: 296–308.

Laird, N.M. and Ware, J.H. (1982) 'Random-effects Models for Longitudinal Data', *Biometrics*, 38: 963–74.

Lange, N. and Laird, N.M. (1989) 'The Effect of Covariance Structure on Variance Estimation in Balanced Growth-curve Models with Random Parameters', *Journal of the American Statistical Association*, 84: 241–7.

Louis, T.A. (1988) 'General Methods for Analysing Repeated Measures'. *Statistics in Medicine*', 7: 29–45.

Mardia, K.V., Kent, J.T., and Bibby, J.M. (1979) *Multivariate Analysis*. Orlando: Academic Press, Inc.

McDonald, R.P. (1985) *Factor Analysis and Related Methods*. Hillsdale, NJ: Lawrence Erlbaum Associates.

Mulaik, S.A. (2009) *The Foundations of Factor Analysis*, 2nd Edition. Boca Raton, FL: Chapman & Hall/CRC.

Núñez-Antón, V. and Woodworth, G.G. (1994) 'Analysis of Longitudinal Data with Unequally Spaced Observations and Time-dependent Correlated Errors', *Biometrics*, 50: 445–56.

Pourahmadi, M. (1999) 'Joint Mean-covariance Models with Applications to Longitudinal Data: Unconstrained Parameterisation', *Biometrika*, 86: 677–90.

Pourahmadi, M. (2001) *Foundations of Time Series Analysis and Prediction Theory*. New York: Wiley.

Pourahmadi, M. (2002) 'Graphical Diagnostics for Modeling Unstructured Covariance Matrices', *International Statistical Review*, 70: 395–417.

Rao, C.R. (1959) 'Some Problems Involving Linear Hypotheses in Multivariate Analysis', *Biometrika*, 46: 49–58.

Reinsel, G. (1982) 'Multivariate Repeated-measurement or Growth Curve Models with Multivariate Random-effects Covariance Structure', *Journal of the American Statistical Association*, 77: 190–5.

Ruppert, D., Wand, M.P., and Carroll, R.J. (2003) *Semiparametric Regression*. Cambridge: Cambridge University Press.

Rutter, C.M. and Elashoff, R.M. (1994) 'Analysis of Longitudinal Data: Random Coefficient Regression Modeling', *Statistics in Medicine*, 13: 1211–31.

Schwarz, G.E. (1978) 'Estimating the Dimension of a Model', *Annals of Statistics*, 6: 461–4.

Seber, G.A.F. (1984) *Multivariate Observations*. New York: Wiley.

Segal, M. (1994) 'Representing Curves for Longitudinal Data via Regression Trees', *Journal of Computational and Graphical Statistics*, 3: 214–33.

Tyler, R.S., Abbas, P., Tye-Murray, N., Gantz, B.J., Knutson, J.F., McCabe, B.F., Lansing, C., Brown, C., Woodworth, G.G., Hinrichs, J., and Kuk, F. (1988) 'Evaluation of Five Different Cochlear Implant Designs: Audiologic Assessment and Predictors of Performance', *The Laryngoscope*, 98: 1100–6.

Verbeke, G. and Molenberghs, G. (2001) *Linear Mixed Models for Longitudinal Data*. New York: Springer.

Verbyla, A.P., Cullis, B.R., Kenward, M.G. and Welham, S.J. (1999) 'The Analysis of Designed Experiments and Longitudinal Data by using Smoothing Splines', *Applied Statistics*, 48: 269–311.

Weiss, R.E. (2005) *Modeling Longitudinal Data*. New York: Springer.

Wolfinger, R.D. (1996) 'Heterogeneous Variance-covariance Structures for Repeated Measures', *Journal of Agricultural, Biological, and Environmental Statistics*, 1: 205–30.

Zimmerman, D.L. (2000) 'Viewing the Correlation Structure in Longitudinal Data through a PRISM', *The American Statistician*, 54: 310–18.

Zimmerman, D.L. and Núñez-Antón, V. (1997) 'Structured Antedependence Models for Longitudinal Data', in T.G. Gregoire, D.R. Brillinger, P.J. Diggle, E. Russek-Cohen, W.G. Warren, and R. Wolfinger (eds), *Modeling Longitudinal and Spatially Correlated Data. Methods, Applications, and Future Directions*. New York: Springer-Verlag, 63–76.

Zimmerman, D.L. and Núñez-Antón, V. (2001) 'Parametric Modeling of Growth Curve Data: An Overview (with Discussion)', *Test*, 10: 1–73.

Zimmerman, D.L. and Núñez-Antón, V. (2010) *Antedependence Models for Longitudinal Data*. Boca Raton, FL: Chapman & Hall/CRC.

Zimmerman, D.L., Núñez-Antón, V., and El Barmi, H. (1998) 'Computational Aspects of Likelihood-based Inference of First-order Antedependence Models', *Journal of Statistical Computation and Simulation*, 60: 67–84.

11

Design Considerations in Multilevel Studies

Gerard van Breukelen
Maastricht University, The Netherlands

Mirjam Moerbeek
Utrecht University, The Netherlands

11.1 INTRODUCTION

Multilevel studies are omnipresent in the social and behavioral sciences, in either of two forms: nesting of persons within larger units, such as pupils in schools, employees in companies, or patients in general practices, or nesting of repeated measures within persons, such as the waves of a panel study or follow-ups in a psychotherapeutic trial. The combination of both forms of nesting in a three-level design (organizations, persons, measures) is also encountered, for instance in longitudinal school effectiveness studies and intervention studies in public health and clinical psychology. Questions to be answered by multilevel analysis of such designs typically concern the effects of one or more predictors on a measure of health, behavior, attitude, or knowledge.

An important question in the design stage of a multilevel study is the sample size needed at each level to have sufficient power and precision for the effects of interest. Useful work on this issue was done by Snijders and Bosker (1993) and Maas and Hox (2005), among others. The dilemma in sample size planning is the fact that larger samples give more power and precision, but also cost more money and time, put a burden on more participants, and may not be feasible. Optimal design, a specialization in statistics, solves this dilemma by mathematically searching for the most powerful design under a cost constraint, or the cheapest design under a power constraint. Berger and Wong (2009) give a good introduction into optimal design, and Atkinson, Donev, and Tobias (2007) give a more technical account. In the context of nested and longitudinal studies, optimal design primarily concerns the best sample size per level of the design: how many clusters, how many persons, how many repeated measures?

This chapter gives an introduction to optimal design of multilevel studies with emphasis on nested designs in intervention research, where persons or organizations are randomly assigned to intervention or control, or intervention A or B. Such studies are often encountered in clinical and educational psychology, public health, and medicine, but they may also be conducted in sociology, economics, and law, for instance to evaluate the effects of administrative, legislative, or managerial changes. The outline of the chapter is as follows. First, the general design and multilevel model are presented for a two-level randomized study with a quantitative outcome measured after the intervention. We then discuss the optimality criterion to be satisfied and present the optimal design as a function of sampling costs and outcome variance at each level of the design. Since this design assumes equal sample sizes per organization and requires the intraclass correlation to be known, we show how both limitations can be overcome in a practical way. After that, we consider the case where the outcome measure is not quantitative, but binary, and the case where covariates are present. Subsequently, optimal design of longitudinal studies is addressed. Here the question is how many repeated measures, and at which time points, is the most efficient. We conclude with a discussion of robust design.

11.2 DESIGN OF INTERVENTION STUDIES WITH PERSONS NESTED IN ORGANIZATIONS

In psychology and medicine the effects of interventions to improve health or to change human behavior are frequently evaluated in a nested setting. Examples are smoking prevention in primary school (Ausems et al., 2002), stress management for primary school children (Kraag et al., 2009), and promotion of healthy behaviors among patients of general practices (Van Keulen et al., 2011). However, nesting of persons within organizations also occurs in, for instance, organizational psychology (companies), educational psychology (schools), sociology (cities), or law (court districts). The two classical nested designs for intervention studies are the cluster randomized trial (Donner and Klar, 1994; Raudenbush, 1997; Hayes and Moulton, 2009), also known as the group randomized trial (Murray, 1998), and the multisite or multicenter trial (Senn, 1998; Raudenbush and Liu, 2000). The difference between these two is in the level of randomization, see Figure 11.1. In a cluster randomized trial, organizations (called clusters) are randomly assigned to one of the treatment conditions and all participating persons within a given cluster are given the same treatment. In a multisite or multicenter trial, persons are randomly assigned to a treatment within each of the organizations (called sites or centers) and so we have blocked or prestratified randomization of persons, with center as the blocking factor or stratifier.

Multisite trials are more efficient than cluster randomized trials because only cluster by treatment interaction and not the cluster 'main' effect enters into the variance of the treatment effect in a multisite trial. However, cluster randomized trials may be the only option for logistic reasons or to prevent treatment contamination. For some treatments, assignment of persons instead of clusters is impossible, for instance if the treatments are teaching methods in classrooms, or safety procedures in companies. For other treatments, person assignment can lead to serious contamination or carry-over of effects from treated onto untreated persons, leading to underestimation of the treatment effect and a loss of power (Moerbeek, 2005a; Rhoads, 2011). Both designs will be discussed, including the simplest model for data analysis and the optimal sample size at each level of the design under cost and power constraints. Equal sample sizes per cluster or site will first be assumed, both for simplicity and because they are the most efficient. Later, this assumption is relaxed.

11.3 CLUSTER RANDOMIZED TRIAL: MODEL FOR ANALYSIS AND OPTIMAL DESIGN

Suppose an intervention is evaluated in a cluster randomized trial with K clusters of n persons per cluster, and the effectiveness of the treatment is inferred from the mean difference

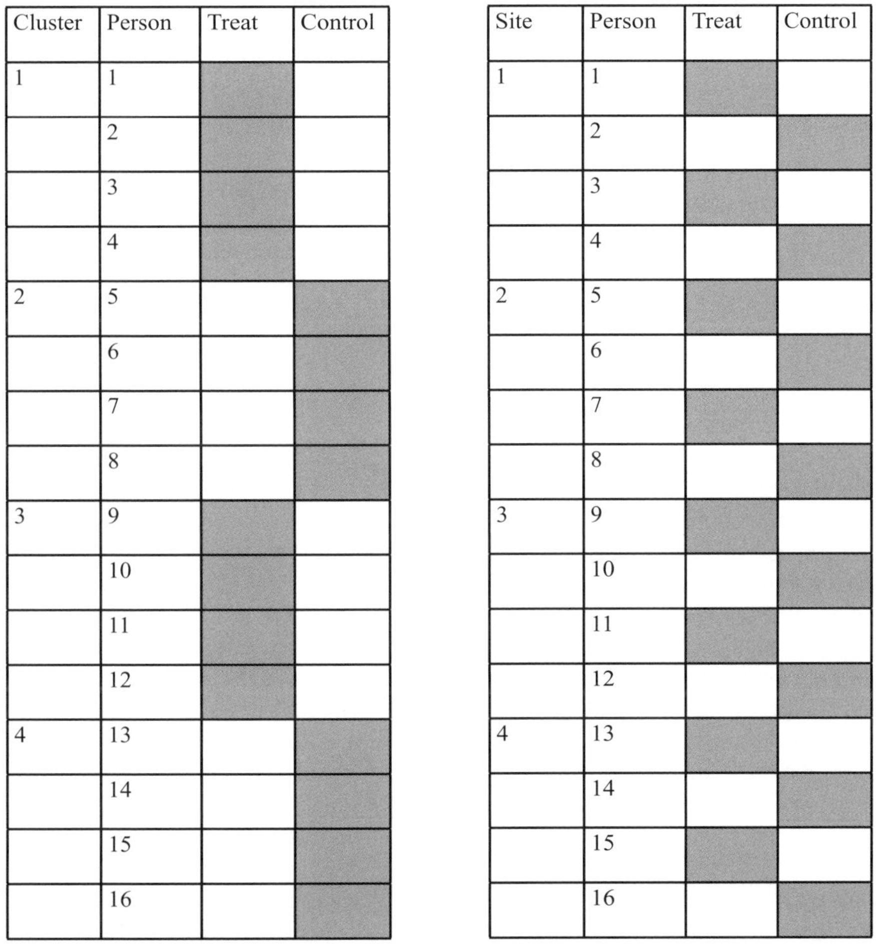

Figure 11.1 Design of a Cluster Randomized Trial (left) and Multisite Trial (right), with Four Clusters/Sites and Four Persons Per Cluster/Site (shaded = observed, blank = missing by design)

between treated and control on a quantitative outcome y. Ignoring covariates for now, the model for data analysis is:

$$y_{ij} = \beta_0 + \beta_1 x_j + u_{0j} + \varepsilon_{ij}, \qquad (11.1)$$

where y_{ij} is the outcome for person i in cluster j, and x_j is the treatment to which cluster j is assigned, coded 1 for treated and 0 for control clusters, or centered as $+0.5$ and -0.5, respectively. Further, u_{0j} is a random cluster effect and ε_{ij} is an error term to accommodate person and measurement error effects, and these two are independently and normally distributed with variances σ^2_{u0} and σ^2_{ε}, respectively. The outcomes y_{ij} and $y_{i'j}$ of two persons i and i' in the same cluster j are correlated, and this intraclass correlation (*ICC*) is

$$\rho = \frac{\sigma^2_{u0}}{\sigma^2_y} = \frac{\sigma^2_{u0}}{\sigma^2_{u0} + \sigma^2_{\varepsilon}}, \qquad (11.2)$$

where σ^2_y is the total outcome variance within each treatment condition.

Ignoring the individual random effects u_{0j} and ε_{ij}, model (11.1) has four parameters: two fixed regression parameters and two variance components, which can be estimated with the maximum likelihood (ML, REML) method. In intervention studies, interest is in the treatment effect β_1, which must be estimated as precisely as possible, thereby also maximizing the test power. So a good optimality criterion is the variance of its estimator, which is (Donner, Birkett, and Buck, 1981; Raudenbush, 1997; Moerbeek, Van Breukelen, and Berger, 2000):

$$Var(\hat{\beta}_1) = [(n-1)\rho + 1] \times \frac{4\sigma_y^2}{nK}, \quad (11.3)$$

where n is the sample size per cluster or site, nK is the total sample size, and $[(n-1)\rho+1]$ is known as the design effect. If $\rho = 0$, the design effect is 1 and $Var(\hat{\beta}_1)$ for cluster randomization is equal to $Var(\hat{\beta}_1)$ for person randomization, since there is no outcome variation at the cluster level in this case. Other optimality criteria than $Var(\hat{\beta}_1)$ exist, notably A- and D-optimality which minimize the trace and determinant of the covariance matrix of estimators, respectively. For the sake of brevity and because intervention studies typically focus on a single treatment effect parameter only, these multivariate criteria will not be discussed in this chapter, but see Moerbeek, Van Breukelen, and Berger (2001a) and Candel and Van Breukelen (2010a) for some applications.

As equation (11.3) shows, $Var(\hat{\beta}_1)$ decreases as the number of persons per cluster (or site) or the number of clusters (or sites) increases, but such increases cost extra money, time, and effort. Also, the effect of increasing n becomes ignorable for large n due to the increasing design effect. Optimal design means finding that sample size at each level that either minimizes $Var(\hat{\beta}_1)$ for a given total sampling cost, or minimizes total sampling cost for a given test power and precision of estimation, that is, for a given value of $Var(\hat{\beta}_1)$. Thus, a cost function is needed to relate sample size to costs. Assume that inclusion of a cluster into the study costs c currency units (e.g., USD, Euro, or GBP), and inclusion of a subject in an included cluster costs s units. The total budget B needed for K clusters of n persons, ignoring study costs that do not depend on the sample size (e.g., general overhead), is then

$$B = cK + snK = K(c + sn), \quad (11.4)$$

where $(c + sn)$ is the total sampling cost per cluster of size n. The optimal design minimizes $Var(\hat{\beta}_1)$ in equation (11.3) as function of n and K given the constraint in equation (11.4). This gives as optimum (Raudenbush, 1997; Cohen, 1998; Moerbeek et al., 2000):

$$K = \frac{B}{\sqrt{\left(\frac{1-\rho}{\rho}\right) sc} + c}, \qquad n = \sqrt{\frac{(1-\rho)c}{\rho s}}. \quad (11.5)$$

As the *ICC*, ρ, increases, the number of clusters K must increase and the number of persons per cluster n must decrease. The first effect follows from the increasing cluster-level variance σ_{u0}^2 which requires a larger sample of clusters to maintain power and precision. The second effect then follows from the budget constraint, as well as from the fact that an increase of σ_{u0}^2 implies a decrease of σ_ε^2 if we keep σ_y^2 constant.

Inserting equation (11.5) in equation (11.3) gives $Var(\hat{\beta}_1)$ for the optimal design, which is the smallest possible $Var(\hat{\beta}_1)$ given our budget, sampling costs, outcome variance, and *ICC*:

$$Var(\hat{\beta}_1) = g(\rho) \times \frac{4\sigma_y^2}{B}, \quad (11.6)$$

where

$$g(\rho) = (\sqrt{\rho c} + \sqrt{(1-\rho)s})^2.$$

Figure 11.2 shows how this minimal $Var(\hat{\beta}_1)$ depends on the *ICC*, ρ, and on cluster size n, and implicitly also on the number of clusters K due to the budget constraint in equation (11.4). The square root of equation (11.6) is the standard error (*SE*) of the treatment effect estimator $\hat{\beta}_1$ under the optimal design, on which the width of the confidence interval and the power of the significance test depend. If the power and precision of the optimal design are still too low, then the budget must be increased, see equation (11.6), and this results in more clusters rather than more persons per cluster, see equation (11.5).

More specifically, the following relation between budget, power, and effect size can be derived from equation (11.6), relying on the asymptotic normality of $\hat{\beta}_1$ and ignoring at first the fact that σ_y^2 is unknown and estimated from the data:

$$\frac{B}{4g(\rho)} = \left(Z_{1-\gamma} + Z_{1-\alpha/2}\right) \times \left(\frac{\sigma_y}{\beta_1}\right)^2, \tag{11.7}$$

where $(1-\gamma)$ is the power, $Z_{1-\gamma}$ is the 100 $(1-\gamma)$th percentile of the standard normal distribution, two-tailed testing with size α is assumed, and the ratio $\frac{\sigma_y}{\beta_1}$ is the inverse of Cohen's effect size d if treatment is coded as (0, 1) or as (-0.5, $+0.5$), so that β_1 is the expected outcome difference between treatments.

11.4 APPLICATION AND PRACTICAL COMPLICATIONS

We illustrate the application of optimal design by taking a sampling budget of $B = 10,000$, a per cluster sampling cost of $c = 200$, a per person sampling cost of $s = 10$, and an *ICC* of $\rho = 0.05$. From equation (11.5) we obtain an optimal design with $K = 26$ clusters and $n = 19$ persons per cluster after rounding K and n to their nearest respective integers (with K even to allow 50:50 allocation), and a total cost of 10,140. If we assume that the outcome variance σ_y^2 is 25 and the smallest clinically meaningful treatment effect β_1 is 2, implying that Cohen's effect size $d = 0.40$, which is a small to medium effect size, then equation (11.6) gives $Var(\hat{\beta}_1) = 0.385$, $SE = 0.62$, and a 95% confidence interval width of 2.43. Using equation (11.7), we find a test power of almost 90% for a two-tailed test with $\alpha = 5\%$, since $Z_{1-\gamma} = 1.26$ given $Z_{1-\alpha/2} = 1.96$. This is a simplified example, embedded within which are four complicating issues, for which we will now discuss simple solutions.

First, the variance components in equation (11.2) must be estimated and the test statistic follows a Student's t-distribution with $df = K - 2$ (Moerbeek et al., 2003), giving slightly less power. Calculations for various values of K from 20 to 100 and power levels 80% and 90% show that this power loss can be compensated by increasing the total number of clusters by 2 if $\alpha = 0.05$ two-tailed, or by 4 if $\alpha = 0.01$ two-tailed. For K from 10 to 20 we need to add 3 or 5 clusters, respectively. So in the example above the number of clusters must be increased from 26 to 28 to have a power of 90% for detecting an effect with size $d = 0.40$, giving a total cost of 10,920.

Second, there are often feasibility boundaries on K or n, like a maximum of 30 general practices due to geographical distance, or a minimum cluster size of 20 per school due to class size. One can then choose, from all feasible designs, the one nearest to the optimal design in terms of K and n and adapt either K or n to approximate the power or budget of the optimal design.

Third, the *ICC* needed in equation (11.5) is usually unknown in the design stage, although at least in primary care it rarely exceeds 0.10 (Adams et al., 2004; Eldridge et al., 2004). Based on equation (11.3), we know that $Var(\hat{\beta}_1)$ increases with the *ICC*, and so one strategy could be to take the largest realistic *ICC* based on published trials. To obtain a less expensive design, one could, instead, use an intermediate *ICC*, like the midpoint of the assumed *ICC* range (e.g., 0.05 if the range is 0 to 0.10). Recent work (Korendijk, Moerbeek, and Maas, 2010; Van Breukelen and Candel, 2013) suggests that if the design is based on an assumed *ICC* value that is half the size of the correct *ICC*, then that design will still have a relative efficiency (RE) above 0.90 or even above 0.95 compared with the optimal design for the correct *ICC*, given the same budget for both designs. The inverse of the RE is the number of clusters or sites in which the design must be replicated in order to give the same information as the optimal

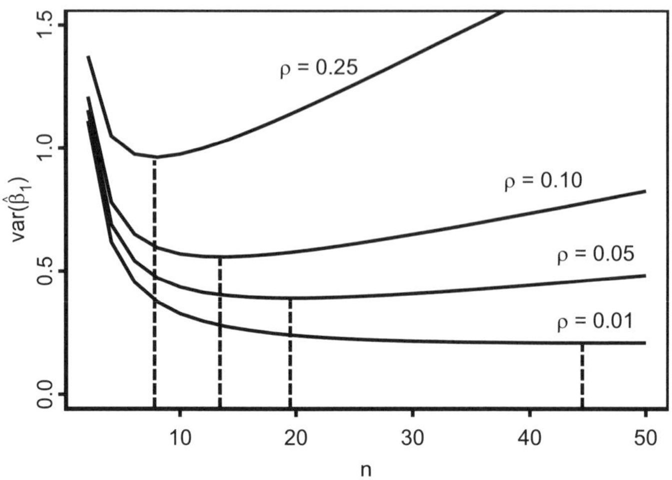

Figure 11.2 Variance of the Treatment Effect Estimator in a Cluster Randomized Trial as a Function of Cluster Size n (and Implicitly Also the Number of Clusters K = B / (c + sn)), for Several Values of the ICC, ρ, with a Sampling Budget B=10.000, Sampling Costs c=200 and s=10, and Outcome Variance σ_Y^2=25. Dotted lines indicate the optimal cluster sizes.

design. So a design with an RE of 0.90 needs $(100/90) = 1.11$ as many clusters as the optimal design, an increase of 11%. It must be noted, however, that, irrespective of the choice between these two designs, the sampling variance of the treatment effect increases as a linear function of the *ICC*; see equation (11.3). So, although an intermediate *ICC* value is a good choice for computing the cluster size and the initial number of clusters and budget needed, estimation of the *ICC* from the data may suggest an increase of the budget and the number of clusters. A recent review by Proschan (2009) suggests that inference on the treatment effect data can then be done on the joint data from the initial and additional clusters without correction for this interim look and sample size recalculation based on the *ICC*.

Fourth and last, the sample size usually varies between clusters, because organizations differ in size and because of recruitment problems and dropout at the person level. Such cluster size variation causes some efficiency loss which can be compensated by sampling more clusters. A simple approximation to this efficiency loss was derived by Manatunga, Hudgens, and Chen (2001) and repeated in various mathematical forms by others (e.g., Eldridge, Ashby, and Kerry, 2006; Taljaard, Donner, and Klar, 2007). Van Breukelen, Candel, and Berger (2007) and Van Breukelen and Candel (2012) showed this approximation becomes very inaccurate if $\bar{n}\rho > 1$, where $\bar{n}$ is the mean cluster size, and combined it with another approximation to show that the minimum relative efficiency (minRE) of unequal versus equal cluster sizes, given the mean cluster size and number of clusters, equals $\frac{2}{2+CV^2}$. Here, *CV* is the coefficient of variation (i.e., *SD*/mean) with respect to cluster size. They also derived a more accurate approximation, supported by simulations (Candel, Van Breukelen, Kotova, and Berger, 2008). According to that approximation, the minRE is $1 - CV^2/4$. Since $\frac{2}{2+CV^2}$ underestimates, and $1 - CV^2/4$ slightly overestimates, the minRE of unequal cluster sizes, we can take their average. In sum, unequal cluster sizes can be accommodated by multiplying the number of clusters K as obtained from equation (11.3) with 1/minRE. In practice, the *CV* of cluster size rarely exceeds 0.7 and is often around 0.5, for instance 0.4 in the trials of Ausems et al. (2002) and Kraag et al. (2009), and 0.6 in Kerry and Bland (2001). The minRE is thus often near 0.90, leading to an increase of the number of clusters needed by 11%. A similar approximation based on the *CV* of cluster size has also been derived for the effect of cluster size variation on variance component estimation (Van Breukelen, Candel, and Berger, 2008).

11.5 MULTISITE/MULTICENTER TRIALS: MODEL FOR ANALYSIS AND OPTIMAL DESIGN

Multisite or multicenter trials (Senn, 1998; Raudenbush and Liu, 2000) are run when person randomization is possible, but a large number of sites or centers (clinics, schools) is needed to recruit a sufficient number of persons or to test for a treatment by center interaction. Center may be treated as fixed or random, depending on several considerations, including the number of centers in the study. If center is treated as random, a simple model for data analysis is:

$$y_{ij} = \beta_0+\beta_1 x_{ij}+u_{0j}+u_{1j}x_{ij}+\varepsilon_{ij}, \quad (11.8)$$

where y_{ij} is again the outcome for person i in cluster (center) j, and x_{ij} is the treatment, which can be coded as 1 for treated and 0 for control, or centered as +0.5 and −0.5. Compared with model (11.1), there is an extra random effect u_{1j} with variance σ^2_{u1} to accommodate treatment by center interaction. Using centered coding of x and assuming no slope-intercept covariance ($\sigma^2_{u0u1} = 0$) implies homogeneity of outcome variance between treatments. More specifically, total outcome variance then equals

$$\sigma^2_y = \sigma^2_{u0} + \frac{1}{4}\sigma^2_{u1} + \sigma^2_\varepsilon.$$

Balanced (50:50) allocation in each center/site is optimal then, and $Var(\hat{\beta}_1)$ satisfies equation (11.3) if we replace σ^2_{u0} with $\frac{1}{4}\sigma^2_{u1}$ in ρ and in σ^2_y, since the random effect u_{0j} drops out of the treatment effect estimator, and so $Var(\hat{\beta}_1)$ does not depend on σ^2_{u0} (Van Breukelen et al., 2007). Consequently, equations (11.5) to (11.7) for the optimal design, $Var(\hat{\beta}_1)$, and power apply to multisite trials if we use the substitution described above. Likewise, the approximation of the relative efficiency of unequal cluster sizes in cluster randomized trials applies to multisite trials. For the more complicated situation in which $\sigma^2_{u0u1} \neq 0$, see Candel et al. (2008, Appendix A2).

Note that the intercept variance σ^2_{u0} in model (11.1) for cluster randomized trials is actually the sum of the cluster and cluster by treatment variance components in equation (11.8), since the two components are not separable in case of cluster randomization. As a result, $Var(\hat{\beta}_1)$ is larger for a cluster randomized trial than for a multisite trial and the latter is more efficient and powerful, assuming absence of treatment contamination, since contamination reduces β_1 and the power; see equation (11.7) and Moerbeek (2005a), Rhoads (2011). Figure 11.3 shows the relative efficiency of a cluster randomized versus multisite trial as a function of the *ICC* and of the ratio of slope-to-intercept variance.

11.6 BINARY OUTCOMES

Some cluster randomized trials have a binary outcome, such as being a smoker at the end of a prevention program. A simple mixed logistic model for this is:

$$\log\left(\frac{P\left(y_{ij}=1\right)}{P\left(y_{ij}=0\right)}\right) = \beta_0 + \beta_1 x_j + u_{0j}, \quad (11.9)$$

where u_{0j} is a random cluster effect with mean zero and variance σ^2_{u0}. Based on first-order Marginal Quasi-Likelihood (MQL1) estimation, Moerbeek, Van Breukelen, and Berger (2001b) derived a closed form for $Var(\hat{\beta}_1)$, which can be rewritten into equation (11.3) if we replace σ^2_ε in ρ and in σ^2_y with δ^2, defined as:

$$\delta^2 = \frac{1}{2}\left(\frac{1}{\pi_c\left(1-\pi_c\right)} + \frac{1}{\pi_t\left(1-\pi_t\right)}\right). \quad (11.10)$$

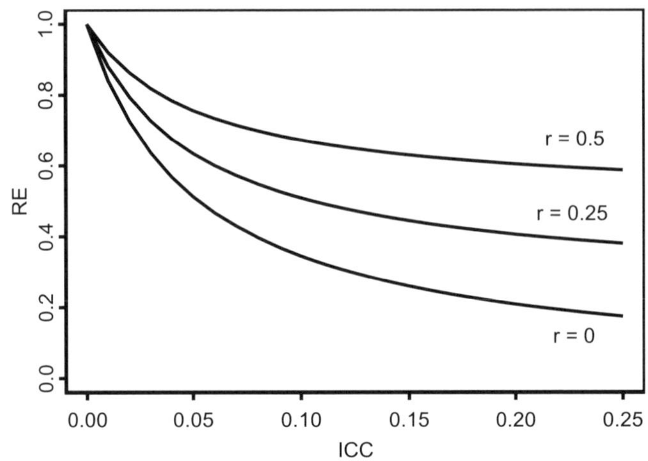

Figure 11.3 Relative Efficiency of Cluster Randomized Versus Multicenter Trial as a Function of the ICC for a Cluster Size n=20 and Various Ratios of Slope-to-Intercept Variance $(ICC = \frac{\sigma_u^2}{\sigma_u^2+\sigma_e^2}, r = \frac{\frac{1}{4}\sigma_{ul}^2}{\sigma_u^2}$, **where** $\sigma_u^2 = \sigma_{u0}^2 + \frac{1}{4}\sigma_{ul}^2)$.

Here, δ^2 is the person-level variance on the log-odds scale, similar to σ_ε^2 in model (11.1), and π_c and π_t are the probabilities $P\left(y_{ij} = 1\right)$ for control and treated if $u_{0j} = 0$. Since MQL1 underestimates $Var(\hat{\beta}_1)$, simulation studies were run to find a correction factor for $Var(\hat{\beta}_1)$ when using second-order Penalized Quasi-Likelihood (PQL2), a fairly unbiased method. These studies show that multiplying $Var(\hat{\beta}_1)$ with 1.25 is a conservative and thus safe correction for sample sizes (Moerbeek et al., 2001b; Candel and Van Breukelen, 2010b).

This means that equations (11.3) to (11.7) can again be used, now by taking $\delta^2 = \sigma_\varepsilon^2$. Likewise, Candel and Van Breukelen (2010b) showed the approximation for the RE of unequal cluster sizes to hold for binary outcomes by letting $\delta^2 = \sigma_\varepsilon^2$. Finally, these results extend to multicenter/multisite trials with a binary outcome by replacing σ_{u0}^2 with $\sigma_{u1}^2/4$ and σ_ε^2 with δ^2. Of course, $Var(\hat{\beta}_1)$ and the number of clusters K and budget B, must be multiplied with the correction factor 1.25 for using PQL2 instead of MQL1. A complication is that δ^2 depends on the unknown fixed effects β_0 and β_1 through π_c and π_t, with a minimum of $\delta^2 = 4$ if $\beta_0 = \beta_1 = 0$ such that π_c and π_t equal 0.50.

There is more literature on nested designs for binary outcomes, based on the logit link (Moerbeek and Maas, 2005; Ouwens, Tan, and Berger, 2006) or the identity link (e.g., Hsieh, 1988; Feng and Grizzle, 1992; Donner, 1998; Zou and Normand, 2001; Jung, Kang, and Ahn, 2001; Thomson, Hayes, and Cousens, 2009). The latter class is not easily extended to include covariates. For more technical details about mixed models for binary data, see Chapter 8 in the present book.

11.7 COVARIATES

Covariates in regression models serve three purposes: to increase power and precision in randomized studies, to adjust for confounding in observational studies, or to evaluate interaction (moderation) in any study. Here, we focus on the first purpose. In a non-nested experimental design, a covariate reduces unexplained outcome variance and thereby the variance of the treatment effect estimator, resulting in more precision and power. This only holds if the covariate is predictive, since otherwise the loss of degree of freedom for estimating the covariate's regression weight, and chance correlation between covariate and treatment, can reduce power and precision. A similar result holds for nested experimental designs, but now we must distinguish between a cluster-level covariate like school type and a person-level covariate like age, noting that a covariate c_{ij} which varies at both levels can be split into a cluster-level part $\bar{c}_j$ (the cluster mean of the covariate) and a person-level part $c_{ij} - \bar{c}_j$ (the cluster-mean centered covariate). Including a cluster-level covariate into the model reduces cluster-level variance σ^2_{u0}, and including a person-level covariate reduces person-level variance σ^2_ε (Moerbeek et al., 2008, p. 188). Both effects improve precision and power for the treatment effect; see equation (11.3).

On the other hand, each covariate costs one degree of freedom, which can only be ignored if the number of clusters K is large, since the degree of freedom for testing the treatment effect is $K - 2$ in a cluster randomized trial, or $K - 1$ in a multicenter trial with random slope (Moerbeek, Van Breukelen, and Berger, 2003). Also, chance correlation between treatment and covariate increases $Var(\hat{\beta}_1)$ in nested just as in non-nested designs. This is especially relevant for cluster randomized trials, where this chance correlation goes to zero only for large K (Moerbeek et al., 2001a). For further details on the costs and benefits of a covariate in a nested experiment, see Moerbeek (2006).

11.8 RANDOMIZED EXPERIMENTS WITH TWO TREATMENTS AND REPEATED MEASURES

Nesting occurs not only of persons in clusters, but also of repeated measures in persons, for instance in field experiments and panel studies. An important question is how often and at which time points persons must be measured. On the one hand, increasing the total follow-up time and frequency of measuring increases the total amount of information per person. On the other hand, it also costs more money that could be spent on sampling more persons. Also, dropout may increase with follow-up time and frequency of measuring. We first consider optimal design of a field experiment with two treatment conditions, fixed follow-up time, no dropout, and no nesting of persons in clusters. After that, we briefly discuss extensions.

Optimal design for a randomized experiment with two treatments and a repeatedly measured quantitative outcome depends on three factors: the expected treatment effect over time (i.e., the treatment by time interaction pattern), the covariance structure of the repeated measures, and the cost per person versus per repeated measure. To clarify this, we start with the following model for the ith repeated measure of the jth person (with $i = 1, 2, \ldots, n$ and $j = 1, 2, \ldots, K$):

$$\begin{aligned} y_{ij} = \beta_{0j} &+ \beta_1 x_j + \sum_{h=2}^{n} \beta_h t_{hj} \\ &+ \beta_{n+1}\, time_{ij}\, x_j + \beta_{n+2} \\ &\quad time^2_{ij}\, x_j + \varepsilon_{ij}. \end{aligned} \tag{11.11}$$

Here, x_j indicates the treatment (1 = treatment, 0 = control) for person j, and the $(n-1)$ t_{hj}s are time indicators, such that $t_{hj} = 1$ if $h = i$, and 0 else, with $i = 1$ being the baseline time point which is the time reference category, and $time_{ij}$ is a linear transformation of real time such that $time = 0$ at baseline ($i = 1$) and $time = 1$ at the end ($i = n$). If time points are equidistant, then $time_{ij} = \frac{i-1}{n-1}$ for all persons $j = 1, 2, \ldots, K$.

This model assumes that the group difference is a second-order polynomial function of time, running from β_1 at the first time point to $\beta_1 + \beta_{n+1} + \beta_{n+2}$ at the last. Note that $\beta_1 = 0$ can be assumed for randomized experiments in this section, but not for nonrandomized studies, where inference about treatment effects is very complicated (Van Breukelen, 2006). Also, note that the time effect in each group according to (11.11) can be of any type due to the $(n - 1)$ time indicators t_{hj}. This model can handle various treatment effect patterns, that is, linear divergence between group means (if $\beta_{n+2} = 0$), nonlinear divergence, and divergence followed by convergence; see Figure 11.4. In a study by Feighner (1980), linear divergence was observed in a randomized trial comparing an antidepressant with placebo in the treatment of primary depression in hospitalized patients. The outcome of interest was the score on the Hamilton Psychiatric Scale for Depression and measured in weeks 0, 1, 2, 3, and 4. Quadratic divergence was observed in a trial on the effect of adenosine triphosphate (ATP) on fat mass in patients with advanced lung cancer. Fat mass and other outcomes were measured at four-week intervals over a period of 28 weeks (Agteresch et al., 2002).

The term ε_{ij} represents error, which may be uncorrelated, autoregressive, or the sum of both. Constraints must be imposed on the covariance matrix of random effects and the covariance matrix of the error term to make the model identifiable, as all covariance parameters must be estimated from the covariance matrix $\boldsymbol{V}$ of the repeated measures. As equation (11.11) suggests, we will first assume fixed slopes. For ε_{ij} we consider uncorrelated measurement error (ME), first-order autoregressive error (AR1), and the sum of both. Combining these with the absence or presence of a random intercept (RI) gives six covariance structures. Of the six, the structure no RI + ME is unrealistic as it implies uncorrelated errors. Accordingly, optimal design for only the other five covariance structures, combined with linear or quadratic divergence, was studied by Winkens et al. (2005, 2006, 2007). The results, based partly on some unpublished analyses (Winkens, 2005), are summarized in Table 11.1.

If the covariance structure resembles compound symmetry, that is, the correlation between repeated measures hardly decreases as the time lag between the measures increases, then about five repeated measures is usually efficient, with the first at baseline ($time = 0$) and the last at the end of follow-up time ($time = 1$). The intermediate measures must be taken as near to the end of the follow-up time as possible for linear divergence, and equidistant for quadratic divergence. The precise optimal number of repeated measures may be larger than five, however, depending on the correlation between repeated measures, and on the cost ratio, that is, the cost per person versus per repeated measure. This is similar to optimal design of a cluster randomized trial, see equation (11.5), which also has a compound symmetric covariance structure. However, the intraclass correlation is usually less than 0.10 in cluster randomized trials and between 0.30 and 0.80 in repeated measures designs.

Conversely, if the covariance structure resembles first-order autoregression, that is, the correlation decreases markedly as the time lag increases, then the optimal number of repeated measures is two for linear divergence and three for quadratic divergence. This is because with an AR1 structure the stronger correlations between intermediate measures reduce their contribution to the total information about the treatment effect. It is then more efficient to spend money on extra persons rather than repeated measures. Further, absence or presence of a random intercept (RI) hardly affects the optimal design under AR1 error.

Finally, if the covariance structure is unknown, then assuming AR1+ME gives the most robust design, that is, a design with a high relative efficiency compared with the optimal design, whether the covariance structure turns out to be like CS or like AR1. The AR1+ME assumption in turn leads to three repeated measures for linear divergence (one at $time = 0$, and two as near $time = 1$ as possible), and for quadratic divergence (at $time = 0, 0.50$, and 1). Again, absence or presence of a RI hardly affects the optimal design once there is autocorrelated error.

Some results were also obtained for model (11.11) with RI+ME plus a random slope of $time$ (by adding a term $b_j time_{ij}$, with b_j random with zero mean to prevent collinearity with the fixed nonlinear time effect $\sum_{h=2}^{n} \beta_h t_{hj}$). This structure implies nonstationarity and the optimal number of repeated measures resembled those for the CS (RI+ME) structure, due to the assumed absence of autocorrelation. Some other authors have come to

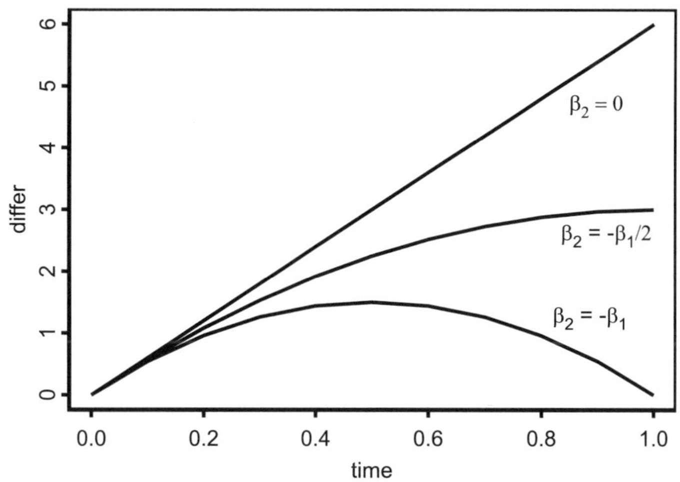

Figure 11.4 Second-Order Polynomial Divergence Between Two Groups in a Randomized Trial, for Various Polynomial Parameter Values. β_1 is the linear group by time effect, β_2 is the quadratic group by time effect, on a time scale from 0 to 1

Table 11.1 Optimal Number and Timing of Repeated Measures in a Randomized Experiment with a Quantitative Outcome Variable and Various Covariance Structures and Effect Patterns, Given a Fixed Duration or Follow-Up Time from 0 (baseline) to 1 (end of follow-up)

Covariance structure	Treatment effect pattern	Number of repeated measures	Timing of repeated measures
Compound symmetry (CS)	Linear divergence	5	[0, 1, 1, 1, 1][a]
	Quadratic divergence	5	[0, 0.25, 0.50, 0.75, 1.0]
First-order autoregressive	Linear divergence	2	[0, 1]
(AR1)	Quadratic divergence	3	[0, 0.50, 1]
AR1 + measurement error	Linear divergence	3	[0, 1, 1]
(AR1 + ME)	Quadratic divergence	3	[0, 0.50, 1]

[a] Intermediate measures to be taken as near to the end of the follow-up time as possible.

similar results for special cases of the models above (e.g., Overall and Doyle, 1994; Maxwell, 1998; Schouten, 1999; Galbraith and Marshner, 2002; Venter, Maxwell, and Bolig, 2002).

11.9 LONGITUDINAL GROWTH STUDIES

Several authors have studied sample sizes for testing (group differences in) polynomial change over time in experimental and observational studies, and some take into account one or more of the following factors: study duration, dropout, frequency and time points of measuring, autoregressive error. No optimal designs are given, but the methods can be implemented in software to compute sample sizes for specific models (Hedeker, Gibbons, and Waternaux, 1999; Raudenbush and Liu, 2001; Galbraith and Marshner, 2002; Yi and Panzarella, 2002, Tu, Kowalski, Zhang, Lynch, and Crits-Christoph, 2004; Moerbeek, 2008; Ortega-Azurduy, Tan, and Berger, 2008; Lu, Mehrotra, and Liu, 2009).

Extensive work on the optimal number and timing of repeated measures in polynomial growth curve analysis within a single cohort has been done by Berger, Tan, and colleagues (Tan and Berger, 1999; Ouwens, Tan, and Berger, 2002; Berger and Tan, 2004; Tekle, Tan, and Berger, 2008a,b; Ortega-Azurduy, Tan, and Berger, 2008, 2009). In line with results by Winkens et al. (2005, 2006, 2007) and Moerbeek (2005b), they find that the optimal number of repeated measures is usually equal to or slightly larger than the number of fixed effects in the polynomial function, e.g., two for linear and three for quadratic growth. They also study Maximin designs, which are more robust against misspecification, effects of dropout, and the choice between a cohort and cross-sectional design. Relatedly, Tekle, Tan, and Berger (2008a) and Moerbeek (2011) studied the optimal number of cohorts for polynomial growth. Finally, Candel (2009) studies optimal design for empirical Bayes estimation of the random effects needed for prediction of the future trajectory of individual growth curves.

Some work has also been done on optimal design for a repeatedly measured binary outcome analyzed with logistic regression, either mixed logistic regression (Tekle, Tan, and Berger, 2008b) or generalized estimating equations (Shih, 1997; Pan, 2001). In using the latter work, it must be realized that, while fixed effects in mixed linear and GEE linear models are equivalent, this does not hold for logistic regression (Neuhaus and Kalbfleisch, 1998; Ritz and Spiegelman, 2004). For further reading on longitudinal data modeling, see also Chapter 9 of this book.

11.10 ROBUST DESIGN

Optimal design for fixed effects estimation in mixed linear models depends on the unknown variance and covariance parameters. For instance, equation (11.5) for the optimal sample sizes in a cluster randomized trial depends on the *ICC*. Optimal design for fixed effects in mixed logistic models additionally depends on the unknown fixed effects themselves, as equation (11.10) indicates. This is known as the local optimality problem in optimal design literature, since a design which is optimal for certain parameter values may be far from optimal for other values, and so misspecification of some parameters may lead to inefficient designs. There are three popular methods to cope with this by obtaining designs that are robust against some misspecification: Maximin design, Bayesian optimal design, and sequential and adaptive design. These three will be briefly discussed.

Maximin design maximizes the minimum efficiency in three steps. First, for every possible parameter value (e.g., for every plausible *ICC* in equation (11.5)), one should compute the efficiency of each possible design, using the chosen optimality criterion (e.g., the inverse of $Var(\hat{\beta}_1)$ in model (11.1)). Second, one should find the minimum efficiency for every design over all parameter values. Third, one should choose the design with the highest minimum efficiency. If, on the other hand, the variance of the estimator of interest, rather than efficiency, is one's criterion of choice, one should use a Minimax design rather than a Maximin design. In this case, we would first find the maximum variance for each design (step 2) and use that design with the smallest maximum (step 3).

A drawback of the Maximin design is that the minimum efficiency is often attained for an unlikely parameter value. For example, in equation (11.3) $Var(\hat{\beta}_1)$ increases in the unknown *ICC*, and so the Maximin design is the locally optimal design for the highest possible *ICC* (e.g., a value of 0.20 if the *ICC* is between 0 and 0.20). One solution is to use as the criterion not the efficiency, but the relative efficiency compared with the locally optimal design. Maximizing the minimum relative efficiency gives a design that stays as close to the locally optimal design as possible for all parameter values, and is optimal for some value well within the range, for instance for an *ICC* of 0.05. Examples of Maximin design are Ouwens, Tan, and Berger (2002, 2006) and Tekle, Tan, and Berger (2008b).

Related to Maximin design is Bayesian optimal design. Here, we maximize not the minimum efficiency but the expected efficiency by integrating over a probability distribution for the unknown parameter(s) upon which the efficiency depends. For a cluster randomized trial, this would mean integrating the inverse of $Var(\hat{\beta}_1)$ in equation (11.3) over a distribution for the *ICC*, for instance a uniform distribution over the interval 0, 0.20. Compared with Maximin design, Bayesian optimal design depends less on efficiencies at unlikely parameter values, just as a sample mean is less affected by outliers than the sample minimum or maximum. But Bayesian optimal design also has drawbacks. First of all, the optimality criterion and optimal design depend on the assumed probability distribution. This is very clear for the expected $Var(\hat{\beta}_1)$ in equation (11.3), which is a linear function of the *ICC*. Secondly, minimizing the expected $Var(\hat{\beta}_1)$ or maximizing the expected efficiency (inverse of $Var(\hat{\beta}_1)$) gives different designs, due to the Jensen inequality (Mood, Graybill, and Boes, 1974). This drawback does not hold for Maximin design. Examples of Bayesian design for cluster randomized trials may be found in Spiegelhalter (2001), Turner, Prevost, and Thomson (2004), and Rotondi and Donner (2009).

A third solution to local optimality is to use information from interim analyses as part of the data that have been collected, to modify the design or stop data collection altogether. Methods can be divided into group sequential and adaptive designs. In the group sequential design, the sample size is based upon a best guess of the parameter values and split into batches (e.g., four batches of 25% of the planned sample size each). Interim analyses of all available data after each batch then lead to a decision to continue the trial, to stop because significance has already been obtained, or to stop because it is unlikely that a significant effect will be found upon completion (futility stopping). As this method involves multiple testing, modified test statistics and significance levels are used. Adaptive design differs from group sequential design in that interim analyses can also lead to a re-estimation of the total sample size needed and to other design modifications, for instance dropping one treatment condition. An introduction into these methods for randomized experiments may be found in Chow and Chang (2007). Lake, Kamman, Klar, and Betensky (2002) discuss adaptive design of cluster randomized trials, and Zou, Donner, and Klar (2005) compare various test statistics for group sequential trials in the context of cluster randomization.

11.11 SOME CONCLUDING REMARKS AND SOFTWARE

This chapter is merely an introduction to optimal design and much more is going on in various fields of application. Examples are optimal sample sizes for three-level designs (Feldman and McKinlay, 1994; Moerbeek et al., 2000; Heo and Leon, 2008, 2009; Teerenstra et al., 2008), hybrids of the cluster randomized and multisite trial (Borm, Melis, Teerenstra, and Peer, 2005; Moerbeek and Teerenstra, 2011; Turner, White, and Croudache, 2007; Hayes and Moulton, 2009), optimal design of fMRI experiments in cognitive psychology (Mumford and Nichols, 2008; Maus, Van Breukelen, Goebel, and Berger, 2011), and optimal design of educational tests (Van der Linden, 2005; Limapassos, Berger, and Tan, 2007). Common to all work is the aim to improve cost-effectiveness of empirical studies. Various authors have written software for computing optimal designs or sample sizes, and some programs are freely available on the internet. We mention five sites with free software plus manuals as a first aid:

`http://stat.gamma.rug.nl/multilevel.htm` for Snijders' program PINT which computes approximate standard errors for fixed effects in two-level designs, technical details of which are given in Snijders and Bosker (1993).

`http://www.bristol.ac.uk/cmm/software/mlpowsim` for MLPowSim, a program from the Centre for Multilevel Modeling, which creates R scripts and MLwiN macros for sample size calculations for various random effects models with R or MLwiN, respectively. More information on R and MLwiN is found in Chapter 26 of this book.

http://ppw.kuleuven.be/english/research/mesrg for ML-DEs which allows the setting up of simulation studies in MLwiN to compare the efficiencies of different multilevel designs under budget and cost constraints.

http://optimal-design.biostat.ucla.edu/optimal/ for various web-based optimal design programs by Wong, Dette, Zhu, Pepelyshev, and co-workers. These can handle a wide range of linear and nonlinear fixed models, but apparently no mixed models.

http://sitemaker.umich.edu/group-based/optimal_design_software for software by Raudenbush et al. which allows the user to plot power versus effect size, ICC, and sample size for cluster randomized trials, multisite trials, and longitudinal studies using specific mixed models.

REFERENCES

Adams, G., Gulliford, M.C., Ukoumunne, O.C., Eldridge, S., Chinn, S., and Campbell, M.J. (2004) 'Patterns of Intracluster Correlation From Primary Care Research to Inform Study Design and Analysis', *Journal of Clinical Epidemiology*, 57: 785–94.

Agteresch, H.J., Rietveld, T., Kerkhofs, L.G.M., Van den Berg, J.W.O., Wilson, J.P.H., and Dagnelie, P.C. (2002) 'Beneficial Effects of Adenosine Triphosphate on Nutritional Status in Advanced Lung Cancer Patients: A Randomized Clinical Trial', *Journal of Clinical Oncology*, 20: 371–8.

Atkinson, A.C., Donev, A.N., and Tobias, R.D. (2007) *Optimum Experimental Designs, With SAS*. Oxford: Oxford University Press.

Ausems, M., Mesters, I., Van Breukelen, G., and De Vries, H. (2002) 'Short-Term Effects of a Randomized Computer-Based Out-of-School Smoking Prevention Trial Aimed at Dutch Elementary Schoolchildren', *Preventive Medicine*, 34: 581–9.

Berger, M.P.F. and Tan, F.E.S. (2004) 'Robust Designs for Linear Mixed Effects Models', *Applied Statistics*, 53: 569–81.

Berger, M.P.F. and Wong, W.K. (2009) *An Introduction to Optimal Designs for Social and Biomedical Research*. Chichester: Wiley.

Borm, G.F., Melis, R.J.F., Teerenstra, S., and Peer, P. (2005) 'Pseudo Cluster Randomization: A Treatment Allocation Method to Minimize Contamination and Selection Bias', *Statistics in Medicine*, 24: 3535–47.

Candel, M.J.J.M. (2009) 'Optimal Designs for Empirical Bayes Estimators of Individual Linear and Quadratic Growth Curves in Linear Mixed Models', *Statistical Methods in Medical Research*, 18: 397–419.

Candel, M.J.J.M. and Van Breukelen, G.J.P. (2010a) 'D-Optimality of Unequal Versus Equal Cluster Sizes for Mixed Linear Regression Analysis of Randomized Trials With Clusters in One Treatment Arm', *Computational Statistics and Data Analysis*, 54: 1906–20.

Candel, M.J.J.M. and Van Breukelen, G.J.P. (2010b) 'Sample Size Adjustments for Varying Cluster Sizes in Cluster Randomized Trials With Binary Outcomes Analyzed With Second-Order PQL Mixed Logistic Regression', *Statistics in Medicine*, 29: 1488–1501.

Candel, M.J.J.M., Van Breukelen, G.J.P., Kotova, L., and Berger, M.P.F. (2008) 'Optimality of Unequal Cluster Sizes in Multilevel Studies with Realistic Sample Sizes', *Communications in Statistics – Simulation and Computation*, 37: 222–39.

Chow, S.C. and Chang, M. (2007) *Adaptive Design Methods in Clinical Trials*. Boca Raton, FL: Chapman and Hall/CRC.

Cohen, M.P. (1998) 'Determining Sample Sizes for Surveys With Data Analyzed by Hierarchical Linear Models', *Journal of Official Statistics*, 14: 267–75.

Donner, A. (1998) 'Some Aspects of the Design and Analysis of Cluster Randomization Trials', *Applied Statistics*, 47: 95–113.

Donner, A., Birkett, N., and Buck, C. (1981) 'Randomization by Cluster: Sample Size Requirements and Analysis', *American Journal of Epidemiology*, 114: 906–14.

Donner, A. and Klar, N. (1994) 'Cluster Randomization Trials in Epidemiology: Theory and Application', *Journal of Statistical Planning and Inference*, 42: 37–56.

Eldridge, S.M., Ashby, D., Feder, G.S., Rudnicka, A.R., and Ukoumunne, O.C. (2004) 'Lessons for Cluster Randomized Trials in the Twenty-First Century: A Systematic Review of Trials in Primary Care', *Clinical Trials*, 1: 80–90.

Eldridge, S.M., Ashby, D., and Kerry, S. (2006) 'Sample Size for Cluster Randomized Trials: Effect of Coefficient of Variation of Cluster Size and Analysis Method', *International Journal of Epidemiology*, 35: 1292–300.

Feighner, J.P. (1980) 'Trazodone, a Triazolopyridine Derivative, in Primary Depressive Disorder', *Journal of Clinical Psychiatry*, 41: 250–5.

Feldman, H.A. and McKinlay, S.M. (1994) 'Cohort Versus Cross-Sectional Design in Large Field Trials: Precision, Sample Size and a Unifying Model', *Statistics in Medicine*, 13: 61–78.

Feng, Z. and Grizzle, J.E. (1992) 'Correlated Binomial Variates: Properties of Estimator of Intraclass Correlation and its Effect on Sample Size Calculation', *Statistics in Medicine*, 11: 1607–14.

Galbraith, S. and Marshner, I.C. (2002) 'Guidelines for the Design of Clinical Trials With Longitudinal Outcomes', *Controlled Clinical Trials*, 23: 257–73.

Hayes, R.J. and Moulton, L.H. (2009) *Cluster Randomized Trials*. Boca Raton, FL: Chapman and Hall.

Hedeker, D., Gibbons, R.D., and Waternaux, C. (1999) 'Sample Size Estimation for Longitudinal Designs With Attrition: Comparing Time-Related Contrasts Between Two Groups', *Journal of Educational and Behavioral Statistics*, 24: 70–93.

Heo, M. and Leon, A.C. (2008) 'Statistical Power and Sample Size Requirements for Three Level Hierarchical Cluster Randomized Trials', *Biometrics*, 64: 1256–62.

Heo, M. and Leon, A.C. (2009) 'Sample Size Requirements to Detect an Intervention by Time Interaction in Longitudinal Cluster Randomized Trials', *Statistics in Medicine*, 28: 1017–27.

Hsieh, F.Y. (1988) 'Sample Size Formulae for Intervention Studies With the Cluster as Unit of Randomization', *Statistics in Medicine*, 8: 1195–201.

Jung, S., Kang, S., and Ahn, C.W. (2001) 'Sample Size Calculations for Clustered Binary Data', *Statistics in Medicine*, 20: 1971–82.

Kerry, S.M. and Bland, J.M. (2001) 'Unequal Cluster Sizes for Trials in English and Welsh General Practices: Implications for Sample Size Calculations', *Statistics in Medicine,* 20: 377–90.

Korendijk, E.J.H., Moerbeek, M., and Maas, C.J.M. (2010) 'The Robustness of Designs for Trials With Nested Data Against Incorrect Initial Intracluster Correlation Coefficient Estimates', *Journal of Educational and Behavioral Statistics*, 35: 566–85.

Kraag, G., Van Breukelen, G.J.P., Kok, G.J., and Hosman, C. (2009) ' "Learn Young, Learn Fair", a Stress Management Program for Fifth and Sixth Graders: Longitudinal Results From an Experimental Study', *Journal of Child Psychology and Psychiatry*, 50: 1185–95.

Lake, S., Kammann, E., Klar, N., and Betensky, R. (2002) 'Sample size Re-Estimation in Cluster Randomized Trials', *Statistics in Medicine*, 21: 1337–50.

Limapassos, V., Berger, M.P.F., and Tan, F.E.S. (2007) 'Test Design Optimization in CAT Early Stage With the Nominal Response Model', *Applied Psychological Measurement*, 28: 72–87.

Lu, K., Mehrotra, D.V., and Liu, G. (2009) 'Sample Size Determination for Constrained Longitudinal Data Analysis', *Statistics in Medicine*, 28: 679–99.

Maas, C.J.M. and Hox, J.J. (2005) 'Sufficient Sample Sizes for Multilevel Modeling', *Methodology*, 1: 86–92.

Manatunga, A.K., Hudgens, M.G., and Chen, S. (2001) 'Sample Size Estimation in Cluster Randomized Studies With Varying Cluster Size', *Biometrical Journal*, 43: 75–86.

Maus, B., Van Breukelen, G.J.P., Goebel, R., and Berger, M.P.F. (2011) 'Optimal Design of Multi-Subject Blocked fMRI Experiments', *NeuroImage*, 56: 1338–52.

Maxwell, S.E. (1998) 'Longitudinal Designs in Randomized Group Comparisons: When Will Intermediate Observations Increase Statistical Power?', *Psychological Methods*, 3: 275–90.

Moerbeek, M. (2005a) 'Randomization of Clusters Versus Randomization of Persons Within Clusters: Which is Preferable?', *American Statistician*, 59: 72–8.

Moerbeek, M. (2005b) 'Robustness Properties of A-, D-, and E-Optimal Designs for Polynomial Growth Models With Autocorrelated Errors', *Computational Statistics and Data Analysis*, 48: 765–78.

Moerbeek, M. (2006) 'Power and Money in Cluster Randomized Trials: When is it Worth Measuring a Covariate?', *Statistics in Medicine*, 25: 2607–17.

Moerbeek, M. (2008) 'Powerful and Cost-Efficient Designs for Longitudinal Intervention Studies With Two Treatment Groups', *Journal of Educational and Behavioral Statistics*, 33: 41–61.

Moerbeek, M. (2011) 'The Effects of the Number of Cohorts, Degree of Overlap Among Cohorts, and Frequency of Observation on Power in Accelerated Longitudinal Designs', *Methodology*, 7: 11–24.

Moerbeek, M. and Maas, C.J.M. (2005) 'Optimal Experimental Designs for Multilevel Logistic Models With Two Binary Predictors', *Communications in Statistics – Theory and Methods*, 34: 1151–67.

Moerbeek, M. and Teerenstra, S. (2011) 'Optimal Design in Multilevel Experiments', in J.J. Hox and J.K. Roberts (eds), *Handbook of Advanced Multilevel Analysis*. New York: Routledge, 257–81.

Moerbeek, M., Van Breukelen, G.J.P., and Berger, M.P.F. (2000) 'Design Issues for Experiments in Multilevel Populations', *Journal of Educational and Behavioral Statistics,* 25: 271–84.

Moerbeek, M., Van Breukelen, G.J.P., and Berger, M.P.F. (2001a) 'Optimal Experimental Designs for Multilevel Models With Covariates', *Communications in Statistics – Theory and Methods*, 30: 2683–97.

Moerbeek, M., Van Breukelen, G.J.P., and Berger, M.P.F. (2001b) 'Optimal Experimental Design for Multilevel Logistic Models', *The Statistician*, 50: 17–30.

Moerbeek, M., Van Breukelen, G.J.P., and Berger, M.P.F. (2003) 'A Comparison Between Traditional Methods and Multilevel Regression for the Analysis of Multicentre Intervention Studies', *Journal of Clinical Epidemiology*, 56: 341–50.

Moerbeek, M., Van Breukelen, G.J.P., and Berger, M.P.F. (2008) 'Optimal Designs for Multilevel Studies', in J. DeLeeuw and E. Meijer (eds), *Handbook of Multilevel Analysis*. New York: Springer, 177–205.

Mood, A.M., Graybill, F.A., and Boes, D.C. (1974) *Introduction to the Theory of Statistics*, 3rd edition. Tokyo: McGraw-Hill.

Mumford, J.A. and Nichols, T.E. (2008) 'Power Calculation for Group fMRI Studies for Arbitrary Design and Temporal Correlation. *NeuroImage*, 39: 261–8.

Murray, D.M. (1998) *Design and Analysis of Group-Randomized Trials*. New York: Oxford University Press.

Neuhaus, J.M. and Kalbfleisch, J.D. (1998) 'Between- and Within-Cluster Covariate Effects in the Analysis of Clustered Data', *Biometrics*, 54: 638–45.

Ortega-Azurduy, S.A., Tan, F.E.S., and Berger, M.P.F. (2008) 'The Effect of Drop-Out on the Efficiency of D-Optimal Designs of Linear Mixed Models', *Statistics in Medicine*, 27: 2601–17.

Ortega-Azurduy, S.A., Tan, F.E.S., and Berger, M.P.F. (2009). 'Highly Efficient Designs to Handle the Incorrect Specification of Linear Mixed Models', *Communications in Statistics – Simulation and Computation*, 38: 14–30.

Ouwens, J.N.M., Tan, F.E.S., and Berger, M.P.F. (2002) 'Maximin D-Optimal Design for Longitudinal Mixed Effects Models', *Biometrics,* 58: 735–41.

Ouwens, J.N.M., Tan, F.E.S., and Berger, M.P.F. (2006) 'A Maximin Criterion for the Logistic Random Intercept Model With Covariates', *Journal of Statistical Planning and Inference*, 126: 962–81.

Overall, J.E. and Doyle, S.R. (1994) 'Estimating Sample Sizes for Repeated Measurement Designs', *Controlled Clinical Trials*, 15: 100–23.

Pan, W. (2001) 'Sample Size and Power Calculations With Correlated Binary Data', *Controlled Clinical Trials*, 22: 211–27.

Proschan, M.A. (2009) 'Sample Size Re-Estimation in Clinical Trials', *Biometrical Journal*, 51: 348–57.

Raudenbush, S.W. (1997) 'Statistical Analysis and Optimal Design for Cluster Randomized Trials', *Psychological Methods*, 2: 173–85.

Raudenbush, S.W. and Liu, X. (2000) 'Statistical Power and Optimal Design for Multisite Trials', *Psychological Methods*, 5: 199–213.

Raudenbush, S.W. and Liu, X. (2001) 'Effects of Study Duration, Frequency of Observation, and Sample Size on Power in Studies of Group Differences in Polynomial Change', *Psychological Methods*, 6: 387–401.

Rhoads, C.H. (2011) 'The Implications of "Contamination" for Experimental Design in Education', *Journal of Educational and Behavioral Statistics*, 36: 76–104.

Ritz, J. and Spiegelman, D. (2004) 'Equivalence of Conditional and Marginal Regression Models for Clustered and Longitudinal Data', *Statistical Methods in Medical Research*, 13: 309–23.

Rotondi, M.A. and Donner, A. (2009) 'Sample Size Estimation in Cluster Randomized Educational Trials: An Empirical Bayes Approach', *Journal of Educational and Behavioral Statistics*, 34: 229–37.

Schouten, H.J.A. (1999) 'Planning Group Sizes in Clinical Trials With a Continuous Outcome and Repeated Measures', *Statistics in Medicine*, 18: 255–64.

Senn, S. (1998) 'Some Controversies in Planning and Analyzing Multi-Centre Trials', *Statistics in Medicine*, 17: 1753–65.

Shih, W.J. (1997) 'Sample Size and Power Calculations for Periodontal and Other Studies With Clustered Samples Using the Method of Generalized Estimating Equations', *Biometrical Journal*, 39: 899–908.

Snijders, T.A.B. and Bosker, R.J. (1993) 'Standard Errors and Sample Sizes for Two-Level Research', *Journal of Educational Statistics*, 18: 237–59.

Spiegelhalter, D.J. (2001) 'Bayesian Methods for Cluster Randomized Trials', *Statistics in Medicine*, 20: 435–52.

Taljaard, M., Donner, A., and Klar, N. (2007) 'Accounting for Expected Attrition in the Planning of Community Intervention Trials', *Statistics in Medicine*, 26: 2615–28.

Tan, F.E.S. and Berger, M.P.F. (1999) 'Optimal Allocation of Time Points for the Random Effects Model', *Communications in Statistics – Simulation and Computation*, 28: 517–40.

Teerenstra, S., Moerbeek, M., Van Achterberg, T., Pelzer, B.J., and Borm, G.F. (2008) 'Sample Size Calculations for 3-Level Cluster Randomized Trials', *Clinical Trials*, 5: 486–95.

Tekle, F.B., Tan, F.E.S., and Berger, M.P.F. (2008a) 'D-Optimal Cohort Designs for Linear Mixed Effects Models', *Statistics in Medicine*, 27: 2586–600.

Tekle, F.B., Tan, F.E.S., and Berger, M.P.F. (2008b) 'Maximin D-Optimal Designs for Binary Longitudinal Responses', *Computational Statistics and Data Analysis*, 52: 5253–62.

Thomson, A., Hayes, R., and Cousens, S. (2009) 'Measures of Between-Cluster Variability in Cluster Randomized Trials With Binary Outcomes', *Statistics in Medicine*, 28: 1739–51.

Tu, X.M., Kowalski, J., Zhang, J., Lynch, K.G., and Crits-Christoph, P. (2004) 'Power Analyses for Longitudinal Trials and Other Clustered Designs', *Statistics in Medicine*, 23: 2799–815.

Turner, R.M., Prevost, A.T., and Thompson, S.G. (2004) 'Allowing for Imprecision of the Intracluster Correlation Coefficient in the Design of Cluster Randomized Trials', *Statistics in Medicine*, 23: 1195–214.

Turner, R.M., White, I.R., and Croudache, T. (2007) 'Analysis of Cluster Randomized Cross-Over Trial Data: A Comparison of Methods', *Statistics in Medicine*, 26: 274–89.

Van Breukelen, G.J.P. (2006) 'ANCOVA Versus Change from Baseline: More Power in Randomized Studies, More Bias in Nonrandomized Studies', *Journal of Clinical Epidemiology*, 59: 920–5.

Van Breukelen, G.J.P. and Candel, M.J.J.M. (2012) 'Comments on "Efficiency Loss Because of Varying Cluster Size in Cluster Randomized Trials is Smaller Than Literature Suggests" ', *Statistics in Medicine*, 31: 397–400.

Van Breukelen, G.J.P. and Candel, M.J.J.M. (2013) 'Efficient Design of Cluster Randomized and Multicentre Trials With Unknown Intraclass Correlation', *Statistical Methods in Medical Research*, in press.

Van Breukelen, G.J.P., Candel, M.J.J.M., and Berger, M.P.F. (2007) 'Relative Efficiency of Unequal Versus Equal Cluster Sizes in Cluster Randomized and Multicentre Trials', *Statistics in Medicine*, 26: 2589–603.

Van Breukelen, G.J.P., Candel, M.J.J.M., and Berger, M.P.F. (2008) 'Relative Efficiency of Unequal Cluster Sizes for Variance Component Estimation in Cluster Randomized and Multicentre Trials', *Statistical Methods in Medical Research*, 17: 439–58.

Van der Linden, W.J. (2005) *Linear Models for Optimal Test Design*. New York: Springer.

Van Keulen, H.M., Mesters, I., Ausems, M., Van Breukelen, G.J.P., Campbell, M., Resnicow, K., Brug, J., and DeVries, H. (2011) 'Tailored Print Communication and Telephone Motivational Interviewing are Equally Successful in Improving Multiple Lifestyle Behaviors in an RCT', *Annals of Behavioral Medicine*, 41: 104–18.

Venter, A., Maxwell, S.E., and Bolig, E. (2002) 'Power in Randomized Group Comparisons: The Value of Adding a Single Intermediate Time Point to a Traditional Pretest-Posttest Design', *Psychological Methods*, 7: 194–209.

Winkens, B. (2005) *Optimal Design and Analysis of Clinical Trials With Repeated Measures*. Unpublished doctoral dissertation, Maastricht University, The Netherlands.

Winkens, B., Schouten, H.J.A., Van Breukelen, G.J.P., and Berger, M.P.F. (2005) 'Optimal Time-Points in Clinical Trials With Linearly Divergent Treatment Effects', *Statistics in Medicine*, 24: 3743–56.

Winkens, B., Schouten, H.J.A., Van Breukelen, G.J.P., and Berger, M.P.F. (2006) 'Optimal Number of Repeated Measures and Group Sizes in Clinical Trials With Linearly Divergent Treatment Effects', *Contemporary Clinical Trials*, 27: 57–69.

Winkens, B., Schouten, H.J.A., Van Breukelen, G.J.P., and Berger, M.P.F. (2007) 'Optimal Designs for Clinical Trials With Second-Order Polynomial Treatment Effects', *Statistical Methods in Medical Research*, 16: 523–37.

Yi, Q. and Panzarella, T. (2002) 'Estimating Sample Size for Tests on Trends Across Repeated Measurements With Missing Data Based on the Interaction Term in a Mixed Model', *Controlled Clinical Trials*, 23: 481–96.

Zou, K.H. and Normand, S.T. (2001) 'On Determination of Sample Size in Hierarchical Binomial Models', *Statistics in Medicine*, 20: 2163–82.

Zou, G.Y., Donner, A., and Klar, N. (2005) 'Group Sequential Methods for Cluster Randomization Trials With Binary Outcomes', *Clinical Trials*, 2: 479–87.

12

Multilevel Models and Causal Inference

Jennifer Hill
New York University, USA

12.1 INTRODUCTION

This chapter discusses the use of multilevel models for causal inference. Multilevel models were not designed specifically to address the concerns most salient for causal inference. However, when data that are being used to identify causal effects have hierarchical structures (students nested within schools) or other types of group dependencies (families classified by religious affilliation, workers classified by industry, patients classified by illness) it may make most sense to incorporate a multilevel model in the analysis strategy. Moreover, causal considerations may dictate a different choice of multilevel model than the researcher would use in a non-causal setting. Choices made will have implications not only for the plausibility of the assumptions and the properties of the estimators but also for the types of effects that can be estimated and for whom.

12.2 BRIEF REVIEW OF CAUSAL INFERENCE NOTATION AND ASSUMPTIONS

We first review the basic structure and notation for causal inference in a non-multilevel setting, before extending it to accommodate multilevel dependencies. Consider individuals indexed by $i = 1, \dots, n$ for whom we observe treatment assignments z_i and post-treatment responses y_i. The causal effect of z on y for individual i can be expressed as a comparison of potential outcomes, $y_i(z_i = 1) = y_i(1)$ and $y_i(z_i = 0) = y_i(0)$, that represent the outcome for that individual under treatment and control conditions, respectively. Then the causal effect for individual i might be expressed, for instance, as the difference in potential outcomes, $y_i(1) - y_i(0)$. Unfortunately, we can never observe both $y_i(1)$ and $y_i(0)$ for any given person; this missing data problem represents the most critical challenge in causal inference.

Now consider estimating average treatment effects such as $\mathrm{E}[y_i(1) - y_i(0)] = \mathrm{E}[y_i(1)] - \mathrm{E}[y_i(0)]$. We can make progress if we consider the implications of possible research designs. For instance, in a completely randomized experiment the distribution of these potential outcomes is balanced across treatment groups; formally, $y_i(1), y_i(0) \perp z_i$. In this case, $\mathrm{E}[y_i(1)] = \mathrm{E}[y_i(1) \mid z_i = 1]$. Since $y_i(1) = y_i$ for observations with $z_i = 1$, this is now an identifiable quantity, $\mathrm{E}[y_i \mid z_i = 1]$. The same holds for $\mathrm{E}[y_i(0)] = \mathrm{E}[y_i(0) \mid$

$z = 0] = \mathrm{E}[y_i \mid z_i = 0]$. Randomized designs that are blocked or stratified have similarly nice properties that simply require conditioning on the blocking variables for the necessary assumptions to hold.

12.2.1 Ignorability Assumption

In the absence of a randomized design, a standard approach is to assume that the potential outcomes are independent of the treatment assignment conditional on a vector of observed covariates, x_i; formally, $y_i(1), y_i(0) \perp z_i \mid x_i$. This assumption is commonly referred to as *ignorability of the treatment assignment* (Rubin, 1978) and implies that for groups of observations with identical values for x_i they should be "as good as randomized" in the sense that their potential outcomes are independent of treatment assignment. Under this assumption, $\mathrm{E}[y_i(0) \mid x_i] = \mathrm{E}[y_i(0) \mid z_i = 0, x_i] = \mathrm{E}[y_i \mid z_i = 0, x_i]$, and likewise $\mathrm{E}[y_i(1) \mid x_i] = \mathrm{E}[y_i(1) \mid z_i = 1, x_i] = \mathrm{E}[y_i \mid z_i = 1, x_i]$. Therefore, in theory, we can get unbiased estimates of conditional average treatment effects (such as $\Sigma_i[y_i(1) - y_i(0) \mid x_i]$) simply by appropriately conditioning on x_i. We can get unbiased estimates of more general average treatment effects (such as $\mathrm{E}[y_i(1) - y_i(0)]$) by additionally appropriately averaging these conditional estimates over the distribution for x_i. The best way to achieve such conditioning on x (either for the sample or for the population) has received a great deal of attention in the causal inference literature (see Imbens, 2004, for a review). Strategies include parametric, semiparametric, and nonparametric models of the response surface, matching, inverse probability of treatment weighting, and combinations of these approaches.

12.2.2 Strong Ignorability and the Common Support Assumption

In addition to the ignorability assumption it is typically assumed that there is sufficient overlap in the joint distributions of the covariates needed for ignorability to hold. This is referred to as *common support* and is generally formalized by a statement such as $0 < \Pr(z_i = 1 \mid x_i) < 1$. This assumption is meant to ensure that there is enough empirical evidence in the data to actually identify the effects being estimated so that we are not forced to rely purely on model extrapolation to make inferences. The combination of the ignorability and common support assumptions is referred to as *strong ignorability*. Strong ignorability should be trivially satisfied in a randomized experiment, with the exception of a randomized block experiment in which the probability of being assigned to treatment is close to 0 or 1 for some blocks (while the assumption would still technically hold here it still could lead to estimation issues in practice). Observational studies, on the other hand, do not automatically satisfy either assumption. While the common support assumption is, in theory, empirically verifiable, since we never know which subset of covariates are needed to satisfy ignorability, if the full set is of high dimension, it can be difficult to check in practice (Hill and Su, 2012).

12.2.3 SUTVA Assumption

The causal inference formulation above implicitly makes an assumption known in the statistics literature as the Stable Unit Treatment Value Assumption (SUTVA; Rubin, 1990). Conceptually, this assumption has two main components. The first is what researchers often colloquially refer to as the "no interference" assumption which requires that the treatment assigned to one person does not affect the outcome for another. The second part of the assumption requires that the treatment is "stable", that is, it means the same thing for each person in the study or there are no "versions" of the treatment. Researchers tend to focus on the former requirement since, in theory, the treatment could always be defined broadly enough to accommodate the latter requirement. In the extreme, such practice can lead to treatment effects estimates that, even if well-estimated, are nearly useless because the treatment is so vaguely defined. From this point onwards, however, this chapter will focus on the "no interference" aspect of the SUTVA assumption when it is discussed, since this is the feature of SUTVA most directly tied to the topic at hand.

To see how we can formalize this assumption, we first re-examine our definition of potential outcomes above. These could have been formalized as $y_i(\mathbf{z} = \mathbf{a}) = y_i(\mathbf{a})$ where $\mathbf{z}$ is the vector of treatment assignments for everyone in the study and $\mathbf{a}$ is a vector of zeros and ones with length equal to the total number of study participants. This definition allows the outcome to depend on the treatment assignment of other study participants. With 2^N possible potential outcomes for each person, allowing this level of generality quickly produces an intractable inference problem. The original definition, on the other hand, represents a special case in which the outcome for person i is only a function of the treatment for person i, not the treatment of anyone else. This simplification allows for only two potential outcomes per person (assuming a treatment with only two conditions), which, as noted above, is still a challenge since we can only observe one of these two outcomes. Accordingly, the formalization of the SUTVA assumption can be expressed with the statement $y_i(\mathbf{z}) = y_i(\mathbf{z}')$ if $z_i = z_i'$. Problems in interpretation that occur when this assumption is not satisfied are explored in Sobel (2006).

When treatments are assigned at the group level (that is, the same treatment is assigned to everyone in the same group) something interesting happens. Consider a special case of the most general definition in which the potential outcome for unit i is allowed to depend on Z_i as well as on the treatment assignments for everyone else in unit i's group, j (collectively $\mathbf{z}^j$), but *not* on the treatment assignments for anyone else in the study. If everyone in that group receives the same treatment then there are still only two possibilities for potential outcomes, $y_i(\mathbf{z}^j = \mathbf{1}^j)$ and $y_i(\mathbf{z}^j = \mathbf{0}^j)$ (where $\mathbf{1}^j$ and $\mathbf{0}^j$ are vectors of ones and zeros, respectively, with length equal to the number of units in group j). Therefore, if results are interpreted at the group level the problem of interference "goes away."[1] This reasoning has acted as a justification for randomizing treatments at the group level. Section 12.7 of this chapter discusses analysis options that allow for partial relaxation of the SUTVA assumption.

12.2.4 *Overview of Differing Concerns Between Randomized and Observation Studies in this Setting*

Fitting multilevel models in a causal inference setting serves different purposes depending on the initial study design. I broadly distinguish between two main types of study design: randomized experiments and observational studies. In a randomized experiment, whether randomization takes place at the individual level or group level, obtaining an unbiased estimate of the treatment effect (typically the biggest challenge in a causal analysis) should be straightforward. What may not be straightforward is calculating an appropriate standard error. Analyses that ignore the grouped structure of data are likely to yield an underestimate of this standard error, thus leading to confidence intervals that do not have nominal coverage rates. Moreover, additional thought might go into identification of heterogeneity in treatment effects or in estimating the effect of treatment receipt in situations in which there is noncompliance with treatment assignment.

In observational settings where ignorability is not trivially satisfied then whatever strategy is used will have multiple goals: to satisfy ignorability (and perhaps work to ensure common support) and to address the multilevel nature of the data. The former goal may lead to complications such as inclusion of a large number of covariates or use of a study design that may complicate the required analysis. The latter goal generally requires use of models that accommodate these data structures. Furthermore, when pre-treatment variables are not balanced across treatment groups, analyses become much less robust to deviations from the parametric assumptions of the model (in particular, linearity). Therefore, in the section on observational studies we discuss some semiparametric alternatives.

This chapter has been divided into two separate sections for randomized experiments (standard analyses and those that address noncompliance with assigned treatment) and observational studies (a section for those that assume ignorability simply by conditioning on covariates and a section for study designs that employ different identification strategies). Within these, when appropriate, there will be separate focus on treatments assigned at the individual versus group level. The last section before the conclusion briefly describes some recent work on relaxation of the SUTVA assumption.

A final note on the models used throughout the chapter. All of the models discussed rely on normality assumptions. As with all parametric models, the researcher should take care to check, to the extent possible, that these assumptions are appropriate for the data at hand. One exception is the instrumental variables models discussed in Section 12.4. For these scenarios I follow the norm in econometrics and model the first stage using a linear regression with normal errors even when the treatment variable is binary. This modeling strategy is a mathematical convenience that should not strongly influence the results.[2]

12.3 RANDOMIZED EXPERIMENTS

Why would we fit a multilevel model in the context of a randomized experiment? As discussed above, in a completely randomized experiment in which individuals are randomly assigned to treatments, the design allows for valid causal inference without using a model; a simple difference in means should yield an unbiased estimate of the average treatment effect. However, if the individual observations are clustered in meaningful ways—specifically, if the individual responses (conditional on any covariates included in the model) are not independent of each other—then a traditional difference in means analysis is not appropriate. In fact with some randomized designs even a linear regression can yield biased estimates of the treatment effect (for an example with cluster randomized designs see Middleton, 2008).

The specific nature of the clustering is important. One key issue is whether randomization took place independent of cluster, within cluster, or at the cluster level. Another, often related, issue is whether treatment assignment is associated with unobserved characteristics of the cluster. These variations in the implementation of randomized experiments can have different implications for how each is modeled. This section will discuss both completely randomized and block randomized experiments. It will also separately handle randomization to treatments that occurs at the individual level versus the group level.

12.3.1 *Completely Randomized Experiments with Clustered Data when Treatment is Assigned at the Individual Level*

Consider an experiment in which each participant had the same probability of receiving a new medical treatment, thought to reduce blood pressure, versus a placebo. Although the randomization took place at the individual level and was unrestricted, the patients might still be naturally clustered into meaningful groups. For instance, they might be clustered by hospital, which is important because the response variable of interest, blood pressure, is likely to systematically vary over these groups due to differences in the types of patient who receive care at each hospital and the quality of the medical services at each. In this case it might make sense to fit a model like:

$$\begin{aligned} y_{ij} &= \mu + \alpha_j + \tau z_{ij} + \varepsilon_{ij} \qquad (12.1)\\ \alpha_j &\sim \mathrm{N}(0, \sigma_\alpha^2)\\ \varepsilon_{ij} &\sim \mathrm{N}(0, \sigma_\varepsilon^2), \end{aligned}$$

where y_{ij} and z_{ij} are the response and binary treatment for the ith person in the jth hospital, respectively, μ is the overall mean of the response, α_j is the random intercept for hospital j (or the deviation of that intercept from μ), and τ is the treatment effect for the binary treatment. σ_ε^2 reflects the error variance within hospitals whereas σ_α^2 reflects block to block variability in mean response.[3]

This model allows for hospital to hospital variation in average levels of the response variable. It also allows for within-hospital dependence in the response variable. The model could be extended to include either individual- or group-level predictors to increase the efficiency of the treatment effect estimates. Importantly, however, this model assumes that $E[\alpha_j \mid z_{ij}] = E\alpha_j$; more generally, this requirement is often phrased as the assumption that the "random effects" are uncorrelated with the predictors. We will refer back to this assumption often as it will be salient for many of the studies discussed.

12.3.2 Randomized Block Experiments with Treatment Assigned at the Individual Level

Randomized block experiments encompass any of a variety of designs in which the randomization is (or can be conceived to have been) performed within strata or groups (called blocks). An experiment may be structured in this way for any of several reasons. A researcher may randomize within groups that define categories or levels of an important covariate (for instance race or income strata) to create perfect balance on that variable (this can be extended for multiple such covariates).

On the other hand, it may simply be logistically more feasible to randomize in this manner. For instance the experiment described in Section 12.3.1 might have been implemented as a multi-site trial in which it might have made sense for a key researcher in each hospital to perform the randomization for that hospital. A standard model for such a setup would still look like (12.1), where in this case j indexes the blocks rather than naturally occurring clusters in the data.

Varying probabilities of treatment assignment

We assumed in our hospital example that the probability of being treated did not vary in any systematic way across hospitals. However, another motivation to randomize separately within blocks is to explicitly allow the probability of receiving the treatment to vary across groups. Consider, for instance, a researcher who wants to examine the effect of provision of free prenatal health care on subsequent birth outcomes. For policy or scientific reasons it may be important to estimate this impact for a sample from the population that represents all income levels. On the other hand, ethical considerations might push researchers to want to exclude participation for those with incomes above a certain threshold. A compromise would be to include participants from all income levels but to divide participants into strata based on their income and randomize separately within each stratum (or "block") such that those from the lowest income strata have the highest probability of receiving access to the intervention and those from the highest income strata have the lowest probability.

A standard multilevel model analysis of these data might again follow the model in 12.1. Note that in this setting, however, the random intercepts, α_j, will be correlated with the treatment variable *by design*. This could lead to biased estimates of the treatment effect. Augmented models which include block specific means of the treatment variable in any of several different ways (Mundlak, 1978; Chamberlain, 1982, 1984; Raudenbush, 2009; Bafumi and Gelman, 2011) can be used to address this problem. The interpretation of the key parameter from these augmented models changes only slightly from the original interpretation in that conceptually the treatment effect is now being estimated within groups.

In some cases the "blocks" in a randomized experiment are defined explicitly by design variables, W. For instance, randomization might occur separately within groups defined by all combinations of participants' sex (male/female) and high school graduation status. A simpler option in this scenario, particularly if there is multilevel structure to the data even beyond these groupings, would be to directly control for these indicators (and their interaction) in the model. This avoids the problems of so-called "correlated random effects" encountered with the traditional multilevel model, but may be less efficient. The trade-offs may be small when there are only a handful of blocks or when there are many observations within each block.

Varying treatment effects

The effect of the treatment effect might also be expected to vary across blocks due, for instance, to different intensities of treatment implementation across blocks or different levels of responsiveness to the treatment by the individuals in different blocks. In this case it would make sense to extend the model above to allow for varying treatment effects. One model for this would be:

$$\begin{aligned} y_{ij} &= \mu + \alpha_j + \tau_j z_{ij} + \varepsilon_{ij} \\ \alpha_j &\sim \mathrm{N}(0, \sigma_\alpha^2) \\ \tau_j &\sim \mathrm{N}(\tau_0, \sigma_\tau^2) \\ \varepsilon_{ij} &\sim \mathrm{N}(0, \sigma_\varepsilon^2) \end{aligned}$$

(for a broader class of models see, for example, Gelman, 2004). Including both varying intercepts and varying slope coefficients markedly increases the complexity of the model, however, because it rarely makes sense to allow these terms to be independent of each other (as is implicitly assumed by the model above). Generally, a more complicated version of this model that allows for correlation between α_j and τ_j is more appropriate. We can express this as:

$$\begin{aligned} y_{ij} &= \mu + \alpha_j + \tau_j z_{ij} + \varepsilon_{ij} \\ \varepsilon_{ij} &\sim \mathrm{N}(0, \sigma_\varepsilon^2) \end{aligned}$$

$$\begin{pmatrix} \alpha_j \\ \tau_j \end{pmatrix} \sim \mathrm{N}\left(\begin{pmatrix} 0 \\ \tau_0 \end{pmatrix}, \begin{pmatrix} \sigma_\alpha^2 & \rho_{\alpha\tau}\sigma_\alpha\sigma_\tau \\ \rho_{\alpha\tau}\sigma_\alpha\sigma_\tau & \sigma_\tau^2 \end{pmatrix} \right).$$

The right panels of Figure 12.1 display 16 block-specific treatment effects (and corresponding 95% intervals) using data from an experiment that randomly assigned low-birth-weight, premature babies either to participate in an intervention that provided intensive child care in years two and three of the child's life (as well as some parental supports) or to a follow-up only group (Brooks-Gunn et al., 1994). Randomization took place within all combinations of the eight sites and two birth weight groups ("lighter" babies were less than 1500 grams, "heavier" were between 1500 and 2500 grams). The dotted line in the plots represents the overall treatment effect (where effects were not allowed to vary across sites) in each birth weight group.

Multiple comparisons

One nice "side-effect" of using a multilevel model to explore treatment effect heterogeneity across blocks or other subgroupings is that it obviates the need for subsequent multiple comparisons adjustments. As discussed in more detail in Gelman, Hill, and Yajima (2012), rather than the strategy employed by traditional multiple comparison corrections of adjusting the p-value of the associated tests, multilevel models naturally shrink estimates towards a common mean without having to compromise power. In this case, within-group estimates of the treatment effect are "shrunk" towards the pooled treatment effect estimates. The more information we have within a given group, the less that group's estimates will move toward the overall effect. There were

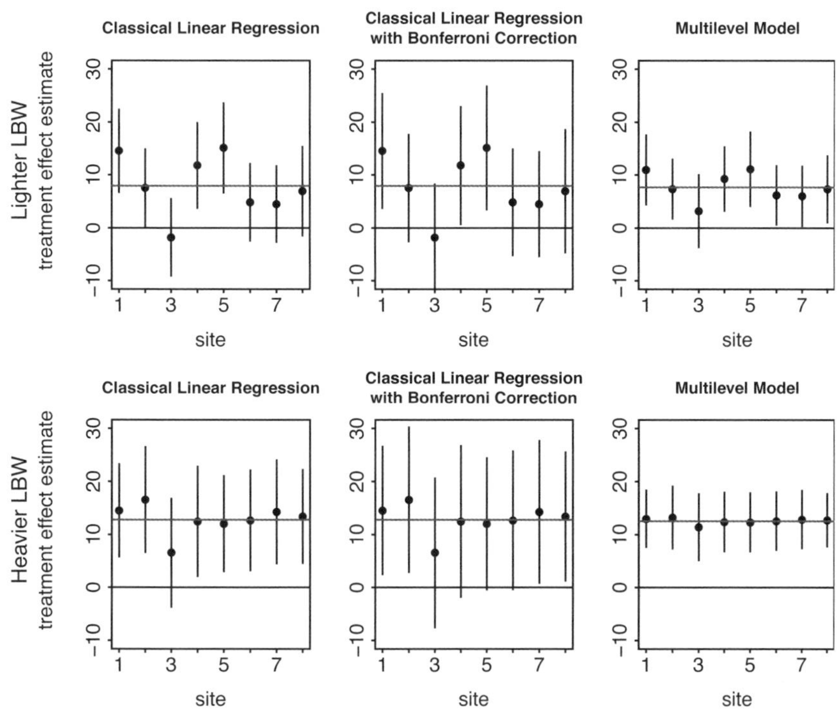

Figure 12.1 Estimates and 95% Uncertainty Intervals for Treatment Effects by Site and Birth Weight Group (Lighter and Heavier) Using Classical Linear Regression, Classical Linear Regression with Bonferroni Corrections, and a Multilevel Model

relatively fewer infants in the "heavier" low birth weight category (bottom row), and thus their estimates were shrunk more strongly than those for the "lighter" low birth weight category (top row).

The left panels of Figure 12.1 provide a comparison. The farthest-left panels display the fit from a more traditional regression model that simply interacts site indicators with the treatment condition. With 16 different blocks (and thus 16 different hypothesis tests against the null hypothesis of zero effect) it is likely that this analysis will suffer from problems with multiple comparisons. The middle panels use the same analysis as a starting point but inflate the uncertainty estimates to reflect a Bonferroni correction; this approach, which assumes independence across tests, is likely to be overly conservative.

The multilevel analysis, on the other hand, capitalizes a more complex model that more accurately reflects the dependence structure within and across groups. In doing so, it retains the power to detect "real" effects and yet appropriately accounts for our uncertainty about the differences between groups.

12.3.3 *Completely Randomized Experiments When Treatment is Assigned at the Group Level*

Studies that randomize interventions or treatments at the group level, sometimes referred to as *cluster randomized experiments*, have experienced a recent rise in popularity. These studies are often motivated by concerns about interference across units (SUTVA violations). Such concerns are heightened in fields like education where the intervention of interest may be thought to work in part through the synergies involved in children's interactions or the dynamics of peer groupings. Consider, for instance, a teen pregnancy prevention program in a high school that relies on an abstinence curriculum that encourages students to make virginity pledges. We could imagine that dynamics within peer groups would be powerful in this setting and thus it would likely be inappropriate to randomize individual students within schools because the control group would almost certainly be contaminated by the information and encouragement received by the treatment group.

A more extreme example is one in which an intervention may be designed to treat the group as a unit. A classic example of this type of intervention is "whole-school reform" (see, for example, Slavin et al., 1996) in which the goal is to transform many different aspects of the school, including the school culture. The philosophy is that, in order for the intervention to work, all the actors in the school community—teachers, administrators, students, and parents—must be involved. Therefore, it is not even possible to randomize this intervention at any unit of aggregation smaller than the school.

Alternatively, randomization may take place at the unit of the group simply due to cost or logistical reasons. For instance, consider a study that examines the effect of a new surgical procedure that requires expensive training. Due to cost considerations it may make sense to randomize at the level of the surgeon rather than the patient because then only half of the surgeons need to be trained. However, there may be no reason to believe that the treatment received by one patient would have an impact on the outcome of another patient.

The key feature shared by all of these studies is that outcomes of the experimental units in the same cluster are likely to be correlated with each other. Teens in the same school who participate in the pregnancy prevention study are likely to have similar probabilities of becoming pregnant while still a teenager simply because these students are likely to be more similar to each other in terms of other characteristics that are associated with teen pregnancy. Similar reasoning applies to the test scores of children participating in the whole-school reform intervention study. Patients with the same surgeon are also likely to have more similar health outcomes.

How should we analyze data arising from such studies? A simple model that allows for this dependence can be expressed similarly to the model in 12.1:

$$\begin{aligned} y_{ij} &= \mu + \alpha_j + \tau^G z_j + \varepsilon_{ij} \\ \alpha_j &\sim \mathrm{N}(0, \sigma_\alpha^2) \\ \varepsilon_{ij} &\sim \mathrm{N}(0, \sigma_\varepsilon^2). \end{aligned}$$

The difference is that the treatment variable is now expressed as a group level covariate. In fact we could re-arrange this model as the following (mathematically equivalent) formulation:

$$\begin{aligned} y_{ij} &= \mu + \alpha_j + \varepsilon_{ij} \\ \alpha_j &\sim \mathrm{N}(\tau^G z_j, \sigma_\alpha^2) \\ \varepsilon_{ij} &\sim \mathrm{N}(0, \sigma_\varepsilon^2). \end{aligned}$$

Interpretation of the causal parameter

In study designs with randomization at the group level, an important choice exists regarding the estimand. The model above implicitly treats the group as the primary unit of analysis and thus focuses on the average effect at that level, $\mathrm{E}[\overline{y}_j(z_j = 1) - \overline{y}_j(z_j = 0)]$, where $\overline{y}_j(z_j = a)$ denotes the average outcome in group j if assigned to treatment a. If cluster size is associated with the treatment effect, however, this estimand will not be equivalent to the estimand that considers the average of effects at the individual level, $\mathrm{E}[y_{ij}(z_j = 1) - y_{ij}(z_j = 0)]$. A relatively simple way of recovering the latter estimate is just to weight each cluster by the cluster size. Conceptually, however, it is worth thinking through whether this individual-level interpretation makes sense in any given setting.

Our experimental manipulation provides evidence about what might happen if an *entire group* were exposed to the treatment versus the result if the *entire group* were exposed to the control. Of the three examples above, the whole-school reform example provides the most extreme illustration of this. Effects from this experiment would provide evidence about what might happen if an entire school received such an intervention or not. It could not possibly provide evidence about a single child in a school receiving the intervention independent of the other children because the intervention was not designed to be implemented for just one child. It also would not provide evidence, for instance, about what might happen to an individual child if moved from a school without the intervention to a school receiving the intervention, in part because that move in and of itself would engender consequences; the child cannot be considered distinct from her self-selected context within the school (peer groups, elective classes and their teachers, after-school activities).

The teen pregnancy prevention program described was intended for a group setting. However, even if we re-imagined this program as a one-on-one counseling intervention it is difficult to imagine that a given teen's outcome would be completely unaffected by whether or not the members of her peer group received or did not receive this counseling as well. Therefore, even this new, individually targeted version of the treatment would not be interpretable at an individual level if it were randomized at the group level. Importantly it would also be difficult to interpret at the individual level if it were randomized at the individual level for the reasons discussed above with regard to SUTVA violations.

The example with the surgeons and patients on the other hand is a more borderline case. It seems quite possible in this scenario that patient-level randomization would not have resulted in a SUTVA violation. Therefore, estimation of the average causal effect of the treatment on an individual might be warranted in this setting despite the cluster randomization.

12.3.4 Randomized Block Experiments with Treatment Assigned at the Group Level

When treatments are randomized at the group level and pre-treatment variables are available that are predictive of the outcome variables, it may be particularly helpful to create blocks in order to increase efficiency. This adds an additional level of hierarchy to the model. Just as when we randomized at the individual level within blocks in order, we need to be aware of the possibility in which unobserved characteristics of the blocks are associated with the probability of being assigned to treatment. This can be avoided with designs in which the probability of treatment is constant across blocks ($k-1$ matched-pairs designs are a simple example of this).

12.4 RANDOMIZED EXPERIMENTS WITH NONCOMPLIANCE

In experiments with human subjects it is not uncommon for study participants to fail to comply with their treatment assignments. When this happens the intention-to-treat effect (the effect of the randomized *assignment*) may differ from the effect of treatment *receipt* in important ways. Under certain conditions we can recover the effect of receiving the treatment using an instrumental variables model (Angrist et al., 1996). We first discuss this model as implemented in a setting absent multilevel data structures and then extend it to handle this complication.

Consider a randomized experiment in which applicants are randomized to receive a scholarship to attend a private school (versus not receiving this scholarship). Let y_i denote the test score for the ith student who participated in the scholarship lottery one year after lottery winners had the opportunity to start attending private school. Let z_i denote whether or not the student was randomized to receive the scholarship; z_i plays the role of the "instrument". Let d_i indicate whether or not the student actually attended private school (versus public) for that year. A simple comparison of the mean scores across groups defined by z should yield an unbiased estimate of the so-called *intention-to-treat* effect, that is, the effect of winning a scholarship (regardless of what type of school was actually attended). However, this may not reflect the effect of attending private (versus public) school. If some people who were randomized to receive a scholarship end up attending public school, or if some students who were randomized to not receive a scholarship end up attending private school, these estimands could be quite different.

Instrumental variables models have been proposed as a strategy to identify the causal effect of the treatment of interest (here, attending private school) in such a setting (Angrist et al., 1996). One way to formulate such a model is:[4]

$$\begin{matrix} y_i = \alpha + \tau d_i + \varepsilon_i \\ d_i = \gamma + \delta z_i + \nu_i \end{matrix}, \quad \text{for } i = 1, \ldots, n,$$

where $\varepsilon_i \sim N(0, \sigma_\varepsilon^2)$ and $\nu_i \sim N(0, \sigma_d^2)$. In this formulation of the model, there are three assumptions that are typically stated in the econometrics literature as required for identification of the causal parameter τ. First, $\mathrm{Cov}(z_i, \nu_i) = 0$. This should be trivially satisfied if z_i is randomized; if it is conditionally randomized then the appropriate design variables can be included in the model to satisfy the assumption conditionally. Second, $\mathrm{Cov}(z_i, \varepsilon_i) = 0$. This assumption corresponds to the so-called exclusion restriction which, informally, says that the instrument (or randomized variable), z_i, can only influence the outcome, y_i, through its impact on the causal variable, d_i. The third assumption is that $\mathrm{Cov}(z_i, d_i) \neq 0$. This is satisfied if the percentage of students in the study who will go to private school if they win the scholarship but not attend private school if they do not is nonzero. This subgroup is sometimes referred to as the "compliers", and this model will identify the effect of attending private school only for these students.[5]

One assumption that is often neglected in discussions of instrumental variables models is the SUTVA assumption. Similar considerations will hold with regard to decisions to randomize at the individual versus group level. However, in this setting SUTVA needs to be extended to address "lack of interference" with respect to both the outcome variable y_i and the other post-randomization variable d_i (see Angrist et al. (1996) for a more thorough discussion). Thus, concern regarding interference affecting either take-up or the outcome might motivate randomization at the group level.

We can rewrite the standard instrumental variables model above as

$$\begin{pmatrix} y_i \\ d_i \end{pmatrix} \sim \mathrm{N}\left(\begin{pmatrix} \alpha + \tau d \\ \gamma + \delta z_i \end{pmatrix}, \begin{pmatrix} \sigma_\varepsilon^2 & \rho\sigma_\varepsilon\sigma_d \\ \rho\sigma_\varepsilon\sigma_d & \sigma_d^2 \end{pmatrix} \right), \quad \text{for } i = 1, \ldots, n.$$

For simplicity, in this section I will not explicitly write out the error term in the model. This formulation will make it easier to generalize this model to multilevel settings. I consider three multilevel settings: 1) treatment is randomized at the individual level, noncompliance occurs at the individual level, but there are multilevel structures to the data; 2) treatment is randomized at the group level but there is individual-level noncompliance; and 3) treatment is randomized at the group level and noncompliance only occurs (or is only observed) at the group level.

All of the models can be extended to incorporate additional pre-randomization variables either to increase efficiency or to appropriately reflect the study design (for instance if randomization occurred within blocks). We omit discussion of such extensions here because they are relatively straightforward. The only particularity worth noting is that it rarely makes sense to include such a variable in just one of the above equations because this would imply a conditional independence that is probably not justifiable.

12.4.1 Treatment Randomization and Noncompliance Occur at the Individual Level

Consider a variation on the example above in which the lottery for scholarships occurs within 10 different cities. When treatments are assigned at the individual level and noncompliance also occurs individual by individual, then the extension of the above model to include varying city-level intercepts for each equation is relatively straightforward but does require addressing the potential association between these two varying intercepts α_j and γ_j:

$$\begin{pmatrix} y_{ij} \\ d_{ij} \end{pmatrix} \sim \mathrm{N}\left(\begin{pmatrix} \mu_{y_{ij}} \\ \mu_{d_{ij}} \end{pmatrix}, \begin{pmatrix} \sigma_\varepsilon^2 & \rho_{\varepsilon d}\sigma_\varepsilon\sigma_d \\ \rho_{\varepsilon d}\sigma_\varepsilon\sigma_d & \sigma_d^2 \end{pmatrix}\right),$$

where $\mu_{y_{ij}} = \alpha_j + \tau d_{ij}$ and $\mu_{d_{ij}} = \gamma_j + \delta z_{ij}$, and

$$\begin{pmatrix} \alpha_j \\ \gamma_j \end{pmatrix} \sim \mathrm{N}\left(\begin{pmatrix} \mu_\alpha \\ \mu_\gamma \end{pmatrix}, \begin{pmatrix} \sigma_\alpha^2 & \rho_{\alpha\gamma}\sigma_\alpha\sigma_\gamma \\ \rho_{\alpha\gamma}\sigma_\alpha\sigma_\gamma & \sigma_\gamma^2 \end{pmatrix}\right).$$

Note that the structural instrumental variables assumptions have not changed in any substantial way in this formulation; the difference is that they now additionally condition on the varying intercepts α_j and γ_j. Therefore, the assumptions only need to hold conditional on group. However, we now need to consider the assumptions regarding the independence of the varying intercept, α_j, and the predictor variable, d_{ij}, as well as the independence between the varying intercept, γ_j, and the predictor variable, z_{ij}. In the current scenario, with binary d_{ij} and z_{ij}, violations of the latter assumption are less likely to cause trouble in practice. If y_{ij} is continuous, violations of the former assumption have greater potential to lead to bias, particularly when σ_ε^2 is large or the within-group sample size is small.

If randomization in this setting was based on separate lotteries for the scholarships and equity considerations dictated that the same number of scholarships be awarded in each city then it is actually extremely unlikely that the probability of treatment would be constant across cities (this would require the same number of lottery applicants in each city). Further, it is unlikely that the probability of attending a private school among lottery participants would be constant across cities; this would likely vary based on the perceived quality of each type of school, tuition, spaces available, and the number of participants who actually won a scholarship, among other factors. If we believe that there are unobserved characteristics of the cities captured by α_j that are correlated with whether students attend private school, we might be concerned about getting a biased estimate of τ. If we believe that there are unobserved characteristics of the cities captured by γ_j that are correlated with the probability of winning a scholarship, we might be concerned about getting a biased estimate of δ (which, in turn, could bias our estimate of τ).

In these settings (as in Section 12.3.2) it may be helpful to extend the model to incorporate this dependence. For instance, if we are worried that α_j is correlated with the d_{ij}, one strategy would be to replace d_{ij} in the model above with $d^*_{ij} = d_{ij} - \overline{d}_j$, where $\overline{d}_j$ represents the block-level means of d_{ij}. This will yield an estimate similar in interpretation to the fixed effects estimate, but it should be more efficient. Another strategy would be to model the α_j directly as a function of $\overline{d}_j$.

SUTVA considerations would require both that 1) student's public versus private attendance may not be influenced by whether another student won a scholarship in the lottery, and 2) her subsequent test scores cannot be influenced by whether other students won a scholarship or whether they attended public versus private school.

12.4.2 Treatment Randomization at the Group Level but Noncompliance at the Individual Level

Now consider an example with treatment assigned at the group level. For instance, consider a flu shot experiment in which doctors who have agreed to participate in the study are randomized to encourage their patients to receive the flu shot or not. Thus, randomization to encourage, z_j, only varies at the level of the doctor (and suppose for simplicity that the doctors cannot refuse to participate after learning their assignments). However, individual patients of a given doctor might choose to receive the flu shot, d_{ij}, or not. The outcome, y_{ij}, might be number of days with flu-related symptoms during the subsequent year.[6] In this case the model changes simply because the assignment can now be considered a group-level variable. This can be expressed in the following way:

$$\begin{pmatrix} y_{ij} \\ d_{ij} \end{pmatrix} \sim \mathrm{N}\left(\begin{pmatrix} \mu_{y_{ij}} \\ \gamma_j \end{pmatrix}, \begin{pmatrix} \sigma_\varepsilon^2 & \rho_{\varepsilon d}\sigma_\varepsilon\sigma_d \\ \rho_{\varepsilon d}\sigma_\varepsilon\sigma_d & \sigma_d^2 \end{pmatrix}\right),$$

where $\mu_{y_{ij}} = \alpha_j + \tau d_{ij}$, and

$$\begin{pmatrix} \alpha_j \\ \gamma_j \end{pmatrix} \sim \mathrm{N}\left(\begin{pmatrix} \mu_\alpha \\ \mu_{\gamma_j} \end{pmatrix}, \begin{pmatrix} \sigma_\alpha^2 & \rho_{\alpha\gamma}\sigma_\alpha\sigma_\gamma \\ \rho_{\alpha\gamma}\sigma_\alpha\sigma_\gamma & \sigma_\gamma^2 \end{pmatrix}\right),$$

where $\mu_{\gamma_j} = \mu_\gamma + \delta z_j$.

The considerations in this model have changed slightly. The exclusion restriction has a similar intuition in that we would need to believe that encouragement does not affect the number of days of flu-related symptoms except through receipt (or not) of a flu shot. Moreover, we no longer need to worry about correlation between γ_j and z_j since the randomization at the group level will ensure their independence.

The SUTVA assumption looks a little different here as potential outcomes can now be defined by both encouragement at the doctor level and flu shot receipt at the individual level. For SUTVA to hold, therefore, we must believe that an individual's decision to receive a flu shot is not influenced by encouragement received by any patient who sees a doctor other than his own. Moreover, his outcome cannot be affected by either the encouragement received by a patient with a different doctor or the flu shot status of any other patient (with the same or different doctor).

12.4.3 Treatment Randomization and Noncompliance at the Group Level

Finally, we consider an example such as an evaluation of a whole-school reform program in which both treatment assignment and noncompliance occur at the group (here school) level (in this setting we will assume that the estimand is interpreted at the group level). In this case, SUTVA will hold as long as there is no interference across schools. The only random variable that we see at the individual level here is the outcome variable (for instance, student-level test scores). Thus, the structural relationships of interest are all happening at the group level:

$$y_{ij} \sim \mathrm{N}(\alpha_j + \tau^G d_j, \sigma_\varepsilon^2)$$

$$\begin{pmatrix} \alpha_j \\ d_j \end{pmatrix} \sim \mathrm{N}\left(\begin{pmatrix} \mu_\alpha \\ \mu_{d_j} \end{pmatrix}, \begin{pmatrix} \sigma_\alpha^2 & \rho_{\alpha\gamma}\sigma_\alpha\sigma_\gamma \\ \rho_{\alpha\gamma}\sigma_\alpha\sigma_\gamma & \sigma_\gamma^2 \end{pmatrix}\right),$$

where $\mu_{d_j} = \mu_\gamma + \delta z_j$.

How does this change the assumptions? Basically, we can just translate all the same concepts from the individual to the group level (i.e., as if the group were the "actor"). Eligibility to participate in the whole-school reform intervention must be randomized (or must be independent of the potential outcomes). Randomization to participate in the whole-school reform intervention must encourage at least some of the schools to participate.[7] And randomization to participate should only be able to influence aggregate school outcomes through its impact on whether schools actually do participate; said another way, schools that would not participate even if randomized to participate should not experience systematically different outcomes depending on their randomized assignment.

As discussed earlier, when treatments are assigned at the group level we can only rarely interpret at the individual level. Thus, once we are making inferences at the group level the most straightforward modeling strategy aggregates the data first. Then we can simply use the model

$$\begin{pmatrix} \overline{y}_j \\ \overline{d}_j \end{pmatrix} \sim \mathrm{N}\left(\begin{pmatrix} \mu_{\overline{y}_j} \\ \mu_{\overline{d}_j} \end{pmatrix}, \begin{pmatrix} \sigma_\varepsilon^2 & \rho_{\varepsilon d}\sigma_\varepsilon\sigma_d \\ \rho_{\varepsilon d}\sigma_\varepsilon\sigma_d & \sigma_{\overline{d}}^2 \end{pmatrix}\right),$$

where $\overline{y}_j = \mu_y + \tau^G \overline{d}_j$ and $\overline{d}_j = \mu_d + \delta \overline{z}_j$. However, if the sizes of the groups vary substantially it can be helpful to incorporate this information into the model (as weights or even sometimes as covariates) to account for the fact that some of the aggregated measures will have been measured with greater accuracy than others.

12.5 OBSERVATIONAL STUDIES IN WHICH IGNORABILITY IS ASSUMED

A more straightforward use of multilevel models occurs in an observational study context in settings where the researcher has made the decision to act as if the ignorability assumption is satisfied conditional on the available pre-treatment covariates. Since unbiased estimation in this setting relies on appropriate conditioning on these pre-treatment values, the focus is typically on the most appropriate model with which to accomplish this conditioning. This can be complicated when there are many covariates or when linear and additive relationships cannot be assumed between the covariates and the response variable. A multilevel data structure adds a further complication. In this section we discuss two different types of scenario—those in which parametric assumptions are assumed to hold and those in which researchers are not comfortable with making traditional parametric assumptions.

12.5.1 *Standard Models for Estimating Causal Effects Under Ignorability*

When ignorability is considered a reasonable assumption conditional on observed covariates in a dataset with a multilevel structure, it may make sense to simply fit a multilevel model that includes these covariates. This is the multilevel analog to estimating causal effects in a simple observational study (that is, without a multilevel structure to the data) using linear regression. The caveat to that analogy is that here we have to think harder about the role played by the groups. Specifically, if we think that the groups are acting as proxies for unobserved covariates that we need to condition on to achieve ignorability then we are in the situation discussed in Section 12.6.2 below. If we truly think that ignorability can be satisfied by the observed covariates then we are fitting the multilevel model for other purposes—for instance, to correctly account for within-group dependence structures or to investigate heterogeneity in treatment effects across groups. In this case, if we think of the confounding covariates as definining blocks within which observations have an equal probability of being treated, the basic ideas from Section 12.3.2 hold here. One caveat is that since the common support assumption is also not guaranteed to hold in this setting, and since problems with failure of the parametric assumptions of the model used for inference will be exacerbated by weak overlap, it may be particularly important to check for common support in this setting.

12.5.2 *Propensity Score Strategies with Grouped Data*

A concern with the approach in Section 12.5.1 that has drawn the attention of many causal researchers is that the assumptions inherent in typical parametric models may not be satisfied. Moreover, as noted above, departures from the parametric assumptions are likely to yield more bias if common support is not strong. These issues have motivated widespread use of causal inference strategies such as those that incorporate estimated propensity scores whose goal is to reduce dependence on parametric assumptions.

The propensity score is defined as the probability of being treated conditional on a set of confounding covariates, $\Pr(z = 1 \mid x)$. It acts as a one-number summary of these covariates and thus more easily allows for less rigidly parametric approaches to conditioning on the covariates through matching, subclassification, or inverse probability weighting (Stuart, 2010). While there is a large literature on propensity score techniques, comparatively little attention has been paid to use of these techniques in multilevel settings. In the past few years, however, several articles have shed light on this topic.

Within a propensity score matching framework, Kim and Seltzer (2007) compare several strategies based on their resulting balance and conclude with a recommendation to estimate the propensity score using a multilevel model with varying intercepts and slopes and then to exact match within group. In related work, Li et al. (2007) note that either fitting a multilevel model for the propensity score or the outcome model will outperform models that ignore the multilevel structure at both stages. Arpino and Mealli (2011) reach similar conclusions. Note, however, that fitting a multilevel model for the response surface in this setting may be particularly vulnerable to violating the assumption that the varying intercepts are uncorrelated with the treatment variable (if ignorability is indeed not satisfied).

12.6 OBSERVATIONAL STUDIES IN WHICH IGNORABILITY IS NOT ASSUMED

The ignorability assumption is not always plausible, which is why randomized experiments are typically preferred. However, sometimes there exists a middle ground in the form of a natural or quasi-experiment. Moreover, sometimes data are structured in such a way that we may be able to condition on what otherwise might be unobservable confounders. We explore these situations in this section.

12.6.1 Instrumental Variables in a Non-Experimental Setting

We discussed use of instrumental variables methods above in the setting of randomized experiments with noncompliance and randomized encouragement designs. However, these strategies are often used outside the context of controlled randomized experiments as well. For instance, consider the goal of estimating the effect of having children on parents labor supply. Angrist and Evans (1998) explored a more limited form of this question, the effect of having two versus three children, by using the sex mix ratio of the first two children as an instrument for having a third child; this works because the sex mix is random and families in which the first two children are of the same sex are marginally more likely on average to have a third child than those who have one child of each sex. These authors fit standard (non-multilevel) models to address this question. However, one could imagine that labor market decisions might be different depending on the specifics of one's local labor market. Therefore, if data had been available on county or state of residence a reasonable extension might have been to embed this instrumental variables framework into a multilevel model as in Section 12.4, with varying county- or state-level intercepts.

The instrumental variables concepts translate directly to these natural experiment settings in that, in principle, the same models can be used. Typically, extra care must be taken, however, when evaluating the plausibility of the requisite assumptions. Furthermore, it is more likely in these settings that the researcher will have to control for additional individual- or group-level covariates to increase the plausibility of the assumptions.

12.6.2 Fixed Versus Random Effects for Causal Inference

Multilevel models provide a means of conditioning on group status and exploring variation across group status. So-called "fixed effects" models, which allow for a separate intercept for the members of each group, have a similar goal.[8] In fact, these models can be thought of as a special case of the multilevel model in (12.1) in which $\sigma_\alpha^2 = \infty$. However, researchers who attempt to identify causal effects using these models have a distinct motivation. In this case, "group membership" is conceptualized as a variable upon which to condition as a means of controlling for group-specific characteristics that are not directly observable. The fixed effects themselves are typically treated as nuisance parameters and as such are generally not directly estimated.

An example can help to elucidate the appeal of the approach. A scientific debate exists regarding whether low birth weight is a cause of future health and developmental problems or merely a proxy variable that is merely associated with these outcomes and the true cause. One way to distinguish these causal pathways was considered by Almond et al. (2005), who examined the impact of lower birth weight through a clever observational design. They focused on a large sample of twins and reasoned that the assignment of twins to have lower or higher birth weight is random (or "haphazard"). Thus, each pair of twins represented a very small natural experiment. A classic econometric approach to analyzing data would be to fit a so-called fixed effects model, such as

$$y_{ij} = \mu + \tau z_j + a_j + \varepsilon_{ij}, \;\; \varepsilon_{ij} \sim \mathrm{N}(0, \sigma_\varepsilon^2),$$

where we use a_j to denote these fixed effects (to distinguish them from the varying intercepts, α_j, in the earlier models).

A more efficient choice in this setting, however, would be to assume a distribution for the twin specific intercepts, such as $a_j \sim \mathrm{N}(0, \sigma_\alpha^2)$; in other words, we could fit a varying intercepts model (in this context this is typically referred to as a "random effects model"). This section will discuss the advantages and disadvantages of using such conditioning methods as a tool for causal inference, as well as the choice between fixed and random effects models in this context. For a more general exposition of the differences between these two types of modes, please refer to Chapter 5 of this handbook.

General motivation for and implications of conditioning on group status to identify causal effects

Fixed effects models have been used as a tool for identifying causal effects in scores of applications. For example, when examining the effect of maternal employment on children's subsequent developmental outcomes using a dataset that had records for multiple children from each household, family fixed effects might be included to control for unobserved family-specific characteristics (see, for example, Waldfogel et al., 2002). The basic idea is similar to the special case above; when we include fixed effects in this model we adjust for family-specific characteristics that do not vary over the members of the family. In essence we can then conceive of each group as a little natural experiment.

Of course, even after conditioning on fixed effects, ignorability still may not hold in any given example; the hope is that the additional conditioning should at least bring us closer to the truth. So is such conditioning always a good thing? Several points are worth noting. First, akin to controlling for an additional confounder in an analysis where ignorability is not guaranteed, we can never know for sure that that particular adjustment leads to a reduction in bias compared to leaving it out. In the absence of any specific knowledge, however, researchers tend to prefer controlling for more rather than fewer sources of potential bias. Second, sometimes fixed effects approaches (and the special case version of these, difference in difference models) are used as a replacement for properly adjusting for observed confounders and this, of course, should be discouraged. Generally, finding a richer dataset with better controls (if possible) is a stronger solution (see, for example, Mincy et al., 2009).

Third, this strategy is often used in settings where the "group" is an individual and the observations vary by time period. For instance, panel data (cross-sectional samples followed over several time periods) have been used in this way to identify the causal effect of marriage on men's earnings (see, for example, Korenman and Neumark, 1991). The thought is that including individual-level fixed effects will control for all unobserved characteristics of each man that do not vary over time. This use of fixed effects models to identify causal effects can be quite problematic specifically because of the time component. To see why, consider in our men's marriage and earnings example what the full model might look like. In addition to individual-level fixed effects we would almost certainly want to additionally condition on other covariates that do change over time—at least those common to the standard wage equation like educational attainment, occupation, industry, years of experience, and perhaps even less common covariates like health and religiosity. However, these variables themselves may be affected by the causing variable of interest, marriage. This means that for some observations we will be estimating the effect of the treatment while conditioning on post-treatment variables, which can bias our treatment effect.[9]

Notice that other settings in which fixed effects models are used may also implicitly have such a time-dependent structure. For instance, in the example where family fixed effects are used to help identify the effect of maternal employment on children's development, effects are estimated in families in which children are born at different time points and different maternal employment decisions are made across children. If other covariates were controlled for in the model as well and these were themselves impacted by the maternal employment decision made at a previous time period, problems could arise.

Choosing between so-called "fixed effects" and "random effects" models for causal inference

In the situation where the problems of variation over time is not an issue, this strategy may make sense as a means of controlling for certain types of additional and otherwise unobserved characteristics. Given that random effects models generally produce more efficient estimates, however, why would one ever choose to fit a fixed effects model? The issue is, in essence, the same as the one that arises in the randomized block experiment in which the block-specific intercepts are correlated with the probability of being treated. Here, the reason we want to control for the "fixed effects" is because we believe that they serve as a proxy for unobserved group-specific characteristics (that don't vary over the members of the group). The reason we would need to control for these characteristics is if we feel they are true confounders, that is, they are associated with both the probability of being treated and the subsequent outcome of interest. But if this dependence indeed exists then the multilevel model assumption of no correlation between the varying intercepts and the predictor variables should not hold (for further discussion see Gelman and Hill, 2007). One compromise, however, might be the types of "correlated random effects" models (and their variants) discussed in Section 12.3.2.

12.7 RELAXING SUTVA

We have discussed strategies for handling potential SUTVA violations through study design. Recently, however, there has been increased discussion in the literature about methods for directly incorporating certain types of SUTVA violation directly into the analysis. Hong and Raudenbush (2006) discuss an example that explores the causal effect of kindergarten grade retention on subsequent achievement using data from the Early Childhood Longitudinal Study Kindergarten Cohort (ECLS-K) in which children were sampled within schools (approximately 20 per school). The "treatment" (being retained in grade) is necessarily assigned at an individual level. These authors relax the SUTVA assumption by defining potential outcomes that depend both on individual treatment assignment and a scalar function of the treatment assignments of everyone else in that school (a binary indicator for "low proportion of children retained" versus "high proportion of children retained"). They dub the relaxed version of SUTVA in their study "no interference between schools". Another example of a modified version of the SUTVA assumption can be found in Hudgens and Halloran (2008). Taking a broader tack, Rosenbaum (2007) describes a nonparametric (randomization-based) approach to inference in the context of randomized experiments that suffer from inference between units.

12.8 CONCLUSION

This chapter provides a brief summary of causal inference designs and analyses in a multilevel context. Key considerations discussed include whether the treatment is randomized or not, the level of hierarchy at which the treatment is assigned, whether the SUTVA assumption is likely to hold, and whether there are unobserved characteristics at the group level that act as confounders of the causal effect. Insufficient space precluded consideration of several other related topics that have been discussed elsewhere. For instance, loss of power when assigning treatments at the group rather than the individual level is described in many places, including Cornfield (1978) and Murray (1998). A helpful summary of the debates that exist regarding the relative benefits of stratifying/blocking versus using matched pairs versus not adjusting at all (in situations where the magnitude of the predictive power of a given blocking/matching variable is a priori unclear) in the context of randomized designs is provided (along with recommendations for best practice) in Imbens (2011). Reardon and Raudenbush (2009) discuss the (heroic) assumptions necessary for multi-site randomized trials to identify multiple causal pathways. Across these settings, researchers need to be aware of the ways in which the models and assumptions commonly used to address the causal inference problem and the multilevel structure of the data change when addressing both issues simultaneously.

NOTES

1 Although arguably this could call into question the "versions of treatment" component of SUTVA again if, for instance, we think that the strength of the treatment is a function of how many people were collectively treated in this group and if group size varies.

2 For further discussion about using linear models with binary outcomes see, for instance, Gail, Wieand, and Piantadosi (1984); Guo and Geng (1995).

3 Readers more familiar with calculating clustered standard errors to accommodate nested data structures might wonder why they should fit a multilevel model instead. In a simple situation such as this the tradeoffs are probably not too dramatic in either direction. One might argue that the multilevel approach relies on normally distributed error components. However, clustered standard errors are often used in models that assume normally distributed residual errors and additionally make unrealistic assumptions about σ_α^2 (typically either that it equals 0 or ∞).

4 While I typically prefer to write out a nonparametric version of the instrumental variables assumptions as in Angrist et al. (1996), it is more straightforward in this context to write out the specific model that I will then generalize to the multilevel setting.

5 Angrist et al. (1996) use this term and provide an alternate framing of the model and assumptions involving potential outcomes that may provide greater intuition. For instance, in their conceptualization the exclusion restriction would focus on students whose d_i is not affected by their z_i. These are students who will go to private school no matter how they are randomized (so-called "always takers") or those who will go to public school no matter how they are randomized (so-called "never takers"). If each of these students would have

the same outcome (or more generally if their outcome would be drawn from the same distribution) no matter the initial random assignment then the exclusion restriction is satisfied. For modeling strategies that explicitly incorporate the potential outcomes in this setting see, for instance, Barnard et al. (2001) and Hirano et al. (2000).

6 We will ignore in this example the fact that this outcome is likely not best modeled using a normal distribution.

7 In this scenario, always takers would not exist since presumably no school would be able to participate in the whole-school reform intervention being evaluated without being randomized to do so.

8 In this context, the term fixed effects is used to refer to models that can be represented by inserting a separate indicator variable for each group (with the exception of the reference group). For an argument against use of this term due to its ambiguous meaning, see Gelman (2005).

9 For a comprehensive discussion of this issue see Sobel, 2012. For a more general discussion of the dangers of controlling for post-treatment outcomes see Rosenbaum (1984). See also Robins and Hernan (2009) for a review of modeling techniques pioneered by Robins and colleagues that may be more appropriate for this setting.

REFERENCES

Almond, D., Chay, K., and Lee, D. (2005), 'The Costs of Low Birth Weight', *The Quarterly Journal of Economics* 120: 1031–83.

Angrist, J. and Evans, W. (1998), 'Children and Their Parents' Labor Supply: Evidence from Exogenous Variation in Family Size', *American Economic Review* 88: 450–77.

Angrist, J.D., Imbens, G.W., and Rubin, D.B. (1996), 'Identification of Causal Effects Using Instrumental Variables', *Journal of the American Statistical Association* 91: 444–72.

Arpino, B. and Mealli, F. (2011), 'The Specification of the Propensity Score in Multilevel Studies', *Computational Statistics and Data Analysis* 55: 1770–80.

Bafumi, J. and Gelman, A. (2011), 'Fitting Multilevel Models when Predictors and Group Effects Correlate', Tech. rep., Columbia University.

Barnard, J., Frangakis, C., Hill, J.L., and Rubin, D.B. (2001), 'School Choice in NY City: A Bayesian Analysis of an Imperfect Randomized Experiment', in *Case Studies in Bayesian Statistics*, eds C. Gatsonis et al., vol. V, New York: Springer-Verlag.

Brooks-Gunn, J., McCarton, C.M., Casey, P.H., McCormick, M.C., Bauer, C.C., Bernbaum, J.C., Tyson, J., Swanson, M., Bennett, F.C., Scott, D.T., Tonascia, J., and Meinert, C.L. (1994), 'Early Intervention in Low-Birth-Weight Premature Infants: Results Through Age 5 from the Infant Health and Development Program', *Journal of the American Medical Association* 272: 1257–62.

Chamberlain, G. (1982), 'Multivariate Regression Models for Panel Data', *Journal of Econometrics* 1: 5–46.

Chamberlain, G. (1984), *Handbook of Econometrics, Volume 2*, chap. Panel Data, pp. 1248–318, Amsterdam: North Holland.

Cornfield, J. (1978), 'Randomization by Group: A Formal Analysis', *American Journal of Epidemiology* 108: 100–102.

Gail, M., Wieand, S., and Piantadosi, S. (1984), 'Biased Estimates of Treatment Effect in Randomized Experiments with Nonlinear Regression and Omitted Covariates', *Biometrika* 71: 431–44.

Gelman, A. (2004), *Applied Bayesian Modeling and Causal Inference from an Incomplete Data Perspective*, chap. Treatment Effects in Before-after Data, pp. 191–8, London: Wiley.

Gelman, A. (2005), 'Analysis of Variance – Why it is More Important Than Ever', *The Annals of Statistics* 33: 1–53.

Gelman, A. and Hill, J. (2007), *Data Analysis Using Regression and Multilevel/Hierarchical Models*, New York: Cambridge University Press.

Gelman, A., Hill, J., and Yajima, M. (2005) 'Why We (Usually) Don't Have to Worry About Multiple Comparisons', *Journal of Research on Educational Effectiveness* 5: 189–211.

Guo, J. and Geng, Z. (1995), 'Collapsibility of Logistic Regression Coefficients', *Journal of the Royal Statistical Society, Series B* 57: 263–7.

Hill, J. and Su, Y.-S. (2012), 'Identifying Areas that Lack Common Support in Observational Studies Using Bayesian Nonparametric Modeling', Tech. rep., New York University.

Hirano, K., Imbens, G.W., Rubin, D.B., and Zhou, A. (2000), 'Assessing the Effect of an Influenza Vaccine in an Encouragement Design', *Biostatistics* 1: 69–88.

Hong, G. and Raudenbush, S. (2006), 'Evaluating kindergarten retention policy: A Case Study of Causal Inference for Multilevel Observational Data', *Journal of the American Statistical Association* 101: 901–10.

Hudgens, M. and Halloran, M. (2008), 'Towards Causal Inference with Interference. PMC2600548', *Journal of the American Statistical Association* 103: 832–42.

Imbens, G. (2004), 'Nonparametric Estimation of Average Treatment Effects Under Exogeneity: A Review', *The Review of Economics and Statistics* 86: 4–29.

Imbens, G.W. (2011), 'Experimental Design for Unit and Cluster Randomized Trials', Tech. rep., prepared for the International Initiative for Impace Evaluation.

Kim, J. and Seltzer, M. (2007), 'Causal Inference in Multilevel Settings in which Selection Processes Vary Across Schools', Tech. rep., Center for Research on Evaluation, Standards and Student Testing (CRESST), UCLA.

Korenman, S. and Neumark, D. (1991), 'Does marriage really make men more productive?' *Journal of Human Resources* 26: 282–307.

Li, F., Zaslavsky, A., and Landrum, M. (2007), 'Propensity Score Analysis with Hierarchical Data' in *Proceeding of Joint Statistical Meetings*, pp. 2474–81, Section on Health Policy Statistics, American Statistical Association.

Middleton, J. (2008), 'Bias of the Regression Estimator for Experiments Using Clustered Random Assignment', *Statistics and Probability Letters* 78: 2654–9.

Mincy, R., Hill, J., and Sinkewicz (2009), 'Marriage: Cause or Mere Indicator of Future Earnings Growth', *Journal of Policy Analysis and Management* 28: 417–39.

Mundlak, Y. (1978), 'On the Pooling of Time Series and Cross Section Data', *Econometrica* 46: 69–85.

Murray, D. (1998), *Design and Analysis of Community Trials*, Oxford Univ. Press, Oxford.

Raudenbush, S.W. (2009), 'Adaptive Centering with Random Effects: An Alternative to the Fixed Effects Model for Studying Time-Varying Treatments in School Settings', *Education Finance and Policy* 4: 468–91.

Reardon, S. and Raudenbush, S.W. (2009), 'Assumptions of Value-added Models for Estimating School Effects', *Journal of Education, Finance and Policy* 4: 492–519.

Robins, J. and Hernan, M.A. (2009), *Longitudinal Data Analysis*, chap. Estimation of the Causal Effects of Time-Varying Exposures, pp. 553–97, Boca Raton, FL: Chapman and Hall.

Rosenbaum, P.R. (1984), 'The Consequences of Adjustment for a Concomitant Variable That Has Been Affected By the Treatment', *Journal of the Royal Statistical Society, Series A, General* 147: 656–66.

Rosenbaum, P.R. (2007), 'Interference Between Units in Randomized Experiments', *Journal of the American Statistical Association* 102: 191–200.

Rubin, D.B. (1978), 'Bayesian Inference for Causal Effects: The Role of Randomization', *The Annals of Statistics* 6: 34–58.

Rubin, D.B. (1990), 'Formal Modes of Statistical Inference For Causal Effects', *Journal of Statistical Planning and Inference* 25: 279–92.

Slavin, R., Madden, N., Dolan, L., and Wasik, B. (1996), *Every Child, Every School: Success for All*, Thousand Oaks, CA: Corwin.

Sobel, M.E. (2006), 'What Do Randomized Studies of Housing Mobility Demonstrate? Causal Inference in the Face of Interference', *Journal of the American Statistical Association* 101: 1398–407.

Sobel, M.E. (2012) 'Does Marriage Boost Men's Wages? Identification of Treatment Effects in Fixed Effects Regression Models for Panel Data', *Journal of the American Statistical Association* 107: 521–9.

Stuart, E.A. (2010), 'Matching Methods for Causal Inference: A Review and a Look Forward', *Statistical Science*, in press.

Waldfogel, J., Han, W., and Brooks-Gunn, J. (2002), 'The Effects of Early Maternal Employment on Child Cognitive Development', *Demography* 39: 369–92.

PART II

Variations and Extensions of the Multilevel Model

To my good friend
Brian, from whom I've learned so
much!

[signature]

13

Multilevel Functional Data Analysis

Ciprian M. Crainiceanu
Johns Hopkins University, USA

Brian S. Caffo
Johns Hopkins University, USA

Jeffrey S. Morris
The University of Texas MD Anderson Cancer Center, USA

13.1 INTRODUCTION

The information explosion and increased access to ever larger and more complex data sets is one of the most important developments of the 21st century. Statistics, the science inspired by and developed for data analysis, is uniquely positioned to take advantage of these unprecedented opportunities. Thus, it has become crucial to reassess the scalability of existing statistical methods and develop new methods to analyze these emerging types of data.

For example, with the advent of new technologies, many studies now collect large-dimensional functional or imaging data on hundreds of subjects at multiple visits. Relevant examples from brain imaging and proteomics are included in Section 13.2. In such settings, collecting, storing, and manipulating the data, even for only one subject at one level (e.g. visit), is challenging. Consequently, statistically principled analysis of such data collected on hundreds or thousands of subjects at multiple levels remains underdeveloped, though some initial approaches are documented in this chapter. While details are different for each particular model, the methods described here borrow heavily from standard multilevel modeling and dimension reduction techniques inspired by functional data analysis (Ramsay and Silverman, 2006). The combination of these two areas of research leads to the new area of multilevel functional data analysis (MFDA). Below we provide a brief introduction to functional data analysis and then describe its natural extensions to multilevel modeling. We do not revisit the basic principles of multilevel modeling as they are covered exhaustively elsewhere in this book.

13.1.1 Functional Data Analysis

Functional data analysis (FDA) is concerned with data obtained by measuring subject-specific functions at a finite number of points. At the most basic level data are an ordered vector representing the observed values of the function, if the function is measured without or with negligible error, or its proxy, if the function is measured with sizeable error. More specifically, if the functional signal $f(s)$ is measured at $s_1, \ldots, s_M$ then the data vector is $(y_1, \ldots, y_M)$, where we assume that $y_m = f(s_m) + \varepsilon_m$, for $m = 1, \ldots, M$, and ε_m is the measurement error, which could be zero. In general, a sample of functional data is available and data at the subject level is of the type $(y_{i1}, \ldots, y_{iM_i})$, where $y_{im} = f(s_{im}) + \varepsilon_{im}$, for $i = 1, \ldots, I$, and $m = 1, \ldots, M_i$, where s_{im} could be the same or different across subjects. Traditionally, FDA referred to functions of one variable, but our definition is much broader and includes functions of multiple variables, such as 2-, 3-, and 4-dimensional images. Unfortunately, extending the definition does not resolve the inferential problems raised by these more complex settings. Instead, FDA provides a conceptual starting point for inferential strategies.

There is a strong connection between functional and multivariate data, as the observed data are vectors in both cases. This connection allows the use of the powerful multivariate analysis inferential machinery to address the new problems raised by functional data. However, some important differences exist. First, most applications involving multivariate data used low-dimensional vectors, with 2 to 100 entries being the norm. In contrast, FDA applications can easily have thousands or hundreds of thousands of observations per subject. Second, FDA assumes that there is a smooth underlying functional process which is sampled at a finite number of points. This is fundamentally different from a vector that contains, for example, a subject's body mass and index (BMI) and systolic blood pressure (SBP). Indeed, the functional process can always be sampled at a finer scale, whereas there are no 'in-between' points for BMI and SBP. The continuous process assumption provides: 1) a natural theoretical foundation; 2) a sound conceptual platform for smoothing of functions, covariance operators, and higher-order moments; and 3) a transparent approach to dealing with missing data and various sampling strategies that could include sparse, quasi-sparse, or dense sampling. Third, functional measurements are on the same scale and are interpretable relative to one another. Thus, smoothing, averaging, ranking, computing principal components is easy to explain to scientific collaborators. Fourth, the argument of the function induces a natural ordering of the data that is crucial to analysis and interpretation. For example, the temporal location of a time series measurement or the voxel (three-dimensional pixel) location in a brain image are some of the most important characteristics of the data. This is a relatively minor difference that can easily be dealt with by labeling and tracking of the various data components. For example, in many imaging applications a three-dimensional image can be unfolded into a long vector where each entry corresponds to a voxel. This is not a problem as long as the mapping from three- to one-dimensional data is preserved together with the original image geometry.

13.1.2 Multilevel FDA

Multilevel functional data occurs in practice for familiar reasons: natural clustering of measured functions. For example, a series of colon carcinogenesis studies with multilevel functional structure was described and analyzed in a number of papers (Morris et al., 2001, 2003; Morris and Carroll, 2006; Li et al., 2007; Baladandayuthapani et al., 2008; Staicu et al., 2010a; Zhou et al., 2010). In one study, several groups of rats were fed different diets and exposed to radiation. Rats were then sacrificed and the concentration of p27 (Sgambato et al., 2000), a cell cycle inhibitor protein, was measured for each cell in multiple colonic crypts. Treating cell-specific measurements as a function of cell location along the crypt wall (Grambsch et al., 1995; Roncucci et al., 2000) transforms the problem to multilevel functional data where functional measurements at the crypt level are nested within rats, who are nested within diet groups. This is an example of functional data where data cannot be treated as being independent even after conditioning on the subject or cluster (a rat in this study). Data structures with multilevel functional structures are becoming increasingly common and include studies of sleep electroencephalograms (EEG) (Crainiceanu et al., 2009b, 2011; Di et al., 2009), brain tractography and morphology (Greven et al., 2010; Zipunnikov et al., 2011), gene expression data (Xiao et al., 2009), proteomics (Morris et al., 2008a, 2011), and calcium ion cellular levels (Martinez et al., 2009).

To better understand the associated structures in MFDA we will use the analogy with standard multilevel modeling. Consider the case when W_{ij} is the jth observation, $j = 1, \ldots, J$ in the ith cluster, $i = 1, \ldots, I$. A standard two-level mixed effects ANOVA for W_{ij} has the structure

$$W_{ij} \sim N(r_i, \sigma_W^2), \quad r_i \sim N(\mu_r, \sigma_r^2),$$

where W_{ij} are conditionally independent of the cluster effects, r_i, and the cluster effects are mutually independent. Thus, the trademark of a multilevel mixed model is the existence of at least two levels of stochastic variables. The main reason for developing multilevel mixed effects is to explicitly decompose the observed variability of the data W_{ij} into simple layers of variability; in this case, within- and between-subject variability. In the functional context, r_i and W_{ij} become random functions and are denoted by $r_i(s)$ and $W_{ij}(s)$, for $s \in [0, 1]$. Thus, the corresponding two-way mixed effect functional ANOVA has the structure (Di et al., 2009)

$$\begin{aligned} W_{ij}(\cdot) &\sim N\{r_i(\cdot), \mathbf{\Sigma}_W\}, \\ r_i(\cdot) &\sim N\{\mu_r(\cdot), \mathbf{\Sigma}_r\}, \end{aligned}$$

where the functional processes $W_{ij}(s)$ are conditionally independent on the cluster mean processes, $r_i(s)$, which, in turn, are assumed to be mutually independent. Including separate mean functions by treatment group, this type of nested multilevel functional data has been termed a 'hierarchical or multilevel functional model' (Baladandayuthapani et al., 2008; Bigelow and Dunson, 2007; Di et al., 2009, 2011; Morris et al., 2003; Staicu et al., 2010a; Zhou et al., 2010). Thus, $W_{ij}(s)$ can be interpreted as: 1) exchangeable functional departures from the cluster mean $r_i(s)$; or 2) replications from a functional measurement error process where the true, unobserved, cluster mean is $r_i(s)$. Note that $\text{Cov}\{W_{ij}(\cdot), W_{ij}(\cdot)\} = \mathbf{\Sigma}_W + \mathbf{\Sigma}_r$, $\text{Cov}\{W_{ij}(\cdot), W_{ij'}(\cdot)\} = \mathbf{\Sigma}_r$ for $j \neq j'$, and $\text{Cov}\{W_{ij}(\cdot), W_{i'j'}(\cdot)\} = 0$ for $i \neq i'$. Thus, $\mathbf{\Sigma}_W$ describes the within-cluster variability and can be viewed as the functional counterpart of the variance component σ_W^2 in the standard two-way mixed model. Similarly, $\mathbf{\Sigma}_r$ controls the between-cluster variability and is the functional counterpart of the variance component σ_r^2. Thus, $\text{trace}(\mathbf{\Sigma}_r)/\{\text{trace}(\mathbf{\Sigma}_r) + \text{trace}(\mathbf{\Sigma}_W)\}$ is a measure of the proportion of between-cluster variability of the total variability. The formal generalization of standard multilevel models to functional multilevel models follows the same recipe used for the two-level mixed ANOVA model: add a functional effect for each cluster effect.

Here, we define MFDA as the *analysis of functional processes with at least two levels of stochastic functional variability*. This definition does not include earlier work in functional modeling of standard longitudinal data (Rao, 1958; Rice, 2004; Rice and Silverman, 1991; Staniswalis and Lee, 1998; Yao et al., 2005; Zhao et al., 2004). These approaches are applications of single-level functional data analysis, or functional ANOVA (FANOVA) (Antoniadis, 2007; Brumback and Rice, 1998; Guo, 2002; Wahba et al., 1995; Wang, 1998), which have only one source of stochastic variability or assume that cluster means are fixed effects functions.

13.1.3 Functional Mixed Models

The general mixed effects model framework (Laird and Ware, 1982) is a useful tool for the analysis of multilevel data. The fixed effects can jointly model the effects of multiple covariates, and the random effects and residuals with various potential covariance structures can be used to account for the correlation among observations due to the clustering or longitudinal nature of the observations. Functional mixed models (FMM), a generalization of the mixed effects model framework, provide a similarly useful framework for multilevel functional data. An early variant of this framework (Guo, 2002) had only one level of stochastic variability, so was not multilevel per se; here, we present a more general version (Morris and Carroll, 2006) that accommodates multiple levels of stochastic variability. Given a sample of observed functions $W_i(s), i = 1, \ldots, n$, fixed effect covariates $X_{ia}, a = 1, \ldots, p$, and random effect covariates $Z^h_{ib}, b = 1, \ldots, q_h, h = 1, \ldots, H$, a general FMM with H hierarchical levels of functional random effects is given by

$$W_i(s) = \sum_{a=1}^{p} X_{ia} B_a(s) + \sum_{h=1}^{H} \sum_{b=1}^{q_h} Z^h_{ib} U^h_b(s) + E_i(s), \tag{13.1}$$

where $U^h_b(s), h = 1, \ldots, H$ and $E_i(s)$ are mutually independent mean zero random processes, and the $B_a(s)$ are fixed effect functions summarizing the partial functional effects of covariate a on the function $W_i(s)$. This framework can also be applied to image data or other higher-dimensional functions by considering vector-valued s (Morris et al., 2011). Various levels of stochastic variability can be handled by the H levels of random effect functions $U^h_b(s)$ and error functions $E_i(s)$, with the Z^h_{ib} indicating cluster membership or longitudinal index.

Many existing models used in multilevel FDA can be written in terms of this model; for example, the two-way mixed effects ANOVA involves $X_{i1} = 1 \forall i$, $B_1(s)$ the overall mean ($\mu_r(s)$ above), $Z^1_{ib} = 1$ if curve i is from cluster b, with cluster b mean given by $B_1(s) + U_b(s)$ ($r_i(s)$ above), and with $U^1_b(\cdot) \sim N(0, \Sigma_r)$ and $E_i(\cdot) \sim N(0, \Sigma_W)$. A key benefit of developing methods that work within this general FMM framework is that they can handle arbitrary fixed and random effect covariates, which gives them the flexibility to model the functional effects of multiple covariates while accounting for the multilevel structure induced by experimental designs.

To develop a usable method in the FMM framework, one must specify distributional assumptions for the random processes in (13.1), covariance assumptions within and between the functional process realizations, and computational and inferential approaches for fitting the model and performing inference. Most work to date has assumed normal distributions, although other alternatives have also been developed (Zhu et al., 2011a; Staicu et al., 2012). Covariance assumptions are tricky here, since the inherently complex nature of many functional data sets frequently precludes use of simple parametric forms, while completely unstructured representations are not feasible for most applications because of their extremely high dimensionality. Many methods involve representation of the functional quantities using basis functions, including pre-specified fixed bases such as splines (Baladandayuthapani et al., 2008; Berhane and Molitor, 2008; Wang et al., 2009), Fourier series, wavelets (Morris and Carroll, 2006; Morris et al., 2008a; Zhu et al., 2011a), as well as empirically determined bases like principal components (Aston et al., 2010; Greven et al., 2010). A suitable choice of basis depends on the application (e.g. wavelets for spiky signals, Fourier for periodic signals, splines or principal components for signals that are consistent across subjects). Depending on the setting, basis function modeling can lead to parsimonious yet flexible modeling of covariance matrices, and together with appropriate penalties or prior distributions on the basis coefficients can lead to effective smoothing of the functional quantities of the model (e.g. fixed effect functions). One general approach to fitting FMM involves a three-step procedure consisting of projecting onto a basis, modeling the basis coefficients, and projecting results back to the original data space for estimation and inference (Morris and Carroll, 2006; Aston et al., 2010; Morris et al., 2011). This has been termed the 'isomorphic modeling approach' (Morris et al., 2011). Both frequentist and Bayesian approaches have been used for modeling and inference.

13.1.4 Inferential Questions in MFDA

Given the richness and complexity of many multilevel functional data sets, there are a large number of different scientific questions that can be addressed by MFDA. The key questions of interest can be roughly grouped into areas: (1) ***group comparison***, (2) ***functional regression/classification***, (3) ***decomposition of variability***, and (4) ***clustering***.

Group comparison involves the regression of multilevel functional responses on sets of scalar predictors, or estimation and inference of fixed effect functions in functional mixed models. Fixed effect function estimates serve as useful summaries, but frequently investigators also have interest in more formal inference, both functional inference, or testing whether an entire functional effect is 0 or not, and pointwise functional inference, which involves identification of which regions of the functions are significantly associated with the predictors. These predictors can be continuous, for example leading to functional linear coefficients, or categorical, for example testing for differences between groups. More general nonlinear models can also be considered.

Functional regression involves the regression of a scalar response on a multilevel functional predictor. Frequently, the scientific goal is prediction of the scalar response based on the functional predictor, although estimation of the functional regression coefficient may also be of interest. When the response is categorical, this type of functional regression amounts to ***classification*** based on multilevel functional predictors.

Decomposition of variability involves estimating and summarizing the forms and relative levels of variability at each level in the multilevel modeling framework. This can be an exploratory exercise to summarize the key modes of variability in the data, but can also be used as building blocks for methods for group comparison, functional regression, and clustering.

Clustering involves the unsupervised grouping of subjects based on their multilevel functional data. This is typically an exploratory exercise, but can also be used as a building block for other inferential statistical methods. We have only encountered one method in the current literature dealing with clustering of multilevel functional data (Rodriguez and Dunson, 2011). This paper presents a Bayesian method for clustering multilevel functional data using mixtures of generalized Dirichlet processes, and is illustrated using hormone profiles from multiple menstrual cycles for women collected in early pregnancy studies. This is clearly a topic of interest, and worthy of further research focus.

Most models and methods for MFDA to date have treated the functions as responses, focusing on group comparison and decomposition of variability; these will be described in Section 13.3, split out based on their use of fixed or empirically estimated basis function modeling. Functional regression methods will be summarized in Section 13.4 along with other extensions.

13.2 DATA EXAMPLES

Driven by new biotechnology and computing advances, high-dimensional, high density, functional multilevel and longitudinal biological signals are becoming commonplace in medical and public health research. These types of signal historically occurred in small clinical or experimental settings. The collection of such biological signals in cohort studies with longitudinal or hierarchical structure represents the next generation of biostatistical problems. Below we provide three examples of such studies.

13.2.1 Brain Morphology

Voxel-based morphometry (VBM) is a whole-brain method for studying localized changes in brain shape (Ashburner and Friston, 2000) that makes fewer anatomical assumptions than traditional region of interest studies. VBM requires that investigators coregister brains to a common anatomical template and retain transformation images in template space. The analysis then considers the now-registered transformation maps across subjects, a common technique in the study of deformational neuroanatomy (Miller et al., 1993). In VBM, typically, the transformation maps are analyzed using voxelwise regression models.

The particular methods of registration we employ were created by Dr Davatzikos and colleagues in a series of papers on deformation models for neuroanatomical study (Davatzikos et al., 2001; Mohamed and Davatzikos, 2004; Xue et al., 2006a,b). The current method creates so called RAVENS (regional analysis of volumes in normalized space maps). The processing pipeline includes: segmentation into gray matter, white matter, and cerebro-spinal fluid, registration to template space, and spatial smoothing. Figure 13.1 displays example gray (left) and white matter (right) RAVENS images from the Former Lead Workers Study described more fully below. The top row displays unsmoothed images, while the bottom images are smoothed with a 10 mm full width at half maximum isotropic Gaussian filter. Here, image intensities correspond to volumes of the associated area in the original brain space.

Because of the registration, every subject's image (either gray or white) will have the same shape. However, intensities will differ across subjects. Hence, a subject with a large head will have an overall brighter image than that of a subject with a smaller head. The benefit of this approach is the ability to study localized changes in brain volume and shape. The template space that we investigate has 1 cubic millimeter resolution for voxels and is of dimension $256 \times 256 \times 198$. Thus, there are roughly 13 million voxels in every image.

The data derive from a study of former organol-lead manufacturing workers (Schwartz et al., 2000b; Stewart et al., 1999; Schwartz et al., 2007, 2000a; Caffo et al., 2008). The subjects in the study are all men, the majority of whom worked in a lead manufacturing plant. The original impetus of the study considered the relation of past peak cumulative lead exposure to cognitive outcomes and brain aging. Two usable MRI scans were collected for roughly 500 subjects five years apart. Using this data, we consider cross-sectional and longitudinal variation in brain shape and volume using the RAVENS images. From the point of view of the data structure, each RAVENS image can be viewed as a function $W_{ij}(v)$, where $i = 1, \ldots, I$ is the subject, $j = 1, 2$ is the visit, and $v = 1, \ldots, V$ is the voxel in the template.

13.2.2 Genomics and Proteomics Data

The emergence of detailed information at the genomic, transcriptomic, proteomic, and epigenetic levels has revolutionized biomedical research. Much current research involves the search for biomarkers at these various levels to use for early diagnosis, patient stratification, and to guide patient treatment. Many of these technologies yield data that can be viewed as functional data on some domain, and since many of these studies obtain multiple measurements for each subject, either replicate functions or repeated functions over time, they give rise to multilevel functional data. In many cases the resulting functionals are complex, irregular, and sampled on very fine grids (thousands to millions), and thus require flexible yet computationally efficient modeling. Here we will provide brief overviews of studies using two of these technologies, mass spectrometry and 2D gel electrophoresis (2DGE) proteomics, as examples of multilevel functional data.

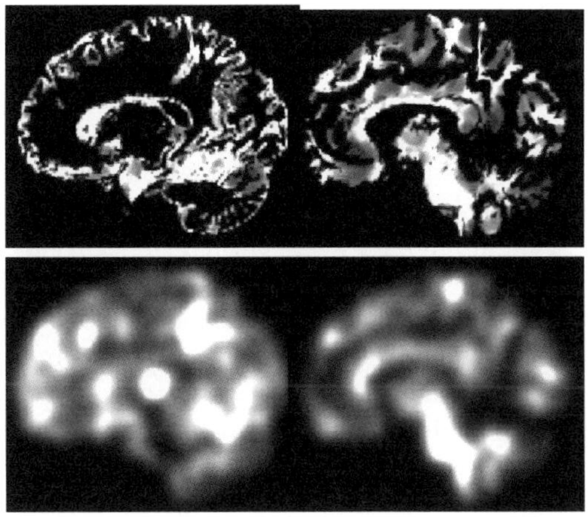

Figure 13.1 RAVENS Images of Gray Matter (left panels) and White Matter (right panels). The top plots are the original data and the bottom panels are obtained from the top panels by applying a 10 mm full width at half maximum isotropic Gaussian filter

Mass spectrometry study

Mass spectrometry methods in proteomics are used to survey the proteomic content of a biological sample by measuring the mass-to-unit charge ratio of charged particles. One commonly-used mass spectrometry method is matrix-assisted laser desorption and ionization time of flight, or MALDI-TOF. In MALDI-TOF, a biological sample of interest is combined with an energy-absorbing matrix substance, and the mixture is placed on a steel plate. The plate is then placed into a vacuum chamber, where a laser strikes the plate, desorbing ionized peptides (protein fragments) from the sample. An electric field accelerates the particles into a flight tube through which they travel at a constant velocity, determined by their molecular mass and charge state, and strike a detector plate that records the number of particles striking it over a series of short, fixed intervals of time, indexed by $\mathbf{t} = (t_1, \ldots, t_T)$, yielding the proteomic spectrum $W(t)$. The spectrum $W(t)$ can be viewed as functional data characterized by many peaks, which correspond to proteins and peptides in the sample. The physics of the instrument allows a mapping of each time t to a mass-to-unit charge ratio (m/z) s. Since most ions have a unit positive charge, the value of a MALDI-TOF spectrum $W(s)$ at a peak roughly measures the abundance of some molecule having a molecular mass of s Daltons. The left panel in Figure 13.2 illustrates a typical MALDI-TOF spectrum.

Consider now the MALDI-TOF data from a cancer biomarker study at MD Anderson Cancer Center (Morris, et al., 2008). In this study, a tumor from one of two cancer cell lines was implanted into either the brain or lungs of 16 nude mice (mice with a genetic mutation that leads to an inhibited immune system). After a period of time, blood serum was extracted from the animals, and two MALDI-TOF spectra were obtained for each mouse, one using a high laser intensity which would more accurately measure lighter molecules, and one using a low laser intensity which more accurately measures the heavier molecules. For each animal $i = 1, \ldots, I = 16$, we observe two MALDI-TOF spectra $W_{ij}(s)$, $j = 1, \ldots, J = 2$ for a grid of m/z values $\mathbf{s}$ of size 7985. These data can be represented in a functional mixed model, with five fixed effect covariates including indicators for the four cell line-by-organ site groups and an indicator of high versus low laser intensity, plus random effect functions for each animal.

In this model, the fixed effect functions represent the mean spectra for the four groups and a contrast spectrum effectively calibrating the high and low laser intensity spectra, while the random effect functions and error functions are independent, mean zero random processes that model the strength and form of the subject-to-subject biological and spectra-to-spectra technical variability. The primary goal was to assess whether differential protein expression was more closely associated with the host organ site or the donor cell line and to search for proteins differentially expressed by organ site, cell line, and their interaction, which involves finding which regions of the domain s were differentially expressed across treatment groups. After preprocessing, standard methods of analysis would involve peak detection and quantification, followed by a surveying of the peaks to determine which are differentially expressed across groups. The methods described here model the entire function using multilevel FDA and thus have potential to find results missed by peak-based methods (Morris et al., 2008a; Morris, 2012).

2DGE cocaine addiction study

2DGE physically separates proteins in a biological sample on a polyacrimide gel by isoelectric point (pH) and molecular mass. In the first step, a pH gradient is applied to the gel and then an electric potential is applied, causing the proteins to migrate across the gels and setting into position based on their pH. Next, the proteins are treated with SDS, which denatures (i.e. unravels) the proteins and attaches negatively charged particles roughly proportional to the protein's length, or molecular mass. Next, an electric potential is applied in the perpendicular direction, causing the proteins to migrate. The friction of the gel acts as a sieve, so lighter proteins will migrate further. Next, some appropriate stain is applied to the gel, binding to the proteins, and the gel is scanned to produce an image $W(s_1, s_2)$, with high intensities in regions of the gel with high protein content. The quantitative interpretation of these image intensities means we can interpret these 2DGE images as functional data, which are typically multilevel as studies usually obtain multiple gels per subject or cluster. The right panel in Figure 13.2 displays an example of a 2DGE image.

Consider the following animal study done to uncover proteomic neurobiological insights into the drug addiction process (Morris et al., 2011). A set of 21 rats were trained to obtain cocaine by pressing a lever for a period of time, after which 6 of the rats were given short durations of drug access (1 hour/day), 7 rats were given long access to the drug (6 hours/day), and 8 control animals were given no access to the drug. After some time, the animals were euthanized, their brain tissue extracted and microdissected to isolate various regions of the brain known to be related to the reward system. The proteomic content of these tissues was then analyzed using 2DGE, with multiple (2–3) gels obtained from each rat for each brain region. The data consists of a total of 53 gels, with $J_i = 2$ or 3 replicate gels from 21 rats $i = 1, \ldots, I = 21$, each gel image consisting of a series of 556,206 pixel intensities observed on a 646×861 grid on the gel image $W_{ij}(s_1, s_2)$. The goal was to search for proteins differentially expressed in these brain regions between animals with long cocaine exposure and control animals, which begins by determining regions on the gel image where the intensities are significantly different across groups. Before analysis, all gel images were aligned using a nonlinear warping algorithm called RAIN (Dowsey et al., 2008), then background corrected and normalized using methods described in (Morris et al., 2008b). Similar to mass spectrometry, typical analysis approaches would involve spot detection and quantification, followed by a determination of which spots were differentially expressed. The methods described in this chapter model the entire image, so have potential to uncover results that could be missed by spot-based methods (Morris et al., 2011).

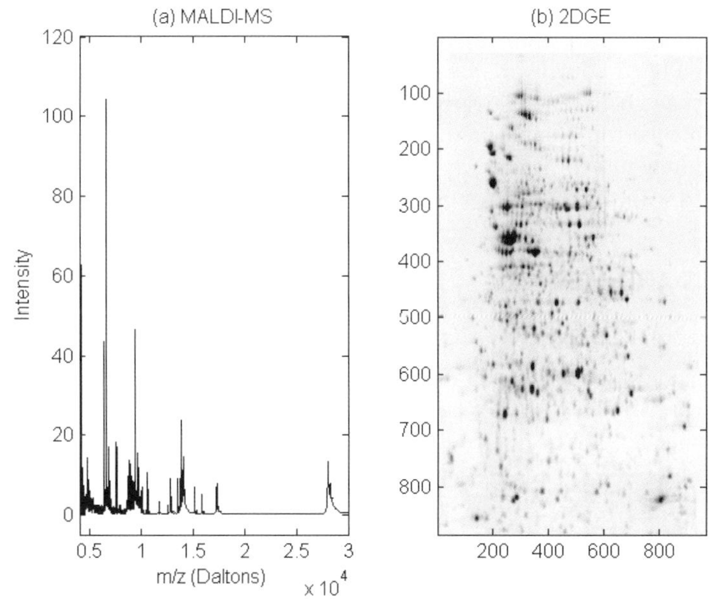

Figure 13.2 Sample Proteomics Data from (a) MALDI-TOF Mass Spectrometry and (b) 2D Gel Electrophoresis

13.2.3 Diffusion Tensor Imaging Tractography

Consider now data from a Diffusion Tensor Imaging (DTI) study comparing the cerebral white matter tracts of multiple sclerosis patients and healthy controls. White matter tracts are combinations of axons, the long projections of nerve cells that carry electrical signals. Axons are surrounded by a fatty substance called myelin, which allows the fast transmission of information. Multiple sclerosis is a demyelinating autoimmune disease that is associated with lesions in the white-matter tracts of affected individuals and results in severe disability. DTI tractography is a magnetic resonance imaging (MRI) technique that allows the study of white-matter tracts by measuring the diffusivity of water in the brain: in white-matter tracts, water diffuses anisotropically in the direction of the tract. Using measurements of diffusivity along several gradients, DTI can provide relatively detailed images of white-matter anatomy in the brain (Basser et al., 1994, 2000; LeBihan et al., 2001; Mori and Barker, 1999).

Here we focus on three measures of water diffusivity along the white matter tracts: fractional anisotropy (FA), parallel diffusivity (L0), and perpendicular diffusivity (LT). The left and right panels in Figure 13.3 display water diffusion measurements at multiple visits for two subjects, respectively. The top panels display FA along the right corticospinal tract (labeled FA RCST), the middle panels display L0 along the corpus callosum (labeled L0 CCA), and the bottom panels display LT along the left optic radiation tract (labeled LT LOPR). For each such measure, functions cluster naturally within subjects and observed data is of the type $\{W_{ij}(d), T_{ij}, \boldsymbol{X}_{ij}\}$, where $W_{ij}(d)$ is the function for subject $i = 1, \ldots, I$ at visit $j = 1, \ldots, J_i$, recorded at distance d along the tract; T_{ij} is the time of the visit, and $\boldsymbol{X}_{ij}$ are other subject/visit-specific covariates.

The basic scientific and statistical questions are similar to the ones in standard longitudinal settings with the difference that at each visit one records a function instead of a scalar. In particular, interest centers around quantifying baseline and longitudinal variability of functions, within- and between-visit variability, and measurement error. Characteristics of the baseline and longitudinal variability can then be associated with various outcomes.

13.3 MODELS AND METHODS

In this section we discuss the common data structures and models for multilevel functional data and cover major related methodological developments.

13.3.1 *Data Structures*

We start by noting that the data structure in what we termed 'next generation observational studies' is of the type $\{y_{ij}, W_{ij1}(\cdot), \ldots, W_{ijL}(\cdot), T_{ij}, \boldsymbol{X}_{ij}, \boldsymbol{Z}_{ij}\}$, for subject $i = 1, \ldots, I$ and visit $j = 1, \ldots, J_i$. Here, y_{ij} denotes the outcome, which can be a scalar or a multivariate low- or high-dimensional vector. The processes $W_{ijl}(\cdot)$ can be low- or high-dimensional with a domain that can be different across modalities, $l = 1, \ldots, L$. However, we assume that for each modality, l, the process $W_{ijl}(\cdot)$ is registered, that is, functional arguments are ordered and the same-index argument is the same for every subject i and visit j. The number of visits for each subject, J_i, can vary across subjects and the times of the visits, T_{ij}, are not required to be equally spaced. The $p \times 1$ dimensional vector $\boldsymbol{X}_{ij}$ contains the fixed effects covariates, while $q \times 1$ dimensional vector $\boldsymbol{Z}_{ij}$ contains the random effects covariates. Alternatively, using the FMM notation, the total set of n observed processes of different modalities can be labeled $W_{il}(\cdot)$, $i = 1, \ldots, n$, and the cluster structure can be indicated through the corresponding random effect covariates Z_i.

This structure of the data is characteristic of all data sets described in Section 13.2. From an inferential perspective, all statistical problems from multilevel data analysis have an immediate counterpart in this context. A major difference in the context of functional data is that most models and inferential methods developed for scalar variables are not naively scalable. Indeed, even loading the analytic data set into memory may not be feasible. A frequently used strategy is to arbitrarily reduce the very high-dimensional data to a set of simple summaries; this is called a 'feature extraction approach'. For example, one could reduce the information contained in millions of voxels in an image to their average over the entire image or several regions of interest (ROIs). These summaries can then be analyzed by classical statistical procedures. While useful in certain hypothesis-driven settings, this approach typically over-reaches in terms of the assumptions and may waste data, as important information that is not contained in the chosen summaries may be lost. We conclude that the most successful analytic methods for multilevel functional data should: 1) be scalable to very large data sets; 2) allow sequential access to the data set and avoid loading the entire data set in the computer memory; and 3) use the information as efficiently as possible.

13.3.2 *Fixed-Basis Modeling Approaches*

A large amount of work done in MFDA involves modeling with pre-determined fixed basis functions. An early paper on nested and crossed curves (Brumback and Rice, 1998) used smoothing splines, and the first paper on MFDA with multiple stochastic levels (Morris et al., 2003) used wavelets. Given a chosen set of L basis functions $\phi_l(s), l = 1, \ldots, L$, a function $W_i(s)$ can be written as a basis expansion $W_i(s) = \sum_{l=1}^{L} \xi_{il}\phi_l(s)$. Common choices of basis include Fourier bases, splines, and wavelets, and should be chosen to match the characteristics of the data. In the MFDA context, these basis expansions can be used in place of all functional quantities in the model, for example in the FMM framework for the observed functions $W_i(s)$, the fixed effect functions $B_a(s)$, and random effect functions $U_b^h(s)$.

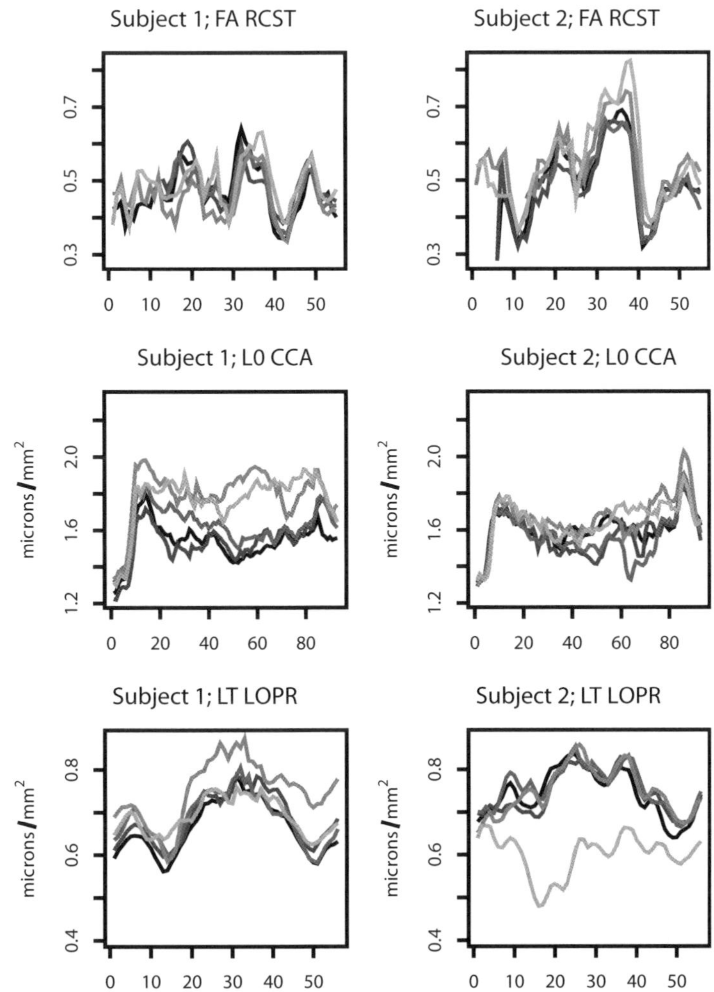

Figure 13.3 Water Diffusion Measurements for Two Subjects at Multiple Visits. Top panels: fractional anisotropy along the right corticospinal tract. Middle panels: parallel diffusivity along the corpus callosum. Bottom panels: perpendicular diffusivity along the left optic radiation tract

The benefits of using a basis function approach are both computational and methodological. A suitably chosen basis can lead to sparse representations of the functional objects, lend itself to more parsimonious modeling structures, and include natural mechanisms to induce effective smoothing of the functional parameters through appropriate penalties or prior distributions. A key issue with these approaches is computational feasibility, as methods that fit the entire data set and all parameters at once are unlikely to scale up. Methods using an isomorphic modeling approach, whereby the observed functional objects are projected to a complete basis space and then modeled there, have the potential to scale up to much larger data sets. In order to do this, one must carefully consider what structure is induced in the original data-space model by the assumptions made in the basis-space model.

Wavelet-based functional mixed models

Here, we provide a brief overview of wavelets. For a more complete introduction to wavelets, see Ogden (1997) and Vidakovic (1999). Wavelets are families of basis functions that decompose the information present in a functional object in both the frequency and time domain. Given a choice of father/mother wavelet pair $\phi(s)$ and $\psi(s)$, a complete orthonormal basis for L^2 can be constructed with basis functions $\phi_{k,l}(s) = 2^{-k/2}\phi(2^{-k}s - l)$ and $\psi_{k,l}(s) = 2^{-k/2}\psi(2^{-k}s - l)$ being dilations and translations of the father/mother pair. A K-level wavelet series for function $W(s)$ is given by $W(s) = \sum_l c_{K,l}\phi_{K,l}(s) + \sum_{k=1}^{K}\sum_{l=1}^{L_k} d_{k,l}\psi_{k,l}(s)$, where the basis coefficients $d_{k,l}$ describe behavior of the functions at frequency $2^{-k/2}$ and location $2^{-k/2}l$, and the $c_{K,l}$ describe smooth behavior at frequency $2^{-K/2}$. Given data sampled on a discrete equally spaced grid of size T, the discrete wavelet transform (DWT) (Mallat, 1989) can be used to obtain the vector of T wavelet coefficients $\mathbf{d} = (c_{K,1}, \ldots, c_{K,L_k}, d_{K,1}, \ldots, d_{K,L_K}, d_{K-1}, 1, \ldots, d_{1,L_1})$ in just $O(T)$ calculations. Natural generalizations exist for higher-dimensional data like images.

Wavelets are derived to have specific properties, some of which make them suitable for modeling functional data, especially data with many local features like change points and peaks. They have a whitening property, in that wavelet coefficients tend to be decorrelated relative to the original functional responses. They have a sparsity property, in that most signals can be reconstructed almost perfectly using just a small subset of wavelet coefficients, yet white noise is distributed equally among all coefficients. For these reasons, their use is ubiquitous in signal processing, denoising, and compression. Also, these properties make them useful for nonparametric regression of spiky functional data using 'wavelet regression' either by thresholding (Donoho and Johnstone, 1995) or shrinking (Abramovich et al., 1998) wavelet coefficients. These same properties make them useful as building blocks for models for multilevel functional data.

One early work in MFDA involved wavelet-space modeling of hierarchical functions in a colon carcinogenesis study (Morris et al., 2003), a Bayesian method which was subsequently generalized (Morris and Carroll, 2006) to the FMM framework described in (13.1). This method is intended for functional data observed on an equally-spaced fine grid, and works well with irregular functional data with local features, but can also be applied to smoother data. Given a sample of functions $W_i(s)$, $i = 1, \ldots, n$, fixed effect covariates X_{ia} and random effect covariates (which can also specify the cluster structure) Z_{ib}^h, recall the FMM (13.1) is given by $W_i(s) = \sum_{a=1}^{p} X_{ia}B_a(s) + \sum_{h=1}^{H}\sum_{b=1}^{q_h} Z_{ib}^h U_b^h(s) + E_i(s)$. In Morris and Carroll (2006) it is assumed that the vector-stacked random effect functions and error functions are $\mathbf{U}^h(s) = \{U_1^h(s), \ldots U_{q_h}^h(s)\}' \sim MGP(\Lambda_U^h, \Sigma_U^h)$ and $\mathbf{E}(s) = \{E_1(s), \ldots, E_n(s)\}' \sim MGP(\Lambda_W, \Sigma_W)$, where a vector of q stacked functions $\mathbf{y}(s) \sim MGP(A, B)$ implies each element of $\mathbf{y}(s)$ is a mean-zero Gaussian Process with separable covariance structure, with A a $q \times q$ between-function covariance matrix and B a within-function covariance surface, with $cov\{y_i(s_j), y_{i'}(s_j')\} = A_{ii'}B(s_j, s_j')$. The between-curve covariances Λ_U^h and Λ_E are specified based on the experimental design, for example identity matrix for cluster designs and AR(1) or some spatial structure, e.g. Baladandayuthapani et al. (2008), for longitudinal functional data. The structure of the within-curve covariances Σ_U^h and Σ_E are determined based on parsimonious assumptions made in the wavelet basis space model; this leads to a flexible set of covariance structures that can accommodate various levels of local non-stationarity.

The wavelet-based functional mixed model method (WFMM, Morris and Carroll, 2006) fits this model using a three-step approach, first projecting the observed functions into the wavelet space by applying the DWT to each observed function, yielding wavelet coefficients $W^*_{ikl}, i = 1, \ldots, n, k = 1, \ldots, K, l = 1, \ldots, L_k$; second, modeling a wavelet-space version of the FMM; and third, projecting the results back to the data space using the inverse DWT (IDWT) to yield estimates and inference on parameters in (13.1). The wavelet-space version of (13.1) is

$$W^*_{ikl} = \sum_{a=1}^{p} X_{ia} B^*_{akl} + \sum_{h=1}^{H} \sum_{b=1}^{q_h} Z^h_{ib} U^{*h}_{bkl} + E^*_{ikl},$$

where B^*_{akl}, U^{*h}_{bkl}, and E^*_{ikl} are the wavelet coefficients for the fixed effect, random effect, and residual error functions, respectively, and, for example for clustered data, we have $U^{*h}_{bkl} \sim N(0, \nu^h_{kl})$ and $E^*_{ikl} \sim N(0, \nu^E_{kl})$. The linearity of the wavelet transform makes it straightforward to investigate the covariance structure for Σ^h_U and Σ_E induced by the wavelet-space assumptions. Here, only the index i refers to subject (group membership), whereas the indexes l and k are used for the wavelet coefficients.

This model is fit using a Bayesian approach, with vague proper priors specified on variance components, and Bayesian variable selection (BVS) priors on the wavelet space fixed effects, $B^*_{akl} = \gamma_{akl} N(0, \tau_{ak}) + (1 - \gamma_{akl}) I_0$, $\gamma_{akl} \sim Bernoulli(\pi_{ak})$, with I_0 a point mass at 0 and τ_{ak} and π_{ak} smoothing parameters than can be estimated from the data or given a hyperprior distribution. This BVS prior distribution induces thresholding or nonlinear shrinkage that leads to adaptively smoothed estimates and inference for the fixed effect functions $B_a(s)$. Model fitting is done using a Markov Chain Monte Carlo (MCMC) sampler that has been coded to run in an automated fashion, and ultimately provides posterior samples of all model parameters in the FMM that can be used to perform any desired Bayesian inference. By fitting separately for each wavelet coefficient, this unified method is highly parallelizable, and being linear in T, the number of observations per curve scales up to extremely large data sets. The software to fit the MCMC and perform some common posterior inferences is freely available for download at `https://biostatistics.mdanderson.org/SoftwareDownload/SingleSoftware.aspx?Software_Id=70`. By default, the method models all wavelet coefficients, but joint wavelet compression approaches (Morris et al., 2011) that take advantage of sparsity in the wavelet space are also available and can be used to further speed calculations, if desired.

One key advantage of the Bayesian modeling approach is that, given the posterior samples, joint or pointwise inference on any function of FMM parameters is immediately available, and this inference automatically accounts for all sources of variability and estimation of nuisance parameters in the model. One key question in MFDA is to assess for which regions of s the predictor X_a is significantly associated with the processes $W(s)$, which involves pointwise functional inference on $B_a(s)$. Given minimum effect size of scientific interest δ, from posterior samples we can compute a posterior probability function $p^\delta_a(s) = Pr\{|B_a(s)| > \delta|W\}$, for which s with large $p^\delta_a(s)$ can be flagged as significance. Following false discovery rate (FDR) principles, we can find a cutpoint on the $p^\delta_a(s)$ corresponding to an expected Bayesian FDR of α.

The WFMM has been extended in various ways, making it able to handle partially observed functional data (Morris et al., 2006), to analyze images and other higher-dimensional functions (Morris et al., 2011), to perform functional discrimination (Zhu et al., 2011b), to apply this same approach using transformations and bases other than wavelets (Morris et al., 2011), and to model nonstationary time series with high-frequency content (Martinez et al., 2011). Additionally, robust methods for MFDA have been developed in this framework (Zhu et al., 2011a) by specifying particular hierarchical distributions in the basis space that lead to robust estimation and inference for fixed and random effect functions $B_a(s)$ and $U^h_b(s)$, respectively, whereby outlying curves or regions of curves are down-weighted. Further, the sparsity distributions implicit to this hierarchical model have excellent variable selection properties, which when applied to wavelet-space modeling lead to improved adaptive regularization of the fixed and random effect functions.

Spline-based approaches

In recent years, a number of methods have been proposed to perform MFDA with models using spline bases to represent the functions. All of these methods involve fitting the entire data set and model in a single model fit, so may not be scalable to the largest modern multilevel functional data sets, but may serve as effective analysis methods for small to moderate sized functional data sets.

One method involves Bayesian adaptive regression splines for hierarchical functional data (Bigelow and Dunson, 2007). The underlying model can be written as a functional mixed model, with the functions represented by piecewise linear models, with variable knot locations stochastically determined using a reversible jump MCMC. Bayesian inference is done by summarizing posterior means for the fixed effect functions along with posterior credible intervals. The method is applied to reproductive hormone profiles in the menstrual cycle, with the goal to characterize differences between conception and nonconception cycles. The data set involves a total of 262 complete menstrual cycles and 199 partial midcycle segments from 173 women. The multilevel aspect comes from the measurement of multiple cycles per woman, and multiple women in the study. The cycles for each woman are modeled as conditionally independent, so involves a clustering covariance structure.

Another method involves modeling of spatially correlated functional data within the hierarchical functional model (Baladandayuthapani et al., 2008). This paper models the p27 colon carcinogenesis data discussed in Section 13.1.2, which involved measurements of p27 along colonic crypts, with crypts nested within rats, rats nested within diet treatment groups, and the spatial orientation of the crypts within the rats measured and accounted for in the analysis. Using the hierarchical functional modeling framework, the authors represent the functions using fixed knot penalized B-splines (P-splines), with the correlation between crypts handled by an isotropic correlation structure. Bayesian posterior means and pointwise credible intervals are used for inference after fitting the model using an MCMC.

Another Bayesian method using fixed knot penalized B-splines was developed for the longitudinal data setting (Berhane and Molitor, 2008). The underlying model can be considered a functional mixed model, with B-spline representations with fixed knots used for the fixed effect functions and random effect functions, and iid errors. The method is applied to growth curve data, where the goal is to relate community-level pollution data to the growth curves of children, while adjusting for various time-varying and fixed covariates. The multilevel aspect of this model comes through the random effect functions for community and for child, with iid errors within the growth curve for each child. The method is fit in WinBugs using a two-stage approach, first estimating the mean and covariance for each community-level curve, followed by a second-stage analysis across communities on nonlinear functionals. A second method also uses B-splines in the functional mixed model framework (Wang et al., 2009). This paper presents two approaches: one a two-step approach that involves modeling each location within the functions as an independent mixed model, followed by a post-smoothing of the results, and the second a single model fit using B-spline representations for the functional quantities in the model, with the resulting model fit using standard mixed models software. Functional ANOVA is done to test whether the fixed effect functions are zero or not. This method is applied to EEG data.

13.3.3 Functional Principal Components Based Approaches

Functional principal components analysis (FPCA) (Ramsay and Silverman, 2005, 2006) is a set of powerful data compression and statistical analysis techniques for single-level functional data analysis. The approach can be viewed as empirically estimating a set of orthonormal basis functions that most efficiently represent the covariance structure of the functional data set. FPCA was developed for and applied to the case when one function or image was recorded for one subject. This includes the case of longitudinal data analysis where *one individual trajectory of one subject* is sampled over time. Some confusion may occur as multilevel modeling is used for this type of classical longitudinal data even though we will call this single-level functional data. Multilevel functional data is defined as data where *multiple individual functions or images are recorded for the same subject.*

We provide a short description of FPCA for single-level functional data. Assume that data are of the type $W_i(s) = \mu(t) + X_i(s) + \varepsilon_i(s)$, where $W_i(t)$ is the observed function, $\mu(t)$ is the population average, $X_i(s)$ is the true process, and $\varepsilon_i(s)$ are independent random variables with mean zero and variance σ_ε^2. The covariance operators of the observed process $K_W(s,r) = \text{Cov}\{W_i(s), W_i(r)\}$ and of the true underlying process $K_X(s,r) = \text{Cov}\{X_i(s), X_i(r)\}$ are linked via the equation $K_W(s,r) = K_X(s,r) + \sigma_\varepsilon^2 \delta_{sr}$, where δ_{sr} is Kronecker's δ-function. These equations suggest a natural solution for estimating the eigenvalues, eigenfunctions, and the nugget variance, σ_ε^2. The first step of the procedure is to estimate the mean function $\mu(t)$ using penalized spline smoothing (Marx and Eilers, 1999; Ruppert et al., 2003; O'Sullivan, 1986; Wood, 2006) under the working independence assumption. The second step is to obtain the method of moment estimates of $K_W(s,r)$, denoted by $\hat{K}_W(s,r)$. The third step is to estimate $\hat{K}_X(s,r)$ by smoothing $\hat{K}_W(s,r)$ for $s \neq r$, as suggested by (Staniswalis and Lee, 1998). Instead of the kernel-based smoothing in Staniswalis and Lee (1998) we use penalized thin-plate spline smoothing (Di et al., 2009; Greven et al., 2010). The fourth step is to predict the diagonal elements, $\hat{K}_X(s,s)$, and estimate the error variance σ_ε^2 as $\hat{\sigma_\varepsilon}^2 = \int\{\hat{K}_W(s,s) - \hat{K}_X(s,s)\}ds$. The fifth step is to estimate the eigenvalues and eigenfunctions of $\hat{K}_X(s,r)$.

Mercer's theorem (see Indritz (1963), Chapter 4) provides the convenient spectral decomposition $K_X(s,r) = \sum_{k=1}^{\infty} \lambda_k \psi_k(s) \psi_k(r)$, where $\lambda_1 \geq \lambda_2 \geq \ldots$ are the ordered eigenvalues and $\psi_k(\cdot)$ are the associated orthonormal eigenfunctions of $K_X(\cdot,\cdot)$ in the L^2 norm. The Karhunen–Loève (KL) decomposition (Karhunen, 1947; Loève, 1945) of the subject level functions is $X_i(s) = \sum_{k=1}^{\infty} \xi_{ik}\psi_k(s)$, where $\xi_{ik} = \int_0^1 X_i(s)\psi_k(s)ds$ are the principal component scores, with $E(\xi_{ik}) = 0$, $\text{Var}(\xi_{ik}) = \lambda_k$ and $\text{Cov}(\xi_{ik}, \xi_{ik'}) = 0$ for every i and $k \neq k'$. Once the eigenfunctions $\psi_k(\cdot)$ and a truncation lag K are fixed, the model for observed functional data can be written as a linear mixed model. Indeed, by construction, ξ_{ik} are mutually uncorrelated with mean 0 and variance λ_k. By assuming a normal shrinkage distribution for scores and errors, the model can be rewritten as a mixed effects model:

$$\begin{cases} W_i(s) & = & \sum_{k=1}^{K} \xi_{ik}\psi_k(s) + \varepsilon_i(s); \\ \xi_{ik} & \overset{iid}{\sim} & (0, \lambda_k); \;\; \varepsilon_i(s) \overset{iid}{\sim} (0, \sigma_\varepsilon^2), \end{cases} \tag{13.2}$$

where $\overset{iid}{\sim} (0, \sigma^2)$ denotes independent identically distributed random variables with mean zero and variance σ^2.

The extension of FPCA to the case of multilevel functional data uses some of these ideas, but requires additional modeling and assumptions. The two current major extensions are: 1) Longitudinal Functional Principal Component Analysis (LFPCA) (Di et al., 2009; Greven et al., 2010), which uses an explicit model for the functional data and principal component analysis to expand the underlying processes; and 2) Functional Principal Component Mixed Effect (FPCME) models

(Aston et al., 2010), which uses a principal component decomposition of the complete data matrix followed by modeling of the principal component scores.

Longitudinal Functional Principal Component Analysis

LFPCA refers to a set of models designed for a wide range of functional data problems. Here, we focus on the particular case when data is of the type $\{W_{ij}(s), T_{ij}, \boldsymbol{X}_{ij}, \boldsymbol{Z}_{ij}\}$, that is when one type of functional measurement, $W_{ij}(s)$, is recorded at multiple visits and there is no outcome, y_{ij}. For example, in the DTI tractography example introduced in Section 13.2.3, $W_{ij}(s)$ could denote the fractional anisotropy along the right corticospinal tract of subject i (top panels in Figure 13.3) at visit j recorded at time T_{ij}.

The general form of the longitudinal functional model is

$$W_{ij}(s) = \eta(s, \boldsymbol{X}_{ij}) + \boldsymbol{Z}'_{ij}\boldsymbol{U}_i^1(s) + U_{ij}^2(s) + \varepsilon_{ij}(s), \quad (13.3)$$

where we assume that:

A.1 $\boldsymbol{U}_i^1(s)$, $U_{ij}^2(s)$, and $\varepsilon_{ij}(s)$ are zero-mean, square-integrable, mutually independent random processes on $[0, 1]$;
A.2 $\boldsymbol{U}_i^1(s)$ is a vector-valued random process with auto-covariance functions for the q components $K_{11}(s, s'), \dots, K_{qq}(s, s')$, and cross-covariance functions $K_{12}(s, s')$ $, \dots, K_{1,q}(s, s'), \dots, K_{q-1,q}(s, s')$;
A.3 $U_{ij}(d)$ is a random process with covariance function $K_U(s, s')$; and
A.4 $\varepsilon_{ij}(d)$ is white noise measurement error with variance σ_ε^2.

The rich class of functional models introduced by (13.3) contains many important particular cases. For example, by setting $p = 1$, $\boldsymbol{X}_{ij} = j, \boldsymbol{Z}_{ij} = 1$, and $\eta(d, X_{ij}) = \mu(d) + \eta_j(d)$, one obtains the two-way functional ANOVA model (Di et al., 2009). By setting $p = 2$, $\boldsymbol{X}_{ij} = T_{ij}$ and $\boldsymbol{Z}_{ij} = (1, T_{ij})'$, the model becomes the functional generalization of the standard random intercept, random slope longitudinal model

$$W_{ij}(s) = \eta(s, T_{ij}) + U_{i,0}^1(s) + U_{i,1}^1(s)T_{ij} + U_{ij}^2(s) + \varepsilon_{ij}(s). \quad (13.4)$$

We focus our discussion on model (13.4) instead of (13.3), because it is easier to explain and because fitting the more complex model requires only minor modifications. The vectors $\boldsymbol{X}_{ij}$ and $\boldsymbol{Z}_{ij}$ contain known covariates for subject i at time T_{ij}. They could, for example, include an additional quadratic term T_{ij}^2, a treatment effect, or other covariates measured for each subject i and/or visit j.

The mean function $\eta(s, \boldsymbol{X}_{ij})$ is the fixed main effect surface, which can depend parametrically, semi-parametrically or nonparametrically on the covariates. The simplest parametric form is a linear mean $\eta(s, \boldsymbol{X}_{ij}) = \alpha_0 + \alpha_1 s + \boldsymbol{X}_{ij}\boldsymbol{\beta}$, while the most complex nonparametric form is a $(p+1)$-dimensional smooth function $\eta(s, \boldsymbol{X}_{ij})$. Intermediate semi-parametric models such as $\eta(d, \boldsymbol{X}_{ij}) = \eta_1(s, X_{ij,1}) + \cdots + \eta_m(s, X_{ij,p})$, where $\boldsymbol{X}'_{ij} = (X_{ij,1}, \dots, X_{ij,p})$, or $\eta(s, \boldsymbol{X}_{ij}) = \eta_0(s) + \boldsymbol{X}'_{ij}\boldsymbol{\beta}$, could also be useful in particular applications. Of particular interest in applications is the case when $p = 1, \boldsymbol{X}_{ij} = X_i$ is a treatment indicator for subject i, and $\eta(s, \boldsymbol{X}_{ij}) = \eta(s, X_i) = \eta_0(s) + X_i\eta_1(s)$, where $\eta_1(s)$ is the functional population effect of treatment. If $\eta_0(s) = \boldsymbol{B}_0(s)\boldsymbol{\beta}_0$ and $\eta_1(s) = \boldsymbol{B}_1(s)\boldsymbol{\beta}_1$ then testing for treatment effect is equivalent to testing the null hypothesis $H_0{:}\boldsymbol{\beta}_1 = \boldsymbol{0}$ versus $H_A{:}\boldsymbol{\beta}_1 \neq \boldsymbol{0}$. This is the familiar case of parametric testing with correlated data. We describe here two strategies for this type of problem. First, one can obtain estimates of $\boldsymbol{\beta}_1$ under the independence assumption and then obtain the joint confidence intervals based on a nonparametric bootstrap of subjects. This is a procedure first discussed in Crainiceanu et al. (2012) and Staicu et al. (2010b). Second, one can estimate $\eta(s, \boldsymbol{X}_{ij})$ under the independence assumption, obtain the covariance of the residuals $W_{ij}(s) - \eta(s, \boldsymbol{X}_{ij})$ under model (13.4) or (13.3), and re-estimate $\eta(s, \boldsymbol{Z}_{ij})$ using the estimated covariance of the residuals. This method was used for a related functional correlated data model in Staicu et al. (2010a).

The functional random intercept, $U^1_{i,0}(s)$, models the cross-sectional variability, whereas the random slope, $U^1_{i,1}(s)$, models the longitudinal transformations to the random intercept as time progresses. One could think about $U^1_{i,1}(s)$ as the shape of the functional transformation applied by the passing of time, T_{ij}, and of $U^1_{i,1}(s)T_{ij}$ as the irreversible, subject-specific transformation of the baseline observation, $U^1_{i,0}(s)$. The two processes $U^1_{i,0}(s)$ and $U^1_{i,1}(s)$ are allowed to be correlated and the subject-specific functional process $U^1_{i,0}(s) + U^1_{i,1}(s)T_{ij}$ can be interpreted as the true unobserved functional process at time T_{ij}. The subject-visit specific process $U^2_{ij}(s)$ is assumed to be exchangeable (across time) and can be interpreted as the time-reversible measurement error. The $U^2_{ij}(s)$ plays a dual role. First, it provides simple model-based registration of functional data by allowing an additive functional noise in the model for longitudinally observed functional data. This procedure is fundamentally different from the functional registration procedures developed for functional data observed at one visit (Kneip et al., 2000; Kneip and Ramsay, 2008). Second, the main directions of variation in the $U^2_{ij}(s)$ process can reveal patterns in the measurement error mechanism. These patterns could provide insights into how to correct or calibrate the functional measurements either using statistical modeling or by changing the measurement tools.

Once an estimator of $\eta(s, T_{ij})$ exists, model (13.4) can be fit using the Karhunen–Loève (KL) decomposition of the processes $\boldsymbol{U}^1_i(s) = \{U^1_{i,0}(s), U^1_{i,1}(s)\}$ and $U_{ij}(s)$. We denote by $K(s, s')$ and $K_U(s, s')$ the covariance operators of $\boldsymbol{U}^1_i(s)$ and $U^2_{ij}(s)$, respectively. With this notation,

$$K(s, s') = \begin{pmatrix} K_0(s, s') & K_{01}(s, s') \\ K_{01}(s', s) & K_1(s, s') \end{pmatrix} = \sum_{k=1}^{\infty} \lambda_k \phi^X_k(s) \phi^X_k(s')',$$

where $\phi^U_k(s) = \{\phi^0_k(s), \phi^1_k(s)\}'$ are the ordered eigenfunctions of $K(s, s')$ corresponding to the eigenvalues $\lambda_1 \geq \lambda_2 \geq \cdots \geq 0$. Similarly, let $K_U(s, s') = \sum_{k=1}^{\infty} \nu_k \phi^U_k(s) \phi^U_k(s')$, where $\phi^U_k(s)$ are the ordered eigenfunctions of $K_U(s, s')$ corresponding to the eigenvalues $\nu_1 \geq \nu_2 \geq \cdots \geq 0$. The eigenfunctions $\{\phi^U_k, k \geq 1\}$ form an orthonormal basis of $L^2[0, 1]$ with respect to the usual $L^2[0, 1]$ scalar product. The eigenfunctions $\{\phi^X_k, k \geq 1\}$ form an orthonormal basis of $L^2[0, 1] \times L^2[0, 1]$ with respect to the additive scalar product

$$\langle (f_0, f_1), (g_0, g_1) \rangle = \int_0^1 (f_0(s) g_0(s) + f_1(s) g_1(s)) ds.$$

The function pairs (ϕ^0_k, ϕ^1_j), (ϕ^0_k, ϕ^U_j), or (ϕ^1_k, ϕ^U_j) are not required to be orthogonal in $L^2[0, 1]$. The Karhunen–Loève expansions of the random processes are $\boldsymbol{U}^1_i(s) = \sum_{k=1}^{\infty} \xi_{ik} \phi^X_k(s)$ and $U^2_{ij}(s) = \sum_{k=1}^{\infty} \zeta_{ijk} \phi^U_k(s)$, where the principal components scores ξ_{ik} and ζ_{ijk} are uncorrelated random variables with mean zero and variances λ_k and ν_k, respectively. Assumption A.1. is ensured by assuming that $\{\xi_{ik}, i = 1, \ldots, I, k \geq 1\}$ and $\{\zeta_{ijk}, j = 1, \ldots, J_i, i = 1, \ldots, I, k \geq 1\}$ are mutually uncorrelated. Because working with infinite expansions is impractical, we consider the finite-dimensional approximations of the U^1 and U^2 processes $\boldsymbol{U}^1_i(d) = \sum_{k=1}^{N_X} \xi_{ik} \phi^X_k(d)$ and $U_{ij}(d) = \sum_{k=1}^{N_U} \zeta_{ijk} \phi^U_k(d)$, where N_X and

N_U are estimated. Conditional on N_X and N_U, the finite approximation to model (13.4) is:

$$\begin{cases} W_{ij}(s) = \eta(s, T_{ij}) \\ \quad + \sum_{k=1}^{N_X} \xi_{ik} \boldsymbol{R}'_{ij} \phi_k^X(s) \\ \quad + \sum_{k=1}^{N_U} \zeta_{ijl} \phi_k^U(s) + \varepsilon_{ij}(s), \\ \xi_{ik} \overset{iid}{\sim} (0, \lambda_k), \zeta_{ijl} \overset{iid}{\sim} (0, \nu_l), \\ \varepsilon_{ij}(s) \overset{iid}{\sim} (0, \sigma^2), \end{cases} \tag{13.5}$$

$\boldsymbol{V}_{ij} = (1, T_{ij})'$, which is a linear mixed model (Di et al., 2009; Crainiceanu et al., 2009b; Greven et al., 2008). We are not assuming normality of the processes in (13.4) or of the scores in (13.5). Extending LFPCA to the more general model (13.3) is similar and estimation is equally fast.

Thus, estimation of model (13.5) becomes relatively easy once consistent estimators of $K(\cdot, \cdot)$ and $K_U(\cdot, \cdot)$ are available. The first step is to obtain centered data $W_{ij}(s) = W_{ij}(s) - \widehat{\eta}(s, T_{ij})$, where $\widehat{\eta}(s, T_{ij})$ is an estimator of the mean function. We continue to denote the zero-mean data as $W_{ij}(s)$ for notational brevity. Under the assumptions of model 13.4,

$$\begin{aligned} \mathrm{E}\{W_{ij}(s) W_{ik}(s')\} &= \mathrm{Cov}\{W_{ij}(s), W_{ik}(s')\} \\ &= K_0(s, s') + T_{ik} K_{01}(s, s') \\ &+ T_{ij} K_{01}(s', s) + T_{ij} T_{ik} K_1(s, s') \\ &+ [K_U(s, s') + \sigma^2 \delta_{ss'}] \delta_{jk}, \end{aligned} \tag{13.6}$$

for all s, s', i, j, and k, where δ_{jk} is Kronecker's delta. Thus, the entries of the covariance operators can be obtained by linear regression of the 'outcome' $W_{ij}(d) W_{ik}(d')$ on the 'covariates' $(1, T_{ik}, T_{ij}, T_{ik} T_{ij}, \delta_{jk})$, where the 'parameters' are $\{\sigma^2,\ K_0(s, s'),\ K_{01}(s, s'),\ K_{01}(s', s),\ K_1(s, s'),\ K_U(s, s')\}$. For an efficient implementation of this method see Greven et al. (2010); the R function implementing LFPCA can be found at `http://www.stat.uni-muenchen.de/~greven/software/LFPCA.r`.

Functional Principal Component Mixed Effect models

An alternative and complementary modeling strategy based on principal components was proposed in Aston et al. (2010). While the methods were developed for general functional mixed model (13.1), not the model described in Section 13.3.1, ideas can be adapted to this context. Using notations from section (13.3.3), a particular case of the model in Aston et al. (2010) can be rewritten as

$$W_{ij}(s) = \eta(s, \boldsymbol{X}_{ij}) + U_{ij}(s) + \varepsilon_{ij}(s), \tag{13.7}$$

where $U_{ij}(s)$ is the subject-/visit-specific from the population mean, $\eta(s, \boldsymbol{X}_{ij})$. While the model in Aston et al. (2010) contained neither $\eta(s, \boldsymbol{X}_{ij})$ nor the white noise process $\varepsilon_{ij}(s)$, adding them to the model does not increase the complexity of the approach. Indeed, one can estimate $\eta(s, \boldsymbol{X}_{ij})$ using methods described, for example, in Section 13.3.3, while 'smoothing on the diagonal' of the covariance operator (Staniswalis and Lee, 1998) can be used to estimate the covariance operator of $U_{ij}^1(s)$, $K(s, s')$. The estimation procedure is then rather simple: 1) estimate $K(s, s')$ under the independence assumption of $W_{ij}(s)$; 2) obtain the eigenfunctions, $\phi_l(s)$, $l \geq 1$, of the covariance estimator $\widehat{K}(s, s')$; and 3) estimate the scores ξ_{ijl} from the functional expansion $X_{ij}(s) = \sum_{l \geq 1} \xi_{ijl} \phi_k(s)$. All these steps are easy to implement. For example, estimators of ξ_{ijl} can be obtained by numerical integration $\widehat{\xi}_{ijl} = \sum_s U_{ij}^1(s) \phi_l(s)$.

Once the scores ξ_{ijl} are estimated, standard mixed effects models can be used to model them conditional on the covariates and random effects for each eigenfunction. For example, in the case where clustering is done at the subject level the models for scores can be written as

$$\xi_{ijl} = \boldsymbol{Z}'_{ij} \boldsymbol{\beta}_l + b_{il} + e_{ijl},$$

where $b_{il} \sim N(0, \sigma_{b,l}^2)$ and $e_{ijl} \sim N(0, \sigma_{e,l}^2)$ are mutually independent. While simple, this example contains the general idea for fitting even the most complex FMMs. Indeed, the motivating application in Aston et al. (2010) is a linguistic study where functional outcomes are clustered by subject and by word. Using slightly more involved notation, one can denote by $W_{ijk}(s)$ the jth measured functional outcome for subject i and word k. Then the models for the scores can easily be augmented to include this additional clustering as

$$\xi_{ijkl} = \boldsymbol{Z}'_{ijk}\boldsymbol{\beta}_l + b_{il} + u_{kl} + e_{ijkl},$$

where $b_{il} \sim N(0, \sigma^2_{b,l}), u_{kl} \sim N(0, \sigma^2_{u,l})$, and $e_{ijl} \sim N(0, \sigma^2_{e,l})$ are mutually independent.

We conclude that the Functional Principal Component Mixed Effect (FPCME) approach introduced in Aston et al. (2010) and the MFPCA and LFPCA approaches introduced in (Di et al., 2009) and Greven et al. (2010) are complementary and provide different perspectives on the data. FPCME is intuitive, easy to explain, and computationally fast, as it requires only to obtain the eigenfunctions of the $K(\cdot, \cdot)$ operator followed by a sequence of standard mixed effects model fits. However, FPCME also has several weaknesses. First, there are no theoretical results about the consistency of the functional principal components. Second, FPCME does not provide inference about the principal components of the within- and between-cluster variability. Third, in most cases FPCME will not provide a parsimonious decomposition of the data because the within- and between-cluster principal components are mixed into population-level principal components. For example, when there are N^1_U within- and N^2_U between-cluster components, an LFPCA approach would fit $N^1_U + N^2_U$ components, whereas a FPCME may require as many as $N^1_U N^2_U$ mixed effects model fits. This is not a serious problem when N^1_U and N^2_U are small, but it may become one in more complex data. LFPCA addresses some of these problems at the expense of currently being less flexible and computationally more expensive.

13.4 OTHER METHODS AND EXTENSIONS

13.4.1 *Multilevel Functional Regression*

There exists a large body of work on functional regression for the case when functional exposures are not correlated (Cardot et al., 1999, 2003; Cardot and Sarda, 2005; Cardot et al., 2007; Crainiceanu and Goldsmith, 2009; Crambes et al., 2009; Goldsmith et al., 2011; James, 2002; James et al., 2009; Marx and Eilers, 1999; Müller and Stadtmüller, 2005; Ramsay and Silverman, 2006; Reiss and Ogden, 2007). Here, we focus on the case when functional exposures are clustered. An example of such data is when the outcome and the functional exposure are observed at multiple visits. In the DTI tractography example introduced in Section 13.2.3, longitudinal cognitive test outcomes were collected in addition to the various tract measurements. The study contains 100 subjects with a median of 3 visits roughly 6 months apart and was analyzed in Goldsmith et al. (2012). We consider the setting where data has a structure of the type $[y_{ij}, W_{ij1}(s), \ldots, W_{ijL}(s), \boldsymbol{X}_{ij}, \boldsymbol{Z}_{ij}]$, where y_{ij} is a scalar outcome, $W_{ijl}(s) \in \mathcal{L}^2[0, 1]$, $1 \le l \le L$, are functional covariates, $\boldsymbol{X}_{ij}$ is a design row vector for fixed effects, and $\boldsymbol{Z}_{ij}$ is a design row vector for random effects. Here, $1 \le i \le I$ are subjects observed at visit $1 \le j \le J_i$. We consider the longitudinal functional regression model

$$\begin{aligned} y_{ij} &\sim \text{EF}(\mu_{ij}, \eta) \\ g(\mu_{ij}) &= \boldsymbol{X}_{ij}\beta + \boldsymbol{Z}_{ij}b_i \\ &\quad + \textstyle\sum_{l=1}^{L} \int_0^1 W_{ijl}(s)\gamma_l(s)ds, \end{aligned}$$

where 'EF(μ_{ij}, η)' denotes an exponential family distribution with mean μ_{ij} and dispersion parameter η. Here, $\boldsymbol{X}_{ij}\beta$ is the standard fixed effects component, $\boldsymbol{Z}_{ij}b_i$ is the standard random effects component, $b_i \sim$ Normal $(0, G)$ are subject-specific random effects, and the integrals $\int_0^1 W_{ijl}(s)\gamma_l(s)ds$ are the subject-visit-specific functional effects. The subject-specific random effects are introduced to account for the correlation between repeated observations at the subject level.

Our proposed method to model fit extends the single-level penalized functional regression approach in Crainiceanu and Goldsmith (2009) and Goldsmith et al. (2011). Specifically, we use functional principal components decomposition to parsimoniously express the $W_{ijl}(s)$ and a large spline basis to estimate the coefficient function $\gamma_l(s)$. Smoothness of $\gamma_l(s)$ is induced via penalization in a mixed effects model. Some strengths of this approach are that: 1) it extends functional regression to model the dynamic association between outcomes and functional predictors in longitudinal studies; 2) it casts a novel functional model in terms of well-understood mixed models; 3) it is applicable in any situation in which the $W_{ijl}(s)$ are observed or can be estimated, including when they are observed sparsely or with error; 4) it can be fit using standard, well-developed statistical software; and 5) it provides confidence or credible intervals for all the parameters of the model, including the functional ones. The `pfr()` and `lpfr()` functions in the R (R Development Core Team, 2009) package `refund` (Crainiceanu et al., 2010) provide implementations of penalized and longitudinal penalized functional regression, respectively.

The choice of penalized spline basis for modeling the functional regression coefficients, $\gamma_l(s)$, is convenient in this and other applications. However, combinations of wavelets, Fourier, splines, and principal components can be used, depending on the particular application. For examples of functional regression using wavelets, see Antoniadis (2007); Brown et al. (2001); Malloy et al. (2010); Wang et al. (2007). The coefficient function, $\gamma_l(s)$, and the functional predictors, $W_{ijl}(s)$, can be expanded in different bases according to the expected type of variation. In many applications, it may even be a good idea not to use any type of expansion for $W_{ijl}(s)$ and use the expansion and smoothing of $\gamma_l(s)$ to control model overfit. This approach can be used with discrete or continuous functional predictors.

Another approach to multilevel functional regression is to perform functional discriminant analysis using functional mixed models (Zhu et al., 2011b). Rather than model the class or predictive variable y_{ij} as the response and the multilevel functional processes $W_{ijl}(s)$ as predictors, one could instead model the functional processes as the responses in FMMs, with the predictive variable as the predictor, and then use posterior predictive probabilities to predict y given the functional responses $W(\cdot)$. For example, after using a Bayesian FMM method to fit the training data with $W(\cdot)$ as the response and y as a predictor, one can compute posterior predictive probabilities to predict unknown y given observed $W(\cdot)$ for a set of validation data. This approach effectively performs a type of functional discriminant analysis. Advantages include the ability to handle multilevel functional data and model covariates of both the class and observed functional data. After fitting the FMM, the computational burden of performing the classification or prediction is quite minor, so that this single analysis workflow allows one to perform both group comparison (e.g. inference on fixed effect functions) and functional regression/classification with the same model fit. This approach has been used with the Gaussian and robust methods for fitting the FMM (Zhu et al., 2011b), with the robust method showing improved classification performance that is due to its downweighting of outliers. While applied using wavelet-based modeling, the approach can be used with FMM using any modeling approach, including other bases like Fourier, spline, or principal components bases.

13.4.2 Correlated Functional Data

The label 'correlated functional data' contains a quasi-tautology, as observations in the same function are inherently correlated. Here, however, we treat the function as the observational unit and 'correlation' refers to correlations between, not within, observational units.

There are a number of papers whose primary goal is estimation of correlation between multiple functions measured for the same individuals. Morris et al. (2001) estimated the correlation function between two functional responses occurring within a nested hierarchical functional design from a colon carcinogenesis experiment, both using parametric and nonparametric representations for the functions. The nonparametric method involved first kernel smoothing each observed function, and then fitting mixed models at each curve location to estimate the correlation function, with parametric bootstrap used for inference. A Bayesian version of this method (Morris et al., 2002) was published that also accounted for measurement error in the functional index measure. Another paper also dealt with correlation between functions, assuming that the correlation was constant but allowing lag effects in the correlation (Dubin and Müller, 2005).

Another series of papers deals with modeling functional data observed on a spatial grid, with part of the emphasis in estimating the spatial correlation of the functions, which is assumed to be stationary across the function. All have been applied to the p27 colon carcinogenesis data described in Section 13.1.2. Modeling approaches have incorporated kernel smoothers (Li et al., 2007), fixed knot splines (Baladandayuthapani et al., 2008), and functional principal components (Staicu et al., 2010a; Zhou et al., 2010), with the correlation function modeled parametrically (Baladandayuthapani et al., 2008; Staicu et al., 2010a; Zhou et al., 2010) or nonparametrically (Li et al., 2007).

Further, this chapter contains many examples of correlated functional data and associated inferential tools. However, we anticipate that the types of correlation and observational units will continue to increase and diversify. The size of these data sets will continue to explode and the need for computational tools and simple models that capture important modes of variation of the data will only increase. This chapter provides a foretaste of what may be possible for statistical research in this area.

13.4.3 Intensive Longitudinal Data

Given the increased number of applications in various areas of application, it should not be surprising that labels also proliferate. Intensive longitudinal data (ILD) (Walls and Shaeffer, 2006) is such an example. ILD refers to the analysis of populations of time series, or functions of time. One example of ILD is the sleep EEG data recorded on thousands of subjects in the Sleep Heart Health Study (SHHS) (Crainiceanu et al., 2009a; Quan et al., 1997). In the SHHS, EEG was recorded at 125 Hz (125 observations per second), which results in 3.6 million observations for 8 hours of sleep for each subject and visit. These data were collected on more than 6000 subjects. Other examples are populations of financial time-series data, Electrocardiograms (ECG), and Blood-oxygen-level dependence (BOLD) in functional MRI (Ogawa et al., 1992). Thus, ILD is a particular case of high-dimensional data.

13.4.4 More Complex Multilevel Models

So far, we have focused on the most general form of FMEs as well as on the particular cases of two-level and longitudinal functional data. However, we anticipate that data sets will have increasingly complex structures including: 1) more than two levels of stochastic functional variability; 2) multilevel cross-classified structures (Goldstein, 1994); and 3) multiple types of functional and imaging data. Several of the ideas presented here can easily be extended to these more complex cases. For example, going from two- to three-level functional principal component analysis can be done by extending the MFPCA ideas in

Di et al. (2009). However, much remains to be done, as the new data types are just now emerging.

13.4.5 High-Dimensional Data

The definition of 'high-dimensional' data is a moving target and has been used for cases where the dimension of the observational unit varies from 10 to 1 billion. If p is the dimension of the data vector we define a data set to be high-dimensional if either: 1) p-dimensional covariance matrices cannot be stored and handled on most computers; or 2) existing implemented analytic methods cannot be used to analyze the data. Thus, a lower limit for defining high-dimensional data would be $p = 10{,}000$, but this will likely increase with increased computational capabilities. Extending even the simplest statistical models from $p = 1$ to $p > 10,000$ typically requires non-trivial methodological and computational developments.

A primary bottleneck in performing functional data analysis on modern data sets is computational viability. For example, consider the RAVENS images. Before preprocessing, there are around 350 subjects with acceptable data at 2 visits where the image array dimensions are $256 \times 256 \times 198$; in other words, nearly ten billion numbers. One possible solution is to project onto Kronecker products of bases, such as in 3-dimensional wavelet, spline, or Fourier transforms. However, to achieve reasonable data reduction, such bases must capture relevant variation in few terms and can make strong assumptions about separability of features of interest with regard to the imaging coordinate system. Nonetheless, these approaches are key tools for data reduction in this area.

In contrast, methods based on singular value decompositions, such as MFPCA, LFPCA, and FPCME are challenging to implement on such data. For example, the covariance matrices of interest in this case are hundreds by tens of millions. Efficiently organizing this data, such as ditching the array structure and only focusing on voxels that overlap with the template brain, reduces the spatial dimension to only three million. However, even loading the seven hundred by three million data matrix into memory is an issue, a problem that will persist under the logical assumption that the density of technological measurement will keep pace with increases in computer memory. Current methods (Zipunnikov et al., 2011) have started to circumvent these problems by focusing on the intrinsic instead of the actual dimension of the data.

Another possible solution is to subsample the data. One could sample points on a grid, randomly sample points, or actually downsample the subject- and visit-specific images to a lower resolution. Note that the latter solution is effectively accomplished by thresholding 3D wavelet transforms. In this case, one compromises fidelity to the data for the sake of computational feasibility.

We note that, in the case of MFPCA and LFPCA, no-compromise solutions exist that yield the exact functional analysis while requiring only blocks of the data matrix at a time. Specifically, it is well known that performing the SVD on a large data matrix can be done by block calculations in the event that one of the matrix dimensions is small (Zipunnikov et al., 2011). Moreover, Zipunnikov et al. (2011) showed that the decomposition of the full data matrix, obtained by stacking all visits, contains all of the relevant information to perform MFPCA. That is, one can obtain eigenvalue decompositions of the within- and between-visit covariances in a computationally feasible manner (linear in the number of non-background voxels). Of course, this solution applies to single-level functional principal components analysis as a special case. Extensions of this method for high-dimensional LFPCA are also available (Zipunnikov et al., 2012).

Expanding and scaling up statistical models to high-dimensional data, failing, and deciding what works and what does not seems to be one of the red hot areas of research in statistics.

ACKNOWLEDGEMENTS

Crainiceanu's and Caffo's research was supported by grant R01NS060910 from the National Institute Of Neurological Disorders And Stroke and by the grant R01 EB012547 from the National Institute of Biomedical Imaging and Bioengineering. Morris's research was supported by grant R01 CA107304 from the National Cancer Institute. The content is solely the responsibility of the authors and does not necessarily represent the official views of the respective National Institutes or of the National Institutes of Health.

REFERENCES

Abramovich, F., Sapatinas, T., and Silverman, B. (1998) 'Wavelet Thresholding via a Bayesian Approach', *Journal of the Royal Statistical Society, Series B*, 60: 725–49.

Antoniadis, A. (2007) 'Wavelet Methods in Statistics: Some Recent Developments and Their Applications', *Statistical Surveys*, 1: 16–55.

Ashburner, J. and Friston, K.J. (2000) 'Voxel-based Morphometry – The Methods', *Neuroimage*, 11: 805–21.

Aston, J.A.D., Chiou, J.-M., and Evans, J.P. (2010) 'Linguistic Pitch Analysis using Functional Principal Component Mixed Effect Models', *Journal of the Royal Statistical Society, Series C*, 59: 297–317.

Baladandayuthapani, V., Mallick, B.K., Hong, M.Y., Lupton, J.R., Turner, N.D., and Carroll, R.J. (2008) 'Bayesian Hierarchical Spatially Correlated Functional Data Analysis with Application to Colon Carcinogenesis', *Biometrics*, 64: 64–73.

Basser, P., Mattiello, J., and LeBihan, D. (1994) 'MR Diffusion Tensor Spectroscopy and Imaging', *Biophysical Journal*, 66: 259–67.

Basser, P., Pajevic, S., Pierpaoli, C., and Duda, J. (2000) 'In Vivo Fiber Tractography using DT-MRI Data', *Magnetic Resonance in Medicine*, 44: 625–32.

Berhane, K. and Molitor, N. (2008) 'A Bayesian Approach to Functional-based Multilevel Modeling of Longitudinal Data: Applications to Environmental Epidemiology', *Biostatistics*, 9: 686–99.

Bigelow, J. and Dunson, D. (2007) 'Bayesian Adaptive Regression Splines for Hierarchical Data', *Biometrics*, 63: 724–32.

Brown, P.J., Fearn, T., and Vannucci, M. (2001) 'Bayesian Wavelet Regression on Curves with Application to a Spectroscopic Calibration Problem', *Journal of the American Statistical Association*, 96: 519–36.

Brumback, B. and Rice, J.A. (1998) 'Smoothing Spline Models for the Analysis of Nested and Crossed Samples of Curves (with discussion)', *Journal of the American Statistical Association*, 93: 961–94.

Caffo, B.S., Chen, S., Stewart, W., Bolla, K., Yousem, D., Davatzikos, C., and Schwartz, B.S. (2008) 'Are Brain Volumes Based on Magnetic Resonance Imaging Mediators of the Associations of Cumulative Lead Dose with Cognitive Function?', *American Journal of Epidemiology*, 167: 429–37.

Cardot, H. and Sarda, P. (2005) 'Estimation in Generalized Linear Models for Functional Data via Penalized Likelihood', *Journal of Multivariate Analysis*, 92: 24–41.

Cardot, H., Ferraty, F., and Sarda, P. (1999) 'Functional Linear Model', *Statistics & Probability Letters*, 45: 11–22.

Cardot, H., Ferraty, F., and Sarda, P. (2003) 'Spline Estimators for the Functional Linear Model', *Statistica Sinica*, 13: 571–91.

Cardot, H., Crambes, C., Kneip, A., and Sarda, P. (2007) 'Smoothing Splines Estimators in Functional Linear Regression with Errors-in-variables', *Computational Statistics and Data Analysis*, 51: 4832–48.

Crainiceanu, C.M. and Goldsmith, A.J. (2009) 'Bayesian Functional Data Analysis using WinBUGS', *Journal of Statistical Software*, 32(11). URL http://www.jstatsoft.org/v32/i11/ paper.

Crainiceanu, C.M., Caffo, B.S., and Punjabi, N.M. (2009) 'Nonparametric Signal Extraction and Measurement Error in the Analysis of Electroencephalographic Activity During Sleep', *Journal of the American Statistical Association*, 104: 541–55.

Crainiceanu, C.M., Staicu, A.M., and Di, C. (2009) 'Generalized Multilevel Functional Regression', *Journal of the American Statistical Association*, 104: 177–94.

Crainiceanu, C.M., Reiss, P., Goldsmith, J., and Huang, L. (2010) *refund: Regression with Functional Data*. URL http://CRAN.R-project.org/package=refund. R package version 0.1-0.

Crainiceanu, C.M., Caffo, B.S., Luo, S., Zipunnikov, V., and Punjabi, N.M. (2011) 'Population Value Decomposition, A Framework for the Analysis of Image Populations (with discussion),' *Journal of the American Statistical Association*, 106: 775–90.

Crainiceanu, C.M., Staicu, A.M., Ray, S., and Punjabi, N.M. (2012) *Bootstrap-based Inference on the Difference in the Means of Two Correlated Functional Processes*. Technical report. URL http://www.bepress.com/jhubiostat/paper225/.

Crambes, C., Kneip, A., and Sarda, P. (2009) 'Smoothing Splines Estimators for Functional Linear Regression', *Annals of Statistics*, 37: 35–72.

Davatzikos, C., Genc, A., Xu, D., and Resnick S.M. (2001) 'Voxel-based Morphometry using the Ravens Maps: Methods and Validation using Simulated Longitudinal Atrophy', *NeuroImage*, 14: 1361–9.

Di, C., Crainiceanu, C.M., Caffo, B.S., and Punjabi, N.M. (2009) 'Multilevel Functional Principal Component Analysis', *Annals of Applied Statistics*, 3: 458–88.

Di, C., Crainiceanu, C.M., and Jank, W. (2011) *Multilevel Sparse Functional Principal Component Analysis*. Technical report. URL `http://www.bepress.com/jhubiostat/paper206/`.

Donoho, D. and Johnstone, I. (1995) 'Adapting to Unknown Smoothness by Wavelet Shrinkage', *Journal of the American Statistical Association*, 90: 1200–24.

Dowsey, A., Dunn, M., and Yang, G. (2008) 'Automated Image Alignment for 2D Gel Electrophoresis in a High-throughput Proteomics Pipeline', *Bioinformatics*, 24: 950–7.

Dubin, J.A. and Müller, H.-G. (2005) 'Dynamical Correlation for Multivariate Longitudinal Data', *Journal of the American Statistical Association*, 100: 872–81.

Goldsmith, A.J., Bobb, J., Crainiceanu, C.M., Caffo, B.S, and Reich, D. (2011) 'Penalized Functional Regression', *Journal of Computational and Graphical Statistics*, 20: 830–51.

Goldsmith, A.J., Crainiceanu, C.M., Caffo, B.S., and Reich, D. (2012) 'Longitudinal Penalized Functional Regression', *Journal of the Royal Statistical Society, Series C*, 61: 453–69.

Goldstein, H. (1994) 'Multilevel Cross-classified Models', *Sociological Methods & Research*, 22: 364–75.

Grambsch, P.M., Randall, B.L., Bostick, R.M., Potter, J.D., and Louis, T.A. (1995) 'Modeling the Labeling Index Distribution: An Application of Functional Data Analysis', *Journal of the American Statistical Association*, 90: 813–21.

Greven, S., Crainiceanu, C.M., Küchenhoff, H., and Peters, A. (2008) 'Restricted Likelihood Ratio Testing for Zero Variance Components in Linear Mixed Models', *Journal of Computational and Graphical Statistics*, 17(4): 870–91.

Greven, S., Crainiceanu, C.M., Caffo, B.S., and Reich, D. (2010) 'Longitudinal Functional Principal Component Analysis', *Electronic Journal of Statistics*, 4: 1022–54.

Guo, W. (2002) 'Functional Mixed Effects Models', *Biometrics*, 58: 121–8.

Indritz, J. (1963) *Methods in Analysis*. New York: Macmillan & Collier-Macmillan.

James, G.M. (2002) 'Generalized Linear Models with Functional Predictors', *Journal Of The Royal Statistical Society, Series B*, 64: 411–32.

James, G.M., Wang, J., and Zhu, J. (2009) 'Functional Linear Regression That's Interpretable', *Annals of Statistics*, 37: 2083–108.

Karhunen, K. (1947) 'Über Lineare Methoden in der Wahrscheinlichkeitsrechnung', *Annales Academiæ Scientiarum Fennicæ, Series A1: Mathematica-Physica, Suomalainen Tiedeakatemia*, 37: 3–79.

Kneip, A. and Ramsay, J.O. (2008) 'Combining Registration and Fitting for Functional Linear Models', *Journal of the American Statistical Association*, 103: 1155–65.

Kneip, A., Li, X., MacGibbon, B., and Ramsay, J.O. (2000) 'Curve Registration by Local Regression', *Canadian Journal of Statistics*, 28: 19–30.

Laird, N.M. and Ware, J.H. (1982) 'Random-effects Models for Longitudinal Data', *Biometrics*, 38: 963–74.

LeBihan, D., Mangin, J., Poupon, C., Clark, C., Pappata, S., Molko, N., and Chabriat, H. (2001) 'Diffusion Tensor Imaging: Concepts and Applications', *Journal of Magnetic Resonance Imaging*, 13: 534–46.

Li, Y., Wang, N., Hong, M., Turner, N.D., Lupton, J.R., and Carroll, R.J. (2007) 'Nonparametric Estimation of Correlation Functions in Longitudinal and Spatial Data, with Application to Colon Carcinogenesis Experiments', *Annals of Statistics*, 35: 1608–43.

Loève, M. (1945) 'Functions Aleatoire de Second Ordre', *Comptes Rendus de l'Acadmie des Sciences*, 220.

Mallat, S. (1989) 'A Theory for Multiresolution Signal Decomposition: The Wavelet Representation', *IEEE Transactions on Pattern Analysis and Machine Intelligence*, 11: 674–93.

Malloy, E.J., Morris, J.S., Adler, S.D., Suh, H., Gold, D.R., and Coll, B.A. (2010) 'Wavelet-based Functional Linear Mixed Models: An Application to Measurement Error-corrected Distributed Lag Models', *Biostatistics*, 11: 432–52.

Martinez, J.G., Huang, J.Z., Burghardt, R.C., Barhoumi, R., and Carroll, R.J. (2009) 'Use of Multiple Singular Value Decompositions to Analyze Complex Intracellular Calcium Ion Signals', *Annals of Applied Statistics*, 3: 1467–92.

Martinez, J.G., Bohn, K. M., Carroll, R. J. and Morris, J. S. (2013). A study of Mexican free-tailed bat syllables: Bayesian functional mixed models for nonstationary acoustic time series. *Journal of the American Statistical Association*, to appear.

Marx, B.D. and Eilers, P.H.C. (1999) 'Generalized Linear Regression on Sampled Signals and Curves: A P-spline Approach', *Technometrics*, 41: 1–13.

Miller, M.I., Christensen, G.E., Amit, Y., and Grenander, U. (1993) 'Mathematical Textbook of Deformable Neuroanatomies', *Proceedings of the*

National Academy of Sciences of the United States of America, 90: 11944.

Mohamed, A. and Davatzikos, C. (2004) 'Shape Representation via Best Orthogonal Basis Selection', *Medical Image Computing and Computer-Assisted Intervention*, 3216: 225–33.

Mori, S. and Barker, P. (1999) 'Diffusion Magnetic Resonance Imaging: Its Principle and Applications', *The Anatomical Record*, 257: 102–9.

Morris, J.S. (2012). Statistical Methods for Proteomic Biomarker Discovery using Feature Extraction or Functional Data Analysis Approaches, *Statistics and its Interface*, 5(1), 117–136.

Morris, J.S. and Carroll, R.J. (2006) 'Wavelet-based Functional Mixed Models', *Journal of the Royal Statistical Society, Series B*, 68: 179–99.

Morris, J.S., Wang, N., Lupton, J., Chapkin, R., Turner, N., Hong, M., and Carroll, R.J. (2001) 'Parametric and Nonparametric Methods for Understanding the Relationship between Carcinogen-induced DNA Adduct Levels in Distal and Proximal Regions of the Colon', *Journal of the American Statistical Association*, 96: 816–26.

Morris, J.S., Wang, N., Lupton, J.R., Chapkin, J.S., Turner, N.D., Hong, M.Y., and Carroll, R.J. (2002) 'A Bayesian Analysis Involving Colonic Crypt Structure and Coordinated Response to Carcinogens Incorporating Missing Crypts', *Biostatistics*, 3: 529–46.

Morris, J.S., Arroyo, C., Coull, B.A., Louise, M.R., Herrick, R.J., and Gortmaker, S.L. (2006) 'Using Wavelet-based Functional Mixed Models to Characterize Population Heterogeneity in Accelerometer Profiles: A Case Study', *Journal of the American Statistical Association*, 101: 1352–64.

Morris, J.S., Brown, P.J., Herrick, R., Baggerly, K., and Coombes, K. (2008) 'Bayesian Analysis of Mass Spectrometry Data using Wavelet-based Functional Mixed Models', *Biometrics*, 12: 479–89.

Morris, J.S., Clark, B., and Gutstein, H. (2008) 'Pinnacle: A Fast, Automatic and Accurate Method for Detecting and Quantifying Protein Spots in 2-dimensional Gel Electrophoresis Data', *Bioinformatics*, 24: 529–36.

Morris, J.S., Baladandayuthapani, V., Herrick, R., Sanna, P., and Gutstein, H. (2011) 'Automated Analysis of Quantitative Image Data using Isomorphic Functional Mixed Models, with Application to Proteomic Data', *Annals of Applied Statistics*, 5: 894–923.

Morris, J.S., Vannucci, M., Brown, P.J., and Carroll, R.J. (2003) 'Wavelet-based Nonparametric Modeling of Hierarchical Functions in Colon Carcinogenesis' (with Discussion), *Journal of the American Statistical Association*, 98: 573–83.

Müller, H.-G. and Stadtmüller, U. (2005) 'Generalized Functional Linear Models', *Annals of Statistics*, 33: 774–805.

Ogawa, S., Tank, D.W., Menon, R., Ellerman, J.M., Kim, S.G., Merkle, H., and Ugurbil, K. (1992) 'Intrinsic Signal Changes Accompanying Sensory Stimulation: Functional Brain Mapping with Magnetic Resonance Imaging', *Proceedings of the National Academy of Sciences*, 89: 5675–9.

Ogden, R.T. (1997) *Essential Wavelets for Statistical Applications and Data Analysis*. Boston: Birkhäuser.

O'Sullivan, F. (1986) 'A Statistical Perspective on Ill-posed Inverse Problems,' *Statistical Science*, 1: 505–27.

Quan, S.F., Howard, B.V., Iber, C., Kiley, J.P., Nieto, F.J., and O'Connor, G.T. (1997) 'The Sleep Heart Health Study: Design, Rationale, and Methods', *Sleep*, 20: 1077–85.

R Development Core Team. *R: A Language and Environment for Statistical Computing*. R Foundation for Statistical Computing, Vienna, Austria, 2009. URL `http://www.R-project.org/`.

Ramsay, J.O. and Silverman, B.W. (2005) *Applied Functional Data Analysis*. New York: Springer-Verlag.

Ramsay, J.O. and Silverman, B.W. (2006) *Functional Data Analysis*. New York: Springer-Verlag.

Rao, C.R. (1958) 'Some Statistical Methods for Comparison of Growth Curves', *Biometrics*, 14: 1–17.

Reiss, P. and Ogden, R. (2007) 'Functional Principal Component Regression and Functional Partial Least Squares', *Journal of the American Statistical Association*, 102: 984–96.

Rice, J. (2004) 'Functional and Longitudinal Data Analysis', *Statistica Sinica*, 14: 631–47.

Rice, J. and Silverman, B. (1991) 'Estimating the Mean and Covariance Structure Nonparametrically when the Data are Curves', *Journal of the Royal Statistical Society, Series B*, 53: 233–43.

Rodriguez, A. and Dunson, D. (2011) *Functional Clustering in Nested Designs: Modeling Variability in Reproductive Epidemiology Studies*. Technical Report. URL `http://ftp.isds.duke.edu/WorkingPapers/10-03.pdf`.

Roncucci, L., Pedroni, M., Vaccina, F., Benatti, P., Marzona, L., and De Pol, A. (2000) Aberrant Crypt Foci in Colorectal Carcinogenesis: Cell and Crypt Dynamics', *Cell Proliferation*, 33: 1–18.

Ruppert, D., Carroll, R.J., and Wand, M.P. (2003) *Semiparametric Regression*. Cambridge, UK: Cambridge University Press.

Schwartz, B.S., Stewart, W.F., Bolla, K.I., Simon, D., Bandeen-Roche, K., Gordon, B., Links, J.M., and Todd, A.C. (2000) 'Past Adult Lead Exposure is Associated with Longitudinal Decline in Cognitive Function', *Neurology*, 55: 1144.

Schwartz, B.S., Stewart, W.F., Kelsey, K.T., Simon, D., Park, S., Links, J.M., and Todd, A.C. (2000) 'Associations of Tibial Lead Levels with BsmI

Polymorphisms in the Vitamin D Receptor in Former Organolead Manufacturing Workers', *Environmental Health Perspectives*, 108: 199.

Schwartz, B.S., Chen, S., Caffo, B., Stewart, W.F., Bolla, K.I., Yousem, D., and Davatzikos, C. (2007) 'Relations of Brain Volumes with Cognitive Function in Males 45 Years and Older with Past Lead Exposure', *Neuroimage*, 37: 633–41.

Sgambato, A., Cittadini, A., Faraglia, B., and Weinstein, I.B. (2000) 'Multiple Functions of p27kip1 and its Alterations in Tumor Cells: A Review', *Journal of Cell Biology*, 183: 18–27.

Staicu, A.-M., Crainiceanu, C.M., and Carroll, R.J. (2010) 'Fast Methods for Spatially Correlated Multilevel Functional Data,' *Biostatistics*, 11: 177–94.

Staicu, A.-M., Lahiri, S.N., and Carroll, R.J. (2010) *Tests of Significance for Spatially Correlated Functional Data*. Technical report.

Staicu, A.-M., Crainiceanu, C.M., Ruppert, D., and Reich, D. (2012) 'Modeling Functional Data with Spatially Heterogeneous Shape Characteristics', *Biometrics*, 68: 331–43.

Staniswalis, J.G. and Lee, J.J. (1998) 'Nonparametric Regression Analysis of Longitudinal Data', *Journal of the American Statistical Association*, 93: 1403–18.

Stewart, W.F., Schwartz, B.S., Simon, D., Bolla, K.I., Todd, A.C., and Links, J. (1999) 'Neurobehavioral Function and Tibial and Chelatable Lead Levels in 543 Former Organolead Workers. *Neurology*, 52: 1610.

Vidakovic, B. (1999) *Statistical Modeling by Wavelets*. Chichester: John Wiley & Sons.

Wahba, G., Wang, Y., Gu, C., Klein, R., and Klein, B. (1995) 'Smoothing Spline ANOVA for Exponential Families, with Application to the Wisconsin Epidemiological Study of Diabetic Retinopathy', *Annals of Statistics*, 23: 1865–95.

Walls, T.A. and Shaeffer, J.L. (eds) (2006) *Models for Intensive Longitudinal Data*. Oxford: Oxford University Press.

Wang, X., Ray, S., and Mallick, B.K. (2007) 'Bayesian Curve Classification using Wavelets', *Journal of the American Statistical Association*, 102: 962–73.

Wang, X., Yang, Q., Fan, Z., Sun, C., and Yue, G. (2009) 'Assessing Time-dependent Association between Scalp EEG and Muscle Activation: A Functional Random-effects Model Approach', *Journal of Neuroscience Methods*, 177: 232–40.

Wang, Y. (1998) 'Mixed Effects Smoothing Spline Analysis of Variance', *Journal of the Royal Statistical Society, Series B*, 60: 159–74.

Wood, S. (2006) *Generalized Additive Models: An Introduction with R*. New York: Chapman & Hall/CRC.

Xiao, G., Reilly, C., and Khodursky, A.B. (2009) 'Improved Detection of Differentially Expressed Genes through Incorporation of Gene Locations', *Biometrics*, 65: 805–14.

Xue, Z., Shen, D., and Davatzikos, C. (2006) 'Statistical Representation of High-dimensional Deformation Fields with Application to Statistically Constrained 3D Warping', *Medical Image Analysis*, 10: 740–51.

Xue, Z., Shen, D., and Davatzikos, C. (2006) 'CLASSIC: Consistent Longitudinal Alignment and Segmentation for Serial Image Computing', *Neuroimage*, 30: 388–99.

Yao, F., Müller, H.-G., and Wang, J.-L. (2005) 'Functional Data Analysis for Sparse Longitudinal Data', *Journal of the American Statistical Association*, 100: 577–90.

Zhao, X., Marrron, J.S., and Wells, M.T. (2004) 'The Functional Data Analysis View of Longitudinal Data', *Statistica Sinica*, 14: 789–808.

Zhou, L., Huang, J.Z., Martinez, J.G., Maity, A., Baladandayuthapani, V., and Carroll, R.J. (2010) 'Reduced Rank Mixed Effects Models for Spatially Correlated Hierarchical Functional Data', *Journal of the American Statistical Association*, 105: 390–400.

Zhu, H., Brown, P.J., and Morris, J.S. (2011a) 'Robust, Adaptive Functional Regression in Functional Mixed Model Framework', *Journal of American Statistical Association*, 106: 1167–79.

Zhu H, Brown PJ, and Morris JS (2012): Robust classification of functional and quantitative image data using functional mixed models. *Biometrics*, 68(4): 1260–1268.

Zipunnikov, V., Caffo, B.S., Yousem, D.M., Davatzikos, C., Schwartz, B.S., and Crainiceanu, C.M. (2011) 'Multilevel Functional Principal Component Analysis for High Dimensional Data', *Journal of Computational and Graphical Statistics*, 20: 852–73.

Zipunnikov, V., Caffo, B.S., Yousem, D.M., Davatzikos, C., Schwartz, B.S., and Crainiceanu, C.M. (2012) *Longitudinal High Dimensional Data Analysis*. Technical report. URL `http://www.bepress.com/jhubiostat/paper234/`.

14

Nonlinear Models

Lang Wu
University of British Columbia, Canada

Wei Liu
York University, Canada

14.1 INTRODUCTION

14.1.1 Linear Models Versus Nonlinear Models

Regression models are widely used in practice. A (univariate) regression model describes a possible relationship between a response variable and a set of *predictors*. The predictors may also be called the *covariates* or *explanatory variables* or *independent variables*. In this chapter, we use the term 'covariates' for convenience. That is, in a regression model covariates are used to partially explain the *systematic variation* in the response. The remaining unexplained variation in the response is treated as a random error. In regression models, a main goal is to understand the dependence of the response on the covariates. However, the *true* relationship between the response and the covariates is typically unknown, may be very complicated, and may not be modeled exactly by a mathematical form. In other words, if there is a relationship between the response and covariates, this relationship is typically nonlinear (in a broad sense) by nature. In practice, our goal is to *approximate* this relationship based on the observed data. The simplest approximation is the linear approximation, which leads to *linear models*.

In a linear model, the relationship between the mean response and covariates is *assumed* to be linear. Linear models are widely used in practice because they are simple, easy to interpret, and their parameter estimates have closed-form mathematical expressions. However, a drawback of linear models is that they often do not represent the true relationship between the responses and the covariates, even if the models fit the observed data well, because linear models only describe the *observed data* rather than the underlying data-generation mechanisms. Linear models are thus called *empirical models*, and predictions based on linear models can be dangerous for values outside the observed data range.

The simplicity of linear models was particularly important in the pre-computer era, since closed-form expressions of parameter estimates for linear models were often available. Given the developments of modern computers and computational tools in the past few decades, however, this advantage is becoming less important. Thus, extensions of linear models, especially nonlinear models, have received great attention in the last few decades. Nonlinear models are typically obtained based on the underlying mechanisms which generate the data. Thus, nonlinear models may better approximate the true relationship between the responses and the covariates and may provide more reliable predictions for values outside the observed data range.

In a nonlinear model, the relationship between the responses and the covariates is allowed to be in a *nonlinear* form. Nonlinear models can be derived in many applications because, based on subject-area knowledge, we may have a good understanding of the *possible* underlying mechanisms which generate the data. For example, in the studies of growth (e.g., animal growth or tree growth), pharmacokinetics, and HIV viral dynamics, nonlinear models can be derived based on such understandings. In these cases, we may be able to better *approximate* the true relationship between a response and covariates using a nonlinear model than using a linear model, based on our understanding of the possible data-generation mechanisms.

Nonlinear models attempt to understand the underlying mechanics of data generation, so they are often called *mechanistic models* or *'scientific' models*. Nonlinear models have the following advantages:

- nonlinear models may better approximate the true regression relationships than linear models;
- nonlinear models often provide better predictions than linear models, especially for values outside the observed-data range;
- parameters in nonlinear models often have natural physical interpretations.

Thus, a nonlinear model may be preferred over a linear model *if* (i) the hypothesized nonlinear form is reasonably correct; (ii) the sample size of the observed data is sufficiently large for a reasonable fit of the nonlinear model; and (iii) the nonlinear relationship is strong enough and nonlinear enough so that a linear approximation may be unsatisfactory. In other words, nonlinear models have the following limitations:

- For many practical problems, nonlinear models may be unavailable because the true relationships are highly complicated or are not known well;
- There may be computational difficulties in parameter estimation, such as slow or non-convergences or sensitivity to choice of starting values, especially when the sample size is not large.
- Sometimes, a simpler linear model may fit the observed data equally as well as a nonlinear model. Parameter estimates from linear models usually have analytic expressions so their finite-sample properties can be well studied, while parameter estimates from nonlinear models usually do not have analytic expressions so one may have to rely on their large-sample asymptotic properties.

Therefore, when choosing a linear or a nonlinear model, one should take the above issues into consideration. Finally, note that both linear and nonlinear models are *hypothesized* models, since the true regression relationships may never be known or may not exactly follow any mathematical forms.

For many real-world problems, the underlying data-generation mechanisms may be described or approximated by a set of differential equations. This is particularly true if we focus on the rates of change of a variable of interest over time, such as growth curve models and dynamic models for longitudinal data. In these cases, the solutions to these differential equations are often nonlinear models. For example, in the analysis of growth curves, the growth rate may be proportional to the remaining size, which can be expressed as a differential equation. In HIV viral dynamic models, the viral elimination and production processes may be expressed as a set of differential equations (Wu and Ding, 1999). The solutions to these differential equations typically lead to nonlinear models. In some cases, the differential equations cannot be solved explicitly, so the nonlinear models are implicit, i.e., the nonlinear models do not have analytic or closed-form expressions (see, e.g., Ramsay et al., 2007; Huang and Lu, 2008).

The developments of nonlinear models require close collaboration between statisticians and subject-area scientists, since subject-area scientists can help statisticians to better understand the scientific problems under consideration and the possible data-generation mechanisms.

Nonlinear models have been widely used in practice, such as in HIV viral dynamics, pharmacokinetics, pharmacodynamics, molecular genetics, and growth or decay. More detailed discussions of nonlinear models can be found in Bates and Watts (1988), Seber and Wild (1989), Lindsey (2001), and Wu (2009), among others.

14.1.2 Models for Multilevel Data

Regression models for iid response data have been well developed. In practice, however, many datasets are not iid. Common examples include *longitudinal data*, or *clustered data*, or *multilevel data*. For example, in a longitudinal study there are multiple *repeated measurements* over time. These longitudinal data may be viewed as multilevel data, where the repeated measurements on each individual are level-1 data and the observations across individuals are the level-2 data (the level-1 data are nested within level-2 data). The repeated measurements on each individual may be viewed as a cluster. Thus, longitudinal data may also be viewed as clustered data. Similarly, in a multicenter study, observations from the same center are nested within this center (cluster, level-1 data), centers may be nested within cities (level-2 data), and cities may be nested within countries (level-3 data). In this chapter, we focus on longitudinal data to illustrate the basic ideas and methods for multilevel data.

A key difference between cross-sectional data (or iid data) and clustered data (or multilevel data) is that the observations within a cluster (or a level) are typically *correlated*, since these observations share similar characteristics. Thus, regression models for clustered response data must incorporate this correlation for valid inference. In other words, a regression model for clustered data must address both the relationship between the response and covariates and the correlation of the response data within each cluster. There are three commonly used models to analyze longitudinal data or clustered data: mixed effects models, marginal GEE models, and transitional models. In this chapter, we mainly focus on mixed effects models, since they are natural models for multilevel data.

In a mixed effects model, random effects are introduced to incorporate the *correlation* within a cluster as well as the *cluster effects*. The random effects may be interpreted as the common (unobserved) characteristics within a cluster, similar to latent variables. Since each cluster shares the same random effects, the observations within a cluster are correlated. For clustered data, there are two sources of variation: the *between-cluster variation* and the *within-cluster variation*. A mixed effects model incorporates both sources of variation: it uses random effects to represent the variation across clusters and uses an additional random error term to represent the within-cluster variation. In other words, random effects are used to accommodate the *heterogeneity* in the data, which may arise from individual or cluster effects. Thus, a mixed effect model allows individual-specific or cluster-specific inference, and they are often called *subject-specific (SS) models*. This is to be contrasted with the population averaged (PA) models or marginal models in which one focuses on population effects.

A mixed effects model can be obtained from the corresponding regression model for cross-sectional data by introducing random effects to appropriate parameters, and thus allowing these parameters to vary across individuals. In regression models for cross-sectional data, all model parameters are fixed, i.e., the parameters are the same for all individuals, so these parameters are also called *fixed effects* or *population parameters*. In longitudinal studies, if data vary greatly across individuals, we may allow each individual to have individual-specific parameters. These individual-specific parameters can be obtained by adding random effects to the population parameters, so the random effects may represent individual deviations from the population averages (plus random errors). Therefore, we can choose random effects informally based on the heterogeneous feature of the data. Formally, we can perform hypothesis testing to test the need of random effects for certain parameters, or we can use formal model selection methods, such as AIC or BIC criteria, to choose between models with and without certain random effects.

14.2 MODELS AND METHODS

In this section, we first consider models for cross-sectional data or iid data, and then we describe how these models can be extended for longitudinal data, clustered data, or multilevel data.

14.2.1 Nonlinear Models for Cross-Sectional Data

We first consider models for cross-sectional data or iid data. Let $\{(y_i, x_{i1}, x_{i2}, \ldots, x_{ir}),\ i = 1, 2, \ldots, n\}$ be an iid sample of n individuals on $r + 1$ variables, where y_i is the response variable and the x_{ij}s are covariates (or predictors) for individual i. The objective is to model a possible relationship between the response and the covariates. In other words, we attempt to use the covariates to partially explain the variation in the response.

The relationship between the response and the covariates, if it exists, is likely to be complicated. However, we may be able to *approximate* this relationship. A linear approximation is the simplest approximation, and it leads to a linear regression model. However, for many practical problems, a *nonlinear* approximation may provide a better approximation since such a nonlinear model is often based on the possible underlying data-generation mechanism for the practical problem under consideration. Note that, in the literature, the term *nonlinear model* often refers to the case where the nonlinear relationship is *known*. When the nonlinear relationship is unknown, the model is often called a *nonparametric model*.

A *nonlinear regression model* for cross-sectional data can be written as

$$y_i = g(\mathbf{x}_i, \boldsymbol{\beta}) + \varepsilon_i, \quad i = 1, 2, \ldots, n, \quad (14.1)$$

where $g(\cdot)$ is a *known nonlinear* function, $\mathbf{x}_i = (x_{i1}, \ldots, x_{ip})'$ is a vector of covariates or predictors, $\boldsymbol{\beta}$ is a vector of unknown regression parameters, and ε_i is a random error. Common assumptions for a nonlinear regression model are the same as those for a linear model, i.e.: (i) the errors ε_i are independent, (ii) the errors ε_i have mean zero and constant variance σ^2, and (iii) the errors ε_i are normally distributed. Some of these assumptions may be relaxed for more general models.

A nonlinear model usually refers to the case where the function $g(\cdot)$ is *nonlinear* in the parameters $\boldsymbol{\beta}$. If the function $g(\cdot)$ is nonlinear in the covariates $\mathbf{x}_i$ but linear in the parameters $\boldsymbol{\beta}$, model (14.1) can be converted to a linear regression model by variable transformations. Thus, the key difference between a linear model and a nonlinear model is whether the regression function is linear or nonlinear in the unknown regression parameters. For a nonlinear model, because $g(\cdot)$ is a nonlinear function, estimates of the parameters $\boldsymbol{\beta}$ typically no longer have closed-form expressions.

By the foregoing definition, a nonlinear model is a parametric model, i.e., the model is controlled by a few unknown parameters (e.g., $\boldsymbol{\beta}$). Once these unknown parameters are estimated from the data, the model is completely known. Other classes of 'nonlinear model' are *nonparametric models* and *semiparametric models*. In a nonparametric model, the function $g(\cdot)$ is completely unspecified, except for the requirement of some smoothness. The goal is then to estimate the functional form of $g(\cdot)$ based on the observed data. Thus, nonparametric models are very flexible in modeling complex data, but nonparametric models should be viewed as empirical models. A semiparametric model, which contains both parametric and nonparametric parts, offers a good compromise between parametric models and nonparametric models.

14.2.2 Nonlinear Models for Multilevel Data

Nonlinear models for cross-sectional data can be extended to model longitudinal data, or clustered data, or multilevel data. A popular approach is to introduce random effects in the models for cross-sectional data to account for the *correlation* among the observations within each individual (or each cluster or each level) and the variation between individuals (or clusters or levels). When random effects are introduced in (parametric) nonlinear models, the resulting models are called *nonlinear mixed effects* (NLME) models. NLME models are useful in many longitudinal studies, such as studies of growth and decay, HIV viral dynamics, and pharmacokinetics analysis.

For simplicity, in the following we focus on two-level data. We can write an NLME model as a two-stage (or two-level) *hierarchical nonlinear model*, which specifically models the intra-individual variation (i.e., variation within a cluster) and the inter-individual variation (i.e., the variation between clusters). Specifically, in the first stage (level 1) we specify the mean and covariance structure for data within an individual or a cluster, and in the second stage (level 2) we model the variation between individuals or clusters through random effects.

In the following, we illustrate the basic ideas with longitudinal data. Let $\mathbf{y}_i = (y_{i1}, \ldots, y_{in_i})^T$ be the n_i observations of the response for individual i or cluster i, where y_{ij} is the response value of individual i measured at time t_{ij}, $i = 1, 2, \ldots, m$; $j = 1, 2, \ldots, n_i$. A general NLME model for longitudinal data can be written as:

$$y_{ij} = g(t_{ij}, \boldsymbol{\beta}_i) + \varepsilon_{ij}, \qquad (14.2)$$
$$\boldsymbol{\beta}_i = h(\mathbf{x}_i, \boldsymbol{\beta}, \mathbf{u}_i), \qquad (14.3)$$
$$i = 1, 2, \ldots, m; \quad j = 1, 2, \ldots, n_i,$$

where $g(\cdot)$ is a known nonlinear function, t_{ij} is the measurement time, ε_{ij} is the random error for within-individual measurement y_{ij}, function $h(\cdot)$ is also known and is often chosen as a linear function, $\boldsymbol{\beta}_i$ and $\boldsymbol{\beta}$ are individual-specific parameters and p fixed-effects parameters respectively, $\mathbf{x}_i$ contains covariates for individual i, and the $\mathbf{u}_i$s are random effects. We often assume that

$$\mathbf{u}_i \sim N_q(\mathbf{0}, \Omega), \qquad \boldsymbol{\varepsilon}_i \sim N_{n_i}(\mathbf{0}, \Sigma_i),$$

where Σ_i is a covariance matrix for the observations within individual i, Ω is a covariance matrix for the random effects, and $\boldsymbol{\varepsilon}_i = (\varepsilon_{i1}, \varepsilon_{i2}, \ldots, \varepsilon_{in_i})^T$ are random errors for observations within individual i. We typically assume that $\boldsymbol{\varepsilon}_i$ and $\mathbf{u}_i$ are independent.

In the NLME model, model (14.2) is chosen based on the longitudinal trajectory of individual i, and model (14.3) is chosen based on the variability of the individual-specific parameter $\boldsymbol{\beta}_i$ across all individuals. The choice of the covariance matrix Σ_i should be guided by practical considerations. Often we can simply choose $\Sigma_i = \sigma^2 I_{n_i}$ for simplicity, which assumes that the within-individual measurements are independent conditioning on that individual. This choice may be reasonable if the observation times t_{ij} are far apart so that the correlation between y_{ij} and y_{ik} is negligible and/or if the between-individual variation (measured by Ω) is dominant. This choice greatly reduces the number of parameters. Other choices may include the AR(1) structure, equal-correlation structure, or other commonly used correlation structures. Note that the accurate specification of Σ_i may be less critical when the between-individual variation is dominant. A large between-individual variation is often a main reason for choosing mixed effects models. The covariance matrix Ω measures the between-individual variation that is not explained by covariates $\mathbf{x}_i$, i.e., the matrix Ω characterizes the variations across individuals or clusters. Typically, Ω is chosen to be unstructured. The function $h(\cdot)$ is often chosen to be simple and parsimonious, such as a linear form.

The random effects $\mathbf{u}_i$ in an NLME model may be viewed as individual effects. They represent the variability of data across individuals, so the choice of random effects can be based on the heterogeneous features of the data. Although the random effects cannot be observed, they can be estimated from the data, and the magnitudes of these estimates reflect the magnitudes of between-individual variations. Thus, individual-specific inference is possible based on the random effects estimates. Since the data within each individual share the same random effects, these within-individual data are thus correlated. In other words, the random effects incorporate both the between-individual variation and the within-individual correlation.

An NLME model is also called a *subject-specific* (SS) or *individual-specific* model since it allows inference at the individual or subject level. The fixed parameters in $\boldsymbol{\beta}$ can be interpreted as the typical values of the individual-specific parameters in $\boldsymbol{\beta}_i$, rather than the typical average response profiles as in a population-averaged (PA) model. In other words, the interpretations of $\boldsymbol{\beta}$ in an SS model and in an PA model are not the same. This distinction is important for NLME models, but not for linear mixed effects (LME) models (Davidian and Giltinan, 1995; Heagerty, 1999). Thus, one should take this interpretation into account when choosing an NLME model in practice.

In the NLME model (14.2) and (14.3), the longitudinal trajectory for *each* individual is modeled *nonlinearly* in (14.2), similar to a nonlinear model for cross-sectional data. The variation across individuals is then incorporated via the random effects in (14.3). Therefore, to build an NLME model, in stage 1 (level 1) we can first assume that the within-individual measurements are iid and derive a nonlinear model as in cross-sectional studies, and then in stage 2 (level 2) we introduce random effects to appropriate parameters in the model.

A simple approach to choosing appropriate random effects in an NLME model is as follows:

- fit a nonlinear model for the data within each individual, as for iid data, and then
- compare the estimates of the parameters across different individuals to check their variabilities.

If the estimates of a parameter from different individuals vary substantially across individuals, we might allow this parameter to be *individual-specific*, i.e., we can introduce a random effect to this parameter; otherwise we may leave this parameter to be fixed across all individuals. More formally, one can consider various models, with or without random effects in certain parameters, and compare these models using standard model selection criteria such as AIC or BIC, and then choose the best model based on these criteria (e.g., the model with the smallest AIC or BIC value). Pinheiro and Bates (2002) provided detailed illustrations of this approach using both S-PLUS and R softwares.

For nonparametric mixed effects models, the procedure for model selection is similar: we can fit the measurements within each individual based on a nonparametric model as for iid data, and then we introduce appropriate random effects to the nonparametric model to account for the between-individual variation and the within-individual correlation. See Wu and Zhang (2006) and Chapters 17 and 18 of this volume for more detailed description of these approaches.

14.2.3 Inference for Nonlinear Models

Parameter estimation and inference for (parametric) nonlinear models are similar to those for linear models. Two approaches are commonly used: the least-squares method and the likelihood method. The least-squares method does not require distributional assumptions for estimation, while the likelihood method does require distributional assumptions. In the following, we briefly describe these two methods.

The *least-squares* method is to find an estimator of $\boldsymbol{\beta}$ to minimize the (squared) differences between the observed response values and the expected response values from the model. That is, for estimation of the parameter $\boldsymbol{\beta}$ in the nonlinear model (14.1), the least-squares method minimizes the quantity

$$Q(\boldsymbol{\beta}) = \sum_{i=1}^{n} (y_i - g(\mathbf{x}_i, \boldsymbol{\beta}))^2.$$

This can be achieved by solving the estimating equation

$$\begin{aligned} h(\boldsymbol{\beta}) &\equiv \frac{\partial Q(\boldsymbol{\beta})}{\partial \boldsymbol{\beta}} \\ &= \sum_{i=1}^{n} \frac{\partial g(\mathbf{x}_i, \boldsymbol{\beta})}{\partial \boldsymbol{\beta}} [y_i - g(\mathbf{x}_i, \boldsymbol{\beta})] = \mathbf{0}. \end{aligned} \tag{14.4}$$

For nonlinear models, a closed-form solution to the above equation typically does not exist, so an iterative algorithm such as the Newton–Raphson method is often required. For example, the Newton–Raphson method iteratively solves the equations

$$\boldsymbol{\beta}_k = \boldsymbol{\beta}_{k-1} - \left(\frac{\partial h(\boldsymbol{\beta})}{\partial \boldsymbol{\beta}} \right)^{-1} \Bigg|_{\boldsymbol{\beta}=\boldsymbol{\beta}_{k-1}} \cdot h(\boldsymbol{\beta}_{k-1}), \quad k = 1, 2, \ldots,$$

where

$$\begin{aligned} \frac{\partial h(\boldsymbol{\beta})}{\partial \boldsymbol{\beta}} &= \left(\frac{\partial h(\boldsymbol{\beta})}{\partial \beta_1}, \ldots, \frac{\partial h(\boldsymbol{\beta})}{\partial \beta_p} \right) \\ &= \left(\frac{\partial^2 Q(\boldsymbol{\beta})}{\partial \beta_i \partial \beta_j} \right)_{p \times p}. \end{aligned}$$

Beginning with an initial value $\boldsymbol{\beta}_0$, the algorithm will usually converge to a (possibly local) maximum or minimum of $Q(\boldsymbol{\beta})$.

The choice of starting value $\boldsymbol{\beta}_0$ in an iterative algorithm is important, if there is a possibility of multiple modes. Many nonlinear models may have multiple modes, so an iterative algorithm may converge to a local maximum closest to the starting value. Also, the iterative algorithm may not always converge. In this case, we can try different starting values. In general, it is always a good idea to try different starting values for an iterative algorithm for nonlinear models to see if they lead to the same solution.

Under the normality assumption for the random error, i.e., under the assumption that ε_i iid $\sim N(0, \sigma^2)$, the maximum likelihood method is to find the *maximum likelihood estimate* (MLE) of $\boldsymbol{\beta}$. Specifically, the MLE of $\boldsymbol{\beta}$ can be obtained by maximizing the likelihood function

$$L(\boldsymbol{\beta}, \sigma^2 | \mathbf{y}) = \prod_{i=1}^{n} \frac{1}{\sqrt{2\pi\sigma^2}} \exp\left[-\frac{(y_i - g(\mathbf{x}_i, \boldsymbol{\beta}))^2}{2\sigma^2} \right].$$

So the MLE of $\boldsymbol{\beta}$ satisfies the likelihood equation

$$\frac{\partial \log L(\boldsymbol{\beta}, \sigma^2 | \mathbf{y})}{\partial \boldsymbol{\beta}} = \mathbf{0},$$

which is identical to the least-squares equation (14.4). Therefore, the least-squares estimator of $\boldsymbol{\beta}$ is the same as the MLE of $\boldsymbol{\beta}$ here. Again, the likelihood equation can be solved using an iterative algorithm such as the Newton–Raphson method, and choice of starting values is important.

A main advantage of the least-squares estimates is that no distributional assumption is needed for the data, while a main advantage of MLEs is that they are asymptotically normal and optimal under usual regularity conditions. Both methods are quite general and can be extended for multilevel data. However, the likelihood method is perhaps more commonly used, since it can also be extended to non-normal response data such as binary data.

For nonlinear models, analytic or closed-form expressions of their parameter estimates are usually unavailable. Thus, unlike linear models where these analytic expressions of parameter estimates are available, it may be difficult to study the properties of parameter estimates from nonlinear models for finite samples. However, when the sample size is large, statistical inference can be based on the asymptotic properties of MLEs from nonlinear models, under the usual regularity conditions. That is, under usual regularity conditions, MLEs of the parameters in nonlinear models are consistent, asymptotically normal, and asymptotically efficient. In other words, when the sample size is large, statistical inference of the model parameters can be based on the asymptotic normal distribution of the MLEs. For example, confidence intervals and hypothesis testing for $\boldsymbol{\beta}$ can be based on the asymptotic normal distributions of the MLEs. When the sample size is small, we can use bootstrap methods or other resample methods for inference.

Although nonlinear models are mechanistic or scientific models, they still may not fit the observed data well, due to the random noise in the data or possibly poor approximations in the derivations of nonlinear models. Thus, it is still important to check if a nonlinear model fits the observed data well. Moreover, the distributional assumptions for nonlinear models should be checked. Model diagnostic methods for nonlinear models are generally similar to those for linear models. Graphical diagnostic tools such as residual plots may be sufficient in many applications. For example, residuals play an important role in model diagnostics. Specifically, we may consider the following diagnostics:

- Goodness of fit of the nonlinear model: a residual plot against the fitted values can usually reveal obvious lack-of-fit and outlying observations.
- Model assumptions: the constant variance assumption can be checked using residual plots, and the normality assumption for the random errors can be checked using QQ plots (quantile-quantile plots).
- Cook's distance plots may be used to check any influential observations.

More details on model diagnostics can be found in Bates and Watts (1988) and Seber and Wild (1989).

For nonparametric models, the objective is to estimate the unknown smooth function $g(\cdot)$, so different methods are used. Commonly used methods include regression splines, kernel estimation, local polynomial method, and smoothing spline method. See Wu and Zhang (2006) and Chapter 17 for more detailed description of these methods.

14.2.4 Inference for NLME Models

For longitudinal data or clustered data or multilevel data, we can introduce appropriate random effects to a nonlinear model for cross-sectional data to account for within-cluster correlation and between-cluster variation. The resulting models are called NLME models. Statistical inference for NLME models is conceptually similar to that of nonlinear models for cross-sectional data, i.e., there are two commonly used methods for estimation and inference: the least-squares method and the likelihood method. A main issue is that NLME models contain random effects, which are unobservable and thus may lead to computational challenges.

For the least-squares method, we can fit a nonlinear model to the data within *each* individual or cluster as for iid data, and obtain the least-squares estimates of the parameters for each individual or cluster. Then, we combine the parameter estimates from all individuals to obtain an overall estimate for each parameter. For this approach, we can compare the estimates of each parameter based on data from different individuals or clusters to check the variability among the estimates of a parameter, which allows us to decide if a random effect is required for this parameter. Specifically, consider the NLME model (14.2) and (14.3). If the number of within-individual data is not small, i.e., if the n_is are large, estimation of parameters can be based on the following *two-step method*: in step 1, the individual-specific parameter $\boldsymbol{\beta}_i$ can be estimated by fitting a nonlinear model to the data within individual i using the least squares method; then in step 2 the individual estimates $\hat{\boldsymbol{\beta}}_i$ for all individuals are used to estimate the fixed parameters $\boldsymbol{\beta}$ (e.g., we can take $\hat{\boldsymbol{\beta}} = \sum_{i=1}^{M} \hat{\boldsymbol{\beta}}_i/M$). Inference can then be based on large-sample asymptotic results. Davidian and Giltinan (1995) provided a detailed description of this two-step method. The two-step method is simple and requires no distributional assumptions, but it requires rich within-individual data (i.e., n_i needs to be large).

In the following we focus on the likelihood method, which is more general and more widely used. Consider NLME model (14.2) and (14.3). Let $\boldsymbol{\theta}$ denote all unknown parameters in the model. The marginal distribution of the response $\mathbf{y}_i$ is given by

$$f(\mathbf{y}_i|\boldsymbol{\theta}) = \int f(\mathbf{y}_i|\mathbf{x}_i, \boldsymbol{\theta}, \mathbf{u}_i) f(\mathbf{u}_i|\boldsymbol{\theta})\, d\mathbf{u}_i, \quad (14.5)$$

i.e., the marginal distribution of the response data is obtained by integrating out the unobservable random effects. Thus, the likelihood for the observed data is given by

$$L(\boldsymbol{\theta}|\mathbf{y}) = \prod_{i=1}^{m} \int f(\mathbf{y}_i|\mathbf{x}_i, \boldsymbol{\theta}, \mathbf{u}_i) f(\mathbf{u}_i|\boldsymbol{\theta}) d\mathbf{u}_i. \quad (14.6)$$

Unlike linear models, the marginal distribution (14.5) and the likelihood (14.6) typically do *not* have closed form expressions (except in special cases), since the NLME model is often nonlinear in the unobservable random effects $\mathbf{u}_i$. This is a key difference between an LME model and an NLME model, and it leads to many computational problems which do not exist for LME models. This difference also leads to the distinction between an SS model and an PA model.

A major challenge of likelihood inference for an NLME model is the evaluation of the intractable integral in the likelihood (14.6). Commonly used methods include numerical or Monte Carlo integration methods, EM algorithms, and approximate methods. Specifically, the following approaches are often used for likelihood inference for an NLME model:

Monte Carlo methods or numerical methods These methods use numerical methods or Monte Carlo integration techniques to approximate the integral in the likelihood (14.6), such as the Gauss–Hermite quadrature method.

EM algorithms These methods use EM algorithms to find the MLEs of model parameters, where the E-step may also be based on numerical or Monte Carlo methods to approximate the expectations.

Approximate methods These methods either use Taylor approximations to *linearize* the NLME model and then iteratively solve the resulting LME models, or use *Laplace approximations* to directly approximate the likelihood (14.6).

Standard software may use one or more of the above methods. For example, in the R library **nlme**, the approximate method based on a Taylor approximation is used, while in the SAS procedure **nlmixed**, both approximate methods and numerical methods (e.g., the Gauss–Hermite quadrature method) are available. In pharmacokinetic analysis, a popular software application is **NONMEM** (Beal and Sheiner, 1989). Other softwares are also available, such as **PKBUGS** for Bayesian approaches.

In the past few decades, linearization methods based on Taylor approximations have been widely used for estimation of NLME models (Beal and Sheiner, 1982; Lindstrom and Bates, 1990), partially because of the availability of softwares. A main advantage of the linearization methods is that they are computationally efficient and they perform reasonably well for NLME models. These linearization methods are iterative algorithms. For example, beginning with reasonable starting values, at each iteration the linearization method of Lindstrom and Bates (1990) approximates an NLME model based on a first-order Taylor series approximation about the current parameter and random effect estimates. This approximation leads to an LME model at each iteration. Thus, this linearization method is equivalent to iteratively solving LME models until convergence. An advantage of the linearization method is that existing methods for LME models can be used at each iteration. Such a first-order Taylor approximation seems to work well for NLME models where the response is assumed to be normally distributed. A popular earlier version of the linearization methods is based on Beal and Sheiner (1982), which estimates the random effects by zeros in the Taylor expansion at each iteration, so it is less accurate than the method of Lindstrom and Bates (1990). Another popular approximate method, implemented in SAS, is based on Laplacian approximations (Wolfinger, 1993).

Other methods for estimation and inference for NLME models have also been developed, such as generalized estimating equation (GEE) methods, moment methods, and Bayesian methods. See Wu (2009, Chapter 2) for a more detailed discussion.

Estimation and inference methods for nonparametric or semiparametric mixed effects models are combinations of nonparametric methods for iid data and methods for parametric mixed effects models. Wu and Zhang (2006) provided an overview of various models and methods. See Chapter 17 for a detailed description.

14.3 CONNECTIONS WITH OTHER MODELS

Standard nonlinear models are used for continuous response data. For other types of data, such as binary data or count data, *generalized linear models* (GLMs) can be used. GLMs for iid data and generalized linear mixed models (GLMMs) for clustered data or multilevel data may be viewed as special nonlinear models, since in these models the responses and predictors are linked nonlinearly. However, in GLMs and GLMMs, predictors enter the models in a linear fashion, as in linear models, and they are often not mechanistic models. On the other hand, due to their nonlinear natures, statistical inferences for GLMMs are similar to those for NLME models. See Chapters 8 and 15 for detailed descriptions of GLMMs.

Although estimation methods for GLMMs and NLME models are similar, the performance of some methods may differ. For example, approximate methods based on Taylor expansions or Laplace approximations may work well for NLME models in which the responses are assumed to be normal, but they may not work well for GLMMs for binary responses, due to the discreteness of the response variables.

Another common type of data is survival data or time-to-event data, for which survival models are used. For clustered or multilevel survival data, frailty models or survival models with random effects are often used. A main feature of survival data is that these data are often *censored*. Commonly used survival models, such as Cox proportional hazards models and accelerated failure time models, may also be viewed as nonlinear models since the responses and predictors in the models are also linked nonlinearly. However, due to censoring in survival data, special statistical methods are often needed. Moreover, in Cox proprotional hazards models, we link the predictors to the hazards rather than the responses. See Chapter 30 for further discussion of these models.

14.4 STRENGTHS AND WEAKNESSES OF THE METHODS

We have discussed various methods for inference for NLME models, including numerical methods, Monte Carlo methods, EM algorithms, and approximate methods. The performance of these methods is similar in many cases, but in some cases it can be different. The Gauss–Hermite quadrature method works well when the random effects have a low dimension (e.g., one or two random effects in the models), but the computational burden increases exponentially with the number of random effects in NLME models. The EM algorithm is stable and general, but it can be computationally intensive and slow to converge when the number of random effects is not small. Other numerical methods and Monte Carlo methods can also be computationally intensive when the number of random effects is not small. An advantage of these methods is that in principle they can be made arbitrarily accurate by increasing the number of quadrature points or the number of Monte Carlo samples, but at the expense of increasing the computational burden.

The approximate methods based on Taylor or Laplace approximations are computationally very efficient and generally work well for NLME models. The performance of these approximate methods improves as the number of within-cluster measurements increases. However, these approximate methods cannot be made arbitrarily accurate, unlike the Gauss–Hermite quadrature method and Monte Carlo methods. Moreover, the standard errors of the parameter estimates sometimes may be under-estimated due to the approximations. The approximate estimates can be excellent starting values for more exact iterative algorithms (e.g., EM algorithms).

For all of these methods for NLME models, a common problem is that sometimes convergence can be quite slow, or even impossible. Choice of starting values is important. Sometimes, reparameterization of the models may help as well. In data analysis, it is desirable to try different methods and compare the results.

14.5 REAL EXAMPLES

In this section, we present several examples of nonlinear models and NLME models, based on motivations from some real problems.

14.5.1 Example 1: HIV Viral Dynamic Models

HIV viral dynamic models describe the viral load trajectories for HIV infected patients during an anti-HIV treatment. These models may help us to better understand the pathogenesis of HIV infection and the anti-HIV treatment. Let T_p, V_I, V_{NI} be the concentrations of the productively infected cells, infectious virus, and non-infectious virus respectively. Wu and Ding (1999) proposed the following differential equations to model HIV dynamics during an anti-HIV treatment:

$$\begin{aligned} dT_p/dt &= k^* T V_I - \delta_p T_p, \\ dV_I/dt &= (1-\eta)P - cV_I, \\ dV_{NI}/dt &= \eta P + P^* + N\delta_p T_p - cV_{NI}, \end{aligned} \tag{14.7}$$

where δ_p is the rate of infected cells T_p killed by HIV, η is the proportion of non-infectious virus produced by infected cells, P^* accounts for virus production from the ignored compartments, k^* is a constant, and N is the number of virions per cell. Under some reasonable assumptions and approximations, an analytic solution to the differential equations (14.7) is the bi-exponential (two-compartments) dynamic model

$$V(t) = P_1 e^{\lambda_1 t} + P_2 e^{\lambda_2 t}, \tag{14.8}$$

where $V(t)$ is the total amount of virus at time t, λ_1 and λ_2 may be interpreted as the turnover rates of productively infected cells and long-lived and/or latently infected cells respectively, and P_1 and P_2 are baseline values. Model (14.8), however, does not consider the dependence of the repeated observations within each individual and the large variation between individuals. Both features may be incorporated by introducing random effects in the model.

We consider a real dataset from a well-known AIDS study (Wu and Zhang, 2002). There were 48 evaluable HIV-1 infected patients who were treated with an anti-HIV treatment. Figure 14.1 displays the viral load trajectories during the first three months. We introduce random effects and random errors into the nonlinear model (14.8) to incorporate the within-patient covariances and between-patient variation, and obtain the NLME model

$$\begin{aligned} y_{ij} &= \log_{10}(P_{1i}e^{-\lambda_{1i}t_{ij}} + P_{2i}e^{-\lambda_{2i}t_{ij}}) + \varepsilon_{ij} \\ P_{1i} &= P_1 + u_{1i}, \qquad \lambda_{1i} = \lambda_1 + u_{2i}, \\ P_{2i} &= P_2 + u_{3i}, \qquad \lambda_{2i} = \lambda_2 + u_{4i}, \end{aligned} \tag{14.9}$$

where $y_{ij} = \log_{10}(V(t_{ij}))$ is the $\log_{10}$-transformation and is used to stabilize the variance and make the data more normally distributed, $\mathbf{u}_i = (u_{1i}, u_{2i}, u_{3i}, u_{4i})^T$ are random effects, and ε_{ij} is a random error. We assume that the random effects $\mathbf{u}_i$ are independent of the random errors ε_{ij}, with $\varepsilon_{ij} \overset{iid}{\sim} N(0, \sigma^2)$ and $\mathbf{u}_i \overset{iid}{\sim} N(\mathbf{0}, \Omega)$, where σ is a precision parameter and Ω is an unstructured variance-covariance matrix.

For comparison, we fit the viral load trajectories based on the nonlinear model (14.8) for iid data, the NLME model (14.9), and an empirical quadratic LME model given by

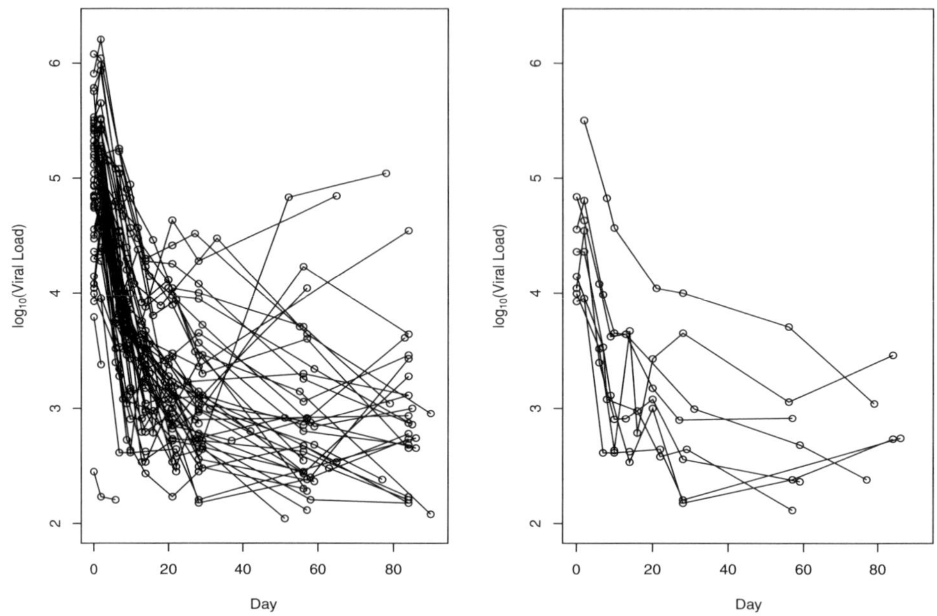

Figure 14.1 Viral Load (in $\log_{10}$ scale) Trajectories During an Anti-HIV Treatment. The open circles are observed values. Left figure: all patients. Right figure: some randomly selected patients

Table 14.1 Parameter Estimates for Nonlinear Models Based on the HIV Dataset

Model		P_1	λ_1	P_2	λ_2	σ
Nonlinear Model (14.8)	Estimate	11.52	28.57	7.50	0.83	0.63
	Standard error	0.17	2.83	0.26	0.47	
NLME Model (14.9)	Estimate	11.60	30.83	7.39	0.89	0.27
	Standard error	0.23	1.38	0.27	0.51	

$y_{ij} = (\beta_0+u_{1i})+(\beta_1+u_{2i})t_{ij}+(\beta_2+u_{3i})t_{ij}^2$. To avoid very small estimates, we rescale the original time t (in days) so that the new time scale is between 0 and 1. Table 14.1 presents the parameter estimates and their standard errors for the corresponding parameters in the NLME model (14.9) and the nonlinear model (14.8) respectively. Although these two models produce similar estimates for the fixed-effects parameters, the NLME model (14.9) should provide a more adequate fit to the longitudinal data than the nonlinear model (14.8) for iid data, since the NLME model (14.9) not only incorporates the correlations of the repeated measurements within each patient but also separately models the variation between patients and the variation within each individual (see Figures 14.2–14.5). In fact, the estimate $\hat{\sigma}$ of the within-individual variation under the NLME model (14.9) is less than half of that under the nonlinear model (14.8). Moreover, the estimate $diag(\hat{\Omega}) = (2.13, 14.44, 2.72, 8.43)$ in NLME model (14.9) provides estimation of the inter-patient variation.

Figure 14.2 shows the fitted viral load trajectories based on model (14.8), model (14.9), and the quadratic LME model for four randomly selected patients. We see the NLME model (14.9) fits the observed data better than

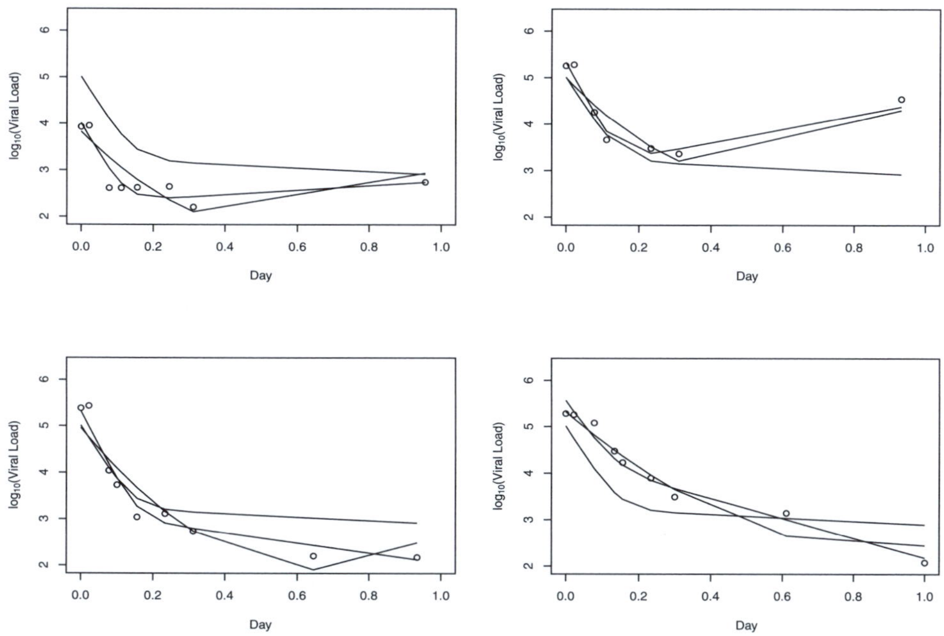

Figure 14.2 Fitted Viral Load Trajectories for Four Randomly Selected Patients in the HIV Study Based on NLME Model (14.9) (solid lines), Nonlinear Model (14.8) (dotted lines), and the Quadratic LME Model (dashed lines). The open circles are observed viral loads

the nonlinear model (14.8) and the quadratic LME model. Figure 14.3 displays plots of the residuals versus the fitted values and the smooth curves fitted by *Loess*. We find that the quadratic LME model fails to capture the nonlinear viral load trajectories over time. Finally, the normality assumptions for the random errors and for the random effects seem to be plausible (see Figures 14.4 and 14.5).

14.5.2 Example 2: Pharmacokinetics Models

Pharmacokinetics plays an important role in the determination of drug action within the living organism. Pharmacokinetics studies the time course of absorption, distribution, metabolism, and elimination of some substance in the body, given a drug dose, i.e., the movement of drugs in the body. Suppose that a substance enters the body via ingestion. Let $y(t)$ be the concentration of the substance in the body at time t (usually measured in the blood), and let $\mu(t) = E(y(t))$. Let $\mu_0(t)$ be the amount at the absorption site (e.g., stomach). A commonly used first-order one-compartment model is based on the differential equations

$$\frac{d\mu(t)}{dt} = \beta_1\mu_0(t) - \beta_2\mu(t),$$
$$\frac{d\mu_0(t)}{dt} = -\beta_1\mu_0(t),$$

where β_1 is the absorption rate and β_2 is the elimination rate. The above differential equations have an analytic solution given by

$$\mu(t) = \frac{\beta_1\beta_2 x}{(\beta_1-\beta_2)\beta_3}\left(e^{-\beta_2 t} - e^{-\beta_1 t}\right), \tag{14.10}$$

where x is the dose of the substance and β_3 is the clearance.

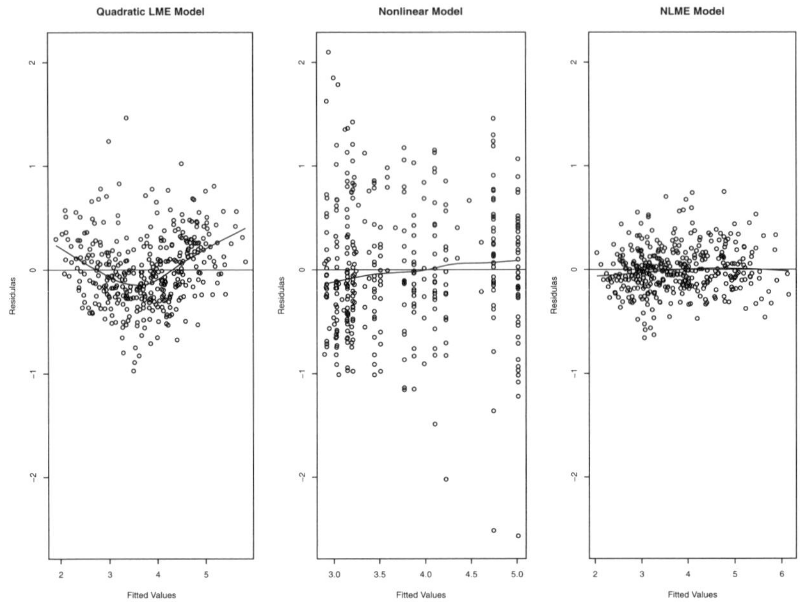

Figure 14.3 Residuals Versus Fitted Values and the Loess Fits Based on NLME Model (14.9), Nonlinear Model (14.8), and the Quadratic LME Model for the HIV Data

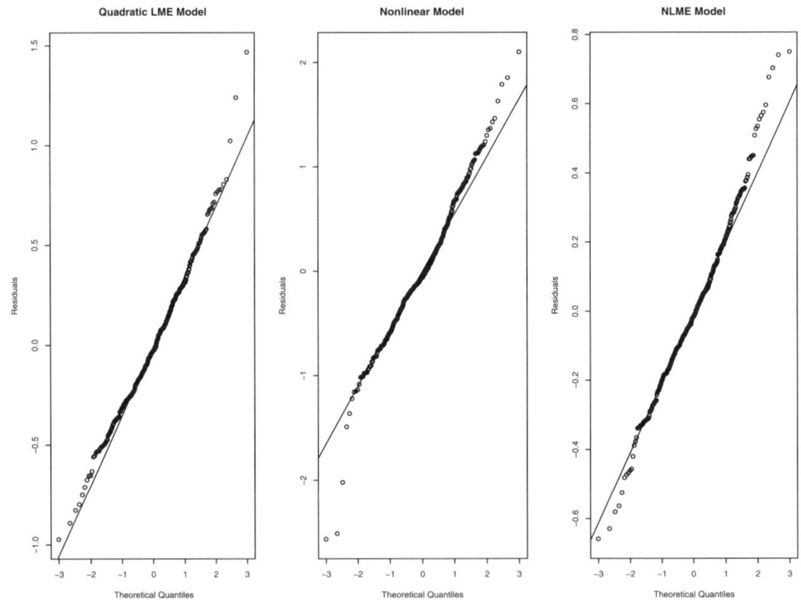

Figure 14.4 QQ Plots of the Residuals Based on the NLME Model (14.9), Nonlinear Model (14.8), and the Quadratic LME Model for the HIV Data

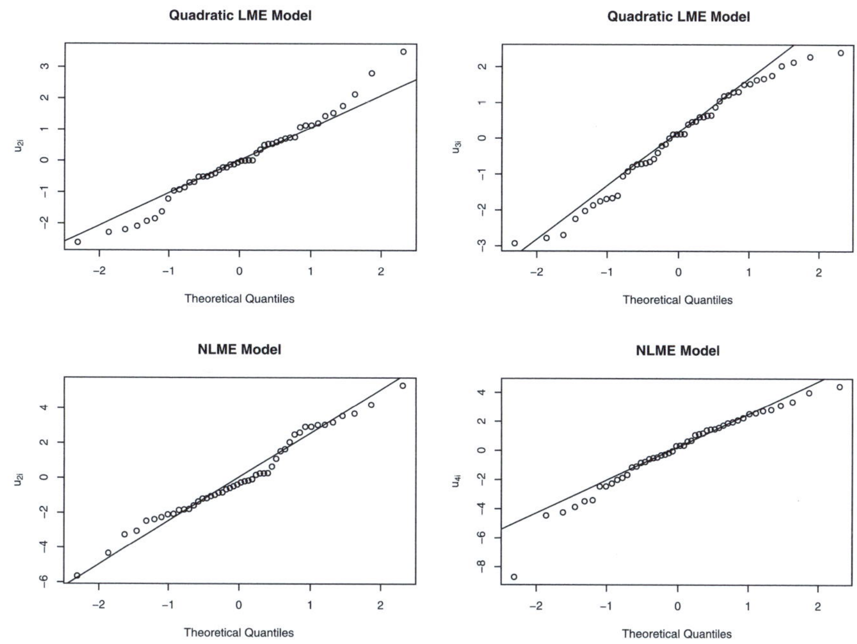

Figure 14.5 QQ Plots of Estimated Random Effects Based on the NLME Model (14.9) and the Quadratic LME Model for the HIV Data

We consider a dataset from an experiment on the pharmacokinetics of theophylline (Boeckmann, Sheiner, and Beal, 1994). In this study, 12 subjects were given oral doses of theophylline, then serum concentrations were measured at 11 time points per subject over the subsequent 25 hours. There are large variations in theophylline concentrations between and within subjects. We therefore introduce random effects and random errors in model (14.10) and obtain the NLME model

$$y_{ij} = \frac{\beta_{1i}\beta_{2i}x_i}{(\beta_{1i}-\beta_{2i})\beta_{3i}}\left(e^{-\beta_{2i}t_{ij}} - e^{-\beta_{1i}t_{ij}}\right) + \varepsilon_{ij}, i = 1, \ldots, m, \quad j = 1, \ldots, n_i, \tag{14.11}$$

where y_{ij} is the theophylline concentration at time t_{ij} for subject i, x_i is the dose of theophylline administered orally to subject i, $\boldsymbol{\beta}_i = (\beta_{1i}, \beta_{2i}, \beta_{3i})^T$ are subject-specific parameters, and ε_{ij} is a random error. We assume that $\varepsilon_{ij} \overset{iid}{\sim} N(0, \sigma^2)$. Note that all subject-specific parameters are necessarily positive to be meaningful, so we reparameterize NLME model (14.11) by transforming these parameters to log scale to enforce the positivity of the $\boldsymbol{\beta}_i$s, i.e., the reparameterized parameter $\boldsymbol{\eta}_i = (\eta_{1i}, \eta_{2i}, \eta_{3i})^T = (\log\beta_{1i}, \log\beta_{2i}, \log\beta_{3i})^T$.

Figure 14.6 shows the estimates and 95% confidence intervals for the subject-specific parameters $\boldsymbol{\eta}_i$ by fitting the NLME model (14.11) individually with R function **nlsList()**. We see that random effects may be needed for the fixed parameters η_1 and η_3, since not all confidence intervals for these two fixed parameters overlap. Thus, we consider the model

$$\eta_{1i} = \eta_1 + u_{1i}, \eta_{2i} = \eta_2, \eta_{3i} = \eta_3 + u_{2i}, \tag{14.12}$$

where $\boldsymbol{\eta} = (\eta_1, \eta_2, \eta_3)^T$ are the fixed effects parameters, and the random effects $\mathbf{u}_i = (u_{1i}, u_{2i})^T \overset{iid}{\sim} N(\mathbf{0}, \Omega)$, where Ω is an

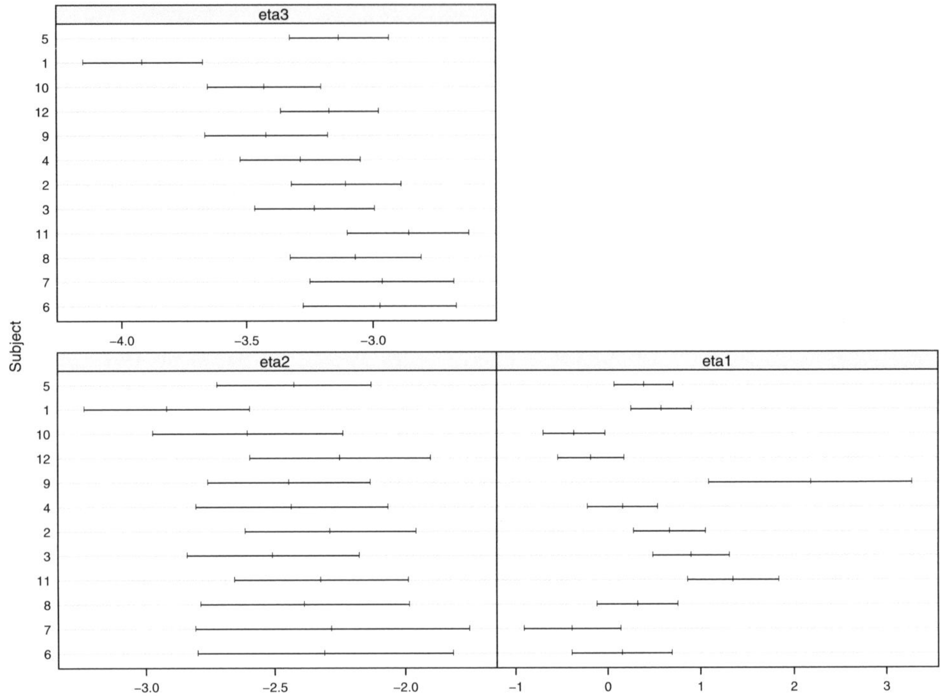

Figure 14.6 Estimates and 95% Confidence Intervals for Parameters η_i

unstructured variance-covariance matrix. We assume that the random effects $\mathbf{u}_i$ are independent of the random errors ε_{ij}.

To explore whether the time-independent covariate **weight** (Wt) is predictive for theophylline concentrations, we consider standard variable selection methods. Table 14.2 shows the values of AIC/BIC and log-likelihoods for various models, as well as the likelihood ratio test statistics and P-values for nested models. We can see that the NLME model (14.11)–(14.12) is substantially improved by incorporating the time-independent covariate weight (Wt) in η_{1i}. Hence, for the pharmacokinetics of theophylline we consider the NLME model (14.11) with

$$\begin{aligned} \eta_{1i} &= \eta_1 + \eta_w Wt_i + u_{1i}, \eta_{2i} = \eta_2, \\ \eta_{3i} &= \eta_3 + u_{2i}, \end{aligned} \tag{14.13}$$

where η_w is the coefficient of the time-independent covariate weight.

We fit NLME model (14.11) and (14.13) to the theophylline data. Table 14.3 presents parameter estimates and standard errors. We conclude that the effect of weight is significant. Figure 14.7 shows the fitted concentration trajectories based on the NLME model (14.11) and (14.13) for four subjects randomly selected from the theophylline data. We see that this NLME model fits the observed data well. Note that the estimate $\hat{\Omega}_{11}$ is equal to 0.41 and 0.30 respectively before and after the covariate weight is introduced in the model, suggesting that the covariate weight may partially explain the inter-subject variation.

14.5.3 Example 3: Growth Curve Models

Growth curve models describe the evolution of a quantity over time. Here we briefly describe a simple monomolecular growth curve model. Let $y(t)$ be the size at time t (e.g., size of an animal), and let $\mu(t) = E(y(t))$. Suppose that the growth rate β_1 is proportional to the remaining size (i.e., the difference

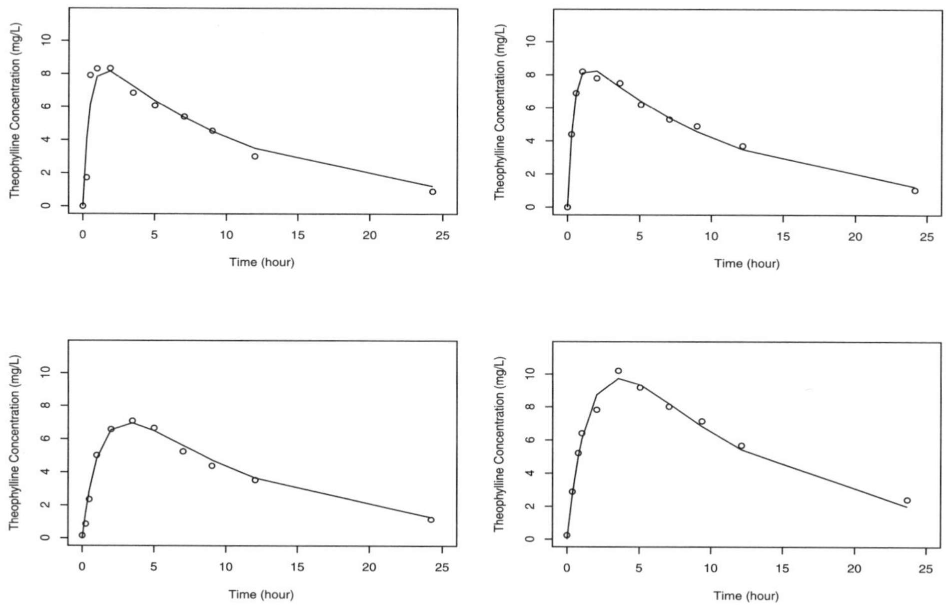

Figure 14.7 Observed (open circles) and Fitted (solid lines) Concentration Trajectories for Four Randomly Selected Subjects in the Theophylline Data

Table 14.2 Model Selection for NLME Model (14.11)–(14.12) with Covariate 'weight (Wt)'

Model	Weight (Wt)	AIC	BIC	loglik.	Test	Lik. Ratio	P-value
1	No Wt	368.04	388.22	−177.02	1 vs 2	4.48	0.03
2	Wt in η_{1i}	**365.57**	**388.63**	**−174.78**			
3	Wt in η_2	370.05	393.11	−177.02			
4	Wt in η_{3i}	367.45	390.52	−175.73			
5	Wt in η_{1i} and η_2	367.49	393.44	−174.75	5 vs 2	0.08	0.78
6	Wt in η_{1i} and η_{3i}	365.82	391.76	−173.91	6 vs 2	1.75	0.19
7	Wt in η_2 and η_{3i}	368.78	394.72	−175.39			
8	Wt in η_{1i}, η_2, and η_{3i}	366.63	395.46	−173.32	8 vs 2	2.93	0.23

between the limiting size β_0 and the current size $\mu(t)$). Then $\mu(t)$ satisfies the differential equation

$$\frac{d\mu(t)}{dt} = \beta_1(\beta_0 - \mu(t)), \qquad \beta_1 > 0.$$

The solution to this differential equation is the following monomolecular growth function

$$\mu(t) = \beta_0 + \beta_2 e^{-\beta_1 t}, \qquad (14.14)$$

where $\beta_2 = \beta_0 - \mu_0$, with μ_0 being the initial size at time $t = 0$.

Next, we apply the monomolecular growth model (14.14) to the cattle weight dataset from an experiment on the control of intestinal parasites in cattle. In an experiment to compare two treatments A and B for controlling the diseases, 60 animals were randomly assigned to treatment groups of size 30. The weight of each animal was recorded 11 times to the nearest kilogram. The detailed data description can be found in Kenward (1987). Figure 14.8 shows the data. To analyze the data, we introduce random effects to parameters β_0 and β_2 to account for the between-animal

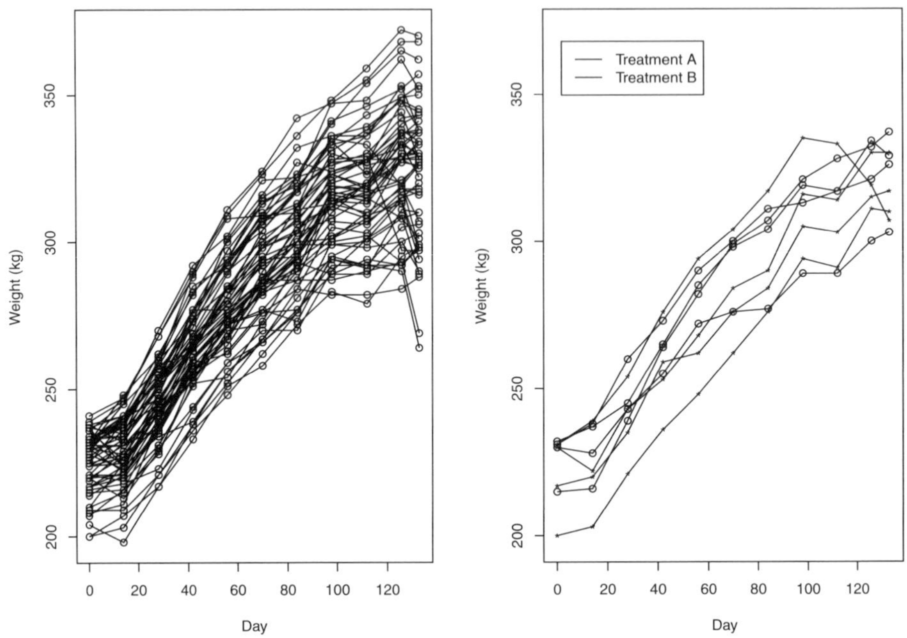

Figure 14.8 Weight Measurements from the Cattle Weight Data. Left figure: all animals. Right figure: four randomly selected animals

Table 14.3 Parameter Estimates for the Theophylline Data

Parameter	Estimate (standard error)
η_1	−2.56 (1.27)
η_w	0.044 (0.018)
η_2	−2.45 (0.05)
η_3	−3.23 (0.06)
σ	0.71
Ω_{11}	0.30
Ω_{12}	0.02
Ω_{22}	0.03

variation and consider the treatment effect on the growth rate.The resulting NLME model is given by

$$y_{ij} = \beta_{0i} + \beta_{2i} e^{-\beta_{1i} t_{ij}} + \varepsilon_{ij}, \quad i = 1, \ldots, m, \quad j = 1, \ldots, n_i, \tag{14.15}$$

$$\beta_{0i} = \beta_0 + u_{1i}, \quad \beta_{1i} = \gamma_0 + \gamma_1 I_{Bi}, \quad \beta_{2i} = \beta_2 + u_{2i}, \tag{14.16}$$

where y_{ij} is the weight for animal i at measurement time t_{ij}, ε_{ij} is the corresponding measurement error, I_{Bi} is an indicator function such that $I_{Bi} = 1$ if animal i is from the group with treatment B or $I_{Bi} = 0$ otherwise, the parameters $(\beta_0, \gamma_0, \gamma_1, \beta_2)^T$ are fixed, and $\mathbf{u}_i = (u_{1i}, u_{2i})^T$ are random effects. We assume that the measurement errors ε_{ij} are independent of the random effects $\mathbf{u}_i$ with $\varepsilon_{ij} \overset{iid}{\sim} N(0, \sigma^2)$ and $\mathbf{u}_i \overset{iid}{\sim} N(0, \Omega)$.

We fit the NLME model (14.15) and (14.16) to the cattle weight data. For comparison, we also fit the quadratic LME model

$$y_{ij} = \beta_{0i} + \beta_1 t_{ij} + \beta_{2i} t_{ij}^2 + \varepsilon_{ij}, \quad i = 1, \ldots, m, \quad j = 1, \ldots, n_i, \tag{14.17}$$

$$\beta_{0i} = \beta_0 + u_{1i}, \quad \beta_{2i} = \gamma_0 + \gamma_1 I_{Bi}. \tag{14.18}$$

Table 14.4 presents the parameter estimates, their standard errors, and the corresponding P-values based on the NLME model

Table 14.4 Parameter Estimates, Standard Errors (SE), and *P*-values for the NLME Model (14.15)–(14.16) and the Quadratic LME (QLME) Model (14.17)–(14.18)

NLME	Parameter	β_0	γ_0	γ_1	β_2	σ	Ω_{11}	Ω_{12}	Ω_{22}
	Estimate	385.64	0.0083	−0.0004	−168.25	7.59	734.96	−637.27	637.51
	SE	6.62	0.0005	0.0005	6.17				
	P-value	0.0000	0.0000	0.3888	0.0000				
QLME	**Parameter**	β_0	β_1	γ_0	γ_1	σ	Ω_{11}		
	Estimate	217.62	1.28	−0.0036	0.0002	9.14	196.50		
	SE	2.03	0.03	0.0002	0.0001				
	P-value	0.0000	0.0000	0.0000	0.061				

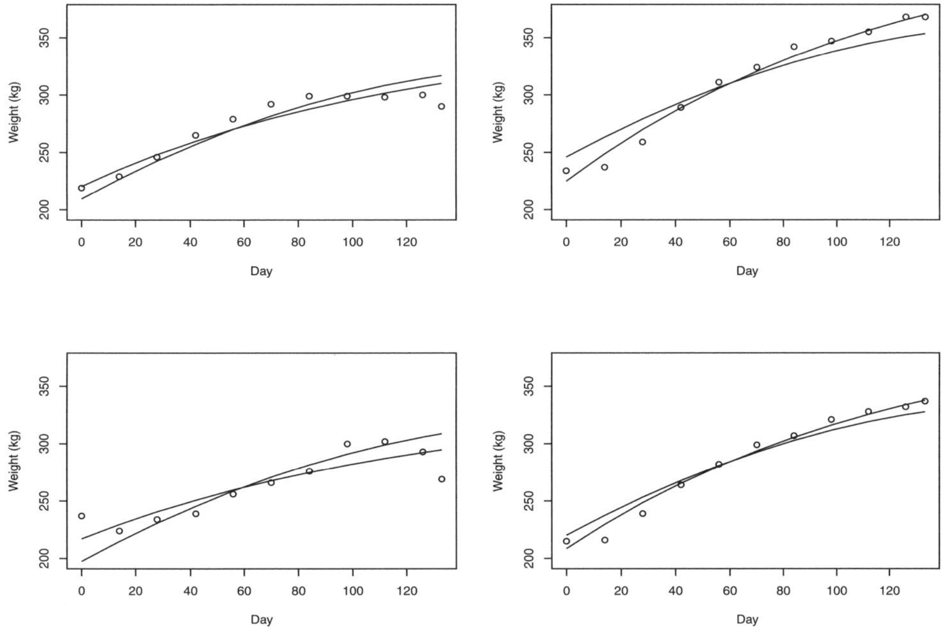

Figure 14.9 Observed (open circles) and Fitted Weight Trajectories Based on the NLME Model (14.15)–(14.16) (solid lines) and the Quadratic LME Model (14.17)–(14.18) (dashed lines) for the Four Cows Randomly Selected from the Cattle Weight Data

(14.15)–(14.16) and the quadratic LME model (14.17)–(14.18). Although the parameters in these two models are not comparable, we conclude that the treatment effect (γ_1) is not significant, which is consistent with the observation in Figure 14.8. Figure 14.9 shows the fitted weight trajectories based on the NLME model (14.15)–(14.16) and the quadratic LME model (14.17)–(14.18) for four cows randomly selected from the cattle weight data. We can conclude that the NLME model fits the observed data better

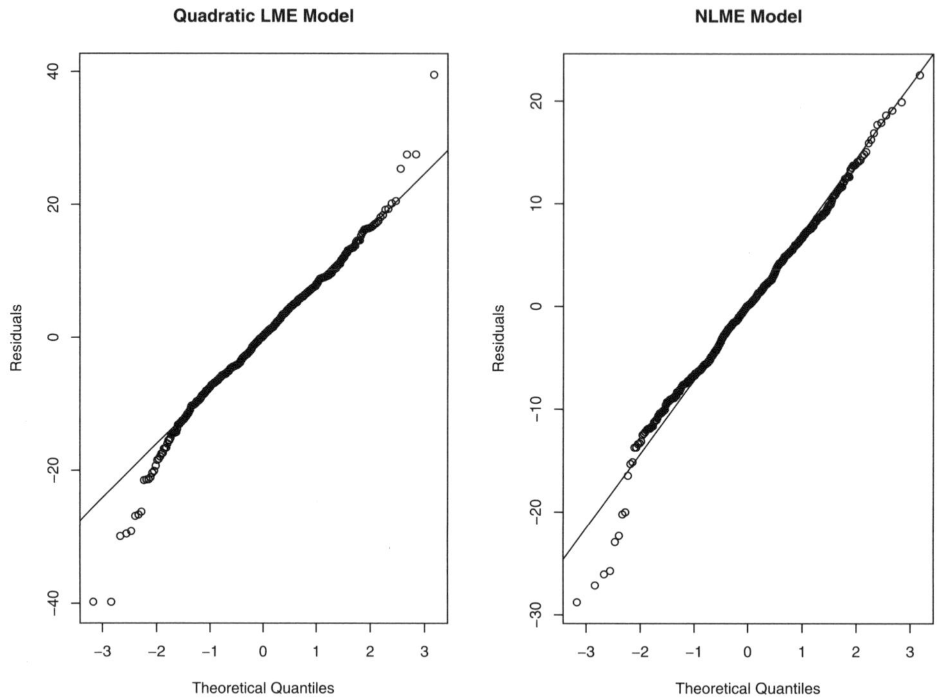

Figure 14.10 QQ Plots of the Residuals Based on the NLME Model (14.15)–(14.16) and the Quadratic LME Model (14.17)–(14.18)

than the quadratic LME model. The normality assumptions of the errors and the random effects appear to be reasonable, as can be seen from the QQ plots in Figures 14.10 and 14.11.

14.6 CONCLUSIONS AND EXTENSIONS

Nonlinear models are typically based on the underlying mechanisms that generate the data, so they may provide better understanding of the scientific problems and may provide more reliable predictions than linear models for values outside the observed-data range. With the development of modern computers and computational tools, which greatly reduce the computational difficulties associated with nonlinear models, nonlinear models are becoming increasingly attractive. However, nonlinear models may not be easily obtained for many real-world problems, and even if a nonlinear model can be derived from a real-world problem, it does not always provide a better fit to the observed data than a simpler linear model. Moreover, statistical inference for nonlinear models may sometimes offer computational problems such as non-convergence, especially when the sample size is not large.

For multilevel data, clustered data, or longitudinal data, we can introduce random effects in nonlinear models for cross-sectional data to specifically model the data in each level or each cluster. These random effects incorporate the correlation within each level or each cluster and the variation across different levels or clusters. The resulting NLME models allow us to make cluster-specific inference, which borrows information across clusters.

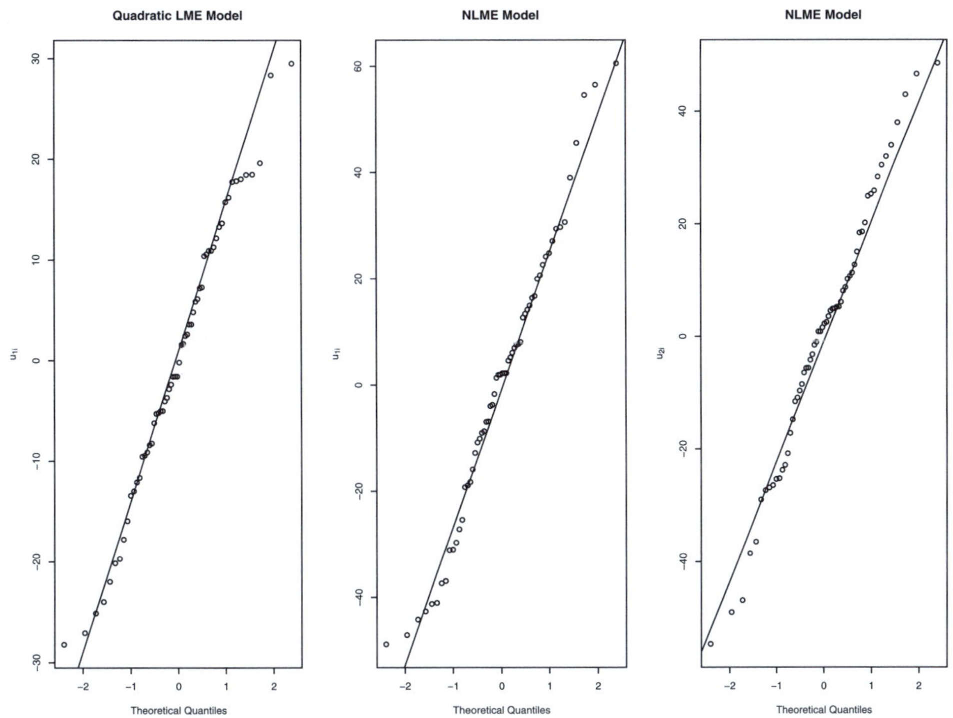

Figure 14.11 QQ Plots of the Random Effects Based on the NLME Model (14.15)–(14.16) and the Quadratic LME Model (14.17)–(14.18)

We have focused on models with two levels. These models can be extended to models with more than two levels. One approach is to introduce random effects at each level to incorporate the within-level correlations and between-level variations. These models can also be extended to multivariate responses. In this case, the covariance structure of the multivariate responses should be incorporated, as in multivariate linear regressions. Another extension is to model non-normal data, similar to GLMMs. Some of these extensions were discussed in Vonesh and Chinchilli (1997).

For cross-sectional data, standard software can be used to fit nonlinear models, such as the **nl()** function in R. A function for fitting multilevel models in R is **bugs()**. For NLME models, available softwares include the **nlme** package in R, the **nlmixed** procedure in SAS, the **mixed** command in SPSS, and the **xt** set of commands such as **xtmixed** in STATA. In pharmacokinetics and pharmacodynamics (PK/PD) modeling, the **NONMEM** package for fitting NLME models is popular. For beginners, an easy software package for fitting NLME models is **WinNonmix**.

REFERENCES

Bates, D.M. and Watts, D.G. (1988) *Nonlinear Regression Analysis and Its Applications*. New York: John Wiley.

Beal, S.L. and Sheiner, L.B. (1982) 'Estimating Population Kinetics', *CRC Critical Reviews in Biomedical Engineering*, 8: 195–222.

Beal, S.L. and Sheiner, L.B. (1989) *NONMEM Users Guide*. San Francisco, CA: University of California.

Boeckmann, A.J., Sheiner, L.B., and Beal, S.L. (1994) *NONMEM Users Guide – Part V, NONMEM Project Group*. San Francisco: University of California.

Davidian, M. and Giltinan, D.M. (1995) *Nonlinear Models for Repeated Measurements Data*. London: Chapman and Hall.

Heagerty, P. (1999) 'Marginally Specified Logistic-Normal Models for Longitudinal Binary Data', *Biometrics*, 55: 688–98.

Huang, Y. and Lu, T. (2008) 'Modeling Long-Term Longitudinal HIV Dynamics with Application to an AIDS Clinical Study', *The Annals of Applied Statistics*, 2: 1384–408.

Kenward, M.G. (1987) 'A Method for Comparing Profiles of Repeated Measurements', *Applied Statistics* 36: 296–308.

Lindsey, J.K. (2001) *Nonlinear Models in Medical Statistics*. Oxford: Oxford University Press.

Lindstrom, M.J. and Bates, D.M. (1990) 'Nonlinear Mixed Effects Models for Repeated Measures Data', *Biometrics*, 46: 673–87.

Pinheiro, J.C. and Bates, D.M. (2002) *Mixed-Effects Models in S and S-PLUS*. New York: Springer.

Ramsay, J.O., Hooker, G., Campbell, D., and Cao, J. (2007) 'Parameter Estimation for Differential Equations: A Generalized Smoothing Approach (with discussion)', *Journal of the Royal Statistical Society, Series B*, 69: 741–96.

Seber, G.A.F. and Wild, C.J. (1989) *Nonlinear Regression*. New York: John Wiley.

Vonesh, E.F. and Chinchilli, V.M. (1997) *Linear and Nonlinear Models for the Analysis of Repeated Measurements*. New York: Marcel Dekker.

Wolfinger, R. (1993) 'Laplace's Approximation for Nonlinear Mixed Models', *Biometrika*, 80: 791–5.

Wu, H. and Ding, A. (1999) 'Population HIV-1 Dynamics in Vivo: Applicable Models and Inferential Tools for Virological Data From AIDS Clinical Trials', *Biometrics*, 55: 410–18.

Wu, H. and Zhang, J.-T. (2002) 'The Study of Long-Term HIV Dynamics Using Semi-Parametric Models', *Statistics in Medicine*, 21: 3655–75.

Wu, H. and Zhang, J.-T. (2006) *Nonparametric Regression Methods for Longitudinal Data Analysis*. New York: Wiley-Interscience.

Wu, L. (2009) *Mixed Effects Models For Complex Data*. Boca Raton, FL: Chapman and Hall/CRC.

15

Generalized Linear Mixed Models: Estimation and Inference

Charles E. McCulloch and John M. Neuhaus

University of California, San Francisco, USA

15.1 INTRODUCTION

Statistical investigations often focus on the analysis of responses gathered in clusters or measured in longitudinal studies. For example, the OAI study (`www.oai.ucsf.edu`) enrolled approximately 5,000 participants and assessed them yearly for osteoarthritis, a degenerative joint disease, which is the major cause of physical disability in older persons. It is characterized by pain, stiffness, and loss of cartilage (a tissue that creates a cushion between bones in joints). The OAI obtained clinical information as well as X-rays and magnetic resonance images (MRIs) of both knees (the most common location of osteoarthritis). Consider studying the outcome of severe pain in a knee with a predictor of body mass index (BMI), a numerical measure of whether a person is overweight. The outcome is binary (yes/no), so a common approach would be to use a logistic regression model. But because the outcome is measured repeatedly over time on a single participant and for both knees, the data are likely correlated within a person, both at a single time (across knees) and across time. So logistic regression, which assumes that observations are independent, will not be appropriate for this multilevel data set.

Studies of osteoarthritis measure a variety of outcomes and predictors. BMI is measured at the participant level at each visit, not at the knee level, while other possible predictors might be measured at the participant level (e.g., gender or race), at the knee level (e.g., measures of early joint degeneration from the MRI taken at the first visit), at the participant and visit level (e.g., their age), or at the participant, knee, and visit level (e.g., X-ray information about remaining cartilage on each knee). In addition, the OAI collected a large amount of information from each participant on each visit. Other possible *outcomes* of interest include: number of days of work missed during the past three months due to knee pain (a count variable), joint space width (the width between the knee bones, which is a measure of the remaining cartilage, a numerical outcome), and how much difficulty participants had with their knees (measured on an ordinal scale of none, mild, moderate, and extreme).

While these outcomes are quite different, we might be interested in whether BMI is associated with each of them and, for each, there are correlated data issues. For work days missed or difficulty with knees there are likely correlations over time within a person, while joint space width, like pain, is measured at the knee level for each visit.

In this chapter we describe a class of models known as *generalized linear mixed models* (GLMMs) to analyze clustered and longitudinal data. These models are able to accommodate correlated data structures, as well as predictors measured at different levels within the data structure (e.g., at the person, person/knee, person/visit, or person/visit/knee level). They can also flexibly handle different types of outcomes by allowing a variety of distributional assumptions and by creating models for different aspects of the distribution, e.g., the mean, as in linear regression, or the log odds of the outcome, as in logistic regression.

We begin by describing in more detail the inferential settings in which GLMMs can be used. Then we discuss how they are specified, contrasting their construction with that of an alternative technique, *generalized estimating equations*. Next we cover methods for fitting them to data, using techniques such as maximum likelihood or modern Bayesian computation. We briefly describe how to diagnose assumptions and assess robustness. We close with a brief summary.

15.2 INFERENTIAL GOALS AND EXAMPLES

Generalized linear mixed models allow the usual statistical inferences available from regression modeling: testing the association of an outcome with a predictor, quantifying the degree of association, or generating estimates of the mean value for given values of the predictors. In the context of the OAI, we might be interested in testing if BMI is associated with the participants progressing more rapidly towards arthritis of the knees, or understanding, on average, the amount of joint space width decrease with increasing age of the participant, or we could obtain predicted probabilities of knee pain for participants with and without evidence of knee degeneration on their MRIs. If the data were not correlated, these questions could be answered, respectively, with ordinal logistic regression, linear regression, and logistic regression.

Because GLMMs accommodate correlated data explicitly by incorporating random effects (as described in detail in the next section), they can be used to perform two types of statistical inference not possible with more standard regression models: prediction of realized random effects and variance components estimation. For example, in OAI, it might be of interest to generate participant-specific predictions of the likelihood of knee pain at the next visit, on the basis of measured predictors (e.g., BMI and gender) *and* on the basis of pain measures from the current and previous visits.

The goal in variance components estimation is to quantify the variability attributable to various levels of a multilevel data set. For example, we might describe the variability in joint space width across participants, within participants across time, or within participants across knees. A study of diagnostic procedures in the Northern California Kaiser healthcare system (Selby et al., 1996) analyzed use of coronary angiography (a diagnostic procedure using x-rays and a special dye to diagnose coronary artery disease) as a binary outcome (yes/no) in patients across 16 hospitals. A number of patient-level predictors were also included to adjust for possible differences in the patient mix across the 16 hospitals. A large variance across hospitals in the model, after adjustment for patient characteristics, could well indicate over- or under-use of the procedure. If the procedure is over-used it indicates poor use of resources. If the procedure is under-used, it could lead to poor clinical outcomes for the patients. In this example, the primary goal was to test for a non-zero variance component and estimate its magnitude.

15.3 SPECIFICATION OF GLMMS

Generalized linear mixed models accommodate correlated data and allow prediction of random effects and quantification of variability by incorporating random terms into the linear predictor portion of a generalized linear model. As outlined in Chapter 8, constructing a GLMM involves four steps:

1. Specify the distribution of the outcome Y for a fixed value of the predictors and conditional on the random effects of Step 4 below. For example, we might choose binary, normal, or Gamma.
2. Specify which predictors of the mean outcome to include. In a generalized linear mixed model the predictors are incorporated through what is known as the *linear predictor*, typically denoted as η, and defined as $\eta = \beta_0 + \beta_1 x_1 + \cdots + \beta_p x_p$, where the βs are regression parameters to be estimated from the data and the x_i are the predictors.
3. Decide what function of the mean to model as a function of the predictors. In a GLMM, this is of the form $\mu = g^{-1}(\eta)$ or $g(\mu) = \eta$, where g is a known monotonic function, called the link function.
4. Declare one or more of the categorical predictors to be random factor(s) by assigning a distribution to the effects associated with the levels of that factor.

In the OAI example, we could model pain as a binary outcome, so its distribution would necessarily be Bernoulli, and we might choose to incorporate the predictors of BMI, gender, age, visit, person, and knee within a person into the model. With $Y = 1$ when a person reports pain (and 0 otherwise) and p the probability of pain, we have $\mathrm{E}[Y] = \mu = p$, and we could model the log odds of the outcome by choosing $g(\mu) = g(p) = \log[p/(1-p)]$ and thus utilizing a form of logistic regression. And we could declare the person effects and the knees-per-person effects to be random. This last step is an extremely important one, because, as illustrated in the very first chapter, declaring a factor to be a random effect induces a correlation among observations that share the same level of the random effect.

15.3.1 The Model

Putting together the ingredients above, we are now in a position to formally write down the generalized linear mixed model. For the ith observation, let Y_i denote the outcome with mean μ_i conditional on the random effects, $\mathbf{x}_i$ be the vector of (non-random) predictors, also called the fixed effects, $\mathbf{z_i}$ be the model matrix for the random effects, $g(\cdot)$ be the link function, $\boldsymbol{\beta}$ be the regression coefficients, and $\mathbf{u}$ be the vector of random effects. Our GLMM is then

$$\begin{aligned} Y_i|\mathbf{u} &\sim \text{indep. } f_Y(Y_i|\mathbf{u}) \\ g(\mu_i) &= \mathbf{x}_i'\boldsymbol{\beta} + \mathbf{z}_i'\mathbf{u} \\ \mathbf{u} &\sim h(\mathbf{u}), \end{aligned} \tag{15.1}$$

where h is a specified density function. For example, for a binary outcome, we might posit a logit link and assume a normal distribution for a model with random intercepts:

$$\begin{aligned} &Y_{ij}|u_i \sim \text{indep.Bernoulli}(p_{ij},), \\ &i = 1, \ldots, m;\ j = 1, \ldots, n_i \\ &\log[p_{ij}/(1-p_{ij})] = \beta_0 + \beta_1 x_i + u_i \\ &\text{or } \ p_{ij} = 1/[1 + \exp\{-(\beta_0 + \beta_1 x_i + u_i)\}] \\ &u_i \sim \text{indep. } \mathcal{N}(0, \sigma_u^2). \end{aligned} \tag{15.2}$$

15.3.2 More General Models

Although the class of models encompassed by the specification in (15.1) is very broad, there are two key restrictions. First, the predictors are all assumed to enter as a linear combination, and through the linear predictor, η. This would not be able to accommodate models in which other aspects of the distribution (e.g., the variance) are functionally dependent on predictors. Second, in its basic form, GLMMs are restricted to covariance structures that can be represented as random effects in the linear predictor. When the distribution is a flexible one, like a normal distribution, more complicated variance-covariance structures can be built into the distribution of Y itself, but this is not as easy with distributions like the Binomial, which do not have natural multivariate analogs.

15.3.3 Conditional Versus Marginal Approaches

A feature of the specification of the GLMM in Section 15.3.1 is that the distribution of Y is specified conditional on the random effects. An alternate approach is to directly specify a regression model for the marginal mean of Y. To contrast this with (15.2) and using $\text{logit}(p) = \log[p/(1-p)]$, we would have

$$\begin{aligned} \text{logit}(\text{E}[Y_i]) &= \beta_0^* + \beta_1^* x_i \\ \text{logit}(\text{E}[Y_i|b_i]) &= \beta_0 + \beta_1 x_i + u_i \qquad (15.3) \end{aligned}$$

for the marginal approach and GLMM approach, respectively. The most popular of the marginal approaches is the method of *generalized estimating equations*, or GEEs. See Chapter 25 for more detail on GEEs.

GEEs do not specify the full distribution of the responses but instead model the marginal distribution of a single observation, along with some working assumptions about the variance-covariance structure of the responses. With the marginal approach, one estimates model parameters as the solution of a set of equations known as the generalized estimating equations and estimates the variance-covariance structure empirically, rather than using likelihood-based methods. It is well known (e.g., Zeger et al., 1988; Neuhaus et al., 1991) that the marginal and conditional approaches do not estimate the same effects of predictors on the response, and the asterisks on the βs in (15.3) emphasize this fact. Unlike conditional approaches, the interpretation of the regression parameters of marginal approaches does not involve the cluster structure of the data and the approaches measure the same covariate effects that one could estimate with a single observation per subject.

Neuhaus and Jewell (1993) assessed differences in the magnitudes of marginal and conditional covariate effects in random intercept settings and presented an approximation to relate them. They showed that

$$\begin{aligned} \beta^* &\approx \beta\, g'[E\{g^{-1}(b)\}]E[1/g'\{g^{-1}(b)\}] \\ &\equiv \beta\, H'(0). \end{aligned}$$

The direction and magnitude of the term $H'(0)$ depends on the curvature of the link function g. For example, $0 < H'(0) < 1$ for $1/g'$ concave and for these links the marginal covariate effects are closer to the null value of 0 than conditional covariate effects. Such links include the logistic, probit, and complementary log-log. $H'(0) = 1$ for identity link models, indicating that marginal and conditional approaches for clustered Gaussian data measure the same effects of covariates on response.

Which method is to be preferred in which instances? Marginal models are most helpful when scientific interest focuses on assessing differences in the marginal expected value of the response between groups defined by levels of a covariate and the assessment of within-cluster differences is not of interest. The approach accommodates the clustering of responses into variance estimation but the cluster structure plays no role in the interpretation of covariate effects. This has sometimes been called a 'public health' focus. In the OAI analysis mentioned earlier, the effect of race, which is constant for a participant over time, might be a logical target of such an approach. On the other hand, within-person trends in pain would not be since the model is built for the 'average' person, not an individual. In contrast to marginal approaches, the interpretation of covariate effects in generalized linear mixed models and conditional approaches is conditional on the random effects in the models, and these approaches directly relate within-cluster changes in covariates to changes in the response. For example, with longitudinal data conditional approaches would assess within-subject changes in the response over time and relate these changes to within-subject changes in predictors. Such associations are the focus of scientific interest in longitudinal studies. See Chapter 9 for more detail on longitudinal analyses.

Needless to say, since the marginal approach does not have person-specific terms, nor a way to describe the random effects variance, it is not useful for generating predictions of the realized values of the random effects, nor for variance components estimation. For a more detailed critique of marginal modeling and discussion of the differences in the marginal and conditional approaches to inference, see Lindsey and Lambert (1998). For an interesting hybrid approach see Heagerty and Zeger (2000).

15.4 FITTING AND INFERENTIAL METHODS

In this section we consider methods for fitting GLMMs to data. These include maximum likelihood, Bayesian methods, and penalized likelihood, as well as methods based on Laplace approximations and h-likelihood. We begin with the most commonly used method, maximum likelihood.

15.4.1 *Maximum Likelihood*

Maximum likelihood, or variants of maximum likelihood such as restricted maximum likelihood, are standard methods of estimation for linear mixed models and generalized linear models (e.g., logistic regression). Maximum likelihood is also widely used for fitting GLMMs and has many attractive properties:

- ML estimators are asymptotically normal, with standard errors available from second derivatives of the log likelihood.
- Under broad conditions, maximum likelihood estimators are asymptotically efficient.
- Hypothesis testing can be carried out using likelihood ratio, score, or Wald tests.

Use of maximum likelihood for GLMMs is no different than for other statistical models, with one possible exception. As noted above, a feature of GLMMs is the ability to quantify variances of random effects. A common hypothesis of interest is testing if the variance component is zero, either because it is of direct inferential interest, as in the Kaiser coronary angiography example, or because the associated random effect can be dropped from the model if the variance component is zero. The likelihood ratio test can still be used in this scenario, but care must be taken with the large sample distribution since the hypothesis is on the boundary of the parameter space. The test is then sometimes known as the $\bar{\chi}^2$ test and its large sample null distribution is a mixture of chi-square distributions (Self and Liang, 1987). In the simple case of a model with a single random effect, the use of the $\bar{\chi}^2$ test is straightforward and is achieved by halving the p-value from a standard, 1 degree of freedom chi-square test. This is akin to adjusting the p-value from a two-sided t-test for a one-sided hypothesis.

Although use of maximum likelihood for GLMMs is conceptually straightforward, it is computationally difficult even to calculate the likelihood, let alone to maximize it. This is because the model is specified conditional on the random effects, and calculation of the likelihood requires integrating over the random effects. Consider the simple logistic mixed model (15.2), with normally distributed random intercepts and a between-cluster covariate, x_i. The contribution to the likelihood from the data associated with the ith cluster is given by $\mathrm{P}\{\mathbf{Y}_i = \mathbf{y}_i | \mathbf{x}_i\}$, or

$$\begin{aligned}
&\int \mathrm{P}\{\mathbf{Y}_i = \mathbf{y}_i | u_i = u, \mathbf{x}_i\} h_{u_i}(u)du \\
&= \int \prod_{j=1}^{n_i} \mathrm{P}\{Y_{ij} = y_{ij} | u_i = u, x_i\} h_{u_i}(u)du \\
&= \int \prod_{j=1}^{n_i} p_{ij}^{y_{ij}} (1 - p_{ij})^{1-y_{ij}} h_{u_i}(u)du \\
&= \int \prod_{j=1}^{n_i} p_{i}^{y_{ij}} (1 - p_{i})^{1-y_{ij}} h_{u_i}(u)du \quad (15.4) \\
&= \int \left(\frac{p_i}{1 - p_i}\right)^{y_{i\cdot}} (1 - p_i)^{n_i} h_{u_i}(u)du \\
&= \int \frac{e^{y_{i\cdot}(\beta_0+\beta_1 x_i+u)}}{(1 + e^{\beta_0+\beta_1 x_i+u})^{n_i}} \frac{e^{-u^2/2\sigma_u^2}}{\sqrt{2\pi\sigma_u^2}} du, \quad (15.5)
\end{aligned}$$

where $y_{i\cdot} = \sum_j y_{ij}$ and the simplification in (15.4) occurs because, for model (15.2), $p_{ij} = p_i$. Unfortunately, (15.5) cannot be evaluated in closed form, so the likelihood cannot be maximized with a closed form solution. In the next section we consider numerical approaches to this problem.

15.4.2 Computation for Maximum Likelihood

As displayed in (15.5), generalized linear mixed models create computational challenges to the use of maximum likelihood. Even the contribution to the likelihood of a single cluster to a very simple logistic model, (15.2), cannot be evaluated in closed form. For more complicated models the likelihood computations become more problematic and, beyond that, there is still the consideration of how to maximize the likelihood.

With a single, normally distributed random effect, the likelihood is the product of one-dimensional integrals of the form

$$\int_{-\infty}^{\infty} s(u)\frac{e^{-u^2/(2\sigma_u^2)}}{\sqrt{2\pi\sigma_u^2}}\,du. \tag{15.6}$$

Using the change of variables $u = \sqrt{2}\sigma_u v$, (15.6) can be written as

$$\int_{-\infty}^{\infty} s(\sqrt{2}\sigma_u v)\frac{e^{-v^2}}{\sqrt{\pi}}\,dv \equiv \int_{-\infty}^{\infty} s^*(v)e^{-v^2}dv, \tag{15.7}$$

where $s^*(\cdot) \equiv s(\sqrt{2}\,\sigma_u\cdot)/\sqrt{\pi}$.

Gauss–Hermite quadrature

Numerical integration over the range from $-\infty$ to ∞ can be difficult. But for the special case where a smooth function is being integrated against e^{-v^2}, the method of Gauss–Hermite quadrature can be applied. In Gauss–Hermite quadrature, the integral in (15.7) is approximated by an easily computed weighted sum:

$$\int_{-\infty}^{\infty} s^*(v)e^{-v^2}\,dv \doteq \sum_{k=1}^{d} s^*(x_k)w_k, \tag{15.8}$$

where standard references, such as Abramowitz and Stegun (1964), provide the weights, w_k, and the evaluation points, x_k.

Gauss–Hermite quadrature has the property that, when the summation in (15.8) extends to d points, Gauss–Hermite quadrature gives the exact answer for all polynomials up to degree $2d - 1$. By using large values of d, (15.8) will yield accurate approximations for integration of likelihood functions as long as they can be well approximated by a high-degree polynomial. Numerical evaluation of these integrals is not automatic. To improve the approximations, the integrands are often centered and scaled to make sure that the calculations are used in the region where (15.5) has the most non-zero contribution.

There is, of course, a tradeoff of degree of computational approximation against longer computational times with large values of d. As a practical matter, we have found that quadrature (especially using a centered calculation) usually gives a good degree of approximation with $d = 20$ quadrature points. Difficult problems can sometimes yield inaccurate fits when using values less than 10.

Derivatives of the log likelihood and predicted values of realized random effects require similar approximations for numerical evaluation. Derivatives of the log likelihood are needed both for calculation of standard errors and Wald tests, and some algorithms to numerically maximize the log likelihood.

Beyond mere calculation of the likelihood, to obtain model fits we would still need to maximize it (numerically) to get estimates of (in our simple example) β_0, β_1, and σ_u^2. To avoid calculation of derivatives (which also cannot be evaluated in closed form and require integral approximations), quasi–Newton algorithms or derivative-free maximization methods are often used.

Laplace approximation

Another approach proposed to approximate (15.5) is the Laplace approximation (e.g., Wolfinger, 1993; Raudenbush et al., 2000).

It is applicable to integrals of the form $\int_{-\infty}^{+\infty} e^{-\kappa(u)}du$, where $\kappa(u)$ has a mode, u_0, such that $\kappa'(u_0) = 0$. A Laplace approximation starts with a Taylor series expansion around u_0:

$$\begin{aligned}\kappa(u) &\approx \kappa(u_0) + \kappa'(u_0)(u - u_0) + \\ &\quad \tfrac{1}{2}\kappa''(u_0)(u - u_0)^2 \\ &= \kappa(u_0) + \tfrac{1}{2}\kappa''(u_0)(u - u_0)^2. \quad (15.9)\end{aligned}$$

Using (15.9) yields

$$\begin{aligned}\int_{-\infty}^{+\infty} e^{-\kappa(u)}du &\approx \int_{-\infty}^{+\infty} \exp(-\kappa(u_0) \\ &\quad - \tfrac{1}{2}\kappa''(u_0)(u - u_0)^2)du \\ &= e^{-\kappa(u_0)}\sqrt{2\pi/\kappa''(u_0)}.\end{aligned}$$

For example, to use the Laplace approximation for (15.5), excepting constants, we would set

$$\begin{aligned}\kappa(u) &= -y_{i\cdot}(\beta_0 + \beta_1 x_i + u) \\ &\quad + \log(1 + e^{\beta_0+\beta_1 x_i+u})^{n_i} + u^2/2\sigma_u^2.\end{aligned}$$

A natural question is whether direct numerical maximization of the likelihood evaluated by use of the Laplace approximation leads to approximately unbiased estimators. Pinheiro and Chao (2006), via simulation, and Joe (2008), via evaluation of the asymptotic limit of maximizers of the Laplace-approximated likelihood, show that it can work well in some cases, but suffers from some degree of bias, especially with binary outcomes and with small cluster sizes. To overcome this drawback, some authors have explored using higher-order Laplace approximations (Raudenbush et al., 2000). General indications are that direct numerical maximization of a Laplace-approximated likelihood is somewhat less biased than quasi-likelihood approaches (see Section 15.4.3).

Other computational methods

The calculations described above are restricted to normally distributed random effects and are for a very simple model. Because the numerical calculations are quite difficult, alternative methods for calculating the maximum likelihood estimates have been explored. Simulation-based evaluation of the integrals coupled with the expectation maximization algorithm, the Newton–Raphson or other approaches to maximize the likelihood have been explored by a number of authors (Booth and Hobert, 1999; McCulloch, 1997; Bellio and Varin, 2005; Steele, 1997). Lele et al. (2010) proposed a novel alternative approach to use the Bayesian computational machinery to evaluate maximum likelihood estimates.

15.4.3 Penalized Quasi-likelihood and h-likelihood

A different approach to fitting GLMMs starts by extending an approach to fitting generalized linear models. Generalized linear models are often fit by maximizing the log of the *quasi-likelihood* (QL) defined as

$$\sum_i Q_i = \sum_i \int_{y_i}^{\mu_i} \frac{y_i - t}{\tau^2 v(t)}dt, \qquad (15.10)$$

where Q_i is the contribution to log QL from the ith observation and $v(\mu)$ is the so-called variance function defined by $\text{var}(Y) = \tau^2 v(\mu)$. Quasi-likelihood is attractive because it does not explicitly make distributional assumptions, but instead rests only on the assumed relationship of variance to mean. By constructing the quasi-likelihood as in (15.10), it retains many of the features of a log likelihood and can form the basis of valid asymptotic inference (McCullagh and Nelder, 1989).

To solve for the QL estimates, it is common to solve the score equation

$$\begin{aligned}\mathbf{0} &= \frac{\partial}{\partial\boldsymbol{\beta}}\sum Q_i \\ &= \sum \frac{y_i - \mu_i}{\tau^2 v(\mu_i)}\frac{\partial\mu_i}{\partial\boldsymbol{\beta}} \\ &= \sum \frac{y_i - \mu_i}{\tau^2 v(\mu_i) g_\mu(\mu_i)}\mathbf{x}_i'.\end{aligned}$$

To see how this might generalize to GLMMs, consider a simple model with the variance of Y proportional to its mean, but not otherwise specifying a distribution for Y:

$$\begin{aligned} \mathrm{E}[Y_{ij}|u_i] &= \mu_{ij} \\ \log \mu_{ij} &= \mathbf{x}'_{ij}\boldsymbol{\beta} + u_i \\ \mathrm{var}(Y_{ij}|u_i) &= \tau^2 \mu_{ij} \\ u_i &\sim \text{indep. } \mathcal{N}(0, \sigma_u^2). \end{aligned} \tag{15.11}$$

Proceeding in a similar vein for GLMMs, the log of the quasi-likelihood for (15.11) is given by

$$\begin{aligned} &\log QL(\boldsymbol{\beta}, \mathbf{u}) \\ &= \Big(\sum_{i,j} y_{ij}\mathbf{x}'_{ij}\boldsymbol{\beta} + \sum_i y_{i\cdot}u_i \\ &\quad - \sum_i e^{u_i} \sum_j e^{\mathbf{x}'_{ij}\boldsymbol{\beta}}\Big)/\tau^2. \end{aligned} \tag{15.12}$$

There is immediately a problem because (15.12) involves the unknown and unobserved random effects u_i. In a full likelihood approach, we would simply integrate over the random effects. Following a similar path for the quasi-likelihood suggests maximizing

$$\int \cdots \int \exp\left\{\log QL(\boldsymbol{\beta}, \mathbf{u}) - \Big(\sum_i u_i^2\Big)/2\right\} (2\pi\sigma_u^2)^{-m/2} du_1 \ldots du_m. \tag{15.13}$$

A primary motivation for considering QL is to avoid the computational difficulties associated with maximum likelihood. But the integrated quasi-likelihood (15.13), is no easier to evaluate than the actual likelihood. Two approaches to this problem have been investigated: iterative, weighted least squares and use of the Laplace approximation.

It is common to fit generalized linear models, such as logistic or Poisson regression, using an iteratively reweighted least squares algorithm. This can be motivated by considering a Taylor series approximation to the outcome, Y, transformed by the link function

$$\begin{aligned} g(Y_i) &\approx g(\mu_i) + (y_i - \mu_i)g'(\mu_i) \\ &= \eta_i + (y_i - \mu_i)g'(\mu_i) \\ &\equiv \mathbf{x}'_i\boldsymbol{\beta} + \varepsilon_i g'(\mu_i) \\ &\equiv w_i. \end{aligned} \tag{15.14}$$

Denote by $\boldsymbol{\beta}^{(k)}$ a current estimate of $\boldsymbol{\beta}$. Then (15.14) can be used to develop the 'working' outcome, $w_i^{(k)} = \mathbf{x}'_i\boldsymbol{\beta}^{(k)} + (y - \mu_i^{(k)})g'(\mu_i^{(k)})$, that can be fit by a linear regression of $w_i^{(k)}$ on the predictors $\mathbf{x}$ using weighted least squares. From this fit, we can update parameter estimates, get a new working outcome, and iterate the process until convergence.

The same idea has been used to fit GLMMs. Starting from (15.1) we can perform the same expansion:

$$\begin{aligned} g(Y_i) &\approx g(\mu_i) + (y_i - \mu_i)g'(\mu_i) \\ &= \eta_i + (y_i - \mu_i)g'(\mu_i) \\ &\equiv \mathbf{x}'_i\boldsymbol{\beta} + \mathbf{z}'_i\mathbf{u} + \varepsilon_i g'(\mu_i) \\ &\equiv w_i. \end{aligned}$$

This is not quite so convenient to fit, because of the random effects and, instead of a linear regression, we would need to fit a linear mixed model at each step, which is typically an iterative process in itself. One way to fit the linear mixed model is via what are called the *mixed model equations* (Searle et al., 1992). These equations use an explicit formula that relates the predicted random effects to the estimate of the random effects variance. The combination of using working variates and the mixed model equations was proposed by Schall (1991) as a way to fit GLMMs.

Another alternative to approximate (15.13) is to use a Laplace approximation (Section 15.4.2). For the model (15.11) we would define κ as

$$\kappa(u) = \left(-y_{i\cdot}u_i + e^{u_i}\sum_j e^{x'_{ij}\beta}\right)/\tau^2 + u_i^2/2\sigma_u^2,$$

so that

$$\kappa'(u) = \left(-y_{i\cdot} + e^{u_i}\sum_j e^{x'_{ij}\beta}\right)/\tau^2 + u_i/\sigma_u^2$$

and

$$\kappa''(u) = \left(e^{b_i} \sum_j e^{x'_{ij}\beta}\right) / \tau^2 + 1/\sigma_u^2.$$

Solving for u_i^0 would thus require

$$0 = \left(-y_{i\cdot} + e^{u_i^0} \sum_j e^{x'_{ij}\beta}\right) / \tau^2 + u_i^0/\sigma_u^2$$

or

$$u_i^0 = \sigma_u^2 \left(y_{i\cdot} - \sum_j \mu_{ij}\right) / \tau^2.$$

We can finally get an approximation to the log quasi-likelihood as

$$\begin{aligned}\log QL \;\; = \;\; & \sum_{ij} y_{ij}(\mathbf{x}'_{ij}\boldsymbol{\beta} + \mathbf{z}'_{ij}\mathbf{u}^0)/\tau^2 \\ & - \sum_i e^{u_i^0} \sum_j e^{x'_{ij}\beta} \sigma_u^2/\tau^2 \\ & - \frac{1}{2} \sum_i \log \\ & \left(e^{u_i^0} \sum_j e^{x'_{ij}\beta} \sigma_u^2/\tau^2 + 1\right) \\ & - \sum_i u_i^{0^2}/2\sigma_u^2. \qquad (15.15)\end{aligned}$$

Ignoring the fact that the term with the logarithm contains β, the remainder of (15.15) is the log quasi-likelihood with the additional 'penalty' term of $\sum_i u_i^{0^2}/2\sigma_u^2$ added on (hence the name penalized quasi-likelihood). We can think of this either as arising from a normal distribution assumption for the random effects, as in (15.11), or merely as a penalty function that keeps the individual random effects u_i relatively 'small'. We can maximize (15.15) as a function of $\boldsymbol{\beta}$. This could be embedded in an iterative algorithm along with a calculation to get an estimate of σ_u^2 from the u_i (e.g., the mixed model equations) to find estimates for the GLMM.

Because the justification is ad hoc and involves approximations, the performance of estimates from this approach must be evaluated afresh. Unfortunately, the method does not work well when the distributions are far from normal, e.g., binary or Poisson with small mean counts (Breslow and Clayton, 1993; Breslow and Lin, 1995; Lin and Breslow, 1996; Neuhaus and Segal, 1997). This is arguably when GLMMs are most needed because we cannot get away with fitting a linear mixed model that assumes approximately normal outcomes.

Lee and Nelder (1996, 2001) developed what can be viewed as a variation of quasi-likelihood, called *h-likelihood*, short for hierarchical likelihood. They form the joint distribution of the outcomes and the random effects, $f_{\mathbf{Y},\mathbf{u}}$, and simultaneously maximize it with respect to $\mathbf{u}$ and $\boldsymbol{\beta}$. This 'joint maximization' method thus finds the predicted value of $\mathbf{u}$ that is needed to solve for the MLE of $\boldsymbol{\beta}$. For the normal distribution linear mixed model it is equivalent to the mixed model equations and works quite well to find the MLEs.

Early implementations of the h-likelihood method suffered from many of the same drawbacks as penalized quasi-likelihood (Engel and Keen, 1996), with badly biased estimators for highly non-normal distributions. By revisiting the estimation equations for the random effects variance, Lee and Nelder (2004) improved the small-sample performance of the method. As a practical analysis method, h-likelihood performs adequately in many circumstances, but there are lingering questions about its theoretical basis (see, for example, the discussions of Lee and Nelder, 2009).

15.4.4 Bayesian Method

The Bayesian paradigm offers an alternative to the frequentist approaches described above for fitting GLMMs. Both frequentist and Bayesian inference begin with a GLMM such as (15.1) to relate unobservable parameters to the measured outcomes. However, in Bayesian modeling further assumptions are made about distributions for the parameters themselves, such as $\boldsymbol{\beta}$, rather than regarding them as fixed, unknown constants, as does the frequentist approach. In the frequentist paradigm, random effects are modeling tangible physical distributions, while prior distributions for parameters are not. For example, in the context of the OAI, the individual propensities to pain (and hence their distribution) could be measured fairly accurately with sufficient data. But the distribution for the overall propensity to pain is a conceptual quantity. See Chapter 4 for more detail on Bayesian approaches.

15.4.5 Other Fitting Methods

There are several other methods of fitting that are less commonly used than those described above, including conditional likelihood and nonparametric fits. Conditional likelihood is a technique that is readily available only for random intercept models, and then only when interest focuses on a predictor that changes within a cluster. It proceeds by constructing the likelihood conditional on the sufficient statistics for the intercept terms, thus removing them completely from the likelihood and obviating any need to specify a parametric distribution for the random effects (see, for example, McCulloch et al., 2008, Section 12.4a). The existence of sufficient statistics for this purpose is restricted to a few distribution and link function combinations. For example, it works for logistic/binomial, log/Poisson and identity/normal but not for other combinations, such as probit/binomial. Neuhaus and McCulloch (2006) suggest between- and within-cluster partitioning methods as a way to approximate conditional inference in situations for which conditional likelihoods cannot be constructed. Conditional likelihood methods, unlike maximum likelihood methods, can be biased under missing data that is 'missing at random' (Rathouz, 2004; Roy et al., 2006). Under the somewhat restrictive situations in which conditional likelihood methods are applicable, they are almost always a good technique to try, at the very least to understand sensitivity to parametric assumptions. Verbeke et al. (2001) explore extensions to conditional maximum likelihood that allow combining features of conditional and random effects inference.

Another method whose goal is to avoid making parametric specification of a random effects distribution is to fit a nonparametric mixing distribution (Laird, 1978; Böhning, 1995; Aitkin, 1999; Lesperance and Kalbfleisch, 1992). The nonparametric fit yields a discrete distribution, which has been criticized as unrealistic by some (Magder and Zeger, 1996). While it is relatively straightforward to fit a discrete mixing distribution with a known number of support points, it is a difficult statistical problem to decide on the number of points. Nonparametric fits typically require large data sets and so may be precluded for complicated models with multiple random effects.

Finally, we mention another approach to fitting that is an 'add-on' to other methods, such as Laplace approximation, which can be biased. Kuk (1995) developed an iterative bias correction method that takes a biased method like penalized quasi-likelihood and makes it asymptotically unbiased. This method uses an iterative method based on a bootstrap to achieve the bias reduction.

15.4.6 Fitting Methods: Recommendations

For fitting GLMMs, our primary recommendation is to use maximum likelihood or conditional maximum likelihood methods. When applicable and computationally feasible, they generally perform well and are robust to many misspecifications (see Section 15.6). In situations in which maximum likelihood methods are not computationally feasible, effective alternatives include Bayesian MCMC methods with objective priors, the data cloning method of Lele et al. (2010), and iterative bias correction (Kuk, 1995) of a Laplace approximation method. We caution against the use of Laplace, h-likelihood, and especially penalized quasi-likelihood methods in situations where the cluster sizes are small and the distribution of the outcome highly non-normal.

15.4.7 *Software*

In recent years, a number of commercial software packages have incorporated and improved analysis methods for GLMMs. See Chapter 26 for guidance on software.

15.5 BEST PREDICTION

An advantage of using GLMMs is the incorporation of random effects and the ability to generate predicted values for them. For example, Zhang et al. (2008) fit a linear mixed model with clustering by patient in order to predict the realized value of the patient-specific slope in kidney function over time. How are 'best' predicted values for random effects defined and derived? The usual criterion is to define the best predicted value as the minimizer of the mean squared error of prediction. That is, choose the best predictor, $\tilde{u}$, of the random effect u according to

$$\tilde{u} \text{ minimizes } \mathrm{E}[(\tilde{u} - u)^2].$$

It is straightforward to show that the minimizer is given by $\tilde{u} = \mathrm{E}[u|\mathbf{Y}]$. In words, the best predicted value is the average of the random effect consistent with the observed data. These best predicted values are described as 'shrinkage' estimators, because they are attenuated in absolute value compared to an estimate that treated each level of the random effect as a fixed effect. They have strong connections to Bayesian estimation because they are the means of the posterior distribution of the random effects. Some software packages report posterior modes rather than means. This is computationally less intensive, and may be a better summary with highly skewed random effects distributions.

15.6 DIAGNOSTICS

Specification of a GLMM involves the steps outlined in Section 15.3.1: decide on the distribution of Y, choose which predictors to include in the model and in what form, pick a link function, and choose a random effects distribution. Are these assumptions crucial and how can violations of the assumptions be diagnosed? We consider each of these choices in turn. Additional information on diagnostics can be found in Chapter 24.

15.6.1 *Distribution of* Y

Choice of a distribution for Y is sometimes easy, a key example being binary data, which must follow a Bernoulli distribution. But other situations are much more difficult. Consider, for example, the count variable: number of days of work missed in the last three months because of knee pain. This is unlikely to be distributed as Poisson, even within a person. It is likely to have a variance quite a bit larger than that assumed by the Poisson distribution and, if so, an analysis assuming a Poisson distribution would produce highly incorrect standard errors, as well as p-values and confidence intervals.

What guidance is available to decide on a distribution? A key diagnostic approach for linear models, residual plots, is not straightforward for GLMMs. Even for a linear mixed model, there are two types of residuals, marginal ($r_{i,m}$) or conditional ($r_{i,c}$), that can be calculated:

$$\begin{aligned} r_{i,m} &= y_i - \mathbf{x}_i'\hat{\boldsymbol{\beta}} \qquad \text{or} \\ r_{i,c} &= y_i - (\mathbf{x}_i'\hat{\boldsymbol{\beta}} + \mathbf{z}_i'\tilde{\mathbf{u}}), \end{aligned}$$

where $\tilde{\mathbf{u}}$ is the predicted value of $\mathbf{u}$. Neither of these have the same properties as residuals from linear models. They are not homoscedastic and are correlated with the predicted values. This makes it more difficult to interpret residual plots. For distributions other than the normal and for nonlinear link functions, residuals tend to be of even less use. Nevertheless, it is sometimes useful to fit a crude model to get estimates of the mean value, calculate residuals and look at plots of residuals versus predicted values or absolute residuals versus predicted values to get an indication of the relationship of variance to mean. This can provide important guidance as to distributional choice.

Other possibilities are to try a distribution that generalizes the one under consideration or try alternate distributions. For example, the negative binomial distribution can be viewed as a gamma mixture of Poisson distributions and, as such, can be used to check the Poisson assumption and, if needed, model situations in which the variability is larger than that arising from a Poisson distribution (Booth et al., 2003). For distributions that do not have ready generalizations, alternate distributions can be tried and roughly compared using log likelihood values. Software such as SAS Proc NLMIXED, which allows a wide variety of distributions for the outcome, including user-specified densities, can be useful in this regard.

There are some theoretical and simulation results for the linear mixed model. Even under non-normality of the outcome, maximum likelihood fits of a linear mixed model are consistent (Richardson and Welsh, 1994) and robust to non-normality (Jacqmin-Gadda et al., 2007). There are less comprehensive results in the literature for GLMMs.

15.6.2 Omitted Predictors

There have been a number of investigations of the consequences of omitting important predictors from the model for generalized linear models (e.g., Gail et al., 1984; Neuhaus and Jewell, 1993), but little work for GLMMs. However, if there are consequences for GLMs, then the results are sure to apply to GLMMs to some degree, since GLMs are special cases of GLMMs as the variance of the random effects approaches zero. The easiest case to consider is when the omitted predictor is uncorrelated with the predictors that *are* included in the model. In the case of a log or linear link, the omission does not bias the results for the included predictors but can decrease efficiency (Neuhaus, 1998). When the link is nonlinear (and other than the log link), the omission typically biases and makes inconsistent the estimators of the other predictors. The direction of the bias depends on the shape of the link function (Neuhaus and Jewell, 1993). When $1/g'(\mu)$ is convex, which is true of many common link functions including the logit, probit, and complementary log-log, then regression parameters are attenuated by the omission of an uncorrelated predictor. When the omitted predictors are correlated with the included predictors, the results are more complicated and the bias can be in either direction.

Results on bias for GLMs are likely to carry over to GLMMs. For example, consider a probit model with a single predictor, x, random intercepts, u_i, and an omitted predictor, α. For simplicity we assume both the random effects and omitted predictors are normally distributed:

$$\begin{aligned} Y_{ij} &\sim \text{Bernoulli}(p_{ij}) \\ p_{ij} &= \Phi\left(\beta_0 + u_i + \beta x_{ij} + \alpha\right) \\ u_i &\sim \text{indep. } \mathcal{N}(0, \sigma_u^2) \\ \alpha &\sim \text{indep. } \mathcal{N}(\mu_\alpha, \sigma_a^2). \end{aligned}$$

From this it is straightforward to work out the marginal mean of Y_{ij}:

$$\text{E}[Y_{ij}] = \Phi\left(\frac{\beta_0 + \beta x_{ij} + \mu_\alpha}{\sqrt{1 + \sigma_u^2 + \sigma_a^2}}\right).$$

When there are no omitted predictors, and considering the probit scale, the regression parameter is $\beta/\sqrt{1+\sigma_u^2}$. So, with an omitted predictor, the regression parameter will be attenuated by $\sqrt{1+\sigma_u^2}/\sqrt{1+\sigma_u^2+\sigma_a^2}$, similar to the situation for a GLM. Straightforward calculations for identity and log links show that regression parameters are not biased by omission of an uncorrelated predictor in GLMMs.

15.6.3 Choice of Link Function

As with omitted predictors, although there is some work on the effects of a misspecified link function for GLMs, there is little work for GLMMs. For the case of binary regression GLMs, Czado and Santner (1992) show that an incorrect link function leads to bias and loss of efficiency. As above, when results have been derived for GLMs, it will have implications for GLMMs. Neuhaus (1999) shows that misclassified binary outcomes lead to misspecified link functions and uses a result from Li and Duan (1989) to understand the resulting bias. Pregibon (1980) developed tests of the correctness of a link function that several computer packages now implement, and Lee and Zhao (1997) consider the influence of individual points on link function diagnosis. To our knowledge, extensions of such tests to GLMMs have not been made.

15.6.4 Distribution of Random Effects

McCulloch and Neuhaus (2011b) provide a detailed review of the impact of random effect distribution misspecification with a focus on the random intercepts case. With some exceptions, the basic message is that GLMMs are highly robust to this form of misspecification. The primary exceptions are a) when the mean of the random effects distribution is associated with a predictor (Neuhaus and McCulloch, 2006), and b) when the variance of the random effects distribution depends on a predictor (Heagerty and Kurland, 2001), either of which can cause significant bias.

There is a considerable body of literature devoted to the issue of misspecification of the *shape* of the random effects distribution, e.g., specifying it as normal when it is actually heavy-tailed. Most inferences are quite robust to this type of misspecification, including estimation of covariate effects, estimation of the random effects variance, and the mean square error of prediction of the realized value of random effects. Minor exceptions are estimation of the intercept when the random intercept distribution is badly misspecified, or estimation of slopes when a corresponding random slope is badly misspecified, in which case bias may arise. Although the mean square error of prediction is little affected, the predicted values themselves and the distribution of the predicted values are sensitive to the assumed form of the random effects distribution.

Particularly robust in the face of a misspecified random intercepts distribution was estimation of the effects of covariates that mainly vary within clusters. This is especially important because evaluation of such covariates is a primary reason for conducting longitudinal or clustered data studies.

Diagnosis of random effects shape misspecification is possible by trying flexible distributional fits (Zhang and Davidian, 2001; Piepho and McCulloch, 2004; McCulloch and Neuhaus, 2011a), nonparametric fits (Aitkin, 1999), or trying alternate distributions. See also Chapter 22 on robust methods. Plotting of the predicted values of the random effects as suggested by Lange and Ryan (1989) (e.g., with a histogram) is not a reliable way to diagnose the shape of the random effects distribution, except with large cluster sizes (McCulloch and Neuhaus, 2011a). Association between the mean of the random effects distribution and a predictor can be diagnosed by decomposing a predictor into its within- and between-cluster components (Neuhaus and McCulloch, 2006).

15.6.5 Missing Data

Missing data are a common occurrence in practice. People refuse to answer sensitive questions on surveys, participants drop out of studies, and measuring instruments break down. Software for analyzing GLMMs typically drops an observation from the analysis if it is missing the outcome or any of the predictors. This is called a *complete case analysis*. When the fact that an observation is missing is totally unrelated to the variables being studied then such a strategy may throw away some data, but will not tend to lead to erroneous conclusions. However, this is often not the case: people who have high incomes tend not to report them, sicker patients tend to drop out more frequently, and measuring instruments tend to break down under extreme situations. When the variables under study are related to the mechanism that creates the missing data, complete case analysis can be biased. Detail is available in the Missing Data section of this book, Chapter 23.

15.7 SUMMARY

We began this chapter by describing the class of inferential scenarios in which GLMMs are applicable and some motivating examples. These include prediction using the fixed effects portion of the model, inference about within- and between-cluster covariates, inference about variances of random effects, and prediction of random effects. Next, we emphasized the conditional specification of the model and contrasted it with direct marginal specification of a regression model, as in generalized estimating equations approaches. Following the descriptions of inferential scenarios and modeling, we detailed methods of fitting GLMMs to data, including maximum and conditional likelihood, as well as some alternate methods. Another brief section described how to check the assumptions involved in specifying GLMMs.

Our goal throughout has been to emphasize the wide applicability of GLMMs, the maturing software enabling even quite complicated models to be fit to data and the types of inference possible from such models.

REFERENCES

Abramowitz, M. and Stegun, I. (eds) (1964) *Handbook of Mathematical Functions.* National Bureau of Standards, Washington, D.C.

Aitkin, M. (1999) 'A General Maximum Likelihood Analysis of Variance Components in Generalized Linear Models', *Biometrics*, 55: 117–28.

Bellio, R. and Varin, C. (2005) 'A Pairwise Likelihood Approach to Generalized Linear Models with Crossed Random Effects', *Statistical Modeling*, 5: 217–27.

Böhning, D. (1995) 'A Review of Reliable Algorithms for the Semi-parametric Maximum Likelihood Estimator of a Mixture Distribution', *Journal of Statistical Planning and Inference*, 47: 5–28.

Booth, J.G. and Hobert, J.P. (1999) 'Maximizing Generalized Linear Mixed Model Likelihoods with an Automated Monte Carlo EM Algorithm', *Journal of the Royal Statistical Society, Series B*, 61: 265–85.

Booth, J.G., Casella, G., Friedl, H., and Hobert, J.P. (2003) 'Negative Binomial Loglinear Mixed Models', *Statistical Modeling*, 3: 179–91.

Breslow, N.E. and Clayton, D.G. (1993) 'Approximate Inference in Generalized Linear Mixed Models', *Journal of the American Statistical Association*, 88: 9–25.

Breslow, N.E. and Lin, X. (1995) 'Bias Correction in Generalized Linear Mixed Models with a Single Component of Dispersion. *Biometrika*, 82: 81–91.

Czado, C. and Santner, T.J. (1992) 'The Effect of Link Misspecification in Binary Regression Inference', *Journal of Statistical Planning and Inference*, 33: 213–31.

Engel, B. and Keen, A. (1996) 'Hierarchical Generalized Linear Models (with discussion)', *Journal of the Royal Statistical Society, Series B*, 56: 619–78.

Gail, M.H., Wieand, S., and Piantadosi, S. (1984) 'Biased Estimates of Treatment Effect in Randomized Experiments with Non-linear Regressions and Omitted Covariates', *Biometrika*, 71: 807–15.

Heagerty, P.J. and Kurland, B.F. (2001) 'Misspecified Maximum Likelihood Estimates and Generalised Linear Mixed Models', *Biometrika*, 88: 973–85.

Heagerty, P.J. and Zeger, S.L. (2000) 'Marginalized Multilevel Models and Likelihood Inference (with comments and a rejoinder)', *Statistical Science*, 15: 1–26.

Jacqmin-Gadda, H., Sibillot, S., Proust, C., Molinac, J.-M., and Thibaut, R. (2007) 'Robustness of the Linear Mixed Model to Misspecified Error Distribution', *Computational Statistics and Data Analysis*, 51: 5142–54.

Joe, H. (2008) 'Accuracy of Laplace Approximation for Discrete Response Mixed Models', *Computational Statistics and Data Analysis*, 52: 5066–74.

Kuk, A.Y.C. (1995) 'Asymptotically Unbiased Estimation in Generalized Linear Models with Random Effects', *Journal of the Royal Statistical Society, Series B*, 57: 395–407.

Laird, N. (1978) 'Nonparametric Maximum Likelihood Estimation of a Mixing Distribution', *Journal of the American Statistical Association*, 73: 805–11.

Lange, N. and Ryan, L. (1989) 'Assessing Normality in Random Effects Models', *Annals of Statistics*, 17: 624–42.

Lee, A.H. and Zhao, Y. (1997) 'Assessing Influence on the Goodness-of-Link in Generalized Linear Models', *Statistics and Probability Letters*, 31: 351–8.

Lee, Y. and Nelder, J.A. (1996) 'Hierarchical Generalized Linear Models (with discussion)', *Journal of the Royal Statistical Society, Series B*, 58: 619–78.

Lee, Y. and Nelder, J.A. (2001) 'Hierarchical Generalised Linear Models: A Synthesis of Generalised Linear Models, Random-effect Models and Structured Dispersions', *Biometrika*, 88: 987–1006.

Lee, Y. and Nelder, J.A. (2004) 'Conditional and Marginal Models: Another View', *Statistical Science*, 19: 219–38.

Lee, Y. and Nelder, J.A. (2009) 'Likelihood Inference for Models with Unobservables: Another View', *Statistical Science*, 24: 255–302.

Lele, S.R., Nadeem, K., and Schumland, B. (2010) 'Estimability and Likelihood Inference for Generalized Linear Mixed Models using Data Cloning', *Journal of the American Statistical Association*, 105: 1617–25.

Lesperance, M.L. and Kalbfleisch, J.D. (1992) 'An Algorithm for Computing the Nonparametric MLE of a Mixing Distribution', *Journal of the American Statistical Association*, 87: 120–6.

Li, K.-C. and Duan, N. (1989) 'Regression Analysis under Link Violation', *Annals of Statistics*, 17: 1009–52.

Lin, X. and Breslow, N.E. (1996) 'Bias Correction in Generalized Linear Mixed Models with Multiple Components of Dispersion', *Journal of the American Statistical Association*, 91: 1007–16.

Lindsey, J.K. and Lambert, P. (1998) 'On the Appropriateness of Marginal Models for Repeated Measurements in Clinical Trials', *Statistics in Medicine*, 17: 447–69.

Magder, L.S. and Zeger, S.L. (1996) 'A Smooth Nonparametric Estimate of a Mixing Distribution using Mixtures of Gaussians', *Journal of the American Statistical Association*, 91: 1141–51.

McCullagh, P. and Nelder, J.A. (1989) *Generalized Linear Models*, 2nd edition. London: Chapman & Hall.

McCulloch, C.E. (1997) 'Maximum Likelihood Algorithms for Generalized Linear Mixed Models', *Journal of the American Statistical Association*, 92: 162–70.

McCulloch, C.E. and Neuhaus, J.M. (2011a) 'Prediction of Random Effects in Linear and Generalized Linear Models under Model Misspecification', *Biometrics*, 67: 270–9.

McCulloch, C.E. and Neuhaus, J.M. (2011b) 'Misspecifying the Shape of a Random Effects Distribution: Why Getting it Wrong May Not Matter', *Statistical Science*, 3: 388–402.

McCulloch, C.E., Searle, S.R., and Neuhaus, J.M. (2008) *Generalized, Linear and Mixed Models*, 2nd edition. New York: Wiley.

Neuhaus, J.M. (1998) 'Estimation Efficiency with Omitted Covariates in Generalized Linear Models', *Journal of the American Statistical Association*, 93: 1124–9.

Neuhaus, J.M. (1999) 'Bias and Efficiency Loss Due to Misclassified Responses in Binary Regression', *Biometrika*, 86: 843–55.

Neuhaus, J.M. and Jewell, N.P. (1993) 'A Geometric Approach to Assess Bias due to Omitted Covariates in Generalized Linear Models', *Biometrika*, 80: 807–15.

Neuhaus, J.M., Kalbfleisch, J.D., and Hauck, W.W. (1991) 'A Comparison of Cluster-specific and Population-averaged Approaches for Analyzing Correlated Binary Data', *International Statistical Review*, 59: 25–35.

Neuhaus, J.M. and McCulloch, C.E. (2006) 'Separating Between and Within-cluster Covariate Effects using Conditional and Partitioning Methods', *Journal of the Royal Statistical Society, Series B*, 68: 859–72.

Neuhaus, J.M. and Segal, M.R. (1997) 'An Assessment of Approximate Maximum Likelihood Estimators in Generalized Linear Mixed Models', in: Gregoire, T.G., Brillinger, D.R., Diggle, P.J., Russek-Cohen, E., Warren, W.G., and Wolfinger, R.D. (eds), *Modeling Longitudinal and Spatially Correlated Data*, 11–22. New York: Springer.

Piepho, H.-P. and McCulloch, C.E. (2004) 'Transformations in Mixed Models: Application to Risk Analysis for a Multienvironment Trial', *Journal of Agricultural, Biological and Environmental Statistics*, 9: 123–37.

Pinheiro, J.C. and Chao, E.C. (2006) 'Efficient Laplacian and Adaptive Gaussian Quadrature Algorithms for Multilevel Generalized Linear Mixed Models', *Journal of Computational and Graphical Statistics*, 15: 58–81.

Pregibon, D. (1980) 'Goodness of Link Tests for Generalized Linear Models', *Applied Statistics*, 29: 15–24.

Rathouz, P.J. (2004) 'Fixed Effects Models for Longitudinal Binary Data with Drop-outs Missing at Random,' *Statistica Sinica*, 14: 969–88.

Raudenbush, S.W., Yang, M.-L., and Yosef, M. (2000) 'Maximum Likelihood for Generalized Linear Models with Nested Random Effects via High-order, Multivariate Laplace Approximation', *Journal of Computational and Graphical Statistics*, 9: 141–57.

Richardson, A.M. and Welsh, A.H. (1994) 'Asymptotic Properties of Restricted Maximum Likelihood (REML) Estimates for Hierarchical Mixed Linear Models', *The Australian Journal of Statistics*, 36: 31–43.

Roy, J., Alderson, D., Hogan, J.W., and Tashima, K.T. (2006) 'Conditional Inference Methods for Incomplete Poisson Data with Endogenous Time-varying Covariates: Emergency Department Use among HIV-infected Women', *Journal of the American Statistical Association*, 101: 424–34.

Schall, R. (1991) 'Estimation in Generalized Linear Models with Random Effects', *Biometrika*, 78: 719–27.

Searle, S.R., Casella, G., and McCulloch, C.E. (1992) *Variance Components*. New York: Wiley.

Selby, J.V., Fireman, B.H., Lundstrom, R.J., Swain, B.E., Truman, A.F., Wong, C.C., Froelicher, E.S., Barron, H.V., and Hlatky, M.A. (1996) 'Variation among Hospitals in Coronary-angiography Practices and

Outcomes after Myocardial Infarction in a Large Health Maintenance Organization', *New England Journal of Medicine*, 335: 1888–96.

Self, S. and Liang, K.-Y. (1987) 'Asymptotic Properties of Maximum Likelihood Estimators and Likelihood Ratio Tests under Nonstandard Conditions', *Journal of the American Statistical Association*, 82: 605–10.

Steele, B.M. (1997) 'A Modified EM Algorithm for Estimation in Generalized Mixed Models', *Biometrics*, 52: 1295–310.

Verbeke, G., Spiessens, B., and Lesaffre, E. (2001) 'Conditional Linear Mixed Models', *American Statistician*, 55: 24–34.

Wolfinger, R.W. (1993) 'Laplace's Approximation for Nonlinear Mixed Models', *Biometrika*, 80: 791–5.

Zeger, S.L., Liang, K.-Y., and Albert, P.S. (1988) 'Models for Longitudinal Data: A Generalized Estimating Equation Approach (Corr: V45 p347)', *Biometrics*, 44: 1049–60.

Zhang, D. and Davidian, M. (2001) 'Linear Mixed Models with Flexible Distribution of Random Effects for Longitudinal Data', *Biometrics*, 57: 795–802.

Zhang, P., Song, P.X.-K., Qu, A., and Greene, T. (2008) 'Efficient Estimation for Patient-specific Rates of Disease Progression using Nonnormal Linear Mixed Models', *Biometrics*, 64: 29–38.

16

Categorical Response Data

Jeroen K. Vermnt
Tilburg University, The Netherlands

16.1 INTRODUCTION

Earlier chapters in this volume have discussed linear mixed models for continuous responses and generalized linear mixed models for dichotomous variables and counts with binomial and Poisson errors at the lower level, respectively. This chapter deals with multilevel models for discrete response variables with more than two categories; that is, with the situation where these errors can be assumed to come from a multinomial distribution, which can be seen as either a multivariate extension of the binomial distribution or a restricted version of the multivariate Poisson distribution. The dependent variables of interest can have ordered response categories modeled using an ordinal regression model, or unordered response categories modeled using a multinomial logit, a probit, or a discrete choice model for first choices, rankings, or other types of choice formats.

While multilevel ordinal regression models are rather similar to models for dichotomous responses, models for nominal responses are somewhat more complex because they typically contain category-specific random effects. The latter makes them computationally much more demanding. Possible ways out of this problem are the use of factor analytic structures for the random effects yielding models similar to item response theory (IRT) models, the use of discrete approximations of the random effects using finite mixture models, or combinations of these. These options will be discussed in more detail below.

An issue that is getting more attention in the analysis of categorical responses is the distinction between heterogeneity in location or preference versus heterogeneity in scale. The latter concerns the error variance in an underlying latent variable specification of the ordinal or nominal regression model concerned. Taking into account heterogeneity in scale will usually yield much simpler explanations for group differences. Below, I will pay attention to this issue.

The remainder of this chapter is organized as follows. The theory is presented by first providing a generic model formulation, and subsequently discussing models for ordinal responses and models for nominal responses. Then attention is paid to factor analytic and to discrete specifications for the random effects. After presenting the theory, I will illustrate multilevel analysis for categorical responses with a rather simple three-level random intercept model for an ordinal response variable and with a more complex model for a combination of ordinal and nominal responses (ratings and best-worst choices).

16.2 GENERIC MODEL FORMULATION

16.2.1 GLMMs for Categorical Response Variables

Using the notation introduced in the previous chapters, I denote the response of subject i by y_i and the vector of predictors (which includes the constant) by $\mathbf{x}_i$. The total number of response categories is denoted by M, implying $1 \leq y_i \leq M$. The probability that person i's response equals m, $P(y_i = m|\mathbf{x}_i)$, is denoted by π_{im}, and the vector of response probabilities by $\boldsymbol{\pi}_i$. Note that $\sum_{m=1}^{M} \pi_{im} = 1$.

Generalized Linear Models (GLMs) for categorical response variables have the following form (McCullagh and Nelder, 1989):

$$h_m(\boldsymbol{\pi}_i) = \mathbf{x}_i' \boldsymbol{\beta}_m, \qquad (16.1)$$

for $2 \leq m \leq M$. That is, $M - 1$ transformations $h_m(\cdot)$ are defined for the response probabilities $\boldsymbol{\pi}_i$ and each of these is modeled via a linear predictor. Each linear predictor may have its own parameter vector $\boldsymbol{\beta}_m$. The main difference compared to a univariate GLM is thus that we have $M - 1$ such linear equations instead of one. As is shown in more detail below, in GLMs for ordinal response variables, predictor effects may be assumed to be equal across equations, eliminating the index m. Moreover, in discrete choice models for nominal responses, the predictor values vary across responses instead of the regression coefficients; that is, $h_m(\boldsymbol{\pi}_i) = \mathbf{x}_{im}' \boldsymbol{\beta}$.

A two-level variant of the model defined in equation (16.1) is obtained as follows (Agresti, Booth, Hobert, and Caffo, 2000):

$$h_m(\boldsymbol{\pi}_{ij}) = \mathbf{x}_{ij}' \boldsymbol{\beta}_m + \mathbf{z}_{ij}' \mathbf{u}_{jm}, \qquad (16.2)$$

where the index i refers to a lower (or level-1) unit and j to a higher (or level-2) unit, and where $\mathbf{u}_{jm}$ is the vector of random effects and $\mathbf{z}_{ij}$ the associated design vector. Note that different from univariate Generalized Linear Mixed Models (GLMMs; Verbeke and Molenberghs, Chapter 8 of this volume), $\mathbf{u}_{jm}$ contains an index m, which indicates that the random effects may vary across response categories. Extension to a third level is rather straightforward:

$$h_m(\boldsymbol{\pi}_{ijk}) = \mathbf{x}_{ijk}' \boldsymbol{\beta}_m + \mathbf{z}_{ijk}' \mathbf{u}_{jkm},$$

where k refers to a unit of a third level, and where $\mathbf{u}_{jkm} = (\mathbf{u}_{jkm}^{(2)\prime}, \mathbf{u}_{km}^{(3)\prime})'$ or $\mathbf{u}_{jkm} = (\mathbf{u}_{jm}^{(a)\prime}, \mathbf{u}_{km}^{(b)\prime})'$ for nested and crossed random effects, respectively.[1]

The above equations show the form of the $M - 1$ linear predictors. More details about the link functions are provided below, where the models for the different types of categorical response are described. Besides the link function and the linear predictor, a GLMM requires the specification of the distribution of the random effects and the distribution of the responses conditional on the random effects. The latter is modeled using a multinomial distribution. The $M - 1$ sets of random effects are typically assumed to come from a multivariate normal distribution. Using these components, we can define the likelihood for the GLMM interest, which, contrary to the one for Linear Mixed Models, contains intractable integrals. Similar to other types of GLMM, approximations are typically used, such as linear approximations yielding analytically solvable integrals, numerical integration using quadrature, or Monte Carlo integration methods (McCulloch and Neuhaus, Chapter 15 of this volume; Skrondal and Rabe-Hesketh, 2004). Another option is to solve the parameter estimation problem within a Bayesian MCMC framework (Train, 2003).

16.3 MODELS FOR ORDINAL RESPONSES

16.3.1 Underlying Latent Variable Approach

A popular approach to the regression analysis of ordinal response variables is to treat the ordinal responses as discretized continuous responses (Agresti, 2002). The continuous responses y_{ij}^* are unobserved or latent variables for which a linear regression model is defined. That is,

$$y_{ij}^{*} = \mathbf{x}_{ij}'\boldsymbol{\beta} + \mathbf{z}_{ij}'\mathbf{u}_j + \varepsilon_{ij}.$$

The relationship between y_{ij}^{*} and the observed discrete response y_{ij} is that $y_{ij} = m$ if $\delta_m < y_{ij}^{*} < \delta_{m+1}$. The δ_m are the so-called thresholds, which are parameters to be estimated for $2 \leq m \leq M$, and with $\delta_1 = -\infty$ and $\delta_{M+1} = \infty$. Depending on the distributional assumptions about the error term ε_{ij}, one obtains a specific type of ordinal regression model. The most common specifications for ε_{ij} are standard normal, logistic, and extreme value, yielding a probit, logit, and complementary log-log model, respectively.

16.3.2 Cumulative Link Functions

The same models that can be defined using an underlying latent variable approach can also be defined using a cumulative link function combined with the assumption that predictor effects are equal across the $M - 1$ linear equations. Let $\pi_{ijm}^{+} = \sum_{k=m}^{M} \pi_{ijm}$, or the probability that person i in group j gives response m or higher. Note that this implies that $\pi_{ijm} = \pi_{ijm}^{+} - \pi_{ij,m-1}^{+}$.

The random-effects cumulative logit or proportional odds model is obtained using a logit transformation for π_{ijm}^{+} (Hedeker and Gibbons, 1994):

$$\log \frac{\pi_{ijm}^{+}}{1 - \pi_{ijm}^{+}} = \alpha_m + \mathbf{x}_{ij}'\boldsymbol{\beta} + \mathbf{z}_{ij}'\mathbf{u}_j,$$

or

$$\pi_{ijm}^{+} = \frac{\exp(\alpha_m + \mathbf{x}_{ij}'\boldsymbol{\beta} + \mathbf{z}_{ij}'\mathbf{u}_j)}{1 + \exp(\alpha_m + \mathbf{x}_{ij}'\boldsymbol{\beta} + \mathbf{z}_{ij}'\mathbf{u}_j)}.$$

Here, α_m are fixed intercept terms which depend on the response category. These are, in fact, the same as the thresholds defined above, but with a reversed sign: $\alpha_m = -\delta_m$. The vector $\boldsymbol{\beta}$ contains the fixed slopes and $\mathbf{u}_j$ the random intercept and slopes. Note that these effects are constant across response categories, a restriction that is usually referred to as the proportional odds assumption when using a model with a logit link. Sometimes the proportionality assumption is relaxed, yielding regression coefficients $\boldsymbol{\beta}_m$ and $\mathbf{u}_{jm}$ which depend on the response category.

Alternatives to the logit transformation are, among others, inverse standard normal and inverse extreme value transformations, yielding a probit and a complementary log-log model, respectively.

16.3.3 Scale Heterogeneity

The latent variable approach described above can be extended by allowing the error terms ε_{ij} to be heteroscedastic (Hedeker, Berbaum, and Mermelstein, 2006). This is usually achieved by defining a log-linear model for the standard deviation of ε_{ij}; that is,

$$\sigma_{\varepsilon_{ij}} \propto \exp(-\mathbf{w}_{ij}'\boldsymbol{\gamma}).$$

This amounts to expanding the cumulative logit model as follows:

$$\begin{aligned} h_m(\boldsymbol{\pi}_{ij}) &= \log \frac{\pi_{ijm}^{+}}{1 - \pi_{ijm}^{+}} \\ &= (\alpha_m + \mathbf{x}_{ij}'\boldsymbol{\beta} + \mathbf{z}_{ij}'\mathbf{u}_j)\exp(\mathbf{w}_{ij}'\boldsymbol{\gamma}). \end{aligned}$$

The term $\exp(\mathbf{w}_{ij}'\boldsymbol{\gamma})$ is sometimes referred to as the scale factor. Though not shown explicitly here, it is also possible to include random terms in the log-linear model for $\sigma_{\varepsilon_{ij}}$.

16.3.4 Other Link Functions

Instead of a cumulative link function, it is also possible to use other types of link function, the most popular of which are the adjacent-category and the continuation-ratio logit models (Agresti, 2002). As in cumulative logit models, the linear term equals $\alpha_m + \mathbf{x}_{ij}'\boldsymbol{\beta} + \mathbf{z}_{ij}'\mathbf{u}_j$. The adjacent-category model is a logit model for $P(y_{ij} = m | m - 1 \leq y_{ij} \leq m)$, the probability of selecting category m out of adjacent categories $m - 1$ and m. The two possible variants of the continuation-ratio logit model concern $P(y_{ij} \geq m | y_{ij} \geq m - 1)$ and $P(y_{ij} \leq m - 1 | y_{ij} \leq m)$, respectively.

16.4 MODELS FOR NOMINAL RESPONSES

16.4.1 First Choice

Also for the case of unordered or nominal responses, it is possible to define the regression model based on an underlying latent variable approach (Grilli and Rampachini, 2007; Skrondal and Rabe-Hesketh, 2004; Swait and Louviere, 1993; Train, 2003). The utilities y^*_{ijm} of the M categories (or alternatives) are unobserved or latent variables for which linear regression models are defined. That is,

$$y^*_{ijm} = \mathbf{x}'_{ij}\boldsymbol{\beta}_m + \mathbf{z}'_{ij}\mathbf{u}_{jm} + \varepsilon_{ijm}.$$

The relationship between y^*_{ijm} and the observed discrete response y_{ij} is as follows: $y_{ij} = m$ if $y^*_{ijm} > y^*_{ijr}$ for all $r \neq m$. In other words, the response corresponds with the category for which the utility is largest. Depending on the assumed distribution of the error terms ε_{ijm}, one obtains another type of regression model. Assuming ε_{ijm} to be extreme value distributed and independent across alternatives yields the multinomial logit model. The multinomial probit model follows from the assumption that the ε_{ijm} come from a multivariate normal distribution. Here, I will use only the more popular multinomial logit specification in which

$$\pi_{ijm} = \frac{\exp(\mathbf{x}'_{ij}\boldsymbol{\beta}_m + \mathbf{z}'_{ij}\mathbf{u}_{jm})}{\sum_{r=1}^{M}\exp(\mathbf{x}'_{ij}\boldsymbol{\beta}_r + \mathbf{z}'_{ij}\mathbf{u}_{jr})}.$$

As in models for ordinal responses, it may be useful to include a scale factor model, which is a model for capturing heterogeneity in the standard deviation of ε_{ijm} (Swait and Louviere, 1993); that is,

$$\sigma_{\varepsilon_{ijm}} \propto \exp(-\mathbf{w}'_{ij}\boldsymbol{\gamma}).$$

This modifies the multinomial logit model into

$$\pi_{ijm} = \frac{\exp[(\mathbf{x}'_{ij}\boldsymbol{\beta}_m + \mathbf{z}'_{ij}\mathbf{u}_{jm})\exp(\mathbf{w}'_{ij}\boldsymbol{\gamma})]}{\sum_{r=1}^{M}[\exp(\mathbf{x}'_{ij}\boldsymbol{\beta}_r + \mathbf{z}'_{ij}\mathbf{u}_{jr})\exp(\mathbf{w}'_{ij}\boldsymbol{\gamma})]}.$$

Below, an example is provided in which the scale factor varies randomly across (latent classes) of respondents.

The nominal response models presented above can be used when predictor values vary across level-1 or level-2 units. Another related class of model has been developed for the situation in which predictor values vary across response categories or choice alternatives. These models are usually referred to as discrete choice models or conditional logit models (McFadden, 1974; Swait and Louviere, 1993). An example application is the analysis of transportation mode choices, which may depend on characteristics of the mode, such as the required time to reach the destination and the price. A model with alternative-specific predictor values for both fixed and random effects has the following form:

$$\pi_{ijm} = \frac{\exp(\mathbf{x}'_{ijm}\boldsymbol{\beta} + \mathbf{z}'_{ijm}\mathbf{u}_j)}{\sum_{r=1}^{M}\exp(\mathbf{x}'_{ijr}\boldsymbol{\beta} + \mathbf{z}'_{ijr}\mathbf{u}_j)}.$$

Also in this model one can incorporate a regression model for scale heterogeneity. Note also that it is straightforward to define hybrids of the multinomial and conditional logit model. This can be achieved by including alternative-specific constants and/or interactions with individual-level predictors in the vectors $\mathbf{x}_{ijm}$ and $\mathbf{z}_{ijm}$.

16.4.2 Other Choice Formats

When confronting individuals with choice tasks, it is often useful to obtain more information than on just the first choice. For example, one may ask for the best and second best alternative, the best and the worst alternative, the complete ranking of the alternatives, or the best two alternatives (without making a distinction between best and second best). These are all variants of ranking tasks, which can be modeled using logistic regression models similar to those for first choices (Skrondal and Rabe-Hesketh, 2003).

The simplest type of ranking model is obtained when it is assumed that the rankings are obtained sequentially, that is, one first selects the best alternative and subsequently selects the second best (or the worst) from the remaining $M - 1$ alternatives. The probability of selecting alternative m_1 as the best and alternative m_2 as the second best (worst) can be written as $\pi_{ij,m_1,m_2} = \pi_{ij,m_1}\pi_{ij,m_2|m_1}$, where the conditioning on m_1 implies that this alternative can no longer be selected. Generalization to three choices yields $\pi_{ij,m_1,m_2,m_3} = \pi_{ij,m_1}\pi_{ij,m_2|m_1}\pi_{ij,m_3|m_1,m_2}$, etc. The logit model for the second best alternative has the following form:

$$\pi_{ijm_2|m_1} = \frac{\exp(\mathbf{x}'_{ijm_2}\boldsymbol{\beta} + \mathbf{z}'_{ijm_2}\mathbf{u}_j)}{\sum_{r \neq m_1} \exp(\mathbf{x}'_{ijr}\boldsymbol{\beta} + \mathbf{z}'_{ijr}\mathbf{u}_j)}, \tag{16.3}$$

which is the same as a first choice model except for the sum in the denominator which excludes the previous choice. In a best-worst choice design, one first selects the most preferred alternative and subsequently the worst alternative. The model for the worst alternative has the same form as the model described in equation (16.3), but with the difference that the sign of the linear term changes since the alternative with the lowest instead of the highest utility is selected (Magidson, Thomas, and Vermunt, 2009).

Rather than a sequential ranking mechanism, it is also possible to define ranking models based on a simultaneous ranking mechanism: best and second best, best and worst, or the two best are selected simultaneously after evaluation of all possible simultaneous choices. The model will be one for a single choice out of an expanded set of joint choices and where the design matrices are weighted to account for the mutual ranking (first choice gets a weight of 2 and second choice of 1), if known. For example,

$$\pi_{ij,m_1,m_2} = \frac{\exp(2 \cdot \mathbf{x}'_{ijm_1}\boldsymbol{\beta} + 2 \cdot \mathbf{z}'_{ijm_1}\mathbf{u}_j + \mathbf{x}'_{ijm_2}\boldsymbol{\beta} + \mathbf{z}'_{ijm_2}\mathbf{u}_j)}{\sum_{r_1=1}^{M} \sum_{r_2 \neq r_1} \exp(2 \cdot \mathbf{x}'_{ijr_1}\boldsymbol{\beta} + 2 \cdot \mathbf{z}'_{ijr_1}\mathbf{u}_j + \mathbf{x}'_{ijr_2}\boldsymbol{\beta} + \mathbf{z}'_{ijr_2}\mathbf{u}_j)}.$$

16.5 FACTOR ANALYTIC RESTRICTIONS ON RANDOM EFFECTS

Random effects models for categorical responses are strongly related to item response theory (IRT) models, which are factor analytic models for multivariate response data (Van der Linden and Hambleton, 1997). The latter can, however, also be seen as two-level data with item responses nested within persons. A simple unidimensional IRT model for nominal responses has the following form:

$$h_m(\boldsymbol{\pi}_{ij}) = \lambda_{im}(\theta_j - \delta_{im}),$$

where θ_j represents the latent trait of person j, δ_{im} the difficulty parameter for category m of item i, and λ_{im} the discrimination parameter (or factor loading) for category m of item i.[2]

The link function is typically logit. Models for ordinal responses impose the restriction that the discrimination parameter is the same across response categories. Another possible restriction used in IRT is that the discrimination parameter is the same across items.

Using the multilevel notation introduced above, the same model can also be formulated as follows:

$$h_m(\boldsymbol{\pi}_{ij}) = \beta_{im} + \lambda_{im}u_j,$$

where $u_j = \theta_j$ and $\beta_{im} = -\delta_{im}\lambda_{im}$. What can be observed is that a single random term u_j is used for modeling the random intercepts of all items and all response categories, whereas a separate scaling term λ_{im} is estimated for each item category. In fact, the random intercept for item i and category m is specified to have a variance which is proportional to $(\lambda_{im})^2$ and the correlation between the random intercepts equals either 1 or -1 depending on whether the two λ parameters concerned have the same sign or not. A multidimensional extension of an IRT model for nominal data can be defined as follows:

$$h_m(\boldsymbol{\pi}_{ij}) = \beta_{im} + \boldsymbol{\lambda}'_{im}\mathbf{u}_j,$$

where $\boldsymbol{\lambda}_{im}$ is a vector for discrimination parameters or loadings, and $\mathbf{u}_j$ a vector of latent variables.

Hedeker (2003) proposed using a lower-dimensional approximation of the random effects distribution similar to IRT models for nominal data. For this purpose, the general model defined in equation (16.2) is reformulated as follows:

$$h_m(\boldsymbol{\pi}_{ij}) = \mathbf{x}'_{ij}\boldsymbol{\beta}_m + \mathbf{z}'_{ij}\boldsymbol{\Lambda}_m\mathbf{u}_j,$$

where $\boldsymbol{\Lambda}_m$ is a kind of factor loadings matrix scaling the random effects variances (Skrondal and Rabe-Hesketh, 2004). The main advantage of this approach is that the dimensionality of the integrals in the likelihood function is reduced by a factor $M - 1$. Factor analytic structures can not only be used to impose simple structures for the random effects across response categories, but also to impose simple structures across the random slopes of different predictors, as is in fact done in the above IRT models.

16.6 FINITE MIXTURE SPECIFICATION OF RANDOM EFFECTS

Rather than using a continuous random effects distribution, it is also possible to use a discrete distribution for the random effects, either as an approximation of a continuous distribution or because one wishes to cluster level-2 units based on differences in the regression coefficients (Aitkin, 1999; Browne and McNicholas, Chapter 20 of this volume; Vermunt and Van Dijk, 2001; Wedel and DeSarbo, 1994). Let C denote the number of latent classes and c a particular latent class; thus, $1 \leq c \leq C$. The resulting two-level mixture regression model can be formulated as follows:

$$h_m(\boldsymbol{\pi}_{ij|c}) = \mathbf{x}'_{ij}\boldsymbol{\beta}_m + \mathbf{z}'_{ij}\mathbf{u}_{cm},$$

where $\mathbf{u}_{cm}$ are free parameters to be estimated. These parameters indicate the location of the effects of latent class c relative to the average effects $\boldsymbol{\beta}_m$. The class membership is assumed to follow a multinomial distribution with probabilities π_c. The $\mathbf{u}_{cm}$ and π_c are sometimes referred to as mass point locations and mass point weights. It should be noted that the computational burden for such a finite mixture two-level regression model is much lower than with continuous random effects since the likelihood contains a sum over C classes (or mass points) instead of a multidimensional integral.

When using the model for clustering purposes, one will usually interpret the terms $\beta_{pm} + u_{pcm}$, which are the class-specific regression parameters for the pth predictor. When using the model as an approximation for a continuous random effects model, one will typically be interested in the fixed effects and the random effects (co)variances. The β_{pm} have the usual interpretation if the u_{pcm} terms are centered across classes; that is, if $\sum_c u_{pcm}\pi_c = 0$. The covariance between two random effects can be obtained as follows: $\sum_c u_{pcm}u_{p'cm'}\pi_c$.

As is shown in the second application below, it sometimes makes sense to use a hybrid model combining discrete and continuous random effects. Such an approach is rather popular in the field of longitudinal data analysis, where these models are referred to as mixture growth models (Muthén, 2004; Vermunt, 2007). The aim is to identify latent classes with different change trajectories, while taking into account within-class residual heterogeneity in these trajectories.

16.7 APPLICATIONS

16.7.1 *A Three-Level Random Intercept Regression Model for an Ordinal Response Variable*

The first application concerns the analysis of a three-level data set using an ordinal regression model. The data come from the Television School and Family Smoking Preventions and Cessation Project (TVSFP) and were analyzed using a linear mixed model by Hedeker, Gibbons, and Flay (1994). The multilevel data structure arises from the fact that the 1600 children are nested within 135 classes which are themselves nested within 28 schools. Consistent with the notation introduced above, the indexes i, j, and k are used to refer to a child, a class, and a school, respectively.

Schools were randomized into four conditions obtained by crossing two experimental factors: media intervention present or absent (x_{1k}) and social-resistance classroom curriculum present or absent (x_{2k}). The outcome is the child's tobacco health knowledge (y_{ijk}) measured on an ordinal scale with 8 categories, $1 \leq m \leq 8$. Tobacco health knowledge was also measured before the intervention (x_{3ijk}). Taking into account the nesting of children within classes and classes within schools, we define the following three-level ordinal regression model:

$$\log \frac{\pi^+_{ijkm}}{1-\pi^+_{ijkm}} = \alpha_m + \beta_1 x_{1k} + \beta_2 x_{2k} + \beta_3 x_{1k} x_{2k} + \beta_4 x_{3ijk} + u^{(2)}_{jk} + u^{(3)}_{k}.$$

The question of interest is whether the treatment effects and their interaction are significant. This was investigated using different specifications for the random intercepts; that is, the full model containing both school and class random effects, models with either class effects or school effects, and a model without random effects. Table 16.1 reports the fit measures and parameter estimates for these four models. The parameters were obtained with the Latent GOLD software (Vermunt and Magidson, 2008) using its default settings for numerical integration. Other software can also be used, such as the GLLAMM routine written for Stata (Skrondal and Rabe-Hesketh, 2003, 2004).

As can be seen from Table 16.1, the model with only a random intercept at the class level should be preferred according to the BIC and AIC statistics. In this model, the main effect of social-resistance classroom curriculum (β_2) is significant, but the main effect of media intervention (β_1) and the interaction (β_3) are not. This is clearly different from the results of the model that ignores the multilevel structure, in which all fixed effects are significant.

16.7.2 A Two-Level Model for a Choice Experiment Combining Best-Worst Choices with Ratings

The second application concerns the analysis of data from an experimental study done for Roche Diagnostics Corporation, a supplier of lab equipment, with the aim to construct a 'needs based segmentation' of the company's clients. The segmentation is based on the importance 305 lab managers attach to 36 different features of lab equipment. The results of this study were used in the training of sales people; that is, when selling to a lab manager, they could get a feel as to which segment a manager belongs and therefore emphasize which lab equipment best filled these needs.[3]

This study used a combination of best-worst choices and ratings. For the choice task, 64 different choice sets were created, each consisting of 5 alternatives ($M = 5$). More specifically, each set contained 5 randomly selected features out of the list of 36 lab equipment features. Each of the 305 lab managers participating in this study was confronted with 16 of the 64 choice sets and requested to pick the best (most important) and the worst (least important) feature. The question asked was 'Which of the following criteria is most important and which is least important to you in choosing a supplier of lab equipment?'. In addition, each lab manager provided an 'importance' rating on a 5-point scale for each of the 36 features ($1 =$ not important, ..., $5 =$ extremely important). The hybrid choice-rating model described below was proposed by Magidson, Thomas, and Vermunt (2009).

Note that the data set has a two-level structure: the 16 best-worst choices and the 36 ratings are nested within persons. The question of main interest concerns the importance lab managers attach to the features and, more specifically, whether segments can be

Table 16.1 Fit Measures and Parameter Estimates (z values between braces) for the Tobacco Health Knowledge Intervention Study

	Full model		Model with only class effect		Model with only school effect		Model without random effects	
α_2	2.03	(10.33)	2.01	(11.21)	2.00	(10.42)	1.93	(12.24)
α_3	0.05	(0.29)	0.03	(0.17)	0.05	(0.31)	−0.01	(−0.06)
α_4	−1.23	(−7.32)	−1.25	(−8.51)	−1.19	(−7.37)	−1.23	(−10.16)
α_5	−2.43	(−13.85)	−2.46	(−15.72)	−2.37	(−13.98)	−2.39	(−18.18)
α_6	−3.87	(−19.95)	−3.89	(−22.04)	−3.78	(−20.20)	−3.79	(−24.65)
α_7	−5.57	(−22.00)	−5.59	(−23.29)	−5.47	(−22.13)	−5.47	(−24.57)
α_8	−7.69	(−14.44)	−7.71	(−14.65)	−7.58	(−14.33)	−7.59	(−14.64)
β_1	0.23	(1.14)	0.20	(1.21)	0.27	(1.38)	0.26	(2.09)
β_2	0.89	(4.35)	0.88	(5.13)	0.91	(4.63)	0.88	(6.92)
β_3	−0.38	(−1.30)	−0.32	(−1.32)	−0.46	(−1.68)	−0.39	(−2.21)
β_4	0.43	(11.20)	0.44	(11.41)	0.43	(11.23)	0.45	(11.90)
σ_{u3}	0.20	(2.07)			0.26	(3.86)		
σ_{u2}	0.39	(4.80)	0.44	(6.01)				
loglik	−2652		−2653		−2658		−2663	
# of par.	13		12		12		11	
BIC($N = 28$)	5348		5346		5356		5363	
AIC	5331		5330		5340		5349	
pseudo R^2	0.19		0.19		0.14		0.12	

identified with different 'needs'. This means that it is most natural to model variation in feature effects using a mixture regression specification with C latent 'needs' classes, which we index by c. The class-specific preference parameters are denoted by β_{pc}, for $1 \leq p \leq 36$ and $1 \leq c \leq C$.

Various other issues need to be taken into account in the model specification:

- Different types of regression model are needed for the 16 best-worst choices and for the 36 ratings. For the former ($1 \leq i \leq 16$), a sequential ranking model was used, and for the latter ($17 \leq i \leq 52$) a proportional odds model.
- The preference parameters can only be assumed to be proportional to one another in the ranking and rating models. This is achieved by using a different log-linear scale parameter in the ranking and the rating model (γ_{1d} and γ_{2d}).
- Respondents may not only differ in preference but also in how certain they are about their preferences. This is the same as saying that the level-1 variances of the underlying latent variables are heteroscedastic. This heterogeneity is modeled by allowing the log-linear scale parameters to vary across D latent scale classes, indexed by d (γ_{1d} and γ_{2d}). An alternative would be to model scale heterogeneity using continuous random effects.
- Ratings are affected by the fact that some persons tend to give higher ratings than others irrespective of the specific content of the question (the feature). We correct for such individual differences in scale use which is unrelated to the importance attached to the features by a random intercept (u_j). These differences in response style affecting the ratings could also be modeled using a latent class specification (Moors, 2003; Morren, Gelissen, Vermunt, 2011).

The model used is formulated as follows:

$$\pi_{ijm_1|cd} = \frac{\exp\left[\left\{\sum_{p=1}^{36}(\beta_p + u_{pc})x_{pim_1}\right\}\exp(\gamma_{1d})\right]}{\sum_{r=1}^{M}\exp\left[\left\{\sum_{p=1}^{36}(\beta_p + u_{pc})x_{pir}\right\}\exp(\gamma_{1d})\right]},$$

$$\pi_{ijm_2|m_1cd} = \frac{\exp\left[\left\{-\sum_{p=1}^{36}(\beta_p + u_{pc})x_{pim_2}\right\}\exp(\gamma_{1d})\right]}{\sum_{r\neq m_1}\exp\left[\left\{-\sum_{p=1}^{36}(\beta_p + u_{pc})x_{pir}\right\}\exp(\gamma_{1d})\right]}$$

for $1 \leq i \leq 16$, and

$$\pi^{+}_{ijm|cd} = \frac{\exp\left[\left\{\alpha_{0m} + \alpha_{1c} + \sum_{p=1}^{36}(\beta_p + u_{pc})x_{pi} + u_j\right\}\exp(\gamma_{2d})\right]}{1 + \exp\left[\left\{\alpha_{0m} + \alpha_{1c} + \sum_{p=1}^{36}(\beta_p + u_{pc})x_{pi} + u_j\right\}\exp(\gamma_{2d})\right]}$$

for $17 \leq i \leq 52$.

Note that x_{pim} equals 1 if alternative m of set i of person j corresponds to attribute p and otherwise 0, and x_{pi} equals 1 if rating i concerns attribute p (if $p = i - 16$) and otherwise 0. The parameters of main interest, the class-specific utilities u_{pc}, appear in both the best-worst and the rating model. Note that these indicate how much class c deviates from the average utility β_p. The class-specific utilities β_{pc}, which are the actual parameters estimated in the mixture regression analysis, equal $\beta_p + u_{pc}$.

Running a series of models for different values of C and D showed that a model with five preference classes ($C = 5$) and two scale classes ($D = 2$) performed best according to the BIC. To illustrate the impact of ignoring differences in the use of the rating scale, in error variances, and in preference, Table 16.2 also shows the fit of restricted models in which either $u_j = 0$, $D = 1$, or $C = 1$, and the model in which all three restrictions are imposed simultaneously. For parameter estimation, the Latent GOLD software was used, including its Choice module (Vermunt and Magidson, 2005, 2008). The model can also be estimated using LEM (Vermunt, 1997).

As can be seen from the BIC and AIC values, ignoring any of the three multilevel components deteriorates the fit significantly. However, the pseudo R^2 values for the choice and ranking variables show that ignoring the heteroscedasticity in the error variances ($D = 1$) does not deteriorate within-sample prediction.

Table 16.3 reports the β_{pc} parameters, the class-specific utilities (importances) of the 36 features. As can be seen, there are large differences in importance across features and across latent classes. Some of the features are considered to be important in all 5 segments, such as features 11 and 17, whereas others are specific for one particular class. For example, features 24, 7, 12, and 16 are important only for sales managers belonging to classes 1, 3, 4, and 5, respectively. Both results are of interest; that is, which features are important in all segments and which are important in specific segments only.

16.8 DISCUSSION

This chapter discussed multilevel regression models for ordinal and nominal response variables. Special attention was paid to issues such as the distinction between location and scale heterogeneity, the specification of restricted random effects using factor analytic structures, and the use of discrete specifications for the random effects, possibly combined with continuous random effects.

This chapter did not discuss models for multivariate categorical data. However, recently various interesting multilevel models for multivariate categorical responses have been proposed. These models fit within the generalized latent variable modeling framework described by Skrondal and Rabe-Hesketh (2004). The most important special cases are multilevel latent class and mixture models (Vermunt, 2003, 2008), multilevel IRT and factor analytic models for categorical responses (Fox and Glas, 2001; Varriale and Vermunt, 2012), and multilevel mixture growth models (Palardy and Vermunt, 2010).

NOTES

1 The superscript denotes the level of the random effect to remove any ambiguity and provide direct reference in this vector notation.
2 The term difficulty comes from the application of IRT models to dichotomous responses from educational tests: the higher δ the more difficult it is to answer the question concerned correctly. The term discrimination refers to the fact that the higher λ the better the item discriminates (distinguishes) persons with different θ values.
3 For another multilevel market research application using preference data, see Sagan (Chapter 32 of this volume).

Table 16.2 Fit Measures for the Segmentation of Lab Managers Study

Model	Loglik.	# of par.	BIC	AIC	R^2 choice	R^2 rating
Unrestricted	−23157	192	47412	46698	0.23	0.46
$\sigma_u = 0$	−23618	191	48329	47618	0.22	0.39
$D = 1$	−23289	189	47659	46956	0.23	0.47
$C = 1$	−24170	44	48591	48427	0.17	0.39
Restricted	−25080	40	50389	50240	0.17	0.24

Table 16.3 Estimates of the Importance of the 36 Features for the 5 Latent Classes of Lab Managers (β_{pc} Parameters), their Average (β_p), and their Standard Deviation. Rows are sorted by decreasing average importance.

Feature number	Class 1	Class 2	Class 3	Class 4	Class 5	Average	Std Dev.
11	2.80	1.94	3.05	2.45	2.35	2.42	0.41
17	2.60	1.81	2.44	3.24	2.16	2.34	0.50
26	2.76	1.96	2.31	2.28	1.75	2.15	0.31
13	2.36	1.13	0.75	1.22	2.43	1.47	0.64
10	1.65	0.75	3.25	1.54	0.63	1.44	0.94
15	1.87	1.11	1.25	1.51	1.63	1.40	0.27
36	0.35	1.28	0.17	1.41	1.62	1.04	0.55
28	1.52	0.13	1.85	0.71	1.30	0.93	0.67
21	0.82	0.62	0.73	1.69	0.47	0.83	0.41
9	0.67	1.04	−0.50	1.00	0.55	0.62	0.56
3	0.60	0.86	−0.02	1.00	0.11	0.56	0.40
33	0.57	0.84	0.16	0.55	0.17	0.51	0.28
24	1.05	0.15	0.59	0.26	−0.08	0.33	0.35
22	0.37	−0.30	−0.65	0.80	1.58	0.24	0.78
32	−0.23	0.53	−0.34	0.13	−0.09	0.09	0.34
30	−0.84	0.50	−0.65	0.16	0.52	0.06	0.55
1	−0.83	−0.15	0.89	−0.45	−0.24	−0.12	0.53
8	0.31	−0.39	0.45	0.67	−1.58	−0.17	0.77
19	0.25	−0.60	0.38	−0.83	0.05	−0.24	0.47
7	−1.34	−0.28	1.06	−0.33	−0.71	−0.26	0.71
16	0.44	−1.16	−0.95	−0.38	1.50	−0.31	0.98
2	−0.78	0.00	−1.03	−0.34	−0.25	−0.39	0.39
23	−0.44	−0.12	−1.06	−0.32	−0.58	−0.45	0.33
12	−0.97	−1.27	0.34	1.67	−1.51	−0.47	1.16
27	−0.88	−0.56	−0.53	−0.84	0.35	−0.48	0.41
25	−1.61	−0.19	−0.40	−0.70	−0.63	−0.58	0.45
6	−1.48	−0.19	−0.15	−0.50	−1.34	−0.61	0.54
18	0.40	−0.54	−1.67	−2.41	−0.19	−0.89	0.92
4	−1.78	−0.81	−0.04	−2.04	−0.23	−0.91	0.74
5	−0.20	−0.37	−1.30	−1.99	−1.36	−0.97	0.65
29	−1.99	−0.59	−1.13	−1.34	−0.91	−1.06	0.45
20	0.35	−1.32	−0.65	−1.65	−1.63	−1.09	0.65
14	−1.20	−1.43	−2.34	−1.20	−2.14	−1.65	0.45
31	−2.20	−1.13	−1.95	−2.01	−1.58	−1.65	0.42
34	−2.17	−1.39	−1.72	−2.41	−1.91	−1.82	0.37
35	−2.81	−1.85	−2.60	−2.58	−2.20	−2.30	0.37
Class proportion	0.13	0.34	0.18	0.17	0.18		

REFERENCES

Aitkin, M. (1999) 'A General Maximum Likelihood Analysis of Variance Components in Generalized Linear Models', *Biometrics*, 55: 117–28.

Agresti, A. (2002) *Categorical Data Analysis*, 2nd edition. Hoboken, NJ: John Wiley & Sons.

Agresti, A., Booth, J.G., Hobert, J.P., and Caffo, B. (2000) 'Random-effects Modeling of Categorical Response Data', *Sociological Methodology*, 30: 27–80.

Fox, J.P. and Glas, C.A.W. (2001) 'Bayesian Estimation of a Multilevel IRT Model using Gibbs Sampling', *Psychometrika*, 66: 269–86.

Grilli, L. and Rampichini, C. (2007) 'A multilevel Multinomial Logit Model for the Analysis of Graduates' Skills', *Statistical Methods and Applications*, 16: 381–93.

Hedeker, D. (2003) 'A Mixed-effects Multinomial Logistic Regression Model', *Statistics in Medicine*, 22: 1433–46.

Hedeker, D., Berbaum, M. and Mermelstein, R. (2006) 'Location-scale Models for Multilevel Ordinal Data: Between- and Within-subjects Variance Modeling', *Journal of Probability and Statistical Science*, 4: 1–20.

Hedeker, D. and Gibbons, R.D. (1994) 'A Random-Effects Ordinal Regression Model for Multilevel Analysis', *Biometrics*, 50: 933–44.

Hedeker, D., Gibbons, R.D., and Flay, B.R. (1994) 'Random-effects Regression Models for Clustered Data: With an Example from Smoking Prevention Research', *Journal of Consulting and Clinical Psychology*, 62: 757–65.

Magidson, J., Thomas, D., and Vermunt, J.K. (2009) 'A New Model for the Fusion of MaxDiff Scaling and Ratings Data', *2009 Sawtooth Software Conference Proceedings*, 83–103.

McCullagh, P. and Nelder, J.A. (1989) *Generalized Linear Models*. London: Chapman and Hall.

McFadden, D. (1974) 'Conditional Logit Analysis of Qualitative Choice Behavior', in *Frontiers in Econometrics*, P. Zarembka (ed.). New York: Academic Press, 105–42.

Moors, G. (2003) 'Diagnosing Response Style Behavior by Means of a Latent-class Factor Approach. Socio-demographic Correlates of Gender Role Attitudes and Perceptions of Ethnic Discrimination Reexamined', *Quality and Quantity*, 37: 277–302.

Morren, M., Gelissen, J.P.T.M., and Vermunt, J.K. (2011) 'Dealing with Extreme Response Style in Cross-cultural Research: A Restricted Latent Class Factor Analysis Approach', *Sociological Methodology*, 41: 13–47.

Muthén, B. (2004) 'Latent Variable Analysis. Growth Mixture Modeling and Related Techniques for Longitudinal Data', in *The Sage Handbook of Quantitative Methodology for the Social Sciences*, D. Kaplan (ed.). Thousand Oaks: Sage Publications, 345–68.

Palardy, G. and Vermunt, J.K. (2010) 'Multilevel Growth Mixture Models for Classifying Groups', *Journal of Educational and Behavioral Statistics*, 35: 532–65.

Skrondal, A. and Rabe-Hesketh, S. (2003) 'Multilevel Logistic Regression for Polytomous Data and Rankings', *Psychometrika*, 68: 267–87.

Skrondal, A. and Rabe-Hesketh, S. (2004) *Generalized Latent Variable Modeling: Multilevel, Longitudinal and Structural Equation Models*. London: Chapman & Hall/CRC.

Swait, J. and Louviere, J. (1993) 'The Role of the Scale Parameter in the Estimation and Comparison of Multinomial Logit Models', *Journal of Marketing Research*, 30, 305–14.

Train, K. (2003) *Discrete Choice Methods with Simulation*. Cambridge: Cambridge University Press.

Van der Linden, W.J. and Hambleton, R.K. (1997) *Handbook of Modern Item Response Theory*. New York: Springer.

Varriale, R. and Vermunt, J.K. (2012) 'Multilevel Mixture Factor Models', *Multivariate Behavioral Research*, 47: 247–75.

Vermunt, J.K. (1997) *LEM 1.0: A General Program for the Analysis of Categorical Data*. Tilburg, NL: Tilburg University.

Vermunt, J.K. (2003) 'Multilevel Latent Class Models', *Sociological Methodology*, 33: 213–39.

Vermunt, J.K. (2007) 'Growth Models for Categorical Response Variables: Standard, Latent-class, and Hybrid Approaches', in *Longitudinal Models in the Behavioral and Related Sciences*, K. van Montfort, H. Oud, and A. Satorra (eds). Mahwah, NJ: Erlbaum, 139–58.

Vermunt, J.K. (2008) 'Latent Class and Finite Mixture Models for Multilevel Data Sets', *Statistical Methods in Medical Research*, 17: 33–51.

Vermunt, J.K. and Magidson, J. (2005) *Technical Guide for Latent GOLD Choice 4.0: Basic and Advanced*. Belmont, MA: Statistical Innovations.

Vermunt, J.K. and Magidson, J. (2008) *LG-Syntax User's Guide: Manual for Latent GOLD 4.5 Syntax Module*. Belmont, MA: Statistical Innovations.

Vermunt, J.K. and Van Dijk, L. (2001) 'A Nonparametric Random-coefficients Approach: The Latent Class Regression Model', *Multilevel Modeling Newsletter*, 13: 6–13.

Wedel, M. and DeSarbo, W.S. (1994) 'A Review of Recent Developments in Latent Class Regression Models, in *Advanced Methods Marketing Research*, R.P. Bagozzi (ed.). Cambridge: Blackwell Publishers, 352–88.

17

Smoothing and Semiparametric Models

Jin-Ting Zhang

National University of Singapore, Singapore[1]

17.1 INTRODUCTION

Longitudinal responses with a number of covariates are frequently observed in various fields. To explore the effects of the covariates on the longitudinal responses, various parametric models such as linear/nonlinear regression models and linear/nonlinear mixed-effects models have been proposed; see Chapters 5, 8, 9, 14, and 15 for interesting discussions and real data examples, and see Davidian and Giltinan (1995), Vonesh and Chinchilli (1997), Diggle, Heagerty, Liang, and Zeger (2002), Pinheiro and Bates (2009), and Verbeke and Molenberghs (2009) among others for more ideas and methodologies for longitudinal data analysis using parametric modeling. Parametric models are very useful for longitudinal data analysis since they provide a parsimonious description of the relationship between the response variable and its covariates. Moreover, they are efficient when the parametric models are correctly specified. However, parametric models are usually restrictive and less robust against the model assumptions; when the parametric model does not fit the data well, the resulting estimates may be biased. To relax the assumptions on parametric forms, various nonparametric models such as nonparametric mixed-effects models have been proposed for longitudinal data analysis; see, for example, Wu and Zhang (2006), among others. Although nonparametric models are more robust against the model assumptions, they are usually more complex and less efficient. Semiparametric models are a good compromise and retain nice features of both parametric and nonparametric models. In semiparametric models, the parametric components are often used to model important factors that affect the responses parametrically, and the nonparametric components are often used for nuisance factors which are usually less important.

In this chapter, we study semiparametric mixed-effects (SPME) models for longitudinal data analysis. For this purpose, in Section 17.2, we first present two longitudinal data sets which are used to motivate and illustrate the methodologies presented in this chapter. A general SPME model is introduced in Section 17.3. Three fitting methods, based on regression splines (Eubank, 1999), smoothing splines (Green and Silverman, 1994), and local polynomial kernel smoothing (Fan and

Gijbels, 1996), are presented in Sections 17.4, 17.5, and 17.6 respectively. We conclude the chapter with some remarks and bibliographical notes in Section 17.7.

17.2 MOTIVATING LONGITUDINAL DATA

17.2.1 CD4 Data

It is well known that human immune-deficiency virus (HIV) destroys CD4 cells (T-lymphocytes, a vital component of the immune system). Thus the number or percentage of CD4 cells in the blood of an HIV patient will reduce after he is infected with HIV. The CD4 cell level is one of the important biomarkers for evaluating the disease progression of HIV-infected subjects. To use the CD4 marker effectively in studies of new antiviral therapies or for monitoring the health status of individual subjects, it is important to build statistical models for CD4 cell count or percentage. For CD4 cell count, Lange et al. (1992) proposed Bayesian models while Zeger and Diggle (1994) employed a semiparametric model, fitted by a backfitting algorithm. For further related references, see Lange et al. (1992).

A subset of HIV monitoring data from the Multi-center AIDS Cohort Study (CD4) contains the HIV status of 283 homosexual men who were infected with HIV during the follow-up period between 1984 and 1991. Kaslow et al. (1987) presented the details for the related design, methods, and medical implications of this study. The response variable is the CD4 cell percentage of a subject at a number of design time points after HIV infection. Three covariates were assessed in this study. The first one, 'Smoking', takes the values of 1 or 0, according to whether a subject is a smoker or nonsmoker. The second covariate, 'Age', is the age of a subject at the time of HIV infection. The third covariate, 'PreCD4', is the last measured CD4 cell percentage level prior to HIV infection. All three covariates are time-independent and subject-specific. All subjects were scheduled to have clinical visits semi-annually for taking measurements of CD4 cell percentage and other clinical status measures, but many subjects frequently missed their scheduled visits which resulted in unequal numbers of measurements and different measurement time points from different subjects in this longitudinal data set. The raw data for six subjects from the CD4 data are presented in Figure 17.1.

The aim of this study is to assess the effects of cigarette smoking, age at seroconversion, and baseline CD4 cell percentage on the CD4 cell percentage depletion after HIV infection among the homosexual male population. From Figure 17.1, we can see that: (1) some subjects, e.g., Subject 1265, only have a few observations, indicating that one needs to borrow information from other subjects to fit the individual curves, i.e., some mixed-effects model should be applied; and (2) the shapes of the individual curves are very complicated, indicating that some nonparametric or semiparametric model should be used to fit the individual curves of the CD4 cell percentage data. In fact, Wu and Zhang (2006, Chapter 8) show that the following SPME model is appropriate:

$$\begin{aligned} y_{ij} &= \beta_0(t_{ij}) + x_{i1}\beta_1(t_{ij}) + x_{i2}\beta_2(t_{ij}) \\ &\quad + x_{i3}\beta_3(t_{ij}) + v_i(t_{ij}) + \varepsilon_{ij}, \\ &\quad j = 1, \dots, n_i;\ i = 1, 2, \dots, n, \end{aligned} \tag{17.1}$$

where x_{i1}, x_{i2}, and x_{i3} denote the observed values of Smoking, Age, and PreCD4 for subject i, respectively; the unknown coefficient functions $\beta_0(t), \beta_1(t), \beta_2(t)$, and $\beta_3(t)$ denote the Baseline, Smoking, Age, and PreCD4 effects, respectively; the random function $v_i(t)$ denotes the ith random-effects function; and ε_{ij} are the measurement errors at time t_{ij}. The coefficient functions $\beta_0(t)$, $\beta_1(t)$, $\beta_2(t)$, $\beta_3(t)$ can be specified parametrically as

$$\begin{aligned} \beta_0(t) &= \beta_{00} + \beta_{01}t + \beta_{02}t^2, \\ \beta_1(t) &= \beta_{10} + \beta_{11}t, \\ \beta_2(t) &= \beta_{20} + \beta_{21}t, \\ \beta_3(t) &= \beta_{30} + \beta_{31}t + \beta_{32}t^2, \end{aligned} \tag{17.2}$$

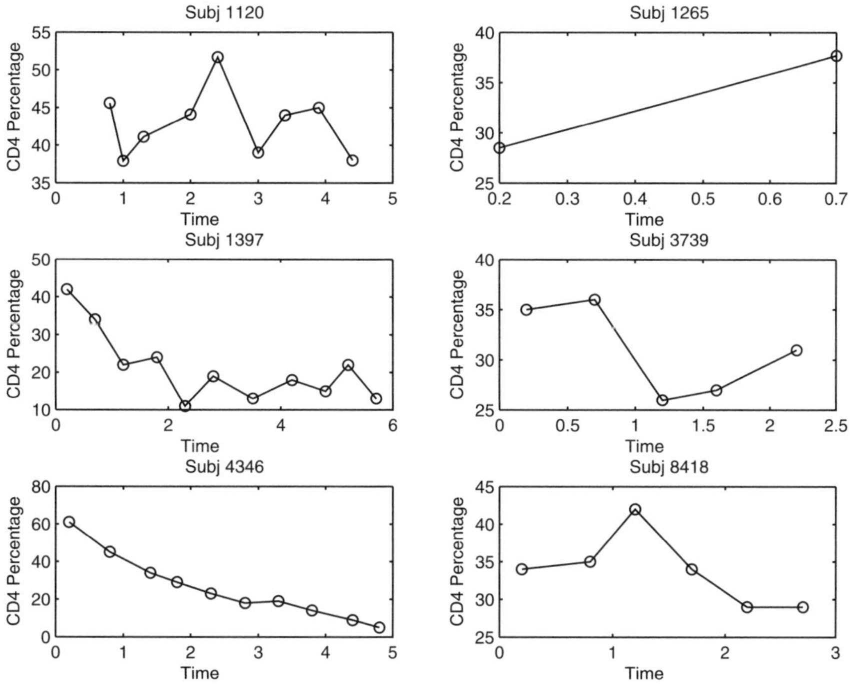

Figure 17.1 Raw Data for Six Subjects from the CD4 Data

while the random-effect functions $v_i(t)$ are assumed to be the realizations of an underlying Gaussian process with mean function 0 and covariance function $\gamma(s, t)$, and the errors ε_{ij} are assumed to be iid with variance σ^2. To fit the nonparametric random-effects function $v_i(t)$ in the SPME model (17.1), some nonparametric smoothing technique is needed.

17.2.2 ACTG 388 Data

The ACTG 388 data were collected in an AIDS clinical trial study conducted by the AIDS Clinical Trials Group (ACTG). This study randomized 517 HIV-1 infected patients to three antiviral treatment arms. The data from one treatment arm will be used for illustration of some of the methodologies proposed in this chapter. This treatment arm includes 166 patients treated with highly active antiretroviral therapy (HAART) for 120 weeks, during which CD4 cell counts were monitored at baseline and at weeks 4, 8, and every 8 weeks thereafter (up to 120 weeks). However, each individual patient might not exactly follow the designed schedule for measurements, and missing clinical visits for CD4 cell measurements frequently occurred. As mentioned previously, CD4 cell count can be used as an important marker for assessing the immunologic response of an antiviral regimen. In this study, of interest are CD4 cell count trajectories over the treatment period for individual patients and for the whole treatment arm. More details about this study and its scientific findings can be found in Fischl et al. (2003) and Park and Wu (2006).

The CD4 cell count data for six subjects from the ACTG 388 data are displayed in Figure 17.2. Again, we can see that: (1) some subjects, e.g., Subject 8, only have a few observations, indicating that borrowing information from other subjects to fit the individual curves is needed; and (2) the shapes of the individual curves are very complicated, indicating that some nonparametric or semiparametric

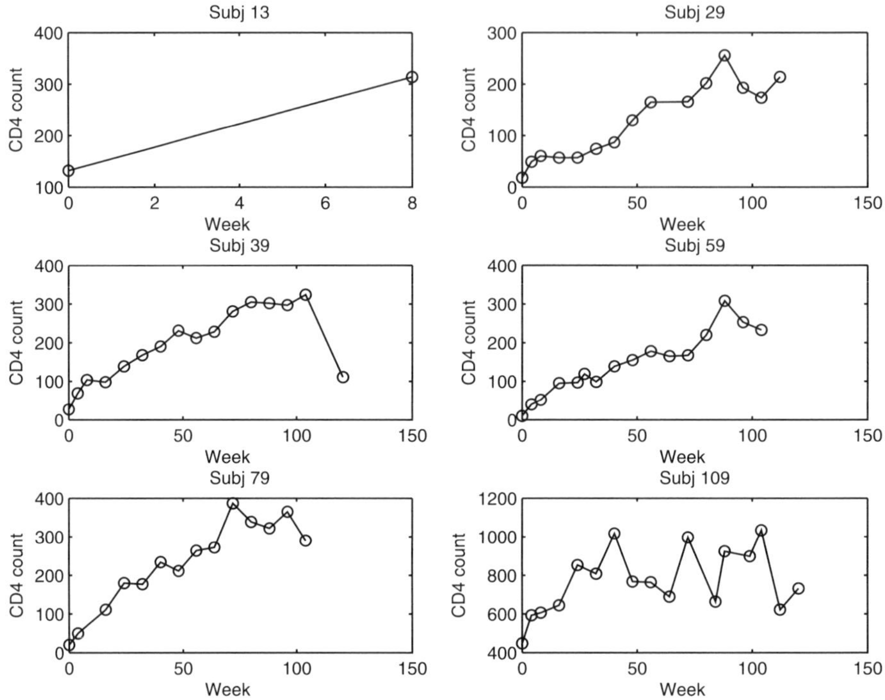

Figure 17.2 Raw Data of Six Subjects from the ACTG 388 Data

models should be used to fit the individual curves of the ACTG 388 data. In fact, Wu and Zhang (2006, Chapter 8) suggested that the following SPME model be used to fit the ACTG 388 data:

$$\begin{aligned} y_{ij} &= \eta(t_{ij}) + z_{ij}'\boldsymbol{u}_i + \varepsilon_{ij}, \\ z_{ij} &= (1, t_{ij}, t_{ij}^2)', \quad \boldsymbol{u}_i = (u_{i1}, u_{i2}, u_{i3})' \\ &\sim N_3(0, \boldsymbol{\Omega}), \end{aligned} \tag{17.3}$$

where $\eta(t)$ is a smooth fixed-effect function and $\boldsymbol{\Omega}$ is an unstructured covariance matrix for the random-effects $\boldsymbol{u}_i$, $i = 1, 2, \ldots, n$. The parametric random-effects component $z_{ij}'\boldsymbol{u}_i = u_{i1} + u_{i2}t_{ij} + u_{i3}t_{ij}^2$ denotes a quadratic polynomial model for the random-effect functions $v_i(t)$, $i = 1, 2, \ldots, n$.

It can be seen that although both the CD4 data and the ACTG 388 data introduced above are from HIV studies, they motivate different SPME models for this chapter. In next section, we shall introduce a general SPME model.

17.3 A GENERAL SEMIPARAMETRIC MIXED-EFFECTS MODEL

Wu and Zhang (2006) discussed the following general SPME model:

$$\begin{aligned} y_{ij} &= \boldsymbol{x}_{ij}'\boldsymbol{\beta} + \eta(t_{ij}) + z_{ij}'\boldsymbol{u}_i + v_i(t_{ij}) + \varepsilon_{ij}, \\ & j = 1, 2, \ldots, n_i; \; i = 1, 2, \ldots, n, \end{aligned} \tag{17.4}$$

where $\boldsymbol{\beta}$ and $\boldsymbol{u}_i$ are a p_0-dimensional parametric fixed-effects component and a q_0-dimensional parametric random-effects component, respectively; $\eta(t)$ and $v_i(t)$ are the nonparametric fixed-effects and random-effects components; $\boldsymbol{x}_{ij}$ and z_{ij} are the associated covariates; and ε_{ij} is the measurement error at time t_{ij} that is not explained by either the fixed-effects component $\boldsymbol{x}_{ij}'\boldsymbol{\beta} + \eta(t_{ij})$ or the random-effects component $z_{ij}'\boldsymbol{u}_i + v_i(t_{ij})$.

In the above SPME model, for simplicity, we assume that

$$\begin{gathered} \boldsymbol{u}_i \sim N_{q_0}(0, \boldsymbol{\Omega}_u), \quad v_i(t) \sim \mathrm{GP}(0, \gamma), \\ \mathrm{E}[\boldsymbol{u}_i v_i(t)] = \boldsymbol{\gamma}_u(t), \\ \boldsymbol{\varepsilon}_i = (\varepsilon_{i1}, \ldots, \varepsilon_{in_i})' \sim N_{n_i}(0, \boldsymbol{\Sigma}_i), \\ i = 1, 2, \ldots, n, \end{gathered} \tag{17.5}$$

where $\mathrm{GP}(\mu, \gamma)$ denotes a Gaussian process with mean function $\mu(t)$ and covariance function $\gamma(s, t)$. That is, $\boldsymbol{u}_i$, $i = 1, 2, \ldots, n$ are independently normally distributed with mean 0 and covariance matrix $\boldsymbol{\Omega}_u$, $v_i(t)$, $i = 1, 2, \ldots, n$ are independent and identical copies of a Gaussian process with mean function 0 and covariance function $\gamma(s, t)$, and $\boldsymbol{\varepsilon}_i$ are independently normally distributed.

When the covariance function $\gamma(s, t)$ is known except for an unknown parameter vector, the above SPME model has been discussed by Zhang et al. (1998) using a smoothing spline method. They used a connection between a smoothing spline and a linear mixed-effects (LME) model for estimating $\eta(t)$ and the variance components. Jacqmin-Gadda et al. (2002) fit the above SPME model (17.4), also using smoothing splines but with smoothing parameters selected by a likelihood-based cross-validation method.

The SPME model (17.4) is very general. It consists of four components: parametric fixed-effects $\boldsymbol{x}'_{ij}\boldsymbol{\beta}$, nonparametric fixed-effects $\eta(t_{ij})$, parametric random-effects $\boldsymbol{z}'_{ij}\boldsymbol{u}_i$, and nonparametric random-effects $v_i(t_{ij})$. When some components are dropped, the SPME model (17.4) reduces to some special SPME models. For example, when both the nonparametric fixed-effects and random-effects components are dropped from the model, it reduces to the usual LME model (Laird and Ware, 1982). When both the parametric fixed-effects and random-effects components are dropped, it reduces to the usual nonparametric mixed-effects model (Wu and Zhang, 2006, Chapter 4). When both the parametric and nonparametric random-effects components are dropped from the model, it reduces to the usual semiparametric regression model. In fact, Wu and Zhang (2006) listed the following six types of special SPME model which are obtained when one or two SPME components are dropped from the general SPME model (17.4). Some of these special SPME models have been discussed in the literature.

Type 1 SPME model: When the parametric fixed-effects component and the nonparametric random-effects component are dropped, the SPME model (17.4) reduces to the following Type 1 SPME model: $y_{ij} = \eta(t_{ij}) + \boldsymbol{z}'_{ij}\boldsymbol{u}_i + \varepsilon_{ij}$, $j = 1, 2, \ldots, n_i$; $i = 1, 2, \ldots, n$. Notice that this SPME model is exactly the same as (17.3), which was proposed for fitting the ACTG 388 data presented in Section 17.2.2. The random-effects component is purely parametric so that when $\boldsymbol{\Sigma}_i = \sigma^2 \boldsymbol{I}_{n_i}$, $i = 1, 2, \ldots, n$, the covariance matrices $\mathbf{V}_i = \mathrm{Cov}(\boldsymbol{y}_i) = z_i \boldsymbol{\Omega}_u \mathbf{z}'_i + \sigma^2 \boldsymbol{I}_{n_i}$ just depend on a parameter vector where $\mathbf{z}_i = (z_{i1}, \ldots, z_{in_i})'$. This allows incorporating the within-subject correlation to estimate $\eta(t)$ using some smoothing method. In fact, Wang (1998) estimated $\eta(t)$ in the above model using a generalized smoothing spline method. Following Wang (1998), Guo (2002) considered likelihood ratio tests for the above Type 1 SPME model. Recently, this model was also considered by Durban et al. (2005) using P-splines.

Type 2 SPME model: When only the parametric fixed-effects component is dropped, the SPME model (17.4) reduces to the following Type 2 SPME model: $y_{ij} = \eta(t_{ij}) + \boldsymbol{z}'_{ij}\boldsymbol{u}_i + v_i(t_{ij}) + \varepsilon_{ij}$, $j = 1, 2, \ldots, n_i$; $i = 1, 2, \ldots, n$.

Type 3 SPME model: When the nonparametric fixed-effects and the parametric random-effects components are dropped, the SPME model (17.4) reduces to the following Type 3 SPME model: $y_{ij} = \boldsymbol{x}'_{ij}\boldsymbol{\beta} + v_i(t_{ij}) + \varepsilon_{ij}$, $j = 1, 2, \ldots, n_i$; $i = 1, 2, \ldots, n$. This model is used when a known parametric model is adequate for the fixed-effects component but a parametric model for the random-effects component is not available. In Section 17.2.1, we mentioned this SPME model for the CD4 data.

Type 4 SPME model: When only the nonparametric fixed-effects component is dropped, the SPME model (17.4) reduces to the following Type 4 SPME model: $y_{ij} = \boldsymbol{x}_{ij}'\boldsymbol{\beta} + \boldsymbol{z}_{ij}'\boldsymbol{u}_i + v_i(t_{ij}) + \varepsilon_{ij}, j = 1, 2, \ldots, n_i;\ i = 1, 2, \ldots, n$. Diggle (1988) considered a special case of the above model when the number of random-effects covariates $q_0 = 1$, and when $v_i(t)$ has a parametric covariance structure such as an auto-regression AR(1) model. This special case was also discussed by Donnelly, Laird, and Ware (1995).

Type 5 SPME model: When only the nonparametric random-effects component is dropped, the SPME model (17.4) reduces to the following Type 5 SPME model: $y_{ij} = \boldsymbol{x}_{ij}'\boldsymbol{\beta} + \eta(t_{ij}) + \boldsymbol{z}_{ij}'\boldsymbol{u}_i + \varepsilon_{ij}, j = 1, 2, \ldots, n_i;\ i = 1, 2, \ldots, n$. The Type 5 SPME model is slightly more complicated than the Type 1 SPME model since it involves an extra term, i.e., the parametric fixed-effects component. Ruppert, Wand, and Carroll (2003) dealt with a simple version (with $q_0 = 1$) of this type of SPME model using P-splines.

Type 6 SPME model: When only the parametric random-effects component is dropped, the SPME model (17.4) reduces to the following Type 6 SPME model: $y_{ij} = \boldsymbol{x}_{ij}'\boldsymbol{\beta} + \eta(t_{ij}) + v_i(t_{ij}) + \varepsilon_{ij},\ j = 1, 2, \ldots, n_i;\ i = 1, 2, \ldots, n$. In Zeger and Diggle (1994), backfitting is used to estimate $\boldsymbol{\beta}$, $\eta(t)$, and $v_i(t)$ separately. In particular, the kernel method is used for estimating the nonparametric fixed-effects component $\eta(t)$ with the bandwidth selected by leaving-one-subject-out cross-validation. However, they did not take the smoothness of $v_i(t)$ into account. Guo (2002) considered the above SPME model for his likelihood ratio test.

17.4 REGRESSION SPLINE METHOD

In this section, we show how to fit the SPME model (17.4) by regression splines. The key idea of the regression spline method is to approximate the nonparametric fixed-effects function $\eta(t)$ and the nonparametric random-effects functions $v_i(t)$ using regression splines so that the SPME model (17.4) reduces to a usual LME model which can be fitted using some existing statistical software such as SAS or S-PLUS (among others). For some basic ideas of regression splines, the reader is referred to Eubank (1999, Chapter 6) and Wu and Zhang (2006, Chapter 3, Section 3.3), among others.

17.4.1 The Regression Spline Estimators

Since $\eta(t)$ is common for all the subjects, while $v_i(t),\ i = 1, 2, \ldots, n$ are regarded as n realizations from a Gaussian process, it is natural to use two regression spline basis vectors $\boldsymbol{\Phi}_{p_1}(t)$ and $\boldsymbol{\Psi}_{q_1}(t)$. These two basis vectors can be taken as a kth degree truncated power basis vector with K knots $\tau_1, \tau_2, \ldots, \tau_K$ and a k_vth degree truncated power basis vector with K_v knots $\delta_1, \ldots, \delta_{K_v}$. Given a sequence of knots $\tau_1, \tau_2, \ldots, \tau_K$, a truncated power basis vector $\boldsymbol{\Phi}_p(t)$ can be defined as

$$\boldsymbol{\Phi}_p(t) = (1, t, \ldots, t^k, (t-\tau_1)_+^k, \ldots, (t-\tau_K)_+^k)', \tag{17.6}$$

where $p = K+k+1$, k is the order of the truncated power basis, and $a_+ = \max(a, 0)$. In particular, when $k = 2$, we obtain a quadratic truncated power basis vector.

With $\boldsymbol{\Phi}_{p_1}(t)$ and $\boldsymbol{\Psi}_{q_1}(t)$, we can approximately express $\eta(t)$ and $v_i(t),\ i = 1, 2, \ldots, n$ as regression splines:

$$\eta(t) = \boldsymbol{\Phi}_{p_1}(t)'\boldsymbol{\eta},\ v_i(t) = \boldsymbol{\Psi}_{q_1}(t)'\boldsymbol{v}_i, \quad i = 1, 2, \ldots, n,$$

where $\boldsymbol{\eta} = (\eta_1, \ldots, \eta_{p_1})'$ and $\boldsymbol{v}_i = (v_{i1}, \ldots, v_{iq_1})',\ i = 1, 2, \ldots, n$, are the associated coefficient vectors. Since $v_i(t)$, $i = 1, 2, \ldots, n$ are assumed to be independent and identical copies of a Gaussian process $v(t)$ with mean function 0 and covariance function $\gamma(s, t)$, the coefficient vectors $\boldsymbol{v}_i$ are independently and identically distributed with mean 0 and covariance matrix, say, $\boldsymbol{\Omega}_v$. Let $\mathrm{E}[\boldsymbol{u}_i \boldsymbol{v}_i'] = \boldsymbol{\Omega}_{uv}$. We then have

$$\begin{pmatrix} \boldsymbol{u}_i \\ \boldsymbol{v}_i \end{pmatrix} \sim N_{q_0+q_1}(\mathbf{0}, \boldsymbol{\Omega}), \text{ where}$$

$$\boldsymbol{\Omega} = \begin{pmatrix} \boldsymbol{\Omega}_u & \boldsymbol{\Omega}_{uv} \\ \boldsymbol{\Omega}_{uv}' & \boldsymbol{\Omega}_v \end{pmatrix}.$$

By using (17.5), we approximately have the following relationships:

$$\begin{aligned} \gamma(s,t) &= \boldsymbol{\Psi}_{q_1}(s)'\boldsymbol{\Omega}_v \boldsymbol{\Psi}_{q_1}(t), \text{ and} \\ \boldsymbol{\gamma}_u(t) &= \boldsymbol{\Omega}_{uv}\boldsymbol{\Psi}_{q_1}(t). \end{aligned}$$

Thus, we can approximately express the SPME model (17.4) as an LME model:

$$\begin{gathered} y_{ij} = \boldsymbol{x}_{ij}'\boldsymbol{\beta} + \boldsymbol{c}_{ij}'\boldsymbol{\eta} + \boldsymbol{z}_{ij}'\boldsymbol{u}_i + \boldsymbol{d}_{ij}'\boldsymbol{v}_i + \varepsilon_{ij}, \\ j = 1, \ldots, n_i, \\ (\boldsymbol{u}_i', \boldsymbol{v}_i')' \sim N_{q_0+q_1}(\mathbf{0}, \boldsymbol{\Omega}), \ \boldsymbol{\varepsilon}_i \sim N_{n_i}(0, \boldsymbol{\Sigma}_i), \\ i = 1, 2, \ldots, n, \end{gathered}$$

where $\boldsymbol{c}_{ij} = \boldsymbol{\Phi}_{p_1}(t_{ij})$ and $\boldsymbol{d}_{ij} = \boldsymbol{\Psi}_{q_1}(t_{ij})$. The above LME model can be written in a standard form:

$$\begin{gathered} y_{ij} = \tilde{\boldsymbol{x}}_{ij}'\tilde{\boldsymbol{\beta}} + \tilde{\boldsymbol{z}}_{ij}'\tilde{\mathbf{u}}_i + \varepsilon_{ij}, \\ j = 1, \ldots, n_i, \\ \tilde{\mathbf{u}}_i \sim N_{q_0+q_1}(\mathbf{0}, \boldsymbol{\Omega}), \ \boldsymbol{\varepsilon}_i \sim N_{n_i}(0, \boldsymbol{\Sigma}_i), \\ i = 1, 2, \ldots, n, \end{gathered} \tag{17.7}$$

where $\tilde{\boldsymbol{x}}_{ij} = (\boldsymbol{x}_{ij}', \boldsymbol{c}_{ij}')'$, $\tilde{\boldsymbol{\beta}} = (\boldsymbol{\beta}', \boldsymbol{\eta}')'$, $\tilde{z}_{ij} = (\boldsymbol{z}_{ij}', \boldsymbol{d}_{ij}')'$, $\tilde{\mathbf{u}}_i = (\boldsymbol{u}_i', \boldsymbol{v}_i')'$, and $\boldsymbol{\varepsilon}_i = (\varepsilon_{i1}, \ldots, \varepsilon_{in_i})'$. In vector form, the above LME model can be expressed as

$$\begin{gathered} \boldsymbol{y}_i = \tilde{\mathbf{X}}_i\tilde{\boldsymbol{\beta}} + \tilde{\mathbf{Z}}_i\tilde{\mathbf{u}}_i + \boldsymbol{\varepsilon}_i, \ \tilde{\mathbf{u}}_i \sim N_{n(q_0+q_1)}(\mathbf{0}, \boldsymbol{\Omega}), \\ \boldsymbol{\varepsilon}_i \sim N_{n_i}(0, \boldsymbol{\Sigma}_i), \ i = 1, 2, \ldots, n, \end{gathered} \tag{17.8}$$

where

$$\begin{aligned} \boldsymbol{y}_i &= (y_{i1}, y_{i2}, \ldots, y_{in_i})', \\ \tilde{\mathbf{X}}_i &= (\tilde{\boldsymbol{x}}_{i1}, \tilde{\boldsymbol{x}}_{i2}, \ldots, \tilde{\boldsymbol{x}}_{i1})', \end{aligned}$$

and

$$\tilde{\mathbf{Z}}_i = (\tilde{z}_{i1}, \tilde{z}_{i2}, \ldots, \tilde{z}_{in_i})'.$$

In matrix form we have

$$\begin{aligned} \boldsymbol{y} &= \tilde{\mathbf{X}}\tilde{\boldsymbol{\beta}} + \tilde{\mathbf{Z}}\tilde{\mathbf{u}} + \boldsymbol{\varepsilon}, \\ \tilde{\mathbf{u}} &\sim N_{n(q_0+q_1)}(\mathbf{0}, \bar{\boldsymbol{\Omega}}), \\ \boldsymbol{\varepsilon} &\sim N_N(0, \boldsymbol{\Sigma}), \end{aligned}$$

where $\boldsymbol{y} = (\boldsymbol{y}_1', \ldots, \boldsymbol{y}_n')'$, $\boldsymbol{\varepsilon} = (\boldsymbol{\varepsilon}_1', \ldots, \boldsymbol{\varepsilon}_n')'$, $\tilde{\mathbf{X}} = (\tilde{\mathbf{X}}_1', \tilde{\mathbf{X}}_2', \ldots, \tilde{\mathbf{X}}_n')'$, $\tilde{\mathbf{Z}} = \oplus_{i=1}^n \tilde{\mathbf{Z}}_i$, $\bar{\boldsymbol{\Omega}} = \oplus_{i=1}^n \boldsymbol{\Omega}$, $\boldsymbol{\Sigma} = \oplus_{i=1}^n \boldsymbol{\Sigma}_i$, with $\oplus$ being the direct sum (block diagonal) operator, and $N = \sum_{i=1}^n n_i$.

Methods for fitting the above standard LME model (17.7) can be found in several books including Davidian and Giltinan (1995), Vonesh and Chinchilli (1997), Pinheiro and Bates (2009), and Verbeke and Molenberghs (2009), among others. See Chapters 3 and 9 for some details.

17.4.2 Fixed-Effects Estimation and Random-Effects Predictions

The estimation of the fixed-effects components $\boldsymbol{x}_{ij}'\boldsymbol{\beta} + \eta(t_{ij})$ and the predictions of the random-effects components $\boldsymbol{z}_{ij}'\boldsymbol{u}_i + v_i(t_{ij})$ at the design time points are important for assessing the fits and also for smoothing parameter selection. This latter topic will be discussed in Section 17.4.3.

By some simple algebra, the estimates of the fixed-effects components at the design time points can be expressed as

$$\hat{\mathbf{y}} = \tilde{\mathbf{X}}\widehat{\tilde{\boldsymbol{\beta}}} = \mathbf{A}\boldsymbol{y}, \tag{17.9}$$

where

$$\begin{aligned} \mathbf{A} &= \tilde{\mathbf{X}}(\tilde{\mathbf{X}}'\mathbf{V}^{-1}\tilde{\mathbf{X}})^{-1}\tilde{\mathbf{X}}'\mathbf{V}^{-1}, \\ &\quad \text{with } \mathbf{V} = \tilde{\mathbf{Z}}\bar{\boldsymbol{\Omega}}\tilde{\mathbf{Z}}' + \boldsymbol{\Sigma}, \end{aligned} \tag{17.10}$$

is the associated smoother matrix for estimating the fixed-effects components. When $\mathbf{V}$ is known, expression (17.9) suggests that the estimator (17.9) is a linear estimator of the fixed-effects components, evaluated at the design time points. The trace of $\mathbf{A}$ are the degrees of freedom of the estimator (17.9), which indicates the model complexity of fitting the fixed-effects components. It is easy to show that $\mathrm{tr}(\mathbf{A}) = p_0 + p_1$ for any $\mathbf{V}$, reflecting the fact that we use p_0 covariates for parametric fixed-effects and p_1 basis functions for approximating the nonparametric fixed-effects curve $\eta(t)$.

The predictions of the random-effects components at the design time points can be expressed as

$$\hat{\mathbf{w}} = \tilde{\mathbf{Z}}\hat{\tilde{\mathbf{u}}} = \tilde{\mathbf{Z}}\bar{\boldsymbol{\Omega}}\tilde{\mathbf{Z}}'\mathbf{V}^{-1}(\boldsymbol{y} - \tilde{\mathbf{X}}\hat{\tilde{\boldsymbol{\beta}}}) = \mathbf{A}_v\tilde{\mathbf{w}}, \tag{17.11}$$

where $\hat{\mathbf{w}} = (\hat{\mathbf{w}}_1', \hat{\mathbf{w}}_2', \ldots, \hat{\mathbf{w}}_n')'$ with $\hat{\mathbf{w}}_i = (z_{i1}'\hat{\mathbf{u}}_i + \hat{v}_i(t_{i1}), \ldots, z_{in_i}'\hat{\mathbf{u}}_i + \hat{v}_i(t_{in_i}))'$, $i = 1, 2, \ldots, n$, and $\tilde{\mathbf{w}} = \boldsymbol{y} - \tilde{\mathbf{X}}\hat{\tilde{\boldsymbol{\beta}}} = (\boldsymbol{I}_N - \mathbf{A})\boldsymbol{y}$ is the residual vector after subtracting the fixed-effects components at the design time points. From (17.11), the associated smoother matrix for estimating the random-effects components at the design time points is then

$$\mathbf{A}_v = \tilde{\mathbf{Z}}\bar{\boldsymbol{\Omega}}\tilde{\mathbf{Z}}'\mathbf{V}^{-1}. \tag{17.12}$$

17.4.3 Knot Locating and Knot Number Selection

In the previous discussion about the regression spline method, we fixed the truncated power basis vectors $\boldsymbol{\Phi}_{p_1}(t)$ and $\boldsymbol{\Psi}_{q_1}(t)$. In practice, for good performance of fitting the nonparametric fixed-effects and random-effects curves, we need to choose these two basis vectors carefully via choosing well their orders k and k_v, locating well their knots $\tau_1, \tau_2, \ldots, \tau_K$ and $\delta_1, \ldots, \delta_{K_v}$, and in particular choosing well the number of knots K and K_v. The knots $\tau_1, \tau_2, \ldots, \tau_K$ and $\delta_1, \delta_2, \ldots, \delta_{K_v}$ for $\boldsymbol{\Phi}_{p_1}(t)$ and $\boldsymbol{\Psi}_{q_1}(t)$ can be specified using the uniform knot spacing method or the quantiles as knots method using all of the design time points from all of the subjects. The details of these two knot placing methods are given in Wu and Zhang (2006, Chapter 3, Section 3.3). When the quantiles as knots method is used, some knots may be tied and only one of the tied knots should be selected so that the associated basis vector is not degenerate. As for k and k_v, we often use lower orders, say, 1, 2, and 3, for easy computation. In practice, linear or quadratic regression splines are widely used.

When the orders k and k_v are fixed, the number of knots, K and K_v will determine how complicated the model is and how closely the model fits the data. When both K and K_v are small, simple models are applied to fit the fixed-effects function $\eta(t)$ and the random-effects function $v_i(t)$, $i = 1, 2, \ldots, n$ but the data may not be fitted closely. In particular, when $K = 0$, no knots are involved to fit $\eta(t)$ so that $\eta(t)$ is fitted using a polynomial of order k. Similarly, $K_v = 0$ means a polynomial of order k_v is fitted to $v_i(t)$, $i = 1, 2, \ldots, n$. On the other hand, when K and K_v are large, a complicated model is fitted to the data so that the data can be fitted closely or even interpolated. Good choice of K and K_v will trade off these two aspects: goodness of fit and model complexity.

Notice that while the choice of the number and location of knots for the regression spline method is a very challenging nonlinear optimization problem, it can be circumvented using penalized splines via choosing a tuning parameter instead. The reader is referred to Chapter 18 for details.

Goodness of fit

Goodness of fit is a concept concerning how close a model fits the data. In the context of longitudinal data analysis, it is often measured by the likelihood or the log-likelihood of the resulting model. The larger the likelihood, the better the goodness of fit is.

Let $\boldsymbol{\mu}_i = (\boldsymbol{x}_{i1}'\boldsymbol{\beta} + \eta(t_{i1}), \ldots, \boldsymbol{x}_{in_i}'\boldsymbol{\beta} + \eta(t_{in_i}))'$ be the fixed-effects at the design time points for the ith subject, and $\mathbf{V}_i = \mathrm{Cov}(\boldsymbol{y}_i)$, $i = 1, 2, \ldots, n$ as defined before. Then under the Gaussian assumption, we have $\boldsymbol{y}_i \sim N_{n_i}(\boldsymbol{\mu}_i, \mathbf{V}_i)$, $i = 1, 2, \ldots, n$. Therefore, the joint probability density of $\boldsymbol{y}$ is

$\text{Prob}(\boldsymbol{y}) = \prod_{i=1}^{n} \frac{\exp\{-\frac{1}{2}(\boldsymbol{y}_i-\boldsymbol{\mu}_i)'\mathbf{V}_i^{-1}(\boldsymbol{y}_i-\boldsymbol{\mu}_i)\}}{(2\pi)^{n_i/2}|\mathbf{V}_i|^{1/2}}$. Taking the logarithm of the above expression, we have $\text{Loglik} = -\frac{1}{2}\sum_{i=1}^{n}\Big\{(\boldsymbol{y}_i - \boldsymbol{\mu}_i)'\mathbf{V}_i^{-1}(\boldsymbol{y}_i-\boldsymbol{\mu}_i)+\log(|\mathbf{V}_i|)\Big\} - \frac{N}{2}\log(2\pi)$, where again $N = \sum_{i=1}^{n} n_i$. In practice, the Loglik is computed with the unknowns $\boldsymbol{\mu}_i$ and $\mathbf{V}_i$ replaced with their estimates. Under the approximation LME model (17.8), $\boldsymbol{\mu}_i$ and $\mathbf{V}_i$ are estimated as $\hat{\boldsymbol{\mu}}_i = \tilde{\mathbf{X}}_i\widehat{\boldsymbol{\beta}}$, $\hat{\mathbf{V}}_i = \tilde{\mathbf{Z}}_i\widehat{\boldsymbol{\Omega}}\tilde{\mathbf{Z}}_i' + \hat{\boldsymbol{\Sigma}}_i$, $i = 1, 2, \ldots, n$. It is expected that when K and K_v increase, Loglik will generally increase due to more basis-based covariates being involved so that the data are fitted more closely.

Model complexity

Model complexity is generally measured by the trace of the associated smoother matrix, or the degrees of freedom of the regression spline fit. It indicates approximately how many parameters are needed if an equivalent parametric model is used to fit the same data set.

For estimating the fixed-effects components at all the design time points, the smoother matrix is $\mathbf{A}$ as defined in (17.10); for estimating the random-effects components at the design time points, the smoother matrix is $\mathbf{A}_v$ as defined in (17.12). It follows that the degrees of freedom are respectively $\text{df} = \text{tr}(\mathbf{A})$ and $\text{df}_v = \text{tr}(\mathbf{A}_v)$, the traces of $\mathbf{A}$ and $\mathbf{A}_v$ respectively.

AIC and BIC

From the above discussions, we can see that the goodness of fit, as indicated by Loglik, will increase as K or K_v or both increase, while at the same time the model complexity, as indicated by $\text{df} + \text{df}_v$, will increase. Therefore, good knot number selection criteria should involve both terms in a proper manner. Wu and Zhang (2006) proposed the following AIC and BIC rules:

$$\text{AIC}(K, K_v) = -2\text{Loglik} + 2[\text{df} + \text{df}_v],$$
$$\text{BIC}(K, K_v) = -2\text{Loglik} + \log(N)[\text{df} + \text{df}_v],$$

where again N is the number of the total measurements in the longitudinal data set. It is seen that the BIC rule is obtained via replacing 2 in the expression of AIC with $\log(N)$ to penalize more seriously the model complexity so that the BIC rule often results in a simpler model, i.e., smaller $\text{df} + \text{df}_v$, compared with the one selected by AIC.

Notice that when the number of possible combinations of (K, K_v) is large, direct minimization of AIC or BIC over (K, K_v) is computationally intensive. A simple trick for reducing the computation is to minimize AIC or BIC on just one argument while keeping the other unchanged. For example, letting $K_v = 1$, we find K^* to minimize AIC. Then letting $K = K^*$, we find K_v^* to minimize AIC. The (K^*, K_v^*) obtained in this way may not minimize AIC globally but it is often satisfactory for practical applications or simulation studies. Otherwise, one may repeat the above process a number of times.

17.4.4 Application to the CD4 Data

As an example, we fit the Type 3 SPME model defined in (17.1) and (17.2) to the CD4 data introduced in Section 17.2.1. In this Type 3 SPME model, we have one parametric fixed-effects component and one nonparametric random-effects component. Therefore, we only need one basis vector to fit the nonparametric random-effects functions $v_i(t)$, $i = 1, 2, \ldots, n$.

We used the quadratic truncated power basis (17.6) with $k = 2$. The knots were located using the quantiles as knots method since the CD4 cell percentages were measured more frequently at the beginning than at the end of the CD4 data collection period. We need to choose a proper number of knots K_v for fitting $v_i(t)$, $i = 1, 2, \ldots, n$. We used $K_v = 0, 1, 2, \ldots, 10$ possible knots. Notice that when $K_v = 0$, a quadratic polynomial model was used to fit the random-effects functions.

Figure 17.3 shows the Loglik, df_v, AIC, and BIC curves in panels (a), (b), (c), and

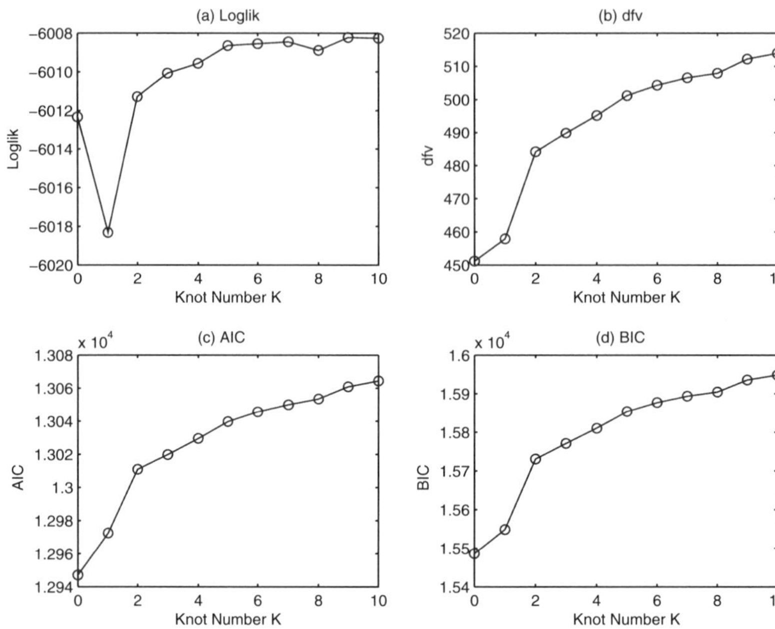

Figure 17.3 Type 3 SPME Modeling of the CD4 Data: Knot Number Selection

(d) respectively. Both the AIC and BIC rules suggest that $K_v = 0$ is the best choice, indicating that a quadratic polynomial model may be adequate for fitting the nonparametric random-effects functions $v_i(t)$, $i = 1, 2, \ldots, n$. Further study is warranted to test if this is the case.

The parametric component estimation results of the Type 3 SPME model fit to the CD4 data are presented in Table 17.1. The parameters for the intercept CD4 percentage level curve are highly significant, with a negative slope, indicating that the intercept CD4 percentage level decreases over time after HIV infection. The parameters for the Smoking and Age effect coefficient functions are insignificant. Thus, the Type 3 SPME model fit suggests that after accounting for the within-subject correlation, the smoking habit and the age at HIV infection have little predictive power on the CD4 percentage levels after HIV infection, but the pre-infection CD4 percentage levels have a positive effect, although the effect decreases over time.

The above conclusions can be seen more clearly from Figure 17.4, where the fitted coefficient functions (solid curves), together with the 95% pointwise standard deviation bands (dashed curves) of the intercept CD4 percentage (Baseline) levels, Smoking, Age, and PreCD4 covariates are displayed. It can be seen that the standard deviation bands of the Smoking and Age coefficient functions contain 0 all the time, indicating that a patient's smoking status and age have limited predictive power over the CD4 percentage levels after HIV infection. The standard deviation bands of the PreCD4 coefficient function are above 0 most of the time, indicating that the PreCD4 levels have a positive effect on CD4 percentage levels after HIV infection.

Figure 17.5 displays the population fits (solid curves) and the individual fits (dashed curves), superimposed by the raw individual data (circles) for six randomly selected subjects from the CD4 data. These individual fits provide good predictions to the raw individual data. The individual plots show that the Type 3 SPME model fit well the raw individual data.

The appropriateness of the SPME model as defined in (17.1) and (17.2) is also partially

Table 17.1 Type 3 SPME Model Fit to the CD4 Data: Estimated Coefficients, Standard Deviations (SD) and Approximate z-test Values and the Approximate P-values for the Parametric Fixed-Effects Components. The quadratic truncated power basis was used with K_v = 0 knots, selected by AIC and BIC

Covariate	Coef.	Estimated Coef.	SD	z-test value	P-value
intercept	β_{00}	36.834	0.685	53.744	0
	β_{01}	−4.848	0.456	−10.635	0
	β_{02}	0.373	0.073	5.102	1.678e-7
Smoking	β_{10}	0.479	1.114	0.430	0.333
	β_{11}	0.110	0.487	0.226	0.410
Age	β_{20}	0.045	0.069	0.655	0.256
	β_{21}	−0.024	0.030	−0.812	0.208
PreCD4	β_{30}	0.566	0.070	8.048	4.441e-16
	β_{31}	−0.138	0.051	−2.717	0.00329
	β_{32}	0.013	0.008	1.498	0.06705

verified by the residual plots presented in Figure 17.6. The standardized residuals are plotted against the SPME fit and time in panels (a) and (b) respectively. The residual plots show that the data structure has been largely explained by the SPME model. Panel (c) presents the histogram of the standardized residuals. It says that the measurement errors are nearly normally distributed. The three plots jointly show that the Type 3 SPME model is indeed an adequate model for the CD4 data.

17.5 SMOOTHING SPLINE METHOD

In the previous section, we investigated the regression spline method for fitting the SPME model (17.4). In this section, we investigate the smoothing spline method instead. We start this section with an LME model which is obtained via evaluating the SPME model (17.4) at all the distinct design time points for all the subjects of a longitudinal data set. Let

$$\tau_1, \tau_2, \cdots, \tau_M \tag{17.13}$$

denote these distinct design time points from the pooled design time points for all the subjects:

$$\{t_{ij}, j = 1, 2, \ldots, n_i, \;\; i = 1, 2, \ldots, n\}. \tag{17.14}$$

It is obvious that the number of distinct design time points, M, is usually larger than the maximum number of measurements for a subject, $\max_{1\le i\le n} n_i$, but is often much smaller than the total number of observations for all the subjects, $N = \sum_{i=1}^n n_i$. Set

$$\begin{aligned} \boldsymbol{\eta} &= (\eta_1, \ldots, \eta_M)', & \eta_r &= \eta(\tau_r), \\ \boldsymbol{v}_i &= (v_{i1}, \ldots, v_{iM})', & v_{ir} &= v_i(\tau_r), \\ & & r &= 1, 2, \ldots, M. \end{aligned}$$

They are obtained by evaluating the nonparametric fixed-effects function $\eta(t)$ and the nonparametric random-effects functions $v_i(t)$ in the SPME model (17.4) at the M distinct design time points (17.13). Moreover, set

$$\begin{aligned} \boldsymbol{c}_{ij} &= (c_{ij1}, \ldots, c_{ijM})', \; j = 1, 2, \ldots, n_i; \\ i &= 1, 2, \ldots, n, \end{aligned}$$

where c_{ijr} equals 1 when $t_{ij} = \tau_r$ and 0 otherwise. Then the SPME model (17.4) evaluated at all the distinct design time points (17.13) can be expressed as

$$\begin{aligned} y_{ij} &= \boldsymbol{x}_{ij}'\boldsymbol{\beta} + \boldsymbol{c}_{ij}'\boldsymbol{\eta} + \boldsymbol{z}_{ij}'\boldsymbol{u}_i \\ &\quad + \boldsymbol{c}_{ij}'\boldsymbol{v}_i + \varepsilon_{ij}, \\ \boldsymbol{u}_i &\sim N_{q_0}(\mathbf{0}, \boldsymbol{\Omega}_u), \quad \boldsymbol{v}_i \sim N_M(\mathbf{0}, \boldsymbol{\Omega}_v), \\ \boldsymbol{\varepsilon}_i &\sim N_{n_i}(\mathbf{0}, \boldsymbol{\Sigma}_i), \\ j &= 1, 2, \ldots, n_i; \; i = 1, 2, \ldots, n, \end{aligned}$$

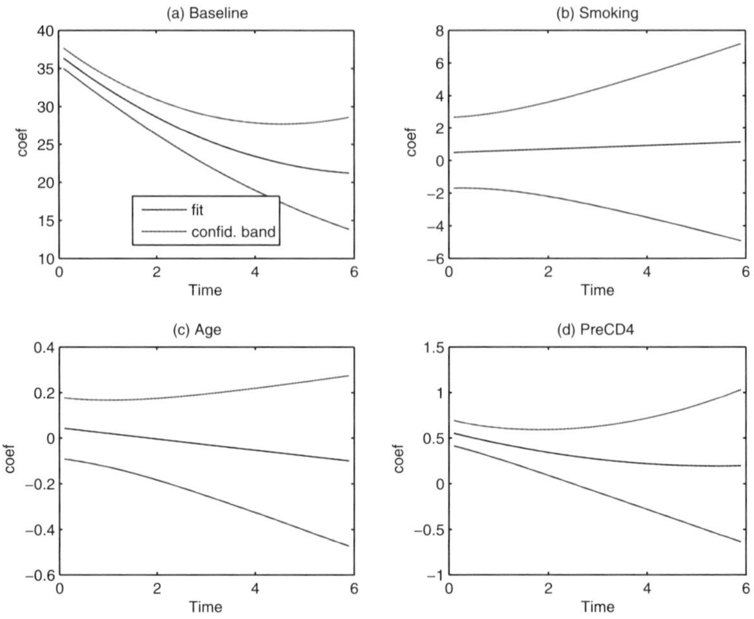

Figure 17.4 Type 3 SPME Modeling of the CD4 Data: Fitted Coefficient Functions (solid curves) with 95% Pointwise Standard Deviation Bands (dashed curves)

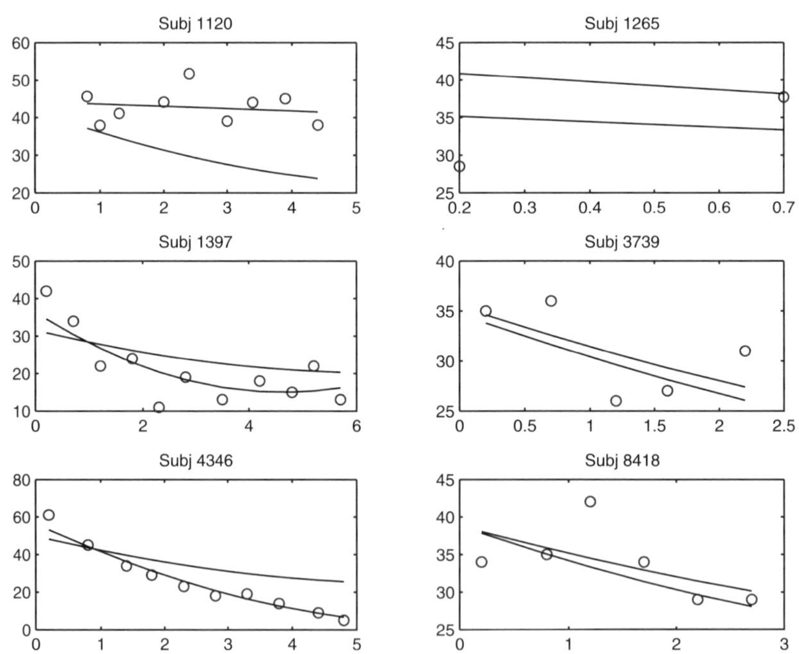

Figure 17.5 Type 3 SPME Modeling of the CD4 Data: Raw Data (circles), Population Fits (solid curves), and Individual Fits (dashed curves) for Six Selected Subjects

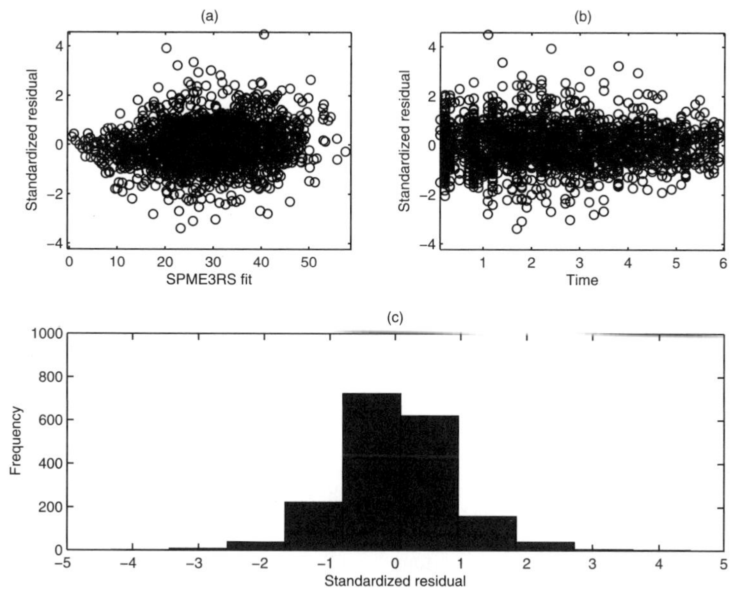

Figure 17.6 Type 3 SPME Modeling of the CD4 Data: Model Checking

where $\boldsymbol{\varepsilon}_i = (\varepsilon_{i1}, \ldots, \boldsymbol{\varepsilon}_{in_i})'$, and the entries of $\boldsymbol{\Omega}_v$ are defined as $\boldsymbol{\Omega}_v(r, s) = \gamma_v(\tau_r, \tau_s)$, $r, s = 1, 2, \ldots, M$. Further, set $\mathbf{X}_i = (\boldsymbol{x}_{i1}, \ldots, \boldsymbol{x}_{in_i})'$, $\mathbf{C}_i = (\boldsymbol{c}_{i1}, \boldsymbol{c}_{i2}, \ldots, \boldsymbol{c}_{in_i})'$, and $\mathbf{Z}_i = (z_{i1}, z_{i2}, \ldots, z_{in_i})'$. In addition, set $\boldsymbol{y}_i = (y_{i1}, \ldots, y_{in_i})'$. Then we have

$$\begin{aligned} \boldsymbol{y}_i &= \boldsymbol{X}_i\boldsymbol{\beta} + \boldsymbol{C}_i\boldsymbol{\eta} + \mathbf{Z}_i\boldsymbol{u}_i + \mathbf{C}_i\boldsymbol{v}_i + \boldsymbol{\varepsilon}_i, \\ \boldsymbol{u}_i &\sim N_{q_0}(\mathbf{0}, \boldsymbol{\Omega}_u), \\ \boldsymbol{v}_i &\sim N_M(\mathbf{0}, \boldsymbol{\Omega}_v), \\ \boldsymbol{\varepsilon}_i &\sim N_{n_i}(\mathbf{0}, \boldsymbol{\Sigma}_i),\ i = 1, 2, \ldots, n. \end{aligned} \tag{17.15}$$

For discussion convenience, we assume that the noise variance is time-independent so that $\boldsymbol{\Sigma}_i = \sigma^2\boldsymbol{I}_{n_i}$, with σ^2 being a noise variance component.

Notice that the model (17.15) is obtained via evaluating the SPME model (17.4) at the distinct design time points (17.13) for all the subjects. It is an LME model with $p_0 + M$ fixed-effects parameters and $n(q_0 + M)$ random-effects parameters. However, direct fitting of this LME model using the usual LME algorithm does not take the roughness of $\eta(t)$ and $v_i(t)$ into account. In fact, when $\boldsymbol{\Omega}$ and $\boldsymbol{\Sigma}_i$ are known, the usual LME solutions are actually obtained via minimizing the following twice negative marginal log-likelihood (up to a constant):

$$\sum_{i=1}^{n}\Big\{(\boldsymbol{y}_i - \tilde{\mathbf{X}}_i\tilde{\boldsymbol{\beta}} - \tilde{\mathbf{Z}}_i\tilde{\mathbf{u}}_i)'\boldsymbol{\Sigma}_i^{-1}(\boldsymbol{y}_i - \tilde{\mathbf{X}}_i\tilde{\boldsymbol{\beta}} - \tilde{\mathbf{Z}}_i\tilde{\mathbf{u}}_i) + \log|\boldsymbol{\Omega}| + \tilde{\mathbf{u}}_i'\boldsymbol{\Omega}^{-1}\tilde{\mathbf{u}}_i + \log|\boldsymbol{\Sigma}_i|\Big\}, \tag{17.16}$$

with respect to $\tilde{\boldsymbol{\beta}}$ and $\tilde{\mathbf{u}}_i$, $i = 1, 2, \ldots, n$, where $\tilde{\mathbf{X}}_i = (\boldsymbol{X}_i, \boldsymbol{C}_i)$, $\tilde{\boldsymbol{\beta}} = (\boldsymbol{\beta}', \boldsymbol{\eta}')'$, $\tilde{\mathbf{Z}}_i = (\boldsymbol{Z}_i, \boldsymbol{c}_i)$, $\tilde{\boldsymbol{u}}_i = (\boldsymbol{u}_i', \boldsymbol{v}_i')'$, and $\boldsymbol{\Omega} = \mathrm{Cov}(\tilde{\mathbf{u}}_i)$, $i = 1, 2, \ldots, n$. The solutions are

$$\begin{aligned} \widehat{\tilde{\boldsymbol{\beta}}} &= \Big(\sum\nolimits_{i=1}^{n}\tilde{\mathbf{X}}_i'\mathbf{V}_i^{-1}\tilde{\mathbf{X}}_i\Big)^{-1}\Big(\sum\nolimits_{i=1}^{n}\tilde{\mathbf{X}}_i'\mathbf{V}_i^{-1}\boldsymbol{y}_i\Big), \\ \widehat{\tilde{\mathbf{u}}}_i &= \boldsymbol{\Omega}\tilde{\mathbf{Z}}_i'\mathbf{V}_i^{-1}(\boldsymbol{y}_i - \tilde{\mathbf{X}}_i\widehat{\tilde{\boldsymbol{\beta}}}),\ i = 1, 2, \ldots, n, \end{aligned}$$

where $\mathbf{V}_i = \tilde{\mathbf{Z}}_i\boldsymbol{\Omega}\tilde{\mathbf{Z}}_i' + \boldsymbol{\Sigma}_i$, $i = 1, 2, \ldots, n$. In matrix notation, they can be written as

$$\begin{aligned} \widehat{\tilde{\boldsymbol{\beta}}} &= (\tilde{\mathbf{X}}'\mathbf{V}^{-1}\tilde{\mathbf{X}})^{-1}\tilde{\mathbf{X}}'\mathbf{V}^{-1}\boldsymbol{y}, \\ \widehat{\tilde{\mathbf{u}}} &= \bar{\boldsymbol{\Omega}}\tilde{\mathbf{Z}}'\mathbf{V}^{-1}(\boldsymbol{y} - \tilde{\mathbf{X}}\widehat{\tilde{\boldsymbol{\beta}}}), \end{aligned} \tag{17.17}$$

where

$$\tilde{\mathbf{X}} = (\tilde{\mathbf{X}}_1', \tilde{\mathbf{X}}_2', \ldots, \tilde{\mathbf{X}}_n')', \mathbf{V} = \oplus_{i=1}^n \mathbf{V}_i,$$
$$\tilde{\mathbf{Z}} = \oplus_{i=1}^n \tilde{\mathbf{Z}}_i, \bar{\mathbf{\Omega}} = \oplus_{i=1}^n \mathbf{\Omega}.$$

To extract the estimators, set

$$\mathbf{E}_1 = (\boldsymbol{I}_{p_0}, \mathbf{0}_{p_0 \times M}), \mathbf{E}_2 = (\mathbf{0}_{M \times p_0}, \boldsymbol{I}_{M \times M}),$$
$$\mathbf{F}_1 = (\boldsymbol{I}_{q_0}, \mathbf{0}_{q_0 \times M}), \mathbf{F}_2 = (\mathbf{0}_{M \times q_0}, \boldsymbol{I}_{M \times M}). \tag{17.18}$$

Then the estimated components can be expressed as

$$\hat{\boldsymbol{\beta}} = \mathbf{E}_1 \widehat{\tilde{\boldsymbol{\beta}}}, \hat{\boldsymbol{\eta}} = \mathbf{E}_2 \widehat{\tilde{\boldsymbol{\beta}}},$$
$$\hat{\mathbf{u}}_i = \mathbf{F}_1 \widehat{\tilde{\mathbf{u}}}_i, \hat{\mathbf{v}}_i = \mathbf{F}_2 \widehat{\tilde{\mathbf{u}}}_i, \quad i = 1, 2, \ldots, n.$$

Since $\boldsymbol{\eta}$ and $\boldsymbol{v}_i$ are obtained via evaluating the nonparametric fixed-effects function $\eta(t)$ and the nonparametric random-effects functions $v_i(t), \ i = 1, 2, \ldots, n$, estimates of $\eta(t)$ and $v_i(t), \ i = 1, 2, \ldots, n$, can be obtained via interpolation based on $\hat{\boldsymbol{\eta}}$ and $\hat{\mathbf{v}}_i, \ i = 1, 2, \ldots, n$ respectively. The resulting estimates $\hat{\eta}(t)$ and $\hat{v}_i(t), \ i = 1, 2, \ldots, n$ are usually not smooth since fitting the LME model (17.15) does not take into account any information about the roughness of $\eta(t)$ and $v_i(t), \ i = 1, 2, \ldots, n$. To overcome this problem, we only need to add the roughness measures of $\eta(t)$ and $v_i(t)$ into the fitting criterion (17.16), leading to the following penalized twice negative marginal log-likelihood (up to a constant):

$$\sum_{i=1}^{n} \Big\{ (\boldsymbol{y}_i - \tilde{\mathbf{X}}_i \tilde{\boldsymbol{\beta}} - \tilde{\mathbf{Z}}_i \tilde{\mathbf{u}}_i)' \boldsymbol{\Sigma}_i^{-1} (\boldsymbol{y}_i - \tilde{\mathbf{X}}_i \tilde{\boldsymbol{\beta}} - \tilde{\mathbf{Z}}_i \tilde{\mathbf{u}}_i)$$
$$+ \log |\mathbf{\Omega}| + \tilde{\mathbf{u}}_i' \mathbf{\Omega}^{-1} \tilde{\mathbf{u}}_i + \log |\boldsymbol{\Sigma}_i|$$
$$+ \lambda_v \int [v_i^{(k_v)}(t)]^2 dt \Big\} + \lambda \int [\eta^{(k)}(t)]^2 dt, \tag{17.19}$$

where λ is the smoothing parameter for the nonparametric fixed-effects function $\eta(t)$, trading off the goodness of fit and the smoothness of $\eta(t)$, and λ_v the smoothing parameter for the nonparametric random-effects functions $v_i(t), \ i = 1, 2, \ldots, n$, trading off the goodness of fit and the smoothness of these random-effects functions, and k and k_v are the times of derivatives used to define the roughness of $\eta(t)$ and $v_i(t)$ respectively. It can be seen that the main difference between the twice negative marginal log-likelihood (17.16) and the penalized twice negative marginal log-likelihood (17.19) is that the latter has penalty terms representing the incorporation of the roughness of $\eta(t)$ and $v_i(t), \ i = 1, 2, \ldots, n$.

In (17.19), we use the same smoothing parameter λ_v for all the nonparametric random-effects functions $v_i(t), \ i = 1, 2, \ldots, n$. We believe this is reasonable since these random-effects functions $v_i(t), \ i = 1, 2, \ldots, n$ are assumed to be the realizations of an underlying zero-mean stochastic process, and hence they should admit similar smoothness, which should be indicated by the same smoothing parameter. We also believe that this is sufficient. Conceptually, however, they may use different smoothing parameters. A typical disadvantage of this latter strategy is that it needs a large computational effort for smoothing parameter selection.

17.5.1 Estimation Using Cubic Smoothing Splines

It is usually not easy to directly compute the integrals involved in (17.19) for general k and k_v. However, when $k = k_v = 2$, i.e., when cubic smoothing splines are used, the integrals can be easily calculated. In fact, we have

$$\int [\eta''(t)]^2 dt = \boldsymbol{\eta}' \mathbf{G} \boldsymbol{\eta},$$
$$\int [v_i''(t)]^2 dt = \boldsymbol{v}_i' \mathbf{G} \boldsymbol{v}_i, \qquad i = 1, 2, \ldots, n,$$

where $\mathbf{G}$ is the cubic spline roughness matrix that can be calculated easily using the formula in Green and Silverman (1994) based on the given distinct design time points (17.13). Thus, (17.19) can now be written as

$$\sum_{i=1}^{n}\Big\{(y_i - \tilde{\mathbf{X}}_i\tilde{\boldsymbol{\beta}} - \tilde{\mathbf{Z}}_i\tilde{\mathbf{u}}_i)'\boldsymbol{\Sigma}_i^{-1}(y_i - \tilde{\mathbf{X}}_i\tilde{\boldsymbol{\beta}} - \tilde{\mathbf{Z}}_i\tilde{\mathbf{u}}_i) + \log|\boldsymbol{\Omega}| + \tilde{\mathbf{u}}_i'\boldsymbol{\Omega}^{-1}\tilde{\mathbf{u}}_i + \log|\boldsymbol{\Sigma}_i| + \lambda_v \boldsymbol{v}_i'\mathbf{G}\boldsymbol{v}_i\Big\} + \lambda\boldsymbol{\eta}'\mathbf{G}\boldsymbol{\eta}. \quad (17.20)$$

Compared with (17.16), the above expression has two extra terms representing the roughness penalties of $\eta(t)$ and $v_i(t)$, $i = 1, 2, \ldots, n$. It can be further written as

$$\sum_{i=1}^{n}\Big\{(y_i - \tilde{\mathbf{X}}_i\tilde{\boldsymbol{\beta}} - \tilde{\mathbf{Z}}_i\tilde{\mathbf{u}}_i)'\boldsymbol{\Sigma}_i^{-1}(y_i - \tilde{\mathbf{X}}_i\tilde{\boldsymbol{\beta}} - \tilde{\mathbf{Z}}_i\tilde{\mathbf{u}}_i) + \log|\tilde{\boldsymbol{\Omega}}| + \tilde{\mathbf{u}}_i'\tilde{\boldsymbol{\Omega}}^{-1}\tilde{\mathbf{u}}_i + \log|\boldsymbol{\Sigma}_i|\Big\} + \tilde{\boldsymbol{\beta}}'\tilde{\mathbf{G}}\tilde{\boldsymbol{\beta}}, \quad (17.21)$$

where $\tilde{\boldsymbol{\Omega}} = (\boldsymbol{\Omega}^{-1} + \tilde{\mathbf{G}}_v)^{-1}$, $\tilde{\mathbf{G}}_v = \mathbf{0}_{q_0\times q_0} \oplus (\lambda_v\mathbf{G})$, and $\tilde{\mathbf{G}} = \mathbf{0}_{p_0\times p_0} \oplus (\lambda\mathbf{G})$. Further, set

$$\tilde{\mathbf{V}}_i = \tilde{\mathbf{Z}}_i\tilde{\boldsymbol{\Omega}}\tilde{\mathbf{Z}}_i' + \boldsymbol{\Sigma}_i,\ i = 1, 2, \ldots, n.$$

Then, assuming that the variance components $\boldsymbol{\Omega}$ and $\boldsymbol{\Sigma}_i$, $i = 1, 2, \ldots, n$, and the smoothing parameters λ and λ_v are known, the minimizers of the penalized marginal log-likelihood (17.21) are

$$\begin{aligned}\widehat{\tilde{\boldsymbol{\beta}}} &= \Big(\sum_{i=1}^{n}\tilde{\mathbf{X}}_i'\tilde{\mathbf{V}}_i^{-1}\tilde{\mathbf{X}}_i + \tilde{\mathbf{G}}\Big)^{-1}\Big(\sum_{i=1}^{n}\tilde{\mathbf{X}}_i'\tilde{\mathbf{V}}_i^{-1}y_i\Big),\\ \widehat{\tilde{\mathbf{u}}}_i &= \tilde{\boldsymbol{\Omega}}\tilde{\mathbf{Z}}_i'\tilde{\mathbf{V}}_i^{-1}(y_i - \tilde{\mathbf{X}}_i\widehat{\tilde{\boldsymbol{\beta}}}),\ i = 1, 2, \ldots, n.\end{aligned} \quad (17.22)$$

In matrix form, the solutions (17.22) can be written as

$$\widehat{\tilde{\boldsymbol{\beta}}} = (\tilde{\mathbf{X}}'\tilde{\mathbf{V}}^{-1}\tilde{\mathbf{X}} + \tilde{\mathbf{G}})^{-1}\tilde{\mathbf{X}}'\tilde{\mathbf{V}}^{-1}y, \quad (17.23)$$

$$\widehat{\tilde{\mathbf{u}}} = \bar{\tilde{\boldsymbol{\Omega}}}\tilde{\mathbf{Z}}'\tilde{\mathbf{V}}^{-1}(y - \tilde{\mathbf{X}}\widehat{\tilde{\boldsymbol{\beta}}}),$$

where

$$\tilde{\mathbf{V}} = \oplus_{i=1}^{n}\tilde{\mathbf{V}}_i,\quad \bar{\tilde{\boldsymbol{\Omega}}} = \oplus_{i=1}^{n}\tilde{\boldsymbol{\Omega}}. \quad (17.24)$$

Compared with the solutions of the LME model (17.15) presented in (17.17), the solutions above are complicated by the incorporation of the roughness of $\eta(t)$ and $v_i(t)$, $i = 1, 2, \ldots, n$. The complexity is mainly determined by the smoothing parameters λ and λ_v. On the one hand, when $\lambda_v = 0$, so that no roughness penalty is imposed for estimating $v_i(t)$, $i = 1, 2, \ldots, n$, we have $\tilde{\boldsymbol{\Omega}} = \boldsymbol{\Omega}$ and $\tilde{\mathbf{V}}_i = \mathbf{V}_i$, $i = 1, 2, \ldots, n$. On the other hand, when $\lambda = 0$, the term $\lambda\mathbf{G}$ disappears in the expression of $\widehat{\boldsymbol{\beta}}$, indicating no roughness penalty imposed for estimating $\eta(t)$. Moreover, when both λ and λ_v are 0 so that no roughness of $\eta(t)$ and $v_i(t)$, $i = 1, 2, \ldots, n$ are imposed, then the penalized marginal log-likelihood criterion (17.19) reduces to the marginal log-likelihood criterion (17.16).

17.5.2 Bayesian Interpretation of the Estimators

For some basic ideas about Bayesian analysis and its applications in multilevel data analysis, the reader is referred to Chapter 4. In this subsection, we show that the solutions (17.23) can be interpreted as the posterior expectations of the parameters of a Bayesian problem. This property will be used to estimate the variance components for the SPME model (17.4). With $\boldsymbol{\Sigma} = \oplus_{i=1}^{n}\boldsymbol{\Sigma}_i$, the Bayesian problem can be defined as follows:

$$y|\tilde{\boldsymbol{\beta}}, \tilde{\mathbf{u}} \sim N_N(\tilde{\mathbf{X}}\tilde{\boldsymbol{\beta}} + \tilde{\mathbf{Z}}\tilde{\mathbf{u}}, \boldsymbol{\Sigma}), \quad (17.25)$$

with priors for $\tilde{\boldsymbol{\beta}}$ and $\tilde{\mathbf{u}}$:

$$\tilde{\boldsymbol{\beta}} \sim N_{p_0+M}(0, \tilde{\mathbf{G}}^{-}),\quad \tilde{\mathbf{u}} \sim N_{n(q_0+M)}(0, \bar{\tilde{\boldsymbol{\Omega}}}), \quad (17.26)$$

where $\tilde{\boldsymbol{\beta}}$, $\tilde{\mathbf{u}}$, and $\boldsymbol{\varepsilon} = y - \tilde{\mathbf{X}}\tilde{\boldsymbol{\beta}} - \tilde{\mathbf{Z}}\tilde{\mathbf{u}}$ are all independent of each other, $\tilde{\mathbf{G}}^{-}$ denotes one of the generalized inverses of $\tilde{\mathbf{G}}$, and $\bar{\tilde{\boldsymbol{\Omega}}}$ is defined as in (17.24). From (17.26), it can be seen that $\tilde{\boldsymbol{\beta}}$ is now a 'random vector' with its variation coming from the roughness of the nonparametric fixed-effects function $\eta(t)$ which itself is not random, while the variation of the random vector $\tilde{\mathbf{u}}$ comes not only from the randomness of the parametric random-effects $\mathbf{u}_i$

and the nonparametric random-effects functions $v_i(t),\ i = 1, 2, \ldots, n$, but also from the roughness of $v_i(t),\ i = 1, 2, \ldots, n$. We can show that the estimates (17.23) that minimize the penalized marginal log-likelihood (17.20) are the same as the posterior expectations of the Bayesian problem defined in (17.25) and (17.26). That is, $\widehat{\tilde{\beta}} = \mathrm{E}(\tilde{\beta}|y),\ \widehat{\tilde{\mathbf{u}}} = \mathrm{E}(\tilde{\mathbf{u}}|y)$. Moreover,

$$\begin{aligned}\tilde{\beta}|y &\sim N_{p_0+M}\left[\widehat{\tilde{\beta}}, (\tilde{\mathbf{G}} + \tilde{\mathbf{X}}'\tilde{\mathbf{V}}^{-1}\tilde{\mathbf{X}})^{-1}\right],\\ \tilde{\mathbf{u}}|y &\sim N_{n(q_0+M)}\left[\widehat{\tilde{\mathbf{u}}}, \bar{\tilde{\Omega}} - \bar{\tilde{\Omega}}\tilde{\mathbf{Z}}'\mathbf{W}\tilde{\mathbf{Z}}\bar{\tilde{\Omega}}\right],\\ \varepsilon|y &\sim N_N\left[\hat{\varepsilon}, \boldsymbol{\Sigma} - \boldsymbol{\Sigma}\mathbf{W}\boldsymbol{\Sigma}\right],\end{aligned}$$

where $\hat{\varepsilon} = y - \tilde{\mathbf{X}}\widehat{\tilde{\beta}} - \tilde{\mathbf{Z}}\widehat{\tilde{\mathbf{u}}}$, and $\mathbf{W} = \tilde{\mathbf{V}}^{-1} - \tilde{\mathbf{V}}^{-1}\tilde{\mathbf{X}}(\tilde{\mathbf{G}} + \tilde{\mathbf{X}}'\tilde{\mathbf{V}}^{-1}\tilde{\mathbf{X}})^{-1}\tilde{\mathbf{X}}'\tilde{\mathbf{V}}^{-1}$. Let $\mathbf{W}_i = \tilde{\mathbf{V}}_i^{-1} - \tilde{\mathbf{V}}_i^{-1}\tilde{\mathbf{X}}_i(\tilde{\mathbf{G}} + \tilde{\mathbf{X}}'\tilde{\mathbf{V}}^{-1}\tilde{\mathbf{X}})^{-1}\tilde{\mathbf{X}}_i'\tilde{\mathbf{V}}_i^{-1}$. Then it is easy to show that

$$\begin{aligned}&\mathrm{E}(\tilde{\mathbf{u}}_i|y) = \widehat{\tilde{\mathbf{u}}}_i,\ \mathrm{Cov}(\tilde{\mathbf{u}}_i|y) = \tilde{\Omega} - \tilde{\Omega}\tilde{\mathbf{Z}}_i'\mathbf{W}_i\tilde{\mathbf{Z}}_i\tilde{\Omega},\\ &\mathrm{E}(\varepsilon_i|y) = \hat{\varepsilon}_i,\ \mathrm{Cov}(\varepsilon_i|y) = \boldsymbol{\Sigma}_i - \boldsymbol{\Sigma}_i\mathbf{W}_i\boldsymbol{\Sigma}_i,\\ &\qquad i = 1, 2, \ldots, n. \end{aligned} \tag{17.27}$$

In next subsection, we shall discuss the restricted maximum likelihood (REML)-based and maximum likelihood (ML)-based EM algorithms for estimating the variance components for the smoothing spline method. The posterior means and covariances of $\tilde{\mathbf{u}}_i$ and ε_i given in (17.27) are closely related to the REML-based EM algorithm for estimating the variance components. For the ML-based EM algorithm, the posterior means and covariances of $\tilde{\mathbf{u}}_i$ and ε_i given y and $\tilde{\beta}$ as in (17.29) are more related. We can show that under the Bayesian framework (17.25) and (17.26), we have

$$\begin{aligned}\tilde{\mathbf{u}}|y, \tilde{\beta} &\sim N_{n(q_0+M)}\Big[\bar{\tilde{\Omega}}\tilde{\mathbf{Z}}'\tilde{\mathbf{V}}^{-1}(y - \tilde{\mathbf{X}}\tilde{\beta}),\\ &\qquad \bar{\tilde{\Omega}} - \bar{\tilde{\Omega}}\tilde{\mathbf{Z}}'\tilde{\mathbf{V}}^{-1}\tilde{\mathbf{Z}}\bar{\tilde{\Omega}}\Big],\\ \varepsilon|y, \tilde{\beta} &\sim N_N\Big[\boldsymbol{\Sigma}\tilde{\mathbf{V}}^{-1}(y - \tilde{\mathbf{X}}\tilde{\beta}),\\ &\qquad \boldsymbol{\Sigma} - \boldsymbol{\Sigma}\tilde{\mathbf{V}}^{-1}\boldsymbol{\Sigma}\Big].\end{aligned} \tag{17.28}$$

From the above results, we actually have $\mathrm{E}(\tilde{\mathbf{u}}|y, \tilde{\beta} = \widehat{\tilde{\beta}}) = \bar{\tilde{\Omega}}\tilde{\mathbf{Z}}'\tilde{\mathbf{V}}^{-1}(y - \tilde{\mathbf{X}}\widehat{\tilde{\beta}}) = \widehat{\tilde{\mathbf{u}}} = \mathrm{E}(\tilde{\mathbf{u}}|y)$, and $\mathrm{E}(\varepsilon|y, \tilde{\beta} = \widehat{\tilde{\beta}}) = \boldsymbol{\Sigma}\tilde{\mathbf{V}}^{-1}(y - \tilde{\mathbf{X}}\widehat{\tilde{\beta}}) = \hat{\varepsilon} = \mathrm{E}(\varepsilon|y)$. Based on these and (17.28), we have

$$\begin{aligned}\mathrm{E}(\tilde{\mathbf{u}}_i|y, \tilde{\beta} = \widehat{\tilde{\beta}}) &= \widehat{\tilde{\mathbf{u}}}_i,\\ \mathrm{Cov}(\tilde{\mathbf{u}}_i|y, \tilde{\beta} = \widehat{\tilde{\beta}}) &= \tilde{\Omega} - \tilde{\Omega}\tilde{\mathbf{Z}}_i'\tilde{\mathbf{V}}_i^{-1}\tilde{\mathbf{Z}}_i\tilde{\Omega},\\ \mathrm{E}(\varepsilon_i|y, \tilde{\beta} = \widehat{\tilde{\beta}}) &= \hat{\varepsilon}_i,\\ \mathrm{Cov}(\varepsilon_i|y, \tilde{\beta} = \widehat{\tilde{\beta}}) &= \boldsymbol{\Sigma}_i - \boldsymbol{\Sigma}_i\tilde{\mathbf{V}}_i^{-1}\boldsymbol{\Sigma}_i,\\ i &= 1, 2, \ldots, n.\end{aligned} \tag{17.29}$$

Comparing (17.29) with (17.27), it is easy to see that $\mathrm{Cov}(\tilde{\mathbf{u}}_i|y, \tilde{\beta} = \widehat{\tilde{\beta}})$ and $\mathrm{Cov}(\varepsilon_i|y, \tilde{\beta} = \widehat{\tilde{\beta}})$ can be obtained respectively from $\mathrm{Cov}(\tilde{\mathbf{u}}_i|y)$ and $\mathrm{Cov}(\varepsilon_i|y)$ by replacing $\tilde{\mathbf{V}}_i^{-1}$ with $\mathbf{W}_i$.

17.5.3 Estimating the Variance Components

In expression (17.22), we assume the variance components $\boldsymbol{\Omega}$ and

$$\boldsymbol{\Sigma}_i = \sigma^2 \boldsymbol{I}_{n_i},\quad i = 1, 2, \ldots, n, \tag{17.30}$$

are known. In practice, both $\boldsymbol{\Omega}$ and σ^2 need to be estimated. In the general LME model, the variance components $\boldsymbol{\Omega}$ and σ^2 are estimated by the ML-based or REML-based EM algorithm (Laird, Lange, and Stram, 1987). These algorithms cannot be directly applied to the variance component estimation of the smoothing spline method. In this subsection, we develop EM algorithms for the estimators using cubic smoothing splines.

The EM algorithms that we shall develop aim to estimate the variance components $\boldsymbol{\Omega}$ and σ^2. When $\tilde{\mathbf{u}}_i$ and ε_i are known, under the Gaussian assumption, both $\tilde{\mathbf{u}}_i$ and ε_i are normal. Therefore, the maximum likelihood estimates (MLEs) of σ^2 and $\boldsymbol{\Omega}$ would be

$$\hat{\sigma}^2 = N^{-1}\sum_{i=1}^{n}\varepsilon_i'\varepsilon_i,\quad \hat{\boldsymbol{\Omega}} = n^{-1}\sum_{i=1}^{n}\tilde{\mathbf{u}}_i\tilde{\mathbf{u}}_i'. \tag{17.31}$$

Because ε_i and $\tilde{\mathbf{u}}_i$ are actually unknown, the above MLEs are not computable. There are two methods that may be used to overcome this problem.

Method 1 Replace $\hat{\sigma}^2$ and $\hat{\mathbf{\Omega}}$ by their posterior expectations $\mathrm{E}(\hat{\sigma}^2|y)$ and $\mathrm{E}(\hat{\mathbf{\Omega}}|y)$ respectively.

Method 2 Replace $\hat{\sigma}^2$ and $\hat{\mathbf{\Omega}}$ by their posterior expectations $\mathrm{E}(\hat{\sigma}^2|y, \tilde{\beta} = \widehat{\tilde{\beta}})$ and $\mathrm{E}(\hat{\mathbf{\Omega}}|y, \tilde{\beta} = \widehat{\tilde{\beta}})$ respectively.

The posterior expectations are computed under the Bayesian framework given in (17.25) and (17.26). They are computed under the assumption that both $\mathbf{\Omega}$ and σ^2 are known, which is not true since, after all, they are the targets of the estimation. An EM algorithm aims to overcome this problem using iterations after initial values are given for $\mathbf{\Omega}$ and σ^2. The maximum likelihood step (M-step) is based on (17.31), while the expectation step (E-step) is based on Method 1 or Method 2 above. The Method-1-based EM algorithm is traditionally called a REML-based EM algorithm since it takes the variation of $\widehat{\tilde{\beta}}$ into account, while the Method-2-based EM algorithm is called an ML-based EM-algorithm, since it does not.

Using the previous results, we can show that, assuming $\mathbf{\Sigma}_i,\ i = 1, 2, \ldots, n$ satisfy (17.30), and under the Bayesian framework defined in (17.25) and (17.26), the posterior expectations of $\hat{\sigma}^2$ and $\hat{\mathbf{\Omega}}$ given y are

$$\mathrm{E}(\hat{\sigma}^2|y) = N^{-1}\sum_{i=1}^{n}\{\hat{\varepsilon}'\hat{\varepsilon} + \sigma^2[n_i - \sigma^2\mathrm{tr}(\mathbf{W}_i)]\}, \quad (17.32)$$

$$\mathrm{E}(\hat{\mathbf{\Omega}}|y) = n^{-1}\sum_{i=1}^{n}\left\{\widehat{\tilde{\mathbf{u}}}_i\widehat{\tilde{\mathbf{u}}}_i' + [\tilde{\mathbf{\Omega}} - \tilde{\mathbf{\Omega}}\tilde{\mathbf{X}}_i'\mathbf{W}_i\tilde{\mathbf{X}}_i\tilde{\mathbf{\Omega}}]\right\}, \quad (17.33)$$

and the posterior expectations of $\hat{\sigma}^2$ and $\hat{\mathbf{\Omega}}$ given y and $\tilde{\beta} = \widehat{\tilde{\beta}}$ are the same as those of $\hat{\sigma}^2$ and $\hat{\mathbf{\Omega}}$ given y after $\mathbf{W}_i$ are replaced with $\tilde{\mathbf{V}}_i^{-1}$ respectively.

The above ideas for the EM algorithms can be iteratively operated using some given initial values for $\mathbf{\Omega}$ and σ^2 until convergence. For simplicity, the initial values can be taken as $\hat{\mathbf{\Omega}} = \hat{\sigma}^2 \boldsymbol{I}_{q_0+M}$ and $\hat{\sigma}^2 = 1$. As an illustration, we state the REML-based EM algorithm in detail. The ML-based EM algorithm can be obtained from the REML-based EM algorithm by replacing those $\mathbf{W}_{i(\cdot)}$ by those $\tilde{\mathbf{V}}_{i(\cdot)}^{-1}$ respectively. The major cycle for the REML-based EM algorithm is as follows:

(a) Given $\hat{\mathbf{\Omega}}$ and $\hat{\sigma}^2$, update $\widehat{\tilde{\beta}}$ and $\widehat{\tilde{\mathbf{u}}}_i$ using (17.22).

(b) Given $\widehat{\tilde{\beta}}$ and $\widehat{\tilde{\mathbf{u}}}_i$, update $\hat{\mathbf{\Omega}}$ and $\hat{\sigma}^2$ using (17.32) and (17.33).

Let $r = 0, 1, 2, \ldots$ index the sequence of iterations, and $\widehat{\tilde{\beta}}_{(r)}$, $\widehat{\tilde{\mathbf{u}}}_{i(r)}$ be the estimated values for $\widehat{\tilde{\beta}}$ and $\widehat{\tilde{\mathbf{u}}}_i$ at iteration r. Other notations are similarly defined. Moreover, let $\delta > 0$ be a number to indicate the required accuracy of the outputs. Examples for δ can be .001 or .00001. Set $\|\mathbf{A}\|$ to be the largest entry of a matrix $\mathbf{A}$. Then, more formally, the REML-based EM algorithm may be written as follows:

The REML-based EM-Algorithm

Step 0. Set r=0; initializing, $\hat{\sigma}^2_{(r)} = 1$, and $\hat{\mathbf{\Omega}}_{(r)} = \hat{\sigma}^2_{(r)}\mathsf{I}_{q_0+M}$.

Step 1. Set $r = r + 1$. Compute

$$\tilde{\mathbf{\Omega}}_{(r-1)} = (\hat{\mathbf{\Omega}}_{(r-1)}^{-1} + \tilde{\mathbf{G}}_v)^{-1},$$
$$\tilde{\mathbf{V}}_{i(r-1)} = \tilde{\mathbf{Z}}_i\tilde{\mathbf{\Omega}}_{(r-1)}\tilde{\mathbf{Z}}_i' + \hat{\sigma}^2_{(r-1)}\mathsf{I}_{n_i},$$
$$i = 1, 2, \ldots, n.$$

Then compute $\widehat{\tilde{\beta}}_{(r)}$, $\widehat{\tilde{\mathbf{u}}}_{i(r)}$, and $\hat{\mathbf{W}}_{ir}$ using

$$\widehat{\tilde{\beta}}_{(r)} = \left(\sum_{i=1}^{n}\tilde{\mathbf{X}}_i'\tilde{\mathbf{V}}_{i(r-1)}^{-1}\tilde{\mathbf{X}}_i + \tilde{\mathbf{G}}\right)^{-1} \times \left(\sum_{i=1}^{n}\tilde{\mathbf{X}}_i'\tilde{\mathbf{V}}_{i(r-1)}^{-1}y_i\right),$$

$$\widehat{\tilde{\mathbf{u}}}_{i(r)} = \tilde{\mathbf{\Omega}}_{(r-1)}\tilde{\mathbf{Z}}_i'\tilde{\mathbf{V}}_{i(r-1)}^{-1}(y_i - \tilde{\mathbf{X}}_i\widehat{\tilde{\beta}}_{(r)}),$$
$$i = 1, 2, \ldots, n,$$

$$\hat{\mathbf{W}}_{ir} = \tilde{\mathbf{V}}_{i(r-1)}^{-1} - \tilde{\mathbf{V}}_{i(r-1)}^{-1}\tilde{\mathbf{X}}_i (\tilde{\mathbf{G}} + \tilde{\mathbf{X}}'\tilde{\mathbf{V}}_{(r-1)}^{-1}\tilde{\mathbf{X}})^{-1} \times \tilde{\mathbf{X}}_i'\tilde{\mathbf{V}}_{i(r-1)}^{-1},$$

where $\tilde{\mathbf{V}}_{(r-1)} = \oplus_{i=1}^{n}\tilde{\mathbf{V}}_{i(r-1)}$.

Step 2. Compute the residuals $\hat{\boldsymbol{\varepsilon}}_{i(r)} = \mathbf{y}_i - \tilde{\mathbf{X}}_i\widehat{\tilde{\boldsymbol{\beta}}}_{(r)} - \tilde{\mathbf{Z}}_i\widehat{\tilde{\mathbf{u}}}_{i(r)}$. Then update $\hat{\sigma}^2_{(r)}$ and $\hat{\boldsymbol{\Omega}}_{(r)}$ using

$$\hat{\sigma}^2_{(r)} = N^{-1}\sum_{i=1}^{n}\Big\{\hat{\boldsymbol{\varepsilon}}'_{i(r)}\hat{\boldsymbol{\varepsilon}}_{i(r)} + \hat{\sigma}^2_{(r-1)} \times[n_i - \hat{\sigma}^2_{(r-1)}\mathrm{tr}(\hat{\mathbf{W}}_{i(r-1)})]\Big\},$$

$$\hat{\boldsymbol{\Omega}}_{(r)} = n^{-1}\sum_{i=1}^{n}\Big\{\widehat{\tilde{\mathbf{u}}}_{i(r)}\widehat{\tilde{\mathbf{u}}}'_{i(r)} + [\tilde{\boldsymbol{\Omega}}_{(r-1)} - \tilde{\boldsymbol{\Omega}}_{(r-1)}\tilde{\mathbf{Z}}'_i\hat{\mathbf{W}}_{i(r-1)}\tilde{\mathbf{Z}}_i\tilde{\boldsymbol{\Omega}}_{(r-1)}]\Big\}.$$

Step 3. When $\|\hat{\boldsymbol{\Omega}}_{(r)} - \hat{\boldsymbol{\Omega}}_{(r-1)}\| < \delta$, stop; otherwise, go back to Step 1.

Notice that the smoothing parameters λ, λ_v and the roughness matrix $\mathbf{G}$ do not change during the entire iteration process.

17.5.4 Statistical Estimation and Inferences

In this subsection, we outline some statistical inferences about the parametric fixed-effects $\boldsymbol{\beta}$, the nonparametric fixed-effects function $\eta(t)$, the parametric random-effects $\mathbf{u}_i$, and the random-effects functions $v_i(t)$, $i = 1, 2, \ldots, n$ of the SPME model (17.4).

Component estimation

Using the matrices defined in (17.18), the estimated components can be expressed as

$$\hat{\boldsymbol{\beta}} = \mathbf{E}_1\widehat{\tilde{\boldsymbol{\beta}}},\ \hat{\boldsymbol{\eta}} = \mathbf{E}_2\widehat{\tilde{\boldsymbol{\beta}}},\ \hat{\mathbf{u}}_i = \mathbf{F}_1\widehat{\tilde{\mathbf{u}}}_i, \quad \hat{\mathbf{v}}_i = \mathbf{F}_2\widehat{\tilde{\mathbf{u}}}_i,\quad i = 1, 2, \ldots, n. \quad (17.34)$$

Let $\mathbf{e}_{r,p}$ denote a p-dimensional column vector with the rth entry being 1 and otherwise 0. Then the estimates of $\eta(t)$ and $v_i(t)$ at τ_r are

$$\hat{\eta}(\tau_r) = \mathbf{e}'_{r,M}\hat{\boldsymbol{\eta}},$$
$$\hat{v}_i(\tau_r) = \mathbf{e}'_{r,M}\hat{\mathbf{v}}_i,\ r = 1, 2, \ldots, M.$$

In addition, we have $\hat{\boldsymbol{\Omega}}_u = \mathbf{F}_1\hat{\boldsymbol{\Omega}}\mathbf{F}'_1$ and $\hat{\boldsymbol{\Omega}}_v = \mathbf{F}_2\hat{\boldsymbol{\Omega}}\mathbf{F}'_2$. The estimates of $\gamma_v(s,t)$ and $\gamma_\varepsilon(s,t)$ at (τ_r, τ_s) are then given by $\hat{\gamma}_v(\tau_r, \tau_s) = \hat{\boldsymbol{\Omega}}_v(r, s)$ and $\hat{\gamma}_\varepsilon(\tau_r, \tau_s) = \hat{\sigma}^2 1_{\{r=s\}}$.

Fixed-effects estimation and random-effects predictions

The estimation of the fixed-effects components $\boldsymbol{x}'_{ij}\boldsymbol{\beta} + \eta(t_{ij})$ at all of the design time points (17.14) can be expressed as the product of a smoother matrix and the response vector $\boldsymbol{y}$. Similarly, the predictions of the random-effects components $z'_{ij}\boldsymbol{u}_i + v_i(t_{ij})$ at all of the design time points can be expressed as the product of a smoother matrix and the residual vector $\tilde{\mathbf{w}} = \boldsymbol{y} - \tilde{\mathbf{X}}\widehat{\tilde{\boldsymbol{\beta}}}$, obtained via removing the fixed effects from $\boldsymbol{y}$. The traces of the associated smoother matrices can be used to measure how complicated the models are.

Using (17.23), the predictions of the fixed-effects components at all of the design time points (17.14) can be expressed as

$$\hat{\mathbf{y}} = \tilde{\mathbf{X}}\widehat{\tilde{\boldsymbol{\beta}}} = \mathbf{A}\boldsymbol{y},$$

where

$$\mathbf{A} = \tilde{\mathbf{X}}(\tilde{\mathbf{X}}'\tilde{\mathbf{V}}^{-1}\tilde{\mathbf{X}} + \tilde{\mathbf{G}})^{-1}\tilde{\mathbf{X}}'\tilde{\mathbf{V}}^{-1} \quad (17.35)$$

is the associated smoother matrix and $\tilde{\mathbf{V}} = \tilde{\mathbf{Z}}\tilde{\boldsymbol{\Omega}}\tilde{\mathbf{Z}}' + \boldsymbol{\Sigma}$. The trace of $\mathbf{A}$ can be used to measure the complexity of the model fitting the fixed-effects components. Similarly, the predictions of the random-effects components at the design time points can be expressed as

$$\hat{\mathbf{w}} = \tilde{\mathbf{Z}}\widehat{\tilde{\mathbf{u}}} = \tilde{\mathbf{Z}}\bar{\tilde{\boldsymbol{\Omega}}}\tilde{\mathbf{Z}}'\tilde{\mathbf{V}}^{-1}(\boldsymbol{y} - \tilde{\mathbf{X}}\widehat{\tilde{\boldsymbol{\beta}}}) = \mathbf{A}_v\tilde{\mathbf{w}},$$

where

$$\mathbf{A}_v = \tilde{\mathbf{Z}}\bar{\tilde{\boldsymbol{\Omega}}}\tilde{\mathbf{Z}}'\tilde{\mathbf{V}}^{-1} \quad (17.36)$$

is the smoother matrix for estimating the random-effects components at the design time points.

Variation of the fixed-effects estimators

Based on (17.23), it is easy to see that $\mathrm{Cov}(\widehat{\tilde{\boldsymbol{\beta}}}) = (\tilde{\mathbf{X}}'\tilde{\mathbf{V}}^{-1}\tilde{\mathbf{X}} + \tilde{\mathbf{G}})^{-1}\tilde{\mathbf{V}}(\tilde{\mathbf{X}}'\tilde{\mathbf{V}}^{-1}\tilde{\mathbf{X}} + \tilde{\mathbf{G}})^{-1}$. It follows from (17.34) that

$$\mathrm{Cov}(\hat{\boldsymbol{\beta}}) = \mathbf{E}_1\mathrm{Cov}(\widehat{\tilde{\boldsymbol{\beta}}})\mathbf{E}'_1,$$
$$\mathrm{Cov}(\hat{\boldsymbol{\eta}}) = \mathbf{E}_2\mathrm{Cov}(\widehat{\tilde{\boldsymbol{\beta}}})\mathbf{E}'_2. \quad (17.37)$$

Using (17.37), we can construct the confidence intervals of each component of $\boldsymbol{\beta}$ and the standard deviation bands for $\eta(t)$ at all of the distinct design time points, i.e., the 95% pointwise standard deviation intervals for $\hat{\boldsymbol{\eta}}$. In fact, we have

$$\begin{aligned}\mathrm{Var}(\hat{\beta}_r) &= \mathbf{e}'_{r,p_0}\mathrm{Cov}(\hat{\boldsymbol{\beta}})\mathbf{e}_{r,p_0}, \\ r &= 1, 2, \ldots, p_0, \\ \mathrm{Var}(\hat{\eta}(\tau_r)) &= \mathbf{e}'_{r,M}\mathrm{Cov}(\hat{\boldsymbol{\eta}})\mathbf{e}_{r,M}, \\ r &= 1, 2, \ldots, M. \end{aligned} \tag{17.38}$$

Then the 95% approximate confidence interval for β_r is

$$\hat{\beta}_r \pm 1.96\widehat{\mathrm{Var}}^{1/2}(\hat{\beta}_r),\ r = 1, 2, \ldots, p_0.$$

Similarly, the approximate standard deviation bands for $\eta(t)$ at the distinct time points (17.13) are

$$\hat{\eta}(\tau_r) \pm 1.96\widehat{\mathrm{Var}}^{1/2}[\hat{\eta}(\tau_r)],\ r = 1, 2, \ldots, M,$$

where $\widehat{\mathrm{Var}}^{1/2}(\hat{\beta}_r)$ and $\widehat{\mathrm{Var}}[\hat{\eta}(\tau_r)]$ are obtained via replacing all the unknowns in (17.37) and (17.38) by their associated estimates.

17.5.5 Selection of Smoothing Parameters

So far, we have assumed that the smoothing parameters λ and λ_v are fixed. In practice, they should be chosen to get good estimators for the component estimations of the SPME model (17.4). For the regression spline method studied in the previous section, we defined the AIC and BIC rules for knot number selection. In this subsection, we shall define the AIC and BIC rules for selecting λ and λ_v. These two rules work by trading off the goodness of fit and the model complexity of the SPME fit.

Goodness of fit

As for the regression spline method, the goodness of fit can be defined by the log-likelihood of the smoothing spline model fitted to the data:

$$\begin{aligned}\mathrm{Loglik} = -\frac{1}{2}\sum_{i=1}^{n}\Big\{&(\boldsymbol{y}_i - \boldsymbol{\mu}_i)'\mathbf{V}_i^{-1}(\boldsymbol{y}_i - \boldsymbol{\mu}_i) \\ &-\frac{1}{2}\log(|\mathbf{V}_i|)\Big\} - \frac{N}{2}\log(2\pi),\end{aligned}$$

with the unknowns $\boldsymbol{\mu}_i = (\boldsymbol{x}'_{i1}\boldsymbol{\beta} + \eta(t_{i1}), \ldots, \boldsymbol{x}'_{in_i}\boldsymbol{\beta} + \eta(t_{in_i}))'$ and $\mathbf{V}_i$ replaced by their estimates. Under (17.20), $\boldsymbol{\mu}_i$ and $\mathbf{V}_i$ are estimated by $\hat{\boldsymbol{\mu}}_i = \tilde{\mathbf{X}}_i\hat{\tilde{\boldsymbol{\beta}}}$ and $\hat{\tilde{\mathbf{V}}}_i = \tilde{\mathbf{Z}}_i\hat{\tilde{\boldsymbol{\Omega}}}\tilde{\mathbf{Z}}'_i + \hat{\boldsymbol{\Sigma}}_i$, $i = 1, 2, \ldots, n$, using some smoothing parameters λ and λ_v. When the smoothing parameter λ (or λ_v) decreases, less penalty is imposed for the roughness of $\eta(t)$ (or $v_i(t)$, $i = 1, 2, \ldots, n$), implying that the model will fit the data closely or even interpolate the data. In this case, it is expected that the Loglik will increase as λ (or λ_v) decreases; alternatively, the Loglik will decrease as λ (or λ_v) increases.

Model complexity

As was the case for the regression spline method, we define the model complexity for estimating the fixed-effects component $\boldsymbol{x}'_{ij}\boldsymbol{\beta} + \eta(t_{ij})$ and the random-effects component $\boldsymbol{z}'_{ij}\boldsymbol{u}_i + v_i(t_{ij})$ separately. First of all, the model complexity for estimating the fixed-effects components is defined as the trace of the smoother matrix $\mathbf{A}$ (17.35):

$$\mathrm{df} = \mathrm{tr}(\mathbf{A}) = \mathrm{tr}\left[(\tilde{\mathbf{X}}'\tilde{\mathbf{V}}^{-1}\tilde{\mathbf{X}} + \tilde{\mathbf{G}})^{-1}\tilde{\mathbf{X}}'\tilde{\mathbf{V}}^{-1}\tilde{\mathbf{X}}\right].$$

Secondly, the model complexity for estimating the random-effects components is defined as the trace of the smoother matrix $\mathbf{A}_v$ (17.36), $\mathrm{df}_v = \mathrm{tr}(\mathbf{A}_v) = \mathrm{tr}(\tilde{\mathbf{Z}}\bar{\tilde{\boldsymbol{\Omega}}}\tilde{\mathbf{Z}}'\tilde{\mathbf{V}}^{-1})$.

AIC and BIC

As for the regression spline method, we can define the AIC and BIC rules for selecting the smoothing parameters λ and λ_v for the smoothing spline method. The AIC rule is defined as $\text{AIC}(\lambda, \lambda_v) = -2\text{Loglik} + 2(\text{df} + \text{df}_v)$, which is obtained via correcting the twice negative log-likelihood by twice the total model complexity; λ and λ_v are then chosen to minimize the above AIC rule. The BIC rule is defined as $\text{BIC}(\lambda, \lambda_v) = -2\text{Loglik} + \log(N)(\text{df} + \text{df}_v)$, where N is the number of all the measurements in the longitudinal data set. When $N \geq 8$, $\log(N) > 2$. It follows that in the BIC rule, the twice negative log-likelihood is corrected more heavily by the model complexity. Thus, the BIC rule is more conservative.

It can be seen that for any fixed λ_v, we can choose λ to minimize the AIC rule. This is because when λ_v is fixed, if we let λ increase, -2Loglik will increase but df will decrease. Similarly, for any fixed λ, we can choose λ_v to minimize $\text{AIC}(\lambda, \lambda_v)$. The same statements are true for the BIC rule.

17.5.6 Application to the ACTG 388 Data

In this subsection, as an example, we applied the smoothing spline method to fit the Type 1 SPME model (17.3) introduced in Section 17.2.2 to the ACTG 388 data. Notice that in the SPME model (17.3), the parametric random-effects component $z_{ij}'\boldsymbol{u}_i = u_{i1} + u_{i2}t_{ij} + u_{i3}t_{ij}^2$ denotes a quadratic polynomial model for the random effects. To fit the nonparametric fixed-effect function $\eta(t)$, we applied the smoothing spline method with only one smoothing parameter λ. To select λ, we computed the Loglik, df, AIC, and BIC values for a grid of smoothing parameters selected properly so that they represent a wide range of possible smoothing parameters. Figure 17.7 shows the Loglik, df, AIC, and BIC curves against λ on a $\log_{10}$ scale. It can be seen that both Loglik and df decrease as λ increases, indicating that as λ increases, the goodness of fit gets worse while the Type 1 SPME model (17.3) becomes simpler. It can be seen that both AIC and BIC curves indicate that there exist proper smoothing parameters to trade off these two aspects well. As expected, the smoothing parameter $\lambda = .0003$ selected by AIC is much smaller than the smoothing parameter $\lambda = 17.241$ selected by BIC. As an illustration, we present just the results using the smoothing parameter selected by BIC.

The variance components estimated by the EM algorithm are

$$
\begin{aligned}
\hat{\sigma}^2 &= 4507, \text{ and } \hat{\boldsymbol{\Omega}} \\
&= \begin{bmatrix} 24297 & 47.843 & -.64092 \\ 47.843 & 8.7093 & -.0523 \\ -.64092 & -.0523 & .0004 \end{bmatrix}.
\end{aligned}
$$

It can be seen that the variation of the intercept term of the random-effects component is much larger than the noise variance, indicating that the between-subject variations dominate the within-subject variations.

Figure 17.8 shows the results of fitting the Type 1 SPME model (17.3) to the ACTG 388 data. Panel (a) shows all the individual fits. Panel (b) shows the population mean function fit (solid curve) with 95% pointwise standard deviation bands of $\eta(t)$. The fitted covariance and correlation functions are presented in panels (c) and (d) respectively.

Figure 17.9 shows the raw data (circles), population fits (dashed curves), and the individual fits (solid curves) for the six subjects presented in Figure 17.2. It seems that the SPME model (17.3) is indeed adequate to fit the ACTG 388 data. To further show the appropriateness of the SPME model (17.3) to the ACTG 388 data, we present the residual plots in Figure 17.10. The standardized residuals are plotted against the SPME fit and time in panels (a) and (b) respectively. Panel (c) presents the histogram of the standard residuals. The histogram indicates that the measurement errors are nearly normally distributed. These plots further show that the Type 1 SPME model (17.3) is indeed an adequate model for the ACTG 388 data.

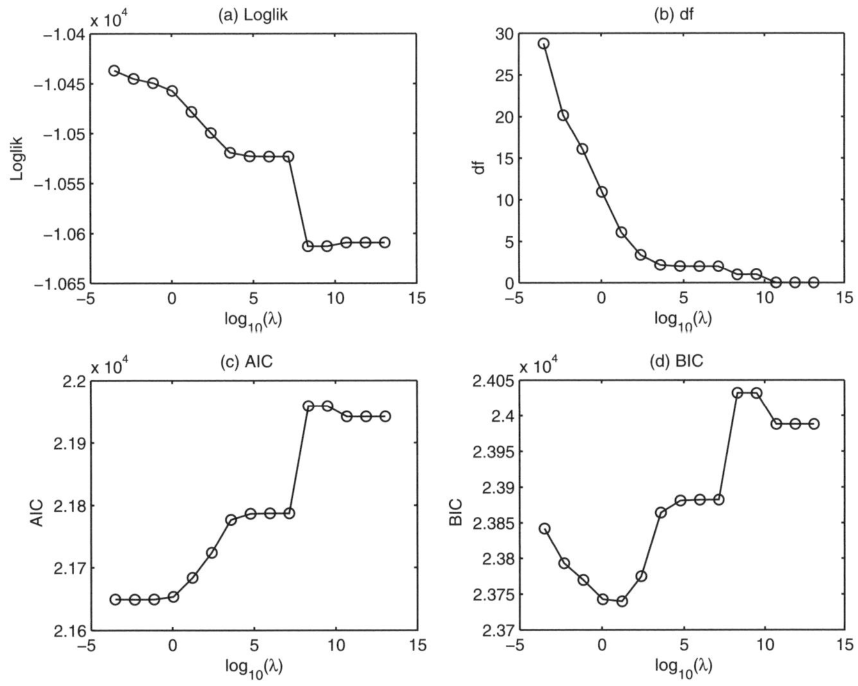

Figure 17.7 Type 1 SPME Modeling of the ACTG 388 Data: Smoothing Parameter Selection

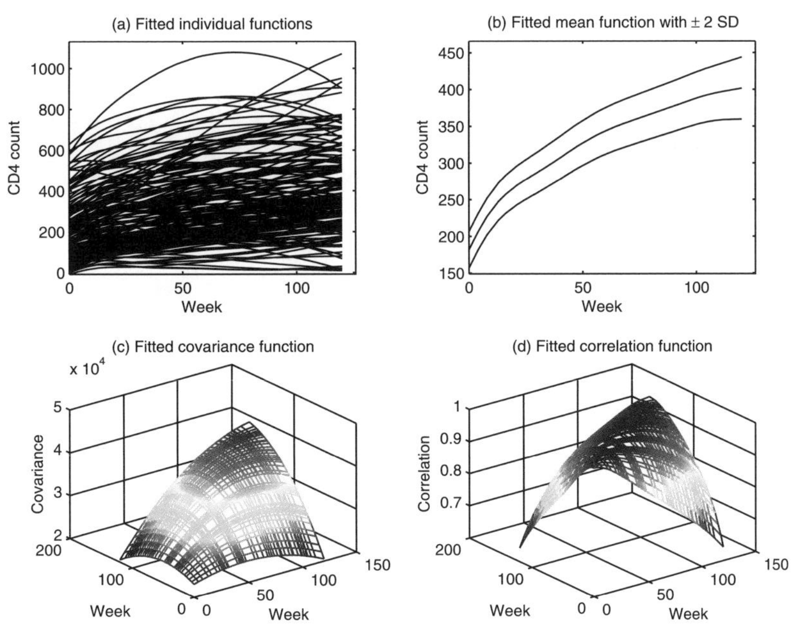

Figure 17.8 Type 1 SPME Modeling of the ACTG 388 Data: Overall Fits

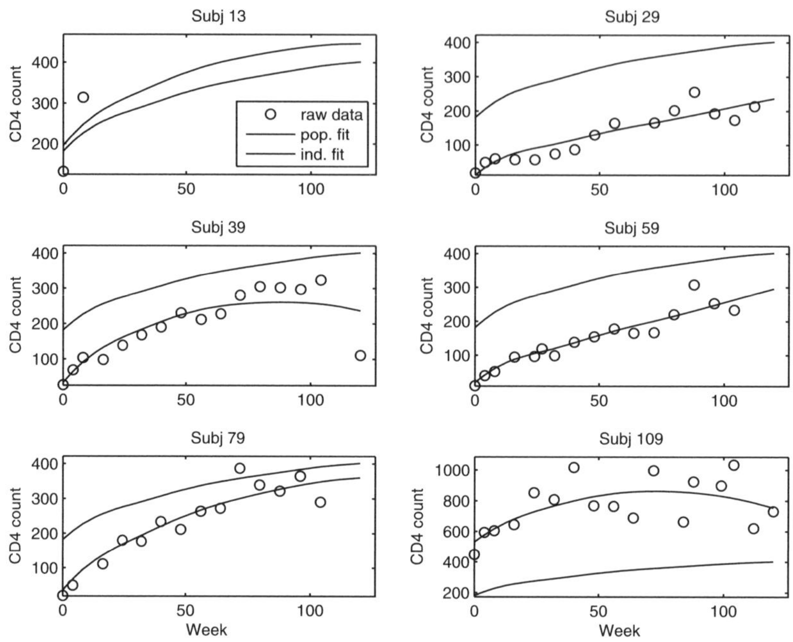

Figure 17.9 Type 1 SPME Modeling of the ACTG 388 Data: Raw Data (circles), Population Fits (dashed curves) and Individual Fits (solid curves) for Six Selected Subjects

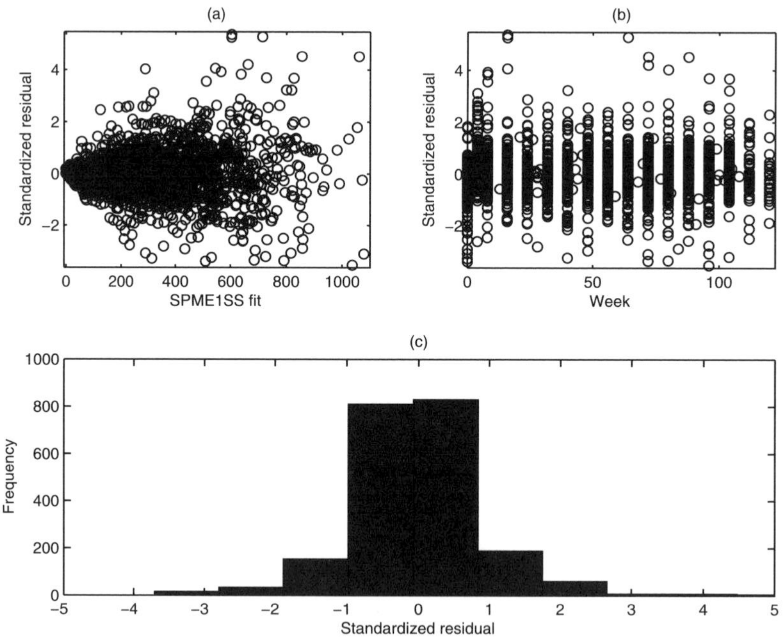

Figure 17.10 Type 1 SPME Modeling of the ACTG 388 Data: Model Checking

17.6 LOCAL POLYNOMIAL KERNEL METHOD

In this section, we investigate the local polynomial kernel (LPK) method for fitting the SPME model (17.4). We describe two methods presented in Wu and Zhang (2006): an LPK backfitting method and a profile LPK method.

17.6.1 *The LPK Backfitting Method*

The LPK backfitting method presented in Wu and Zhang (2006) is a generalization of the Zeger and Diggle (1994) kernel-based backfitting method. Notice that, on the one hand, when the parametric fixed-effects component $\boldsymbol{x}_{ij}'\boldsymbol{\beta}$ and the random-effects component $\boldsymbol{z}_{ij}'\boldsymbol{u}_i$ are given, the SPME model (17.4) essentially reduces to the nonparametric mixed-effects (NPME) model discussed in Wu and Zhang (2006, Chapter 4):

$$\tilde{y}_{ij} = \eta(t_{ij}) + v_i(t_{ij}) + \varepsilon_{ij}, \quad j = 1, 2, \ldots, n_i;\ i = 1, 2, \ldots, n, \tag{17.39}$$

where $\tilde{y}_{ij} = y_{ij} - \boldsymbol{x}_{ij}'\boldsymbol{\beta} - \boldsymbol{z}_{ij}'\boldsymbol{u}_i$, $j = 1, 2, \ldots, n_i$; $i = 1, 2, \ldots, n$. On the other hand, when the nonparametric fixed-effects $\eta(t_{ij})$ and random-effects components $v_i(t_{ij})$ are given, the SPME model (17.4) essentially reduces to the standard LME model reviewed in Wu and Zhang (2006, Chapter 2):

$$r_{ij} = \boldsymbol{x}_{ij}'\boldsymbol{\beta} + \boldsymbol{z}_{ij}'\boldsymbol{u}_i + \varepsilon_{ij}, \quad j = 1, 2, \ldots, n_i;\ i = 1, 2, \ldots, n, \tag{17.40}$$

where $r_{ij} = y_{ij} - \eta(t_{ij}) - v_i(t_{ij})$, $j = 1, 2, \ldots, n_i$; $i = 1, 2, \ldots, n$. Therefore, we can employ the backfitting strategies (Park and Wu, 2006) to fit the SPME model (17.4) via iteratively fitting the NPME model (17.39) and the LME model (17.40). The LPK backfitting algorithm for fitting the SPME model (17.4) can be written as:

(a) Given the current estimates of $\boldsymbol{\beta}$ and $\boldsymbol{u}_i$, $i = 1, 2, \ldots, n$, fit (17.39) using the LPK method described in Wu and Zhang (2006, Chapter 4), and obtain the estimates of $\eta(t)$ and $v_i(t)$, $i = 1, 2, \ldots, n$.

(b) Given the current LPK estimates of $\eta(t)$ and $v_i(t)$, $i = 1, 2, \ldots, n$, fit (17.40) using a standard lme fitting routine such as the S-PLUS function lme or the SAS procedure PROC MIXED, and obtain the estimates of $\boldsymbol{\beta}$ and $\boldsymbol{u}_i$, $i = 1, 2, \ldots, n$.

Intuitively, the above LPK backfitting algorithm is simple. However, like all other backfitting-based approaches, inferences about the SPME model (17.4) are often challenging. More research in this direction is definitely warranted.

17.6.2 *The Profile LPK Method*

We now describe the profile LPK method for fitting the SPME model (17.4). To this end, we first write the SPME model (17.4) in vector and matrix notation. In vector notation, the SPME model (17.4) can be written as

$$\begin{aligned} \boldsymbol{y}_i &= \mathbf{X}_i\boldsymbol{\beta} + \boldsymbol{\eta}_i + \mathbf{Z}_i\boldsymbol{u}_i + \boldsymbol{v}_i + \boldsymbol{\varepsilon}_i, \\ \boldsymbol{u}_i &\sim N_{q_0}(0, \boldsymbol{\Omega}_u),\ \boldsymbol{v}_i \sim N_{n_i}(0, \boldsymbol{\Gamma}_i), \\ &\quad \mathrm{E}(\boldsymbol{u}_i\boldsymbol{v}_i') = \boldsymbol{\Gamma}_{iu}, \\ \boldsymbol{\varepsilon}_i &\sim N_{n_i}(0, \boldsymbol{\Sigma}_i),\ i = 1, 2, \ldots, n, \end{aligned} \tag{17.41}$$

where $\boldsymbol{y}_i = (y_{i1}, \ldots, y_{in_i})'$, $\boldsymbol{\eta}_i = (\eta(t_{i1}), \ldots, \eta(t_{in_i}))'$, and $\boldsymbol{v}_i$ and $\boldsymbol{\varepsilon}_i$ are similarly defined, $\mathbf{X}_i = (\boldsymbol{x}_{i1}, \ldots, \boldsymbol{x}_{in_i})'$, $\mathbf{Z}_i = (z_{i1}, \ldots, z_{in_i})'$, the (k, l)th entry of $\boldsymbol{\Gamma}_i$ is $\gamma(t_{ik}, t_{il})$ and the lth column of $\boldsymbol{\Gamma}_{iu}$ is $\boldsymbol{\gamma}_u(t_{il})$. In matrix notation, the SPME model (17.41) can be further written as

$$\begin{aligned} \boldsymbol{y} &= \mathbf{X}\boldsymbol{\beta} + \boldsymbol{\eta} + \mathbf{Z}\boldsymbol{u} + \boldsymbol{v} + \boldsymbol{\varepsilon}, \\ \boldsymbol{u} &\sim N_{nq_0}(0, \bar{\boldsymbol{\Omega}}_u),\ \boldsymbol{v} \sim N_N(0, \boldsymbol{\Gamma}), \\ &\quad \mathrm{E}(\boldsymbol{u}\boldsymbol{v}') = \boldsymbol{\Gamma}_u, \\ \boldsymbol{\varepsilon} &\sim N_N(0, \boldsymbol{\Sigma}), \end{aligned} \tag{17.42}$$

where $\boldsymbol{y} = (\boldsymbol{y}_1', \ldots, \boldsymbol{y}_n')'$, $\boldsymbol{\eta} = (\boldsymbol{\eta}_1', \ldots, \boldsymbol{\eta}_n')'$, and $\boldsymbol{v}$ and $\boldsymbol{\varepsilon}$ are similarly defined; $\mathbf{X} = (\mathbf{X}_1', \ldots, \mathbf{X}_n')'$, $\mathbf{Z} = \oplus_{i=1}^n \mathbf{Z}_i$, $\boldsymbol{\Gamma} = \oplus_{i=1}^n \boldsymbol{\Gamma}_i$, $\boldsymbol{\Gamma}_u = \oplus_{i=1}^n \boldsymbol{\Gamma}_{iu}$, and $\boldsymbol{\Sigma} = \oplus_{i=1}^n \boldsymbol{\Sigma}_i$. Then in matrix notation the NPME model (17.39) can be written as

$$\tilde{\boldsymbol{y}} = \boldsymbol{\eta} + \boldsymbol{v} + \boldsymbol{\varepsilon}, \tag{17.43}$$

where $\tilde{\mathbf{y}} = \boldsymbol{y} - \mathbf{X}\boldsymbol{\beta} - \boldsymbol{Z}\boldsymbol{u}$. Let $\mathbf{A}$ and $\mathbf{A}_v$ be the LPK smoother matrices for fitting the above NPME model; see Wu and Zhang (2006, Chapter 4) for details for constructing $\mathbf{A}$ and $\mathbf{A}_v$. Then $\hat{\eta} = \mathbf{A}\boldsymbol{y}$ and $\hat{\mathbf{v}} = \mathbf{A}_v\mathbf{Q}\boldsymbol{y}$ where $\hat{\eta}$ and $\hat{\mathbf{v}}$ are the LPK estimators of η and $\boldsymbol{v}$, and $\mathbf{Q} = \boldsymbol{I}_N - \mathbf{A}$ with $\boldsymbol{I}_N$ being the identity matrix of size N. It follows that when $\boldsymbol{\beta}$ and $\boldsymbol{u}$ are known, by fitting (17.43) using LPK smoothing, we have the following profile smoothers of η and $\boldsymbol{v}$:

$$\hat{\eta} = \mathbf{A}(\boldsymbol{y} - \mathbf{X}\boldsymbol{\beta} - \boldsymbol{Z}\boldsymbol{u}), \hat{\mathbf{v}} = \mathbf{A}_v\mathbf{Q}(\boldsymbol{y} - \mathbf{X}\boldsymbol{\beta} - \boldsymbol{Z}\boldsymbol{u}). \quad (17.44)$$

Notice that both smoother matrices $\mathbf{A}$ and $\mathbf{A}_v$ depend on the covariance matrix $\mathbf{V}_v = \mathrm{Cov}(\boldsymbol{v} + \boldsymbol{\varepsilon}) = \boldsymbol{\Gamma} + \boldsymbol{\Sigma}$.

Replacing η and $\boldsymbol{v}$ in (17.42) with their profile smoothers, letting $\mathbf{P} = \boldsymbol{I}_N - \mathbf{A} - \mathbf{A}_v\mathbf{Q} = (\boldsymbol{I}_N - \mathbf{A}_v)\mathbf{Q}$, and after simplification, we have

$$\mathbf{P}\boldsymbol{y} = \mathbf{P}\mathbf{X}\boldsymbol{\beta} + \mathbf{P}\mathbf{Z}\boldsymbol{u} + \boldsymbol{\varepsilon}, \quad (17.45)$$

where again $\boldsymbol{u} \sim N_{nq_0}(0, \bar{\boldsymbol{\Omega}}_u)$ and $\boldsymbol{\varepsilon} \sim N_N(0, \boldsymbol{\Sigma})$. For given $\mathbf{A}$ and $\mathbf{A}_v$, (17.45) is a standard LME model. Let $\mathbf{V}_u = \mathbf{P}\mathbf{Z}\bar{\boldsymbol{\Omega}}_u\mathbf{Z}'\mathbf{P}' + \boldsymbol{\Sigma}$. Solving the above LME model (17.45) leads to the following estimators of $\boldsymbol{\beta}$ and $\boldsymbol{u}$:

$$\hat{\boldsymbol{\beta}} = (\mathbf{X}'\mathbf{P}'\mathbf{V}_u^{-1}\mathbf{P}\mathbf{X})^{-1}\mathbf{X}'\mathbf{P}'\mathbf{V}_u^{-1}\mathbf{P}\boldsymbol{y},$$
$$\hat{\mathbf{u}} = \bar{\boldsymbol{\Omega}}_u\mathbf{Z}'\mathbf{P}'\mathbf{V}_u^{-1}\mathbf{P}(\boldsymbol{y} - \mathbf{X}\hat{\boldsymbol{\beta}}).$$

Plugging the above estimators into the profile LPK smoothers in (17.44), we have

$$\hat{\eta} = \mathbf{A}(\boldsymbol{y} - \mathbf{X}\hat{\boldsymbol{\beta}} - \mathbf{Z}\hat{\mathbf{u}}), \hat{\mathbf{v}} = \mathbf{A}_v\mathbf{Q}(\boldsymbol{y} - \mathbf{X}\hat{\boldsymbol{\beta}} - \mathbf{Z}\hat{\mathbf{u}}). \quad (17.46)$$

Note that the LPK method used in the LPK backfitting algorithm and in the profile LPK method is not crucial and it can be replaced by any smoothing technique such as smoothing splines (Green and Silverman, 1994), regression splines (Eubank, 1999), and penalized splines (Ruppert, Wand, and Carroll, 2003) among others. See Chapters 5, 6, and 7 of Wu and Zhang (2006) for details on these techniques.

17.7 CONCLUDING REMARKS AND BIBLIOGRAPHICAL NOTES

In this chapter, we described three smoothing methods for fitting the general SPME model (17.4) for longitudinal data where the within-subject correlations are properly modeled using random-effects parametrically or nonparametrically. However, the data we used to illustrate the methodologies are only for two special SPME models where only one parametric component and one nonparametric component are involved. These two special models can be fit using only one smoothing parameter, and hence it allows us to simplify the methodologies we discussed in this chapter. When an SPME model contains all four components, we can still apply our methodologies but more computation may be involved since we now may need two smoothing parameters. Penalized splines can also be used to fit the general SPME model (17.4). For some basic ideas of penalized splines and their applications in multilevel data analysis, the reader is referred to Chapter 18.

We now give some bibliographical notes about SPME models. In the last decade, a number of methods have been developed for SPME models. Most of the literature focuses on some special SPME models as listed in Section 17.3 of this chapter. For example, Diggle (1988) considered a special case of the Type 4 SPME model when the number of random-effects covariates $q_0 = 1$, and $v_i(t)$ has a parametric covariance structure such as the AR(1) model. This special case was also discussed by Donnelly, Laird, and Ware (1995). Zeger and Diggle (1994) considered fitting the Type 6 SPME model using kernel smoothing and a backfitting algorithm. Moyeed and Diggle (1994) investigated the rate of convergence for such estimators. Wang (1998) considered fitting the Type 1 SPME model using the reproducing kernel-based smoothing splines of Wahba (1990) and a connection of a smoothing spline to an LME model. Following Wang (1998), Guo (2002) considered likelihood ratio tests about the Type 1 SPME model. Zhang et al. (1998) extended the Zeger and Diggle (1994) SPME model to the general SPME model (17.4) that we considered in this chapter, but they assumed a parametric covariance structure for the nonparametric random-effects component $v_i(t),\ i = 1, 2, \ldots, n$. They employed smoothing splines to fit the nonparametric fixed-effects component $\eta(t)$ and used a connection of a smoothing spline to an LME model. Jacqmin-Gadda et al. (2002) investigated the same SPME model but they computed the roughness term using a basis of cubic M-splines. Recently, Ruppert, Wand, and Carroll (2003) dealt with a simple version (with $q_0 = 1$) of the Type 5 SPME model using P-splines and a connection of a P-spline to an LME model, and Durban et al. (2005) considered fitting the general SPME model using P-spline smoothing and a connection of a P-spline to an LME model. Tao et al. (1999) investigated a semiparametric mixed-effects model for analysis of clustered or longitudinal data with a continuous, ordinal, or binary outcome. Lin and Zhang (1999) studied generalized semiparametric additive mixed models using smoothing splines.

It is more challenging to fit and conduct statistical inference about semiparametric nonlinear mixed-effects (SNLME) models. More work is needed on this topic. Key references on SNLME models include Ladd and Lindstrom (2000), Ke and Wang (2001), and Wu and Zhang (2002), among others. Ladd and Lindstrom (2000) considered semiparametric self-modeling for two-dimensional curves. Ke and Wang (2001) employed smoothing splines to fit a special case of the SNLME model. In Ke and Wang's SNLME model, the only nonparametric component is the nonparametric fixed-effect function. Wu and Zhang (2002) fit a general SNLME model using a basis-based approach. This approach essentially transforms the original SNLME model into a series of standard NLME models indexed by the bases, which can be solved by some existing NLME software. Wu and Zhang (2002) proposed to choose the bases using the AIC and BIC rules for standard NLME models defined in Davidian and Giltinan (1995).

NOTE

1 The work was supported by the National University of Singapore Academic Research Grant R-155-000-085-112. The author thanks the editors for their kind invitation to write this chapter. He also thanks the editors and a reviewer for their detailed comments and suggestions on the first draft. Thanks also go to his research assistant Miss Wu Hao for her help in editing the tables and figures used in this chapter.

REFERENCES

Davidian, M. and Giltinan, D.M. (1995) *Nonlinear Models for Repeated Measurement Data*. New York: Chapman and Hall/CRC.

Diggle, P.J. (1988) 'An approach to the analysis of repeated measurements', *Biometrics*, 44: 959–71.

Diggle, P., Heagerty, P., Liang, K.Y., and Zeger, S. (2002) *Analysis of Longitudinal Data*. Oxford: Oxford University Press.

Donnelly, C.A., Laird, N.M., and Ware, J.H. (1995) 'Prediction and Creation of Smooth Curves for Temporally Correlated Longitudinal Data', *Journal of the American Statistical Association*, 90: 984–9.

Durban, M., Harezlak, J., Wand, M.P., and Carroll, R.J. (2005) 'Simple Fitting of Subject-Specific Curves for Longitudinal Data', *Statistics in Medicine*, 24: 1153–67.

Eubank, R.L. (1999) *Nonparametric Regression and Spline Smoothing*. New York: Marcel Dekker.

Fan, J. and Gijbels, I. (1996) *Local Polynomial Modeling and Its Applications*. London: Chapman and Hall.

Fischl, M.A., Ribaudo, H.J., Collier, A.C., Erice, A., Giuliano, M., Dehlinger, M., Eron, J.J. Jr., Saag, M.S., Hammer, S.M., Vella, S., et al. (2003) 'A Randomized Trial of 2 Different 4-drug Antiretroviral Regimens Versus a 3-drug Regimen, in Advanced Human Immunodeficiency Virus Disease', *The Journal of Infectious Diseases*, 188: 625–34.

Green, P.J. and Silverman, B.W. (1994) *Nonparametric Regression and Generalized Linear Models: A Roughness Penalty Approach*. London: Chapman and Hall.

Guo, W. (2002) 'Inference in Smoothing Spline Analysis of Variance', *Journal of the Royal Statistical Society, Series B*, 64: 887–98.

Jacqmin-Gadda, H., Joly, P., Commenges, D., Binquet, C., and Chene, G. (2002) 'Penalized Likelihood Approach to Estimate a Smooth Mean Curve on Longitudinal Data', *Statistics in Medicine*, 21: 2391–402.

Kaslow, R.A., Ostrow, D.G., Detels, R., Phair, J.P., Polk, B.F., Rinaldo, C.R. Jr., et al. (1987) 'The Multicenter AIDS Cohort Study: Rationale, Organization, and Selected Characteristics of the Participants', *American Journal of Epidemiology*, 126: 310–18.

Ke, C. and Wang, Y. (2001) 'Semiparametric Nonlinear Mixed-Effects Models and Their Applications (with discussion)', *Journal of the American Statistical Association*, 96 1272–98.

Ladd, M.J. and Lindstrom, W.M. (2000) 'Self Modeling for Two-Dimensional Response Curves', *Biometrics*, 56: 89–97.

Laird, N., Lange, N., and Stram, D. (1987) 'Maximum Likelihood Computations With Repeated Measures: Application of the EM Algorithm', *Journal of the American Statistical Association*, 82: 97–105.

Laird, N.M. and Ware, J.H. (1982) 'Random-Effects Models for Longitudinal Data', *Biometrics*, 38: 963–74.

Lange, N., Carlin, B.P., and Gelfand, A.E. (1992) 'Hierarchical Bayes Models for the Progression of HIV Infection Using Longitudinal CD4 T-cell Numbers', *Journal of the American Statistical Association*, 87: 615–26.

Lin, X. and Zhang, D. (1999) 'Inference in Generalized Additive Mixed Models by Using Smoothing Splines', *Journal of Royal Statistical Society, Series B*, 61: 381–400.

Moyeed, R.A. and Diggle, P.J. (1994) 'Rates of Convergence in Semiparametric Modeling of Longitudinal Data', *Australian Journal of Statistics*, 36: 75–93.

Park, J.G. and Wu, H. (2006) 'Backfitting and Local Likelihood Methods for Nonparametric Mixed-Effects Models With Longitudinal Data', *Journal of Statistical Planning and Inference*, 136: 3760–82.

Pinheiro, J.C. and Bates, D.M. (2009) *Mixed-Effects Models in S and S-PLUS*. New York: Springer.

Ruppert, D., Wand, M.P., and Carroll, R.J. (2003) *Semiparametric Regression*. Cambridge: Cambridge University Press.

Tao, H., Palta, M., Yandell, B.S., and Newton, M.A. (1999) 'An Estimation Method for the Semiparametric Mixed-Effects Model', *Biometrics*, 55: 102–10.

Verbeke, G. and Molenberghs, G. (2009) *Linear Mixed Models for Longitudinal Data*. New York: Springer.

Vonesh, E.F. and Chinchilli, V.M. (1997) *Linear and Nonlinear Models for the Analysis of Repeated Measurements*. New York: Marcel Dekker.

Wahba, G. (1990) *Spline Models for Observational Data*. Philadelphia: SIAM.

Wang, Y. (1998) 'Mixed-Effects Smoothing Spline Anova', *Journal of the Royal Statistical Society, Series B*, 60: 159–74.

Wu, H. and Zhang, J.T. (2002) 'The Study of Long-Term HIV Dynamics Using Semiparametric Nonlinear Mixed-Effects Models', *Statistics in Medicine*, 21: 3655–75.

Wu, H. and Zhang, J.T. (2006) *Nonparametric Regression Methods for Longitudinal Data Analysis*. New York: Wiley.

Zeger, S.L. and Diggle, P.J. (1994) 'Semiparametric Models for Longitudinal Data With Application to CD4 Cell Numbers in HIV Seroconverters', *Biometrics*, 50: 689–99.

Zhang, D., Lin, X., Raz, J., and Sowers, M.F. (1998) 'Semiparametric Stochastic Mixed Models for Longitudinal Data', *Journal of the American Statistical Association*, 93: 710–19.

18

Penalized Splines and Multilevel Models

Göran Kauermann
University of Munich, Germany
Torben Kuhlenkasper
Goethe University Frankfurt, Germany

18.1 MOTIVATION

In this chapter we will extend multilevel models to include smooth, functional components. This means we will move forward from a parametric model as discussed in Chapters 3 and 8 towards non- and semiparametric modeling. To motivate the need for such a model class let us look at the following example. We record with y the wage of a person and intend to model how the wage depends on maturity and working experience recorded with x as years of professional experience. It is natural that y does depend non-linearly on x and we therefore use the smoothing model (see Chapter 17) and set

$$y = m(x) + \varepsilon \tag{18.1}$$

with $m(.)$ as a smooth but otherwise unspecified function and ε as unstructured error. It is helpful to extract the general level of $m(x)$ as intercept and rewrite (18.1) to

$$y = \beta_0 + m(x) + \varepsilon$$

where β_0 is the intercept.

For reasons that become clear later we explicitly list the intercept β_0, and therefore to achieve identifiability we additionally impose the constraint

$$\int_{-\infty}^{\infty} m(x)\, \mathrm{d}F(x) = 0, \tag{18.2}$$

where $F(.)$ is the cumulative distribution function of x, sometimes also called the design of x. If $F(.)$ is unknown, which is most often the case unless one is in a purely experimental setting, one replaces $F(.)$ in (18.2) by its empirical version $F_n(.)$, say.

With y denoting wage we observe that data on income and wages are often made available through survey panels, so that y_{it} is the wage of individual i at panel wave t and x_{it} the corresponding covariate, the years of professional experience. Following the multilevel modeling framework introduced in Chapter 1 we set

$$y_{it} = \beta_{0i} + m(x_{it}) + \varepsilon_{it} \tag{18.3}$$
$$\beta_{0i} = \beta_0 + u_i, \tag{18.4}$$

with $u_i \sim N(0, \sigma_u^2)$ iid. For estimation of the multilevel components we will combine spline smoothing with linear mixed model techniques. To do so we will first demonstrate how to estimate $m(x)$ with penalized splines and mixed models. This is then extended to accommodate multilevel effects. As will be seen, the estimation framework is surprisingly coherent and we can fit model (18.3) with results from the previous sections and with available software quite easily.

18.2 MODEL AND METHODS

18.2.1 Penalized Spline Smoothing

Before discussing the estimation framework for the model with both smooth and level-specific effects we leave the level component u_i in (18.4) aside and focus on the smooth component only. That is, we look at model (18.1). The underlying idea of penalized splines is to replace function $m(x)$ by some flexible, high-dimensional basis representation. This is available numerically in a quite stable manner using B-splines (de Boor, 1978). For clarity and ease of presentation it is easier to demonstrate the subsequent idea with so-called truncated polynomials. B-splines and truncated polynomials are in fact analytically equivalent, though numerically the first have advantages. Let $\tau_1 < \tau_2 < \cdots < \tau_K$ be a set of (not necessarily equidistant) knots covering the range of observed x values. We then approximate $m(x)$ through

$$m(x) = x\beta_1 + x^2\beta_2 + \sum_{k=1}^{K}(x - \tau_k)_+^2 v_k, \tag{18.5}$$

where $(x)_+^2 = \max(0, \text{sign}(x)x^2)$. As an example we provide in Figure 18.1, left plot, simulated data being fitted with a truncated quadratic basis using ordinary least squares to fit coefficients $\boldsymbol{\beta} = (\beta_0, \beta_1, \beta_2)$ and $\mathbf{v} = (v_1, \ldots, v_K)'$. It is clear that the least squares fit is inappropriate. Due to the high-dimensional parameterization, estimation variability is dominant, leading to a wiggled estimate. This is shown in the bottom part of the plot where we show the truncated polynomials $(x - \tau_k)_+^2$ weighted with their least squares coefficients v_k. The wiggliness can be controlled by imposing a penalty on the spline coefficient vector $\mathbf{v}$. This was originally suggested by O'Sullivan (1986) and coined as P-spline smoothing in Eilers and Marx (1996). The idea has received extensive consideration in the statistical literature since then as demonstrated in the survey article by Ruppert et al. (2009). The principle of penalized splines is therefore as follows. Instead of minimizing ordinary least squares, one minimizes the penalized version

$$\sum_{i=1}^{n}(y_i - X_i\boldsymbol{\beta} - Z_i\mathbf{v})^2 + \lambda\mathbf{v}'D\mathbf{v} \tag{18.6}$$

based on an iid sample y_i, x_i, $i = 1, \ldots, n$. X_i is the row vector $X(x_i) = (1, x_i, x_i^2)$, and $Z_i(x_i) = ((x_i - \tau_1)_+^2, \ldots, (x_i - \tau_K)_+^2)$. Matrix D in (18.6) is called the penalty matrix, which for truncated polynomials is suitably chosen as the identity matrix. In this case one penalizes changes in the quadratic slope of the fitted function. Finally, λ in (18.6) plays the role of the penalty parameter. Choosing $\lambda = 0$ yields an unpenalized fit (see Figure 18.1, left plot), while choosing $\lambda \equiv \infty$ and D as identity gives a quadratic fit with $\mathbf{v} \equiv \mathbf{0}$. Apparently, λ needs to be chosen in a data-driven way, and when doing so one achieves the fit visualized in the right-hand-side plot of Figure 18.1. As can be seen, the true function is reproduced by shrinking the fitted coefficients v_k appropriately to zero. We give more details on how to choose λ in the next section.

We have not yet raised the question of how the knots τ_k should be located and how many should be selected, i.e. how large K should be. Ruppert (2002) suggests choosing K based on the rule of thumb $K = \min(40, n/4)$, which performs uniformly well in a wide range of simulations and applications. Kauermann and Opsomer (2011) propose a procedure to select K based on the linear mixed model framework discussed in the next section. Asymptotic results are found in Claeskens et al. (2009) and references given there. From a

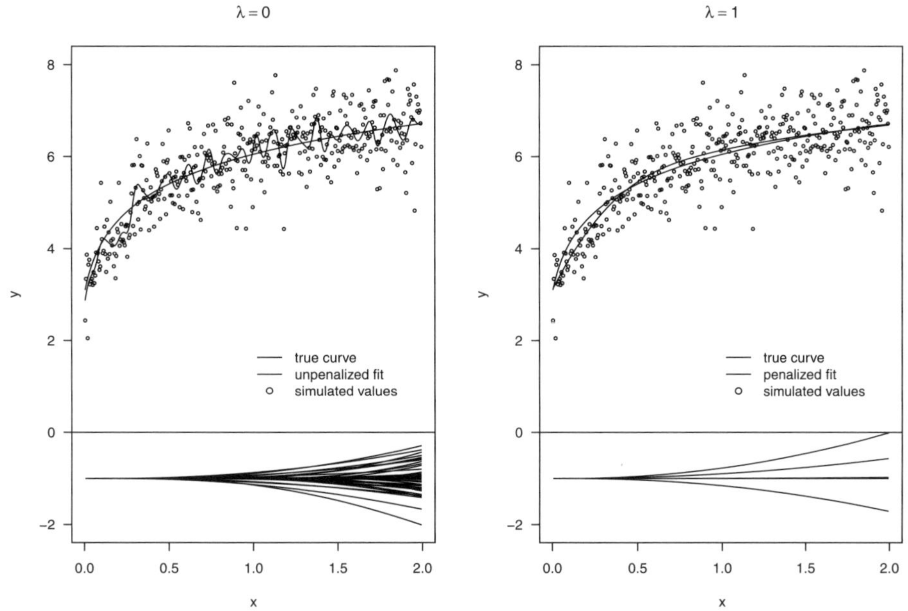

Figure 18.1 Unpenalized Fit (left) and Penalized Fit (right) for Simulated Data

practical point of view, one can recommend that selecting K large, e.g. based on Ruppert's 2002 rule of thumb, performs satisfactorily well. For knot location, equidistant or quantile-based allocation has proven to provide stable and convincing results. Generally, experiments have shown that the location of knots is not very influential on the properties of the estimator.

Finally, the fitted curve in Figure 18.1 is not normalized to fulfill constraint (18.2). This can be easily done after fitting by replacing (18.2) with its empirical version. Since this is a technical, rather than modeling, issue it is not discussed further here.

18.2.2 Penalized Spline Smoothing and Linear Mixed Models

We will now motivate (18.6) as a log likelihood coming from a linear mixed model. We therefore view the penalty term in (18.6) as a normal prior following a Bayesian viewpoint. That is, we impose a prior distribution on some parameters. We will subsequently assume the model

$$\begin{aligned} \mathbf{y}|\mathbf{v} &\sim N(X\boldsymbol{\beta} + Z\mathbf{v}, \sigma_\varepsilon^2 I_n) \\ \mathbf{v} &\sim N(0, \sigma_v^2 D^-), \end{aligned} \tag{18.7}$$

where $\mathbf{y} = (y_1, \ldots, y_n)$ and X and Z are matrices with row X_i and Z_i, respectively, and I_n is the n-dimensional identity matrix. Matrix D^- is the (generalized) inverse of D and if D is not of full rank, one has linear constraints on vector $\mathbf{v}$. In the case where $D = I_K$ we obtain, with (18.7), a standard linear mixed model. In general, for any spline basis, we can always decompose the spline approach to obtain a form like (18.7); see Ngo and Wand (2004). To relate the mixed model (18.7) to the penalized least squares (18.6) we observe that $\sigma_v^2 = \sigma_\varepsilon^2/\lambda$ plays the role of the penalization parameter. Note that the linear mixed model (18.7) can be rewritten by integrating out $\mathbf{v}$, which yields

$$\mathbf{y} \sim N\big(X\boldsymbol{\beta}, \sigma_\varepsilon^2(I_n + \lambda^{-1} Z D^- Z')\big).$$

The penalty parameter λ is now a regular parameter in a purely parametric model.

Hence, one can use Maximum Likelihood or Restricted Maximum Likelihood (REML) to estimate λ (see Chapter 3).

It is also not difficult to show that

$$\hat{\beta} = (X'\hat{\Omega}X)^{-1}X'\hat{\Omega}y$$

is the Maximum Likelihood estimate, with $\hat{\Omega}^{-1} = I + \hat{\lambda}^{-1}ZD^{-}Z$. Moreover, the Best Linear Unbiased Predictor (BLUP) for $\mathbf{v}$ is

$$\hat{\mathbf{v}} = (Z'Z + \hat{\lambda}D)^{-1}Z'(y - X\hat{\beta}),$$

and with

$$\hat{m}(x) = X(x)\hat{\boldsymbol{\beta}} + Z(x)\hat{\mathbf{v}}$$

we get the penalized estimate that would arise from (18.6). In other words, the formulation of the penalty as a prior normal distribution (18.7) provides a data-driven estimate of the penalty parameter simply as a variance component estimate; see Kauermann (2005), Wood (2011), and Reiss and Ogden (2009). This formulation even allows for more, namely the combination of smoothing with multilevel models that is demonstrated in the next section.

18.2.3 Penalized Splines and Multilevel Models

Assume now that observations are grouped by levels and let y_{ij} denote the ith observation in the jth group with $j = 1, \ldots, J$ and $i = 1, \ldots, n_j$. We record with x_{ij} the corresponding covariate and assume a linear influence of x_{ij}. The standard two-level model is

$$y_{ij} = \beta_{0j} + x_{ij}\beta_1 + \varepsilon_{ij} \tag{18.8}$$

$$\beta_{0j} = \beta_0 + u_{0j}, \tag{18.9}$$

where $u_{0j} \sim N(0, \sigma^2_{u0})$ are iid and independent of $\varepsilon_{ij} \sim N(0, \sigma^2_{\varepsilon})$ iid. Assuming now that x_{ij} is metrically scaled we can generalize (18.8) by replacing the linear influence by a functional form

$$y_{ij} = \beta_{0j} + m(x_{ij}) + \varepsilon_{ij},$$

with $m(.)$, as above, as a smooth but otherwise unspecified function. The next step is to replace $m(x_{ij})$ by a spline representation (18.5) and impose a penalty on the spline coefficients. Interpreting the penalty as a normal prior we end up with the structured multilevel model

$$\begin{aligned} y_{ij} &= \beta_{0j} + X_1(x_{ij})\boldsymbol{\beta}_1 + Z_1(x_{ij})\mathbf{v} + \varepsilon_{ij} \\ \beta_{0j} &= \beta_0 + u_{0j} \end{aligned} \tag{18.10}$$

and mutually independent random components

$$\begin{aligned} \varepsilon_{ij} &\sim N(0, \sigma^2_{\varepsilon}) \text{ iid}, \ \mathbf{v} \sim N(0, \sigma^2_v D^{-}), \\ u_{0j} &\sim N(0, \sigma^2_{u0}) \text{iid}. \end{aligned} \tag{18.11}$$

In (18.10), matrix $X_1(x_{ij})$ gives the row vector (x_{ij}, x^2_{ij}), assuming quadratic polynomials for the spline fit, and accordingly $\boldsymbol{\beta}_1 = (\beta_{11}, \beta_{12})'$, while $Z_1(x_{ij}) = ((x_{ij} - \tau_1)^2_+, \ldots, (x_{ij} - \tau_K)^2_+)$. We recognize that smooth and multilevel modeling can be carried out coherently in one model framework and parameters can be estimated through maximum likelihood methodology for the marginalized model, i.e. after integrating out $\mathbf{v}$ and $u_{0j}, j = 1, \ldots, J$. Therefore, let X denote the matrix with rows $(1, X_1(x_{ij}))$. Moreover, let Z_1 have rows $Z_1(x_{ij})$ and let Z_0 be the indicator matrix with J columns and entry 1 in the jth column for rows matching the $y_{ij}, i = 1, \ldots, n_i$ and entries 0 otherwise. Defining $\mathbf{y} = (y_{11}, y_{21}, \ldots, y_{12}, \ldots)'$, we obtain, with model (18.10) and (18.11),

$$\mathbf{y} \sim N(X\beta, \sigma^2_{\varepsilon}I + \sigma^2_{u0}Z_0Z_0' + \sigma^2_v Z_1 D^{-} Z_1).$$

Letting $\hat{\beta}, \hat{\sigma}^2_{\varepsilon}, \hat{\sigma}^2_{u0}$, and $\hat{\sigma}^2_v$ denote the resulting maximum likelihood estimates it is not difficult to explicitly derive the BLUP for spline coefficient vector $\mathbf{v}$ through

$$\hat{\mathbf{v}} = (Z_1'Z_1 + \lambda D)^{-1}Z'(\mathbf{y} - X\boldsymbol{\beta} - Z_0\hat{\mathbf{u}}_0),$$

with $\hat{\mathbf{u}}_0 = (\hat{u}_{01}, \ldots, \hat{u}_{0J})'$ as BLUP for the level effects and $\lambda = \sigma^2_{\varepsilon}/\sigma^2_v$.

It becomes obvious that the modeling strategy is now easily extendable to accommodate more complex models with parametric components and higher-level components, respectively. We will demonstrate this with analysis of the income data referred to in the first section.

As stated above, the idea of combining penalized splines and multilevel models combines the multilevel approach with smooth and semiparametric modeling. The approach can also be generalized to a non-normal response model as discussed in Chapter 8.

The obvious advantage of the combination of penalized splines and multilevel models is that the resulting model is coherent, which allows for numerically stable and feasible estimation. The weakness is that numerically the models can be burdensome if the number of observations is large or the complexity of the function is strong.

18.3 REAL EXAMPLE

18.3.1 Smooth Multilevel Model

To illustrate the penalized splines introduced above in the context of a socio-economic application we will analyze the wages of females as one major element within human capital theory. Although many covariates are likely to influence the wage of an individual, we first focus on the full-time work experience in years as the major covariate affecting the wage of a woman. According to Becker (1993), the ongoing experience an employee has gained while working full time is assumed to influence the wage in a positive, but not necessarily linear, way. As full-time work experience in the labour market continues, wages are assumed to increase with slower rates. We therefore model

$$\log wage_{it} = \beta_{0i} + m(fulltime_{it}) + \varepsilon_{it} \qquad (18.12)$$

$$\beta_{0i} = \beta_0 + u_{0i},$$

with $wage_{it}$ as wage of female i at time point t and $fulltime_{it}$ as full-time work experience up to time point t. Errors ε_{it} and u_{i0} are distributed as specified subsequent to (18.8) and (18.9). We fit the model using data from the German Socio Economic Panel (GSOEP, `www.diw.de/soep`). For a detailed introduction to GSOEP, we refer to Haisken-DeNew and Frick (2005) and Wagner et al. (2007). Model (18.12) is a simplified relationship which traces back to Mincer (1974), and will be specified by using the inflation-adjusted, contract-based hourly wage for a German woman. Due to assumed major structural differences in the labour market for differently educated females, we divide the data into three strata defined with respect to three different categories of educational attainment. Using the International Standard Classification of Education (ISCED), the first group consists of women with low educational achievements, represented by the levels 1 and 2 in the ISCED. The second strata is defined by averagely educated females, having achieved the ISCED levels 3 and 4. Highly educated women have at least achieved ISCED level 5 and define the third strata. The data base is further restricted to German females without migration background aged between 18 and 55 years who have not given birth to a child (yet). Due to the unbalanced panel structure of our database, we end up with multiple observations per individual ranging from $n_i = 1$ to $n_i = 25$. The multilevel component u_{i0} captures the individual latent effects and can be interpreted as the deviation from the population effect β_0. We estimate the model defined in (18.12) for each strata with 3532 observations from 1230 low-educated females; with 10542 observations from 2450 women with average educational attainment; and with 4203 observations from 964 highly educated females. As an estimation tool we use the open-source statistical software R (R Development Core Team, 2010) with package GAMM4 (Wood, 2010). Further details about R can be found in Crawley (2007), while the package employed for multilevel modeling traces back to Wood (2006). Alternatives to the GAMM4 routine can be found in the R packages AMER (Scheipl, 2011) and GAM (Hastie, 2011).

The fitted variances for the error term and the individual random effects are listed in Table 18.1. Looking at the variances first, we see that low-educated females are faced with

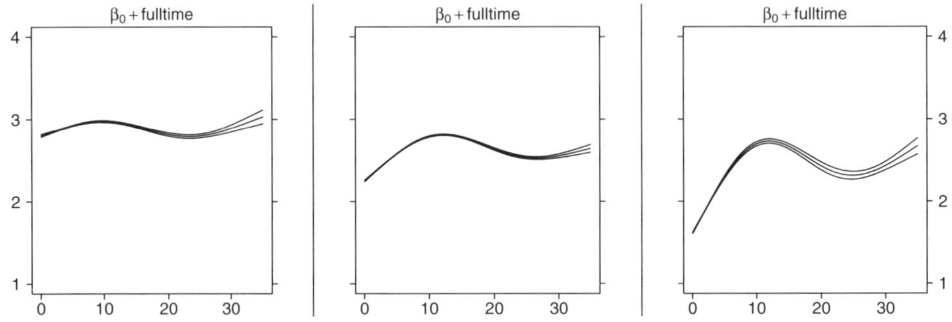

Figure 18.2 Fitted Dynamic Effect of Full-Time Work Experience for High, Middle and Low Educated Women (from left to right) Based on Model (18.12)

Table 18.1 Parametric Estimation Results for Model (18.12)

	$\hat{\beta}_j$ (p-value)		
effect	high ISCED	middle ISCED	low ISCED
(Intercept)	2.89 (< 0.01)	2.55 (< 0.01)	1.95 (< 0.01)
$\hat{\sigma}_u$	0.19	0.21	0.28
$\hat{\sigma}_\varepsilon$	0.08	0.1	0.14

high variability, both on the residual as well as on the individual level. The resulting fits of $\hat{m}(fulltime_{it})$ for each ISCED stratum are given in Figure 18.3. In the displayed graphical results we add the estimated population effect $\hat{\beta}_0$ to the smooth effect of $\hat{m}(\cdot)$ to make the levels of the fitted function comparable between the strata.

While the effect of full-time work experience indicates, for all three levels of educational attainment, a clear increasing shape to the threshold of 10 years, the effect decreases between 10 and 20 years and starts rising again towards the end of the working careers. Although it is not surprising to see an overall higher wage level for higher-educated women, the dynamic pattern of the effects, with a decreasing effect after having 10 years of full-time experience, can hardly be explained with classical economic theory. This therefore motivates the extension of the models, appending additional effects and thus gaining a more comprehensive insight into the estimated relationship between work experience and individual wages.

18.4 EXTENSION TO ADDITIVE MODELS

Including more than one covariate in the model leads to a semiparametric model. We exemplify this by extending model (18.12) to analyze the relationship between human capital theory and the individual wage more adequately. Besides the construction of different strata capturing the importance of educational achievements and the major covariate of full-time work experience, many other covariates are likely to influence an individual's wage. Mincer (1974) and Becker (1993) give further details for additional covariates with respect to human capital theory. Analogous to $fulltime_{it}$, the (metrical) covariates $parttime_{it}$ and $unemployed_{it}$ capture the time or experience woman i has up to the year t

working with reduced weekly working hours and being officially registered unemployed, respectively. In addition, $firmtime_{it}$ is defined as the time of attachment to the current firm or employer. While an ongoing attachment to the current employer is assumed to be rewarded with higher wages, the time spent in unemployment is likely to have negative effects on wages. However, there is no clear theoretical evidence about the effect of part-time experience on wages. Finally, it is reasonable to assume that wages do not exclusively depend on the individual stock of human capital. We therefore include an additional (metrical) covariate in a functional form capturing a possible calendar effect of the business cycles between the years of 1984 and 2008 in Germany denoted with $Year_{it}$. In addition to all of the above-described metrical covariates with corresponding functional effects, we control further for the size of the company with dummy variables indicating if woman i is working in a firm with fewer than 20 employees ($small.firm_{it}$) in year t or with more than 2000 employees ($large.firm_{it}$) in year t, respectively. We finally add the dummy variable $south.germany_{it}$ to the models, indicating if the female is living in the states of southern Germany, which are assumed to perform more strongly economically than other states and might therefore reward the individual stock of human capital differently. The resulting semi-parametric multilevel model for each stratum is therefore defined as:

$$\begin{aligned}\log wage_{it} &= \beta_{0i} + m_1(fulltime_{it}) \\ &\quad + m_2(parttime_{it}) \\ &\quad + m_3(unemployed_{it}) \\ &\quad + m_4(firmtime_{it}) \\ &\quad + m_5(Year_{it}) + \beta_1\, small.firm_{it} \\ &\quad + \beta_2\, large.firm_{it} \\ &\quad + \beta_3\, south.germany_{it} \\ &\quad + \varepsilon_{it} \qquad (18.13) \\ \beta_{0i} &= \beta_0 + u_{i0}.\end{aligned}$$

The resulting fits are shown in Figure 18.3, and the estimation results of the parametric part of (18.13) are listed in Table 18.2. In Figure 18.3, the new covariates with their effects of (18.13) in comparison to (18.12) can easily be interpreted as deviations from the major effect of full-time work experience. However, this major effect changes its dynamic pattern in the context of the semi-parametric model (18.13), being consistent now with the general economic theory. While experience in part-time employment is likely to have negative effects on wages, the attachment to the current employer is less important when analyzing females' earnings, at least for average and low-educated women. The effects of the binary coded covariates in Table 18.2 indicate a positive effect of being employed in large companies and underlines the more dynamic labour market situation in the three southern states of Germany. The dependence of wages on the business cycle, however, is striking through all three strata, attenuating the dominance of human capital theory. It is worth noting that all of the estimated variances of the random terms in model (18.13) are smaller than their counterparts coming from model (18.12), indicating a more reliable estimation with respect to the data. These extensions motivate the use of penalized splines for routine estimation in non- and semi-parametric multilevel models for any kind of empirical research.

18.5 CONCLUSIONS

The combination of multilevel modeling with penalized spline smoothing techniques provides advanced and sophisticated models which can be fitted in a coherent framework. This does not apply in the same way to alternative smoothing methods like local likelihood based approaches or classical smoothing spline techniques. This makes the use of penalized splines, such as those presented here, attractive and appealing. We demonstrated how the models provide new insights into the analysis of empirical panel data.

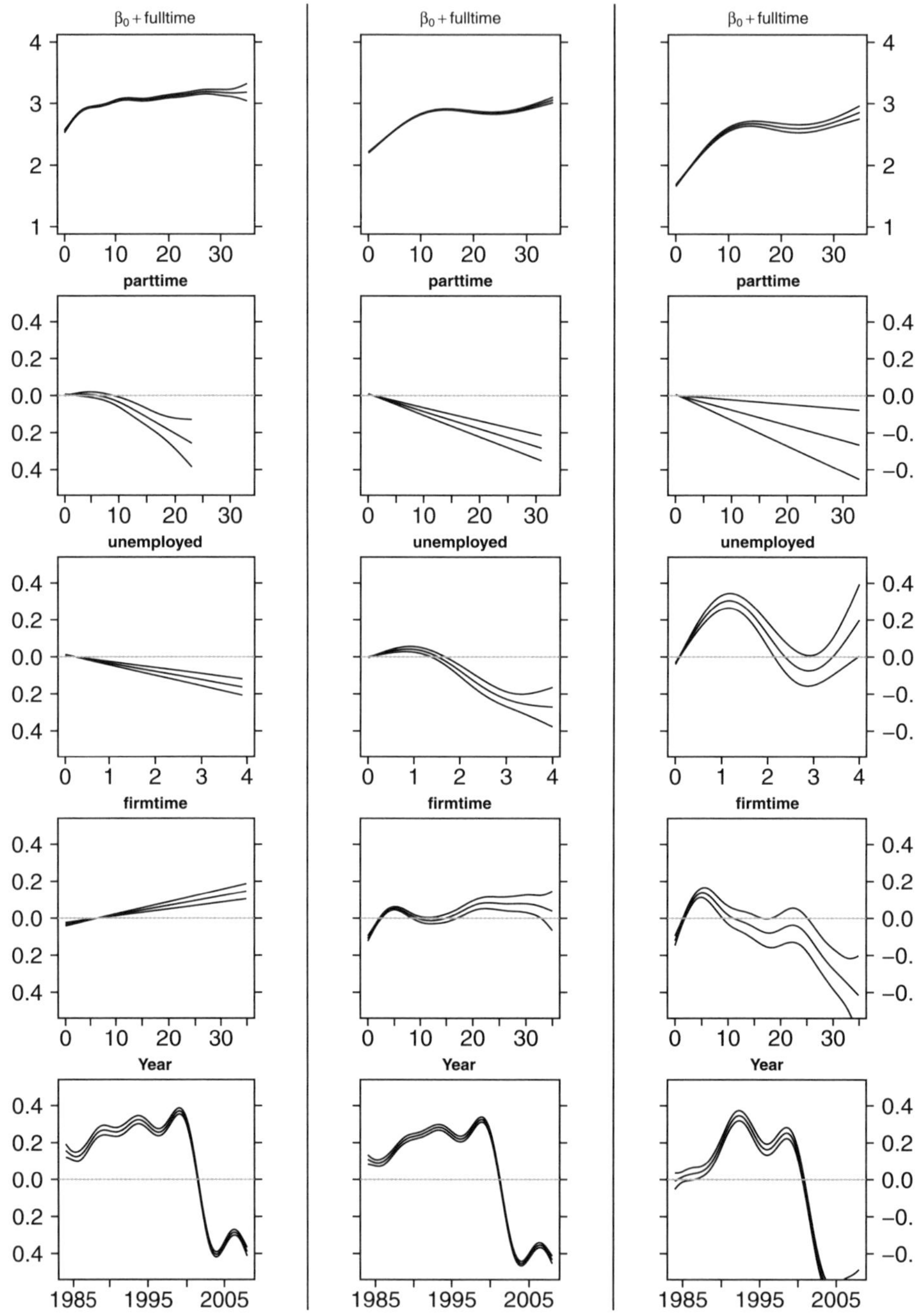

Figure 18.3 Fitted Dynamic Effects for High, Middle, and Low Educated Women (first, second, and third column respectively) Based on Model (18.13)

Table 18.2 Parametric Estimation Results for Model (18.13)

	$\hat{\beta}_j$ (p-value)		
effect	high ISCED	middle ISCED	low ISCED
(Intercept)	2.93 (< 0.01)	2.6 (< 0.01)	1.99 (< 0.01)
small.firm	−0.04 (< 0.01)	−0.07 (< 0.01)	−0.17 (< 0.01)
large.firm	0.04 (< 0.01)	0.07 (< 0.01)	0.15 (< 0.01)
south.germany	0.06 (< 0.01)	0.04 (< 0.01)	0.11 (< 0.01)
$\hat{\sigma}_u$	0.12	0.11	0.18
$\hat{\sigma}_\varepsilon$	0.04	0.07	0.1

REFERENCES

Becker, G.S. (1993) *Human Capital*, 3rd edition. Chicago: University of Chicago Press.

Claeskens, G., Krivobokova, T., and Opsomer, J. (2009) 'Asymptotic properties of penalized spline estimators', *Biometrika*, 96: 529–44.

Crawley, M.J. (2007) *The R Book*. Chichester: Wiley.

de Boor, C. (1978) *A Practical Guide to Splines*. Berlin: Springer.

Eilers, P.H.C. and Marx, B.D. (1996) 'Flexible Smoothing With B-Splines and Penalties', *Statistical Science*, 11: 89–121.

Haisken-DeNew, J.P. and Frick, J.R. (2005) *DTC – Desktop Companion to the German Socio-Economic Panel (SOEP)*, 8th edition. Essen, Berlin: RWI Essen & DIW Berlin.

Hastie, T. (2011) *gam: Generalized Additive Models*. R package version 1.04.1.

Kauermann, G. (2005) 'Penalised Spline Fitting in Multivariable Survival Models With Varying Coefficients', *Computational Statistics and Data Analysis*, 49: 169–86.

Kauermann, G. and Opsomer, J. (2011) 'Data-Driven Selection of the Spline Dimension in Penalized Spline Regression', *Biometrika*, 98: 225–30.

Mincer, J. (1974) *Schooling, Experience, and Earnings*. New York: National Bureau of Economic Research.

Ngo, L. and Wand, M.P. (2004) 'Smoothing With Mixed Model Software', *Journal of Statistical Software*, 9: 1–54.

O'Sullivan, F. (1986) 'A Statistical Perspective on Ill-Posed Inverse Problems (with discussion)', *Statistical Science*, 1: 502–18.

R Development Core Team (2010) *R: A Language and Environment for Statistical Computing*. Vienna, Austria: R Foundation for Statistical Computing.

Reiss, P. and Ogden, R. (2009) 'Smoothing Parameter Selection for a Class of Semiparametric Linear Models', *Journal of the Royal Statistical Society, Series B*, 71: 505–23.

Ruppert, D. (2002) 'Selecting the Number of Knots for Penalized Splines', *Journal of Computational and Graphical Statistics*, 11: 735–57.

Ruppert, D., Wand, M.P., and Carroll, R.J. (2009) 'Semiparametric Regression During 2003–2007', *Electronic Journal of Statistics*, 3: 1193–256.

Scheipl, F. (2011) *amer: Additive Mixed Models With lme4*. R package version 0.6.10.

Wagner, G.G., Frick, J., and Schupp, J. (2007) 'The German Socio-Economic Panel Study (SOEP) – Scope, Evolution and Enhancements', *Schmollers Jahrbuch (Journal of Applied Social Science Studies)*, 127: 139–69.

Wood, S.N. (2006) *Generalized Additive Models: An Introduction With R*. Boca Raton, FL: Chapman & Hall/CRC.

Wood, S.N. (2010) *gamm4: Generalized Additive Mixed Models Using mgcv and lme4*. R package version 0.0-4.

Wood, S.N. (2011) 'Fast Stable Resctricted Maximum Likelihood and Marginal Likelihood Estimation of Semiparametric Generalized Linear Models', *Journal of the Royal Statistical Society, Series B*, 73: 3–36.

19

Hierarchical Dynamic Models

Marina Silva Paez and Dani Gamerman

Universidade Federal do Rio de Janeiro, Brazil

19.1 INTRODUCTION

Consider a population of students divided into groups (schools or classes), for which we believe there are similarities (about a certain aspect of interest) within students in the same group. More specifically, suppose we are interested in analyzing the proficiency in maths of students attending at the same level at different schools, and all of them are submitted to the same test. It is reasonable to expect that the average score obtained by the students is not the same at every school, as characteristics of the school could possibly be affecting this average. For this specific example, the teaching skills of the maths teacher could be an important aspect. Other variables could also be important, such as the average social-economic background of the parents. This problem can be expressed in a hierarchical framework written in two hierarchical levels: first, a regression equation is specified at the student level, possibly including explanatory variables such as gender or age. The intercept and possibly some regression coefficients of this regression are allowed to vary between schools, according to equations specified in the second level of the hierarchy. These equations can be specified, for example, as regressions at the school level, taking into account explanatory variables at this level. By allowing the regression coefficients of certain explanatory variables to vary between schools, the model is taking into account the fact that the effects of these variables in the maths score of the students can vary by school.

Suppose now that the scores are obtained yearly for a certain period of time. It is reasonable to expect that the conditions of each school could be changing through time. Therefore, a modification can be made in the model described above to allow the regression intercept to vary between schools and also through time. In a more general framework, all the regression coefficients can be allowed to vary within these dimensions. Hierarchical models that consider a time variation for the parameters through dynamic linear models (Harrison and Stevens, 1979) are named Dynamic Hierarchical Models. These models are well documented in Gamerman and Migon (1993).

Dynamic hierarchical models can be applied to data in many fields of interest. Specifically, many applications are made to environmental data due to the usual requirement for modeling their variation in time, and often in space as well. Some motivating examples and applications to real datasets are presented throughout this chapter. The remainder of the chapter is organized as follows: in Section 19.2, the dynamic hierarchical models are presented, and inference for these models is presented in Section 19.3. Section 19.4 presents extensions of these models for observations in time and space. Finally, Section 19.5 presents some concluding remarks.

19.2 DYNAMIC HIERARCHICAL MODELS FOR UNIVARIATE OBSERVATIONS

In this section, dynamic hierarchical models are introduced through Example 1 below, which illustrates the problem presented in the introduction of this chapter. After the methodology is presented and its specific notation is introduced, two other examples will be given.

19.2.1 Example 1: Maths Scores of Pupils from Different Schools Over Time

This example illustrates the problem presented in the introduction of this chapter, where the interest lies in modeling the scores obtained in maths tests by students of different schools.

First, suppose we have a total of J schools, with n_j students being tested in the jth school, $j = 1, \ldots, J$. Let y_{ij} be the maths score for the ith student of the jth school, and suppose his/her age (x_{1ij}) and gender (x_{2ij}) were also observed. A linear regression model could be specified to explain the maths scores using age and gender as explanatory variables. A possibly better specification, however, would take into consideration that the students coming from the same school tend to have similarities. This means that students with the same characteristics (gender and age) should not be expected to achieve the same scores if they study at different schools. To accommodate this, one could allow the intercept of the original regression to vary by school. Variables observed at the school level could help to explain this variation. In this example, we suppose that the following school-related variables were also observed: years of experience of the maths teacher (x_{3j}), number of maths classes per week (x_{4j}), and an indicator of whether the students received any support from a tutor (x_{5j}), $j = 1, \ldots, J$. It is reasonable to expect that the regression coefficients related to these explanatory variables will be all positive—meaning that the scores of the students of a certain school tend to be higher if the maths teacher is more experienced, they have a higher number of maths classes, and received the support of a tutor. A possible hierarchical model for this data would be:

$$\begin{aligned}
y_{ij} =& \beta_{0,j} + \beta_1 x_{1ij} + \beta_2 x_{2ij} + \varepsilon_{ij}, \\
&\varepsilon_{ij} \sim N(0, \sigma^2), \\
\beta_{0,j} =& \beta_0 + \beta_3 x_{3j} + \beta_4 x_{4j} + \beta_5 x_{5j} + u_j, \\
&u_j \sim N(0, \sigma_u^2).
\end{aligned}$$

Now suppose tests are applied yearly to students of a certain level over a period of T years. For simplification purposes, we suppose that the number of students tested in each school does not vary through the years, under the hypothesis that no student will change schools over this period, or fail a year. It is now reasonable to assume that the intercept of the regression at the student level is changing over time, as well as by school, if you take into account that characteristics of the schools could be changing over time. We will also suppose that the intercept of the regression at the school level is changing over time through an autoregressive structure, which reflects that other characteristics of the school, not specified in the model, can also be affecting the scores of the students—and these characteristics may vary smoothly in time. The proposed model becomes:

$$\begin{aligned}
y_{ijt} &= \beta_{0,jt} + \beta_1 x_{1ij} + \beta_2 x_{2ij} + \varepsilon_{ijt}, \\
&\quad \varepsilon_{ijt} \sim N(0, \sigma^2), \\
\beta_{0,jt} &= \beta_{0,t} + \beta_3 x_{3jt} + \beta_4 x_{4jt} \\
&\quad + \beta_5 x_{5jt} + u_{jt}, \\
&\quad u_{jt} \sim N(0, \sigma_u^2),
\end{aligned}$$

$$\beta_{0,t} = \beta_{0,t-1} + w_t, \quad w_t \sim N(0, \sigma_w^2), \quad (19.1)$$

where y_{ijt} denotes the score obtained by the ith student of the jth school at the tth year. Note also that the variables measured at the school level are now indexed by time. The age of the students can be considered fixed if we work, for example, with the age given at the beginning of the experiment.

For this example, the coefficients related to gender and age are fixed. Note, however, that they could also be varying by school, time, or both. Similarly, the slope coefficients at the second level of the hierarchy, β_3, β_4, and β_5, could also be varying over time.

In matrix notation, the above model can be written as

$$\begin{aligned} \boldsymbol{y}_t &= \boldsymbol{X}_{1t}\boldsymbol{\beta}_{1,t} + \boldsymbol{v}_{1t}, \\ & \quad \boldsymbol{v}_{1t} \sim N_n(\boldsymbol{0}, \boldsymbol{V}_1), \quad (19.2) \\ \boldsymbol{\beta}_{1,t} &= \boldsymbol{X}_{2t}\boldsymbol{\beta}_{2,t} + \boldsymbol{v}_{2t}, \\ & \quad \boldsymbol{v}_{2t} \sim N_{3J}(\boldsymbol{0}, \boldsymbol{V}_2), \quad (19.3) \\ \boldsymbol{\beta}_{2,t} &= \boldsymbol{I}_4\boldsymbol{\beta}_{2,t-1} + \boldsymbol{w}_t, \\ & \quad \boldsymbol{w}_t \sim N_4(\boldsymbol{0}, \boldsymbol{W}), \quad (19.4) \end{aligned}$$

$t \geq 1$, where $\boldsymbol{I}_k$ denotes a $k \times k$ identity matrix. $\boldsymbol{y}_t$ is a vector of size n where the first n_1 elements represent the scores of the students of the 1st school, the following n_2 elements represent the scores of the students of the 2nd school, and so on, for a fixed time t. That way, $n = n_1 + n_2 + \cdots + n_J$. $\boldsymbol{X}_{1t}$ is a matrix of dimension $n \times 3J$, obtained by the direct sum of the matrices $\boldsymbol{\theta}_{1j}$, $j = 1, \ldots, J$, where $\boldsymbol{\theta}_{1j}$ is an $n_j \times 3$ matrix with all the elements in the first column being equal to one, and the ith elements of the second and third columns being the gender and the age of the ith student of the jth school. $\boldsymbol{\beta}_{1,t}$ is a vector of order $3J \times 1$ of coefficients, where the first n_1 sequences of three elements are given by $\beta_{0,1t}$, β_1, and β_2 (the coefficients corresponding to the first school); the following n_2 sequences of three elements are given by $\beta_{0,2t}$, β_1, and β_2 (the coefficients corresponding to the second school); and so forth. $\boldsymbol{v}_{1t}$ is an $n \times 1$ vector of errors, and $\boldsymbol{V}_1 = \boldsymbol{I}_n\sigma^2$. Analogously, $\boldsymbol{X}_{2t}$ is a $3J \times 4$ matrix, obtained by the direct sum of the matrices $\boldsymbol{\theta}_{2j}$, where $\boldsymbol{\theta}_{2j}$ is a 1×4 matrix with unitary elements in the first column and the other three columns being the values of the covariates at the second level of the hierarchy for school j; $\boldsymbol{\beta}_{2,t}$ and $\boldsymbol{v}_{2t}$ are vectors of coefficients and errors, respectively, of dimension 4×1, and $\boldsymbol{V}_2 = \boldsymbol{I}_J\sigma_u^2$. $\boldsymbol{w}_t$ is a vector of size 4, with the first element representing the error term in equation (19.1) and the other elements being constant and equal to zero (as β_3, β_4, and β_5 do not vary in time). That way, $\boldsymbol{W}$ is a 4×4 matrix with element $[1, 1]$ equal to σ_w^2 and the other elements are equal to zero.

Now we are ready to present a general formulation for the dynamic hierarchical models, as introduced by Gamerman and Migon (1993). The model is composed of three parts: an observation equation, structural equations, and a system equation. The observation equation describes the distribution of the observations through a regression model, as in (19.2); the structural equation describes the structure of hierarchy of the regression parameters, as in (19.3); and the system equation describes the evolution of the parameters through time, as in (19.4). For a linear hierarchical model of three levels, these equations can be written, respectively, as:

Observation equation:

$$\boldsymbol{y}_t = \boldsymbol{X}_{1t}\boldsymbol{\beta}_{1,t} + \boldsymbol{v}_{1t}, \quad \boldsymbol{v}_{1t} \sim N_n(\boldsymbol{0}, \boldsymbol{V}_{1t}), \quad (19.5)$$

Structural equations:

$$\boldsymbol{\beta}_{1,t} = \boldsymbol{X}_{2t}\boldsymbol{\beta}_{2,t} + \boldsymbol{v}_{2t}, \quad \boldsymbol{v}_{2t} \sim N_{r_1}(\boldsymbol{0}, \boldsymbol{V}_{2t}), \quad (19.6)$$

$$\boldsymbol{\beta}_{2,t} = \boldsymbol{X}_{3t}\boldsymbol{\beta}_{3,t} + \boldsymbol{v}_{3t}, \quad \boldsymbol{v}_{3t} \sim N_{r_2}(\boldsymbol{0}, \boldsymbol{V}_{3t}), \quad (19.7)$$

System equation:

$$\boldsymbol{\beta}_{3,t} = \boldsymbol{G}_t\boldsymbol{\beta}_{3,t-1} + \boldsymbol{w}_t, \quad \boldsymbol{w}_t \sim N_{r_3}(\boldsymbol{0}, \boldsymbol{W}_t), \quad (19.8)$$

where n is the total number of observations, $\boldsymbol{v}_{1t}$, $\boldsymbol{v}_{2t}$, $\boldsymbol{v}_{3t}$, and $\boldsymbol{w}_t$ are disturbance terms which are independent; $\boldsymbol{X}_{1t}$, $\boldsymbol{X}_{2t}$, $\boldsymbol{X}_{3t}$, and $\boldsymbol{G}_t$ are known matrices possibly incorporating explanatory variables; $\boldsymbol{V}_{1t}$, $\boldsymbol{V}_{2t}$, $\boldsymbol{V}_{3t}$, and

$\boldsymbol{W}_t$ are variance-covariance matrices that can be allowed to vary over time; $\boldsymbol{\beta}_i$ is a vector of coefficients of size r_i, $i = 1, \ldots, 3$, with $r_1 > r_2 > r_3$.

Hierarchical models of higher levels can easily be obtained by adding extra levels in the structural equations. The two-level model can be obtained by setting $\boldsymbol{X}_{3t}$ to be the identity matrix, and $\boldsymbol{V}_{3t}$ to be a matrix of null elements.

Other simple hypothetical examples are presented below to illustrate the use of these models.

19.2.2 Example 2: Weight Measurements in a Population of Patients Under Treatment

Suppose that the variation of weight in a population of patients under the same kind of experimental treatment is being investigated. Since the beginning of the experiment, a different sample of patients is selected every week from the population and weighted, for a total of T weeks. Suppose that at the tth week, a sample of size n_t is selected. To model the variation of weight through time, we can use a simple two-stage model, given by:
Observation equation:

$$y_{it} = \beta_{i,t} + \varepsilon_{it},\ \varepsilon_{it} \sim N(0, \sigma^2),\ i = 1, \ldots, n_t,$$

Structural equation:

$$\beta_{i,t} = \mu_t + u_{it},\ u_{it} \sim N(0, \sigma_u^2),\ i = 1, \ldots, n_t,$$

System equation:

$$\mu_t = \mu_{t-1} + w_t,\ w_t \sim N(0, \sigma_w^2).$$

This model is a collection of the steady models of West and Harrison (1997), which are related through similar mean levels. The $\beta_{i,t}$s are the observation levels assumed to form an exchangeable (with respect to index i) sample of means with common mean μ_t. Note that the mean μ_t is allowed to vary over time: the treatment can cause average weight loss or weight gain through time. This model can be written in matrix notation as in (19.5)–(19.8):

$$\begin{aligned} \boldsymbol{y}_t &= \boldsymbol{X}_{1t}\boldsymbol{\beta}_{1,t} + \boldsymbol{v}_{1t}, \\ &\quad \boldsymbol{v}_{1t} \sim N_{n_t}(\boldsymbol{0}, \boldsymbol{V}_{1t}), \quad (19.9) \\ \boldsymbol{\beta}_{1,t} &= \boldsymbol{X}_{2t}\beta_{2,t} + \boldsymbol{v}_{2t}, \\ &\quad \boldsymbol{v}_{2t} \sim N_{n_t}(\boldsymbol{0}, \boldsymbol{V}_{2t}), (19.10) \\ \beta_{2,t} &= G_t\beta_{2,t-1} + w_t, \\ &\quad w_t \sim N(0, W_t), \quad (19.11) \end{aligned}$$

where $\boldsymbol{y}_t = (y_{1t}, \ldots, y_{n_t t})^T$, $\boldsymbol{\beta}_{1,t} = (\beta_{1,t}, \ldots, \beta_{n_t,t})^T$, $\boldsymbol{X}_{1t} = \boldsymbol{I}_{n_t}$, $\boldsymbol{V}_{1t} = \boldsymbol{I}_{n_t}\sigma^2$, $\boldsymbol{X}_{2t} = \mathbf{1}_{n_t}$, $\beta_{2,t} = \mu_t$, $\boldsymbol{V}_{2t} = \boldsymbol{I}_{n_t}\sigma_u^2$, $G_t = 1$, and $W_t = \sigma_w^2$. $\mathbf{1}_n$ represents an n-dimensional vector of ones.

19.2.3 Example 3: Weight Measurements in a Population of Children With Malnutrition

The second example can be illustrated by the following experiment: A population of children with malnutrition is being treated with a caloric diet. As in Example 2, a different sample (of size n_t) of children is selected from the population at the tth week of the experiment and weighed. Differently from the previous example, however, the children are expected to gain weight through time. The model proposed in Example 1 can therefore be modified to accommodate this expected growth in the average weight. The proposed model for this example can be written as:
Observation equation:

$$y_{it} = \beta_{it} + \varepsilon_{it},\ \ \varepsilon_{it} \sim N(0, \sigma^2),\ i = 1, \ldots, n_t,$$

Structural equation:

$$\beta_{i,t} = \mu_t + u_{it},\ \ u_{it} \sim N(0, \sigma_u^2),\ i = 1, \ldots, n_t,$$

System equations:

$$\mu_t = \rho_1(\mu_{t-1} + \delta_{t-1}) + w_{1t},\ \ w_{1t} \sim N(0, \sigma_{w_1}^2),$$
$$\delta_t = \rho_2\delta_{t-1} + w_{2t},\ \ w_{2t} \sim N(0, \sigma_{w_2}^2).$$

Note that the expected growth of the mean μ_t is specified by the system equations. This model can be represented in matrix notation as in (19.9)–(19.11), where $\boldsymbol{y}_t$ and $\boldsymbol{\beta}_{1,t}$ are defined as before and $\boldsymbol{X}_{1t} = \boldsymbol{I}_{n_t}$, $\boldsymbol{V}_{1t} = \boldsymbol{I}_{n_t}\sigma^2$, $\boldsymbol{\beta}_{2,t} = (\mu_t, \delta_t)^T$, $\boldsymbol{X}_{2t} = (\boldsymbol{1}_{n_t}, \boldsymbol{0}_{n_t})$, $V_{2t} = \boldsymbol{I}_{n_t}\sigma_u^2$,

$$\boldsymbol{G}_t = \begin{pmatrix} \rho_1 & \rho_1 \\ 0 & \rho_2 \end{pmatrix}, \text{ and } \boldsymbol{W}_t = \begin{pmatrix} \sigma_{w1}^2 & 0 \\ 0 & \sigma_{w2}^2 \end{pmatrix},$$

where $\boldsymbol{0}_n$ denotes an n-dimensional vector of zeros.

19.3 INFERENCE FOR DYNAMIC HIERARCHICAL MODELS

Inference will be presented here from the Bayesian point of view. To do so, it will be necessary to specify prior distributions for all the unknown parameters of the model. More details about the Bayesian approach can be seen in Chapter 4. Our aim for this kind of modeling is usually to obtain the posterior distribution of these parameters and perform forecasting for future observations. In this section, some basic results which were presented in Gamerman and Migon (1993) will be reviewed.

Let us define D_t as all the information obtained up to time t, including the observations $\boldsymbol{y} = \{\boldsymbol{y}_1, \ldots, \boldsymbol{y}_t\}$, and the prior information, represented by D_0. We assume, therefore, that $D_t = \{\boldsymbol{y}_t, D_{t-1}\}$. We can represent the ($k$-stage) dynamic hierarchical model as:

$$\begin{aligned} \boldsymbol{y}_t|\boldsymbol{\beta}_{1,t} &\sim N_n(\boldsymbol{X}_{1t}\boldsymbol{\beta}_{1,t}, \boldsymbol{V}_{1t}), \quad (19.12) \\ \boldsymbol{\beta}_{i,t}|\boldsymbol{\beta}_{i+1,t} &\sim N_{r_i}(\boldsymbol{X}_{i+1,t}\boldsymbol{\beta}_{i+1}, \boldsymbol{V}_{i+1,t}), \\ & \qquad i = 1, \ldots, k-1, \quad (19.13) \\ \boldsymbol{\beta}_{k,t}|\boldsymbol{\beta}_{k,t-1} &\sim N_{r_k}(\boldsymbol{G}_t\boldsymbol{\beta}_{k,t-1}, \boldsymbol{W}_t), (19.14) \end{aligned}$$

with initial prior $\boldsymbol{\beta}_{k,0}|D_0 \sim N(\boldsymbol{m}_{k,0}, \boldsymbol{C}_{k,0})$. Note that the matrices $\boldsymbol{X}_{it}, i = 1, \ldots, k-1$, and $\boldsymbol{G}_t$ are assumed known.

19.3.1 Updating and Forecasting: Known Variances Case

At this point, let us also assume that the variances $\boldsymbol{V}_{it}$ are known. Even though it is in most cases an unrealistic assumption, it enables us to obtain the basic updating and forecasting operations. Under that assumption, Gamerman and Migon (1993) showed that the model specified by equations (19.12)–(19.14), considering the prior $\boldsymbol{\beta}_{k,0}|D_0 \sim N(\boldsymbol{m}_{k,0}, \boldsymbol{C}_{k,0})$, give the following prior, posterior, and predictive distributions:

- the prior distribution at time t is given by
$$\boldsymbol{\beta}_{i,t}|D_{t-1} \sim N(\boldsymbol{a}_{it}, \boldsymbol{R}_{it}), \quad i = 1, \ldots, k, \quad (19.15)$$
where $\boldsymbol{a}_{it} = \boldsymbol{X}_{i+1,t}\boldsymbol{a}_{i+1,t}$, $\boldsymbol{R}_{it} = \boldsymbol{X}_{i+1,t}\boldsymbol{R}_{i+1,t}\boldsymbol{X}_{i+1,t}^T + \boldsymbol{V}_{i+1,t}$, $i = 1, \ldots, k+1$, $\boldsymbol{a}_{kt} = \boldsymbol{G}_t\boldsymbol{m}_{k,t-1}$, and $\boldsymbol{R}_{kt} = \boldsymbol{G}_t\boldsymbol{C}_{k,t-1}\boldsymbol{G}_t^T + \boldsymbol{W}_t$;
- the predictive distribution one step ahead is given by
$$\boldsymbol{y}_t|D_{t-1} \sim N(\boldsymbol{f}_t, \boldsymbol{Q}_t), \quad (19.16)$$
where $\boldsymbol{f}_t = \boldsymbol{X}_{1t}\boldsymbol{a}_{1t}$ and $\boldsymbol{Q}_t = \boldsymbol{X}_{1t}\boldsymbol{R}_{1t}\boldsymbol{X}_{1t}^T + \boldsymbol{V}_{1t}$;
- the posterior distribution at time t is given by
$$\boldsymbol{\beta}_{i,t}|D_t \sim N(\boldsymbol{m}_{it}, \boldsymbol{C}_{it}), \quad i = 1, \ldots, k, \quad (19.17)$$
where $\boldsymbol{m}_{it} = \boldsymbol{a}_{it} + \boldsymbol{S}_{it}\boldsymbol{Q}_t^{-1}(\boldsymbol{y}_t - \boldsymbol{f}_t)$, $\boldsymbol{C}_{it} = \boldsymbol{R}_{it} - \boldsymbol{S}_{it}\boldsymbol{Q}_t^{-1}\boldsymbol{S}_{it}^T$, $\boldsymbol{S}_{it} = \boldsymbol{R}_{it}\boldsymbol{E}_{0it}^T$, $\boldsymbol{E}_{ijt} = \prod_{l=i+1}^{j}\boldsymbol{X}_{lt}$, $0 \leq i < j \leq k$, and $\boldsymbol{E}_{iit} = \boldsymbol{I}_{r_i}$ for $1 \leq i < k$.

Gamerman and Migon (1993) also showed that h-steps-ahead forecasts, $h > 1$, can easily be obtained from this result. Suppose we are interested in predicting $\boldsymbol{y}_{t+h}$ given D_t or, in other words, we want to obtain the distribution of $\boldsymbol{y}_{t+h}|D_t$. Note that

$$\begin{aligned} \boldsymbol{y}_{t+h}|\boldsymbol{\beta}_{1,t+h} &\sim N(\boldsymbol{X}_{1,t+h}\boldsymbol{\beta}_{1,t+h}, \boldsymbol{V}_{1,t+h}), \text{and} \\ \boldsymbol{\beta}_{k,t}|D_t &\sim N(\boldsymbol{m}_{kt}, \boldsymbol{C}_{kt}). \end{aligned}$$

Then the distribution of $\boldsymbol{\beta}_{k,t+h}|D_t \sim N(\boldsymbol{a}_{kt}(h), \boldsymbol{R}_{kt}(h))$ can be recursively obtained (West and Harrison, 1997) with $\boldsymbol{a}_{kt}(h) = \boldsymbol{G}_{t+h}\boldsymbol{a}_{kt}(h-1)$ and $\boldsymbol{R}_{kt}(h) = \boldsymbol{G}_{t+h}\boldsymbol{R}_{kt}(h-1)\boldsymbol{G}_{t+h}^T + \boldsymbol{W}_{t+h}$ with starting values $\boldsymbol{a}_{kt}(0) = \boldsymbol{m}_{kt}$ and $\boldsymbol{R}_{kt}(0) = \boldsymbol{C}_{kt}$. Successive integrations give:

$$\begin{aligned} \boldsymbol{\beta}_{i,t+1}|D_t &\sim N(\boldsymbol{a}_{it}(h), \boldsymbol{R}_{it}(h)), \\ & \qquad i = 1, \ldots, k-1 \\ \boldsymbol{y}_{t+h}|D_t &\sim N(\boldsymbol{f}_t(h), \boldsymbol{Q}_t(h)), \end{aligned}$$

where $\boldsymbol{f}_t(h)$ and $\boldsymbol{Q}_t(h)$ are analogous to $\boldsymbol{f}_t$ and $\boldsymbol{Q}_t$ as defined previously, substituting $\boldsymbol{a}_{it}$ and $\boldsymbol{R}_{it}$ by $\boldsymbol{a}_{it}(h)$ and $\boldsymbol{R}_{it}(h)$.

When a more accurate description of the model parameters is required, or if events in the past are still of interest, a procedure called *smoothing* or *filtering* can be applied at any given time point. This procedure filters back the information smoothly via a recursive algorithm, allowing for parametric estimation of the system at previous times given all available information at a given period of time. Thus, the smoothed distributions of state parameters are denoted by $[\boldsymbol{\beta}_{i,t}|D_n]$, as opposed to their online posterior distributions, denoted by $[\boldsymbol{\beta}_{i,t}|D_t]$. Gamerman and Migon (1993) show that the smoothed distribution for a dynamic hierarchical model is $\boldsymbol{\beta}_{i,t}|D_n \sim N(\boldsymbol{m}^n_{it}, \boldsymbol{C}^n_{it}), t = 1, \ldots, n, i = 1, \ldots, k$, with moments recursively defined as

$$\boldsymbol{m}^n_{it} = \boldsymbol{m}_{it} + \boldsymbol{A}^T_{it}\boldsymbol{R}^{-1}_{i,t+1}(\boldsymbol{m}^n_{i,t+1} - \boldsymbol{a}_{i,t+1}),$$
$$\boldsymbol{C}^n_{it} = \boldsymbol{C}_{it} - \boldsymbol{A}^T_{it}\boldsymbol{R}^{-1}_{i,t+1}(\boldsymbol{R}_{i,t+1} - \boldsymbol{C}^n_{i,t+1})\boldsymbol{R}^{-1}_{i,t+1}\boldsymbol{A}_{it},$$

and initialized at $t = n$ with $\boldsymbol{m}^n_{in} = \boldsymbol{m}_{in}$ and $\boldsymbol{C}^n_{in} = \boldsymbol{C}_{in}$, where

$$\boldsymbol{A}_{it} = \boldsymbol{E}_{ik,t+1}\boldsymbol{G}_{t+1}\boldsymbol{C}_{kt}\{(\boldsymbol{I}_{r_i} - \boldsymbol{V}^*_{ikt}\boldsymbol{E}^T_{0it}\boldsymbol{V}^{*-1}_{0kt}\boldsymbol{E}_{0it})\boldsymbol{E}_{ikt}\}^T,$$

and $\boldsymbol{V}^*_{ijt} = \sum_{l=i+1}^{j} \boldsymbol{E}_{i,l-1,t}\boldsymbol{V}_{lt}\boldsymbol{E}^T_{i,l-1,t}$.

19.3.2 Updating and Forecasting: Unknown Variances Case

As stated before, the hypothesis that the variances in the model are known is hardly ever realistic in a real application. In many applications, however, it is reasonable to suppose independence of the errors in the observation equation, such that $\boldsymbol{V}_{it} = \sigma^2_{it}\boldsymbol{I}_N$, where N is the dimension of $\boldsymbol{y}_t$. A conjugate analysis is possible when all the variances $\boldsymbol{V}_{it}$, $\boldsymbol{W}_t$, and $\boldsymbol{C}_{k0}$ are scaled by σ^2, an unknown factor, with an Inverted Gamma prior distribution: $\sigma^2|D_0 \sim IG(n_0/2, d_0/2)$. The model can be written as in (19.12)–(19.14) with the variances multiplied by the factor σ^2.

The distribution of σ^2 can be updated at time t by $IG(n_t/2, d_t/2)$ where $n_t = n_{t-1}+n$ and $d_t = d_{t-1} + (\boldsymbol{y}_t - \boldsymbol{f}_t)^T\boldsymbol{Q}_t^{-1}(\boldsymbol{y}_t - \boldsymbol{f}_t)$. The results presented by equations (19.15), (19.16), and (19.17) remain valid except that all variances should be multiplied by σ^2. The predictive distribution (one step ahead) in this case (after integrating σ^2 out) is obtained by replacing the Normal distributions by Student T distributions with n_t degrees of freedom and substituting σ^2 by its estimate d_t/n_t. The posterior distribution is also obtained by substituting the Normal by a Student T distribution, but in this case with n_{t-1} degrees of freedom, and the estimate for σ^2 is given by d_{t-1}/n_{t-1}. We denote the Student T distribution with mean $\boldsymbol{m}$, ν degrees of freedom, and variance-covariance matrix $\boldsymbol{\Sigma}$ by $T(\boldsymbol{m}, \nu, \boldsymbol{\Sigma})$.

In a more realistic approach, however, just one unknown factor σ^2 is not enough to handle all of the uncertainty at every level of the hierarchy. In general, a more realistic representation considers at least one unknown parameter for the variance of the errors at each hierarchical level—with the single parameter being the constant variance under the assumption of independent errors. Models of this kind are not analytically tractable and therefore it is necessary to make use of computational methods to simulate from the distributions of interest (posterior and/or predictive), or make use of deterministic approximation methods.

Numerical methods are a very important tool for applied Bayesian analysis. When the posterior (or predictive) distribution of the unknown parameters is not analytically tractable, samples can hardly ever be obtained directly from this distribution, and, as an alternative, simulation methods that sample indirectly from the posterior distribution can be applied. One of the most used methods to obtain these samples is Markov Chain Monte Carlo (MCMC), a computer-intensive method where the idea is to sample from a Markov chain whose equilibrium distribution is the posterior/predictive distribution of interest. That way, after convergence is reached, the simulated values can be seen as a sample of the distribution of interest, and inference can be performed based on this sample. Different algorithms can be proposed in MCMC, with Gibbs Sampling and Metropolis–Hastings being the most used. A broad discussion about these and other algorithms can be found in Gamerman and Lopes (2006).

More recently, a promising alternative to inference via MCMC in latent Gaussian models, the Integrated Nested Laplace Approximations (INLA)—was proposed by Rue et al. (2009). INLA is an approximation method for the posterior/predictive distributions of interest that, unlike the empirical Bayes approaches (Fahrmeir et al., 2004), incorporates posterior uncertainty with respect to hyperparameters. Other alternative methods that have been successfully used for inference in problems of this kind are Particle Filters, which are sequential Monte Carlo methods based on point mass (or 'particle') representations of probability densities, and are very attractive computationally. Examples of the use of these methods can be found in Liu and West (2001), Johannes and Polson (2007), and Carvalho et al. (2010), amongst others. Both these methods (INLA and Particle Filters) can be applied to the hierarchical dynamic models presented here and extensions of these models that will be presented in Section 19.4.

19.4 MODEL EXTENSIONS

In this section, extensions to the dynamic hierarchical models will be presented, including examples illustrating applications to real data-sets. In Section 19.4.1, Matrix-Variate Dynamic Hierarchical Models will be presented as a multivariate extension to the univariate case already introduced. In Section 19.4.2, a particular case of this model will be considered, imposing a parametric structure to account for the spatial correlation between observations made in different locations in space. In Section 19.4.3, an extension to observations in the exponential family will be presented.

19.4.1 Matrix-Variate Dynamic Hierarchical Models

Suppose q observations are made through time in r different locations in space such that $\boldsymbol{y}_{1t}, \ldots, \boldsymbol{y}_{qt}$ represent q vectors of observations of dimension $r \times 1$. Data with this structure can easily be found in the literature for the study of environmental processes. As an example, in many locations around the world, monitoring stations are implemented to monitor the air quality, measuring concentrations of all kinds of pollutants at a certain periodicity. q observations of certain pollutants made through time, and registered at r different locations, can be seen as q vectors of observations of dimension $r \times 1$. In this section, the hierarchical dynamic models for univariate responses will be extended to responses of this kind, leading to the Matrix-Variate Hierarchical Dynamic Model. To introduce the notation, this section presents the Matrix-Variate Dynamic Model, then presents Matrix-Variate Hierarchical Dynamic Models as a simple extension. Inference to these models is then presented, and finally an application is given.

Matrix-variate dynamic models

Let $\boldsymbol{y}_{1t}, \ldots, \boldsymbol{y}_{qt}$ represent q vectors of observations of dimension $r \times 1$, with the ith element of each vector representing an observation made at location i. If we denote

$$\boldsymbol{y}_t = [\boldsymbol{y}_{1t}, \ldots, \boldsymbol{y}_{qt}],$$

we have a matrix of observations through time. West and Harrison (1989) define a Matrix-Variate Dynamic Model by the equations:

$$\begin{aligned} \boldsymbol{y}_t &= \boldsymbol{X}_t\boldsymbol{\beta}_t + \boldsymbol{v}_t, \\ \boldsymbol{\beta}_t &= \boldsymbol{G}_t\boldsymbol{\beta}_{t-1} + \boldsymbol{w}_t, \end{aligned}$$

where

- $\boldsymbol{y}_t$ is the $r \times q$ observations matrix at time t;
- $\boldsymbol{X}_t$ is a known matrix of regressors $r \times p$;
- $\boldsymbol{\beta}_t$ is the matrix of unknown parameters;

- v_t is a matrix of errors;
- G_t is a known evolution matrix $r \times p$;
- w_t is the matrix of evolutions of errors $p \times q$.

It is assumed that β_t and v_t are independent, β_{t-1} and w_t are independent, and v_t and w_t are independent. Under the hypotheses of normality, the errors v_t and w_t, which are matrices, will follow a matrix-variate normal distribution. If Z is a random matrix of dimension $r \times q$, we say that Z has a matrix-variate normal distribution with right covariance matrix Σ and left covariance matrix C, denoted by $Z \sim N(M, C, \Sigma)$, if $vec(Z) \sim N_{rq}(vec(M), \Sigma \otimes C)$, where $vec(Z)$ represents the vectorization of the matrix Z in columns.

As an example, suppose that q types of pollutants are observed in r different locations in space, resulting in an $r \times q$ matrix of observations, for a fixed period of time t. If this matrix follows an $N(M, C, \Sigma)$ distribution, then M is an $r \times q$ matrix of means, the left matrix C is $r \times r$, and represents the correlation of the observations in space, and the right matrix Σ is $q \times q$ and represents the correlation between the different kinds of pollutants. Note that in this model, C and Σ are not identifiable, and $N(M, C, \Sigma)$ is equivalent to $N(M, C/k, k\Sigma), k \in \mathbb{R}$. If the observations of each variable of interest are not correlated in space at a given period of time t, then $C = I_r$. If the observations are independent at a fixed period of time and fixed location in space, then $\Sigma = I_q$.

Hierarchical formulation of the model

Matrix-Variate Hierarchical Dynamic Models are an extension of the model described above; an extension where, for each period of time, we observe a matrix of observations instead of a vector. This model can be written, as defined by Landim and Gamerman (2000), as follows:

$$y_t = X_{1t}\beta_{1,t} + v_{1t},\quad v_{1t}|\Sigma \sim N(0, V_1, \Sigma), \tag{19.18}$$

$$\beta_{1,t} = X_{2t}\beta_{2,t} + v_{2t},\quad v_{2t}|\Sigma \sim N(0, V_2, \Sigma), \tag{19.19}$$

$$\vdots$$

$$\beta_{k-1,t} = X_{kt}\beta_{k,t} + v_{kt},\quad v_{kt}|\Sigma \sim N(0, V_k, \Sigma), \tag{19.20}$$

$$\beta_{k,t} = G_t\beta_{k,t-1} + w_t,\quad w_t|\Sigma \sim N(0, W, \Sigma), \tag{19.21}$$

where, for a fixed period of time t,

- y_t is the $r \times q$ observations matrix;
- $\beta_{i,t}, i = 1, \ldots, k$ are regression coefficients matrices of order $r_i \times q$;
- X_{it} are known matrices of regressors of order $r_{i-1} \times r_i$, with $r_0 = r, i = 1, \ldots, k$;
- G_t is the known evolution matrix $r_k \times r_k$;
- w_t is the matrix of evolutions of errors $p \times q$.

We assume that $\beta_{i,t}$ and v_{it} are independent, $i = 1, \ldots, k$, $\beta_{k,t-1}$ and w_t are independent, and $v_{1t}, v_{2t}, \ldots, v_{kt}$ and w_t are independent. Matrices $V_1, V_2, \ldots, V_k$ and W are considered to be known and, for simplicity, are considered to be fixed over time.

Note that a restriction is made, stating that part of the conditional covariance between elements of y_t, which account for the different responses, is specified in Σ. The same restriction is imposed on the conditional covariances of $\beta_{1,t}, \beta_{2,t}, \ldots, \beta_{k,t}$, all sharing the same structure Σ to account for the covariance between elements coming from different responses. This assumption is usually made in multivariate dynamic models (see, for example, West and Harrison, 1997, Chapter 16, and Quintana, 1987), and is reasonable since the magnitude of the covariance between elements at different hierarchical levels is not imposed to be the same. The models above could be specified and operated without this restriction but that would substantially increase parameter dimensionality.

To complete the specification of the model presented above, it is necessary to specify prior distributions for the unknown parameters of the model. Usual choices of priors, used by West and Harrison (1997), Quintana

(1987), and Landim and Gamerman (2000), are the following:

$$\boldsymbol{\beta}_{k,0}|\boldsymbol{\Sigma}, D_0 \sim N(\boldsymbol{M}_{k,0}, \boldsymbol{C}_{k,0}, \boldsymbol{\Sigma}), \text{ and}$$
$$\boldsymbol{\Sigma}|D_0 \sim WI(n_0, \boldsymbol{S}_0).$$

Inference

As in the beginning of Section 19.3, we will start this section presenting results of inference based on the assumption that $\boldsymbol{V}_1, \boldsymbol{V}_2, \ldots, \boldsymbol{V}_k$ and $\boldsymbol{W}$ are known. Conditionally on the value of $\boldsymbol{\Sigma}$, it is possible to obtain the distributions of $\boldsymbol{\beta}_{i,t}|D_{t-1}, \boldsymbol{y}_t|D_{t-1}$, and $\boldsymbol{\beta}_{i,t}|D_t$ analytically. These results are given below, and the demonstration can be found in Landim and Gamerman (2000). For $t = 1, 2, \ldots$ and $i = 1, \ldots, k$,

a) $\boldsymbol{\beta}_{i,t}|D_{t-1} \sim N(\boldsymbol{a}_{it}, \boldsymbol{R}_{it}, \boldsymbol{\Sigma})$,

$$\boldsymbol{a}_{it} = \boldsymbol{X}_{i+1}\boldsymbol{a}_{i+1,t}, \quad i = 1, \ldots, k-1,$$
$$\boldsymbol{a}_{kt} = \boldsymbol{G}_t\boldsymbol{M}_{k,t-1},$$
$$\boldsymbol{R}_{it} = \boldsymbol{X}_{i+1,t}\boldsymbol{R}_{i+1,t}\boldsymbol{X}_{i+1,t}^T + \boldsymbol{V}_{i+1,t},$$
$$i = 1, \ldots, k-1,$$
$$\boldsymbol{R}_{kt} = \boldsymbol{G}_t\boldsymbol{C}_{k,t-1}\boldsymbol{G}_t^T + \boldsymbol{W}_t;$$

b) $\boldsymbol{y}_t|D_{t-1} \sim N(\boldsymbol{f}_t, \boldsymbol{Q}_t, \boldsymbol{\Sigma})$,

$$\boldsymbol{f}_t = \boldsymbol{X}_{1t}\boldsymbol{a}_{1t},$$
$$\boldsymbol{Q}_t = \boldsymbol{X}_{1t}\boldsymbol{R}_{1t}\boldsymbol{F}_{1t}^T + \boldsymbol{V}_{1t};$$

c) $\boldsymbol{\theta}_{i,t}|D_t \sim N(\boldsymbol{M}_{it}, \boldsymbol{C}_{it}, \boldsymbol{\Sigma})$,

$$\boldsymbol{M}_{it} = \boldsymbol{a}_{it} + \boldsymbol{S}_{it}\boldsymbol{Q}_t^{-1}\boldsymbol{E}_t,$$
$$\boldsymbol{S}_{it} = \boldsymbol{R}_{it}\boldsymbol{E}_0\boldsymbol{R}_{it}^T,$$
$$\boldsymbol{E}_t = \boldsymbol{y}_t - \boldsymbol{f}_t,$$
$$\boldsymbol{C}_{it} = \boldsymbol{R}_{it} - \boldsymbol{S}_{it}\boldsymbol{Q}_t^{-1}\boldsymbol{S}_{it}^T.$$

The posterior distribution of $\boldsymbol{\Sigma}$ can also be obtained analytically, and is given by

$$\boldsymbol{\Sigma}|D_t \sim WI(n_t, \boldsymbol{S}_t),$$

with $n_t = n_{t-1} + N$ and $\boldsymbol{S}_t = \boldsymbol{S}_{t-1} + \boldsymbol{E}_t^T\boldsymbol{Q}_t^{-1}\boldsymbol{E}_t$. Unconditionally on $\boldsymbol{\Sigma}$, the distributions of $\boldsymbol{\Theta}_{it}|D_{t-1}, \boldsymbol{y}_t|D_{t-1}$, and $\boldsymbol{\beta}_{i,t}|D_t$ become Student T, given by:

1 $\boldsymbol{\beta}_{i,t}|D_{t-1} \sim T(\boldsymbol{a}_{it}, \boldsymbol{R}_{it}, n_t, \boldsymbol{S}_t)$,
2 $\boldsymbol{y}_t|D_{t-1} \sim T(\boldsymbol{f}_t, \boldsymbol{Q}_t, n_t, \boldsymbol{S}_t)$,
3 $\boldsymbol{\beta}_{i,t}|D_t \sim T(\boldsymbol{M}_{it}, \boldsymbol{C}_{it}, n_t, \boldsymbol{S}_t)$,

where $T(\boldsymbol{a}, \boldsymbol{R}, n, \boldsymbol{S})$ denote the Matrix Variate Student T distribution with mean $\boldsymbol{a}$, left side covariance matrix $\boldsymbol{R}$, n degrees of freedom and right side covariance matrix $\boldsymbol{S}$. If $\boldsymbol{Z}$ is a random matrix such that $\boldsymbol{Z} \sim T(\boldsymbol{a}, \boldsymbol{R}, n, \boldsymbol{S})$ then $vec(\boldsymbol{X}) \sim T(vec(\boldsymbol{a}), n, \boldsymbol{R} \otimes \boldsymbol{S})$.

Equivalently to the case of Univariate Hierarchical Dynamic Linear Models, forecasts h steps ahead, or the distribution of $(\boldsymbol{y}_{t+h}|D_t, \boldsymbol{\Sigma})$, can be obtained.

When the variance matrices $\boldsymbol{V}_1, \boldsymbol{V}_2, \ldots, \boldsymbol{V}_k$, $\boldsymbol{W}$, and $\boldsymbol{\Sigma}$ are unknown, it is not possible to obtain the distributions of $\boldsymbol{\beta}_{i,t}|D_{t-1}$, $\boldsymbol{y}_t|D_{t-1}$, and $\boldsymbol{\beta}_{i,t}|D_t$ analytically. In that case, numerical methods can help in obtaining approximations for these distributions.

Under the Bayesian point of view, prior distributions must be specified for these quantities. Landim and Gamerman (2000) considered independent Inverted Wishart prior distributions for all of the variance-covariance matrices, and a Normal prior distribution for $\boldsymbol{\beta}_{k,0}|\boldsymbol{\Sigma}, D_0$, given by $\boldsymbol{\beta}_{k,0}|\boldsymbol{\Sigma}, D_0 \sim N(\boldsymbol{M}_{k0}, \boldsymbol{C}_{k0}, \boldsymbol{\Sigma})$.

The joint posterior distribution of $(\{\boldsymbol{\beta}_1\}, \ldots, \{\boldsymbol{\beta}_k\}, \boldsymbol{V}_1, \ldots, \boldsymbol{V}_k, \boldsymbol{W}, \boldsymbol{\Sigma})$ is given by

$$\begin{aligned} p(\{\boldsymbol{\beta}_1\}, \ldots, \{\boldsymbol{\beta}_k\}, \boldsymbol{V}_1, \ldots, \boldsymbol{V}_k, \boldsymbol{W}, \boldsymbol{\Sigma}|\boldsymbol{y}) &\propto \prod_{t=1}^{T} p(\boldsymbol{y}_t|\boldsymbol{\beta}_{1,t}, \boldsymbol{V}_1, \boldsymbol{\Sigma}) \\ &\times p(\boldsymbol{W})p(\boldsymbol{\Sigma})p(\boldsymbol{\beta}_{k,0})|\boldsymbol{\Sigma})\prod_{i=1}^{k} p(\boldsymbol{V}_i) \\ &\prod_{t=1}^{T}\prod_{i=1}^{k-1} p(\boldsymbol{\beta}_{i,t}|\boldsymbol{\beta}_{i+1,t}, \boldsymbol{V}_{i+1}, \boldsymbol{\Sigma}) \\ &\prod_{t=2}^{T} p(\boldsymbol{\beta}_{k,t}|\boldsymbol{\beta}_{k,t-1}, \boldsymbol{W}, \boldsymbol{\Sigma}). \end{aligned}$$

Even though it is not possible to obtain the posterior distribution of the model parameters analytically, it is possible to obtain their full conditional distributions. Landim and Gamerman (2000) showed that the full conditional distribution for each of the parameters $\boldsymbol{\beta}_{i,t}, t = 1, \ldots, n, i = 1, \ldots, k$, is Normal and the full conditional distribution for the variance parameters $\boldsymbol{V}_1, \boldsymbol{V}_2, \ldots, \boldsymbol{V}_k, \boldsymbol{W}$, and $\boldsymbol{\Sigma}$ is Inverted Wishart. That is, it is possible to obtain samples from the posterior distribution of these parameters using Gibbs Sampling, which is an MCMC algorithm. Details of how to obtain the full conditional distributions and the proposed algorithm can be seen in Landim and Gamerman (2000).

Application to the occupied population and average salary in Brazil

Landim and Gamerman (2000) presented an application to jointly model the population at work and the average salary for the biggest metropolitan regions in Brazil: Rio de Janeiro (RJ), Salvador (SAL), São Paulo (SP), Belo Horizonte (BH), Porto Alegre (PA), and Recife (RE). The data was made available by Instituto de Pesquisa Econômica Apliacada (IPEA), in Brazil.

Observations were made monthly from May, 1982 to December, 1996. Preliminary analysis showed a seasonal pattern for both variables in time. Landim and Gamerman (2000) worked with the series aggregated at every three months, having a total of $T = 56$ periods of times of observation.

The original data was transformed via a logarithmic transformation which made the observations closer to those coming from a Normal distribution. Also, the series of salaries was transformed using the INPC index, which is a national index of prices for consumers, using December 1996 as a base. That way, the salaries were all to the same scale of reals (R$) of December 1996. Figure 19.1 shows the aggregated series of the logarithm of people at work and average salary for the six metropolitan regions, showing that both variables change significantly and smoothly in time.

Note that this example concerns two response variables which are varying not only in time but also in space. Preliminary analysis showed that it is reasonable to consider that the observations follow an autoregressive model in time and that the coefficients of this model vary in time and space.

Let $\boldsymbol{y}_t$ be a 6×2 matrix with element $[i, 1]$ representing the logarithm of the population size in the ith region, and $[i, 2]$ representing the logarithm of the average salary in the ith region for a fixed period of time t. That way, the response variable in the model can be represented by

$$\boldsymbol{y}_t = \begin{bmatrix} y_{11t} & y_{12t} \\ & \vdots & \\ y_{61t} & y_{62t} \end{bmatrix}.$$

Landim and Gamerman (2000) suggest a dynamic autoregressive model of order 1 to model this response matrix, where the observation equation for the element $[i, j]$ of $\boldsymbol{y}_t$ is given by:

$$\begin{aligned} y_{ijt} &= \beta_{1,ijt} + \beta_{2,ijt} y_{i1,t-1} \\ &\quad + \beta_{3,ijt} y_{i2,t-1} + \varepsilon_{ijt}, \end{aligned}$$

$i = 1, \ldots, 6$, $j = 1, 2$ and $t = 1, \ldots, 56$. In matrix notation, we can write:

$$\boldsymbol{y}_t = \boldsymbol{X}_{1t} \boldsymbol{\beta}_{1,t} + \boldsymbol{v}_{1t}, \quad \boldsymbol{v}_{1t} | \boldsymbol{\Sigma} \sim N(\boldsymbol{0}, \boldsymbol{V}_1, \boldsymbol{\Sigma}),$$

where $\boldsymbol{X}_{1t}$ is given by:

$$\boldsymbol{X}_{1t} = \begin{bmatrix} 1 & y_{11,t-1} & y_{12,t-1} & 0 & \cdots & 0 & 0 & 0 \\ & & & & \vdots & & & \\ 0 & 0 & 0 & 0 & \cdots & 1 & y_{61,t-1} & y_{62,t-1} \end{bmatrix},$$

and the matrix of autoregressive coefficients is given by:

$$\boldsymbol{\beta}_{1,t}^T = \begin{bmatrix} \beta_{1,11t} \beta_{2,11t}\ \beta_{3,11t} & \cdots & \beta_{1,61t}\ \beta_{2,61t}\ \beta_{3,61t} \\ \beta_{1,12t} \beta_{2,12t}\ \beta_{3,12t} & \cdots & \beta_{1,62t}\ \beta_{2,62t}\ \beta_{3,62t} \end{bmatrix}.$$

For each $j = 1, 2$, and fixed period of time t, the parameters β_1, β_2, and β_3 are supposed to

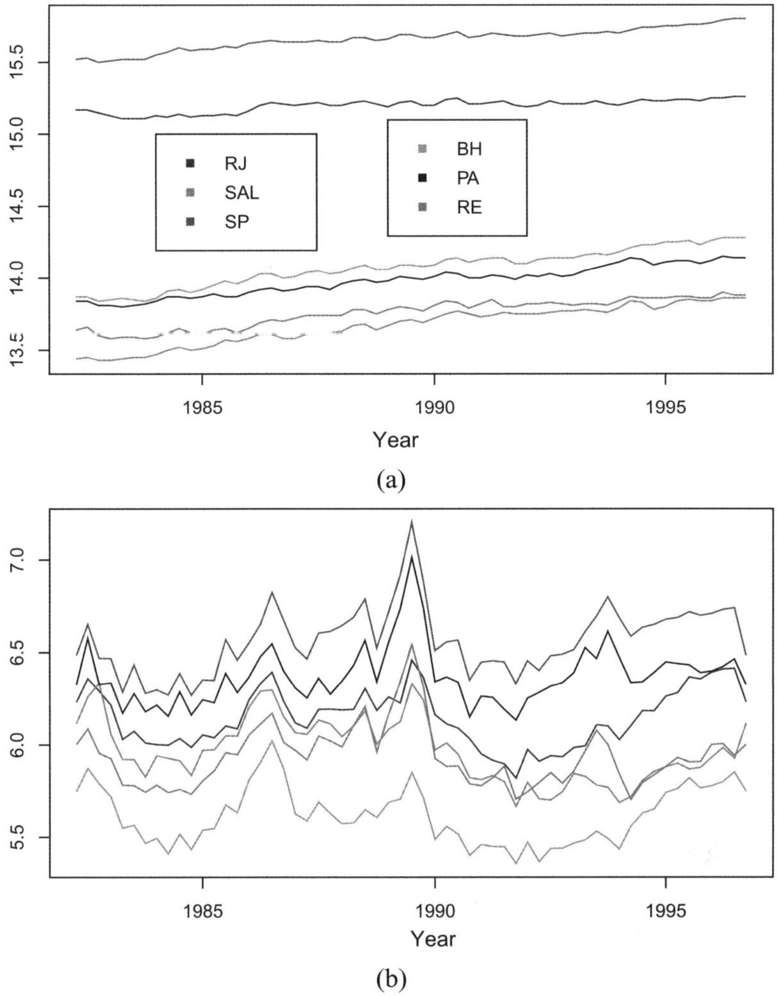

Figure 19.1 (a) Logarithm of People at Work; and (b) Logarithm of Average Salary, for Each One of the Six Regions Through Time

have the same mean for every location i, as specified by the equations below:

$$\beta_{1,ijt} = \beta_{1,jt} + v_{21,ijt},$$
$$\beta_{2,ijt} = \beta_{2,jt} + v_{22,ijt},$$
$$\beta_{3,ijt} = \beta_{3,jt} + v_{23,ijt},$$

where $i = 1, \ldots, 6$. That way, we can write the matrix $\boldsymbol{\beta}_{1,t}$ in the form:

$$\boldsymbol{\beta}_{1,t} = \boldsymbol{X}_{2t}\boldsymbol{\beta}_{2,t} + \boldsymbol{v}_{2t}, \quad \boldsymbol{v}_{2t}|\boldsymbol{\Sigma} \sim N(\boldsymbol{0}, \boldsymbol{V}_2, \boldsymbol{\Sigma}),$$

where $\boldsymbol{X}_{2t}$ is 18×3 and is given by

$$\boldsymbol{X}_{2t}^T = \begin{bmatrix} 1 & 0 & 0 & \cdots & 1 & 0 & 0 \\ 0 & 1 & 0 & \cdots & 0 & 1 & 0 \\ 0 & 0 & 1 & \cdots & 0 & 0 & 1 \end{bmatrix} = 1_6^T \otimes I_3,$$

the hyperparameters matrix is given by

$$\boldsymbol{\beta}_{2,t} = \begin{bmatrix} \beta_{1,1t} & \beta_{1,1t} \\ \beta_{2,1t} & \beta_{2,2t} \\ \beta_{3,1t} & \beta_{3,2t} \end{bmatrix},$$

and

$$v_{2t}^T = \begin{bmatrix} v_{21,11t} & \cdots & v_{23,11t} & \cdots & v_{21,61t} & \cdots & v_{23,61t} \\ v_{21,12t} & \cdots & v_{23,12t} & \cdots & v_{21,62t} & \cdots & v_{23,62t} \end{bmatrix}.$$

The evolution matrix is given by:

$$\boldsymbol{\beta}_{2,t} = \boldsymbol{\beta}_{2,t-1} + \boldsymbol{w}_t, \qquad \boldsymbol{w}_t \sim N(\mathbf{0}, \boldsymbol{W}, \boldsymbol{\Sigma}).$$

To complete the model, Inverted Wishart distributions were set as the prior distributions for the parameters $\boldsymbol{V}_1$, $\boldsymbol{V}_2$, $\boldsymbol{W}$, and $\boldsymbol{\Sigma}$, and a Normal distribution was set as the prior distribution for $\boldsymbol{\beta}_{2,0}|\boldsymbol{\Sigma}$.

Landim and Gamerman (2000) used an MCMC algorithm to obtain samples from the posterior distribution of the unknown model parameters. After 59500 iterations, convergence was achieved and the last 5000 iterations were used as samples of the posterior distribution. They obtained credibility intervals for each unknown parameter of the model. One important finding was that the correlations in matrix $\boldsymbol{V}_1$ were not significantly different from zero, which indicates independence between observations made in different regions. Besides that, there is a strong indication that matrix $\boldsymbol{V}_2$ could be written as a block diagonal matrix. These hypotheses were incorporated into the model, and it was re-estimated. Convergence for this case was faster. The posterior distributions of coefficients $\boldsymbol{\beta}_2$ showed that past observations of population size are a good explanatory variable for the present population size, but do not explain much of the average salary. In the same way, past observations of average salary can explain part of the variability of the present average salary, but do not explain much of the population size. Matrix $\boldsymbol{\Sigma}$ showed small but positive correlation between the response variables.

19.4.2 Spatially Structured Matrix-Variate Dynamic Hierarchical Models

In this section we describe a class of models proposed by Paez et al. (2008) which are a particular case of the Matrix-Variate Dynamic Hierarchical Models of Landim and Gamerman (2000), but with a parametric structure imposed to account for the spatial correlation between observations made in different locations in space. With that restriction, the spatial correlation can be captured without the need to completely estimate unknown covariance matrices. A spatially structured matrix can be specified with a small number of parameters and, as a consequence, interpolation of the response variable can be made, for any fixed period of time, to any set of non-observed locations in space.

Gelfand et al. (2005) use a similar approach to deal with this same kind of problem, where the temporal variance is described by dynamic components which capture the association between measurements made for a fixed location in space and period of time, as well as the dependence between space and time themselves. The multivariate space processes described in that paper were developed through the use of coregionalization techniques and non-separable space-time covariance structures. They applied the proposed methodology to model the variation of monthly maximum temperature in the State of Colorado, U.S.A., using the monthly mean precipitation as an explanatory variable.

Other applications of multilevel modeling taking spatial structures into consideration can be seen in Chapter 31.

General model framework

The model presented in this section is a special case of the model described by equations (19.18)–(19.21). Without loss of generality, the model will be described for a two-level hierarchical dynamic model. A simple extension can be made to obtain models of higher hierarchical levels. Suppose we observe q $(q > 1)$ response variables in a discrete set of periods of time $t = 1, \ldots, T$ and set of locations $\{s_1, \ldots, s_r\}$ in a continuous space S. Analogously to equations (19.18)–(19.21), the Matrix-Variate

Hierarchical Dynamic Space-Time Model can be written as:

$$y_t = X_{1t}\beta_{1,t} + v_{1t},\ \ v_{1t} \sim N(0, V_1, \Sigma), \tag{19.22}$$

$$\beta_{1,t} = X_{2t}\beta_{2,t} + v_{2t},\ \ v_{2t} \sim N(0, V_2, \Sigma), \tag{19.23}$$

$$\beta_{2,t} = G_t\beta_{2,t-1} + w_t,\ \ w_t \sim N(0, W, \Sigma), \tag{19.24}$$

for $t = 1, \ldots, T$. We assume that the matrices X_{1t}, X_{2t}, and G_t are known, with X_{1t} and X_{2t} possibly incorporating the values of explanatory variables. The dimensions of the matrices in equation (19.22)–(19.24) are the same as specified previously for equations (19.18)–(19.21).

A spatial structure can be incorporated in the variance matrices V_1 and V_2, being specified through parametric structures describing spatial dependency. In particular, one can define $V_i = C \otimes V$, which corresponds to $v_{it} \sim N(0, V_i\rho_i(\lambda_i, \cdot), \Sigma_i)$, where C is the matrix specified through the correlation function $\rho_i(\lambda_i, \cdot)$, $i = 1, 2$. For a fixed period of time t and given the parameter λ_i, the correlation function $\rho_i(\lambda_i, \cdot)$ gives the correlation between the elements of $v_{it}, i = 1, 2$. This function can be appropriately specified in order to model spatial dependency, giving higher correlation between elements which correspond to observations made closer in space.

Paez et al. (2008) suggest independent Inverted Wishart prior distributions for the covariance matrices and a matrix-variate Normal prior distribution for $(\beta_{2,0}|\Sigma)$. A prior distribution must also be specified for the parameters $\lambda_i, i = 1, 2$, which define the spatial correlation matrices. Inference for this model, including the estimation of the model parameters, time forecasting and spatial interpolation, can be performed using MCMC methods as described in Paez et al. (2008).

Pollution in the northeast of the United States

This example was presented in Paez et al. (2008), and refers to a data-set made available by the Clean Air Status and Trends Network (CASTNet) at the web site `www.epa.gov/castnet`. CASTNet is the primary source for data on dry acidic deposition and rural ground-level ozone in the United States. It consists of over 70 sites across the eastern and western United States and is cooperatively operated and funded by the United States Environmental Protection Agency (EPA) with the National Park Service.

When dealing with pollutants (and most environmental data-sets), the main interest is usually to be able to explain some possible causes of variation as well as to be able to perform predictions. In this particular example, the interest is to model and make predictions of the levels of two pollutants: SO_2 and NO_3, which were measured in micrograms per cubic meter of air (μg/m^3). In this example, Paez et al. (2008) work with observations made in 24 monitoring stations from the 1st of January 1998 to the 30th of December 2004, in a total of 342 periods of time. The data, however, is not completely available, and about 4% of it is missing. Under the Bayesian approach, handling missing observations is straightforward, as they can be considered as unknown parameters of the model and can be estimated if needed.

Figure 19.2 shows a map of the 24 monitoring stations in the northeast United States and Figure 19.3 shows the trajectory of the time series observed at five randomly selected locations in space, after a logarithmic transformation. No explanatory variables are available to explain the variation of these pollutants. However, it is clear from Figure 19.3 that part of the time variation can be explained by seasonal components, such as sine and cosine waves.

Exploratory analysis clearly shows that the levels of the time series vary with the location where they were measured. Also, it can be noticed that the amplitude of their oscillation varies as well. The mean level of the

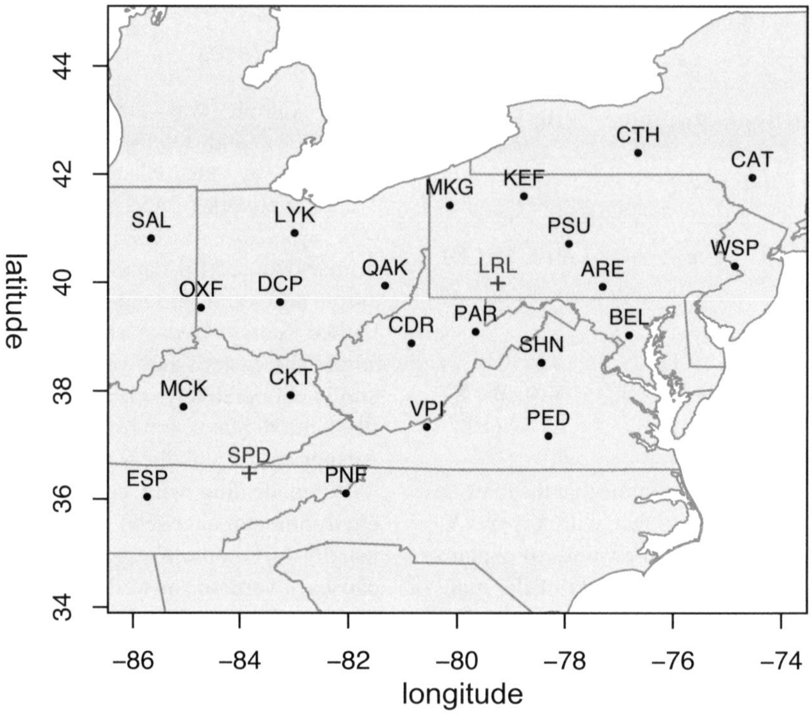

Figure 19.2 Map of the 24 Monitoring Stations in the Northeast United States. The coo

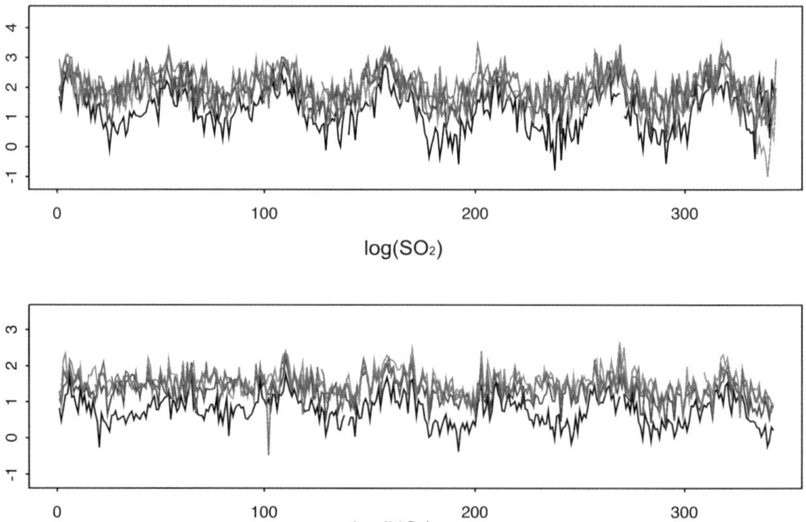

Figure 19.3 Series of $log(SO_2)$ and $log(NO_3)$ Through Time for Five Randomly Selected Monitoring Stations

series is also varying in time (it is slowly decreasing), as well as their amplitude (which is slowly increasing). These variations suggest the use of a hierarchical model with varying intercept and also varying coefficients of sine and cosine waves. It is intuitive, however, that these coefficients are not independent for each period of time or location in space, but rather vary smoothly in these dimensions. This should also be considered in the model specification.

A logarithmic transformation was applied to the observed levels of both pollutants. Preliminary analysis showed strong correlation between the transformed response variables $log(SO_2)$ and $log(NO_3)$. Paez et al. (2008) worked with observations from 22 monitoring stations and left the other 2 to validate the models (stations LRL and SPD, which can be seen in Figure 19.2).

The authors compared the interpolation performance between two univariate models (one for each response variable) and one bivariate model (where the two response variables are jointly modeled). They showed an advantage of working under the multivariate model, which was described by the authors through the following equations:

$$\begin{aligned}
\boldsymbol{y}_t &= \boldsymbol{X}_{1t}\boldsymbol{\beta}_{1,t} + \boldsymbol{v}_{1t}, \quad \boldsymbol{v}_{1t} \sim N(\boldsymbol{0}, \boldsymbol{I}_r, \boldsymbol{\Sigma}),\\
\boldsymbol{\beta}_{1,t} &= \boldsymbol{\beta}_{2,t} + \boldsymbol{v}_{2t}, \quad \boldsymbol{v}_{2t} \sim N(\boldsymbol{0}, \boldsymbol{C} \otimes \boldsymbol{V}, \boldsymbol{\Sigma}),\\
\boldsymbol{\beta}_{2,t} &= \boldsymbol{\beta}_{2,t-1} + \boldsymbol{w}_t, \quad \boldsymbol{w}_t \sim N(\boldsymbol{0}, \boldsymbol{W}, \boldsymbol{\Sigma}),
\end{aligned}$$

where $\boldsymbol{y}_t = (y_{1t}, y_{2t})$, with $y_{1t} = log(SO_2)_t$ and $y_{2t} = log(NO_3)_t$. Sine and cosine waves were used as explanatory variables to explain the seasonality present in the observations, so for each location s_i, $\boldsymbol{X}_{1t}(s_i) = (1, \sin(2\pi t/52), \cos(2\pi t/52)), i = 1, \ldots, 22$, with $\boldsymbol{X}_{1t} = (\boldsymbol{X}_{1t}(s_1), \ldots, \boldsymbol{X}_{1t}(s_r))^T$. Note that this model is a special case of the model specified by equations (19.22)–(19.24), where $\boldsymbol{V}_1 = \boldsymbol{I}_r$, $\boldsymbol{V}_2 = \boldsymbol{C} \otimes \boldsymbol{V}$, $\boldsymbol{X}_{2t} = \boldsymbol{I}_3$, and $\boldsymbol{G}_t = \boldsymbol{I}_3$. The spatial structure is specified through a spatial correlation function which defines matrix $\boldsymbol{C}$.

In this application, the spatial correlation function was specified in the Matérn family (Matérn, 1986; Handcock and Stein, 1993), which is given by

$$\begin{aligned}
C(i, j) &= \frac{1}{2^{\kappa-1}\Gamma(\kappa)}\left(\frac{d_{ij}}{\phi}\right)^{\kappa} K_\kappa\left(\frac{d_{ij}}{\phi}\right),\\
&\ \phi, \kappa > 0,
\end{aligned}$$

where K_κ is the modified Bessel function of order κ. This is a flexible family, containing the exponential function ($\phi = 0.5$), and the squared exponential function, which is the limiting case when $\kappa \rightarrow \infty$. κ is a range parameter that controls how fast the correlation decays with distance, and ϕ is a smoothness (or roughness) parameter that controls geometrical properties of the random field. d_{ij} is a measure of the distance between the locations of observation.

As the scale and range parameters in the Matérn family cannot be estimated consistently (Zhang, 2004), Paez et al. (2008) worked with a fixed value of $\kappa = 2.5$, and gave to ϕ a prior distribution based on the reference prior of Berger, de Oliveira, and Sansó (2001). The choice of $\kappa = 2.5$ was made based on forecast performances comparing a set of different values for this parameter. Non-informative priors were specified for the other unknown parameters of the model.

Samples from the posterior distributions of the parameters were obtained through an MCMC algorithm. The elements of matrix $\boldsymbol{\beta}_2$ were shown to change significantly in time, as can be seen in Figure 19.4, for both $log(SO_2)$ and $log(NO_3)$. Thus, the use of temporally varying coefficients seems to be justified in this application.

An interpolation of $\boldsymbol{\beta}_1$ was performed at an equally spaced grid of points, for a fixed period of time $t = 342$. Figure 19.5 shows a significant spatial variation of the elements of $\boldsymbol{\beta}_1$, which supports the importance of allowing these parameters to vary in this dimension. Thus, the use of spatially varying coefficients also seems to be justified in this application. The variation of these elements is smooth in space, especially for the coefficients of sine and cosine of $log(NO_3)$. Based on the interpolated coefficients $\boldsymbol{\beta}_1$, the response process $\boldsymbol{y}_t$ is also interpolated at $t = 342$. Each value

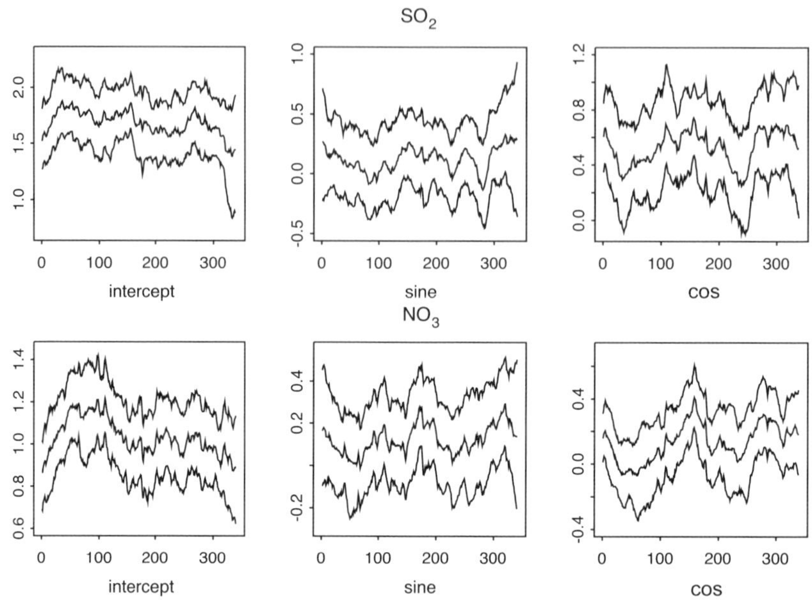

Figure 19.4 Posterior Median and 95% Credibility Intervals for the Elements of Matrix β_2 Through Time

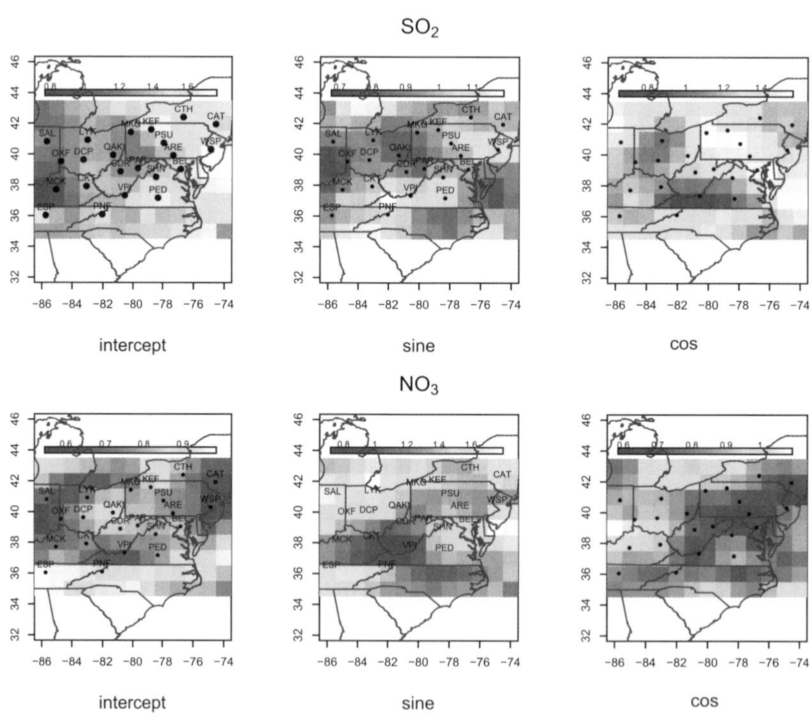

Figure 19.5 Posterior Median of the Elements of Matrix β_1 Through Space

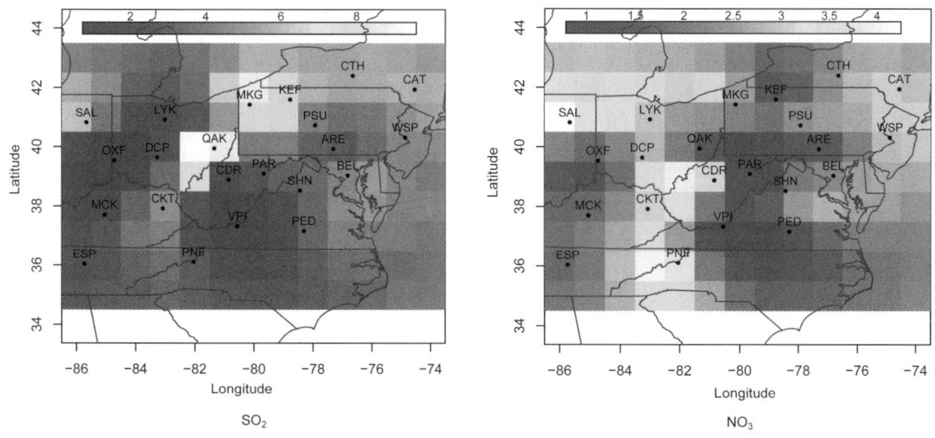

Figure 19.6 Predicted Surface of SO_2 and NO_3 Levels ($\mu g/m^3$) for $t = 342$

sampled from $y_{1,342}$ and $y_{2,342}$ received an exponential transformation, so that a sample of values in the original scale of the pollutants SO_2 and NO_3 were obtained. The spatial variation of the posterior median of these pollutants is shown, to the original scale, in Figure 19.6. Finally, Figures 19.7 and 19.8 show the predictions (95% credibility intervals and median of the posterior distribution) made for SO_2 (Figure 19.7) and NO_3 (Figure 19.8), for the two stations (SPD and LRL) which were used for validation. The true observed trajectories are also shown, for comparison. It can be seen that the obtained predictions are good, with most of the real observations following inside the posterior credibility intervals.

19.4.3 Dynamic Hierarchical Models: Exponential Family Observations

The methodology which will be introduced here was presented by Hansen (2009), and is an extension of previous models to observations in the exponential family.

The motivation for this extension is that it is sometimes not possible to work under the hypothesis of normality of the observations that otherwise could be modeled through the hierarchical dynamic models with structure in space proposed by Paez et al. (2008). Measurements of rainfall are an example of observations that could benefit from this new approach, as they cannot be assumed to be normally distributed. It is reasonable, however, to assume that they come from another distribution belonging to the exponential family.

General model framework

Consider a set of discrete periods of time, $t = 1, \ldots, T$, where, for every t, q variables are observed in r different locations in space $s_1, \ldots, s_r$. Let $\boldsymbol{y}_t$ be the $r \times q$ observation matrix. Suppose that the distribution of $\boldsymbol{y}_t$ belongs to the exponential family. A family of distributions is said to belong to the exponential family if the probability density function (or probability mass function, for discrete distributions) can be written as

$$f(x|\lambda) = h(x) \exp\left(\sum_{i=1}^{s} \eta_i(\lambda) T_i(x) - A(\lambda)\right),$$

where $h(x)$, $A(\lambda)$, $T_i(x)$, and $\eta_i(\lambda)$, $i = 1, \ldots, s$ are known functions. We will use the notation $X \sim EF(\mu)$ to denote that the random variable (either a scalar, vector, or matrix) X follows a distribution coming from the exponential family with mean μ.

Suppose that $\boldsymbol{y}_t$ has mean $\boldsymbol{\phi}_t$ and that $\boldsymbol{\phi}_t$ can be modeled through a function of a regression equation in which the covariate effects vary smoothly through time and space. This function is called a link function and it links the

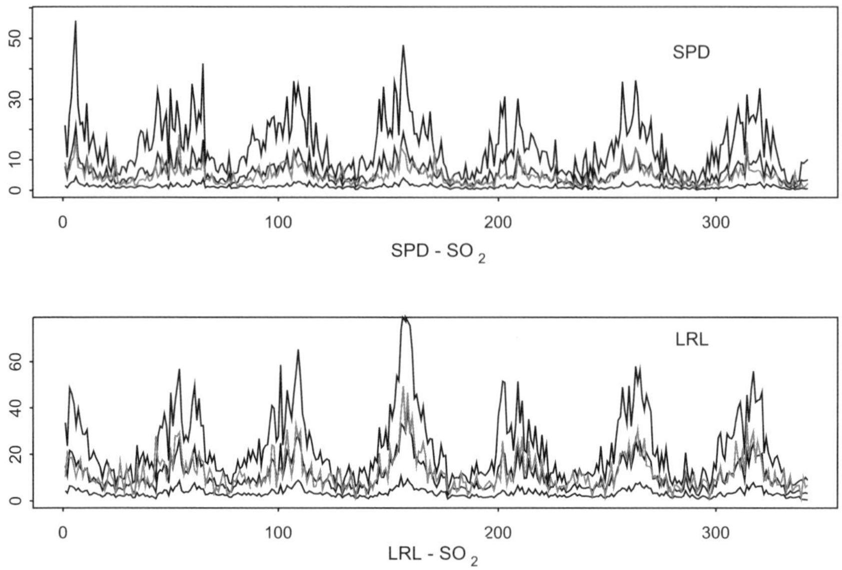

Figure 19.7 SO_2 Levels ($\mu g/m^3$) Observed at Stations SPD and LRL, for $t=1,\ldots,342$, and the Posterior Median and 95% Credibility Intervals Obtained for SO_2. The true trajectory is in gray

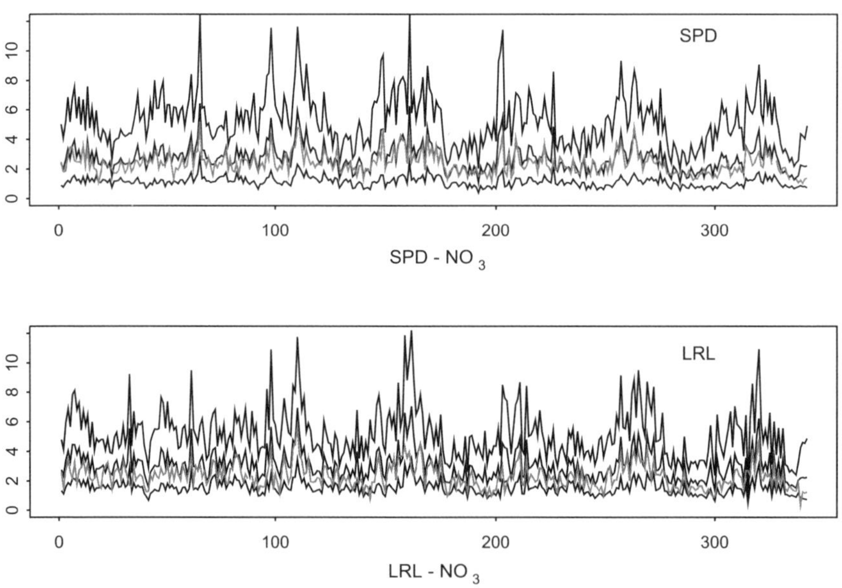

Figure 19.8 NO_3 Levels ($\mu g/m^3$) Observed at Stations SPD and LRL, for $t=1,\ldots,342$, and the Posterior Median and 95% Credibility Intervals Obtained for NO_3. The true trajectory is in gray

linear predictor to the mean of the distribution function as in the Generalized Linear Models (Nelder and Weddenburn, 1972). The model can be specified as below:

$$\begin{aligned} y_t &\sim EF(\mu_t), && (19.25)\\ g(\mu_t) &= X_{1t}\beta_{1,t} + v_{1t}, \\ & \quad v_{1t} \sim N(0, V_{1t}, \Sigma), && (19.26)\\ \beta_{1,t} &= X_{2t}\beta_{2,t} + v_{2t}, \\ & \quad v_{2t} \sim N(0, V_{2t}, \Sigma), && (19.27)\\ \beta_{2,t} &= G_t\Theta_{2,t-1} + w_t, \\ & \quad w_t \sim N(0, W, \Sigma), && (19.28) \end{aligned}$$

where $g(\mu_t)$ is a known link function and the other quantities are defined as in (19.22)–(19.24).

Application: Rainfall data-set in Australia

In this section we present an application of the models presented above to a single response variable consisting of measurements of rainfall in $r = 15$ monitoring stations in Australia (Hansen, 2009). Here, the observations can be assumed as coming from a Gamma distribution, which belongs to the exponential family. We denote by $y_t(s_i)$ the amount of rainfall observed in time t and location s_i. We assume that $y_t(s_i)$ follows a Gamma distribution with mean $\phi_t(s_i)$ and coefficient of variation η denoted by $y_t(s_i) \sim G(\phi_t(s_i), \eta)$.

Exploratory analysis showed that the observations present a clear seasonal cycle of one year. Hansen (2009) worked with rainfall amounts aggregated by year to eliminate the seasonal effect. No explanatory variables were considered in the model. Another possibility would be to consider seasonal covariates. To model the rainfall observations, capturing time and space dependencies, Hansen (2009) proposed the use of a particular case of the model in (19.25)–(19.28) for a univariate response, where $X_{1t} = I_r$, $X_{2t} = 1_r$, $G_t = 1$, $V_{1t} = 0 \times I_r$, $V_{2t} = C$, and $W = \sigma_w^2$. A spatial structure is incorporated through the specification of the elements of matrix C. In this application, these elements are specified by the exponential family, where $C[i, j] = \rho \exp\{-\lambda d_{i,j}\}$. Note that this is an isotropic covariance function, which depends on the locations of observation s_i and s_j only through the distance between them ($d_{i,j}$), and depends on two unknown parameters ρ and λ. Hansen (2009) also works with a logarithmic link function. The model can be written as follows:

$$\begin{aligned} &y_t(s_i) \sim G(\phi_t(s_i), \eta), \\ &\log(\phi_t(s_i)) = \beta_{1,t}(s_i), \\ &\beta_{1,t} = \beta_{2,t} + v_{2t}, \quad v_{2t} \sim N(0, C), \\ &\beta_{2,t} = \beta_{2,t-1} + w_t, \quad w_t \sim N(0, \sigma_w^2), \end{aligned}$$

where $\beta_{1,t} = (\beta_{1,t}(s_1), \ldots, \beta_{1,t}(s_r))$. To complete the model specifications, prior distributions for the unknown model parameters must be specified. Hansen (2009) works with vaguely informative priors. As in the previous applications, $\beta_{2,0}$ follows a Normal distribution and W follows an Inverted Gamma distribution. The prior distribution for ρ, which is also a variance parameter, was set to be an Inverted Gamma as well. Gamma priors were specified for η and λ. Samples of the posterior distribution of the unknown parameters were obtained through MCMC, using Gibbs Sampling and Metropolis–Hastings. Table 19.1 shows statistics of the posterior distribution of the fixed parameters of the model.

It can be noticed that λ is centered around 0.08 and ρ is centered around 0.8. Taking into account the distances between the monitoring stations, these estimated values define a covariance matrix C with values that vary from 0.16 to 0.70. In this way, we conclude that the model captured some high correlations between observations made in different locations in space. The posterior samples obtained for σ_w^2 show high probability density around 0.08, meaning that the parameters $\beta_{2,t}$ do not vary much in time. The spatial variation of these parameters, however, is higher, with the estimated value for the spatial variance parameter ρ being around ten times larger than the estimated value for σ_w^2.

Table 19.1 Statistics from the Posterior Samples Obtained for the Fixed Parameters

parameter	mean	2.5%	median	97.5%
η	0.00	0.09	0.18	0.22
ρ	0.80	0.67	0.79	0.94
λ	0.08	0.07	0.08	0.10
σ_w^2	0.08	0.05	0.08	0.14

19.5 CONCLUDING REMARKS

In this chapter we presented a methodological review of hierarchical dynamic models and generalizations of these models to accommodate multivariate responses, spatial variation of regression coefficients, and observations in the exponential family.

The models presented here are very flexible, permitting the smooth variation of regression coefficients in time and/or space, and they can be applied to model data in many areas of interest. In this chapter we presented examples and applications made with real data-sets focusing mainly on environmental problems. The interest in this kind of application is usually to do time forecasting and interpolation in space, which can be easily done under the proposed methodology.

Inference is made under the Bayesian point of view. Usually, for models like the ones presented in this chapter, the posterior distributions of the unknown parameters are not analytically tractable, and numerical methods must be used to approximate these distributions. In the applications presented here, MCMC methods were used.

ACKNOWLEDGMENTS

The authors would like to thank Dr Aline Maia for providing us with the Australian rainfall data-set and Dr Flávia Landim for providing us with the occupied population and mean salary data-set for the metropolitan regions in Brazil. They would also like to thank CNPq and FAPERJ for their financial support.

REFERENCES

Berger, J.O., De Oliveira, V., and Sanso, B. (2001) 'Objective Bayesian Analysis of Spatially Correlated Data', *Journal of the American Statistical Association*, 96: 1361–74.

Carvalho, C.M., Johannes, M.S., Lopes, H.F., and Polson, N.G. (2010) 'Particle Learning and Smoothing', *Statistical Science*, 25: 88–106.

Fahrmeir, L., Kneib, T., and Lang, S. (2004) 'Penalized Structured Additive Regression for Space-time Data: A Bayesian Perspective', *Statistica Sinica*, 14: 731–61.

Gamerman, D. and Lopes, H.F. (2006) *Monte Carlo Markov Chain: Stochastic Simulation for Bayesian Inference*, 2nd edition. London: Chapman and Hall.

Gamerman, D. and Migon, H.S. (1993) 'Dynamic Hierarchical Models', *Journal of the Royal Statistical Society, Series B*, 55: 629–42.

Gelfand, A.E., Banerjee, S., and Gamerman, D. (2005) 'Spatial Process Modeling for Univariate and Multivariate Dynamic Spatial Data', *Environmetrics*, 16: 465–79.

Handcock, M.S. and Stein, M.L. (1993) 'A Bayesian Analysis of Kriging', *Technometrics*, 35: 403–10.

Hansen, N. (2009) 'Models with Dynamic Coefficients Varying in Space for Data in the Exponential Family', unpublished MSc dissertation, IM, Universidade Federal do Rio de Janeiro (in Portuguese).

Harrison, P.J. and Stevens, C.F. (1979) 'Bayesian Forecasting', *Journal of the Royal Statistical Society, Series B*, 38: 205–47.

Johannes, M. and Polson, N.G. (2007) 'Exact Particle Filtering and Learning', *Working Paper.* The University of Chicago Booth School of Business.

Landim, F.M.P.F. and Gamerman, D. (2000) 'Dynamic Hierarchical Models – An Extension to Matrix Variate Observations', *Computational Statistics and Data Analysis*, 35: 11–42.

Liu, J. and West, M. (2001) 'Combined Parameter and State Estimation in Simulation-based Filtering', in Doucet, A., de Freitas, N., and Gordon, N. (eds) *Sequential Monte Carlo Methods in Practice*. New York: Springer-Verlag.

Matérn, B. (1986) *Spatial Variation*, 2nd edition. Berlin: Springer-Verlag.

Nelder, J.A. and Wedderburn, R.W.M. (1972) 'Generalized Linear Models', *Journal of the Royal Statistical Society, Series A*, 135: 370–84.

Paez, M.S., Gamerman, D., Landim, F.M.P.F., and Salazar, E. (2008) 'Spatially Varying Dynamic Coefficient Models,' *Journal of Statistical Planning and Inference*, 138: 1038–58.

Quintana, J. (1987) 'Multivariate Bayesian Forecasting Models', unpublished PhD Thesis, University of Warwick.

Rue, H., Martino, S., and Chopin, N. (2009) 'Approximate Bayesian Inference for Latent Gaussian Models by using Integrated Nested Laplace Approximations (with Discussion)', *Journal of the Royal Statistical Society, Series B*, 71: 319–92.

West, M. and Harrison, P.J. (1989) *Bayesian Forecasting and Dynamic Model*. New York: Springer.

West, M. and Harrison, P.J. (1997) *Bayesian Forecasting and Dynamic Model*, 2nd edition. London: Springer.

Zhang, H. (2004) 'Inconsistent Estimation and Asymptotically Equal Interpolations in Model-based Geostatistics', *Journal of the American Association*, 99: 250–61.

20

Mixture and Latent Class Models in Longitudinal and Other Settings

Ryan P. Browne and
Paul D. McNicholas
Department of Mathematics and Statistics
University of Guelph, Canada

20.1 INTRODUCTION

In this chapter, we see how mixture models can be an important tool for multilevel models. We place some focus on longitudinal data while also trying to give a fairly broad coverage of the literature. The structure of this chapter is as follows. We begin by looking at how mixture models and latent class analysis fit into the multilevel modeling framework. Then we examine and discuss the relationship between mixture models and latent class models. In Section 20.2.2, latent variable models are reviewed and linked with finite mixture model-based clustering and classification. In Section 20.2.3, we review finite mixture models. We then review three families of Gaussian mixture models, one of which is designed explicitly for the analysis of longitudinal data (Section 20.3). In Section 20.4, non-Gaussian approaches are considered, while covariates and missing data are discussed in Section 20.5. In Section 20.6, a longitudinal data analysis is presented. We then conclude with a summary and discussion (Section 20.7).

20.2 BACKGROUND

20.2.1 Latent Class Models in Multilevel Analysis

In multilevel modeling, we typically have information that groups observations together, e.g., school for student scores or treatment group for panel data. The number of groups can be large or small, depending on the application. A large number of groups might occur when considering test scores from classes within a sizable school district. A small number of groups might occur when we have panel or longitudinal data from a treatment/control study.

Considering the longitudinal setting, the goal is to compare data from each group. We might begin by fitting an intercept and slope (linear trend) to each group. We can then use latent class models to add flexibility to the fitted model within each group. That is, we would assume that each group can be described by a mixture model. Because we apply a mixture model to each level of the data, we need to be careful about the number of parameters, especially in the covariance term. Furthermore, the necessity for more parsimonious mixture models will also arise when using a single mixture model. Parsimony can be introduced into mixture models through various decompositions of the covariance matrix (cf. Section 20.3).

When considering test scores from many classes or longitudinal data from many groups, the data typically appear heterogeneous. However, fitting a separate intercept and slope for each group may not be feasible due to a small number of observations within each group. To circumvent this problem, researchers sometimes use random effects models. In such models, we assume the intercepts and/or slopes are not fixed but random variables that commonly follow a Gaussian distribution. As an alternative to a random effects model, we can use the latent class framework to add flexibility into the model to deal with heterogeneous data. We can accomplish this by ignoring the given groupings (classes) and fitting a finite mixture model. This clustering approach will give a sort of 'natural' grouping within the data.

20.2.2 Latent Class Analysis

Many variables or objects are often unobservable based on the data that manifest. For example, we cannot measure happiness, pain, religious faith, or health. However, each of these unobservable variables has many observable surrogate measures that are related to the variable of interest. For example, the athletic ability of an individual cannot be measured directly but we might argue that those who are more athletic tend to have a gym membership, walk more, run more, run faster, jump higher, and so on. Because we believe that these manifest, or observable, variables are caused by a latent variable, we can study the interrelationship between the manifest variables to estimate the latent variable; that is, the correlation between manifest variables can be largely explained by the latent variable.

In latent class modeling, the latent variable is assumed to be a discrete random variable with levels $1, \ldots, G$ and associated probabilities $\pi_1, \ldots, \pi_G$. The goal is typically to group similar observations so that the within-group variability is smaller than the between-group variability. However, the group allocation and the number of groups are unknown; that is, group or cluster membership is unknown for each observation, whereas this information is known in a multilevel model. From this viewpoint, the goal in latent class modeling might be to find hidden or unknown groupings within the data. When viewed in this context, latent class modeling is analogous to clustering based on mixture models. In certain applications, however, we might want to predict the unknown labels for a subset of observations given that the others have known labels. When latent class modeling is viewed within this context, the appropriate analogy is classification based on mixture models.

Under the latent model to determine $h(\mathbf{x})$, the marginal distribution of the observed random variable $\mathbf{X}$, we use the law of total probability to sum over the discrete latent variable Y:

$$\begin{aligned} h(\mathbf{x}) &= \sum_{g=1}^{G} f_g\left(\mathbf{x}, Y=g\right) \\ &= \sum_{g=1}^{G} f_g\left(\mathbf{x} \mid Y=g\right) P(Y=g) \\ &= \sum_{i=1}^{G} \pi_g f_g\left(\mathbf{x} \mid Y=g\right), \qquad (20.1) \end{aligned}$$

where $f_g\left(\mathbf{x} \mid Y=g\right)$ is the conditional distribution of $\mathbf{X}$ given the latent variable Y in group g. The conditional distribution can depend on the level of Y in different ways. However, a common assumption is to restrict the conditional distribution to have the same parametric form but allow the parameters from the conditional distribution to vary. This assumption can be written as $f_g(\mathbf{x} \mid Y = g) = f(\mathbf{x} \mid \boldsymbol{\theta}_g)$. Equation 20.1 then becomes

$$f(\mathbf{x} \mid \boldsymbol{\vartheta}) = \sum_{g=1}^{G} \pi_g f_g(\mathbf{x} \mid \boldsymbol{\theta}_g), \qquad (20.2)$$

where $\boldsymbol{\vartheta} = (\pi_1, \ldots, \pi_G, \boldsymbol{\theta}_1, \ldots, \boldsymbol{\theta}_G)$. Equation 20.2 is a finite mixture model, and so we see that latent class models and finite mixture models can be considered equivalent.

20.2.3 Model-Based Clustering and Classification

We say that a random vector $\mathbf{X}$ arises from a parametric finite mixture distribution if, for all $\mathbf{x} \subset \mathbf{X}$, we can write its density in the form given in (20.2), where $\pi_g > 0$, such that $\sum_{g=1}^{G} \pi_g = 1$ are the mixing proportions, $f_1(\mathbf{x} \mid \boldsymbol{\theta}_1), \ldots, f_G(\mathbf{x} \mid \boldsymbol{\theta}_G)$ are called component densities, and $\boldsymbol{\vartheta} = (\pi_1, \ldots, \pi_G, \boldsymbol{\theta}_1, \ldots, \boldsymbol{\theta}_G)$ is the vector of parameters. Formally, we can refer to the density specified in (20.2) as a G-component finite mixture density. The component densities $f_g(\mathbf{x} \mid \boldsymbol{\theta}_g)$ are usually taken to be of the same type, with the Gaussian distribution being the most popular choice due to its mathematical malleability. The Gaussian mixture model density is

$$f(\mathbf{x} \mid \boldsymbol{\vartheta}) = \sum_{g=1}^{G} \pi_g \phi(\mathbf{x}_i \mid \boldsymbol{\mu}_g, \boldsymbol{\Sigma}_g), \qquad (20.3)$$

where $\phi(\mathbf{x}_i \mid \boldsymbol{\mu}_g, \boldsymbol{\Sigma}_g)$ is the density of a multivariate Gaussian random variable with mean $\boldsymbol{\mu}_g$ and covariance matrix $\boldsymbol{\Sigma}_g$. Alternatives to the multivariate Gaussian component density continue to be explored and examples are given in Section 20.4. Extensive details on finite mixture models and their applications are given by Titterington et al. (1985), McLachlan and Basford (1988), McLachlan and Peel (2000b), and Frühwirth-Schnatter (2006).

In practice, the number of parameters in the component covariance matrices $\boldsymbol{\Sigma}_1, \ldots, \boldsymbol{\Sigma}_G$ often poses a model fitting issue: there are $Gp(p+1)/2$ parameters in the component covariance matrices alone and this is in addition to the other $(G-1) + Gp$ parameters in (20.3). In clustering and classification applications, parsimony is typically introduced into these models via decomposition of the covariance structure followed by the imposition of constraints upon one or more of the constituent elements. Of course, the same approach can taken to the component means $\boldsymbol{\mu}_g$, but the opportunity to gain parsimony is usually far less. When parsimony is introduced through decomposition and constraining of component covariance structures, the result is several models that are said to form a 'family' of mixture models, wherein each member is a special case of the unconstrained covariance decomposition.

The idiom 'model-based clustering' is used to describe the application of a family of finite mixture models to data for the purpose of clustering. In such cases, all members of a family are typically applied and the best is then selected via a criterion such as the Bayesian information criterion (BIC; Schwarz, 1978). Suppose that n p-dimensional data vectors $\mathbf{x}_1, \ldots, \mathbf{x}_n$ are observed, all having unknown group memberships. The Gaussian model-based clustering likelihood is then given by

$$\mathcal{L}(\boldsymbol{\vartheta} \mid \mathbf{x}) = \prod_{i=1}^{n} \sum_{g=1}^{G} \pi_g \phi(\mathbf{x}_i \mid \boldsymbol{\mu}_g, \boldsymbol{\Sigma}_g).$$

Model parameters $\boldsymbol{\vartheta}$ can be estimated using the expectation-maximization (EM) algorithm (Dempster et al., 1977) and the resulting cluster memberships are given by the maximum a posteriori (MAP) values of the estimated component membership probabilities. That is, $\text{MAP}\{\hat{z}_{ig}\} = 1$ if $\max_g\{\hat{z}_{ig}\}$ occurs at component g, and $\text{MAP}\{\hat{z}_{ig}\} = 0$ otherwise, where $\mathbf{z}_i$ denotes the component membership of observation i, so that $z_{ig} = 1$ if observation i belongs to component g and $z_{ig} = 0$ otherwise. We will not go into any detail on the EM algorithm in this chapter, instead referring the reader to McLachlan and Krishnan (2008) for extensive details.

The locutions 'partial classification' or 'model-based classification' (cf. McLachlan,

1992, Section 2.7) refer to a semi-supervised version of model-based clustering. Now, k of our n observations have known group memberships and we want to use these k to help classify the remaining $n - k$. Without loss of generality, we order the n observations so that the first k have known group memberships and the Gaussian model-based classification likelihood can be written

$$\mathcal{L}(\boldsymbol{\vartheta} \mid \mathbf{x}, \mathbf{z}) = \prod_{i=1}^{k}\prod_{g=1}^{G}\left[\pi_g \phi(\mathbf{x}_i \mid \boldsymbol{\mu}_g, \boldsymbol{\Sigma}_g)\right]^{z_{ig}} \times \prod_{j=k+1}^{n}\sum_{h=1}^{H}\pi_h \phi(\mathbf{x}_j \mid \boldsymbol{\mu}_h, \boldsymbol{\Sigma}_h), \tag{20.4}$$

for $H \geq G$. The flexibility inherent in allowing more groups to exist (H) than have been observed (G) is important and yet heretofore under-utilized within the literature. The EM algorithm can be used for parameter estimation and the classifications are given by $\text{MAP}\{\hat{z}_{jh}\}$, for $j = k + 1, \ldots, n$ and $h = 1, \ldots, H$.

The term 'model-based discriminant analysis' describes a supervised version of model-based clustering, wherein density estimation is carried out based on the k observations with known labels before the remaining $n-k$ observations are classified into groups based on these density estimates (cf. Hastie and Tibshirani, 1996). Again, we order our n observations so that the first k have known group memberships. First, we form the likelihood

$$\mathcal{L}(\boldsymbol{\vartheta} \mid \mathbf{x}, \mathbf{z}) = \prod_{i=1}^{k}\prod_{g=1}^{G}\left[\pi_g \phi(\mathbf{x}_i \mid \boldsymbol{\mu}_g, \boldsymbol{\Sigma}_g)\right]^{z_{ig}},$$

then we estimate parameters and compute the posterior expected values

$$\hat{z}_{jg} = \frac{\hat{\pi}_g \phi(\mathbf{x}_j \mid \hat{\boldsymbol{\mu}}_g, \hat{\boldsymbol{\Sigma}}_g)}{\sum_{h=1}^{G}\hat{\pi}_h \phi(\mathbf{x}_j \mid \hat{\boldsymbol{\mu}}_h, \hat{\boldsymbol{\Sigma}}_h)},$$

for $j = k + 1, \ldots, n$. The MAP classifications associated with these expected values then serve as the discriminant rule.

In model-based classification, we use all n observations to estimate our model parameters; when performing model-based discriminant analysis, parameter estimation is based only on the k observations with known group memberships. A further distinction is that model-based classification offers the option of estimating $H \geq G$ components. One should note that neither model-based classification nor discriminant analysis consistently outperforms the other; their relative performance is heavily dependent on the data (see Browne and McNicholas, 2012, for an interesting example).

20.3 THREE FAMILIES OF GAUSSIAN MIXTURE MODELS

20.3.1 The MCLUST Family

The MCLUST family (Banfield and Raftery, 1993; Celeux and Govaert, 1995; Fraley and Raftery, 2002) is currently the most prevalent family of mixture models within the literature. The members (Table 20.1) arise from constraints upon eigen-decomposed component covariance matrices. In the entirely unconstrained case, the VVV case, we have $\boldsymbol{\Sigma}_g = \lambda_g \mathbf{D}_g \mathbf{A}_g \mathbf{D}'_g$, where $\mathbf{D}_g$ is the matrix of eigenvectors, $\mathbf{A}_g$ is the diagonal matrix with entries proportional to the eigenvalues, and λ_g is the constant of proportionality. The MCLUST family is available through the `mclust` package (Fraley and Raftery, 2006) for the R software (R Development Core Team, 2012). Note that the MCLUST family is a subset of the family of Gaussian parsimonious clustering models (GPCM) introduced by Celeux and Govaert (1995) and available through the `mixture` package (Browne and McNicholas, 2013) for R.

MCLUST has been used for discriminant analysis and model-based classification. MCLUST discriminant analysis (MCLUST DA; Fraley and Raftery, 2002) is available within `mclust` and allows multiple components per known group by using the BIC to choose the number of components as

Table 20.1 Nomenclature, Covariance Structure, and Number of Free Covariance Parameters for each Member of the MCLUST Family

Model	Volume	Shape	Orientation	Σ_g	Free cov. parameters
EII	Equal	Spherical	—	$\lambda\mathbf{I}$	1
VII	Variable	Spherical	—	$\lambda_g\mathbf{I}$	G
EEI	Equal	Equal	Axis-Aligned	$\lambda\mathbf{A}$	p
VEI	Variable	Equal	Axis-Aligned	$\lambda_g\mathbf{A}$	$p+G-1$
EVI	Equal	Variable	Axis-Aligned	$\lambda\mathbf{A}_g$	$pG-G+1$
VVI	Variable	Variable	Axis-Aligned	$\lambda_g\mathbf{A}_g$	pG
EEE	Equal	Equal	Equal	$\lambda\mathbf{DAD}'$	$p(p+1)/2$
EEV	Equal	Equal	Variable	$\lambda\mathbf{D}_g\mathbf{AD}'_g$	$Gp(p+1)/2-(G-1)p$
VEV	Variable	Equal	Variable	$\lambda_g\mathbf{D}_g\mathbf{AD}'_g$	$Gp(p+1)/2-(G-1)(p-1)$
VVV	Variable	Variable	Variable	$\lambda_g\mathbf{D}_g\mathbf{A}_g\mathbf{D}'_g$	$Gp(p+1)/2$

well as the best model for each known cluster. Dean et al. (2006) used MCLUST for classification and discriminant analysis, also comparing the relative performance of both approaches. The non-diagonal members of the MCLUST family (EEE, EEV, VEV, and VVV) have $\mathcal{O}(p^2)$ covariance parameters, which can render the analysis of high-dimensional data impractical. Parameter estimation for each member of the MCLUST family is carried out using an EM algorithm, and starting values for these EM algorithms are generated within `mclust` via a Gaussian model-based agglomerative hierarchical clustering procedure (cf. Murtagh and Raftery, 1984; Banfield and Raftery, 1993).

20.3.2 Mixtures of Factor Analyzers and the PGMM Family

Factor analysis (Spearman, 1904; Bartlett, 1953) is a multivariate statistical technique that is popular in applications in several fields, including psychology and sociology. Given a p-dimensional random vector $\mathbf{X}$, the factor analysis model uses a q-dimensional vector of latent factors $\mathbf{U}$, where $q \ll p$, to model $\mathbf{X}$. The factor analysis model is $\mathbf{X} = \boldsymbol{\mu}+\boldsymbol{\Lambda}\mathbf{U}_i+\boldsymbol{\varepsilon}_i$, where $\boldsymbol{\Lambda}$ is a $p\times q$ matrix of factor loadings, the latent factors $\mathbf{U} \sim \mathcal{N}(\mathbf{0}, \mathbf{I}_q)$, and $\boldsymbol{\varepsilon}_i \sim \mathcal{N}(\mathbf{0}, \boldsymbol{\Psi})$, where $\boldsymbol{\Psi}$ is a $p \times p$ diagonal matrix. The resulting marginal distribution of $\mathbf{X}$ is $\mathcal{N}(\boldsymbol{\mu}, \boldsymbol{\Lambda\Lambda}' + \boldsymbol{\Psi})$. In an interesting and classic paper, Lawley and Maxwell (1962) discuss how factor analysis 'became the black sheep of statistical theory'. One clear problem with the factor analysis model is that the matrix $\boldsymbol{\Lambda}$ is not uniquely defined, because if $\boldsymbol{\Lambda}$ is replaced by $\boldsymbol{\Lambda}^* = \boldsymbol{\Lambda}\mathbf{D}$, where $\mathbf{D}$ is an orthonormal matrix, then $\boldsymbol{\Lambda\Lambda}' + \boldsymbol{\Psi} = (\boldsymbol{\Lambda}^*)(\boldsymbol{\Lambda}^*)' + \boldsymbol{\Psi}$.

The mixture of factor analyzers model has the Gaussian mixture model density (20.3) with component covariance structure $\boldsymbol{\Sigma}_g = \boldsymbol{\Lambda}_g\boldsymbol{\Lambda}'_g + \boldsymbol{\Psi}_g$. Ghahramani and Hinton (1997) actually used $\boldsymbol{\Sigma}_g = \boldsymbol{\Lambda}_g\boldsymbol{\Lambda}'_g + \boldsymbol{\Psi}$; Tipping and Bishop (1997, 1999) imposed the isotropic constraint $\boldsymbol{\Psi}_g = \psi_g\mathbf{I}_p$ to give $\boldsymbol{\Sigma}_g = \boldsymbol{\Lambda}_g\boldsymbol{\Lambda}'_g + \psi_g\mathbf{I}_p$, which is known as the mixture of probabilistic principal component analyzers model. McLachlan and Peel (2000a) used the most general form $\boldsymbol{\Sigma}_g = \boldsymbol{\Lambda}_g\boldsymbol{\Lambda}'_g + \boldsymbol{\Psi}_g$. These three mixture models were further developed by McNicholas and Murphy (2005, 2008, 2010b) to give a family of twelve parsimonious Gaussian mixture models (PGMMs, Table 20.2). They consider the constraint $\boldsymbol{\Lambda}_g = \boldsymbol{\Lambda}$, which together with the constraints $\boldsymbol{\Psi}_g = \psi_g\mathbf{I}_p$ and $\boldsymbol{\Psi}_g = \boldsymbol{\Psi}_g$ gives a family of eight models. Four extra models arise by setting $\boldsymbol{\Psi}_g = \omega_g\boldsymbol{\Delta}_g$, where $\omega_g \in \mathbb{R}^+$ and $\boldsymbol{\Delta}_g$ is a diagonal matrix such that $|\boldsymbol{\Delta}_g| = 1$, and imposing all valid combinations of the constraints $\omega_g = \omega$, $\boldsymbol{\Delta}_g = \boldsymbol{\Delta}$, and $\boldsymbol{\Delta}_g = \mathbf{I}_p$.

All 12 members of the PGMM family have $\mathcal{O}(p)$ covariance parameters, which is an advantage in the analysis of high-dimensional data. Another such advantage arises from the Woodbury identity (Woodbury, 1950), which

Table 20.2 The Covariance Structure and Number of Free Covariance Parameters for each Member of the PGMM Family

Covariance Structure	Number of Free Covariance Parameters
$\boldsymbol{\Sigma}_g = \boldsymbol{\Lambda}\boldsymbol{\Lambda}' + \omega\mathbf{I}_p$	$[pq - q(q-1)/2] + 1$
$\boldsymbol{\Sigma}_g = \boldsymbol{\Lambda}\boldsymbol{\Lambda}' + \omega_g\mathbf{I}_p$	$[pq - q(q-1)/2] + G$
$\boldsymbol{\Sigma}_g = \boldsymbol{\Lambda}_g\boldsymbol{\Lambda}_g' + \omega\mathbf{I}_p$	$G[pq - q(q-1)/2] + 1$
$\boldsymbol{\Sigma}_g = \boldsymbol{\Lambda}_g\boldsymbol{\Lambda}_g' + \omega_g\mathbf{I}_p$	$G[pq - q(q-1)/2] + G$
$\boldsymbol{\Sigma}_g = \boldsymbol{\Lambda}\boldsymbol{\Lambda}' + \omega\boldsymbol{\Delta}$	$[pq - q(q-1)/2] + p$
$\boldsymbol{\Sigma}_g = \boldsymbol{\Lambda}\boldsymbol{\Lambda}' + \omega_g\boldsymbol{\Delta}$	$[pq - q(q-1)/2] + [G + (p-1)]$
$\boldsymbol{\Sigma}_g = \boldsymbol{\Lambda}_g\boldsymbol{\Lambda}_g' + \omega\boldsymbol{\Delta}$	$G[pq - q(q-1)/2] + p$
$\boldsymbol{\Sigma}_g = \boldsymbol{\Lambda}_g\boldsymbol{\Lambda}_g' + \omega\boldsymbol{\Delta}_g$	$G[pq - q(q-1)/2] + [G + (p-1)]$
$\boldsymbol{\Sigma}_g = \boldsymbol{\Lambda}\boldsymbol{\Lambda}' + \omega\boldsymbol{\Delta}_g$	$[pq - q(q-1)/2] + [1 + G(p-1)]$
$\boldsymbol{\Sigma}_g = \boldsymbol{\Lambda}\boldsymbol{\Lambda}' + \omega_g\boldsymbol{\Delta}_g$	$[pq - q(q-1)/2] + Gp$
$\boldsymbol{\Sigma}_g = \boldsymbol{\Lambda}_g\boldsymbol{\Lambda}_g' + \omega\boldsymbol{\Delta}_g$	$G[pq - q(q-1)/2] + [1 + G(p-1)]$
$\boldsymbol{\Sigma}_g = \boldsymbol{\Lambda}_g\boldsymbol{\Lambda}_g' + \omega_g\boldsymbol{\Delta}_g$	$G[pq - q(q-1)/2] + Gp$

can be used to completely avoid the inversion of non-diagonal $p \times p$ matrices. In addition to model-based clustering, the PGMM family has been used for model-based classification (McNicholas, 2010) and discriminant analysis (Andrews and McNicholas, 2011b). The PGMM family is supported by the `pgmm` package (McNicholas et al., 2011) for R.

The reader should note that there has been other work on extending mixtures of factor analyzers. Yoshida et al. (2004) developed a mixed factors model and Montanari and Viroli (2010) proposed heteroscedastic factor mixture analysis. Baek et al. (2010) developed a variant of the mixture of factor analyzers model, known as the mixture of common factors model.

20.3.3 *Model-Based Clustering and Classification of Longitudinal Data*

Longitudinal data typically arise from prospective studies, but there are important cases where clustering or classification techniques need to be applied to longitudinal data. One such case arises with gene expression time course data, where clustering is necessary to assess which groups of genes behave in a similar fashion throughout the time course. To facilitate clustering of longitudinal data, McNicholas and Murphy (2010a) decompose the component covariance matrix $\boldsymbol{\Sigma}_g$ using the modified Cholesky decomposition, $\mathbf{T}_g\boldsymbol{\Sigma}_g\mathbf{T}_g' = \mathbf{D}_g$, where $\mathbf{T}_g$ is a unique lower unitriangular matrix and $\mathbf{D}_g$ is a unique diagonal matrix with strictly positive diagonal entries (Pourahmadi, 1999). This decomposition may also be expressed in the form $\boldsymbol{\Sigma}_g^{-1} = \mathbf{T}_g'\mathbf{D}_g^{-1}\mathbf{T}_g$, where it is very useful for model-based clustering and classification of longitudinal data. The values of $\mathbf{T}_g$ and $\mathbf{D}_g$ can be interpreted as generalized autoregressive parameters and innovation variances, respectively (cf. Pourahmadi, 1999).

McNicholas and Murphy (2010a) impose constraints on the component precision matrices $\boldsymbol{\Sigma}_g^{-1} = \mathbf{T}_g'\mathbf{D}_g^{-1}\mathbf{T}_g$ to give a family of eight mixture models each with $\mathcal{O}(p^2)$ covariance parameters (Table 20.3). These eight models arise from the option to have $\mathbf{T}_g$ and $\mathbf{D}_g$ equal across components as well as the isotropic constraint $\mathbf{D}_g = \delta_g\mathbf{I}_p$. For this family of models, the fact that $\mathcal{O}(p^2)$ covariance parameters exist is not usually a problem because longitudinal data, such as gene expression time course data, usually have relatively few time points.

When using finite mixtures, the question regarding the number of components is central. Depending on the model, we may also need to select the parametric structure as well as, perhaps, the number of latent variables.

Table 20.3 The Nomenclature, Covariance Structure, and Number of Covariance Parameters for each Model Used by McNicholas and Murphy (2010a)

Model	$\mathbf{T}_g$	$\mathbf{D}_g$	$\mathbf{D}_g$	Number of Covariance Parameters
EEA	Equal	Equal	Anisotropic	$p(p-1)/2+p$
VVA	Variable	Variable	Anisotropic	$G[p(p-1)/2]+Gp$
VEA	Variable	Equal	Anisotropic	$G[p(p-1)/2]+p$
EVA	Equal	Variable	Anisotropic	$p(p-1)/2+Gp$
VVI	Variable	Variable	Isotropic	$G[p(p-1)/2]+G$
VEI	Variable	Equal	Isotropic	$G[p(p-1)/2]+1$
EVI	Equal	Variable	Isotropic	$p(p-1)/2+G$
EEI	Equal	Equal	Isotropic	$p(p-1)/2+1$

The BIC is used to select the best member of this family of Gaussian mixture models. The BIC can be written as BIC $= 2l(x, \widehat{\theta}) - \rho \log N$, where $l(x, \widehat{\theta})$ is the maximized log-likelihood, $\widehat{\theta}$ is the maximum likelihood estimate of $\widehat{\theta}$, ρ is the number of free parameters in the model, and N is the number of observations. The use of the BIC for mixture model selection is justified by Keribin (1998, 2000), who shows that it gives consistent estimates of the number of components in a mixture model, under certain regularity conditions. Furthermore, Fraley and Raftery (1998, 2002) and McNicholas and Murphy (2008) give practical evidence that the BIC is effective as a model selection criterion for Gaussian mixture models.

20.4 NON-GAUSSIAN MIXTURE MODEL-BASED APPROACHES

20.4.1 Introduction

The three families of Gaussian mixture models reviewed in Section 20.3 are just a small selection of mixture models that have been used for clustering, classification, and discriminant analysis. In this section, non-Gaussian approaches to model-based clustering, classification, and discriminant analysis are discussed, starting with multivariate t-analogues of the families discussed in Section 20.3. Then we consider models that incorporate skewness, as well as models for rank data and data of mixed type.

20.4.2 Multivariate t-Distributions

Model-based clustering using mixtures of multivariate t-distributions has been around for some time (McLachlan and Peel, 1998; Peel and McLachlan, 2000). More recently, Zhao and Jiang (2006) developed a t-analogue of the probabilistic principal components analysis model and considered mixtures thereof. McLachlan et al. (2007) introduced a mixtures of t-factor analyzers model. Andrews and McNicholas (2011a,b) built on this work to introduce a t-analogue of the PGMM family for model-based clustering, classification, and discriminant analysis. Andrews et al. (2011) illustrated the performance of a four-member family of mixtures of multivariate t-distributions for model-based classification, and Steane et al. (2012) studied two specific mixtures of t-factors models for model-based classification. Baek and McLachlan (2011) considered a t-analogue of the mixture of common factor analyzers model.

Andrews and McNicholas (2012a) developed a t-analogue of the MCLUST family of models, including two additional covariance structures (i.e., they use 12 of the 14 models of Celeux and Govaert, 1995), for model-based clustering, classification, and discriminant analysis. This family of models is supported by the `teigen` package for R (Andrews and McNicholas, 2012b). McNicholas and Subedi (2012) develop a t-analogue of the approach of McNicholas and Murphy (2010a) to cluster and classify longitudinal data; they also considered modeling the component means. The `longclust` package (McNicholas et al., 2012) for R implements the approach of McNicholas and Subedi (2012) while also allowing the user to select a Gaussian mixture model.

20.4.3 Selected Other Non-Gaussian Approaches

In recent years, work has burgeoned on skew distributions, including that by Lin (2009, 2010) on skew-normal and skew-t distributions, respectively, and that by Karlis and Santourian (2009) on non-elliptically contoured distributions. Lee and McLachlan (2011) and Vrbik and McNicholas (2012) discuss model fitting issues around skew-t mixtures. Karlis and Meligkotsidou (2007) considered mixtures of multivariate Poisson distributions, along with a variety of applications. Handcock et al. (2007) discuss an interesting application of model-based clustering to social networks data, while Gormley and Murphy (2008) develop an approach for rank data. Browne et al. (2012) develop a mixture model wherein each mixture component is itself a mixture of a Gaussian and a uniform distribution. Browne and McNicholas (2012) introduce a method for clustering and classification of data of mixed type.

20.5 COVARIATES AND MISSING DATA

20.5.1 Covariates

Vermunt and Magidson (2002) generalized latent class models and finite mixture models by allowing the inclusion of covariates $\mathbf{z}$. In this model, covariates can affect the inclusion probability and/or conditional distribution of the observed data. The model density can be written

$$f(\mathbf{x}_i \mid \boldsymbol{\vartheta}, \mathbf{z}_i) = \sum_{g=1}^{G} \pi_{g|\mathbf{z}_i} f_g(\mathbf{x} \mid \boldsymbol{\theta}_g, \mathbf{z}_i).$$

Vermunt and Magidson (2002) suggest restricting $\pi_{g|\mathbf{z}_i}$ to a multinomial logit model, i.e., a logit model with 'linear effects' and no higher-order interactions. Vermunt and Magidson (2002) present a less general model because they assume the observed data are independent given the latent class and the covariates. To predict the class or group to which observation $\mathbf{x}_i$ belongs, we use the MAP classifications. In doing this, we are assigning an observation to the group with the highest conditional posterior group membership expectation or probability.

20.5.2 Missing Data

The model-based clustering framework is set up for balanced data with no missing elements. However, longitudinal or panel data commonly have missing data and/or data that are otherwise incomplete. In most cases, incomplete data can be dealt with within the EM algorithm framework (cf. McLachlan and Krishnan, 2008). Shaikh et al. (2010) present a method for clustering incomplete longitudinal data using a 'pseudo-EM' (PEM) algorithm. Their PEM algorithm only accounts for one of the two sufficient statistics for the missing data, which are taken to be Gaussian, during each E-step. In essence, Shaikh et al. (2010) are accounting for the mean of the missing data but not the variance. Accounting for the other sufficient statistic would lead to a *bona fide* EM algorithm. Their method, where just one sufficient statistic was accounted for, did, however, prove very effective for the gene expression time course they considered.

Other techniques, such as Buck's method (Buck, 1960) and extensions thereof (e.g., Beale and Little, 1975), can be used to account for incomplete data in this setting. An interesting review of methods for handling missing data is given by Little (1992). Little and Rubin (2002) describe several interesting approaches to handling missing data within the EM algorithm framework.

When carrying out model-based clustering, we are of course proceeding as if the group membership labels are missing within the multilevel setting. This type of missing data fits easily within the model-based clustering framework because prediction of the group memberships is typically the principal goal. Furthermore, model-based classification and discriminant analysis can be considered from the multilevel viewpoint where the group memberships are partially observed; i.e., some are observed and the rest are missing.

20.6 LONGITUDINAL DATA EXAMPLE

20.6.1 Introduction

A longitudinal data clustering example is considered. The `longclust` package in R is used to fit a mixture of multivariate t-distributions with a modified Cholesky decomposition for each component covariance structure and the option of a linear model for each component mean (cf. McNicholas and Subedi, 2012, for extensive details). The data, which are simulated using the `mvtnorm` package, are very similar to the data used in the help file for the `longclustEM()` function (cf. McNicholas et al., 2012).

20.6.2 The Data

The data were generated in R as follows.

```
library("mvtnorm")
m1 <- c(23, 34, 39, 45, 51, 56)

S1 <- matrix(c(1.00, -0.90, 0.18,
-0.13, 0.10, -0.05, -0.90, 1.31,
-0.26, 0.18, -0.15, 0.07, 0.18,
-0.26, 4.05, -2.84, 2.27, -1.13,
-0.13, 0.18, -2.84, 2.29, -1.83,
0.91, 0.10, -0.15, 2.27, -1.83,
3.46, -1.73, -0.05, 0.07, -1.13,
0.91, -1.73, 1.57), 6, 6)

m2 <- c(16, 18, 15, 17, 21, 17)

S2 <- matrix(c(1.00, 0.00,
-0.50, -0.20, -0.20, 0.19,
0.00, 2.00, 0.00, -1.20, -0.80,
-0.36,-0.50, 0.00, 1.25, 0.10,
-0.10, -0.39, -0.20, -1.20, 0.10,
2.76, 0.52, -1.22,-0.20, -0.80,
-0.10, 0.52, 1.40, 0.17, 0.19,
-0.36, -0.39, -1.22, 0.17, 3.17),
6, 6)

m3 <- c(8, 11, 16, 22, 25, 28)

S3 <- matrix(c(1.00, 0.00, 0.00,
0.00, 0.00, 0.00, 0.00, 1.00,
-0.20, -0.64, 0.26, 0.00, 0.00,
-0.20, 1.04, -0.17, -0.10, 0.00,
0.00, -0.64, -0.17, 1.50, -0.65,
0.00, 0.00, 0.26, -0.10, -0.65,
1.32, 0.00, 0.00, 0.00, 0.00,
0.00, 0.00, 1.00), 6, 6)

data <- matrix(0, 30, 6)
set.seed(12345)
data[1:10,] <- rmvnorm(10, m1,
S1)
data[11:20,] <- rmvnorm(10, m2,
S2)
data[21:30,] <- rmvnorm(10, m3,
S3)
```

20.6.3 Model-Based Clustering—Linear Means

First, we run model-based clustering with a linear model for each component mean, using the following R code.

```
library("longclust")
clus <- longclustEM(data, 2, 6,
        linearMeans=TRUE)
summary(clus)
plot(clus,data)
```

The group structure seems to have been well captured by the selected model, a $G = 3$ component mixture of multivariate t-distributions with the EVI covariance structure (cf. Table 20.3) and a linear model for the mean (Figure 20.1). The red and green

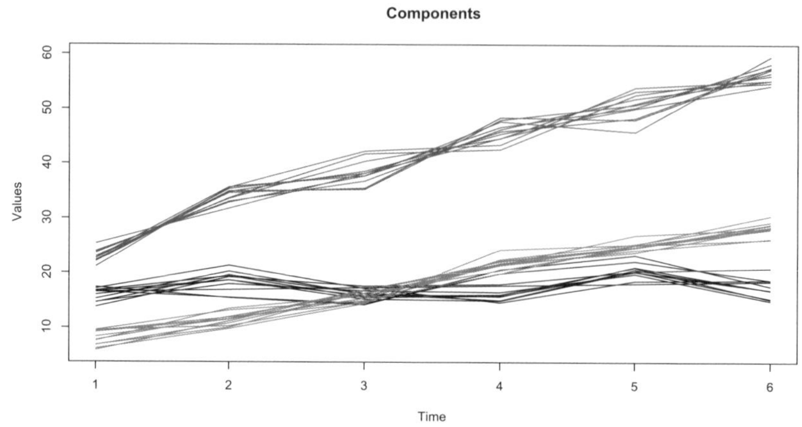

Figure 20.1 Fitted Model Using `longclustEM()` **with Linear Means on the Simulated Data from Section 20.6.2**

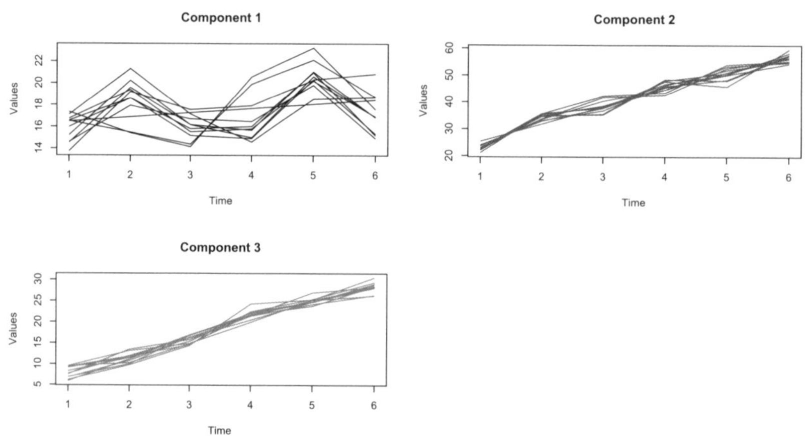

Figure 20.2 The Three Components, Shown Separately, for the Fitted Model Using `longclustEM()` **with Linear Means on the Simulated Data from Section 20.6.2**

components lend themselves rather well to a linear model for the mean, but this seems to be less so in the case of the black component (Figure 20.2).

20.6.4 Model-Based Clustering—Raw Means

Next, we repeat the analysis of Section 20.6.3 without using a linear model for each component mean. Therefore, Gp parameters are estimated for each component mean. We use the following R code.

```
clus2 <- longclustEM(data, 2, 6,
         linearMeans=FALSE)
summary(clus2)
plot(clus2,data)
```

This time, a $G = 4$ component mixture of multivariate t-distributions with the EVI covariance structure was selected. Comparing Figures 20.1 and 20.3, we can see that the component that was not well served by a

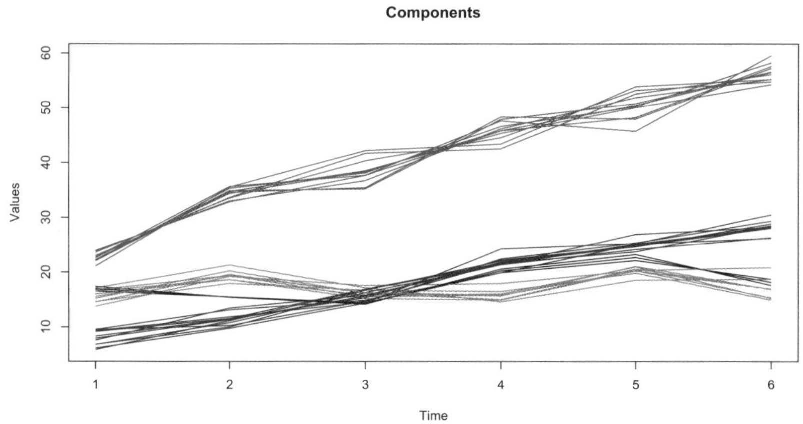

Figure 20.3 Fitted Model Using `longclustEM()` with Raw Means on the Simulated Data from Section 20.6.2

linear mean has been split in two when the linearity restriction is lifted. Looking closely at the two components (black and green in Figure 20.4), the shapes of the time courses do appear different. Notably, the BIC for the $G = 4$ component model with raw means is greater than the BIC for the $G = 3$ component model with linear means (-790.90 and -862.12, respectively). This lends some support for use of the former model ($G = 4$ components, raw means), even more so considering that it is less parsimonious than the latter.

Whether the $G = 3$ or $G = 4$ component model is preferable might depend on the particular application. If this were a gene expression time course analysis, for example, then the difference between the $G = 3$ and $G = 4$ component models would almost certainly be considered important. Looking at the results from the $G = 4$ component model (Figure 20.4), we can see that the two genes in the black component are initially down-regulated followed by up-regulation after time 3, whereas genes in the green component are initially up-regulated, followed by down-regulation after time 2 and up-regulation again after time 4. Given this distinction, one might wonder why the $G = 3$ component model was selected in Section 20.6.3. The answer is likely a combination of two factors: that the linear mean was not appropriate for the genes in question (component 1 in Figure 20.4)—a linear mean is not suited to the genes in component 3 (Figure 20.4) either—and that the BIC over-penalized the addition of a fourth component in the analysis in Section 20.6.3.

20.7 SUMMARY AND DISCUSSION

In this chapter, we have discussed the relationship between mixture models, latent class analysis, and multilevel models. Work on the application of families of mixture models for clustering, classification, and discriminant was reviewed, starting with a focus on three families of Gaussian mixture models. One of these families was specifically designed for longitudinal data and became the focus of our data analysis example, which also served as an illustration of the `longclust` package for R. The reader should note that the three families of Gaussian mixture models reviewed, by no means constitute an exhaustive list of families of Gaussian mixture models; a particularly noteworthy example of other work is the methodology developed by Bouveyron et al. (2007) for clustering high-dimensional data.

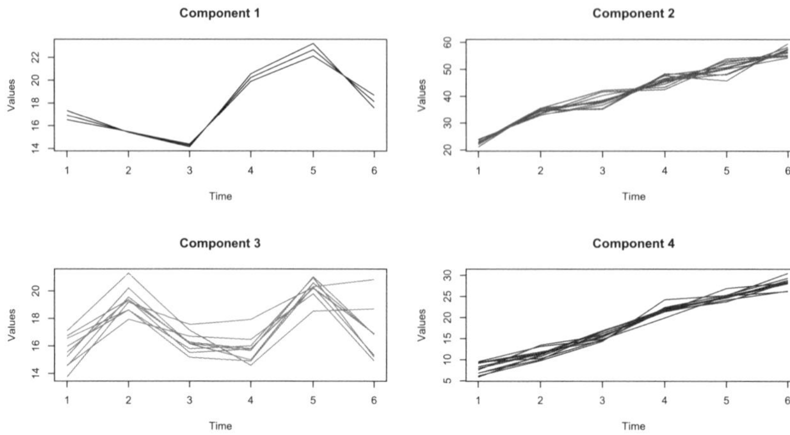

Figure 20.4 The Three Components, Shown Separately, for the Fitted Model Using `longclustEM()` **with Raw Means on the Simulated Data from Section 20.6.2**

Model-based approaches using the multivariate t-distribution were also discussed, along with a selection of non-Gaussian approaches. The incorporation of covariates and accounting for missing data were also considered. Almost the entire corpus of work on parameter estimation for these models was omitted from this chapter; work on variational approximations (cf. McGrory and Titterington, 2007) is perhaps of particular interest and there is a rich vein of work on specific problems related to the EM algorithm, including Greselin and Ingrassia (2010) and Ingrassia and Rocci (2011).

The literature on model-based learning, which as explained herein is generally of direct relevance to multilevel modeling problems, has grown rapidly in the past decade and there is no sign of that changing in the near future. Moreover, this growth is multi-faceted. Certainly, work on non-Gaussian approaches is ongoing and is sure to yield many developments over the coming years. Work on parameter estimation, both to improve and to replace the EM algorithm, continues to occupy a good portion of published work. There are, however, many roads to be investigated. As we approach what many consider the 50th anniversary of the birth of model-based clustering (Wolfe, 1963), all that is certain is that there is much more work ahead.

REFERENCES

Andrews, J.L. and McNicholas, P.D. (2011a) 'Extending Mixtures of Multivariate t-Factor Analyzers', *Statistics and Computing*, 21: 361–73.

Andrews, J.L. and McNicholas, P.D. (2011b) 'Mixtures of Modified t-Factor Analyzers for Model-Based Clustering, Classification, and Discriminant Analysis', *Journal of Statistical Planning and Inference*, 141: 1479–86.

Andrews, J.L. and McNicholas, P.D. (2012a) 'Model-Based Clustering, Classification, and Discriminant Analysis via Mixtures of Multivariate t-Distributions', *Statistics and Computing*, 22: 1021–1029.

Andrews, J.L. and McNicholas, P.D. (2012b) *teigen: Model-Based Clustering and Classification with the Multivariate t-Distribution*, R package version 1.0.

Andrews, J.L., McNicholas, P.D., and Subedi, S. (2011) 'Model-Based Classification via Mixtures of Multivariate t-Distributions', *Computational Statistics and Data Analysis*, 55: 520–9.

Baek, J. and McLachlan, G.J. (2011) 'Mixtures of Common t-Factor Analyzers for Clustering High-Dimensional Microarray Data', *Bioinformatics*, 27: 1269–76.

Baek, J., McLachlan, G.J., and Flack, L.K. (2010) 'Mixtures of Factor Analyzers with Common Factor Loadings: Applications to the Clustering and

Visualization of High-Dimensional Data', *IEEE Transactions on Pattern Analysis and Machine Intelligence*, 32: 1298–309.

Banfield, J.D. and Raftery, A.E. (1993) 'Model-Based Gaussian and non-Gaussian Clustering', *Biometrics*, 49: 803–21.

Bartlett, M.S. (1953) 'Factor Analysis in Psychology as a Statistician Sees It', in *Uppsala Symposium on Psychological Factor Analysis*, Nordisk Psykologi's Monograph Series, 3, 23–43.

Beale, E.M.L. and Little, R.J.A. (1975) 'Missing Values in Multivariate Analysis', *Journal of the Royal Statistical Society, Series B*, 37: 129–45.

Bouveyron, C., Girard, S., and Schmid, C. (2007) 'High-Dimensional Data Clustering', *Computational Statistics and Data Analysis*, 52: 502–19.

Browne, R.P. and McNicholas, P.D. (2012) 'Model-Based Clustering, Classification, and Discriminant Analysis of Data with Mixed Type', *Journal of Statistical Planning and Inference*, 142: 2976–2984.

Browne, R.P., McNicholas, P.D., and Sparling, M.D. (2012) 'Model-Based Learning Using a Mixture of Mixtures of Gaussian and Uniform Distributions', *IEEE Transactions on Pattern Analysis and Machine Intelligence*, 34: 814–17.

Buck, S.F. (1960) 'A Method of Estimation of Missing Values in Multivariate Data Suitable for Use with an Electronic Computer', *Journal of the Royal Statistical Society, Series B*, 22: 302–6.

Celeux, G. and Govaert, G. (1995) 'Gaussian Parsimonious Clustering Models', *Pattern Recognition*, 28: 781–93.

Dean, N., Murphy, T.B., and Downey, G. (2006) 'Using Unlabelled Data to Update Classification Rules with Applications in Food Authenticity Studies', *Journal of the Royal Statistical Society, Series C*, 55: 1–14.

Dempster, A.P., Laird, N.M., and Rubin, D.B. (1977) 'Maximum Likelihood from Incomplete Data via the EM Algorithm', *Journal of the Royal Statistical Society, Series B*, 39: 1–38.

Fraley, C. and Raftery, A.E. (1998) 'How Many Clusters? Which Clustering Methods? Answers via Model-Based Cluster Analysis', *The Computer Journal*, 41: 578–88.

Fraley, C. and Raftery, A.E. (2002) 'Model-Based Clustering, Discriminant Analysis, and Density Estimation', *Journal of the American Statistical Association*, 97: 611–31.

Fraley, C. and Raftery, A.E. (2006) 'MCLUST Version 3 for R: Normal Mixture Modeling and Model-Based Clustering', Technical Report 504, Department of Statistics, University of Washington, September 2006. Minor revisions January 2007 and November 2007.

Frühwirth-Schnatter, S. (2006) *Finite Mixture and Markov Switching Models*. New York: Springer-Verlag.

Ghahramani, Z. and Hinton, G.E. (1997) 'The EM Algorithm for Factor Analyzers', Technical Report CRG-TR-96-1, University of Toronto.

Gormley, I.C. and Murphy, T.B. (2008) 'Exploring Voting Blocs within the Irish Electorate: A Mixture Modeling Approach', *Journal of the American Statistical Association*, 103: 1014–27.

Greselin, F. and Ingrassia, S. (2010) 'Constrained Monotone EM Algorithms for Mixtures of Multivariate t-Distributions', *Statistics and Computing*, 20: 9–22.

Handcock, M.S., Raftery, A.E., and Tantrum., J.M. (2007) 'Model-Based Clustering for Social Networks', *Journal of the Royal Statistical Society, Series A*, 170: 301–54.

Hastie, T. and Tibshirani, R. (1996) 'Discriminant Analysis by Gaussian Mixtures', *Journal of the Royal Statistical Society, Series B*, 58: 155–76.

Ingrassia, S. and Rocci, R. (2011) 'Degeneracy of the EM Algorithm for the MLE of Multivariate Gaussian Mixtures and Dynamic Constraints', *Computational Statistics and Data Analysis*, 55: 1715–25.

Karlis, D. and Meligkotsidou, L. (2007) 'Finite Mixtures of Multivariate Poisson Distributions with Application', *Journal of Statistical Planning and Inference*, 137: 1942–60.

Karlis, D. and Santourian, A. (2009) 'Model-Based Clustering with non-Elliptically Contoured Distributions', *Statistics and Computing*, 19: 73–83.

Keribin, C. (1998) 'Estimation Consistante de l'Ordre de Modèles de Mélange', *Comptes Rendus de l'Académie des Sciences. Série I. Mathématique*, 326: 243–8.

Keribin, C. (2000) 'Consistent Estimation of the Order of Mixture Models', *Sankhyā. The Indian Journal of Statistics, Series A*, 62: 49–66.

Lawley, D.N. and Maxwell, A.E. (1962) 'Factor Analysis as a Statistical Method', *Journal of the Royal Statistical Society, Series D*, 12: 209–29.

Lee, S. and McLachlan, G.J. (2011) 'On the Fitting of Mixtures of Multivariate Skew t-Distributions via the EM Algorithm', arXiv:1109.4706.

Lin, T-I. (2009) 'Maximum Likelihood Estimation for Multivariate Skew Normal Mixture Models', *Journal of Multivariate Analysis*, 100: 257–65.

Lin, T.-I. (2010) 'Robust Mixture Modeling using Multivariate Skew t Distributions', *Statistics and Computing*, 20: 343–56.

Little, R.J.A. (1992) 'Regression with Missing X's: A Review', *Journal of the American Statistical Association*, 87: 1227–37.

Little, R.J.A. and Rubin, D.B. (2002) *Statistical Analysis With Missing Data*, 2nd edition. New Jersey: John Wiley & Sons.

McGrory, C.A. and Titterington, D.M. (2007) 'Variational Approximations in Bayesian Model Selection for Finite Mixture Distributions', *Computational Statistics and Data Analysis*, 51: 5352–67.

McLachlan, G.J. (1992) *Discriminant Analysis and Statistical Pattern Recognition*. New Jersey: John Wiley & Sons.

McLachlan, G.J. and Basford, K.E. (1988) *Mixture Models: Inference and Applications to Clustering*. New York: Marcel Dekker Inc.

McLachlan, G.J. and Krishnan, T. (2008) *The EM Algorithm and Extensions*, 2nd edition. New York: Wiley.

McLachlan, G.J. and Peel, D. (1998) 'Robust Cluster Analysis via Mixtures of Multivariate t-Distributions', in *Lecture Notes in Computer Science*, 1451: 658–66. Berlin: Springer-Verlag.

McLachlan, G.J. and Peel, D. (2000a) 'Mixtures of Factor Analyzers', in *Proceedings of the Seventh International Conference on Machine Learning*, San Francisco: Morgan Kaufmann, 599–606.

McLachlan, G.J. and Peel, D. (2000b) *Finite Mixture Models*. New York: John Wiley & Sons.

McLachlan, G.J., Bean, R.W., and Jones, L.B.-T. (2007) 'Extension of the Mixture of Factor Analyzers Model to Incorporate the Multivariate t-Distribution', *Computational Statistics and Data Analysis*, 51: 5327–38.

McNicholas, P.D. (2010) 'Model-Based Classification using Latent Gaussian Mixture Models', *Journal of Statistical Planning and Inference*, 140: 1175–81.

McNicholas, P.D. and Murphy, T.B. (2005) 'Parsimonious Gaussian Mixture Models', Technical Report 05/11, Department of Statistics, Trinity College Dublin, Ireland.

McNicholas, P.D. and Murphy, T.B. (2008) 'Parsimonious Gaussian Mixture Models', *Statistics and Computing*, 18: 285–96.

McNicholas, P.D. and Murphy, T.B. (2010a) 'Model-Based Clustering of Longitudinal Data', *The Canadian Journal of Statistics*, 38: 153–68.

McNicholas, P.D. and Murphy, T.B. (2010b) 'Model-Based Clustering of Microarray Expression Data via Latent Gaussian Mixture Models', *Bioinformatics*, 26: 2705–12.

McNicholas, P.D. and Subedi, S. (2012) 'Clustering Gene Expression Time Course Data Using Mixtures of Multivariate t-Distributions', *Journal of Statistical Planning and Inference*, 142: 1114–27.

McNicholas, P.D., Jampani, K.R., and Subedi, S. (2012) *longclust: Model-Based Clustering and Classification for Longitudinal Data*, R package version 1.1.

McNicholas, P.D., Jampani, K.R., McDaid, A.F., Murphy, T.B., and Banks, L. (2011) *pgmm: Parsimonious Gaussian Mixture Models*, R package version 1.0.

Montanari, A. and Viroli, C. (2010) 'Heteroscedastic Factor Mixture Analysis', *Statistical Modeling*, 10: 441–60.

Murtagh, F. and Raftery, A.E. (1984) 'Fitting Straight Lines to Point Patterns', *Pattern Recognition*, 17: 479–83.

Peel, D. and McLachlan, G.J. (2000) 'Robust Mixture Modeling Using the t Distribution', *Statistics and Computing*, 10: 339–48.

Pourahmadi, M. (1999) 'Joint Mean-Covariance Models with Applications to Longitudinal Data: Unconstrained Parameterisation', *Biometrika*, 86: 677–90.

R Development Core Team (2012) *R: A Language and Environment for Statistical Computing*, R Foundation for Statistical Computing, Vienna, Austria. URL `http://www.R-project.org`.

Schwarz, G. (1978) 'Estimating the Dimension of a Model', *The Annals of Statistics*, 6: 461–4.

Shaikh, M., McNicholas, P.D., and Desmond, A.F. (2010) 'A Pseudo-EM Algorithm for Clustering Incomplete Longitudinal Data', *The International Journal of Biostatistics*, 6: 8.

Spearman, C. (1904) 'The Proof and Measurement of Association between Two Things', *American Journal of Psychology*, 15: 72–101.

Steane, M.A., McNicholas, P.D., and Yada, R.Y. (2012) 'Model-Based Classification via Mixtures of Multivariate t-Factor Analyzers', *Communications in Statistics – Simulation and Computation*, 41: 510–23.

Tipping, T.E. and Bishop, C.M. (1997) 'Mixtures of Probabilistic Principal Component Analysers', Technical Report NCRG/97/003, Aston University (Neural Computing Research Group), Birmingham, UK.

Tipping, T.E. and Bishop, C.M. (1999) 'Mixtures of Probabilistic Principal Component Analysers', *Neural Computation*, 11: 443–82.

Titterington, D.M., Smith, A.F.M., and Makov, U.E. (1985) *Statistical Analysis of Finite Mixture Distributions*. Chichester: John Wiley & Sons.

Vermunt, J.K. and Magidson, J. (2002) 'Latent Class Cluster Analysis', in J. Hagenaars and A. McCutcheon (eds), *Applied Latent Class Analysis*, Cambridge: Cambridge University Press, 89–106.

Vrbik, I. and McNicholas, P.D. (2012) 'Analytic Calculations for the EM Algorithm for Multivariate Skew-t Mixture Models', *Statistics and Probability Letters*, 82: 1169–74.

Wolfe, J.H. (1963) 'Object Cluster Analysis of Social Areas', Master's Thesis, University of California, Berkeley.

Woodbury, M.A. (1950) *Inverting Modified Matrices*, Statistical Research Group, Memorandum Report 42. Princeton University, NJ.

Yoshida, R., Higuchi, T., and Imoto, S. (2004) 'A Mixed Factors Model for Dimension Reduction and Extraction of a Group Structure in Gene Expression Data', in *Proceedings of IEEE 3rd Computational Systems Bioinformatics (CSB2004)*, 161–72.

Zhao, J. and Jiang, Q. (2006) 'Probabilistic PCA for t Distributions', *Neurocomputing*, 69: 2217–26.

21

Multivariate Response Data

Helena Geys
Janssen Pharmaceutica, Johnson and Johnson, Belgium

Christel Faes
Universiteit Hasselt, Belgium

21.1 INTRODUCTION

The term *multivariate* refers to the setting where different measurements are taken from the same individual, e.g. the weight and height of a person. *Hierarchical* or *multilevel* refers to the setting where the dependent variables correspond to the same measure, e.g. in a clustered setting where measurements are taken from groups of individuals. Methods for multilevel or multivariate data are widely available. The combination of these two, e.g. involving multiple measurements taken repeatedly on an individual, is more challenging. For continuous data, the normal distribution with its elegant properties plays a prominent role. However, when some or all of the outcome variables are non-normal, techniques are less standard, because of the lack of an analogue to the multivariate normal distribution. In this chapter we will first introduce methods for multivariate outcomes, then extend these to the hierarchical setting, and finally describe the application of the proposed methodologies to two different datasets.

21.2 MULTIVARIATE MODELS

Whereas methods for multivariate normal models are well established, joint models for non-normal outcomes are less well known and often cumbersome. In practice, there are situations where one is interested in the joint modeling of continuous, categorical, time-to-event, longitudinal, ... endpoints and combinations thereof.

Random effects models are probably the most frequently used models to analyze multivariate data, in particular when they are of a longitudinal or otherwise hierarchical nature. The generalized linear mixed model (GLMM) is the most frequently used random-effects model for non-normal outcomes. For a multivariate response vector $\boldsymbol{y}_i = (y_{1i}, y_{2i}, \ldots, y_{mi})'$ with m different endpoints measured for each unit i, conditionally on random effects $\boldsymbol{u}_i$, the elements y_{ij} follow a distribution that belongs to the exponential family:

$$\begin{aligned} f_i(y_{ij}) &\equiv f_i(y_{ij}|\theta_{ij}, \phi) \\ &= \exp\{\phi^{-1}[y_{ij}\theta_{ij} - \psi(\theta_{ij})] \\ &\quad + c(y_{ij}, \phi)\}, \end{aligned}$$

where the mean is modeled through a linear predictor containing fixed regression parameters $\boldsymbol{\beta}$ as well as subject-specific parameters $\boldsymbol{u}$. Such random-effects models can be fitted by maximization of the marginal likelihood, obtained by integrating out the random effects. As stated by Molenberghs and Verbeke (2005), no analytic expressions are available for the integrals in general and approximations are needed. One approach that will be adopted here is to decompose the data into the mean $\boldsymbol{\mu}_i$ and an appropriate error term $\boldsymbol{\varepsilon}_i$.

The mean $\boldsymbol{\mu}_i$ is specified in terms of fixed and random effects:

$$\boldsymbol{\mu}_i = \boldsymbol{\mu}_i(\boldsymbol{\eta}_i) = \boldsymbol{h}(\boldsymbol{X}_i\boldsymbol{\beta} + \boldsymbol{Z}_i\boldsymbol{u}_i),$$

where $\boldsymbol{X}_i$ and $\boldsymbol{Z}_i$ are $(m \times p)$- and $(m \times q)$-dimensional matrices of known covariate values, and $\boldsymbol{\beta}$ a p-dimensional vector of unknown fixed regression coefficients. Further, $\boldsymbol{u}_i \sim N(\boldsymbol{0}, \boldsymbol{D})$ are the q-dimensional random effects. The error terms have the appropriate distribution, depending on the type of the endpoint, with variance equal to $\mathrm{Var}(y_{ij}|\boldsymbol{u}_i) = \phi v(\mu_{ij})$, and $v(\cdot)$ is the variance function corresponding with the exponential family distribution. For continuous endpoints, the decomposition is straightforward, with the inverse link function $\boldsymbol{h}(\cdot)$ and variance function $v(\cdot)$ corresponding to the identity function and overdispersion parameter ρ equal to the variance σ^2. For binary endpoints with a logistic natural link function, the inverse link function $\boldsymbol{h}(\cdot)$ is given by the expit function, and the error terms ε_{ij} equal $1 - \pi_{ij}$ with probability π_{ij} and $-\pi_{ij}$ with probability $1 - \pi_{ij}$. It is also straightforward to use this mixed model representation in situations where various outcomes of a different nature are observed (e.g. Fitzmaurice et al., 2009) by allowing the components of the inverse link function $\boldsymbol{h}(\cdot)$ and the components of the error term $\boldsymbol{\varepsilon}_i$ to change with the nature of the various outcomes in $\boldsymbol{y}_i$.

Based on this general formulation, one can derive approximate expressions for the variance-covariance matrix of $\boldsymbol{y}_i$ (Molenberghs and Verbeke, 2005), which are useful for investigating the association between the endpoints. The variance-covariance matrix is given by

$$V_i = \mathrm{Var}(\boldsymbol{y}_i) \simeq \Delta_i Z_i D Z_i' \Delta_i' + \Sigma_i, \quad (21.1)$$

with

$$\Delta_i = \left(\frac{\partial \boldsymbol{\mu}_i}{\partial \boldsymbol{\eta}_i}\right)\Big|_{\boldsymbol{u}_i=\boldsymbol{0}}$$

and

$$\Sigma_i \simeq \Xi_i^{1/2} A_i^{1/2} R_i(\alpha) A_i^{1/2} \Xi_i^{1/2},$$

where A_i is a diagonal matrix containing the variances following from the generalized linear model specification of y_{ij} given the random effects $\boldsymbol{u}_i = \boldsymbol{0}$, i.e., with diagonal elements $v(\mu_{ij}|\boldsymbol{u}_i = \boldsymbol{0})$. Likewise, Ξ_i is a diagonal matrix with the overdispersion parameters along the diagonal. Finally, R_i is a correlation matrix.

From expression (21.1) it is clear that the correlation between the multivariate outcomes can be induced by either assuming the errors to be correlated or by specification of a random effects structure. In a standard GLMM one oftentimes assumes no correlations in $R_i(\alpha)$, leading to a conditional independence model or purely random-effects model. When the model does not contain random effects, but $\boldsymbol{\varepsilon}_i$ is assumed to be correlated, a marginal model is obtained, which we refer to as a marginal generalized linear model (MGLM).

As an illustration, these multivariate models will be explained for the special case of two endpoints, one of which is continuous and the other binary.

21.2.1 Marginal GLMMs

For the special case of a continuous and binary endpoint, a model with no random effects (MGLM) is of the form

$$\begin{pmatrix} y_{1i} \\ y_{2i} \end{pmatrix} = \begin{pmatrix} \alpha_0 + \alpha_1 X_i \\ \frac{\exp(\beta_0+\beta_1 X_i)}{1+\exp(\beta_0+\beta_1 X_i)} \end{pmatrix} + \begin{pmatrix} \varepsilon_{1i} \\ \varepsilon_{2i} \end{pmatrix},$$

in which the first component of the inverse link function $\boldsymbol{h}$ equals the identity link, and the second component corresponds with the logit link. The correlation among the endpoints y_{1i} and y_{2i}, denoted by ρ_{12}, is specified via the error variance

$$\text{Var}\begin{pmatrix} \varepsilon_{1i} \\ \varepsilon_{2i} \end{pmatrix} = \begin{pmatrix} \sigma^2 & \rho\sigma\sqrt{v_{2i}} \\ \rho\sigma\sqrt{v_{2i}} & v_{2i} \end{pmatrix},$$

where $v_{2i} = \pi_{2i}[1-\pi_{2i}]$ and where ρ denotes the correlation between ε_{1i} and ε_{2i}. Since the variance of $\boldsymbol{y}_i$ in the case of no random effects reduces to

$$V_i = \text{Var}(\boldsymbol{y}_i) \simeq \Sigma_i,$$

ρ is exactly the correlation between the two endpoints y_{1i} and y_{2i}, i.e. $\rho_{12} = \rho$.

21.2.2 Shared Random Effects

Another approach is to use a shared random parameter model. More specifically, for a mixed continuous/binary setting, the general linear mixed model can be written in the form

$$\begin{pmatrix} y_{1i} \\ y_{2i} \end{pmatrix} = \begin{pmatrix} \alpha_0 + \alpha_1 X_i + \lambda u_i \\ \frac{\exp(\beta_0+\beta_1 X_i+u_i)}{1+\exp(\beta_0+\beta_1 X_i+u_i)} \end{pmatrix} + \begin{pmatrix} \varepsilon_{1i} \\ \varepsilon_{2i} \end{pmatrix},$$

where u_i denotes a shared random effect of the two outcomes. The errors are assumed to be uncorrelated, and hence do not lead to additional parameters. Note that a scale parameter λ in the continuous component of the model is included, given that the continuous and binary outcomes are measured on different scales. For this model, the variance of $\boldsymbol{y}_i$ can be derived from equation (21.1), in which

$$Z_i = \begin{pmatrix} \lambda \\ 1 \end{pmatrix}, \Delta_i = A_i = \begin{pmatrix} 1 & 0 \\ 0 & v_{i2} \end{pmatrix},$$

and

$$\Xi_i = \begin{pmatrix} \sigma^2 & 0 \\ 0 & 1 \end{pmatrix}.$$

This implies that the variance of $\boldsymbol{y}_i$ is approximately equal to

$$\begin{aligned} V_i &= \begin{pmatrix} \lambda^2 & v_{i2}\lambda \\ v_{i2}\lambda & v_{i2}^2 \end{pmatrix}\tau^2 + \begin{pmatrix} \sigma^2 & 0 \\ 0 & v_{i2} \end{pmatrix} \\ &= \begin{pmatrix} \lambda^2\tau^2 + \sigma^2 & v_{i2}\lambda\tau^2 \\ v_{i2}\lambda\tau^2 & v_{i2}^2\tau^2 + v_{i2} \end{pmatrix}. \end{aligned}$$

As a result, the approximate correlation among the two outcomes y_{1i} and y_{2i} is equal to

$$\rho_{12} = \frac{v_{i2}\lambda\tau^2}{\sqrt{\lambda^2\tau^2 + \sigma^2}\sqrt{v_{i2}^2\tau^2 + v_{i2}}}. \quad (21.2)$$

This correlation depends on the fixed effects through v_{i2}.

21.3 HIERARCHICAL MULTIVARIATE MODELS

In many practical applications, joint outcomes are observed within a hierarchical context. For example, in a longitudinal study, bivariate (or multivariate) endpoints are often collected repeatedly over time, introducing another form of hierarchy on top of the bivariate responses. Another specific hierarchy stems from clustered data where, e.g., a continuous and a categorical endpoint are measured on each member in a cluster.

Extension of the multivariate GLMM-based models towards inclusion of repeated measurement, meta-analysis, clustered data, correlated data, etc. can be performed by appropriate formulation of the random effects and error structure in this model. In this section, three possible models are described which take into account the multilevel structure of bivariate endpoints. We will formulate methods in terms of a continuous and binary endpoint, but other extensions of the model are possible as well. In the application, the settings of two continuous endpoints, two binary endpoints or mixed binary and continuous endpoints are described.

21.3.1 Marginal GLMMs

A first method is to incorporate both the association between the two endpoints at each time point as well as the association due to the longitudinal structure of the data in the error structure, by the specification of a general (i.e. unstructured) correlation matrix $R_i(\alpha)$, allowing each pair of outcomes to have its own correlation coefficient. Correlations follow in a straightforward fashion when such a purely marginal GLMM is used. While such a marginal model, with fully unstructured $2n \times 2n$ variance-covariance matrix, is very appealing because of ease of interpretation, it can become computationally very demanding, especially when the number of measurements per subject (n) gets larger. Other specifications of the correlation structure of interest could be a product of an unstructured (2×2) correlation matrix and a structured ($n \times n$) correlation matrix (e.g. AR(1), CS), where the first correlation matrix corresponds to the multivariate structure in the data and the second corresponds to the multilevel structure. However, these covariance structures are not readily available in current software packages.

21.3.2 Independent Random Effects

A second approach is to use a random-intercepts model for each endpoint with a correlated error structure to account for the association among the endpoints. In this model, a random effect is introduced at the level of the linear predictor after application of the link function to account for the longitudinal structure in the data, while the residual correlation is introduced at the level of ε_i to accommodate the association among the binary and continuous measurements. The model is also of the form (21.1), but where u_{1i} and u_{2i} are independent zero-mean normally distributed random effects with variance τ_1^2 and τ_2^2 respectively. The error is assumed to have variance-covariance matrix

$$\begin{pmatrix} \sigma^2 & \rho\sigma\sqrt{\upsilon_{2ij}} \\ \rho\sigma\sqrt{\upsilon_{2ij}} & \upsilon_{2ij} \end{pmatrix}.$$

Here, ρ denotes the correlation between ε_1 and ε_2. The variance among the continuous and binary endpoints at time point j, for subject i, can be derived from the variance-covariance matrix (21.1), and is approximately equal to

$$\begin{aligned} V_{ij} &= \begin{pmatrix} \tau_1^2 & 0 \\ 0 & \tau_2^2\upsilon_{2ij}^2 \end{pmatrix} \\ &\quad + \begin{pmatrix} \sigma^2 & \rho\sigma\sqrt{\upsilon_{2ij}} \\ \rho\sigma\sqrt{\upsilon_{i2}} & \upsilon_{2ij} \end{pmatrix} \\ &= \begin{pmatrix} \tau_1^2+\sigma^2 & \rho\sigma\sqrt{\upsilon_{2ij}} \\ \rho\sigma\sqrt{\upsilon_{2ij}} & \upsilon_{2ij}^2\tau_2^2+\upsilon_{2ij} \end{pmatrix}. \end{aligned}$$

As a result, the approximate correlation among the two endpoints is equal to

$$\rho_{12} = \frac{\rho\sigma\sqrt{\upsilon_{2ij}}}{\sqrt{\tau_1^2+\sigma^2}\sqrt{\upsilon_{2ij}^2\tau_2^2+\upsilon_{2ij}}}.$$

In the special case of no random effects, the model is a MGLM as introduced in the previous section, and the correlation reduces to ρ. Further, ρ_{12} equals zero in the case of conditional independence.

21.3.3 Correlated Random Effects

A third approach is to specify a correlated random-intercepts model, with a general variance-covariance matrix D for the random effects. This model can be written in the following form:

$$\begin{pmatrix} y_{1ij} \\ y_{2ij} \end{pmatrix} = \begin{pmatrix} \alpha_0+\alpha_1 X_{ij}+u_{1i} \\ \frac{\exp(\beta_0+\beta_1 X_{ij}+u_{2i})}{1+\exp(\beta_0+\beta_1 X_{ij}+u_{2i})} \end{pmatrix} + \begin{pmatrix} \varepsilon_{1ij} \\ \varepsilon_{2ij} \end{pmatrix}, \qquad (21.3)$$

where the random effect u_{1i} and u_{2i} are normally distributed as

$$\begin{pmatrix} u_{1i} \\ u_{2i} \end{pmatrix} \sim N\left(\begin{pmatrix} 0 \\ 0 \end{pmatrix}, \begin{pmatrix} \tau_1^2 & \rho\tau_1\tau_2 \\ \rho\tau_1\tau_2 & \tau_2^2 \end{pmatrix}\right)$$

and where ε_{1ij} and ε_{2ij} are independent. The random effects u_{1i} and u_{2i} are used to accommodate the longitudinal structure in the data. The correlation ρ_{12} among the continuous and binary endpoints is induced by the incorporation of a correlation ρ among the two random effects. The variance of y_{1ij} and y_{2ij} can be derived from equation (21.1), in which

$$Z_i = \begin{pmatrix} 1 & 0 \\ 0 & 1 \end{pmatrix}, \Delta_i = A_i = \begin{pmatrix} 1 & 0 \\ 0 & v_{i2} \end{pmatrix},$$

$$D = \begin{pmatrix} \tau_1^2 & \rho\tau_1\tau_2 \\ \rho\tau_1\tau_2 & \tau_2^2 \end{pmatrix}, \Xi_i = \begin{pmatrix} \sigma^2 & 0 \\ 0 & 1 \end{pmatrix},$$

and where $v_{2ij} = \pi_{2ij}(\boldsymbol{u}_i = \mathbf{0})[1 - \pi_{2ij}(\boldsymbol{u}_i = \mathbf{0})]$. As a result, the approximate variance-covariance matrix of the two measurements for subject i at time point j is equal to:

$$\begin{aligned} V_{ij} &= \begin{pmatrix} \tau_1^2 & \rho\tau_1\tau_2 v_{2ij} \\ \rho\tau_1\tau_2 v_{2ij} & v_{2ij}^2\tau_2^2 \end{pmatrix} \\ &\quad + \begin{pmatrix} \sigma^2 & 0 \\ 0 & v_{2ij} \end{pmatrix} \\ &= \begin{pmatrix} \tau_1^2 + \sigma^2 & \rho\tau_1\tau_2 v_{2ij} \\ \rho\tau_1\tau_2 v_{2ij} & v_{2ij}^2\tau_2^2 + v_{2ij} \end{pmatrix}. \end{aligned}$$

Thus, the correlation ρ_{12} among the binary and continuous outcome follows from the fixed effects and variance components, and is approximately equal to

$$\rho_{12} = \frac{\rho\tau_1\tau_2 v_{2ij}}{\sqrt{\tau_1^2 + \sigma^2}\sqrt{v_{2ij}^2\tau_2^2 + v_{2ij}}}.$$

In the case of conditional independence ($\rho \equiv 0$), the approximate marginal correlation function ρ_{12} also equals zero. In case $\rho \equiv 1$, this model reduces to the shared parameter model as presented before. The correlation between the two endpoints y_{1ij} and y_{2ij} is as given by expression (21.2) in which the scale factor λ is equal to τ_1/τ_2.

21.4 APPLICATION

21.4.1 Irwin Study

Data description

Irwin's toxicity test is designed for primary screening of the safety of psychotropic compounds. The goal of the test is to study possible effects on the general behavior and activity of a rat. The data considered here come from a single-dose oral safety pharmacology study on rats. In this study, a series of non-invasive observational and interactive measurements are evaluated to study the behavioral and neurofunctional integrity of the rat. Male rats are exposed to a single dose of 0, 2.5, 10, or 40 mg/kg body weight. Autonomic, neurologic, and behavioral observations were recorded 24h before the exposure, and also 0.5h, 2h, 4h, and 7 days after the exposure. There were five rats in each dose group.

For this chapter, three outcomes are considered: wire maneuver, body temperature, and pupil size. The first outcome is a neurologic endpoint. If the rat is able to actively grip a wire with hind legs then the outcome equals 0; if, however, the rat has difficulties or is unable to grasp the wire, the outcome is 1. The second and third outcomes are autonomic observations measured on a continuous scale. Figure 21.1 (left panel) shows the rat-specific evolutions of body temperature per dose group. We can clearly see that there is a decrease in body temperature at the higher dose levels between 2 and 4 hours after dosage. After 7 days the temperature has come to its normal level again. Figure 21.1 (right panel) summarizes the proportion of unsuccessful wire maneuvers at each time point per dose level, aggregated over the rats. Failures are most common for animals in the highest dose group, especially at 2 and 4 hours after the exposure.

Analysis per time point

First, we investigate the effects of the test substance at 0.5h, 2h, and 4h after dose administration. For each time point, a marginal generalized linear mixed model is used. Parameter estimates (standard errors) are displayed in Table 21.1. The treatment seems to result in a significant decrease of body temperature at 2h and 4h after dosing. Also, increasing dosage leads to an increase in difficulties in performing the wire maneuver, with a significant effect at 4h

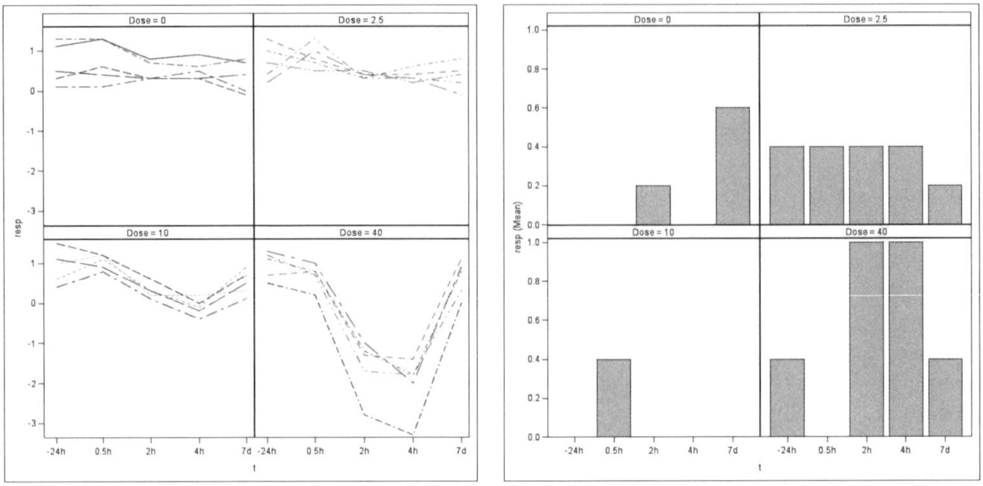

(a) Body Temperature

(b) Wire Maneuver

Figure 21.1 Time Trend Per Dose Level

after dosing. There is no significant effect on the response to temperature or wire maneuver at 0.5h post-dosing. The correlation between the two endpoints is negative. At 2h after dosing, the correlation, estimated as $\widehat{\rho} = -0.511$, is strongest. At none of the other time points is the correlation significant.

A shared random effects model (GLMM) was also fitted to the data. However, in this case, the instability of the model due to the non-linearity induced by the inflation factor resulted in non-convergence. The inflation factor often appears to diverge. Careful selection of initial values is important, as is monitoring of the convergence process and fine tuning of the optimization method. This shows the difficulty in the practical use of this model.

Joint longitudinal model

We now focus on the full hierarchical model, accounting for all time points. It is of interest to test whether there is a significant treatment effect on body temperature or wire maneuver at any time point. While a separate analysis at each time point is useful, it is not an efficient method and possible effects present in the data might be missed. For the dose toxicity study, a longitudinal analysis accounting for possible association among time points is considered.

We consider a model without random intercepts, and correlated and uncorrelated random-intercepts models. Table 21.2 shows the estimates and standard errors corresponding to each model. The first column shows the results of the MGLM model, with no random effects but a general residual variance-covariance matrix. This model estimates the fully unstructured 10×10 variance-covariance matrix given in Table 21.3, where only the lower-triangular is presented. The diagonal presents the estimated overdispersion (for the binary endpoint) and variance (for the continuous endpoint) of the endpoints. The upper left quadrant corresponds to the variance-covariance matrix of the binary endpoint, the lower right quadrant to the continuous endpoint, and the lower left quadrant with the binary-continuous association. In line with the analyses per time point, it can be seen that the strongest association between body temperature and wire maneuver occurs at 2h after dosing. The intra-class correlations between any two time points for the binary endpoint tend to be much stronger than the ones for the continuous endpoint.

Table 21.1 Toxicity at 0.5h, 2h, and 4h Post-Dosing: Parameter Estimates (Standard Errors) of the Bivariate Analysis for Temperature (Continuous Endpoint) and Wire Maneuver (Binary Endpoint)

Effect	Par.	0.5h	2h	4h
		Binary Endpoint		
Intercept	β_0	−1.367 (0.981)	−1.825 (1.020)	−3.297 (1.393)
log(dose)	β_1	−0.011 (0.429)	0.760 (0.425)	1.274 (0.541)
overdisp. par.	ϕ	0.133 (0.044)	0.394 (0.131)	0.376 (0.125)
		Continuous Endpoint		
Intercept	α_0	0.830 (0.136)	0.842 (0.234)	0.911 (0.229)
log(dose)	α_1	0.003 (0.059)	−0.516 (0.102)	−0.667 (0.100)
Variance	σ^2	1.111 (0.370)	1.095 (0.366)	0.934 (0.312)
		Association		
Correlation	ρ	−0.043 (0.258)	−0.511 (0.175)	−0.183 (0.236)

Table 21.2 Toxicity Post-Dosing: Parameter Estimates (Standard Errors) of the Bivariate Analysis for Temperature (Continuous Endpoint) and Wire Maneuver (Binary Endpoint)

Effect	time	MGLM	uncorr R.I.	corr R.I.
		Binary Endpoint		
Intercept	−24h	−1.433 (0.448)	−1.768 (0.663)	−2.035 (0.839)
	0.5h	−1.270 (0.713)	−1.460 (1.035)	−1.768 (1.248)
	2h	−1.692 (0.749)	−1.973 (1.105)	−2.490 (1.356)
	4h	−3.064 (0.986)	−3.683 (1.541)	−4.578 (1.988)
	7d	−0.178 (0.612)	−0.080 (0.906)	−0.298 (1.099)
log(dose)	0.5h	−0.083 (0.323)	−0.168 (0.421)	−0.138 (0.517)
	2h	0.620 (0.305)	0.783 (0.463)	0.973 (0.568)
	4h	1.121 (0.371)	1.406 (0.630)	1.736 (0.804)
	7d	−0.408 (0.298)	−0.509 (0.387)	−0.524 (0.489)
overdisp. par.	ϕ		2.535 (1.367)	
R.I. Var	τ_1		0.124 (0.021)	0.112 (0.043)
		Continuous Endpoint		
Intercept	−24h	0.815 (0.178)	0.813 (0.110)	0.815 (0.106)
	0.5h	0.848 (0.297)	0.914 (0.169)	0.912 (0.163)
	2h	0.900 (0.310)	0.930 (0.169)	0.931 (0.163)
	4h	0.978 (0.286)	0.996 (0.169)	1.001 (0.163)
	7d	0.378 (0.298)	0.404 (0.169)	0.404 (0.163)
log(dose)	0.5h	−0.007 (0.129)	−0.042 (0.070)	−0.042 (0.067)
	2h	−0.549 (0.133)	−0.565 (0.070)	−0.566 (0.067)
	4h	−0.705 (0.122)	−0.716 (0.070)	−0.717 (0.067)
	7d	0.045 (0.127)	0.028 (0.070)	0.030 (0.067)
Variance	σ^2		0.118 (0.048)	0.112 (0.018)
R.I. Var	τ_2		0.725 (0.123)	2.934 (2.094)
		Association		
Correlation	ρ		−0.135 (0.118)	0.818 (2.218)

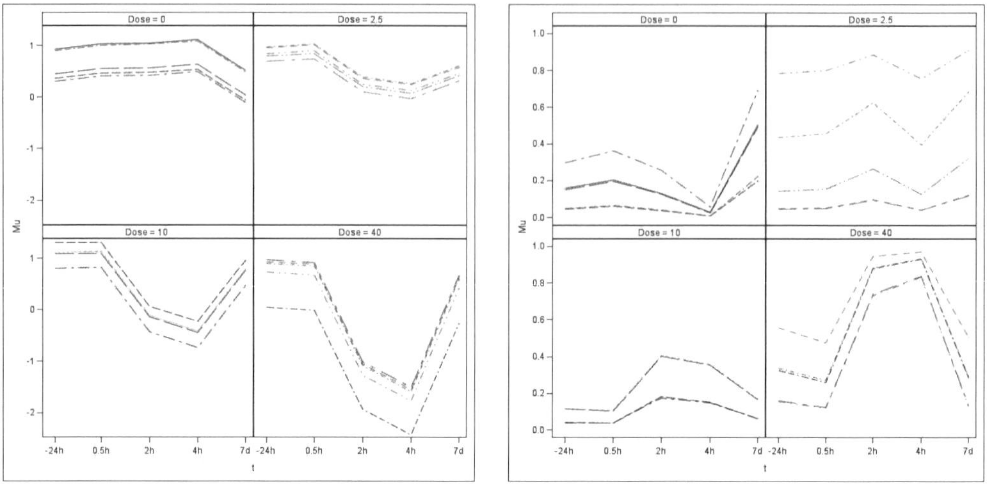

(a) Body Temperature (b) Wire Maneuver

Figure 21.2 Predicted Time Trend Per Dose Level Based on Uncorrelated Random-Intercepts Model

Table 21.3 Lower-Triangular Portion of the Unstructured Variance-Covariance Matrix Based on the MGLM Model

$$\left(\begin{array}{rrrrr|rrrrr}
0.18 & & & & & & & & & \\
0.51 & 0.14 & & & & & & & & \\
0.35 & 0.64 & 0.39 & & & & & & & \\
0.31 & 0.60 & 0.92 & 0.38 & & & & & & \\
0.64 & 0.45 & 0.36 & 0.34 & 0.14 & & & & & \\
\hline
-0.11 & 0.06 & 0.04 & 0.09 & 0.23 & 1.04 & & & & \\
-0.04 & 0.24 & 0.35 & 0.40 & 0.21 & 0.06 & 1.09 & & & \\
-0.23 & -0.38 & -0.51 & -0.35 & -0.035 & 0.27 & 0.12 & 1.05 & & \\
-0.02 & -0.04 & -0.04 & -0.22 & 0.14 & 0.39 & 0.32 & 0.78 & 0.80 & \\
-0.39 & -0.29 & -0.57 & -0.46 & -0.06 & 0.26 & -0.07 & 0.05 & 0.34 & 1.09
\end{array}\right)$$

The second column shows the parameter estimates of the uncorrelated random-intercepts model. The results are also presented in Figure 21.2. The model fit seems to closely follow the data. The correlation parameter represents the correlation in the error structure. A negative but insignificant correlation between the binary and continuous endpoints is estimated. The third column presents the results of the correlated random-intercepts model. The fixed effects parameter estimates and standard errors for the continuous endpoints are approximately the same for all considered models, while the fixed effects estimates for the binary endpoints differ somewhat more. This follows from the fact that interpretation of the parameters in a marginal and in a random-effect model is different in the case of binary endpoints. Note that the correlation parameter in this model represents the correlation between the random intercepts. This estimate can be used to calculate the correlation among the continuous and discrete outcomes.

The models are consistent in estimation of the effect of dose on the endpoints. In all models, the dose significantly reduces the ability to perform the wire maneuver at 4h after dosing. A reduction is estimated 2h after dosing, but this effect is not significant. After 7 days the effect of the treatment on the performance of the wire maneuver has disappeared. The dose effect on the body temperature is not significant at 0.5h after exposure, but is significant at 2h and 4h after dosing, with highest drop in temperature 4h after dosing. After 7 days, there are no effects of the treatment on body temperature.

21.4.2 HET-CAM

Data description

The modified HET-CAM assay is an assay in which the chorioalantoic membrane (CAM) of a fertilized chicken egg is topically treated with some test compound until the complete CAM is covered. After application, the CAM (i.e. the blood vessels, capillaries, and albumen) is examined microscopically for major irritation effects for 5 minutes and then after 15, 30, 45, and 60 minutes. All experiments are performed in triplicate. For the purpose of this chapter, only two endpoints are considered. The first endpoint records whether there is bleeding of at least one of the CAM vessels (binary). The second endpoint is a time-to-event endpoint which investigates the time to a first decrease or complete block of the blood flow inside the CAM vessels. Observations for which this lasts longer than 60 minutes are censored. Data are available from 111 different experiments with 3 replicate eggs each. Therefore, it is likely that some correlation among replicate measurements within a study may be present. Table 21.4 shows the distribution of the number of studies over different dose levels (Vehicle (V), Low (L), Medium (M), High (H), and Super-High (SH)) of a test compound.

Table 21.4 Number of Studies Per Dose Level

Dose Level	V	L	M	H	SH
# Experiments	41	5	25	19	21

Figure 21.3 shows the evolutions of (a) the observed average probabilities of haemorrhages, and (b) the average time to coagulation for each dose level. With increasing dose levels, there is a clear increase in haemorrhage probability and a decrease in time to coagulation.

Data analysis

In any data analysis it is always a good idea to do some univariate analyses before proceeding to more complicated models. Therefore, we first propose a model to describe the relationship between the percentage of haemorrhage and treatment. Next, a dose–response model is proposed for the time to coagulation. And finally, the two models are combined in a joint model.

Haemorrhage

To account for possible correlations among replicates within the same experiment, a random effects logistic model is fitted to the haemorrhage data with treatment group as a fixed factor and experiment as a random factor. As expected, there is a clear dose–response relationship. The percentage of haemorrhages increases with dose ($F_{(4,230)} = 17.01$; $p < .0001$). Table 21.5 tabulates the odds ratio for each of the dose groups versus vehicle. From the medium dose onwards there is evidence of a significant difference versus vehicle.

The variance of the random effects (τ^2) equals 1.60 (s.e. 0.52). The marginalized intraclass correlation coefficient therefore equals 0.34 ($\tau^2/(\tau^2+\pi^2/3)$), indicating some non-ignorable degree of correlation among replicates within an experiment.

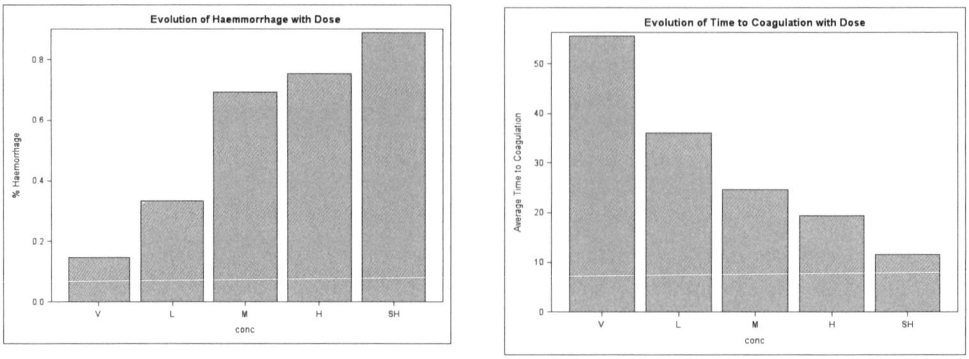

(a) Percent Haemorrhages

(b) Time to Coagulation

Figure 21.3 Average Response Evolutions

Table 21.5 Haemorrhage: Odds Ratios Per Dose Group

Treatment	Odds Ratio	95% Confidence Limits
Low	3.67	(0.62, 21.62)
Medium	20.25	(7.29, 56.25)
High	26.80	(8.97, 80.07)
Super-High	82.15	(23.32, 289.36)

Table 21.6 Time to Coagulation: Percentage of Censored Observations Per Dose Group

Treatment	% censored
Vehicle	73.02
Low	6.67
Medium	6.67
High	1.64
Super-High	0.00

Time to coagulation (TCC)

The time to coagulation (TTC) is a so-called "time-to-event" variable. There are certain aspects of time-to-event data, such as censoring and non-normality, that generate great difficulty when trying to analyze the data using traditional statistical models such as regression and ANOVA. If there is no censoring, standard regression procedures could be used. However, these may be inadequate because time-to-event is restricted to be positive and has a skewed distribution, the probability of surviving past a certain point in time may be of more interest than the expected time of the event, and the hazard function, used for regression in survival analysis, can lend more insight into the failure mechanism than linear regression.

Table 21.6 shows the percentage of censored observations within each treatment group. The number of censored observations decreases drastically with increasing concentration. Whereas the percentage of censored observations is as high as 73% in the vehicle group, all eggs show coagulation within the first 60 minutes in the highest concentration group. Ignoring censoring would therefore lead to severe underestimation of the true TTC in the vehicle group.

Here, a parametric dose–response model based on the Weibull distribution is proposed to describe the relationship between the TTC and treatment. Let t_{ij} be the TTC for egg j ($j = 1, \ldots, 3$) in experiment i ($i = 1, \ldots, 111$) and δ_{ij} a censoring indicator which equals 1 if the TTC was censored (at 60 minutes) and 0 otherwise. First, assume that the replicates within an experiment are independent. The likelihood over all observation then equals

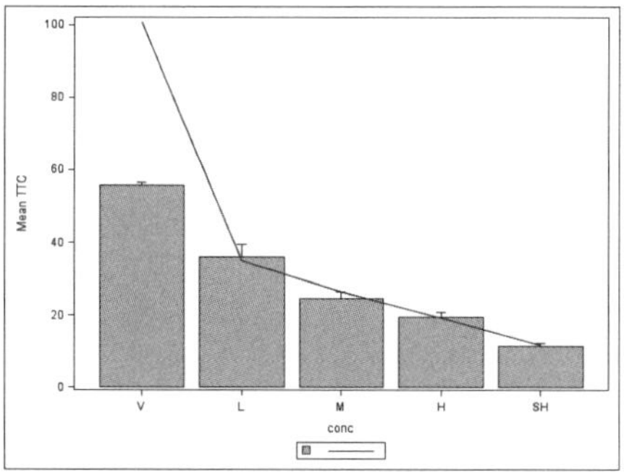

Figure 21.4 Observed and Fitted Average Time to Coagulation

Table 21.7 Parameter Estimates for the Univariate Weibull Model on the TTC

Parameter	Estimate	s.e.	p-value
Scale Parameter			
Intercept	−14.32	0.954	<0.0001
Low	3.17	0.829	0.0002
Medium	4.58	0.547	<0.0001
High	5.12	0.580	<0.0001
Super-High	6.83	0.638	<0.0001
Shape Parameter			
κ	3.02	0.197	<0.0001
Random Effect Variance			
τ^2	2.290	0.569	<0.0001

$$\ell = \prod_i \prod_j \ell_{ij} = \prod_i \prod_j (\kappa \lambda t_{ij}^{\kappa-1})^{1-\delta_{ij}} e^{-\lambda t_{ij}^{\kappa}}. \tag{21.4}$$

Note that the Weibull model reduces to the exponential model when the shape parameter κ equals 1. The mean TTC equals $E[T] = \lambda^{(-1/\kappa)}\Gamma[\frac{1}{\kappa} + 1]$. Hence, the scale parameter λ is proportional to the mean. In order to relate the TTC with treatment (and other possible covariates), a regression model can be specified on λ:

$$\lambda = \exp(X_i \boldsymbol{\beta}),$$

with X_i a design matrix of known covariate values for experiment i and $\boldsymbol{\beta}$ a vector of unknown fixed effects regression coefficients. Figure 21.4 shows the observed average TTC values (bar chart) and the fitted average TTC values (line plot) superimposed on top of each other. Due to the large amount of censoring in the vehicle group, there is quite some discrepancy between the observed and fitted average TTC.

In order to further account for possible correlations among replicates within a study, a study-specific random intercept u_i can be included:

$$\lambda = \exp(X_i \boldsymbol{\beta} + u_i),$$

where u_i is assumed to follow a normal distribution with mean 0 and variance τ^2. Of course, this could easily be extended towards multiple random effects should such be of interest.

Upon fitting a random intercept model to the TTC data, all treatment groups are significantly different from the vehicle group. Table 21.7 shows the parameter estimates of the random intercept model corresponding to the TTC.

Joint model (GLMM approach)

Table 21.8 shows the parameter estimates of the final joint model following the GLMM approach. As before, there is evidence of a significant difference versus vehicle for all but the low concentration group on the haemorrhage response. For time to coagulation, all treatment groups are significantly different from the vehicle group. Furthermore, haemorrhage and time to coagulation are highly correlated. The correlation coefficient equals 0.924 (s.e. 0.270). The more bleedings observed, the longer also the time to decrease or complete block of the blood flow. A further advantage of the joint modeling approach is that one can investigate the dependence of this correlation upon covariates. In this case there was no evidence of a dependence upon (for example) concentration.

Table 21.8 Parameter Estimates for the Joint Models

Parameter	GLMM			Copula		
	Estimate	St. Err.	*p*-value	Estimate	St. Err.	*p*-value
Haemorrhage						
Intercept	−3.015	0.600	<0.0001	−3.944	0.689	<0.0001
Low	1.909	1.245	0.126	2.798	1.308	0.035
Medium	4.296	0.848	<0.0001	5.211	0.929	<0.0001
High	4.648	0.894	<0.0001	5.580	0.971	<0.0001
Super-High	6.325	1.122	<0.0001	7.328	1.209	<0.0001
Coagulation						
Intercept	−14.376	0.944	<0.0001	−14.133	0.946	<0.0001
Low	3.167	0.838	0.0002	3.029	0.800	0.0002
Medium	4.588	0.546	<0.0001	4.460	0.530	<0.0001
High	5.123	0.580	<0.0001	4.991	0.562	<0.0001
Super-High	6.847	0.635	<0.0001	6.692	0.621	<0.0001
κ	3.021	0.197	<0.0001	3.001	0.197	<0.0001
Random Effect Variances						
τ_1^2	3.745	1.422	<0.0001	4.110	1.495	<0.0001
τ_2^2	2.371	0.574	<0.0001	2.088	0.532	<0.0001
Associations						
ρ	0.924	0.270	0.0008			
ψ				0.315	0.136	0.023

Alternative model (copula approach)

While the focus in this chapter is on GLMM-based models for the joint analysis of endpoints of different natures, alternative models do exist. In recent years, copula-based models have gained an increasing amount of popularity both in actuarial sciences and biostatistics. They have recently been used in many biostatistical applications where interest lies in, for example, the joint modeling of time-to-event outcomes (Burzykowski et al., 2001), mixed categorical and time-to-event endpoints (Burzykoski, Molenberghs, and Buyse, 2004), and mixed discrete and continuous outcomes (e.g. Geys et al., 2001; Faes et al., 2004; and de Leon and Wu, 2011).

As an example, we will concentrate here on the use of the Plackett copula (Dale, 1986). The Plackett family of copulas is given, for $\theta \neq 1$, by:

$$C_\theta(u, v) = \Big[1 + (u + v)(\theta - 1) - \sqrt{[1 + (\theta - 1)(u + v)]^2 + 4\theta(1 - \theta)uv}\Big] / [2(\theta - 1)],$$

and for $\theta = 1$:

$$C_\theta(u, v) = uv.$$

From Mardia (1970) it follows that $C_\theta(u, v)$ is always a bivariate copula with θ in $[0, \infty]$. This copula method provides an alternative to the frequently used multivariate normal latent variable models. The main advantages are the flexibility with which the marginal densities can be chosen (normal, logistic, complementary log-log, etc.) and the familiarity of the odds ratio which is used as a measure of association. Another advantage of this approach is that it allows direct modeling of the bivariate association θ as a function of the covariates of interest. Also, extensions towards other types of endpoint are possible.

Extending the above-described copula models to a hierarchical setting can be done relatively easily by introducing subject-specific (random) effects into the marginal distributions, which also results in a GLMM. The association between the responses is then dealt with through the copula, whereas the hierarchical structure is accounted for through (for example normally distributed) random effects.

Table 21.8 shows the parameter estimates of the copula approach. As before, there is evidence of a significant difference versus vehicle for all concentrations. Whereas the difference between the low concentration group and the vehicle group was not significant based on the GLMM approach, it is now borderline significant. For time to coagulation, all treatment groups are significantly different from the vehicle group. Furthermore, note again that (even after adjustment for treatment) haemorrhage and time to coagulation are highly associated. The odds ratio equals 0.315, with 95% confidence interval $(0.13, 0.74)$. The more bleedings observed, the longer also the time to decrease or complete block of the blood flow. Note that the methodology described above for joint models can easily be extended to other non-standard joint models, such as two time-to-events, categorical-continuous, and so on.

21.5 DISCUSSION

Methods where the normal distribution plays a prominent role for multilevel, multivariate, continuous data are widely available. When the outcome variables are of a mixed type, techniques are less standard due to the lack of a discrete analog to the multivariate distribution. In this chapter we have focused on the use of a generalized linear mixed model, allowing the link function and residual error structure to differ for the various outcomes and assuming correlated random effects to account for the correlation between the outcomes. This general framework makes it possible to consider arbitrary combinations of outcome types, for example combined discrete and continuous outcomes, survival and discrete outcomes, etc. Extensions to higher dimensions are possible by specifying a joint (higher-dimensional) distribution for the random effects. When many outcomes are of interest, this results in a high-dimensional integration problem, which is computationally no longer feasible. In that case, a pseudo-likelihood function can be used instead of the full likelihood (Fieuws and Verbeke, 2006; Faes et al., 2008). The principal idea of the pseudo-likelihood methodology is to replace the computationally cumbersome likelihood by a function set up as the product of easy-to-compute functions, but which is no longer equal to the full likelihood. For example, the pseudo-likelihood function could be defined as the product of all pairwise bivariate models. This pseudo-likelihood yields unbiased estimates and valid standard errors. One should be careful, however, with the analysis of incomplete data (see Chapter 23) since the behavior of the pseudo-likelihood under the assumption of missing at random is presently not fully understood from a methodological standpoint, and is being investigated. All statistical analyses in this chapter were performed with standard available software procedures, such as PROC GLIMMIX and PROC NLMIXED in SAS. Code for the models considered in this chapter is available at the web site for this book.

REFERENCES

Burzykowski, T., Molenberghs, G., and Buyse, M. (2004) 'The Validation of Surrogate Endpoints using Data from Randomized Clinical Trials: A Case Study in Advanced Colorectal Cancer', *Journal of The Royal Statistical Society, Series A*, 167: 103–24.

Burzykowski, T., Molenberghs, G., Buyse, M., Renard, D., and Geys, H. (2001) 'Validation of Surrogate Endpoints in Multiple Randomized Clinical Trials with Failure-time Endpoints', *Applied Statistics*, 50: 405–22.

Dale, J.R. (1986) 'Global Cross-Ratio Models for Bivariate, Discrete, Ordered Responses', *Biometrics*, 42: 909–17.

de Leon, A.R. and Wu, B. (2011) 'Copula-Based Regression Models for a Bivariate Mixed Discrete and Continuous Outcome', *Statistics in Medicine*, 30: 175–85.

Faes, C., Geys, H., Aerts, M., Molenberghs, G., and Catalano, P. (2004) 'Modeling Combined Continuous and Ordinal Outcomes in a Clustered Setting', *Journal of Agricultural, Biological and Environmental Statistics*, 9: 515–30.

Faes, C., Aerts, M., Molenberghs, G., Geys, H., Teuns, G., and Bijnens, L. (2008) 'A High Dimensional Joint Model for Longitudinal Endpoints of Different Nature', *Statistics in Medicine*, 27: 4408–27.

Fieuws, S. and Verbeke, G. (2006) 'Pairwise Fitting of Mixed Models for the Joint Modeling of Multivariate Longitudinal Profiles', *Biometrics*, 62: 424–31.

Fitzmaurice, G., Davidian, M., Verbeke, G., and Molenberghs, G. (2009) *Longitudinal Data Analysis: A Handbook of Modern Statistical Methods*. Boca Raton, FL: Chapman and Hall.

Geys, H., Regan, M., Catalano, P., and Molenberghs, G. (2001) 'Two Latent Variable Risk Assessment Approaches for Mixed Continuous and Discrete Outcomes From Developmental Toxicity Data', *Journal of Agricultural Biological and Environmental Statistics*, 6: 340–55.

Mardia, K.V. (1970) *Families of Bivariate Distributions*. London: Griffin.

Molenberghs, G. and Verbeke, G. (2005) *Models for Discrete Longitudinal Data*. New York: Springer.

PART III

Practical Considerations in Model Fit and Specification

22

Robust Methods for Multilevel Analysis

Joop Hox, Rens van de Schoot
Utrecht University, The Netherlands

22.1 INTRODUCTION

What are robust statistical methods? Robust methods are statistical procedures for estimation and establishing confidence intervals that are not very sensitive to violations of the assumptions of the underlying statistical methodology. Examples from standard univariate statistics are the finding that, in general, parametric tests are not very sensitive to violations of the normality assumption, provided the sample size is not too small. If the violation is extreme, as in having a single extreme outlier in the data, robustness does not apply. Obviously, if we define robustness as insensitivity to violations of assumptions, the next question is: which assumptions? This chapter examines the common assumptions of multilevel models, and discusses to what extent violations of these assumptions threaten the validity of statistical conclusions. First, a short discussion follows of the problem of how to diagnose whether specific assumptions are violated, and solutions involving transformation of the data. Subsequently, three approaches are discussed that are useful when the assumptions are known or suspected to be violated. These approaches are using robust standard errors, bootstrapping, and Bayesian estimation (see also Chapter 4). The different approaches are illustrated with an example data set from a longitudinal study on child deviant behavior, a field where skewed outcome measures and outliers are commonly found. The chapter ends with a discussion of the strengths and weaknesses of the different methods.

22.2 WHICH ASSUMPTIONS?

We examine the multilevel structural equation model and the multilevel regression model, which is, in fact, a special case of the multilevel structural equation model. First, the assumptions of multilevel structural equation models are the same as the assumptions of single-level structural equation models, that is, sufficient sample size, linear relationships, absence of multicollinearity, and multivariate normality for the dependent variables, which implies homoscedasticity and absence of outliers (cf. Tabachnick and Fiddell, 2007). Errors are assumed to be independent and identically distributed (iid). The assumption of independent observations is lessened in multilevel modeling, because multilevel modeling includes specific forms of dependence. However, there is still an issue with unmodeled dependencies, which may result from omitting a level or ignoring dependence over time in longitudinal models. Finally, multilevel models generally assume that the errors at different levels are independent from each other.

In multilevel modeling, the requirement that the sample size is adequate applies most forcefully to the highest level, since that generally has the smallest sample size. Eliason (1993) states that a sample size of at least 60 is needed for small models. This coincides well with simulation results from Hox et al. (2010), who report that for accurate variance estimates and standard errors 100 groups are needed, but also conclude that 50 groups provide sufficient accuracy to be workable in practice. When the interest is solely in the regression coefficients, even 20 groups may be sufficient (Maas and Hox, 2005); however, at such small sample sizes the accuracy of variance estimates is low, and estimates are sensitive to violations of assumptions (Maas and Hox, 2004). However, this applies mostly to maximum likelihood estimation, which is asymptotic and therefore must assume large samples. Bayesian estimation is valid at small sample sizes. A comparison of Bayesian and maximum likelihood estimation is provided by Stegmueller (2011), who reports simulations of a simple multilevel regression model, and by Hox et al. (2012), who report simulations of a small structural model containing one latent variable and a country-level exogenous variable. Both simulations support the claim that Bayesian estimation works well with far smaller sample sizes than maximum likelihood estimation.

In addition to distributional assumptions, the variances of the parameter estimates also depend on the correctness of the model. If the model is misspecified, this will in general not only result in biased parameter estimates, but also in incorrect estimates of their variability. To some extent, robust standard errors and Bayesian estimation ameliorate this. Bayesian estimation methods in particular can be modified in ways to contend with model misspecification; this type of robustness is discussed in more detail in Section 22.7. For a general description of Bayesian inference we refer to Chapter 4 of this volume.

22.3 DIAGNOSIS AND TRANSFORMATIONS

As an example, we use a data set compiled by Patrick Curran from a large longitudinal data set. The data are a sample of 405 children who were within the first two years of entry to elementary school. The data consist of four repeated measures of both the child's antisocial behavior and the child's reading recognition skills. In addition, at the first measurement occasion, measures were collected of emotional support and cognitive stimulation provided by the mother, the child's gender, and the ages of the mother and the child. The data were collected using face-to-face interviews of both the child and the mother at two-year intervals between 1986 and 1992.[1]

These data can be analyzed in several ways, including a growth curve model using structural equation analysis. As provided, the data are in the multivariate (wide) format where each child is a separate case, with measures across time represented as different variables. For the multilevel analyses, the data are restructured to a stacked (long) format where each measurement occasion is a separate row in the data matrix, with measurement occasions nested within children. The age variables are included in the file as time-varying predictors, and the child's gender, emotional support, and cognitive stimulation are included as time invariant predictors. The measurement occasion is indicated as 0–3, and missing measurement occasions are omitted from the long data file.

Measures of deviant behavior often have a skewed distribution and outliers, and the example data are no exception. Figure 22.1a shows boxplots for the four measurements of antisocial behavior, plus a boxplot for the stacked antisocial behavior variable. Figure 22.1b clearly shows skewness and several outliers for each variable (the circles indicate outliers, the stars extreme outliers).

Transformations of the dependent variable are often used to achieve linearity, homogeneity of variance around the regression line, or normality. If data follow a multivariate normal distribution, linearity and homogeneity follow. Thus, a transformation that approaches (multivariate) normality often improves linearity and homogeneity as well. Since the effect of transformations in multilevel regression do not really differ from the effects in single-level regression, we will not go into transformations, but refer to, e.g., Tabachnick and Fiddell (2007) instead.

For single-level regression a number of diagnostics have been developed to assess how well assumptions are met. Multivariate normality is often assessed by plotting residuals. Indicators for multicollinearity are the tolerance (or the Variance Inflation Factor—VIF) defined by $tolerance = 1 - R_j^2$ and $VIF = tolerance^{-1}$, where R_j^2 is the squared multiple correlation of variable j with all other predictor variables. A VIF larger than 10 (or equivalently a tolerance lower than 0.1) indicates a serious multicollinearity problem (Field, 2009). Multicollinearity is also indicated by the condition number of the matrix of predictors. The condition number of a matrix is the square root of the largest eigenvalue divided by the smallest eigenvalue. If there is perfect multicollinearity, the smallest eigenvalue equals zero. Thus, with increasing collinearity, the condition number tends to infinity. A condition number higher than 30 is usually taken as indicating significant multicollinearity.

Residuals are often plotted to assess normality and to look for outliers. In addition, several diagnostic indices exist to inspect leverage and influence. Leverage indicates an observation with an extreme value on a predictor variable. Influence indicates that parameter estimates substantially change when that observation is removed from the data. Influence is a combination of leverage and outlyingness. One index for influence is Cook's distance (CD—Field, 2009), which indicates the combined influence of a case. A CD larger than $(2k + 2)\,n^{-1}$ (k is the number of predictors and n is the sample size) is often interpreted as indicating very influential cases. Finally, a useful overall summary index is the Mahalanobis distance, which indicates the distance of a case to the mean of all cases in the multivariate space. Critical values for the Mahalanobis distance depend on the number of variables and the sample size; see Field (2009) for suggestions for cut-off values.

All of these indices can be generalized to multilevel data; for a discussion of the statistical techniques see Langford and Lewis (1998) and Chapter 24 of this volume. A problem that complicates diagnostic procedures in multilevel analysis is that most of the usual indicators can be assessed at each separate level. Hence, we can inspect residuals, distinguish between influential cases and influential groups, and test for multicollinearity at separate levels. In addition, multilevel regression modeling often involves regression slopes that are assumed to vary across groups, which leads to the additional possibility of strikingly different slopes that should be inspected too.

Diagnosing violations of assumptions in multilevel models suffers from software limitations. In general, statistical packages for multiple regression can calculate a large variety of diagnostics to assess potential violations of assumptions. Multilevel analysis software typically incorporates only a few of these procedures, or none. This may be the reason why reports of multilevel analyses seldom discuss how the assumptions were examined (Ferron et al., 2008). For example, the mixed model module in the popular statistical package SPSS calculates only a single residual, which makes it difficult to assess violations of assumptions at separate levels (Leyland, 2006, describes a procedure to calculate separate residuals for two levels in SPSS). The program HLM can output a file with various residuals, Mahalanobis distances, variance estimates and other variables that are helpful in diagnosing possible violations, but

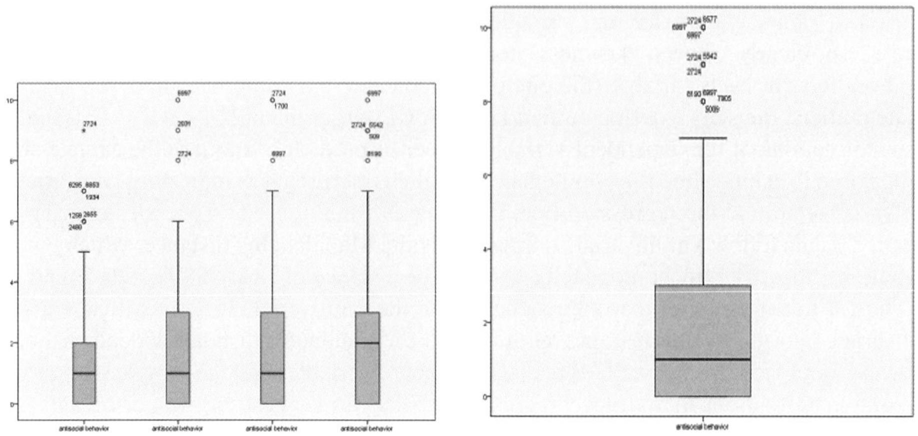

(a) Four Time Points (b) Total

Figure 22.1 Boxplots of Antisocial Behavior at Four Time Points and in Total

the analysis details depend on the imagination of the analyst. MLwiN allows direct inspection of multilevel residuals, multilevel indicators of influence and leverage, and a variety of plots of these indicators.

Given that the ordinary least squares (OLS) estimators for regression coefficients in multilevel models are unbiased and consistent (Goldstein, 2011a), we can carry out a single-level regression analysis, and use the available regression diagnostics. The significance tests produced by this procedure are invalid, but we can inspect linearity plots, Mahalanobis distances, VIF values, and apply other descriptive procedures. If our multilevel analysis software produces residuals at separate levels, we can plot these and look for indications of problematic cases or groups. If our multilevel analysis software does not produce residuals, we can at least calculate the group-level scores $\mathbf{Y}_{\mathrm{B}} = \bar{\mathbf{Y}}_j$ and the individual-level deviation scores $\mathbf{Y}_{\mathrm{W}} = \mathbf{Y}_{ij} - \bar{\mathbf{Y}}_j$ (subscript i for individuals, j for groups), and examine the distribution of these variables.

22.4 EXAMPLE

As an example we estimate a model on our example data with outcome variable *Antisocial behavior* and predictors the time-varying predictor *child's age* centered on its median value of six and the time-invariant predictor *emotional support at home*. The slope of child age is allowed to vary across children. Using REML estimation in HLM 7, we obtain the following estimated equation:

$$Antisocial_{ti} = 1.65 + 0.08 * Childage - 0.17 * HomeEmo. \tag{22.1}$$

All regression coefficients are highly significant ($p < 0.001$). The child-level variances of the intercept and slope are also highly significant ($p < 0.001$). The value of 0.08 for the effect of age indicates that antisocial behavior increases when the children become older. However, the standard deviation of the distribution of the slopes is 0.15, which implies that for some children the individual slope is downwards, while for other children the increase is considerably larger than indicated by the average value of 0.08.

HLM produces a residuals file for each distinct level. Figure 22.2 is a boxplot for the child-level Empirical Bayes (EB) residuals for the intercept and slope in this model. The boxplots indicate some skewness in the residuals, and one clear outlier for the intercept, marked

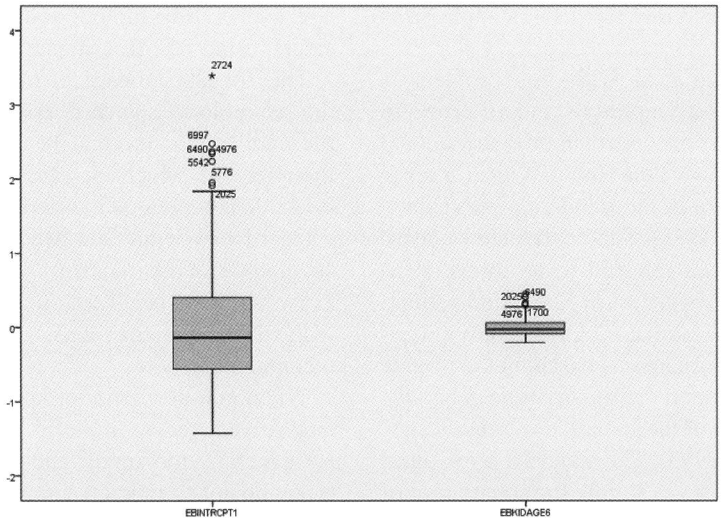

Figure 22.2 Boxplots for Child-Level Residuals of Intercept and Slope

by identification number 2724. Interestingly enough, in Figure 22.1 child 2724 also is an extreme outlier on antisocial behavior on the first measurement occasion, and among the ordinary outliers at all three other measurements. Even more interesting, a boxplot of the single residual that SPSS can produce does not show up case 2724 as problematic. Clearly, the residual that simply reflects how well the entire regression equation predicts the lowest-level outcomes is not very useful.

However, some other single-level indicators applied to the multilevel data are useful. For example, the diagnostics in the single-level regression analysis yield a VIF of 1.0 for both predictors, and a condition number of 3.4, both indicating no multicollinearity. A scatter plot of antisocial behavior against child age and home emotional support shows no outliers, and indicates that the relationship is approximately linear.

Since child 2724 appears to be an outlier in several diagnostic analyses, we carry out a sensitivity analysis by estimating our model after removing this child from the data set. The regression coefficients remain mostly unchanged; the largest change is the slope for home emotional support which changes from −0.17 to −0.15. The largest change is the child-level intercept variance, which changes from 0.83 to 0.70. As a result, the correlation between intercept and slope also changes, from 0.66 to 0.75. There is a tendency that children who have a high average level of antisocial behavior also show a larger increase as they grow older, and removing this outlier makes this relationship more apparent.

As the example analysis shows, multilevel diagnostics are useful, even if we need to work with software that has no special provisions for multilevel diagnostics. Summing up, we strongly recommend that researchers carry out *some* kind of diagnostic analysis before embarking on a complex multilevel analysis.

22.5 ROBUST STANDARD ERRORS

When the response variables do not have a normal distribution, the parameter estimates produced by the maximum likelihood method are generally still consistent and asymptotically unbiased (Eliason, 1993). However, the asymptotic standard errors are incorrect, and they cannot be trusted to produce accurate significance tests or confidence intervals (Goldstein, 2011a, p. 93). This problem does *not* always decrease when the sample sizes increase.

One way to deal with this problem is to correct the asymptotic standard errors to make them less dependent upon distributional assumptions, using the Huber/White (or sandwich) estimator of the standard error (Huber, 1967; White, 1982). Such corrected standard errors are often referred to as robust standard errors. In maximum likelihood estimation, the usual estimator of the sampling variances and covariances is based on the inverse of the Information matrix, or, more generally, on the inverse of the so-called Hessian matrix (cf. Eliason, 1993). The standard errors used in the Wald test are simply the square root of the sampling variances that are found on the diagonal of this inverse. Thus, using matrix notation, the asymptotic variance-covariance matrix of the estimates can be written as:

$$\mathbf{V}_{\mathrm{A}}(\hat{\boldsymbol{\theta}}) = \mathbf{H}^{-1},$$

where $\mathbf{V}_{\mathrm{A}}$ is the asymptotic covariance matrix of the regression coefficients, and $\mathbf{H}$ is the Hessian matrix. The Huber/White estimator is given as

$$\mathbf{V}_{\mathrm{R}}(\hat{\boldsymbol{\theta}}) = \mathbf{H}^{-1}\mathbf{C}\mathbf{H}^{-1},$$

where $\mathbf{V}_{\mathrm{R}}$ is the robust covariance matrix of the regression coefficients, and $\mathbf{C}$ is a correction matrix based on the cross product of the raw residuals. If these residuals follow a normal distribution, $\mathbf{V}_{\mathrm{A}}$ and $\mathbf{V}_{\mathrm{R}}$ are both consistent estimators of the covariances of the regression coefficients, but the model-based asymptotic covariance matrix $\mathbf{V}_{\mathrm{A}}$ is more efficient if the distributional assumptions are correct. However, if the model is misspecified, or we cannot rely on normality assumptions, the asymptotic standard errors tend to be too small, and robust standard errors better reflect the true variability.

Table 22.1 presents the results of an analysis of the antisocial behavior example data. The table presents the maximum likelihood estimates of the regression coefficients and the variances of the intercept and slope of child age, with both asymptotic and robust standard errors, using REML estimation in MLwiN.

The robust standard errors are close to the asymptotic standard errors, except for the individual- (second-)level variance of the intercept, which increases from 0.098 to 0.143. When there is a difference, the robust standard errors are always larger than the asymptotic standard errors. Given that we know that there is at least one extreme outlier in the data, using the robust standard errors is recommended here.

When non-normality or heteroscedasticity is involved, the asymptotic standard errors are generally too small. The robust standard errors do not completely correct this, but they do result in more accurate significance tests and confidence intervals. Since the robust standard errors are based on the observed residuals, they do need a reasonable level-two sample size to be accurate (Kauermann and Carroll, 2001); single-level simulation results suggest a sample size of at least 100. In multilevel analysis, this would translate to a minimal second-level sample size of 100 for the robust standard errors to work well. Multilevel simulations with strongly non-normal two-level data (Maas and Hox, 2004) confirm these recommendations. Cheong et al. (2001) find that robust standard errors even provide reasonable protection against omitting an entire level from the analysis. Raudenbush and Bryk (2002, p. 278) suggest comparing the robust to the asymptotic standard errors to appraise the possible effect of model misspecification and non-normality. Thus, robust standard errors are used as an indicator for misspecifications of the model or its assumptions. If the robust standard errors are much different from the asymptotic standard errors, this should be interpreted as a warning sign that some assumption is violated, and an indicator to look into the problem. Multilevel structural equation modeling often also produces a chi-square goodness of fit test. Several robust chi-square variants have been proposed, based on procedures to correct the asymptotic chi-square to deal with excessive skewness or kurtosis in the

Table 22.1 Estimates with Asymptotic and Robust Standard Errors

Parameter	Estimate with (asymptotic s.e.)	Estimate with (robust s.e.)
Constant	1.65 (0.082)	1.65 (0.082)
Child age	0.08 (0.019)	0.08 (0.019)
Emotional support	−0.17 (0.031)	−0.17 (0.033)
σ^2_{occ}	1.76 (0.098)	1.76 (0.143)
σ^2_{ind}	0.83 (0.211)	0.83 (0.255)
σ^2_{age}	0.023 (0.010)	0.023 (0.011)

observed data. The Satorra–Bentler (Satorra and Bentler, 2001) and the Yuan–Bentler (Yuan and Bentler, 2000) corrections are often used, and appear to work well in practice.

22.6 BOOTSTRAPPING

The *bootstrap* method (Efron, 1982; Efron and Tibshirani, 1993) estimates the parameters of a model and their standard errors strictly from the sample, without reference to a theoretical sampling distribution. In classical statistical inference, the sampling variation is modeled by a theoretical sampling distribution, for instance a normal distribution, and estimates of the expected value and the variability are taken from this distribution. In bootstrapping, we draw b times a sample (with replacement) from the observed sample at hand. In each bootstrap sample, we calculate the statistic(s) of interest, and the observed distribution of the b statistics is used for the sampling distribution. Thus, estimates of the expected value and the variability of the statistics are taken from this empirical sampling distribution (Mooney and Duvall, 1993).

The number of bootstrap iterations b is typically large, with b between 1000 and 2000 (Booth and Sarkar, 1998; Carpenter and Bithell, 2000). Increasing the number of bootstrap samples will, of course, not increase the amount of information available from the original sample. It will, however, decrease the amount of random error generated by the bootstrap procedure itself. If the interest is in establishing a very accurate confidence interval, we need an accurate estimate of percentiles close to zero or one hundred, which requires an even larger number of bootstrap replicates, such as $b > 5000$.

The bootstrap is not without its own assumptions. A key assumption of the bootstrap is that the *resampling* properties of the statistic resemble its *sampling* properties. A second assumption is that the resampling scheme used in the bootstrap reflects the actual sampling mechanism used to collect the data (Carpenter and Bithell, 2000). This assumption is important in multilevel modeling, since in multilevel problems we have a hierarchical sampling mechanism, which must be simulated correctly in the bootstrapping procedure.

Bootstrapping for standard errors is straightforward. The mean of the bootstrapped sampling distribution is taken as the point estimate for the parameter. The confidence interval can be established in two different ways. One approach is using the standard deviation of the bootstrapped sampling distribution as an estimate of the standard error, and establishing the usual 95% confidence interval as the mean bootstrap value ± 1.96 times this standard error. This, of course, brings us back to the assumption that the sampling distribution of that parameter is normal. Since bootstrapping is often used to overcome problems of non-normality, this is evidently not the preferred approach. Closer to the non-parametric spirit of the bootstrap is to estimate the 2.5th and 97.5th percentiles of the bootstrap distribution, and using these as the endpoints of the bootstrapped 95% confidence interval.

An interesting application of the bootstrap is bootstrapping for bias correction. The bias-corrected bootstrap (usually shortened to bc bootstrap) assumes that the difference between the mean of the bootstrap replications and the parameter estimate in the actual sample indicates the estimation bias. The bias in a parameter estimate $\hat{\boldsymbol{\theta}}$ is defined as the difference between the expected value of the parameter estimate and the population value,

$$Bias(\hat{\boldsymbol{\theta}}) = \boldsymbol{\theta} - E(\hat{\boldsymbol{\theta}}).$$

Many statistics, for example the sample mean, are, under very general conditions, an unbiased estimate of the corresponding population value. Other statistics, for example the sample correlation, are known to be biased estimators of the corresponding population value (the correlation coefficient has a small negative bias, a fact already known to Fisher, 1915). For such statistics, the amount of bias can be estimated using

$$Bias_{\mathrm{B}}(\hat{\boldsymbol{\theta}}) = \hat{\boldsymbol{\theta}} - \bar{\boldsymbol{\theta}}_{\mathrm{B}},$$

which estimates the bias as the difference between the asymptotic parameter estimate and the mean of the bootstrapped estimates. A similar bias correction is applied to the percentiles, which leads to the bias-corrected percentile confidence interval (Efron and Tibshirani, 1993; Mooney and Duval, 1993). The bias-corrected bootstrap can be repeated, with each subsequent bootstrap including the bias correction estimated previously. This leads to the iterated bootstrap, which in combination with a large number of bootstrap samples for accuracy can be computationally demanding. The simplest form of bootstrapping resamples entire cases. In multilevel analysis this is problematic, because the resampling scheme for multilevel data should follow the assumed multilevel data-generating mechanism. The obvious scheme to bootstrap groups and subsequently bootstrap individuals within groups does not work adequately. For instance, if the group sizes are not equal, bootstrapping groups changes the individual-level sample size. Bootstrapping individuals, followed by determining which groups they are in, is even more problematic because it leads to varying numbers of groups and does not preserve the within-groups correlation structure. Van der Leeden et al. (2008, p. 413) discuss examples where simple case- or group-based resampling may actually be optimal. For example, when the multilevel structure involves repeated measures within individuals, resampling individuals (i.e., level-two units) appears natural. Similarly, for cross-cultural research involving respondents sampled within countries, resampling only individuals within countries (i.e., level-one units) is again an acceptable resampling scheme. In general, however, one would prefer a resampling scheme that preserves the structure of the original sampling design.

One solution is to use parametric bootstrapping, where the bootstrap samples are generated by adding to the fixed model predictions random errors from the appropriate error distributions. Again, this reintroduces the assumption of specific distributions such as multivariate normality for the residual errors. The preferred approach to multilevel bootstrapping is nonparametric residual bootstrapping, which bootstraps residuals. In the nonparametric bootstrap, multilevel estimation is carried out once on the total data set. The regression coefficients from this estimation are used to produce predicted values, and the residuals from this analysis are resampled at each appropriate level in the bootstrap replications. Van der Leeden et al. (2008) discuss the appropriateness of using either raw residuals or shrunken (empirical Bayes) residuals for bootstrapping. As Goldstein (2011b) makes clear, this discussion neglects the fact that either set of residuals generally underestimates the residual variances, and therefore must be adjusted to reflect the estimated covariance matrix exactly. Computational details are provided by Goldstein (2011a, 2011b); for an example, see Carpenter et al. (2003).

The multilevel nonparametric residuals bootstrap is currently implemented only in the software MLwiN, which contains both the bootstrap-based bias correction and the iterated bootstrap described earlier.

Table 22.2 presents the asymptotic and bootstrapped estimates for the antisocial behavior model. The asymptotic results are based on REML estimation in MLwiN, and the bootstrap results are bias-corrected estimates using the nonparametric residuals method with 5000 replications. As Table 22.2 makes clear, the differences from the asymptotic estimates are very small. The only visible difference is in the confidence intervals for the variances. These (small) differences do not indicate that distributional assumptions are violated. Rather, they indicate that the traditional way of establishing a confidence interval using the standard error is not very precise if the sampling distribution is not normal. Variances have a chi-square distribution, and the bootstrap distribution will correctly reflect that in an asymmetric percentile-based confidence interval.

Since bootstrapping takes the observed data as the sole information about the population, it needs a reasonable original sample size. Good (1999, p. 107) suggests a minimum sample size of 50 when the underlying distribution is not symmetric. Yung and Chan (1999) review the evidence on the use of bootstrapping with small samples. They conclude that it is not possible to give a simple recommendation for the minimal sample size for the bootstrap method. However, in general the bootstrap appears to compare favorably with asymptotic methods. A large simulation study involving complex structural equation models (Nevitt and Hancock, 2001) suggests that for accurate results in the presence of large violations of normality assumptions, the bootstrap needs an observed sample of more than 150. Given such results, the bootstrap is not the best approach when the major problem is a small sample size. When the problem is violations of assumptions, or establishing bias-corrected estimates and valid confidence intervals for variance components, the bootstrap appears to be a viable alternative to asymptotic estimation methods.

22.7 BAYESIAN ESTIMATION METHODS

As put forward by Walker et al. (2007, p. 366), 'the Bayesian approach offers innovative solutions to some challenging analytical problems that plague research in [...] psychology' (see also Howard et al., 2000; Lee and Wagenmakers, 2005; van de Schoot et al., 2011). In this section we will discuss the advantages of using Bayesian multilevel analyses in the context of robustness to violations of assumptions. Two types of robustness will be discussed here: (1) the sensitivity of Bayesian estimators to the prior as a statement about the robustness of the approach, and (2) misspecification of the likelihood function. For a detailed introduction to Bayesian statistics we refer to Chapter 4, but see also Lynch (2007), or for more technical details Gelman et al. (2004), for Bayesian model selection see Hoijtink et al. (2008), for Bayesian structural equation modeling see Lee (2007), and for Bayesian multilevel modeling see Jedidi and Ansari (2001). There is a variety of software packages available for Bayesian multilevel analyses (R packages, WinBUGS, OpenBUGS, MLwiN, Mplus v6.x), but for the current chapter we use the software Mplus (Muthén and Muthén, 1998–2010) because it is often used by applied researchers. Bayesian estimation became available in Mplus in version 6. For a technical implementation of Bayesian statistics in Mplus, see Asparouhov and Muthén (2010) and Muthén and Asparouhov (2012).

The core idea of Bayesian inferences is that a priori beliefs are updated with observed evidence, the likelihood function, and both are combined in a so-called posterior distribution. Bayesian estimation uses Markov Chain Monte Carlo (MCMC) algorithms to obtain estimates for the posterior distribution. The idea behind MCMC is that the conditional distribution of one set of parameters given other sets can be used to make random draws of parameter values. This results in an approximation of the joint distribution

Table 22.2 Estimates with Asymptotic and Bootstrapped Standard Errors

Parameter	Asymptotic estimate	Asymptotic 95% CI	Bootstrapped bias-corrected estimate	Bootstrapped 95% CI
Constant	1.65	(1.48, 1.81)	1.65	(1.49, 1.81)
Child age	0.08	(0.04, 0.11)	0.08	(0.04, 0.11)
Emotional support	−0.17	(−0.23, −0.11)	−0.17	(−0.23, −0.11)
σ^2_{occ}	1.76	(1.57, 1.96)	1.76	(1.57, 1.94)
σ^2_{ind}	0.83	(0.42, 1.25)	0.82	(0.36, 1.22)
σ^2_{age}	0.023	(0.003, 0.043)	0.023	(0.002, 0.042)

of all the parameters. The default MCMC sequence used in most software packages is a Gibbs sampler where t iterations are used to obtain new values for the parameters in each step drawing from a conditional posterior parameter distribution. Typically, several MCMC chains are used and the first part of each chain is discarded as a burn-in phase. Mplus automatically monitors convergence using the Gelman–Rubin convergence criterion, which considers within- and between-chain variability of the parameter estimates (Gelman et al., 2004, pp. 296–8). After the chains have converged and the estimates from the burn-in phase have been omitted, the sample of MCMC draws is used to approximate the posterior distribution. From the posterior distribution, the mean, mode, or median of the desired parameters can be computed.

Consider the example of antisocial behavior and the regression equation in (22.1). This model was analyzed using Mplus v6.11 using the default settings. To illustrate the convergence properties, consider Figure 22.3, a trace plot of the parameter σ^2_{ind}. On the x-axis, the iterations of the MCMC procedures are displayed (400 iterations were used), and on the y-axis, the parameter value of each iteration. For the first iteration the starting value 1.85 has been used, which is far away from the final posterior mean. Therefore, the burn-in phase (i.e., the part on the left-hand side of the vertical line) must be discarded. The posterior distribution of the variance around the slope is shown in Figure 22.4. As can be seen, the posterior is not normally distributed, and this is taken into account in the calculation of the 95% credibility interval (.04–.12), which is allowed to be asymmetric in Bayesian analyses.

Maximum likelihood (ML) estimation methods are often used in multilevel modeling, and one might wonder what the advantages are of a Bayesian multilevel approach relative to the maximum likelihood approach. Jedidi and Ansari (2001, pp. 130–31) list some advantages: (1) prior information can be used (see also van de Schoot et al., 2011); (2) it allows estimating individual-specific parameters (such as missing observations) while accounting for the uncertainty; (3) Bayesian procedures do not rely on asymptotic inference; (4) by using MCMC procedures, simulation-based estimates of the parameters are obtained, which circumvents the need for evaluating the complex multidimensional integrals that are often required to implement ML methods on multilevel data. We have already mentioned the advantage of Bayesian estimation when sample sizes are small. Other advantages are the robustness of Bayesian estimation with respect to (1) the robustness of the approach, and (2) the robustness of misspecifying the likelihood function.

First we discuss the sensitivity of Bayesian estimators to the prior as a statement about the robustness of the approach. Bayesian estimation always results in admissible values, and this feature is exactly why the Bayesian toolbox helps in the context of robustness. Let us explain why.

In our example about antisocial behavior there are two sets of unknown parameters $\boldsymbol{\theta}$:

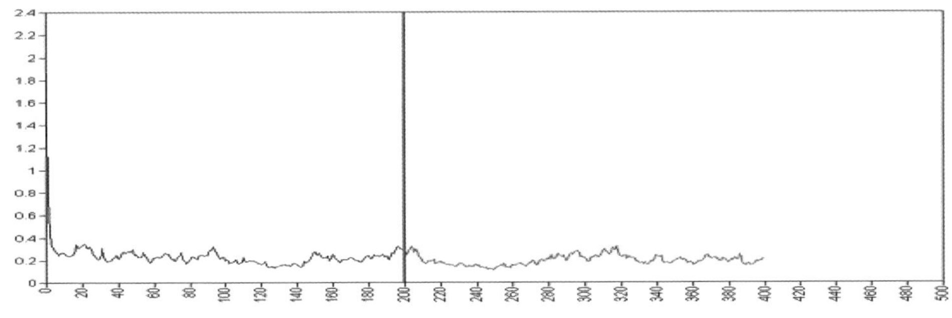

Figure 22.3 Trace Plot of the Parameter σ^2_{ind} with 400 Iterations and the Burn-In Phase on the Left-Hand Side of the Vertical Line

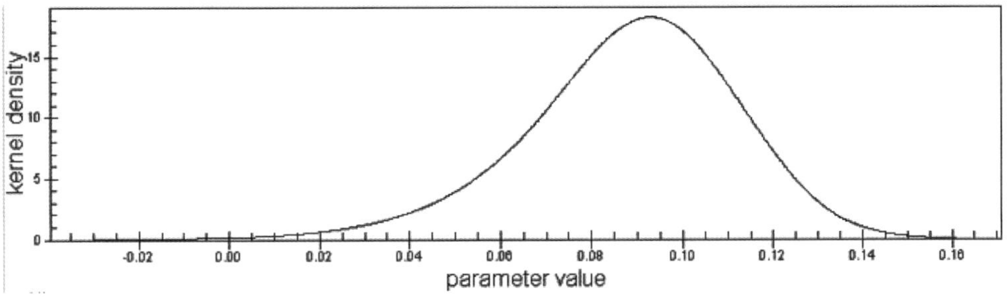

Figure 22.4 Posterior Distribution of the Parameter σ^2_{ind}

(1) fixed parameters (denoted by $\boldsymbol{\theta}_1$), and (2) the random error terms (denoted by $\boldsymbol{\theta}_2$). The prior distributions for the fixed part, $p(\boldsymbol{\theta}_1)$, are often normal distributions with a prior mean and a prior variance. The defaults in, for example, Mplus are a prior mean of zero and a very large prior variance. The prior distributions for the random part, $p(\boldsymbol{\theta}_2)$, are often gamma distributions or inverse chi-square distributions. The default in Mplus for variances/residual variances is a vague gamma distribution. Because the prior distributions of the random parameters are gamma or chi-square distributed, the posterior parameters obtained can never be negative. This feature makes Bayesian estimation robust in the sense of always obtaining admissible parameters, in contrast to the ML approach where negative variance/residual variance terms may be obtained.

In Table 22.3, the ML estimates, as well as the Bayesian estimates using the default settings in Mplus, are displayed. Because σ^2_{age} shows some skewness, see Figure 22.4, differences in the 95% CI can be observed between the asymptotic and Bayesian estimates. The Bayesian posterior 95% CI takes the skewness of σ^2_{age} into account, whereas the asymptotic results do not.

A second feature of the Bayesian toolbox is that one can experiment with the distributions that are used, which makes the method robust to the misspecification of the likelihood function and the prior distribution. By using robust methods, unusual observations do not exert an undue influence on the parameters obtained. That is, if normal distributions are specified for the likelihood or prior distribution, most statistical models are notoriously nonrobust to outliers.

Table 22.3 Asymptotic (ML) and Bayesian Estimates

Parameter	Asymptotic estimate	Asymptotic 95% CI	Posterior Mode	Posterior 95% CI
Constant	1.65	(1.48, 1.81)	1.64	(1.48, 1.81)
Child age	0.08	(0.04, 0.11)	0.08	(0.04, 0.11)
Emotional support	−0.17	(−0.23, −0.11)	−0.17	(−0.23, −0.11)
σ^2_{occ}	1.76	(1.57, 1.96)	1.76	(1.60, 1.95)
σ^2_{ind}	0.83	(0.42, 1.25)	0.83	(0.57, 1.20)
σ^2_{age}	0.023	(0.003, 0.043)	0.09	(0.04, 0.12)

In the multilevel framework, normal distributions are often specified for the random effects and for the within-subject errors. Both, however, are sensitive to outliers. An extreme case can have a strong effect on the posterior distribution, which might be unwanted. In the Bayesian framework, we can reduce the influence of outliers by replacing the normal distribution used for the likelihood and prior distribution by a longer-tailed family of distributions. Such a longer-tailed distribution allows for the possibility of outliers. By 'long-tailed' we mean a distribution with relatively high probabilities far away from the center (e.g., Gelman et al., 2004, pp. 443–59). Gelman et al. suggest switching to these distributions when 10% of the observations lie a distance more than 1.5 times the interquartile range away from the median (p. 445). Long-tailed distributions can be specified for the likelihood distribution, or they can be specified for the prior distribution.

An example of a long-tailed distribution is the t-distribution, or Cauchy distribution. As can be seen in Figure 22.5, the t-distribution allows for a higher probability of extreme cases. The t-distribution is defined by three parameters: the center, the scale, and the degrees of freedom. If the number of degrees of freedom is equal to one, the t-distribution is equivalent to the Cauchy distribution. Such a distribution is rather uninformative since it has infinite mean and variance. If the number of degrees of freedom is large, the t-distribution approaches the normal distribution. In applications where one wants to adjust for outliers it would make no sense to specify many degrees of freedom, because one could use the normal distribution in that case. It also makes no sense to specify just one or two degrees of freedom, because an infinite variance, which is the result, is usually not realistic in practice.

Another example is the use of (finite) mixture models. In these types of model, a normal distribution is used for the bulk of the observations, and a second distribution, for example a scaled chi-square distribution, is specified for the outliers. See Figure 22.6 for an example of a two-component mixture distribution. A complication of drawing from a mixture distribution is that such models have multimodal posterior densities (i.e., multiple modes). The Metropolis algorithm is often used instead of the Gibbs sampler to deal with this complication.

Staudenmayer et al. (2009) and Lin and Lee (2007) demonstrate how the t-distribution can be used for multilevel modeling. Other alternatives have also been studied in the context of multilevel modeling. For example, Rosa, Gianola, and Padovani (2004) and Rosa, Padovani, and Gianola (2003) used the slash distribution and the contaminated normal. Lachos, Cabral, and Abanto-Valle (2012) showed that, for random effects in a multilevel framework multivariate, thick-tailed distributions can be used to obtain robust parameter estimates. Lachos, Ghosh, and Arellano-Valle (2010) used multivariate skew-normal/independent distributions for the random effects in a longitudinal model and symmetric normal/independent distributions for the random errors. Software for Bayesian analysis using long-tailed distributions is available for R (e.g., Lachos

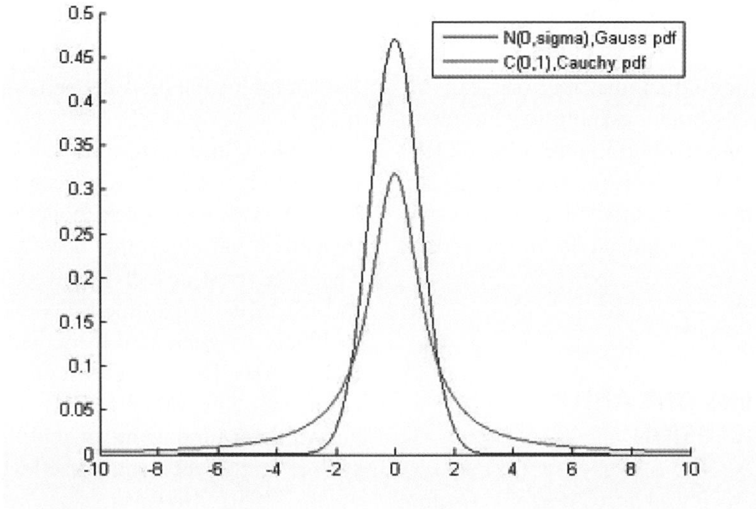

Figure 22.5 An Example of a Normal Distribution Versus a Cauchy /*t*-Distribution

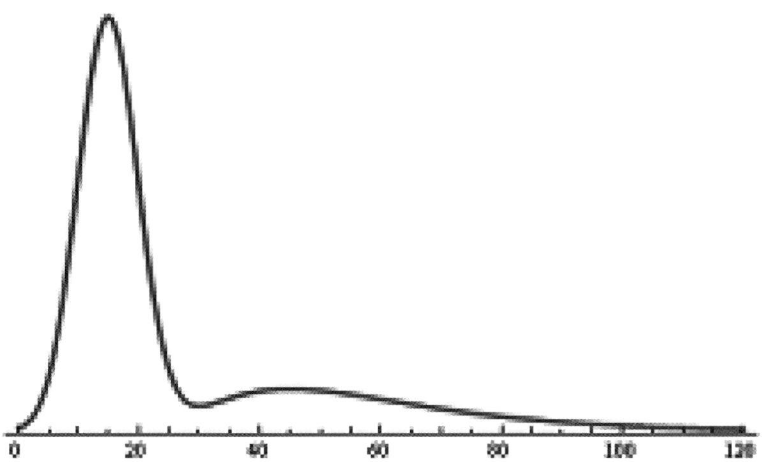

Figure 22.6 An Example of a Mixture of Distributions

et al., 2012) and BRUGS/WinBugs (Marley and Wand, 2010), but not yet for Mplus or MLwiN.

In addition to compensating for outliers, robust distributions, such as the *t*-distribution or the two-component mixture distribution, can be used to assess the sensitivity of the posterior distribution to model assumptions (see, for example, Gelman et al., 2004, and Lachos et al., 2012). This can be done by 'playing around' with the specification of the prior distribution using a set of priors with different degrees of freedom. By specifying more degrees of freedom, the prior gains more influence on the posterior distribution. One can inspect whether major differences occur in posterior estimates between the different prior specifications. The bigger the differences, the less robust the model is to assumptions like the influence of outliers.

In sum, long-tailed distributions can be used to evaluate model assumptions, to identify outliers, and to obtain more robust inferences. Long-tailed distributions provide an appealing robust alternative to the usual symmetric normal distribution in linear mixed models.

22.8 EXTENSIONS AND CONCLUSION

The methods reviewed here generally assume multivariate normality and are focused mostly on multilevel regression analysis. In multilevel structural equation modeling, the issues are the same, but the increased complexity of the models leads to additional complications. Structural equation modeling (SEM) is an inherently multivariate method, which makes diagnostic procedures to assess the effect of non-normality of outliers more cumbersome. In multilevel SEM (MSEM), we have the problem that violations of assumptions or misspecifications of the model at one level may propagate into the other level. Just as in multilevel regression, influential cases may include both influential individuals and influential groups. So far, the field of diagnosing violations of assumptions in MSEM is largely unexplored. Robust standard errors and Bayesian methods can be applied in MSEM in a manner similar to their application in multilevel regression analysis. However, bootstrapping in MSEM is more difficult. As outlined above, a naïve casewise resampling procedure does not work well for multilevel data. When cases are sampled, the number of groups can change, and when groups are resampled, the total sample size can change. Timmerman et al. (2009) discuss several different resampling schemes in the context of multilevel component analysis. They conclude that the appropriate resampling scheme depends on the design, i.e., which levels are considered random and which levels are considered fixed. Van der Leeden et al. (2008) also conclude that the choice for a specific bootstrapping method depends on the study design.

In the context of multilevel regression, resampling residuals was recommended. In MSEM, with a complex model and multiple dependent variables, this becomes unwieldy. Thus, the most appropriate way to bootstrap in MSEM is the parametric bootstrap, where residuals are generated from their theoretical distribution. This can be useful to determine confidence intervals for effects that are difficult to determine using asymptotic methods, such as multilevel mediation effects (Preacher et al., 2010). However, since residuals are generated from their theoretical distributions rather than resampled, parametric bootstrapping does not address the problem of non-normal distributions.

The approaches for robust multilevel estimation described above generalize easily to nonlinear models such as models for categorical variables or count data. Again, the bootstrap is the exception. Nonlinear models are generally analyzed as generalized linear models, where a linear regression equation predicts a latent outcome, and a nonlinear link function links the latent variable to the observed non-normal outcome. As a result, residuals would be latent, and resampling observed residuals is not possible.

Given the problems with multilevel bootstrapping when the models become more complex, good general approaches to robust estimation appear using robust standard errors and chi-squareds, and Bayesian estimation. Robust standard errors and chi-squares work well if the sample size is not too small, recommendations ranging from 50 to 100 as a lower limit, with 100 preferred. Bayesian estimation as incorporated currently in standard multilevel analysis software does not incorporate provisions for dealing with outliers or unduly influential data points, but they will always produce parameter estimates that are permissible (e.g., no negative variances). As a result, the impact of outliers will be lessened, and

Bayesian estimation deals better with small samples than asymptotic maximum likelihood estimation. Bayesian estimation can also be extended to address violations of assumptions, but here one should be wary. Our suggestion is to carefully inspect trace plots to assess convergence, and if default convergence criteria are used we suggest setting the cut-off criteria to a stricter value. This requires expert knowledge of both multilevel modeling and Bayesian statistics, and is beyond the scope of this chapter.

NOTE

1 The example data are a data set constructed by Patrick Curran for a symposium *Comparing Three Modern Approaches to Longitudinal Data Analysis: An Examination of a Single Developmental Sample* conducted at the 1997 Biennial Meeting of the Society for Research in Child Development. In this symposium, several different approaches to longitudinal modeling (latent growth curves, multilevel analysis, and mixture modeling) were compared and contrasted by letting experts analyze a single shared data set. This data set, herein called the *Curran data*, was compiled from a large longitudinal data set. A copy of the data in SPSS format is available from `http://www.joophox.net/mlbook2/MLbook.htm`.

REFERENCES

Asparouhov, T. and Muthén, B. (2010) 'Bayesian Analysis using Mplus: Technical implementation', Technical Report. Version 3. `http://statmodel.com/papers.shtml`.

Booth, J.G. and Sarkar, S. (1998) 'Monte Carlo Approximation of Bootstrap Variances', *American Statistician*, 52: 354–7.

Carpenter, J.M. and Bithell, J. (2000) 'Bootstrap Confidence Intervals: When, Which, What? A Practical Guide for Medical Statisticians', *Statistics in Medicine*, 19: 1141–64.

Carpenter, J.M., Goldstein, H., and Rasbash, J. (2003) 'A Novel Bootstrap Procedure for Assessing the Relationship between Class Size and Achievement', *Applied Statistics*, 52: 431–43.

Cheong, Y.F., Fotiu, R.P., and Raudenbush, S.W. (2001) 'Efficiency and Robustness of Alternative Estimators for Two- and Three-level Models: The Case of NAEP', *Journal of Educational and Behavioral Statistics*, 26: 411–29.

Efron, B. (1982) *The Jackknife, the Bootstrap and Other Resampling Plans*. Philadelphia: Society for Industrial and Applied Mathematics.

Efron, B. and Tibshirani, R.J. (1993) *An Introduction to the Bootstrap*. New York: Chapman and Hall.

Eliason, S.R. (1993) *Maximum Likelihood Estimation*. Newbury Park, CA: Sage.

Ferron, J.M., Hogarty, K.Y., Dedrick, R.F., Hess, M.R., Niles, J.D., and Kromrey, J.D. (2008) 'Reporting results from multilevel analyses', in A.A. O'Connell and D.B. McCoach (eds), *Multilevel Modeling of Educational Data*. Charlotte, NC: Information Age Publishing, Inc., 391–426.

Field, A. (2009) *Discovering Statistics Using SPSS*. Los Angeles: Sage.

Fisher, R.A. (1915) 'Frequency Distribution of the Values of the Correlation Coefficient in Samples from an Indefinitely Large Population', *Biometrika*, 10: 507–21.

Gelman, A., Carlin, J.B., Stern, H.S., and Rubin, D.B. (2004) *Bayesian Data Analysis*, 2nd edition. London: Chapman and Hall.

Goldstein, H. (2011a) *Multilevel Statistical Models*, 4th edition. London: Wiley.

Goldstein, H. (2011b) 'Bootstrapping in Multilevel Models', in J.J. Hox and J.K. Roberts (eds), *Handbook of Advanced Multilevel Analysis*. New York: Routledge, 163–72.

Good, P.I. (1999) *Resampling Methods: A Practical Guide to Data Analysis*. Boston/Berlin: Birkhäuser.

Hoijtink, H., Klugkist, I., and Boelen, P.A. (eds) (2008) *Bayesian Evaluation of Informative Hypotheses*. New York: Springer.

Howard, G.S., Maxwell, S.E., and Fleming, K. (2000) 'The Proof of the Pudding: An Illustration of the Relative Strengths of Null Hypothesis, Meta-analysis, and Bayesian Analysis', *Psychological Methods*, 5: 315–32.

Hox, J.J., Maas, C.J.M., and Brinkhuis, M.J.S. (2010) 'The Effect of Estimation Method and Sample Size in Multilevel Structural Equation Modeling', *Statistica Neerlandica*, 64: 157–70.

Hox, J.J., van de Schoot, R., and Matthijsse, S. (2012) 'How Few Countries Will Do? Comparative Survey Analysis from a Bayesian Perspective', *Survey Research Methods*, 6, 87–93.

Huber, P.J. (1967) 'The Behavior of Maximum Likelihood Estimates under Non-standard Conditions', in *Proceedings of the Fifth Berkeley Symposium on Mathematical Statistics and Probability*. Berkeley, CA: University of California Press, 221–33.

Jedidi, K. and Ansari, A. (2001) 'Bayesian Structural Equation Models for Multilevel Data', in G.A. Marcoulides and R.E. Schumacker (eds), *New Developments and Techniques in Structural Equation Modeling*. New York: Erlbaum, 129–57.

Kauermann, G. and Carroll, R.J. (2001) 'A Note on the Efficiency of Sandwich Covariance Matrix Estimation', *Journal of the American Statistical Association*, 96: 1387–96.

Lachos, V.H., Cabral, C.R.B., and Abanto-Valle, C.A. (2012) 'A Noniterative Sampling Bayesian Method for Linear Mixed Models with Normal Independent Distributions', *Journal of Applied Statistics*, 39: 531–49.

Lachos, V.H., Ghosh, P., and Arellano-Valle, R.B. (2010) 'Likelihood Based Inference for Skew-normal Independent Linear Mixed Models', *Statistica Sinica*, 20: 303–22.

Langford, I.H. and Lewis, T. (1998) 'Outliers in Multilevel Models (with discussion)', *Journal of the Royal Statistical Society, Series A*, 161: 121–60.

Lee, M.D. and Wagenmakers, E.-J. (2005) 'Bayesian Statistical Inference in Psychology: Comment on Trafimow (2003)', *Psychological Review*, 112: 662–8.

Lee, S.-Y. (2007) *Structural Equation Modeling: A Bayesian Approach*. West Sussex: John Wiley and Sons Ltd.

Leyland, A.H. (2006) 'A Review of Multilevel Modelling in SPSS', Centre for Multilevel Modelling Software Reviews of Multilevel Analysis Packages, `http://www.bristol.ac.uk/cmm/learning/mmsoftware/reviewspss.pdf`.

Lin, T.I. and Lee, J.C. (2007) 'Bayesian Analysis of Hierarchical Linear Mixed Modeling using the Multivariate t-distribution', *Journal of Statistical Planning and Inference*, 137: 484–95.

Lynch, S. (2007) *Introduction to Applied Bayesian Statistics and Estimation for Social Scientists*. New York: Springer.

Maas, C.J.M. and Hox, J.J. (2004) 'Robustness Issues in Multilevel Regression Analysis', *Statistica Neerlandica*, 58: 127–37.

Maas, C.J.M. and Hox, J.J. (2005) 'Sufficient Sample Sizes for Multilevel Modeling', *Methodology*, 1: 85–91.

Marley, J.K. and Wand, M.P. (2010) 'Non-Standard Semiparametric Regression via BRugs', *Journal of Statistical Software*, 37: 1–30.

Mooney, C.Z. and Duval, R.D. (1993) *Bootstrapping. A Nonparametric Approach to Statistical Inference*. Newbury Park, CA: Sage.

Muthén, B. and Asparouhov, T. (2012) 'Bayesian SEM: A More Flexible Representation of Substantive Theory', *Psychological Methods*, to appear.

Muthén, L.K. and Muthén, B.O. (1998–2010). *Mplus User's Guide*, 6th edition. Los Angeles, CA: Muthén and Muthén.

Nevitt, J. and Hancock, G.R. (2001) 'Performance of Bootstrapping Approaches to Model Test Statistics and Parameter Standard Error Estimation in Structural Equation Modeling', *Structural Equation Modeling*, 8: 353–77.

Preacher, K.J., Zyphur, M.J., and Zhang, Z. (2010) 'A General Multilevel SEM Framework for Assessing Multilevel Mediation', *Psychological Methods*, 15: 209–33.

Raudenbush, S.W. and Bryk, A.S. (2002) *Hierarchical Linear Models*. Thousand Oaks, CA: Sage.

Rosa, G.J.M., Gianola, D., and Padovani, C.R. (2004) 'Bayesian Longitudinal Data Analysis with Mixed Models and Thick-tailed Distributions using MCMC', *Journal of Applied Statistics*, 31, 855–73.

Rosa, G.J.M., Padovani, C.R., and Gianola, D. (2003) 'Robust Linear Mixed Models with Normal/ Independent Distributions and Bayesian MCMC Implementation', *Biometrical Journal*, 45: 573–90.

Satorra, A. and Bentler, P.M. (2001) 'A Scaled Difference Chi-square Test Statistic for Moment Structure Analysis', *Psychometrika*, 66: 507–14.

Staudenmayer, J., Lake, E.E., and Wand, M.P. (2009) 'Robustness for General Design Mixed Models Using the t-distribution', *Statistical Modelling*, 9: 235–55.

Stegmueller, D. (2013) 'How Many Countries for Multilevel Modeling? A Comparison of frequentist Bayesian and Approaches'. *American Journal of Political Science*, DOI: 10.1111/aips.120001.

Tabachnick, B.G. and Fidell, L.S. (2007) *Using Multivariate Statistics*. New York: Pearson.

Timmerman, M.E., Kiers, H.A.L., Smilde, A.K., Ceulemans, E., and Stouten, J. (2009) 'Bootstrap Confidence Intervals in Multi-level Simultaneous Component Analysis', *British Journal of Mathematical and Statistical Psychology*, 62: 299–318.

van de Schoot, R., Hoijtink, H., Mulder, J., Van Aken, M.A.G., Orobio de Castro, B., Meeus, W., and Romeijn, J.-W. (2011) 'Evaluating Expectations about Negative Emotional States of Aggressive Boys using Bayesian Model Selection', *Developmental Psychology*, 47: 203–12.

van der Leeden, R., Meijer, E., and Busing, F.M.T.A. (2008) 'Resampling Multilevel Models', in J. de Leeuw and E. Meijer (eds), *Handbook of Multilevel Analysis*. New York: Springer, 401–33.

Walker, L.J., Gustafson, P. and Frimer, J.A. (2007) 'The Application of Bayesian Analysis to Issues in Developmental Research', *International Journal of Behavioral Development*, 4: 366–73.

White, H. (1982) 'Maximum Likelihood Estimation of Misspecified Models', *Econometrica*, 50: 1–25.

Yuan, K.H. and Bentler, P.M. (2000) 'Three Likelihood-based Methods for Mean and Covariance Structure Analysis with Nonnormal Missing Data', *Sociological Methodology*, 30: 165–200.

Yung, Y.-F. and Chan, W. (1999) 'Statistical Analyses using Bootstrapping: Concepts and Implementation', in R. Hoyle (ed.), *Statistical Strategies for Small Sample Research*. Thousand Oaks, CA: Sage, 82–108.

23

Missing Data

Geert Molenberghs and Geert Verbeke
Interuniversity Institute for Biostatistics and Statistical Bioinformatics
Universiteit Hasselt and Katholieke Universiteit Leuven, Belgium

23.1 INTRODUCTION

The problem of dealing with missing values is common throughout statistical work and is particularly present whenever human subjects are enrolled. Respondents may refuse participation or may be unreachable. Patients in clinical and epidemiological studies may withdraw their initial consent without further explanation. Early work on missing values was largely concerned with algorithmic and computational solutions to the induced lack of balance or deviations from the intended study design (Afifi and Elashoff, 1966; Hartley and Hocking, 1971). More recently, general algorithms such as Expectation–Maximization (EM) (Dempster et al., 1977) and data imputation and augmentation procedures (Rubin, 1987; Tanner and Wong, 1987), combined with powerful computing resources, have largely provided a solution to this aspect of the problem. There remains the very difficult and important question of assessing the impact of missing data on subsequent statistical inference. Conditions can be formulated under which an analysis that proceeds as if the missing data are missing by design, that is, ignoring the missing value process, can provide valid answers to study questions. While such an approach is attractive from a pragmatic point of view, the difficulty is that such conditions can rarely be assumed to hold with full certainty. Indeed, assumptions will be required that cannot be assessed from the data under analysis. Hence, in this setting there cannot be anything that could be termed a definitive analysis, and hence any analysis of preference is ideally to be supplemented with a so-called sensitivity analysis.

This chapter is organized as follows. First, a couple of illustrative examples are introduced. Second, attention is devoted to an overview of the fundamental concepts of incomplete hierarchical data, including a general framework within which missing data ideas can be developed as well as a modeling framework for both Gaussian and non-Gaussian outcomes, useful to further develop and illustrate practical strategies for dealing with incomplete data. Third, simple methods that are often used but carry quite a bit of danger in them are discussed. Fourth, an overview is provided of methods which are valid when missingness is missing at random (MAR), that is, when the probability of dropout is allowed to depend on the observed measurements. Such methods can be used as a primary analysis. In this chapter, the focus will be primarily on incomplete outcomes. However, incomplete covariate data, perhaps in conjunction with incomplete outcome data, are relatively common too. It should be noted, therefore, that multiple imputation, discussed in the fifth instance, can nicely handle this type of missingness as well. Sixth, two approaches to modeling data with non-random dropout are considered: the selection model framework and the pattern-mixture modeling family. The optimal place of such missing not at random (MNAR) models is within a sensitivity analysis, the final topic of this chapter.

23.2 ILLUSTRATIVE EXAMPLES

23.2.1 *The Vorozole Study*

This study was an open-label, multicenter, parallel group design conducted at 67 North American centers. Patients were randomized to either the new drug Vorozole (2.5 mg taken once daily) or the standard drug megestrol acetate (40 mg four times daily). The patient population consisted of postmenopausal patients with histologically confirmed estrogen-receptor-positive metastatic breast carcinoma. All 452 randomized patients were followed until disease progression or death. The main objective was to compare the treatment groups with respect to response rate, whereas secondary objectives included a comparison relative to duration of response, time to progression, survival, safety, pain relief, performance status, and quality of life. Full details of this study are reported in Goss et al. (1999). In this chapter, we will focus on overall quality of life, measured by the total Functional Living Index: Cancer (FLIC; Schipper, Clinch, and McMurray, 1984); a higher FLIC score is the more desirable outcome. Even though this outcome is, strictly speaking, of the ordinal type, the total number of categories encountered exceeds 70, justifying the use of continuous-outcome methods.

Patients underwent screening and, for those deemed eligible, a detailed examination at baseline (occasion 0) took place. Further measurement occasions were months 1, then from months 2 at bimonthly intervals until month 44.

Goss et al. (1999) analyzed FLIC using a two-way ANOVA model with effects for treatment and disease status, as well as their interaction. No significant difference was found. Apart from treatment, important covariates are dominant site of the disease as well as clinical stage.

23.2.2 *Orthodontic Growth Data*

These data, introduced by Pothoff and Roy (1964), contain growth measurements for 11 girls and 16 boys. For each subject, the distance from the center of the pituitary to the maxillary fissure was recorded at ages 8, 10, 12, and 14. The data were used by Jennrich and Schluchter (1986) to illustrate estimation methods for unbalanced data, where unbalancedness is now to be interpreted in the sense of an unequal number of boys and girls. Individual profiles and sex group by age means are plotted in Figure 23.1.

Little and Rubin (2002) deleted 9 of the $[(11 + 16) \times 4]$ measurements at age 10, thereby producing 9 incomplete subjects; they described the mechanism to be such that subjects with a low value at age 8 are more likely to have a missing value at age 10. The data are presented in Table 23.1. The measurements that were deleted are marked with an asterisk.

23.3 FUNDAMENTAL CONCEPTS

When referring to the missing-value, or non-response, process we will use the terminology of Little and Rubin (2002). A non-response process is said to be *missing completely at random* (MCAR) if the missingness is independent of both unobserved and observed data, and *missing at random* (MAR) if, conditional on the observed data, the missingness is independent of the unobserved measurements. A process that is neither MCAR nor MAR is termed *missing not at random* (MNAR). In the context of likelihood inference, and when the parameters describing the measurement

Table 23.1 The Orthodontic Growth Data: Data for 11 Girls and 16 Boys. Measurements marked with * were deleted by Little and Rubin (2002)

	Age (in years)					Age (in years)			
Girl	8	10	12	14	Boy	8	10	12	14
1	21.0	20.0	21.5	23.0	1	26.0	25.0	29.0	31.0
2	21.0	21.5	24.0	25.5	2	21.5	22.5*	23.0	26.5
3	20.5	24.0*	24.5	26.0	3	23.0	22.5	24.0	27.5
4	23.5	24.5	25.0	26.5	4	25.5	27.5	26.5	27.0
5	21.5	23.0	22.5	23.5	5	20.0	23.5*	22.5	26.0
6	20.0	21.0*	21.0	22.5	6	24.5	25.5	27.0	28.5
7	21.5	22.5	23.0	25.0	7	22.0	22.0	24.5	26.5
8	23.0	23.0	23.5	24.0	8	24.0	21.5	24.5	25.5
9	20.0	21.0*	22.0	21.5	9	23.0	20.5	31.0	26.0
10	16.5	19.0*	19.0	19.5	10	27.5	28.0	31.0	31.5
11	24.5	25.0	28.0	28.0	11	23.0	23.0	23.5	25.0
					12	21.5	23.5*	24.0	28.0
					13	17.0	24.5*	26.0	29.5
					14	22.5	25.5	25.5	26.0
					15	23.0	24.5	26.0	30.0
					16	22.0	21.5*	23.5	25.0

Source: Pothoff and Roy (1964), Jennrich and Schluchter (1986).

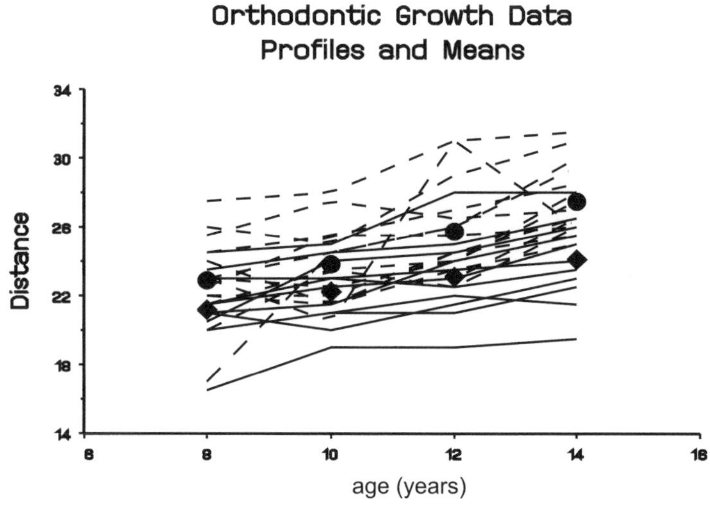

Figure 23.1 The Orthodontic Growth Data: Observed Profiles and Group by Age Means. Solid lines and diamonds are for girls, dashed lines and bullets are for boys

process are functionally independent of the parameters describing the missingness process, MCAR and MAR are *ignorable*, while a non-random process is non-ignorable. Thus, under ignorable dropout, one can literally ignore the missingness process and nevertheless obtain valid estimates of, say, the treatment. The definitions above are conditional on including the correct set of covariates in the model. An overview of the various mechanisms, and their (non-) ignorability under likelihood, Bayesian, or frequentist inference, is given in Table 23.2.

Let us first consider the case where one follow-up measurement per subject is made. When dropout occurs, and hence there are no follow-up measurements, one is usually forced to discard such a subject from analysis, thereby violating the intention to treat (ITT) principle which stipulates that all randomized subjects should be included in the primary analysis and according to the randomization scheme. Of course, the effect of treatment can be investigated under extreme assumptions, such as, for example, a worst-case and a best-case scenario, but such scenarios are most often not really helpful.

Early work regarding missingness focused on the consequences of the induced lack of balance of deviations from the study design (Afifi and Elashoff, 1966; Hartley and Hocking, 1971). Later, algorithmic developments took place such as the expectation–maximization algorithm (EM; Dempster, Laird, and Rubin, 1977) and multiple imputation (Rubin, 1987). These have brought likelihood-based ignorable analysis within reach of a large class of designs and models.

In the meantime, however, clinical trial practice in particular has put a strong emphasis on methods such as *complete case analysis* (CC) and *last observation carried forward* (LOCF), or other simple forms of imputation. Claimed advantages include computational simplicity, no need for a full longitudinal model analysis (see Chapter 9; e.g., when the scientific question is in terms of the last planned measurement occasion only), and, for LOCF, compatibility with the ITT principle. However, a CC analysis assumes MCAR and the LOCF analysis makes peculiar assumptions on the (unobserved) evolution of the response, underestimates the variability of the response, and ignores the fact that imputed values are not real data.

On the other hand, a likelihood-based longitudinal analysis (Chapter 9) requires only MAR, uses all data (obviating the need for both deleting and filling in data), and is also consistent with the ITT principle. Further, it can also be shown that the incomplete sequences contribute to estimands of interest (treatment effect at the end of the study), even for early dropouts. For continuous responses, the linear mixed model is quite popular and is a direct extension of ANOVA and MANOVA approaches, but more broadly valid in incomplete data settings. For categorical responses and count data, so-called marginal (e.g., generalized estimating equations, GEE: Chapters 15 and 25) and random-effects (e.g., generalized linear mixed-effects models, GLMM) approaches are in use. While GLMM parameters can be fitted using maximum likelihood, the same is not true for the frequentist GEE method (Chapters 15 and 25), but modifications have been proposed to accommodate the MAR assumption (Robins, Rotnitzky, and Zhao, 1995).

Finally, MNAR missingness can never be fully ruled out based on the observed data only. It is argued that, rather than going either for discarding MNAR models entirely or for placing full faith in them, a sensible compromise is to make them a component of a sensitivity analysis.

23.4 SIMPLE METHODS

We will focus on two relatively simple methods that have been and still are in extensive use. A detailed account of simple methods to handle missingness is given in Verbeke and Molenberghs (1997, 2000).

23.4.1 Complete Case Analysis

A *complete case analysis* includes only those cases for analysis for which all measurements were recorded. This method has obvious advantages. It is very simple to describe and, since the data structure is as would have resulted from a complete experiment, standard statistical software can be used without additional work. Further, since the entire estimation is done on the same subset of

Table 23.2 Overview of Missing Data Mechanisms

Acronym	Description	Likelihood/Bayesian	Frequentist
MCAR	missing completely at random	ignorable	ignorable
MAR	missing at random	ignorable	non-ignorable
MNAR	missing not at random	non-ignorable	non-ignorable

completers, there is a common basis for inference. Unfortunately, the method suffers from severe drawbacks. First, there is nearly always a substantial loss of information. The impact on precision and power is dramatic. Further, such an analysis will only be representative for subjects who remain in the study. Of course, a complete case analysis could have a role as an auxiliary analysis, especially if a scientific question relates to it. A final important issue about a complete case analysis is that it is only valid when the missingness mechanism is MCAR. However, severe bias can result when the missingness mechanism is MAR but not MCAR. This bias can go both ways, i.e., either overestimating or underestimating the true effect.

23.4.2 Last Observation Carried Forward

A method that has received a lot of attention (Siddiqui and Ali, 1998; Mallinckrodt et al., 2003a,b) is *last observation carried forward* (LOCF). As already noted before, whenever a value is missing, in the LOCF method the last observed value is substituted. For the LOCF approach, the MCAR assumption is necessary but not sufficient for an unbiased estimate. Indeed, it further assumes that subjects' responses would have been constant from the last observed value to the endpoint of the trial. These conditions seldom hold (Verbeke and Molenberghs, 2000). In a clinical trial setting, one might believe that the response profile *changes* as soon as a subject goes off treatment, and even that it would flatten. However, the constant profile assumption is even stronger. Therefore, carrying observations forward may bias estimates of treatment effects and underestimate the associated standard errors (Verbeke and Molenberghs, 2000; Gibbons et al., 1993; Heyting, Tolboom, and Essers, 1992; Lavori, Dawson, and Shera, 1995; Mallinckrodt et al., 2001a,b). Further, this method artificially increases the amount of information in the data, by treating imputed and actually observed values on an equal footing.

Despite its shortcomings, LOCF has been the longstanding method of choice for primary analysis in clinical trials because of its simplicity, ease of implementation, and the belief that the potential bias from carrying observations forward leads to a "conservative" analysis in comparative trials. An analysis is called conservative when it leads to no treatment difference, while in fact there is treatment difference. However, reports of anti-conservative or liberal behavior of LOCF are common (Beunckens et al., 2005; Molenberghs et al., 2004; Mallinckrodt et al., 2004; Little and Yau, 1996; Liu and Gould, 2002). This means that a LOCF analysis can create treatment effect when none exists. Thus, the statement that LOCF analysis has been used to provide a conservative estimate of treatment effect is unacceptable.

Historically, an important motivation behind the simpler methods was their simplicity. Indeed, the main advantage, shared with complete case analysis, is that complete data software can be used. However, with the availability of commercial software tools such as, for example, the SAS procedures MIXED and NLMIXED and the SPlus and R nlme libraries, this motivation no longer applies.

It is often quoted that LOCF or CC, while problematic for parameter estimation, produce randomization-valid hypothesis testing, but this is questionable. First, in a CC analysis partially observed data are selected out, with probabilities that may depend on post-randomization outcomes, thereby undermining any randomization justification. Second, if the focus is on one particular time point, e.g., the last one scheduled, then LOCF plugs in data. Such imputations, apart from artificially inflating the information content, may deviate in complicated ways from the underlying data. Third, although the size of a randomization-based LOCF test may reach its nominal size under the null hypothesis of no difference in treatment profiles, there will be other regions of the alternative space where the power of the LOCF test procedure is equal to its size, which is completely unacceptable.

23.5 METHODS VALID UNDER MISSING AT RANDOM

We will first provide a graphical illustration, using an artificial example, of the various simple methods we have considered and then turn to so-called direct likelihood analysis.

23.5.1 Illustration of Simple Methods

Let us take a look at an artificial but insightful example, depicted in Figure 23.2, which displays the results of the traditional methods, CC, and LOCF, next to the result of an MAR method. In this example, the mean response is supposed to be linear. For both groups (completers and dropouts), the slope is the same but intercepts differ. Patients with incomplete observations dropped out halfway through the study, e.g., because they reached a certain level of the outcome. This is obviously an MAR missingness mechanism. Using a method valid under the MAR assumption yields the correct mean profile, being a straight line centered between the mean profiles of the completers and incompleters. If one performed a CC analysis, the fitted profile would coincide with the mean profile of the complete cases (bold line). Next, under LOCF, data are imputed (dashed line). The resulting fitted profile will be the bold dashed line. Clearly, both traditional methods produce an incorrect result.

Further, in a traditional available case analysis (AC), one makes use of the information actually available. One such set of estimators could be the treatment-specific mean at a number of designed measurement occasions. With a decreasing sample size over time, means later in time would be calculated using fewer subjects than means earlier in time. Very importantly, without correcting for this the set of subjects available at one time may differ in important characteristics from sets of subjects at other times. Figure 23.2 shows a dramatic instance of this, due to the rather extreme nature of this illustrative example. The key message is that such an approach is unable to remove major sources of bias.

23.5.2 Direct Likelihood Analysis

For continuous outcomes, Verbeke and Molenberghs (2000) describe likelihood-based mixed-effects models, which are valid under the MAR assumption. Indeed, for longitudinal studies, such as described in Chapter 9, where missing data are involved, a mixed model only requires that the missing data are MAR. In contrast to the traditional techniques, mixed-effects models permit the inclusion of subjects with missing values at some time points (both dropout and intermittent missingness).

This likelihood-based MAR analysis is also termed likelihood-based ignorable analysis, or, as we will be using in the remainder of this chapter, a *direct likelihood analysis*. In such a direct likelihood analysis, the observed data are used without deletion or imputation. In doing so, appropriate adjustments are made to parameters at times when data are incomplete, due to the within-subject correlation.

Thus, even when interest lies, for example, in a comparison between the two treatment groups at the last occasion, such a full longitudinal analysis (Chapter 9) is a good approach,

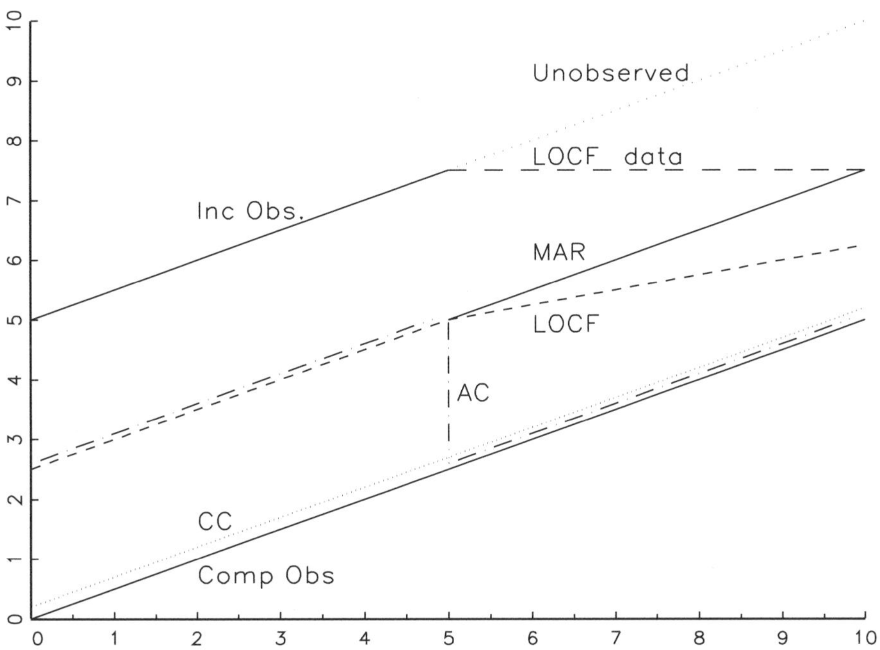

Figure 23.2 Artificial Situation, Illustrating the Results of the Traditional MCAR Methods (CC and LOCF) Next to the Result of the Direct Likelihood Method

since the fitted model can be used as the basis for inference at the last occasion.

In many clinical trials the repeated measures are balanced in the sense that a common (and often limited) set of measurement times is considered for all subjects, which allows the a priori specification of a "saturated" model. An example of this is when a full group-by-time interaction for the fixed effects combines with an unstructured covariance matrix (Chapter 10). Such a model specification is sometimes termed Mixed-effects Model Repeated-Measures analysis (MMRM; Mallinckrodt et al., 2003b). Thus, MMRM is a particular form of a linear mixed model, relevant for acute-phase confirmatory clinical trials fitting within the direct likelihood paradigm. Moreover, this direct likelihood MMRM analysis is equivalent to a classical MANOVA analysis when the data are complete, but more generally valid when they are incomplete. This is a strong answer to the common criticism that a direct likelihood method is making strong assumptions. Indeed, its coincidence with MANOVA for data without missingness shows that the assumptions made are very mild. Therefore, it constitutes a very promising alternative for CC and LOCF. When a relatively large number of measurements is made within a single subject, the full power of random effects modeling can be used (Verbeke and Molenberghs, 2000).

The practical implication is that a software module with likelihood estimation facilities, and with the ability to handle incompletely observed subjects, manipulates the correct likelihood, providing valid parameter estimates and likelihood ratio values.

A few cautionary remarks are warranted. First, when at least part of the scientific interest is directed towards the non-response process, both processes obviously need to be considered. Under MAR, both questions can be answered separately. This implies that a conventional method can be used to study questions in terms of the outcomes of interest, such as treatment effect and time trend, whereafter a separate model can be considered to study missingness. Second, likelihood inference is often surrounded with references to the sampling distribution (e.g., to construct measures of precision for estimators and for statistical hypothesis tests; Kenward and Molenberghs, 1998). However, the practical implication is that standard errors and associated tests, when based on the observed rather than the expected information matrix, and given that the parametric assumptions are correct, are valid. Thirdly, it may be hard to rule out the operation of an MNAR mechanism. This point was brought up in the introduction and will be discussed further in Section 23.8.

23.5.3 Illustration: Orthodontic Growth Data

As an example, we use the orthodontic growth data introduced in Section 23.2.2.

The simple methods and direct likelihood method from Sections 23.4 and 23.5 are now compared using the growth data. A linear mixed model is used for this purpose, assuming an unstructured mean, i.e., assuming a separate mean for each of the eight age $\times$ sex combinations, together with an unstructured covariance structure, and using maximum likelihood (ML) as well as restricted maximum likelihood (REML). The mean profiles of the linear mixed model using maximum likelihood for all four data sets, for boys, are given in Figure 23.3. The girls' profiles are similar and hence not shown.

Next to this longitudinal (Chapter 9) approach, we will consider a full MANOVA analysis and a univariate ANOVA analysis, i.e., one per time point. For all of these analyses, Table 23.3 shows the estimates and standard errors for boys at ages 8 and 10, for the original data and all available incomplete data, as well as for the CC and the LOCF data.

First, we consider the group means for the boys in the original data set in Figure 23.3, i.e., we observe a relatively straight line. Clearly, there seems to be a linear trend in the mean profile.

In a complete case analysis of the growth data, the 9 subjects which lack one measurement are deleted, resulting in a working data set with 18 subjects. This implies that 27 available measurements will not be used for analysis, a quite severe penalty on a relatively small data set. Observing the profiles for the CC data set in Figure 23.3, all group means increased relative to the original data set, but mostly so at age 8. The net effect is that the profiles overestimate the average length.

For the LOCF data set, the 9 subjects that lack a measurement at age 10 are completed by imputing the age 8 value. It is clear that this procedure will affect the apparently increasing linear trend found for the original data set. Indeed, the imputation procedure forces the means at ages 8 and 10 to be more similar, thereby destroying the linear relationship. Hence, a simple, intuitively appealing interpretation of the trends is made impossible.

In the case of direct likelihood, we now see two profiles, one for the observed means and one for the fitted means. These two coincide at all ages except age 10. As mentioned earlier, the complete observations at age 10 are those with a higher measurement at age 8. Due to the within-subject correlation, they are the ones with a higher measurement at age 10 as well, and therefore the fitted model corrects in the appropriate direction. The consequences of this are very important. While we are inclined to believe that the fitted means do not follow the observed means all that well, this nevertheless is precisely what we should observe. Indeed, since the observed means are based on a non-random subset of the data, the fitted means take into account all observed data points, as well as information on the observed data at age 8, through the measurements that have been taken for such children at different time points.

As an aside to this, note that, in case of direct likelihood, the observed average at age 10 coincides with the CC average, while the fitted average does not coincide with anything else. Indeed, if the model specification is correct, then a direct likelihood analysis produces a consistent estimator for the average profile, as if nobody had dropped out. Of course, this effect might be blurred in relatively small data sets due to small-sample variability.

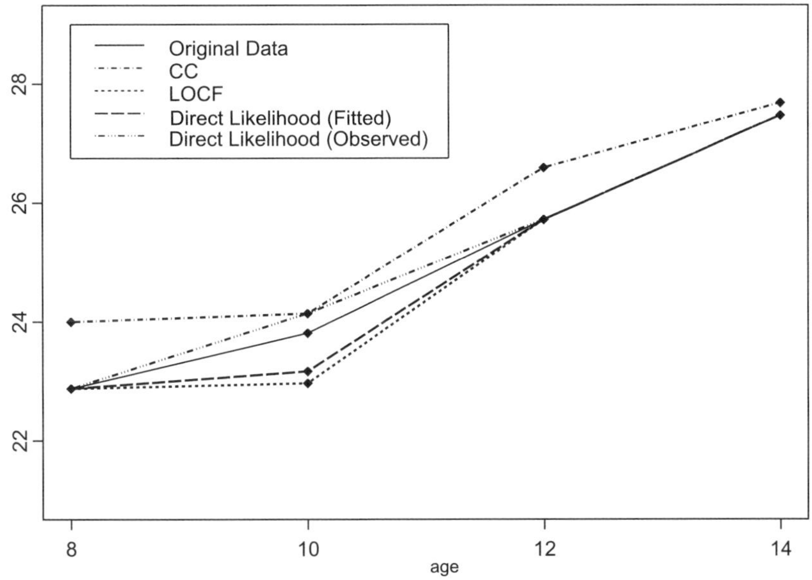

Figure 23.3 Orthodontic Growth Data: Profiles for the Original Data, CC, LOCF, and Direct Likelihood for Boys

Irrespective of the small-sample behavior encountered here, the validity under MAR and the ease of implementation are good arguments that favor this direct likelihood analysis over other techniques.

Let us now compare the different methods by means of Table 23.3, which shows the estimates and standard errors for boys at age 8 and 10, for the original data and all available incomplete data, as well as for the CC data and the LOCF data.

Table 23.3 shows some interesting features. In all four cases, a CC analysis gives an upward-biased estimate for both age groups. This is obvious, since the complete observations at age 10 are those with a higher measurement at age 8, as we have seen before. The LOCF analysis gives a correct estimate for the average outcome for boys at age 8. This is not surprising since there were no missing observations at this age. As noted before, the estimate for boys of age 10 is biased downwards. When the incomplete data are analyzed, we see from Table 23.3 that direct likelihood produces good estimates. The MANOVA and ANOVA per time point analyses give an overestimation of the average for age 10, as in the CC analysis. Further, the MANOVA analysis also yields an overestimation of the average at age 8, again the same as in the CC analysis. Note that the ANOVA per time point is exactly equal to the AC analysis, mentioned earlier. It is clear that the average at the age of 8 is correct, because there is no missingness at this point in time, while there is an overestimation at age 10. This is not surprising, because the subgroup available at age 10 consists of the children with higher measurements, a fact that cannot be corrected with this method.

Thus, direct likelihood shares the elegant and appealing features of ANOVA and MANOVA for fully observed data, but is superior with incompletely observed profiles.

23.6 MULTIPLE IMPUTATION

A well-known alternative method valid under the MAR assumption is multiple imputation (MI), which was introduced by Rubin (1978)

Table 23.3 Orthodontic Growth Data: Comparison of Analyses Based on Means at (Completely Observed Age 8 and Incompletely Observed Age 10 Measurement)

Method	Boys at Age 8	Boys at Age 10
	Original Data	
Direct likelihood, ML	22.88 (0.56)	23.81 (0.49)
Direct likelihood, REML	22.88 (0.58)	23.81 (0.51)
MANOVA	22.88 (0.58)	23.81 (0.51)
ANOVA per time point	22.88 (0.61)	23.81 (0.53)
	All Available Incomplete Data	
Direct likelihood, ML	22.88 (0.56)	23.17 (0.68)
Direct likelihood, REML	22.88 (0.58)	23.17 (0.71)
MANOVA	24.00 (0.48)	24.14 (0.66)
ANOVA per time point	22.88 (0.61)	24.14 (0.74)
	Complete Case Analysis	
Direct likelihood, ML	24.00 (0.45)	24.14 (0.62)
Direct likelihood, REML	24.00 (0.48)	24.14 (0.66)
MANOVA	24.00 (0.48)	24.14 (0.66)
ANOVA per time point	24.00 (0.51)	24.14 (0.74)
	Last Observation Carried Forward Analysis	
Direct likelihood, ML	22.88 (0.56)	22.97 (0.65)
Direct likelihood, REML	22.88 (0.58)	22.97 (0.68)
MANOVA	22.88 (0.58)	22.97 (0.68)
ANOVA per time point	22.88 (0.61)	22.97 (0.72)

and further discussed in Schafer (1997) and Little and Rubin (2002). The key idea of the procedure is to first replace each missing value with a set of M plausible values drawn from the conditional distribution of the unobserved outcomes, given the observed ones. This conditional distribution represents the uncertainty about the right value to impute. In this way, M imputed data sets are generated (imputation stage), which are then analyzed using standard complete data methods (analysis stage). Finally, the results from the M analyses have to be combined into a single one (pooling stage) by means of the method laid out in Rubin (1978), to produce inferences. To be valid, multiple imputation in its basic form requires the missingness mechanism to be MAR, even though versions under MNAR have been proposed (Rubin, 1987; Molenberghs et al., 1997).

Thus, multiple imputation is valid under the same conditions as direct likelihood, and therefore does not suffer from the problems encountered in most single-imputation methods. However, there are a number of situations where multiple imputation is particularly useful. An example of this is when outcomes as well as covariates are missing; then multiple imputation is a sensible route. The method is also useful when several analyses, perhaps conducted by different analysts, have to be done on the same set of incomplete data. In such a case, all analyses could start from the same set of multiply imputed sets of data and enhance comparability. Generally speaking, multiple imputation can flexibly incorporate additional information, such as covariates not present in the actual analysis model. The same is not at all straightforward with direct likelihood.

More precisely, assume that the parameter vector of the distribution of the response $\boldsymbol{y}_i = (\boldsymbol{y}_i{}^o, \boldsymbol{y}_i{}^m)$ is denoted by $\boldsymbol{\theta} = (\boldsymbol{\beta}, \boldsymbol{\alpha})'$, in which $\boldsymbol{\beta}$ denotes the vector of fixed-effects parameters and $\boldsymbol{\alpha}$ the vector of covariance parameters. Multiple imputation uses the observed data $\boldsymbol{Y}^o$ to estimate the conditional distribution of $\boldsymbol{Y}^m$ given $\boldsymbol{Y}^o$. The missing data are sampled several times from this conditional distribution and augmented to the observed data. The resulting completed data are then used to estimate $\boldsymbol{\theta}$. If the distribution of $\boldsymbol{y}_i = (\boldsymbol{y}_i{}^o, \boldsymbol{y}_i{}^m)$ were known, with parameter vector $\boldsymbol{\theta}$, then $\boldsymbol{y}_i{}^m$ could be imputed by drawing a value of $\boldsymbol{y}_i{}^m$ from the conditional distribution $f(\boldsymbol{y}_i{}^m|\boldsymbol{y}_i{}^o, \boldsymbol{\theta})$. The objective of the imputation phase is to sample from this true predictive distribution. However, $\boldsymbol{\theta}$ in the imputation model is unknown, and therefore needs to be estimated from the data first, say $\widehat{\boldsymbol{\theta}}$, after which $f(\boldsymbol{y}_i{}^m|\boldsymbol{y}_i{}^o, \widehat{\boldsymbol{\theta}})$ is used to impute the missing data. Precisely, this implies one first generates draws from the distribution of $\widehat{\boldsymbol{\theta}}$, thereby taking sampling uncertainty into account. Generally, the parameter vector in the imputation model differs from the parameter vector that governs the analysis model. Alternatively, a Bayesian approach, in which uncertainty about $\boldsymbol{\theta}$ is incorporated by means of some prior distribution for $\boldsymbol{\theta}$, can also be taken.

In the last phase of multiple imputation, the results of the analyses for the M imputed data sets are pooled into a single inference. The combined point estimate for the parameter of interest from the multiple imputation is simply the average of the M complete-data point estimates (Schafer, 1997). Let $\boldsymbol{\theta}$ denote the parameter of interest. Then the estimate and its estimated variance are given by:

$$\widehat{\boldsymbol{\theta}} \equiv \frac{1}{M}\sum_{m=1}^{M}\widehat{\boldsymbol{\theta}}^m \quad \text{and}$$

$$\widehat{\text{Var}}(\widehat{\boldsymbol{\theta}}) \equiv \boldsymbol{v} = W + \left(\frac{M+1}{M}\right)B,$$

where

$$W = \frac{\sum_{m=1}^{M}\boldsymbol{u}^m}{M} \quad \text{and}$$

$$B = \frac{\sum_{m=1}^{M}(\widehat{\boldsymbol{\theta}}^m - \widehat{\boldsymbol{\theta}})(\widehat{\boldsymbol{\theta}}^m - \widehat{\boldsymbol{\theta}})'}{M-1},$$

with W denoting the average *within-imputation* variance and B the *between-imputation* variance (Rubin, 1987).

As mentioned before, one can choose any complete data method to analyze the imputed data sets in the analysis stage of the MI method. This is particularly useful when interest lies in a marginal model for non-Gaussian outcomes. WGEE has been proposed as an alternative for GEE (Chapters 15 and 25) to enable valid inferences based on incomplete data under the MAR assumption. However, in WGEE, all subjects are given weights, which are calculated using the hypothesized dropout model. As a consequence, any misspecification of this dropout model will affect all subjects and thus all results. Therefore, we can consider MI combined with GEE at the analysis phase (MI-GEE). In essence, this method comes down to first using the predictive distribution of the unobserved outcomes given the observed ones, and perhaps covariates. After this step, the missingness mechanism can be further ignored, provided we assume MAR. In these MI cases, a misspecification made in the imputation step will only affect the unobserved (that is, imputed) but not the observed part of the data. Meng's (1994) results show that, as long as the imputation model is not grossly misspecified, this approach will perform well. Moreover, Beunckens et al. (2008) have proven that, even though WGEE is asymptotically unbiased, MI-GEE is less biased and more precise in small and moderate samples and MI-GEE is more robust to misspecification in either the imputation or measurement model.

Returning to the specific case of hierarchical models, it is important to note that one can take various views on the use of multiple imputation. Suppose, for illustration's sake, that one wants to consider a linear mixed model $\boldsymbol{y}_i = X_i\boldsymbol{\beta} + Z_i\boldsymbol{u}_i + \boldsymbol{\varepsilon}_i$ in the analysis stage, where X_i and Z_i are $n_i \times p$ and $n_i \times q_i$ design matrices, respectively, $\boldsymbol{\beta}$ a vector of fixed effects, $\boldsymbol{u}_i \sim N_{q_i}(0, \Omega_i)$ a vector of random effects, and $\boldsymbol{\varepsilon}_i \sim N_{n_i}(0, \Sigma_i)$ a vector of errors. As indicated in Chapter 9, the corresponding marginal model is specified by

$$\boldsymbol{y}_i \sim N_{n_i}(\boldsymbol{\mu}_i = X_i\boldsymbol{\beta}, V_i = Z_i\Omega_i Z_i' + \Sigma_i). \tag{23.1}$$

In case $\boldsymbol{y}_i = (\boldsymbol{y}_i^{o'}, \boldsymbol{y}_i^{m'})'$, split over t_i observed and $n_i - t_i$ missing components, one can impute in at least three different ways. First, imputation can be done based on a multivariate normal model, where $\boldsymbol{\mu}_i$ and V_i are not as in (23.1) but rather fully general, or at least more general. This is fine, as long as the imputation model does not contain restrictions that would not be used in the analysis model. Second, (23.1) itself can be chosen for imputation. Third, in some cases, for example when $\boldsymbol{y}_i$ is non-normally distributed (Chapters 8 and 15), deriving the marginal model is not straightforward, and yet one would like to maintain this particular model in the imputation stage (see also Chapter 5). It is then possible to first draw random effects $\boldsymbol{u}_i$ and, given these, missing outcomes $\boldsymbol{y}_i^m|\boldsymbol{y}_i^o, \boldsymbol{u}_i$. For illustration, in the first and second cases, partition (23.1) as:

$$\left(\begin{array}{c}\boldsymbol{y}_i^o\\ \hline \boldsymbol{y}_i^m\end{array}\right) \sim N_{n_i}\left[\left(\begin{array}{c}\boldsymbol{\mu}_i^o\\ \hline \boldsymbol{\mu}_i^m\end{array}\right), \left(\begin{array}{c|c}V_i^{oo} & V_i^{om}\\ \hline V_i^{mo} & V_i^{mm}\end{array}\right)\right]. \tag{23.2}$$

(23.2) produces

$$\boldsymbol{y}_i^m|\boldsymbol{y}_i^o \sim N_{n_i - t_i}[\boldsymbol{\mu}_i^m + V_i^{mo}(V_i^{oo})^{-1}(\boldsymbol{y}_i^o - \boldsymbol{\mu}_i^o), \Sigma_i^\star],$$

where $\Sigma_i^\star = V_i^{mm} - V_i^{mo}(V_i^{oo})^{-1}V_i^{om}$, from which one can draw without any trouble.

23.7 MISSING NOT AT RANDOM

23.7.1 *Selection Models*

Much of the early development of, and debate about, selection models appeared in the econometrics literature in which the tobit model (Heckman, 1976) played a central role. This combines a marginal Gaussian regression model for the response, as might be used in the absence of missing data, with a Gaussian-based threshold model for the probability of a value being missing. For simplicity, consider a single Gaussian distributed response variable $Z \sim N(\mu, \sigma^2)$. The probability of Z being missing is assumed to depend on a second Gaussian variable $Z_m \sim N(\mu_m, \sigma_m^2)$, where $\mathrm{P}(R = 0) = \mathrm{P}(Z_m < 0)$. Dependence of missingness on the response Z is induced by introducing a correlation between Z and Z_m. To avoid some of the complications of direct likelihood maximization, a two-stage estimation procedure was proposed by Heckman (1976) for this type of model. The use of the tobit model and associated two-stage procedure was the subject of considerable debate in the econometrics literature, much of it focusing on the issues of identifiability and sensitivity (Amemiya, 1984; Little, 1986). We shall see the same issues arising in developments of these ideas in the biometric and epidemiologic settings.

At first sight the tobit model does not appear to have the selection model structure commonly encountered, in that there is no conditional partition of $f(\boldsymbol{y}, \boldsymbol{r})$. However, it is simple to show, from the joint Gaussian distribution of z and z_m, that in the tobit model $\mathrm{P}(r = 0 \mid z)$ equals $\Phi(\beta_0 + \beta_1 z)$ for suitably chosen parameters β_0 and β_1, with $\Phi(\cdot)$ the Gaussian distribution function. This can be seen as a probit regression model for the (binary) missing value process. This basic structure underlies the simplest form of selection model that has been proposed for longitudinal data in the biometric and epidemiologic setting. A suitable response model, such as the multivariate Gaussian, is combined with a binary regression model for dropout. At each time point the occurrence of dropout can be regressed on previous and current values of the response, as well as covariates. We now explore such models in more detail.

Although modeling a continuous response is generally more straightforward than a categorical response, we begin by considering the latter case. By reducing the problem to a very simple setting, that of two repeated binary responses, we are able to illustrate some key points in a very explicit way. With both measurements (Y_1, Y_2) observed, a subject will generate one of four responses, with associated probabilities π_{ij}:

	Time 2		
Time 1	0	1	Total
0	π_{00}	π_{01}	$\pi_{0\cdot}$
1	π_{10}	π_{11}	$\pi_{1\cdot}$
Total	$\pi_{\cdot 0}$	$\pi_{\cdot 1}$	1

These probabilities can be parameterized in many different ways, for example using marginal or Markov representations. For the moment, we are not concerned with the particular parameterization chosen and will work with the joint probabilities, noting that three of these determine the fourth. It is assumed at first that missing values are restricted to dropout and in the most general form the probability of dropout may depend on the outcome at either time, so we write, for $r = 1$, the event of being observed on the second occasion:

$$P(r = 1 \mid y_1 = s_1, y_2 = s_2) = \phi_{s_1 s_2},$$

for $s_1 = 0, 1;\ s_2 = 0, 1$. There are four probabilities, making a total of seven degrees of freedom for the model that combines the response and dropout components. If we could observe the data from those who drop out, the full data could be represented as a $2 \times 2 \times 2$ contingency table, classified by the Time 1 and Time 2 outcomes and dropout status. The model described saturates the degrees of freedom in this table. In practice, two tables of data are observed; a 2×2 table from the completers and a 2×1 table of Time 1 outcomes from the dropouts, making a total of five degrees of freedom. Clearly, some parameters in the full model cannot be identified, and if a model is to be estimated from the observed data some constraints must be applied. This reflects the information lost with the dropouts. The MCAR and MAR cases correspond to the simple constraints. Indeed, MCAR: $\phi_{s_1 s_2} = \phi$ and MAR: $\phi_{s_1 s_2} = \phi_{s_1}$. Allowing $\phi_{s_1 s_2}$ to depend on s_2 makes the dropout non-random. We can relate this dropout model in a simple way to the logistic regression models that have often been used in the selection model (Diggle and Kenward, 1994; Baker, 1995; Fitzmaurice et al., 1996a,b; Molenberghs et al., 1997) as follows:

$$\text{logit}(\phi_{ij}) = \beta_0 + \beta_1 y_1 + \beta_2 y_2.$$

Note that this does not saturate the dropout model; the introduction of the interaction term in $y_1 y_2$ would be required for this.

23.7.2 Selection Models Applied to the Vorozole Study

Let us consider Diggle and Kenward's (1994) selection model as introduced in the previous section. For the measurement model, we start by ignoring the dropout mechanism. Since we are modeling change versus baseline, all models are forced to pass through the origin. This is done by allowing the main covariate effects, but only through their interactions with time. The following covariates were considered for the measurement model: baseline value, treatment, dominant site, and time in months (up to a cubic time trend). Second-order interactions were considered as well. Then, a backwards selection procedure was performed. For design reasons, treatment was kept in the model in spite of its non-significance. An F-test for treatment effect produces a p-value of 0.5822. Apart from baseline, no other time-stationary covariates

Table 23.4 Vorozole Study: Selection Model

Effect	Estimate (s.e.)
Fixed-Effect Parameters:	
time	7.78 (1.05)
time × baseline	−0.065 (0.009)
time × treatment	0.086 (0.157)
time^2	−0.30 (0.06)
time^2 × baseline	0.0024 (0.0005)
Variance Parameters:	
random intercept (δ^2)	105.42
serial variance (τ^2)	77.96
serial association (ϕ)	7.22
measurement error (σ^2)	77.83

were kept. A quadratic time effect provided an adequate description of the time trend. Based on the variogram, we confined the random-effects structure to random intercepts, and supplemented this with a spatial Gaussian process and measurement error (Chapter 31). The final model is presented in Table 23.4. The total correlation between two measurements, one month apart, equals 0.696. The residual correlation, which remains after accounting for the random effects, is still equal to 0.491. The serial correlation, obtained by further ignoring the measurement error, equals $\rho = \exp(-1/7.22^2) = 0.981$.

Fitted profiles are displayed in Figure 23.4 and Figure 23.5. In Figure 23.5, empirical Bayes estimates of the random effects are included, whereas in Figure 23.4 the purely marginal mean is used. For each treatment group, we obtain three sets of profiles. The fitted complete profile is the average curve that would be obtained, had all individuals been completely observed. If we use only those predicted values that correspond to occasions at which an observation was made, then the fitted incomplete profiles are obtained. The latter are somewhat above the former when the random effects are included, and somewhat below when they are not, suggesting that individuals with lower measurements are more likely to disappear from the study. In addition, while the fitted complete curves are very close (the treatment effect was not significant), the fitted incomplete curves are not, suggesting that there is more dropout in the standard arm than in the treatment arm. This is in agreement with the observed dropout rate and should not be seen as evidence of a bad fit. Finally, the observed curves, based on the measurements available at each time point, are displayed. These are higher than the fitted ones, but this should be viewed with the standard errors of the observed means in mind.

Next, we will study factors which influence dropout. A logistic regression model for the probability of dropping out at a given time, given that one is still at risk for dropping out, is used. To start, we restrict attention to MAR processes, whence $\psi_d = 0$. The first model includes treatment, dominant site, baseline value, and the previous measurement, but only the last two are significant, producing

$$\begin{aligned}\text{logit}[g(\boldsymbol{h}_{ij})] &= 0.080(0.341)\\ &-0.014(0.003)\text{base}_i\\ &-0.033(0.004)y_{i,j-1}.\end{aligned} \quad (23.3)$$

Diggle and Kenward (1994) and Molenberghs et al. (1997) considered non-random versions of this model by including the current, possible unobserved measurement. This requires more elaborate fitting algorithms, since the missing data process is then non-ignorable, and hence numerical integration needs to be used. Diggle and Kenward used the simplex algorithm (Nelder and Mead, 1965), while Molenberghs et al. (1997) fitted their models with the EM algorithm. With larger datasets such as this one, convergence can be painstakingly difficult and one has to worry about apparent convergence. Therefore, we first proceed in an alternative way. Both Diggle and Kenward and Molenberghs et al. observed that, in informative models, dropout tends to depend on the increment, i.e., the difference between the current and previous measurements $Y_{ij} - Y_{i,j-1}$. Clearly, a very similar quantity is obtained as $Y_{i,j-1} - Y_{i,j-2}$, but a major advantage of such a model is that

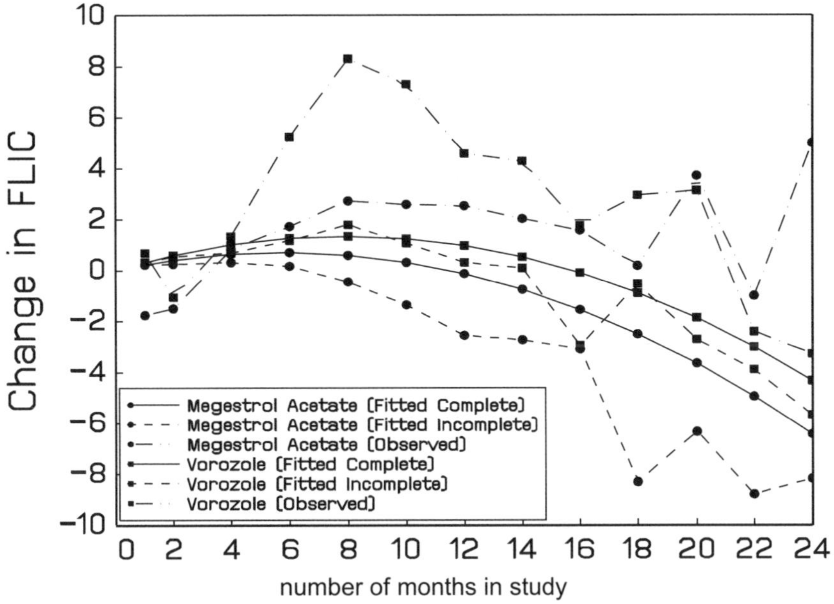

Figure 23.4 Vorozole Study: Fitted Profiles (averaging the predicted means for the incomplete and complete measurement sequences, without the random effects)

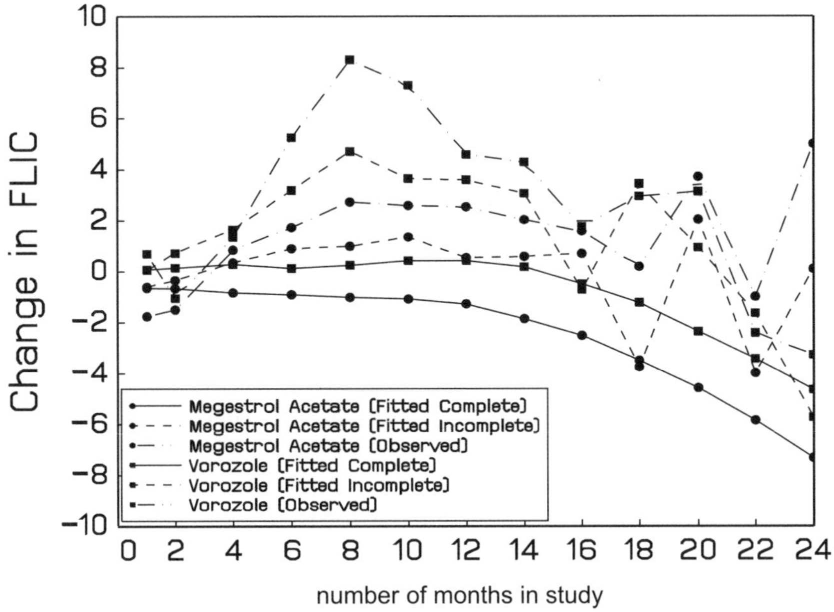

Figure 23.5 Vorozole Study: Fitted Profiles (averaging the predicted means for the incomplete and complete measurement sequences, including the random effects)

it fits within the MAR framework. In our case, we obtain

$$\begin{aligned}\text{logit}[g(\boldsymbol{h}_{ij})] &= 0.033(0.401)\\ &\quad -0.013(0.003)\text{base}_i\\ &\quad +0.012(0.006)y_{i,j-2}\\ &\quad -0.035(0.005)y_{i,j-1} \qquad (23.4)\\ &= 0.033(0.401)\\ &\quad -0.013(0.003)\text{base}_i\\ &\quad -0.023(0.005)\frac{y_{i,j-2}+y_{i,j-1}}{2}\\ &\quad -0.047(0.010)\frac{y_{i,j-1}-y_{i,j-2}}{2},\end{aligned}$$

indicating that both size and increment are significant predictors for dropout. We conclude that dropout increases with a decrease in baseline, in overall level of the outcome variable, as well as with a decreasing evolution in the outcome.

Both dropout models (23.3) and (23.4) can be compared with their non-random counterparts, where y_{ij} is added to the linear predictor. The first one becomes

$$\begin{aligned}\text{logit}[g(\boldsymbol{h}_{ij}, y_{ij})] &= 0.53 - 0.015\text{base}_i\\ &\quad -0.076y_{i,j-1} + 0.057y_{ij}, \qquad (23.5)\end{aligned}$$

while the second one becomes

$$\begin{aligned}&\text{logit}[g(\boldsymbol{h}_{ij}, y_{ij})]\\ &\quad = 1.38 - 0.021\text{base}_i - 0.0027y_{i,j-2}\\ &\qquad -0.064y_{i,j-1} + 0.035y_{ij}. \qquad (23.6)\end{aligned}$$

Formal testing of dropout models (23.5) versus (23.3) and (23.6) versus (23.4) are possible in principle, but will not be carried out for two reasons. First, the likelihood function tends to be very flat for non-random dropout models and therefore the determination of the likelihood ratio is often computationally nontrivial. More fundamentally, Rubin (1994), Little (1994), Laird (1994), and Molenberghs et al. (1997) pointed out that formal testing for non-random dropout faces philosophical objections. Indeed, non-random dropout models are identified only due to strong but unverifiable assumptions. Hogan and Laird (1997) suggested pattern-mixture models as a viable alternative.

23.7.3 Pattern-Mixture Models

As a simple illustration consider a continuous response at three times of measurement which will be modeled using a trivariate Gaussian distribution. Assume that there may be dropout at time 2 or 3, and let the dropout indicator R take the values 1 and 2 to indicate that the last observation occurred at these times and 3 to indicate no dropout. Then, in the first instance, the model implies a different distribution for each time of dropout. We can write:

$$\boldsymbol{y} \mid r \sim N(\boldsymbol{\mu}^{(r)}; \Sigma^{(r)}),$$

where

$$\boldsymbol{\mu}^{(r)} = \begin{pmatrix}\mu_1^{(r)}\\ \mu_2^{(r)}\\ \mu_3^{(r)}\end{pmatrix} \text{ and}$$

$$\Sigma^{(r)} = \begin{pmatrix}\sigma_{11}^{(r)} & \sigma_{21}^{(r)} & \sigma_{31}^{(r)}\\ \sigma_{21}^{(r)} & \sigma_{22}^{(r)} & \sigma_{32}^{(r)}\\ \sigma_{31}^{(r)} & \sigma_{32}^{(r)} & \sigma_{33}^{(r)}\end{pmatrix},$$

for $r = 1, 2, 3$. Let $\text{P}(r) = \pi_r$. Then the marginal distribution of the response is a mixture of normals with, for example, mean

$$\boldsymbol{\mu} = \sum_{r=1}^{3} \pi_r \boldsymbol{\mu}^{(r)}.$$

However, although the π_r can be simply estimated from the observed proportions in each dropout group, only 16 of the 27 response parameters can be identified from the data without making further assumptions. These 16 comprise all the parameters from the completers plus those from the following two submodels. For $r = 2$:

$$N\left[\begin{pmatrix}\mu_1^{(2)}\\ \mu_2^{(2)}\end{pmatrix}; \begin{pmatrix}\sigma_{11}^{(2)} & \sigma_{21}^{(2)}\\ \sigma_{31}^{(3)} & \sigma_{32}^{(3)}\end{pmatrix}\right],$$

and for $r = 1$: $N(\mu_1^{(2)}; \sigma_{11}^{(1)})$. This is a *saturated* pattern-mixture model and the representation makes it very clear what information each dropout group provides, and consequently the assumptions that need to be made if we are to predict the behavior of the unobserved responses, and so obtain marginal models for the response. If the three sets of parameters $\boldsymbol{\mu}^{(r)}$ are simply equated, this implies MCAR. Progress can be made with less stringent restrictions, however. In practice, choice of restrictions will need to be guided by the context. In addition, the form of the data will typically be more complex, requiring, for example, a more structured model for the response with the incorporation of covariates. Hence, such models can be constructed in many ways.

23.7.4 Pattern-Mixture Models Applied to the Vorozole Study

First, we analyse the data using basic pattern-mixture models. Later in this chapter we will apply pattern-mixture models based on identifying restrictions.

Initial pattern-mixture models

The dropout process simplifies to $f(d_i|\boldsymbol{\psi})$, which is a (possibly covariate-corrected) model for the probability of belonging to a particular pattern. Its components describe the dropout rate at each occasion.

The measurement model has to reflect dependence on dropout. In its most general form, this implies

$$\begin{cases} y_i = X_i\boldsymbol{\beta}(d_i) + Z_i\boldsymbol{u}_i + \boldsymbol{\varepsilon}_i, \\ \boldsymbol{u}_i \sim N[\mathbf{0}, D(d_i)], \\ \boldsymbol{\varepsilon}_i \sim N[\mathbf{0}, \Sigma_i(d_i)]. \end{cases} \quad (23.7)$$

Thus, the fixed effects as well as the covariance parameters are allowed to change with dropout pattern, and a priori no restrictions are placed on the structure of this change.

As discussed in Section 23.7.3, model family (23.7) contains underidentified members since it describes the full set of measurements in pattern d_i, even though there are no measurements after occasion $d_i - 1$. To avoid this problem, simplified (identified) models can be considered. The advantage is that the number of parameters decreases, which is generally an issue with pattern-mixture models. Hogan and Laird (1997) noted that in order to estimate the large number of parameters in general models, one has to make the awkward requirement that each dropout pattern is sufficiently "filled", in other words one has to require large numbers of dropouts.

In analogy with our selection model, the profiles are forced to pass through the origin. This is done by allowing only the time main effect and interactions of other covariates with time in the model.

The most complex pattern-mixture model we consider includes a different parameter vector for each of the observed patterns. This is done by including the interaction of all effects in the model with *pattern*, a factor variable calculated as 2 plus the number of observations after baseline. We then proceed by backward selection in order to simplify the model. First, we found that the covariance structure is common to all patterns, encompassing random intercept, a serial exponential process, and measurement error.

For the fixed effects we proceeded as follows. A backward selection procedure, starting from a model that includes a main effect of time and time^2, as well as interactions of time with baseline value, treatment effect, dominant site, and pattern, and the interaction of pattern with time^2. This procedure revealed significant effects for nearly all included factors. The only exception was treatment effect, as was the case with the selection model in Table 23.4. Indeed, a single degree of freedom F-test yields a p-value of 0.6868. Note that such a test is possible since treatment effect does not interact with pattern, in contrast to the model which we will describe later. The fitted profiles are displayed in Figure 23.6. We observe that the profiles for both arms are very similar. This is due to the fact that treatment effect is not significant, but perhaps also because we did not allow a more complex treatment effect. For example, we might consider an interaction of treatment with the square of time and, more importantly, a treatment effect which is pattern-specific. Our second, expanded, model allowed for up to cubic time effects, the interaction of time with dropout pattern, dominant site, baseline value, and treatment, as well as their two- and three-way interactions. After a backward selection procedure, the effects included are time and time^2, the two-way interaction of time and dropout pattern, as well as three-factor interactions of time and dropout pattern with (1) baseline, (2) group, and (3) dominant site. Finally, time^2 interacts with dropout pattern and with the interaction of baseline and dropout pattern. No cubic time effects were necessary. The model is graphically represented in Figure 23.7.

Contrasting Figures 23.6 and 23.7, they look similar at first sight, even though the latter allows for, and shows, a much richer set of profiles than the former. This difference occurs because in the former, while the effect of pattern on outcome is present, it comes in a restrictive way: only the time by pattern interaction is present. In the second model, a much richer structure is allowed for, with pattern interacting with all other covariates. Note that, when pattern interacts with all effects, and when also the covariance structure would be different for different patterns, then a completely different model per pattern would result. Thus, our parametric family of pattern-mixture models compromises between the selection model's mean description on the one hand (i.e., with pattern absent from the outcome model) and the fully unstructured pattern-mixture model on the other. Of course, in a fully specified selection, there is also an explicit model for dropout, and it is there that the relationship between outcomes and dropout is modeled. Nevertheless, the parametric pattern-mixture model may give insightful indications as to how, say, early, middle, and late dropouts differ from each other. That said, one should be very careful when extrapolating such a mean profile beyond the time over which observations are collected. We expand on this and related cautionary remarks in the sensitivity analysis section.

In the latter model, because a pattern-specific parameter has been included, we have several options for the assessment of treatment. Since there are 13 patterns (remember we cut off the patterns at 2 years), one can test the global hypothesis, based on 13 degrees of freedom, of no treatment effect. We obtain $F = 1.25$, producing $p = 0.2403$, indicating that there is no evidence for an overall treatment effect. Each of the treatment effects separately is at a non-significant level. Alternatively, the *marginal* effect of treatment can be calculated, which is the weighted average of the pattern-specific treatment effects, with weights given by the probability of occurrence of the various patterns. Its standard error is calculated using a straightforward application of the delta method. This effect is equal to $-0.286(0.288)$, producing a p-value of 0.3206, which is even more non-significant.

In summary, we obtain a non-significant treatment effect from all our different models, which gives more weight to this conclusion. Further, the pattern-mixture model formulation has led to an important complementary insight, which in a selection model would go unnoticed.

23.8 SENSITIVITY ANALYSIS

When there is residual doubt about the plausibility of MAR, one can conduct a sensitivity analysis. While many proposals have been made, this is still a very active area of research. Obviously, a number of MNAR models can be fitted, provided one is prepared to approach formal aspects of model comparison with due caution. Such analyses can be complemented with appropriate (global and/or local) influence analyses (Verbeke et al., 2001). Another route is to construct pattern-mixture models, where the measurement model is considered, conditional upon the observed dropout pattern, and to compare the conclusions with

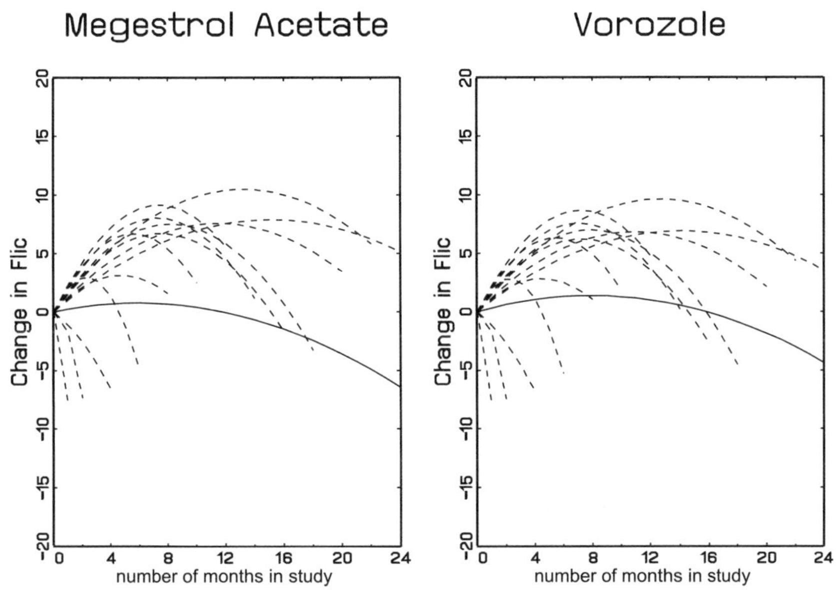

Figure 23.6 Vorozole Study: Fitted Selection and First Pattern-Mixture Model

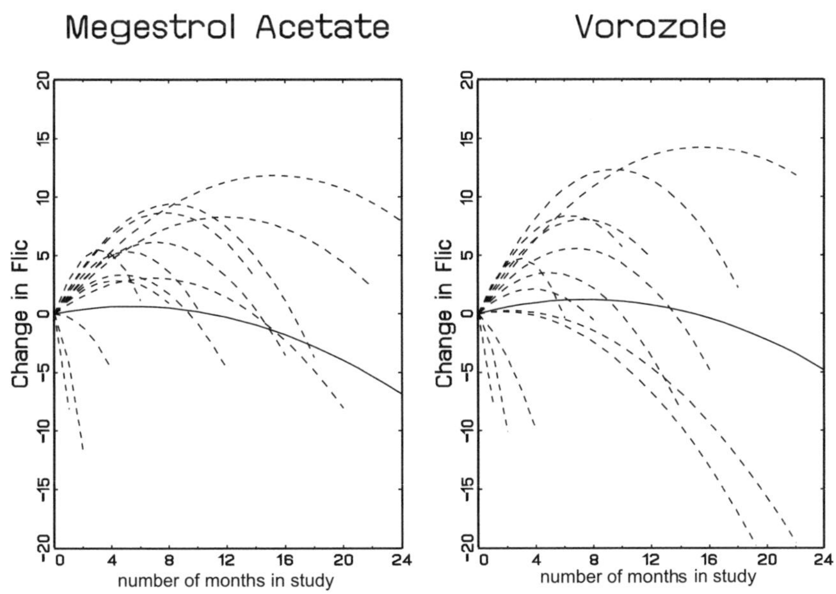

Figure 23.7 Vorozole Study: Fitted Selection and Second Pattern-Mixture Models

those obtained from the selection model framework, where the reverse factorization is used (Michiels et al., 2002; Thijs et al., 2002). Alternative sensitivity analysis frameworks are provided by Robins, Rotnitzky, and Scharfstein (1998) and Forster and Smith (1998), who present a Bayesian sensitivity analysis, and Raab and Donnelly (1999). A further paradigm, useful for sensitivity analysis, are so-called shared parameter models, where common latent or random effects drive both the measurement process as well as the process governing missingness.

Nevertheless, ignorable analyses may provide reasonably stable results, even when the assumption of MAR is violated, in the sense that such analyses constrain the behavior of the unseen data to be similar to that of the observed data. A discussion of this phenomenon in the survey context has been given in Rubin, Stern, and Vehovar (1995). These authors first argue that, in well-conducted experiments (some surveys and many confirmatory clinical trials), the assumption of MAR is often to be regarded as a realistic one. Secondly, and very important for confirmatory trials, an MAR analysis can be specified a priori without additional work relative to a situation with complete data. Thirdly, while MNAR models are more general and explicitly incorporate the dropout mechanism, the inferences they produce are typically highly dependent on the untestable and often implicit assumptions built in regarding the distribution of the unobserved measurements given the observed ones. The quality of the fit to the observed data need not reflect at all the appropriateness of the implied structure governing the unobserved data. Based on these considerations, we recommend, for primary analysis purposes, the use of ignorable likelihood-based methods or appropriately modified frequentist methods. To explore the impact of deviations from the MAR assumption on the conclusions, one should ideally conduct a sensitivity analysis (Verbeke and Molenberghs, 2000). Various routes are discussed in Molenberghs and Kenward (2007).

23.9 CONCLUDING REMARKS

In conclusion, it is sensible to set out assuming MAR, and then choosing for one or a combination of the following methods: direct likelihood analysis, direct Bayesian analysis, multiple imputation, and/or weighted estimating equations. In addition, it is prudent, if not imperative, to additionally conduct a sensitivity analysis, for which an ever-expanding toolkit is being developed.

The statistical analysis done in this chapter made use of the SAS System. It is particularly useful to note that the direct likelihood method does not require any additional software, other than the longitudinal analysis tools, such as the SAS procedures MIXED, NLMIXED, or GLIMMIX, for example, that one would use had the data been complete. Evidently, also LOCF, CC, and AC analyses are easy to conduct, although less advisable. Multiple imputation has been implemented in the SAS procedures MI and MIANALYZE. For the MNAR and sensitivity analysis tools, dedicated macros and functions are needed. Programs, output, and datasets used are available through the authors' web pages: `www.ibiostat.be/software`. Note also that Verbeke and Molenberghs (2000), Molenberghs and Verbeke (2005), and Molenberghs and Kenward (2007) provide ample detail on SAS implementations of various techniques mentioned here.

ACKNOWLEDGMENTS

The authors gratefully acknowledge support from IAP research Network P6/03 of the Belgian Government (Belgian Science Policy).

REFERENCES

Afifi, A. and Elashoff, R. (1966) 'Missing Observations in Multivariate Statistics I: Review of the Literature', *Journal of the American Statistical Association*, 61: 595–604.

Amemiya, T. (1984) 'Tobit Models: A Survey', *Journal of Econometrics*, 24: 3–61.

Baker, S.G. (1995) 'Marginal Regression for Repeated Binary Data with Outcome Subject to Mon-ignorable Non-response', *Biometrics*, 51: 1042–52.

Beunckens, C., Molenberghs, G., and Kenward, M.G. (2005) 'Tutorial: Direct Likelihood Analysis versus Simple Forms of Imputation for Missing Data in Randomized Clinical Trials', *Clinical Trials*, 2: 379–86.

Beunckens, C., Molenberghs, G., Verbeke, G., and Mallinckrodt, C. (2008) 'A Latent-class Mixture

Model for Incomplete Longitudinal Gaussian Data', *Biometrics*, 64: 96–105.

Dempster, A.P., Laird, N.M., and Rubin, D.B. (1977) 'Maximum Likelihood from Incomplete Data via the EM Algorithm (with Discussion)', *Journal of the Royal Statistical Society, Series B*, 39: 1–38.

Diggle, P.J. and Kenward, M.G. (1994) 'Informative Drop-out in Longitudinal Data Analysis (with Discussion)', *Applied Statistics*, 43: 49–93.

Fitzmaurice, G.M., Heath, G., and Clifford, P. (1996a) 'Logistic Regression Models for Binary Data Panel Data with Attrition', *Journal of the Royal Statistical Society Series A*, 159: 249–64.

Fitzmaurice, G.M., Laird, N.M., and Zahner, G.E.P. (1996b) 'Multivariate Logistic Models for Incomplete Binary Response', *Journal of the American Statistical Association*, 91: 99–108.

Forster, J.J. and Smith, P.W. (1998) 'Model-based Inference for Categorical Survey Data Subject to Non-ignorable Non-response', *Journal of the Royal Statistical Society, Series B*, 60: 57–70.

Gibbons, R.D., Hedeker, D., Elkin, I., Waternaux, D., Kraemer, H.C., Greenhouse, J.B., Shea, M.T., Imber, S.D., Sotsky, S.M., and Watkins, J.T. (1993) 'Some Conceptual and Statistical Issues in Analysis of Longitudinal Psychiatric Data', *Archives of General Psychiatry*, 50: 739–50.

Goss, P.E., Winer, E.P., Tannock, I.F., and Schwartz, L.H. (1999) 'Breast Cancer: Randomized Phase III Trial Comparing the New Potent and Selective Third-generation Aromatase Inhibitor Vorozole with Megestrol Acetate in Postmenopausal Advanced Breast Cancer Patients', *Journal of Clinical Oncology*, 17: 52–63.

Hartley, H.O. and Hocking, R. (1971) 'The Analysis of Incomplete Data', *Biometrics*, 27: 7783–808.

Heckman, J.J. (1976) 'The Common Structure of Statistical Models of Truncation, Sample Selection and Limited Dependent Variables and a Simple Estimator for such Models', *Annals of Economic and Social Measurement*, 5: 475–92.

Heyting, A., Tolboom, J., and Essers, J. (1992) 'Statistical Handling of Dropouts in Longitudinal Clinical Trials', *Statistics in Medicine*, 11: 2043–61.

Hogan, J.W. and Laird, N.M. (1997) 'Mixture Models for the Joint Distribution of Repeated Measures and Event Times', *Statistics in Medicine*, 16, 239–58.

Jennrich, R.I. and Schluchter, M.D. (1986) 'Unbalanced Repeated Measures Models with Structured Covariance Matrices', *Biometrics*, 42: 805–20.

Kenward, M.G. and Molenberghs, G. (1998) 'Likelihood Based Frequentist Inference when Data are Missing at Random', *Statistical Science*, 12: 236–47.

Laird, N.M. (1994) 'Discussion of Diggle, P.J. and Kenward, M.G.: Informative Dropout in Longitudinal Data Analysis', *Applied Statistics*, 43: 84.

Lavori, P.W., Dawson, R., and Shera, D. (1995) 'A Multiple Imputation Strategy for Clinical Trials with Truncation of Patient Data', *Statistics in Medicine*, 14, 1913–25.

Little, R.J.A. (1986) 'A Note about Models for Selectivity Bias', *Econometrika*, 53: 1469–74.

Little, R.J.A. (1994) 'A Class of Pattern-mixture Models for Normal Incomplete Data', *Biometrika*, 81: 471–83.

Little, R.J.A. and Rubin, D.B. (2002) *Statistical Analysis with Missing Data*. New York: John Wiley & Sons.

Little, R.J.A. and Yau, L. (1996) 'Intent-to-Treat Analysis in Longitudinal Studies with Drop-Outs.' *Biometrics*, 52: 1324–33.

Liu, G. and Gould, A.L. (2002) 'Comparison of Alternative Strategies for Analysis of Longitudinal Trials with Dropouts', *Journal of Biopharmaceutical Statistics*, 12: 207–26.

Mallinckrodt, C.H., Clark, W.S., Carroll, R.J., and Molenberghs, G. (2003a) 'Assessing Response Profiles from Incomplete Longitudinal Clinical Trial Data under Regulatory Considerations', *Journal of Biopharmaceutical Statistics*, 13: 179–90.

Mallinckrodt, C.H., Clark, W.S., and Stacy, R.D. (2001a) 'Type I Error Rates from Mixed-effects Model Repeated Measures versus Fixed Effects Analysis of Variance with Missing Values Imputed via Last Observation Carried Forward', *Drug Information Journal*, 35, 4: 1215–25.

Mallinckrodt, C.H., Clark, W.S., and Stacy, R.D. (2001b) 'Accounting for Dropout Bias using Mixed-effects Models', *Journal of Biopharmaceutical Statistics*, 11, (1 & 2): 9–21.

Mallinckrodt, C.H., Sanger, T.M., Dube, S., Debrota, D.J., Molenberghs, G., Carroll, R.J., Zeigler Potter, W.M., and Tollefson, G.D. (2003b) 'Assessing and Interpreting Treatment Effects in Longitudinal Clinical Trials with Missing Data', *Biological Psychiatry*, 53: 754–60.

Mallinckrodt, C.H., Watkin, J.G., Molenberghs, G., and Carroll, R.J. (2004) 'Choice of the Primary Analysis in Longitudinal Clinical Trials', *Pharmaceutical Statistics*, 3: 161–9.

Meng, X.-L. (1994) 'Multiple-imputation Inferences with Uncongenial Sources of Input', *Statistical Science*, 9: 538–58.

Michiels, B., Molenberghs, G., Bijnens, L., Vangeneugden, T., and Thijs, H. (2002) 'Selection Models and Pattern-mixture Models to Analyze Longitudinal Quality of Life Data Subject to Dropout', *Statistics in Medicine*, 21: 1023–41.

Molenberghs, G. and Kenward, M.G. (2007) *Missing Data in Clinical Studies*. Chichester: John Wiley & Sons.

Molenberghs, G., Kenward, M.G., and Lesaffre, E. (1997) 'The Analysis of Longitudinal Ordinal Data with Non-random Dropout', *Biometrika*, 84: 33–44.

Molenberghs, G., Thijs, H., Jansen, I., Beunckens, C., Kenward, M.G., Mallinckrodt, C., and Carroll, R.J. (2004) 'Analyzing Incomplete Longitudinal Clinical Trial Data', *Biostatistics*, 5: 445–64.

Molenberghs, G. and Verbeke, G. (2005) *Models for Discrete Longitudinal Data*. New York: Springer.

Nelder, J.A. and Mead, R. (1965) 'A Simplex Method for Function Minimisation', *The Computer Journal*, 7: 303–13.

Potthoff, R.F. and Roy, S.N. (1964) 'A Generalized Multivariate Analysis of Variance Model useful Especially for Growth Curve Problems', *Biometrika*, 51: 313–26.

Raab, G.M. and Donnelly, C.A. (1999) 'Information on Sexual Behaviour when Some Data are Missing', *Applied Statistics*, 48: 117–33.

Robins, J.M., Rotnitzky, A., and Scharfstein, D.O. (1998) 'Semiparametric Regression for Repeated Outcomes with Non-ignorable Non-response', *Journal of the American Statistical Association*, 93: 1321–39.

Robins, J.M., Rotnitzky, A., and Zhao, L.P. (1995) 'Analysis of Semiparametric Regression Models for Repeated Outcomes in the Presence of Missing Data', *Journal of the American Statistical Association*, 90: 106–21.

Rubin, D.B. (1978) 'Multiple Imputations in Sample Surveys – A Phenomenological Bayesian Approach to Nonresponse', in *Imputation and Editing of Faulty or Missing Survey Data*. Washington, DC: U.S. Department of Commerce, 1–23.

Rubin, D.B. (1987) *Multiple Imputation for Nonresponse in Surveys*. New York: John Wiley & Sons.

Rubin, D.B. (1994) 'Discussion of Diggle, P.J. and Kenward, M.G.: Informative Dropout in Longitudinal Data Analysis', *Applied Statistics*, 43: 80–2.

Rubin, D.B., Stern, H.S., and Vehovar, V. (1995) 'Handling "Don't Know" Survey Responses: The Case of the Slovenian Plebiscite', *Journal of the American Statistical Association*, 90: 822–8.

Schafer, J.L. (1997) *Analysis of Incomplete Multivariate Data*. London: Chapman & Hall.

Schipper, H., Clinch, J., and McMurray, A. (1984) 'Measuring the Quality of Life of Cancer Patients: The Functional-Living Index-Cancer: Development and Validation', *Journal of Clinical Oncology*, 2: 472–83.

Siddiqui, O. and Ali, M.W. (1998) 'A Comparison of the Random-effects Pattern Mixture Model with Last Observation Carried Forward (LOCF) Analysis in Longitudinal Clinical Trials with Dropouts', *Journal of Biopharmaceutical Statistics*, 8: 545–63.

Tanner, M.A. and Wong, W.H. (1987) 'The Calculation of Posterior Distributions by Data Augmentation', *Journal of the American Statistical Association*, 82: 528–50.

Thijs, H., Molenberghs, G., Michiels, B., Verbeke, G., and Curran, D. (2002) 'Strategies to Fit Pattern-mixture Models', *Biostatistics*, 3: 245–65.

Verbeke, G. and Molenberghs, G. (1997) *Linear Mixed Models in Practice: A SAS-Oriented Approach*. Lecture Notes in Statistics 126. New York: Springer-Verlag.

Verbeke, G. and Molenberghs, G. (2000) *Linear Mixed Models for Longitudinal Data*. New York: Springer-Verlag.

Verbeke, G., Molenberghs, G., Thijs, H., Lesaffre, E., and Kenward, M.G. (2001) 'Sensitivity Analysis for Non-random Dropout: A Local Influence Approach', *Biometrics*, 57: 7–14.

24

Lack of Fit, Graphics, and Multilevel Model Diagnostics

Gerda Claeskens
Katholieke Universiteit Leuven, Belgium

24.1 MODEL ASSUMPTIONS

Multilevel models rely on several assumptions which we can group into the following three types:

1. Assumptions about the distribution of the error terms and about the distribution of the random effects. For linear mixed models it is quite common to assume normality for the errors and the random effects; see Chapter 13.
2. Assumptions stating that all of the observations obey the specified model and hence that there are no aberrant or outlying observations. In cases where this assumption does not hold, and robustness of the estimators is deemed important, robust estimation methods may be used; see Chapter 22.
3. Assumptions on the form of the mean response, which is often taken to be a parametric function. In the linear mixed model, such assumptions commonly translate to linear effects of the covariates X, that is, it is assumed that $E(y) = X\beta$; see Chapters 14, 17, and 18.

In this chapter, methods will be discussed to verify and check such assumptions.

24.2 DATA EXAMPLE 1: THE SLEEP STUDY

The first data example reports, in a sleep deprivation study, the average reaction times for 18 people, measured on 10 consecutive days. The data are available under the name 'sleepstudy' (see also Belenky et al., 2003) in the software R, within the library lme4. The individuals who participated in the study slept normally the night before the start of the study (that is, day 0). During the nine following days these people were allowed to only have three hours of sleep each night. On each day of the study a series of tests were taken and the persons' reaction times were recorded. This set of data is balanced, with 10 observations for each subject; there are no missing values. The data matrix consists of three columns: the average reaction times on the tests (taken to be the response variable), a column with values 0 through 9 indicating the day on which the measurement was taken, and a column with a subject identification number.

Visual representations of the data are helpful as a guide to starting the model building step. Of particular interest are Trellis plots to visualize effects in pairs of variables. For the sleep study, we wish to see, for each subject, whether and how the response variable 'reaction time' varies over the different days. Figure 24.1 is obtained by using the command 'xyplot' in R, which is available in library(lattice).

Rather than showing profiles (lines connecting the response values over the different days for each individual separately), this figure displays in each panel a simple least squares fit assuming a linear regression model. While the same information might also be contained in a single graph showing these linear regression fits, it is now immediately clear that there are apparently different intercept values for the different subjects, as well as different slopes. This Trellis display motivates modeling the data with a mixed-effects model, with a subject-specific random intercept and a subject-specific random slope. The regression lines in each panel are drawn with only 10 observations, and are not expected to be highly accurate. Not all points in all of the panels follow the fitted straight line closely, though the deviations do not seem to be too large (by sight, we do not perform a test at this early model building stage), not warranting anything more complicated than a global linear fixed effect for the variable day. A first model for these data can take the following form. Denote by y_{ij} the reaction time of individual i ($i = 1, \ldots, 18$) on day j ($j = 0, \ldots, 9$). Let variable x_j indicate day number j. We take a bivariate random effects variable $u_i = (u_{0i}, u_{1i})^t$ representing the random intercept and slope; the error term is denoted by ε_{ij}. We assume that there is independence between the error terms and the random effects, that there is independence amongst the error terms, and independence amongst the set of 18 bivariate random effects. For both random effects and error terms, we assume a normal distribution such that $u_i \sim N_2(0, \Omega)$, with

$$\Omega = \begin{pmatrix} \sigma_{u0}^2 & \sigma_{u01} \\ \sigma_{u01} & \sigma_{u1}^2 \end{pmatrix}$$

and $\varepsilon_{ij} \sim N(0, \sigma_\varepsilon^2)$ for all $i = 1, \ldots, 18$ and $j = 0, \ldots, 9$. This leads to the model

$$y_{ij} = \beta_0 + \beta_1 x_j + u_{0i} + u_{1i} x_j + \varepsilon_{ij}. \tag{24.1}$$

When fitting this model with the method of restricted maximum likelihood, we obtain that $\hat{\beta}_0 = 251.4$ (standard error 6.8), $\hat{\beta}_1 = 10.5$ (standard error 1.5), $\hat{\sigma}_\varepsilon^2 = 654.9$, $\hat{\sigma}_{u0}^2 = 612.1$, $\hat{\sigma}_{u1}^2 = 35.1$, and the estimated correlation between u_{0i} and u_{1i} is equal to 0.07 or, equivalently, a covariance equal to 9.6.

While the structure in this data example is not that complicated, and the number of variables as well as the number of subjects is small, Trellis displays are particularly useful for visualizing structure in datasets with a large number of variables. In addition, observations that are in some sense different from the other observations can be detected by inspecting such graphs. Becker et al. (1996) clearly explain the use and construction of such displays and they illustrate their use with various examples. Trellis displays can help to visualize interactions between variables. By putting the graphs in several panels, the method is also able to deal with many explanatory variables (these variables may be discrete or continuous) by showing possible relationships between two or three variables conditional on other variables. While in this example we have used scatter plots with added least squares lines, a Trellis display can also consist of dot plots, box plots, normal probability plots, estimated surfaces, etc. Nonparametric estimates may be added in each subplot, or profiles can be added by connecting the points. Many options are available and it is worth spending some time on such graphical displays to help build a good model.

24.3 DATA EXAMPLE 2: THE FRAMINGHAM CHOLESTEROL STUDY

We use the Framingham cholesterol data as described in Zhang and Davidian (2001),

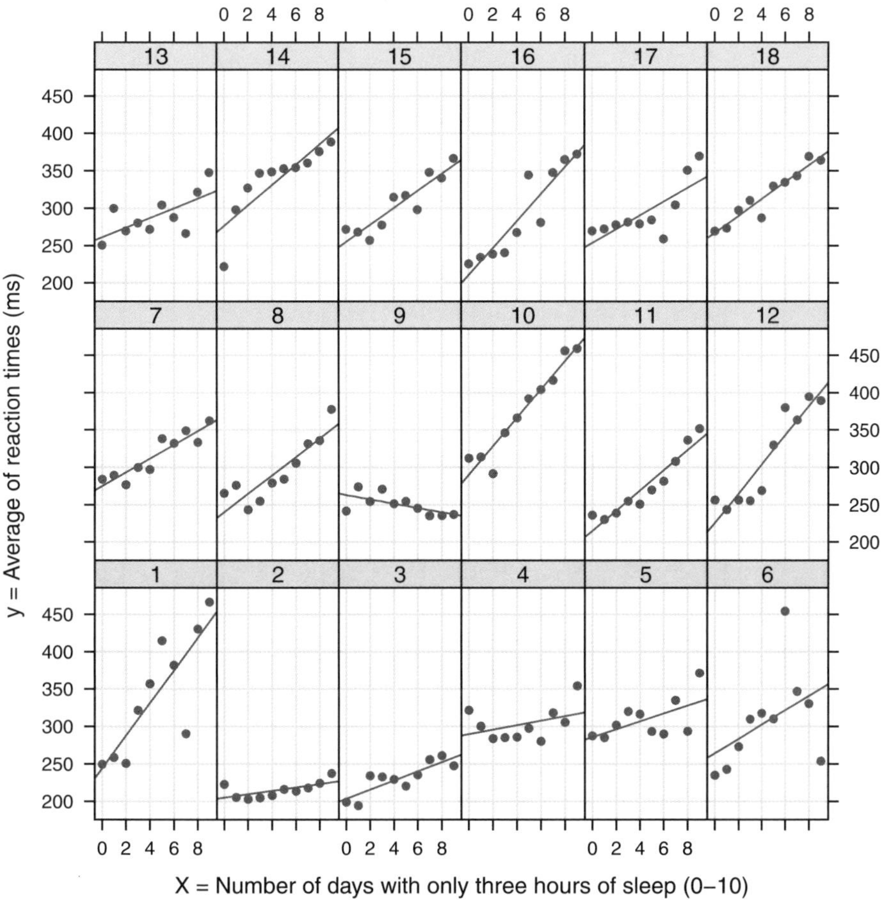

Figure 24.1 Trellis Plot for the Sleep Study. Average reaction times (in ms) versus day, per subject. Fitted linear regression lines are added to each panel

where we restrict analysis to the subset of 73 women and 60 men for which complete data were recorded. For each participant in the study, cholesterol levels were measured at baseline, and further at two-year intervals, resulting in 6 measurements in total. The age at baseline and gender is also available. Because of the size of the dataset, we plot Trellis displays separately for men and women, see Figures 24.2 and 24.3, where the cholesterol level (vertically) is plotted versus the measurement times (at years 0, 2, 4, 6, 8, 10 from baseline) per subject.

The difference in intercepts and slopes of the added linear regression fits per panel motivate the use of a random intercept and slope in the time effect, as well as of a separate gender variable. Including age in the model leads to the mixed effect model

$$y_{jk} = \beta_0 + \beta_1 x_{1j} + \beta_2 x_{2j} + \beta_3 x_{3jk} + u_{0j} + u_{1j} x_{3jk} + \varepsilon_{jk}, \quad (24.2)$$

where x_{1j} denotes the age of the jth subject, x_{2j} is a binary variable indicating the gender, with 0 for females and 1 for males, x_{3jk} is the time (taking values in $\{0, 2, 4, 6, 8, 10\}$ for subject j at the kth measurement period, with $k = 1, \ldots, 6$ and $j = 1, \ldots, 133$. There are two subject-specific unobserved

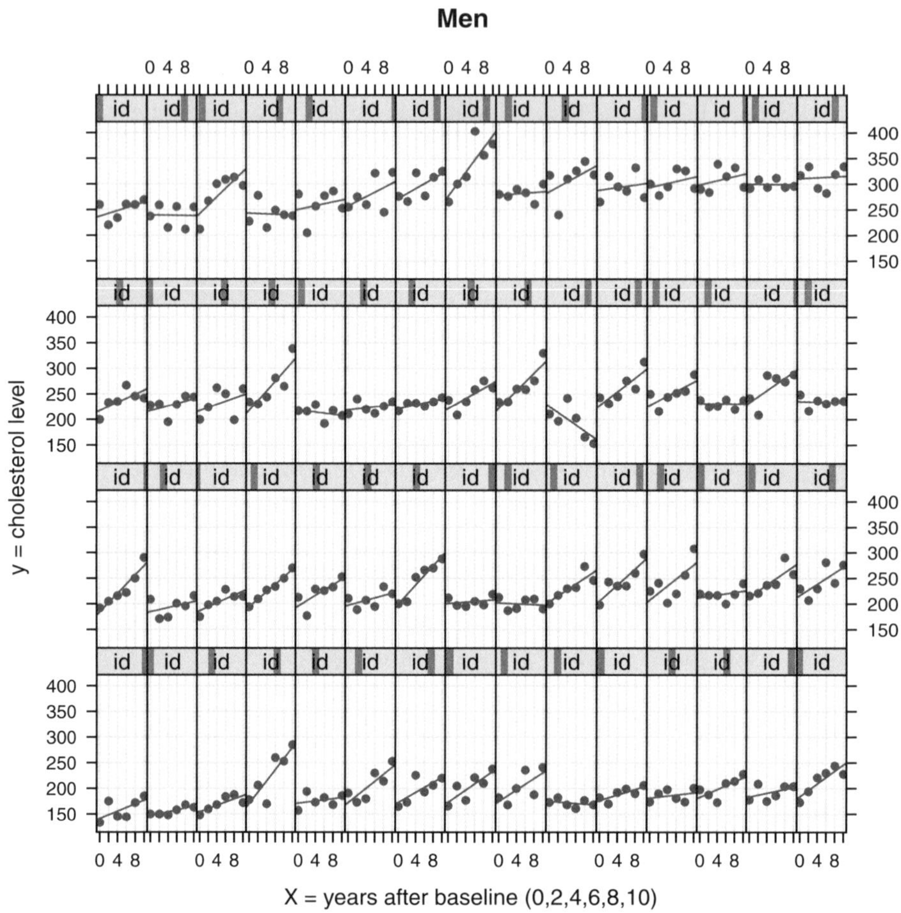

Figure 24.2 Trellis Plot for the Framingham Cholesterol Study for the 60 Men Separately. Cholesterol level versus years after baseline, per subject. Linear regression fits are added to each panel

random effects, u_{0j} and u_{1j}, the coefficients $\beta_0, \ldots, \beta_3$ are fixed but unknown, and ε_{jk} is the unobserved error term.

24.4 TESTS FOR THE DISTRIBUTION OF RANDOM EFFECTS

24.4.1 Are Residuals Useful in Graphical Displays?

In a linear regression model with only fixed effects of the form $y_i = \boldsymbol{X_i}\boldsymbol{\beta} + \varepsilon_i$ and with independent observations, normality of the error terms ε_i $(i = 1, \ldots, n)$ may be checked through tests on the distribution of the fitted residuals $e_i = y_i - \boldsymbol{X_i}\hat{\boldsymbol{\beta}}$. In this model, residual plots are helpful to detect model misspecifications, and a normal probability plot is a graphical way of inspecting the normality assumption.

A direct generalization towards hierarchical models of the form $y_i = \boldsymbol{X_i}\boldsymbol{\beta} + \boldsymbol{Z_i}\boldsymbol{u_i} + \varepsilon_i$ is difficult for several reasons. First, how should one construct residuals? A first type of residual is obtained by mimicking the situation

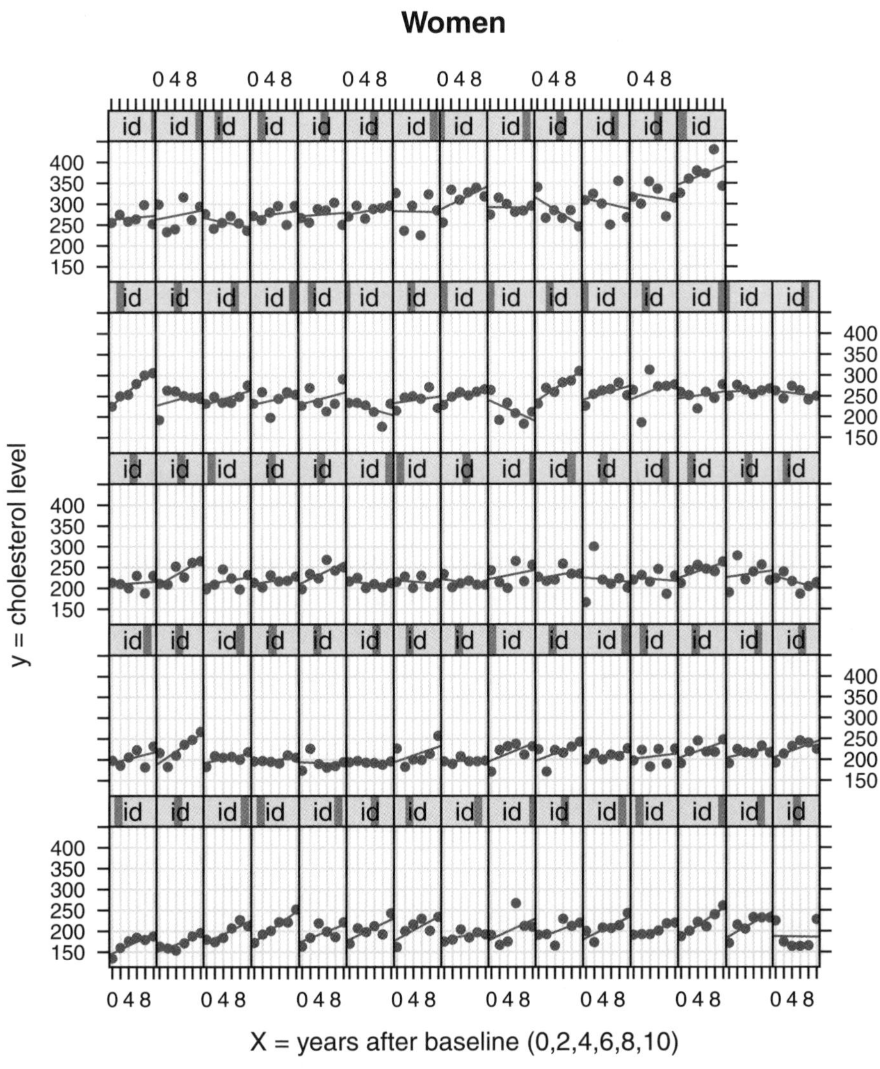

Figure 24.3 Trellis Plot for the Framingham Cholesterol Study for the 73 Women Separately. Cholesterol level versus years after baseline, per subject. Linear regression fits are added to each panel

in the fixed effects model, namely, by taking $e_i^{(1)} = y_i - \boldsymbol{X}_i\hat{\boldsymbol{\beta}}$. When conditioning on the random effects, another type of residual is $e_i^{(2)} = y_i - \boldsymbol{X}_i\hat{\boldsymbol{\beta}} - \boldsymbol{Z}_i\hat{\boldsymbol{u}}_i$. For $e_i^{(2)}$ it would not be possible to check for the distribution of the random effects since one has conditioned on these effects. For $e_i^{(1)}$ this conditioning does not take place, but with these residuals it does not immediately seem possible to separate the random effects from the 'pure' errors ε_i. A second difficulty of directly working with these residuals is that the within-subject residuals are correlated, while most tests about the error terms are valid for uncorrelated residuals. Panels (a) and (b) of Figure 24.4 show the raw residuals $e_i^{(1)}$ and $e_i^{(2)}$ for the sleep study, using the subject number as an index for plotting. While the subjects numbered 1 and

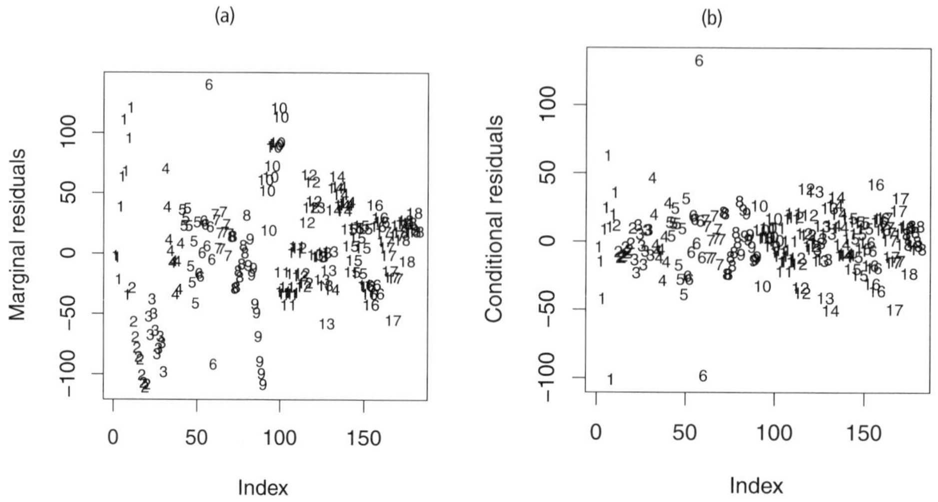

Figure 24.4 Residual Plots for the Sleep Study. (a) $e_i^{(1)}$, reaction times y_{ij} minus fitted fixed effects $\hat{\beta}_0 + \hat{\beta}_1 x_j$ versus the observation index; (b) $e_i^{(2)}$, reaction times minus predicted values $\hat{\beta}_0 + \hat{\beta}_1 x_j + \hat{u}_{0i} + \hat{u}_{1i} x_j$ versus the observation index. The subject number is used as a label

6 seem to have both a large and a small conditional residual, it is not immediately clear from this plot whether this is a source of doubt for the validity of the model assumptions. Further plots (not shown here) of robust versus classical Mahalanobis distances identify these two points as possible outliers.

The unstandardized marginal residual plots in Figure 24.5 show for the Framingham cholesterol study some large (in absolute value) residuals for several of the subjects (e.g. 4, 19, 85, 94, 106, 108, 113, 116). This plot does not allow the analyst to make conclusions about whether there might be problems with the normality assumption of the error terms or of the random effects.

Several attempts have been undertaken to make the residuals of linear mixed models uncorrelated. Calvin and Sedransk (1991) suggest working with the BLUP residuals $e_i^{(2)}$ and to plot a standardized version of $e_i^{(2)}$ versus $\hat{y}_i = \boldsymbol{X}_i\hat{\boldsymbol{\beta}} - \boldsymbol{Z}_i\hat{\boldsymbol{u}}_i$. A problem with such an approach, that is acknowledged and explained in that paper, is that for a correct standardization, the vector of $e_i^{(2)}$ values (for $i = 1, \ldots, n$) is premultiplied with an estimated variance matrix of the conditional residuals. Thus, the standardization is performed on the whole vector, not on individual observations. In the case of abnormal observations after standardization, it is difficult, if not impossible in most cases, to identify the original (untransformed) observations that caused the standardized observations to become extreme in value. Indeed, if a linear combination of values is considered much larger than the remaining standardized values, this might have happened because a few values happen to be simultaneously somewhat large (while not extreme) and this effect is magnified by taking the linear combination. On the other hand, taking a linear combination of values might cause extreme values to be downweighted resulting in a standardized value completely in the range of the other values, and hence the outlying unstandardized value might not get noticed. Moreover, since estimated variances and covariances are used for the estimation of the variance matrix instead of population values, this additional uncertainty should be taken into account,

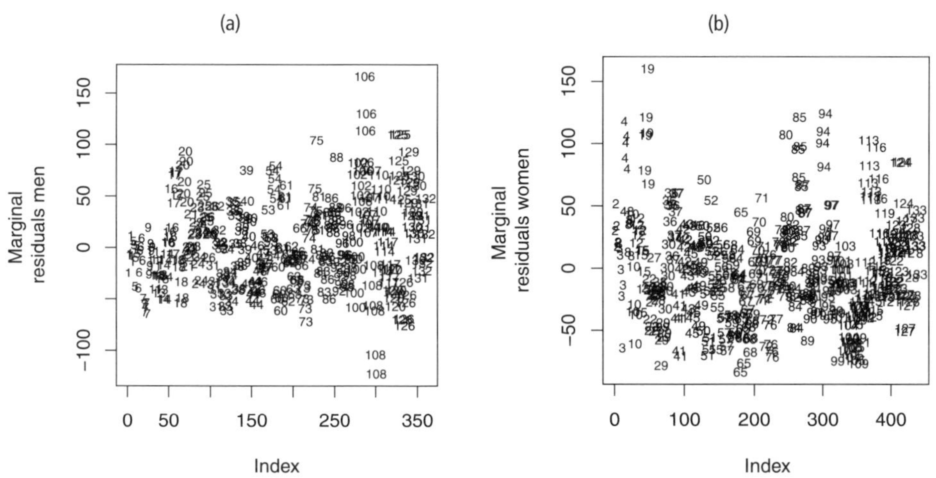

Figure 24.5 Marginal Residual Plots of $e_i^{(1)}$ Versus the Observation Index for the Framingham Cholesterol Study When Fitting Model (24.2) to the Data Set of 133 Subjects with Complete Measurements. (a) for men, (b) for women, using the subject number as a label

though the effect of the estimation is difficult to measure.

For the same reasons, Proust and Jacqmin-Gadda (2005) acknowledge that plots of the standardized residuals versus the predicted values or covariates are not useful. Instead, they ignore the effect of the estimation of the model parameters, and use for the standardization of the marginal residuals $e_i^{(1)}$ a Cholesky square root matrix $(\hat{V}^{-1})^{1/2}$ that is constructed such that $\hat{V}^{-1} = \{(\hat{V}^{-1})^{1/2}\}^t(\hat{V}^{-1})^{1/2}$. By ignoring the effect of the estimation of the unknown parameters, this leads to considering a normal probability plot for the standardized residuals. Such a normal probability plot for the sleep study is shown in Figure 24.6 panel (a), and for the Framingham cholesterol data in panel (b), where some deviation from normality seems visible from the plot. For the interpretation of such a graph, we should pay attention to the facts that (i) the effect of the estimation of unknown parameters has not been taken into account, and (ii) that we work with the marginal residuals, hence this plot cannot give information about the random effects distribution and about the error distribution separately.

A further problem with the use of normal probability plots, and Q-Q plots in general, is explained in McCulloch and Neuhaus (2011). The shape of the distribution of the predicted random effect values depends quite a bit on the shape of the assumed distribution. In other words, when normality has been assumed for the random effects distribution, this will be reflected in the normal probability plot, whether or not this distribution is actually normal. In that paper it is hence advised against using histograms or Q-Q plots for diagnostics about the shape of the true distribution of the random effects. McCulloch and Neuhaus (2011) argue that a wrong specification of the shape of the distribution introduces a possibly large bias for the intercept estimator, which can cause a further bias for the estimation of other quantities, such as the expected outcome.

An alternative graphical method for assessing the assumed random effects distribution is based on directional derivatives of the log likelihood function of the data, constructed from the marginal density of the response values; see Verbeke and Molenberghs (2013). When the random effects distribution function G is replaced by a mixture $(1 - a)G + aH$, for

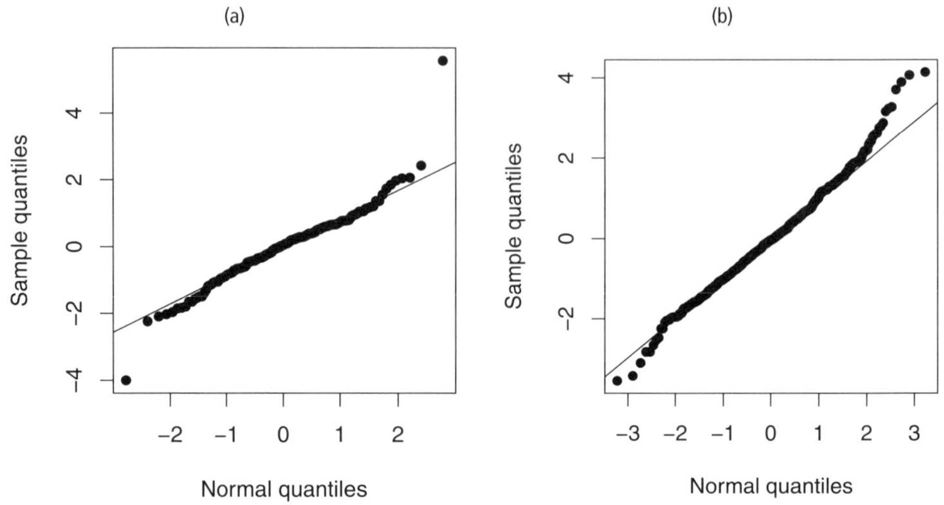

Figure 24.6 Normal Probability Plot Using the Standardized Marginal Residuals for (a) the Sleep Study and (b) the Framingham Cholesterol Study

an infinitesimal value a and another distribution function H, the gradient, or directional derivative, measures the change in the log likelihood when G is replaced by $(1 - a)G + aH$. The gradient function $\Delta(G, \boldsymbol{u})$ is defined as the average over the number of cases $(i = 1, \ldots, n)$ of the ratios of the conditional density functions $f_i(y_i|\boldsymbol{u})$ over the marginal densities $f_i(y_i|G)$ assuming the random effects distribution to be equal to G, thus $\Delta(G, \boldsymbol{u}) = n^{-1} \sum_{i=1}^{n} \{f_i(y_i|\boldsymbol{u})/f_i(y_i|G)\}$. In Verbeke and Molenberghs (2013) it is argued that the gradient stays below the value 1 in the case where no distribution H can be found that gives a better fit than the (estimate of the) distribution G. A graphical representation of such a gradient may thus indicate deviations from the assumed distribution G. This graphical method is useful for one- or two-dimensional random effects distributions.

A misspecification of the shape of the random effect distribution does not always cause harm, as the effect on the estimation of the random effect variances seems to be minor, as is the effect on the expected squared prediction error. For more information and references, see McCulloch and Neuhaus (2011).

24.4.2 Testing the Normality of the Random Effects in a Linear Mixed Model

The null hypothesis and alternatives

We specifically focus attention on testing for normality since this is the most commonly assumed random effects distribution. The null hypothesis that we are interested in is

$$H_0: \boldsymbol{u} \sim N_q(\boldsymbol{0}, \Omega).$$

Parametric tests compare the distribution under the null hypothesis with another prespecified distribution. In the current context, that would be difficult to specify since the distribution can deviate in so many ways from the specified normal distribution. For this reason we rather rely on nonparametric tests. In this case we do not precisely specify the distribution under the alternative hypothesis, but rather say that we test against

$$H_a: \boldsymbol{u} \sim \text{a distribution other than a normal one.}$$

Such a test is called nonparametric, as indeed the alternative hypothesis is not specified via a parametric description. Often, such tests make use of nonparametric estimators to estimate under the alternative hypothesis. For the testing problem on the distribution of the random effects this means that we will estimate this distribution in a way that avoids saying beforehand precisely which distribution we will use.

For ease of exposition we will work with a standardized random effect $\boldsymbol{\gamma}$ such that $\boldsymbol{u} = \boldsymbol{G}\boldsymbol{\gamma}$, with $\Omega = \boldsymbol{G}\boldsymbol{G}^t$. The null hypothesis that we wish to test is equivalent to

$$H_0\colon \boldsymbol{\gamma} \sim N_q(\mathbf{0}, \boldsymbol{I}),$$

a q-variate standard normal distribution. The nonparametric estimator for the density function of $\boldsymbol{\gamma}$ is constructed via a series expansion around the standard normal density function, also called an Edgeworth expansion. For a univariate variable γ (with dimension $q = 1$), the expansion uses Hermite polynomials and is

$$f_\gamma(u) = \phi(u)\{1+k_3H_3(u)+k_4H_4(u)+\cdots\}.$$

The reason for choosing Hermite polynomials, which satisfy $H_j(u)\phi(u) = (-1)^j \frac{d^j\phi(u)}{du^j}$, is that they are orthogonal with respect to the standard normal density function ϕ. This means that $\int H_j(u)H_k(u)\phi(u)du = 0$ when $j \neq k$. Hermite polynomials are polynomial functions in u. One may compute that $H_3(u) = u^3 - 3u$, $H_4(u) = u^4 - 6u^2 + 3$, $H_5(u) = u^5 - 10u^3 + 15u$, and $H_6(u) = u^6 - 15u^4 + 45u^2 - 15$ (see Olver et al., 2010, Section 18.5). The formulae for a vector-valued $\boldsymbol{u}$ can be computed following the approach of Withers (2000).

By taking this expansion for a large number of terms, and letting the coefficients k_j vary, we construct a large set of density functions. Importantly, the standard normal density function is included as a special case where all coefficients k_j, $j \geq 3$, are equal to zero. Such a series expansion allows us to consider a whole class of density functions which are all alternatives to the normal density function. In practice, the series is truncated after M terms, and the corresponding density is denoted by $f_{\gamma,M}$.

This approach of constructing random effects distributions that are more complicated than the normal distribution is studied by Zhang and Davidian (2001) for linear mixed models, and there named semi-nonparametric estimation. They have implemented this using SAS. Chen et al. (2002) discuss the estimation of generalized linear mixed models with such distributions for the random effects.

Order selection tests

Once we know how to construct a series expansion about the normal density function, we can try fitting models to the data with several such expansions for the random effect density. Each time, we use a truncated expansion with, for example, a growing truncation point to construct models that are more and more complex.

The idea behind order selection testing is to try several approximations to the underlying density function by letting the series truncation point M vary between 0 (the null model) and some upper bound M_{up}. The truncation point M is also called the 'order'. Next, a model selection method is applied to select the best value of M. The idea of the test is then quite simple: the null hypothesis is rejected whenever the selected value of M does not correspond to the value of the model under the null hypothesis.

Originally, Eubank and Hart (1992) used a modified Mallows' C_p criterion to determine M, and Aerts et al. (1999, 2000) used a modified Akaike information criterion (AIC) for regression models in a likelihood setting or with more than one covariate; see also Hart (1997) for more details on the order selection tests for linear regression models, and Claeskens and Hjort (2008, Chapter 8) for a comparison with related approaches. For random effects models, such tests using a modified AIC are described in Claeskens and Hart (2009a).

It is clear that the characteristics of the model selection method that is used to select the order determine the asymptotic distribution of the performed test. We here explain how the asymptotic distribution may be derived when working with the AIC. Akaike's 1973 information criterion is defined as minus twice the value of the log likelihood, where unknown parameters are replaced by maximum likelihood estimators, penalized with twice the number of estimated parameters in the model. It is readily obtained that when working with the AIC, there is an equivalent formulation of the test in terms of maxima of weighted likelihood ratio statistics. Indeed, the null model S_0 is selected by the AIC if and only if its AIC value is smaller than the AIC values of all other models S_m. Let p_j denote the number of unknown parameters in the model S_j, for $j = 0, \ldots, J$, with J the number of models under consideration.

$$
\begin{aligned}
&\text{Model } S_0 \text{ is selected}\\
&\Leftrightarrow \mathrm{AIC}(S_0) < \mathrm{AIC}(S_m), \text{ for all } m \geq 1,\\
&\Leftrightarrow -2\log \mathrm{Lik}(S_0) + 2p_0 < -2\log \mathrm{Lik}(S_m)\\
&\qquad\qquad + 2p_m, \text{ for all } m \geq 1,\\
&\Leftrightarrow 2\log\left(\frac{\mathrm{Lik}(S_m)}{\mathrm{Lik}(S_0)}\right) < 2(p_m - p_0),\\
&\qquad\qquad \text{for all } m \geq 1,\\
&\Leftrightarrow T_{n,OS} = \max_{m\geq 1} \frac{\mathrm{LRT}(S_m, S_0)}{p_m - p_0} < 2,
\end{aligned}
$$

where we assumed that $p_m > p_0$ for all $m \geq 1$. For the likelihood ratio statistic $\mathrm{LRT}(S_m, S_0) = 2\log\{\mathrm{Lik}(S_m)/\mathrm{Lik}(S_0)\}$, it holds that under the null hypothesis $\mathrm{LRT}(S_m, S_0) \xrightarrow{d} \chi^2_{p_m - p_0}$, and it then follows that the order selection statistic has a well-defined limit in distribution,

$$T_{n,OS} \xrightarrow{d} T_{OS} = \sup_{m\geq 1} \frac{1}{p_m - p_0} \sum_{j=1}^{p_m - p_0} Z_j^2,$$

with $Z_j \sim N(0, 1)$ and $Z_1, Z_2, \ldots$ independent. From this follows an explicit expression to obtain critical values for the test statistic. For the situation where the number of variables increases by one when going from model S_m to model S_{m+1}, and thus $p_m - p_0 = m$, it holds that (Eubank and Hart, 1992)

$$P(T_{OS} \leq x) = \exp\left\{-\sum_{j=1}^{\infty} \frac{P(\chi_j^2 > jx)}{j}\right\}.$$

The null hypothesis is rejected for large values of T_{OS}. Applying this formula gives that when we use the AIC for selecting the model in the sequence, we use as a critical value 2. This '2' comes from using as a penalty twice the number of parameters in the AIC. When using the above formula (e.g., using a truncated series of 500 terms rather than summing the series all the way to infinity) we find that $P(T_{OS} \geq 2) \approx 0.29$, which is too large for common testing problems. Hence, for the probability of a type I error equal to α (typical values are 10%, 5%, 1%), we wish to find a value C_α such that $P(T_{OS} \geq C_\alpha) = \alpha$. Working with the T_{OS} form of the test, there is no obligation to use 2 as a critical value. For example, when $p_m - p_0 = m$, the value $C_{0.05} = 4.18$ corresponds to $\alpha = 5\%$ and may be used instead. Also, for the equivalent form of the test phrased in terms of the AIC, this problem with the large type I error is easily solved. For the purpose of performing this test, we modify the AIC with a different penalty constant as in

$$
\begin{aligned}
\widetilde{\mathrm{AIC}}(S_m; C_\alpha) &= -2\log \mathrm{Lik}(S_m)\\
&\quad + C_\alpha(p_m - p_0),
\end{aligned}
$$

with C_α that value such that $P(T_{OS} \leq C_\alpha) = 1-\alpha$. Since the models are nested for growing values of m and all models contain the normal model as a special case, we subtract from the penalty in the $\widetilde{\mathrm{AIC}}$ value the number of parameters estimated in the normal model (p_0). This is a constant value and will not change the results of the minimization process to find the best value of m. Denote $\widehat{M}_{\widetilde{\mathrm{aic}}}$ the (random) value of M that is selected by the modified AIC. The following tests are then equivalent:

$$
\begin{aligned}
&\text{Reject } H_0 \text{ if and only if}\\
&\widehat{M}_{\widetilde{\mathrm{aic}}} = \arg\min_m\{\widetilde{\mathrm{AIC}}(S_m; C_\alpha); m \geq 0\} > 0,
\end{aligned}
\qquad (24.3)
$$

and

Reject H_0 if and only if

$$T_{n,OS} = \max_{m \geq 1} \frac{\text{LRT}(S_m, S_0)}{p_m - p_0} > C_\alpha. \tag{24.4}$$

While at first the test (24.3) does not resemble the usual form of a test statistic, the equivalent form in (24.4) makes it immediately clear that the nonparametric test that we have constructed is in fact a maximum of weighted likelihood ratio test statistics. We indeed do not perform a single likelihood ratio test, and hence avoid the restrictions of parametric testing whereby it is necessary to specify the type of the alternative model beforehand. On the other hand, we also do not fall in the trap of performing multiple tests, where it is difficult to control the level of the test. All separate likelihood ratio values are combined into one single test statistic $T_{n,OS}$, of which the asymptotic distribution function is known.

Normality tests for the random effects in the sleep study

In the model for the sleep study, see (24.1) in Section 24.2, there is a bivariate vector of random effects representing a subject-specific intercept and slope. We wish to test for bivariate normality

$$(u_0, u_1) \sim N_2(0, \Omega),$$

with Ω unspecified. The order selection test using Hermite polynomials is applied with an upper bound of $M = 9$ terms in the series expansion. This leads to the results in Table 24.1. For each model with different truncation points m in the series expansion of the random effects density function, we give the value of the maximized log likelihood, together with the added number of estimated parameters as compared to the normal model (this is the model with $m = 0$). These numbers are used to form an adjusted AIC where, instead of the usual value 2 in the penalty, we use the value $C_\alpha = 3.084$, with $\alpha = 0.05$. Other values of interest may be $C_{0.1} = 2.474$ and $C_{0.01} = 4.584$. These values are obtained through simulation of the asymptotic distribution (see also Claeskens and Hart, 2009a, Section 6). The value of m that is selected by the modified $\widetilde{\text{AIC}}(\cdot; 3.084)$ is zero for this data set. At the 5% level of significance we cannot reject the null hypothesis of bivariate normality for the random effects of the sleep study. Equivalently, we can compute the order selection statistic $T_{n,OS}$ with an observed value of 2.627. This value does not exceed the simulated critical value of 3.084 at the 5% level of significance, and we—of course—come to the same conclusion.

Normality tests for the random effects in the Framingham cholesterol study

The specified random effects model for the cholesterol study in Section 24.3 contains a bivariate vector of random effects consisting of a subject-specific intercept and slope. We wish to test for bivariate normality

$$(u_0, u_1) \sim N_2(0, \Omega),$$

with Ω unspecified.

The order selection test is applied to this dataset with $M = 5$. Table 24.2 gives the observed values of the log likelihood with estimated parameter values inserted, the differences in number of parameters for the considered extended models, as well as the adjusted AIC values using the constant 3.084 instead of the usual value 2 in the penalty part of the criterion. This choice of penalty reflects testing at a 5% level of significance. The resulting $\widetilde{\text{AIC}}(\cdot; 3.084)$ selects the value $m = 1$ as the best, followed by $m = 2$, $m = 3$, and only at the fourth place comes the null model with $m = 0$. Based on this result we have evidence to reject the null hypothesis of bivariate normality for this set of data. The equivalent form of the test statistic is obtained by taking the maximum of the values in the last column in Table 24.2, resulting in the value 12.39, which has a corresponding P-value equal to 10^{-5}, confirming the previous

Table 24.1 Sleep Study. Values of the maximized log likelihood for different truncation points m in the series expansion of the random effects density, difference in degrees of freedom $p_m - p_0$, modified AIC values with penalty constant 3.084 selecting the model with m=0, the values of the log likelihood ratio statistic LRT and the scaled LRT values with the difference in degrees of freedom $p_m - p_0$. The maximum of the last column gives the value of the order selection statistic $T_{n,OS}$, here equal to 2.627

m	log Lik.	$p_m - p_0$	$\widetilde{\text{AIC}}(\cdot; 3.084)$	LRT	LRT/$(p_m - p_0)$
0	−875.970	—	1751.940	—	—
1	−873.343	2	1752.854	5.254	2.627
2	−872.346	5	1760.112	7.248	1.450
3	−869.198	9	1766.152	13.543	1.505
4	−869.278	14	1781.732	13.384	0.956
5	−868.133	20	1797.946	15.673	0.784
6	−868.468	27	1820.204	15.003	0.556
7	−866.511	35	1840.962	18.917	0.540
8	−866.018	44	1867.732	19.904	0.452
9	−865.742	54	1898.020	20.455	0.379

conclusion about rejecting the null hypothesis of bivariate normality.

Contours of a nonparametric estimate of the bivariate density of the random intercept and slope using a penalized Gaussian mixture linear mixed model (Ghidey et al., 2004) is shown in Figure 24.7. This figure also confirms the previous conclusion. In particular, the distribution of the random intercept seems to deviate from normality and could possibly be better represented by a mixture of two normal distributions.

24.4.3 *Simultaneous Tests on the Error and Random Effects Densities*

When testing for the normality of the distribution of the error terms, a similar approach to the previous section can be taken. Using Hermite polynomials, a series expansion of the error density is given by

$$f_{\varepsilon,M_\varepsilon}(\boldsymbol{v}) = \phi(\boldsymbol{v}) \left\{ \sum_{j \le M_\varepsilon} b_j H_j(\boldsymbol{v}) \right\}.$$

This expansion of the error term can be used in combination with that of the random effects. To fit a model with such more complicated density functions, we write the marginal density of the response $\boldsymbol{y}$ as

$$f_{M,M_\varepsilon}(\boldsymbol{y};\theta) = \int f_{\varepsilon,M_\varepsilon}(\boldsymbol{y} - \boldsymbol{X\beta} - \boldsymbol{Zu}|\boldsymbol{u},\theta) f_{U,M}(\boldsymbol{u})d\boldsymbol{u}.$$

Maximum likelihood estimators of the model parameters and of the series coefficients that are introduced by the expansion of the density functions are obtained by maximizing the likelihood of the data. The null hypothesis that we wish to test is normality of both the random effects and the error term, thus $M = M_\varepsilon = 0$. The order selection test approach uses a model selection method such as a modified AIC, where the penalty constant is appropriately chosen, to obtain data-driven values for M and M_ε. A particular advantage of this test is that, in case of rejection, it can indicate where the discrepancy with the null model is. If, for example, only $M_\varepsilon > 0$, there might be a problem with the normality assumption of the error terms, though not of the random effects.

24.4.4 *Other Test Statistics*

Score statistics in the order selection test

The order selection test statistic $T_{n,OS}$ in (24.4) is phrased as a maximum of weighted

Table 24.2 Framingham Cholesterol Study. Values of the maximized log likelihood for different truncation points m in the series expansion of the random effects density, difference in degrees of freedom $p_m - p_0$, modified AIC values with penalty constant 3.084 here selecting the model with $m = 1$, values of the log likelihood ratio statistic LRT and the scaled LRT values with the difference in degrees of freedom $p_m - p_0$. The maximum of the last column gives the value of the order selection statistic $T_{n,OS}$, here equal to 12.389

m	log Lik.	$p_m - p_0$	$\widetilde{\text{AIC}}(\cdot; 3.084)$	LRT	LRT$/(p_m - p_0)$
0	−160.986	—	321.972	—	—
1	−148.597	2	303.362	24.778	12.389
2	−146.891	5	309.202	28.190	5.638
3	−145.470	9	318.696	31.032	3.448
4	−142.917	14	329.010	36.138	2.581
5	−142.606	20	346.892	36.760	1.838

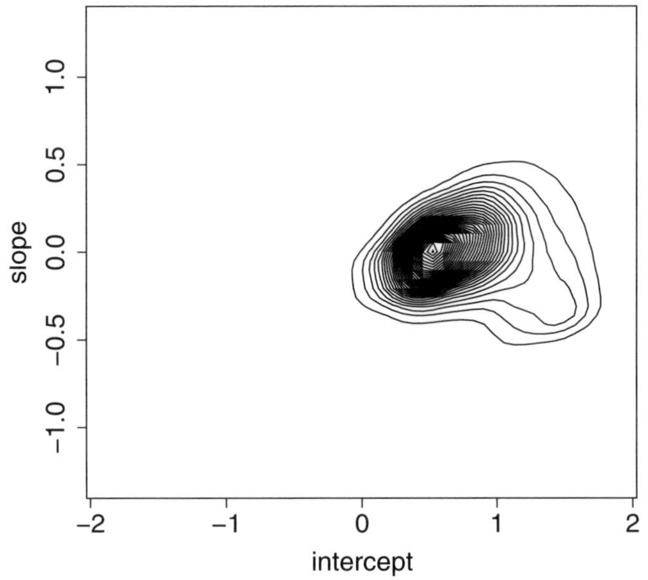

Figure 24.7 Framingham Cholesterol Study: Contours of a Nonparametric Estimate of the Bivariate Density of the Random Intercept and Slope Using a Penalized Gaussian Mixture Linear Mixed Model. A formal hypothesis test (see main text) rejects the null hypothesis of bivariate normality

likelihood ratio statistics. While this is statistically a good idea, in practice it might lead to computational efforts to find the maximum likelihood estimators of all parameters in all of the considered models. The bigger the truncation point m, the more parameters need to be estimated. A more attractive substitute for the likelihood ratio statistic is the score statistic, since such score statistics only need maximum likelihood estimates computed for the null model. Since a single score statistic has the same asymptotic distribution as the corresponding likelihood ratio statistic, the limiting distribution of the order selection test statistic is also the same for both versions of the test. Score tests for order selection testing have been studied for linear multiple regression models by Aerts et al. (1999, 2000). For linear mixed models, see Thas (2009) and Claeskens and Hart (2009b).

Pearson χ^2 test

Jiang (2007, pp. 91–2) constructed a type of Pearson χ^2 test for jointly testing for the distributions of random effects and errors. Similar to the classical Pearson χ^2 tests, this test starts with dividing the real line into a number of classes C_j ($j = 1, \ldots, M$). Let N_j be the count of how many of the response values belong to class j. This is the observed 'cell' (or class) count. An expected cell count E_j for the jth class is computed as a sum over all cases n of the probability that a normal random variable with mean the corresponding row of $X\hat{\beta}$ and with variance the corresponding diagonal element of $\widehat{V}$ (with estimators inserted for unknowns) belongs to class C_j. Unlike the traditional use of Pearson's χ^2 statistic, the current application, where observations are correlated due to the presence of random effects, implies that N_j is no longer a sum of independent variables, and hence the distribution of $\frac{1}{a}\sum_{j=1}^{M}(N_j - E_j)^2$ is not χ^2. The constant a in this expression is a 'normalizing' constant, though is not uniquely determined. For the special case of a balanced one-way random effects model with n cases with r replicated observations each, Jiang (2001) advised using $a = nr^2$. An explicit formula for use in general, more complicated models is not available, but could possibly be derived using the theoretical results of Jiang (2001). The choice of the number of classes M, as well as their construction (equal in size, or equal in probability) is up to the user to determine, but will have consequences for the power of the test.

Minimum distance tests

Hart and Cañette (2012) develop and study nonparametric estimation methods for random effect models that are particularly applicable when there is only a small number of cases in the data set (n small) in combination with a large number of observations per case. Such situations typically arise in microarray studies. In that method, no parametric model is required for estimation of either the random effects distribution or the error distribution. The starting point is a nonparametric estimator of the characteristic function, which by Fourier inversion yields an estimator of the corresponding density function. Parameter estimators are found by minimizing a certain distance function (hence the name of the method), that in this case is different from a usual likelihood or likelihood ratio. In Claeskens and Hart (2009a, Section 7), this kind of nonparametric estimator is used as the basis of a minimum distance test for the distribution of the random effects and the error terms in linear mixed models. The estimated discrepancy measure itself is used as a test statistic. In that paper, a bootstrap has been applied to provide p-values for the test statistics. A study of similar such tests and their asymptotic behavior for specific multilevel models might be worthwhile.

24.5 DEALING WITH OUTLIERS

Dealing with 'abnormal' values, outliers, in multilevel models is a very complicated task, since observations that do not seem to have been generated by the same probability model as the rest of the data can occur at every level of the model.

Rather than identifying outlying observations, estimation methods or distributional assumptions may be changed to encompass all observations in a single model fit.

Staudenmayer et al. (2009) stay close to the mixed model with normal random effects and advocate the use of t-distributions instead, both for the random effects as well as for the error term. While the normal distribution does not accommodate very large or very small values, a t-distribution (especially with low degrees of freedom) is much more capable of fitting such data.

When outliers are thought to be present, robust estimators could be used. Copt and Victoria-Feser (2006) use robust S-estimation in mixed linear models. Chapter 22 contains more information about robust estimation methods in multilevel models.

24.5.1 Testing for Larger Variances in Variance Shift Models

Gumedze et al. (2010) develop an approach for outlier detection which is based on variance shift outlier models. This approach is especially developed to deal with observations arising from a distribution with a larger variance than that for the other observations, which is their definition of an outlier. This makes sense intuitively since a larger variance allows for smaller or larger values to be observed. This outlier detection procedure works for models with independent random effects and errors, thus $\Omega = \sigma_u^2 I_q$ and $\Sigma = \sigma_\varepsilon^2 I_n$. A variance shift is incorporated for individual observations, resulting in the model formulation

$$y = X\beta + Zu + \delta_i e_i + \varepsilon,$$

where e_i is a unit vector of length n, with a single value one at position i, and $\delta_i \sim N(0, \sigma_{\delta_i}^2)$. Rewriting $\sigma_{\delta_i}^2 = \omega_i \sigma_\varepsilon^2$, with $\omega_i \geq 0$, results in an increased variance for the ith observation.

The procedure continues by computing a test statistic such as the restricted likelihood ratio statistic or the score statistic for the one-sided testing problem

$$H_0: \omega_i = 0 \text{ versus } H_a \quad \omega_i > 0.$$

Since negative values of ω_i are not allowed, both test statistics have a non-trivial asymptotic distribution because the value of ω_i under the null hypothesis belongs to the boundary of its parameter space, namely $(0, \infty)$. An additional issue is the multiple testing problem, since such a variance shift model may (and often will) be written down for each of the n observations, resulting in n tests. To deal with both problems simultaneously, a parametric bootstrap may be used. For a bootstrap test to be valid, data have to be generated under the null hypothesis, thus with $\omega_i = 0$. Parameter estimates are inserted for the unknown fixed effects and variance components. A sample of the same size as the original data is generated from the estimated model $X\hat{\beta}+Z\tilde{u}+\tilde{\varepsilon}$, where $\tilde{u} \sim N_q(0, \hat{\sigma}_u^2 I_q)$ and $\tilde{\varepsilon} \sim N_n(0, \hat{\sigma}_\varepsilon^2 I_n)$. With these generated data, the restricted likelihood ratio or score statistics are computed, in precisely the same way as for the original data. When repeating the data generation and the computation of the test statistics a large number of times, say $B = 5000$, all the results of the test statistics are taken together to compute quantiles of the resulting empirical distribution. These are then used to provide a bootstrap P-value for the test statistic values computed from the original data.

Outliers may occur at the level of the random effects, meaning that a whole group of observations (as defined by the random effect) may be considered outlying. In a similar way, a variance shift model is constructed where now

$$y = X\beta + Z(u + \delta_j e_j) + \varepsilon,$$

with e_j a unit vector of length q with a single value one at position j, and $\delta_j \sim N(0, \omega_j \sigma_u^2)$, with $\omega_j \geq 0$. A similar testing procedure allows the detection of observations with large values of ω_j $(j = 1, \ldots, q)$.

One possible drawback of this method is that it is computationally intensive because of the use of the bootstrap. The pointwise testing approach in combination with bootstrap resampling might be too time consuming for large datasets. Extensions of the variance shift outlier models to the general linear mixed model with correlated errors are discussed in Gumedze et al. (2010).

24.5.2 Cook's Distance and Case/Subject Deletion Diagnostics

For linear models, the influence of a particular observation on the parameter estimates is examined by leaving out each observation in turn and by recomputing the estimates. When there are no random effects in the model, exact formulae exist for leverage values (defined as $\partial \hat{y}_i / \partial y_i$), as well as for Cook's distance. Cook's distance is defined as the squared standardized difference between the estimates with and without the left-out observation. Large values indicate observations that might need to be considered in more detail. Generalizations of these concepts for investigating the influence of an observation are presented in Demidenko and Stukel (2005); more references may be found there.

While leaving out one observation might be natural for linear models without random effects, for linear mixed models one often takes out a whole subject (or cluster, group) rather than a single observation.

Demidenko and Stukel (2005) define a Cook's distance for linear mixed models, including non-balanced models for two situations: (1) for a known variance matrix Ω of the variance components, and (2) for Ω unknown. Only the first case leads to explicit formulae. For case (1), denote the leverage matrix, or 'hat' matrix, for the fixed effects by

$$H_{i,\text{fixed}} = X_i \left(\sum_{j=1}^{m} X_j^t V_j^{-1} X_j \right)^{-1} X_i^t V_i^{-1},$$

with $i = 1, \ldots, m$. When a common error variance may be separated from the variance matrices such that $V_i = \sigma_\varepsilon^2 \tilde{V}_i$, with $\tilde{V}_i$ assumed to be known, Cook's distance for subject number i is defined (with p the number of fixed effect parameters) as:

$$C_i = \frac{1}{p\hat{\sigma}_\varepsilon^2} (\boldsymbol{y}_i - X_i \hat{\boldsymbol{\beta}})(I_n - H_{i,\text{fixed}})^{-1} \tilde{V}_i^{-1} \times X_i \left(\sum_{j=1}^{m} X_j \tilde{V}_j X_j \right)^{-1} X_i^t \tilde{V}_i^{-1} \times (I_n - H_{i,\text{fixed}})^{-1} (\boldsymbol{y}_i - X_i \hat{\boldsymbol{\beta}}).$$

A plot of the C_i values versus the subject index i may show particular values that are much larger than the others, pointing towards influential observations. No well-defined critical value of the Cook's distance values is available at this time. In the case of a large value for Cook's distance C_i, it cannot immediately be determined from the plot or size of C_i whether this is due to a large residual $\boldsymbol{y}_i - X_i \hat{\boldsymbol{\beta}}$ (pointing towards an outlying observation), or to a large leverage $H_{i,\text{fixed}}$ (pointing towards a potentially influential observation).

While Cook's distance uses only the part of the leverage that is concerned with estimation of the fixed effects, the part of the leverage matrix relating to the prediction of random effects may also be used to identify influential observations. In general, for linear mixed models one defines the leverage matrix as

$$H_i = \frac{\partial \hat{\boldsymbol{y}}_i}{\partial \boldsymbol{y}_i} = \frac{\partial (X_i \hat{\boldsymbol{\beta}} + Z_i \hat{\boldsymbol{u}}_i)}{\partial \boldsymbol{y}_i} = H_{i,\text{fixed}} + H_{i,\text{random}},$$

with (Demidenko and Stukel, 2005)

$$H_{i,\text{random}} = Z_i \Omega_i Z_i^t \tilde{V}_i^{-1} (I_n - H_{i,\text{fixed}}).$$

It is informative to make two plots: $H_{i,\text{fixed}}$ versus i and $H_{i,\text{random}}$ versus i. Since the response values are only indirectly included in the leverage matrices through the residuals, such plots are mostly used to determine influential observations in either the observed X_i values or the observed Z_i values.

24.5.3 Local Influence

Influence measures indicate how sensitive estimators are to small changes in the data or in the model.

Verbeke and Molenberghs (2000, Chapter 11) explain, based on differential geometry arguments, how a measure of influence is obtained that explains, for each observation, how a small perturbation changes the estimator. In this approach, each case's contribution to the log likelihood is given a case-specific weight. Estimators computed using these weights are similar to the original estimators in the case of no or just a small influence. Twice the difference between the maximized log likelihood value with and without the weights is called the likelihood displacement. For a detailed study of this local influence measure, including graphical representations and examples, see Verbeke and Molenberghs (2000, Chapter 11).

Demidenko (2004, Section 9.9) stays closer to the influence measures for linear regression models. This is possible in cases where the effect of the estimation of variance components is small. For details, see the mentioned chapter.

24.6 TESTING THE PARAMETRIC FORM OF THE FIXED EFFECTS

In this section we consider the question of checking the parametric form of the fixed effects in the model, thus whether $\boldsymbol{X\beta}$ is the correct functional form.

We will not address checking whether the random effects need to be present or might be omitted from the model. Such a testing approach is complicated by the fact that under a null hypothesis of zero values for the variance components, the value zero is on the left boundary of the allowable parameter space for variance components, which are always nonnegative. The fact that the parameters under the null hypothesis are not at the interior of their parameter space implies that the asymptotic distributions of test statistics such as the likelihood ratio statistic or the score statistic are more complicated. Often this results in mixture distributions with a point mass at zero. For more details on testing under constraints, see Silvapulle and Sen (2005). Questions about the inclusion of random effects can also be addressed via model selection methods; see Vaida and Blanchard (2005), Greven and Kneib (2010), and Chapter 7.

Zhang and Lin (2003), as well as Lombardía and Sperlich (2008), construct a semiparametric generalized linear mixed effects model, where to the fixed and random effects is added an unspecified function $m(t_{ij})$ that is assumed smooth, but is otherwise unknown. Their model takes the form

$$g\{E(y_{ij}|\boldsymbol{u})\} = m(t_{ij}) + X_{ij}\beta + Z_{ij}u_i,$$

for a specified and known link function g that is used similarly in generalized linear models (see Chapters 8 and 17). There are several options to estimate the function m.

Zhang and Lin (2003) use smoothing splines, which include in the criterion for estimation a penalty that is a constant (the smoothing parameter) times the integral of the estimated, often squared order, derivative of the function m. Through a connection with Bayesian estimation they rewrite the smoothing spline model as a random effects model, as is done similarly in Ruppert et al. (2003) for regression spline models (see Chapter 18). Testing whether the parametric part of the model suffices then boils down to testing whether a variance component is equal to zero. This gives the same aforementioned complications with the asymptotic distribution of the test, which in this case is a mixture of χ^2 distributions. In Zhang and Lin (2003) an approximation to this asymptotic distribution is applied.

Lombardía and Sperlich (2008) use kernel estimators for m instead. By integrating out the random effects density, they obtain the marginal density of the response values $\boldsymbol{y}$, which leads to the marginal likelihood $\sum_{i=1}^{n} \log f(\boldsymbol{y}_i|\boldsymbol{u}_i)$, which contains as unknowns the fixed effects parameters $\boldsymbol{\beta}$, the variance components for the random effects and the function m. Adding kernel weights to the log likelihood function gives a function that is to be maximized for the unknown parameters. A profile likelihood estimator is studied, both theoretically as well as in practice. Their test statistic compares the nonparametric estimator of the mean of $g\{E(y_{ij}|\boldsymbol{u})\}$

to that of a fully parametric estimator. Since kernel estimation methods, as for most other nonparametric estimators, introduce a bias, the use of a bootstrap procedure that takes this bias into account is advocated instead of using an asymptotic distribution.

Henderson et al. (2008) likewise use kernel estimators in their panel data models and use a bootstrap for comparing the nonparametric to the parametric estimates.

24.7 CONCLUSIONS

As with any other model, multilevel models rely on model assumptions. Detecting deviations from the assumptions is not always straightforward due to the intrinsic nature of the model, consisting of several levels and having correlated outcomes on certain of these levels. Existing simple graphical methods, such as normal probability plots, might be misleading. Formal tests and diagnostic tools developed for mixed models should be applied instead. The research area of diagnostics, lack-of-fit, and goodness-of-fit testing for multilevel models is still in full development; more results are welcome.

REFERENCES

Aerts, M., Claeskens, G., and Hart, J.D. (1999) 'Testing the Fit of a Parametric Function', *Journal of the American Statistical Association*, 94: 869–79.

Aerts, M., Claeskens, G., and Hart, J.D. (2000) 'Testing Lack of Fit in Multiple Regression', *Biometrika*, 87: 405–24.

Akaike, H. (1973) 'Information Theory and an Extension of the Maximum Likelihood Principle', in B. Petrov and F. Csáki (eds), *Second International Symposium on Information Theory*. Budapest: Akadémiai Kiadó, 267–81.

Becker, R., Cleveland, W., and Shyu, M.-J. (1996) 'The Design and Control of Trellis Display', *Journal of Computational and Graphical Statistics*, 5: 123–55.

Belenky, G., Wesensten, N.J., Thorne, D.R., Thomas, M.L., Sing, H.C., Redmond, D.P., Russo, M.B., and Balkin, T.J. (2003) 'Patterns of Performance Degradation and Restoration During Sleep Restriction and Subsequent Recovery: A Sleep Dose-Response Study', *Journal of Sleep Research*, 12: 1–12.

Calvin, J.A. and Sedransk, J. (1991) 'Bayesian and Frequentist Predictive Inference for the Patterns of Care Studies', *Journal of the American Statistical Association*, 86: 36–48.

Chen, J., Zhang, D., and Davidian, M. (2002) 'A Monte Carlo EM Algorithm for Generalized Linear Mixed Models with Flexible Random Effects Distribution', *Biostatistics*, 3: 347–60.

Claeskens, G. and Hart, J.D. (2009a) 'Goodness-of-Fit Tests in Mixed Models', *Test*, 18: 213–39.

Claeskens, G. and Hart, J.D. (2009b) 'Rejoinder on: Goodness-of-Fit Tests in Mixed Models', *Test*, 18: 265–70.

Claeskens, G. and Hjort, N.L. (2008) *Model Selection and Model Averaging*. Cambridge: Cambridge University Press.

Copt, S. and Victoria-Feser, M. (2006) 'High-Breakdown Inference for Mixed Linear Models', *Journal of the American Statistical Association*, 101: 292–300.

Demidenko, E. (2004) *Mixed Models*. Hoboken, NJ: Wiley-Interscience.

Demidenko, E. and Stukel, T.A. (2005) 'Influence Analysis for Linear Mixed-Effects Models', *Statistics in Medicine*, 24: 893–909.

Eubank, R.L. and Hart, J.D. (1992) 'Testing Goodness-of-Fit in Regression Via Order Selection Criteria', *The Annals of Statistics*, 20: 1412–25.

Ghidey, W., Lesaffre, E., and Eilers, P. (2004) 'Smooth Random Effects Distribution in a Linear Mixed Model', *Biometrics*, 60: 945–53.

Greven, S. and Kneib, T. (2010) 'On the Behavior of Marginal and Conditional Akaike Information Criteria in Linear Mixed Models', *Biometrika*, 97: 773–89.

Gumedze, F.N., Welham, S.J., Gogel, B.J., and Thompson, R. (2010) 'A Variance Shift Model for Detection of Outliers in the Linear Mixed Model', *Computational Statistics and Data Analysis*, 54: 2128–44.

Hart, J.D. (1997) *Nonparametric Smoothing and Lack-of-fit Tests*. New York: Springer-Verlag.

Hart, J.D. and Cañette, I. (2012) 'Nonparametric Estimation of Distributions in Random Effects Models', *Journal of Computational and Graphical Statistics*, 20: 461–78.

Henderson, D.J., Carroll, R.J., and Li, Q. (2008) 'Nonparametric Estimation and Testing of Fixed Effects Panel Data Models', *Journal of Econometrics*, 144: 257–75.

Jiang, J. (2001) 'Goodness-of-Fit Tests for Mixed Model Diagnostics', *The Annals of Statistics*, 29: 1137–64.

Jiang, J. (2007) *Linear and Generalized Linear Mixed Models and Their Applications*. New York: Springer.

Lombardía, M.J. and Sperlich, S. (2008) 'Semiparametric Inference in Generalized Mixed Effects

Models', *Journal of the Royal Statistical Society, Series B*, 70: 913–30.

McCulloch, C.E. and Neuhaus, J.M. (2011) 'Misspecifying the Shape of a Random Effects Distribution: Why Getting It Wrong May Not Matter', *Statistical Science*, 26: 388–402.

Olver, F. W.J., Lozier, D.W., Boisvert, R.F., and Clark, C.W. (eds) (2010) *NIST Handbook of Mathematical Functions*. Washington, DC: U.S. Department of Commerce National Institute of Standards and Technology.

Proust, C. and Jacqmin-Gadda, H. (2005) 'Estimation of Linear Mixed Models With a Mixture of Distribution for the Random Effects', *Computer Methods and Programs in Biomedicine*, 78: 165–73.

Ruppert, D., Wand, M., and Carroll, R. (2003) *Semiparametric Regression*. Cambridge: Cambridge University Press.

Silvapulle, M.J. and Sen, P.K. (2005) *Constrained Statistical Inference*. Hoboken, NJ: Wiley-Interscience.

Staudenmayer, J., Lake, E.E., and Wand, M.P. (2009) 'Robustness for General Design Mixed Models Using the t-Distribution', *Statistical Modelling*, 9: 235–55.

Thas, O. (2009) 'Comments on: Goodness-of-Fit Tests in Mixed Models. Smooth Tests of Goodness-of-Fit for the Random Effects Distribution in Linear Mixed Models', *Test*, 18: 260–4.

Vaida, F. and Blanchard, S. (2005) 'Conditional Akaike Information for Mixed-Effects Models', *Biometrika*, 92: 351–70.

Verbeke, G. and Molenberghs, G. (2000) *Linear Mixed Models for Longitudinal Data*. New York: Springer-Verlag.

Verbeke, G. and Molenberghs, G. (2013) 'The Gradient Function as an Exploratory Goodness-of-Fit Assessment of the Random Effects Distribution in Mixed Models', *Biostatistics*, to appear.

Withers, C.S. (2000) 'A Simple Expression for the Multivariate Hermite Polynomials', *Statistics and Probability Letters*, 47: 165–9.

Zhang, D. and Davidian, M. (2001) 'Linear Mixed Models With Flexible Distributions of Random Effects for Longitudinal Data', *Biometrics*, 57: 795–802.

Zhang, D. and Lin, X. (2003) 'Hypothesis Testing in Semiparametric Additive Mixed Models', *Biostatistics*, 4: 57–74.

25

Multilevel Models: Is GEE a Robust Alternative in the Presence of Binary Endogenous Regressors?

Robert Crouchley

University of Lancaster, United Kingdom

25.1 INTRODUCTION

Generalized Estimating Equations (GEE) have been used to analyze clustered univariate response data in its many varied forms. This includes panel data on a sequence of outcomes for different subjects and clinical trials data of different subjects obtained from a clustered sample design. For nested 2-level clustered data, we typically have $j = 1, \ldots, J$ clusters of $i = 1, \ldots, N_j$ responses (y_{ij}), with the associated covariate vectors, $\mathbf{x}_{ij} = [x_{ijk}]$, $k = 1, \ldots, K$. For longitudinal data, j typically denotes a subject, and $i = 1, \ldots, N_j$ the time-ordered sequence of responses for subject j. GEE models treat the intra-cluster correlation as a nuisance and focus on the marginal mean of the responses (y_{ij}). By using an appropriately weighted score function, and the sandwich form of the standard errors, the GEE approach can enable the intra-cluster correlation between responses to be allowed for in the inference about the covariate effects on the marginal means.

Diggle, Liang, and Zeger (1994) suggest that the objectives (or the type of inferences) of a study should determine the type of statistical model to be used, i.e. use conditional models, such as multilevel models (MLMs), for inferences about individual responses and GEE marginal mean models for inferences about margins. Lee and Nelder (2004) argue that this is the wrong emphasis. They regard statistical analysis as being made up from two activities: the first activity is model selection, which aims to find parsimonious and well-fitting models for the responses under investigation; the second activity is model prediction, where the estimates from the selected models of the first activity are used to predict quantities of interest, such as marginal means.

In both GEE and MLMs for clustered univariate responses, the analysis is directed at distinguishing systematic variation from random variation due to other effects. To simplify the discussion, we will assume that there are no cluster-specific random covariate effects present in the MLMs, just random intercepts or effects. The cluster-specific random effects in MLMs account for the intra-cluster correlation in the responses. The GEE models provide a characterization of a marginal response to changing covariates. For example, a marginal gender comparison compares the marginal mean among men to the marginal mean among women, holding the other covariates constant. The MLMs explicitly account for the unobserved or random variation across subjects and responses and describe how a subject's response varies as a function of the covariates. There are two ways of obtaining an MLM gender comparison. We can compare the mean among men to the mean among women, for the same value of the random effect, holding the other covariates constant. The problem with this comparison is that it is typically only relevant for a small subset of the sampled population. MLMs can also provide a marginal mean gender comparison. This comparison is obtained from the difference between the marginal MLM model mean for males and the marginal MLM model mean for females, conditional on the other covariates being the same in both marginal models. See Chamberlain (1984, Section 3.1) and Galbraith et al. (1991) for further comments. However, in either GEE or MLM comparisons it may not always be appropriate at a particular ij instance to simply condition on only the values of the other covariates at the same ij.

Comparison of GEE and MLMs can lead to some stark differences, e.g., Lindsey and Lambert (1998) provide an illustration of Simpson's paradox, in which the GEE marginal model suggests an average positive treatment effect while the MLM model suggests a negative treatment effect for each sampled individual. Crouchley and Davies (2001) compared the GEE and MLM model-building procedures on a one-year panel study of depression and help-seeking behaviour in Los Angeles. The GEE model suggested the presence of a lagged depression effect on current depression, little, if any, intra-cluster correlation, and no seasonal effects. The MLM random effects model indicated no lagged depression effect, substantial intra-cluster correlation, and seasonal effects. That is, the MLM found depression to be a cross-sectional process consistent with a GEE population average model, while the GEE independent working model suggested that depression is a higher-order process, i.e. something less parsimonious.

There are many alternative perspectives on the differences between GEE models and MLMs. In an attempt to reconcile some aspects of this debate, we distinguish between the population average model $E\left(y_{ij} \mid \mathbf{x}_{ij}\right)$, Zeger et al. (1988), and the marginal mean $E\left(y_{ij} \mid \mathbf{x}_{1j}, \ldots, \mathbf{x}_{N_j j}\right)$, see, e.g. Gourieroux et al. (1984), Prentice (1988), Pepe and Anderson (1994), Emond et al. (1997). That is, the population average model only conditions on the covariates present for response (time) i, while the marginal model conditions on all the covariates in the cluster. There are also the 2-level model conditional means $E\left(y_{ij} \mid \mathbf{x}_{ij}, u_j\right)$ and $E\left(y_{ij} \mid \mathbf{x}_{1j}, \ldots, \mathbf{x}_{N_j j}, \mu_j\right)$, see Chamberlain (1984). In the conditional models, the u_j are unobserved 2-level cluster-specific effects. In MLMs, the conditional models are estimated by integrating the u_j out of the cluster-specific log likelihoods.

In longitudinal data it is important to distinguish between the different types of covariate that may be present. The simplest covariates are time-constant, e.g. ethnic origin or gender. There are several types of time-varying covariate. Kalbfleisch and Prentice (1980) make a distinction between defined external and internal time-varying covariates. Defined external covariates can be treated in the same way as time-invariant covariates and conditioned on in a model. This is because the values of defined external covariates do not typically change as a consequence of the subject's behaviour or response, e.g. smoking behaviour for the level of atmospheric air pollution. On the other hand, internal covariates have a time path that is jointly determined with the responses of interest. Thus, conditioning on internal covariates can create specification bias. Internal covariates are one type of endogenous regressor. Furthermore, there are time-varying covariates that are not defined externally to the subject's responses but which are not necessarily internal either, e.g. maternal job tenure for a child's illness. For a related definition of time-varying covariates, see Lai and Small (2006).

Having clustered univariate responses helps the identification of within-cluster varying endogenous effects without the need to use instrumental variables as an identifying restriction. The study of endogeneity is important in many substantive fields, see, e.g., Chamberlain (1984) in economics; Winship and Mare (1988), Crouchley and Pickles (1989) in sociology; Duncan et al. (2004) in developmental research; and Bartels (1991) in political science. Standard scientific procedures such as model interpretation, replication, and inference are prejudiced if the role of endogenous variables is not properly addressed. An endogenous explanatory variable is one that is correlated with the omitted variables of the model. Simple examples could include: maternal smoking during pregnancy for a child's birth weight and parental socio-economic status for a child's educational achievement. Within-cluster (time) constant covariates are exogenous by definition, internal covariates tend to be endogenous. Treating the baseline values of time-varying endogenous covariates as if they are time-constant exogenous covariates in GEE models and MLMs can also create misspecification, see Crouchley and Davies (1999). We study the performance of a test that helps distinguish between the different types of within-cluster (time) varying covariate in both GEE marginal models and MLMs. If any endogenous regressors are present we could get a misleading impression of the role of exogenous covariates by, e.g. comparing the mean of men to that of women, holding all the other covariates, including the endogenous regressor, constant at a particular ij. Gender could have two different impacts, a main effect and an additional effect through the endogenous regressor.

There are a range of issues that can affect the relative robustness of GEE models and MLMs. For instance, the classical GEE models can only generally ignore missing data if the dropout or missing data mechanism does not depend on either the covariates or any of the subjects' responses. The likelihood-based MLMs can be considered less restrictive as they can ignore missing data even if the dropout or missing data mechanism is related to the covariates and to pre-dropout responses; see Kenward and Molenberghs (1998). Thus the GEE model's attempt at relaxing assumptions is not without risk as it requires stronger assumptions about the sample selection and missing data mechanisms. However, neither approach is robust to a dropout or sample selection mechanism where the probability of dropout is related to responses at the time of dropout and possibly afterwards. Lindsay and Lambert (1998) compared GEE and MLMs on sample selection, the plausibility of the implied data-generating mechanism, and model interpretation. They concluded that GEE models may sometimes be appropriate for descriptive epidemiological studies but that they should be used with great care in clinical trials.

Statistical theory has been developed for making inferences when some forms of misspecification are present; see, e.g. Cox (1961), Huber (1967), White (1982), and Gourieroux et al. (1984). In the analysis of clustered data, the main results from this work apply to univariate response models in which the mean of each response is correctly specified, i.e. $E\left(y_{ij} \mid \mathbf{x}_{1j}, \ldots, \mathbf{x}_{N_j j}\right) = E\left(y_{ij} \mid \mathbf{x}_{ij}\right)$, but the intra-cluster correlation is either ignored or not correctly modeled. The estimated parameters from these models have a covariance matrix given by the appropriate observed information form of the sandwich estimator, Liang and Zeger (1986).

Many other forms of misspecification in GEE for clustered univariate response models have been investigated. For instance, Park and Weisberg (1998) investigate the conditions under which consistent estimates of regression coefficients can be obtained even if the link function is misspecified. Wall, Dai, and Eberly (2005) study the use of misspecified time-varying covariates. Chao et al. (1997) examine the consequences of an omitted confounder. We will consider the relationship between the various means when the data are generated by an MLM and present specification tests of

$$E(y_{ij} \mid \mathbf{x}_{1j}, \ldots, \mathbf{x}_{N_j j}) \\ = E(y_{ij} \mid \mathbf{x}_{ij}),$$

$$E\{E_u(y_{ij} \mid \mathbf{x}_{1j}, \ldots, \mathbf{x}_{N_j j}, u_j)\} \\ = E\{E_u(y_{ij} \mid \mathbf{x}_{ij}, u_j)\}.$$

These equalities do not hold if endogenous regressors are present; see Chamberlain (1984).

This chapter considers binary endogenous regressors. It assumes that they are generated by continuous latent variables that cross thresholds; this type of regressor inhibits the estimation of reduced forms. A simulation study is presented comparing the robustness of clustered cross-sectional probit GEE and MLMs to the misspecification that occurs by not properly addressing contemporaneous endogeneity of a binary regressor in multilevel data. Simulations are used as it is not possible to obtain simple analytic forms for the misspecification. This work differs from that of Pepe and Anderson (1994) and Emond et al. (1997) as the endogeneity in these models arose from the lagged value of the response variable in the absence of multilevel effects. We also study the behaviour of the Mundalk (1978) and Chamberlain (1984) specification tests for marginal means and MLMs in multilevel data. The simulations show that GEE and MLMs which ignore contemporaneous endogeneity in clustered univariate responses are equally misspecified. An empirical illustration is also provided using data taken from the Mother's Stress and Children's Morbidity (MSCM) Study (Zeger and Liang, 1986).

While the main focus is on fitting clustered univariate binary probit models to data generated by clustered bivariate ($h = 1, 2$) binary probit models in which the binary response y_2 is endogenous for y_1, the ideas readily generalize to more dimensions with more endogenous covariates, more levels, and other members of the exponential family of regression models (Nelder and Wedderburn, 1972).

25.2 CASE STUDY

The illustrative data we use are taken from the Mother's Stress and Children's Morbidity (MSCM) Study (Zeger and Liang, 1986). The data are on 167 pre-school children who were: (1) living with their mothers, (2) free from chronic illness, and (3) attending an inner-city pediatric clinic. The child's mother kept a daily record for over 28 days of their child's illness (1 ill, 0 otherwise) and their level of maternal stress (1 stressed, 0 otherwise). Like earlier researchers, we ignore the possible selective nature of the data. Days with missing `illness` or `stress` were dropped from the analysis; the data was reduced from $167 \times 28 = 4676$ to 4612 days of data. The daily binary data on illness and stress for 6 randomly selected child–mother pairs is presented in Figure 25.1.

Subject 69 is missing day 11 and Subject 37 is missing day 14. These plots illustrate some of the variety in the daily sequences of child illness and maternal stress. For instance, the 69th subject's child has 1 occurrence of illness on day 1, the mother has no occurrences of stress throughout. The 164th subject's child is ill on 3x2=6 days, and the mother is stressed on 3 days. There is also some suggestion of dependence between child illness and maternal stress at the start of the sequence.

The MSCM data also contains a range of baseline covariates, Diggle et al. (2002, Table 12.1). These include employed outside the home (binary), marital status (binary), self-reported maternal and child health (ordered with 5 levels), child race (binary), maternal educational attainment (binary), and household size (binary). Amongst other things, the summary data in Table 25.1 below shows that of the 112 unemployed mothers, 77 were in households of 3 or more and 64 were not married, and that of the 55 employed mothers, 46 were high school graduates.

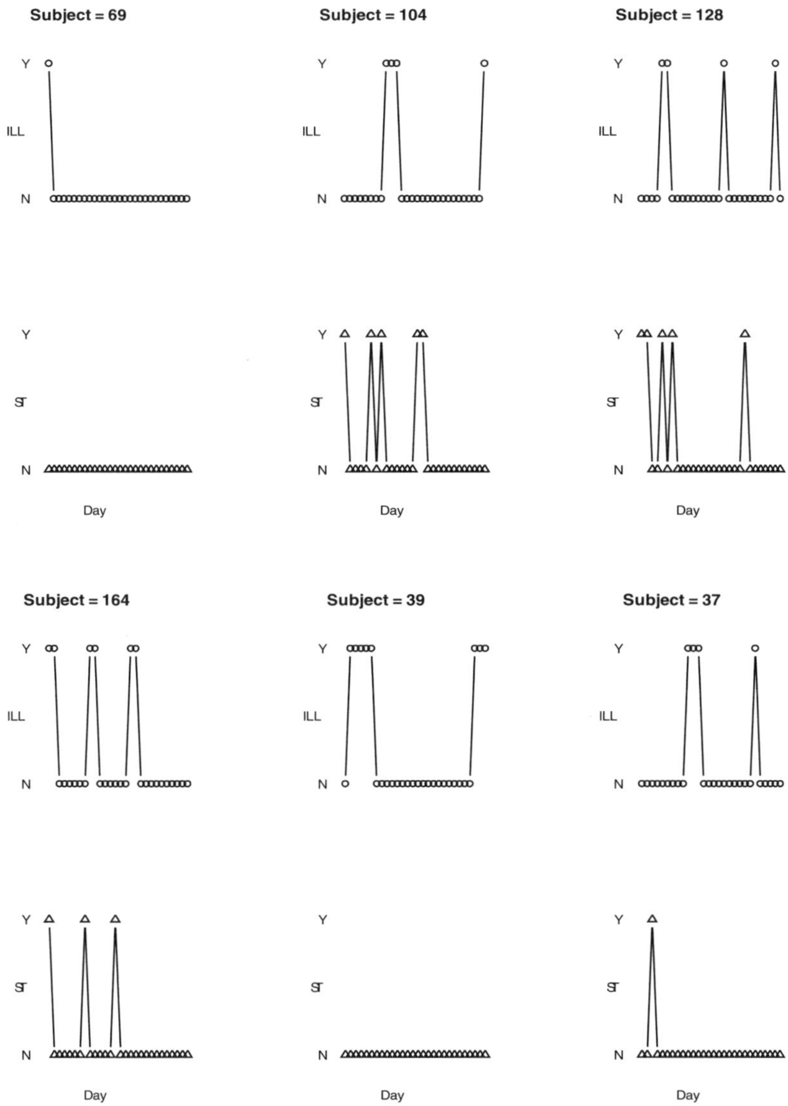

Figure 25.1 Plots of Child Illness (ILL) and Maternal Stress (ST) by Day, for 6 Illustrative Child–Parent Pairs from the Diggle et al. (2002) MSCM Study

The baseline measures are treated as time-constant covariates, Diggle et al. (2002). Results for a clustered cross-sectional model for the contemporaneous effect of maternal stress on child illness are presented. See also Diggle et al. (2002, Table 12.4). In the analysis of Section 25.6, we use the probit link instead of the logit and we do not use lagged values of illness or stress as in the more complex model of Diggle et al. (2002, Table 12.7).

Table 25.1 Summary Data for the MSCM Study, Disaggregated by Employment Status

Variable (variable label)	Employed (employed=1)	Not Employed (employed=0)
Married (married)		
No=0	23	64
Yes=1	32	48
Maternal Health (mhlth)		
fair/poor=1,2	5	19
good=3	18	38
very good=4	26	36
excellent=5	6	19
Child Health (chlth)		
fair/poor=1,2	4	6
good=3	4	18
very good=4	30	51
excellent=5	17	37
Race (race)		
White=0	34	41
Black=1	21	71
Education (educ_bin)		
up to High School=0	9	66
Graduate=1	46	46
Household Size (housize)		
less than 3=0	21	35
3 or more=1	34	77

25.3 MLMS AND GEE

It is often assumed that a 2-level model for the hth-dimensional binary response of the ith subject in the jth cluster is given by

$$y^*_{hij} = \eta_{hij} + v_{hij},$$

where $v_{hij} = u_{hj} + \varepsilon_{hij}$, such that

$$\begin{aligned}\Pr(y_{hij} &= 1 \mid x_{ij}, u_{hj}) \\ &= \Pr(y^*_{hij} > 0 \mid x_{ij}, u_{hj}),\end{aligned}$$

where $-\infty \leq y^*_{hij} \leq \infty$; η_{hij} is a linear predictor of regressors $\mathbf{x}_{ij}$; u_{hj} is a cluster-specific random effect for the hth dimension, which for the moment we assume to be independent of $\mathbf{x}_{ij}$; ε_{hij} is an iid stochastic disturbance term, i.e. $E\left(\varepsilon_{hij} \mid \mathbf{x}_{ij}, u_{hj}\right) = 0$. This implies that

$$E\left(y_{hij} \mid \mathbf{x}_{ij},\ u_{hj}\right) = \Pr\left(y_{hij} = 1 \mid \mathbf{x}_{ij}, u_{hj}\right)$$
$$= \int_{-\eta_{hij}-u_{hj}}^{\infty} g\left(\varepsilon_{hij} \mid \mathbf{x}_{ij}, u_{hj}\right) d\varepsilon_{hij},$$

which gives the MLM marginal mean

$$\begin{aligned}\mu_{hij} &= E_R\left(y_{hij} \mid \mathbf{x}_{ij}\right) \\ &= E\left\{E_u\left(y_{hij} \mid \mathbf{x}_{ij},\ u_{hj}\right)\right\} \\ &= \Pr\left(y_{hij} = 1 \mid \mathbf{x}_{ij}\right) \\ &= \int_{-\infty}^{\infty} \Pr\left(y_{hij} = 1 \mid \mathbf{x}_{ij}, u_{hj}\right) \\ &\quad dF\left(u_{hj}; \alpha_h\right).\end{aligned}$$

The subscript R is used for the MLM marginal mean to distinguish it from the GEE marginal mean. This formulation assumes $E\left(\varepsilon_{hij} \mid \mathbf{x}_{ij}, u_{hj}\right) = 0, E\left(u_{hj} \mid \mathbf{x}_{ij}\right) = 0$, but because an MLM undertakes the joint analysis of all the responses in a cluster we actually require the stronger conditions

$$E\left(\varepsilon_{hij} \mid \mathbf{x}_{1j}, \ldots, \mathbf{x}_{N_j j}, u_{hj}\right) = 0$$

and

$$E\left(u_{hj} \mid \mathbf{x}_{1j}, \ldots, \mathbf{x}_{N_j j}\right) = 0.$$

Endogeneity can be said to arise in an MLM if $E\left(u_{hj} \mid \mathbf{x}_{1j}, \ldots, \mathbf{x}_{N_j j}\right) \neq 0$. In this chapter we concentrate on endogeneity due to simultaneity, but endogeneity can also be thought to arise from measurement error in the covariates or the inclusion of proxy regressors. If endogeneity is present, u_{hj} will be dependent on the covariates in the cluster, and the correctly specified MLM marginal mean is

$$\begin{aligned} &\int_{-\infty}^{\infty} \Pr\left(y_{hij} = 1 \mid \mathbf{x}_{ij}, u_{hj}\right) \\ &dF\left(u_{hj} \mid \mathbf{x}_{1j}, \ldots, \mathbf{x}_{N_j j}; \alpha_h\right) \\ &= E_R\left(y_{hij} \mid \mathbf{x}_{1j}, \ldots, \mathbf{x}_{N_j j}\right). \end{aligned}$$

We return to endogeneity in the next section.

There are a range of possible choices for the distribution (g) of the stochastic errors and for the random effects (u_{hj}). In this chapter, we assume for convenience that both the MLM random effects u_{hj} and ε_{hij} are normally distributed, i.e. $u_{hj} \sim N(0, \sigma_h^2)$, $\varepsilon_{hij} \sim N(0, \sigma_h^2)$, where $\sigma_\varepsilon^2 = 1$, so that the implied MLM marginal means take probit form. As the MLM probit marginal mean is formed by integrating out both u_{hj} and ε_{hij}, it will have a larger scale factor than the standard probit model. Thus the MLM probit marginal mean regressor parameter estimates will tend to have larger absolute values than those of the corresponding standard normal probit. Probit model regressor parameter estimates are only identified up to a multiplying scale factor. For a discussion on interpreting covariate effects in marginal means from probit models, see Chamberlain (1984, Section 3.1). If the u_{hj} are from some other distribution besides $N(0, \sigma_h^2)$, we can obtain the MLM marginal means $E_R\left(y_{hij} \mid \mathbf{x}_{ij}\right)$ and $E_R\left(y_{hij} \mid \mathbf{x}_{1j}, \ldots, \mathbf{x}_{N_j j}\right)$ using numerical methods. If ε_{hij} is from a logistic or extreme value distribution the 2-level marginal means do not have a closed form, but they can be approximated by reference to the probit–logit relationship. There are other choices for u_{hj} and ε_{hij} which do have analytically tractable 2-level model marginal means; see, for example, Conaway (1990).

25.3.1 Generalized Estimating Equations (GEE)

In the derivation of the GEE approach, Liang and Zeger (1986) appeal to a multivariate generalization of quasi likelihood (QL). GEE estimates a model for a vector of univariate correlated responses $\mathbf{y}_{1ij}$ (say) in cluster j by means of an adjusted score equation:

$$\begin{aligned} S\left(\beta_{G1}, \alpha_{G1}\right) &= \sum_j S_j\left(\beta_{G1}, \alpha_{G1}\right) \\ &= \sum_j \left(\frac{\partial \mu_{1j}}{\partial \beta_{G1}}\right)' \left[Var\left(\mathbf{y}_{1j} \mid \mathbf{x}_{1j}\right)\right]^{-1} \\ &\quad \left(\mathbf{y}_{1j} - \mu_{1j}\right) \\ &= \mathbf{0}, \end{aligned}$$

where $\mathbf{y}'_{1j}$ is the vector of the univariate responses for subject i of the jth cluster. The vector of means μ_{1j} is modeled with parameter vector β_{G1} for covariates $\mathbf{x}_{1j}$. The score equation depends not only upon β_{G1} but also on any over-dispersion parameters α_{G1}. As Lee and Nelder (2004) and Crowder (1995) note, there is in general no QL for which these QL-type estimating equations are score equations, implying that GEE models do not necessarily have a probabilistic or likelihood basis.

The GEE estimates will approach the limit vector β_{G1}^* given by the solution to the set of equations

$$E_c\left[N^{-1} S\left(\beta_{G1}, \alpha_{G1}\right)\right] = 0$$

as the sample size (N) approaches infinity. E_c is the expectation under the correct model, i.e.

the process that generates the observed data; see Cox (1961) and Huber (1967).

Let $r_j(st)$ denote the stth element of $\left[Var\left(\mathbf{y}_{1j} \mid \mathbf{x}_{1j}\right)\right]^{-1}$, and suppose we decide to select a model for $E\left(y_{1tj}\right)$ which is conditional only on the covariates present for intra-cluster response (time) t, i.e. $E\left(y_{1tj}\right) = \Phi\left(\beta_{G1}\mathbf{x}_{1tj}\right)$. This implies

$$E_c\left(r_j(tt)\frac{\partial\Phi\left(\beta_{G1}\mathbf{x}_{1tj}\right)}{\partial\beta_{G1k}}\right)\left(y_{1tj} - \Phi\left(\beta_{G1}\mathbf{x}_{1tj}\right)\right) = 0,$$

for $t = 1, \ldots, N_j$, but $E\left(y_{1tj}\right) = \Phi\left(\beta_{G1}\mathbf{x}_{1tj}\right)$ does not necessarily imply

$$E_c\left(r_j(st)\frac{\partial\Phi\left(\beta_{G1}\mathbf{x}_{1sj}\right)}{\partial\beta_{G1k}}\right)\left(y_{1tj} - \Phi\left(\beta_{G1}\mathbf{x}_{1tj}\right)\right) = 0,$$

for $s \neq t$, and thus the limit $E_c[N^{-1}S(\beta_{G1}, \alpha_{G1})] = 0$ may not hold, suggesting that the marginal mean is not correctly specified, see also Pepe and Anderson (1994), Lai and Small (2006). In Sections 25.3.3 and 25.4 we explore some of the issues that can arise in the specification of μ_{1j} when the correct model, i.e. the model generating the data, is a MLM.

Liang and Zeger (1986) proposed that a working covariance matrix be formed from

$$\begin{aligned}Var\left(\mathbf{y}_{1j} \mid \mathbf{x}_{1j}\right) &= Var(\mathbf{y}_{1j}; \beta_{G1}, \alpha_{G1} \mid \mathbf{x}_{1j}) \\ &= \mathbf{A}_{1j}^{1/2} Corr\left(\mathbf{y}_{1j} \mid \mathbf{x}_{1j}\right)\mathbf{A}_{1j}^{1/2},\end{aligned}$$

where $\mathbf{A}_{1j} = diag\left[Var\left(y_{1ij} \mid \mathbf{x}_{1ij}\right)\right]$ is specified by the marginal distributions. Essentially, GEE is using a working correlation matrix $Corr\left(\mathbf{y}_{1j} \mid \mathbf{x}_{1j}\right)$ to weight the standard generalized linear model (GLM) score equations by cluster. Various forms of the working correlation matrix are used, e.g. independent working model (IWM), Exchangeable (E), AR(1), and Unstructured (U). Estimation consists of specifying a model for $Corr\left(\mathbf{y}_{1j} \mid \mathbf{x}_{1j}\right)$ and cycling between the estimation of β_{G1} via $S(\beta_{G1}, \alpha_{G1})$ and the estimation of $Corr\left(\mathbf{y}_{1j} \mid \mathbf{x}_{1j}\right)$ by the method of moments; see, e.g. Prentice (1988), Thall and Vail (1990). The GEE procedure is available in many statistical software packages, including R, SAS, Splus, and Stata, e.g. `xtgee`.

GEE software also produces the sandwich variance estimate for the regressor parameter estimates. Liang and Zeger (1986, Theorem 2) note that if $\widehat{\beta}_{G1}$ is the solution to say, $S(\beta_{G1}, \alpha_{G1}) = 0$, then $N^{1/2}\left(\widehat{\beta}_{G1} - \beta_{G1}\right)$ is asymptotically multivariate Gaussian with zero mean and covariance matrix V_G where

$$V_{G1} = \lim_{N\to\infty} N\ \mathbf{Q}^{-1}\mathbf{M}\mathbf{Q}^{-1},$$

where

$$\mathbf{Q} = \left[\sum_j \left(\frac{\partial\mu_{1j}}{\partial\beta_{G1}}\right)'\left[Var\left(\mathbf{y}_{1j} \mid \mathbf{x}_{1j}\right)\right]^{-1}\left(\frac{\partial\mu_{1j}}{\partial\beta_{G1}}\right)\right]$$

and

$$\mathbf{M} = \sum_j \mathbf{B}Cov(\mathbf{y}_{1j} \mid \mathbf{x}_{1j})\mathbf{B}',$$

with

$$\mathbf{B} = \left(\frac{\partial\mu_{1j}}{\partial\beta_{G1}}\right)'\left[Var\left(\mathbf{y}_{1j} \mid \mathbf{x}_{1j}\right)\right]^{-1}.$$

The variance estimate $\widehat{V}_{G1}$ of $\widehat{\beta}_{G1}$ can be obtained by replacing $Cov(\mathbf{y}_{1j} \mid \mathbf{x}_{1j})$ by $\left(\mathbf{y}_{1j} - \mu_{1j}\right)\left(\mathbf{y}_{1j} - \mu_{1j}\right)'$ and the parameters β_{G1}, α_{G1} by their estimates.

If the GEE mean is correctly specified and the working intra-cluster correlation is of the correct form but missing data are present and the missing data mechanism is related to the covariates and to pre-dropout responses, the estimates will be consistent but the sandwich estimator may not be; see Kenward and Molenberghs (1998).

Pepe and Anderson (1994) note that the GEE parameter estimates will be sensitive to the specification of $Var\left(\mathbf{y}_{1j} \mid \mathbf{x}_{1j}\right)$ if the covariates in the jth longitudinal cluster do not satisfy the condition

$$E\left(y_{1ij} \mid \mathbf{x}_{11j}, \ldots, \mathbf{x}_{1N_j j}\right) = E\left(y_{1ij} \mid \mathbf{x}_{1ij}\right) = \mu_{1ij}.$$

Pepe and Anderson (1994, p. 493) recommend either (1) using $\mu_{1ij} = E\left(y_{1ij} \mid \mathbf{x}_{1ij}\right)$ with a diagonal covariance matrix (IWM), or (2) validating $E\left(y_{1ij} \mid \mathbf{x}_{11j}, \ldots, \mathbf{x}_{1N_j j}\right) = E\left(y_{1ij} \mid \mathbf{x}_{1ij}\right)$. Chamberlain (1984) proposed testing this equality with the IWM, but his interest was in detecting endogenous variables in $\mathbf{x}_{1ij}$. He did not propose using $\mu_{1ij} = E\left(y_{1ij} \mid \mathbf{x}_{1ij}\right)$ if this condition had not been confirmed. To simplify the notation we often drop the distinction on h in the covariates.

Pepe and Couper (1997) give recommendations about using GEE models for lagged endogenous variables. This involves application of the GEE method to the conditional mean $E\left(y_{tj} \mid y_{t-1,j}, \mathbf{x}_{tj}\right)$ for a sequence $\left(t = 2, \ldots, N_j\right)$ of responses y_{tj}, with a diagonal working covariance matrix (IWM). Pepe and Couper (1997) also suggest that inference on the structural parameters be based upon the robust or sandwich standard errors. Diggle et al. (1994, p. 199) note that the robust standard errors have valid coverage in large samples even when the Markov assumption is incorrect. Diggle et al. (2002) elaborate on this position but make essentially the same point.

For examples of GEE being applied to more complex levels of clustering, see e.g. Betensky et al. (2000), Chao (2006), and Teerenstra et al. (2010).

25.3.2 Multilevel Models

In multivariate MLMs the individual-level unobserved factors u_{hj} are integrated out of the joint distribution of the N_j responses in cluster j to obtain the marginal likelihood function. For bivariate clustered responses $(h = 1, 2)$,

$$\begin{aligned} L_j\left(\beta_R, \alpha_R; \mathbf{y}_j\right) &= \int_{-\infty}^{\infty}\int_{-\infty}^{\infty} \Pr(y_{11j}, y_{21j}, \ldots, y_{1N_j j}, y_{2N_j j} \mid \mathbf{x}_j, \mathbf{u}_j; \beta_R, \alpha_R) dF(\mathbf{u}_j; \alpha_R) \\ &= \int_{-\infty}^{\infty}\int_{-\infty}^{\infty} \prod_i \Pr(y_{1ij}, y_{2ij} \mid \mathbf{x}_{ij}, \mathbf{u}_j; \beta_R, \alpha_R)\, dF(\mathbf{u}_j; \alpha_R), \end{aligned}$$

where $\beta_R = [\beta_{R1}, \beta_{R2}]$, $\alpha_R = [\alpha_{R1}, \alpha_{R2}]$, $\mathbf{y}_j = \left[\mathbf{y}_{1j}, \mathbf{y}_{2j}\right]$, $\mathbf{x}_j = \left[\mathbf{x}_{1j}, \mathbf{x}_{2j}\right]$, $\mathbf{u}_j = \left[u_{1j}, u_{2j}\right]$. With endogeneity in which y_2 is endogenous for y_1, this becomes

$$L_j\left(\beta_R, \alpha_R; \mathbf{y}_j\right) = \int_{-\infty}^{\infty}\int_{-\infty}^{\infty} \prod_i \Pr\left(y_{1ij} \mid y_{2ij}, \mathbf{x}_{ij}, u_{1j}; \beta_{R1}, \alpha_{R1}\right) \times \Pr\left(y_{2ij} \mid \mathbf{x}_{ij}, u_{2j}; \beta_{R2}, \alpha_{R2}\right) dF\left(\mathbf{u}_j; \alpha_R\right).$$

The integrals in the bivariate probit MLM likelihoods can be estimated by various methods, including simulated maximum likelihood. For Gaussian quadrature there is the Stata library `gllamm` (`http://www.gllamm.org/`) and the SabreR package for R (Berridge and Crouchley, 2011, Chapter 11). If

$$L\left(\beta_R, \alpha_R; \mathbf{y}\right) = \prod_j L_j\left(\beta_R, \alpha_R; \mathbf{y}_j\right),$$

then the maximum likelihood estimates will approach the limit vector β^*_{R1} given by the solution to the set of equations

$$E_c\left[N^{-1}\frac{\partial L\left(\beta_R, \alpha_R; \mathbf{y}\right)}{\partial \beta_{Rk}}\right] = 0$$

as the sample size $N = \sum_j N_j$ approaches infinity. E_c is the same expectation used for GEE, i.e. it is the model that generated the observed data; see Cox (1961) and Huber (1967).

If robustness against the wrong choice of parametric distribution for $F(\cdot)$ is required, the nonparametric maximum likelihood estimator of Laird (1978) can be used. GEE can also be used to estimate MLMs so long as the implied MLM marginal mean and intracluster covariance structure is used; for some MLMs this will require the use of numerical integration or the use of Taylor expansions.

If we assume u_{1ij}, u_{2ij} are independent, the likelihood partitions are as follows:

$$L_j(\beta_{R1}, \alpha_{R1}; \mathbf{y}_j) = \int_{-\infty}^{\infty} \prod_i \Pr(y_{1ij} \mid y_{2ij}, \mathbf{x}_{ij}, u_{1j}; \beta_{R1}, \alpha_{R1})\, t(u_1; \alpha_{R1}) \times \int_{-\infty}^{\infty} \prod_i \Pr(y_{2ij} \mid \mathbf{x}_{ij}, u_{2j}; \beta_{R2}, \alpha_{R2})\, dF(u_2; \alpha_{R2})$$
$$= L_{1j}(\beta_{R1}, \alpha_{R1}; \mathbf{y}_{1j} | \mathbf{y}_{2j}).L_{2j}(\beta_{R2}, \alpha_{R2}; \mathbf{y}_{2j}).$$

We compare the use of $\prod_j L_{1j}(\beta_{R1}, \alpha_{R1}; \mathbf{y}_{1j} \mid \mathbf{y}_{2j})$ with $S(\beta_{G1}, \alpha_{G1})$ for inference about the y_1 regressors when $[u_{1ij}, u_{2ij}]$ are correlated in the process generating the data.

25.3.3 Comparing GEE and MLMs

Many papers compare MLMs and GEE models for clustered responses, e.g. Carriere and Bouyer (2002), Gardiner et al. (2009), Subramanian and O'Malley (2010), and Ritz and Spiegelman (2004). Though as Lee and Nelder (2004) note, comparisons of MLMs with some GEE marginal means models, e.g. Heagerty and Zeger (2000), are often meaningless because differences are mainly caused by preimposed unidentifiable constraints on the MLM random effects. Lee and Nelder (2004) also regard the MLM model as providing the more fundamental building block from which marginal predictions can be made. Senn (2004), while being less adherent to MLMs, comments on Lee and Nelder (2004), noting that the desire to issue marginal predictions is not in itself a reason for not using MLMs and that inferences will usually be superior if MLMs are used.

The supporters of GEE models, e.g. Hubbard et al. (2010), tend to argue that because GEEs make fewer explicit assumptions they are less vulnerable to misspecification. But as Pepe and Anderson (1994) and Lee and Nelder (2004) show, it is the implicit assumptions we need to worry about. One of the other advantages claimed for GEE models over MLMs is that MLMs can have computationally demanding marginal likelihoods. This argument implicitly assumes that the GEE μ_{1ij} and $Var(\mathbf{y}_{1j} \mid \mathbf{x}_{1j})$ have a simple closed form while marginal likelihoods do not, but even this apparent advantage is disappearing with the increasing availability of algorithms that can use MPI and OpenMP for numeric integration; see, e.g. Stata/MP (`http://www.stata.com/statamp/`).

Misspecification will occur in both MLM and GEE models for y_1 if we ignore the endogeneity of y_2 for y_1, i.e. treat y_2 as exogenous. If $[u_{1ij}, u_{2ij}]$ are positively correlated then both the MLM and IWM models for y_1 which ignore this stochastic dependence will overestimate the y_2 coefficient. The increased dependence between y_1 and y_2 induced by the correlation between $[u_{1ij}, u_{2ij}]$ will be attributed to the endogenous variable y_2, and the y_2 parameter in η_1 will be overestimated. This misspecification also creates estimation bias in the other entirely exogenous variables. This effect has been noted by Hsiao (1986), amongst others, in the context of lagged responses in univariate MLMs.

For univariate clustered responses, the probit MLM and IWM approaches usually adopt a different standardization. The relationship between the estimated parameters is, if we take the models for y_1 as an example,

$$\widehat{\beta}_{G1} \simeq \widehat{\beta}_{R1} \left(1 - \frac{\widehat{\sigma}_1^2}{1 + \widehat{\sigma}_1^2}\right)^{1/2},$$

where $\widehat{\beta}_{R1}$ is the MLM covariate estimate, $\widehat{\sigma}_1^2$ is the estimated variance of the random effect u_{1j} from $\prod_j L_{1j}(\beta_{R1}, \alpha_{R1}; \mathbf{y}_1 \mid \mathbf{y}_2)$, and $\widehat{\beta}_{G1}$ is the IWM covariate estimates. See also Zeger et al. (1988).

It is only because we have assumed the presence of normally distributed 2-level random effects in a probit model that the implied MLM marginal means also take probit form. This was done to simplify the comparison of GEE and MLM models. For GEE and MLM logit models (Zeger et al., 1988), we would have

$$\widehat{\beta}_{G1} \simeq \widehat{\beta}_{R1}\left(1 - \frac{c^2\widehat{\sigma}_1^2}{1 + c^2\widehat{\sigma}_1^2}\right)^{1/2},$$

where $c = 16\sqrt{3}/(15\pi)$. For an empirical comparison of the implied 2-level logit model marginal means with those of a GEE logit model with an unstructured working correlation matrix, see Chapter 8, Table 8.2 and Figure 8.3.

25.4 MUNDALK AND CHAMBERLAIN MODELS FOR ENDOGENOUS EFFECTS

To allow for the possibility of endogeneity, Mundalk (1978) and Chamberlain (1980, 1984) proposed making the distribution of the random effects (u_{1j}) conditional on all regressor variables in the cluster. If y_2 is endogenous for y_1, the MLM likelihood for the y_1 responses conditional on y_2 in cluster j becomes

$$L_{1j}\left(\beta_{R1}^*, \alpha_{R1}^*; \mathbf{y}_{1j} \mid \mathbf{y}_{2j}\right) = \int_{-\infty}^{\infty} \prod_i \Pr\left(y_{1ij} \mid y_{2ij}, \mathbf{x}_{ij}, u_{1j}; \beta_{R1}^*, \alpha_{R1}^*\right) dF\left(u_{1j} \mid \mathbf{y}_{2j}, \mathbf{x}_j; \alpha_{R1}^*\right).$$

Mundalk and Chamberlain (MC) consider several ways of specifying the random effect distribution $F\left(u_{1j} \mid \mathbf{y}_{2j}, \mathbf{x}_j; \alpha_{R1}^*\right)$. One approximation they propose involves making the u_{1j} a linear function of the covariate cluster averages $\left[\overline{\mathbf{x}}_j, \overline{y}_{2j}\right]$, i.e.

$$u_{1j} = \pi_{R1}\overline{\mathbf{x}}_j + \theta_{R1}\overline{y}_{2j} + w_{1j},$$

where w_{1j} is iid. Thus, the MLM likelihood becomes

$$L_1\left(\beta_{R1}^*, \alpha_{R1}^*; \mathbf{y}_1 \mid \mathbf{y}_2\right) = \int_{-\infty}^{\infty} \prod_i \Pr(y_{1ij} \mid y_{2ij}, \overline{y}_{2j}, \mathbf{x}_{ij}, \overline{\mathbf{x}}_j, w_{1j}; \beta_{R1}^*, \alpha_{R1}^*)\, dF(w_{1j}; \alpha_{R1}^*),$$

with the marginal means

$$\begin{aligned}\mu_{1ij} &= \Pr\left(y_{1ij} = 1 \mid y_{2ij}, \overline{y}_{2j}\mathbf{x}_{ij}, \overline{\mathbf{x}}_j\right) \\ &= \int_{-\infty}^{\infty} \Pr\left(y_{hij} = 1 \mid y_{2ij}, \overline{y}_{2j}, \mathbf{x}_{ij}, \overline{\mathbf{x}}_j, w_{1j}\right) dF\left(w_{1j}; \alpha_{R1}^*\right) \\ &= \Phi\left(\frac{\beta_{R1}\mathbf{x}_{ij} + \delta_{R1}y_{2ij} + \pi_{R1}\overline{\mathbf{x}}_j + \theta_{R1}\overline{y}_{2j}}{\sqrt{1 + \sigma_{w1}^2}}\right).\end{aligned}$$

We have included the cluster means of all the regressors in the linear approximation $u_{1j} = \pi_{R1}\overline{\mathbf{x}}_j + \theta_{R1}\overline{y}_{2j} + w_{1j}$, as the endogeneity in y_2 could induce some endogeneity in $\mathbf{x}_{ij}$ when y_2 is determined by $\mathbf{x}_{ij}$ in a nonlinear model like the probit. Endogeneity will be present if any of the elements of $[\pi_{R1}, \theta_{R1}]$, or equivalently $[\pi_{G1}, \theta_{G1}]$, are significantly different from zero. If u_{1j} is positively correlated with u_{2j}, θ_{G1} will be positive. Those covariates that do not vary within the cluster will, by definition, have zero elements in $[\pi_{R1}, \theta_{R1}]$ and $[\pi_{G1}, \theta_{G1}]$.

These GEE IWMs and MLMs can also be written in terms of the within-cluster components $\mathbf{x}_{ij} - \overline{\mathbf{x}}_j$, $y_{2ij} - \overline{y}_{2j}$ and cluster means $\overline{\mathbf{x}}_j, \overline{y}_{2j}$, i.e.

$$\begin{aligned}\mu_{1ij} &= \Phi(\beta_{G1}\mathbf{x}_{ij} + \delta_{G1}y_{2ij} + \pi_{G1}\overline{\mathbf{x}}_j + \theta_{G1}\overline{y}_{2j}) \\ &= \Phi(q),\end{aligned}$$

with

$$
\begin{aligned}
q &= \beta_{G1}\left(\left[\mathbf{x}_{ij}-\overline{\mathbf{x}}_j\right]+\overline{\mathbf{x}}_j\right) \\
&\quad +\delta_{G1}\left(\left[y_{2ij}-\overline{y}_{2j}\right]+\overline{y}_{2j}\right) \\
&\quad +\pi_{G1}\overline{\mathbf{x}}_j+\theta_{G1}\overline{y}_{2j} \\
&= \beta_{G1}\left[\mathbf{x}_{ij}-\overline{\mathbf{x}}_j\right]+\delta_{G1}\left[y_{2ij}-\overline{y}_{2j}\right] \\
&\quad +\left[\beta_{G1}+\pi_{G1}\right]\overline{\mathbf{x}}_j+\left[\delta_{G1}+\theta_{G1}\right]\overline{y}_{2j}.
\end{aligned}
$$

Thus, the MC models imply that many of the papers that consider GEE and MLM models with separate within-cluster and between-cluster covariate components, e.g. Desai and Begg (2008), Galecki (1994), Goetgeluk and Vansteelandt (2008), Neuhaus and Kalbfleisch (1998), Neuhaus and McCulloch (2006), Shen et al. (2008), Palta and Yao (1991), Preacher et al. (2010), and Ware et al. (1990), could be more broadly interpreted in terms of models that allow for the endogeneity of covariates.

Using the covariate means of time-varying covariates is not the only way to explore endogeneity. Chamberlain (1984) tested for endogeneity by imposing restrictions on the coefficients of $\mu_{1ij} = \Phi\left(\sum_t \beta_{G1ti}\mathbf{x}_{tj}\right)$, where β_{G1ti} is a vector of coefficients for response occasion i for $\mathbf{x}_{tj}$, where $\mathbf{x}_{tj}$ is the vector of covariates from response occasion $t = 1,\ldots,N_j$. The coefficients for each response occasion i were estimated using an IWM. Lai and Small (2006) test for endogeneity using a moment condition on a function of the score equations.

25.5 SIMULATION

We performed a simulation experiment to study the sampling behaviour of the regressor parameter estimates and the MC tests for different levels of correlation between u_{1ij} and u_{2ij} in a bivariate binary probit model for (y_1, y_2) in which y_2 is also endogenous for y_1. There are two continuous exogenous explanatory variables: x_1, which is common over responses h but varying by observation pair i, and x_2, which is constant for all responses h and observations i in a given cluster j. Both x_1 and x_2 are present in the linear predictors of y_1 and y_2. For the ith observation from the jth cluster we have the following:

$$y_{1ij}^* = \beta_{10}+\beta_{11}x_{1ij}+\beta_{12}x_{2j}+\delta y_{2ij}+u_{1j}+\varepsilon_{1ij},$$

with $y_{1ij} = 1$ if $y_{1ij}^* > 0$, 0 otherwise, and

$$y_{2ij}^* = \beta_{20}+\beta_{21}x_{1ij}+\beta_{22}x_{2j}+u_{2j}+\varepsilon_{2ij},$$

with $y_{2ij} = 1$ if $y_{2ij}^* > 0$, 0 otherwise. The random effects $u \sim N(0, \Omega)$,

$$\Omega = \begin{bmatrix} \sigma_1^2 & \sigma_1\sigma_2\rho \\ \sigma_1\sigma_2\rho & \sigma_2^2 \end{bmatrix},$$

where $\sigma_1 = 0.5$, $\sigma_2 = 0.8$, and with ρ taking a range of values. For $\rho = 0$, y_2 is exogenous for y_1 but it is endogenous for all other values of ρ. Both x_1 and x_2 are $N(0, 1)$, also $\varepsilon_{kij} \sim N(0, 1)$, $k = 1, 2$. The regression coefficients have the following values: $\beta_{10} = \beta_{20} = 0.5$, $\beta_{11} = 2$, $\beta_{12} = -2$, $\delta = 0.2$, $\beta_{21} = 1$, $\beta_{22} = -1$. We simulated 1000 data sets using Stata 11, for a range of correlations ($\rho = 0$ (0.2) 0.8), each data set having $J = 1000$ clusters, with each cluster having $N_j = 4$ sets of bivariate responses.

The first model we estimated on the simulated clustered y_1 responses was for $\mu_{1ij} = \Phi\left(\beta_{G1}\mathbf{x}_{ij}+\delta_{G1}y_{2ij}\right)$. We call this the naive IWM. These results indicate the bias that occurs by ignoring the endogeneity of y_2. The naive IWM model for y_1 was estimated using the Stata procedure `probit`; the results are presented in Table 25.2.

Judging by the means, the naive IWM probit model only gets β_{G11}, δ_{G1}, β_{G12} right for $\rho = 0$. As ρ increases, the bias in the mean of δ_{G1} increases, by $\rho = 0.8$, there has been +213% change in the mean of δ_{G1}. As ρ increases; the role of $\overline{y_2}$ becomes more important in both the naive IWM (Table 25.2) and naive MLM (Table 25.3). $\overline{y_2}$ becomes more of a surrogate for u_{1j}, leaving less of the remaining correlation between the clustered responses $\left[y_{1ij}, y_{1i'j}\right]$ to be accounted for by u_{1j}, thus making $\widehat{\sigma_1}$ smaller, as judged by its means (Table 25.2).

Table 25.2 Descriptive Statistics for the Naive IWM Probit Model Estimates on the Simulated Data

	Values of ρ									
	0.0		0.2		0.4		0.6		0.8	
Parameter	Mean	SD	Mean	SD	Mean	SD	Mean	SD	Mean	SD
β_{G10}	0.450	0.051	0.392	0.052	0.333	0.051	0.275	0.050	0.216	0.051
β_{G11}	1.793	0.059	1.769	0.058	1.748	0.058	1.730	0.058	1.714	0.058
δ_{G1}	0.176	0.059	0.267	0.059	0.360	0.058	0.454	0.057	0.550	0.058
β_{G12}	-1.794	0.060	-1.770	0.060	-1.749	0.059	-1.730	0.059	-1.714	0.059

Table 25.2 suggests that the parameter estimates for the exogenous covariates are much less biased; by $\rho = 0.8$, the means of $\beta_{G11,}$ and β_{G12} have a -4.4% and -4.5% change. These naive IWM parameters stay reasonably close to the implied β_{G11}, δ_{G1}, β_{G12} as seen comparing the Table 25.2 values with the bottom three rows of the naive MLM model of Table 25.3. As mentioned above, the implied β_{G11}, δ_{G1}, β_{G12} of Table 25.3 are obtained from the MLM by $\widehat{\beta}_{G1} = \widehat{\beta}_{R1}\left(1 - \widehat{\sigma}_1^2/\left[1+\widehat{\sigma}_1^2\right]\right)^{1/2}$.

The naive MLM model for y_1, i.e. with $\mu_{1ij} = \Phi\left(\beta_{R1}\mathbf{x}_{ij} + \delta_{R1}y_{2ij} + u_{1j}\right)$, was estimated using the Stata procedure `xtprobit` on the clustered responses. The results are presented in Table 25.3.

Table 25.3 shows that the naive MLM only really gets the means of β_{R11}, δ_{R1}, β_{R12}, σ_1 right for $\rho = 0$. As ρ increases, the bias in the mean of δ_{R1} increases; when $\rho = 0.8$, there has been $+184\%$ change in the mean of δ_{R1}. The naive MLM probit, like the naive IWM, clearly overestimates the parameter on the endogenous regressor. Like the naive IWM, Table 25.3 suggests that the parameter estimates for the exogenous covariates are much less biased; by $\rho = 0.8$, the means of β_{R11} and β_{R12} have a -6.3% and -6.5% change. Tables 25.2 and 25.3 suggest that the exogenous covariates are, suprisingly, slightly vulnerable to ignoring the endogeneity in y_2. The bias of σ_1 increases with increasing ρ, as implied by its mean; by $\rho = 0.8$, it has a -14.9% change. The misspecification of the naive MLM is confirmed by the Wald tests of the regressor parameter estimates against their true values. By $\rho = 0.2$ we reject 16% of the tests at the 5% level, and by $\rho = 0.8$ we reject every test at the 5% level. There is nothing in the descriptive statistics of Tables 25.2 and 25.3 which suggests that the naive IWM is either more or less robust than the naive MLM.

We estimate the Mundalk (1978) and Chamberlain (1984) suggested augmentations by including the cluster means of the regressors that vary by j (i.e. x_1 and y_2) in the linear predictor. The significance of the regressor cluster means indicate the extent of the misspecification present. The augmented IWM results for the model $\mu_{1ij} = \Phi\left(\beta_{G1}\mathbf{x}_{ij} + \delta_{G1}y_{2ij} + \pi_{G1}\overline{\mathbf{x}}_j + \theta_{G1}\overline{y}_{2j}\right)$ are presented in Table 25.4.

The means of the regressor parameter estimates, β_{G11}, δ_{G1}, β_{G12}, of Table 25.4 change as ρ increases. However, they do stay close to the implied β_{G11}, δ_{G1}, β_{G12} as judged by the means from the augmented MLM model of Table 25.5. The reduction in $\widehat{\sigma}_1$ also changes the MLM probit scaling. This effect is seen in Tables 25.4 and 25.5, with the mean of δ_{G1} increasing with increasing ρ; also note that the mean of π_{G11}, the coefficient on $\overline{x_1}$, an exogenous covariate, has an increasingly negative mean with increasing ρ. This is a manifestation of the MC linear approximation as x_1 is in both the y_1 and y_2 linear predictors. If x_1 was not in the linear predictor for y_2 this would not happen. The mean of β_{G12} gets further away

Table 25.3 Descriptive Statistics for the Naive MLM Probit Model Estimates on the Simulated Data

	Values of ρ									
	0.0		0.2		0.4		0.6		0.8	
Parameter (true vales)	Mean	SD	Mean	SD	Mean	SD	Mean	SD	Mean	SD
β_{R10} (0.5)	0.503	0.058	0.447	0.059	0.389	0.058	0.329	0.057	0.266	0.058
β_{R11} (2.0)	2.003	0.081	1.977	0.081	1.948	0.080	1.915	0.079	1.878	0.077
δ_{R1} (0.2)	0.197	0.065	0.283	0.065	0.370	0.064	0.459	0.063	0.551	0.063
β_{R12} (-2.0)	-2.004	0.084	-1.978	0.084	-1.948	0.083	-1.915	0.082	-1.877	0.080
σ_1 (0.5)	0.495	0.068	0.490	0.067	0.477	0.069	0.455	0.071	0.422	0.074
Wald (β_{R11}=2, δ_{R1}=0.2, β_{R12}=-2)	0.065		0.155		0.588		0.932		1.000	
β_{G11}	1.793	0.058	1.773	0.058	1.756	0.057	1.740	0.057	1.726	0.058
δ_{G1}	0.176	0.058	0.254	0.058	0.334	0.058	0.417	0.058	0.507	0.059
β_{G12}	-1.794	0.060	-1.774	0.059	-1.756	0.059	-1.740	0.058	-1.725	0.059

Table 25.4 Descriptive Statistics for the Augmented IWM Estimates on the Simulated Data

	Values of ρ									
	0.0		0.2		0.4		0.6		0.8	
Parameter	Mean	SD	Mean	SD	Mean	SD	Mean	SD	Mean	SD
β_{G10}	0.453	0.083	0.295	0.082	0.137	0.081	-0.024	0.081	-0.188	0.081
β_{G11}	1.794	0.061	1.794	0.061	1.801	0.060	1.816	0.062	1.840	0.064
π_{G11}	0.001	0.082	-0.054	0.081	-0.110	0.081	-0.168	0.082	-0.227	0.084
δ_{G1}	0.178	0.075	0.182	0.074	0.187	0.073	0.194	0.072	0.203	0.074
θ_{G1}	-0.006	0.158	0.247	0.154	0.504	0.153	0.767	0.155	1.037	0.157
β_{G12}	-1.796	0.065	-1.741	0.064	-1.691	0.062	-1.648	0.061	-1.611	0.061

from its true value as ρ increases. The augmented MLM probit results are presented in Table 25.5.

The augmented MLM probit models get close to the true values of β_{R11}, δ_{R1} for all values of ρ, even though there appears to be some slight drift in the means. The number of rejections of the Wald test against their true values at the 5% level are very close to 5% for all values of ρ. Just like the IWM results, as ρ increases θ_{R1} $(\overline{y_2})$ becomes more positive, while π_{R11} $(\overline{x_1})$ becomes more negative. Table 25.5 also includes the % rejections at the 5% level of a Wald test of π_{R11} $(\overline{x_1}) = 0$. The percent rejections at the 5% level increase with ρ, and when $\rho = 0.8$ we reject the hypothesis that $\pi_{R11} = 0$ 79% of the time. The MC linear approximation seems to have made x_1 endogenous. Judging by its mean, the time-constant covariate effect β_{R12} in the MLM of Table 25.5 becomes increasingly biased with increasing ρ. There is nothing in the descriptive statistics of Table 25.5 which suggests that the augmented IWM is more or less robust than the augmented MLM.

25.6 MSCM STUDY

In this section we compare the inference from the naive and augmented IWMs and MLMs for the effect of daily maternal stress (1=present, 0=absent) on a child's daily illness (1=ill, 0=not ill) while controlling for the baseline covariates of Table 25.1. Like

Table 25.5 Descriptive Statistics for the Augmented MLM Estimates on the Simulated Data

	Values of ρ									
	0.0		0.2		0.4		0.6		0.8	
Parameter (true values)	Mean	SD	Mean	SD	Mean	SD	Mean	SD	Mean	SD
β_{R10} (0.5)	0.506	0.092	0.328	0.091	0.150	0.089	-0.028	0.088	-0.204	0.087
β_{R11} (2.0)	2.003	0.084	1.999	0.084	1.994	0.083	1.988	0.082	1.982	0.082
π_{R11}	0.001	0.091	-0.060	0.090	-0.123	0.089	-0.184	0.090	-0.245	0.091
δ_{R1} (0.2)	0.199	0.083	0.202	0.083	0.206	0.081	0.210	0.079	0.216	0.080
θ_{R1}	-0.006	0.175	0.278	0.172	0.563	0.170	0.845	0.170	1.123	0.172
β_{R12} (-2.0)	-2.005	0.087	-1.939	0.086	-1.871	0.083	-1.803	0.081	-1.733	0.077
σ_1 (0.5)	0.493	0.068	0.487	0.068	0.468	0.069	0.437	0.073	0.389	0.077
Wald (β_{R11}=2, δ_{R1}=0.2)	0.055		0.049		0.045		0.047		0.054	
Wald (π_{R11}=0)	0.048		0.081		0.257		0.511		0.786	
β_{G11}	1.794	0.061	1.794	0.060	1.803	0.060	1.819	0.061	1.843	0.064
δ_{G1}	0.178	0.075	0.182	0.074	0.186	0.073	0.192	0.072	0.201	0.074
β_{G12}	-1.796	0.064	-1.741	0.063	-1.692	0.062	-1.649	0.061	-1.612	0.060

Diggle et al. (2002, Table 12.4), we treat the self-reported maternal and child health at baseline as interval-level covariates. Time is also treated as an interval-level covariate in the form `wk` = (`day` − `14`)/`7`. All the other covariates are binary dummy variables. There are no within-cluster varying exogenous covariates in the published data. The naive (IWM1) and augmented (IWM2) IWM results of child illness were obtained using the Stata `probit` routine and are presented in Table 25.6.

Table 25.6 shows that the contemporaneous effect of maternal stress (`stress`), 0.368 (0.077), in IWM1 is significant, and both `stress`, 0.240 (0.075), and `stress_mean`, 1.088 (0.388), are significant in IWM2. By comparing the IWM1 and IWM2 estimates of `stress`, we see that IWM1 has overestimated the role of `stress` by over 50%. The positive sign on `stress_mean` in IWM2 suggests that there is a positive correlation between $\left[u_{1ij}, u_{2ij}\right]$. The baseline covariates (i.e. `wk`, `married`, `chlth`, `race`, `housize`) remain significant in both IWM1 and IWM2. The naive (MLM1) and augmented (MLM2) MLM child illness results were obtained using the Stata `xtprobit` routine and are presented in Table 25.7.

The same general pattern in the IWM1 and IWM2 results was also obtained for the univariate MLMs, though there is some slight difference in regressor significance between them. For instance, `wk` is more significant in the MLM analyses than it is in the IWM analyses.

Comparing the IWM and MLM log likelihoods gives a chi square (1 df) improvement of 134 for MLM1 over IWM1 and 116 for MLM2 over IWM2, suggesting that the random effects are very important. Both IWM2 and MLM2 confirm the significance of the maternal stress mean, which can be interpreted as indicating the endogeneity of stress in the model for illness. A joint illness–stress simultaneous MLM was estimated using the Stata plug-in `SabreStata` (`http://sabre.lancs.ac.uk/sabreStatause_intro.html`), and these results are presented in Table 25.8.

In Table 25.8, both $\widehat{\sigma_1}$, 0.467 (0.045), and $\widehat{\sigma_2}$, 0.538 (0.049), are more significant than any regressor. The estimate of the correlation ($\widehat{\rho}$) between $\left[u_{Ij}, u_{Sj}\right]$, 0.323 (0.113), is also significant, implying that maternal stress (`stress`) is endogenous for child illness (`illness`). The estimates of $\widehat{\sigma_1}$ from Tables 25.7 and 25.8 are very close.

Table 25.6 Univariate IWMs of the MSCM Data

	IWM1			IWM2		
Covariate	Estimate	Robust SE	Z score	Estimate	Robust SE	Z score
_cons	-0.510	0.220	-2.32	-0.842	0.259	-3.25
employed	-0.065	0.092	-0.7	-0.044	0.092	-0.48
wk	-0.097	0.029	-3.29	-0.100	0.030	-3.36
married	0.275	0.081	3.41	0.244	0.084	2.92
mhlth	-0.013	0.051	-0.25	0.021	0.048	0.45
chlth	-0.158	0.048	-3.33	-0.131	0.051	-2.59
race	0.240	0.088	2.74	0.208	0.086	2.41
educ_bin	-0.030	0.102	-0.3	-0.088	0.097	-0.91
housize	-0.381	0.083	-4.57	-0.333	0.086	-3.87
stress	0.368	0.077	4.8	0.240	0.075	3.21
stress_mean				1.088	0.388	2.8
Pseudo LogL	-1683.75			-1670.58		

Table 25.7 Univariate MLMs of the MSCM Data

	MLM1			MLM2		
Covariate	Estimate	SE	Z score	Estimate	SE	Z score
_cons	-0.572	0.261	-2.19	-0.935	0.281	-3.320
employed	-0.021	0.109	-0.19	0.003	0.106	0.030
wk	-0.110	0.022	-4.95	-0.111	0.022	-4.990
married	0.283	0.099	2.87	0.242	0.097	2.510
mhlth	0.011	0.056	0.2	0.046	0.055	0.820
chlth	-0.187	0.061	-3.07	-0.156	0.060	-2.610
race	0.255	0.101	2.53	0.223	0.098	2.280
educ_bin	-0.060	0.106	-0.56	-0.114	0.104	-1.100
housize	-0.422	0.099	-4.25	-0.366	0.098	-3.740
stress	0.309	0.072	4.31	0.264	0.073	3.610
stress_mean				1.146	0.384	2.990
σ_1	0.467	0.045		0.446	0.044	
LogL	-1616.808			-1612.453		

The simulations suggest that the estimates of the cluster-constant (baseline) covariates in Tables 25.6 and 25.7 are biased. However, Tables 25.6, 25.7, and 25.8 are in agreement about which baseline covariates are significant for illness, perhaps because $\widehat{\rho}$ is quite

Table 25.8 Bivariate MLM of the MSCM Data

	Illness Model			Stress Model		
Covariate	Estimate	SE	Z score	Estimate	SE	Z score
_cons	-0.552	0.261	-2.11	-0.241	0.289	-0.83
employed	-0.026	0.109	-0.24	-0.178	0.121	-1.47
wk	-0.111	0.022	-4.98	-0.128	0.022	-5.70
married	0.281	0.099	2.85	0.177	0.109	1.63
mhlth	0.010	0.056	0.18	-0.156	0.063	-2.47
chlth	-0.189	0.061	-3.10	-0.146	0.068	-2.15
race	0.249	0.101	2.48	0.108	0.111	0.97
educ_bin	-0.055	0.106	-0.52	0.273	0.120	2.28
housize	-0.424	0.099	-4.27	-0.288	0.110	-2.62
stress	0.268	0.073	3.68			
σ	0.467	0.045	10.42	0.538	0.049	10.89
ρ	0.323	0.113	2.86			
LogL	-3205.04					

small. The Stress Model of Table 25.8 shows that the better maternal health `mhlth`, −0.156 (0.063), or the child's health `chlth`, −0.146 (0.068), at the baseline, the lower the probability of the mother experiencing stress during the study. Also, the higher the mother's education `educ_bin`, 0.273 (0.120), the higher the probability that the mother feels stress. Some covariates are significant in both the `illness` and maternal stress (`stress`) models, e.g. `wk`, `chlth`, and `housize`; these covariates have both a direct effect on `illness` and an indirect effect on `illness` though the endogenous regressor `stress`.

For illustrative purposes we have used a simple recursive model in which contemporaneous maternal stress has a direct effect on child illness. We are not saying that there are no reciprocal effects or lagged endogenous covariates present.

25.7 CONCLUSION

This paper compared the robustness of IWMs and MLMs to ignoring the contemporaneous endogeneity of a within-cluster varying regressor. This can occur in both cross-sectional and longitudinal clustered univariate responses. Ignoring endogeneity is a form of misspecification that creates a difference in the between- and within-cluster covariate effects. We have shown that the MLM and IWM GEE analyses of this are equivalent; they are just conditional and marginal formulations with respect to the cluster-specific random effect. The Mundalk (1978) and Chamberlain (1980, 1984) misspecification test, which uses a linear function of the within-cluster regressor means in the linear predictor, is able to detect misspecification due to endogeneity and has similar sampling properties in both approaches. A re-analysis of the MSCM data suggests that contemporaneous maternal stress is endogenous for child illness.

This paper has studied some of the issues that arise in GEE marginal models and MLMs when we have assumed that the model generating the data is from a multivariate MLM. It would also be worth investigating the misspecification issues that arise in MLMs when the model generating the data is consistent with that of a population average model with a range of different GEE working covariance structures for the within-cluster responses.

ACKNOWLEDGEMENTS

I would like to thank David Stott for his help in writing the Stata code for the simulations and for the analysis of the MSCM data. I would also like to thank Jim Taylor for his comments on a draft of this chapter.

REFERENCES

Bartels, L.M. (1991) 'Instrumental and "Quasi-Instrumental" Variables', *American Journal of Political Science*, 35: 777–800.

Berridge, D.M. and Crouchley, R. (2011) *Multivariate Generalized Linear Mixed Models Using R*. Boca Raton, FL: CRC Press.

Betensky, R.A., Talcott, J.A., and Weeks, J.C. (2000) 'Binary Data with Two, Non-Nested Sources of Clustering: an Analysis of Physician Recommendations for Early Prostate Cancer Treatment', *Biostatistics*, 1: 219–30.

Carriere, I. and Bouyer, J. (2002) 'Choosing Marginal or Random-effects Models for Longitudinal Binary Responses: Application to Self-Reported Disability among Older Persons', *BMC Medical Research Methodology*, 2: 15.

Chamberlain, G. (1980) 'Analysis of Covariance with Qualitative Data', *The Review of Economic Studies*, 47: 225–38.

Chamberlain, G. (1984) 'Panel Data', in Z. Griliches, M.D. Intriligator (eds), *Handbook of Econometrics*. Amsterdam: Elsevier, Chapter 22.

Chao, E.C. (2006) 'Structured Correlation in Models for Clustered Data', *Statistics in Medicine*, 25: 2450–68.

Chao, W.-H., Palta, M. and Young, T. (1997) 'Effects of Omitted Confounders on the Analysis of Correlated Binary Data', *Biometrics*, 53: 678–89.

Conaway, M.R. (1990) 'A Random Effects Model for Binary Data', *Biometrics*, 46: 317–28.

Cox, D. R. (1961) 'Tests of Separate Families of Hypotheses', in J. Neyman (ed.), *Proceedings of the 4th Berkeley Symposium on Mathematical Statistics and Probability*. Berkeley: University of California Press, 105–23.

Crouchley, R. and Davies, R.B. (1999) 'A Comparison of Population Average and Random Effects Models for the Analysis of Longitudinal Count Data with Baseline Information', *Journal of the Royal Statistical Society, Ser. A*, 162: 331–47.

Crouchley, R. and Davies, R.B. (2001) 'A Comparison of GEE and Random Effects Models for Distinguishing Heterogeneity, Nonstationarity and State Dependence in a Collection of Short Binary Event Series', *Statistical Modeling*, 1: 271–85.

Crouchley, R. and Pickles, A. (1989) 'An Empirical Comparison of Conditional and Marginal Likelihood Methods in a Longitudinal Study', *Sociological Methodology*, 19: 161–81.

Crowder, M.J. (1995) 'On the Use of a Working Correlation Matrix in Using Generalised Linear Models for Repeated Measures', *Biometrika*, 82: 407–10.

Desai, M.D. and Begg, M.D. (2008) 'A Comparison of Regression Approaches for Analyzing Clustered Data', *American Journal of Public Health*, 98: 1425–9.

Diggle, P.J., Heagerty, P., Liang, K-Y., and Zeger, S.L. (2002) *Analysis of Longitudinal Data*, 2nd edition. Oxford: Oxford University Press.

Diggle, P.J., Liang, K.-Y., and Zeger, S.L. (1994) *Analysis of Longitudinal Data*. Oxford: Clarendon Press.

Duncan, G.J., Magnuson, K., and Ludwig, J. (2004) 'The Endogeneity Problem in Developmental Studies', *Research in Human Development*, 1: 59–80.

Emond, M.J., Ritz, J., and Oakes, D. (1997) 'Bias in GEE Estimates from Misspecified Models for Longitudinal Data', *Communications in Statistics – Theory and Methods*, 26: 15–32.

Galbraith, J.I., Zeger, S.L., Liang, K.-Y., and Albert, P.S. (1991) 'The Interpretation of a Regression Coefficient', *Biometrics*, 47: 1593–6.

Galecki, A.T. (1994) 'General Class of Covariance Structures for Two or More Repeated Factors in Longitudinal Data Analysis', *Communications in Statistics – Theory and Methods*, 23: 3105–19.

Gardiner, J.C., Luo, Z., and Roman, L.A. (2009) 'Fixed Effects, Random Effects and GEE: What are the Differences?', *Statistics in Medicine*, 28: 221–39.

Goetgeluk, S. and Vansteelandt, S. (2008) 'Conditional Generalized Estimating Equations for the Analysis of Clustered Longitudinal Data', *Biometrics*, 64: 772–80.

Gourieroux, C., Monfort A., and Trognon, A. (1984) 'Pseudo Maximum Likelihood Methods: Applications to Poisson Models,' *Econometrica*, 52: 701–20.

Heagerty, P.J. and Zeger, S.L. (2000) 'Marginalized Multilevel Models and Likelihood Inference (with Discussion)' *Statistical Science*, 15: 1–26.

Hsiao, G. (1986) *Analysis of Panel Data*. Cambridge, UK: Cambridge University Press.

Hubbard, A.E., Ahern, J., Fleischer, N.L., van del Laan, M., Lippman, S.A., Jewell, N., Bruckner, T., and Satariano, W.A. (2010) 'To GEE or Not to GEE. Comparing Population Average and Mixed Models for Estimating the Associations Between Neighborhood Risk Factors and Health', *Epidemiology*, 21: 467–74.

Huber P.J. (1967) 'The Behaviour of Maximum Likelihood Estimates Under Nonstandard Conditions', *Proceedings of the 5th Berkeley Symposium on Mathematical Statistics and Probability*, 1: 221–33.

Kalbfleisch, J.D. and Prentice, R.L. (1980) *The Statistical Analysis of Failure Time Data*. New York: John Wiley and Sons.

Kenward, M.G. and Molenberghs, G. (1998) 'Likelihood Based Frequentist Inference When Data are Missing at Random', *Statistical Science*, 13: 236–47.

Lai, T.Z. and Small, D. (2006) 'Marginal Regression Analysis of Longitudinal Data with Time-Dependent Covariates: A Generalized Method-of-Moments Approach', *Journal of the Royal Statistical Society, Ser. B*, 69: 79–99.

Laird, N.M. (1978) 'Nonparametric Maximum Likelihood Estimation of a Mixing Distribution', *Journal of the American Statistical Association* 73: 805–11.

Lee, Y. and Nelder, J.A. (2004) 'Conditional and Marginal Models: Another View', *Statistical Science*, 19: 219–28.

Liang, K.-L. and Zeger, S.L. (1986) 'Longitudinal Data Analysis Using Generalised Linear Models', *Biometrika*, 73: 13–22.

Lindsey, J.K. and Lambert, P. (1998) 'On the Appropriateness of Marginal Models for Repeated Measurements in Clinical Trials', *Statistics in Medicine*, 17: 447–69.

Mundalk, Y. (1978) 'On the Pooling of Time Series and Cross Section Data', *Econometrica*, 46: 69–85.

Nelder, J.A. and Wedderburn, R.W.M. (1972) 'Generalized Linear Models', *Journal of the Royal Statistical Society, Ser. A*, 135: 370–84.

Neuhaus, J.M. and Kalbfleisch, J.D. (1998) 'Between- and Within-Cluster Covariate Effects in the Analysis of Clustered Data', *Biometrics*, 54: 638–45.

Neuhaus, J.M. and McCulloch, C.E. (2006) 'Separating Between- and Within-Cluster Covariate Effects by Using Conditional and Partitioning Methods', *Journal of the Royal Statistical Society, Ser. B*, 68: 859–72.

Palta, M. and Yao, T.-J. (1991) 'Analysis of Longitudinal Data with Unmeasured Confounders', *Biometrics*, 47: 1355–69.

Park, C. and Weisberg, S. (1998) 'Fisher Consistency of GEE Models Under Link Misspecification', *Computational Statistics and Data Analysis*, 27: 229–35.

Pepe, M.S. and Anderson, G.L. (1994) 'A Cautionary Note on Inference for Marginal Regression Models with Longitudinal Data and General Correlated Response Data', *Communications in Statistics – Simulation and Computation*, 23: 939–51.

Pepe, S. and Couper, D. (1997) 'Modeling Partly Conditional Means with Longitudinal Data', *Journal of the American Statistical Association*, 92: 991–8.

Preacher, K.J., Zyphur, M.J., and Zhang, Z. (2010) 'A General Multilevel SEM Framework for Assessing Multilevel Mediation', *Psychological Methods*, 15: 209–33.

Prentice, R.L. (1988) 'Correlated Binary Regressions with Covariates Specific to Each Binary Observation', *Biometrics*, 44: 1033–48.

Ritz, J. and Spiegelman, D. (2004) 'Equivalence of Conditional and Marginal Regression Models for Clustered and Longitudinal Data', *Statistical Methods in Medical Research*, 13: 309–23.

Senn, S. (2004) 'Conditional and Marginal Models: Another View: Comment', *Statistical Science*, 19: 228–30.

Shen, L., Shao, J., Park, S., and Palta, M. (2008) 'Between and Within-Cluster Covariate Effects and Model Misspecification in the Analysis of Clustered Data', *Statistica Sinica*, 18: 731–48.

Subramanian, S.V. and O'Malley, J.O. (2010) 'Modeling Neighborhood Effects: The Futility of Comparing Mixed and Marginal Approaches', *Epidemiology*, 21: 475–8.

Teerenstra, S., Lu, B., Preisser, J.S., Van Achterberg, T., and Borm, G.F. (2010) 'Sample Size Considerations for GEE Analyses of Three-Level Cluster Randomized Trials', *Biometrics*, 66: 1230–7.

Thall, P.F. and Vail, S. (1990) 'Some Covariance Models for Longitudinal Count Data with Overdispersion', *Biometrics*, 46: 657–71.

Wall, M.M., Dai, Y., and Eberly, L.E. (2005) 'GEE Estimation of a Misspecified Time-Varying Covariate: An Example with the Effect of Alcoholism Treatment on Medical Utilization', *Statistics in Medicine* 24: 925–39.

Ware, J.H., Dockerty, D.W., Louis, T.A., Xu, X., Ferris (Jr), B.G., and Speizer, F.E. (1990) 'Longitudinal and Cross-Sectional Estimates of Pulmonary Function Decline in Never Smoking Adults', *American Journal of Epidemiology*, 132, 685–700.

White, H. (1982) 'Maximum Likelihood Estimation of Misspecified Models' *Econometrica*, 50: 1–25.

Winship, C. and Mare, R.D., (1988) 'Endogenous Switching Regression Models for the Causes and Effects of Discrete Variables', in J. Scott Long (ed.), *Common Problems in Quantitative Social Research*. London: Sage Press.

Zeger, S.L. and Liang, K.-Y. (1986) 'Longitudinal Data Analysis for Discrete and Continuous Outcomes', *Biometrics*, 42: 121–30.

Zeger, S.L., Liang, K.-Y., and Albert, P.A. (1988) 'Models for Longitudinal Data: A Generalized Estimating Equation Approach', *Biometrics*, 44: 1049–60.

26

Software for Fitting Multilevel Models

Andrzej T. Gałecki and Brady T. West

University of Michigan, USA

26.1 INTRODUCTION

Today's applied statistician or research analyst has the luxury of working with a variety of powerful statistical software procedures that enable users to efficiently fit many of the multilevel models covered in this volume. Developments in software for multilevel models, including linear (LMMs), generalized (GLMMs), and non-linear (NLMMs) mixed-effects models, are propelled both by advances in statistical methodology and technological progress in meeting computational demands.

With this chapter, we aim to provide readers with some history and an overview of the current software procedures that are available for fitting these models, and the abilities of the software to accommodate the many analytic aspects that accompany this broad class of models.

This chapter begins in Section 26.2, with some background information and an overview of important historical developments leading to the development of the powerful software available today. An overview of currently available software is given in Section 26.3. Section 26.4 discusses implementation details for the various procedures. Sections 26.5 and 26.6 conclude the chapter by discussing available software for fitting more advanced models, and directions for future developments.

26.2 BACKGROUND

For many years, software for the analysis of clustered and/or longitudinal data relied on simplifying model assumptions and employed straightforward computational algorithms. In the case of linear models, software implementing repeated measures analysis of variance (rmANOVA) models was the only option for longitudinal data. Fitting these models required data sets with very rigid structures: balanced counts of observations between groups and over time, no missing data, homogeneity of variance components, and no time-dependent covariates. For non-linear models, a very popular two-stage approach could be implemented using standard non-linear regression software, followed by simple ANOVA-type analysis of subject-specific estimates obtained in the first stage. This overly simplistic approach was based on the premise that subject-specific estimates obtained from individual non-linear fits were exactly equal to true values of subject-specific parameters. This approach suffered from similar shortcomings to those exhibited by rmANOVA; for example, time-varying covariates were not allowed. Moreover, overall variability of the dependent variable was not properly partitioned into between- and within-cluster/subject variation, leading to biased estimates of variance components.

A noteworthy exception in this regard was the NONMEM procedure, developed at the University of California for non-linear mixed-effects models dating back to 1980 (Beal and Sheiner, 1980). At its early stages, NONMEM employed an estimation method referred to as extended least squares (Sheiner et al., 1972; Beal and Sheiner, 1982), which was not widely accepted outside of the pharmaceutical community.

With the advent of personal computers, the progress in developing new software was greatly accelerated. In the mid 1980s, software programs for fitting mixed-effects models diverged into three broad groups, roughly ordered with respect to increasing levels of generality and computational difficulty: LMMs, GLMMs, and NLMMs.

In this section, we first focus on frequentist approaches to model fitting, followed by Bayesian approaches.

26.2.1 Frequentist Approaches

From the estimation point of view, we will first focus on the likelihood-based approach for mixed-effects models. The common feature of procedures using this approach is that the objective function, i.e., *marginal likelihood*, is expressed in the form of an integral over random effects, which has important implications for how the computations are performed.

In what follows, we briefly present discussions of approaches for fitting LMMs, GLMMs, and NLMMs, with references to methodological literature highlighting important computational and software developments.

Linear mixed-effects models

Over the years, the list of available software for fitting LMMs has become very impressive. We start our presentation with software for fitting LMMs, as specified by Laird and Ware (1982), for at least three reasons. First, these are the simplest types of mixed-effects models. Second, in many cases, they are the primary choice for the analysis of longitudinal and/or clustered data because of their numerical stability, substantive theoretical advances, and the availability of the software. Third, several aspects of LMMs are shared by and, to varying degrees, often adopted by more advanced models. By considering LMMs first, we set the stage for the more advanced classes of models described in the remainder of this chapter.

In their seminal paper, Laird and Ware (1982) introduced the specification of LMMs for longitudinal data and demonstrated how to obtain likelihood-based estimates using the EM algorithm (Dempster et al., 1977). In contrast to previous approaches, Laird–Ware's specification allowed for time-dependent covariates, missing values of the dependent variable, and structured variance-covariance matrices.

In the late 1980s, the 5V function (Schluchter, 1988) for two-level models for longitudinal and repeated measures data was introduced in BMDP, a general purpose statistical package. Origins of the two major software packages for fitting multilevel models, i.e., HLM (Bryk et al., 1988) and MLwiN, with ML2 (Rasbash et al., 1989), ML3 (Prosser et al., 1991), and MLn being predecessors of the latter, also date back to that time. These programs became the software of choice for analysts for many years to come.

The popularity of BMDP/5V diminished after SAS/PROC MIXED was introduced in 1992. Originally, the stand-alone software VARCL (Longford, 1990) was also one of the major packages, but its development, similar to that of BMDP/5V, has been discontinued. Additional software packages developed in later years are described in Section 26.3.

Nowadays, it is difficult to imagine a general purpose statistical software package without a module or function for fitting LMMs.

Other developments have involved faster algorithms based on sparse matrices (George and Liu, 1981) and are designed for large-scale problems encountered in fields such as genetics. Software employing this type of methodology includes ASReml (under development since 1992), SAS/PROC HPMIXED (SAS Institute Inc., 2010) and the `lmer` function in the R/**lme4** package (Bates et al., 2011), first released in 2005. We note that for the latter two programs the increased execution speed comes at the expense of a limited choice of variance-covariance structures for random effects and errors.

With respect to estimation, the key feature of LMMs is that the marginal likelihood can be expressed in a closed form, which essentially allows the software to directly employ standard optimization algorithms, such as Newton–Raphson (N–R) and Fisher scoring (Jennrich and Schluchter, 1986), to obtain maximum likelihood estimates.

Generalized linear mixed-effects models

In contrast to LMMs, the key computational difficulty for GLMMs is that, in general, the integral used in the marginal likelihood representation does not have a closed-form solution. Consequently, computational methods need to circumvent this difficulty, most often by approximating the likelihood before proceeding to the likelihood optimization. The choice of the approximation method may have considerable implications for the properties of the parameter estimates, and we therefore elaborate on this issue in more detail.

First, we note that a variety of approximations of the marginal likelihood have been proposed in the literature. One group of approximations, referred to as *deterministic*, provide an analytic formula for approximating the likelihood. Selected examples of deterministic approximations include:

- Non-adaptive Gaussian-Hermite quadrature (Stroud and Secrest, 1966);
- Laplace (Tierney and Kadane, 1986) approximation of the integrand function;
- Quasi-likelihood approaches, e.g., marginal (MQL) in Goldstein (1991) and Breslow and Clayton (1993), and penalized (PQL) in Breslow and Clayton (1993) and Wolfinger and O'Connell (1993);
- Adaptive Gaussian quadrature (AGQ) in Liu and Pierce (1994) and Pinheiro and Bates (1995a).

Approximations based on non-adaptive Gauss-Hermite quadrature are computationally intensive. This type of approximation was implemented for the first time for selected examples of two-level models in EGRET (Mauritsen, 1990).

Approximations of the likelihood based on the Laplace and quasi-likelihood methods are simple to implement but, at the same time, their early versions were fairly crude. Rodriguez and Goldman (1995) report that MQL-based estimates of fixed effects and variance components are subject to substantial downward bias when the random effects are relatively large. A similar observation was made for the PQL method in Breslow and Clayton (1993).

The Laplace approximation in Tierney and Kadane (1986) was implemented in early versions of the HLM software. Quasi-likelihood methods have been implemented in several software procedures, including earlier versions of SAS/PROC GLIMMIX, HLM, MLwiN, and others.

To alleviate the shortcomings of the aforementioned early implementations, the Laplace approximation was refined in Raudenbush et al. (2000). Similarly, the quasi-likelihood methods were also improved by using second-order terms in the Taylor expansion (Rodriguez and Goldman, 1995; Goldstein and Rasbash, 1996). Subsequently, these refinements were implemented in the HLM and MLwiN software.

AGQ, a substantial improvement over regular non-adaptive Gaussian-Hermite, was implemented for the first time in SAS/PROC NLMIXED (Wolfinger, 1999) and later in the `gllamm` command in Stata (Rabe-Hesketh et al., 2004). It is well known that by choosing one quadrature point AGQ becomes Laplace's approximation. By increasing the number of quadrature points, the accuracy of the approximation can be increased at the expense of longer execution times.

Alternative likelihood approximations are based on Monte Carlo integration methods and are referred to as *stochastic approximations* or, equivalently, *simulated maximum likelihood methods*. They employ, for example, Monte Carlo type or importance sampling based algorithms to simulate the likelihood function (Pinheiro and Bates, 1995a). On the one hand, stochastic approximations introduce another source of error, i.e., error due to sampling variance; on the other hand, with a large number of simulations, this error becomes negligible. Moreover, sampling variance can be used to assess the accuracy of the stochastic approximation.

When fitting GLMMs, approximation of the marginal likelihood is followed by an optimization step, most commonly the Expectation–Maximization (EM) algorithm (Dempster et al., 1977), N–R, or Fisher scoring. Each of the aforementioned approximation methods of the marginal likelihood can be combined with one of the optimization methods, leading to a large number of choices in finding optimal values for parameter estimates.

Other approaches, such as hierarchical likelihood, or h-likelihood (Lee and Nelder, 1996, 2001), attempt to avoid integration of the likelihood over random effects altogether by employing an objective function that is different from the marginal likelihood. This approach has been implemented, for example, in the **hglm** package (Alam et al., 2010) in the R software.

Nonlinear mixed-effects models

With respect to likelihood optimization, NLMMs with Gaussian errors share a similar challenge to that encountered for GLMMs, i.e., an intractable integral in the definition of the marginal likelihood. Approximations similar to those used for GLMMs have been implemented in modern software procedures.

The most common approximations are based on a linear Taylor expansion of the model mean function around current estimates of the fixed effects and/or empirical Bayesian (EB) estimates of the random effects. This linearization-based approach is conceptually similar to the one used in the quasi-likelihood approach for GLMMs.

The first-order (FO) linearization proposed in Beal and Sheiner (1982) was implemented, for example, in NONMEM. It performs in a satisfactory way if random effects enter the model in a linear fashion. In this approximation, the Taylor expansion is performed with EB estimates of the random effects set to zero.

Conditional first-order (FOCE) linearization, drawing on a method presented by Lindstrom and Bates (1990), is an improvement over the FO method in a similar way that PQL is an enhancement of MQL. In contrast to the FO method, it involves expansion of the model function around parameter estimates, including the current values of the EB estimates. This method is carried out in NONMEM. It is also the basis of the estimation algorithm implemented in the `nlme` function of the R/**nlme** package, and in the SAS NLINMIX macro (Littell et al., 1996, 2006).

Similar to developments for GLMMs, more recent approaches involve stochastic approximations of the likelihood function and are implemented, for example, in the Monolix software (Lavielle, 2008).

26.2.2 Bayesian Approaches

The Bayesian framework is especially well suited for multilevel models due to their hierarchical structure. The key feature in model specification, distinguishing from the frequentist approach, is the need for a prior distribution for the model parameters. Bayesian inference itself is based on the posterior distribution of the parameters given data. In general, the posterior distribution is analytically intractable for multilevel models. Thanks to Markov Chain Monte Carlo (MCMC) methods, this distribution can be stochastically and efficiently estimated based on generated samples. A useful reference in this context is Congdon (2005).

In the mid 1990s, the WinBUGS software (Spiegelhalter et al., 1996a,b,c) was developed for Bayesian analysis with MCMC chains constructed using the Gibbs sampler. The beginnings of this program date back to 1989, when its predecessor BUGS (Bayesian inference Using Gibbs Sampling) was introduced. Currently, the development of the program is focused on the OpenBUGS project, drawing on the open source concept. In addition, users of the R software can call WinBUGS and OpenBUGS from R, and Gelman and Hill (2007) provide a good reference on this practice.

WinBUGS, available at `http://www.mrc-bsu.cam.ac.uk/bugs/winbugs/contents.shtml`, can be used for a wide range of multilevel models. Bayesian analysis using the MCMC approach was also implemented for fitting GLMMs in the MLwiN software (Browne, 2009) and in the **MCMCglmm** (Hadfield, 2010) R package. Other software, for example MPlus (Muthén and Muthén, 1998–2010) and JAGS (Plummer, 2003), can also be used for Bayesian analysis of multilevel models. More recently, alternatives to the MCMC approach based on analytic approximations have been developed in Rue et al. (2009) and implemented in the R package **inla**.

26.2.3 Recap

Interestingly, many different groups have worked on developing software for multilevel models. Apart from general-purpose statistical software packages, such as SAS, SPSS, BMDP, Stata, and R, stand-alone software has been extensively developed. This software, dedicated specifically to multilevel models, was geared toward various audiences and became popular within specific disciplines. Examples include the HLM software's popularity in education and social science research, and aML's use by econometricians. NONMEM is a leading software tool in the pharmaceutical community for analysis of data from population pharmacokinetics/pharmacodynamics (PK/PD) studies. ASReml is generally dedicated to analytical problems in agriculture and genetics. There are also examples of applications of the **lme4** package in R to ecological and linguistic research.

Due to dynamic progress, stand-alone software dedicated to fitting multilevel models (e.g., MLWin and HLM) had a fairly complicated history, including changing names, alternating between research or production phases, and some of them being commercialized. We note similarities in developmental paths, however. Initially, this software was developed for models with two levels for a linear case. The next step was adding levels to a model's hierarchy, so multilevel linear models could be addressed. In the mid 1990s, with the advent of the Windows operating system, these software packages added graphing capabilities and some of them added "Win" to the software names.

In general, software for fitting LMMs came first, followed by software for fitting GLMMs, and possibly other classes of models. Recent advances have also resulted in procedures capable of fitting both LMMs and GLMMs in general-purpose statistical computing packages like SAS, SPSS, Stata, and R.

Software procedures for fitting NLMMs were developed in parallel, often as separate products. Interfaces between the different software were developed to gain flexibility and take advantage of other software environments.

Marginal likelihood approximations for GLMMs and NLMMs using deterministic and stochastic methods are critical components of computational algorithms and can be performed in a variety of ways. Unfortunately, in the case of many software products information about the approximation method being used is not readily available.

The Bayesian approach to model fitting and inference was limited in early development stages by computer speed, but this is not currently an issue. This is partially due to increased computing speed, but also due to the advent and widespread use of MCMC methods. More recently, analytical approximations of the posterior marginal distribution have been proposed. Consequently, the Bayesian approach provides a very attractive, unified framework, which allows analysts to systematically address a variety of multilevel and more advanced models.

Our presentation of historical developments is largely based on the work of Kreft et al. (1994), de Leeuw and Kreft (2001), Pillai et al. (2005), and de Leeuw and Meijer (2008).

At present, there are various venues of support for users thanks to internet resources. Examples of relevant web pages include `http://www.bristol.ac.uk/cmm/` and `http://statcomp.ats.ucla.edu/mlm/` at Bristol University and UCLA, respectively. Discussion lists, such as `r-sig-mixed-models@r-project.org` for R users, provide a vivid forum for discussing issues related to multilevel modeling.

In our personal view, the most influential software, besides NONMEM, developed over the years, includes HLM, MLwiN, SAS/PROC MIXED, and SAS/PROC NLMIXED. The `gllamm` command in Stata has proven widely useful, due to its versatility and ability to address models other than LMMs and GLMMs (see Section 26.5). These developments will certainly continue in the future, and the next section addresses the different software procedures that are currently available.

26.3 AVAILABLE SOFTWARE

In this section, we present a broad range of currently available software that can be used to fit many examples of LMMs, GLMMs, and NLMMs that frequently arise in a variety of applied fields. For the sake of simplicity, we concentrate on parametric models with normally distributed random effects. Extensions of this basic modeling framework and a discussion of related software procedures are deferred to Section 26.5.

Also for the sake of simplicity, we tentatively divide the available software into three groups: general-purpose software, standalone software dedicated for multilevel models, and software designed for a wider class of models encompassing mixed-effects models as a special case.

26.3.1 General-Purpose Software

In Table 26.1, we present examples of functions and modules available for fitting mixed-effects models in general-purpose statistical software. Typically, as reflected in the columns' labels, the available software contains functions/modules designed for a specific class of models.

Major general statistical software packages include SAS, SPSS, Stata, and R. Relevant references in this context are Littell et al. (2006), SAS Institute Inc. (2008), SPSS Inc. (2010), Rasbash (2004), West (2009), StataCorp. (2009), R Development Core Team (2010), and Pinheiro and Bates (2000). GENSTAT (Payne et al., 2009), SYSTAT, and STATISTICA are other general-purpose statistical software packages not included in the table that allow users to fit similar classes of models.

We note that the R/**nlme** package with the `lme` and `nlme` functions was available in S-plus (Pinheiro and Bates, 1995b) before being implemented in R. Also, the R/**lme4** package is under ongoing development. We also note that Stata has developed two specialized commands for selected GLMMs, namely `xtmelogit` and `xtmepoisson`.

It is also worth noting that although modules/functions are often limited to a particular class of models, their strength comes from being immersed in a rich software/package environment. Several examples of LMM syntax for HLM software and appropriate functions listed in Table 26.1 are given in West et al. (2007).

Table 26.1 General-Purpose Statistical Software Packages Containing Modules/Functions Designed To Fit Various Classes of Mixed-Effects Models. Software version/approximate year of the corresponding first release are given in parentheses

Software/	Class of mixed-effects models		
Package	LMM	GLMM	NLMM
SAS	`PROC MIXED` (6.07/1992)	`PROC GLIMMIX` (9.1/2006)	`PROC NLMIXED` (8.0/1999)
SPSS	`MIXED` (11.0/2001)	`GENLINMIXED` (19.0/2010)	
Stata	`xtmixed` (9/2005)	`gllamm` (8/2003)	
R/**nlme**	`lme` (1999)		`nlme` (1999)
R/**lme4**	`lmer` (2005)	`glmer` (2006)	`nlmer` (2007)

26.3.2 Software Dedicated to Multilevel Models

In Table 26.2, we include examples of stand-alone software dedicated to fitting multilevel models, or at least selected classes of these models.

In the upper part of Table 26.2, we include stand-alone software applicable to a fairly wide range of multilevel models. HLM and MLwiN are major software products in this group. Both are commercially distributed by Scientific Software International (SSI) and the Centre for Multilevel Modeling based at the University of Bristol, respectively. We note that MLwiN functionality is substantially enhanced by the presence of macro capability, allowing users to write their own functions. EGRET stands for Epidemiological GRaphics Estimation Testing, originally developed at the University of Washington. This software product is currently available from Cytel.

The Mixed-Up Suite was originally developed at the University of Illinois at Chicago. It is also referred to as MIXFOO, and contains the MIXOR, MIXREG, MIXNO, MIXPREG functions written in FORTRAN for ordinal, normally distributed, nominal, and count response data, respectively. More recently, the aforementioned functions were combined into a single application SuperMix (Hedeker et al., 2008) distributed by SSI.

The name aML stands for applied Multilevel Modeling, and this is powerful statistical software for multilevel and multi-process models. In general, software outlined in the upper part of this table tends to address LMMs and GLMMs, but does not cover NLMMs.

In the lower part of Table 26.2, we include software applicable to a fairly narrow range of multilevel models. MLA (Busing et al., 2005) is research software designed to fit two-level models. An interesting feature of MLA, distinguishing it from many other software packages, is the implementation of bootstrap and jackknife techniques for variance estimation.

ASReml is specialized software designed for fitting linear mixed-effects models using REML. It was developed jointly by statisticians from the Biometrics Program of the NSW Department of Primary Industries in Australia and the Biomathematics Unit of Rothamsted Research in Britain. Initially it was stand-alone software; later, in 1997, its core was linked into Genstat (Release 4.1). At present it is distributed by VSN International Ltd (VSNi). SAS/PROC HPMIXED is a new procedure developed by SAS Institute, Inc. As we have already mentioned, both ASReml and SAS/PROC HPMIXED use techniques based on sparse matrices and are designed for efficiently fitting LMMs to large and complex clustered or longitudinal data sets.

Table 26.2 Software Dedicated to Fitting Multilevel Models

Software	First release	Reference(s)
Software for a wide range of models		
HLM	1988	Bryk et al. (1988)
		Raudenbush et al. (2004a)
MLwiN	1997	Rasbash et al. (2004)
EGRET	1990	Mauritsen (1990)
Mixed-Up Suite	1996	Hedeker and Gibbons (1996a,b)
		Hedeker (1998, 1999)
Supermix		Hedeker et al. (2008)
aML	2000	Lillard and Panis (2003)
Software for selected classes of models		
MLA	1994	Busing et al. (2005)
ASReml	1992	Butler et al. (2007)
		Gilmour et al. (2009)
SAS/HPMIXED	2010	SAS Institute Inc. (2010)
NONMEM	1980	Beal and Sheiner (2004)
Monolix	2008	Lavielle (2008)

NONMEM is a leading software package for fitting NLMMs in research communities conducting analysis of data from population PK/PD studies. Starting with Version 7 of NONMEM, ICON Developments Solutions has exclusive rights to this software. Monolix is a fairly new software package also designed for NLMMs, employing similar advanced estimation techniques to those recently implemented in NONMEM.

In stand-alone software dedicated to multilevel models, we anticipate more specialized options compared to the available software in general-purpose statistical packages. Compared to general-purpose software, the environment, e.g., graphics capabilities, may be less developed. We note that HLM, aML, and MLA are true multilevel software, in the sense that they allow for the specification of regression models at multiple levels, making them an attractive choice for these types of analyses.

26.3.3 Software for More General Models

In this section, we present software that can be used for fitting more general models than mixed-effects models. This software can also be used to fit mixed-effects models as a special case of the general class. Examples of the software are given in Table 26.3.

Interestingly, the `gllammm` command in Stata, included in Table 26.1, is an example of such a software procedure. More specifically, it allows the user to fit a more general class of models referred to as Generalized Linear Latent And Mixed Models (GLLAMM), described in Rabe-Hesketh et al. (2001) and Skrondal and Rabe-Hesketh (2004), with GLMMs obviously included as a special case.

The Mplus and LatentGold software packages were originally designed primarily for latent class models. More recent releases have also enabled analysts to fit GLMMs. Mplus and LatentGold are commercial products distributed by Muthén & Muthén and Statistical Innovations, respectively.

LISREL is designed for structural equation modeling, and is a commercial product distributed by Scientific Software International (SSI). Procedures for fitting GLMMs were implemented in Release 8.8 (2006).

The Automatic Differentiation Model Builder (ADMB) (Fournier, 1993–2011) is a very influential software package used in ecology and the field of fisheries.

Table 26.3 Software for Fitting a More General Class of Models

Software	First release	Reference(s)
MPlus	1998	Muthén and Muthén (1998–2010)
LatentGold	2000	Vermunt and Magidson (2008)
LISREL	1974	Jöreskog and Sörbom (2006)
ADMB	2004	Fournier (1993–2011)
WinBUGS	1996	Spiegelhalter et al. (1996a,b,c)

The distinguishing and attractive feature of ADMB is that the automatic differentiation (also known as algorithmic differentiation) capability is implemented. More specifically, the code to evaluate derivatives used for likelihood optimization is generated automatically by the computer program. ADMB-RE (Skaug and Fournier, 2004–2011), first introduced in 2004, is a software package that adds mixed-effects modeling capabilities to ADMB. Consequently, ADMB-RE enables users to include random effects in a wide range of non-linear models tackled by ADMB. Both ADMB and ADMB-RE are accurate, flexible, and computationally efficient programs.

WinBUGS is an MCMC-based package for Bayesian analysis developed at the Medical Research Council Biostatistics Research Unit in Cambridge, and can be used for a wide range of models, including multilevel models. As mentioned in Section 26.2.2, recent developments have also enabled users of R to call BUGS routines from R.

26.3.4 Recap

There are many software products currently available allowing analysts to fit multilevel models. For the sake of simplicity, we have classified the available software into three groups, with examples presented in Tables 26.1, 26.2, and 26.3.

The most popular software procedures available today for LMMs are SAS/PROC MIXED, the `lme` function in the R/**nlme** package, the `lmer` function in the R/ **lme4** package, HLM, the `xtmixed` command in STATA, and the MIXED procedure in SPSS. An extensive overview of the options currently available in software designed to fit LMMs can be found in West and Gałecki (2011). For GLMMs, the leading software procedures are SAS/PROC GLIMMIX, HLM, MlwiN, and the `gllamm` command in Stata. For NLMMs, the most important software procedures are SAS/PROC NLMIXED and NONMEM.

Over the last decade, we have observed an emerging trend in the development of software, initiated by the `gllamm` command in Stata, employing a unified framework for different but related classes of models, including multilevel, latent class, and/or structural equation models. We note, for example, that Mplus, LatentGold, LISREL, and ADMB, presented in Table 26.3 and originally designed for a different class of models, i.e., latent class or SEMs, have, over time, added the capability to fit multilevel models with random effects.

Examples of non-commercial products include R, the MIXED-Up suite, ADMB, aML, MLA, Monolix, and WinBUGS. It is also worth noting that in this group of software, R, ADMB, and OpenBUGS are developed as open source projects.

We also note that auxiliary software has been developed to address specific issues associated with mixed-effects modeling, such as sample size calculations (see Section 26.5).

26.4 IMPLEMENTATION DETAILS

In this section, we review some aspects of the implementation of procedures for fitting

LMMs, GLMMs, and NLMMs. Some of these aspects were already mentioned in Section 26.2 in a broader context. Here, we present them in more detail. We will stress issues/tasks common to LMMs, GLMMs, and NLMMs, and point out some differences.

26.4.1 Input Information

We consider two aspects of input information, namely data organization and model specification.

Data organization

The leading approach used by the majority of available software procedures is to use data in a "long" format, with multiple records per subject or cluster. Another very natural method, implemented in the HLM software, is to have data organized essentially in the form of a relational database, with separate tables of data for variables measured at different levels of the data hierarchy (e.g., for subjects at Level 2 and individual measurements on subjects at Level 1). In contrast to the aforementioned data formats, the "wide" format, used for example in the 5V BMDP software (one of the very first software tools for fitting LMMs), poses substantial difficulties in specifying models for longitudinal data with irregular time observations and is essentially no longer supported by today's software for fitting mixed-effects models.

Model specification

As stated in Chapter 2 of this volume, multilevel models can be specified in a variety of ways. In most software, syntax typically follows a general matrix formulation. For example, PROC MIXED employs a single-equation specification of mixed-effects models. In contrast, the HLM, MLwiN, MLA, and aML software packages implement model specification based on "level notation", which naturally corresponds to the hierarchy of multilevel models.

Current software procedures for fitting multilevel models offer a variety of ways for users to specify different components of hierarchical models. Some software procedures (e.g., HLM, MLwiN) offer users a graphical interface for specifying models in a multilevel manner, while others rely on syntax (e.g., SAS/PROC MIXED) or possibly selection of choices from a menu (e.g., SPSS MIXED) for specifications more in the spirit of a "single equation". In general, in a given software package, the specification of design matrices for fixed and random effects in LMMs is similar to specifications used for classical linear models. A desirable feature that future enhancements of these software procedures might include is to have well-integrated polynomials and regression splines, so that specification of the design matrices may be enhanced.

Crossed (as opposed to nested) random effects are increasingly implemented in many software packages. Examples include SAS/PROC MIXED, SPSS MIXED, HLM, and R/`lmer`. Goldstein (1999, 2010) describes a trick for setting up the crossed random-effects model for a hierarchical software package such as MLwiN or `gllamm` in Stata.

Software procedures have different syntax or different options allowing users to specify alternative covariance structures in these models, which is especially important for errors in the context of longitudinal continuous data. Although the choices of variance-covariance matrices may be impressive (e.g., SAS/PROC MIXED), it may also be desirable to have an option to allow for user-defined structures, as is done in the R/**nlme** package. Readers can refer to West et al. (2007) for more details about specifying alternative covariance structures in a variety of software packages capable of fitting LMMs.

In the context of GLMMs, it may be desirable to allow for user-defined link functions, and this feature of GLMMs has not yet been widely implemented. For NLMMs, an important feature is to allow for the variance function to depend on the mean value of the dependent variable. This feature is available, for example, in the NONMEM software, but it is not available in SAS/PROC NLMIXED.

Caution needs to be exercised when using some software. For example, in aML, a potential source of confusion for some users is that the labeling of hierarchical levels is the opposite of that used in most of the multilevel literature and software.

26.4.2 Estimation

In most of the software for multilevel models, estimation is primarily based on optimization of the likelihood function, but other methods such as iterative generalized least squares (IGLS) are also available.

In the *linear* case, likelihood-based software typically relies on optimization of a marginal likelihood for covariance parameters with fixed effects being profiled out. Optimization of the likelihood in its general form is complicated by underlying non-linear constraints, assuring that covariance matrices remain positive-definite. Some remedies, such as transformation of the variance-covariance parameters (resulting in an unconstrained optimization) have been proposed (Pinheiro and Bates, 1996). The most common computational approach for ML optimization is based on the N–R algorithm. In some cases, e.g., HLM or the `xtmixed` command in Stata, the EM algorithm is used to refine initial values. Other methods such as Ridge-stabilized N–R are also often available.

Most current software procedures provide users with a choice of REML or ML estimation, depending on the test of interest. Typically, restricted maximum likelihood (REML) is used as a default, due to its property of producing unbiased estimates of the covariance parameters. Software procedures may vary in terms of methods used to derive starting values for estimates of covariance parameters, and some software procedures (e.g., SAS/PROC MIXED) allow the user to specify starting values for the covariance parameter estimates. Most current software procedures, in addition to having default methods for optimizing likelihood functions, use various combinations of these methods to arrive at the final estimates. Current computing power generally makes differences between these methods trivial for models with nested random effects. Models with crossed random effects require more advanced methods that are becoming more widely implemented in LMM software procedures (e.g., R/**lmer**).

As mentioned in Section 26.2.1 in a *non-linear case*, i.e. GLMMs and NLMMs, optimization is typically preceded by approximation of the marginal likelihood. In this section, we elaborate on this issue in more detail in the context of implementation in a variety of software.

Due to its simplicity, a basic version of the Laplace approximation is implemented in many software packages. More advanced versions of this approximation introduced in Raudenbush et al. (2000) have been implemented in the HLM software.

Approximations based on non-adaptive Gaussian quadrature have been implemented, for example, in MIXOR, MIXNO, aML, and EGRET.

Quasi-likelihood methods are implemented by iteratively executing two steps, namely creating pseudo-data and fitting a working LMM. An example of such an approach is PQL, as implemented in the `glmmPQL` function from the **MASS** package (Venables and Ripley, 2002) in R.

AGQ is currently implemented in several popular software packages that are aimed at fitting GLMMs, e.g., SAS/PROC GLIMMIX, SAS/PROC NLMIXED, and the `gllamm` command in Stata. In software packages implementing both methods, one can choose between non-adaptive and adaptive Gaussian quadrature. Interestingly, SAS/PROC NLMIXED allows the user to modify the number of quadrature points during the iteration process, so that better approximation can be accomplished. Additionally, SAS/PROC NLMIXED allows for the importance sampling method, an example of stochastic approximation, to be used.

Another example of software implementing importance sampling is ADMB.

The Mplus software can provide users with both frequentist and Bayesian inference. The numerical integration of the marginal likelihood can be carried out with or without adaptive quadrature. Bayesian analyses conducted using Mplus employ MCMC methods.

Assuming that the method of marginal likelihood approximation is selected, the next step is to optimize the likelihood. The most commonly used techniques are the EM, N–R, and Fisher scoring algorithms. We briefly describe the EM algorithm only. At first glance, it appears that the EM algorithm manages to bypass the likelihood approximation step and allows the user to go directly to finding optimum values of parameter estimates. Unfortunately, at the E-step, the EM algorithm also requires the calculation of an intractable integral, and therefore the E-step needs to employ a deterministic or, alternatively, a stochastic approximation. An example of the latter approach, called stochastic approximation EM (SAEM), was introduced in Delyon et al. (1999) and implemented in the Monolix software. More specifically, this method replaces the analytically intractable integral in the E-step of the EM algorithm with one iteration of a stochastic approximation procedure.

A different estimation technique known as IGLS, presented in Goldstein (1986) and Davidian and Giltinan (1995), is referred to in Chapter 3 and is used less often in software. The IGLS method is implemented, for example, in the `lme` function from R/**nlme**, and can prove useful for fitting models where the variance depends on the mean.

26.4.3 Tools for Inference and Model Selection

In this section, we present selected tools for inference and model selection, focusing primarily on hypothesis tests. We conclude with a short note on information criteria, and more recent developments.

Although tests of hypotheses for both fixed effects and covariance parameters are often performed using approximate Wald-type tests, we describe them separately due to important differences.

We first discuss hypothesis tests for *fixed effects*. In the LMM case, software procedures will typically automatically compute F- or t-test statistics, approximate degrees of freedom for the underlying null distributions, and corresponding p-values. Software procedures will vary in terms of the approximations used (or available for use) for the degrees of freedom, and some software procedures (e.g., the `lmer` function in R) will not propose any degrees of freedom approximations and p-values will not be reported. Denominator numbers of degrees of freedom for these tests are currently a topic of vigorous debate.

GLMM estimation is also based on maximum likelihood principles, and therefore Wald-type tests for fixed effects can also be performed. A word of caution is needed when quasi-likelihood approximations of the likelihood are used, because the output is based on the results of fitting a working LMM to pseudo-data, not to observed data. We encounter this type of situation, for example, when fitting the models using the `glmmPQL` function from the **MASS** package in R.

Since analytical forms of null distributions for the test statistics are merely approximations in most cases when fitting mixed-effects models, it is often desirable to derive them empirically using bootstrapping, jackknife procedures (for example in MLA), or by applying Bayesian methods (i.e., generating draws from posterior distributions based on a fitted model, for example in the **lme4** R package).

Hypothesis tests for *covariance* parameters can also be performed using Wald tests. In the case where the null hypotheses for covariance parameters do not define the parameters to be on the boundary of a parameter space (e.g., a null hypothesis that the covariance between two random errors is equal to zero), asymptotic likelihood ratio tests for the parameters can be performed in a standard way by referring likelihood ratio test statistics to a χ^2 distribution with an appropriate number of degrees of freedom (typically the difference in the number of parameters between a null model and an alternative model). Difficulties arise, for example, when testing whether variance components are equal to 0 (which places the variance under the null hypothesis on the boundary of its parameter space) and consequently classical testing procedures are no longer valid. Some software procedures will present the asymptotic Wald test (i.e., the estimated variance component divided by its estimated standard error) for covariance parameters automatically; others might report confidence intervals for covariance parameters (see Bottai and Orsini, 2004, for a method implemented in Stata). The HLM software uses a χ^2 test for variance components explained by Bryk and Raudenbush (2002, pp. 63–4). Selected procedures for fitting LMMs, e.g., SAS/PROC GLIMMIX and the `xtmixed` command in Stata, have currently implemented asymptotic likelihood ratio tests for variance components based on mixtures of chi-square distributions (Gutierrez et al., 2001).

The exact null distribution of the likelihood ratio test statistic for a variance component under more general conditions (including small samples) has been defined (Crainiceanu and Ruppert, 2004), and the R/**RLRsim** package (Scheipl, 2010) implements likelihood ratio tests based on simulations from this distribution. Appropriate null distributions of likelihood ratio test statistics for multiple covariance parameters have not been derived to date; classical likelihood ratio tests comparing nested models with multiple variance components constrained to be 0 in the reduced model should be considered conservative. The `xtmixed` command in Stata, for example, makes explicit note of this when users fit models with multiple random effects. MCMC methods based on draws from posterior distributions defined by a given model are generally considered more appropriate for making inferences about covariance parameters; however, readily available software procedures for this type of inference (e.g., the `mcmcsamp` generic function from the **lme4** package in R) are not as common.

Akaike's (AIC) and Bayesian (BIC) information criteria are often included as a part of the default output after fitting models. Also, the **lmmlasso** package (Schelldorfer, 2011) in R allows for model selection using LASSO penalties. Methodology related to information criteria and LASSO penalties is discussed in Chapter 7 of this volume.

26.4.4 Model Diagnostics

Assessment of model diagnostics is more complicated in the case of multilevel models as opposed to a "one-level" model, given that there are more underlying assumptions. Most multilevel software can readily compute empirical Bayesian (EB) estimates of random effects and model-based residuals for assessing simple model diagnostics. Work by Schabenberger (2004) has allowed SAS users to examine influence diagnostics when using SAS/PROC MIXED and SAS/PROC GLIMMIX, but diagnostic procedures still tend to be fairly limited in most software procedures. See the case studies in West et al. (2007) for examples of simpler model diagnostics for LMMs that can be implemented in most current software packages, including plots of residuals and plots of EB estimates. The ability of a given software procedure to generate relevant diagnostic statistics for LMMs is an important consideration for model assessment, and future software developments are certainly needed in this area. See Chapter 24 for further discussion.

26.4.5 Recap

At present, desirable features for multilevel modeling software include hierarchical specifications of the model, supported by graphical representation (e.g., HLM). Efficient estimation algorithms, including accurate approximations of the marginal likelihood in a nonlinear case, need to be implemented. Availability of various structured covariance matrices at different levels of the model hierarchy (both for the random effects and errors), dealing with crossed random effects, finding starting values for variance-covariance parameters, assuring that variance-covariance matrices remain positive-definite during the iteration process, imposing user-defined constraints on variance parameters, allowing for multiple levels of model hierarchy, and utilities for testing for the need for random effects are all features that we feel should be included in "ideal" software for fitting mixed-effects models. Methods for empirically deriving the null distributions of test statistics also need to be more widely implemented in the available software. Finally, graphical facilities for visualizing model fit and diagnostics also tend to be lacking in much of the available software, and we hope that future developments will simplify generation of these types of graphics, especially in the spirit of Cleveland (1993). Several aspects of the multilevel model implementation in various software are compared by Ben Bolker at `http://glmm.wikidot.com/pkg-comparison`.

26.5 MORE ADVANCED MODELS

So far, this chapter has focused on fairly standard examples of multilevel models, namely LMMs, GLMMs, and NLMMs, with random effects being normally distributed. Here, we briefly discuss extensions of these models, going in a variety of directions. When relaxing some of the assumptions underlying these models, the models become even more flexible, and stand-alone and specialized software procedures are more likely to address these advanced issues.

Accounting for complex sample design features

Data sets having hierarchical structures are often produced from large complex samples, where several units of analysis are sampled from randomly sampled clusters of units, and the clusters of units may be stratified in some way. In addition, sampled units may not have equal probabilities of selection in complex multi-stage sample designs, leading to the use of weights to offset the unequal probabilities of selection (and also possibly account for nonresponse and calibration of estimates to known population totals). Pfeffermann et al. (1998), Asparouhov and Muthén (2006), and Rabe-Hesketh and Skrondal (2006) have developed theory for estimating multilevel models in a way that incorporates the weights, and Rabe-Hesketh and Skrondal (2006), Carle (2009), and Heeringa et al. (2010) have presented applications of these approaches using currently available software procedures. See Gelman (2007) for a summary of the issues surrounding the use of weights when fitting multilevel and mixed-effects models. In particular, the Mplus, HLM, and MLwiN software packages allow users to address complex sample design features when fitting a wide class of models. More research in this area is very likely in the near future.

Different types of the dependent variables

Dependent variables measured on an ordinal scale or with censored observations lead to models that are not examples of GLMMs. Examples of software implementing models for these types of dependent variables include MLWin and aML. Also, specific functions, such as the MIXOR and MIXSUR functions incorporated in the Mixed-Up suite or in the Supermix software, can be used.

In the context of NLMMs, such as those discussed in Chapter 14, an important role is being played by models expressed as a solution of ordinary differential equations (ODE). These can be fitted, for example, using the NONMEM software. In the general-purpose statistical package SAS, the NLMEM macro (Gałecki, 1998), which builds on the SAS macro NLINMIX, is a possible choice.

Similarly, the R/**nlmeODE** package (Tornøe et al., 2004; Tornøe, 2010), which builds on the `nlme` function in R/**nlme** package, can be used.

Joint/multi-process modeling

Multi-process or joint modeling refers to the capability of jointly fitting models to more than one dependent variable, measured over time or for clustered data.

Initial approaches to the problem of specifying a joint density of outcomes of the same type go back to Gałecki (1994) and were implemented for LMMs in SAS/PROC MIXED. Examples of joint modeling for outcomes of different types, for example continuous and discrete data, using standard software (e.g., SAS/PROC GLIMMIX and SAS/NLMIXED) are given in Molenberghs and Verbeke (2005). See Chapter 21 for further discussion of data of this type. For joint modeling of longitudinal and time-to-event data the **JM** (Rizopoulos, 2010) package in R can be used. The aML software appears to be especially well-suited for joint (or multi-process) modeling of correlated outcomes of *any* type.

Distributions for random effects

In some cases, the assumption of a normal distribution for the random effects may be too restrictive. Relaxing this assumption can be accomplished in a variety of ways.

In a nonparametric approach, the researcher may leave this (possibly continuous) distribution entirely unspecified. In such a case, the distribution function is estimated with a step function having a finite number of steps (Laird, 1978). The method of finding such a step function is called nonparametric maximum likelihood estimation (NPMLE) or fully semi-parametric estimation, and is implemented in the `gllamm` command in Stata.

Other choices of distributions for the random effects are possible. For example, the random effects might follow a mixture of two or more normal distributions, as presented in Verbeke and Lesaffre (1996). Both Mplus and LatentGold are well suited to fit these types of model.

Another way to relax assumptions about random effects is introduced in Hierarchical Generalized Linear Models, with generalized linear models specified at every level of the model hierarchy. Estimation based on the aforementioned h-likelihood method (see Section 26.2.1) is implemented in the HG system of facilities, called from GENSTAT, and in the R/**hglm** package (Rönnegård et al., 2010).

Another group of extensions aims to build on existing classes of models, and incorporate random effects into them. We discuss examples of these types of models below.

Generalized Additive Mixed-effects Models (GAMM)

This class of models builds on Generalized Additive Models (GAM) and allows flexible functional dependence of an outcome variable on covariates by using nonparametric regression. This class of models can be fitted for example using the R/**amer** package, which extends the range of models fitted by R/**lme4**. Implementation is based on the observation that the penalized least squares problem for smooth terms can be reformulated as a likelihood for mixed-effects models, with the smoothing parameter becoming a variance component (see Chapter 18).

Semi-parametric nonlinear mixed-effects models (SNMM) have also been proposed by Ke and Wang (2001). The function `snm` in the R/**assist** package (Wang and Ke, 2011) can be used for fitting models from this class. Similar models are considered in Chapter 17.

Random effects incorporated in Structural Equation Models (SEM)

These models build on SEMs by incorporating random effects (possibly to account for clustering of observations) and are implemented in programs such as LISREL and Mplus.

Social relations analysis

Analysis of social relations, which takes into account correlations of observations in dyads or groups of subjects, requires specialized software. Examples include the **TripleR** package (Schönbrodt et al., 2012) and StOCNET (Huisman and Van Duijn, 2003). The **TripleR** package is used for Social Relation Model (SRM) analyses for single or multiple round-robin groups. On the other hand, StOCNET is a project that builds on a software system developed for the advanced analysis of social networks and has been written in collaboration between software engineers of Science Plus and other researchers. Data of this type are discussed in Chapter 33.

Sample size considerations

Specialized software for power calculations includes, but it is not limited to, Raudenbush et al. (2004b) and Browne et al. (2009). Sample size calculations are discussed in the context of study design considerations in Chapter 11 of this volume.

26.6 CONCLUSION

The recent trend discussed in this chapter of developing software encompassing a wide range of models, in a unified framework enabling both likelihood-based and Bayesian approaches, will be continued in the future. Likelihood-based approaches are valid for fitting these models under the assumption that missing data are missing at random (MAR), or equivalently that missingness depends on the observed data, but not on the unobserved data. Future developments for mixed-effects models might consider models for data sets with data exhibiting a not missing at random (NMAR) mechanism (Little and Rubin, 1987, 2002).

With more powerful hardware, certain options for fitting these models will become more attractive and will (hopefully) be widely implemented (e.g., empirically deriving null distributions of test statistics and sampling from these distributions). This is especially the case with resampling methods such as the bootstrap, multiple imputation, and MCMC.

Given dynamic developments and the rapidly changing software landscape, we have merely touched on features of the software procedures without going into too much detail. In many places we have provided our personal impressions of the currently available software. This is a wide-ranging topic encompassing many disciplines, and we have indicated areas where we think that additional developments are needed. We take responsibility for any inaccuracies or omissions in this chapter.

ACKNOWLEDGEMENTS

We gratefully acknowledge valuable comments provided by Ben Bolker from McMaster University and Sue Welham from Rothamsted Research. We are thankful to the editors and anonymous reviewers of this chapter for their patience and feedback. We thank Marta Gałecki for editing early drafts of this chapter. We acknowledge support from the Claude Pepper Center Grants AG08808 and AG024824 from the National Institutes of Aging.

REFERENCES

Alam, M., Ronnegard, L., and Shen, X. (2010) *hglm: Hierarchical Generalized Linear Models*. R package version 1.1.1.

Asparouhov, T. and Muthén, B. (2006) 'Multilevel modeling of complex survey data', in *Proceedings of the American Statistical Association*, Seattle, WA: American Statistical Association.

Bates, D., Maechler, M., and Bolker, B. (2011) *lme4: Linear mixed-effects models using S4 classes*. R package version 0.999375-42.

Beal, S. and Sheiner, L. (1980) 'The NONMEM system', *American Statistician*, 34: 118–19.

Beal, S. and Sheiner, L. (1982) 'Estimating population kinetics', *Critical Reviews in Biomedical Engineering*, 8: 195–222.

Beal, S. and Sheiner, L. (2004) *NONMEM User's Guide*.

Bottai, M. and Orsini, N. (2004) 'XTVC: Stata module to compute confidence intervals for the variance component of random-intercept linear models',

Statistical Software Components, Boston College Department of Economics.

Breslow, N. and Clayton, D. (1993) 'Approximate inference in generalized linear mixed models', *Journal of the American Statistical Association*, 88: 9–25.

Browne, W. (2009) *MCMC Estimation in MLwiN, v2.13*. Centre for Multilevel Modeling, University of Bristol.

Browne, W., Golalizadeh Lahi, M., and Parker, R. (2009) *A guide to sample size calculations for random effect models via simulation and the MLPowSim software package*. Centre for Multilevel Modeling, University of Bristol.

Bryk, A. and Raudenbush, S. (2002) *Hierarchical Linear Models: Applications and Data Analysis Methods*. Thousand Oaks, CA: Sage.

Bryk, A., Raudenbush, S., Seltzer, M., and Congdon, R. (1988) *An Introduction to HLM: Computer Program and Users Manual*. Chicago: University of Chicago Dept. of Education.

Busing, F., Meijer, E., and van der Leeden, R. (2005) *MLA: Software for MultiLevel Analysis of Data with Two Levels. User's Guide for Version 4.1*. Universiteit Leiden.

Butler, D., Cullis, B., Gilmour, A., and Gogel, B. (2007) *ASReml-R reference manual*. State of Queensland Department of Primary Industries and Fisheries.

Carle, A. (2009) 'Fitting multilevel models in complex survey data with design weights: Recommendations', *BMC Medical Research Methodology*, 9: 49.

Cleveland, W. (1993) *Visualizing Data*. Summit, NJ: Hobart Press.

Congdon, P. (2005) *Bayesian Statistical Modeling*, 2nd edition. New York: Wiley.

Crainiceanu, C. and Ruppert, D. (2004) 'Likelihood ratio tests in linear mixed models with one variance component', *Journal of the Royal Statistical Society: Series B (Statistical Methodology)*, 66: 165–85.

Davidian, M. and Giltinan, D. (1995) *Nonlinear Models for Repeated Measurement Data*. Boca Raton, FL: Chapman and Hall/CRC.

de Leeuw, J. and Kreft, I. (2001) 'Software for Multilevel Analysis', in A.H. Leyland and H. Goldstein (eds), *Multilevel Modeling of Health Statistics*. New York: Wiley, 187–204.

de Leeuw, J. and Meijer, E. (2008) *Handbook of Multilevel Analysis*. New York: Springer Verlag.

Delyon, B., Lavielle, M., and Moulines, E. (1999) 'Convergence of a Stochastic Approximation Version of the EM Algorithm', *Annals of Statistics*, 27: 94–128.

Dempster, A., Laird, N., and Rubin, D. (1977) 'Maximum Likelihood from Incomplete Data Via the EM Algorithm', *Journal of the Royal Statistical Society. Series B (Methodological)*, 39: 1–38.

Fournier, D. (1993–2011) *An Introduction to AD Model Builder for Use in Nonlinear Modeling and Statistics Version 10.0*.

Gałecki, A. (1994) 'General Class of Covariance Structures for Two or More Repeated Factors in Longitudinal Data Analysis', *Communications in Statistics – Theory and Methods*, 23: 3105–19.

Gałecki, A. (1998) 'NLMEM: A New SAS/IML Macro for Hierarchical Nonlinear Models', *Computer Methods and Programs in Biomedicine*, 55: 207–16.

Gelman, A. (2007) 'Struggles With Survey Weighting and Regression Modeling', *Statistical Science*, 22: 153–64.

Gelman, A. and Hill, J. (2007) *Data Analysis Using Regression and Multilevel/Hierarchical Models*. New York: Cambridge University Press.

George, A. and Liu, J. (1981) *Computer Solution of Large Sparse Positive Definite Systems*. Englewood Cliffs, NJ: Prentice-Hall.

Gilmour, A., Gogel, B., Cullis, B., and Thompson, R. (2009) *ASReml User Guide Release 3.0*.

Goldstein, H. (1986) 'Multilevel Mixed Linear Model Analysis Using Iterative Generalized Least Squares', *Biometrika*, 73: 43–56.

Goldstein, H. (1991) 'Nonlinear Multilevel Models, With an Application to Discrete Response Data', *Biometrika*, 78: 45–51.

Goldstein, H. (1999, 2010) *Multilevel Statistical Models*. New York: Wiley.

Goldstein, H. and Rasbash, J. (1996) 'Improved Approximations for Multilevel Models With Binary Responses', *Journal of the Royal Statistical Society. Series A (Statistics in Society)*, 159: 505–13.

Gutierrez, R., Carter, S., and Drukker, D. (2001) 'On boundary-value likelihood-ratio tests', *Stata Technical Bulletin*, 10(60): 15–18.

Hadfield, J. (2010) 'MCMC Methods for Multi-Response Generalised Linear Mixed Models: The MCMCglmm R Package', *Journal of Statistical Software*, 33(2): 1–22.

Hedeker, D. (1998) *MIXPREG: A Computer Program for Mixed-Effects Poisson Regression*. Chicago: School of Public Health, University of Illinois at Chicago.

Hedeker, D. (1999) 'MIXNO: A Computer Program for Mixed-Effects Nominal Logistic Regression', *Journal of Statistical Software*, 4(5): 1–92.

Hedeker, D. and Gibbons, R. (1996a) 'MIXOR: A Computer Program for Mixed-Effects Ordinal Regression Analysis', *Computer Methods and Programs in Biomedicine*, 49: 157–76.

Hedeker, D. and Gibbons, R. (1996b) 'MIXREG: A Computer Program for Mixed-Effects Regression Analysis With Autocorrelated Errors', *Computer Methods and Programs in Biomedicine*, 49: 229–52.

Hedeker, D., Gibbons, R., du Toit, M., and Cheng, Y. (2008) *SuperMix: Mixed Effects Models*. Scientific Software International.

Heeringa, S., West, B., and Berglund, P. (2010) *Applied Survey Data Analysis*. Boca Raton, FL: Chapman and Hall/CRC.

Huisman, M. and Van Duijn, M. (2003) 'StOCNET: Software for the Statistical Analysis of Social Networks', *Connections*, 25: 7–26.

Jennrich, R. and Schluchter, M. (1986) 'Unbalanced Repeated-Measures Models With Structured Covariance Matrices', *Biometrics*, 42: 805–20.

Jöreskog, K. and Sörbom, D. (2006) *LISREL 8.8 for Windows*.

Ke, C. and Wang, Y. (2001) 'Semiparametric Nonlinear Mixed-Effects Models and Their Applications', *Journal of the American Statistical Association*, 96: 1272–98.

Kreft, I., de Leeuw, J., and van der Leeden, R. (1994) 'Review of Five Different Statistical Packages for Hierarchical Linear Regression: BMDP-5V, GENMOD, HLM, ML3 and VARCL', *American Statistician*, 48: 324–35.

Laird, N. (1978) 'Nonparametric Maximum Likelihood Estimation of a Mixing Distribution', *Journal of the American Statistical Association*, 73: 805–11.

Laird, N. and Ware, J. (1982) 'Random-Effects Models for Longitudinal Data', *Biometrics*, 38: 963–74.

Lavielle, M. (2008) *Monolix Users Guide V.2.4*.

Lee, Y. and Nelder, J. (1996) 'Hierarchical Generalized Linear Models', *Journal of the Royal Statistical Society. Series B (Methodological)*, 58: 619–78.

Lee, Y. and Nelder, J. (2001) 'Hierarchical Generalised Linear Models: A Synthesis of Generalised Linear Models, Random-Effect Models and Structured Dispersions', *Biometrika*, 88: 987–1006.

Lillard, L. and Panis, C. (2003) *aML Multilevel Multiprocess Statistical Software, version 2.0*. Los Angeles, CA: EconWare.

Lindstrom, M. and Bates, D. (1990) 'Nonlinear Mixed Effects Models for Repeated Measures Data', *Biometrics*, 46: 673–87.

Littell, R., Milliken, G., Stroup, W., and Wolfinger, R. (1996) *SAS® System for Mixed Models*. Cary, NC: SAS Institute.

Littell, R., Milliken, G., Stroup, W., Wolfinger, R., and Schabenberger, O. (2006) *SAS® for Mixed Models*, 2nd edition. Cary, NC: SAS Institute.

Little, R. and Rubin, D. (1987, 2002) *Statistical Analysis With Missing Data*. New York: Wiley.

Liu, Q. and Pierce, D. (1994) 'A Note on Gauss-Hermite Quadrature', *Biometrika*, 81: 624–9.

Longford, N. (1990) *VARCL. Software for Variance Component Analysis of Data With Nested Random Effects (Maximum Likelihood)*. Princeton, NJ: Educational Testing Service.

Mauritsen, R. (1990) *EGRET Reference Manual*. Seattle: Statistics and Epidemiology Research Corporation.

Molenberghs, G. and Verbeke, G. (2005) *Models for Discrete Longitudinal Data*. New York: Springer.

Muthén, L. and Muthén, B. (1998–2010) *Mplus: Statistical Analysis With Latent Variables: User's guide*, 6th edition.

Payne, R., Harding, S., Murray, D., Soutar, D., Baird, D., Glaser, A., Channing, I., Welham, S., Gilmour, A., Thompson, R., et al. (2009) *The Guide to GenStat Release 12, Part 2: Statistics*.

Pfeffermann, D., Skinner, C., Holmes, D., Goldstein, H., and Rasbash, J. (1998) 'Weighting for unequal selection probabilities in multilevel models', *Journal of the Royal Statistical Society. Series B (Statistical Methodology)*, 60: 23–40.

Pillai, G., Mentré, F., and Steimer, J. (2005) 'Non-Linear Mixed Effects Modeling – From Methodology and Software Development to Driving Implementation in Drug Development Science', *Journal of Pharmacokinetics and Pharmacodynamics*, 32: 161–83.

Pinheiro, J. and Bates, D. (1995a) 'Approximations to the Log-Likelihood Function in the Nonlinear Mixed-Effects Model', *Journal of Computational and Graphical Statistics*, 4: 12–35.

Pinheiro, J. and Bates, D. (1995b) *lme and nlme: Mixed-Effects Methods and Classes for S and S-Plus*.

Pinheiro, J. and Bates, D. (1996) 'Unconstrained Parameterizations for Variance-Covariance Matrices', *Statistics and Computing*, 6: 289–96.

Pinheiro, J. and Bates, D. (2000) *Mixed-Effects Models in S and S-PLUS*. New York: Springer.

Plummer, M. (2003) 'JAGS: A Program for Analysis of Bayesian Graphical Models Using Gibbs Sampling', in *Proceedings of the 3rd International Workshop on Distributed Statistical Computing (DSC 2003)*, 20–22.

Prosser, R., Rasbash, J., Goldstein, H., and University of London Institute of Education (1991) *ML3, Software for Three-level Analysis: User's Guide*.

R Development Core Team (2010) *R: A Language and Environment for Statistical Computing*. Vienna: R Foundation for Statistical Computing.

Rabe-Hesketh, S., Pickles, A., and Skrondal, A. (2001) 'Gllamm: A Class of Models and a Stata Program', *Multilevel Modeling Newsletter*, 13(1): 17–23.

Rabe-Hesketh, S. and Skrondal, A. (2006) 'Multilevel Modeling of Complex Survey Data', *Journal of the Royal Statistical Society. Series A (Statistics in Society)*, 169: 805–27.

Rabe-Hesketh, S., Skrondal, A., and Pickles, A. (2004) *GLLAMM manual*.

Rasbash, J. (2004) 'Statistical Software Review: Mixed Models in SPSS', *British Journal of Mathematical and Statistical Psychology*, 57: 189–90.

Rasbash, J., Prosser, R., and Goldstein, H. (1989) *ML2: Software for Two-Level Analysis, User's Guide*.

Rasbash, J., Steele, F., Browne, W., and Prosser, B. (2004) *A User's Guide to MLwiN, Version 2*.

Raudenbush, S., Bryk, A., Cheong, Y., and Congdon, R. (2004a) *HLM 6: Hierarchical Linear and Nonlinear Modeling*. Scientific Software International.

Raudenbush, S., Spybrook, J., Liu, X., and Congdon, R. (2004b) *Optimal Design for Longitudinal and Multilevel Research: Documentation for the Optimal Design Software*.

Raudenbush, S., Yang, M., and Yosef, M. (2000) 'Maximum Likelihood for Generalized Linear Models With Nested Random Effects Via High-Order, Multivariate Laplace Approximation', *Journal of Computational and Graphical Statistics*, 9: 141–57.

Rizopoulos, D. (2010) 'JM: An R Package for the Joint Modeling of Longitudinal and Time-to-Event Data', *Journal of Statistical Software*, 35(9), 1–33.

Rodriguez, G. and Goldman, N. (1995) 'An Assessment of Estimation Procedures for Multilevel Models With Binary Responses', *Journal of the Royal Statistical Society. Series A (Statistics in Society)*, 158: 73–89.

Rönnegård, L., Shen, X., and Alam, M. (2010) 'hglm: A Package for Fitting Hierarchical Generalized Linear Models', *R Journal*, 2(2): 20–28.

Rue, H., Martino, S., and Chopin, N. (2009) 'Approximate Bayesian Inference for Latent Gaussian Models by Using Integrated Nested Laplace Approximations', *Journal of the Royal Statistical Society. Series B (Statistical Methodology)*, 71: 319–92.

SAS Institute Inc. (2008) *SAS/STAT® 9.2 Users Guide*.

SAS Institute Inc. (2010) *SAS/STAT® 9.22 Users Guide*.

Schabenberger, O. (2004) 'Mixed Model Influence Diagnostics. Paper 189–29', in *Proceedings of the Twenty-Ninth Annual SAS® Users Group International (SUGI) Conference*. Cary, NC: SAS Institute Inc.

Scheipl, F. (2010) *RLRsim: Exact (Restricted) Likelihood Ratio Tests for Mixed and Additive Models*. R package version 2.0–5.

Schelldorfer, J. (2011) *lmmlasso: Linear Mixed-Effects Models With Lasso*. R package version 0.1–2.

Schluchter, M. (1988) *BMDP 5V: Unbalanced Repeated Measures: Models with Structured Covariance Matrices*. BMDP Statistical Software.

Schönbrodt, F., Back, M., and Schmukle, S. (2012) 'TripleR: An R Package for Social Relations Analyses Based on Round-Robin Designs', *Behavior Research Methods*, 44: to appear.

Sheiner, L., Rosenberg, B., and Melmon, K. (1972) 'Modeling of Individual Pharmacokinetics for Computer-Aided Drug Dosage', *Computers and Biomedical Research*, 5: 441–59.

Skaug, H. and Fournier, D. (2004–2011) *Random Effects in AD Model Builder. ADMB-RE User Guide. Version 10.0*.

Skrondal, A. and Rabe-Hesketh, S. (2004) *Generalized Latent Variable Modeling: Multilevel, Longitudinal, and Structural Equation Models*. Boca Raton, FL: CRC Press.

Spiegelhalter, D., Thomas, A., Best, N., and Gilks, W. (1996a) *BUGS 0.5: Bayesian Inference Using Gibbs Sampling Manual (Version ii)*.

Spiegelhalter, D., Thomas, A., Best, N., and Gilks, W. (1996b) *BUGS 0.5 Examples*, Volume 1.

Spiegelhalter, D., Thomas, A., Best, N., and Gilks, W. (1996c) *BUGS 0.5 Examples*, Volume 2.

SPSS Inc. (2010) *SPSS 19.0 Advanced Models*. Englewood Cliffs, NJ: Prentice Hall.

StataCorp (2009) *Stata Statistical Software: Release 11*. College Station, TX: Stata Corporation.

Stroud, A. and Secrest, D. (1966) *Gaussian Quadrature Formulas*. Upper Saddle River, NJ: Prentice-Hall.

Tierney, L. and Kadane, J. (1986) 'Accurate Approximations for Posterior Moments and Marginal Densities', *Journal of the American Statistical Association*, 81: 82–6.

Tornøe, C., Agersø, H., Jonsson, E., Madsen, H., and Nielsen, H. (2004) 'Non-Linear Mixed-Effects Pharmacokinetic/Pharmacodynamic Modeling in NLME Using Differential Equations', *Computer Methods and Programs in Biomedicine*, 76: 31–40.

Tornøe, C.W. (2010) *nlmeODE: Non-Linear Mixed-Effects Modeling in nlme Using Differential Equations*. R package version 12.2.2.

Venables, W.N. and Ripley, B.D. (2002) *Modern Applied Statistics with S*, 4th edition. New York: Springer.

Verbeke, G. and Lesaffre, E. (1996) 'A Linear Mixed-Effects Model With Heterogeneity in the Random-Effects Population', *Journal of the American Statistical Association*, 91: 217–21.

Vermunt, J. and Magidson, J. (2008) *LG-Syntax User's Guide: Manual for Latent GOLD 4.5 Syntax Module*.

Wang, Y. and Ke, C. (2011) *assist: A Suite of S-Plus Functions Implementing Smoothing Splines*. R package version 3.1.2.

West, B. (2009) 'Analyzing Longitudinal Data With the Linear Mixed Models Procedure in SPSS', *Evaluation and the Health Professions*, 32: 207–28.

West, B. and Gałecki, A. (2011) 'An Overview of Current Software Procedures for Fitting Linear Mixed Models', *The American Statistician*, 65: 274–82.

West, B., Welch, K., and Gałecki, A.T., with contributions from Gillespie B. (2007) *Linear Mixed Models: A Practical Guide Using Statistical Software*. Boca Raton, FL: Chapman and Hall/CRC.

Wolfinger, R. (1999) 'Fitting Nonlinear Mixed Models With the New NLMIXED Procedure', in *Proceedings of the 24th Annual SAS Users Group International Conference (SUGI 24)*, 1–10.

Wolfinger, R. and O'Connell, M. (1993) 'Generalized Linear Mixed Models: A Pseudo-Likelihood Approach', *Journal of Statistical Computation and Simulation*, 48: 233–43.

PART IV

Selected Applications

27

Meta-Analysis

Larry V. Hedges
Northwestern University, USA
Kimberly S. Maier
Michigan State University, USA

27.1 INTRODUCTION

In many fields, the growth of the empirical research enterprise has led to an accumulation of empirical research findings that must be organized and synthesized to clearly identify what is known, and to identify critical areas where research has been insufficient, inconclusive, or even contradictory. Examples of areas in which quantitative syntheses of research have become routine include physics (e.g., Birge, 1932 or Rosenfeld, 1975), agriculture (e.g., Cochran, 1937), and chemistry (e.g., Zwolinski and Chao, 1972). Other sciences such as medicine (where an early example, Pearson, 1904, predates routine use in that science by more than 60 years), ecology (e.g., Gurevitch and Hedges, 1993), and software engineering (e.g., Kampenes, Dybå, Hannay, and Sjøberg, 2007) have embraced quantitative synthesis of research more recently. The contemporary term "meta-analysis" describes a more longstanding activity of the use of statistical methods to combine the results of analyses in studies for the purposes of drawing general conclusions. Glass coined the term meta-analysis in 1976 in the context of combining research results across studies in educational psychology (Glass, 1976). Since that time, statistical methods to combine information from different research studies has been extensively used in education, psychology, and other social sciences (see Cooper, Hedges, and Valentine, 2009).

Meta-analysis uses a numerical index that describes the results of a study, the effect size (forms include a correlation coefficient, a standardized mean difference, or an odds ratio). Each study provides an estimate of the effect size and an uncertainty (standard error) of that estimate. The estimates are then combined across studies to obtain a summary. The specific analytic model used to combine the effect size estimates will depend on the question the meta-analytic summary is intended to address. Sometimes the question of interest concerns the typical or average study result. For example, in studies that measure the effect of some treatment or intervention, the average effect of the treatment is often of interest (see, e.g., Smith and Glass, 1977). In other cases, the degree of variation in results across studies will be of primary interest. For instance, meta-analysis is often used to study the generalizability of employment test validities across situations (see, e.g., Schmidt and Hunter, 1977). In yet other cases, the primary interest is in the factors that are related to study results. For example, meta-analysis is often used to identify the contexts in which a treatment or intervention is most successful or has the largest effect (see, e.g., Cooper, 1989).

For any of these kinds of question, the natural way to describe the data is via a two-level hierarchical model with a level-1 model for within-study (estimation error) variation and a level-2 model for between-study variation in effects. Hierarchical models (see Raudenbush and Bryk, 2002, for an introduction) allow the natural hierarchical structure of data to be formally modeled. Also known as a mixed or random coefficient model, a hierarchical linear model consists of at least two levels of hierarchy. The first level of the hierarchy models the effect within a single group or context (in meta-analysis the level-1 unit is a study). Each of the coefficients of the first level are modeled at the second level with an intercept and, if appropriate, additional covariates. Usually, at least the intercept is assumed to vary across studies, and one or more of the coefficients may also have corresponding random components.

Sometimes the term meta-analysis is used to connote the entire process of quantitative research synthesis. More recently, the term has also been used to refer specifically to the statistical component of research synthesis. This chapter deals exclusively with the more narrow meaning of the term, as a description of statistical methods only. However, it is crucial to understand that in research synthesis, as in any research, statistical methods are only one part of the enterprise. Contemporary treatments of research synthesis (often called systematic reviewing in fields such as medicine and public health) emphasize that systematic reviews include many other parts (including problem formulation, data collection, data evaluation, and presentation of results), all of which have implications for the validity of the review. The data analysis component is important, but cannot alone assure the validity of conclusions. Treatments of the non-statistical aspects of research synthesis are available in Cooper (2010); Cooper, Hedges, and Valentine (2009); and Lipsey and Wilson (2001).

27.2 EFFECT SIZES

Effect sizes are numerical indexes used to summarize the results of the studies in a meta-analysis. Effect sizes reflect the magnitude of the association between the independent and dependent variables of interest in each study. There are many different measures of effect size, and choosing the appropriate effect size often requires considerable thought. In general, the choice of an effect size is governed by two, often conflicting, goals. On the one hand, the effect size should be as "natural" for each study as possible, involving as little departure from the metric of the original data analysis as possible. On the other hand, since many studies will be combined, the effect size should also be interpretable in the same way in every study. An index of effect size that meets these two goals represents the results of a study in a way that is both easily interpretable and comparable across studies. Ideally, effect sizes should express the results of all studies "using a common metric" so that they can be readily interpreted, compared, and combined.

The choice of an effect size index will depend on the design of the studies, the way in which the outcome is measured, and the statistical analysis used in each study. One consideration that often affects the choice of effect size indexes in the social sciences is that many outcome measures (e.g., mental test scores, attitude measures, measures of psychological functioning) do not have natural scales of measurement. Moreover, there are many alternative measures of the same construct, so that most research syntheses will involve studies using a variety of different measurement instruments that are not on the same numerical scale. Therefore, some form of standardization is often relied upon when creating effect size measures so that they are more comparable across studies. It is important to recognize that there is nothing inherent in meta-analysis that requires or privileges standardized effect sizes. Standardized measures are widely used simply because they are needed to create effect size indices that are more comparable across studies.

We distinguish the effect size estimate from the effect size parameter (the true effect size) in a study. In principle, the effect size estimate will vary somewhat between the possible samples that might be obtained in a particular study. The effect size parameter for a particular study is, in principle, a fixed quantity. The effect size parameter might be conceived as the estimate that would be obtained if the study had a very large (essentially infinite) sample, so that the estimation error is negligible.

Effect size estimates used in meta-analysis typically have approximately normal distributions about their respective effect size parameters. The conditional variance of the effect size estimate (the variance of the estimate given the corresponding parameter) is typically a function largely of sample sizes and the effect size. Often the dependence of the variance on the effect size is slight and consequently the conditional variances are taken to be known. In these instances, essentially the same methods for meta-analysis can be used regardless of which effect size measure is chosen. For example, an approximate $100\,(1-\alpha)$ percent confidence interval for the effect size parameter θ based on an estimate T with conditional variance σ^2 is

$$T - z_{\alpha/2}\sigma \leq \theta \leq T + z_{\alpha/2}\sigma,$$

where $z_{\alpha/2}$ is the $100\,(1-\alpha)$ percentile of the standard normal distribution, regardless of which effect size index T and θ represent.

The effect size indices most commonly used in the social sciences will fall into one of three families of effect sizes: the standardized mean difference family, the odds ratio family, and the standardized regression coefficient family (of which the correlation coefficient, the standardized regression coefficient with a single predictor, is by far the most widely used).

27.3 THE STANDARDIZED MEAN DIFFERENCE

In many studies of the effects of a treatment or intervention that measure the outcome on a continuous scale, a natural effect size is the standardized mean difference. The standardized mean difference is the difference between the mean outcome in the treatment group and the mean outcome in the control group divided by the within-group standard deviation. The standardized mean difference parameter δ is

$$\delta = \frac{\mu^T - \mu^C}{\sigma},$$

where μ^T is the population mean in the treatment group, μ^C is the population mean outcome in the control group, and σ is the population within-group standard deviation of the outcome. This effect size is easy to interpret since it is just the treatment effect in standard deviation units. It can also be interpreted as having the same meaning across studies (see Hedges and Olkin, 1985).

The sample estimate of the standardized mean difference d is

$$d = \frac{\bar{Y}^T - \bar{Y}^C}{S},$$

where $\bar{Y}^T$ is the sample mean of the outcome in the treatment group, $\bar{Y}^C$ is the sample mean of the outcome in the control group, and S is the within-group standard deviation of the outcome. Sometimes d is modified slightly to adjust for the small sample bias of d (Hedges, 1981), namely $g = dJ\left(n^T + n^C - 2\right)$, where $J(m) = 1 - 3/(4m-3)$.

The usual estimate of the conditional variance (the variance of d given δ) is

$$v = \frac{n^T + n^C}{n^T n^C} + \frac{d^2}{2\left(n^T + n^C\right)},$$

where n^T and n^C are the treatment and control group sample sizes, respectively. Note that this variance can be computed from a single effect size estimate if the sample sizes of the two groups within a study are known.

Several variations of the standardized mean difference are also sometimes used as effect sizes (see Borenstein, 2009).

27.4 THE LOG ODDS RATIO

In many studies of the effects of a treatment or intervention that measure the outcome on a dichotomous scale, a natural effect size is the log odds ratio; this is readily interpretable by researchers who frequently encounter this kind of data. This effect size measure also has the same meaning across studies, making it suitable for combining (see Fleiss and Berlin, 2009).

The log odds ratio is just the log of the ratio of the odds of a particular one of the two possible outcomes (the target outcome) in the treatment group to the odds of that particular outcome in the control group. That is, the log odds ratio parameter is

$$\begin{aligned}\omega &= \log\left(\frac{\pi^T/(1-\pi^T)}{\pi^C/(1-\pi^C)}\right)\\ &= \log\left(\frac{\pi^T(1-\pi^C)}{\pi^C(1-\pi^T)}\right),\end{aligned}$$

where π^T and π^C are the population proportions in the treatment and control groups, respectively, that have the target outcome. The usual sample estimate of the log odds ratio is

$$\begin{aligned}\log(OR) &= \log\left(\frac{\hat{\pi}^T/\left(1-\hat{\pi}^T\right)}{\hat{\pi}^C/\left(1-\hat{\pi}^C\right)}\right)\\ &= \log\left(\frac{\hat{\pi}^T\left(1-\hat{\pi}^C\right)}{\hat{\pi}^C\left(1-\hat{\pi}^T\right)}\right),\end{aligned}$$

where $\hat{\pi}^T$ and $\hat{\pi}^C$ are the sample proportions of the treatment and control groups, respectively, that have the target outcome.

The estimate of the conditional variance of the log odds ratio sample estimate is

$$\begin{aligned}v = {} & \frac{1}{n^T\hat{\pi}^T} + \frac{1}{n^T\left(1-\hat{\pi}^T\right)} + \frac{1}{n^C\hat{\pi}^C}\\ & + \frac{1}{n^C\left(1-\hat{\pi}^C\right)},\end{aligned}$$

where n^T and n^C are the treatment and control group sample sizes, respectively.

There are several other indexes in the odds ratio family, including the risk ratio (the ratio of proportion having the target outcome in the treatment group to that in the control group, or π^T/π^C) and the risk difference (the difference between the proportion having a particular one of the two possible outcomes in the treatment group and that in the control group, $\pi^T - \pi^C$). For a discussion of effect size measures for studies with dichotomous outcomes, including the odds ratio family of effect sizes, see Fleiss and Berlin (2009).

27.5 THE CORRELATION COEFFICIENT

The correlation coefficient has a long history as a measure of association between variables, but in this context it is useful to recognize that the correlation coefficient is the standardized regression coefficient when there is a single predictor variable. Thus, the correlation coefficient is the number of standard deviations of change in the dependent variable that are associated with a one standard deviation change in the independent variable. In many studies of the relation between two continuous variables, the correlation coefficient is a natural measure of effect size.

Let ρ be the correlation parameter and let r be its sample estimate. Often, the statistical analysis is carried out in the metric of the Fisher z-transform of the correlations rather than in the metric of the correlations, so that the effect size parameter is

$$\zeta = \frac{1}{2}\log\left(\frac{1+\rho}{1-\rho}\right)$$

and the sample estimate of ζ is

$$z = \frac{1}{2} \log \left(\frac{1+r}{1-r} \right).$$

The sampling uncertainty of the z-transformed correlation z is characterized by its (approximate) variance

$$v = \frac{1}{n-3},$$

where n is the sample size of the study.

27.5.1 *Estimating the Average Effect Across Studies*

Consider now the principal consideration of meta-analysis that was raised above, namely combining the effect size estimates across studies to estimate the average effect size.

First, consider the simplest meta-analytic problem: estimating a common effect size across studies. We consider this problem in detail to demonstrate how meta-analytic approaches, which are often not explicitly described in terms of multilevel statistical models, can be addressed using the multilevel modeling framework. Please note that some of the following notation differs from that defined earlier in this book in order to follow conventions used in meta-analysis.

Let θ_i be the (unobserved) effect size parameter in the ith study, let T_i be the corresponding observed effect size estimate, and let v_i be its conditional variance. Thus, the data from a set of k studies are the effect size estimates $T_1, \ldots, T_k$ and their variances $v_1, \ldots, v_k$.

In the level-one (within-study) model, the effect size estimate T_i is just the effect size parameter plus an estimation error ε_i:

$$T_i = \theta_i + \varepsilon_i, \; i = 1, \ldots, k \;\; \varepsilon_i \sim N(0, v_i). \qquad (27.1)$$

In the level-two (between-study) model, the effect size parameters are modeled with a mean effect size β_0 (a fixed effect) plus a study-specific random effect η_i. That is,

$$\theta_i = \beta_0 + \eta_i, \; i = 1, \ldots, k \eta_i \sim N(0, \tau^2). \qquad (27.2)$$

Here, the η_i represent differences between the effect size parameters from study to study. The parameter τ^2, often called the between-studies variance component, describes the amount of variation across studies in the random effects (η_i), and therefore in the effect size parameters (θ_i).

Note that this meta-analytic model is very similar to the hierarchical linear model often used in the primary analysis of social science data. However, it differs from that model in three ways. First, in the usual model, the level-1 variance is identical across level-1 units. In the meta-analytic model, the level-1 variances (v_i) are different for each of the level-1 units (in this case, studies). That is, each study has a *different* estimation error variance at level 1. Second, in the usual model, the level-1 variance is unknown and must be estimated from the data. In the meta-analytic model, the level-1 variances, which differ across the studies, are known. Third, there is only one level-1 observation within each level-2 unit.

This model, with level-1 variances constrained to known values, can be easily estimated using multilevel modeling software, which will provide estimates of β_0, its standard error, and τ^2, along with a test of its statistical significance. Constraining the level-1 variances is done somewhat differently in different software. For example, it is accomplished via the PARMS/EQCONS option in SAS Proc MIXED (a useful exposition about the use of SAS Proc MIXED in this context is given in Sheu and Suzuki, 2001). The maximum likelihood estimate of β_0 is a weighted linear combination of the T_i with weights inversely proportional to the (unconditional) variance, $v_i + \hat{\tau}^2$.

The model (27.1)–(27.2) is often called the random effects model in meta-analysis because the effect size parameters (the θ_i) are random effects. A variant of this model would posit that the effect size parameters are identical across all k studies, that is $\theta_1 = \cdots = \theta_k$, or, equivalently, $\tau^2 = 0$. This is often called the fixed effects (or constant effect) model in meta-analysis because, in this model, the effect size parameters are fixed effects. The question of whether the fixed or the random effects model is more appropriate is often difficult to answer on a purely empirical basis, but analysis of the heterogeneity of effect sizes is often informative.

27.6 CHARACTERIZING HETEROGENEITY OF EFFECT SIZES

In multilevel models, estimates of the variance component τ^2 using maximum likelihood estimation or restricted maximum likelihood estimation are standard features of the analysis. The estimate of τ^2 (or its square root) is a pure measure of the "amount" of heterogeneity of effect sizes across studies. However, τ^2 is often not very intuitive to users of meta-analysis. Consequently, there have been attempts to provide indexes that characterize the magnitude of τ^2 compared to the estimation variances. Perhaps the most widely used is the I^2 index,

$$I^2 = \left(\frac{\tau^2}{v + \tau^2}\right) \times 100\%,$$

where v is an "average" of the v_i (Higgins and Thompson, 2002). This index was inspired by the idea of intraclass correlation (the ratio of between-studies variance to total variance of effect estimates) which, in the multilevel model context, Raudenbush and Bryk (2002) characterize as the reliability of the individual effects. One difference between this reliability coefficient and I^2 is that the latter uses what might appear to be an unusual average v, namely

$$v = \frac{(k-1)\sum w_i}{\left(\sum w_i\right)^2 - \sum w_i^2},$$

where $w_i = 1/v_i$, which allows an important connection to be drawn between I^2 and the (likelihood ratio) test statistic Q used to test the hypothesis that $\tau^2 = 0$.[1] With this definition of v, it can be shown that $I^2 = 100\% \times [Q - (k-1)]/Q$.

In the multilevel context, tests (e.g., a likelihood ratio test) of the hypothesis that the variance component $\tau^2 > 0$ arise naturally. A simple test (which turns out to be equivalent to the likelihood ratio test under this model) of the hypothesis that $\tau^2 = 0$ uses the weighted sum of squares about the weighted mean that would be obtained if $\tau^2 = 0$. Specifically, it uses the statistic

$$Q = \sum_{i=1}^{k} (T_i - \hat{\beta}_0)^2 / v_i,$$

where $\hat{\beta}_0$ is the maximum likelihood estimate of β_0 that would be obtained if $\tau^2 = 0$ (the weighted mean using inverse conditional variance weights). The statistic Q has the chi-squared distribution with $(k-1)$ degrees of freedom if $\tau^2 = 0$. Therefore, a test of the null hypothesis that $\tau^2 = 0$ at significance level α rejects the hypothesis if Q exceeds the $100\,(1-\alpha)$ percent point of the chi-square distribution with $(k-1)$ degrees of freedom.

This (or any other statistical hypothesis test) should not be interpreted too literally. The test is not very powerful if the number of studies is small or if the estimation variances (the v_i) are large (see Hedges and Pigott, 2001). Consequently, even if the test does not reject the hypothesis that $\tau^2 = 0$, the actual variation in effects across studies may be consistent with a substantial range of nonzero values of τ^2, some of which may be rather large. Even when the maximum likelihood estimate of τ^2 is equal to zero, there may still be many nonzero values of τ^2 that are quite compatible with the effect size data (see Raudenbush and Bryk, 1985). This has led many investigators to consider the use of Bayesian estimators that compute the average of the entire meta-analysis over a range of plausible values of τ^2 (see, e.g., Hedges, 1998).

27.7 EXAMPLE

Raudenbush (1984) carried out a review of 19 experimental studies of the effects of teacher expectancy on pupil IQ. The results of these 19 studies are given in Table 27.1. Contact is estimated weeks of teacher–student contact prior to expectancy induction. These data are from Raudenbush and Bryk (1985), but see also Raudenbush (1984). The standardized mean difference and standard error of that estimate are given in the table.

Computing the test statistic Q, we obtain a value of $Q = 35.83$ with $19 - 1 = 18$ degrees of freedom, which is significant at the $p = 0.01$ level of significance. An analysis using SAS Proc MIXED, estimating τ^2 via restricted maximum likelihood, yields $\hat{\tau}^2 = 0.018$ and the estimate of the average effect size of $\hat{\beta}_0 = 0.08$. The standard error of the estimate is 0.052, which is less than two standard errors greater than zero, so the hypothesis that the average effect size is different from 0 cannot be rejected at the $\alpha = 0.05$ level of significance. We note that, in the fixed effects analysis, constraining τ^2 to be zero (which the significance test based on the Q statistic tells us is quite inconsistent with these data), the estimate of the average effect size would have been $\hat{\beta}_0 = 0.06$, with a standard error of 0.036, and again the average effect size is not significantly different from zero.

27.8 THE META-REGRESSION MODEL

One of the fundamental issues facing meta-analysts is how to model the association between characteristics of studies and their effect sizes. One widely used strategy involves the use of a hierarchical linear model.

Suppose that we have the same level-1 (within-study) model as before, namely

$$T_i = \theta_i + \varepsilon_i, \; i = 1, \ldots, k \quad \varepsilon_i \sim N(0, v_i), \tag{27.3}$$

where the v_i are known.

Suppose now that there are p known predictor variables for the fixed effects, $X_1, \ldots, X_p$, and assume that they are related to the effect sizes via a linear model. In this case the level-2 model for the ith effect size parameter becomes

$$\theta_i = \beta_0 + \beta_1 x_{i1} + \beta_2 x_{i2} + \cdots + \beta_p x_{ip} + \eta_i, i = 1, \ldots, k, \tag{27.4}$$

where $x_{i1}, \ldots, x_{ip}$ are the values of the predictor variables $X_1, \ldots, X_p$ for the ith study (that is, x_{ij} is the value of predictor variable X_j for study i), the regression coefficients $\beta_0, \ldots, \beta_p$ are fixed effects, and the η_i are mutually independent study-specific random effects, normally distributed with zero expectation and variance τ^2. Note that the variance τ^2 is the conditional variance of the effect size parameters given the covariates.

This model, with the level-1 variances constrained to known values, can easily be estimated using multilevel modeling software, which will provide an estimate of $\boldsymbol{\beta} = (\beta_0, \ldots, \beta_p)'$, the covariance matrix of the estimates, and an estimate of τ^2 along with a test of its statistical significance. The maximum likelihood estimate of $\boldsymbol{\beta}$ is the weighted least squares estimate of $\boldsymbol{\beta}$ with weights inversely proportional to the (unconditional) variance, $v_i + \hat{\tau}^2$, where $\hat{\tau}^2$ is the maximum likelihood estimate of τ^2. The restricted maximum likelihood estimate of τ^2 is often used in place of the unrestricted (full) maximum likelihood estimate to improve estimator properties in small samples.

The model (27.3)–(27.4) is often called the random effects meta-regression model in meta-analysis because the effect size parameters (the θ_i) are random effects (although it would be more appropriate to call it a mixed model, because $\beta_0, \ldots, \beta_p$ are fixed effects). A variant of this model would posit that all of the effect size parameters are completely determined by the covariates, so that $\theta_i = \beta_0 + \beta_1 x_{i1} + \beta_2 x_{i2} + \cdots + \beta_p x_{ip}$, or, equivalently, $\tau^2 = 0$. This is often called the fixed effects meta-regression model because the effect size parameters are fixed effects.

Table 27.1 Studies of the Effects of Teacher Expectancy on Pupil IQ

Study	Contact	d	se	Study	Contact	d	se
1	2	.03	.125	11	0	.54	.302
2	21	.12	.147	12	0	.18	.223
3	19	−.14	.167	13	1	−.02	.289
4	0	1.18	.373	14	2	.23	.290
5	0	.26	.369	15	17	−.18	.159
6	3	−.06	.103	16	5	−.06	.167
7	17	−.02	.103	17	1	.30	.139
8	24	−.32	.220	18	2	.07	.094
9	0	.27	.164	19	7	−.07	.174
10	1	.80	.251				

27.9 EXAMPLE

Returning to the data from Raudenbush (1984), we consider a study-level covariate that may explain some of the variation in effects. The teacher expectancy manipulation depends on its credibility to the teachers involved, and it was hypothesized that teachers who had more prior contact with students would find the teacher expectancy manipulation less credible, and it would therefore have a smaller effect. To examine the evidence for this hypothesis, the covariate X was assigned the value 0, 1, 2, or 3, depending on whether the teachers in the study had 0, 1, 2, or 3 (or more) weeks of prior contact with the students.

Fitting the level-2 model $\theta_i = \beta_0 + \beta_1 x_i + \eta_i$, using SAS Proc MIXED for the analysis, and estimating τ^2 via restricted maximum likelihood yields $\hat{\tau}^2 = 0.000$ $(p > 0.50)$. The estimate of the intercept was $\hat{\beta}_0 = 0.41$ (with a standard error of 0.087) and the estimate of the slope coefficient was $\hat{\beta}_1 = -0.16$ (with a standard error of 0.036), so that both coefficients are significantly different from zero at the $\alpha = 0.05$ level of significance. Therefore, the analysis suggests that there is a statistically reliable teacher expectancy effect if the expectancy manipulation is used with teachers who have had no prior contact with the students involved, but that the effect is substantially reduced as a function of the teachers' prior contact with the students involved.

27.10 ROBUST ESTIMATION OF STANDARD ERRORS OF FIXED EFFECTS

In situations where the number of studies is large and there is concern that the distribution of the between-study random effects may be highly non-normal, the use of robust estimates of the standard errors of the fixed effects has sometimes been suggested (see, e.g., Raudenbush, 2009). These robust estimates usually use the assumption of normally distributed between-study random effects to compute variance component estimates used in the weights which, in turn, are used to compute the fixed effects. The standard errors of the fixed effect estimates are computed from the square roots of the diagonal elements of the robust estimate of the covariance matrix

$$\frac{k}{k-p}\left(\sum_{i=1}^{k} w_i \mathbf{x}_i \mathbf{x}_i'\right)^{-1}\left(\sum_{j=1}^{k} w_i^2 e_i^2 \mathbf{x}_i \mathbf{x}_i'\right) \cdot \left(\sum_{i=1}^{k} w_i \mathbf{x}_i \mathbf{x}_i'\right)^{-1},$$

where $w_i = 1/(v_i + \hat{\tau}^2)$ is the weight, $\mathbf{x}_i = (x_{i1}, \ldots, x_{ip})'$ is the vector of covariate values, and $e_i = T_i - \mathbf{x}'_i \hat{\boldsymbol{\beta}}$ is the residual associated with the ith study (see, e.g., Raudenbush, 2009). Although the theory justifying robust estimation of the standard errors of the fixed effects depends on a large sample size (here, a large number of effect size estimates), there is some limited evidence from simulation studies that the method can work well in meta-analysis with modest sample sizes (see Hedges, Tipton, and Johnson, 2010).

27.11 META-ANALYTIC MODELS INVOLVING DEPENDENCE AMONG EFFECT SIZE ESTIMATES

The simplest statistical models for meta-analysis assume that all of the effect size estimates are statistically independent. These models are appropriate when there is a single effect size (and effect size estimate) per study and the studies are independent. The assumption of independent estimates can be violated in ways that can lead to dependency among the (random) effect size parameters themselves or (if there are multiple estimates per study) among the estimation errors, even if the effect size parameters remain independent.

It is possible that groups of studies have been conducted in the same laboratory, by the same investigators, or reported in the same papers. If the effect size parameters generated by the same laboratory, investigator, or research report are more alike than those generated by different laboratories, investigators, or research reports, the effect size parameters will not be independent. In meta-analysis, this phenomenon has been referred to as hierarchical dependence (see Gurevitch and Hedges, 1999, or Stevens and Taylor, 2009).

Alternatively, it is possible that effect size parameters and estimates that arise from the group of studies are independent, but there is more than one effect size estimate per study. This may occur if more than one measure of the outcome variable is given to each individual. Alternatively, it may occur if there is one outcome measure and several independent treatment groups, and effect sizes are generated by comparing each treatment group with a common control group. In either case, the effect size estimates generated from the same study are not independent. In this case, however, the dependence occurs through correlations among the estimation errors.

27.12 THE HIERARCHICAL DEPENDENCE MODEL

Consider a situation in which there are $n = k_1 + k_2 + \cdots + k_m$ effect size parameters and estimates are obtained within m clusters of estimates (e.g., estimates from the same groups of studies), with $k_i \geq 1$ estimates within the ith cluster for $j = 1, \ldots, m$. Denote the jth effect size parameters, estimates, and estimation errors, respectively, in the ith cluster by θ_{ij}, T_{ij}, and v_{ij}. Then it is natural to consider this data in terms of a three-level hierarchical model where the level-1 model is

$$T_{ij} = \theta_{ij} + \varepsilon_{ij},\ i = 1, \ldots m;\ j = 1, \ldots, k_i,$$

where $\varepsilon_{ij} \sim N\left(0, v_{ij}\right)$ and the v_{ij} are known. The level-2 model might be

$$\theta_{ij} = \gamma_i + \eta_{ij},\ i = 1, \ldots, m;\ j = 1, \ldots, k_i,$$

where $\eta_{ij} \sim N\left(0, \tau_i^2\right)$, so that the γ_i represent the mean effect size for the ith cluster of studies and η_{ij} is a random effect that represents the difference between the effect size in the jth study and the mean effect size in the ith cluster, and τ_i^2 is the between-studies (within-cluster) variance of effect sizes in the ith cluster. The τ_i^2 are often constrained to have the same value across clusters, but this is not essential if there is a sufficiently large dataset to support estimation of separate values for the τ_i^2. The level-3 model might be

$$\gamma_i = \beta_0 + \xi_i,\ i = 1, \ldots, m, \xi \sim N(0, \omega^2),$$

so that the fixed effect β_0 is the overall mean effect size, and ξ_i is a random effect that represents the difference between the mean effect size in the ith cluster and the overall mean effect size, and ω^2 is the between-cluster variance of cluster mean effect sizes.

In hierarchical dependence models, attention often focuses on the overall average effect size β_0, and the multilevel modeling primarily serves to obtain correct standard errors for estimates of β_0 and estimates of within- and between-cluster variation in effect sizes. However, variations on the hierarchical dependence model might include covariates in level-2 or level-3 models to represent important between-study or between-cluster differences.

27.13 MODELS WITH CORRELATED ESTIMATION ERRORS

The classic multivariate meta-analysis model is one in which there are $k = mq$ effect size parameters and estimates obtained within m independent studies (e.g., estimates from the same groups of studies), with $q \geq 1$ estimates within each study. Denote the jth effect size parameters and estimates, respectively, in the ith study by θ_{ij} and T_{ij}, and the vector of effect size parameters and estimates in the ith study by $\boldsymbol{\theta}_i = (\theta_{i1}, \ldots, \theta_{iq})'$and $\mathbf{T}_i = (T_{i1}, \ldots, T_{iq})'$. Denote the $q \times q$ (conditional) covariance matrix of T_i given θ_i (the estimation error covariance matrix) by Σ_i. The classical multivariate meta-analysis model assumes that the estimation error covariance matrices (the Σ_i) are known, which is the natural generalization of the assumption that the estimation error variances (the v_i) are known in the univariate meta-analysis model (see Jackson, Riley, and White, 2011).

It is natural to consider a two-level hierarchical model in which the level-1 model is

$$\mathbf{T}_i{=}\boldsymbol{\theta}_i + \boldsymbol{\varepsilon}_i,\ i = 1, \ldots, m\ \boldsymbol{\varepsilon}_i \sim N(\boldsymbol{\theta}_i, \Sigma_i),$$

where $\boldsymbol{\varepsilon}$ is a vector of estimation errors and the estimation error covariance matrix Σ_i is known. A natural level-2 meta-regression model might be

$$\boldsymbol{\theta}_i = X_i\boldsymbol{\beta} + \boldsymbol{\eta}_i,\ i = 1, \ldots, m \boldsymbol{\eta}_i \sim N(\mathbf{0}, \mathrm{T}),$$

where X_i is a $q \times p$ design submatrix containing the covariate values for the estimates in the ith study:

$$X_i = \begin{pmatrix} x_{i11} & \cdots & x_{i1p} \\ \vdots & \ddots & \vdots \\ x_{iq1} & \cdots & x_{iqp} \end{pmatrix},$$

x_{ija} is the value of the ath covariate for the jth effect size in the ith study, $\boldsymbol{\beta} = (\beta_1, \ldots, \beta_p)'$ is a vector of regression coefficients (fixed effects), $\boldsymbol{\eta} = (\eta_1, \ldots, \eta_q)'$ is a vector of study-specific random effects, and T is a $q \times q$ matrix of covariance components.

Because the estimation error covariance matrix Σ_i is known, it is possible to transform the within-study model so that the estimation errors are independent. For each study, perform the Cholesky factorization of each estimation error covariance matrix Σ_i so that $\Sigma_i = \mathrm{F}_i\mathrm{F}_i'$, where F_i is a known matrix (since Σ_i is a known matrix). Define the transformed effect size vector $\mathbf{Z}_i$ by

$$\mathbf{Z}_i \equiv \mathrm{F}_i^{-1}\mathbf{T}_i$$

and the transformed estimation error vector by $\tilde{\boldsymbol{\varepsilon}}_i \equiv \mathrm{F}_i^{-1}\boldsymbol{\varepsilon}_i$, so that the covariance matrix of $\tilde{\boldsymbol{\varepsilon}}_i$ is a $q \times q$ identity matrix. The new within-study (level-1) model becomes

$$\mathbf{Z}_i = \mathrm{F}_i^{-1}\mathbf{T}_i = \mathrm{F}_i^{-1}\boldsymbol{\theta}_i + \mathrm{F}_i^{-1}\boldsymbol{\varepsilon}_i = \mathrm{F}_i^{-1}\boldsymbol{\theta}_i + \tilde{\boldsymbol{\varepsilon}}_i,$$

where the transformed effect size estimates $\mathbf{Z}_i$ are independent, but the effect size parameter vector $\boldsymbol{\theta}_i$ is the same as in the original model.

Thus, the within-study model (level-1) along with the between-study (level-2) model is a conventional two-level hierarchical linear model with independent estimation errors (and unit variances) at the first level. Therefore, conventional software can be used to estimate $\boldsymbol{\beta}$ and $\mathbf{T}$, constraining the level-1 variances to unity (see Kalaian and Raudenbush, 1996).

Unfortunately, implementing this procedure requires knowledge of the covariance structure of the effect size estimation errors (which in turn requires knowledge of the dependence structure of the raw data). While the variances of the estimates are typically known, the covariances are not, so the information needed to implement this strategy is rarely available. Moreover, unless the number of studies is small, these methods can be cumbersome to use, even if the estimation error covariances *are* known. For both of these reasons, this strategy is rarely used.

A strategy that does not require the availability of the within-study covariance structure of the estimates is based on robust estimation of the variance of the fixed effects, as introduced by Hedges, Tipton, and Johnson (2010). The ideas are an extension of methods proposed by Huber (1967) and White (1980). This strategy makes it easier to accommodate a model in which different studies have different numbers of effect size estimates. To illustrate that point, suppose that the sample of effect size estimates consists of $k = q_1 + \cdots + q_m$ estimates, with $q_i \geq 1$ estimates from the ith study for $i = 1, \ldots, m$. The vector of fixed effects $\boldsymbol{\beta}$ is estimated via weighted least squares:

$$\hat{\boldsymbol{\beta}} = \left(\sum_{i=1}^{m} X_i' W_i X_i\right)^{-1} \left(\sum_{i=1}^{m} X_i' W_i \mathbf{T}_i\right), \tag{27.5}$$

where $\mathrm{W} = \mathrm{diag}\,(\mathrm{W}_1, \ldots, W_k)$ and W_i is a $q_i \times q_i$ diagonal matrix of weights for the elements of $\mathbf{T}_i$. We discuss the choice of weights below. When the number m of studies is large, a robust estimate of the covariance matrix of the estimate $\hat{\boldsymbol{\beta}}$ calculated via (27.5) is

$$\frac{k}{k-p}\Sigma^{-1}\left(\sum_{i=1}^{m} X_i' W_i \mathbf{e}_i \mathbf{e}_i' W_i X_i\right)\Sigma^{-1},$$

where

$$\Sigma = \sum_{i=1}^{m} X_i' W_i X_i$$

and $\mathbf{e}_i = \mathrm{T}_i - \mathrm{X}_i \hat{\boldsymbol{\beta}}$ is the $(q_i \times 1)$ residual vector in the ith study. This estimate of the fixed effects and its standard error has the considerable advantage that it can be calculated without knowledge of the covariance structure of the estimation errors (the Σ_is).

While this method is completely general, it provides no direct method of computing the weight matrix W. Optimal weights would be $\mathrm{W}_i = (\mathrm{T} + \Sigma_i)^{-1}$, but construction of these weights requires the Σ_is to be known since estimation of the error covariance matrix using available data is typically not feasible in meta-analyses.

One way to proceed is to approximate the optimal weights in a way that does not involve the within-study covariances at all. An alternative is to use a simplification of the estimation error covariance structure to estimate between-study variance components as described below (see Hedges, Tipton, and Johnson, 2010). In deriving approximations, it is useful to remember that any weights that are reasonably close to optimal will give estimates that are nearly as efficient as the optimal weights.

The variance of all effect size estimates depends strongly on sample size. Sample sizes typically vary substantially across studies. Correlated effects within a study arising from multiple measurements on the same individuals are likely to be based on similar (if not identical) sample sizes. It follows that there is likely to be much more variation in precision of effect size estimates between studies than within studies. Thus, a sensible approach might be to give equal weights to all of the correlated estimates within a study, assigning each estimate in a study a weight proportional to the inverse of the average of the estimation error variances of estimates in that study.

Alternatively, one might estimate a between-studies variance component τ^2 based on a specified covariance structure for the estimates. Assume, for the purposes of computing weights, a convenient choice of covariance structure among the correlated estimates (imputing values of those correlations), such as one in which the effect size estimates within studies have correlation ρ, and the covariance matrix of the $\boldsymbol{\theta}_i$ is τ^2 times an identity matrix, so that the total (unconditional) covariance matrix of $\mathbf{T}_i$ is

$$\tau^2 (1-\rho) I + \rho\tau^2 J + V_i,$$

where I is a $q \times q$ identity matrix, J is a $q \times q$ matrix of ones, and $V_i = \text{diag}(v_{i1}, \ldots, v_{iq})$. This specification permits estimation of τ^2 given a fixed ρ, or even joint estimation of τ^2 and ρ. In either case, it would be desirable to conduct sensitivity analyses to ensure that estimates are not highly sensitive to the choice of the covariance structure. Ishak et al. (2008) found, in simulation studies, that the results of multivariate meta-regressions were relatively insensitive to incorrect values of the within-study correlations. Our experience is also that these estimates are not highly sensitive to choices of covariance structure in practical situations, but it does seem reasonable that estimates of between-study variance components will be most sensitive to within-study correlation structure when between-study covariance components are small compared to within-study covariances (see Riley, 2009). For an extensive discussion of estimating approximately optimal weights in this situation, see Hedges, Tipton, and Johnson (2010).

27.14 BAYESIAN APPROACHES TO META-ANALYSIS

Random effects meta-analyses can be seen as essentially using the data to estimate τ^2 and then carrying out the weighted analysis by substituting a single value of τ^2 (e.g., the maximum likelihood estimate) into the weights for estimating the fixed effects and their uncertainties (e.g., standard errors). The problem with this estimation approach hinges on the practice of treating the estimated value of τ^2 as a known quantity without error. However, this estimate has an associated uncertainty that arises from variation between studies and, in the case of a small number of studies, this uncertainty can be quite large. Therefore, an analysis (such as the conventional random effects analysis) that treats an estimated value of the variance component as if it were known with certainty, is problematic.

Bayesian analyses address this problem by recognizing that there is a family of random effects analyses, one for each value of the variance component. The Bayesian analyses can be seen as essentially averaging over the family of results, assigning each one the weight that is appropriate given the posterior probability of each variance component value conditional on the observed data. Some approaches to Bayesian inference for meta-analysis do this directly (see, e.g., DuMouchel and Harris, 1983; DuMouchel, 1994; Hedges, 1998; or Rubin, 1981). They compute the summary statistics of the posterior distribution (such as the mean and variance) conditionally given τ^2, sometimes plotting the estimates of the fixed effects in the meta-analysis as a function of (that is conditional on) τ^2, then average (integrate over) these conditional values, weighting by the posterior distribution of τ^2 given the data. These approaches are transparent and provide a direct approach to obtaining estimates of parameters of the posterior distribution. While they require numerical integrations, they are not particularly challenging to accomplish.

An alternative is the use of Markov chain Monte Carlo methods, which provide the posterior distribution directly without requiring numerical integration. These methods provide not only the mean and variance of the posterior distribution, but the entire posterior distribution, making it possible to compute a variety of descriptive statistics for the distribution. Another major advantage of these methods is that they permit models in which the study-specific random effects are not normally distributed, but have heavier-tailed distributions such as a Student's t or gamma distribution (see Seltzer, 1993; Smith, Spiegelhalter, and Thomas, 1995; Seltzer, Wong, and Bryk, 1996).

27.15 PUBLICATION BIAS IN META-ANALYSIS

A pervasive concern in the interpretation of scientific research is that the research results observed (e.g., in the published literature) may be selected in ways that influence the findings that are observed. For example, there is evidence that studies finding statistically or clinically significant results are more likely to be published (Dickerson, 2005), that journal reviewers and editors tend to view statistical significance as a criterion in selecting papers for publication (Greenwald, 1975), and that the published literature in many fields has an unexpectedly high proportion of statistically significant results given the likely statistical power of that research (Sterling, Rosenbaum, and Weinkam, 1995; Seidmeier and Gigerenzer, 1989). While selection favoring statistically significant results is not the only possible form of publication selection, it is perhaps the most widely discussed and also one that has the potential to introduce substantial bias into the set of observed (that is, published) effect size estimates from that research.

Publication selection has the potential to create very substantial bias among the observed (e.g., published) effects. Consider a two group (treatment and control group) experiment and the extreme selection mechanism in which only statistically significant results and no statistically insignificant results are observed. Hedges (1984) showed that under this extreme selection model the expected value of the observed effect sizes could be over 200% of the true value. While this model of selection is too extreme to be realistic, it does illustrate that the effects of publication selection can be large enough to seriously affect interpretation of research findings. Note that bias due to publication selection is not just a problem for meta-analysis, but for the use of published scientific evidence more generally. For example, Ioannidis (2005) observed that selection favoring statistically significant effects makes it much more likely that treatments identified as having large effects will later be found to have smaller or negligible effects upon replication.

While there is little that can be done to detect publication selection from a single finding or to adjust that single finding for the potential effects of publication bias, the replications inherent in meta-analysis make it possible to detect potential publication bias in a series of studies and to adjust for its potential effects (see Rothstein, Sutton, and Borenstein, 2005). In large datasets, the most satisfactory approaches involve positing a selection model and a multilevel model for the variation of effect sizes and simultaneously estimating parameters in the selection model and the effect size (multilevel) model (see, e.g., Duval, 2005; Hedges and Vevea, 2005). Of course, not all selection is undesirable. Selection based on quality of the research is an essential part of the scientific publication system. However, eliminating publication selection that is correlated with effect estimates but is not otherwise related to study quality should be a goal of the scientific publication system.

27.16 CONCLUSION

This chapter presented the most widely used models for meta-analysis, situating these methods within the context of hierarchical linear models. Three different approaches to estimation were discussed, and the use of both random (mixed) and fixed effects models was illustrated with an example of data from 19 experimental studies of the effects of teacher expectancy on pupil IQ. Traditional statistical software packages such as HLM (using the V Known option), MLwiN, and Proc MIXED in SAS can carry out mixed models analyses for meta-analytic data with nested structure. While the emphasis in this chapter is on two- and three-level models, the mixed effects models presented here can easily be extended to three or more levels of hierarchy, capturing random variation at higher levels.

NOTE

1 The average variance v arises naturally in conjunction with the test statistic Q, which has expected value $(k-1)$ under the null hypothesis that $\tau^2 = 0$. The ratio $Q/(k-1)$, called the Birge ratio, is sometimes used as an index of heterogeneity. The expected value of $Q/(k-1)$ is $1+\tau^2/v$. Because τ^2/v determines the average value of $Q/(k-1)$, it is natural to see v as a kind of "average" variance.

REFERENCES

Birge, R.T. (1932) 'The Calculation of Errors by the Method of Least Squares', *Physical Review*, 40: 207–27.

Borenstein, M. (2009) 'Effect Sizes for Continuous Data', in *The Handbook of Research Synthesis and Meta-analysis*, 2nd edition, H. Cooper, L.V. Hedges, and J.C. Valentine (eds). New York: Russell Sage Foundation, 221–36.

Cochran, W.G. (1937) 'Problems Arising in the Analysis of a Series Of Similar Experiments', *Journal of the Royal Statistical Society (Suppl.)*, 4: 102–18.

Cooper, H. (1989) *Homework*. New York: Longman.

Cooper, H.C., Hedges, L.V., and Valentine, J.C. (eds) (2009) *The Handbook of Research Synthesis and Meta-analysis*, 2nd edition. New York: The Russell Sage Foundation.

Cooper, H.M. (2010) *Research Synthesis and Meta-analysis: A Step By Step Approach*. Thousand Oaks, CA: Sage Publications.

Dickerson, K. (2005) 'Publication Bias: Recognizing the Problem, Understanding Its Origins and Scope, and Preventing Harm', in *Publication Bias in Meta-analysis*, H.R. Rothstein, A.J. Sutton, and M. Borenstein (eds). Chichester (UK): Wiley, 11–34.

DuMouchel, W.H. (1994) *Hierarchical Bayes Linear Models for Meta-analysis*. Technical Report Number 27, National Institute of Statistical Sciences, Research Triangle Park, NC. http://www.niss.org/sites/default/files/pdfs/technicalreports/tr27.pdf.

DuMouchel, W.H. and Harris, J.E. (1983) 'Bayes Method for Combining the results of Cancer Studies in Humans and Other Species', *Journal of the American Statistical Association*, 78: 293–315.

Duval, S. (2005) 'The trim and fill method', in *Publication Bias in Meta-analysis*, H.R. Rothstein, A.J. Sutton, and M. Borenstein (eds). Chichester (UK): Wiley, 127–44.

Fleiss, J.L. and Berlin, J.A. (2009) 'Effect Sizes for Dichotomous Data', in *The Handbook of Research Synthesis and Meta-analysis*, 2nd edition, H. Cooper, L.V. Hedges, and J.C. Valentine (eds). New York: Russell Sage Foundation, 237–55.

Glass, G.V. (1976) 'Primary, Secondary, and Meta-analysis of Research', *Educational Researcher*, 5: 3–8.

Greenwald, A.G. (1975) 'Consequences of Prejudice Against the Null Hypothesis', *Psychological Bulletin*, 82: 1–20.

Gurevitch, J. and Hedges, L.V. (1993) 'Meta-Analysis: Combining the Results of Independent Experiments', in *Design and Analysis of Ecological Experiments*, S.M. Scheiner and J. Gurevitch (eds). New York: Chapman & Hall, 378–98.

Gurevitch, J. and Hedges, L.V. (1999) 'Statistical Issues in Ecological Meta-analyses', *Ecology*, 80: 1142–9.

Hedges, L.V. (1981) 'Distribution Theory for Glass's Estimator of Effect Size and Related Estimators', *Journal of Educational Statistics*, 6: 107–28.

Hedges, L.V. (1984) 'Estimation of Effect Size Under Nonrandom Sampling: The effects of Censoring Studies Yielding Statistically Insignificant Mean Differences', *Journal of Educational Statistics*, 9: 61–85.

Hedges, L.V. (1998) 'Bayesian Approaches to Meta-Analysis', in *Recent advances in the Statistical Analysis of Medical Data*, B. Everitt and G. Dunn (eds). London: Edward Arnold, 251–75.

Hedges, L.V. and Olkin, I. (1985) *Statistical Methods for Meta-analysis*. New York: Academic Press.

Hedges, L.V. and Pigott, T.D. (2001) 'The Power of Statistical Test in Meta-analysis', *Psychological Methods*, 6: 203–17.

Hedges, L.V., Tipton, E., and Johnson, M. (2010) 'Robust Variance Estimation in Meta-regression with Dependent Effect Size Estimates', *Research Synthesis Methods*, 1: 39–65. (Erratum, 1: 164–5.)

Hedges, L.V. and Vevea, J. (2005) 'Selection Model Approaches', in *Publication Bias in Meta-Analysis*, H.R. Rothstein, A.J. Sutton, and M. Borenstein (eds). Chichester (UK): Wiley, 145–74.

Higgins, J.P.T. and Thompson, S.G. (2002) 'Quantifying Heterogeneity in a Meta-analysis', *Statistics in Medicine*, 21: 1539–58.

Huber, P. (1967) 'The Behavior of Maximum-likelihood Estimates Under Nonstandard Conditions' in *Proceedings of the Fifth Berkeley Symposium On Mathematical Statistics and Probability*, L.M. LeCam and J. Neyman (eds). Berkeley: University of California Press, 221–33.

Ioannidis, J.P.A. (2005) 'Why Most Published Research Findings Are False', *PLoS Medicine*, 2(8), e124. doi:10.1371/journal.pmed.0020124.

Ishak, K.J., Platt, R.W., Joseph, L., and Hanley, J.A. (2008) 'Impact of Approximating or Ignoring Within-study Covariances in Multivariate Meta-analyses', *Statistics in Medicine*, 27: 670–86.

Jackson, D., Riley, R., and White, I.R. (2011) 'Multivariate Meta-analysis: Potential and Promise', *Statistics in Medicine*, 30: 2481–98.

Kalaian, H.A. and Raudenbush, S.W. (1996) 'A Multivariate Mixed Linear Model for Meta-analysis', *Psychological Methods*, 1: 227–35.

Kampenes, V.B., Dybå, T., Hannay, J., and Sjøberg, D. (2007) 'A Systematic Review of Effect Size in Software Engineering Experiments', *Information and Software Technology*, 49: 1073–86.

Lipsey, M.L. and Wilson, D.L. (2001) *Practical Meta-analysis*. Thousand Oaks, CA: Sage Publications.

Pearson, K. (1904) 'Report on Certain Enteric Fever Inoculation Statistics', *British Medical Journal*, 2: 1243–6.

Raudenbush, S.W. (1984) 'Magnitude of Teacher Expectancy Effects on Pupil IQ as a Function of the Credibility of Expectancy Induction: A Synthesis Of Findings From 18 Experiments', *Journal of Educational Psychology*, 76: 85–97.

Raudenbush, S.W. (2009) 'Analyzing Effect Sizes: Random-effects Models', in *The Handbook of Research Synthesis and Meta-analysis*, 2nd edition, H. Cooper, L.V. Hedges, and J.C. Valentine (eds). New York: Russell Sage Foundation, 295–315.

Raudenbush, S.W. and Bryk, A.S. (1985) 'Empirical Bayes Meta-analysis', *Journal of Educational Statistics*, 10: 75–98.

Raudenbush, S.W. and Bryk, A.S. (2002) *Hierarchical Linear Models: Applications and Data Analysis Methods*. Thousand Oaks, CA: Sage.

Riley, R.D. (2009) 'Multivariate Meta-analysis: The Effect of Ignoring Within-study Correlation', *Journal of the Royal Statistical Society, Series A*, 172: 789–811.

Rosenfield, A.H. (1975) 'The Particle Data Group: Growth and Operations', *Annual Review of Nuclear Science*, 25: 555–99.

Rothstein, H.R., Sutton, A.J., and Borenstein, M. (2005) *Publication Bias in Meta-analysis*. Chichester (UK): Wiley.

Rubin, D.B. (1981) 'Estimation in Parallel Randomized Experiments', *Journal of Educational Statistics*, 6: 377–401.

Schmidt, F.L. and Hunter, J. (1977) 'Development of a General Solution to the Problem of Validity Generalization', *Journal of Applied Psychology*, 62: 529–40.

Seidmeier, P. and Gigerenzer, G. (1989) 'Do Studies of Statistical Power Have an Effect on the Power of Studies?', *Psychological Bulletin*, 105: 309–16.

Seltzer, M. (1993) 'Sensitivity Analysis for Fixed Effects in the Hierarchical Model: A Gibbs Sampling Approach'. *Journal of Educational Statistics*, 18: 207–35.

Seltzer, M.H., Wong, W.H., and Bryk, A.S. (1996) 'Bayesian Analysis in Applications of Hierarchical Models: Issues and Methods', *Journal of Educational and Behavioral Statistics*, 21: 131–67.

Sheu, C. and Suzuki, S. (2001) 'Meta-analysis Using Linear Mixed Models', *Behavior Research Methods, Instruments, and Computers*, 33: 102–7.

Smith, M.L. and Glass, G.V. (1977) 'Meta-analysis of Psychotherapy Outcome Studies', *American Psychologist*, 32: 752–60.

Smith, T.C., Spiegelhalter, D.J., and Thomas, A. (1995) 'Bayesian Approaches To Random-effects Meta-analysis: A Comparative Study', *Statistics in Medicine*, 14: 2685–99.

Sterling, T.D., Rosenbaum, W.I., and Weinkam, J.J. (1995) 'Publication Decisions Revisited: The Effect of the Outcome of Statistical Tests on the Decision To Publish', *American Statistician*, 49: 108–12.

Stevens, J.R. and Taylor, A.M. (2009) 'Hierarchical Dependence in Meta-analysis', *Journal of Educational and Behavioral Statistics*, 34: 46–73.

White, H. (1980) 'A Heteroscedasticity-consistent Covariance Matrix and a Direct Test for Heteroscedasticity', *Econometrica*, 48: 817–38.

Zwolinski, B.J. and Chao, J. (1972) 'Critically Evaluated Tables Of Thermodynamic Data', in *International Review of Science–Physical Chemistry: Vol 10. Thermochemistry and Thermodynamics*, H.A. Skinner (ed.). London: Butterworth, 93–120.

28

Modeling Policy Adoption and Impact with Multilevel Methods

James E. Monogan III[1]
University of Georgia, USA

Policy analysts study both the causes leading to the adoption of specific policies and the consequences of policies once they are enacted. When investigating either question, it is essential to recognize that policies are dynamic over time and are often developed and implemented through a hierarchical structure. For example, policy might be carried out through clinics or schools, each of which have unique qualities as they fulfill their jobs. Further, legislation in many cases, including the American case, is decided within a federal structure. Hence, policies adopted by counties should be thought of in the context of their respective states, and state policy should be thought of within the context of national policy. Thus, clustering is a pervasive feature in the design and administration of programs.

Additionally, policy is not static. Rather, policy constantly changes over time. In fact, some programs are designed explicitly to be staggered over years, as they are phased in or out. Plus, the social factors lawmakers try to influence (such as unemployment, smoking rates, or workforce education levels) are also in continuous flux, whether a policy treatment is present or not. Thus, researchers should also be prepared to analyze longitudinal data to account for the dynamic nature of lawmaking. To ignore any of these aspects of policy data is to ignore sources of variance and to potentially draw false inferences.

Intriguingly, policy studies in the field of political science, which often focus on the questions of the origins and factors behind enacting legislation, have underutilized multilevel methods, thereby failing to account for many important attributes of policy data.[2] By contrast, the field of education presents many more examples of multilevel research, more often on questions regarding the effectiveness of programs on outcomes. Regardless of the varying degrees to which multilevel methods are used in public policy research, this chapter offers the reasons such methods are important for both nested and longitudinal policy data.

The chapter begins by laying out the means by which multilevel modeling can be applied to policy data. Second, it describes how modeling policy data compares to other applications of hierarchical models and makes the case that the spatial element of policy data should be more fully exploited. Third, it discusses the difficulties that can emerge when trying to model policy data using multilevel methods and contrasts alternative methods. Fourth, the chapter offers an applied example that analyzes state actions in enforcing the Clean Air Act from 2001–2009. Fifth, the chapter lists some of the software options available for conducting policy analysis with multilevel data. Finally, the chapter offers advice to the applied policy researcher.

28.1 MODELS AND METHODS

Since programs are rarely implemented through independent units, but rather hierarchies, and since laws constantly change over time, how should models be specified for policy data? This section offers several illustrations of how multilevel models can be used to account for these data structures. Taking this approach is important because failure to account for unit heterogeneity can produce overconfidence. Further, regressors may have heterogeneous effects across units, and this section will show how such variation can be considered. The section first explains how to account for hierarchical structures, then gives an example with real data, and finally offers techniques for dealing with longitudinal policy data.[3]

28.1.1 Nested Data

As an example of analyzing the impact of a program which is implemented in a hierarchical manner, suppose that an outcome of interest is fifth-grade student test scores in mathematics. In this case, the program of instructing fifth-grade students in math is implemented by placing several students within classrooms, which are part of their respective schools, which are organized into school districts. Variations in how policy is implemented occur at each level: The teacher of the classroom will have unique instructional techniques and qualifications. The school will have different facilities available to students and will implement unique disciplinary policies. The districts will impose varying curriculum requirements and purchase different textbooks. A researcher may be interested primarily in, for example, whether a teacher's holding of a master's degree leads to better math test scores among students, which would have implications relevant to teacher certification requirements. Naturally, the researcher will want to control for as many observable attributes of the classroom, school, and district as possible. The classroom size, school budget, textbook used, and a variety of other factors should be included in a full model of student performance. But what about the nested nature of instruction?

The fact is, there are probably classroom, school, and district effects that even comprehensive models will not explain away. (Alternatively, there may be a comprehensive covariate specification that would explain all of the unit effects, were it not for the curse of dimensionality.) The presence of school-level unit effects, for example, implies that some schools have a tendency to produce better outcomes—in the form of higher average math scores—than would be expected from covariate values, while other schools perform worse relative to expectations. If a researcher failed to account for this in the model, perhaps by estimating a linear model with ordinary least squares (OLS), the estimates would be inefficient, and the standard errors would be artificially small. This is because the hierarchical structure of the data produces autocorrelation between observations within the same upper-level group. Specifically, the fact that students in the same school are influenced by the same unobserved school qualities means that their error terms will correlate with each other. The presence of intraclass correlation generally causes standard errors to be biased downward, raising the values of t- and z-ratios and raising the tendency for researchers to report false positive results.

To avoid inefficient estimation and biased standard errors, models of programs that are implemented in a clustered way should incorporate random effects to account for differing means among higher units. Using our working example, suppose we are modeling the test score of student i, who is in classroom j, at school k, in district ℓ. We might specify the following model:

$$y_{ijk\ell} = \mathbf{x}_{ijk\ell}\boldsymbol{\beta} + u_\ell + \delta_{k\ell} + \nu_{jk\ell} + \varepsilon_{ijk\ell}. \quad (28.1)$$

Here, $\mathbf{x}_{ijk\ell}$ is a row vector of covariates (including a constant) for each observation and $\boldsymbol{\beta}$ is a column vector of fixed coefficients. In this case, the unexplained variance would be broken into four components: district-level (u_ℓ), school-level ($\delta_{k\ell}$), classroom-level ($\nu_{jk\ell}$), and student-level ($\varepsilon_{ijk\ell}$) variance. Since the estimator accounts for the non-independence of schools within a district or classrooms within a school, the aforementioned inefficiency and tendency for added false positive results can be avoided.

Besides being more efficient, a second advantage of multilevel models is that they can entertain unit-specific effects. For example, if the impact of teacher credentials differs across schools, this can be informative when requirements for teacher training are set. Suppose again that we wish to model fifth-grade math scores. Assume there is a similar data setup to that of (28.1), but with only two input variables: students' math scores in the fourth grade and whether the teacher holds a master's degree. In this situation, however, we also believe that the consequences of a teacher's credentials will vary across schools. This model can now be written as

$$\begin{aligned} y_{ijk\ell} = \beta_0 + x_{1jk\ell}\beta_1 + x_{2ijk\ell}\beta_2 + u_\ell + \delta_{0k\ell} \\ + x_{1jk\ell}\delta_{1k\ell} + \nu_{jk\ell} + \varepsilon_{ijk\ell}. \end{aligned}$$

Here, $x_{1jk\ell}$ is the indicator for whether a teacher has a master's degree, β_1 is the fixed coefficient for teacher credentials, $\delta_{1k\ell}$ is the random component of the effect of teacher credentials (varying by school), and $x_{2ijk\ell}$ is the student's fourth-grade test score.

The term $\delta_{1k\ell}$ is assumed to have mean zero, implying that, on average, the effect of having a master's degree is β_1. If $\delta_{1k\ell}$ has a variance close to zero, then the researcher could conclude that the effect of a master's degree is close to β_1 across all schools. Further, if β_1 is large and positive, a lawmaker might even conclude that requiring teachers to hold a master's degree will raise statewide test scores.[4] If $\delta_{1k\ell}$ has a large variance, though, then the consequence of holding a master's degree varies substantially from school to school. Thus, imposing a universal requirement would not have a uniform effect across the state.[5] In all, for the ability to account for unit heterogeneity in intercepts and in the slopes of predictors, multilevel models are key to analyzing hierarchical programs.

Nesting example: Smoking prevention through education

An illustration using real data can quickly reveal how serious the problems of ignoring clustering can be. Flay et al. (1995) designed an experimental study which considered various possible programs for smoking prevention among teenagers, with a focus on learning about the health hazards of smoking. Two independent treatments were assigned to students: a classroom-based social resistance curriculum and a television-based intervention. Treatment assignment was done at the school level. Using a 2×2 factorial design, schools may have been randomly assigned to neither, one, or both of these treatments. Following Fitzmaurice et al. (2004, pages 453–5), who more explicitly modeled these data in a multilevel structure, this example focuses on a subset of 1600 observations from the original data.

Equation (28.2) shows the model specification for this question:

$$\begin{aligned} y_{ijk} &= \mathbf{x}_{ijk}\boldsymbol{\beta} + u_k + \delta_{jk} + \varepsilon_{ijk} \\ u_k &\sim \mathcal{N}(0, \sigma_u^2) \\ \delta_{jk} &\sim \mathcal{N}(0, \sigma_\delta^2) \\ \varepsilon_{ijk} &\sim \mathcal{N}(0, \sigma_\varepsilon^2). \end{aligned} \quad (28.2)$$

The outcome variable (y_{ijk}) is how students scored on a tobacco and health knowledge questionnaire after treatments were administered. The input variables (included in the vector $\mathbf{x}_{ijk}$) are how students scored on the knowledge questionnaire before treatments were administered, whether the student received the classroom curriculum treatment, whether the student received the television treatment, and whether the student received both the classroom and television treatments. The coefficient $\boldsymbol{\beta}$ represents the fixed coefficients of this equation, while u_k, δ_{jk}, and ε_{ijk} represent the school-, classroom-, and student-level random effects, respectively. Each of these terms is assumed to come from a Gaussian distribution with mean zero and a unique homoscedastic variance.

Table 28.1 presents the results of this analysis. The column labeled OLS shows the estimates of coefficients computed from a simple linear model. The column labeled MLM shows the coefficient estimates from the model of (28.2), which accounts for the three levels of the data: student, classroom, and school. These results also show the estimated components of unexplained variance associated with schools, classes, and individual students, respectively.

Table 28.1 clearly demonstrates how OLS can produce overconfidence. Since simple linear models ignore the nested nature of the data, the column labeled OLS does not account for correlation among observations within the same group. Therefore, the reported standard error on every coefficient is smaller than the reported standard error for a multilevel model (seen in the MLM column). Despite the fact that a multilevel model is more efficient with clustered data such as these, the standard errors under OLS are underestimated so substantially that one would conclude that the main effect of the TV treatment and the interaction of the TV and classroom treatments are statistically different from zero. However, under the multilevel model, the standard errors better account for the structure of the data, and neither of these coefficients differs significantly from zero. For these data, then, a linear model estimated with OLS would have produced two false positive errors that a multilevel model would not have. Thus, this smoking prevention study should illustrate how essential it is that the model specification properly consider the hierarchical nature of a program's implementation.

28.1.2 Longitudinal Data

Besides nested data, analyzing policy can often require the study of panels, or data in which a cross-section of individuals is observed at several waves over time.[6] When studying the crafting or implementation of programs, the cross-sectional "individuals" may be schools, clinics, counties, or states. The "waves," or repeated measurement occasions, could be school years, calendar years, months, or legislative sessions. The key advantage of using panels when studying the significance of programs is that such data offer the opportunity to better evaluate causal propositions about a program than strictly cross-sectional data. Whereas cross-sectional data only allow the researcher to contrast regimes with different policies, panel data further allow the researcher to observe whether a change in an input precedes a change in the outcome. In other words, since longitudinal data consist of the same individuals over time, the analyst can observe a shift in responses as a reaction to a policy treatment.

There are two classes of panel data sets: long panels, which have many measurement occasions relative to the size of the cross-section, and short panels, which have many individuals in the cross-section relative to the number of waves. A potential short panel in policy analysis might be a field experiment that follows the employment status of persons who did or did not receive a treatment of job training.[7] This kind of study could be useful in evaluating the effectiveness of such training as a possible workforce education program. While such an investigation might recruit hundreds of participants, it may only have funding to reinterview these individuals three or four times.

A long panel, by contrast, may study the annual level of foreign direct investment (FDI) in developing countries, with the treatment variable being the tax regime. In this case, there may be twenty to fifty countries

Table 28.1 Tobacco and Health Knowledge, Modeled Using OLS and a Multilevel Model

Variable	OLS	MLM
Intercept	1.6613*	1.7020*
	(0.0844)	(0.1254)
Pre-Intervention Knowledge	0.3252*	0.3054*
	(0.0258)	(0.0259)
Classroom	0.6406*	0.6413*
	(0.0921)	(0.1609)
TV	0.1987*	0.1821
	(0.0900)	(0.1572)
Classroom×TV	−0.3216*	−0.3309
	(0.1303)	(0.2245)
School-Level Variance (σ_u^2)	—	0.0386
Class-Level Variance (σ_δ^2)	—	0.0647
Student-Level Variance (σ_ε^2)	—	1.6023

Notes: $N = 1600$. Cell entries are parameter estimates.
Standard errors in parentheses. * $p < .05$
Estimates computed with R 2.12.1.

from a particular global region, such as Latin America or Africa. Yet, if information on tax policy and FDI is publicly available, the researcher may have observations for as many as sixty years. Hence, it can be seen how questions about policy and its impact can focus primarily on cross-sectional variance or on time-dependent variance.

Laying out the model specification in a situation like this, suppose the outcome of interest is annual FDI in country i in year t. A common model for this situation is what econometricians call a random effects model—so called for having a random intercept for each unit:

$$y_{it} = \mathbf{x}_{it}\boldsymbol{\beta} + u_i + \varepsilon_{it}. \tag{28.3}$$

Here, $\mathbf{x}_{it}$ is a row vector of covariates for a given nation-year (including a constant), $\boldsymbol{\beta}$ is a column vector of estimated coefficients, u_i is the random intercept for each nation, and ε_{it} is the disturbance for each nation-year.

In this example, if some nations tend to draw higher FDI year-in, year-out than others for reasons not explained by the covariates, then it is essential to account for this unit heterogeneity somehow. Just as in the hierarchical case, failure to do so can produce inefficiency and underestimated standard errors, thereby raising the specter of false inference. The random effects model is one cure for this inefficiency. (See the section on "strengths and weaknesses" for more alternatives on this issue.)

A generalization of the "random effects model" would be a traditional linear mixed effects model, as is easily specified in a multilevel framework. Using the example of modeling FDI, a researcher may have reason to believe that the effect of some time-varying factor, such as the tax regime or per capita GNP, differs from country to country. For time-variant inputs, the researcher has the option of allowing the coefficient to randomly vary by country. In this case, a mixed effects model might be specified as

$$y_{it} = \mathbf{x}_{it}\boldsymbol{\beta} + \mathbf{z}_{it}\mathbf{u}_i + \varepsilon_{it}. \tag{28.4}$$

In (28.4), the terms have the same meaning as in the random effects model of (28.3), except that now $\mathbf{z}_{it}$ is a row vector that is a subset of $\mathbf{x}_{it}$, and $\mathbf{z}_{it}$ includes a constant and possibly some of the time-varying covariates (such as per capita GNP) found in $\mathbf{x}_{it}$. The variable $\mathbf{u}_i$ now includes the random intercept for each unit and random coefficients for any time-varying covariates included in $\mathbf{z}_{it}$. Just as with a mixed model for hierarchical data, the terms in $\mathbf{u}_i$ typically are not directly estimated with frequentist methods, but can be predicted after model estimation using best linear unbiased predictions or empirical Bayes predictions. (More information on predicting random effects can be found in Chapter 3 of this book, or in Fitzmaurice et al. 2004, pages 206–10.) As an example of analyzing longitudinal policy data, see Section 28.4 on Clean Air Act enforcements later in this chapter.

28.2 CONNECTIONS TO OTHER MODELS

Having had a look at the basic layout of hierarchical and longitudinal models when studying policy, what complexities might arise in real research? To start, the examples of the previous section all assumed continuous outcomes with Gaussian disturbances. They also assumed that, other than correlation caused from holding a common upper-level unit, disturbances were independent of each other. For this reason, several of the methods described in other chapters of this volume are relevant for the study of programs.

One worthwhile case to consider is when the dependent variable is a duration time. In other words, the outcome variable may be the length of time until a political unit adopts a new policy. As Chapter 30 points out, duration times like this are not Normally distributed. Even more importantly, duration data can be misleading due to censoring—meaning that some observations may have adopted the policy before the start of the study and others may have not yet adopted the policy by the time data collection is complete. The method of event history analysis is designed to deal with this non-Normality and censoring in data (see Box-Steffensmeier and Jones, 2004, for a full treatment of these models). Event history models have found substantial application in the study of legislation in the American states: Berry and Berry (1990) studied the factors shaping the time it takes for a state to adopt an education lottery, and Grossback et al. (2004) looked at the time it takes for states to adopt academic bankruptcy laws and sentencing guidelines, just to name a few examples. While there are several methods within the event history framework, researchers often structure the data as consisting of several time points nested within the political unit; hence, the logic of multilevel methods can be useful.

Another consideration is that a major part of understanding when and why laws will be enacted is understanding the behavior of policy makers. A common approach is to develop a theory of what motivates a legislator, executive, or bureaucrat. Then, the researcher can empirically model how outcomes—such as these actors' voting patterns, official requests, or regulatory behavior—vary based on inputs like electoral context, personal factors, or policy needs. To evaluate how policy will come to pass in this way is a behaviorally motivated undertaking, and as such will parallel many of the considerations that enter into social and behavioral research (which is laid-out much more broadly in Chapter 29).

Third, policy data, as has been mentioned, are often longitudinal, so the considerations that arise in modeling longitudinal economic data may also be relevant. Economic data are often long panels, as measures in many cases are public records, making repeated observation easier. For example, research on how monetary policy affects indicators such as inflation and unemployment could draw from cross-national data which are measured quarterly over several years. Short panel studies, though, such as those that recruit participants, will put much more emphasis on cross-sectional variance. Thus, the structure of the study should weigh substantially in the kinds of techniques used.

28.2.1 The Relevance of Spatial Modeling

An additional complication of policy analysis that is overlooked entirely in many literatures is geographic dependency. In the study of laws or programs, observations can often be placed in space in one of two ways: One possibility is that observations may be towns, schools, clinics, or something that can be located at points on a map. By recording, for instance, the latitude and longitude of every clinic being studied, the analyst can gauge how near or far any one observation is from the other. The other possibility is that observations are nations, counties, school districts, or some governing body that can be defined by a border. These data are called areal or lattice data, and the analyst can easily classify whether two observations are geographic neighbors, neighbors of neighbors, or how far they are removed. Since policies are often adopted by governing regions with borders and are often implemented through stations at precise points in space, geography is a pervasive feature of policy data.

With observations that can be located in space, a common issue is the fact that some unobserved or omitted input variable may be spatially dependent. (Or, alternately, the outcome variable simply has a quality that neighbors tend to take more similar values than non-neighbors for reasons that cannot be explained.) For instance, a study of policy liberalism in the fifty states may not have a good measure of cultural idiosyncrasies that shape outcomes. In the absence of a good measure, the effect of this variable is captured by the error term. It is also quite possible that neighboring states, or those that are more proximate to each other, share similar cultural tendencies. Because this variable takes similar values among more proximate observations and because it is omitted, errors will be spatially autocorrelated, with the correlation rising as proximity increases.[8] Ignoring this attribute of the data creates yet another circumstance where inefficiency and false positive errors are a potential problem.

Regarding how to handle this, there is an extensive literature in political science about policy diffusion over space and time (Berry and Berry, 2007; Gray, 1973), where many researchers consider geographic diffusion by including the behavior of neighbor states in the structural portion of the model. As a second solution, depending on the structure of the data, spatial autocorrelation also may diminish merely by accounting for the multilevel nature of a data set. For instance, if clinics are nested within states, then accounting for state unit effects may capture most of the similarities among proximate clinics. A third approach to dealing with geographic dependence is to explicitly account for it in the model's assumptions. Much of the heaviest development in model-based methods for the study of spatial data have come from the field of epidemiology, meaning that some of the best policy applications of spatial methods are in the field of public health. For example, Banerjee et al. (2004, pages 335–8) reviews a study originally reported in O'Hara et al. (1998) on smoking cessation. Here, the focus is a duration model of time until a subject quit smoking, and the key treatment variable is whether a participant was randomly placed into a smoking intervention group or received usual care. In this study, the best-fitting models specify a conditionally autoregressive (CAR) error structure, which assumes that patients from the same zip codes are likely to exhibit the most similar behavior, but that patients from neighboring zip codes are also likely to exhibit somewhat similar behavior.[9] Hence, better results can be obtained from considering geographic dependency. Chapter 31 from this book describes many of the methods developed by epidemiologists, both for data located at points and lattice data. In fact, these models are largely extensions of hierarchical models designed to represent dependencies between the hierarchical units. Therefore, spatial modeling fits easily in the multilevel modeling framework. To account for similarity of more proximate entities, policy analysis would be well served to consider using these methods. As an illustration, the example in this chapter on Clean Air Act enforcements in the fifty states (Section 28.4) illustrates how spatial conditionally autoregressive models can be applied with real data.

28.3 STRENGTHS AND WEAKNESSES OF MULTILEVEL METHODS

The applications of multilevel models laid out in this chapter have several tradeoffs compared with other methods that might be used for policy data. Specifically, fixed effects models and experimental methods offer several plausible alternatives in the study of policy. The major problems to consider are the number of upper-level units that are observed and whether or not random intercepts are independent of covariates. This section first entertains the concerns and alternative methods associated with hierarchical data and then moves to panel studies.

28.3.1 Attributes of Models of Nested Data

Much of the dialogue in this chapter has presumed that the researcher is interested in field data. To analyze these data, the empirical model must use statistical control to account for alternate explanatory factors and model-based corrections for features like intraclass correlation. Depending on the question, field data may be the only sensible alternative in a study. However, in some cases, experimental data can also be of use. In policy analysis, the experiments that will find the most use are natural experiments and field experiments.

Natural experiments are studies wherein similar locales are studied, with at least one that adopts a policy treatment and at least one that does not. (Alternatively, one locale may be observed over time, adopting an intervention at some point.) An example of this might be a study of state pension plans and how well they are funded, with the treatment being the manner in which revenue is managed. Field experiments, on the other hand, are studies where the investigator introduces a treatment in one locale, but not in another. The article replicated earlier in this chapter by Flay et al. (1995) actually took this approach by introducing a smoking awareness treatment in some television viewing areas, but not others. In either kind of experiment, structuring the study effectively sometimes can make the analysis as simple as doing difference-of-means tests or contingency tables.

As the reference to the Flay et al. (1995) article likely makes clear, though, even in an experimental setup, data often are nested or longitudinal in their nature. Thus, even though these researchers randomized which schools received treatments, the structure of the data still needs to be considered in the analysis, as Table 28.1 demonstrates. Further, many outcomes are fundamentally temporal in nature. For instance, even if a researcher randomizes which participants receive job training, there is still the need to reinterview the individual multiple times to assess the impact of this treatment. Therefore, longitudinal methods would be necessary in this case as well. Overall, experimental design can be useful for understanding policy, but even clean data such as these will often require multilevel models.[10]

Micronumerosity

A common problem that can emerge with clustered data in the study of policy is micronumerosity of upper-level units. When political entities form a higher-level unit, then there are often only a handful of observations at the highest level. To wit, there are only fifty American states and twenty countries in Latin America. Yet, within each of these upper-level units, there might be hundreds of municipalities or thousands of eligible voters. This can pose an issue if the treatment only varies at the highest level. The statistical power for inference on top-level inputs is driven heavily by the number of upper-level units. In fact Bryk and Raudenbush (1987, page 211) state that, as a rule of thumb, a data set should have ten upper-level observations per upper-level covariate. The upshot of all of this is that researchers who are studying a small population of governing bodies will have to recognize that they have less power than might be expected from the total number of lower-level observations. Thus, the need to carefully weigh parsimony against necessary covariates becomes heightened.

All of this raises the relevance of Bayesian methods in the analysis of multilevel policy data. Specifically, Gill (2001) makes the case that, when analyzing a full population rather than a sample, Bayesian methods are a more natural approach for interpreting the uncertainty in results. Thus, policy researchers who are limited to the study of a small population could benefit from analyzing their data using Bayesian techniques, as Chapter 4 lays out. Beyond matters of inference, though, Bayesian methods allow a greater array of model specifications in that the researcher must specify priors for every parameter. For example, in a Bayesian model, the researcher has a lot of flexibility in choosing the distributional assumption of random effects and is not limited to the assumption of a Normal distribution. Many of these advantages are illustrated in Section 28.4 about Clean Air Act enforcement later in this chapter.

Limits to random effects

As a final point, random intercepts associated with each upper-level unit are not a sufficient correction for heterogeneity if the unit effects correlate with any of the regressors. Consider once more the hypothetical example of modeling fifth-grade students' math test scores with the predictor of whether a teacher holds a master's degree. Again, suppose each school has a unique mean for student test scores, even beyond what would be expected from the covariates. If the schools that tended to perform better than expected also had more teachers with master's degrees, then an endogeneity bias comparable to omitted variable bias will emerge because a variable in the model covaries with the error term. In this case, the estimate of the effect of a master's degree will also capture school effects. Regardless of whether the error term is broken down into school-level variance and lower-level variance (as a random effects model would do), this bias would emerge because nothing in the estimation algorithm partials out unit heterogeneity from the variance in the input variables.

In this example, one alternative would be to include an indicator variable for each school, save a reference unit. This model uses the same technology as the fixed effects model frequently used for panel data (see next section). The advantage of doing this is that it would account for unit heterogeneity and would partial out the effect of heterogeneity from the effect of teacher credentials. However, there are major disadvantages of this approach as well: With a large number of upper-level units relative to the sample, this would be an inefficient technique as a degree of freedom has to be used to estimate each school's intercept. Further, if the variable of interest was measured at the school level (i.e., number of administrators on campus) instead of the classroom level, this method would preclude estimation of that effect. This is because the indicator variables will be perfectly collinear with any input that varies only at the school level. Thus, the nested fixed effects approach might be useful in cases where the researcher is not confident about capturing all of the essential input variables at a certain level. Random effects, though, are the most feasible approach if the predictor of interest varies only at a higher level.[11]

It should be noted that there are alternatives to the fixed effects and random effects model specifications. These alternative models have the advantage of accounting for covariation between unmeasured upper-level factors and observed lower-level factors without estimating a lot of extra parameters. Gelman (2006) describes an approach in which the group mean of a lower-level predictor is also used to model the group-level heterogeneity. This serves to curb bias from covariance between an individual-level predictor and group heterogeneity. Similarly, Neuhaus and Kalbfleisch (1998) explain how a conditional likelihood solution can be useful for estimating the unique effect of a predictor, independent of group-level heterogeneity. For a more extensive description of "hybrid" models of unit effects, see Allison (2005). The reader also may want to consult Chapter 6 for more information on centering predictors.

28.3.2 Alternative Longitudinal Approaches

Regarding longitudinal data, this chapter has entertained mixed effects models and their special case of random intercepts only (i.e., a random effects model). Yet these models have shortcomings for which alternate methods are often more applicable. In particular, the fixed effects model and models with more sophisticated autocorrelation structures are often useful.

The biggest potential drawback of the random effects model of (28.3) is that it assumes that unit effects are independent of all covariates. Much like the case of non-independent unit heterogeneity in clustered data, if the unit effects in longitudinal data are correlated with any of the input variables, then the random effects model is biased and inconsistent. Using the earlier example of modeling FDI, if countries that have a lower tax rate tend to draw more investments, even beyond the true effect of taxes, then this term's coefficient will be overstated. This problem, akin to omitted variable bias, can be corrected by instead specifying what econometricians call a fixed effects model, typically estimated with least squares dummy variables (LSDV).[12] This approach estimates the desired model using OLS, including dummy variables for each individual, save a reference individual. This approach has the advantages of being computationally simple and correcting for unit heterogeneity that correlates with regressors.

Yet the fixed effects model is not without its limitations. Chief among these is that individual dummies are perfectly collinear with any variable which varies only cross-sectionally. Hence, LSDV precludes the inclusion of time-invariant variables in a model, whereas multilevel models do not. This can be a fundamental problem in many policy analyses because many policy treatments only vary cross-sectionally. Another drawback of the fixed effects model is that it can be inefficient due to the loss of information from estimating a unique intercept for each individual (whereas the random effects model often does not directly estimate this). Particularly in the case of short panels where the number of individuals is much bigger than the number of waves, the inefficiency issue can become substantial. Thus, fixed effects often are a more sensible choice for long panels, while random effects are often reasonable for short panels—although diagnostics and the purposes of the analysis should certainly be considered when weighing this decision. For more on the decision between fixed and random effects for panel data, including the use of Hausman tests to weigh the validity of random effects, see Chapter 5 of this volume (or, alternatively, Cameron and Trivedi, 2005, pages 271–4).

A smaller issue than the choice of fixed or random effects is that there are alternative, often more realistic, techniques for specifying intra-individual correlation besides random effects. The model of (28.3) simply includes an intercept that randomly varies by individual. This model assumes that any two repeated observations of the same individual will correlate at the same level: it does not matter whether they are two observations that are one wave apart or if they are the first and last observation taken. While this assumption is substantially better than ignoring serial correlation completely, the truth is usually more nuanced. In particular, a researcher can specify a variety of autocorrelation models with the idea that observations from the same individual should correlate, but at a level that diminishes with temporal separation.[13]

28.4 APPLICATION: CLEAN AIR ACT ENFORCEMENT

Given the wide array of considerations that must go into the analysis of public policy, this chapter concludes by working through the complex enforcement of the Clean Air Act (CAA). Clean air regulation requires activity from several levels of American government. The United States Environmental Protection Agency (EPA) and environmental agencies from each of the fifty states all play prominent roles in the process. Given that each state will likely have a unique preference on how much it should regulate business in the interest of curbing pollution, it stands to reason that enforcement will not be even-handed across the nation. Hence, this study asks: How do state-level factors influence implementation of the Clean Air Act?

Some background on clean air law is key to understanding the way government activity is divided. Originally, Congress viewed pollution abatement as a policy falling entirely within the domain of the states. In fact, early laws continued to recognize states' rights to enforce pollution control as they saw fit. For example, the original Clean Air Act of 1963 was fairly limited, and its main provisions included, among other items, a call for the development of air quality criteria, matching funds to state and local agencies for regulatory control, and the expansion of federal authority to intervene when pollution in one state might endanger the health of citizens in another state (Wark et al., 1998, page 63).

Power over environmental policy gradually shifted to the federal level, though. In the early 1960s, several states began developing emissions requirements for automobiles, and car manufacturers became concerned about developing vehicles that could meet fifty specialized sets of requirements (Zimmerman, 2005, page 75). As a result, the CAA was amended by the Motor Vehicle Air Pollution Control Act of 1965, which adopted on a national basis the automobile standards California had at the time, effective starting with 1968 model cars (Wark et al., 1998, page 64).

Yet, federal clean air regulations truly gained muscle with the Clean Air Act Amendments of 1970. This act established national standards for two general types of pollutants, criteria air pollutants and hazardous air pollutants. States were to develop state implementation plans (SIPs) by which they would meet emissions requirements. Though the states held the primary responsibility for regulating emissions, the EPA had oversight for whether their SIPs would actually meet federal timetables for air quality standards. Besides establishing standards and requiring SIPs, the 1970 amendments also established additional grants to the states, granted the EPA the right of entry to examine industry emission records, imposed fines and criminal penalties for violating implementation plans, and substantially expanded citizens' rights to sue for violation of emission standards. Overall, then, the EPA gained substantial authority in both its oversight of SIPs and the many other enforcement avenues created by these amendments (Wark et al., 1998, pages 66–8).

Further revisions to the law occurred with the Clean Air Act Amendments of 1990. This law allows the EPA to require states to make further progress in curbing air pollution when reviewing SIPs (Zimmerman, 2005, pages 171–2). In this review process, states are required to show through modeling how regulations can bring nonattainment areas into compliance with federal standards. Additionally, it requires states to develop comprehensive emission inventory tracking systems to determine progress toward air standards in nonattainment areas (Wark et al., 1998, pages 94–7).[14] Congress, therefore, strengthened federal authority in 1990.

To understand how these various provisions in a hierarchical federal structure were implemented in various locales, Wood (1992) studied data from 1977–1985 and found that a variety of centralizing and decentralizing factors influenced outcomes. Coming from the national level, prior enforcement activity by the Environmental Protection Agency (EPA) and EPA grants-in-aid to the states for clean air purposes both prompted the states to raise their level of enforcement activity. Local variation, though, did provide unique outcomes in the level of enforcement. For example, state-level political pressure and institutional control shaped the number of state enforcement actions, meaning that Democratic control of elected offices raised the level of enforcement, higher unemployment diminished enforcement, and a greater number of CAA-based lawsuits originating in a state raised the number of actions. Further, as a state confronted larger pollution problems, in the forms of higher levels of emissions or higher levels of manufacturing output, the state made greater efforts to enforce the CAA. Finally, state activity in a given year tended to be similar to the previous year's behavior, potentially due to bureaucratic inertia.

The previous study by Wood offers clear priors of what should shape state behavior. Since this time frame, though, the 1990 CAA amendments have substantially changed the law, so there is a strong possibility that the manner in which states behave has substantially changed. In particular, since the 1990 amendments gave greater power to the EPA, the upshot may be that clean air enforcement is more responsive to centralizing factors and less responsive to local context. Therefore, following the covariate specification found in Wood's study, this study analyzes data from a more recent time frame, 2001–2009.

28.4.1 Data and Methods

In this study, the policy outcome of interest is the number of enforcement actions each state took against air polluters.[15] These data are longitudinal because the individual states are observed every year from 2001 to 2009. The advantage of these data is that they will allow us to assess how temporal changes in factors such as partisanship influence state implementation. The input variables are: lagged state-level actions, lagged federal-level actions, partisan control of the state government, the state unemployment rate, state manufacturing outputs, EPA grants to the state, and environmental lawsuits originating from a state.[16]

As the discussion of this chapter indicates, analyzing policy data properly can be complex and requires attention to many facets of the data. In this case, a critical attribute of these data is the fact that they consist of repeated observations of the same states. Figure 28.1 shows line plots of the number of enforcement actions taken in the period 2001–2009 by each of the states. The horizontal axis is the year of activity, and the vertical axis is the number of acts on a logged scale. Each line represents a state. As the figure illustrates, much of the variance is between states, rather than within them. Therefore, addressing unit heterogeneity in some way is critical.

Second, the outcome variable is a count. Thus, a standard linear model is inappropriate for this outcome. It makes more sense to assume this outcome has a Poisson distribution, rather than a Normal distribution. As can be seen, though, a dynamic model is more difficult for a generalized linear model.

Third, a state's activity is unlikely to be independent of the actions of other states. Rather, neighboring states are likely to share similar unobserved qualities, thereby making their behavior in clean air enforcement similar, even in the context of a broad model. Therefore, it would be useful to account for such spatial correlation in the model.

Finally, and perhaps most importantly, the causal flow from the predictors to the outcome is likely not to be static, but rather a dynamic relationship. In other words, a change in an input at one time is likely to affect the outcome for several years rather than only concurrently. If, in fact, the causal processes are dynamic, then failure to properly specify this can be problematic. Wood (1992) crafted a dynamic linear model, though he did so within a linear framework, meaning that including a lagged dependent variable was sufficient for such a specification. As Brandt and Williams (2001) point out, however, to specify

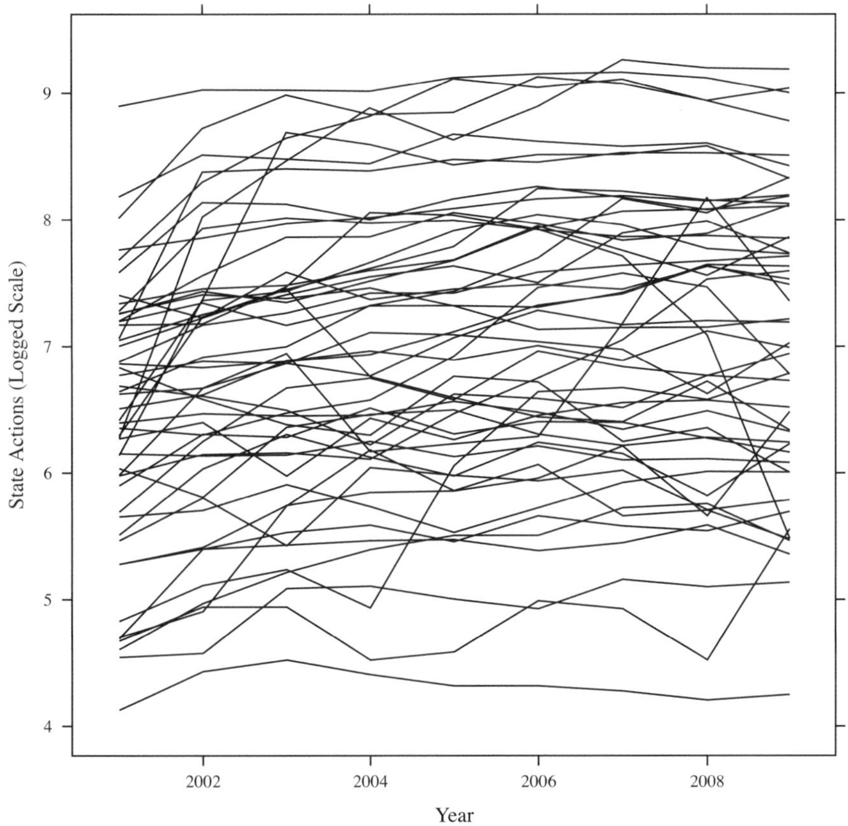

Figure 28.1 Clean Air Act Enforcement Actions by State and Year

a dynamic model in the context of count data requires the complexities of a state-space model.

As a solution to the issues of unit heterogeneity, a count outcome, spatial correlation, and a dynamic functional form, I estimate a state-space model for panel data that is designed in Blundell et al. (2002) and similar to the model Brandt and Williams (2001) use for a single time series.[17] The model is specified as follows:

$$\begin{aligned} y_{it} &\sim \mathcal{P}(\lambda_{it}) \\ \lambda_{it} &= \rho y_{i,t-1} + \exp\{\mathbf{x}_{it}\boldsymbol{\beta} + \phi_i\} \\ \beta_j &\sim \mathcal{N}(0, 100) \\ \rho &\sim \mathcal{B}(2, 5) \\ \phi_i &\sim CAR(1/\tau_c) \\ 1/\tau_c &\sim \mathcal{G}(.1, .1). \end{aligned} \quad (28.5)$$

Here, y_{it} is the number of enforcement actions taken by state i in year t, λ_{it} is the Poisson parameter, ρ is the first-order time-autoregressive parameter, $\boldsymbol{\beta}$ is a vector of population parameters, β_j for $j = 0, \ldots, 8$ refers to the individual elements of $\boldsymbol{\beta}$, and ϕ_i is the state-level spatial random effect. $\mathbf{x}_{it}$ is the vector of all covariates for an observation including a lag of standardized federal enforcement actions, an indicator for unified Democratic control of state government, an indicator for unified Republican control of state government, the unemployment rate, manufacturing outputs, EPA grants, the number of lawsuits related to clean air originating in a state, and a constant. Equation (28.5) also shows that the coefficients have vague Normal priors, the autoregressive parameter has a Beta prior to constrain it to the interval $\rho \in (0, 1)$, the spatial random effects have a conditionally autoregressive (CAR) prior, and the hyperprior on the precision of the random effects has a vague Gamma distribution.

In choosing the model specified in (28.5), one clear alternative would be a simpler generalized linear mixed model—following the approach of Chapter 8—that did not incorporate an autoregressive model of the Poisson parameter. Either approach is valid, but I chose to take this route because it allows for a dynamic specification of the model, which corresponds to what Wood (1992) did. If I had fewer time points, or was less inclined to believe there was a dynamic effect, a more standard generalized linear mixed model would have seemed a more sensible solution.

By specifying that the random effects are conditionally autoregressive, states are assumed to have similar average values beyond what would be expected from the broader model. Additionally, the inclusion of the time lag in the model for the mean parameter produces dynamic causality. This happens because, as the values of covariates shape the outcome at one time point, they will have an indirect effect on the next time point through their effect on the contemporaneous value. Thus, this model proposes one solution for these four issues: repeated observations of the same states, dynamic functional form, spatial correlation, and a count process as the outcome variable.[18]

28.4.2 Results

Table 28.2 reports the results of the model presented in (28.5). Many of the results align well with what Wood (1992) reports. Just as then, both enforcement actions taken by the federal EPA one time period ago and EPA grants to states lead states to take greater action in the enforcement of the Clean Air Act. Therefore, the EPA still has a centralizing effect on clean air policy. Further, clean air enforcements tend to be more common in states governed by the Democratic party, just as Wood found. Noteworthy, though, is that unified governments tend to enforce the Clean Air Act more rigorously than divided governments, with even unified Republican government enforcing the law more strongly than a divided government. Regarding the local economy, the consequence of having a larger manufacturing output continues to raise the level of activity by a state agency, serving as a particularly powerful predictor of the level of enforcement actions. In a bit of a contrast since the prior time frame, neither the number of lawsuits initiated by groups nor the level of unemployment exhibit a robust effect. With these two small exceptions, the results that Wood reported seem to be robust overall across time and with new methods.

Besides the effect of covariates, the model also resembles Wood's results in that it shows a strong autoregressive tendency over time. The mean autoregressive parameter in the model from Table 28.2 is $\hat{\rho} = 0.699$, meaning that, just as in the 1977–1985 time frame, a change in an independent variable continues to shape a state's clean air policy for years to come. Adding the spatial CAR specification also adds new information to analyze, though. In this case, the variance of the random effects in the spatial model has a posterior mean of 3.1133.[19] This result implies that there is clustering in state outcomes, and a substantial proportion of the unexplained variance can be accounted for by recognizing that, holding the input variables equal, neighboring states are likely to take similar actions on clean air policy.

28.5 SOFTWARE FOR MULTILEVEL POLICY ANALYSIS

The policy analyst wishing to use the methods described in this volume has a variety of software tools available in order to do so.

Table 28.2 Model of State Clean Air Act Enforcement, 2001–2009 (Posterior Summaries)

Input Variable	Mean	Std Dev.	[90% Cred. Int.]
Lagged EPA action	0.071	0.006	[0.061, 0.080]
Democratic unified government	0.057	0.004	[0.050, 0.064]
Republican unified government	0.022	0.005	[0.014, 0.030]
% unemployment	0.004	0.004	[−0.002, 0.010]
Manufacturing outputs	0.185	0.012	[0.166, 0.205]
EPA grants	0.028	0.003	[0.023, 0.033]
Environmental lawsuits	0.002	0.004	[−0.005, 0.008]
Intercept	5.902	0.012	[5.881, 5.922]
Autoregressive term ($\hat{\rho}$)	0.699	0.004	[0.692, 0.706]
Precision for spatial random effects ($\hat{\tau}_c$)	0.321	0.065	[0.222, 0.436]

Notes: $N = 50$. $T = 9$. $DIC = 40358.6$. Estimates computed in WinBUGS 1.4.3.
Results based on a post burn-in MCMC sample of 270,000.
The burn-in period was 30,000 iterations. All input variables were standardized.

R, Stata, SAS, and WinBUGS are just a few of the available options. This chapter includes an example of a model fit using maximum likelihood methods in R (the smoking prevention study reported in Table 28.1) as well as a model fit using Bayesian methods in WinBUGS (the Clean Air Act analysis reported in Table 28.2). Scholars wishing to try this code are encouraged to download the data and software code from the author's dataverse, `http://hdl.handle.net/1902.1/17800`.

Each of the available programs has its benefits. R is a freely available program for data analysis that has the added benefit of easily incorporating user-contributed libraries. Within R, the `lme4` library, for instance, provides an extremely versatile array of multilevel methods. By contrast, Stata and SAS are not free options, but they have found wide use in many circles. Commands such as `xtreg` in Stata and `proc mixed` in SAS (among others) allow estimation of multilevel models for analysts preferring to use these programs. Finally, WinBUGS is a freely available program that allows immense user flexibility in model specification. By simply specifying the distribution of random effects and appropriately placing them in a model, the user can easily take a Bayesian approach to hierarchical modeling (as is further described in Chapter 4).[20]

28.6 IMPLICATIONS FOR THE PRACTICAL RESEARCHER

This chapter has aimed to illustrate the usefulness of multilevel models in order to yield more accurate inferences in the analysis of policy data. In doing so, hopefully a few clear ideas stand out for the applied policy analyst. First and foremost, whenever data are clustered or longitudinal in nature, unit heterogeneity absolutely must be addressed in some way. Including random intercepts can correct for the intraclass correlation that results in inefficiency and false inferences. In cases where a fixed effect model would not impede evaluation of a treatment effect, it also is a good idea to weigh whether the random effects model is appropriate with a Hausman test (Cameron and Trivedi, 2005, pages 271–4).

While unit effects are often going to be the biggest threat to accurate inference, empirical models should also reflect the complex aspects of policy data. In other words, the applied researcher should carefully consider what assumptions are fair to make and how the model can be built to reflect the data accurately. This may require the specification of a generalized linear mixed model for a limited dependent variable (i.e., binomial or count) or perhaps specifying something like a duration model which entertains censoring in measures. Further, if the data have additional complications beyond the autocorrelation of clustering or repeated observations, such as geographic dependence, then entertaining these can not only yield more accurate inferences in substantive conclusions, but can offer new quantities that are interesting to report (such as the level of neighbor similarity).

As a final point, policy analysts may want to consistently ask: "Are my data a random sample from a broader population?" If they are, then the techniques of classical inference are more suitable for analyzing the data. However, data on policy outcomes often are based on facts, not a sample of independent events. Further, policy studies often can exhaust an entire population of governing bodies in data collection, undermining the assumptions behind traditional inference techniques such as significance testing. Therefore, the researcher answering "no" and concluding that the data are not a sample in the traditional sense should consider shifting to the Bayesian paradigm. Using these techniques, the data are assumed fixed and the parameters are random, which is sensible for policy scholars who have full information about programs and social outcomes but would rather make probabilistic statements about model parameters.

The key point is that policy data have unique aspects that require careful attention and can often lead to complex models. However, sources such as this volume lay out important techniques for remedying these issues, and how most issues can be resolved with thoughtful modeling. Of course, clustered, time-dependent, and geographically oriented data pose difficulties, but also offer unique opportunities for assessing causality and entertaining unique kinds of relationships. Thus, the complexities of policy data can also be a benefit in the development of deeper understanding of the development and impact of programs.

NOTES

1 I would like to thank Jeff Gill, Christopher Faricy, Christopher Witko, Bernd Weiss, the editors, and an anonymous reviewer for their gracious assistance. For sharing data, I thank Brian Flay and Garrett M. Fitzmaurice. For coding data used in the Clean Air Act analysis, I thank Dustin Elliott and Ryan Williamson. Replication code and data are available at the author's dataverse: `http://hdl.handle.net/1902.1/17800`.

2 However, other substantive areas of political science make more use of these techniques. Steenbergen and Jones (2002) offer an excellent exposition into how multilevel models can be usefully applied in political science.

3 A more comprehensive explanation of the multilevel model framework can be found at the start of this book and in Chapter 1.

4 Conditions and assumptions for causal inference are explored in detail in Chapter 12. In this chapter, I may refer to interpretations of model parameters that are suggestive of cause and thus potentially useful in forming policy. The reader interested in causal interpretations, however, should consult Chapter 12 for more details.

5 Note that frequentist methods typically do not estimate random effects directly. Bayesian estimation, however, can generate posterior estimates of random effects. Even when using a technique that does not directly estimate random effects, researchers wishing to know the values of random effects can predict them with a variety of techniques. The best linear unbiased predictor (BLUP) is one predictor of note, which is described along with other methods in Chapter 3 of this volume.

6 Chapter 9 offers a more thorough description of longitudinal data modeling.

7 For the analysis of experimental data, it is important to consider how best to conduct causal inference in a multilevel framework. Chapter 12 describes how to do this.

8 Naturally, in any case where variables are omitted, omitted variable bias is a possible concern. Thus, any proposed remedy for spatial dependence is designed to be considered above and beyond all efforts to include essential variables.

9 Also, because responses from neighboring zip codes correlate at a certain rate, neighbors of neighbors implicitly correlate at a lower rate (and third-degree neighbors at a still lower rate), which seems plausible.

10 Although the replication of Flay et al. (1995) did not explicitly incorporate the logic of causal modeling, this can be useful in the analysis of experimental data. See Chapter 12.

11 Chapter 5 considers the choice between fixed and random effects in depth.

12 It should be noted that the terms fixed and random effects have multiple meanings, which can be a source of confusion. Econometricians typically call an LSDV model a fixed effects model and a multilevel, longitudinal model with random intercepts

a random effects model. These terms can take a different meaning within the context of using multilevel models. Specifically, in a multilevel model a fixed effect refers to any model quantity estimated in the fitting of a model (i.e., obtained via least squares or maximum likelihood), while a random effect refers to any parameter that is unique to the individual but can be predicted separately.

13 Additional details on complexities such as these can be found in Chapter 10. The random effects model of (28.3) can also be referred to as a "compound symmetry" error structure, whereas first-order autoregressive or Toeplitz error structures can more easily account for temporal decay in serial correlation. Models with error structures like these can be estimated using techniques such as feasible Generalized Least Squares (fGLS, of which iterative algorithms include Prais–Winsten and Cochrane–Orcutt), maximum likelihood, or restricted maximum likelihood (REML).

14 A novel feature of these amendments was that they authorized the trading of emission allowances, meaning that plants which keep emissions below the required maximum can sell their additional allowance to another company (Zimmerman, 2005, pages 171–2).

15 State and federal clean air enforcement actions were downloaded from `http://www.epa-echo.gov/echo/idea_download.html` on 16 January 2011. Only 2001–2009 are studied because the EPA website states "the data quality before November 2000 has not been assessed and should be considered unknown." This source also provides the annual number of federal-level actions in each state.

16 These covariates were measured from the following sources: Partisan control of state government, Klarner (2003), `http://www.indstate.edu/polisci/klarnerpolitics.htm`, accessed 3 March 2012. Manufacturing outputs by state (in chained 2000 dollars), Bureau of Economic Analysis, Regional Economic Accounts, `http://www.bea.gov/regional/` accessed 17 July 2010. Unemployment by state, Bureau of Labor Statistics, `http://data.bls.gov/`, accessed 18 July 2010. Non-construction grant allotments, Environmental Protection Agency, `http://yosemite.epa.gov/oarm/igms_egf.nsf/HomePage?ReadForm`, accessed 12 April 2011. Lawsuits under the Clean Air Act, Bureau of National Affairs (2010).

17 This model uses one of several possible techniques to deal with a dynamic functional form wherein an input shapes the outcome over time. There are many other approaches that specifically focus on causal inference in a dynamic setting. Robins et al. (2000), for example, describes how marginal structural models can be used whenever there are time-dependent confounders that are also affected by previous treatments. Blackwell (2010) more specifically illustrates how to use inverse probability of treatment weighting to estimate a marginal structural model.

18 Again, alternatives such as those proposed by Robins et al. (2000) and Blackwell (2010) may be more appropriate in many cases.

19 This is simply the posterior mean of $1/\hat{\tau}_c$. The 90% credible interval for the variance of the random effects is [2.296,4.501].

20 The primary drawback of WinBUGS, however, is that it only runs on the Windows operating system. JAGS is a similar alternative that runs on other operating systems, though there are some differences in the programs. JAGS, for example, currently does not have a module for spatial data analysis quite like GeoBUGS in WinBUGS.

REFERENCES

Allison, P. (2005) *Fixed Effects Regression Models for Longitudinal Data Using SAS*. Cary, NC: SAS Publishing.

Banerjee, S., Carlin, B.P., and Gelfand, A.E. (2004) *Hierarchical Modeling and Analysis for Spatial Data*. New York: Chapman and Hall.

Berry, F.S. and Berry, W.D. (1990) 'State Lottery Adoptions as Policy Innovations: An Event History Analysis', *The American Political Science Review*, 84: 395–415.

Berry, F.S. and Berry, W.D. (2007) 'Innovation and Diffusion Models in Policy Research', in *Theories of the Policy Process*, Sabatier, P.A. (ed.). Boulder, CO: Westview Press.

Blackwell, M.L. (2010) 'All Politics is Dynamic: How Time and Strategy Mask the Importance of Negative Advertising'. Paper presented at the Annual Meeting of the American Political Science Association, Washington, DC.

Blundell, R., Griffith, R., and Windmeijer, F. (2002) 'Individual Effects and Dynamics in Count Data Models', *Journal of Econometrics*, 108: 113–31.

Box-Steffensmeier, J.M. and Jones, B.S. (2004) *Event History Modeling: A Guide for Social Scientists*. New York: Cambridge University Press.

Brandt, P.T. and Williams, J.T. (2001) 'A Linear Poisson Autoregressive Model: The Poisson AR(p) Model', *Political Analysis*, 9: 164–84.

Bryk, A.S. and Raudenbush, S.W. (1987) 'Application of Hierarchical Linear Models to Assessing Change', *Psychological Bulletin*, 101: 147–58.

Bureau of National Affairs (2001–2010) *Environment Reporter—Cases*, volume 51–70. Washington, DC: The Bureau of National Affairs, Inc.

Cameron, A.C. and Trivedi, P.K. (2005) *Microeconometrics*. New York: Cambridge University Press.

Fitzmaurice, G.M., Laird, N.M., and Ware, J.H. (2004) *Applied Longitudinal Analysis*. Hoboken, NJ: Wiley-Interscience.

Flay, B.R., Miller, T.Q., Hedeker, D., Siddiqui, O., Britton, C.F., Brannon, B.R., Johnson, C.A., Hansen, W.B., Sussman, S., and Dent, C. (1995) 'The Television, School And Family Smoking Prevention And Cessation Project: VIII Student Outcomes And Mediating Variables', *Preventive Medicine*, 24: 29–40.

Gelman, A. (2006) 'Multilevel (Hierarchical) Modeling: What It Can and Cannot Do', *Technometrics*, 48: 432–5.

Gill, J. (2001) 'Whose Variance is It Anyway? Interpreting Empirical Models with State-level Data', *State Politics and Policy Quarterly*, 1: 318–38.

Gray, V. (1973) 'Innovation in the States: A Diffusion Study', *The American Political Science Review*, 67: 1174–85.

Grossback, L.J., Nicholson-Crotty, S., and Peterson, D.A.M. (2004) 'Ideology and Learning in Policy Diffusion', *American Politics Research*, 32: 521–45.

Klarner, C.E. (2003) 'Measurement of the Partisan Balance of State Government', *State Politics and Policy Quarterly*, 3: 309–19.

Neuhaus, J.M. and Kalbfleisch, J.D. (1998) 'Between- and Within-cluster Covariate Effects in the Analysis of Clustered Data', *Biometrics*, 54: 638–45.

O'Hara, P., Connett, J.E., Lee, W.W., Nides, M., Murray, R., and Wise, R. (1998) 'Early and Late Weight Gain Following Smoking Cessation in the Lung Health Study', *American Journal of Epidemiology*, 148: 821–30.

Robins, J.M., Hernan, M.A., and Brumback, B. (2000) 'Marginal Structural Models and Causal Inference in Epidemiology', *Epidemiology*, 11: 550–60.

Steenbergen, M.R. and Jones, B.S. (2002) 'Modeling Multilevel Data Structures', *American Journal of Political Science*, 46: 218–37.

Wark, K., Warner, C.F., and Davis, W.T. (1998) *Air Pollution: Its Origin and Control*, 3rd edition. Reading, MA: Addison-Wesley.

Wood, B.D. (1992) 'Modeling Federal Implementation as a System: The Clean Air Case', *American Journal of Political Science*, 36: 40–67.

Zimmerman, J.F. (2005) *Congressional Preemption*. Albany: State University of New York Press.

29

Multilevel Models in the Social and Behavioral Sciences

David Rindskopf
CUNY Graduate Center, USA

29.1 OVERVIEW

Multilevel models were developed and adopted very quickly in the social sciences during the 1980s and 1990s. Two factors seem to have caused this rapid change. First, social scientists generally knew that nested data should not be treated as if the observations were independent, but had been stymied by the lack of algorithms or software to treat the data correctly. Secondly, new ways of thinking about these models contributed to their quick adoption: rather than being just statistical necessities, the models could be expressed in ways that gave the parameters interesting and useful subject-matter interpretations. This interpretation arose in sociology through the work of William Mason and collaborators (e.g., Mason, Wong, and Entwisle, 1983), and education and psychology through the work of Anthony Bryk and Stephen Raudenbush (e.g., Raudenbush and Bryk, 1986), Harvey Goldstein (e.g., Goldstein, 1986), and Murray Aitkin (e.g., Aitkin and Longford, 1986), and derived from the conceptualization of Lindley and Smith (1972).

These interpretations arose in two general areas: individuals nested within higher-level units (students within classes or schools, residents within geographical or political areas, etc.), and longitudinal data (observations nested within individuals). In the first case, the interpretation was that (i) each higher-level unit would have a regression equation, with its own intercept and slopes; (ii) the intercepts and slopes might vary across these higher-level units; and (iii) characteristics of the higher-level units might help explain variation in the intercepts and/or slopes. In the second case, the procedure would be similar: (i) the growth of each individual would be modeled; (ii) the parameters of the model might vary across individuals; and (iii) individual (time-invariant) characteristics of the individuals might explain some of the variation across individuals in the parameters.

These models were originally developed with linear regression as the model and with two levels. Over the years, extensions were instituted in every aspect of these models. Most obviously, many models would have more than two levels. For example, we could have students nested within classes nested within schools nested within districts (and so on). Or one could combine growth models with nesting and have observations nested within individuals nested within classes (and so on). The dependent variable might not be normal or continuous, so models were extended to include many varieties of nonlinear models, mostly in the class of generalized linear models. These models were soon seen to be interpretable as latent variable models, and this led to further extensions of latent growth and other latent variable applications of nested models. This chapter explores the use of these models, their expansion to many areas not originally considered when the models were first developed, and some possibilities for additional uses of the models.

Another area of application that would not be obvious is meta-analysis, which is the quantitative summary of studies that all relate to a single subject area. Early applications included the analysis of studies on the benefits of psychotherapy, the effects of small class size on student achievement, and the Pygmalion Effect in psychology. Each study is characterized by an effect size, and possibly other variables that represent different characteristics of the studies. Interest lies in how consistent the studies are in their estimate of an effect and, if they are inconsistent, how much of the inconsistency can be explained by study characteristics. Nesting occurs implicitly, because research participants are nested within the studies; the individual-level data are (usually) not available, so there is only an equation for the aggregate data.

29.2 VARIANCE COMPONENTS

Before the full multilevel model was developed, there were models for separating the total variance among a set of numbers into the contribution due to different influences. One of the most obvious was in education, where it was of considerable interest to know how much variation among test scores was due to variation among schools, and how much was due to variation among students within schools.

The representation of this model in terms of equations at each level is

$$y_{ij} = \beta_{0j} + \varepsilon_{ij}$$

for the first (student) level, and

$$\beta_{0j} = \gamma_{00} + u_{0j}$$

for the second (school) level. In words, a student's score is the sum of a school mean plus a term showing how much the student deviates from the school mean; and the school mean is the sum of an overall mean plus a term showing how much that school's mean differs from the overall mean.

Combining the two equations by substituting the right-hand side of the Level 2 equation into the Level 1 equation gives the combined model

$$y_{ij} = \gamma_{00} + u_{0j} + \varepsilon_{ij},$$

with variance components $Var(u_{0j}) = \tau_{00}$ and $Var(\varepsilon_{ij}) = \sigma^2$. In words, a student's score is the sum of an overall mean, plus a term showing how much the student's school mean differs from the overall mean, plus a term showing how much the student's score differs from the school mean. The variance components represent variation of school means around the overall mean, and variation of student scores around their school means.

The intraclass correlation, which represents the proportion of total variation that is between schools, is

$$\rho = \frac{\tau_{00}}{\tau_{00} + \sigma^2}.$$

In the United States, it is the common belief that there are "good" schools and "bad" schools, which implies that schools are relatively homogeneous (little variation within schools), but that there are large differences among schools. In technical terms, the intraclass correlation would be expected to be large. The reality is quite surprising to most people: the variation among public schools is generally around 20 to 30 percent, and within schools between 70 and 80 percent for standard academic subjects. That is, within most public schools, there is a wide range of performance.

29.3 A SIMPLE EXAMPLE OF A MULTILEVEL MODEL

Consider the example from Chapter 2, where students are nested within schools. The outcome is student achievement on a test of history and related topics; the student-level predictor is Sex (coded using effects coding) and the school-level variable is proportion of students receiving free or reduced-price lunch (FSL). The subscript i will refer to individual students, and j to schools.

For those new to multilevel models, it is useful to think of the data as coming from a collection of studies, one in each school. In each school a researcher has collected data on a sample of students, and is doing a regression to predict test score from the sex of the student. So, in school j, the regression equation (called the Level 1 equation) is

$$y_{ij} = \beta_{0j} + \beta_{1j}\, Sex_{ij} + \varepsilon_{ij}.$$

As the subscripts show, each school could have a different intercept and slope. Then, we imagine a "meta-researcher" who wants to combine the results from the regressions in each school, and is interested in issues such as (i) what is the average intercept and slope? (ii) how much variation is there in the intercepts and slopes, over and above sampling variation? (iii) if the variation is not negligible, how much variation across schools in the intercepts and/or slopes can be explained by school characteristics? In this example, one such school-level variable, FSL, is used to try to account for variation in intercepts and slopes, resulting in the following Level 2 equations:

$$\beta_{0j} = \gamma_{00} + \gamma_{01}\, FSL_j + u_{0j}$$
$$\beta_{1j} = \gamma_{10} + \gamma_{11}\, FSL_j + u_{1j}.$$

The remaining parts of the model are variances and covariances of the error terms in each equation. The level-1 error is usually assumed to be independent of the level-2 errors, so in this case we have three variances, $Var(\varepsilon_{ij}) = \sigma^2$, $Var(u_{0j}) = \tau_{00}$, $Var(u_{1j}) = \tau_{11}$, and one covariance, $Cov(u_{0j}, u_{1j}) = \tau_{01}$. The covariance is seldom important in typical hierarchical models, and is omitted from many computer program outputs. (Note that one additional assumption is often made, that the error variance σ^2 in each school is the same. This assumption is testable, and models can be extended to allow heterogeneous variation in this respect.)

As noted in Chapter 2, multilevel models are expressed in different, often equivalent, ways. Why is the system above so common in the social sciences? I believe it is because of the conceptual interpretation that is natural in this framework: a regression model holds for each group, and although the form of the model is the same across groups, the regression coefficients can vary across groups. Some of this variation in coefficients can be accounted for by characteristics of the groups, while some variation is unaccounted for. Statistically, the nature of the group is irrelevant, but for the social scientist it is very important. The group represents some context or environment; a particular form of model is believed to hold in each such context, but the values of the parameters may vary across contexts, and we want to estimate and account for that variation. The context may be (among other things) classrooms, schools, neighborhoods, work groups, election districts, cultural groups, or countries. In each case we are interested in commonalities and differences among these contexts.

One other advantage of this way of expressing the model is that it makes obvious what the fixed and random effects are. It is obvious that only the coefficients in the Level 1 equation can vary randomly at Level 2. Another advantage is that it is consistent with the way some computer software packages (most notably HLM) represent the model.

One disadvantage with this form of notation is that it is not obvious that some effects represent interactions; in the model above, γ_{11} is the interaction between the level-1 variable Sex and the level-2 variable FSL. This would only be obvious in the combined-equation forms of the model presented next.

The combined-equation form results from substituting the right-hand sides of the Level 2 equations into the Level 1 equation. For our example, this becomes

$$\begin{aligned} y_{ij} &= \gamma_{00} + \gamma_{01}\, FSL_j + u_{0j} \\ &\quad + (\gamma_{10} + \gamma_{11}\, FSL_j + u_{1j})\, Sex_{ij} \\ &\quad + \varepsilon_{ij}. \end{aligned}$$

This can be rearranged in different ways to emphasize certain important characteristics of the equation. One common way of rearranging the combined equation is

$$\begin{aligned} y_{ij} &= \gamma_{00} + \gamma_{01}\, FSL_j \\ &\quad + \gamma_{10}\, Sex_{ij} + \gamma_{11}\, Sex_{ij}\, FSL_j \\ &\quad + u_{0j} + u_{1j}\, Sex_{ij} \\ &\quad + \varepsilon_{ij}. \end{aligned}$$

This representation emphasizes the difference between the fixed effects (with γ coefficients), the level-2 random effects (u terms), and the level-1 random effect (ε term). In the matrix form discussed in Chapters 1 and 2, but using notation similar to that used above, this way of writing the model would be expressed as

$$\mathbf{y} = X\boldsymbol{\gamma} + Z\mathbf{u} + \boldsymbol{\varepsilon}.$$

Another way to rearrange the equation is

$$\begin{aligned} y_{ij} &= \gamma_{00} + u_{0j} \\ &\quad + (\gamma_{10} + u_{1j})\, Sex_{ij} \\ &\quad + \gamma_{01}\, FSL_j \\ &\quad + \gamma_{11}\, Sex_{ij}\, FSL_j \\ &\quad + \varepsilon_{ij}. \end{aligned}$$

In this representation it is easy to see that the intercept and effect of Sex are random; that is, they vary across schools.

These two representations are useful in part because several computer programs express models in one of these two ways.

29.4 ILLUSTRATIVE EXAMPLES FROM SOCIOLOGY AND POLITICAL SCIENCE

The main illustrative examples I use come from education and psychology; I also mention some applications in other social science areas. One of the first applications of multilevel models was in sociology. Mason, Wong, and Entwisle (1983) discussed multilevel models as contextual models; that is, a different relationship among variables might hold in different contexts. In this case, the contexts were cultures, for which a proxy was the country in which women lived. The outcome to be explained was the number of children ever born to each married woman aged 40–44 years. The individual-level variables used to explain this were education level and whether the woman was raised in an urban versus rural environment. The contextual (country-level) predictors were the gross national product (a measure of economic development) and a measure of the country's family planning effort.

(If these data were fit today, the model used would probably treat the counts as Poisson, possibly with inflation of the number of families with no children.)

In political science, the case is the same as in most other social sciences: nesting is ubiquitous. In addition to obvious repeated measures (surveys that follow the same people over time, for example), political scientists might study voters in election districts (nested within counties and then states), or people nested within countries (for cross-national research). Duch and Stevenson (2005) studied the theory of economic voting across a number of countries and across a number of surveys in each country. The outcome was whether the voter reported voting for the incumbent party; economic voting theory says that voters will reward or punish the incumbent party depending on economic conditions, as well as personal characteristics of the voters (such as their left/right leaning, gender, age, etc.). Duch and Stevenson tested a multilevel model in which ideology and perception of the state of the economy were the principal individual-level predictors, and concentration of authority in the country was the principal explanatory variable at the country level.

29.5 CROSSED RANDOM EFFECTS

Most multilevel models involve nesting of random effects, such as students (randomly sampled) nested within schools (randomly sampled from a population of schools). Some also involve crossed random effects. One researcher might follow students over a number of years, and would have primary school crossed with secondary school attended. Another might have students classified by both the school they attend and the neighborhood within which they live (in cases where school attendance is not determined strictly by neighborhood). Such designs, with primary school crossed by secondary school, or schools crossed by neighborhoods, is often sparse in the sense that many combinations have no students; depending on the pattern of this sparseness there may be problems either of estimation or interpretation of the results.

Methodology and two example analyses of studies involving crossed random effects are described in Raudenbush (1993a). In one of the examples, a reanalysis of data from Garner and Raudenbush (1991), students in Scotland are nested within neighborhoods and schools, with neighborhoods and schools crossed. The design is very sparse, in that most neighborhoods have students from only a few schools, so the estimation of interactions between the random effects was not possible. An interesting result from the analysis was that almost all of the differences among schools and among neighborhoods were accounted for by student characteristics and by the level of deprivation of the neighborhood in which they live (which had a large and significant effect).

29.6 CENTERING PREDICTORS, AND EXTENSIONS

In ordinary regression, little attention is paid to the scaling of predictor variables, because a linear (unlike a nonlinear) transformation changes nothing essential about the nature of the prediction equation. It does change the meaning of the intercept, but in the social sciences the intercept seldom is of interest. With multilevel models, the situation is different, because the intercept of the equations at one level are explained by equations at the next highest level of the model. If the intercept is not interesting, then explaining variation in the intercept across individuals is also not interesting. If the intercept can be made interesting by a transformation, then explaining variation in the intercept also becomes interesting. Centering is discussed in detail in Chapter 6, so here I will only give a few simple examples relevant for situations I will discuss.

Before going to the multilevel case, consider first an example using simple linear regression to predict weight (in pounds) from height (in inches); suppose that the prediction equation from some set of data is

$$\hat{w} = -300 + 7h,$$

where w is weight and h is height.

In regression, the intercept is the predicted value on the outcome (weight) when all of the predictors (here, just height) are 0. In this case, we predict that someone who has a height of 0 will weigh -300 pounds; all regression books warn against making such predictions, which use values outside the observed (or even anticipated) range of values on the predictor. In ordinary regression this causes no harm, because we would not care to interpret the intercept, but in multilevel models the intercept is modeled at the next higher level, so it should have a reasonable interpretation.

One way to give the intercept a reasonable interpretation is to center the predictors by subtracting the mean from all observations; then the intercept is not only interpretable, but it is the mean on the outcome variable. There are other advantages of mean centering, mostly in terms of improved numerical estimation, but these are generally not relevant to interpretation of the results. The literature has generally not recognized that there are other (linear) transformations besides centering that can be useful.

First, in some instances it is more useful to do approximate centering. After all, the mean is usually estimated, not known, so subtracting a round number close to the mean may make interpretation easier in some cases. For example, if the mean of X is 29.2, subtracting 30 from all scores rather than 29.2 would make it easier to interpret scores on the new scale and translate back to the original scale without using a calculator.

In other cases, scales have values that are meaningful for diagnostic purposes, e.g., for diagnosing depression. One might want to subtract the cut score from all values, so that on the transformed scale a negative value means the person is not depressed, and a positive value means that the person is depressed. There is no term more general than centering in current use to describe this type of linear transformation, but three possibilities are *rescaling*, *translation*, and *re-expression*.

In the analysis of single-case designs (discussed later) or other time series data, there is often a time point where treatment phases change. In these cases, it is often useful to transform the variable that measures time (or age) by subtracting the value at the phase change, so that zero on the new scale is the point of change.

Scaling of predictors is also important for categorical predictors. In the example we are discussing from Chapter 2, *Sex* was coded using the values .5 and −.5. This means that β_0 in each school is the value of Y that would be expected on average, if the school consisted of half males and half females (i.e., halfway between .5 and −.5). If, instead, *Sex* was coded using a dummy variable (e.g., 0 for males and 1 for females), then β_0 would represent the average value of Y for males. A number of such coding choices are discussed in Raudenbush and Bryk (2002).

The principle of centering also applies to nonlinear models, and in much the same way. For any generalized linear model, the right-hand side of the model is still linear, and the intercept for the Level 1 equation will still vary (and be modeled by predictors) at Level 2 (and so on for other levels). Therefore, the value of 0 for all predictors should be meaningful, and each predictor should be scaled to ensure this. (An example is discussed in Section 29.10 on nonlinear models.)

29.7 REPEATED MEASURES

29.7.1 Growth Models

In many behavioral studies, participants are measured repeatedly. Multilevel models treat such data as nested, with the observation nested within the individual participant. (There can be additional layers of nesting, such as individuals nested within groups; these models will be discussed below.) The most common situation discussed in the literature is called "growth models", but more generally it applies to situations in which time (or measures of it, such as age) is a primary (or only) predictor variable at the lowest level of the model. For example, if vocabulary size (or a transformation of it, such as square root or logarithm) is a linear function of a child's age, one could write an equation for each child as

$$y_{it} = \pi_{0i} + \pi_{1i}\text{Age}_{it} + \varepsilon_{it},$$

where each child (labelled by the subscript i) has an intercept and slope.

Huttenlocher, Haight, Bryk, and Seltzer (1991) did such a study, following infants from age 14 months until age 26 months. Not all infants were measured at each occasion (t), nor were all measured for the full time period; one of the advantages of multilevel models is that this causes no problem for the conceptualization, analysis, or interpretation of the results. The dependent variable was vocabulary size, and the relationship between age and vocabulary size was not linear over the range of ages observed. Huttenlocher et al. used a quadratic function (i.e., they included the square of age) in their model to allow for the lack of linearity. (They also tried a nonlinear model, the exponential, but it did not fit the data.)

In addition, they observed that at the age of 12 months, no infants knew any words; this led them to scale age by subtracting 12, so that the intercepts reflect vocabulary at one year of age, which was zero in all cases, so the intercept term was no longer needed. Another consequence of this choice was that the linear term was no longer needed in the model, and was dropped. The coefficient of the quadratic (square) term, which estimated growth rate, varied across children; this variation was modeled as a function of gender and the amount of talking the mothers did with the children.

29.7.2 Time-Irrelevant Models

In some repeated measures designs, time is either totally or relatively unimportant. Multilevel models offer a unified framework for analyzing such data, and easily solve problems that are difficult or impossible using traditional repeated measures analyses. In addition, some designs that apparently involve time series do not need to take time into consideration directly in the model.

Consider first a typical one-way repeated measures analysis of variance design. An example Bock (1975) uses from the early literature on the effectiveness of various sleep medicines illustrates the issues. Each subject was measured under four different conditions that, for simplicity, I will describe as Placebo, Drug 1, Drug 2A, or Drug 2B; Drugs 2A and 2B are slightly different chemical forms of the same general compound. (For details on how carryover and other design issues were treated, see Bock or the original source cited in Bock; I will ignore these here.) With this simple design, the usual one-way ANOVA model would be fine, but for illustrative purposes I will show how the model can be represented in its multilevel form. (Raudenbush, 1993b, contains a detailed discussion of the relationship between multilevel models and some typical ANOVA-based analyses of experimental designs.)

The dependent variable is the average number of hours slept when tested under each drug; had the full set of individual observations been available, the analysis could have been even more informative. The model for each observation (an average) is

$$y_{ij} = \beta_{0j} + \beta_{1j} Drug1_{ij} + \beta_{2j} Drug2A_{ij} + \beta_{3j} Drug2B_{ij} + \varepsilon_{ij},$$

where dummy (0/1) variables are used to represent whether or not a particular drug was used for measurement j on person i. Of course, as in any formulation of a (general) linear model, other coding schemes could be used to represent other contrasts of interest among the groups.

The Level 2 equations would contain a random term for the intercept to allow variation among subjects:

$$\beta_{0j} = \gamma_{0j} + u_{0j}$$
$$\beta_{1j} = \gamma_{1j}$$
$$\beta_{2j} = \gamma_{2j}$$
$$\beta_{3j} = \gamma_{3j}.$$

The combined equation for the model is

$$y_{ij} = \gamma_{0j} + \gamma_{1j}\, Drug1_{ij} + \gamma_{2j}\, Drug2A_{ij} + \gamma_{3j}\, Drug2B_{ij} + u_{0j} + \varepsilon_{ij},$$

which shows that observations vary because of the fixed effects of treatments (the γ values), random variation among subjects (the u_{0j} effects), and (random) differences in the effect of treatments across subjects (the ε_{ij} effects).

Using non-orthogonal coding (such as the dummy variables illustrated here) will not produce an overall sum of squares for the treatment effect that a usual ANOVA will. If one wants to be able to reproduce the usual ANOVA, one can use orthogonal contrasts instead of dummy variables. For the example here, Helmert contrasts correspond to theoretically sensible comparisons: First, compare the placebo group to the mean of the three treatment groups; next compare Drug 1 to the the average of Drug 2A and Drug 2B (the type 2 drugs); finally, compare Drug 2A to Drug 2B. The sets of contrast codes (non-normalized) for these comparisons are $(3, -1, -1, -1)$, $(0, 2, -1, -1)$, and $(0, 0, 1, -1)$.

An example adapted from linguistics illustrates the use of a similar design for a nonrandomized study, as well as showing how easily multilevel models can handle some additional complications. A researcher was studying the speech of students who were just learning Spanish, and was interested in how frequently errors were made in four different conditions. The frequency with which these conditions occurred in the language sample varied from person to person, and were generally unequal across the four conditions. In general, there were at least several replications for each condition for each person, unlike in the previous example. Furthermore, the dependent variable was dichotomous: whether there was an error or not. Therefore, a model for a binomial outcome was used, in which the logarithm of the odds of an error was modeled (as in logistic regression). One can easily add person-level predictors, such as gender, the age at which language learning started, or the length of time the student has been studying the language.

Next, consider a study in which the investigator wished to see the degree to which mood in females was affected by hormone changes; all variables were measured every day (mood by questionnaire, hormones by blood tests). At first it seems as if a time series model would be appropriate, but time did not enter into the research theory. The only issue was whether mood was affected by today's hormone levels (and possibly yesterday's hormone levels if there were carryover effects), so there was no need for a time variable in the model. (If the study lasted for a very long period of time, one might need to consider whether there was a time trend up or down in mood, and whether this varied across participants.)

29.7.3 Intensive Longitudinal Data

Behavioral researchers are making use of both paper-and-pencil as well as electronic means to collect large amounts of data on each individual. A person might push a button on a small hand-carried electronic recorder whenever he or she smokes a cigarette, for example. Or they might keep a diary that records certain events, as well as the circumstances surrounding the events, as they occur through each day. Such data sets will be much larger than the usual longitudinal studies where participants might be measured once a year or even less frequently, and have a total of fewer than 10 times of observation. The more frequent, longer-term testing is called intensive longitudinal data (ILD) or ecological momentary assessment (EMA). Multilevel models of various sorts (both linear and nonlinear) are major tools in the analysis of such data. For theory and information on applications, see Walls and Schafer (2006) and Chapter 13.

29.7.4 Single-Case Designs

Some psychological research involves the intense study of a small number of individuals, often over an extended period of time. This research typically originates from a behavioral orientation (deriving from the work of B.F. Skinner and others), and is called single-case research to emphasize the importance of establishing causality by within-subject comparisons (rather than traditional between-group comparisons). Because behaviorists expect to see extremely large effects, they have traditionally rejected the need for statistical analysis; the only test they respect is the intra-ocular traumatic test (i.e., the effect is so big it hits you between the eyes). For a number of reasons, this attitude is slowly beginning to change. (In part, the change is driven by government policies: in education, the What Works Clearinghouse requires a quantifiable effect size to include studies in research reviews.)

Most single-case designs involve repeated observations of a case under different conditions, so that changes in treatment create phases of observation. In an AB design, for example, the participants are first observed for several sessions in condition A (usually a baseline, control, or "treatment as usual", condition), and then are switched to condition

B (typically a new treatment) for several sessions. More typically, there are more phases, and the simplest generally used design is an ABAB design. If effects were immediate, and if there were no carryover from one session (or phase) to another, the observations should go from one flat straight line immediately to a higher or lower level (depending on whether the treatment should increase or decrease the behavior), with little variation within a phase. This ideal, unfortunately, is seldom observed in practice, thus necessitating the use of statistical models.

The single-case design is a natural candidate for multilevel modeling, as it is a repeated measures design, with observations nested within individuals. A level-1 model is constructed for each individual; then similarities among and differences between individuals are described by a level-2 model. The first to suggest this approach were Van den Noortgate and Onghena (2003a, 2003b, 2007, 2008).

The analysis of data from single-case designs presents a number of problems. First, most such data come from time series, so that time trends may be present (and such trends may vary across phases of the design.) As with any time series data, autocorrelation is a possibility, but generally there are too few observations (the time series is too short) to estimate this well. Most behavioral studies do not have continuous outcomes, but rather counts (sometimes translated into rates or percentages), which require nonlinear models. In many cases, behavior does not change suddenly after a phase change, but somewhat slowly rises or falls to a new level; more complicated nonlinear models are needed in such cases, such as floor and ceiling effect models (see, e.g., Rindskopf, 2002) Finally, although single-case designs are generally not single cases, the number of cases is generally small, making fully Bayesian estimation more appealing than empirical Bayes estimation.

29.8 SHRINKAGE ESTIMATES

If one wanted to make predictions for a higher-level unit (such as a school, election district, or country) the most obvious way to proceed would be to use the data from that unit alone. However, if there are other units that are similar, why not use some information from those other units to aid in prediction for the unit in question? The opposite extreme from analyzing data from each unit separately is to pool data from all units and fit one equation. This has the opposite problem from ignoring other units: it assumes that all units are identical. A compromise is to "borrow strength" from other units to the extent that they are similar to the unit for which a prediction must be made.

A simple example from the Bayesian literature shows why such a method would be desired. Suppose that a randomly selected major league baseball player has batted two times, and got a hit both times, so the observed proportion of successes is 2/2 = 1.000. If you had to predict the proportion of successes in this player's next 100 times at bat, what would you predict? Suppose you know (or are told) that major-league baseball players generally have long-run proportions of success in hitting that average about .270, and that the standard deviation among players is about .15. It is therefore most likely that the player who got 2 hits in 2 at bats has an average that might be above .270, but certainly below 1.000; a Bayesian estimate shrinks the estimate down from the observed value of 1.000 towards the average of .270. The amount of shrinkage in this case would be great because of the small amount of data on this player; for players with more data, shrinkage would not be as great.

In many social science areas, variable scaling is arbitrary. (For example, why should a GRE [Graduate Record Examination] score have a mean of 500 and standard deviation of 100? Answer: no reason at all, and in fact it will have changed by the time you read this.) Therefore, if everyone's score shrunk by the same amount, one might not learn much of interest (unless we were predicting to another test administration), because the rank ordering (and relative distances between individuals) would remain the same. However, if we have more information about some individuals than about others, shrinkage will be greater for those about whom we know least, and the rank ordering of shrunken estimates could be very different from the rank ordering of observed scores. This could be taken into account when equating scores from different tests; an observed z-score of 2 on a reliable test is more impressive than an observed z-score of 2 (or perhaps even 2.5 or 3) on an unreliable test.

In the social sciences, shrinkage estimates are often used in Value Added Models (VAM) in education (Braun, 2005). These models are used to assess the progress made by students, relative to what they would be expected to gain; the results are often used to evaluate teachers, principals, schools, and school districts. (In some countries, so-called "league tables", named after similar ranking tables for sports teams, are used to rank schools.) Although these models are better than using unadjusted test score averages to produce rankings, VAMs have many pitfalls due to model misspecification and problems with scaling of the test scores used as outcomes. Similar models are used in other areas, such as medicine, to rank the performance of doctors or hospitals.

In all such cases, one is trying to make causal inferences with nonexperimental data, with all the dangers inherent in any such project. In order for this to be plausible, you must account for all effects other than the one to which you wish to attribute causality; furthermore, these must be measured without error, and the functional form (linear, quadratic, other; possibly with interactions) must be correct. In many educational applications in the US, for example, SES (socioeconomic status) is measured by whether the child receives free or reduced-price school lunches; this is a very crude measure (2 or 3 categories), and is almost certainly inadequate to control for all SES effects.

Although shrinkage estimators have, in general, excellent statistical properties, those applying them should be cautious about unintended consequences that may have adverse ethical (or even legal) effects. Suppose one wants to rank students, for example, and that the statistical model includes race as a predictor (because it typically is significant, at least in the United States). Then estimates for an individual are shrunk (at least in part) towards the mean of his or her racial group. If that group mean is lower than the mean of another group, then two individuals with the same observed test score would be estimated to have different true scores, with one person's score lowered and the other's raised just because of the racial group to which they belong, resulting in (presumably unintentional) racial discrimination.

29.9 BAYESIAN PHILOSOPHY AND METHODS

Some data analysts in the social sciences are drawn to Bayesian methods because of their theoretical appeal, others by necessity (because of Bayesian software that is very flexible). As I noted earlier, several strands of research and software packages in the behavioral sciences came directly from Lindley and Smith's (1972) Bayesian approach to linear models.

One major advantage of Bayes is in the intuitive interpretation of parameter estimates and intervals. To a Bayesian, parameters are random quantities and have distributions; therefore, inference is in accord with the usual naive interpretation given by beginning statistics students: e.g., there is a 95 percent probability that the parameter is in the (95 percent) interval. Other areas under the posterior distribution of parameters (i.e., the distribution after observing the data) have similar interpretations, so one could easily estimate the probability that an effect is positive, or that it is larger than some minimally important value.

Bayesian statistics also allows the incorporation of prior information about parameters in an analysis, but in many analyses there is little such prior information, in which case Bayesians use what is called a noninformative prior (or sometimes a very minimally informative prior). In this way, the posterior distribution is determined totally (or almost totally for minimally informative priors) by the information in the data. The parameter estimates and standard errors are then similar (or in some cases the same) as for frequentists using maximum likelihood estimation, but the interpretation is different, as indicated above.

Even if one does not adopt a Bayesian interpretation of parameters, the software (such as WinBUGS) can be so flexible, including allowance for missing data, that many frequentists find themselves using Bayesian software and then interpreting the results in classical ways. In addition to missing data problems, this is also true for many nonlinear models, which can be easily estimated (and produce nearly unbiased estimates of parameters and standard errors). For more detailed information on the application of Bayesian methods to multilevel data, see Chapter 4.

29.10 NONLINEAR MODELS

As with non-nested data, linear models are not always sufficient. One reason for requiring nonlinear models is that the nature of some dependent variables makes linear models impossible or unreasonable. The most obvious example is for dichotomous outcomes, because the expected value of such a variable is a proportion. Because proportions are restricted to lie between 0 and 1, any straight line predicting from a continuous X will result in predictions less than 0 and greater than 1, which are impossible. The logit transform (logarithm of the odds) solves this problem. Other dependent variables, such as counts, generally follow distributions that have variances increase with the mean, which is reasonable in most cases. Many models for this type of problem fall into the category of generalized linear models (McCullagh and Nelder, 1989; Dobson and Barnett, 2008).

In other cases it is the theoretical relationship that requires nonlinear models, even with normally distributed dependent variables. Most growth models, for example, would have a maximum value (i.e., growth does not continue forever), requiring an asymptote in the model. These models are not yet very common in the social sciences, but are widely used in many areas of biology and medicine, such as pharmacokinetics (which describes how drug levels in the system change over time).

In expanding these models to the multilevel case, the primary additional considerations are the same as before. First, the predictors should be scaled appropriately, whether this involves centering or some other transformation. Second, the model should be expressed in a way that ensures parameters will have meaningful interpretations. As an example, some growth patterns could be adequately described by a quadratic model (i.e., a model with a linear and a squared term) rather than a nonlinear model. However, polynomial terms are generally not as meaningful in a growth context as are parameters in a nonlinear growth model.

Alternative parameterizations

When fitting nonlinear models, the model can frequently be written in more than one way, resulting in parameters with different interpretations. Each way of writing the model may result in parameters that are conceptually meaningful; which one is *correct* is then not the appropriate question, but rather which meaning is important in a particular context. For example, the usual logistic regression model can be written

$$\ln\left(\frac{\pi}{1-\pi}\right) = \beta_0 + \beta_1 X.$$

In this equation, β_0 is the intercept (the predicted logit when $X = 0$), and β_1 is the slope (rate of change in the predicted logit with each unit increase in X). But in some contexts it might be more important to have a parameter that describes the location of the logistic curve; in both item response theory (IRT) and toxicology this is usually the point at which the response rate is .5 (the logit is 0). One can then compare curves for different individuals (in growth models) or contexts/groups (in a typical multilevel model) in terms of where they are located (to the left, middle, or right). This is done by rewriting the equation as follows:

$$\ln\left(\frac{\pi}{1-\pi}\right) = \beta_1(X - H),$$

where β_1 is the same slope as before, and H is the halfway point; that is, the value of X where the predicted proportion rises halfway between 0 and 1. (In IRT, H would be the item difficulty; in toxicology, H is the LD50, which is the amount of drug that kills 50 percent of the sample.) Software for general nonlinear models (for example, WinBUGS) will allow this method of writing the model, but standard multilevel programs do not.

29.11 SURVEY SAMPLING

Most surveys are constructed so that they produce multilevel data. Often, geographic areas are randomly sampled, and then individuals (or households, schools, etc.) are randomly sampled from the selected geographic areas. (Further levels, such as teachers within schools, and students within teachers, are also possible.) Additional complications of some surveys not directly incorporated in traditional multilevel model formulation include stratification and oversampling. Geographic areas may first be stratified by size or population density (urban/suburban/rural), and then units randomly selected within each stratum. Some units may be oversampled if they are rare but important; such is frequently the case for certain racial/ethnic minority groups. These provide additional complications for estimation. Most standard software for survey analysis (and many for multilevel models) can now handle all of these issues. See Pfeffermann et al. (1998) for some of the technical details involved in weighting.

29.12 AGGREGATION BIAS AND ECOLOGICAL INFERENCE: MODERN VIEWS ON OLD ISSUES

Multilevel statistical models can be viewed as a solution (in whole or in part) to several problems that have a long and contentious history in the social sciences. The unit-of-analysis problem refers to the desire of researchers to analyze individual-level data even when the individuals are nested in larger units. For example, educational studies often deal with students given instruction in whole classroom units. Frequently, only a small number of classrooms are available. Before multilevel modeling was developed, it was known that ignoring nesting was wrong, but only for the simplest designs was proper estimation feasible (and even then, equal sample sizes in groups was frequently needed). Additionally, researchers were loath to use the correct analysis even when it was possible, because statistical power was so low. The educational literature is filled with studies for which the analysis was incorrect due to neglect of nesting. Now, even though multilevel modeling is more widely known, incorrect analyses are still too common, again most likely due to researchers not wanting to sacrifice statistical power when they have few classrooms or schools.

Another issue that plagued social research was called *aggregation bias*. Frequently, information collected on individuals is aggregated to higher levels; for example, the average age or education of individuals, or the proportion minority, in a city, county, state, or country. Robinson (1950) drew attention to this issue; he showed that the relationship between two variables when measured at the individual level can be very different (including a change of sign) than the relationship measured by variables aggregated to higher levels. (When the aggregation is over a fixed rather than random variable, the reversal of direction of relationships is called Simpson's Paradox.) In Robinson's example, the proportion of immigrants in a state was positively related to the proportion of the state's population that was literate, even though at the individual level immigrant status was negatively related to literacy (that is, immigrants were less likely to be literate, but states with many immigrants were likely to have higher literacy rates).

After Robinson's demonstration, researchers attempted to determine whether they could estimate relationships on an individual level even when only aggregate data were available; in other words, they attempted to duplicate a multilevel model when only data on the higher level were known. This is known as the problem of ecological inference; the most detailed recent approach was developed by Gary King (1997). King was motivated by the large number of problems in political science and related areas for which individual-level data were not available. One such problem was the enforcement of the Voting Rights Act, in which one needed to know the voting patterns of blacks and whites, but due to the secret ballot only aggregate data by precinct were available on race and voting.

Critiques of this area (see, e.g., Freedman, 1999, and references therein) have demonstrated that, in many instances, recovery of the desired individual-level relationships is either impossible or can be done only very inaccurately. In addition, as with any missing data problem, whether the estimates are believable depends on whether certain (often untestable) assumptions are believable.

29.13 META-ANALYSIS

I have already described the basic idea of meta-analysis, which is to provide a quantitative analysis of the results of a group of studies on the same topic. Chapter 27 of this book is devoted to this topic, so here I will just describe how meta-analysis fits into the multilevel framework.

Suppose a number of studies of some topic all have the same design (a generalization to the case of different designs is conceptually straightforward, though not always simple in practice), which is a two-group experiment comparing a new treatment to an old treatment. Studies may use different dependent variables; in order to make results comparable across studies, for each study divide the dependent variable by the standard deviation of the control group (or pooled standard deviation for both groups) of that study. The mean difference between groups is now standardized within each study, so effect sizes are comparable across studies.

The Level 1 equation for each study is

$$y_{ij} = \beta_{0j} + \beta_{1j}\,\mathrm{Trt}_{ij} + \varepsilon_{ij},$$

where Trt_{ij} is a dummy variable representing whether person i in study j did (Trt $= 1$) or did not (Trt $= 0$) get the new treatment. The Level 2 equations are

$$\begin{aligned}\beta_{0j} &= \gamma_{00} + u_{0j}\\ \beta_{1j} &= \gamma_{10} + u_{1j}.\end{aligned}$$

Here, we are only interested in the second of these equations, because it represents the average effect size (γ_{10}). We are also interested in the variance of the error in the second equation (i.e., $\tau_{11} = Var(u_{1j})$), because it shows how much the effect size varies across studies. If τ_{11} is large, then we would add study-level predictors to try to explain variation across studies in the effect size. For one potential predictor the equation can be represented as

$$\beta_{1j} = \gamma_{10} + \gamma_{11}\,\mathrm{Rand}_j + u_{1j},$$

where Rand_j is a dummy variable equal to 1 if the study implemented random assignment to groups, and 0 otherwise. Such an equation would assess whether randomization affects the treatment effect size. In general, a number of such variables might be plausible in a meta-analysis, ranging from study design characteristics, to participant characteristics (age, education, geographic region, for example), to the nature of the outcome variable.

Because the level-1 data are seldom available, and also because the nature of the equation may vary because of different study designs, the Level 1 equation is seldom directly mentioned in statistical treatments of meta-analysis. As noted above, the only equation in which there is interest in meta-analysis is the second Level 2 equation, and that is the one usually used. The dependent variable is the effect size from a study; sometimes it is a standardized difference in means, as described here, other times a correlation (or a transform of a correlation) or other effect estimate rescaled to be comparable to the standardized difference between means.

The multilevel software most commonly used in the social sciences (HLM and MLWin) contains provisions for conducting meta-analyses. A package for the Splus language, hblm, using the approach described in DuMouchel (1994), can do fully Bayesian (rather than empirical Bayesian) estimation; as pointed out in Chapter 27, this is especially important when the number of studies is small. WinBUGS can also be used for Bayesian estimation, and has the additional advantage over hblm that it can be used when there are more levels of nesting.

29.14 POWER, PRECISION, AND SAMPLE SIZE

In any statistical analysis one would like sufficient power for hypothesis tests or, equivalently, reasonable precision (small standard errors) for interval estimation. Power and precision are generally influenced by study design factors and sample size, and in the case of multilevel models one must consider sample size at all levels of the design. There are some general rules; one generally would prefer more level-2 units than a large number of level-1 units per level-2 unit, for example. However, tradeoffs depend in part on the relative size of the variance within level-2 units and the variance between level-2 units (measured by the intraclass correlation), and which parameters are most important to estimate accurately. For more details, see Chapter 11.

In the social sciences, one of the most important design considerations is made when conducting program evaluations. Some programs are administered at the contextual level; for example, a whole-school reform by definition applies to a whole school, and cannot be given just to some students within a school. This limits the design choices, and if there is randomization (the optimal choice for program evaluation), schools, not students, must be randomized to treatment conditions. This severely limits power (see, e.g., Schochet, 2011, and the accompanying detailed technical report). In other instances, however, there is a choice of randomization at the student level or the school (or classroom or other higher-level unit) level; in this case, the optimal choice for increased power and precision is to randomize at the lowest possible level.

For growth studies, choices must be made not only about how many respondents to include in a study, but how many times to measure each, and the spacing of the observation times (equal spacing is not always optimal). The timing of observations depends on the nature of the growth model; different choices would be made for linear than for nonlinear models, and each nonlinear model would have its own optimal spacing.

29.15 DYADIC AND OTHER SMALL-GROUP DATA

People often come in pairs, married couples being the obvious example, and social scientists often study the social dynamics of couples. More broadly, they study small groups such as families. In some cases, these small groups have members that inhabit distinguishable roles (husband and wife, for example), and in other cases they do not (siblings). In some cases, the relationships are mutual (e.g. two people are married to each other), and in other cases they are directional (person A may like person B, while person B may not like A; person A may speak to person B or perform some other action of which B is the recipient).

Difficulties in the analysis of this type of data arise in part merely because the groups are so small. In addition, some of the types of relationships provide conceptual difficulties beyond those experienced in the general multilevel model, and require special modeling techniques. An authoritative source on dyadic data is Kenny, Kashy, and Cook (2006).

29.16 MULTILEVEL LATENT VARIABLE MODELS

A number of models in the social sciences involve latent variables; that is, variables that are not directly observable, but are measured with error. These include factor analysis, structural equation models, latent class models (see Chapter 20), and a variety of item response theory models for measurement. These models differ mainly in the nature of the variables; the latent variables can be either continuous or categorical, and the same is true for the observed variables. Originally, factor analysis involved continuous observed variables that measured (with error) continuous latent variables (factors); more recent work extended factor analysis to include categorical observed variables. Item response theory (IRT) is a group of psychometric models which began as one-factor models with dichotomous observed measures (typically, test items that are scored correct or incorrect), and have evolved to include ordered and unordered polytomous observed measures, as well as multiple latent variables, thus making IRT and factor analysis very similar in many regards. Some IRT models included additional item parameters to account for guessing or similar problems that do not occur in traditional factor analysis models. Latent class models (see Chapter 20) posit categorical latent and observed variables, and latent profile models generally involve categorical latent but continuous observed variables.

The theory for many of these models can be found in Muthén (1994), Skrondal and Rabe-Hesketh (2004), and Vermunt (2003). Software for these models includes Mplus (Muthén and Muthén, 2010), LatentGOLD (Vermunt and Magidson, 2008), LISREL (Jöreskog and Sörbom, 2005), and GLLAMM (Rabe-Hesketh, Skrondal, and Pickles, 2004).

29.17 MISCELLANEOUS ISSUES IN MULTILEVEL BEHAVIORAL RESEARCH

Nesting: Fixed vs random effects

One frequent conceptual error is to mistakenly believe that any nesting makes a model multilevel. In a typical one-way analysis of variance, subjects are nested within treatment groups, but the model is not a multilevel model because treatment group is a fixed factor. A common definition of multilevel models requires that one random factor be nested in another random factor. If, for example, subjects were nested within therapists instead of type of therapy, the model would be multilevel because therapists would usually be treated as randomly selected from some population of therapists. However, this requirement is not absolute (see Gelman and Hill, 2007, for some discussion); the distinction between fixed and random effects has a long history and is discussed in Chapter 5.

Sampling assumptions

As in most statistical models, random effects are generally assumed to arise from sampling from a population, though the Bayesian framework offers an alternative approach. However, with most data sets this assumption is not valid, as (except with designed surveys) most data sets use whatever sample (of both individuals and higher-level units) happen to be available. One could legitimately ask whether inference can (or should) be done under such circumstances. As in models that are not multilevel, most people imagine a universe or population from which the one that is observed have been drawn. Informally, analysts make an appeal to generalize to “other units like these”; such an appeal may not appeal to everyone. The Bayesian approach is rather subtle, but ends up in the same place (in this case) as the classical approach: the Bayesian assumes that observations (in most cases, this really refers to the errors) are *exchangeable*, meaning that there is no reason to believe that errors from one unit could not just as easily have come from any other unit. One could, in other words, move errors to other cases at random and the model would still be valid. (Note the similarity to permutation tests in classical statistics.) If one believes this, then an analysis proceeds by assuming that all errors are drawn from some common distribution, which duplicates what a frequentist statistician would do.

A related issue arises when all units of a particular type are used in a study (e.g., all classes in a school, all schools in a district, all states/provinces in a country, all countries in the world). If all units are included, how is sampling theory relevant? A number of theoretical solutions are available in this case. One is to imagine the observed units as a sample from a hyperpopulation of all possible ways the world could have turned out. A related conceptualization that is appropriate in some cases is to think of the observed data as a sample from an ongoing process, and the desire is to make inferences about this process and possible future observations of it. These conceptualizations are very contentious, as some people will easily imagine such cases, while others will have great difficulty doing so.

Small sample sizes

Some social science applications involve extremely small sample sizes. For example, one might have six classrooms, three of which are randomly assigned to each of two instructional methods. Even if there are many students in each class, the effective sample size could be very small. With six classrooms, variation among classroom means is not well-estimated. The major issue here is design; one has six units randomly assigned, so even if students were randomly assigned to classrooms, there is little power. The same is often true of single-case designs, in which there are not only few cases, but sometimes few observations per subject. In such cases, one can fit models and describe the (potentially large) degree of imprecision of estimates.

Measurement issues: Generalizability theory

An area not generally considered in the multilevel model literature, perhaps because it deals mostly with measurement issues, is generalizability theory (g-theory; Cronbach, Gleser, Nanda, and Rajaratnam, 1972). G-theory is a measurement theory that considers very general designs, many of which involve nesting of random factors, and others of which involve crossing of random factors. G-theory includes a number of reliability coefficients, some of which will be familiar to those who reference Raudenbush and Bryk (2002) or use HLM software.

Issues from regression

Because multilevel models can be complicated, researchers sometimes forget lessons they learned about ordinary least squares (OLS) regression. A multilevel model is still regression, and can be plagued by outliers, unusual distributions, deviations from linearity, hard-to-detect interactions, heterogeneity of variance, multicollinearity, and other problems. Sometimes researchers hope to make causal inferences without checking the plausibility of those inferences, which they would do if using ordinary regression. (Such inferences are unfortunately facilitated by the language of regression, which often describes "effects" of independent variables, even though these merely represent prediction, not estimates of causal effects.)

Omitted factors

In some instances, unusual results occur because of an omitted (and possibly unknown or even unknowable) factor. Suppose that a school district deliberately puts students from varying ability or achievement levels in each class, rather than the more usual grouping of students of similar ability. If the variable used as a basis for assignment were available, this would be similar to a randomized blocks design, with each class being a block, and "treatments" being prior achievement level. However, by construction, the variance within classes will be greater than expected by chance (if random assignment had taken place), and some formulas (based on sums of squares) will yield a negative estimate of the between-class variance.[1] For random sampling, a negative population variance would be impossible, but in the case of deliberate sampling (with the responsible variable omitted from the analysis) such a result is expected and not an anomaly.

29.18 SOFTWARE

Two stand-alone packages to fit the usual range of multilevel models have developed a wide audience among social scientists: HLM (Raudenbush et al., 2011) and MLWin (Rasbash, Steele, Browne, and Goldstein, 2009). Most general statistical software packages, including SAS, Stata, and SPSS, also have facilities to fit at least some multilevel models. Advanced researchers are also moving in two new directions for software: R, a free general statistical package, has a library for multilevel models called lme4 (Bates, Maechler, and Bolker, 2011), and WinBUGS/OpenBUGS (Spiegelhalter et al., 2002a,b) does Bayesian modeling that includes the ability to construct multilevel models. As mentioned in Section 29.16, there are specialty programs that are constructed to fit multilevel latent variable models. Some of the above packages will handle the complexities of survey sampling, and there are additional packages especially constructed for the analysis of surveys, including AM (2012), SUDAAN (2012), and WesVar (2012).

29.19 CONCLUDING COMMENTS

The speed with which multilevel models have been adopted in the behavioral sciences indicates the power of these models to conceptualize and solve problems that we typically encountered, as well as an array of new ones; these problems were shown to have great commonality. Who would have thought that repeated measures, meta-analysis, and nested data had so much in common? In some instances it was the lack of software that prevented the application of these methods; in other instances, it was the lack of theory. Whatever the reason for the prior logjam, few doing quantitative research today in these areas can afford not to know about multilevel models.

NOTE

1 Editors' note: Software for model-based estimates of variance typically reparameterizes those variance components so that negative variance estimates are not possible.

REFERENCES

Aitkin, M. and Longford, N. (1986) 'Statistical Modeling Issues in School Effectiveness Studies (with Discussion)', *Journal of the Royal Statistical Society A*, 149: 1–43.

AM Statistical Software (2012) Retrieved April 25, 2012 from url `http://am.air.org/default.asp`.

Bates, D., Maechler, M., and Bolker, B. (2011) *lme4: Linear Mixed-effects Models Using S4 Classes*. R package.

Bock, R.D. (1975) *Multivariate Statistical Methods in Behavioral Research*. New York: McGraw Hill.

Braun, H.I. (2005) *Using Student Progress to Evaluate Teachers: A Primer on Value-Added Models*. Princeton, NJ: Educational Testing Service. url `http://www.ets.org/Media/Research/pdf/PICVAM.pdf`.

Cronbach, L.J., Gleser, G.C., Nanda, H., and Rajaratnam, N. (1972) *The Dependability of Behavioral Measurements: Theory of Generalizability for Scores and Profiles*. New York: John Wiley.

Dobson, A.J. and Barnett, A.G. (2008) *Introduction to Generalized Linear Models*, 3rd edition. Boca Raton, FL: Chapman and Hall/CRC.

Duch, R.M. and Stevenson, R. (2005) 'Context and the Economic Vote: A Multilevel Analysis', *Political Analysis*, 13: 387–409.

DuMouchel, W. (1994) 'Hierarchical Bayes Linear Models for Meta-analysis', Technical Report 27, National Institute of Statistical Sciences.

Freedman, D.A. (1999) 'Ecological Inference and the Ecological Fallacy', Technical Report No. 549, University of California, Berkeley, Department of Statistics. url `http://www.stat.berkeley.edu/tech-reports/549.pdf`.

Garner, C.L. and Raudenbush, S.W. (1991) 'Neighbourhood Effects on Educational Attainment: A Multilevel Analysis of the Influence of Pupil Ability, Family, School and Neighbourhood', *Sociology of Education*, 64: 252–62.

Gelman, A. and Hill, J. (2007) *Data Analysis Using Regression and Multilevel/Hierarchical Models*. New York: Cambridge.

Goldstein, H. (1986) 'Multilevel Mixed Linear Model Analysis Using Iterative Generalized Least Squares', *Biometrika*, 74: 43–56.

Huttenlocher, J.E., Haight, W., Bryk, A.S., and Seltzer, M. (1991) 'Early Vocabulary Growth: Relations to Language Input and Gender', *Developmental Psychology*, 22: 236–49.

Jöreskog, K.G. and Sörbom, D. (2005) *LISREL for Windows [Computer Software]*. Lincolnwood, IL: Scientific Software International, Inc.

Kenny, D.A., Kashy, D.A., and Cook, W.L. (2006) *Dyadic Data Analysis*. New York: Guilford Press.

King, G. (1997) *A Solution to the Ecological Inference Problem: Reconstructing Individual Behavior from Aggregate Data*. Princeton, NJ: Princeton University Press.

Lindley, D.V. and Smith, A.F.M. (1972) 'Bayes Estimates for The Linear Model', *Journal of the Royal Statistical Society B*, 34: 1–41.

Mason, W.M., Wong, G.Y., and Entwisle, B. (1983) 'Contextual Analysis Through the Multilevel Linear Model', *Sociological Methodology*, 14: 72–103.

McCullagh, P. and Nelder, J. (1989) *Generalized Linear Models*, 2nd edition. Boca Raton, FL: Chapman and Hall/CRC.

Muthén, B.O. (1994) 'Multilevel Covariance Structure Analysis', in J. Hox and I. Kreft (eds), *Multilevel Modeling*, a special issue of *Sociological Methods and Research*, 22: 376–98.

Muthén, L.K. and Muthén, B.O. (2010) *Mplus User's Guide*, 6th edition. Los Angeles, CA: Muthén and Muthén.

Pfeffermann, D., Skinner, C., Goldstein, H., Holmes, D., and Rasbash, J. (1998) 'Weighting for Unequal Selection Probabilities in Multilevel Models (with discussion)', *Journal of the Royal Statistical Society, Series B*, 60: 23–56.

Rabe-Hesketh, S., Skrondal, A., and Pickles, A. (2004) *GLLAMM Manual*. U.C. Berkeley Division of Biostatistics Working Paper Series. Working Paper 160.

Rasbash, J., Steele, F., Browne, W.J., and Goldstein, H. (2009) *A User's Guide to MLwiN, v2.10*. Centre for Multilevel Modeling, University of Bristol.

Raudenbush, S.W. (1993a) 'A Crossed Random Effects Model for Unbalanced Data with Applications in Cross-sectional and Longitudinal Research', *Journal of Educational Statistics*, 18: 321–49.

Raudenbush, S.W. (1993b) 'Hierarchical Linear Models and Experimental Design', in Lynne K. Edwards (ed.), *Applied Analysis of Variance in Behavioral Science*. New York: Marcel Dekker.

Raudenbush, S.W. and Bryk, A.S. (1986) 'A Hierarchical Model for Studying School Effects', *Sociology of Education*, 59: 1–17.

Raudenbush, S.W. and Bryk, A.S. (2002) *Hierarchical Linear Models: Applications and Data Analysis Methods*, 2nd edition. Newbury Park, CA: Sage.

Raudenbush, S.W., Bryk, A.S., Cheong, Y.F., Congdon, R.T., and du Toit, M. (2011) *PHLM7: Linear and Nonlinear Modeling*. Lincolnwood, IL: Scientific Software.

Rindskopf, D. (2002) 'The Use Of Latent Class Analysis in Medical Diagnosis', in *American Statistical Association Proceedings of the Social Statistics Section*, 2912–16. url `http://www.amstat.org/sections/srms/Proceedings/y2002/files/JSM2002000332.pdf`.

Robinson, W.S. (1950) 'Ecological Correlations and the Behavior of Individuals', *American Sociological Review*, 15: 351–7.

Schochet, P.Z. (2011) 'Do Typical RCTs of Education Interventions Have Sufficient Statistical Power for Linking Impacts on Teacher Practice and Student Achievement Outcomes?', *Journal of Educational and Behavioral Statistics*, 36: 441–71. Technical report available online at url `http://ies.ed.gov/ncee/pdf/20094065.pdf`.

Skrondal, A. and Rabe-Hesketh, S. (2004) *Generalized Latent Variable Modeling: Multilevel, Longitudinal and Structural Equation Models*. Boca Raton, FL: Chapman and Hall/CRC.

Spiegelhalter, D., Thomas, A., Best, N., Gilks, W., and Lunn, S. (2002a) *Bugs: Bayesian Inference Using Gibbs Sampling*. Cambridge, UK: MRC Biostatistics Unit.

Spiegelhalter, D., Thomas, A., Best, N. and Lunn, D. (2002b) *WinBUGS User Manual Version 1.4*. Cambridge, UK: MRC Biostatistics Unit.

SUDAAN (2012) Retrieved on April 27, 2012 from url `http://www.rti.org/sudaan/index.cfm`.

Van den Noortgate, W. and Onghena, P. (2003a) 'Combining Single-case Experimental Data Using Hierarchical Linear Models', *School Psychology Quarterly*, 18:325–46.

Van den Noortgate, W. and Onghena, P. (2003b) 'Hierarchical Linear Models for the Quantitative

Integration of Effects Sizes in Single-case Research', *Behavior Research Methods, Instruments and Computers*, 35: 1–10.

Van den Noortgate, W. and Onghena, P. (2007) 'The Aggregation of Single-case Results Using Hierarchical Linear Models', *The Behavior Analyst Today*, 8: 196–209.

Van den Noortgate, W. and Onghena, P. (2008) 'A Multilevel Meta-analysis of Single Subject Experimental Design Studies', *Evidence-Based Communication Assessment and Intervention*, 3: 142–51.

Vermunt, J.K. (2003) 'Multilevel Latent Class Models', *Sociological Methodology*, 33: 213–39.

Vermunt, J.K. and Magidson, J. (2008) *LG-Syntax User's Guide: Manual for Latent GOLD 4.5 Syntax Module*. Belmont, MA: Statistical Innovations Inc.

Walls, T.A. and Schafer, J.L. (2006) *Models for Intensive Longitudinal Data*. New York: Oxford University Press.

WesVar(2012) Retrieved from url http://www.westat.com/expertise/information_systems/WesVar/.

30

Survival Analysis and the Frailty Model

Ardo van den Hout
University College London, United Kingdom

Brian D.M. Tom
MRC Biostatistics Unit, Institute of Public Health, United Kingdom

30.1 INTRODUCTION

Survival analysis is a statistical investigation of time to an event. Applications can be found, for instance, in medical science where the event may be defined as death, or in engineering with regard to the failure of a machine component. The distinguishing feature of survival analysis is that the time-to-event data may be subject to censoring, in which case not all the events under investigation are observed. A typical example of this is right-censoring in a medical study, where individuals are followed up but not all experience the event of interest during the study period.

The term *survival model* will be used in this chapter to denote a model for data where the event is a one-off occurrence. There are many instances of this classical survival model and the basic features will be discussed. The term *multi-state model* will be used for a generalization of the survival model that is defined for more than one event. That is, a multi-state model is an extension of the survival model which allows movements amongst a finite number of states of a stochastic process, as compared to the survival model which allows only one possible transition from an 'alive' state to an absorbing state, 'death'. A typical example of this model is an illness–death model, where intermediate living states before the death state are distinguished. The most basic illness–death model has a healthy state, an ill-health state, and the death state. There are fundamental links between survival models and multi-state models, and we make these links explicit by working within one encompassing modeling framework.

Frailty models are multilevel models for survival where random effects are linked to the risk of an event. The terminology originates from medical science, where frailty denotes a general susceptibility to ill-health. As a random effect, a frailty term can be specified for an individual, but can also be cluster-specific, in which case individuals within clusters share the same frailty. Applications of frailty models are not limited to medical science but can be found in other areas where survival analysis is undertaken.

This chapter discusses an application of survival analysis and frailty models to data from the Medical Research Council Cognitive Function and Ageing Study (CFAS, `www.cfas.ac.uk`, Brayne *et al.*, 2006). This is a population-based longitudinal study of health in the older population (i.e., 65 years and older) of England and Wales with data collected from five centres. The baseline interviews took place in 1991–1993. Here, we are primarily interested in the association of education with survival and disability, and in long-term prediction. Matthews *et al.* (2006) showed that there are regional differences in the UK in aspects of health, and we will formulate frailty models to take these into account by modeling the centres as clusters in a multilevel model.

In the application, the time scale is age and the covariate education is a binary variable with zero indicating less than 10 years of schooling, and one 10 or more years of schooling. CFAS is a population-based study, and as a consequence age and background at baseline varies across individuals (although everyone is 65 years of age or older at the start of the study). To take into account a possible interaction between education and age, extensions of standard proportional hazards regression survival models are investigated.

Specific literature references will be given in Section 30.3, but books that provide a good starting point for survival analysis and/or frailty models are Cox and Oakes (1984), Kalbfleisch and Prentice (2002), Collett (2003), Duchateau and Janssen (2008), and Aalen *et al.* (2008).

The outline of the rest of the chapter is as follows. Section 30.2 provides an introductory example of a survival model with a cluster-specific frailty term. Section 30.3 gives an overview of basic terminology in survival models and multi-state models with references to the literature. Section 30.4 formulates the survival models and multi-state models that are used in the application, whereas statistical inference is discussed briefly in Section 30.5. Data analysis is presented in Section 30.6. Section 30.7 discusses advantages and disadvantages of the methods that are used in this chapter. The conclusion is set out in Section 30.8.

30.2 GOMPERTZ FRAILTY MODEL FOR SURVIVAL

As an illustrative introduction to survival analysis and frailty models, the Gompertz regression model will be discussed. We consider a random subset ($N = 2000$) of the men in CFAS. This data set will be detailed in Section 30.6.1. Death times are known exactly for those who died in the period 1991 up to 2005. The centres in CFAS will be called clusters from here on. Cluster-specific sample sizes are 369, 420, 419, 412, and 380. Of interest is how education is associated with mortality.

In survival analysis, models are specified for the hazard. The hazard is the instantaneous risk of an event. In the survival model for the CFAS data, the event of interest is death. Given that no event has occurred at age t, a high hazard at age t means that the probability of an event in a small time interval $(t, t + \epsilon)$ is high.

Typical of survival data is that the event may not always be observed. CFAS is a study with a fixed follow-up, and not every participant dies during the study. Deaths that have not yet occurred have survival times that are right-censored. In the data, there are 1581 (79%) observed deaths and 419 right-censored survival times (21%). Another aspect of cohort data such as CFAS is that individuals who die before the baseline survey are not in the sample. This is called left-truncation. The hazard can be unbiasedly estimated from censored and truncated cohort data when account is made of these observation and selection schemes; see, e.g., Kalbfleisch and Prentice (2002, Chapter 3).

The following model specifies a log-linear relationship between the (always positive) hazard $h_{ij}(t)$ at age t for individual i in cluster j, and covariates given by the vector $\boldsymbol{x}_{ij}(t)$. The relation is

$$h_{ij}(t) = h_0(t)\exp[\boldsymbol{x}_{ij}(t)\boldsymbol{\beta} + u_j],$$

where $h_0(t) > 0$ is the baseline hazard at age t, and u_j is the cluster-specific frailty. Note that contrary to notation often adopted in models for longitudinal data, index j denotes the cluster and is *not* an index for measurements within a cluster. Our notation is in line with the multilevel notation used by Scott *et al.* in Chapter 2. The covariate vector $\boldsymbol{x}_{ij}(t)$ does *not* include an intercept. Including the frailty takes into account a possible cluster-specific effect on the hazard. We assume that these effects can be described by a normal distribution with mean zero and unknown variance σ^2, i.e., $u_j \sim N(0, \sigma^2)$. Further specification of the baseline hazard yields the Gompertz regression model given by

$$h_{ij}(t) = \lambda \exp(\gamma t) \exp[\boldsymbol{x}_{ij}(t)\boldsymbol{\beta} + u_j], \tag{30.1}$$

where $\lambda > 0$ is called the scale parameter. In this model, the baseline hazard follows a Gompertz distribution. The baseline hazard specifies how the hazard changes over time when $\boldsymbol{x}_{ij}(t)$ is kept constant.

To fit the model, we redefine the time scale t to be equal to age minus 60 years. The covariates are the binary variable *Education* (0 for less than ten years of education, 1 for ten or more years), the variable *Year of Birth* (year of birth minus 1900), and the interaction between *Education* and t. The first two covariates are time-independent. We will model the scale parameter by $\lambda = \exp(\beta_0)$. Thus, the parameter vector for model (30.1) is given by $(\gamma, \beta_0, \beta_1, \beta_2, \beta_3, \sigma)$, where β_1, β_2, and β_3 correspond to the effects associated with education, year of birth, and the interaction between education and t, respectively.

CFAS is a population-based study with an over-sampling of 75-year-olds. One way to deal with this is by using the sampling weights of the study in the analysis. Another way, which is applied here, is to add year of birth to the model as a survey-design variable, see, e.g., Korn and Graubard (1999, Section 4.5).

Table 30.1 presents the results for a series of Gompertz survival models. Details of the estimation will be presented in Section 30.4. Model 1 is the model without any covariates and $-2\times$ the value of the maximized log-likelihood (-2Loglik) is 18124. Model 1 is readily improved upon by adding *Education* (Model 2, with -2Loglik $= 18116$). Further improvement is achieved by adding *Year of Birth* (Model 3, -2Loglik $= 18099$).

Model 3 implies proportionality of hazards with regard to education when we condition on year of birth. That is, the hazard when $Education = 1$ divided by the hazard when $Education = 0$, given same year of birth, is $\exp(\beta_1)$. This is an appealing property and leads to a straightforward interpretation of the effect of education.

Figure 30.1 depicts Kaplan–Meier curves for the two education groups in the CFAS data. A Kaplan–Meier curve is a non-parametric estimator of the survival function, see, e.g., Collett (2003, Section 2.1). The graph suggests that the effect of education on the hazard may change with age, which would be a violation of the proportional hazards assumption. This trend is confirmed by the graph of the log-cumulative hazards, also shown in Figure 30.1. The hazards in this graph are derived from the Kaplan–Meier estimators. The fact that the curves in the graph are not parallel indicates non-proportionality of the hazards. However, the evidence for non-proportionality is not strong. Typically, crossing survival curves are considered to be strong evidence and this is not the case here. However, since we are also interested in long-term prediction, a minor violation of the proportional hazards assumption may still have a meaningful impact.

To investigate the proportionality of the hazards further, the interaction between *Education* and t is added to the model (cf. Kalbfleisch and Prentice 2002, p. 111), yielding Model 4 with -2Loglik $=$ 18097. The likelihood ratio test of Model 3 versus Model 4 has p-value 0.12. Although this provides weak statistical evidence against the proportional hazards assumption, we decide to forgo this assumption.

Table 30.1 Gompertz Survival Models Fitted to the CFAS Data for Men. Point estimates and estimated standard errors in parentheses for the effects of the covariates

Year Model	Education	Year of Birth	Educ. × Time	−2×Log-Likelihood
1				18124
2	−0.15 (0.05)			18116
3	−0.14 (0.05)	−0.02 (0.01)		18099
4	−0.35 (0.15)	−0.03 (0.01)	0.01 (0.01)	18097
5	−0.41 (0.19)	−0.03 (0.01)	0.01 (0.01)	18097

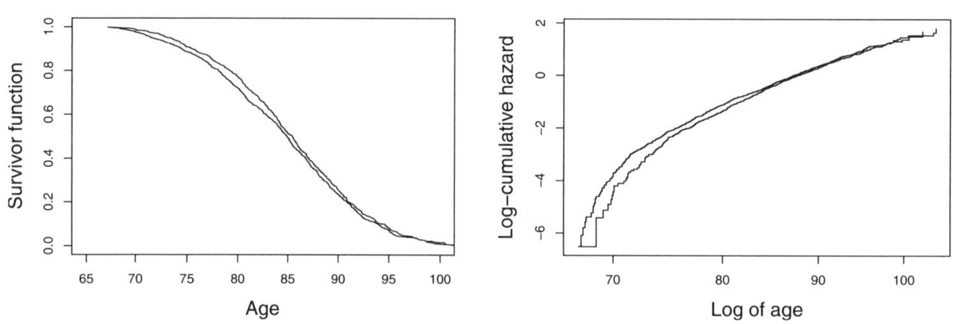

Figure 30.1 Kaplan-Meier Curves on the left-hand side, and Cumulative Hazard Curves on the right-hand side. Solid line for men with ten or more years of education, dashed line for those with less than ten years

Next, we estimate the frailty model (30.1) by maximizing the marginal likelihood. This is Model 5, with $-2\text{Loglik} = 18097$. In terms of -2Loglik, introducing the frailty term to the model does not lead to a statistically significant improvement. The standard deviation σ of the random effect distribution is small and estimated as 0.024. The corresponding small random effects u_j may not be statistically significant with regard to mortality, but these effects may still be relevant and indicative in a comparison of the clusters. We return to this issue later.

For Model 5, the estimates of the baseline hazard parameters are $\widehat{\beta_0} = -3.18$ (0.22), and $\widehat{\gamma} = 0.06$ (0.01), where estimated standard errors are in parentheses. The estimate of γ clearly shows the effect of age, leading to an increased hazard with increasing age.

The negative estimate of the effect of year of birth means that, for instance, those who are 85 in 2001 have a lower hazard at that age than those who are 85 in 1991. This is consistent with the fact that cohort life expectancy in the UK is increasing (UK Office for Statistics, `www.statistics.gov.uk`).

The effect of education is of primary interest. The negative effect on the hazard implies a positive effect of more education on survival. Although there is a change of the main effect of education when going from Model 4 to Model 5, this change is minimal compared with the estimated standard errors. The positive coefficient for the interaction between education and age means that the positive effect of more education weakens with increasing age. This is in line with information from the Kaplan–Meier curves in Figure 30.1, where the two curves are very similar for the older ages.

In Section 30.5, the Gompertz frailty model will be estimated within the Bayesian framework and further interpretation of results and estimated frailties will be given.

30.3 OVERVIEW OF SURVIVAL ANALYSIS AND FRAILTY MODELS

This section provides a brief overview of multilevel survival analysis. The discussion and the literature references are not exhaustive (nor historic), but aim to convey basic terminology and important issues.

Various names are used for data for the time to an event: *duration data*, *time-to-event data*, *failure time data*, *survival data*, and *event history data*, where the latter is often used to denote processes with recurrent events.

Censoring is the distinguishing feature of survival analysis. When an event is right-censored it did not occur during the follow-up. When the event is interval-censored it occurred in a time interval with known lower and upper bounds but at an unknown time within the interval. Left-censoring is a specific form of interval-censoring, namely the case where the lower bound is zero. The models in this chapter assume that the censoring is non-informative, i.e., the censoring mechanism is independent of the stochastic time-to-event process. An example of informative censoring is a survival time that is censored because the treatment stopped due to a deterioration of the physical condition of the patient (Collett, 2003).

When an event is left-censored, it is at least known that the event took place. Left-truncation refers to the situation where processes with events are unknown because the event time is less than some threshold. The following example is from Kalbfleisch and Prentice (2002). The time origin is time of birth, and of interest is whether people were exposed to some environmental risk. If individuals are selected for study after responding to an advertisement, then individuals who died prior to the advertisement are not observed. Those who are in the study are subject to left-truncation.

The time scale in a survival model is important. The time origin specifies from which moment survival is measured. Typical choices of the time scale are time from randomization in a clinical trial, time from exposure in a study of infectious disease, or age in an observational study (Bull and Spiegelhalter, 1997).

Model formulation in this chapter is with respect to the hazard, and inference is within the framework of Cox and Oakes (1984), Collett (2003), Kalbfleisch and Prentice (2002). A more rigorous mathematical approach for inference is based on counting processes and martingale theory (Andersen *et al.*, 1993; Aalen *et al.*, 2008). An alternative to modeling the hazard is using direct modeling of the survival time. For such a model with random effects see Ha *et al.* (2002). When the logarithm of the survival time is used, this is called an accelerated failure time model; see, e.g., Wei (1992) or Collett (2003).

The notation in this chapter is conventional with regard to survival models. Stochastic variable T denotes time to the event. The hazard is $h(t) = \lim_{\Delta \downarrow 0} P(t \leq T < t + \Delta | T \geq t)/\Delta$, and the survival function is $S(t) = P(T > t) = \exp(-\int_0^t h(u)du)$.

A multi-state model where one of the states is a death state (or, more generally, an absorbing state) can be seen as an extension of a survival model. Statistical analysis of multi-state survival processes can be found in many areas. Research in medical statistics that is related to work in this chapter is presented in Kay (1986), Kalbfleisch and Lawless (1985), Jackson *et al.* (2003), and Putter *et al.* (2007). Theory for multi-state models is derived from theory on Markov processes (Norris, 1997). A first-order Markov assumption simplifies model formulation but may not always be realistic, in which case extensions such as a semi-Markov model may be required; see, e.g., Kalbfleisch and Prentice (2002, Chapter 8) and the references therein.

Transition between living states in a multi-state model are often interval-censored since constant monitoring of the process is too expensive, or just not feasible. Data sparseness with respect to some of the transitions in the model can be a problem, especially when the effect of covariates on transition hazards are of interest. A multilevel structure may help to obtain a parsimonious model by replacing covariate information by random effects.

A *frailty model* is a type of multilevel survival model. The term frailty is typically used in medical statistics when random effects describe cluster-specific susceptibility for ill health, but as a multilevel model the approach can, of course, be applicable to other areas. Specific to frailty models is the choice of the frailty distribution. Duchateau and Janssen (2008) and Aalen *et al.* (2008) discuss several choices for the frailty distribution. Wu (2010) discusses frailty survival models within the wider framework of mixed-effects models. Information about possible frailty in survival data differs from the information in multi-state data. Because of the repeated measurements in multi-state data, frailty may be easier to detect than in the univariate survival data.

30.4 MODELS

In this chapter, a survival model is a model for data where the event is a one-off occurrence, and a multi-state model is a generalization of the survival model that is defined for more than one event. This section specifies survival models for right-censored, left-truncated, continuous-time time-to-event data. In addition, continuous-time multi-state models are specified for data, where death is one of the states, times of death are known, and transitions between living states are interval-censored (i.e., exact transition times are unknown, but lie within known time intervals), or right-censored. Frailty terms are included in the models to take clustering of data into account.

30.4.1 *Piecewise-Constant Hazard Model*

To deal with time-dependent covariates in the regression equations for the hazard, piecewise-constant hazards are used both in the survival models and in the multi-state models. The following explains the approach for a survival model, but the multi-state model can be formulated analogously. The basic idea is to approximate the survival function by a product of survival functions derived from exponential distributions.

The survival function for time t is given by

$$S(t) = \exp\left[- \int_0^t h(u)du \right], \qquad (30.2)$$

where h is the hazard function. Define time grid $t_1 = 0$, $t_k = t_{k-1} + \Delta$, for $k = 2, \ldots, K$ and $\Delta > 0$, such that $t_K = t$. Approximating the integral in (30.2) using this grid yields

$$\begin{aligned} S(t) &\approx \exp\left[- \sum_{k=1}^{K-1} \Delta h(t_k) \right] \qquad (30.3) \\ &= \prod_{k=1}^{K-1} \exp\Big[- \Delta h(t_k) \Big] = \prod_{k=0}^{K-1} S_k^E(\Delta), \end{aligned}$$

where S_k^E is the survival function for an exponential distribution with constant hazard given by $h(t_k)$. In this approximation, the hazard is kept constant in between two time points in the grid, but is allowed to change at the grid points according to a specified distribution such as the Gompertz or the Weibull.

For the models in this chapter, the piecewise-constant hazards are defined with respect to the time scale age. Because the change of age is deterministic given age at baseline, these values are known at all the grid points.

In the survival model, we will impose a time grid as defined by Δ above. In case t is not a multiple of Δ, we allow the last interval of the grid to be shorter than Δ. For example, if $\Delta = 1$ and $t = 2.1$, then the imposed grid is t_1, t_2, t_3, t_4 equal to 0, 1, 2, 2.1.

In the multi-state model, we used the times of observation as the points in the grid. For example, if living states are observed at times t_1, t_2, t_3, and death is observed at t_4, then the grid for the piecewise-constant hazard is t_1, t_2, t_3, t_4. A finer grid can be imposed, but we use the times of observation in this case to limit computations.

30.4.2 Parametric Survival Models

For individual i, $i = 1, \ldots, n_j$, in cluster j, $j = 1, \ldots, J$, we use two times: $t_{b.ij}$ is the time at baseline, and t_{ij}, which is either the time to the event or the time of right-censoring. Define $\delta_{ij} = 1$ if the event time is observed, $\delta_{ij} = 0$ if the event time is right-censored.

The conditional likelihood for the jth cluster is

$$L_j(\boldsymbol{\theta}|u_j) = \prod_{i=1}^{n_j} \Big(P(\text{death at } t_{ij} | \text{alive at } t_{ij} - \epsilon) \, S(t_{ij} - \epsilon | t_{b.ij}) \Big)^{\delta_i} \times S(t_{ij}|t_{b.ij})^{(1-\delta_i)}, \quad (30.4)$$

where the probabilities on the right-hand side are conditional on $\boldsymbol{\theta}$ and u_j, and $\epsilon > 0$ is a small value. The function $S(t|t_b)$ is the conditional survival probability $P(T > t | T > t_b)$ which is equal to $S(t)/S(t_b) = \exp(-\int_{t_b}^{t} h(u)du)$. This takes left-truncation into account. Note that $S(t|t_b = 0) = S(t)$.

The model for the hazard is given by

$$h_{ij}(t) = h_0(t) \exp[\boldsymbol{x}_{ij}(t)\boldsymbol{\beta} + u_j],$$

where the distribution of the frailty is normal, i.e., $u_j \sim N(0, \sigma^2)$. The baseline hazard can be chosen from

$$\begin{aligned} \text{Exponential: } h_0(t) &= \lambda \\ \text{Weibull: } h_0(t) &= \lambda \rho t^{\rho-1} \\ \text{Gompertz: } h_0(t) &= \lambda \exp(\gamma t). \end{aligned}$$

For the survival functions in (30.4), we use piecewise-constant approximations as in (30.3). The chosen distribution for $h_0(t)$ determines how the hazard changes over time in a piecewise-constant fashion. The probability $P(\text{death at } t_{ij} | \text{alive at } t_{ij} - \epsilon)$ in (30.4) is approximated by $1 - S^E(\epsilon)$, where S^E is the survival function of an exponential distribution with hazard $h(t_{ij} - \epsilon)$. By replacing the exact death time by an interval-censored death time in a small time interval $(t_{ij} - \epsilon, t_{ij})$, it is ensured that the likelihood value is a probability.

In the frequentist framework, the marginal log-likelihood is maximized. The marginal likelihood for the jth cluster is $L_{marg,j}(\boldsymbol{\theta}) = \int L_j(\boldsymbol{\theta}|u_j) p(u_j|\boldsymbol{\theta}) du_j$, where the density $p(u_j|\boldsymbol{\theta})$ is derived from the normal density for u_j. The marginal likelihood for all clusters is given by $L_{marg}(\boldsymbol{\theta}) = \prod_{j=1}^{J} L_{marg,j}(\boldsymbol{\theta})$.

In the Bayesian framework, the inference concerns the posterior density of $\boldsymbol{\theta}$ and the random effects $\boldsymbol{u}$. This density is given by

$$\begin{aligned} p(\boldsymbol{\theta}, \boldsymbol{u}|\boldsymbol{y}) &= \frac{p(\boldsymbol{y}|\boldsymbol{\theta}, \boldsymbol{u}) p(\boldsymbol{u}|\boldsymbol{\theta}) p(\boldsymbol{\theta})}{p(\boldsymbol{y})} \\ &\propto p(\boldsymbol{y}|\boldsymbol{\theta}, \boldsymbol{u}) p(\boldsymbol{u}|\boldsymbol{\theta}) p(\boldsymbol{\theta}), \end{aligned}$$

where $p(\boldsymbol{\theta})$ is the prior density of $\boldsymbol{\theta}$, and $p(\boldsymbol{y})$ is the marginal likelihood of the data. Note that $p(\boldsymbol{y}|\boldsymbol{\theta}, \boldsymbol{u}) = \prod_{j=1}^{J} L_j(\boldsymbol{\theta}|u_j)$. The density of the random effects is $p(\boldsymbol{u}|\boldsymbol{\theta}) = \prod_{j=1}^{J} p(u_j|\boldsymbol{\theta})$.

30.4.3 Parametric Multi-State Models

For individual i, $i = 1, \ldots, n_j$, in cluster j, $j = 1, \ldots, J$, we observe state $z_{ij.k}$ at time $t_{ij.k}$, $k = 1, \ldots, K_i$. The state at $t_{ij.K_i}$ is allowed to be right-censored, in which case the state is not observed. Note that k in this notation does not refer to a subgroup in a multilevel structure (cf. Scott *et al.* in Chapter 2), but refers to subsequent observation times for individual i in cluster j.

The model for the hazard of transition at time t from state r to state s is given by

$$h_{ij.rs}(t) = h_{rs.0}(t) \exp(\boldsymbol{x}_{ij}(t)\boldsymbol{\beta}_{rs} + u_{rs.j}),$$

where the transition-specific frailties are assumed to be normally distributed, i.e., $u_{rs.j} \sim N(0, \sigma^2_{rs})$. The baseline hazards can be chosen from

$$\begin{aligned} \text{Exponential: } h_{rs.0}(t) &= \lambda_{rs} \\ \text{Weibull: } h_{rs.0}(t) &= \lambda_{rs} \rho_{rs} t^{\rho_{rs}-1} \\ \text{Gompertz: } h_{rs.0}(t) &= \lambda_{rs} \exp(\gamma_{rs} t). \end{aligned}$$

The conditional likelihood for the jth cluster is

$$\begin{aligned} L_j(\boldsymbol{\theta}|\boldsymbol{u}_j, z_1) &= \prod_{i=1}^{n_j} P(z_{ij.K_i}, z_{ij.K_i-1}, \ldots, z_{ij.2}|\boldsymbol{u}_j, z_{ij.1}) \qquad (30.5) \\ &= \prod_{i=1}^{n_j} P(z_{ij.K_i}|z_{ij.K_i-1}, \boldsymbol{u}_j) \times \cdots \\ &\quad \times P(z_{ij.2}|\boldsymbol{u}_j, z_{ij.1}), \end{aligned}$$

where $\boldsymbol{u}_j$ is the vector with the transition-specific frailties $u_{rs.j}$, and z_1 is the baseline state data. The second equation follows from the first-order Markov assumption. If the state at $t_{ij.K_i}$ is the death state D, then

$$\begin{aligned} &P(z_{ij.K_i}|z_{ij.K_i-1}, \boldsymbol{u}_j) \\ &\quad = \sum_s P(z_{ij.K_i} = D|s \text{ at } t_{ij.K_i} - \epsilon, \boldsymbol{u}_j) \\ &\quad \times P(s \text{ at } t_{ij.K_i} - \epsilon|z_{ij.K_i-1}, \boldsymbol{u}_j), \end{aligned}$$

where the summation is over all living states (cf. Kay, 1986). This implies an unknown state just before death and then death within a small time interval ϵ. If the state at $t_{ij.K_i}$ is censored, then

$$\begin{aligned} &P(z_{ij.K_i}|z_{ij.K_i-1}, \boldsymbol{u}_j) \\ &\quad = \sum_s P(z_{ij.K_i} = s|z_{ij.K_i-1}, \boldsymbol{u}_j), \end{aligned}$$

where again the summation is over all living states.

Transition probabilities such as $P(z_{ij.k}|z_{ij.k-1}, \boldsymbol{u}_j)$ for time interval $(t_{ij.k-1}, t_{ij.k})$ are computed by assuming that the transition-specific hazards during the interval are constant and equal to $h_{ij.rs}(t_{ij.k-1})$. Theory on continuous-time Markov models (Norris, 1997) describes how transition probabilities can be computed for a fixed time interval given a constant hazard.

If there are two states, alive and death, then the multi-state model becomes a survival model with likelihood (30.4).

In the frequentist framework, the marginal likelihood for all clusters is $L_{marg}(\boldsymbol{\theta}|z_1) = \prod_{j=1}^{J} \int L_j(\boldsymbol{\theta}|\boldsymbol{u}_j, z_1) p(\boldsymbol{u}_j|\boldsymbol{\theta}) d\boldsymbol{u}_j$. In the Bayesian framework, the posterior density is given by

$$\begin{aligned} p(\boldsymbol{\theta}, \boldsymbol{u}|z) \quad &\propto \quad p(z_{\text{follow-up}}|\boldsymbol{\theta}, \boldsymbol{u}, z_1) \\ &\qquad p(\boldsymbol{u}|\boldsymbol{\theta}) p(\boldsymbol{\theta}), \end{aligned}$$

where $p(z_{\text{follow-up}}|\boldsymbol{\theta}, \boldsymbol{u}, z_1) = \prod_{j=1}^{J} L_j (\boldsymbol{\theta}|\boldsymbol{u}_j, z_1)$.

30.4.4 Extending the Multilevel Structure

For ease of exposition, the survival models and the multi-state models are presented with one random effect only (the frailty term for a cluster). Other multilevel structures can be formulated in a similar way. We give two examples.

First, fixed effects can be turned into random effects for individuals. In the survival model this means that $h_{ij}(t) = h_0(t) \exp[\boldsymbol{x}_{ij}(t)\boldsymbol{\beta}_i + u_j]$, where $\boldsymbol{\beta}_i$ follows a multivariate distribution. This distribution is often chosen to be a multivariate normal such that $\boldsymbol{\beta}_i \sim MVN(\boldsymbol{\beta}_0, \boldsymbol{\Sigma})$, with mean $\boldsymbol{\beta}_0$ and variance matrix $\boldsymbol{\Sigma}$ to be estimated from the data.

Second, in the three-state illness–death model in the application, we use both a cluster-specific frailty and a subject-specific frailty. The latter is to capture heterogeneity with regard to transition hazards for the death state (state 3) and is used for reasons of parsimony. The cluster-specific frailty is used to further investigate the process of interest which is the hazard of moving from the healthy state (state 1) to the disability state (state 2).

30.5 INFERENCE

Both frequentist methods and Bayesian methods will be used in this chapter. Since the models that we use in the data analysis have a simple random-effects structure, we use the maximization of marginal likelihoods as the starting point for statistical inference. Inference for multilevel models using marginal likelihood is briefly alluded to by Goldstein

(Chapter 3). Gaussian quadrature is used to approximate the integrals in the likelihoods, and the maximization is undertaken in R which is a free programming language and software environment for statistical computing (`cran.r-project.org`). The programming language C is used within R to speed up computations within the likelihood. The R function for the maximization is the multipurpose optimizer optim with the Nelder–Mead algorithm (Nelder and Mead, 1965). The models estimated using the marginal likelihood are compared by looking at $-2 \times$ the value of the maximized log-likelihood. The chosen model is further investigated using Bayesian inference.

The Bayesian framework for multilevel modeling is introduced by Fahrmeir *et al.* (Chapter 4). For the Bayesian inference in this chapter, we use Markov chain Monte Carlo (MCMC) to obtain the posterior distribution of the model parameters and the frailties. The MCMC is undertaken in OpenBUGS (Thomas *et al.,* 2006). OpenBUGS is free computer software for the Bayesian analysis of statistical models using MCMC.

The implementation in OpenBUGS of the three-state model in the data analysis is based upon the methods presented in Van den Hout and Matthews (2009), which extends the methods in Welton and Ades (2005) and Pan *et al.* (2007). Van den Hout and Matthews present a progressive illness–death model whereas, in this chapter, the model for the data analysis allows for reversible transitions, i.e., transitions from the ill-health state back to the healthy state. The possibility of recovery complicates the formulas needed to compute transition probabilities, but the basic approach stays the same. Transition probabilities for a specified time interval are used in the formulation of the likelihood and can be computed using the regression equations for the transition intensities. For a three-state illness–death model with recovery, the closed-form formulas are presented in Welton and Ades (2005).

The estimation using OpenBUGS is time consuming, and this is the reason why we use the frequentist methods as the starting point for statistical inference.

The data, R code, and OpenBUGS code that are used for the analysis in this chapter are available on the web site of this book or can be obtained by contacting the first author.

30.6 ANALYSIS

30.6.1 Data

To recap, the Medical Research Council Cognitive Function and Ageing study (CFAS, Brayne *et al.,* 2006) is a population-based longitudinal study of cognition and health in the older population of England and Wales. Interviews were conducted in 5 centres (clusters). The study was designed to have 2500 individuals in each cluster. There were 13004 individuals who undertook the baseline interview in 1991–1993, with follow-up interviews up to 2004. All individuals were aged 65 years and above at baseline, and all deaths up to the end of 2005 have been included.

For the analysis in this chapter we use a random subset from the men in CFAS. The sample size of these data is $N = 2000$. The data can be obtained from the web site. To prevent disclosure of identities of individuals in the data, the clusters are anonymized, and dates are rounded to the first of the month.

For the survival models, the event is death, and deaths after 2005 are right-censored. For the multi-state models it is also of importance what happens before death. The living states are defined with regard to disability. State one is the disability-free state and state two is the disability state. State three is the death state. The definition of disability is based on activities of daily living. Individuals were classified as disabled at time of interview if they were unable to perform at least one of the following five activities without human help: transfer to and from a chair (from interviewer assessment), put on shoes and socks, prepare a hot meal, get around outside, and have a bath or an all-over wash (Spiers *et al.,* 2005; Jagger *et al.*, 2007). People in this group require help at least several times per week.

The time between CFAS interviews varies between and within individuals and the number of interviews is not fixed. The last recorded state of an individual at the end of December 2005 is either death or right-censored. Death times are known, but transitions between the living states are interval-censored. Table 30.2 presents frequencies for the number of times each pair of states were observed at successive observation times. Note that the total of the two columns for death and right-censoring is equal to the sample size $N = 2000$.

For the living states, the frequencies of the number of observations per individual are 624, 757, 453, and 166, for 1, 2, 3, and 4 observation(s), respectively. At baseline, there happens to be one 64-year-old man, but the other men are 65 or older. The oldest man at baseline is 99. There are 1320 men with less than ten years of education, and 680 with ten or more years of education.

30.6.2 Survival Models

For the event death, we fitted survival models where the baseline hazard follows both a Gompertz distribution and a Weibull distribution. The models were fitted using an imposed time grid with Δ equal to two years (see Section 30.4.1), and with ϵ in the log-likelihood equal to one month (see Section 30.4.2). The time scale t is age minus 60 years. The covariates are *Education*, *Year of Birth*, and the interaction between *Education* and t (see Section 30.2). The left-truncation in the CFAS data is taken into account. We will report $-2 \times$ the value of the maximized log-likelihood (-2Loglik).

The fixed-effects Gompertz model (Model 3 in Section 30.2) has -2 Loglik $=$ 18097. The corresponding fixed-effects Weibull model has -2Loglik $=$ 18105, where the shape parameter ρ is estimated as 2.10 with estimated standard error 0.12. The Weibull model has the same number of parameters as the Gompertz model. Because the latter yields a better fit in terms of the likelihood, we do not pursue the analysis with the Weibull model any further. The graph of the log-cumulative hazard in Figure 30.1 is additional evidence that the Weibull model is not a good choice: a Weibull distribution for the survival times would have implied straight lines for the log-cumulative hazard.

Notice that since the exponential model is a restricted version of the Weibull model, there is no reason to pursue the exponential model either. Instead, we will discuss Bayesian inference of the Gompertz frailty model (30.1) which takes a cluster-specific effect on the hazard into account (Model 5 in Section 30.2).

Even though Model 5 with the frailty does not lead to a better fit in terms of -2Loglik when compared to a model without the frailty (see Section 30.2), we are still interested in a further investigation of Model 5 in terms of the effect of the frailty on the prediction of survival. Even though the frailty term may not be significant with respect to the fit, small differences between the clusters may be relevant when it comes to long-term prediction.

For Bayesian inference, prior densities have to be specified for the parameters. For the frailty model (30.1), the parameters vector is $(\gamma, \beta_0, \beta_1, \beta_2, \beta_3, \sigma)$. For γ and each of the β-coefficients, we specify independent vague univariate priors given by $N(0, 10)$. For the standard deviation σ of the frailty distribution, we use the prior $U(0, 5)$, i.e., a uniform distribution on the interval $(0, 5)$.

The posterior means of the regression coefficients in the frailty model are very similar to the point estimates in Table 30.1. The main effect of *Education*, for instance, has a posterior mean equal to -0.40, with posterior standard deviation 0.19. The 95% credible interval (CI) is $(-0.76, -0.05)$. The posterior mean of standard deviation σ is 0.08 and the 95% CI for this parameter is $(0.01, 0.26)$. The result for the standard deviation of the frailty term reflects the more general result that using marginal likelihood leads to an underestimation of the variance of the random-effects distribution. Additional analyses were performed to investigate the sensitivity of the results with

Table 30.2 Frequencies of Observed Transitions Between the States in the CFAS Subset with 2000 Men

From	To No disability	Disability	Death	Right-censored
No disability	1820	176	1205	396
Disability	42	123	376	23

respect to the choice of the priors. For example, specifications $N(0, 20)$ and $U(0, 1)$ were used for the priors for the regression coefficients and the standard deviation, respectively. The effect of these alternative specifications on the results was very minor.

Figure 30.2 presents boxplots of the posterior distributions of the frailties in the Gompertz survival model. As already mentioned in Section 30.2, the frailty term is not statistically significant: the fixed-effects model performs equally well with respect to −2Loglik. However, differences as depicted in Figure 30.2 are still relevant when cluster-specific frailty is of interest.

We consider prediction with regard to the effect of education on survival for men at a specified age, with a specified year of birth. Consider a man aged 70 who was born in 1920. First, assume that the frailty of this man is the mean of the frailty distribution, which is zero. If this man has less than ten years of education, then the posterior mean of the predicted survival probability at age 80 is 0.53 with 95% CI (0.49, 0.58). If this man has ten or more years of education, then the posterior mean is 0.60 (0.56, 0.65). Although there is an overlap of the CIs, there is a clear and relevant difference in survival: more education is associated with better survival in older age.

Figure 30.3 presents survival curves for men aged 70 and born in 1920, with separate curves for the five clusters, and for years of education. This is a long-term prediction over a time which exceeds the follow-up of the study. There is a clear positive effect of more education, but the cluster effects are rather minimal.

30.6.3 Three-State Models

In this section we analyze the three-state model for disability and death. Frailty terms are used to investigate cluster-specific susceptibility to disability, and to take into account individual-specific heterogeneity with regard to hazard of death.

We use the same time-independent covariate vector $\boldsymbol{x} = (x_1, x_2, x_3)$ as in the survival model: x_1 is *Education*, x_2 is *Year of Birth*, and x_3 is the interaction between *Education* and t. The time scale t is age minus 60 years. The four regression equations for the transition intensities are given by

$$\begin{aligned} h_{ij.12}(t) &= h_{12.0}(t)\exp(\boldsymbol{x}_{ij}\boldsymbol{\beta}_{12} + u_j) \\ h_{ij.13}(t) &= h_{13.0}(t)\exp(v_i) \\ h_{ij.21}(t) &= h_{21.0}(t) \\ h_{ij.23}(t) &= h_{23.0}(t)\exp(v_i), \end{aligned} \tag{30.6}$$

where u_j is a cluster-specific frailty and v_i is a subject-specific frailty. The baseline hazards of the transition intensities follow a Gompertz model, and are given by

$$h_{rs.0}(t) = \lambda_{rs.0}\exp(\gamma_{rs}t),$$
$$\text{for } (r, s) \in \{(1, 2), (1, 3), (2, 1), (2, 3)\}.$$

As in the survival model, we replace positive $\lambda_{rs.0}$ in the baseline hazards by $\exp(\beta_{rs.0})$. The random-effects distributions are given by

$$u_j \sim N(0, \sigma^2) \quad \text{and} \quad v_i \sim N(0, \tau^2).$$

The specification of this frailty model is driven by data considerations, computational issues, and the research question. The methodology for a multi-state model does not change when the number of states change. Likewise, the number of covariates does not affect

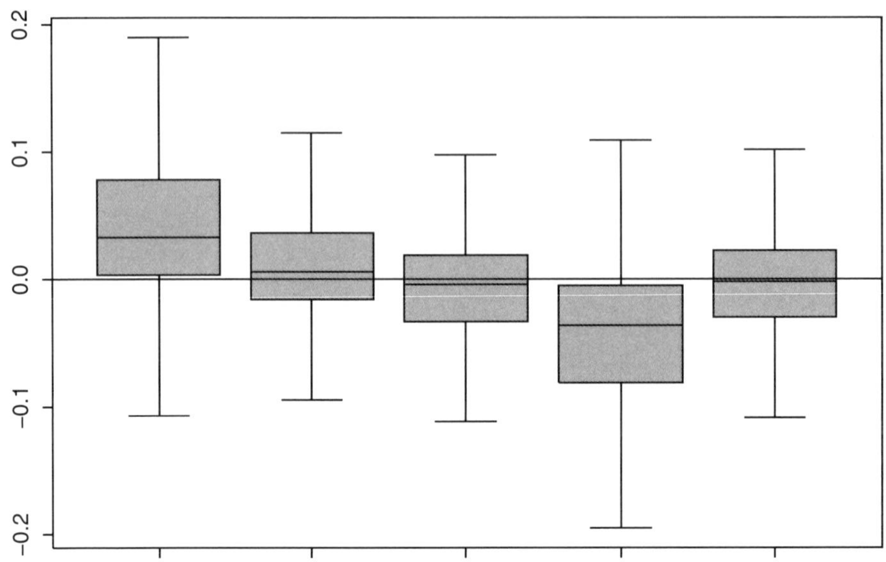

Figure 30.2 Posterior Distribution of Anonymized Cluster Frailties in the Gompertz Survival Model

the basic structure of the model. However, estimation becomes rapidly computationally intensive when the number of states or the number of covariates is increased. We make a few remarks with regard to our model choice. First, there are not many transitions observed from state 2 back to state 1, which means that there is limited information in the data on covariate effects for this transition. Second, the research question is about the effect of education on the transition from state 1 to 2. Hence we link education and the other relevant covariates to the hazard for this transition. Third, for the transitions into the death state we aim for parsimony but at the same time we want to acknowledge that there is heterogeneity with regard to hazard of death. Using individual-specific frailty will take that heterogeneity into account without the need for additional covariates.

As stated before, the grid for the piecewise-constant hazards is defined by the observation times (see Section 30.4.1). As in the survival models, ϵ in the log-likelihood is taken to be equal to one month (see Section 30.4.3).

We start by estimating the fixed-effects model which is derived by restricting $\sigma = 0$ and $\tau = 0$. This model has -2Loglik $=$ 19262. We compared this model with the fixed-effects model where the baseline hazards of the transition intensities follow a Weibull model. The latter has -2Loglik $=$ 19273. This is the reason why we use the Gompertz baseline hazards in the frailty model.

The marginal likelihood of the Gompertz frailty model as specified by (30.6) has -2Loglik $=$ 19260. This is only a slight improvement compared with the fixed-effects model. The estimated standard deviation of the distribution of the individual-specific frailties is very small, but the estimated standard deviation of the distribution of the cluster-specific frailties seems to indicate some differences between the clusters. We will use this model for Bayesian inference. Since we are interested in prediction, small differences in cluster-specific hazards may be relevant.

For Bayesian inference, prior densities have to be specified for the parameters. For

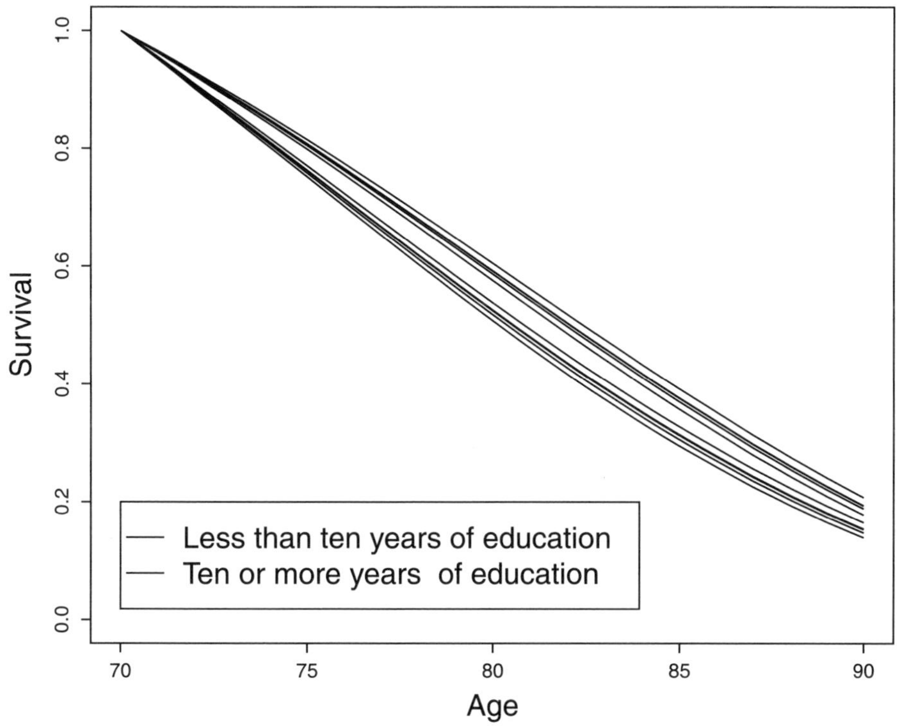

Figure 30.3 For Men Aged 70 and Born in 1920, Posterior Means of Survival Derived from the Gompertz Frailty Model. For both education groups, five survival curves corresponding to the five clusters in the data

model (30.6), the parameter vectors for the baseline hazards are $\boldsymbol{\gamma} = (\gamma_{12}, \gamma_{21}, \gamma_{21}, \gamma_{23})$, and $\boldsymbol{\beta}_0 = (\beta_{12.0}, \beta_{13.0}, \beta_{21.0}, \beta_{23.0})$. For the covariates effects we have $(\beta_{12.1}, \beta_{12.2}, \beta_{12.3})$, and for the frailty terms we have (σ, τ). Analogous to the survival frailty model, for the entries of $\boldsymbol{\gamma}$, $\boldsymbol{\beta}_0$, and $(\beta_{12.1}, \beta_{12.2}, \beta_{12.3})$, we specify vague priors given by $N(0, 10)$. For the standard deviations of the frailty distributions, we use $U(0, 5)$.

We also investigated a model where the individual frailty for the transition from the disability state to the death state was specified as θv_i, with θ a parameter to be estimated from the data, i.e., $h_{ij.23}(t)$ in (30.6) is replaced by $h_{ij.23}(t) = h_{23.0}(t)\exp(\theta v_i)$. This model takes into account the fact that the frailty for individual i may be transition-specific with respect to death. However, the parameter θ could not be identified using the data at hand. For θ to be identified, multi-state data is needed where many individuals are observed in both state 1 and state 2 several times.

Table 30.3 presents posterior means and 95% CIs for the model parameters in the three-state model. The effect of age parameterized by γ_{rs} is as expected. For the transitions $1 \rightarrow 2$, $1 \rightarrow 3$, and $2 \rightarrow 3$ the effect is positive, showing the increase of risk of a transition when men get older. The negative effect for transitions $2 \rightarrow 1$ is also understandable, as recovery from disability becomes less likely with increasing age.

For the hazard of transition $1 \rightarrow 2$, the effect of more education is negative (the posterior mean of $\beta_{12.1}$ is -0.34). This effect becomes less negative with increasing age (the posterior mean of $\beta_{12.3}$ is 0.01). However, the 95% CIs for these effects are quite wide. The hazards of the transition $1 \rightarrow 2$ seem proportional with respect to education (within each cluster), since $\beta_{12.3}$ does not seem different from zero in a statistically significant way. This is an interesting finding when compared to the survival model where there was more evidence that the effect of education with respect to mortality changes over time.

The individual frailties for the transitions into the death state do not seem to play an important role: the posterior mean of the standard deviation τ is 0.06 with 95% CI equal to $(0.01, 0.16)$. With regard to the cluster frailties there seems to be more heterogeneity: the posterior mean of the standard deviation σ is 0.25 with 95% CI equal to $(0.02, 0.76)$. Figure 30.4 depicts the posterior distribution of the cluster frailties. One cluster is clearly different from the other clusters (lower frailty, so less risk of disability).

The sensitivity of the results with respect to the choice of the priors was investigated with additional analyses. For example, we compared specifications $N(0, 10)$ and $N(0, 20)$ for the independent priors for the entries of $\boldsymbol{\gamma}$ and $\boldsymbol{\beta}$. The effect of these alternative specifications was minor for most of the parameters. The posterior means of γ_{12} and $\beta_{12.0}$ varied slightly. There is a high posterior correlation between these parameters, and both parameters have wide 95% credible intervals. However, the effect of this uncertainty on the inference for the other parameters is small. For example, the posterior distribution of the frailties in Figure 30.4 does not really change with different priors.

For a man born in 1920, and aged 70 years old, Figure 30.5 shows the change of the probability to be healthy (not disabled) at $t > 10$ conditional on being alive at t, where the scale for t in the model is age minus 60. This is the probability $P(Z_t = 1|Z_{10} = 1, Z_t \neq 3)$, where we assume that the individual-specific frailty for a transition into death is zero. So, for example, after ten years (at 80 years old), the conditional probability is about 0.9 for a man with more than ten years of education when he is in the best-performing cluster. Interestingly, one of the cluster effects is stronger than the education effect. The effect of being in the corresponding cluster outweighs the education effect after about 15 years.

30.7 DISCUSSION

We briefly discuss the choice of the distribution for the baseline hazard in the survival model, the approximation using piecewise-constant hazards, the approximation of the exact death time by interval censoring, and the choice of the frailty distribution.

In the absence of censoring and truncation, the estimation of the covariate effects in the survival model is consistent across the choices for the distribution of the baseline hazard (Kalbfleisch and Prentice 2002, Section 3.10). In the presence of right-censoring and left-truncation, inconsistency due to an incorrect choice of distribution is expected to be relatively small. For the CFAS data, this is illustrated by fitting the fixed-effects Cox proportional hazards model with covariates *Education* and *Year of Birth*. The Cox model is a regression model for the hazard where the baseline hazard is non-parametric; see, for details and software, Therneau and Grambsch (2000). Taking left-truncation into account, the effect of *Education* in the Cox model is estimated as -0.14 (0.05), and the effect of *Year of Birth* is -0.02 (0.01). These estimates are similar to the estimates for Model 3 in Section 30.2. This illustrates, to a certain extent, that the estimated effect of education is robust against misspecification of the distribution of the baseline hazard. The robustness holds with respect to the point estimates – not for the estimated standard errors. The advantage of the parametric Gompertz model is that the estimation of the effects is, in general, more efficient (although this is not apparent in the application), and that we can use the model for prediction.

Table 30.3 Gompertz Three-State Model for the Men in the CFAS Data. Posterior means and 95% credible intervals for the model parameters

Model parameters					
Baseline hazards			*Covariate effect*		
$\beta_{12.0}$	−1.99	(−3.94, −0.21)	$\beta_{12.1}$	−0.34	(−1.07, 0.42)
$\beta_{13.0}$	−4.01	(−4.37, −3.70)	$\beta_{12.2}$	−0.08	(−0.13, −0.02)
$\beta_{21.0}$	−2.12	(−3.01, −1.24)	$\beta_{12.3}$	0.01	(−0.03, 0.05)
$\beta_{23.0}$	−1.72	(−2.06, −1.41)			
γ_{12}	0.04	(−0.02, 0.10)			
γ_{13}	0.06	(0.05, 0.08)	*Frailty*		
γ_{21}	−0.03	(−0.07, 0.02)	σ	0.25	(0.02, 0.76)
γ_{23}	0.02	(0.01, 0.04)	τ	0.06	(0.01, 0.16)

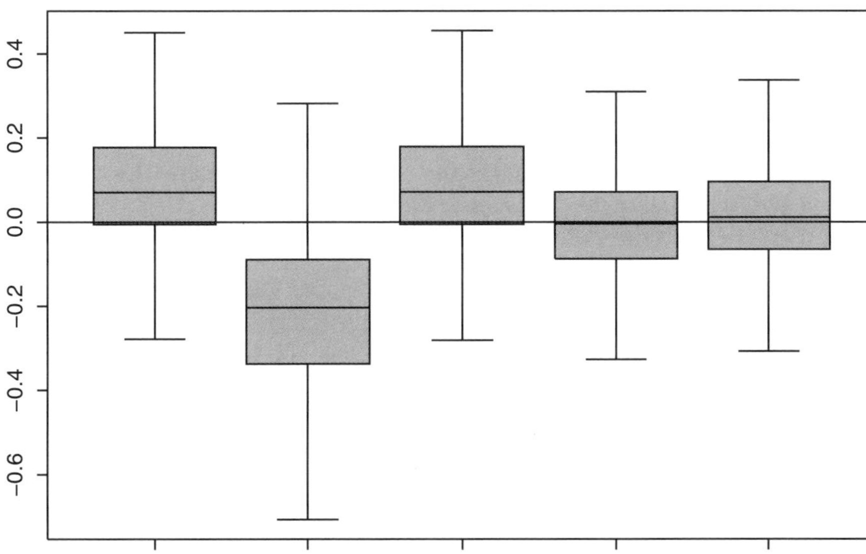

Figure 30.4 For the Gompertz Three-State Model, Posterior Distribution of Anonymized Cluster Frailties for the Hazard of Moving from the Healthy State to the Disability State

The exact death time is approximated by a short time interval ϵ in the likelihood (30.4) for the survival model and in the likelihood for the multi-state model (30.5). We have used this formulation because it ensures that the likelihood is a probability which enables a straightforward implementation in the software that was used for the MCMC. The approximation may not be a common choice, but produces similar results when compared with the maximum likelihood estimation where the hazard is included directly. We illustrate this for fixed-effects survival Model 3 in Section 30.2. Instead of $P(\text{death at } t_i|\text{alive at } t_i - \epsilon)S(t_i - \epsilon|t_{b.i})$ in (30.4) for time to event t_i, we use $h(t_i)S(t_i|t_{b.i})$. For γ we obtain 0.07 (0.01), and for the effects of *Education* and *Year of Birth* we get −0.14 (0.05) and −0.03 (0.01), respectively. This illustrates that the estimated effect of education in Model 3 is robust to the ϵ-approximation of the exact death time. The

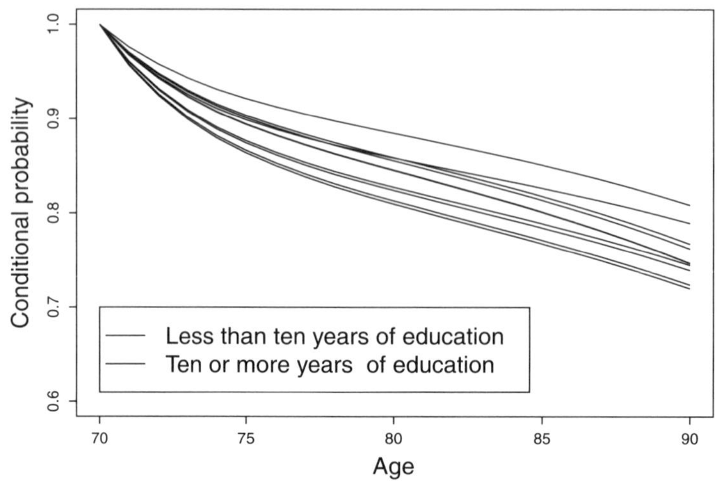

Figure 30.5 For Men Born in 1920, Posterior Means of the Probability of Not Being Disabled Conditional on Being Alive, and no Disability at Age 70. For both education groups, five survival curves corresponding to the five clusters in the data. The top two curves at age 90 correspond to the cluster with the lowest frailty

advantage of the ϵ-approximation is that we have exactly the same likelihood for the frequentist inference and the Bayesian inference.

Approximating the survival function using piecewise-constant hazards is an approximation of an integral. The chosen grid will affect the performance of the approximation. For the fixed-effects survival model 3 with the ϵ-approximation, we replaced the two-yearly grid with a series of finer grids. The effect of the specification of the grid seems limited. Given a one-yearly grid, for γ we obtain the estimate (and estimated standard error) 0.07 (0.01), and for the effects of *Education* and *Year of Birth* we get -0.14 (0.05) and -0.02 (0.01), respectively. The advantage of the grid is that time-dependent covariates can be taken into account easily. Using a relatively coarse grid reduces the computations needed to maximize the likelihood.

In this chapter, the normal distribution was chosen for the frailty term. Another common choice is the gamma distribution. In the regression equation for the survival model this would mean $h_{ij}(t) = h_0(t)w_j \exp(\boldsymbol{x}_{ij}\boldsymbol{\beta})$, where the frailty w_j follows a gamma distribution. For the Weibull model with the gamma frailty, the likelihood has a closed form, but for survival models with normal frailty there is no such closed form and numerical methods are needed to approximate the integrals in the likelihood (Duchateau and Janssen, 2008, Sections 2.2 and 4.7). We did not investigate other distributions besides the normal. The disadvantage of using the normal frailty is that taking the mean of the frailty distribution does not produce the mean trend in the population due to the non-linearity of the model for the hazard. However, the advantages are that the modeling structure resembles those of the multilevel models described elsewhere in this book, and users have, in general, a basic understanding of the normal distribution.

30.8 CONCLUSION

In this chapter, frailty models for survival are applied to longitudinal data from a population-based multi-centre study of older men in England and Wales. With regard to mortality, the effect of education was investigated, taking into account the clustering of men within centres. There is a clear association between more education and a better overall survival.

An illness–death frailty model with three states was also investigated for the same data. In this model, state one is a state without disability, state two indicates disability, and state three is the death state. In this multi-state survival process, there is also a clear effect of more education in the sense that more education is associated with a lower risk of disability. The risk of disability is confounded by centre effects, and using a multilevel structure for centre-specific frailty is essential to control for this.

That more education is associated in the UK with better survival and better health in older age is not a new finding. But this chapter has shown that by using multilevel modeling, we can distinguish centre effects from the effect of education. This provides a better insight into the role of education, and also improves prediction. For one of the centres in the study, the long-term centre effect on the risk of disability was stronger than the education effect.

The analysis in this chapter should not be considered as the final analysis of the CFAS data. To make the data used publicly available, we have taken a subset of the full data and centres have been anonymized. Moreover, other models can be formulated, but have not been investigated in this chapter. We have used the normal distribution for the frailty terms, but other choices such as the gamma distribution are possible. Furthermore, we plan to investigate in future research work whether introduction of transition-specific choices for the baseline hazards in the multi-state model would be worthwhile pursuing. Likewise, other distributions for the baseline hazard in the survival model could be investigated as well. Despite the limitations of the current data analysis, we think that the work in this chapter provides a useful insight into the effect of education on survival and disability in older age. In addition, we hope that we have illustrated the possibilities and the advantages of multilevel modeling.

ACKNOWLEDGEMENTS

The Medical Research Council CFAS is supported by major awards from the Medical Research Council [grant number G99001400] and the Department of Health. This work was supported by the Medical Research Council [grant numbers G0400077, U105261167].

REFERENCES

Aalen, O.O., Borgan, O., and Gjessing, H.K. (2008) *Survival and Event History Analysis. A Process Point of View*. New York: Springer-Verlag.

Andersen, P.K., Borgan, O., Gill, R.D., and Keiding, N. (1993) *Statistical Models Based on Counting Processes*. New York: Springer-Verlag.

Brayne, C., McCracken, C., and Matthews, F.E. (2006) 'Cohort Profile: The Medical Research Council Cognitive Function and Ageing Study (CFAS)', *International Journal of Epidemiology*, 35: 1140–45.

Bull, K. and Spiegelhalter, D.J. (1997) 'Survival Analysis in Observational Studies', *Statistics in Medicine*, 16: 1041–74.

Collett, D. (2003) *Modeling Survival Data in Medical Research*, 2nd edition. London: Chapman and Hall/CRC.

Cox, D.R. and Oakes, D. (1984) *Analysis of Survival Data*. London: Chapman and Hall/CRC.

Duchateau, L. and Janssen, P. (2008) *The Frailty Model*. New York: Springer.

Ha, I.D., Lee, Y., and Song, J.-K. (2002) 'Hierarchical-likelihood Approach for Mixed Linear Models with Censored Data', *Lifetime Data Analysis*, 8: 163–76.

Jackson, C.H., Sharples, L.D., Thompson, S.G., Duffy, S.W., and Couto, E. (2003) 'Multi-state Markov Models for Disease Progression with Classification Error', *Statistician*, 52: 193–209.

Jagger, C., Matthews, R.J., Matthews, F.E., Spiers, N.A., Nickson, J., Paykel, E.S., Huppert, F.A., Brayne, C., and the Medical Research Council Cognitive Function and Ageing Study (2007) 'Cohort Differences in Disease and Disability in the Young-old: Findings from the MRC Cognitive Function and Ageing Study (MRC-CFAS)', *BMC Public Health*, 7.

Kalbfleisch, J.D. and Lawless, J.F. (1985) 'The Analysis of Panel Data Under a Markov Assumption',

Journal of the American Statistical Association, 80: 863–71.

Kalbfleisch, J.D. and Prentice, R.L. (2002) *The Statistical Analysis of Failure Time Data*, second edition. New York: Wiley.

Kay, R. (1986) 'A Markov Model for Analysing Cancer Markers and Disease States in Survival Studies', *Biometrics*, 42: 855–65.

Korn, E.L. and Graubard, B.I. (1999) *Analysis of Health Surveys*. New York: Wiley.

Matthews, F.E., Miller, L.L., Brayne, C., and Jagger, C. (2006) 'Regional Differences in Multidimensional Aspects of Health: Findings from the MRC Cognitive Function and Ageing Study', *BMC Public Health*, 6.

Nelder, J.A. and Mead, R. (1965) 'A Simplex Algorithm for Function Minimization', *Computer Journal*, 7: 308–13.

Norris, J.R. (1997) *Markov Chains*. Cambridge: Cambridge University Press.

Pan, S.L., Wu, H.M., Yen, A.M.F., and Chen, T.H.H. (2007) 'A Markov Regression Random-effects Model for Remission of Functional Disability in Patients Following a First Stroke: A Bayesian Approach', *Statistics in Medicine*, 26: 5335–53.

Putter, H., Fiocco, M., and Geskus, R.B. (2007) 'Tutorial in Biostatistics: Competing Risks and Multi-state Models', *Statistics in Medicine*, 26: 2389–430.

Spiers, N.A., Matthews, R.J., Jagger, C., Matthews, F.E., Boult, C., Robinson, T.G., and Brayne, C. (2005) 'Diseases and Impairments as Risk Factors for Onset of Disability in the Older Population in England and Wales: Findings from the Medical Research Council Cognitive Function and Ageing Study', *Journals of Gerontology: Medical Sciences*, 6: 248–54.

Therneau, T.M. and Grambsch, P.M. (2000) *Modeling Survival Data: Extending the Cox Model*. New York: Springer.

Thomas, A., O'Hara, B., Ligges, U., and Sturtz, S. (2006) 'Making BUGS Open.' *R News*, 6: 12–17.

Van den Hout, A. and Matthews, F.E. (2009) 'Estimating Dementia-free Life Expectancy for Parkinson's Patients Using Bayesian Inference and Microsimulation', *Biostatistics*, 10: 729–43.

Wei, L.J. (1992) 'The Accelerated Failure Time Model: A Useful Alternative to the Cox Regression Model in Survival Analysis', *Statistics in Medicine*, 11: 1871–9.

Welton, N.J. and Ades, A.D. (2005) 'Estimation of Markov Chain Transition Probabilities and Rates from Fully and Partially Observed Data: Uncertainty Propagation, Evidence Synthesis, and Model Calibration', *Medical Decision Making*, 25: 633–45.

Wu, L. (2010) *Mixed Effects Models for Complex Data*. Boca Raton, FL: Chapman & Hall/CRC.

31

Point-Referenced Spatial Modeling

Andrew O. Finley
Michigan State University, USA

Sudipto Banerjee
University of Minnesota, USA

31.1 INTRODUCTION

This chapter is inherently different from most of the other chapters in this Handbook. While multilevel modeling (MLM) customarily refers to settings where observations are clustered by subjects within groups, or, perhaps, subjects within time, in the current chapter we envision observing multiple outcomes at *each* spatial location. While this is not entirely dissimilar to longitudinal/within-subject analysis, an important distinction is that the multiple observations in spatial data typically come from different variables that are themselves posited to be associated with each other. In this regard, our approach resembles *multivariate* modeling rather than repeated measures.

Inferential interest requires accounting for two types of association. The first is the association between observations of the *same* variable at different locations. The second is the association among the different variables within each location. Our approach builds joint probability distributions that can distinguish between these different types of association and infer about them.

We should also clarify at the outset that our multilevel models are, in fact, Bayesian hierarchical models (Chapter 4) that incorporate prior information on quantities such as spatial range (i.e., the distance at which the spatial association becomes negligible). We primarily focus upon *point-referenced* data, which means that the locations are indexed by their coordinates. A salient feature of these multilevel models is that they deploy distributions on the spatial random effects that are induced from a *spatial process*. The variables being studied are assumed to exist at any arbitrary location within a geographical region of interest but have, naturally, been observed only at a finite set of locations. Such data are commonly encountered in the natural and environmental sciences such as forestry, ecology, climate science, and so on. We will also discuss, albeit briefly, approaches for multivariate data that arise as regional aggregates or summaries. Such data are more common in public health, where geographical coordinates are not disclosed due to confidentiality issues.

31.2 STATISTICAL MODELS IN SPATIAL DATA ANALYSIS

The scientific community is moving into an era where data-rich environments provide extraordinary opportunities to understand the complexity of spatially indexed datasets. Unprecedented access to spatial data is a result of investments to collect data for regulatory, monitoring, and resource management objectives, and technological advances in spatially enabled sensor networks along with geospatial information storage, analysis, and distribution systems. These data sources are increasingly diverse and specialized, e.g., computer model outputs, monitoring station instruments, remotely located sensors, and georeferenced field measurements. Across scientific fields, researchers face the challenge of coupling these data with imperfect models to better understand variability in their system of interest. The inference garnered through these analyses often supports decisions with important economic and ecological implications; therefore, it is critical to correctly estimate inferential uncertainty. However, developing modeling frameworks capable of accounting for various sources of uncertainty is not a trivial task—massive datasets from multiple sources with complex spatial dependence structures only serve to aggravate the challenges.

Proliferation of spatial data has spurred considerable development in statistical modeling; see, for example, the books by Cressie (1993), Chilés and Delfiner (1999), Møller and Waagepetersen (2004), Schabenberger and Gotway (2004), and Wackernagel (2003). See also Banerjee et al. (2004) and Diggle and Ribeiro (2007) for a variety of methods and applications. The statistical literature acknowledges that spatial and temporal associations are captured most effectively using models that build dependencies in different stages. Multilevel models are especially advantageous with datasets having several lurking sources of uncertainty and dependence, where they can estimate much richer models with less stringent assumptions than traditional modeling paradigms. These models follow the Bayesian paradigm of statistical inference (see, e.g., Carlin and Louis, 2000; Gelman et al., 2004), where analysis uses sampling from the posterior distributions of model parameters.

Computational advances with regard to Markov chain Monte Carlo (MCMC) methods have contributed enormously to the popularity of multilevel models in a wide array of disciplines (e.g., Gilks et al., 1996; Robert and Casella, 2005), and spatial modeling is no exception (see, e.g., Banerjee et al., 2004). In the realm of spatial statistics, multilevel models have been widely applied to analyze both areally referenced as well as point-referenced or geostatistical data. For the former, a class of models known as Conditionally Autoregressive (CAR) models have become very popular as they are easily implemented using MCMC methods such as the Gibbs sampler. In fact, these models are somewhat naturally suited to the Gibbs sampler, which draws samples from conditional distributions that are fully specified by the CAR models. Their popularity has increased in no small measure due to their automated implementation in the OpenBUGS software package, which offers a flexible and user-friendly interface to construct multilevel models that are implemented using a Gibbs sampler (see, e.g., Chapter 4). This is performed by identifying a multilevel model with a directed acyclic graph (DAG) whose nodes form the different components of the model and allowing the language to identify the full conditional distributions that need to be updated. OpenBUGS is an offshoot of the BUGS (Bayesian inference Using Gibbs Sampling) project and the successor of the WinBUGS software.

From an automated implementation perspective, the challenges are somewhat greater for point-referenced models. First, expensive matrix computations are required that can become prohibitive with large datasets.

Second, routines to fit unmarginalized models are less suited for direct updating using a Gibbs sampler in the BUGS paradigm and result in slower convergence of the chains. Third, investigators often encounter multivariate spatial datasets with several spatially dependent outcomes, whose analysis requires multivariate spatial models that involve matrix computations that are poorly implemented in BUGS. These issues have, however, started to wane with the delivery of relatively simpler `R` packages (`www.r-project.org`), including `mcmc`, `MCMCpack`, `geoRglm`, and `spBayes`, that help automate Bayesian methods for point-referenced data and diagnose convergence.

The primary focus of this chapter is on multilevel models for Gaussian univariate and multivariate point-referenced data. Some detail is also provided for modeling non-Gaussian point-referenced data and for data arising over areal units. These multilevel spatial process models, implemented through MCMC methods, offer increased flexibility to fit models that would be infeasible with classical methods within inappropriate asymptotic paradigms. With this increased flexibility comes substantial computational demands. Estimating these models involves expensive matrix decompositions whose computational complexity increases in cubic order with the number of spatial locations, rendering such models infeasible for large spatial datasets. An approach to overcoming this computational hurdle is to consider a representation of the spatial process in terms of lower-dimensional realizations. One such representation, detailed here, leads to a class of *predictive process models* for spatial and spatiotemporal data. The methods presented in this chapter are illustrated using a multivariate forestry inventory dataset collected on the Penobscot Experimental Forest in Maine by the U.S. Forest Service. Here, need for specialized multilevel models is especially great because the data exhibit strongly spatially structured residuals. Special attention is also given to exploration and visualization of spatial data and the practical and accessible implementation of the models.

31.3 MODELS

31.3.1 Univariate Spatial Regression

Geostatistical settings typically assume, at locations $s \in D \subseteq \mathbb{R}^2$, a Gaussian outcome variable $y(s)$ along with a $p \times 1$ vector of spatially referenced predictors $x(s)$ which are associated through a spatial regression model such as

$$y(s) = x(s)'\beta + u(s) + \varepsilon(s). \quad (31.1)$$

In this model, the residual comprises a spatial process, $u(s)$, and an independent white-noise process, $\varepsilon(s)$, that captures measurement error or micro-scale variation. With any collection of n locations, say $\mathcal{S} = \{s_1, \ldots, s_n\}$, we assume that the $\varepsilon(s_i)$s are identically and independently distributed as $N(0, \tau^2)$, where τ^2 is often called the *nugget*.

The $u(s)$ are spatial random effects, providing local adjustment (with structured dependence) to the mean, interpreted as capturing the effect of unmeasured or unobserved covariates with spatial pattern. A popular modeling choice for the spatial random effects is the *Gaussian process*, $u(s) \sim GP(0, C(\cdot, \cdot; \theta))$, specified by a valid covariance function $C(s, s'\theta) = \mathrm{Cov}(u(s), u(s'))$ that models the covariance corresponding to a pair of sites s and s'. The process realizations are collected into an $n \times 1$ vector, say $u = (u(s_1), \ldots, u(s_n))'$, which follows a multivariate normal distribution $N_n(\mathbf{0}, \Sigma_u)$, where Σ_u is the $n \times n$ covariance matrix of u with (i, j)th element given by $C(s_i, s_j; \theta)$.

Clearly, $C(s, s'; \theta)$ cannot be just any function; it must ensure that the resulting Σ_u matrix is symmetric and positive definite. Such functions are known as positive definite functions and are characterized as the characteristic function of a symmetric random variable (due to a famous theorem by Bochner).

Further technical details about positive definite functions can be found in Cressie (1993), Chilés and Delfiner (1999), and Banerjee et al. (2004).

We customarily specify $C(\boldsymbol{s}, \boldsymbol{s}'; \boldsymbol{\theta}) = \sigma^2 \rho(\boldsymbol{s}, \boldsymbol{s}'; \boldsymbol{\phi})$ with $\boldsymbol{\theta} = \{\sigma^2, \boldsymbol{\phi}\}$, where $\rho(\cdot; \boldsymbol{\phi})$ is a *correlation function* while $\boldsymbol{\phi}$ includes parameters quantifying rate of correlation decay and smoothness of the surface $u(\boldsymbol{s})$. Then $\text{Var}(u(\boldsymbol{s})) = \sigma^2$ represents a spatial variance component in the model in (31.1). This specification assumes *stationary* and *isotropic* processes. Stationarity, in spatial modeling contexts, refers to the setting when $C(\boldsymbol{s}, \boldsymbol{s}'; \boldsymbol{\theta}) = C(\boldsymbol{s} - \boldsymbol{s}'; \boldsymbol{\theta})$; that is, the covariance function depends upon the separation of the sites. Isotropy goes further and specifies $C(\boldsymbol{s}, \boldsymbol{s}') = C(\|\boldsymbol{s} - \boldsymbol{s}'\|)$, where $\|\boldsymbol{s} - \boldsymbol{s}'\|$ is the Euclidean distance between the sites $\boldsymbol{s}$ and $\boldsymbol{s}'$. A very versatile class of correlation functions is the Matérn correlation function given by

$$\rho(\|\boldsymbol{s} - \boldsymbol{s}'\|; \boldsymbol{\phi}) = \frac{1}{2^{\phi_2 - 1}\Gamma(\phi_2)} (\|\boldsymbol{s} - \boldsymbol{s}'\|\phi_1)^{\phi_2} \mathcal{K}_{\phi_2}(\|\boldsymbol{s} - \boldsymbol{s}'\|; \phi_1), \tag{31.2}$$

for $\phi_1 > 0$, $\phi_2 > 0$. Here, $\boldsymbol{\phi} = \{\phi_1, \phi_2\}$, with ϕ_1 controlling the decay in spatial correlation and ϕ_2 controlling process smoothness. Specifically, if ϕ_2 lies between positive integers m and $(m + 1)$, then the spatial process $u(\boldsymbol{s})$ is mean-square differentiable m times, but not $m+1$ times. This fact has been used to study the derivatives of spatial surfaces (e.g., Banerjee, Gelfand, and Sirmans, 2003). Also, Γ is the usual Gamma function while $\mathcal{K}_{\phi_2}$ is a modified Bessel function of the third kind with order ϕ_2.

For multilevel models, one assigns prior (hyperprior) distributions to the model parameters (hyperparameters) and inference proceeds by sampling from the posterior distribution of the parameters (see, e.g., Banerjee et al., 2004). Note that the spatial process induces a $N_n(\boldsymbol{0}, \sigma^2 R(\boldsymbol{\phi}))$ distribution on $\boldsymbol{u}$, where $R(\boldsymbol{\phi})$ is the $n \times n$ dispersion matrix whose (i, j)th element is $\rho(\boldsymbol{s}_i, \boldsymbol{s}_j; \boldsymbol{\phi})$. For the remaining parameters, suppose that $\boldsymbol{\beta}$ is assigned a multivariate Gaussian prior, say $N_p(\boldsymbol{\mu}_\beta, \Sigma_\beta)$, while the spatial variance component σ^2 and the measurement error variance τ^2 are assigned inverse-Gamma (IG) priors. The process correlation parameter(s), $\boldsymbol{\phi}$, are usually assigned informative priors (e.g., uniform over a finite range and to perhaps ensure a desired degree of freedom for the spatial surface) based on the underlying spatial domain and the physical processes under study.

Using notations similar to Gelman et al. (2004), we can write the posterior distribution by $p(\boldsymbol{\beta}, \boldsymbol{u}, \sigma^2, \tau^2, \boldsymbol{\phi} \,|\, \boldsymbol{y})$, where $\boldsymbol{y} = (y(\boldsymbol{s}_1), \ldots, y(\boldsymbol{s}_n))'$, as proportional to

$$p(\boldsymbol{\phi}) \times IG(\tau^2 \,|\, a_\tau, b_\tau) \times IG(\sigma^2 \,|\, a_\sigma, b_\sigma) \times N_p(\boldsymbol{\beta} \,|\, \boldsymbol{\mu}_\beta, \Sigma_\beta) \times N_n(\boldsymbol{u} \,|\, \boldsymbol{0}, \sigma^2 R(\boldsymbol{\phi})) \times \prod_{i=1}^{n} N(y(\boldsymbol{s}_i) \,|\, \boldsymbol{x}(\boldsymbol{s}_i)'\boldsymbol{\beta} + u(\boldsymbol{s}_i), \tau^2). \tag{31.3}$$

Estimation of (31.3) customarily proceeds using an MCMC algorithm. Often, a marginalized likelihood is used that is obtained by integrating out the spatial effects $\boldsymbol{u}$. This yields the posterior distribution $p(\boldsymbol{\beta}, \sigma^2, \tau^2, \boldsymbol{\phi} \,|\, \boldsymbol{y})$ that is proportional to

$$p(\boldsymbol{\phi}) \times IG(\tau^2 \,|\, a_\tau, b_\tau) \times IG(\sigma^2 \,|\, a_\sigma, b_\sigma) \times N_p(\boldsymbol{\beta} \,|\, \boldsymbol{\mu}_\beta, \Sigma_\beta) \times N_n(\boldsymbol{y} \,|\, X\boldsymbol{\beta}, \sigma^2 R(\boldsymbol{\phi}) + \tau^2 I_n), \tag{31.4}$$

where X is the matrix of regressors whose ith row is $\boldsymbol{x}(\boldsymbol{s}_i)'$ and $\boldsymbol{I}_n$ is the $n \times n$ identity matrix.

An efficient Markov chain Monte Carlo (MCMC) algorithm is obtained by updating $\boldsymbol{\beta}$ from its full conditional $N_p(\boldsymbol{\mu}_{\beta|\cdot}, \Sigma_{\beta|\cdot})$, where

$$\Sigma_{\beta|\cdot} = [\Sigma_\beta^{-1} + X'\Sigma_y^{-1}X]^{-1} \text{ and } \boldsymbol{\mu}_{\beta|\cdot} = \Sigma_{\beta|\cdot} X'\Sigma_y^{-1}\boldsymbol{y}, \tag{31.5}$$

where $\Sigma_y = \sigma^2 R(\boldsymbol{\phi}) + \tau^2 I_n$. The remaining parameters can be updated using Metropolis steps, where the target distribution for any (set of) parameter(s) is proportional to the product of the terms in (31.4) that involve that (those) parameter(s). These will yield posterior samples of $\Omega = \{\boldsymbol{\beta}, \sigma^2, \boldsymbol{\phi}, \tau^2\}$.

In this sampling scheme, the spatial random effects $\boldsymbol{u}$ are not sampled directly. This reduces the parameter space, which results in a more efficient MCMC algorithm. A key advantage of the first-stage Gaussian model, as in (31.1), is that samples from the posterior distribution of $\boldsymbol{u}$ can be recovered in a posterior predictive fashion. More precisely, we seek to evaluate

$$p(\boldsymbol{u}\,|\,\boldsymbol{y}) \propto \int p(\boldsymbol{u}\,|\,\Omega, \boldsymbol{y})\,p(\Omega\,|\,\boldsymbol{y})\, d\Omega\,. \tag{31.6}$$

Because the full conditional distribution of $\boldsymbol{u}$ in (31.6) is again multivariate normal, (31.6) is easily evaluated using *composition sampling*. To be precise, for each sample $\{\boldsymbol{\beta}^{(l)}, \sigma^{2(l)}, \tau^{2(l)}, \boldsymbol{\phi}^{(l)}\}$, we draw $\boldsymbol{u}^{(l)}$ from $N_n(\boldsymbol{\mu}_{\boldsymbol{u}|\cdot}, \Sigma_{\boldsymbol{u}|\cdot})$, where

$$\begin{aligned}\Sigma_{\boldsymbol{u}|\cdot} &= \left[\frac{R\left(\boldsymbol{\phi}^{(l)}\right)^{-1}}{\sigma^{2(l)}} + \frac{I_n}{\tau^{2(l)}}\right]^{-1} \quad \text{and}\\ \boldsymbol{\mu}_{\boldsymbol{u}|\cdot} &= \Sigma_{\boldsymbol{u}|\cdot}\frac{\left(\boldsymbol{y} - X\boldsymbol{\beta}^{(l)}\right)}{\tau^{2(l)}}.\end{aligned} \tag{31.7}$$

These posterior samples from $\boldsymbol{u}$ can be mapped using contours or interpolation routines to produce plots of the spatial process, see, e.g., Figure 31.2 in Section 31.5.

For predictions, if $\mathcal{S}_0 = \{\boldsymbol{s}_{0,1}, \boldsymbol{s}_{0,2}, \ldots, \boldsymbol{s}_{0,n_0}\}$ is a collection of n_0 *new* locations, the posterior predictive distribution of $p(\boldsymbol{u}_0\,|\,\boldsymbol{y})$, where $\boldsymbol{u}_0 = (u(\boldsymbol{s}_{0,1}), u(\boldsymbol{s}_{0,2}), \ldots, u(\boldsymbol{s}_{0,n_0}))'$, is

$$p(\boldsymbol{u}_0\,|\,\boldsymbol{y}) \propto \int p(\boldsymbol{u}_0\,|\,\boldsymbol{u}, \Omega, \boldsymbol{y})\; p(\boldsymbol{u}\,|\,\Omega, \boldsymbol{y})\,p(\Omega\,|\,\boldsymbol{y})\,d\Omega\boldsymbol{u}.$$

Given posterior samples, $\{\Omega^{(l)}\}_{l=1}^{L}$, this distribution can again be obtained via composition sampling: we first draw $\boldsymbol{u}^{(l)}$ for each l as described in (31.6) and then draw $\boldsymbol{u}_0^{(l)}$ from $p(\boldsymbol{u}_0\,|\,\boldsymbol{u}^{(l)}, \Omega^{(l)}, \boldsymbol{y})$, where this last distribution is derived as a conditional distribution from a multivariate normal and, hence, is again multivariate normal. More precisely, the process realizations over the new locations are conditionally independent of the observed outcomes given the realizations over the observed locations and the process parameters. In other words, $p(\boldsymbol{u}_0\,|\,\boldsymbol{u}, \Omega, \boldsymbol{y}) = p(\boldsymbol{u}_0\,|\,\boldsymbol{u}, \Omega)$, which is a multivariate normal distribution with mean and variance given by

$$\begin{aligned}\mathrm{E}[\boldsymbol{u}_0\,|\,\boldsymbol{u}, \Omega] &= \mathrm{Cov}(\boldsymbol{u}_0, \boldsymbol{u})\mathrm{Var}^{-1}(\boldsymbol{u})\boldsymbol{u}\\ &= R_0(\boldsymbol{\phi})'R(\boldsymbol{\phi})^{-1}\boldsymbol{u}\\ \mathrm{Var}[\boldsymbol{u}_0\,|\,\boldsymbol{u},\Omega] &= \sigma^2\{R(\boldsymbol{\phi}) - R_0(\boldsymbol{\phi})'R(\boldsymbol{\phi})^{-1}R_0(\boldsymbol{\phi})\},\end{aligned}$$

where $R_0(\boldsymbol{\phi})'$ is the $n_0 \times n$ matrix with (i, j)th element given by $\rho(\boldsymbol{s}_{0,i}, \boldsymbol{s}_j; \boldsymbol{\phi})$. Finally, given a set of covariates at a new location $\boldsymbol{s}_0$, samples from the posterior predictive distribution of the outcome variable, $y(\boldsymbol{s}_0)^{(l)}$, are drawn from $N(\boldsymbol{x}(\boldsymbol{s}_0)'\boldsymbol{\beta}^{(l)} + \boldsymbol{u}_0^{(l)}, \tau^{2(l)})$ for $l = 1, 2, \ldots, L$.

Spatial process models for large datasets

Full inference and accurate assessment of (31.1) will involve matrix decompositions of either Σ_y or Σ_u, depending on whether we choose to integrate out $\boldsymbol{u}$ from the likelihood or not. The computational complexity of these factorizations is of the order of $O(n^3)$, where n is the number of locations. Hence, we encounter computational infeasibility or the 'big n' problem for large point-referenced datasets.

Modeling large spatial datasets has received much attention in the statistical literature. Vecchia (1988) proposed approximating the likelihood with a product of appropriate conditional distributions to obtain maximum-likelihood estimates. Stein et al. (2004) adapted this to restricted maximum likelihood estimation. Another possibility is to approximate the likelihood using spectral representations of the spatial process (Fuentes, 2007). These likelihood approximations yield a joint distribution, but not a process that facilitates spatial interpolation. Yet another approach considers compactly supported correlation functions (Furrer et al., 2006; Kaufman et al., 2008; Du et al., 2009) that yield sparse correlation structures. More efficient sparse solvers can then be employed for kriging and variance estimation, but the tapered structures may limit modeling flexibility. Also, full likelihood-based inference still requires determinant computations that may be problematic.

Rather than approximations, one could build models especially geared towards handling large spatial datasets. These are representations of the spatial process in a lower-dimensional subspace and are often referred to as low-rank or reduced-rank spatial models (Higdon, 2002; Kamman and Wand, 2003; Stein, 2007, 2008; Cressie and Johannesson, 2007; Banerjee et al., 2008; Crainiceaniu et al., 2008). Many of these methods are variants of the so-called 'subset of regressors' methods used in Gaussian process regressions for large datasets in machine learning (e.g., Rasmussen and Williams, 2006). The idea here is to consider a smaller set of locations, or 'knots', say $\mathcal{S}^* = \{s_1^*, \ldots, s_{n^*}^*\}$, where the number of knots, n^*, is *fixed* to be much smaller than the number of observed sites, and to express the spatial process realizations over n locations in terms of its realizations over the smaller set of knots. It is reasonable to assume there will be insignificant loss of spatial information in the underlying process from using a smaller set of locations—the knots—with adequate domain coverage.

Recently, Banerjee et al. (2008) proposed a class of knot-based spatial process models for large spatial datasets. These models consider a fixed set of $\mathcal{S}^*$, again with $n^* \ll n$, which may or may not form a subset of the entire collection of observed locations in $\mathcal{S}$. The Gaussian process $u(s)$ yields an n^*-vector of realizations over the knots, say $\boldsymbol{u}^* = (u(s_1^*), \ldots, u(s_{n^*}^*))'$, which follow an $N_{n^*}(\mathbf{0}, \sigma^2 R^*(\boldsymbol{\phi}))$ distribution where $R^*(\boldsymbol{\phi})$ is the corresponding $n^* \times n^*$ dispersion matrix whose (i, j)th element is $\rho(s_i^*, s_j^*; \boldsymbol{\phi})$. Spatial interpolation (or 'kriging') at a generic site $\boldsymbol{s}$ is executed through

$$\tilde{u}(\boldsymbol{s}) = E\{u(\boldsymbol{s}) \mid \boldsymbol{u}^*\} = \boldsymbol{r}(\boldsymbol{s}; \boldsymbol{\phi})' R^{*-1}(\boldsymbol{\phi})\boldsymbol{u}^*.$$

This yields a spatial process $\tilde{u}(\boldsymbol{s}) \sim GP(0, \sigma^2\tilde{\rho}(\cdot))$, where

$$\tilde{\rho}(\boldsymbol{s}, \boldsymbol{s}'; \boldsymbol{\phi}) = \boldsymbol{r}(\boldsymbol{s}; \boldsymbol{\phi})' R^{*-1}(\boldsymbol{\phi})\boldsymbol{r}(\boldsymbol{s}', \boldsymbol{\phi}),$$

and $\boldsymbol{r}(\boldsymbol{s}; \boldsymbol{\phi}) = [\rho(\boldsymbol{s}, s_j^*; \boldsymbol{\phi})]_{j=1}^{n^*}$. We refer to $\tilde{u}(\boldsymbol{s})$ as the *predictive process* derived from the *parent process* $u(\boldsymbol{s})$. The predictive process is a spatially adaptive linear transformation of the realizations of $u(\boldsymbol{s})$ over $\mathcal{S}^*$, with $\boldsymbol{r}(\boldsymbol{s}; \boldsymbol{\phi})' R^{*-1}(\boldsymbol{\phi})$ comprising the coefficients of the transformation. This also implies that $\tilde{u}(\boldsymbol{s})$ is non-stationary, even though $u(\boldsymbol{s})$ is not, and may allow the model to adapt better to fit the data.

Replacing $u(\boldsymbol{s})$ in (31.1) with $\tilde{u}(\boldsymbol{s})$, we obtain the predictive process model

$$y(\boldsymbol{s}) = \boldsymbol{x}(\boldsymbol{s})'\boldsymbol{\beta} + \tilde{u}(\boldsymbol{s}) + \varepsilon(\boldsymbol{s}). \qquad (31.8)$$

Dimension reduction occurs since the computations now involve evaluating the $n^* \times n^*$ matrix $R^{*-1}(\boldsymbol{\phi})$. Unlike other knot-based methods, the predictive process does not introduce any additional parameters nor involves projecting data onto a grid, while enjoying attractive theoretical properties that justify its use as a *best approximation* for the parent process. For example, $\tilde{u}(\boldsymbol{s})$ is an orthogonal projection of $u(\boldsymbol{s})$ on an appropriate linear subspace minimizing $E[\{u(\boldsymbol{s}) - f(\boldsymbol{u}^*)\}^2 \,|\boldsymbol{u}^*]$ over all real-valued functions $f(\boldsymbol{u}^*)$.

Rather than an approximation to the parent process, we consider the predictive process as a dimension-reducing model for large point-referenced datasets. It is crucial, therefore, that its parameters should be interpreted with respect to (31.8) and not (31.1). In fact, being smoother than the parent process, the predictive process tends to have lower variance which, in turn, leads to an upward bias in the nugget. The following inequality reflects, more formally, the shrinkage in variability for the predictive process:

$$0 \le \text{Var}\{u(\boldsymbol{s}) \mid \boldsymbol{u}^*\} = \sigma^2\{1 - \boldsymbol{r}(\boldsymbol{s}; \boldsymbol{\phi})' R^{*-1}\boldsymbol{r}(\boldsymbol{s}; \boldsymbol{\phi})\} = \sigma^2 - \text{Var}\{\tilde{u}(\boldsymbol{s})\},$$

which implies that $\text{Var}\{\tilde{u}(\boldsymbol{s})\} \le \sigma^2 = \text{Var}\{u(\boldsymbol{s})\}$. The diminished variability in $\tilde{u}(\boldsymbol{s})$

is often manifested by an overestimation of the nugget variance τ^2. Banerjee et al. (2010) explore these biases in greater detail.

Finley et al. (2009) consider modifying the predictive process by adding a heteroscedastic white-noise Gaussian process. More specifically, they propose replacing $\tilde{u}(s)$ in (31.8) with $\tilde{u}_{\varepsilon}(s) = \tilde{u}(s) + \tilde{\varepsilon}(s)$, where

$$\tilde{\varepsilon}(s) \overset{iid}{\sim} N\{0, \sigma^2(1 - r(s;\phi)'R^{*-1}(\phi)r(s;\phi))\}.$$

The white-noise term can be regarded as an additional nugget effect in (31.8).

A key issue in predictive process modeling is the number and selection of knots, which is a challenging problem, with choice in two dimensions more difficult than in one. The choice of n^* is governed by computational cost and sensitivity to choice. Customarily, the analysis is implemented over different choices of n^* and knot locations. The issue is not dissimilar to a spatial design problem (e.g., Nychka and Saltzman, 1998; Xia and Gelfand, 2006; Diggle and Lophaven, 2006). The standard method is to experiment with different knot configurations. Using MCMC for such experimentations will, however, be a daunting task and fast, accurate approximation methods will need to be explored.

31.3.2 Multivariate Spatial Regression

Here, we extend the univariate spatial regression to consider m outcome variables observed at each location. Let $y(s) = (y_1(s), \ldots, y_m(s))'$ be the collection of the outcomes measured at each location s. Given such data, our model should account for inherent association between the outcomes within each location and between observations across locations. We consider the multivariate regression model

$$y(s) = X(s)'\beta + u(s) + \varepsilon(s). \qquad (31.9)$$

Here, $X(s)'$ is an $m \times p$ block-diagonal matrix ($p = \sum_{l=1}^{m} p_l$), with lth diagonal element being the $1 \times p_l$ vector $x_l(s)'$, and $\beta = (\beta_1, \ldots, \beta_p)'$ is a $p \times 1$ vector of regression coefficients, with each β_l a $p_l \times 1$ vector of regression coefficients corresponding to $x_l(s)'$. This specifies the mean structure that accounts for large-scale variation in the outcome. Spatial variation in the outcome is modeled using an $m \times 1$ vector of random effects $u(s) = (u_1(s), \ldots, u_m(s))'$. Customarily, we assume the unstructured residuals $\varepsilon(s)$, defined analogously to $u(s)$, follow a zero-centered multivariate normal distribution with zero mean and an $m \times m$ dispersion matrix Ψ.

Spatially structured dependence is introduced in (31.9) through a multivariate ($m \times 1$) spatial process $u(s) \sim GP(\mathbf{0}, C(\cdot,\cdot;\theta))$—see, e.g., Cressie (1993)—where the *cross-covariance function* $C(s, s'; \theta)$ is now defined to be the $m \times m$ matrix with (i, j)th entry $C_{ij}(s, s') = \mathrm{Cov}(u_i(s), u_j(s'))$. As in the univariate case, for stationary processes $C_{ij}(s, s') = C_{ij}(h)$, where $h = s' - s$, and when the process is isotropic, $C_{ij}(s, s') = C_{ij}(d)$, where $d = \|s' - s\|$.

The cross-covariance function completely determines the joint dispersion structure implied by the spatial process. Specifically, for any n and any arbitrary collection of locations $\mathcal{S}$, the $nm \times 1$ vector $u = (u(s_1)', \ldots, u(s_n)')'$ will have the variance-covariance matrix given by $\Sigma_u(\theta)$, an $nm \times nm$ block matrix whose (i, j)th block is the cross-covariance matrix $C(s_i, s_j; \theta)$. Since $\Sigma_u(\theta)$ must be symmetric and positive-definite, it is immediate that the cross-covariance function must satisfy (i) $C(s_i, s_j; \theta) = C(s_j, s_i; \theta)'$ for any pair of locations, and (ii) $\sum_{i=1}^{n}\sum_{j=1}^{n} v_i' C(s_i, s_j; \theta) v_j > 0$ for all $v_i, v_j \in \mathbb{R}^m \setminus \{\mathbf{0}\}$. The first condition asserts that $\Sigma_u(\theta)$ is symmetric, although the cross-covariance function itself need not be. The second condition ensures the positive-definiteness of $\Sigma_u(\theta)$ and is, in fact, quite stringent: this condition must hold for any arbitrary set of distinct locations. The above definition also implies that $C(s, s; \theta)$ is a symmetric and positive definite function.

One possibility for $C(s_1, s_2; \theta)$, where $\theta = \{\Sigma, \phi\}$, is a separable specification, $\rho(s_1, s_2; \phi)\Sigma$, where Σ is an $m \times m$ covariance matrix between the outcomes and $\rho(\boldsymbol{s}_1, \boldsymbol{s}_2; \boldsymbol{\phi})$ is a spatial correlation function such as (31.2). This implies $\text{Var}(\boldsymbol{u}) = R(\boldsymbol{\phi}) \otimes \Sigma$, where $R(\boldsymbol{\phi})$ was defined previously in Section 31.3.1. A separable spatial association structure is rather restrictive here, assigning a single set of spatial correlation parameters $\boldsymbol{\phi}$ to be shared by all the outcomes. This can be inappropriate—why should all the outcomes have, for example, the same strength of spatial associations?

More general cross-covariance functions are not routine to specify since they demand that for any number of locations and any choice of these locations the resulting covariance matrix for the associated data be positive definite. Various constructions are possible. One possibility is the moving average approach of Ver Hoef and Barry (1998). The technique is also called kernel convolution and is a well-known approach for creating rich classes of stationary and non-stationary spatial processes, as discussed in Higdon, Swall, and Kern (1999) and Higdon, Lee, and Holloman (2003). Yet another approach could attempt a multivariate version of local stationarity, extending ideas in Fuentes (2002).

The primary characterization theorem for cross-covariance functions (Cramér, 1940; Yaglom, 1987) applies to stationary processes. It says that real-valued functions, say $C_{ij}(\boldsymbol{h})$, will form the elements of a valid cross-covariance matrix $C_u(\boldsymbol{h}) = [C_{ij}(\boldsymbol{h})]_{i,j=1}^m$ if and only if each $C_{ij}(\boldsymbol{h})$ has the *cross-spectral representation* $C_{ij}(\boldsymbol{h}) = \int \exp(2\pi \iota \boldsymbol{t}'\boldsymbol{h}) d(C_{ij}(\boldsymbol{t}))$ with respect to a positive definite measure $C(\cdot)$, i.e., where the cross-spectral matrix $M(B) = [C_{ij}(B)]_{i,j=1}^m$ is positive definite for any Borel subset $B \subseteq \mathbb{R}^d$ and $\iota = \sqrt{-1}$. The cross-spectral representation is the analogue of Bochner's Theorem for covariance functions and is the most general representation of cross-covariance functions. Essentially, one requires a choice of the $C_{ij}(\boldsymbol{t})$s. Matters simplify when $C_{ij}(\boldsymbol{h})$ is assumed to be square-integrable, ensuring that a spectral density function $c_{ij}(\boldsymbol{t})$ exists such that $d(C_{ij}(\boldsymbol{t})) = c_{ij}(\boldsymbol{t})d\boldsymbol{t}$. Now one simply needs to ensure that $[c_{ij}(\boldsymbol{t})]_{i,j=1}^p$ are positive definite for all $\boldsymbol{t} \in \mathbb{R}^d$. Corollaries of the above representation lead to the approaches proposed by Gaspari and Cohn (1999) and Majumdar and Gelfand (2007) for constructing valid cross-covariance functions as convolutions of covariance functions of stationary random fields. For isotropic settings the development is very similar; customary notations use $C(d_{ij})$, where $d_{ij} = \|\boldsymbol{s}_i - \boldsymbol{s}_j\|$ is the distance between any two sites.

Cross-spectral forms, however, are often computationally infeasible, resulting in integral representations with embedded parameters that are difficult to estimate. We motivate an alternative approach as follows. For the separable model, notice that we can write $C(\boldsymbol{s}_1, \boldsymbol{s}_2; \boldsymbol{\theta}) = A\Theta(\boldsymbol{s}_1, \boldsymbol{s}_2; \boldsymbol{\phi})A'$, where $\Sigma = AA'$ is a Cholesky factorization and $\Theta(\boldsymbol{s}_1, \boldsymbol{s}_2; \boldsymbol{\phi}) = \rho(\boldsymbol{s}_1, \boldsymbol{s}_2; \boldsymbol{\phi})I_m$. More generally, any valid cross-covariance matrix can be expressed as $C(\boldsymbol{s}_1, \boldsymbol{s}_2; \boldsymbol{\theta}) = A(\boldsymbol{s}_1)\Theta(\boldsymbol{s}_1, \boldsymbol{s}_2; \boldsymbol{\phi})A(\boldsymbol{s}_2)'$. We call $\Theta(\boldsymbol{s}_1, \boldsymbol{s}_2; \boldsymbol{\phi})$ the *cross-correlation* function which must satisfy the two conditions for a cross-covariance function and, in addition, must satisfy $\Theta(\boldsymbol{s}, \boldsymbol{s}; \boldsymbol{\phi}) = I_m$ for all $\boldsymbol{\phi}$. This implies that $C(\boldsymbol{s}, \boldsymbol{s}) = A(\boldsymbol{s})A(\boldsymbol{s})'$, and $A(\boldsymbol{s})$ identifies with a matrix square root (e.g., Cholesky) of $C(\boldsymbol{s}, \boldsymbol{s})$.

For modeling $A(\boldsymbol{s})$, without loss of generality one can assume that $A(\boldsymbol{s}) = C^{1/2}(\boldsymbol{s}, \boldsymbol{s})$ is a lower-triangular square root; the one-to-one correspondence between the elements of $A(\boldsymbol{s})$ and $C_{\boldsymbol{u}}(\boldsymbol{s}, \boldsymbol{s})$ is well known (see, e.g., Harville, 1997, p. 229). Therefore, $A(\boldsymbol{s})$ determines the association between the elements of $\boldsymbol{u}(\boldsymbol{s})$ within $\boldsymbol{s}$. Choices for modeling $A(\boldsymbol{s})$ include an inverse spatial Wishart process for $A(\boldsymbol{s})A(\boldsymbol{s})'$ (Gelfand et al., 2004), element-wise modeling with Gaussian and log-Gaussian processes, or using parametric association structures suggested by the design under consideration (Banerjee and Johnson, 2006). Alternatively, instead of a lower-triangular Cholesky factor, we could opt for a spectral square root, setting $A(\boldsymbol{s}) = P(\boldsymbol{s})\Lambda(\boldsymbol{s})^{1/2}$, where $C(\boldsymbol{s}, \boldsymbol{s}) = P(\boldsymbol{s})\Lambda P(\boldsymbol{s})'$ is the spectral decomposition for $C(\boldsymbol{s}, \boldsymbol{s})$ (following Daniels and Kass, 1999).

When stationarity is assumed, so that $C(\boldsymbol{s},\boldsymbol{s}) = C(\mathbf{0})$ and $A(\boldsymbol{s}) = A$ do not vary with location, we could assign a prior, e.g., inverse Wishart, to Σ. Also, we will adopt a flexible and computationally feasible approach that specifies $\Theta(\boldsymbol{s}_1, \boldsymbol{s}_2; \boldsymbol{\phi})$ as a diagonal matrix with $\rho_k(\boldsymbol{s}_1, \boldsymbol{s}_2; \boldsymbol{\phi}_k), k = 1, \ldots, m$, as its diagonal elements and $\boldsymbol{\phi} = \{\boldsymbol{\phi}_1, \boldsymbol{\phi}_2, \ldots, \boldsymbol{\phi}_m\}$. This incorporates a set of m correlation parameters, or even m different correlation functions, offering an attractive, easily interpretable, and flexible approach. This approach resembles the linear model of coregionalization (LMC) as in, e.g., Grzebyk and Wackernagel (1994), Wackernagel (2003), Schmidt and Gelfand (2003), and Gelfand et al. (2004). See, also, Reich and Fuentes (2007) for a Bayesian nonparametric adaptation. We will assume this structure in our subsequent development.

Once a stationary cross-covariance function is specified for a multivariate Gaussian process, the realizations of $\boldsymbol{u}(\boldsymbol{s})$ over the set of observed locations $\mathcal{S}$ is given by $N_p(\mathbf{0}, \Sigma_{\boldsymbol{u}}(\boldsymbol{\theta}))$, where $\boldsymbol{\theta} = \{A, \boldsymbol{\phi}\}$, $\Sigma_{\boldsymbol{u}}(\boldsymbol{\theta})$ is an $mn \times mn$ block matrix whose (i, j)th block is the $m \times m$ cross-covariance $C(\boldsymbol{s}_i, \boldsymbol{s}_j; \boldsymbol{\theta}), i, j = 1, \ldots, n$. Without further specifications, estimating (31.9) will involve factorizations that yield the inverse and determinant of the $mn \times mn$ matrix $\Sigma_{\boldsymbol{u}}(\boldsymbol{\theta})$. Such computations invoke linear solvers or Cholesky decompositions of complexity $O(n^3m^3)$, once every MCMC iteration, to produce estimates of $\boldsymbol{\theta}$. With large n, this is computationally infeasible and we will have to again resort to multivariate predictive process models (Banerjee et al., 2008, 2010).

Collecting observations from the n locations into an $mn \times 1$ vector $\boldsymbol{y} = (\boldsymbol{y}(\boldsymbol{s}_1)', \ldots, \boldsymbol{y}(\boldsymbol{s}_n)')'$, we can write down the multivariate analogue of (31.3) as

$$\prod_{k=1}^{m} p(\boldsymbol{\phi}_k) \times p(A) \times N_p(\boldsymbol{\beta} \mid \boldsymbol{\mu}_\beta, \Sigma_\beta)$$

$$\times \prod_{i=1}^{m} IG(\Psi_i | a_i, b_i) \times N_{mn}(\boldsymbol{u} \mid \mathbf{0}, \Sigma_{\boldsymbol{u}}(\boldsymbol{\theta}))$$

$$\times \prod_{i=1}^{n} N_m(\boldsymbol{y}(\boldsymbol{s}_i) \mid X(\boldsymbol{s}_i)'\boldsymbol{\beta} + \boldsymbol{u}(\boldsymbol{s}_i), \boldsymbol{\Psi}), \tag{31.10}$$

while the marginalized version, analogous to (31.4), is obtained as

$$\prod_{k=1}^{m} p(\boldsymbol{\phi}_k) \times p(A) \times N_p(\boldsymbol{\beta} \mid \boldsymbol{\mu}_\beta, \Sigma_\beta) \tag{31.11}$$

$$\times \prod_{i=1}^{m} IG(\Psi_i \mid a_i, b_i) \times N_{mn}(\boldsymbol{y} \mid X\boldsymbol{\beta}, \Sigma_{\boldsymbol{y}}(\boldsymbol{\theta})),$$

where $X = [X(\boldsymbol{s}_1) \quad X(\boldsymbol{s}_2) \quad \cdots \quad X(\boldsymbol{s}_n)]'$ is now the $nm \times p$ matrix of covariates, $\Sigma_{\boldsymbol{y}} = (I_n \otimes A)\Sigma_\Theta(\boldsymbol{\phi})(I_n \otimes A') + I_n \otimes \Psi$, and $\Sigma_\Theta(\boldsymbol{\phi})$ is an $mn \times mn$ block matrix whose (i, j)th block is the cross-correlation matrix $\Theta(\boldsymbol{s}_i, \boldsymbol{s}_j; \boldsymbol{\phi})$. The set of parameters that are to be updated in the marginalized model are $\Omega = \{\boldsymbol{\beta}, \Psi, A, \boldsymbol{\phi}\}$.

The full conditional for $\boldsymbol{\beta}$, $N_p(\boldsymbol{\mu}_{\boldsymbol{\beta}|\cdot}, \Sigma_{\boldsymbol{\beta}|\cdot})$, is multivariate normal and analogous to (31.5), with the appropriate adjustments to the dimensions of the vectors and matrices. The remaining parameters are updated by using Metropolis steps, possibly using block updates. Typically, random-walk Metropolis steps with (multivariate) normal proposals are adopted; since all parameters with positive support are converted to their logarithms, some Jacobian computation is needed. Upon convergence, the MCMC output generates L samples, say $\{\Omega^{(l)}\}_{l=1}^{L}$, from the posterior distribution in (31.11). Then, given these posterior samples of Ω, recovery of $\boldsymbol{u}$ is analogous to (31.7), but with

$$\Sigma_{\boldsymbol{u}|\cdot} = \left[(I_n \otimes A^{-1(l)}) \Sigma_\Theta \left(\boldsymbol{\phi}^{(l)}\right)^{-1} (I_n \otimes A^{-1(l)\prime}) + I_n \otimes \Psi^{-1(l)} \right]^{-1}$$

$$\boldsymbol{\mu}_{\boldsymbol{u}|\cdot} = \Sigma_{\boldsymbol{u}|\cdot} (I_n \otimes \Psi^{-1(l)}) \left(\boldsymbol{y} - X\boldsymbol{\beta}^{(l)}\right).$$

Subsequent prediction of the spatial effects $\boldsymbol{u}(\boldsymbol{s}_0)$ and the outcome $\boldsymbol{y}(\boldsymbol{s}_0)$ is analogous to the univariate setting.

For large n, the univariate predictive process in Section 31.3.1 is easily extended to a multivariate analogue. Letting $\boldsymbol{u}^* = (\boldsymbol{u}(\boldsymbol{s}_1^*)', \ldots, \boldsymbol{u}(\boldsymbol{s}_n^*)')$ be the realizations of the parent process over $\mathcal{S}^*$, we formulate a *multivariate predictive process* defined as

$$\begin{aligned}\tilde{\boldsymbol{u}}(\boldsymbol{s}) &= E\{u(\boldsymbol{s}) \mid \boldsymbol{u}^*\} \\ &= \text{Cov}(\boldsymbol{u}(\boldsymbol{s}), \boldsymbol{u}^*)\text{Var}^{-1}(\boldsymbol{u}^*)\boldsymbol{u}^* \\ &= \mathcal{C}(\boldsymbol{s}; \boldsymbol{\theta})'\Sigma^*(\boldsymbol{\theta})^{-1}\boldsymbol{u}^*, \qquad (31.12)\end{aligned}$$

where $\mathcal{C}(\boldsymbol{s}; \boldsymbol{\theta})'$ is an $m \times mn^*$ block matrix with the $m \times m$ matrix $C(\boldsymbol{s}, \boldsymbol{s}_j^*; \boldsymbol{\theta})$ being the jth block, and $\Sigma^*(\boldsymbol{\theta})$ is the $n^*m \times n^*m$ dispersion matrix of $\boldsymbol{u}^*$, i.e., with $C(\boldsymbol{s}_i^*, \boldsymbol{s}_j^*; \boldsymbol{\theta})$ as its (i, j)th block. Equation (31.12) reveals $\tilde{\boldsymbol{u}}(\boldsymbol{s})$ to be a zero mean $m \times 1$ spatial process with cross-covariance matrix given by $C_{\tilde{\boldsymbol{u}}}(\boldsymbol{s}_i, \boldsymbol{s}_j; \boldsymbol{\theta}) = \mathcal{C}(\boldsymbol{s}_i; \boldsymbol{\theta})'\Sigma^*(\boldsymbol{\theta})^{-1}\mathcal{C}(\boldsymbol{s}_j; \boldsymbol{\theta})$. Again, the appeal behind this formulation is that every spatial process (parent) model produces a corresponding predictive process version. Thus, the predictive process counterpart of the parent model in (31.9), and consequently in (31.10) and (31.11), is obtained by replacing $\boldsymbol{u}(\boldsymbol{s})$ with $\tilde{\boldsymbol{u}}(\boldsymbol{s})$. No new parameters are introduced and the cross-covariance function of the predictive process derives directly from that of the parent process.

As in the univariate case, bias in the covariance components of the multivariate predictive process model are corrected by adding a spatially adaptive nugget process to $\tilde{\boldsymbol{u}}(\boldsymbol{s})$. More precisely, we form $\tilde{\boldsymbol{u}}_\varepsilon(\boldsymbol{s}) = \tilde{\boldsymbol{u}}(\boldsymbol{s}) + \tilde{\boldsymbol{\varepsilon}}(\boldsymbol{s})$, where $\tilde{\boldsymbol{\varepsilon}}(\boldsymbol{s}) \overset{iid}{\sim} N_m\left\{\boldsymbol{0}, C(\boldsymbol{s}, \boldsymbol{s}; \boldsymbol{\theta}) - \mathcal{C}(\boldsymbol{s}; \boldsymbol{\theta})'\Sigma^*(\boldsymbol{\theta})^{-1}\mathcal{C}(\boldsymbol{s}; \boldsymbol{\theta})\right\}$. Details on the algorithms to estimate these bias-corrected multivariate predictive process models can be found in Finley et al. (2009) and Banerjee et al. (2010).

31.3.3 Model Selection

There are several approaches to assessing model performance and subsequent comparison; see, e.g., DIC in Chapter 24. Like its predecessors, the AIC and BIC, the DIC also uses two times the negative log-likelihood as a measure of deviance. What is different is how DIC penalizes model complexity. The approach here is to estimate the effective number of parameters in the model and use that as a measure of model complexity (Spiegelhalter et al., 2002).

Here, in addition to DIC, we also consider a particularly convenient approach that is based on simulating *independent* replicates for each observed outcome. For example, for the full multivariate spatial model (31.9), and each location $\boldsymbol{s}_i$, we compute, for each $i = 1, 2, \ldots, n$,

$$p(\boldsymbol{y}_{rep}(\boldsymbol{s}_i) \mid \boldsymbol{y}) = \int N(\boldsymbol{y}_{rep}(\boldsymbol{s}_i) \mid X\boldsymbol{\beta} + \boldsymbol{u}(\boldsymbol{s}_i), \Psi) \times p(\Omega, \boldsymbol{u}(\boldsymbol{s}_i) \mid \boldsymbol{y})\, d\Omega\, d\boldsymbol{u}(\boldsymbol{s}_i),$$

where the conditional distribution of the replicate $\boldsymbol{y}_{rep}(\boldsymbol{s}_i)$ given the parameters is simply the likelihood component corresponding to $\boldsymbol{y}(\boldsymbol{s}_i)$. Simulation from other candidate models, such as the nonspatial ($\boldsymbol{u} = \boldsymbol{0}$) or the predictive process models, follow this same approach. Then, letting $\boldsymbol{\mu}_{rep,i}$ and $\boldsymbol{\Sigma}_{rep,i}$ be the posterior predictive mean and variance for each $\boldsymbol{y}_{rep}(\boldsymbol{s}_i)$, we will prefer models that will perform well under a decision-theoretic balanced loss function, penalizing both departure of replicated means from their observed values (lack of fit) and excessive uncertainty in the replicated data. Using a squared error loss function (e.g., Gelfand and Ghosh, 1998), the measures for these two criteria are evaluated as $G = \sum_{i=1}^n \|\boldsymbol{y}(\boldsymbol{s}_i) - \boldsymbol{\mu}_{rep,i}\|^2$, where $\|\cdot\|$ is the standard Euclidean norm, and $P = \sum_{i=1}^n \text{tr}(\boldsymbol{\Sigma}_{rep,i})$. We will use the score $D = G + P$ as a model selection criterion, with lower values of D indicating better models.

31.3.4 Non-Gaussian First Stage Models

There are two typical non-Gaussian first stage settings: (i) binary response at locations modeled using logit or probit regression, and (ii) count data at locations modeled using Poisson regression. Diggle, Tawn, and Moyeed (1998) unify the use of generalized linear models in spatial data contexts. See also Lin et al. (2000), Kammann and Wand (2003), and Banerjee et al. (2004). Essentially, we replace (31.1) with the assumption that $E[y(\mathbf{s})]$ is linear on a transformed scale, i.e., $\eta(\mathbf{s}) \equiv g(E(y(\mathbf{s}))) = \mathbf{x}(\mathbf{s})'\boldsymbol{\beta} + u(\mathbf{s})$, where $g(\cdot)$ is a suitable link function. With the Gaussian first stage, we can integrate out the $u(\mathbf{s}_i)$s and work on a reduced parameter space. With, say, a binary or Poisson first stage, such marginalization is precluded; we have to update the $u(\mathbf{s}_i)$s in running our Gibbs sampler. Convergence with a large number of locations may become problematic and matters can be improved by using the predictive process models. Using the predictive process, we only have to update the $n^* \times 1$ vector $\mathbf{u}^*$.

Some additional clarification may be useful. As described in the previous paragraph, the resulting predictive process model would take the form $\prod_{i=1}^{n} p(y(\mathbf{s}_i) \mid \boldsymbol{\beta}, \mathbf{u}^*, \phi) p(\mathbf{u}^* | \sigma^2, \phi) p(\boldsymbol{\beta}, \phi, \sigma^2)$. Though $\mathbf{u}^*$ is only $m \times 1$, updating this vector through its full conditional distribution may be awkward due to the way that $\mathbf{u}^*$ enters the likelihood. Computations can be simplified by introducing the bias correction. The full model will now take the form

$$\prod_{i=1}^{n} p(y(\mathbf{s}_i) \mid \eta(\mathbf{s}_i)) \times \prod_{i=1}^{n} p(\eta(\mathbf{s}_i) \mid \boldsymbol{\beta}, \mathbf{u}^*, \phi) p(\mathbf{u}^* \mid \phi, \sigma^2) \, p(\boldsymbol{\beta}, \phi, \sigma^2).$$

The full conditional distribution for $\mathbf{u}^*$ is now multivariate normal and the $\eta(\mathbf{s}_i)$s are conditionally independent. General categorical data settings can be treated using these ideas.

31.4 AREAL MODELS

In public health services, to protect patient privacy, spatial data are usually available as case counts or rates aggregated over *areal* regions (e.g., counties, census tracts, or ZIP codes) rather than the geographical location of the individual residences. For areally referenced data, the association structures are built upon adjacencies or neighborhood structures of the regions.

There are several key differences between how we model areally referenced (or regionally aggregated) data and point-referenced data (Sections 31.3.1 and 31.3.2). One key distinction is that point-referenced models are inherently continuous over space, while areal models are discrete. Point-referenced models are more suitable for outcomes that can be envisioned to exist at every arbitrary point in the geographical domain of interest. Examples include temperature, precipitation, biomass, pollutant levels, and so on. In other words, one can envision a continuous underlying surface for the outcome. This surface is modeled using a stochastic process customarily characterized by a covariance or cross-covariance function. Spatial association is also customarily modeled as a continuous function of the locations.

The models for areal data, on the other hand, are inherently discrete. Spatial association is based upon the number of neighbors a region has. There is no concept of spatial association decaying continuously as a function of distance. In somewhat more technical terms, the underlying topology used is the discrete topology of a *graph*: the regions are nodes and there is an 'edge' between any two nodes if they are neighbors. These models are, therefore, more suitable when the outcome is aggregated over regions and cannot be easily envisioned to exist at locations. While spatial process models can be used for areal data by assuming the outcome is observed at the centroid of a region, this is less desirable and less interpretable. For example, what would it mean to assign a single geographical coordinate to the number of lung cancer cases in a particular county of Minnesota? In fact, the inferential objectives pursued by these models are also quite different. While a major goal of spatial process models is to estimate the underlying surface by interpolating or predicting at arbitrary locations, areal models are usually less suited for predicting at new regions. Instead, they focus upon smoothing the data and capturing residual spatial patterns after adjusting for explanatory variables or confounders.

The statistical models for areal data regard observations from a region to be more similar to those from its neighboring regions than those arising from regions farther away. Two such models are the *spatial lag model* and the *spatial error model*. They have been widely employed for smoothing maps and evincing spatial trends and clusters in econometrics (e.g., Anselin, 1988; Le Sage and Pace, 2009) and in public health (e.g., Banerjee et al., 2004).

The spatial lag model takes the form $\boldsymbol{y} = \rho W\boldsymbol{y} + X\boldsymbol{\beta} + \boldsymbol{\varepsilon}$, where $\boldsymbol{y}$ is a vector of observations on the outcome variable, W is a spatial weights matrix, $W\boldsymbol{y}$ is a spatially lagged outcome variable, ρ is a spatial autoregressive coefficient, X is a matrix of observations on the explanatory variables, $\boldsymbol{\beta}$ is a vector of parameters (regression coefficients), and $\boldsymbol{\varepsilon}$ is a vector of independent normally distributed random errors, as in the standard linear regression model. This error could be representative of variability from a number of sources including unobserved explanatory variables, sampling error, and so on. In contrast to the spatial lag model, the spatial error model incorporates spatial dependence into the error term, which is defined differently from that in the spatial lag model. The spatial error model can also be expressed as $\boldsymbol{y} = X\boldsymbol{\beta} + \boldsymbol{\varepsilon}$, where $\boldsymbol{\varepsilon} = \lambda W\boldsymbol{\varepsilon} + \boldsymbol{\zeta}$ is a vector of error terms, $W\boldsymbol{\varepsilon}$ are spatially lagged errors, λ is an autoregressive coefficient, and $\boldsymbol{\zeta}$ is a vector of independent random errors.

The spatial lag and error models belong to the class of Simultaneous Autoregressive (SAR) models. They are both appropriate for continuous outcomes where the normal distribution of errors is reasonable (e.g., log-transformed rates). However, health service data often do not yield continuous outcomes. For example, we might want to directly model the number of patients opting for services in each county. Alternatively, we might encounter a binary outcome: whether a patient has opted for service or not. Finally, even when the outcome is continuous, normal error models can be inappropriate for outcomes such as time to seek care. Transforming outcomes to render them closer to normality, while occasionally helpful, is often inadequate. Bayesian hierarchical models (Carlin and Louis, 2000; Banerjee et al., 2004) offer a richer and more flexible option in this regard by providing a framework for combining data from multiple sources, and providing an uncertainty assessment that takes into account all sources of variation. Depending upon the data type, a likelihood or joint probability distribution (e.g., Poisson for count data, binary for spatial data, hazard models for time to event data, etc.) is specified for the outcome, conditional on a set of parameters and unknown random effects. Next, these random effects are modeled using spatially informative distributions, and all remaining parameters are modeled using *prior* distributions that may quantify strength of expert opinion available about these parameters. A popular class of spatial statistical models on areal units is the *Markov Random Field* (MRF); see, e.g., Cressie (1993, Section 7.4), Banerjee et al. (2004, Chapter 3). Here, local spatial dependence between the observed areal unit characteristics is captured by the neighborhood structure of the units, where a unit is assumed to be independent of others *conditional* upon its neighbors. Various neighborhood structures are possible (e.g., Banerjee et al., 2004, pp. 70–71), but all propose higher statistical dependence between data values from areas that are spatially closer, thus inducing local smoothing. These models build joint distributions from a set of full conditional distributions assuming Markovian dependence structures and are known as Conditional Autoregressive (CAR) models.

To exemplify such models with count data, let y_i be the observed number of patients who underwent a specific preventive or clinical outcome in areal unit i, $i = 1, \ldots, n$, and let E_i be the expected number of outcomes for that unit. We treat the y_i as random variables, while the E_i are assumed fixed and known (and are often simply taken as proportional to the number of potential outcomes in the region). For rare outcomes, a Poisson model approximation to a binomial sampling distribution for outcome counts is often used. Thus, a commonly used likelihood when mapping a single outcome is

$$y_i \overset{ind}{\sim} Poisson(E_i e^{\mu_i}) \ , \ \ i = 1, \ldots, n,$$

where $\mu_i = \boldsymbol{x}_i' \boldsymbol{\beta} + \phi_i$ represents the log-relative risk, estimates of which are often based on the departures of observed from expected counts. The $\boldsymbol{x}_i$ will include explanatory, region-level covariates or predictors, having parameter coefficients $\boldsymbol{\beta}$. Bayesian inference will deliver *exact inference* through the posterior distribution of $\boldsymbol{\beta}$ conditional upon the data. Each ϕ_i represents the *spatial random effect* associated with region i. Often, the outcome variable is appropriately transformed into a rate or an areal average (perhaps subsequently log-transformed). Now we can work with a Gaussian likelihood:

$$y_i = \mu_i + \phi_i + \varepsilon_i, \ i = 1, \ldots, n,$$

where μ_i is as above and $\boldsymbol{\varepsilon} = \{\varepsilon_i\}$ represents random unstructured errors of the outcome across the map. We can now model the spatial random effects as $\boldsymbol{\phi} = \lambda W \boldsymbol{\phi} + \boldsymbol{\zeta}$, where $\boldsymbol{\phi} = (\phi_1, \ldots, \phi_n)'$ is the vector of spatial random effects, $W\boldsymbol{\phi}$ are spatially lagged random effects, λ is an autoregressive coefficient, and $\boldsymbol{\zeta}$ is a vector of independent random errors. This is a SAR model, as above. An alternative approach builds a joint distribution on the spatial random effects using conditional distributions. More precisely, we specify, for each ϕ_i,

$$\phi_i \mid \phi_j, j \neq i, \ \sim \ N\left(\frac{\rho}{m_i}\sum_{i \sim j} \phi_j, \ \frac{1}{\tau m_i}\right), \tag{31.13}$$

for $i, j = 1, \ldots, n$, where $i \sim j$ denotes that region j is a *neighbor* (typically defined in terms of spatial adjacency) of region i. Using MRF theory (as above; see also Besag, 1974; Besag et al., 1991), it can be shown that (31.13) implies the following joint distribution for the spatial effects:

$$\boldsymbol{\phi} \ \sim \ N_n\left(\boldsymbol{0}\, , \ \sigma^2 \left(D - \rho W\right)^{-1}\right) , \tag{31.14}$$

where N_n denotes the n-dimensional normal distribution, D is a $n \times n$ diagonal matrix with diagonal elements m_i that denote the number of neighbors of area i, and $W = \{w_{ij}\}$ is the adjacency matrix of the map (i.e., $w_{ii} = 0$, and $w_{ij} = 1$ if i is adjacent to j and 0 otherwise). In the joint distribution (31.14), σ^2 is the spatial dispersion parameter, and ρ is a spatial autocorrelation parameter. We denote this distribution concisely as $CAR(\rho, \sigma^2)$. Further technical details, including the positive-definiteness (hence nonsingularity) of the matrix $D - \rho W$ in (31.14), can be found in Banerjee et al. (2004, Chapter 3). The CAR structure (31.14) reduces to the well-known intrinsic conditionally autoregressive (ICAR) model if $\rho = 1$, or an independence model if $\rho = 0$. The ICAR model induces 'local' smoothing by borrowing strength from the neighbors, while the independence model assumes independence of spatial rates and induces 'global' smoothing. The smoothing parameter ρ in the CAR prior (31.14) controls the strength of spatial dependence among regions, though it has long been appreciated that a fairly large ρ may be required to deliver significant spatial correlation; see Wall (2004) for recent discussion and exemplification.

Models for analyzing multiple areally referenced outcomes have also been proposed. Mardia (1988), Carlin and Banerjee (2003), and Gelfand and Vounatsou (2002) proposed multivariate versions of the CAR model. They make an assumption of separability, which permits modeling of correlations among the p diseases while maintaining spatial dependence across space. Separability assumes that the association structure separates into a non-spatial and spatial component, and is similar to the point-referenced models in Section 31.3.2. Jin, Banerjee, and Carlin (2007) proposed non-separable models for areal data by modeling areal cross-covariances. Zhang, Hodges, and Banerjee (2009) offered a smoothed ANOVA version of multivariate areal models.

In the next section we return to point-referenced multivariate spatial modeling and illustrate with an example from forestry.

31.5 APPLICATION

Spatially explicit data layers of forest variables are important inputs to large-scale assessments of forest sustainability, biodiversity, timber biomass, carbon cycle, and forest health monitoring. Because complete enumeration of these variables across a forest is prohibitively expensive, foresters must sample the domain of interest. Then, given valid sampling designs and corresponding estimators, design-based inference is commonly used to produce asymptotically unbiased estimates of the forest variables (see, e.g., Gregoire and Valentine, 2008). The resulting coarse spatial scale estimates of the forest variables are often not at a sufficiently fine spatial scale to support modern ecosystem and economic assessments. Further, design-based estimators do not easily incorporate ancillary data or complex spatial dependence structures and cannot leverage the among-forest variable covariation to improve the accuracy and precision of parameter estimates and/or prediction. These shortcomings are addressed within a model-based inferential framework as detailed in Section 31.3, and illustrated using an analysis of forest inventory data collected on the Penobscot Experimental Forest (PEF), Maine.[1]

For the analysis, we focused on point-referenced data observed at 589 forest inventory plots across PEF, Figure 31.1(a). Sendak et al. (2003) describe the PEF sampling design and multiple outcome variables measured at each inventory plot location. Here, we used only mean tree diameter at breast height (MDBH) (breast height is measured at 1.37 m above the forest floor) and tree density measured as number of trees per hectare (TPH). These variables are commonly used, in combination with other information, to assess a forest's economic and ecological value. The relationship between MDBH and TPH is influenced by many individual and environmental factors. Individual factors include age, species, and genetics, whereas environmental factors include quality of soil, quantity of water and light, and competition for these resources. Forest disturbance history (e.g., timber harvesting, fire, or windthrow) also strongly influence MDBH and TPH. These factors often vary spatially and at different scales across a forest and are too expensive or impossible to measure directly. It is generally expected that as a forest ages, MDBH increases and the number of trees per hectare decreases, which implies a negative relationship between mean DBH and TPH. Interpolated surfaces of MDBH and TPH are illustrated in Figures 31.1(c) and (d), respectively. These non-statistical surfaces clearly show the negative relationship between MDBH and TPH across the PEF.

As suggested above, the goal of this analysis is to predict MDBH and TPH as a continuous surface across the PEF. We will use remotely sensed predictor variables, recorded at a high spatial resolution across the PEF to improve prediction at unobserved locations. In addition to using information from the predictor variables, the multivariate spatial model detailed in Section 31.3.2 informs prediction using proximity to observed inventory plots and the relationship among the multiple outcome variables—captured by the cross-covariance matrix. The predictor variables are derived from NASA's Laser Vegetation Imaging Sensor (LVIS; `http://lvis.gsfc.nasa.gov`) airborne waveform Light Detection and Ranging (LiDAR) sensor. Figure 31.1(b) illustrates the waveform signals recorded at eight forest inventory plots identified in the inset in Figure 31.1(a).

These high-dimensional waveform LiDAR data contain information about the vertical and horizontal structure of the forest canopy and explain some of the variability in MDBH and TPH. LiDAR signals were measured at $q = 63,370$ locations across the PEF (589 of these coincide with the observed inventory plots). A signal measured at any given location comprises $r = 100$ values (over which we interpolated to create the canopy profiles in Figure 31.1(b)). Obviously, we cannot include the r highly collinear predictors in our regression model, and thus must consider some way to reduce the signals' dimension.

One approach is to form a $q \times r$ matrix M, with each row corresponding to the signal at a given location. Then decompose M using a Singular Value Decomposition (SVD) into $U\Lambda V'$, where U and V are orthogonal matrices of dimension $q \times q$ and $r \times r$ respectively, and Λ is a diagonal matrix comprising the non-negative singular values of M. The number of positive entries in Λ equals the rank of M. Linear dependencies in M usually arise among the columns corresponding to small singular values. Therefore, we order the column vectors of U, which are precisely the orthonormal eigenvectors of MM', from largest corresponding eigenvalue to the least (these vectors are referred to as *left-singular vectors* or LSVs), and take a subset thereof as the covariates in our regression analysis.

The data were analyzed using four candidate submodels of (31.9), including the non-spatial (i.e., $\boldsymbol{u} = \mathbf{0}$), full spatial, and predictive process version based on 50 and 200 knot intensities. The locations of the knots were based on a *k-means* clustering algorithm. The candidate models' performance was evaluated using DIC and GP model fit criteria, and root mean squared prediction error (RMSPE) calculated using a 10% hold-out set of randomly selected observations. Those observations used for parameter estimation ($n = 529$) and subsequent prediction evaluation ($n_0 = 60$) are indicated in Figure 31.1(a) with $\circ$ and $+$, respectively.

Both TPH and MDBH were log transformed to more closely approximate normal distributions. In addition to an intercept, the first four LSVs were used as predictor variables. This decision was based on an exploratory analysis using the non-spatial regression model, which suggested there was marginal information gain beyond the first four LSVs. Obviously, we might use a formal Bayesian variable selection method to choose which columns of U should serve as covariates (see, e.g., a review of popular methods by O'Hara and Sillanpää, 2009).

To complete the model specification we assigned flat priors to each of the 10 β parameters (i.e., $p = \sum_l^2 p_l = 10$, with each p_l including an intercept and first four columns of U). The covariance matrices Ψ and $\Sigma = AA'$ each received an inverse-Wishart, $IW(df,\ S)$, with $df = q + 1 = 3$ degrees of freedom. Empirical semivariograms fit to the TPH and MDBA non-spatial model residuals provided reasonable estimates for the diagonal elements of S for Σ and Ψ, respectively. These semivariograms are plotted in Figure 31.1(e) and (f). Here, the difference between the upper and lower horizontal line is σ^2, the lower line is τ^2, and the vertical line is the effective spatial range, which, assuming an Exponential semivariogram model, is equal to $-\log(0.05)/\phi \approx 3/\phi$, where ϕ is the spatial decay parameter (here, choice of a correlation of 0.05 is simply a convention). The Exponential semivariogram model estimates for TPH are ($\sigma^2 = 1.14$, $\tau^2 = 0.18$, $\phi = 0.004$) and MDBH are ($\sigma^2 = 0.11$, $\tau^2 = 0.07$, $\phi = 0.005$). These variance estimates were used to define the IW scale matrix hyperparameter for Σ as $\text{Diag}(S) = (1.14, 0.11)$ and for Ψ as $\text{Diag}(S) = (0.18, 0.07)$. The estimates from the semivariograms also help us define the prior on the ϕs. The spatial decay parameter is often poorly defined by the data and therefore requires a mildly informative prior. We assume ϕ_1 and ϕ_2 follow a Uniform $U(0.002,\ 3)$, which corresponds to support between 1 and 1,500 (m).

Posterior inference was based on three chains of 10,000 iterations (the first 1,000 iterations were discarded as burn-in). Samples were collected using the `spMvLM` function

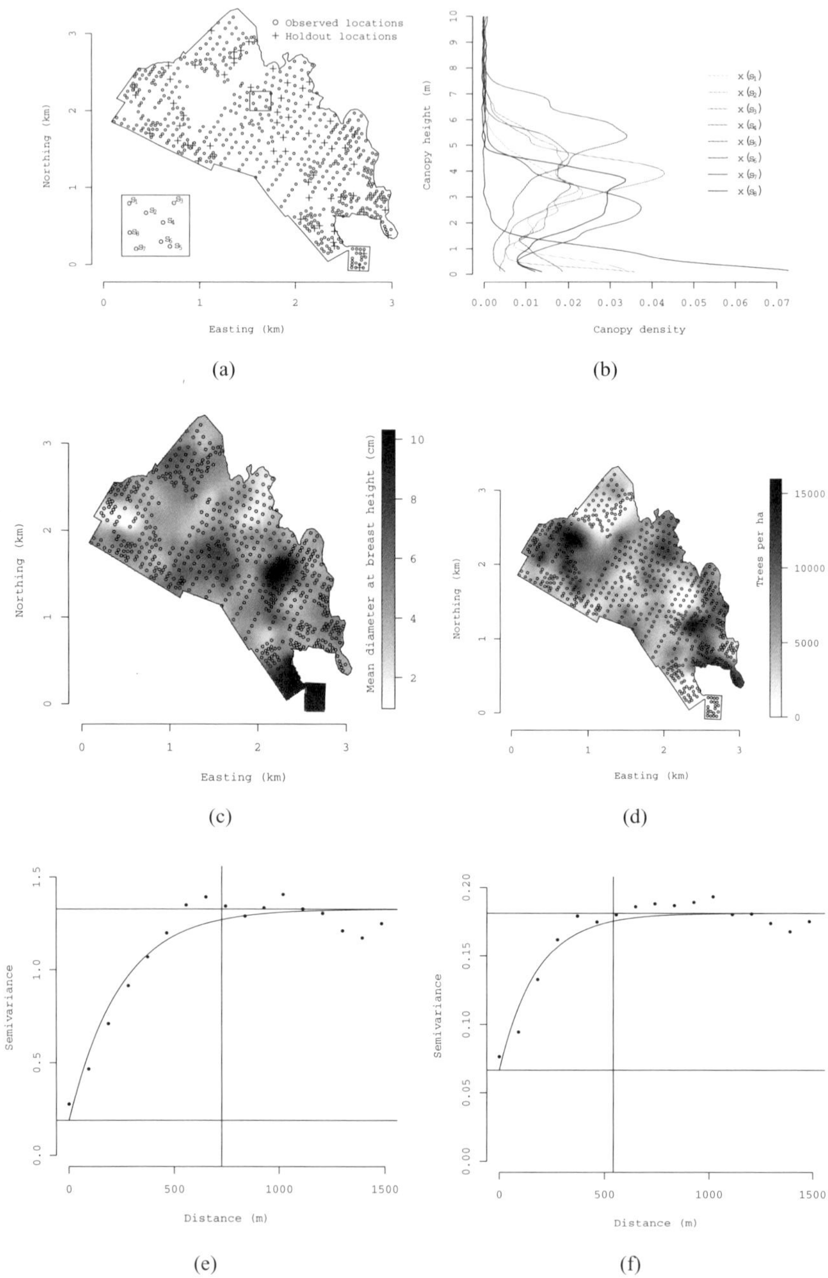

Figure 31.1 Penobscot Experimental Forest Data Analysis: (a) forest inventory plot locations; (b) LiDAR signals measured at eight locations illustrated in (a) inset; (c) interpolated surface of the mean tree diameter at breast height; (d) inventory plots' tree count expressed on a per hectare basis; (e) semivariogram estimated from the non-spatial candidate model's residuals for log transformed trees per hectare; (f) log transformed mean tree diameter at breast height

in the `spBayes` `R` package. This function calls lower-level `C++` and `Fortran` `BLAS` and `LAPACK` routines for matrix computations. All analyses were conducted on a Linux workstation using an Intel Nehalem quad-Xeon processor.

31.5.1 Results and Discussion

Parameter estimates and model fit criteria for the candidate models are presented in Table 31.1. Here, we see that the regression slope parameters estimated by the three spatial models are similar in sign, magnitude, and significance. These models also produce similar estimates of the spatial cross-covariance, residual, and spatial decay parameters. Likewise, the estimated spatial cross-correlation, $\rho(\Sigma_{TPH,MDBH}) = \Sigma_{TPH,MDBH}/(\Sigma_{TPH,TPH}\Sigma_{MDBH,MDBH})^{1/2}$, is nearly identical among the three models and identifies a negative correlation that differs significantly from zero. This reflects the biological relationship noted earlier, which does not seem to be captured by the models' mean. The similarity among these parameter estimates suggest that reducing the dimension of the full model's 529 observations to the predictive process model's 50 knots has little effect on inference. Interestingly, the

Table 31.1 Summary of the Candidate Models' Parameter Estimates, Model Fit, and Holdout Set Prediction Performance. Posterior credible intervals summarized using the 50 (2.5, 97.5) quantiles. Parameter posterior credible intervals that do not include 0 at the 0.05 level are in bold

	Non-spatial	Full	Predictive process	
			50	100
$\beta_{TPH,0}$	**8.70 (7.63, 9.78)**	**9.12 (8.23, 9.97)**	**8.82 (7.92, 9.74)**	**9.00 (8.11, 9.84)**
β_{TPH,LSV_1}	−15.89 (−41.10, 9.42)	**−27.94 (−43.91, −12.57)**	**−23.81 (−41.14, −6.66)**	**−26.56 (−42.47, −9.60)**
β_{TPH,LSV_2}	−0.31 (−3.83, 2.99)	−0.65 (−2.61, 1.44)	−0.77 (−2.95, 1.44)	−0.71 (−2.79, 1.33)
β_{TPH,LSV_3}	**−5.41 (−10.01, −0.82)**	**−7.21 (−10.47, −3.92)**	**−6.97 (−10.52, −3.42)**	**−7.03 (−10.52, −3.68)**
β_{TPH,LSV_4}	**3.99 (1.06, 7.01)**	**2.82 (0.90, 4.76)**	**2.45 (0.32, 4.56)**	**2.75 (0.63, 4.71)**
$\beta_{DBH,0}$	**0.72 (0.30, 1.12)**	**0.87 (0.51, 1.25)**	**0.92 (0.51, 1.30)**	**0.92 (0.53, 1.32)**
β_{DBH,LSV_1}	**18.03 (8.51, 27.96)**	**14.23 (6.81, 21.31)**	**13.66 (5.99, 21.31)**	**13.21 (5.29, 21.06)**
β_{DBH,LSV_2}	−1.00 (−2.23, 0.33)	−0.37 (−1.27, 0.51)	−0.26 (−1.23, 0.75)	−0.30 (−1.27, 0.69)
β_{DBH,LSV_3}	**7.28 (5.51, 9.13)**	**5.55 (3.96, 7.00)**	**5.88 (4.26, 7.49)**	**5.51 (3.89, 7.13)**
β_{DBH,LSV_4}	**−1.75 (−2.89, −0.67)**	**−1.51 (−2.35, −0.60)**	**−1.43 (−2.35, −0.52)**	**−1.46 (−2.38, −0.49)**
$\Sigma_{TPH,TPH}$	—	1.10 (0.84, 1.35)	1.19 (0.92, 1.49)	1.05 (0.81, 1.29)
$\Sigma_{TPH,MDBH}$	—	**−0.34 (−0.43, −0.26)**	**−0.39 (−0.51, −0.29)**	**−0.33 (−0.43, −0.24)**
$\Sigma_{MDBH,MDBH}$	—	0.17 (0.14, 0.21)	0.18 (0.14, 0.24)	0.18 (0.13, 0.23)
$\rho(\Sigma_{TPH,MDBH})$	—	**−0.78 (−0.86, −0.71)**	**−0.83 (−0.88, −0.78)**	**−0.78 (−0.84, −0.71)**
$\Psi_{TPH,TPH}$	1.39 (1.23, 1.55)	0.20 (0.17, 0.27)	0.11 (0.08, 0.16)	0.18 (0.11, 0.21)
$\Psi_{TPH,MDBH}$	−0.41 (−0.46, −0.36)	−0.09 (−0.11, −0.07)	−0.05 (−0.07, −0.02)	−0.07 (−0.09, −0.04)
$\Psi_{MDBH,MDBH}$	0.19 (0.17, 0.22)	0.06 (0.05, 0.07)	0.04 (0.03, 0.05)	0.05 (0.04, 0.07)
Eff. range$_1$ (m)	—	955.20 (746.09, 1105.57)	1039.05 (751.44, 1179.16)	955.01 (686.14, 1066.99)
Eff. range$_2$ (m)	—	1088.75 (756.07, 1158.12)	1055.67 (814.01, 1186.96)	1186.73 (1109.56, 1197.00)
p_D	12	498	89	137
DIC	−162	−1401	−991	−1092
G	857	75	297	248
P	863	213	26	36
D	1720	288	322	284
RMSPE$_{TPH}$	4572.94	3858.83	4279.19	3891.93
RMSPE$_{MDBH}$	2.17	1.51	1.61	1.55

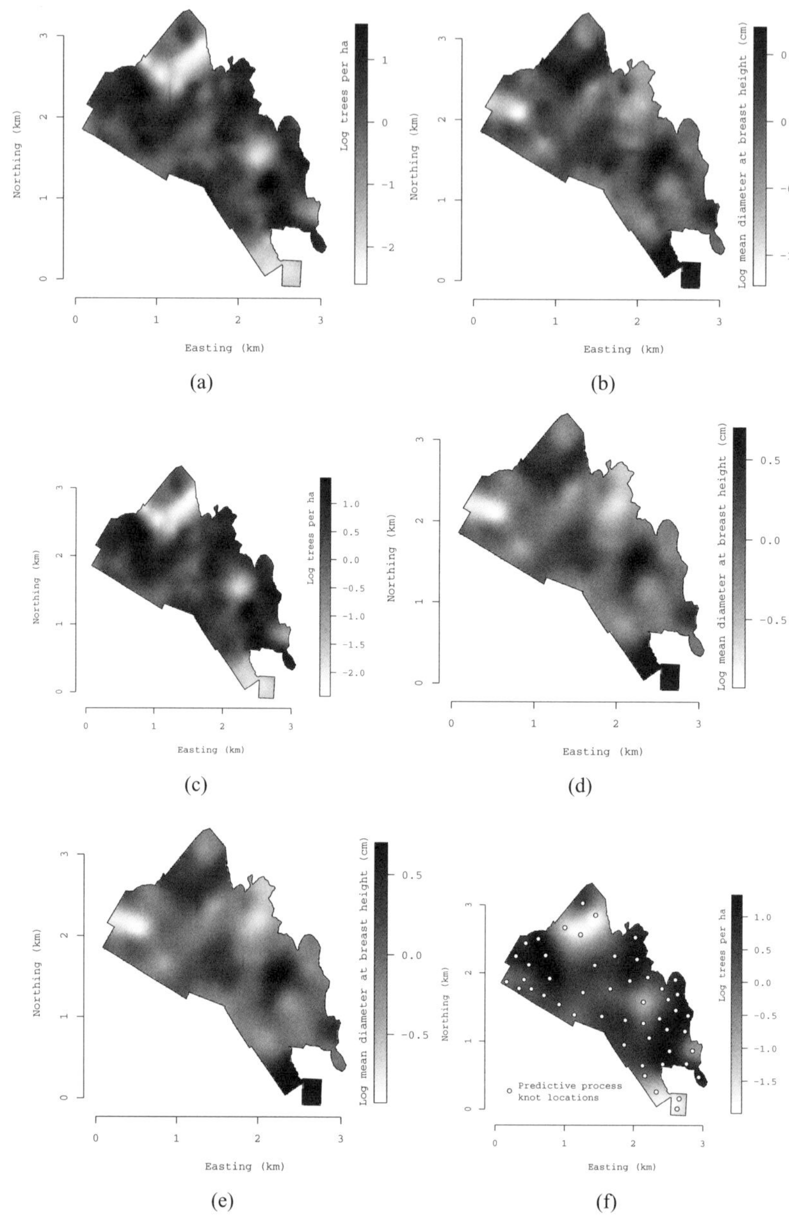

Figure 31.2 Penobscot Experimental Forest Data Analysis: (a) interpolated surface of the non-spatial candidate model's residuals for log transformed trees per hectare; (b) mean tree diameter at breast height; (c) interpolated surface of the full candidate model's mean random spatial effects for log transformed trees per hectare; (d) mean tree diameter at breast height; (e) interpolated surface of the 100-knot predictive process candidate model's mean random spatial effects for log transformed trees per hectare; (f) mean tree diameter at breast height

reapportioning of the residual spatial correlation between the outcome variables is clearly seen by the decrease of Ψ's off-diagonal element when moving from the non-spatial to spatial models.

Turning to model fit and predictive performance, compared to the full spatial model, the 50-knot predictive process model has marginally larger DIC and D, which suggest some information is lost by reducing the dimensionality of the spatial random effects. However, the predictive process models' competitive DIC and D values are the result of smaller P_D and P *penalty* terms in these fit criteria and not comparable fit terms, e.g., the G for the full and 50-knot predictive process models is 75 and 295, respectively. Compared to the non-spatial model, all three spatial models have lower RMSPEs, which supports the addition of the spatial random effects. When comparing the RMSPEs among the spatial models, it is clear that some predictive inference is lost to smoothing as the number of knots is reduced from 100 to 50. This smoothing effect is seen to some degree in Figure 31.2. Here, interpolated surfaces of the non-spatial model residuals for TPH (a) and MDBH (b) can be compared to the corresponding mean spatial random effects from the full and 50-knot predictive process models.

The full model required approximately 90 minutes to collect a single chain of 10,000 MCMC samples. The 100 and 50 knot predictive process models required approximately 23 and 10 minutes, respectively, to collect the same number of samples. Obviously, with only 529 observations and two outcome variables, it is computationally feasible to fit this model without resorting to the predictive process. However, this exercise provides some insight into the trade-off between inference and computational demand that might be expected when the number of observations becomes too large and we are forced to consider a lower-rank approximation.

NOTE

1 Data for this study were provided by a unit of the Northern Research Station, U.S. Forest Service, located at the Penobscot Experimental Forest in Maine. Significant funding for collection of these data was provided by the U.S. Forest Service.

REFERENCES

Anselin, L. (1988) *Spatial Econometrics: Methods and Models*. Boston: Kluwer Academic Publishers.

Banerjee, S., Carlin, B.P., and Gelfand, A.E. (2004) *Hierarchical Modeling and Analysis for Spatial Data*. Boca Raton, FL: Chapman and Hall/CRC Press.

Banerjee, S., Finley, A.O., Waldmann, P., and Ericsson, T. (2010) 'Hierarchical Spatial Process Models for Multiple Traits in Large Genetic Trials', *Journal of the American Statistical Association*, 105: 506–21.

Banerjee, S., Gelfand, A.E., Finley, A.O., and Sang, H. (2008) 'Gaussian Predictive Process Models for Large Spatial Datasets', *Journal of the Royal Statistical Society, Series B*, 70: 825–48.

Banerjee, S., Gelfand, A.E., and Sirmans, C.F (2003) 'Directional Rates of Change Under Spatial Process Models', *Journal of the American Statistical Association*, 98: 946–54.

Banerjee, S. and Johnson, G.A. (2006) 'Coregionalized Single- and Multi-resolution Spatially-varying Growth Curve Modeling with Application to Weed Growth', *Biometrics*, 61: 617–25.

Besag, J. (1974) 'Spatial Interaction and the Statistical Analysis of Lattice Systems (with Discussion)', *Journal of the Royal Statistical Society, Series B*, 36: 192–236.

Besag, J., York, J., and Mollie, A. (1991) 'Bayesian Image Restoration, with Two Applications in Spatial Statistics (with Discussion)', *Annals of the Institute of Statistical Mathematics*. 43: 1–59.

Carlin, B.P. and Banerjee, S. (2003) 'Hierarchical Multivariate CAR Models for Spatially Correlated Survival Data', in *Bayesian Statistics 7*. Oxford: Oxford University Press, 45–64.

Carlin, B.P. and Louis, T.A. (2000) *Bayes and Empirical Bayes Methods for Data Analysis*. New York: Chapman and Hall/CRC.

Chilés, J.P. and Delfiner, P. (1999) *Geostatistics: Modeling Spatial Uncertainty*. New York: Wiley.

Crainiceaniu, C.M., Diggle, P.J., and Rowlingson, B. (2008) 'Bivariate Binomial Spatial Modeling of Loa Loa Prevalence in Tropical Africa', *Journal of the American Statistical Association*, 103: 21–37.

Cramér, H. (1940) 'On the Theory of Stationary Random Processes', *Annals of Mathematics*, 41: 215–30.

Cressie, N.A.C. (1993) *Statistics for Spatial Data*, 2nd edition. New York: Wiley.

Cressie, N.A.C. and Johannesson, G. (2007) 'Spatial Prediction for Massive Datasets', *Journal of the Royal Statistical Society, Series B*, 70: 209–26.

Daniels, M.J. and Kass, R.E. (1999) 'Nonconjugate Bayesian Estimation of Covariance Matrices and its Use in Hierarchical Models', *Journal of the American Statistical Association*, 94: 1254–63.

Diggle, P.J. and Lophaven, S. (2006) 'Bayesian Geostatistical Design', *Scandinavian Journal of Statistics*, 33: 53–64.

Diggle, P.J. and Ribeiro, P.J. (2007) *Model-based Geostatistics*. New York: Springer.

Diggle, P.J., Tawn, J.A., and Moyeed, R.A. (1998) 'Model-based Geostatistics (with Discussion)', *Journal of the Royal Statistical Society, Series C (Applied Statistics)*, 47: 299–350.

Du, J., Zhang, H., and Mandrekarm, V.S. (2009) 'Fixed-domain Asymptotic Properties of Tapered Maximum Likelihood Estimators', *The Annals of Statistics*, 37: 3330–61.

Finley, A.O., Sang, H., Banerjee, S., and Gelfand, A.E. (2009) 'Improving the Performance of Predictive Process Modeling for Large Datasets', *Computational Statistics and Data Analysis*', 53: 2873–84.

Fuentes, M. (2002) 'A New Class of Nonstationary Spatial Models', *Biometrika*, 89: 197–210.

Fuentes, M. (2007) 'Approximate Likelihood for Large Irregularly Spaced Spatial Data', *Journal of the American Statistical Association*, 102: 321–31.

Furrer, R., Genton, M.G., and Nychka, D. (2006) 'Covariance Tapering for Interpolation of Large Spatial Datasets', *Journal of Computational and Graphical Statistics*, 15: 502–23.

Gaspari, G. and Cohn, S.E. (1999) 'Construction of Correlation Functions in Two and Three Dimensions', *The Quarterly Journal of the Royal Meteorological Society*, 125: 723–57.

Gelfand, A.E. and Ghosh, S.K. (1998) 'Model Choice: A Minimum Posterior Predictive Loss Approach', *Biometrika*, 85: 1–11.

Gelfand, A.E., Schmidt, A., Banerjee S., and Sirmans, C.F. (2004) 'Nonstationary Multivariate Process Modeling through Spatially Varying Coregionalization (with Discussion)', *Test*, 13: 1–50.

Gelfand, A.E. and Vounatsou, P. (2002) 'Proper Multivariate Conditional Autoregressive Models for Spatial Data Analysis', *Biostatistics*, 4: 11–25.

Gelman, A., Carlin, J.B., Stern, H.S., and Rubin, D.B. (2004) *Bayesian Data Analysis*, 2nd edition. Boca Raton, FL: Chapman and Hall/CRC Press.

Gilks, W.R., Richardson, S., and Spiegelhalter, D.J. (1996) *Markov Chain Monte Carlo in Practice*. Boca Raton, FL: Chapman and Hall/CRC.

Gregoire, T.G. and Valentine, H.T. (2008) *Sampling Strategies for Natural Resources and Environment*. Boca Raton, FL: Chapman-Hall/CRC.

Grzebyk, M. and Wackernagel, H. (1994) 'Multivariate Analysis and Spatial/Temporal Scales: Real and Complex Models', in *Proceedings of the XVIIth International Biometrics Conference*, Hamilton, Ontario, 19–33.

Harville, D.A. (1997) *Matrix Algebra from a Statistician's Perspective*. New York: Springer.

Higdon, D.M. (2002) 'Space and Space-Time Modeling Using Process Convolutions', in Anderson, C., Barnett, V., Chatwin, P.C., and El-Shaarawi A.H. (eds), *Quantitative Methods for Current Environmental Issues*. New York: Springer-Verlag, 37–56.

Higdon, D.M., Lee, H., and Holloman, C. (2003) 'Markov Chain Monte Carlo-based Approaches for Inference in Computationally Intensive Inverse Problems (with Discussion)', *Bayesian Statistics 7. Proceedings of the Seventh Valencia International Meeting*, 181–97.

Higdon, D.M., Swall, J., and Kern, J. (1999) 'Non-Stationary Spatial Modeling', in Bernardo, J.M., Berger, J.O., Dawid, A.P., and Smith, A.F.M. (eds), *Bayesian Statistics 6*. Oxford: Oxford University Press, 761–8.

Jin, X., Banerjee, S., and Carlin, B.P. (2007) 'Order-free Coregionalized Lattice Models with Application to Multiple Disease Mapping', *Journal of the Royal Statistical Society, Series B*, 69: 817–38.

Kammann, E.E. and Wand, M.P. (2003) 'Geoadditive Models', *Applied Statistics*, 52: 1–18.

Kaufman, C.G., Schervish, M.J., and Nychka, D.W. (2008) 'Covariance Tapering for Likelihood-based Estimation in Large Spatial Data Sets', *Journal of the American Statistical Association*, 103: 1545–55.

Le Sage, J. and Pace, K. (2009) *Introduction to Spatial Econometrics*. Boca Raton, FL: Chapman and Hall/CRC.

Lin, X., Wahba, G., Xiang, D., Gao, F., Klein, R., and Klein, B. (2000) 'Smoothing Spline ANOVA Models for Large Data Sets with Bernoulli Observations and the Randomized GACV', *Annals of Statistics* 28: 1570–600.

Majumdar, A. and Gelfand, A.E. (2007) 'Multivariate Spatial Modeling for Geostatistical Data using Convolved Covariance Functions', *Mathematical Geology*, 39: 225–45.

Mardia, K.V. (1988) 'Multi-dimensional Multivariate Gaussian Markov Random Fields with Application to Image Processing', *Journal of Multivariate Analysis*, 24: 265–84.

Møller, J. and Waagepetersen, R.P. (2004) *Statistical Inference and Simulation for Spatial Point Processes*. Boca Raton, FL: Chapman and Hall/CRC.

Nychka, D. and Saltzman, N. (1998) 'Design of Air-quality Monitoring Networks', in Nychka, D., Cox, L., and Piegorsch, W. (eds), *Case Studies in Environmental Statistics, Lecture Notes in Statistics*. New York: Springer Verlag, 51–76.

O'Hara, R.B. and Sillanpää, M.J. (2009) 'Review of Bayesian Variable Selection Methods: What, How and Which', *Bayesian Analysis*, 4: 85–118.

Rasmussen, C.E. and Williams, C.K.I. (2006) *Gaussian Processes for Machine Learning*. Cambridge, MA: MIT Press.

Reich, B.J. and Fuentes, M. (2007) 'A Multivariate Nonparametric Bayesian Spatial Framework for Hurricane Surface Wind Fields', *The Annals of Applied Statistics*, 1: 249–64.

Robert, C.P. and Casella, G. (2005) *Monte Carlo Statistical Methods*, 2nd edition. New York: Springer.

Schabenberger, O. and Gotway, C.A. (2004) *Statistical Methods for Spatial Data Analysis. Texts in Statistical Science Series*. Boca Raton, FL: Chapman and Hall/CRC.

Schmidt, A. and Gelfand, A.E. (2003) 'A Bayesian Coregionalization Model for Multivariate Pollutant Data', *Journal of Geophysics Research-Atmospheres*, 108: 8783.

Sendak, P.E., Brissette, J.C., and Frank, R.M. (2003) 'Silviculture Affects Composition, Growth, and Yield in Mixed Northern Conifers: 40-year Results from the Penobscot Experimental Forest', *Canadian Journal of Forest Research*, 33: 2116–28.

Spiegelhalter, D.J., Best, N., Carlin, B.P., and van der Linde, A. (2002) 'Bayesian Measures of Model Complexity and Fit (with Discussion)', *Journal of the Royal Statistical Society, Series B*, 64: 583–639.

Stein, M.L. (2007) 'Spatial Variation of Total Column Ozone on a Global Scale', *Annals of Applied Statistics*, 1: 191–210.

Stein, M.L. (2008) 'A Modeling Approach for Large Spatial Datasets', *Journal of the Korean Statistical Society*, 37: 3–10.

Stein, M.L., Chi, Z., and Welty, L.J. (2004) 'Approximating Likelihoods for Large Spatial Datasets', *Journal of the Royal Statistical Society, Series B*, 66: 275–96.

Vecchia, A.V. (1988) 'Estimation and Model Identification for Continuous Spatial Processes', *Journal of the Royal Statistical Society Series B*, 50: 297–312.

Ver Hoef, J.M. and Barry, R.D. (1998) 'Modeling Crossvariograms for Cokriging and Multivariable Spatial Prediction', *Journal of Statistical Planning and Inference*, 69: 275–94.

Wackernagel, H. (2003) *Multivariate Geostatistics: An Introduction with Applications*. Berlin: Springer-Verlag.

Wall, M. (2004) 'A Close Look at the Spatial Structure Implied by the CAR and SAR Models', *Journal of Statistical Planning and Inference*, 121: 311–24.

Xia, G. and Gelfand, A.E. (2006) 'Stationary Process Approximation for the Analysis of Large Spatial Datasets', Technical Report, ISDS, Duke University.

Yaglom, A.M. (1987) *Correlation Theory of Stationary and Related Random Functions I*. New York: Springer.

Zhang, Y., Hodges, J.S., and Banerjee, S. (2009) 'Smoothed ANOVA with Spatial Effects as a Competitor to MCAR in Multivariate Spatial Smoothing', *Annals of Applied Statistics*, 3: 1805–30.

32

Market Research and Preference Data

Adam Sagan
Cracow University of Economics, Poland

32.1 INTRODUCTION

In contemporary marketing, research perspectives take a more holistic view on customer research, where one underlines interdependencies among market partners, focusing on value co-creation rather than value delivery and interactive processes in relationships that create dyadic and network structures in particular situations. According to this view, customers act in a highly institutionalized context, as they are nested in various market settings and clustered in social structures. Therefore, marketing variables have to reveal the contextual and relational nature of interactions and exchanges between business partners. They may describe the states and processes of contact episodes like frequencies of interactions, the strength of relationship and relational bonds, the asymmetry of information and market power, loyalty toward partners, mutual confidence in each other, commitment, and trust.

Marketing data today are predominantly longitudinal and dynamic, and describe relational characteristic of partners that can be dyadic (e.g. partnership relations in distribution channels in the business-to-business (B2B) market) or network-based (e.g. word-of-mouth diffusion of information on business-to-consumer (B2C)/consumer-to-business (C2B) markets)—Iacobucci and Hopkins (1992). This view on customer behavior has become popular in those modern marketing orientations that stress the dynamic nature of marketing relationships, situational and contextual aspects of buyer–seller (buyer–firm) interactions, and interdependencies among marketing actors.

32.2 MULTILEVEL ANALYSIS IN MARKETING

Both marketing strategies and research perspectives must take into account the hierarchical nature of marketing data. Marketing variables may reflect the institutional, relational, contextual, situational, interactive, and processual market factors. This analytical framework opens a wide perspective for introducing multilevel models in various marketing applications. These challenges open up new trends and opportunities to develop multilevel modeling in marketing (Wieseke et al., 2008).

In recent years, multilevel models with latent variables have grown in importance in marketing (Macintosh and Lockshin, 1997). This involves using the theoretical constructs represented by latent variables. This unified approach includes confirmatory factor analysis and path models that form structural equation models (SEM) that are based on covariance structure analysis. The integration of multilevel analysis within the SEM methodology provides the possibility of accounting for nested effects and higher-level variability.

These trends may be classified into the following groups:

1. Preference models and conjoint analysis in a multilevel framework. Conjoint analysis and discrete choice models are among the best-known approaches in preference measurement for positioning strategies and new product development. They can be reformulated as multilevel models of nested data with repeated measures, where measurements are nested in the individuals. Additionally, latent or manifest variables can be used to predict stated or revealed preferences that enable the integration of conjoint or discrete choice with latent variable models (Temme et al., 2008).
2. Measurement models of marketing constructs. These involve the development of multilevel confirmatory factor analysis and multilevel item response theory models (Fox, 2005). These models are used to assess individual differences in response behavior (persons' response functions) due to response inconsistencies and multidimensionality in the measurement of consumer preferences (Romaniuk et al., 1999; Gabrielsen, 2001).
3. Analysis of the interaction between and within institutional levels. This involves the analysis of the relationships between individuals and institutions, and interactions between them and the social context to which they belong. In marketing studies, examples of this framework may be the buyer–store interaction in place strategy developed to identify hierarchical interactions between individual consumer-level and store-level variables (Ray and Ray, 2008; Malthouse et al., 2004; Macintosh and Lockshin, 1997). Another example is in highly institutional professional service markets, such as the analysis of patient–service provider relations in the health care market (Rudawska and Sagan, 2007). In addition, three-level hierarchical interactions within business relations can be investigated this way, such as client–salesperson–firm interactions in explaining the return on investment (ROI) in relationship marketing (Palmatier et al., 2006).
4. Modeling latent states. Consumer behavior in contemporary markets is more sensitive to situational and contextual factors than was previously the case. Marketing offers, especially from a Service Dominant Logic perspective, are defined using a process-based, service-oriented series of pre-transactional and post-transactional episodes in buyer–firm interactions. Therefore, flexible consumers' moods, satisfaction and dissatisfaction, anger and regret can result from person–situation interaction in a given service encounter (Gardner, 1985). Within this context, multilevel latent state-trait models (LS-T) are used for the analysis of variability and fluctuations of temporary consumer states (Courvoisier et al., 2007).
5. Analysis of mutual interdependencies between a buyer and seller in a long-term relationship. In partnership marketing in B2B and B2C markets, relations are based on formal contracts, trust, or specific investments of both partners. In a long-term relationship they may form more or less stable dyadic or network relations. Models that can explain mutual relational effects and influences are developed in multilevel settings. These models are rooted in the multilevel actor–partner interdependence models (APIM) as a special case of multilevel social relations models (SRM) (Snijders and Kenny, 1999; Little and Card, 2005; Cook and Kenny, 2005). See Chapter 33 for a general description of network models.
6. International market segmentation analysis. Firms can operate in a global environment, so segmentation studies take into account institutional or spatial dimensions of the customers in segmentation strategies. Multilevel latent class models or multilevel mixture models have become popular in segmentation, where customer-level segmentation variables are integrated with country-level characteristics (Ter Hofstede et al., 2002; Bijmolt et al., 2004).

Multilevel models take into account various nested data structures. These can appear as the effects of given sampling schemes (e.g. cluster or multistage complex sampling, where individuals are nested in primary sampling units (PSUs), or repeated measures, i.e. longitudinal and conjoint studies, where multiple observations are nested within individuals), non-independence of observations and individual responses due to interactions between subjects (e.g. the mutual influence of family members, a dyadic relationship in couples such as buyer–seller interactions), and organizational/institutional or situational reinforcement (e.g. retail institutions or buying situations in which the clients are nested).

In the modeling process of relational and time-oriented data, especially in an institutional context, assumptions regarding the independence of observations are often violated. Also, the variability of intercepts and slopes of response variables may be due to specific sample design (e.g. cluster sampling), depending on higher-level factors. In this case, multilevel analysis is more suitable than traditional approaches such as ordinary least squares multiple regression or analysis of variance because the latter methods can produce large Type I errors and biased parameter estimates (Peugh, 2010).

Of course, the presence of nested data sets is not a sufficient condition to use the multilevel perspective. If there is no variability in response variables across higher-level units, then the data can be analyzed using traditional methods exclusively on the individual level. The need for the multilevel approach involves the calculation of intraclass correlation coefficients (*ICC*) and design effect statistics (*deff*). The hypothesis of the multilevel data structure is used to check the intraclass correlation coefficients (*ICC*). *ICC* is the estimate of the proportion of level-2 variance compared to the total variance (the sum of within- and between-cluster variances),

$$\frac{\delta_\alpha^2}{\delta_\alpha^2 + \delta_\varepsilon^2}.$$

Usually, *ICC* values between 0.05 and 0.2 and a *deff* value of 2 or more suggest the need to consider multilevel modeling to separate additional (level-2) sources of variation (Muthén and Satorra, 1995).

Moreover, hierarchical and nested data require larger sample sizes in comparison to traditional analysis. In the estimation procedures (i.e. maximum likelihood, Muthén's maximum likelihood, and weighted least squares), the sample size should generally meet the 30/30 rule, that is, it should include at least 30 groups/classes, each of which should have at least 30 observations. For the analysis of cross-level interactions this rule is a more restrictive 50/20 one; for the analysis of data with random effects (variance and covariance components), some authors even suggest a 100/10 rule (100 classes, each of at least 10 observations per class); see Hox (2010), pp. 233–5.

Multilevel SEM can utilize two general approaches (Hox, 2010, p. 290). The first, and the more often used, is Muthén's (1997) decompositional approach, which separates the individual-level ('within') and group-level ('between') covariance matrices. In this approach, an individual data value under study Y_{ij} is decomposed into between-cluster variability ($Y_B = Y_j$) and within-cluster variability ($Y_W = Y_{ij} - Y_j$). Both composites are orthogonal and additive, so $Y_T = Y_B + Y_W$ (Hox and Maas, 2004, p. 137). This kind of decomposition is used to calculate the interclass covariance matrix S_B (the covariance matrix of the disaggregated group means) and the intragroup covariance matrix (the covariance matrix of the deviations of individual observations from the appropriate group mean) S_W. These matrices are also additive and orthogonal: $S_T = S_B + S_W$. The second and more computationally demanding approach is based on direct estimation for multilevel data, including variables at each level, using full maximum likelihood estimation.

The process of estimating the multilevel SEM model in the framework of 'within and between matrices' consists of six basic steps:

1 Estimating intraclass and interclass correlation coefficients and between and within covariance matrices.
2 Exploratory factor analysis (without taking into account the clustered nature of the data) for preliminary identification of the factor structure.
3 Confirmatory factor analysis (on both basic and complex sample data).
4 Confirmatory factor analysis on the intraclass covariance matrix ('within' covariance matrix).

5 Confirmatory factor analysis on the interclass covariance matrix ('between' covariance matrix) and diagnosis of ecological fallacy.
6 The multilevel model, estimated simultaneously on the two levels of analysis.

To sum up, nested data structures can be the result of complex sampling design, repeated measures within individuals, or the institutional and hierarchical data context. These problems require the adoption of multilevel analysis for solving research problems in the field of consumer and marketing research.

In the following sections, two multilevel models are presented. In the first model we consider a multilevel regression with demographic and attitudinal variables in the framework of conjoint analysis of preference data on the health care service market. The hierarchical structure of the data is based on repeated measures ('within' level) nested in the individuals ('between' level).

The second model is a multilevel model of buyer–seller interactions on the mobile phone market. In this model the multilevel data consists of dyadic data, where individual customers ('within' level) are nested in the service providers ('between' level).

32.3 MULTILEVEL MODELS OF PREFERENCES

Modeling preference data is the core problem in consumer response analysis and market behavioral segmentation. Preferences are usually treated as latent non-observable theoretical constructs, directly resulting from the adoption of the theory of consumer behavior. Preference measurement is done on the basis of certain responses expressed on the particular choices, ranking or rating scales (stated preferences), or by directly observing the results of actual market behavior expressed in historical data (revealed preferences).

In general, three basic types of research approaches and associated methods for preference measurement can be distinguished. The first family of methods are called decompositional, the second compositional, and the third are mixed methods.

In decompositional methods the preference measurement is based on the rule of simultaneous rating, ranking, or choice of a comparable set of brands or product categories that are described by a bundle of attributes with a certain number of levels. On this basis, the total utilities are estimated and decomposed (hence the name of the approach) on partial utilities of particular levels of the products' attributes (part-worth functions). The basic method used to analyze the structure of preferences in this area is conjoint analysis.

Compositional methods rely on independent evaluations or comparisons of different products across particular attributes. The overall map of preference structure is composed and revealed using principal component analysis, correspondence analysis or multidimensional scaling on the matrix of ratings or rankings of individual attributes (hence the name of the approach). The basic method of analysis in this area is a multidimensional preference scaling that provides preference maps in a reduced low-dimensional space. Mixed methods are a combination of these approaches. This area usually contains hybrid and adaptive conjoint analysis.

Integrated preference models with latent variables are in fact the combination of discrete choice or conjoint (with ordinal and interval preference measurement) models with latent variable models. They allow for such latent factors as values and attitudes that are measured by the number of manifest indicators to be introduced into traditional conjoint analysis (e.g. items of Likert scales)—Temme et al. (2008). These two types of analysis can also be combined in a sequential, two-step procedure. In the first step, the model with latent variables is estimated and the factor scores are calculated for the localization of individual respondents on the latent continuum. In the second step, the previously calculated factor scores serve as predictors of the partial utilities that are calculated in a separate conjoint analysis. As Temme et al. stress, the popularity of integrated choice and latent variable models (ICLV) is rather limited in the marketing literature. These models combine classic discrete choice or conjoint models with the SEM approach.

The problems in incorporating latent variables that underlie unobservable attitudinal constructs is thoroughly covered in the literature. At least four approaches can be adopted in dealing with this issue. The easiest is to simply include specific indicators of attitudes and values as manifest variables directly in utility functions (Koppelman and Hauser, 1979; Green, 1984; Harris and Keane, 1979). In this approach, however, it is assumed that measuring latent constructs is done without error, which should be dealt with in the estimation process. The second frequently used approach is a sequential one. In the first step, factor analysis is performed on the indicators, and in the second step the fitted latent variables are used in the utility function (Prashker, 1979; Madanat et al., 1995). The fitted values contain measurement error and consistent estimates are obtained by integrating the choice probability over the distribution of the latent variables (Morikawa, 1989). A third approach, which is very popular in internal market structure analysis, is based on inference of the latent variables and preferences from preference (choice) data (Elrod, 1991; Elrod and Keane, 1995). Recently, several extended approaches were developed to include latent variables into preference models in one simultaneous estimation. In contrast to sequential methods, this results in consistent and efficient estimates of model parameters (Ben-Akiva et al., 2002; Ashok et al., 2002).

Sequential approaches, although relatively easy to apply, have numerous drawbacks (Temme et al., 2008). They lead to inconsistent and biased partial utilities and they do not allow the more complex relationships between values or attitudes and the partial utilities of product attributes to be tested. Usually, the integrated approach does not suffer these disadvantages. It both enables the estimation of the preference structure and explains the preferences with the help of additional predictors in the model.

32.3.1 Multilevel Model of Preference Data

As an illustration, conjoint analysis with covariates was done using a two-level regression model with repeated measures. The application of multilevel modeling stems from the fact that in a conjoint study the measures are nested in the individual respondents, and as a result they form the 'within' level of analysis. Latent and observed characteristics of the respondents are consequently used to explain the consumers' preference ratings on the 'between' level.

In the study, 506 respondents were requested to rate on a 0–100 scale, the eight predefined health care service profiles based on the following characteristics.[1] The characteristics of health care sub-products are as follows:

1 Cost of health service: (1) free, (2) paid
2 Accessibility: (1) poor, (2) good, (3) average
3 Recommendation: (1) not recommended, (2) moderately recommended, (3) highly recommended
4 Service complexity: (1) low, (2) high
5 Image: (1) poor, (2) average, (3) good

The combination of these characteristics was used to generate eight profiles of health care sub-products on the basis of an orthogonal structure:

Profile 1: free care, poor accessibility, poor image, moderately recommended, low complexity
Profile 2: paid care, poor accessibility, average image, highly recommended, low complexity
Profile 3: free care, good accessibility, poor image, not recommended, high complexity
Profile 4: paid care, good accessibility, good image, moderately recommended, low complexity
Profile 5: free care, average accessibility, good image, highly recommended, high complexity
Profile 6: free care, good accessibility, average image, not recommended, high complexity

Profile 7: free care, poor accessibility, average image, moderately recommended, high complexity
Profile 8: paid care, average accessibility, poor image, moderately recommended, high complexity

The relations between combinations of attributes' levels (expressed by dummy variables) and profile ratings were given on 0–100 probability scales and evaluated at the 'within-unit' level. The dependent variable is the preference rating and the independent variables are dummy coded variables that apply to the combinations of the attributes. Each respondent had to evaluate eight profiles and the 'within' level consists of eight repeated measures nested within the individuals.

In the 'between-unit' level the means of preference ratings on the individual level are explained by the selected demographic characteristics of the individuals, namely gender and age classes and two relational latent variables: formal bond with service provider (measured by information quality, level of specialization, and location of the health-care unit—HCU) and structural bond (measured by the reputation of the HCU, ethics, and patient co-deciding). This two-level regression model is equivalent to a classical regression-based conjoint analysis with $n - 1$ dummy variables.

The estimated intraclass correlation coefficient for the dependent variable was 0.164. Figure 32.1 shows the 'within' and 'between' parts of the conjoint model with two latent variables (formal and structural bonds), gender, and age classes as demographic characteristics.

The 'within' part consists of variables that describe levels of health service attributes: P – paid care, MA – moderate accessibility, WA – weak accessibility, HI – high image, AI – average image, HR – highly recommended, MR – moderately recommended, LC – low complexity. The boxes represent dummy (manifest) variables that indicate the dimensions of health service offer. A black dot at the end of an arrow indicates a random intercept that varies across individuals. The regression coefficients show that paid care (−19.99) is less preferred than free care (0.00), moderate accessibility (10.24) is better than weak (−11.13) and good accessibility (0.00), and also the higher the image of a health care unit, the stronger the preference. The location and reputation of HCU have insignificant path coefficients.

In the 'between' part of the model are variables that measure the attitudinal and demographic characteristics of the consumers: IQ – information quality, SPE – specialization, LOC – localization, REP – reputation, ETH – ethics, COD – patient's co-deciding, FB – formal bond, SB – social bond, BPREF – ratings on between-level, gender (female), age classes (25 – less than 25 years old, 35 – less than 35 years old, 45 – less than 45 years old, 55 – less than 55 years old, 65 – less than 65 years old).

The model has a good fit to the data. The value of the chi-square statistic is 64.14 on 48 degrees of freedom, with a p-value of 0.059. The chi-square statistic shows marginally adequate model fit, but because it is very sensitive to sample size and represents the 'exact fit' statistic, the chi-square test forces even quite good models to be rejected. For example, the population-based RMSEA value of 0.01 is also acceptable (is less than 0.05). Additionally, the incremental fit indices CFI and TLI show good fit in comparison with the baseline model (both CFI and TLI indices equal 0.99). The standardized root mean square residuals (SRMR) for the 'within' model are 0.00, and 0.019 for the 'between' model. A comparison of parameter estimates and standard errors (in parenthesis) for the 'within' model shows that all but one of the parameter estimates are statistically significant. The attributes with the highest (most preferred) coefficients are free care, good accessibility, good image, and high complexity of health care services. Patients recommendations have an insignificant effect on the overall consumer ratings. The R^2 coefficient of determination on the 'within' level equals 0.451 (s.e. 0.016).

The visualization of utility coefficients in conjoint analysis is based on the plot of the

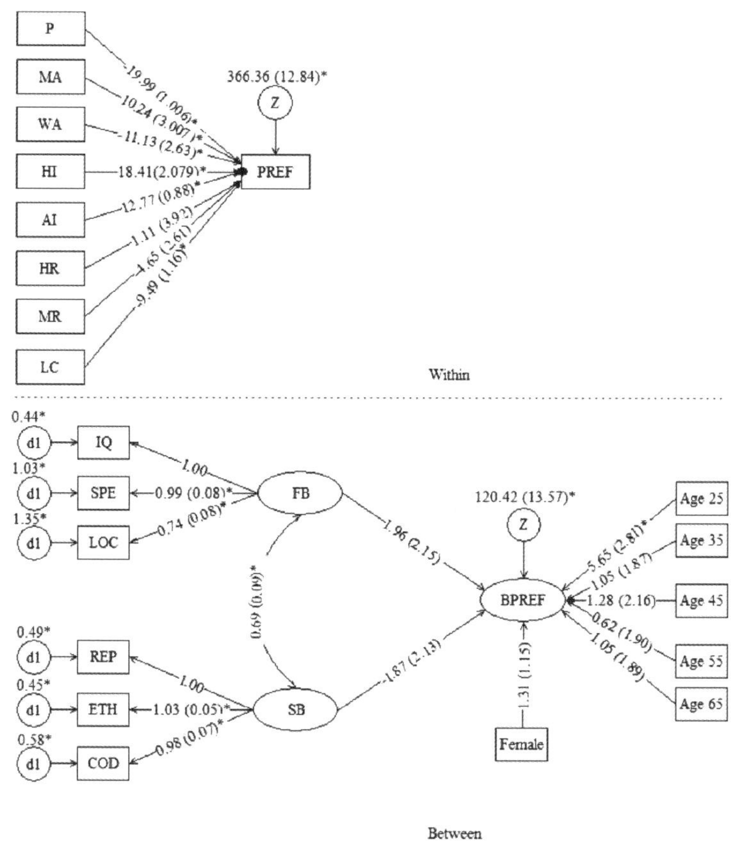

* significant parameters ($p < 0.05$)

Figure 32.1 Integrated Conjoint Model for Health Care Performance Data

utility for each of the attribute levels on a graph. The part-worth function plots are given in Figure 32.2.

In the 'between' part of the model, the preference ratings intercepts (intercepts are random effects that vary across the individuals) are explained by gender, age, and formal and structural bonds. The parameter estimates are significant only for the extreme age classes (lowest and highest age group—above 65 years old served as reference group). The impact of formal and structural bonds and gender on overall ratings is negligible. Also, R^2 for the 'between' part is very low and insignificant (0.023).

Conjoint analysis with latent variables is a promising area to explore for the development of preference models, as it allows the limitations of the two-step approach in estimation procedure to be overcome, and the integration of the consumers' preference data and personal values, lifestyles, and attitudes towards products into one analytical model. Analysis of health care service preference along with dimensions of relational bonds allows for a more detailed explanation of decision-making processes when potential customers are selecting health service providers.

32.4 MULTILEVEL LS-T/API MODEL

In relationship marketing, the modeling of consumer states of satisfaction and loyalty

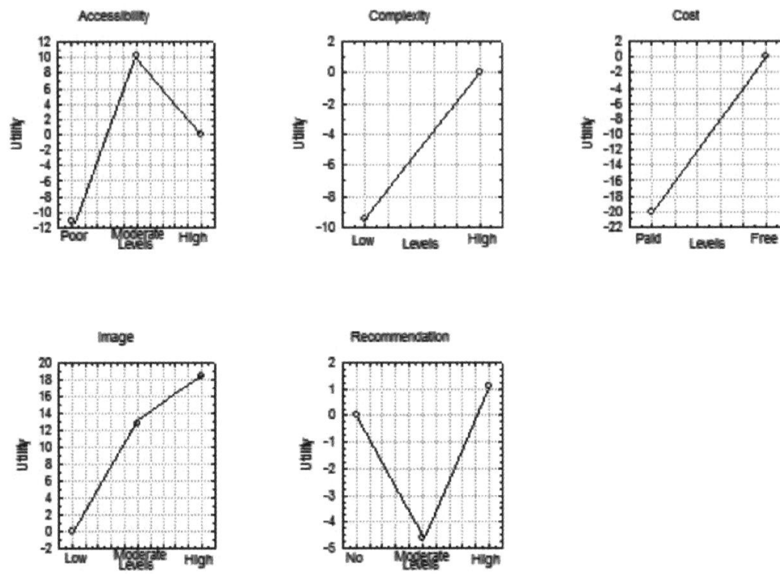

Figure 32.2 Part-Worth Utilities for Health Care Performance Data

toward brands or service providers has become ever more important. Analysis of buyer–seller relationships involves multilevel research settings. Firms' loyalty programs tend to develop stronger links between themselves and clients. These relationship marketing strategies cause the clients to become nested in particular service providers or sellers on the B2C markets. In order to explain this phenomenon, two models are combined and integrated: a latent state-trait model and an actor–partner interdependence model.

Latent state-trait models (LS-T) are a latent class model extension of classic Markov chains, and concern the description of short-time fluctuations of temporary customer characteristics. Unlike in the analysis of changes in the levels of manifest or latent variables (as in the latent growth curve models—LGCM), LS-T models relate to the variations and fluctuations of the momentary emotional states and moods of the consumers. These states are characterized by short-term fluctuations. Consumers' happiness, (dis)satisfaction, or anger depend not only on the preceding states, but also on the average mood level as well as situational factors and interactions between situations and consumers' characteristics.

The advantage of using this kind of model is that it provides the opportunity to separate three basic components of the interindividual structure of variability of consumer response in a given situation: (a) a relatively stable component, the latent trait of the consumer (latent trait); (b) specific, situational factors that explain unstable and liable consumer states (occasions); and (c) the error component. It is assumed that latent states and traits as well as error variances are uncorrelated. The variability of indicators is an interaction between more stable traits (e.g. the general level of satisfaction or attitude), situational and specific consumer states and the interaction between traits and states (person–situation interaction). Traits, therefore, are mainly characterized by an inter-object variability, and consumer states are defined as temporary deviations from the level of latent trait, reflecting primarily the influence of situational factors, and, furthermore, are characterized by both intra- and inter-object variability (Courvoisier et al., 2007; Sagan and Kowalska-Musial, 2008). They mainly reflect the situational differences and trait–occasion interaction at a particular measurement time (Steyer and Schmitt, 1990). They are particularly applicable in measuring the variability of both a consumer's mood and that of his/her latent traits, which are sensitive to situational factors.

To sum up, the variables representing latent traits are consumer characteristics that are relatively independent of situational factors and person–situation interactions. Latent states represent the momentary deviations from latent trait in particular situations (Steyer et al., 2004).

In actor–partner interdependence models (APIM), the dyad is assumed to be the basic unit of the analysis. Dependence between partners results from the assumption that the information about the characteristics and behavior of one actor in the dyad allow us to predict the behavior or characteristic of his/her partner. These relations can result from mutual similarities, the dominance of one partner as a result of power asymmetry, mutual interaction, and external factors (common fate) that may affect the mutual adjustments of the partners (Kenny et al., 2006).

In the analysis of this kind of data, two forms of dyad can be identified: (a) a distinguished dyad (where the social roles of the partners are clearly defined, e.g. buyer and seller, or consumer and producer), and (b) undistinguished dyads, where parters may play a similar role (e.g. friends, collaborators in the project, parents, etc.). In dyadic data analysis, the distinguished dyad can be analyzed using standard confirmatory factor analysis or path models, and undistinguished dyads can be analyzed using multilevel models (Olsen and Kenny, 2006).

The sources of variability can be divided into three categories: (a) interdyadic variability, in which values can differ across dyads but are the same for both partners in a particular dyad (e.g. the length of time of relationships, negotiated price level); (b) intradyadic variability, in which values can differ between objects in the same dyad, but are the same across dyads (e.g. the gender of partners in a marriage); and (c) mixed variability, in which the values may vary both within and between dyads (e.g. the age of partners, and the average age of couples)—Kenny et al. (2006).

In the structure of actor–partner interdependence models (APIM) with latent variables, it is assumed that factor loadings across the individuals in the dyad are equal (the same common meaning of constructs), and error variances for similar items and means of latent variables are the same. The covariances between latent variables for an actor and his/her partner's latent traits are also assumed to be significant and relatively high.

The structure of interdependencies between buyers and sellers in dyadic relationships can be divided into two important effects: actor and partner effects. The sources of these interdependencies can be rooted in the common characteristics of partners, the asymmetry of power distribution in dyads, the frequency of mutual interactions, external common fate, and so on (Kenny et al., 2006).

Actor effect means here that characteristics of the buyer (or seller) can influence his/her own characteristics (e.g. the buyer's satisfaction influences his/her loyalty towards a partner, or an actor's satisfaction with the relation in a particular actor's state influences his/her satisfaction with the relation in the next stage of the relationship). The partner effect, on the other hand, occurs when the characteristics of an actor influence those of the partner (e.g. the buyer's satisfaction influences the seller's loyalty towards the buyer, or the actor's satisfaction with the relation in a particular actor's state influences a partner's satisfaction with the relation in the next stage of their relationship). Additionally, various mediators and moderators of the relationship can be added into the model, resulting in several moderation/mediation effects, including actor-moderated actor's effect, actor-moderated partner's effect, partner-moderated partner's effect, and partner-moderated actor's effect.

In consumer behavior models, especially in relationship marketing, the models of satisfaction–loyalty relations are the center of interest. Consumer satisfaction can be regarded in the framework of the latent state-trait as an emotional and experiential consumer state, which can fluctuate in the service consumption processes. It can be assumed that the individual states and traits of both partners in a service encounter can affect the mutual loyalty of both partners, inducing certain actor and partner effects.

Because buyers are usually nested in sellers (service providers or business partners), a level-1 model ('within') also contains information about random intercepts of the indicators and random slopes for the structural part of the model. Random intercepts represent the variability of means for manifest variables that are explained by the characteristics of level-2 units at the 'between' level, and which are characteristics of the service institution the customers are nested in. Random slopes represent the variability of path coefficients, determining the relation between the satisfaction and the loyalty of clients and sellers. This kind of variability is explained by level-2 variables. Buyer–seller interactions are also explained by the situational, contextual, and institutional factors as a frame of reference for the level-2 model. These factors are expressed by so-called global variables (characteristics of the level-2 analysis), and structural variables that constitute aggregated variables from level-1 units (individual respondents).

We now provide an illustration of the multilevel LS-T/API model in the analysis of satisfaction–empathy relations between buyers and service providers in the mobile phone market. The data come from 48 service outlets of mobile services and 146 observations of individual customers. The data were gathered using an observational questionnaire including behavioral indicators of satisfaction and empathy between a buyer and seller.[2]

Two variables were identified: satisfaction (measured two times and having two latent states) and the mutual empathy of partners (regarded as a relational variable). In the presented model, the relationship between satisfaction and the mutual empathy of partners is explained by institutional features of the sales point (the breadth of the offer and location from the center of the town). The intercepts for variables on level 1 were treated as random variables (random intercepts), and therefore on the second level of the analysis these indicators are presented as latent variables (unobserved heterogeneity). Additionally, global exogenous manifest variables on level 2 (length of product line and firm locations) explain the buyer and seller empathy factor on this level, while the variability of the slope coefficients explain the direction and strength of the relationship between partners' satisfaction and empathy.

As the data were based on an observational questionnaire, the indicators of customer and seller satisfaction (CS and SS respectively) reflected the non-verbal behavior of partners in two time points: (1) a smile (CS11, CS12/SS1, SS12) and (2) nodding in agreement (CS2/SS2). Indicators of empathy were commitment to the conversation (CE1/SE1), repeating words (CE2/SE2), and expressing the desire to meet again (CE3/SE3), and were measured at the final, closing stage of the encounter. The observational nature of the data, however, limits the accuracy and generalizability of the model.

Customer satisfaction was measured in two waves by two indicators (CS11, CS12, CS21, CS22), and customer and seller empathy were measured by three indicators (CE1/SE1, CE2/SE2, CE3/SE3). Therefore, satisfaction is treated as a latent state of the customer and seller, which is measured in two 'moments of truth' in the service process: solving the problem and closing the transaction.

The aim of the preliminary step in the multilevel analysis is to check the assumptions and

Table 32.1 Intraclass Correlation Coefficients for Mobile Phone Satisfaction Data

Manifest variables	ICC
CS11	0.13
CS12	0.19
CS21	0.19
CS22	0.09
CE1	0.13
CE2	0.07
CE3	0.14

whether there is a need to introduce a two-level framework for the analysis. At this stage the intraclass coefficients (*ICC*) or design effects coefficients (*deff*) are examined before the estimation of higher-level models.

Table 32.1 shows the *ICC*s used to assess the proportion of between-group variance compared with total variance, and thus provides the acceptance of a multilevel model for further analysis. Intraclass correlation coefficients are predominantly greater than 0.1 (except CS22 and CE2), indicating the usefulness of the multilevel model. On average, more than 3.5 customers were observed at each point of sale.

32.4.1 Exploratory Multilevel Factor Analysis

A preliminary step in multilevel analysis with latent variables is to check the factor structure of constructs on the individual ('within') and cluster ('between') levels. An individual data value Y_{ij} is decomposed into a between-group component Y_B (the disaggregated group mean) and a within-group component Y_W (the individual deviation from the group means). The factor structure is identified using exploratory factor analysis on the individual-level ('within') and the group-level ('between') covariance matrices.

The total covariance matrix is decomposed into additive and orthogonal 'within' and 'between' covariance matrices using the formula

$$\Sigma_T = \Sigma_B + \Sigma_W.$$

In two-level factor analysis, the within part of the factor analysis model is

$$y_W = \Lambda_W \cdot \eta_W + \varepsilon_W,$$

while the between part of the factor analysis model is

$$\mu_B = \mu_B + \Lambda_B \cdot \eta_B + \varepsilon_B,$$

where μ_B are random intercepts that vary across groups, Λ_W is the factor loadings matrix at the 'within' level, Λ_B is the factor loadings matrix at the 'between' level, and ε_W and ε_B are errors at the 'within' and 'between' levels. The final unified multilevel CFA model is then given as

$$y_{ij} = \mu_B + \Lambda_W \cdot \eta_W + \Lambda_B \cdot \eta_B + \varepsilon_B + \varepsilon_W.$$

Tables 32.2 and 32.3 show the results of an exploratory factor analysis carried out on 'within' (at the individual level) and 'between' (at the service provider level) covariance matrices. The goodness of fit indices of the EFA model with Geomin rotation for the 'within' covariance matrix of the clients are low. The chi-square statistic has a value of 33.91 with 8 degrees of freedom and a p-value < 0.01. The population-based index Root Mean Square Error of Approximation also shows the bad fit of the model (RMSEA = 0.14 with 90 percent C.I. 0.09–0.19). The standardized Root Mean Square Residual (SRMR) is acceptable (0.03). The first factor (F1) has higher factor loadings across the satisfaction items and the second (F2) has higher loadings on empathy indicators. The Geomin correlation between factors is 0.69.

The second model is an exploratory factor analysis on the covariance matrix for the sellers (Table 32.3). Model fit is unacceptable, with the chi-square statistic at a level of 44.34 and 8 degrees of freedom. The p-value is less than 0.01, the RMSEA is 0.16 (90 percent C.I. = 0.119–0.213) and SRMR is acceptable (0.04). The factor loadings structure is similar to that of the EFA models for customers. Factor F1 has higher loadings across satisfaction

Table 32.2 Exploratory Factor Analysis: Model 'Within' for the Customers for Mobile Phone Satisfaction Data

Manifest variables	F1	S.E.	F2	S.E.
CS11	0.67*	0.17	0.17	0.29
CS12	0.45*	0.18	0.34	0.19
CS21	0.91*	0.03	0.01	0.06
CS22	0.64*	0.15	0.20	0.18
CE1	–0.07	0.02	0.79*	0.07
CE2	0.29	0.18	0.52*	0.18
CE3	0.11	0.22	0.57*	0.21

* significant loadings ($p < 0.05$)

Table 32.3 Exploratory Factor Analysis: Model for Sellers for Mobile Phone Satisfaction Data

Manifest variables	F1	S.E.	F2	S.E.
SS11	0.69*	0.06	0.11	0.03
SS12	0.94*	0.11	–0.08	0.13
SS21	0.52*	0.11	0.19	0.11
SS22	0.61*	0.08	0.13	0.11
SE1	0.28*	0.11	0.54*	0.11
SE2	–0.05	0.01	0.81*	0.10
SE3	0.09	0.11	0.42*	0.11

* significant loadings ($p < 0.05$)

items and factor F2 across empathy indicators. The Geomin factor F1–F2 correlation is 0.47.

In a subsequent analysis, the exploratory factor analyses were done on 'between' level covariance matrices. The 'between' level corresponds to level 2 of the service providers. The results are presented in Table 32.4. The goodness of fit of the EFA model is acceptable. The chi-square statistic has a value of 0.34 with 8 degrees of freedom and the p-value is 0.99. The RMSEA index is 0.00 while SRMR is 0.01. The Geomin factor correlation is 0.46. However, the factor loading structure is unclear with respect to clients' empathy and satisfaction measurements. Additionally, the 3-factor structure suggests a correlation between satisfaction in the second measurement occasion and the empathy of clients.

32.4.2 Within and Between LS-T/API Models

The integrated multilevel model of latent state-trait and actor–partner interdependence (LS-T/APIM) explains the relationships between the states of buyers' and sellers' satisfaction and the mutual empathy of both partners. In the process of multilevel model development, several models were identified and compared.

The first model was the LS-T/APIM general model based on the total covariance matrix. The second model was LS-T/APIM based on the total covariance matrix with complex sampling (where individual customers are nested in service providers). The third model was a two-level LS-T/APIM with random intercept as an outcome. The fourth model was

Table 32.4 Exploratory Factor Analysis: Model 'Between' for the Clients for Mobile Phone Satisfaction Data

Manifest variables	F1	S.E.	F2	S.E.
CS11	0.57*	0.15	–0.01	0.16
CS12	0.40*	0.11	0.20	0.17
CS21	0.17	0.11	0.46*	0.11
CS22	0.08	0.10	0.45*	0.11
CE1	0.13	0.11	0.34*	0.13
CE2	–0.04	0.09	0.32*	0.12
CE3	0.22*	0.11	0.22*	0.11

* significant loadings ($p < 0.05$)

a two-level LS-T/APIM with random slopes as an outcome. In the estimation of the last two models the two-step approach was used because of the numerical problems that occur in model estimation. In the first step, the separate LS-T and APIM multilevel submodels were estimated, then the final LST/API integrated model was developed. The estimation of the models was done on the factor scores estimated in the first step of the analysis.

Table 32.5 presents the goodness-of-fit indices of four multilevel LS-T/API models: (1) a general model based on total covariance matrix, (2) a general model based on total covariance matrix with complex sampling, (3) a two-stage two-level LS-T/API model (TSTL), and (4) a two-stage two-level LS-T/API model with random slopes (TSTLRS).

As a final version of the multilevel LS-T/API model, the model with random slopes was selected on the basis of the Akaike Information Criterion and Bayesian Information Criterion. Figure 32.3 shows the integrated LS-T/API 'within' and 'between' models for individual customers nested within the sellers. The variables on the 'within' part of model are: CS, the overall customer satisfaction; CS1, the customer satisfaction at solving problem stage; CS2, the customer satisfaction at closing transaction; and CE, the overall customer empathy.

The variables situated on the 'between' part are: CSB – the overall customer satisfaction on the 'between' level, CSB1 – customer satisfaction at solving problem stage on the 'between' level, CSB2 – customer satisfaction at closing the transaction on the 'between' level, CEB – the overall customer empathy on the 'between' level, SS – the overall seller satisfaction, SS1 – the seller satisfaction at solving problem stage, SS2 – seller satisfaction at closing the transaction, SE – the overall seller empathy, LOC – the localization of distributor, PLINE – the length of product line, S1–S3 – the random slopes, and Z1–Z5 – the disturbance terms.

In the 'within' part of the model, the regression parameters indicate that the only significant path is between overall customer satisfaction and empathy toward the seller (CS – CE = 0.60). Temporary latent states of the customers' satisfaction in two 'moments of truth' (problem solving and closing the transaction of the selling process) have insignificant estimates. The correlation between the states is also insignificant. The black dots depicted at the end of the arrows represent the random intercepts, and black dots lying on the regression paths indicate the random slope model.

The 'between' model explains the customers'–sellers' satisfaction and empathy relationship on the service providers' ('between') level of analysis (CSB and CEB variables respectively). For the sellers, the satisfaction and empathy variables are regarded as global variables that exist only on the 'between' level. Additionally, it is assumed

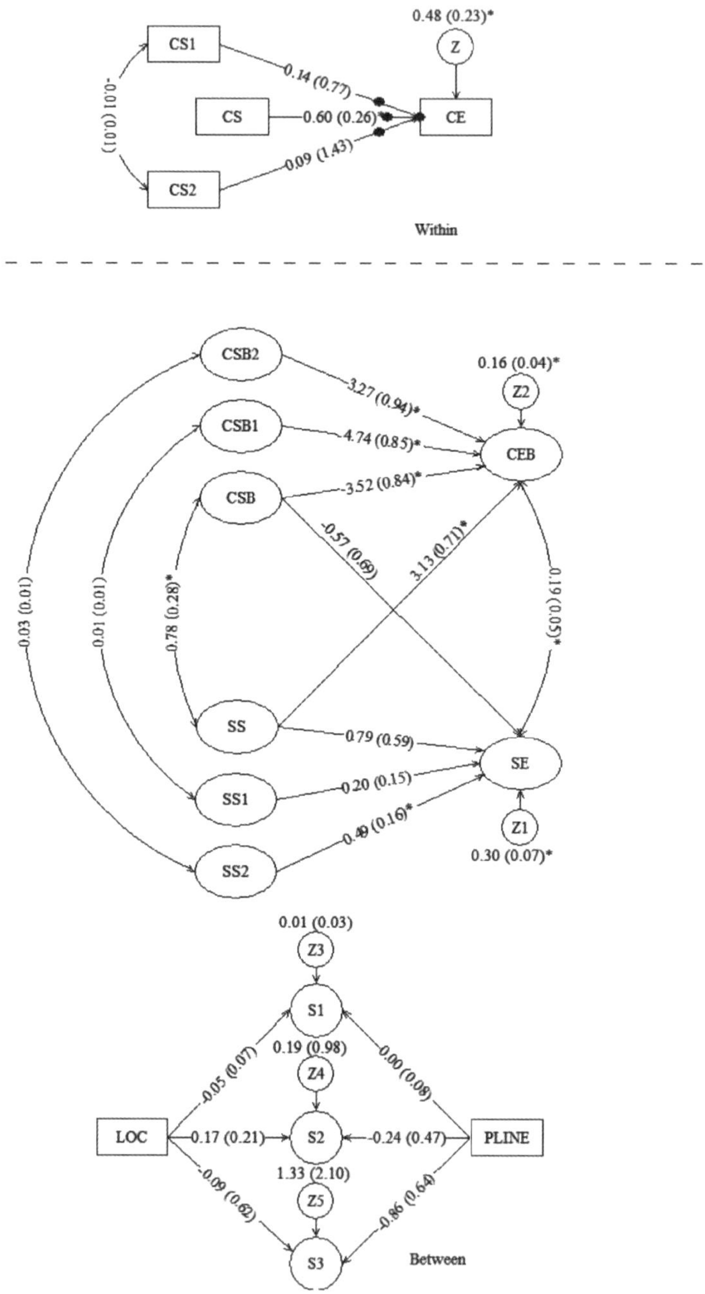

* significant parameters ($p < 0.05$)

Figure 32.3 Multilevel 'Within' and 'Between' LS-T/API Model for Mobile Phone Satisfaction Data

Table 32.5 Comparison of LS-T/API Models for Mobile Phone Satisfaction Data

Model	Chisq	LL	Df	*p*-value	CFI	TLI	AIC	BIC	RMSEA
General	211.55	—	95	0.00	0.93	0.89	7936.22	8130.98	0.08
General complex	264.57	—	16	0.00	0.90	0.86	8961.34	9214.97	0.09
TSTL	155.26	—	16	0.00	0.77	0.75	311.23	425.22	0.23
TSTLRS	—	−115.36	—	—	—	—	317.524	452.11	—

that an actor's temporary states (CS1/SS1, CS2/SS2) affect only the actor effect (states of consumer or seller satisfaction affect only the empathy on his/her own level), but overall satisfaction (CS/SS) shapes both the actor's and the partner's effects.

The model presented in Figure 32.3 has random slopes, which explains the variability of the slopes between consumer satisfaction and empathy in the 'within' level. Random slopes are represented by the circles (as latent variables) at the bottom of Figure 32.3. The variability of the slopes may be explained by the global variables LOC (location, the distributor's distance from the town center) and PLINE (the length of the product line and service offers).

The results show that the opposite actor's effects can be observed on both the individual and cluster levels of the analysis. This suggests the existence of Simpson's paradox (an ecological error). Both satisfaction states and overall customer satisfaction with the service influence the customer's empathy towards the seller. The relationships can be expressed in the negative: the higher the satisfaction with the 'problem solving' stage of interaction, the lower will be the empathy communicated toward the seller. Also, general customer satisfaction is negatively related to one's own customers' empathy. A customer's partner effect is insignificant—the customer's satisfaction has no impact on the seller's empathy toward the clients on the service providers' level.

On the other hand, the seller's partner effect is positive and significant and it implies that sellers' general feelings of satisfaction influence the customers' empathy toward them. The seller's actor effect is insignificant; the sellers' satisfaction has no impact on their empathy toward clients. Only the sellers' satisfaction with the closing stage interaction has a significant effect on the sellers' empathy towards their customers.

The variability of slopes is explained by two institutional-level variables (LOC and PLINE). Both have an insignificant path on the variability of the relationship between stages of customer satisfaction (CS1, CS2) and empathy (CE), and between overall customers' satisfaction (CS) and empathy (CE).

32.5 CONCLUSIONS

Multilevel models have taken on increased importance in transactional as well as relationship marketing. They help analysts to understand various types of nested data structures like repeated measures nested in individuals, individual customers nested in sellers, stores, or firms, business partners in dyads or networks, etc. Also, models that take into account the interactions inside or between the actors are a promising direction in development of behavioral analysis in service-oriented marketing. They allow correct modeling of both customer and seller behavioral characteristics in the context of situational factors. Moreover, multilevel models allow for inclusion of institutional, contextual, and hierarchical effects for understanding consumer behavior in more complex buying situations.

The popularity of multilevel approaches is strongly related to the accessibility of contextual and hierarchical marketing data. The growth of relational databases and data warehouses, and the integration of data coming from various sources, give an opportunity for multilevel models to be a more popular approach in marketing research and analysis.

NOTES

1 Empirical study has been carried out by Prof. Iga Rudawska, PhD, from the University of Szczecin, to whom I express my gratitude for providing the data.

2 An empirical study has been carried out by Magdalena Kowalska-Musial, PhD, from the Higher School of Finance and Banking in Cracow, to whom I express my gratitude for providing the data.

REFERENCES

Ashok, K., Dillon, W.R., and Yuan, S. (2002) 'Extending Discrete Choice Models to Incorporate Attitudinal and Other Latent Variables', *Journal of Marketing Research*, 39: 31–46.

Ben-Akiva, M., Walker, J., Bernardino, A.T., Gopinath, D.A., Monikawa, T., and Polydoropoulu, A. (2002) 'Integration of Choice and Latent Variable Models', in Mahmassani, H.S. (ed.), *Perpetual Motion: Travel Behaviour Research Opportunities and Application Challenges*. Amsterdam: Elsevier.

Bijmolt, T.H.A., Paas, L.J., and Vermunt, J.K. (2004) 'Country and Consumer Segmentation: Multilevel Latent Class Analysis of Financial Product Ownership', *International Journal of Research in Marketing*, 21: 323–40.

Cook, W.L. and Kenny, D.A. (2005) 'The Actor-Partner Interdependence Model: A Model of Bidirectional Effects in Developmental Studies', *International Journal of Behavioral Development*, 29: 101–9.

Courvoisier, D.S., Eid, M., and Nussbeck, F.W. (2007) 'Mixture Distributions Latent State-Trait Analysis: Basic Ideas and Applications', *Psychological Methods*, 12: 80–104.

Elrod, T. (1991) 'Internal Analysis of Market Structure: Recent Developments and Future Prospects', *Marketing Letters*, 23: 252–66.

Elrod, T. and Keane, M.P. (1995) 'A Factor-Analytic Probit Model for Representing the Market Structure in Panel Data', *Journal of Marketing Research*, 32: 1–16.

Fox, J.P. (2005) 'The Multilevel IRT Using Dichotomous and Polytomous Data', *British Journal of Mathematical and Statistical Psychology*, 58: 145–72.

Gabrielsen, G. (2001) 'A Multilevel Model for Preferences', *Food Quality and Preference*, 12: 337–44.

Gardner, A. (1985) 'Mood States and Consumer Behavior. A Critical Review', *Journal of Consumer Research*, 12: 281–300.

Green, P. (1984) 'Hybrid Models of Conjoint Analysis: an Expository Review', *Journal of Marketing Research*, 21: 155–69.

Harris, K.M. and Keane, M.P. (1979) 'A Model of Health Plan Choice: Inferring Preferences and Perceptions from a Combination of Revealed Preference and Attitudinal Data', *Journal of Econometrics*, 89: 131–57.

Hox, J. (2010) *Multilevel Analysis. Techniques and applications*. New York: Routledge.

Hox, J. and Maas, C. (2004) 'Multilevel Structural Equation Models: the Limited Information Approach and the Multivariate Multilevel Approach', in K. van Monforf, J. Oud, and A. Satorra (eds), *Recent Developments on Structural Equation Models*. Dordrecht: Kluwert Academic Press, 135–49.

Iacobucci, D. and Hopkins, N. (1996) 'Modeling Dyadic Interactions and Networks in Marketing', *Journal of Marketing*, 29: 5–17.

Kenny, D.A., Kashy, D.A., and Cook, W.L. (2006) *The Analysis of Dyadic Data*. New York: Guilford Press.

Koppelman, F. and Hauser, J. (1979) 'Destination Choice for Non-Grocery-Shopping Trips', *Transportation Research Record*, 673: 157–65.

Little, T.D. and Card, N.A. (2005) 'On Use the Social Relations and Actor-Partner Interdependence Models in Developmental Research', *International Journal of Behavioral Development*, 29: 173–9.

Macintosh, G. and Lockshin, L.S. (1997) 'Retail Relationship and Store Loyalty: A Multilevel Perspective', *International Journal of Research in Marketing*, 14: 487–97.

Madanat, S.M., Yang, C.Y.D., and Yen, Y.M. (1995) 'Analysis of Stated Route Diversion Intentions Under Advanced Traveler Information Systems Using Latent Variable Modeling', *Transportation Research Record*, 1485: 10–17.

Malthouse, E.C., Oakley, J.L., Calder, B.J., and Iacobucci, D. (2004) 'Customer Satisfaction Across Organizational Units', *Journal of Service Research*, 6: 231–42.

Morikawa, T. (1989) *Incorporating Stated Preference Data in Travel Demand Analysis*. Unpublished PhD dissertation, Department of Civil Engineering, MIT.

Muthén, B. (1997) 'Latent Variable Modeling with Multilevel and Longitudinal Data', *Sociological Methodology*, 27: 453–80.

Muthén, B. and Satorra, A. (1995) 'Complex Sample Data in Structural Equation Modeling', *Sociological Methodology*, 25: 87–99.

Olsen, J.A. and Kenny, D.A. (2006) ' Structural Equation Modeling With Interchangeable Dyads', *Psychological Methods*, 11: 127–41.

Palmatier, R.W., Gopalakhrishna, S., and Houston, M.B. (2006) 'Returns of Business to Business

Marketing Investments: Strategies for Leveraging Profits Marketing Science', *Marketing Science*, 25: 477–93.

Peugh, J.L. (2010) 'A Practical Guide to Multilevel Modeling', *Journal of School Psychology*, 48: 85–112.

Prashker, J.A. (1979) 'Mode Choice Models with Perceived Reliability Measures', *Transportation Engineering Journal*, 105: 251–62.

Ray, J.C. and Ray, D. (2008) 'Multilevel Modeling in Marketing: A Primer', *Recherche et Applications en Marketing*, 23: 1–24.

Romaniuk, H., Skinner, C.J., and Cooper, P.J. (1999) 'Modeling Consumers' Use of Product', *Journal of Royal Statistical Society, Ser. A*, 162: 407–21.

Rudawska, I. and Sagan, A. (2007) 'Multilevel Dimensions of Consumer Relationships in the Healthcare Service Market. MLSEM vs MLIRT Approach', in Decker, R. and Lenz, H.-J. (eds), *Advances in Data Analysis*. Heidelberg: Springer-Verlag, 481–8.

Sagan, A. and Kowalska-Musial, M. (2008) 'Dyadic Interactions in Service Encounter: Bayesian SEM Approach', in Fink, A., Lausen, B., Seidel, W., and Ultsch, A. (eds), *Advances in Data Analysis, Data Handling and Business Intelligence*. Heidelberg: Springer, 573–82.

Snijders, T.A.B. and Kenny, D. (1999) 'The Social Relations Model for Family Data: A Multilevel Approach', *Personal Relationships*, 6: 471–86.

Steyer, R., Krambeer, S., and Hannover, W. (2004) 'Modeling Latent Trait-Change', in Monfort, K., Oud, J., and Satorra, A. (eds), *Recent Developments on Structural Equation Models*. Dordrecht: Kluwer Academic Press, 337–57.

Steyer, R. and Schmitt, M.J. (1990) 'Latent State-Trait Models in Attitude Research', *Quality and Quantity*, 24: 427–45.

Temme, D., Paulssen, M., and Dannewald, T. (2008) 'Incorporating Latent Variables Into Discrete Choice Models – A Simultaneous Estimation Approach Using SEM Software', *Business Research Official Open Access Journal of VHB*, 1: 220–37.

Ter Hofstede, F., Wedel, M., and Steenkamp, J.-B.E.M. (2002) 'Identifying Spatial Segments in International Markets', *Marketing Science*, 21: 160–77.

Wieseke, J., Lee, N., Broderick, A.J., Dawson, J.F., and Van Dick, R. (2008) 'Multilevel Analyses in Marketing Research: Differentiating Analytical Outcomes', *Journal of Marketing Theory and Practice*, 16: 321–39.

33

Multilevel Modeling of Social Network and Relational Data

Marijtje A.J. van Duijn
Interuniversity Center for Social Science Theory and Methodology, University of Groningen, The Netherlands

33.1 INTRODUCTION

Social network analysis is a large and interdisciplinary research field with its main roots in the social sciences (sociology, psychology, anthropology) and mathematics (graph theory). Its objective can be summarized as understanding the structure of the social network through description, visualization, and mathematical and statistical analysis. Since the late 1970s, statistical models for social network analysis have been developed and applied, aided by the development of statistical models in other areas and by the advancement of computational techniques. The models are part of a larger body of statistical models available for the analysis of social network data, of which recent overviews can be found in Snijders (2011), van Duijn and Huisman (2011), Goldenberg et al. (2009), and O'Malley and Marsden (2008). General introductions to social network analysis are given by Scott and Carrington (2011), Scott (2000), and Wasserman and Faust (1994).

The main outcome variable in models for social network data is a tie variable, measuring the relation or relational tie (*edge* or *link* in graph-theoretical terminology) between two individuals (*nodes*, *vertices*, or *actors* in social network terminology). Edges can be undirected, just recording a link between two nodes, or directed, where information is available on the *sender* or *ego* and the *receiver* or *alter* of the relation, i.e., it is known which actor reports his/her relation with another actor. Thus, two reports are available on the relation between the two nodes, who both act as sender and receiver. Social network data can be represented as graphs or by a so-called adjacency matrix, a square matrix $\boldsymbol{Y}$, where Y_{ij} represents the relation from actor i to actor j, where the diagonal is usually undefined. For undirected network data the adjacency matrix is symmetric. A pair of actors and their relations is called a *dyad*, (Y_{ij}, Y_{ji}). Whether a social network is directed, whether actors contribute more than one reported tie to the data, and whether covariate information on the actors or dyads is recorded as well depends on the type of design or data collection method.

An important distinction in social network data designs is made between *ego(centric)* or *personal* network data and *complete* network or whole-network data. In personal network data (a sample of) individuals report on their relations with one or more alters. In complete network data all directed or undirected relations between a closed group of individuals are collected, in a *round robin* design. Marsden (2011) provides an overview of social network data collection designs and methods. The (survey) sampling of social network data is not straightforward in view of the dependence between nodes and between edges. Estimation of (population) statistics requires special methods dependent on the exact sampling procedure. See Frank (2011) for an introduction to and overview of design-based estimators, and Handcock and Gile (2010) for a model- or likelihood-based approach. The presence of missing data is a related challenge to statistical inference (Handcock and Gile, 2007) for which methods are still being developed, roughly categorized as model-based (Koskinen et al., 2010) and imputation-based (Huisman, 2009).

The aim of statistical models for tie variables can be generally described as *explaining* or *predicting* the observed social network using covariate information on actors and/or dyadic information, taking into account the dependence structure of the tie variables. From a statistical point of view it is much easier to deal with personal network data than with complete network data, due to a simpler dependence structure of tie observations. Likewise, the models are easier from a computational point of view when a normal distribution can be assumed for the tie variables. Typically in social network analysis, however, it is reported whether or not ties are present, leading to binary outcome variables.

This chapter provides an overview of the use and application of the most straightforward multilevel—or random effects—models in social network analysis. It is organized by increasing complexity in dependence structure and computational requirements, starting with egocentric networks for continuous tie variables and ending with models for multiple complete binary networks.

Where appropriate, reference is made to related models and recent applications. At the end of the chapter some comparisons will be made with a suitable selection of models for complete network data.

Two important classes of statistical model for social network analysis are not mentioned further: contagion models and models for longitudinal network data. Contagion models investigate the effect of the social network on actor outcomes. A good overview and illustration of contagion models in social network analysis is given by O'Malley and Marsden (2008); a recent multilevel extension of contagion models for a binary actor-level outcome is proposed by Koskinen and Stenberg (2012). Models for longitudinal data include stochastic actor-based models (Snijders, 2001), investigating network dynamics, with the possibility of distinguishing contagion and selection effects (Steglich et al., 2010), and models for relational events, based on event history models (Butts, 2008), or on an actor-oriented Markov process (Stadtfeld and Geyer-Schulz, 2011).

All models treated in this chapter are illustrated using data from a study on classroom networks (Vermeij, 2006). The data are analyzed using R version 2.12, MLwiN version 2.22 (Rasbash et al., 2010), and STOCNET (Boer et al., 2006).

33.2 DESCRIPTION OF THE CLASSROOM SOCIAL NETWORK DATA

Data on several social relations in 76 classrooms in 33 secondary schools were collected by Lotte Vermeij in 1999/2000 for her dissertation research (Vermeij, 2006). A total of 1521 pupils (52% girls) participated in the study, on average 15.6 years old. Vermeij's main research question concerned the influence of ethnicity and ethnic composition in the classroom and in the neighborhood on the relationships between pupils (see also Vermeij et al., 2009). An even distribution of the sampled schools over urban, suburban, and rural areas, and over the four main tracks in secondary education, was obtained.

Seven social network questions were administered to the students who could nominate up to twelve classmates using a separate list with numbers for all of their fellow pupils. Five items are used to measure the content of the social relationship to measure support, communication, and shared activities.[1] The other two items were about desired or undesired relations, the first asking whom the respondent would like to see more, the second asking the respondent to indicate whom s/he did not like to be around. The median number of nominations is 2 for the support items and shared activities. For communication, it was 1 for the first (talking about norms) and 5 for the second (talking about taste). The desire for seeing each other more is quite low, with a median of 0. The median for not liking to be around is 1.

From the first five items, a variable 'tie strength' was constructed by adding the nominations. The resulting variable has possible outcomes 0–5, where 0 indicates no tie. The distribution of 'tie strength' is far from being normal. The overall mean tie strength of observed ties is 2.67 (s.d. 1.03; range 1–5), the average school mean tie strength 2.66 (s.d. 0.32, range 2.01–3.60). No valid nominations were obtained in two schools with a total of 34 pupils, and for 94 other pupils in other schools. The potential bias due to the (probably not at random) missingness of these pupils is ignored here (cf. Handcock and Gile, 2007).

Various pupil covariates are available, of which three are used: gender, a known important determinant of social relationships among children and adolescents, school marks, as a proxy for cognitive ability and school motivation, and importance of the classroom as a social context. School mark was measured as the average mark obtained at the most recent report card, or, if not available, by the average of the marks for language (Dutch) and mathematics, considered the most important and difficult subjects. To measure the importance of the classroom social environment, a classroom identification scale was constructed consisting of 6 five-point Likert scale items (α=0.78). The first three items tapped on emotions toward being part of the (specific) classroom: positive feelings, pride, and regret (reversed). The second three items measured the importance of the classroom for the respondent (self image and feelings), and the influence of classmates (reversed). Although available, no covariate information on ethnicity is used.

In Table 33.1, an overview is given of the characteristics of the original data set. No special attention or treatment is given to missing data, which is most often present for the marks covariate. See Chapter 23 and Snijders and Bosker (2012, Chapter 9) for methods to deal with missing data in multilevel models.

From the larger original data set, several smaller data sets are extracted to serve as an illustration of the models presented. They all investigate, in some way or another, the effect of pupils' ability and motivation and the effect of the classroom context on the social networks of boys and girls. When the method calls for a continuous outcome variable, tie strength is used. For the models for binary social network data, the negative relation 'not like to be around' is chosen.

33.3 MULTILEVEL MODELING OF PERSONAL NETWORK DATA

The application of the basic multilevel model to personal network data is straightforward, with alters *nested* in egos, assuming that there is no overlap between the alters of different egos. The structure of the data is represented in Figure 33.1. The lines not only indicate the nesting structure, but also represent the actual outcome, the tie from ego to alter. Note that personal network data may be unbalanced, i.e. that egos do not mention the same number of alters. The multilevel model for personal network data was first proposed by Snijders et al. (1995) in an application to a sample of cocaine users and the importance of cocaine in their relationships with alters.[2] The multilevel

Table 33.1 Description of the Classroom Social Network Data

	Gender % girl	Tie Strength	Classroom Identification	Average Mark
Original Data Set				
n=1521				
Mean (SD)	0.55	2.28 (1.02)	3.18 (0.79)	6.70 (0.75)
# Missing	31	128	22	185
Random Personal Networks				
Ego n=74				
Mean (SD)	0.61	2.70 (1.07)	3.24 (0.80)	6.57 (0.84)
Alter n=297				
Mean (SD)	0.65	2.49 (1.48)	3.30 (0.79)	6.73 (0.73)
Random Dyads				
Ego/Alter n=922				
Mean (SD)	0.59	3.14 (1.52)	3.24 (0.79)	6.72 (0.74)
Complete Dyads				
Sender/Receiver n=1131				
Mean (SD)	0.56	2.70 (1.03)	3.20 (0.79)	6.71 (0.73)
Random Binary Personal Networks				
Ego n=74				
Mean (SD)	0.58	0.089 (0.13)	3.14 (0.79)	6.69 (0.83)
Alter n=1206				
Mean (SD)	0.57	0.077	3.39 (0.80)	6.69 (0.75)
Complete Binary Dyads				
Sender/Receiver n=1282				
Mean (SD)	0.53	0.080 (0.11)	3.17 (0.79)	6.69 (0.75)

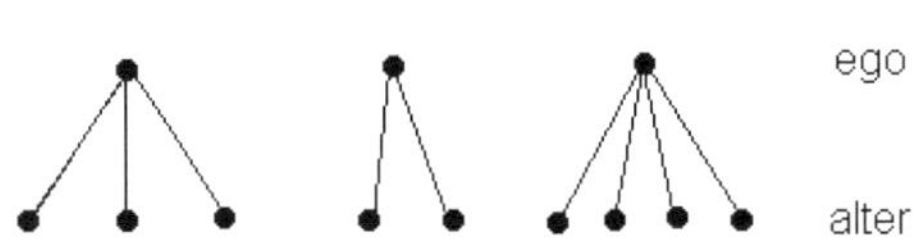

Figure 33.1 Schematic Representation of Personal or Egocentric Network Data

model has proved to be very useful in studying continuous or tie variables; see Roberts and Dunbar (2011) for a recent application.

The basic multilevel model is defined for outcome variable y_{ij}, here tie strength of the relation of respondent or ego j with classmate or alter i. The model contains one explanatory variable at the first level, x_{ij}, e.g. gender of alter, and one at the second level, z_j, e.g. gender of ego. The regression equation at the first level is then

$$y_{ij} = \beta_{0j} + x_{1ij}\beta_{1j} + \varepsilon_{ij}, \qquad (33.1)$$

where regression coefficients β_{0j} and β_{1j} vary over egos, referred to as the *random intercept* and the *random slope*, respectively. ε_{ij} is the level-one error term with mean 0 and variance σ^2.

Two regression equations are defined, one for the random intercept and the second for the random slope at level 2 (ego):

$$\beta_{0j} = \beta_0 + x_{2j}\beta_2 + u_{0j} \qquad (33.2)$$

and

$$\beta_{1j} = \beta_1 + x_{2j}\beta_3 + u_{1j}. \qquad (33.3)$$

The regression coefficients β_0, β_1, β_2, and β_3 are *fixed* effects. They do not vary across egos and can be viewed as the average effects over alters and egos. The random intercept and random slope are regressed on the level-two explanatory variable. The random error terms (or random effects) u_{0j} and u_{1j} represent the unexplained variation that remains among egos. This is analogous to the error ε_{ij}, which expresses unexplained variance of egos' ties with alters. It is assumed that the random effects follow a bivariate normal distribution with expected value 0 and a covariance matrix with elements $\text{var}(u_{0j}) = \sigma^2_{u0}$, $\text{var}(u_{1j}) = \sigma^2_{u_1}$, and $\text{cov}(u_{0j}, u_{1j}) = \sigma_{u_{01}}$.

The composite model is obtained by substituting equations (33.2) and (33.3) into (33.1):

$$y_{ij} = \beta_0 + x_{1ij}\beta_1 + x_{2j}\beta_2 + x_{1ij}x_{2j}\beta_3 + u_{0j} + x_{1ij}u_{1j} + \varepsilon_{ij}, \tag{33.4}$$

containing the fixed effects in the first part and the random effects in the second part. A fixed *cross-level interaction* term regression coefficient β_3 is obtained as the product of the level-one and level-two variables x_{1ij} and x_{2j}. A random 'interaction' term is given as $x_{1ij}u_{1j}$ with variance $x^2_{1ij}\sigma^2_{u_1}$, thus allowing heteroscedasticity.

The cross-level interaction terms are of special interest in the application of the multilevel model to personal network data, as they represent the *combination* of ego and alter, the dyadic *interaction*. Instead of a cross-level interaction term as defined by the product, other dyadic definitions are possible, reflecting dyadic proximity, measured for example by the absolute difference score, or a binary variable measuring similarity. Mollenhorst et al. (2008) analyze similarity as outcome in personal network data using the basic multilevel model with several ego and alter characteristics.

The model is estimated with (R)IGLS estimation, as implemented in standard random effects software (see Chapter 3 for details on likelihood estimation). The significance of the fixed effects is tested with t-tests, based on the ratio of parameter estimate to standard error. The significance of random effects is tested with a likelihood ratio or deviance test. The deviance is defined as minus twice the (IGLS) log-likelihood value and never increases when adding parameters to a model. The difference in deviance of two nested models can be used as a test statistic with an approximate χ^2 distribution, with the number of added parameters as the degrees of freedom under the null hypothesis of no model goodness-of-fit improvement. These tests are used as measures in the model selection process. In most cases, a forward selection process is applied in multilevel analysis (see Van Duijn et al., 1999, for an application to personal network data).

33.3.1 Application to the Classroom Social Network Data

From the classroom social network data, a personal network data set is obtained by random sampling one pupil per classroom and recording his/her nominations of classmates. The sampling procedure was set up such as to only include pupils with complete ego and alter covariate information. Thus, a random sample was obtained consisting of 74 pupils (one per school), the egos, with a total of 297 alters 4 (on average/per ego). Most egos indicate three alters, with the possible maximum of 12 alters reached by one pupil. Information on the composition of the data set, and a comparison with the complete data set, is given in Table 33.1. For the egos, the mean and standard deviation of their (individual) classroom identification scores and marks are reported, as well as of the tie strength averaged over their alters. For the alters, (unweighted) statistics are reported. The differences between the random personal networks data sets and the original data set are small, although the random ego data set contains a somewhat larger percentage of girls (61% egos and 65% alters) and seems to consist of pupils with slightly higher classroom identification scores.

For the classroom personal network data, the joint effects of gender, school marks, and classroom identification on tie strength were investigated, either as ego, alter characteristics, and/or as a dyadic similarity effect (*homophily* in social network jargon) or difference effect (*distance* or reversed similarity).

The analysis was carried out in three steps with the resulting models presented in Table 33.2. Starting point is a so-called empty model with no covariates and just the variances at alter (1) and ego (2) level (Model 0). The intercept is 2.6, which is the (weighted) mean tie strength. The total tie strength variance is separated into a between-alters within-ego (level-1) variance and a between-ego (level-2) variance, where the former is larger than the latter. The intraclass correlation, the percentage of variance residing at the ego level,[3] equals 0.21. To this model, gender effects were added, distinguishing the four possible dyadic gender combinations, with main ego and alter effects for girls and their interaction, with the boy–boy ego–alter combination as the reference category (not reported as a distinct model in the table).

As a second step, the effect of average marks and classroom identification were investigated. The best parameterization of the model turned out to be through dyadic effects, a difference score for marks and an absolute difference of ego–alter classroom identification scores. The accompanying alter and ego effects (using centered variables) were not significant, close to zero, and therefore left out of the model, shown as Model 1 in Table 33.2. Disregarding the gender effect for a moment, the negative effect of the marks difference score implies that ties tend to be stronger between egos and alters with higher marks (than ego), and weaker between egos and alters with lower marks. The negative effect of the classroom identification absolute difference score represents a distance effect: the higher the difference between ego and alter in classroom identification, the weaker their tie (as reported by ego), regardless of whether ego or alter finds the classroom social context most important.

This interpretation holds when comparing identical gender dyads. From the gender parameter estimates in Model 1 it becomes clear that the gender effects are strong. The (almost equal) negative ego and alter parameter estimates of gender indicate that boy–girl and girl–boy combinations have a lower tie strength than the ties in boy–boy combinations (with equal marks and classroom identification scores). The positive effect of the girl–girl (interaction) effect is larger than the sum of the two main effects and therefore indicates that the ties with a girl reported by a girl are the strongest, on average.

As a third step, in Model 2 a different parameterization through similarity effects is chosen, distinguishing boy–boy and girl–girl ego–alter combinations. In this model, the main ego and alter effects are left out, thus choosing not to distinguish between the two mixed-gender combinations (now the reference category). The girl–girl similarity parameter estimate is higher than the boy–boy counterpart, although a model with one gender-similarity parameter, with a deviance of 1025.4, is not significantly worse than Model 2 (χ^2 value is 3.1 on 1 df, $p = 0.08$).[4]

As a final step (not reported in the table), the effect of personal network composition was investigated by averaging the alter scores per ego. No significant parameter estimates were found for the ego (level 2) variables constructed in this way.

The model results underline the importance of the ('cross-level interaction') dyadic covariate information.[5] The negative dyadic difference in marks can be viewed as a hierarchical or aspiration effect where the tie strength is stronger in the upward than in the downward direction. On the other hand, the negative dyadic absolute difference in classroom identification scores indicates a homophily effect, where pupils are 'closer' the more equally they assess the importance

Table 33.2 Multilevel Analysis Results for the Random Personal Network Data Set

	Model 0	Model 1	Model 2
(REML)	Parameter Estimate (SE)	Parameter Estimate (SE)	Parameter Estimate (SE)
Fixed Effects			
Intercept	2.60 (0.12)	2.53 (0.21)	2.11 (0.23)
Ego (n=74)			
Girl			−0.61 (0.34)
Alter (n=297)			
Girl			−0.57 (0.34)
Ego–Alter			
Girl–Girl		1.60 (0.44)	1.02 (0.23)
Boy–Boy			0.59 (0.27)
Mark (Ego Alter Difference)		−0.29 (0.10)	−0.29 (0.10)
Classroom			
Identification (Absolute Difference)		−0.33 (0.14)	−0.33 (0.14)
Random Effects			
Ego variance	0.47 (0.68)	0.45 (0.67)	0.45 (0.67)
Alter variance	1.73 (1.32)	1.56 (1.25)	1.56 (1.25)
(ML) Deviance	1056.4	1022.2	1022.3
Difference in Deviance		34.2	0.1
(df; *p*)		(5; < 0.001)	(1; 0.75)

of the classroom as a social context. Note that the gender effects are stronger than the cognitive/ motivational and social context effects. To compensate the same-gender dyadic effect, a score difference of two to three points in average mark and classroom identification score is required, which is a large difference given the standard deviations of these variables. The overall explanatory power of the variables is modest, with an explained variance of approximately 10%.

A logical next step is to take the dyad as the unit of analysis, which is done in the actor–partner interdependence model, the topic of the next section.

33.4 MULTILEVEL MODELING OF DYADIC DATA

In the analysis of personal network data, the outcome variable y_{ij} is defined as tie strength reported by an ego about an alter. If each ego only reports about *one* alter, a simple regression model would suffice. The tie strength as reported by alter about ego (y_{ji}) is not an outcome variable.[6] To use both sources of information, a multilevel model for dyads as unit of analysis and the two (dyadic) directed tie variables is defined. Under the assumption that the dyads are independent, dyadic data can be viewed as nested data, with two ties nested within a dyad, or, alternatively, as two repeated measurements reflecting the dyadic design, where both actors fulfill two roles, as a sender and as a receiver.

The basic model as presented in equation (33.4) is applicable, although defining the first level is not as straightforward as before. The direction of the tie outcome variable is essential. Figure 33.2 gives a graphical presentation of independent dyadic data, where the two crossed lines indicate the two directed ties from the dyad, to and from the two actors forming the dyad, represented by the two sets of dots, as senders and receivers. The index i now does not indicate an actor but the direction of the tie variable (from and to), and j the dyad. In the model defined in this way, actor covariates can be used as sender and/or receiver characteristics. As in the previous model, a combination of the two actors'

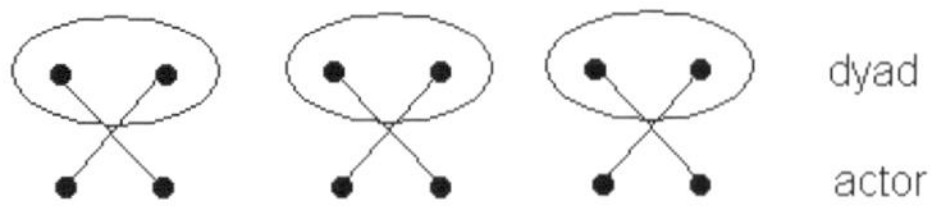

Figure 33.2 Schematic Representation of Independent Dyadic Network Data with Interdependent Actors

covariates is a dyadic variable. In the multilevel model for dyadic data, however, it is not a cross-level interaction term, but an explanatory variable defined at level 1 (tie) if its value depends on the direction of the tie (e.g. the difference) or at level 2 (dyad) if it has the same value for both dyadic observations (e.g. the absolute difference). Usually, no random covariate effects are specified.

The dyadic relations model is well known in social psychology when studying couples or other pairs of interaction partners (e.g. parent/child), as reviewed in the textbook on dyadic data analysis by Kenny et al. (2006). The model has been named the Actor–Partner Interdependence Model (APIM; see, e.g. Cook and Kenny, 2005) after being developed by David Kenny and his coworkers in the 1990s, and has been and still is widely applied, more recently also in organizational research (see Gooty and Yammarino, 2011, for an overview).

In the APIM tradition, the two actors in the dyad are not 'random' (or exchangeable) but have a fixed role (husband and wife, or leader and follower) and therefore often the model is defined with two intercepts, one for each directed tie. In the random effects model this is achieved by defining indicator variables distinguishing the two directed ties. In addition to the two intercepts, two variances can be specified at level 1, for both roles. The APIM can be viewed as the simplest case of the Social Relations Model treated in the next section. The first estimation approach was through (adjusted) analysis of variance methods. The random effects modeling approach is more flexible. Alternatively, the model can be estimated as a structural equation model (Kenny et al., 2006).

33.4.1 Application to the Classroom Social Network Data

From the classroom social network data, a dyadic data set is obtained by randomly sampling independent dyads from the classrooms. The sampling procedure started by randomly selecting one pupil and drawing one of the classmates the pupil had nominated and whom the pupil was nominated by. Thus, the sampling frame was restricted to pairs of students who had nominated each other, and, additionally, who had complete covariate data. After selecting the first dyad, a second pair of pupils was randomly selected in the same way, excluding the pupils already assigned to the first dyad, etc. This was repeated for all classrooms, resulting in a sample of 461 independent dyads in 74 classes. In Table 33.1, the characteristics of the resulting random dyads data set with 922 ties and actors are given. Most likely, the increase in mean tie strength of about 0.5 points is due to the restriction to mutual dyads, i.e. dyads where both pupils have nominated each other. Most dyads are same-gender (for boys 89% and for girls 92%).

In the random dyads data set, the effect of ability and motivation and of importance of classroom as a social context on tie strength is investigated, both within and between dyads, taking into account the gender differences. As with the personal network data, the analysis was carried out in three steps, first assessing the overall variance structure and the exchangeability of the actors. After adding gender effects, the next step was to decide on the best model parameterization to capture the mark and classroom identification effects. In a final step, the gender effects are reduced to a simpler structure if possible. Only the null model and the final model are reported in Table 33.3.

The empty model (Model 0 in Table 33.3), with intercept equal to the simple mean over the 922 tie observations, reflects the nested structure of the random dyads data with three levels: ties within dyads within classrooms.

Table 33.3 Multilevel Analysis Results for the Random Dyads Data Set (n=922)

	Model 0	Final Model
(REML)	Parameter Estimate (SE)	Parameter Estimate (SE)
Fixed Effects		
Intercept	3.14 (0.05)	2.74 (0.23)
Dyadic		
Girl–Girl		0.97 (0.22)
Boy–Boy		0.34 (0.23)
Mark (Absolute Difference)		–0.14 (0.11)
Classroom		
Identification (Absolute Difference)		–0.25 (0.11)
Random Effects		
Classroom variance	0.015 (0.05)	
Dyad variance	1.41 (0.14)	1.28 (1.13)
Tie variance	0.90 (0.06)	0.90 (0.95)
(ML) Deviance	3174.9	3133.3
Difference in Deviance		31.6
(df; p)		(4; $<$ 0.001)

The classroom (level-3) variance turns out to be negligible, however, and is ignored in the further analyses. The intraclass correlation, equal to 0.61, indicates the strong correlation between the two tie outcomes in the dyad, a phenomenon that can be interpreted as reciprocity. Put differently, the relatively large dyad (level-2) variance indicates that there are more differences between dyads than within dyads. Further, it was checked whether indeed the two actors in a dyad could be regarded as exchangeable. This was done by replacing the intercept by two dummy variables and estimating separate variances as well (not shown in the table). The model turned out not to be an improvement over the empty model (deviance equal to 3172.2, the difference of 2.7 not a significant difference on 2 degrees of freedom).

Second, the effects of marks and classroom identification were investigated by including sender and receiver effects, and several dyadic combinations, difference score (at level 1) and absolute difference, mean, maximum, and minimum score. The level-1 effects were all small and non-significant. The dyadic effect of mark was not strong and almost equal for absolute difference score and maximum score, with the absolute difference slightly stronger, although not significant. For the classroom identification score, the absolute difference was clearly the best parameterization, with a significant effect that is slightly smaller than in the personal network data analysis.

In the final model (see Table 33.3), the gender effect is presented through the two same-gender combinations. All-girl dyads report markedly higher tie strengths than all other dyadic combinations, including the boy–boy dyad. The tie variance (level 1) is not lower than in the empty model; the (level-2) dyad variance is slightly lower, with an explained variance of 7%. This is less than in the personal network data, due to the inability of the model to explain within-dyad variability.

The analysis confirms the previous findings of a strong homophily effect with respect to gender and, less strongly, of the importance of the classroom as a social context. The cognitive/motivational dyadic effect is a bit more difficult to interpret. Possibly due to the selection of mutual dyads, the 'hierarchical' difference in marks effect, as found in the personal network analysis, was not significant. Only a weak similarity in marks effect could be reported.

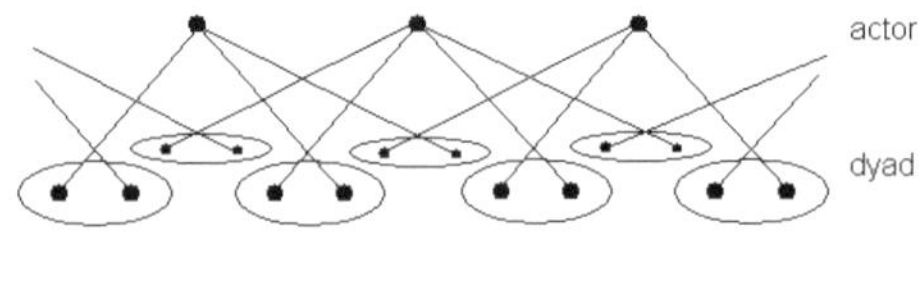

Figure 33.3 Schematic Representation of Complete Network Data Through Interdependent Dyads

33.5 MULTILEVEL MODELING OF COMPLETE SOCIAL NETWORK DATA

In the modeling of all available dyads, i.e. 'complete' social network data, the assumption of independent dyads is usually violated: actors report about relations with all other actors in the social setting. Complete social network data can be modeled by treating the dyads as cross-nested within the actors. The resulting dependence structure, depicted in Figure 33.3, acknowledges the multiple dyadic observations of actors and their double roles as sender and receiver, leading to a 'cross-level' structure. Every dyad is nested within two actors, where all–actors are involved in multiple dyads.

To model complete social network data, the random part of the independent dyads model (33.4), where i indicates the directed tie and j the dyad, is extended with two vectors of random effects $\boldsymbol{u_1}$ and $\boldsymbol{u_2}$, of length N. They are weighted, with indicator variables $\boldsymbol{s}$ and $\boldsymbol{r}$, to indicate the sending and receiving actors for all tie outcomes:

$$y_{ij} = \beta_0 + x_{1ij}\beta_1 + x_{2j}\beta_2 + \sum_{d=1}^{N}(s_{dij}u_{1d} + r_{dij}u_{2d}) + u_{0j} + \varepsilon_{ij}. \quad (33.5)$$

The vectors $\boldsymbol{u_1}$ and $\boldsymbol{u_2}$ can be interpreted as random sender and receiver effects, corresponding to (latent) individual tendencies to send and receive ties, known as outgoingness and popularity in social network analysis. Their inclusion in the model provides an adequate way to handle the cross-nesting of the two dyadic outcomes in the actors as well as the dependence between the two roles of an actor, by assuming equal (co)variances of the random sender and receiver effects of all actors,

$$\text{var}(u_{i1}) = \text{var}(u_{i2}) = \cdots = \text{var}(u_{iN}) = \sigma^2_{u_i}$$

for $i = 1, 2$, and

$$\text{cov}(u_{11}, u_{21}) = \text{cov}(u_{12}, u_{22}) = \cdots = \text{cov}(u_{1N}, u_{2N}) = \sigma_{u_{12}}, \quad (33.6)$$

whereas sender and receiver effects of different actors are assumed to be uncorrelated: $\text{cov}(u_{1j}, u_{2k}) = 0$ for $j \neq k$.

The variances of the sender and receiver random effects can be interpreted as the (residual or unexplained by covariates) individual variability in outgoingness and popularity. It is natural to assume that latent tendencies are correlated within individuals but uncorrelated across individuals. The dyadic error term u_{0j} is the common part of the two ties forming the dyad, and its variance is thus interpreted as a measure of reciprocity.

Equation (33.5) is the random effects parameterization of the Social Relations Model (SRM), allowing dyadic and actor covariates, proposed by Snijders and Kenny (1999). Kenny and La Voie (1984) presented the SRM as an ANOVA model, splitting up the variance in components belonging to senders, receivers, and their interaction (the dyad). The SRM has found many applications.[7]

As a regression model, the SRM handles unbalanced data, here defined as complete social network data with missing observations either at the dyadic level or at the directed tie level (i.e. not all relational ties between all actors in the social network are reported or recorded). Like the APIM, the SRM can deal with non-exchangeable actors, for instance in family networks (see Gerlsma et al., 1997, for an early example). For actors with different roles (such as parents and children), distinct variances and covariance components are specified. The SRM is also easily extended to a three-level random effects model, for the analysis of multiple networks. The SRM can be estimated with (R)IGLS using standard software for random effects models, with careful specification of the random sender and receiver parameters using indicator variables and equality constraints for their (co)variances.

33.5.1 Application to the Classroom Social Network Data

The Social Relations Model is applied to the classroom social network data, selecting dyads of pupils with complete covariate data containing at least one reported tie. Thus, 5136 tie observations are obtained, from 1239 pupils in 74 classrooms, varying from 3 to 28 pupils per classroom. Note that these data encompass the personal network data and the dyads data. Due to the design, the data are not balanced (i.e. it is not a round-robin design where all students have reported their tie strengths with all other students in the classroom).[8] Moreover, a small portion of the students are only mentioned by others (35 exclusive receivers), or never mentioned by others even though they nominate classmates (73 exclusive senders).

The characteristics of the 1131 students who both act as senders and receivers of ties are given in Table 33.1. The difference in characteristics of the exclusive sender and receivers with the larger group are negligible. Except for the number of students eligible for the analysis, there is little variation in (average) classroom characteristics (not reported in the table). The model is estimated in MLwiN (Rasbash et al., 2010), using the macros developed by Tom Snijders.[9]

The tie strength reported in Table 33.1 is the unweighted average over the mean tie strength reported. The descriptives resemble the original data set rather closely, not surprising given the large overlap between the two data sets.

The analysis with the SRM makes it possible to investigate all available data, modeling the dependence between ties going to and coming from the same actors, within and between dyads. With multiple classrooms, the between-classroom variability is also modeled. Again, the analysis is aimed at investigating the influence of ability and motivation as well as of importance attached to the classroom as a social context on tie strength. The analysis proceeded in four general steps, starting with an empty model, adding gender effects, selecting the covariate effects, including classroom-level covariates (constructed through aggregation over pupils within classrooms), and finally leaving out insignificant effects to arrive at the final model. Only the empty model and the final model are reported in Table 33.4.

The average mean strength (Table 33.1) is slightly higher than the intercept in the Social Relations Model analysis using no covariate information, reported as Model 0 in Table 33.4. This was also observed in the random personal network data due to the weighting. The random effects in Model 0 provide some more information on the variance components. As in the random dyads analysis, the classroom variability is small. The sender variance is much higher than the receiver variance, indicating more individual variability in (self-reported) outgoingness than in popularity. The correlation between sender and receiver effects is modest with a value of 0.18, whereas the reciprocity between ties within a dyad is substantial, 0.64. The residual, 'unexplained', tie variance is 1.78 in the null model and in the final model only slightly reduced.

Adding the gender effects improved the model considerably; adding average marks and classroom identification as sender and receiver effects only indicated a significant modest sender effect of classroom identification (not reported in the table). The best covariate parameterization was through the absolute difference of classroom identification and of average mark. In view of this small homophily effect of cognitive ability, the hierarchy effect of cognitive ability earlier found in the analysis of the personal network data set is outweighed by the homophily effect in

Table 33.4 SRM Analysis Results for the Unbalanced Complete Data Set (n=5136)

	Model 0	Final Model
(REML)	Parameter Estimate (SE)	Parameter Estimate (SE)
Fixed Effects		
Intercept	2.46 (0.04)	2.00 (0.07)
Dyadic		
Girl–Girl		1.10 (0.06)
Boy–Boy		0.62 (0.07)
Mark (Absolute Difference)		–0.09 (0.04)
Classroom		
Identification (Absolute Difference)		–0.29 (0.04)
Random Effects		
Classroom variance	0.039 (0.02)	0.041(0.02)
Sender variance	0.28 (0.03)	0.29(0.03)
Receiver variance	0.063 (0.02)	0.019(0.02)
Sender–Receiver Covariance	0.024 (0.02)	0.053(0.02)
Residual tie variance	1.78 (0.05)	1.62(0.04)
Within-dyad covariance	1.14 (0.05)	0.97(0.05)
(ML) Deviance	17607.9	17249.5
Difference in Deviation		358.4
(df; p)		(5; < 0.001)

the complete network data, where 66% of the dyads are mutual. Notably, the effects of gender and classroom identification are as strong or stronger than in the personal network data analysis.

33.6 MULTILEVEL MODELING OF BINARY SOCIAL NETWORK DATA

The models presented thus far all assume normally distributed errors and are therefore not suited for (too) skewed data and binary network data. In social network analysis, binary data (i.e. recording whether or not two actors are tied) are common, however. The basic (composite) model (33.4) can be easily reformulated as a generalized linear mixed model (GLMM; see also Chapters 8 and 15), a multilevel logistic, or Poisson regression model. Recent applications, also using multilevel (polytomous) logistic models for binary tie variables include Völker and Flap (2007), Carrasco et al. (2008), and De Miguel Luken and Tranmer (2010). Vermunt and Kalmijn (2006) proposed a random effects conditional logit latent class model for personal network data.

For the analysis of dyadic data, this GLMM formulation does not work, due to the tie-level (1) variance components that cannot be freely modeled in logistic regression. With binary data, the dyad has four possible outcomes. A polytomous logistic regression model would seem suitable, but needs to reflect the reciprocity structure. The p_1 model (Holland and Leinhardt, 1981), the first probability model for binary dyadic network data, does precisely that,[10] using (fixed) sender and receiver parameters and an explicit reciprocity parameter. Similar to the proposals of Wong (1987) and Gill and Swartz (2004) for Bayesian formulations of the p_1 model, the p_2 model (Van Duijn et al., 2004) has correlated random sender and receiver parameters, for which a normal distribution is assumed. As it features actor and dyadic covariates, also for the reciprocity parameter, the p_2 model can be viewed as the binary counterpart of the SRM (and of the APIM).

The combination of a (polytomous) logistic regression model with normally distributed random effects implies that no closed-form expression of the likelihood can be obtained. MCMC estimation is required to obtain unbiased parameter estimates (Zijlstra et al., 2009). Extensions to multiple (independent) networks and multiple relations are available (Zijlstra et al., 2006; Zijlstra, 2008); for an application, see Vermeij et al. (2009). Recently, Spain et al. (2012) proposed a random effects version of the APIM for binary data, and investigated the estimation properties of standard HLM software. Thomas (2009, Chapter 2) provided a general overview of random effects models for (binary) social network data.

33.6.1 Application to the Binary Classroom Social Network Data

Two analyses are presented for the binary negative tie variable 'not liking to be around' (from now on called 'dislike'). The first is of personal network data, with a random effects logistic regression model (using R), the second is of the 'complete' networks, using a multilevel p_2 model using the software implemented in StOCNET (Boer et al., 2006). In both analyses the effect of gender, of ability and motivation, and of the importance of the classroom as a social context is investigated. It is expected that the results for the negative social relation are different from the positive tie variable used in the analyses thus far.

To construct the personal network data, in each class a pupil (ego) is randomly drawn from the pupils with complete covariate data. The personal network of this pupil consists of the ties with all other pupils (alters with complete covariate data) in the class, with value 1 if ego indicated alter as somebody s/he did not like to be around and 0 if alter was not indicated by ego, thus extending the personal network to all pupils in the classrooms (with complete covariate data). Therefore, the resulting personal networks are larger than for the ties based on the five relations, varying from 3 to 28 pupils, with a total of 1206 alters (see Table 33.1). Dislike is rare, occurring in 8% of the ties, with a slightly higher mean ego dislike percentage. The classroom identification scores and average marks closely resemble the original data set.

The personal network data are analyzed with a random effects logistic regression model, first estimating an empty model and next investigating the ego and alter effects of the covariates and their dyadic effects, either as homophily or distance effects. As a third analysis, a different approach is used to characterize the differences between ego and alter. The percentage of girls, the mean average mark, and mean classroom identification scores are used as ego-level (2) explanatory variables to investigate the effect of 'classroom environment' (from the perspective of ego). Before averaging, the variables are grand-mean centered, except for ego.

The results of the three analyses are presented in Table 33.5. The empty model (Model 0) shows a negative intercept, indicating that the probability of a negative tie is far below 50%. As the alter variance (level 1) is determined by the model, only the ego variance is estimated, showing that there is a sizable difference in relative dislike between egos. Model 1 contains the results of the analysis with ego, alter, and directed dyadic effects. Although similar in parameterization, the parameter estimates are opposite in sign compared to the corresponding analysis of the positive social relation in Table 33.2. Thus, girls tend to dislike boys, and vice versa. Moreover, both sexes tend to indicate more often dislike for fellow pupils who are dissimilar in average marks and classroom identification. As in the positive personal network analysis, the effect of gender is strongest, with the parameter estimate of importance of classroom as a social context slightly larger than the estimated ability effect. Note, however, that the parameter estimates are now to be interpreted on a log-odds scale.

Table 33.5 Multilevel Logistic Regression Results for the Random Personal Binary Network Data Set

(Laplace approx.)	Model 0 Parameter Estimate (SE)	Model 1 Parameter Estimate (SE)	Model 2 Parameter Estimate (SE)
Fixed Effects			
Intercept	−2.95 (0.19)	−3.37 (0.34)	−2.75 (0.31)
Ego (n=74)			
Girl			−0.38 (0.34)
Classroom Identification (Cent.)			0.86 (0.31)
Girl*Cl. Identification (Cent.)			−1.01 (0.42)
Mean Classroom Ident. of Alters (Centered)			−1.77 (0.58)
Alter (n=1206)			
Classroom Identification (Cent.)			−0.42 (0.16)
Ego–Alter			
Same Gender		−0.52 (0.24)	−0.50 (0.24)
Mark (Absolute Difference)		0.33 (0.18)	
Classroom Identification (Absolute Difference)		0.46 (0.19)	
Random Effects			
Ego variance	1.35 (1.16)	1.37 (1.17)	0.68 (0.82)
(Laplace) Deviance	617.7	603.4	579.9
Difference in Deviance (w.r.t. Model 0)		14.3	37.8
(df; p)		(3; 0.002)	(6; $<$ 0.001)

Model 2 in Table 33.5 presents the results of the analysis using (aggregated) 'classroom environment' variables. These are quite different from the previous analysis, except for the negative homophily effect of gender. The distance effect of average mark was not significant. The effect of classroom identification is present in various forms. The largest estimate is for the mean (over alters) classroom identification, the social environment effect. It is negative, -1.77, indicating that for egos in classrooms with a higher mean importance of the classroom as a social context the incidence of dislike ties is lower. This parameter is larger (in absolute value) than the alter classroom identification score parameter estimate, -0.42 (cf. the contextual effect or within/between regression effect mentioned in the multilevel analysis literature; see, e.g. Snijders and Bosker, 2012, Section 4.6). The effect of ego classroom identification, however, was found to be different for boys and girls. The positive main effect of ego classroom identification implies that boys tend to dislike more fellow-pupils the larger their classroom identification, although this individual ego effect is smaller than the social context effect. The negative interaction between classroom identification and gender indicates that this positive ego effect does not hold for girls. It is clear that classroom identification is quite important in the explanation of dislike, both at the ego and alter level and as a classroom environment effect.

For the analysis with the multilevel p_2 model, the complete social network data are used for the 74 classrooms, again including only pupils with complete covariate data. A total of 29,226 observed tie variables (in 14,613 dyads), from 1282 pupils, with 4 through 29 fellow classmates, are available for the analysis, of which many contain no dislike ties. The characteristics of the 1282 pupils closely resemble those of the original data set, with somewhat more girls dropping from the analysis. The incidence of dislike is low (on average 8% per pupil), but rather variable, partly caused by the large differences in classroom size.

As a continuation of the analysis of the dislike personal network data, the focus of the analysis with the multilevel p_2 model is on the social environment effects of classroom identification and gender, discarding the effect of average mark. As a starting point, an empty model is estimated to assess the variability between classrooms, and, within classrooms, between actors in their roles of senders and receivers. Model 0 in Table 33.6 represents the first analysis, showing a negative intercept (density parameter) and a positive reciprocity, indicative of the relative rareness and symmetry of dislike relations between the pupils, respectively. The sender and receiver variances are rather large with a small negative correlation (-0.08), whereas the classroom variance is smaller but still sizable.

Controlling for gender homophily, and sender and receiver classroom identification, the mean classroom identification did not have a significant effect, and is not included in Model 1 in Table 33.6. The percentage of girls in each class, however, accounted for some of the variability between classes: the more girls, the lower the probability of dislike ties. The negative parameter estimates of classroom identification for senders and receivers show that the larger the classroom identification of a pupil, the less likely his/her involvement in a dislike relation, either as sender or receiver. These results correspond with the previous analysis of the personal network data, where it was found that girls have a lower tendency to indicate dislike ties with their classmates. This is in line with the finding in the complete data that there is less dislike in classes with more girls. The variances are slightly reduced, the sender–receiver correlation slightly increased (correlation equal to -0.11). Note that no deviance or other goodness-of-fit measures are reported due to the approximate nature of the MCMC estimation.

33.7 OTHER MODELS FOR COMPLETE SOCIAL NETWORK DATA

An important assumption in all dyadic random effects models for complete network data is that, given the fixed regression part of the model, the dependence of the network data is adequately modeled through the dyadic dependence structure, the random sender and receiver effects. In the class of statistical models for the analysis of (cross-sectional) complete network data, some models deal with more complex dependence structures. For instance, transitivity is a common phenomenon of grouping of relational ties in social networks, where actors tend to have ties with actors having ties with each other. To model transitivity involves (at least) three actors and requires a triadic dependence structure.

The exponential random(-family) graph model (ERGM; Snijders et al., 2006; Robins et al., 2007; see also Robins, 2011, for an overview) builds on the p_1 model (Holland and Leinhardt, 1981) and handles the dependence structure within the network by incorporating appropriate network configurations as statistics in the model. In an ERGM, the outcome variable is the complete $N \times N$ network, whose probability is modeled relative to all possible networks for this set of actors and their covariates, and its parameters are the weights of the network configurations representing the network structure and covariate effects. The ERGM therefore does not need random effects to model dependence between the observations. The ERGM is estimated with MCMC maximum likelihood estimation. Earlier extensions of the p_1 model, known as the Markov random graph model (Frank and Strauss, 1986), assuming independence between ties not having a common actor, and the p^* model (Wasserman and Pattison, 1996), developed with the same goal of having a simpler, approximate, likelihood estimation, proved to be inadequate in empirical applications.

Table 33.6 Multilevel p_2 Analysis Results for the Binary Complete Data Set (n=29226)

(MCMC)	Model 0 Parameter Estimate (SE)	Model 1 Parameter Estimate (SE)
Fixed Effects		
Intercept	−4.13 (0.11)	−3.54 (0.27)
Reciprocity	1.12 (0.13)	1.09 (0.12)
Classroom		
Percentage Girls		−1.15 (0.48)
Sender		
Classroom Identification (Cent.)		−0.47 (0.09)
Receiver		
Classroom Identification (Cent.)		−0.27 (0.06)
Dyadic		
Same Gender		0.13 (0.05)
Random Effects		
Classroom variance	0.28 (0.10)	0.29 (0.12)
Sender variance	2.86 (0.21)	2.70 (0.21)
Receiver variance	2.15 (0.16)	2.09 (0.16)
Sender–Receiver Covariance	−0.21 (0.12)	−0.26 (0.12)

Another extension of the p_1 model is the stochastic block model, first proposed by Wang and Wong (1987), distinguishing groups of actors based on an a priori grouping of the actors, for instance by an observed covariate. This approach was extended to a latent grouping for an undirected (or symmetrized) complete network by Frank (1995, 1996), and for directed complete network data by Nowicki and Snijders (2001). Both models are aimed at identifying an a priori unknown number of groups based on the assumption that actors in the same group interact more with each other and have a similar pattern of interaction with other groups. The mixed membership stochastic block model proposed by Airoldi et al. (2008) relaxes the assumption that every actor belongs to just one (latent) group.

In the latent space models for cross-sectional complete social network data (Hoff et al., 2002; Handcock et al., 2007), the actors are positioned in a latent space with an a priori unknown dimensionality and an unknown number of clusters to account for the higher-order dependence between the dyads, through the distance between the positions.[11] The outcome variable in these models is a single, directed or undirected, binary or count tie variable. Hoff (2005) proposed a related generalized bilinear regression model (having the social relations model as a limiting case). Krivitsky et al. (2009) extended the latent space cluster models with individual random sender and receiver effects, thus also accommodating individual differences in popularity and outgoingness in a GLMM formulation.[12]

Specialized software is available, mostly in R; see Handcock et al. (2008), and Krivitsky and Handcock (2008). ERGMs can also be estimated with PNET (Wang, Robins, and Pattison, 2009). For a general overview of software for social network analysis see Huisman and Van Duijn (2011), who show that the software is rapidly developing. Two important reasons are the availability of new models and methods, and the necessity to analyze very large data sets, such as ones 'harvested' from social media.

33.8 CONCLUDING REMARKS

The analyses, although primarily aimed at illustration, were able to reveal interesting differences between the various multilevel models for social network analysis presented here. To the same basic research question, the models provided different answers, showing the influence of the sampling design, the (resulting) network information available, the measurement of the tie variable, and the parameterization of the effects of interest as individual (actor) and/or dyadic (interaction) effects.

Disregarding any bias due to the sampled data, some tentative substantive conclusions are drawn from the analyses. Over and above the always prominent effect of gender in social relations, the importance of the classroom as a social context as measured by individual classroom identification, turned up as a determinant of positive and negative social relations. Having a similar classroom identification (regardless of its value) tends to bind pupils in a positive relationship, sharing time, opinions, resources. This effect of classroom identification was slightly different for the negative relation 'dislike', where higher scores predict less dislike in all combinations, at least for girls. For boys, the effect depends on the social environment provided by the classroom, in identification and gender composition.

From the many other statistical models available for social network analysis, it is clear that the multilevel models constitute just a small part, with a focus on dyadic outcomes. The dyad forms the basis of social network analysis, and is important for understanding the (complex) structure of the network data. This underlines the usefulness of multilevel models for social network analysis, if only as a starting point. To date, no extensive studies have been carried out to compare the different models in their performance, for instance in estimating covariate effects. It is to be expected, however, that severe misrepresentation of structural effects harms the estimation of the other effects. Thus, multilevel models are not to be preferred for the analysis of social network data aimed at unraveling their structural properties.

ACKNOWLEDGMENTS

The author would like to thank Lotte Vermeij for her permission to use the data, the editors and an anonymous reviewer for their helpful comments on the first draft, and Mark Huisman for his comments on the revision.

NOTES

1 The exact formulation of the five items was (Vermeij, 2006, pp. 110–11):

- Who helps you with practical matters, like homework or finding a job?
- Who helps you when you feel down, for example because of problems at home or because your relationship ended?
- With whom do you talk about what's right and what's wrong, for example about religion and justice?
- With whom do you talk about what you like, for example in music or art?
- With whom do you spend time outside of school?

2 Because no sample frame exists for the population of cocaine users, the data were obtained in a so-called snowball sample, where alters nominated by egos were invited to name their alters (see, e.g., Frank, 2011). Thus, the assumption of independence between egos does not hold exactly. In Van Duijn et al.'s (1999) application of the basic multilevel model to a study on change in personal networks after three types of life events, the sample design permitted the assumption of independence.

3 Or equivalently, the correlation between two randomly drawn alters belonging to the same ego.

4 Note that the models in Table 33.2 seem not to be nested, but they are. Not only for the binary covariate, defining four combinations of ego and alter; the continuous covariates defined as (absolute) difference also imply a direct relation with the ego and alter effects. Next to these ego–alter effects, only one (main) effect can be estimated, either for ego or for alter. Thus, for this type of model, the iron rule that a model should not contain interaction effects without the accompanying main effects seems to be violated, but is in fact just modified by acknowledging that the unit of analysis is a dyadic combination.

5 Estimating the models specifying a Poisson distribution for the tie strength measured as the sum of its components, to check the sensitivity of the

results to the normality assumption, did not lead to substantive differences in results.

6 Although it could be used as a covariate.

7 See, for a long list, `http://www.davidakenny.net/doc/srmbiblio.pdf`.

8 Balanced data could be obtained by filling in zeroes (no tie strength) for the 'empty' dyadic observations. As this makes the tie strength distribution quite skewed, the unbalanced data design is preferred for illustrating the Social Relations Model. The largest changes occur in the estimate of the intercept (much smaller) and the sender variance (smaller and more equal to the receiver variance, due to the almost 20,000 zero tie observations added). Surprisingly, the covariate effects are much less affected, leading to the same substantive conclusions.

9 To be found at `http://www.stats.ox.ac.uk/ snijders/`. R features a package for SRM analysis, TripleR, which does not accommodate covariates.

10 Implicitly, as it is presented and estimated as a log-linear model.

11 Note that the latent stochastic block models can be viewed as latent space models in a one-dimensional discrete space.

12 A similar extension can be found in Schweinberger (2007, Chapter 4) for the analysis of longitudinal social network data with so-called stochastic actor-oriented models (Snijders, 2001).

REFERENCES

Airoldi, E.M., Blei, D.M., Fienberg, S.E., and Xing, E.P. (2008) 'Mixed Membership Stochastic Blockmodels', *Journal of Machine Learning Research*, 9: 1981–2014.

Boer, P., Huisman, M., Snijders, T.A.B., Steglich, C.E.G., Wichers, L.H.Y., and Zeggelink, E.P.H. (2006) *StOCNET: An Open Software System for the Advanced Statistical Analysis of Social Networks*. Version 1.8. Groningen: ICS, University of Groningen/SciencePlus.

Butts, C.T. (2008) 'A Relational Event Framework for Social Action', *Sociological Methodology*, 38: 155–200.

Carrasco, J.A., Hogan, B., Wellman, B., and Miller, E.J. (2008) 'Agency in Social Activity Interactions: The Role of Social Networks in Time and Space', *Tijdschrift voor Economische en Sociale Geografie*, 99: 562–83.

Cook, W.L. and Kenny, D.A. (2005) 'The Actor-Partner Interdependence Model: A Model of Bidirectional Effects in Developmental Studies', *International Journal of Behavioral Development*, 29: 101–9.

De Miguel Luken, V. and Tranmer, M. (2010) 'Personal Support Networks of Immigrants to Spain: A Multilevel Analysis', *Social Networks*, 32: 253–62.

Frank, K.A. (1995) 'Identifying Cohesive Subgroups', *Social Networks*, 17: 27–56.

Frank, K.A. (1996) 'Mapping Interactions Within and Between Cohesive Subgroups', *Social Networks*, 18: 93–119.

Frank, O. (2011) 'Survey Sampling in Networks', in J. Scott and P.J. Carrington (eds), *The SAGE Handbook of Social Network Analysis*. London: Sage, 389–403.

Frank, O. and Strauss, D. (1986) 'Markov Graphs', *Journal of the American Statistical Association*, 81: 832–42.

Gerlsma, C., Snijders, T.A.B., van Duijn, M.A.J., and Emmelkamp, P.M.G. (1997) 'Parenting and Psychopathology: Differences in Family Members Perceptions of Parental Rearing Styles', *Personality and Individual Differences*, 23: 271–82.

Gill, P.S. and Swartz, T.B. (2004) 'Bayesian Analysis of Directed Graphs Data With Applications to Social Networks', *Journal of the Royal Statistical Society, Series C (Applied Statistics)*, 53: 249–60.

Goldenberg, A., Zheng, A.X., Fienberg, S.E., and Airoldi, E.M. (2009) 'A Survey of Statistical Network Models', *Foundations and Trends in Machine Learning*, 2: 129–233.

Gooty, J. and Yammarino, F.J. (2011) 'Dyads in Organizational Research: Conceptual Issues and Multilevel Analyses', *Organizational Research Methods*, 14: 456–83.

Handcock, M.S. and Gile, K. (2007) *Modeling Social Networks With Sampled or Missing Data*, Working Paper 75, Center for Statistics and the Social Sciences, University of Washington.

Handcock, M.S. and Gile, K.J. (2010) 'Modeling Social Networks From Sampled Data', *The Annals of Applied Statistics*, 4: 5–25.

Handcock, M.S., Hunter, D.R., Butts, C.T., Goodreau, S.M., and Morris, M. (2008) 'statnet: Software Tools for the Representation, Visualization, Analysis and Simulation of Network Data', *Journal of Statistical Software*, 24(1).

Handcock, M.S., Raftery, A.E., and Tantrum, J.M. (2007) 'Model-Based Clustering for Social Networks (with discussion)', *Journal of the Royal Statistical Society, Series A*, 170: 301–54.

Hoff, P.D. (2005) 'Bilinear Mixed-Effects Models for Dyadic Data', *Journal of the American Statistical Association*, 100: 286–95.

Hoff, P.D., Raftery, A.E., and Handcock, M.S. (2002) 'Latent Space Approaches to Social Network Analysis', *Journal of the American Statistical Association*, 97: 1090–8.

Holland, P.W. and Leinhardt, S. (1981) 'An Exponential Family of Probability Distributions for Directed

Graphs', *Journal of the American Statistical Association*, 76: 33–50.

Huisman, M. (2009) 'Imputation of Missing Network Data: Some Simple Procedures', *Journal of Social Structure*, 10(1).

Huisman, M. and van Duijn, M.A.J. (2011) 'A Reader's Guide to SNA Software', in J. Scott and P.J. Carrington (eds), *The SAGE Handbook of Social Network Analysis*. London: Sage, 578–600.

Kenny, D.A., Kashy, D.A., and Cook, W.L. (2006) *Dyadic Data Analysis*. New York: Guilford Press.

Kenny, D.A. and La Voie, L. (1984) 'The Social Relations Model', in L. Berkowitz (ed.), *Advances in Experimental Social Psychology Vol. 18*. New York: Academic Press, 141–82.

Koskinen, J.H., Robins, G.L. and Pattison, P.E. (2010) 'Analysing Exponential Random Graph (p-star) Models with Missing Data Using Bayesian Data Augmentation', *Statistical Methodology*, 7: 366–84.

Koskinen, J. and Stenberg, S. (2012) 'Bayesian Analysis of Multilevel Probit Models for Data With Friendship Dependencies', *Journal of Educational and Behavioral Statistics*, 37: 203–30.

Krivitsky, P.N. and Handcock, M.S. (2008) 'Fitting Latent Cluster Models for Networks With latentnet', *Journal of Statistical Software* 24(5).

Krivitsky, P.N., Handcock, M.S., Raftery, A.E., and Hoff, P.D. (2009) 'Representing Degree Distributions, Clustering, and Homophily in Social Networks With Latent Cluster Random Effects Model', *Social Networks*, 31: 204–13.

Marsden, P.V. (2011) 'Survey Methods for Network Data', in J. Scott and P.J. Carrington (eds), *The SAGE Handbook of Social Network Analysis*, London: Sage, 370–88.

Mollenhorst, G.W., Völker, B.G.M., and Flap, H.D. (2008) 'Social Contexts and Personal Relationships: The Effect of Meeting Opportunities on Similarity for Personal Relationships of Different Strength', *Social Networks*, 30: 60–8.

Nowicki, K. and Snijders, T.A.B. (2001) 'Estimation and Prediction for Stochastic Blockstructures', *Journal of the American Statistical Association*, 96: 1077–87.

O'Malley, A.J. and Marsden, P.V. (2008) 'The Analysis of Social Networks', *Health Services and Outcomes Research Methodology*, 8: 222–69.

Rasbash, J., Browne, W.J., Healy, M., Cameron, B. and Charlton, C. (2010) *MLwiN Version 2.22*. Bristol: Centre for Multilevel Modeling, University of Bristol.

Roberts, S.G.B. and Dunbar, R.I.M. (2011) 'Communication in Social Networks: Effects of Kinship, Network Size, and Emotional Closeness', *Personal Relationships*, 18: 439–52.

Robins, G. (2011) 'Exponential Random Graph Models for Social Networks', in J. Scott and P.J. Carrington (eds), *The SAGE Handbook of Social Network Analysis*. London: Sage, 484–500.

Robins, G.L., Snijders, T.A.B., Wang, P., Handcock, M.S., and Pattison, P. (2007) 'Recent Developments in Exponential Random Graph (p^*) Models for Social Networks', *Social Networks*, 29: 192–215.

Schweinberger, M. (2007) *Statistical Methods for Studying the Evolution of Networks and Behavior*. Unpublished PhD thesis, University of Groningen. http://dissertations.ub.rug.nl/faculties/gmw/2007/m.schweinberger

Scott, J. (2000) *Social Network Analysis: A Handbook*. London: Sage.

Scott, J. and Carrington, P.J. (eds) (2011) *The SAGE Handbook of Social Network Analysis*. London: Sage.

Snijders, T.A.B. (2001) 'The Statistical Evaluation of Social Network Dynamics', in M.E. Sobel and M.P. Becker (eds), *Sociological Methodology*, 31: 361–95.

Snijders, T.A.B. (2011) 'Statistical Models for Social Networks', *Annual Review of Sociology*, 37: 129–51.

Snijders, T.A.B. and Bosker, R.J. (2012) *Multilevel Analysis. An Introduction to Basic and Advanced Multilevel Modeling*, 2nd edition. London: Sage.

Snijders, T.A.B. and Kenny, D.A. (1999) 'The Social Relations Model for Family Data: A Multilevel Approach', *Personal Relationships*, 6: 471–86.

Snijders, T.A.B., Pattison, P.E., Robins, G.L., and Handcock, M.S. (2006) 'New Specifications for Exponential Random Graph Models', *Sociological Methodology*, 36: 99–153.

Snijders, T.A.B., Spreen, M., and Zwaagstra, R. (1995) 'The Use of Multilevel Modeling for Analysing Personal Networks: Networks of Cocaine Users in an Urban Area', *Journal of Quantitative Anthropology*, 5: 85–105.

Spain, S.M., Jackson, J.J., and Edmonds, G.W. (2012) 'Extending the Actor-Partner Interdependence Model for Binary Outcomes: A Multilevel Logistic Approach', *Personal Relationships*, 19: 431–44.

Stadtfeld, C. and Geyer-Schulz, A. (2011) 'Analyzing Event Stream Dynamics in Two-Mode Networks: An Exploratory Analysis of Private Communication in a Question and Answer Community', *Social Networks*, 33: 258–72.

Steglich, C., Snijders, T.A.B., and Pearson, M. (2010) 'Dynamic Networks and Behavior: Separating Selection from Influence', *Sociological Methodology*, 40: 329–93.

Thomas, A.C. (2009) *Hierarchical Models for Relational Data*. PhD thesis, Harvard University. http://gradworks.umi.com/33/85/3385417.html

Van Duijn, M.A.J., van Busschbach, J.T., and Snijders, T.A.B. (1999) 'Multilevel Analysis of Personal

Networks as Dependent Variables', *Social Networks* 21: 187–209.

Van Duijn, M.A.J. and Huisman, M. (2011) 'Statistical Models for Ties and Actors', in J. Scott and P.J. Carrington (eds), *The SAGE Handbook of Social Network Analysis*. London: Sage, 459–83.

Van Duijn, M.A.J., Snijders, T.A.B., and Zijlstra, B.J.H. (2004) 'p_2: A Random Effects Model with Covariates for Directed Graphs', *Statistica Neerlandica* 58: 234–54.

Vermeij, L. (2006) *What's Cooking? Cultural Boundaries Among Dutch Teenagers of Different Ethnic Origins in the Context of School*. Unpublished PhD thesis, ICS/University of Utrecht.

Vermeij, L., Van Duijn, M.A.J., and Baerveldt, C. (2009) 'Ethnic Segregation in Context: Social Discrimination Among Native Dutch Pupils and Their Ethnic Minority Classmates', *Social Networks*, 31: 230–9.

Vermunt, J.K. and Kalmijn, M. (2006) 'Random Effects Models for Personal Networks', *Methodology*, 2: 34–41.

Völker, B.G.M. and Flap, H.D. (2007) '16 Million Neighbors. A Multilevel Study of the Role of Neighbors in the Personal Networks of the Dutch', *Urban Affairs Review*, 43: 256–84.

Wang, P., Robins, G., and Pattison, P. (2009) *PNet: Program for the Simulation and Estimation of Exponential Random Graph (p^*) Models. User Manual*. Melbourne: University of Melbourne.

Wang, Y.J. and Wong, G.Y. (1987) 'Stochastic Blockmodels for Directed Graphs', *Journal of the American Statistical Association*, 82: 8–19.

Wasserman, S. and Faust, K. (1994) *Social Network Analysis. Methods and Applications*. Cambridge: Cambridge University Press.

Wasserman, S. and Pattison, P. (1996) 'Logit Models and Logistic Regressions for Social Networks: I. An Introduction to Markov Graphs and p*', *Psychometrika*, 61: 401–25.

Wong, G.Y. (1987) 'Bayesian Models for Directed Graphs', *Journal of the American Statistical Association*, 82: 140–8.

Zijlstra, B.J.H. (2008) *Random Effects Models for Directed Graphs with Covariates*. Unpublished PhD thesis, University of Groningen, the Netherlands.

Zijlstra, B.J.H., van Duijn, M.A.J., and Snijders, T.A.B. (2006) 'The Multilevel p_2 Model: A Random Effects Model for the Analysis of Multiple Social Networks', *Methodology*, 2: 42–7.

Zijlstra, B.J.H., van Duijn, M.A.J., and Snijders, T.A.B. (2009) 'MCMC Estimation of the p_2 Network Regression Model with Crossed Random Effects', *British Journal of Mathematical and Statistical Psychology*, 62: 143–66.

Name Index

Index of names: the index is in word-by-word alphabetical order. Names with prefixes, e.g. De Boor, C., Van Dick, R., have been listed under the first letter of the prefix. Bibliographical information can be found in the References at the end of each chapter.

Subject Index

Index of subjects: Tables and Figures are indicated by page number in bold. Given the volume of topics covered, references are made to keywords that are deemed central on a chapter by chapter basis.